10	11 1B	12 2B	13 3A	14 4A	15 5A	16 6A	17 7A	18 8A
								2 **He** 4.00260
			5 **B** 10.811	6 **C** 12.011	7 **N** 14.0067	8 **O** 15.9994	9 **F** 18.9984	10 **Ne** 20.1797
			13 **Al** 26.9815	14 **Si** 28.0855	15 **P** 30.9738	16 **S** 32.066	17 **Cl** 35.4527	18 **Ar** 39.948
28 **Ni** 58.69	29 **Cu** 63.546	30 **Zn** 65.39	31 **Ga** 69.723	32 **Ge** 72.61	33 **As** 74.9216	34 **Se** 78.96	35 **Br** 79.904	36 **Kr** 83.80
46 **Pd** 106.42	47 **Ag** 107.868	48 **Cd** 112.411	49 **In** 114.818	50 **Sn** 118.710	51 **Sb** 121.75	52 **Te** 127.60	53 **I** 126.904	54 **Xe** 131.29
78 **Pt** 195.08	79 **Au** 196.967	80 **Hg** 200.59	81 **Tl** 204.383	82 **Pb** 207.2	83 **Bi** 208.980	84 **Po** (209)	85 **At** (210)	86 **Rn** (222)
110 (269)	111 (272)	112 (277)						

64 **Gd** 157.25	65 **Tb** 158.925	66 **Dy** 162.50	67 **Ho** 164.930	68 **Er** 167.26	69 **Tm** 168.934	70 **Yb** 173.04	71 **Lu** 174.967
96 **Cm** (247)	97 **Bk** (247)	98 **Cf** (251)	99 **Es** (252)	100 **Fm** (257)	101 **Md** (258)	102 **No** (259)	103 **Lr** (260)

Temoshok, Lydia.

The type C connection.

$22.50

DATE			

H EDITION

...tion to Gene... ...logical ...stry

...Ouellette

...hio State University

PRENTICE HALL Upper Saddle River, New Jersey 07458

Library of Congress Cataloging-in-Publication Data

Ouellette, Robert J.
Introduction to general, organic, and biological chemistry/Robert J. Ouellette.—4th ed.
p. cm.
Includes index.
ISBN 0–02–389670–1
1. Chemistry. I. Title
QD31.2.084 1997
540—dc20 95–32247
CIP

Editorial Director: Tim Bozik
Editor-in-Chief: Paul F. Corey
Acquisitions Editor: Ben Roberts
Associate Editor: Mary Hornby
Director of Production and Manufacturing: David W. Riccardi
Executive Managing Editor: Kathleen Schiaparelli
Production Supervisor: Rose Kernan
Creative Director: Paula Maylahn
Interior Designer: Wanda Kossak
Art Director: Joe Sengotta
Cover Designer: Wanda Kossak
Cover Credit: Neo Vision/IMA US
Photo Researcher: Rona Tuccillo
Manufacturing Manager: Trudy Pisciotti

Printed in the United States of America

10 9 8 7 6 5 4 3 2 1

ISBN 0-02-389670-1

Prentice-Hall International (UK) Limited, London
Prentice-Hall of Australia Pty. Limited, Sydney
Prentice-Hall Canada, Inc., Toronto
Prentice-Hall Hispanoamericana, S.A., Mexico
Prentice-Hall of India Private Limited, New Delhi
Prentice-Hall of Japan, Inc., Tokyo
Simon & Schuster Asia Pte. Ltd., Singapore
Editora Prentice-Hall do Brasil, Ltda., Rio de Janeiro

Contents

Preface xiii

1
Chemistry 1

1.1 The Human Condition: Past and Present 1
1.2 What Is Science? 2
1.3 How Do We Do Science? 3
1.4 What Is Chemistry? 5
1.5 Matter Exists in Three States 11
1.6 Physical and Chemical Properties 11
1.7 Mixtures and Pure Substances 14
1.8 Elements 16
1.9 Energy Is a Property 18
1.10 Spontaneity in Chemistry 18

ESSAY
Serendipity, Chemistry, and Medicine 10

Summary 19 • Exercises 20

2
Measurements 23

2.1 What Is a Measurement? 23
2.2 How Reliable Are Measurements? 24
2.3 Significant Figures in Measurements 25
2.4 Expressing Large and Small Numbers 27
2.5 Using Units in Problem Solving 28
2.6 The Metric System 29
2.7 Temperature Units 34
2.8 Extensive and Intensive Properties 37

ESSAYS
The Scale of Our Universe 31
Homeostasis, Hyperthermia, and Hypothermia 36
Calories and the Basal Metabolic Rate 41

Summary 42 • Exercises 42

3
Atomic Structure and the Periodic Table 45

3.1 What Is the Smallest Unit of Matter? 45
3.2 The Atom Is Very Small 47
3.3 Atoms Contain Subatomic Particles 48
3.4 Composition of Atoms 49
3.5 Isotopes 50
3.6 Mendeleev's Classification of Elements 53
3.7 The Bohr Atom 56
3.8 Shells, Subshells, and Orbitals 57
3.9 Electron Configuration 61
3.10 Electron-Dot Symbols 65
3.11 The Periodic Table and Electron Configuration 65
3.12 Properties of Atoms and the Periodic Table 67

ESSAYS
Isotopes in Biochemistry 52
Atmospheric Nuclear Tests 55

Summary 72 • Exercises 73

4
Ionic Compounds 77

4.1 Bonds, Structure, and Properties 77
4.2 Types of Bonds 78

4.3 Cations and Anions 78
4.4 Ionic Bonds 81
4.5 Structure of Ionic Compounds 83
4.6 Names of Ions 84
4.7 Predicting Formulas of Ionic Compounds 87
4.8 Names of Ionic Compounds 89

ESSAYS
Ionic Radii Affect Cell Composition and Enzyme Reactions 81
Calcium Ions and Your Diet 84
Phosphates in Agriculture 88

Summary 90 • Exercises 91

5 Molecular Elements and Compounds 93

5.1 Covalent Bonds 93
5.2 Drawing Lewis Structures 96
5.3 Polar Covalent Bonds 98
5.4 Coordinate Covalent Bonds 99
5.5 Resonance Structures 101
5.6 Molecular Geometry 103
5.7 Polarity of Molecules 106
5.8 Orbitals and Molecular Shapes 107
5.9 Structural Formulas of Organic Compounds 109
5.10 Structural Models 113

ESSAY
Coordinate Covalent Bonds and Respiration 100

Summary 113 • Exercises 114

6 Chemical Equations and Reactions 117

6.1 Chemical Equations 117
6.2 Balancing Equations 119
6.3 Chemical Reactions 125
6.4 Combination Reactions 126
6.5 Decomposition Reactions 127
6.6 Single-Replacement Reactions 128
6.7 Double-Replacement Reactions 129
6.8 Neutralization Reactions 132
6.9 Organic Acids and Bases 133
6.10 Oxidation–Reduction Reactions 134
6.11 Redox Reactions in Organic Chemistry 140

ESSAYS
The Chemistry of Rockets 124
Preventing Corrosion, an Unwanted Oxidation 139
Redox Reactions in Biochemistry 141

Summary 142 • Exercises 143

7 Composition and Stoichiometry 146

7.1 Submicroscopic and Macroscopic Views 146
7.2 Molecular and Formula Weights of Compounds 147
7.3 Percent Composition 148
7.4 Avogadro's Number and the Mole 149
7.5 Calculations Using Moles 152
7.6 Empirical Formula and Molecular Formula 154
7.7 Stoichiometry and the Balanced Equation 158
7.8 Solving Stoichiometry Problems 159
7.9 Heat of Reaction 161
7.10 Percent Yield 162
7.11 Limiting Reagent 163

ESSAYS
Products from the Earth, Sea, and Air 153
Limiting Reagents in Life's Processes 164

Summary 165 • Exercises 165

8 Gases, Liquids, and Solids 169

8.1 The Organization of Matter 169
8.2 Pressure 172
8.3 The Gas Laws 174
8.4 Avogadro's Hypothesis and the Ideal Gas Law 180
8.5 Dalton's Law of Partial Pressure 181
8.6 Liquids 183
8.7 Solids 189
8.8 Le Châtelier's Principle and Changes of State 189
8.9 Types of Intermolecular Forces 190

ESSAYS
Breathing and Boyle's Law 176
Partial Pressures in Medicine 182
Water—An Unusual Compound 196

Summary 197 • Exercises 198

9

Solutions 202

9.1 The Nature of Solutions 202
9.2 Types of Solvents and Solutes 203
9.3 Concentration—The Amount of Solute in Solution 204
9.4 Percent Concentration 205
9.5 Molarity 207
9.6 Dilution of Solutions 209
9.7 Electrolytes and Nonelectrolytes 209
9.8 Electrolytes and Body Fluids 210
9.9 The Formation of a Solution 212
9.10 Structure and Solubility 215
9.11 Factors that Affect Solubility 216
9.12 Colligative Properties 219
9.13 Osmosis 221
9.14 Osmosis and Cells 224

ESSAYS
Gout, Kidney Stones, and Solubility 205
Hot Packs and Cold Packs 214
Solubility of Vitamins 216
Gas Solubility and the Bends 217
Design of Medicines 219

Summary 225 • Exercises 225

10

Reactions Rates and Equilbrum 228

10.1 Reactions—An Overview 228
10.2 Energy and Chemical Reactions 229
10.3 Spontaneity and Chemical Reactions 231
10.4 Free Energy 233
10.5 Kinetics 236
10.6 Activation Energy 240
10.7 Chemical Equilibrium 244
10.8 Chemical Equilibrium and Le Châtelier's Principle 247

ESSAYS
Explosions and Entropy 233
Free Energy and Biological Processes 235
The Effect of Temperature on Organisms 238
Catalysis in Living Systems 239
Catalytic Converters 244
Production of Ammonia for Fertilizer Products 251

Summary 251 • Exercises 252

11

Acid–Base Equilibria 255

11.1 Acids and Bases 255
11.2 Properties of Acids and Bases 257
11.3 Reactions of Acids and Bases 260
11.4 Conjugate Acids and Conjugate Bases 262
11.5 Self-Ionization of Water 263
11.6 Strengths of Acids and Bases 266
11.7 The pH Scale 269
11.8 Hydrolysis of Salts 273
11.9 Buffers 276
11.10 Buffers in the Body 278
11.11 Titration 279

ESSAYS
Acids in the Environment 256
Antacids 262
Acids and Bases in Agriculture 276

Summary 283 • Exercises 284

12

Nuclear Chemistry 287

12.1 Chemistry of the Nucleus 287
12.2 Radioactivity and Nuclear Reactions 289
12.3 Balancing Nuclear Equations 291
12.4 Half-Lives 292
12.5 Nuclear Transmutations 295
12.6 Nuclear Fission Reactions 297
12.7 Nuclear Fusion 300
12.8 Detection of Radiation 301
12.9 Biological Effects of Radiation 302
12.10 Radioisotopes in Diagnosis 304
12.11 Radioisotopes in Therapy 306

ESSAYS
Dating Old Rocks and Fossils 295
Chernobyl 299
Positron Emission Tomography 305

Summary 307 • Exercises 307

13

The Organization of Organic Chemistry 310

13.1 Organic Compounds 310
13.2 Bonding and Molecular Shapes 312
13.3 Structural Formulas and Conformations 314
13.4 Functional Groups 316
13.5 Isomers 320
13.6 Organic Reactions and Mechanisms 323
13.7 Acid–Base Reactions in Organic Chemistry 324
13.8 Oxidation–Reduction Reactions 326
13.9 Classification of Organic Reactions 327

ESSAYS
Organic Compounds in Daily Life 312
Communication in the Insect World 320

Summary 329 • Exercises 330

14

Alkanes and Cycloalkanes 333

14.1 Classes of Hydrocarbons 333
14.2 Alkanes 334
14.3 Nomenclature of Alkanes 337
14.4 Cycloalkanes 340
14.5 Physical Properties of Saturated Hydrocarbons 344
14.6 Oxidation of Alkanes and Cycloalkanes 345
14.7 Halogenation of Saturated Hydrocarbons 346

ESSAYS
Physiological Properties of Alkanes 344
Octane Numbers 346
Freon, Radicals, and the Ozone Layer 348

Summary 349 • Exercises 349

15

Unsaturated Hydrocarbons 353

15.1 Types of Unsaturated Hydrocarbons 353
15.2 Geometric Isomerism 356
15.3 Nomenclature of Alkenes 359
15.4 Oxidation and Reduction of Alkenes 360
15.5 Addition Reactions of Alkenes 362
15.6 Polymerization of Alkenes 365
15.7 Nomenclature of Alkynes 366
15.8 Reactions of Alkynes 367
15.9 Aromatic Hydrocarbons 369
15.10 Nomenclature of Aromatic Hydrocarbons 371
15.11 Polycyclic Aromatic Hydrocarbons 374
15.12 Heterocyclic Aromatic Compounds 375

ESSAYS
Physiologically Active Alkynes 356
Geometric Isomers and the Sex Life of Moths 358
Carcinogenic Aromatic Hydrocarbons 374

Summary 376 • Exercises 376

16

Alcohols, Phenols, and Ethers 381

16.1 Compounds of Oxygen 381
16.2 Common Alcohols and Phenols 383
16.3 Nomenclature and Classification of Alcohols 385
16.4 Hydrogen Bonding in Alcohols 388
16.5 Acid–Base Reactions 389
16.6 Substitution Reactions 390
16.7 Dehydration of Alcohols 392
16.8 Oxidation of Alcohols 394
16.9 Ethers 397
16.10 Sulfur Compounds 400

ESSAYS
Fetal Alcohol Syndrome 384
Phosphate and Pyrophosphate Esters 391
Ethanol–An Antidote for Methanol Poisoning 396
Polyethers in the Laboratory and in Medicine 399

Summary 401 • Exercises 401

17 Aldehydes and Ketones 405

17.1 The Carbonyl Group 405
17.2 Naturally Occurring Aldehydes and Ketones 407
17.3 Nomenclature of Aldehydes and Ketones 408
17.4 Physical Properties of Aldehydes and Ketones 410
17.5 Oxidation of Carbonyl Compounds 412
17.6 Reduction of Carbonyl Compounds 414
17.7 Addition Reactions of Carbonyl Compounds 416
17.8 Addition of Oxygen Compounds 417
17.9 Addition of Nitrogen Compounds 420
17.10 Reactivity of the α-Carbon Atom 421
17.11 The Aldol Condensation 423

ESSAYS
Monitoring Blood Sugar and Diabetes 413
Biological Oxidation with Enzymes 415
Isomerism, Addition Reactions, and Vision 421

Summary 425 • Exercises 425

18 Carboxylic Acids and Esters 429

18.1 The Carboxyl and Acyl Groups 429
18.2 Nomenclature of Carboxylic Acids 433
18.3 Physical Properties of Carboxylic Acids 435
18.4 Acidity of Carboxylic Acids 437
18.5 Salts of Carboxylic Acids 439
18.6 Decarboxylation of Carboxylic Acids 440
18.7 Esters 441
18.8 Hydrolysis and Saponification of Esters 445
18.9 Nucleophilic Acyl Substitution 448
18.10 Esters and Anhydrides of Phosphoric Acid 450
18.11 The Claisen Condensation 452

ESSAYS
Polyfunctional Carboxylic Acids in Biochemistry 431
Carboxylate Salts as Food Additives 440
Polyesters 444
Soaps and Detergents 446
Thioesters Are Nature's Active Acyl Compounds 449

Summary 453 • Exercises 453

19 Amines and Amides 457

19.1 Organic Nitrogen Compounds 457
19.2 Bonding and Structure of Amines 459
19.3 Classification and Nomenclature of Amines 459
19.4 Physical Properties of Amines 463
19.5 Basicity of Amines 466
19.6 Quaternary Ammonium Salts 469
19.7 Reactions of Amines with Nitrous Acid 470
19.8 Amides 472
19.9 Formation of Amides 474
19.10 Hydrolysis of Amides 476

ESSAYS
Heterocyclic Compounds From the Ocean 462
Amphetamines 465
Muscle Relaxants 470
Nitrites and Cancer 471
Polyamides 475

Summary 477 • Exercises 478

20 The Organization of Biochemistry 482

20.1 What is Biochemistry? 482
20.2 Composition and Structure of Biochemicals 483
20.3 Energy and Biochemical Reactions 487
20.4 Types of Cells 488
20.5 The Eukaryotic Cell 489
20.6 Stereoisomers 492
20.7 Molecular Chirality 494
20.8 Optical Activity 497
20.9 Fischer Projection Formulas 499
20.10 Multiple Chiral Centers 502
20.11 How Are Stereoisomers Formed? 505

ESSAY
Chirality and the Senses 504

Summary 506 • Exercises 507

21

Carbohydrates 510

21.1 Carbohydrates and Life 510
21.2 Classification of Carbohydrates 511
21.3 Configuration of Monosaccharides 512
21.4 Metabolic Reactions of Glucose 516
21.5 Cyclic Forms of Monosaccharides 517
21.6 Glycosides 521
21.7 Oxidation of Monosaccharides 522
21.8 Reduction of Monosaccharides 524
21.9 Disaccharides 525
21.10 Polysaccharides 528

ESSAYS
Lactose Intolerance 526
Carbohydrates and Human Blood Groups 530

Summary 531 • Exercises 531

22

Lipids 535

22.1 Classification of Lipids 535
22.2 Terpenes 537
22.3 Steroids 537
22.4 Waxes 540
22.5 Fatty Acids 541
22.6 Triacylglycerols 544
22.7 Glycerophospholipids 546
22.8 Sphingophospholipids 548
22.9 Glycosphingolipids 549
22.10 Biological Membranes 550
22.11 Transport Across Membranes 553

ESSAYS
Whale Oil 541
Eicosanoids 543

Summary 554 • Exercises 555

23

Amino Acids and Proteins 558

23.1 The Biological Role of Proteins 558
23.2 Amino Acids 561
23.3 Acid–Base Properties of Amino Acids 563
23.4 Isoionic Point 565
23.5 Peptides 567
23.6 Biological Activity of Peptides 569
23.7 Bonding in Proteins 570
23.8 Structure of Proteins 572
23.9 Denaturation of Proteins 576
23.10 Hydrolysis of Proteins 579

ESSAYS
Lipoproteins and Blood Cholesterol 560
Straight and Curly Hair 578
Complete Proteins in the Diet 580

Summary 581 • Exercises 581

24

Enzymes 584

24.1 Biological Catalysts 584
24.2 Classification of Enzymes 586
24.3 Models of Enzyme Action 588
24.4 Active Sites 590
24.5 Cofactors 590
24.6 Kinetics of Enzyme Reactions 592
24.7 Enzyme Inhibition 596
24.8 Regulatory Enzymes: Feedback Inhibition 599
24.9 Zymogens 600

ESSAYS
Hot Spas for Bacteria 595
Mercury and Lead Poisoning 598

Summary 601 • Exercises 601

25

Biochemical Energy 604

25.1 Food and Free Energy 604
25.2 Free Energy and Oxidation State 608
25.3 Metabolic Reactions and ATP 609
25.4 Coupling of Reactions 612
25.5 Substrate-Level Phosphorylation 613
25.6 Energy and Redox Reactions 615
25.7 Cellular Respiration 617
25.8 Citric Acid Cycle 619
25.9 The Mitochondria 621

Summary 623 • Exercises 624

26

Metabolism of Carbohydrates 626

26.1 Digestion of Carbohydrates 626
26.2 Metabolic Pathways of Glucose 627
26.3 Glycolysis 629
26.4 Metabolism of Pyruvic Acid 633
26.5 The Yield of ATP from Glucose 637
26.6 Gluconeogenesis 638
26.7 The Pentose Phosphate Pathway 641
26.8 Glycogenolysis and Glycogenesis 642
26.9 Regulation of Carbohydrate Metabolism 643

ESSAY
Fermentation—An Anaerobic Process 635

Summary 645 • Exercises 645

27

Metabolism of Triglycerides 647

27.1 Digestion of Triglycerides 647
27.2 Catabolic Reactions of Triglycerides 651
27.3 Energy Production from Fatty Acids 655
27.4 Ketosis 657
27.5 Biosynthesis of Fatty Acids 658

ESSAY
Fats, Exercise, and Diet 650

Summary 663 • Exercises 663

28

Metabolism of Proteins 665

28.1 Digestion of Proteins 665
28.2 Overview of Amino Acid Metabolism 666
28.3 Deamination of Amino Acids in Catabolism 669
28.4 The Urea Cycle 674
28.5 The Fate of the Carbon Skeleton 677
28.6 Decarboxylation of Amino Acids 679
28.7 Biosynthesis of Amino Acids 680

ESSAYS
Diet, Amino Acid Supplements, and Muscles 667
Nitrogen Balance, Starvation, and Kwashiorkor 669
Phenylketonuria 678

Summary 681 • Exercises 682

29

Nucleic Acids 684

29.1 The Control of Life Processes 684
29.2 Nucleic Acids are Polymers 686
29.3 Polynucleotides 689
29.4 The Double Helix of DNA 691
29.5 DNA Replicates Itself 693
29.6 The Flow of Genetic Information 695
29.7 DNA Produces Several Types of RNA 696
29.8 The Genetic Code 698
29.9 Protein Synthesis and Translation 700
29.10 Gene Mutations and Genetic Disease 704
29.11 Recombinant DNA and Genetic Engineering 707
29.12 Viruses 710

ESSAYS
Nucleosides in Medicine 687
A Gene Defect May Cause Osteoarthritis 706
Recombinant DNA and Agriculture 709
Gene Therapy in Cystic Fibrosis 709

Summary 711 • Exercises 712

30

Body Fluids 714

30.1 Solutions in the Body 714
30.2 pH of Body Fluids 715
30.3 Solutes in Body Fluids 716
30.4 Digestive Juices 718
30.5 Blood 721
30.6 Blood Cells 723
30.7 Blood Gases 725
30.8 The Immune System 728
30.9 Blood Clotting 730
30.10 Urine 732
30.11 The Kidneys 734

ESSAYS
Bacteria and Ulcers 720
Bone Marrow Transplants 725

Summary 736 • Exercises 736

APPENDIX A
Scientific Notation and Significant Figures 738

APPENDIX B
Glossary 743

APPENDIX C
Answers to Problems 757

APPENDIX D
Answers to Odd-Numbered Exercises 760

Index 772

Preface

Audience

The approach to chemistry for non-science majors presented in this text has evolved from the author's 33 years of experience in teaching a course covering general, organic, and biological chemistry, his teaching experience in the traditional second-year organic chemistry course, and his long-term interest in biochemistry. The subject of this book is directed toward students with interests in agriculture, health education, home economics, medical technology, nursing, nutrition, physical therapy, and other fields generally described as life or health sciences. For most of these students, this course will be their only exposure to the principles of chemistry, although the applications of chemistry will permeate a number of subjects that they will study as part of their major.

Required Math Background

The text is written for students who have had no previous course in chemistry and whose mathematical background is limited. The emphasis is on essential chemical concepts rather than giving a cursory treatment to a large number of topics related to each of the diverse interests of the students.

Discussions of many topics are qualitative. However, since chemistry is a quantitative science, an appreciation of chemical concepts in areas such as stoichiometry or gas laws is best achieved by working quantitative problems. The students need only arithmetic and basic algebraic skills. All problems are solved using the factor-unit method, which is emphasized from the very beginning of the book. Appendix A summarizes the use of significant figures and exponential notation.

Philosophy

Including two editions of an earlier text on the same subject with Harper and Row followed by three editions with Macmillan, this edition of Introduction to General, Organic, and Biological Chemistry with Prentice Hall is the sixth time that the author has evaluated the current role of a text for the general education of non-science majors. In 1967 this author first put pen to paper (a yellow note pad). In 1996, not only is white paper preferred over colored, but the computer keyboard has long since replaced the pen. The author is reminded of the words of a song, "The times they are a-changing", because the changes in chemistry have been as dramatic as the computers used to write books. Biochemistry, which in the 1960s was only a small component of books for non-science majors, is now the focal point to which much of the other traditional components—general and organic chemistry—are directed. However, these two components of the science have not stagnated either. The principles presented in general chemistry are largely unchanged, but the applications of those principles in a rapidly changing world have made the study of chemistry much more important. Furthermore, the organic chemistry of 30 years ago is hardly recognizable today as the products of this fertile field have come to be so indispensable to us.

The student of today is no more sophisticated or skilled in math and science than in the mid-sixties. The dilemma then is how much of the current advances in chemistry can be covered in an academic year and what topics would be beneficial for the students. Can we cover more material? The author feels that, in a parody of the words describing Kansas City in the play "Oklahoma", we have gone about as far as we can go.

One of the characteristics of all chemistry texts for the last decade, regardless of level or the intended audience, is an increase in the length of the text as well as the number and type of special features. Every author has to question whether more is necessarily better or advantageous as well as what can be deleted. Choices must be made and the material selected must have a focus that will hold the student's attention. This requirement dictates that there must be substantial integration of subjects and explanations for why a particular topic of chemistry should be studied. We can't simply urge students to take their medicine and tell them to wait for something exciting just over the horizon. That might have worked several decades ago but it doesn't today when even events of a few years ago are regarded as ancient history.

Retained Features

Each new term is presented using bold face. In this way students can recognize that the sentence incorporating the new term also provides the definition which they can highlight with a colored marker as seems to be the norm today. A Glossary provided in Appendix B gives a definition similar to that used in the text. In addition, the more important terms are incorporated in the narrative-style summary at the end of each chapter.

This edition retains the feature of solved in-chapter Examples and unsolved Problems paired with the Examples. The Examples still number about 200 but there have been additions and deletions. Answers for the unsolved Problems are in Appendix C.

The Exercises at the end of the chapters still number about 1600. Again as in the case of Examples, there have been some additions and deletions. There are more than enough Exercises for faculty to provide sufficient assignments for students. As in the previous edition all Exercises are arranged in pairs of questions on related material. Answers to the odd-numbered Exercises are in Appendix D of the text; solutions to all Exercises are in the Study Guide.

Essays are set off from the main sections of the chapters. About 25% of the 80 essays of the third edition were dated and hence dropped. However, they have been replaced with new essays, so the essays still total 80. There is no single theme among the essays. They vary in content and style but are generally brief and related to the immediate subject of the chapter. They have been selected to extend the subject of the chapter as well as to provide material of interest to students. Each essay is placed near related subject matter to avoid disturbing the continuity of the textual material. Furthermore they are distributed as evenly as possible over all sections of the text. Comparable material that could just as well have been the subject of an essay is often integrated within the text.

Organization

Although every author feels that his or her book provides the best approach to teaching the subject, the successful book acknowledges the variety of instructors, students, and goals of the many courses offered in this country. Accordingly, care has been taken to write sections of each chapter to be as free-standing as possible. In general, the more difficult concepts or areas that may not be covered are isolated in separate sections or are placed at the ends of the chapters.

The Overall Theme

The unifying theme of the book remains the chemistry of life. The traditional three segments of the book covering general, organic, and biological chemistry have been retained. However, the subjects are more strongly integrated. This is accomplished within the textual material as well as by the use of essays. In addition, many more of the Examples and Problems within the chapter as well as many of the Exercises have been selected to serve the same purpose. For example, organic molecules are often selected for emphasis based on their importance in pharmaceutical drugs and related biochemistry.

The Energy Theme

Chemistry is a study of matter and the changes that it undergoes. However, in texts at this level the role of free energy is ignored and regarded as too difficult for mathphobic students. Thermodynamics is certainly not an appropriate subject for the course for non-science majors, but the qualitative aspects of ΔH° and ΔG° can be understood by them. Without considering thermodynamic concepts, there is no rhyme or reason as to why any of the chemical reactions presented in this book proceed in the indicated direction.

The concept of spontaneity is presented in Chapter 1. It is reinforced in discussions of bonding, phase changes, solution processes, and chemical equilibrium. Only in the discussion of chemical equilibria is ΔG° formally defined. The student should have a better appreciation of the importance of the role of ΔG° when biochemical energy is finally discussed in Chapter 25 and used in some detail in subsequent chapters.

The Molecular Theme

Chemistry is a molecular science and chemists visualize both physical and chemical change at the molecular level. Accordingly in this edition there are many more representations of molecules rather than generic spheres. Molecules are shown to illustrate phase changes, solubility, acid-base reactions, and even in balancing equations. This approach decreases the gap between general chemistry and organic chemistry. It also should meet with approval of students who are more often visual learners and not as skilled at abstract thinking.

General Chemistry

The number of chapters dealing with general chemistry has decreased by one to a total of 12. The subject matter of the first 8 chapters of the third edition has been rearranged and reduced into 7 chapters. This material now places atomic structure, the periodic table, and bonding before chemical equations and stoichiometry. As a result, the quantitative aspects of chemistry are delayed and now appear near chapters on gases and solutions which also require numerical calculations.

The original bonding chapter is now divided into two chapters to emphasize the difference between ionic compounds and molecular compounds and to highlight the latter. Hybridization, which was found in the first "organic" chapter, is now discussed in Chapter 5, as are the fundamentals of the structure of organic compounds. However, this material appears at the end of the chapter and can be passed over and covered later in the course if the instructor chooses to maintain the separation between inorganic and organic compounds. Subsequent general chemistry chapters do contain some examples of organic molecules in simple reactions and in describing some physical properties. However, they are not great in number and only small molecules are considered. The author feels that there are benefits that accrue by using organic molecules as examples but realizes that this opinion may not be shared by all. For those that disagree, there should be little difficulty in bypassing the material. However, those teachers with an interest in avoiding the separation of fields of chemistry should welcome this foundation on which they can build with more interesting examples based on this important branch of chemistry.

Organic Chemistry

As in the third edition, an introductory chapter provides an overview of organic chemistry and discusses acid-base chemistry and oxidation-reduction reactions as encountered in organic chemistry. However, the material is now also previewed in some sections in the general chemistry chapters.

The organic chemistry chapters still present the simpler aspects of organic mechanisms. Without an indication that many organic reactions are just Lewis acid-base reactions, students may feel that organic chemistry is all black magic and is a subject that must simply be memorized. For those who do not favor the use of mechanisms, note that the mechanisms are few in number and are restricted to isolated sections so that they need not be used. For those who choose to use mechanisms, the material will provide a foundation that can be expanded to explain biochemical reactions as well.

As in previous editions, the focus of the organic chemistry chapters is on the structure and properties of molecules—not their synthesis. Although the number of reagents for reactions in organic chemistry continues to increase, the author, who is an organic chemist, has maintained his focus on the goals of the text. Accordingly, the majority of the reactions discussed also occur in biochemical systems. Whenever possible, examples are selected that involve medical or pharmaceutical applications.

Biochemistry

The chapter on nucleic acids has been moved from its former position following the three chapters on the three classes of biomolecules—carbohydrates, lipids, and proteins—to Chapter 29. This shift allows a freer flow from the chapters on the structure of biomolecules to chapters on their metabolic reactions. Thus, function (Chapters 24–28) follows naturally from structure (Chapters 21–23). For many instructors, nucleic acids are a highlight of the course and this placement will be appropriate. This placement is logical because nucleic acids control so many of the fundamental processes that must be first described in earlier chapters. However, for "short" courses the nucleic acid chapter could be presented after Chapter 23 or 24.

Trying to be reasonably up-to-date in the current state of nucleic acid chemistry and genetic engineering is a losing battle. For example, the essay on cystic fibrosis is based on 1995 information which may pave the way for a cure yet may not be successful in 1996. The author has tried to provide detailed information without overwhelming the students. For example, exons and introns as well as the role of Okazaki fragments in the DNA replication are not discussed.

A new chapter on Body Fluids has been added to the biochemistry part of the text. This chapter is essential for some courses but optional in others. In any case, placement as the last chapter should please all. The chapter interrelates concepts such as acid-base reactions, osmotic pressure, enzyme reactions, and transport across membranes presented in earlier chapters.

Essays

Students often ask if they are going to be tested on certain sections of a text such as the essays. Of course that decision is what makes each instructor and the course that they teach different. Many of the essays enhance and broaden the subjects in the main text. However, if an instructor wants to assign only the main text, the continuity will not be disrupted nor will anything be missing that would affect understanding of concepts in later chapters.

The impact of chemistry on life in a modern society has been profound. The essays are used to highlight some of these aspects of chemistry in the world around us. For the most part, the language is not technical and the goal is to provide information in a form readable by students.

There is no single theme for the essays, although medical applications form a substantial percentage of the subjects. A list of essays appears in the Table of Contents arranged by chapter.

Among the medical topics of current interest discussed are Fetal Alcohol Syndrome (Chapter 16), Monitoring Blood Sugar and Diabetes (Chapter 17), Carbohydrates and Blood Groups (Chapter 21), Blood Cholesterol (Chapter 23), Gene Therapy in Cystic Fibrosis (Chapter 29), and Bone Marrow Transplants (Chapter 30). Dietary and nutritional subjects considered include Calcium Ions and Your Diet (Chapter 4), Gout, Kidney Stones, and Solubility (Chapter 9), Antacids (Chapter 11), Carboxylate Salts as Food Additives (Chapter 18), Nitrites and Cancer (Chapter 19), Lactose Intolerance (Chapter 21), Fats, Exercise and Diet (Chapter 27), Nitrogen Balance, Starvation, and Kwashiorkor (Chapter 28), Diet, Amino Acid Supplements, and Muscles (Chapter 28), and Bacteria and Ulcers (Chapter 30).

Although not easily classified, a number of essays present information on phenomena of general interest to the average student population. These include The Chemistry of Rockets (Chapter 6), Hot Packs and Cold Packs (Chapter 9), Catalytic Converters (Chapter 10), Acids in the Environment (Chapter 11), Chernobyl (Chapter 12), Freon, Radicals, and the Ozone Layer (Chapter 14), Amphetamines (Chapter 19), Straight and Curly Hair (Chapter 23), Fermentation–An Anaerobic Process (Chapter 26), and Recombinant DNA and Agriculture (Chapter 29).

Some essays provide a preview of material found in later chapters and help to integrate the various chemical disciplines. These include Serendipity, Chemistry, and Medicine (Chapter 1), Isotopes in Biochemistry (Chapter 3), Ionic Radii Affect Cell Composition and Enzyme Reactions (Chapter 4), Coordinate Covalent Bonds and Respiration (Chapter 5), Redox Reactions in Biochemistry (Chapter 6), Free Energy and Biological Processes (Chapter 10), Chemical Communication in the Insect World (Chapter 13), Phosphate and Pyrophosphate Esters (Chapter 16), Biological Oxidation with Enzymes (Chapter 17), and Thioesters are Nature's Active Acyl Compounds (Chapter 18). All of this material precedes the biochemistry section of the text.

Depending on the instructor's interest it should be possible to select different groups of essays for discussion in class. The list of essays selected illustrates that they are well distributed among the chapters.

Use of Color

Although many figures and tables have been retained throughout the text, a substantial number have been deleted or replaced by new representations. Virtually all have been redesigned to take advantage of color. Molecular structures are highlighted with color to provide a focus on the important parts that are changed in a chemical reaction. This approach is especially helpful to students when applied to biomolecules and their biochemical pathways. In total there has been an increase in the number of figures in this edition. There are more structures, with accompanying legends and arrows pointing to special features, placed within the textual material. These are color coded as well.

Acknowledgments

The helpful comments of the reviewers—John Anderson of Mid-State Technical College, Jerry Bergman of Northwest State University, Gerald Berkowitz of Erie Community College–North Campus, P. S. Braterman of University of North Texas, Donald Campbell of University of Wisconsin–Eau Claire, Morris Fishman of New York University, Robert Fromland of San Diego Mesa College, Pat M. Garvey of Des Moines Area Community College, Robert Gooden of Southern University, Henry Harris of Armstrong State College, Liland Horten of North Park College, Richard Morrison of West Virginia University, Joe Muscarella of Henry Ford Community College, Robert N. Nelson of Georgia Southern University, Elva Mae Nicholson of Eastern Michigan University, Bernard L. Powell of University of Texas at San Antonio, Mona Rampy of Columbia Basin College, James E. Rodgers of Azusa Pacific University, Peter Ronman of Brevard Community College, Guy Rosenthal of University of Vermont, Iden Smith of Edison Community College, Sister Mary Benedict Joseph Watters of Mount Aloysius College, Danny White of American River College—are gratefully acknowledged.

The author appreciates the work of Judy Casey, who read this text in the galley stage while preparing the Teacher's Manual. The author is indebted particularly to Mary Bailey for her work in reading the original manuscript, the galleys, and the page proofs. In addition, she checked the Exercises and prepared the Study Guide/Solutions Manual.

R.J.O.

Our earth, as seen by Apollo 11 astronauts from the moon, and all other parts of the universe are composed of chemicals. (NASA)

1

Chemistry

1.1 The Human Condition: Past and Present
1.2 What Is Science?
1.3 How Do We Do Science?
1.4 What Is Chemistry?
1.5 Matter Exists in Three States
1.6 Physical and Chemical Properties
1.7 Mixtures and Pure Substances
1.8 Elements
1.9 Energy Is a Property
1.10 Spontaneity in Chemistry

Essays
- *Serendipity, Chemistry, and Medicine*

1.1 The Human Condition: Past and Present

Survival has been the foremost concern of all humans throughout recorded history, and it is highly probable that it has also been our primary goal since the beginning of time. The same is true for all other life forms. To some degree, all animals observe their environment and the changes that occur in it; but the extent to which they profit from these observations and thereby improve their adaptability varies considerably. Only humans can transmit the large amounts of accumulated information from one generation to the next.

Once their basic needs are met, humans, in contrast with other animals, try to improve the quality of life both for themselves and for future generations. Our early ancestors made observations to control their lives and improve their adaptability, but they were not concerned with why things happen. Over the centuries we have developed methods to achieve such insights, so we are able to control certain aspects of our environment. Now we can improve the quality and quantity of food, clothing, and shelter—the three fundamental goals that have shaped human existence. Most of the processes used to produce food, process it, and store it involve chemistry. Even clothing and shelter have been changed dramatically by chemistry, which has produced new materials with specially designed properties. As you study

Aluminum is used in the construction of airplanes. The chemical processes used to produce aluminum from ore, make our world seem a little smaller. (Morton Beebe, The Image Bank)

chemistry in this text you will become more aware of these materials and how they are made.

You live in a world of constant chemical change—both extremely slow and in many cases far too fast. Chemical reactions are occurring in the atmosphere and in the oceans as well as in geological formations deep within the earth. All are areas of our world that are important to you.

Methane from the flatulence of cows, the natural oils from trees, and many substances from the eruptions of volcanos all change the air layer closest to earth. In the upper atmosphere, the ozone layer, which protects us from the ultraviolet radiation from the sun, is changing.

The changes in the oceans affect our weather patterns and our oxygen supply. Although photosynthesis is a process normally associated with terrestrial plants, the photosynthesis by phytoplankton in the oceans is the major source of carbohydrates and oxygen on earth.

The petroleum in geological formations has been produced very slowly over long periods of time. We extract petroleum from underground and convert it into products such as synthetic fibers used in clothes and antibiotics used to fight infections. Ores are also extracted from the earth to produce metals such as iron and aluminum for many products, from airplanes to soft-drink cans. In this text you will learn about how these natural materials are converted into products that we use every day.

Humans have long been curious about life, sickness, and death and about what is responsible for these events. Superstition dominated our ancestors' thoughts about why people became sick or died. We now know that chemical processes occur in all life forms, and it is a goal of chemists to understand them. The germination of seeds in the spring, the growth of plants in the summer, the changes in the colors of leaves in the autumn, and the decay of vegetation in the winter all involve chemical reactions. The conception, life, and death of humans and other life forms involve biological and physiological processes that are chemical reactions. Understanding these life processes allows us to intercede to improve health and in some cases to delay death. Medicine relies on drugs now produced by chemists who create new materials rather than relying on only the natural herbs and potions of earlier generations. In addition, chemists have produced new materials for prosthetic devices such as hip and knee replacements and have developed other materials to patch organs as important as the heart.

A major theme of this book is the chemical basis of life. To understand chemistry and its role in our lives, you will need to start with some simple facts and gradually extend your knowledge to the more complex. Let's start this study by first discussing science in general. Then we will consider the science we call chemistry.

1.2 What Is Science?

The word science comes from the Latin *scientia,* meaning "knowledge." **Science** is a method of learning about our universe. Scientists—people who "do science"—learn by making observations, usually about a limited part of our world.

We All Make Observations

What types of observations are made by scientists? The same kinds that you make every day. Scientists use the senses of sight, hearing, smell, taste, and touch. In addition, scientists use instruments to extend their powers of observation. For example, a thermometer is used to determine the temperature of something without having to touch it to find out how hot it is.

Making observations may seem simple, but different people observing the same phenomenon might record different information. Our ability to use our senses depends on many things, including interest, education, prior experiences, prejudices, and special training in observation. Thus, a scientist must develop the skill of making careful observations. Then, if possible, the same event should be repeated and observed again. Scientists have confidence in observations only if they and others can repeat the same observations.

Classification Helps Us Organize Things

In our daily lives we recognize and describe materials by their characteristics or **properties;** then we classify or categorize the materials according to those properties We have learned that it is easier to deal with the general characteristics of categories of items than with a large amount of information about individual items. The classification method allows us to make generalizations about the common properties of the members of each class. These generalizations provide a focus for asking questions about what is responsible for the properties of those materials.

Although you may not have seen this animal before, you should have little difficulty in identifying it as a fish. (Robert Ouellette)

Consider how you might characterize the animals that we call fish. "Fish are animals that live in water" might be a reasonable statement, but upon further consideration you would realize that whereas all fish do live in water, not all species that live in water are fish. Porpoises and whales live in water, but both are mammals. Thus, placing things into categories must be done with care. Biologists use a variety of criteria to categorize plants and animals.

Chemists classify the individual components that make up our world, known as **substances,** by their properties (Section 1.6). For example, substances known as metals have some properties in common. Consider lead and iron, which are both metals. Iron is attracted by a magnet and will rust, whereas lead is not attracted to a magnet and does not rust. Thus, neither magnetic properties nor susceptibility to rust are characteristic of all metals. What other properties can be considered? Both lead and iron are heavy, and both conduct electricity. Is "heaviness" a property common to all metals? No, metals such as aluminum and magnesium are light. What about electrical conductivity? Aluminum and magnesium also conduct electricity; in fact, all metals conduct electricity. Electrical conductivity is a useful criterion for classifying metals.

1.3 How Do We Do Science?

Although a wide variety of procedures are used, the general methodology of science is termed the *scientific method.* It includes each of the following, although not necessarily in the order listed:

1. Making scientific observations
2. Recognizing patterns
3. Developing theories
4. Performing experiments

Making Scientific Observations

Scientific observations include anything that seems relevant to us and perhaps some things that may be important but less obvious. The record of information about observations is commonly called **data.** This information may be qualitative or quantitative. **Qualitative data** are nonnumerical facts such as "water freezes when it is cold." **Quantitative data** are numerical facts such as "water freezes at 0 °C." Quantitative data are preferred in science, but qualitative data can be very important too.

Recognizing Patterns

Scientists search for repeating trends and interrelationships that can be summarized by a general statement. There isn't any specific way to do this. In fact, recognizing patterns requires some creativity and considerable training. You might think that there is a bit of luck involved in discovering patterns; however, Louis Pasteur summed it up well when he said, "Chance favors the prepared mind."

Developing Theories

Although still on the tree, these apples are attracted to the earth by the forces of gravity. (Peter Arnold, Inc.)

The first educated guess to "explain" why things happen is a **hypothesis.** Scientists express hypotheses using **models** that may correspond to what is responsible for a phenomenon. Note the word *may.* A model is never certain. In fact, a model has to be tested to determine if it is reasonable. Does the imagined cause explain all past observations as well as any new ones obtained in additional experiments? If not, a new hypothesis must be proposed. If after continued experiments the hypothesis stands up to tests, it is called a **theory.**

We may describe a model to others by waving our hands about, drawing a picture on paper, or representing it with mathematical equations. Pictures, which are said to be worth a thousand words, are easier to understand than equations and are used frequently in this text.

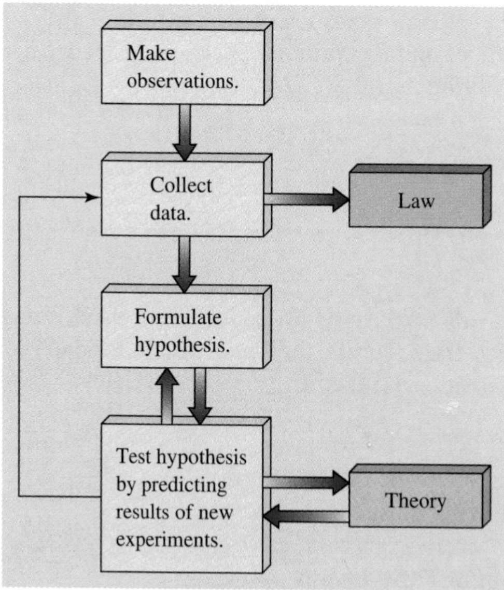

FIGURE 1.1 Outline of the Scientific Method

Performing Experiments

The development of a theory depends not only on the verification of the observed facts but also on the development of new experiments that test the theory. These experiments require intellectual honesty and cannot be biased in any way. The goal is to devise tests that can be verified by any other scientists.

Scientific Laws

Scientists summarize scientific facts in statements called *laws.* A **law** is a conclusion or an explicit statement of fact that has been found to be true for all cases examined. Thus a law represents the present state of our knowledge.

A law does not explain why nature behaves as it does. For example, Sir Isaac Newton observed that released objects fall to earth. We now know that all objects in the universe have an attraction for all other objects that is directly proportional to their masses and inversely proportional to the square of the distance between them. That statement is the law of gravity. The theory developed to explain this law is that gravitational forces exist that are responsible for the phenomenon expressed by the law. The relationships among hypotheses, theories, experiment, and laws are summarized in Figure 1.1.

1.4 What Is Chemistry?

There are many subdivisions of the natural sciences that focus on the universe in different ways. The broadest division of the natural sciences is into the physical and biological sciences (Figure 1.2). Over time each of these areas has been subdivided and continues to be further subdivided as new areas develop that bridge currently defined areas. **Chemistry** is the science of the composition, structure, properties, and reactions of matter. This definition is straightforward, but to understand any definition, you must understand the words used in stating it. Before we deal with the terminology, let's consider again the statement that bridging areas are constantly developing.

Some fields of study bridge disciplines within one of the two broad categories of natural sciences, and some even bridge these two seemingly diverse areas. Even in geology, an area that might appear to be quite restrictive, there are geophysicists and geochemists. Paleobiology, the study of animal fossils, is an area that bridges geology and biology. Chemistry bridges so many areas that it has been called the "central science." For example, there are both physical chemists and chemical physicists who investigate similar phenomena. Biochemistry is an important broad interdisciplinary field that bridges practically all the biological sciences and chemistry. This field deals with the chemistry of life, which is a major focus of this book. But first we have to understand basic chemistry, so we return to the terminology used to define chemistry.

Matter

Any material that you can see, smell, taste, or touch consists of matter. Some examples of the types of substances that constitute matter are the oxygen in the air that you breathe, the water that you drink, and the aspirin that you take for a headache. **Matter** is material that occupies space and has mass. *Mass* is not a term that you commonly use. Normally you describe a quantity of material by its weight, but there is a subtle difference between mass and weight. **Mass** is the quantity of matter in an object. **Weight** is the result of the force of the gravitational attraction between the earth and that object. Without gravity, an object would not have weight, but it would still have mass. For example, although the astronauts in the space shuttle are weightless in space, they still have mass. Although mass and weight are different, two objects that have the same mass also have the same weight under the same gravity. Therefore, mass and weight are often used interchangeably.

Composition

In chemistry, **composition** is defined as the identity and amounts of the components of matter or, in other words, what a substance is made of and in what quantities. To understand the term *composition,* let's consider the components of a bicycle. A bicycle is composed (made) of a set of handlebars, a frame, a seat, two wheels, and two pedals. This description, which is a statement of composition, gives the type and number of items that are contained in a bicycle.

How do chemists describe the composition of the different types of substances of our world? The composition of any substance can be described at the **macroscopic** level, which means in amounts that can be seen and weighed. However, a substance can also be described at the **submicroscopic** level—that is, in amounts that cannot be seen with optical microscopes. The submicroscopic composition of a substance is described by its numbers and types of atoms. **Atoms** are simple units of matter that we will discuss in detail in Chapter 3. Most substances consist of

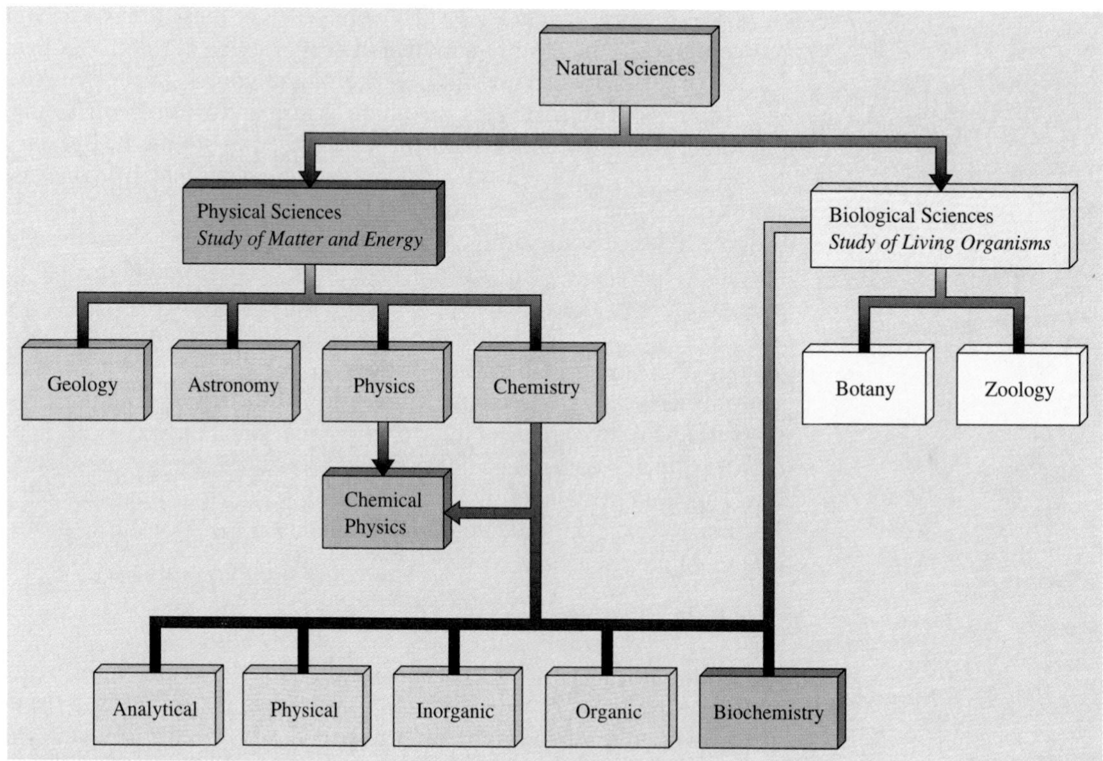

FIGURE 1.2
Areas of the Natural Sciences

FIGURE 1.3 Models of Atoms Found in Biological Molecules

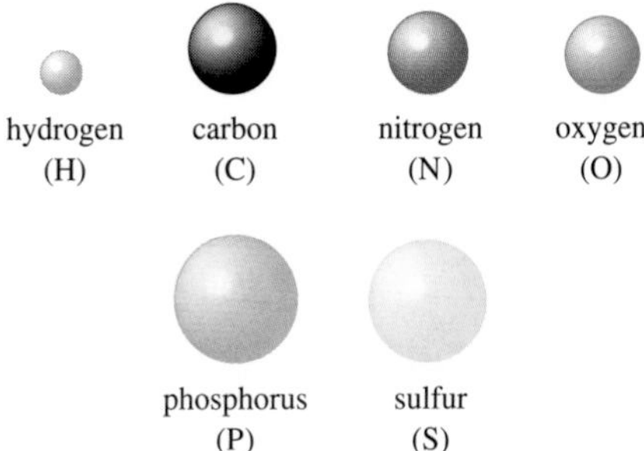

molecules which are combinations of atoms. The helium used to inflate balloons consists of helium atoms; water consists of water molecules.

You have probably heard water described as H_2O ("aitch two oh"). The letter H represents the hydrogen atom; the letter O represents the oxygen atom. The 2 following the H means that there are two hydrogen atoms in one water molecule. No number follows the letter O, and by convention this means that there is only one oxygen atom per molecule. Representations using symbols and numbers to show the composition of a molecule are molecular **formulas.** More details about molecular formulas will be given in later chapters.

We will often represent molecules using colored circles or spheres for the constituent atoms (Figure 1.3). The atoms themselves are not this color; however, you should note that the different-sized circles or spheres represent the different sizes of atoms. The hydrogen atom, which is prevalent in biological molecules, is the smallest. The carbon, nitrogen, and oxygen atoms, which also are found in biological molecules, are larger than hydrogen but decrease slightly in size in the order carbon $>$ nitrogen $>$ oxygen. Later we will consider what atoms weigh.

Structure and Bonds

Composition by itself does not tell us all we need to know about a substance. For example, could someone who had never seen a bicycle visualize it if you provided only a picture of each of the components of the bicycle? The answer is no, because it is necessary to know how the parts of the bicycle fit together.

In chemistry, **structure** indicates which atoms are connected to each other and what shape they form. In this book you will learn that atoms in molecules are bonded to one another. **Bonded** means held or fastened together. The number of bonds formed by atoms and the arrangement of the atoms in space describe a structure. The bonding of atoms in the molecules of some common substances is shown in Figure 1.4. In a water molecule the two hydrogen atoms are bonded to an oxygen atom but not to each other. The lines between the symbols for the atoms represent the bonds connecting the atoms in the molecule. Each atom forms a characteristic number of bonds. The hydrogen atom forms only one bond. Oxygen, nitrogen, and carbon atoms form two, three, and four bonds, respectively. The reasons for these facts will be discussed in Chapters 4 and 5.

Molecules are combinations of atoms that each have the "proper" number of bonds. For simple molecules you can write a structure showing the connections between atoms if you know both the number of atoms and the number of bonds that each atom forms. For more complicated molecules such as sucrose the problem is not so easy. Table sugar consists of sucrose molecules that contain 12 carbon atoms, 22 hydrogen atoms, and 11 oxygen atoms. You probably can't write a structure for sucrose now, but you will be able to before you finish this book.

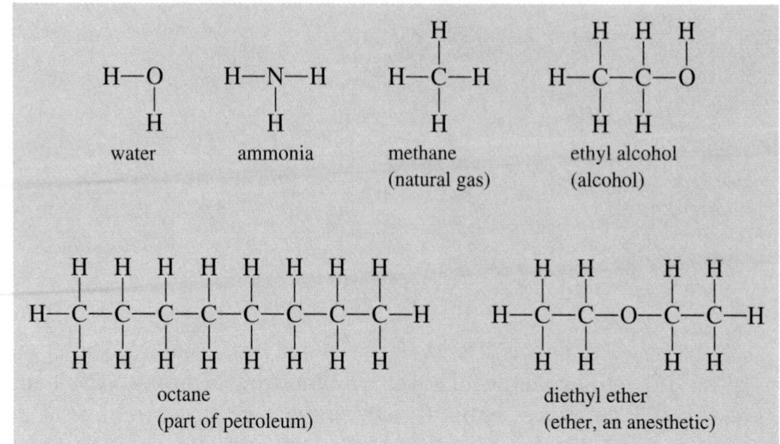

FIGURE 1.4 Structural Representations of Molecules
The dashes between the chemical symbols represent bonds that hold the molecule together and create a structure.

The bonds shown between the symbols representing atoms in water, ammonia, and methane are in two-dimensional representations called **molecular structures** (Figure 1.4). A molecular structure gives the connectivity between atoms, not the actual shape of the molecule. These molecules are shown in Figure 1.5 using three dimensions. The relationship of the atoms to one another in space determines the shape of the molecule. Further details of structure and characteristic shapes of molecules will be given in Chapter 5.

EXAMPLE 1.1

Describe the composition of ethyl alcohol (Figure 1.4) in terms of the atoms in the molecule.

Solution
The representation of ethyl alcohol in Figure 1.4 has two C's, six H's, and one O. This representation informs us that the molecule has two carbon atoms, six hydrogen atoms, and one oxygen atom.

PROBLEM 1.1

Describe the composition of ammonia (Figure 1.4).

EXAMPLE 1.2

Describe the structure of ethyl alcohol (Figure 1.4).

Solution
Lines are used to represent bonds between atoms. From the structure in Figure 1.4 you can see that the carbon atoms, each represented by the letter C, are bonded to each other. One

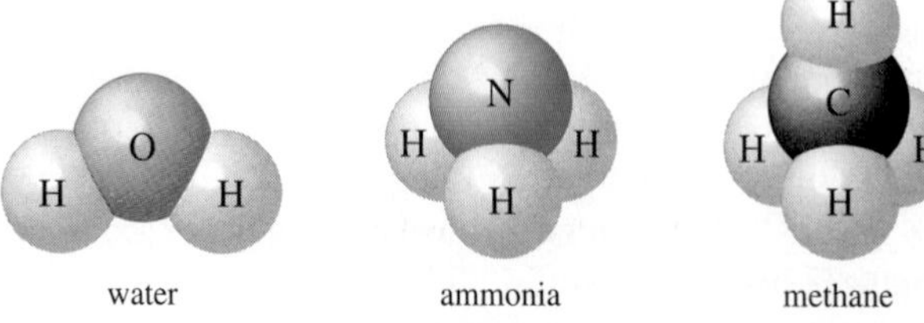

FIGURE 1.5 Shapes of Some Simple Molecules
The hydrogen atoms located about the central atom create a three-dimensional shape.

carbon atom is also bonded to three hydrogen atoms (represented by H's). The other carbon atom is also bonded to two hydrogen atoms and one oxygen atom. The oxygen atom is bonded to both a carbon atom and a hydrogen atom.

PROBLEM 1.2

Describe the structure of diethyl ether (Figure 1.4).

Properties of Matter

The properties of a substance such as the freezing point of water reflect the structure of molecules. We will find that the macroscopic features of a substance give us some clue about its submicroscopic structure. The macroscopic features are the result of interactions between large collections of very tiny molecules. We will discuss this relationship in Chapter 8 after we learn how atoms form molecules and why specific molecular architectures result.

Reactions Change Matter

In a chemical **reaction,** changes occur in the composition and structure of one or more substances. Some substances disappear and others appear, but it turns out that the component atoms are unchanged. They are only redistributed in a new way to form new substances. Because these substances have different compositions and structures, the macroscopic properties that you observe are different. You have seen many examples of chemical reactions. A combustion process such as burning gasoline in a car is a chemical reaction. Chemical reactions that occur in your body make vision, muscle movement, and the metabolism of food possible. Although you might think of these as physiological processes, they are chemical reactions because all involve changes in the composition and structure of matter.

We represent chemical reactions using **chemical equations.** Consider the metabolic processes in your body that convert glucose into carbon dioxide and water. The body uses oxygen in this reaction. For the metabolism of glucose the equation is

$$\underbrace{\text{glucose} + \text{oxygen}}_{\text{These substances are reactants.}} \longrightarrow \underbrace{\text{carbon dioxide} + \text{water}}_{\text{These substances are products.}}$$

The substances to the left of the arrow are **reactants,** the substances that are used in the reaction and undergo change. The substances named to the right of the arrow are **products,** the substances that result from or are produced in a chemical reaction. The arrow means "is converted into."

Reactions of molecules depend on how the molecules "fit" as they encounter each other. If molecules do not have the proper shape, they cannot come close enough to each other to exchange atoms in a reaction. Molecular structure is especially important in chemical reactions in living systems. Molecules known as **enzymes** are involved in many reactions in cells. The shapes of enzymes are complex, and as a result only certain molecules "fit" each enzyme. An enzyme is like a baseball glove, which is designed to catch baseballs. The glove is not suitable for catching basketballs or Ping Pong balls!

EXAMPLE 1.3

Photosynthesis in plants can be represented by the following equation. What are the reactants and the products?

$$\text{carbon dioxide} + \text{water} \longrightarrow \text{glucose} + \text{oxygen}$$

Solution

Carbon dioxide and water, written to the left of the arrow, are the reactants. Glucose and oxygen, written to the right of the arrow, are the products.

PROBLEM 1.3

Nitrogen fixation, as it occurs in the roots of some plants, is a natural process that converts nitrogen into nitrogen-containing compounds. Nitrogen is also "fixed" in an industrial process where nitrogen is combined with hydrogen to give ammonia. What are the reactants and the products in this chemical reaction? Write a word equation for the process.

Serendipity, Chemistry, and Medicine

Experimental observations may be the result of planned experiments, trial-and-error approaches, or serendipity. Serendipity means a fortuitous discovery usually made while searching for something else. Of course, serendipity doesn't occur unless the observer has a trained mind and recognizes that something important has been discovered. It is a case of "chance favoring the prepared mind." Once observed, do we further explore the subject rather than continue with our original goals, or do we forget about it? How many potentially useful products have never been discovered because the observer didn't have the intuition that some serendipitous event had occurred? Two examples of observations that were not overlooked are the discoveries of penicillin, an antibiotic, and cisplatin, an anticancer drug.

Alexander Fleming, a bacteriologist at the University of London, serendipitously discovered penicillin in 1928. He was working with cultures of *Staphylococcus aureus* and noticed that one of the cultures had accidentally been contaminated with a mold. Fleming further observed that in an area around the mold, the bacteria had been destroyed. Using the mold, he prepared a broth and added it to other types of bacterial cultures, which were also destroyed. Fleming was unable to purify and isolate the antibacterial agent, but it was finally obtained by scientists at the University of Oxford in 1940. Fortunately, Fleming didn't discard the contaminated culture, because penicillin saved many lives in World War II, where it was used to treat a variety of infections.

In 1964 Barnett Rosenberg at Michigan State was studying the effect of electric current on the growth of bacteria such as *Escherichia coli.* The apparatus had platinum electrodes, so chosen because of platinum's lack of reactivity. Rosenberg observed that cell division ceased in the cultures. By examining the solution in the culture medium, he discovered that small amounts of a platinum compound had formed from the usually unreactive platinum electrodes. That substance, known as cisplatin, was not new—it had been known since 1845. However, no one had ever considered that it might be a potential medicine. Its medicinal value might have been discovered by someone else through a trial-and-error procedure of adding random samples of known compounds in a study of bacterial growth. However there are 10 million known substances! Furthermore, if choices were made to test substances from among groups of most likely candidates, it is unlikely that platinum-containing substances would have been considered. After the discovery of the effect of cisplatin on cell division, the question was asked if it might also be useful in preventing uncontrolled cell division in cancer. After some years of testing, the U.S. Food and Drug Administration (FDA) approved its use in 1979 for the treatment of some forms of cancer.

1.5 Matter Exists in Three States

This iceberg off Antarctica and the surrounding water represent two of the three states of water. Gaseous water cannot be seen. The clouds are actually small droplets of liquid water. (Peter Arnold, Inc.)

As a child you learned that matter exists as solid, liquid, or gas. These forms of matter are called **states.** You learned about the characteristics of the solid state as you touched your toys. While splashing about in your bath you found that liquid water has different properties from solids. Although you could not see the gaseous state, you might have caught your breath as a sudden wind blew against your face.

We associate most materials with a single state. We think of iron as a solid, gasoline as a liquid, and air as a gas. However, most matter can exist in any of the three states under the proper conditions. You have observed water in all three states. At normal atmospheric pressure, water exists as a solid below 0 °C and as a gas above 100 °C. Although we use the terms *ice* and *steam* to describe the solid and gaseous states of water, its composition and structure have not changed. It is still water.

Solids have definite volume and shape. Salt and sugar are common solids whose shape can be seen with a magnifying glass or a microscope. Solids such as ice, salt, sugar, and even iron melt when heated sufficiently. Wood will not melt; a chemical reaction occurs with the oxygen of the air, and the wood burns.

Liquids have definite volume but no definite shape. You know that a liquid fills a container such as a measuring cup from the bottom up. When the liquid is poured from a cup into a bowl, it flows and assumes the shape of the bowl. However, its volume is unchanged. When a liquid is cooled, it freezes to form a solid. The temperature at which freezing of a liquid occurs is the same temperature at which melting of the corresponding solid occurs when it is heated. You may refer to either the **melting point** or the **freezing point** of a substance—but the temperature is the same. For ice and liquid water that temperature is 0 °C.

Scuba divers can survive under water by using a large volume of air compressed under pressure in an air tank. (Robert Ouellette)

A **gas** has no characteristic shape or volume and completely fills any container into which it is placed. Gases are compressible, and under increased pressure the volume of a gas decreases. Because of this compressibility large amounts of oxygen can be stored in the cylinders used in hospitals. For the same reason, a scuba diver's air tank provides a large volume of air. Gases are converted into the liquid state when they are cooled. For example, steam is converted into liquid water at 100 °C. The reverse of this process, the conversion of liquid water into steam, also occurs at 100 °C. Conversion of a liquid into a gas is **boiling;** conversion of a gas into a liquid is **condensation.** The temperature at which these two processes occur is called the **boiling point.**

1.6 Physical and Chemical Properties

Every substance has a characteristic set of properties, which are divided into two classes. **Physical properties** depend only on the substance itself. **Chemical properties** depend on the changes that a substance undergoes in a chemical reaction.

Physical Properties

Physical properties are determined without altering the composition of the substance. Some examples include boiling point, melting point, electrical conductivity, odor, and color. Pure water is a colorless, odorless, and tasteless liquid; it freezes at 0 °C and boils at 100 °C. Both the qualitative adjectives and the quantitative numbers describe physical properties of water. Similarly, chlorine is a yellow-green gas with a suffocating odor and a sharp, sour taste; it freezes at −101.6 °C and boils at −34.5 °C.

Physical properties can be used to identify substances. For example, aspirin melts at 143 °C, whereas sugar melts at 186 °C. Thus the identity of a white material known to be either aspirin or sugar can be determined by measuring its melting point.

Matter can often change appearance without changing its chemical composition. For example, the tungsten filament in a light bulb glows when the electricity is turned on, but the tungsten filament is chemically the same. When the light is turned off, the filament reverts to its original appearance, and the composition of the filament is still unchanged.

Changes of state, such as melting a solid or boiling a liquid, are the most common examples of physical changes. The melting points and boiling points of some substances are given in Table 1.1.

Although most substances change from solid to liquid and then to a gas as the temperature is increased, there are some exceptions. The change of a substance from a solid to a gas without passing through the liquid state is called **sublimation.** Dry ice, which is solid carbon dioxide, sublimes at −78.5 °C.

EXAMPLE 1.4

Chloroform is no longer used as an anesthetic because it causes liver damage. In what state does chloroform exist at 25 °C? (See Table 1.1.)

Solution

Chloroform melts at −63.5 °C and boils at 61.7 °C. At 25 °C chloroform must be a liquid, because that temperature is above its freezing point but below its boiling point.

PROBLEM 1.4

Natural gas is mostly methane. In what state does methane exist at −158 °C? (See Table 1.1.)

TABLE 1.1 Melting Points and Boiling Points of Some Pure Substances

Substance	Melting Point (°C)	Boiling Point (°C)
ammonia	−77.7	−33.4
carbon monoxide	−199	−191.5
copper	1083	2567
chlorine	−101.0	−34.5
chloroform	−63.5	61.7
methane	−182.5	−164
octane	−56.8	125.7
salt	801	1413

Chemical Properties

Sodium dropped into water reacts violently, releasing hydrogen gas and heat that causes ignition of the hydrogen. (Richard Megna, Fundamental Photographs)

Chemical properties are descriptions of the changes that a substance undergoes in chemical reactions. For example, water reacts violently with sodium metal to produce hydrogen gas, which may ignite, and a caustic substance known as sodium hydroxide. Chlorine reacts with sodium to form sodium chloride, known as table salt. In both examples, the products have different compositions than the reactants.

Let us compare a light bulb and a flashcube, which was once used in photography but has been largely replaced by electronic flash. When a current is passed into the fine wires of magnesium metal in a gaseous atmosphere of oxygen in the flashcube, a flash results. In this a physical change similar to that of turning on a light bulb? The answer is no. That side of the cube will not flash again, and the contents of the cube are changed. A white powdery substance known as magnesium oxide has formed, and its properties are very different from those of the original materials. This is a chemical change, or *reaction,* because the composition or structure of one or more substances is altered.

In the reaction within a flashcube, magnesium and oxygen are the reactants, and magnesium oxide is the product. The word equation is

$$\text{magnesium} + \text{oxygen} \longrightarrow \text{magnesium oxide}$$

EXAMPLE 1.5

Methane gas burns in oxygen and produces carbon dioxide and water. Is this process a physical change or a chemical change?

Solution

The substances change composition; therefore the process is a chemical change. This chemical change is represented by an equation in which the methane and oxygen are reactants, and the carbon dioxide and water are products.

$$\text{methane} + \text{oxygen} \longrightarrow \text{carbon dioxide} + \text{water}$$

PROBLEM 1.5

Electrolysis of water produces hydrogen and oxygen gases. Is this process a physical change or a chemical change?

Mass Is Conserved in Chemical Changes

It has been shown experimentally that the products of a chemical reaction have the same total mass as the sum of the masses of the starting materials. Based on these observations, the **law of conservation of mass** states that matter can neither be created or destroyed. Consider our garbage, which has to go somewhere in some form. We know that its total mass is unchanged no matter what physical process is used to shred or compact it. The fact that the law of conservation of mass applies to chemical changes used to dispose of garbage is less obvious. After all, we burn garbage in a trash-burning power plant and it disappears. Unfortunately, this is not actually the case. The mass of our garbage is still there in the gaseous products, which are not always totally innocuous. Even carbon dioxide can be a problem, as it is one of the "greenhouse gases" that may be contributing to the warming of the earth.

1.7 Mixtures and Pure Substances

Upon closer examination of Half Dome in Yosemite, you could see the components of the mixture making up its edifice. (Tony Stone Images)

Most of the things that you see around you contain two or more different substances. Granite contains the minerals quartz, feldspar, and biotite. Gasoline is a mixture of at least 20 liquids. All foods consist of an incredible number of components. The human body contains hundreds of thousands of substances.

Some mixtures are uniform, and you cannot see the components. Other mixtures have uneven textures, and the components are easily seen. Mixtures are classified according to their uniformity into two categories, homogeneous mixtures and heterogeneous mixtures. **Homogeneous mixtures** have the same physical and chemical properties throughout. A homogeneous mixture is uniform—that is, it has no internal boundaries. Automobile radiator antifreeze is a homogeneous mixture; it contains ethylene glycol and various rust inhibitors and hose conditioners.

Heterogeneous mixtures have physical and chemical properties that are not uniform throughout the sample. Granite is not uniform; it has boundaries of several minerals. You can see that it is a heterogeneous mixture. A pepperoni pizza is also obviously a heterogeneous mixture.

Homogeneous mixtures are called **solutions.** Solutions are possible in all three states. The air we breathe is a gaseous solution. Salt in water is a liquid solution. An alloy such as brass, which contains zinc and copper, is a solid solution. We will discuss solutions further in Chapter 9.

All mixtures have variable compositions. For example, a salt–water mixture might contain a pinch of salt or a teaspoonful of salt. A pepperoni pizza could have a few slices of pepperoni or many slices. A summary of the characteristics of homogeneous and heterogeneous mixtures is given in Figure 1.6.

Separation of Mixtures

A mixture can be separated into pure substances by physical methods that do not change the identity of the components. Many procedures have been developed to separate mixtures by taking advantage of the differences in the physical properties of matter. For example, liquids have different boiling points. The process of **distillation,** a method used to separate liquids, is based on the differences in boiling points of the liquids. When a mixture of liquids is heated, the substance with the lowest boiling point is converted into a gas and "boils" away from the mixture. This method is used to separate various fractions of petroleum into commercial products such as automobile fuel, jet fuel, domestic heating oil, and lubricant oil.

FIGURE 1.6 The Classification of Mixtures

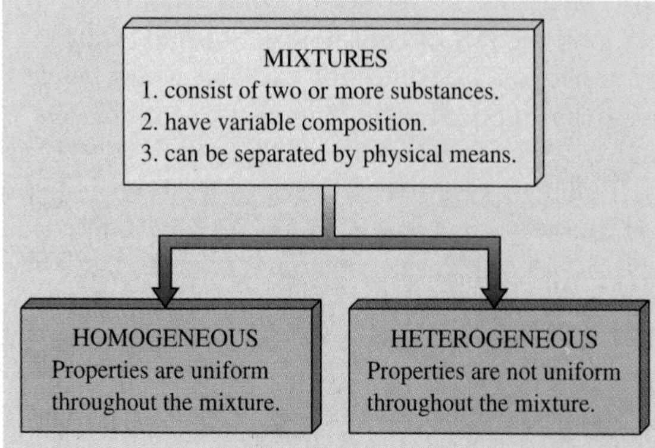

Pure Substances

When a mixture is completely separated into its individual components by a physical process, pure substances result. **Pure substances** have uniform properties, have a definite composition, and cannot be further separated into other components by physical methods. Chemists have isolated about 10 million pure substances, but few pure substances are in common use. Copper used in electrical wiring is pure; however, steel is a mixture of iron and other substances. Sugar is pure, but table salt has additives such as potassium iodide to help prevent goiter and magnesium chloride to prevent caking.

Each pure substance has its own set of physical and chemical properties. For example, helium is a colorless and odorless gas that does not burn in air and does not react with other substances. Ethyl alcohol is a liquid that mixes with water to form solutions and will burn in air. Table salt (sodium chloride) is a white solid that mixes with water to form solutions and will not burn in air.

There are two classes of pure substances: *elements* and *compounds.* Elements form the smaller class; of the 10 million pure substances, only 110 are elements. At one time an element was defined as a substance that cannot be made from or decomposed into simpler substances. This definition was modified when it became known that elements are composed of electrons, protons, and neutrons (Chapter 3). **Elements** are now defined as substances that cannot be constructed from or decomposed into simpler substances by ordinary chemical processes.

A **compound** is composed of two or more elements in fixed proportion by mass. A compound can be decomposed into other compounds or elements only by chemical reactions. Why is a compound different from a homogeneous mixture? Why can't a compound be separated into elements by physical means? Compounds are composed of elements joined together by bonds, which are not destroyed by physical changes; only chemical reactions affect bonds. In a mixture, the components are simply mixed together and are not bonded to one another. A summary of the characteristics of homogeneous matter is given in Figure 1.7.

FIGURE 1.7 The Classification of Homogeneous Matter

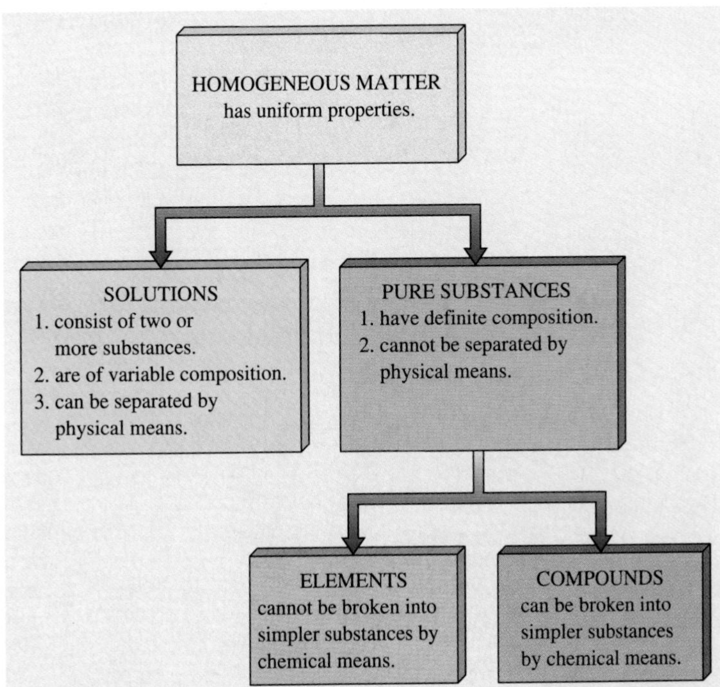

1.8 Elements

A **chemical symbol** is a chemical shorthand for the name of the element. Chemical symbols for most elements consist of one or two letters, but some elements obtained by nuclear reactions were temporarily assigned symbols consisting of three letters. When more than one letter is used, only the first is capitalized. The symbols for 10 of the elements are derived from Latin names (Table 1.2).

The first letter of the name is used for only 12 of the elements. The majority of elements have symbols consisting of two letters. The symbols of many elements consist of the first two letters of the name or the first and third letters of the name. However, there are also symbols in which some other letter is used. Consider carbon, calcium, cadmium, and californium, whose symbols are C, Ca, Cd, and Cf, respectively. A variety of letters are needed for the symbols for these four elements because their first two letters are identical. The symbols of some elements are given in Table 1.3. All the elements and their symbols are listed inside the cover of this book.

TABLE 1.2 Symbols for Elements Derived from Latin Names

Present Name	Symbol	Latin Name
antimony	Sb	stibium
copper	Cu	cuprum
gold	Au	aurum
iron	Fe	ferrum
lead	Pb	plumbum
mercury	Hg	hydrargyrum
potassium	K	kalium
silver	Ag	argentum
sodium	Na	natrium
tin	Sn	stannum

TABLE 1.3 Symbols for Selected Elements

First Letter		First and Second Letter		First and Third Letter		First and Some Other Letter	
H	hydrogen	He	helium	Cl	chlorine	Cm	curium
C	carbon	Li	lithium	Mg	magnesium	Pd	palladium
N	nitrogen	Ca	calcium	Mn	manganese	Pt	platinum
O	oxygen	Al	aluminum	As	arsenic	Nb	niobium
F	fluorine	Br	bromine	Cr	chromium	Rn	radon
P	phosphorus	Ni	nickel	Rb	rubidium		
S	sulfur	Si	silicon	Zn	zinc		

The Periodic Table

You have probably noticed a **periodic table** on the classroom wall or on the inside cover of this book. It contains the symbols for all the elements and some numbers. The periodic table (Figure 1.8) lines up the elements in horizontal rows called **periods** and vertical columns called **groups.** If you "read" the periodic table from left to right and from top to bottom as you would read a page, you will encounter the elements in approximate order of increasing weight of the atoms. This property, called the atomic weight, is presented in Chapter 3. Although the form of the table may not make much sense to you at this time, the elements are arranged in a very logical order. The most important feature is the similarity of properties of elements within a group. The arrangement of the elements in the periodic table will be discussed in detail in Chapter 3. You will discover that the periodic table summarizes and correlates information about the elements and may be used to predict their physical properties and chemical reactions.

Metals and Nonmetals

Various elements have long been classified as either metals or nonmetals. About three-fourths of the elements are metals. With the exception of mercury, metals are solids at room temperature. The metals are located in the left region of the periodic table.

About half of the nonmetals are gases, such as chlorine, nitrogen, and oxygen. Argon, helium, krypton, neon, radon, and xenon—a group of very unreactive gases known as the **noble gases**—are also nonmetals. The only nonmetal that is a liquid at room temperature is bromine. A few solid nonmetals, such as phosphorus and sulfur, have low melting points. Note that the nonmetals are located on the right side and toward the top of the periodic table.

Metals as a class are more dense than nonmetals. In addition, metals are good conductors of heat and electricity. Many metals are ductile (can be drawn into

FIGURE 1.8
The Periodic Table
The atomic weights are given with the symbol of the atom in the periodic table.

Period	IA	IIA	IIIB	IVB	VB	VIB	VIIB	VIIIB	VIIIB	VIIIB	IB	IIB	IIIA	IVA	VA	VIA	VIIA	VIIIA
							transition elements											
1	H 1.0079																	He 4.003
2	Li 6.941	Be 9.012											B 10.81	C 12.011	N 14.007	O 15.999	F 18.998	Ne 20.179
3	Na 22.990	Mg 24.305											Al 26.982	Si 28.086	P 30.974	S 32.06	Cl 35.453	Ar 39.948
4	K 39.098	Ca 40.08	Sc 44.956	Ti 47.88	V 50.942	Cr 51.996	Mn 54.938	Fe 55.847	Co 58.933	Ni 58.69	Cu 63.546	Zn 65.38	Ga 69.72	Ge 72.59	As 74.922	Se 78.96	Br 79.904	Kr 83.80
5	Rb 85.4678	Sr 87.62	Y 88.906	Zr 91.22	Nb 92.906	Mo 95.94	Tc (98)	Ru 101.07	Rh 102.906	Pd 106.42	Ag 107.868	Cd 112.41	In 114.82	Sn 118.69	Sb 121.75	Te 127.60	I 126.904	Xe 131.29
6	Cs 132.905	Ba 137.3	La* 138.906	Hf 178.49	Ta 180.948	W 183.85	Re 186.207	Os 190.2	Ir 192.22	Pt 195.08	Au 196.966	Hg 200.59	Tl 204.383	Pb 207.2	Bi 208.980	Po (209)	At (210)	Rn (222)
7	Fr (223)	Ra 226.025	Ac** 227.028	Rf (261)	Ha (262)	Sg (263)	Ns (262)	Hs (262)	Mt (266)									

lanthanides *	Ce 140.12	Pr 140.9077	Nd 144.24	Pm (145)	Sm 150.36	Eu 151.96	Gd 157.25	Tb 158.925	Dy 162.50	Ho 164.930	Er 167.26	Tm 168.934	Yb 173.04	Lu 174.967
actinides **	Th 232.0381	Pa 231.0359	U 238.02	Np 237.0482	Pu (244)	Am (243)	Cm (247)	Bk (247)	Cf (251)	Es (252)	Fm (257)	Md (258)	No (259)	Lr (260)

wires) and malleable (can be rolled into sheets). These are physical properties; the chemical differences between metals and nonmetals will be examined in later chapters.

There are a few elements whose properties are between those of metals and nonmetals. These elements, known as **semimetals** or metalloids, are antimony, arsenic, boron, germanium, silicon, and tellurium. Their physical properties are closer to those of metals, whereas their chemical properties are closer to those of nonmetals. Semimetals are used in semiconductors.

1.9 Energy Is a Property

Unlike matter, energy cannot be seen, held, collected, or stored in a bottle. However, the effect that energy has on objects can be observed. **Energy** is the ability to do work. Thus, energy makes things happen. There are many forms of energy, each of which can be converted to other forms. For example, electrical energy can be converted into heat energy in a heater or into mechanical energy in an electric motor. In this process none of the energy is destroyed; it only changes form. The **law of conservation of energy** states that energy can be converted from one form to another but can be neither created nor destroyed.

In this text we will discuss energy changes in matter as it undergoes physical changes such as changes of state as well as energy changes in chemical reactions. Chemical reactions release or absorb energy. The energy required or produced by chemical reactions may be electrical, heat, or light energy or a combination of these forms of energy. A reaction that releases heat energy is **exothermic,** whereas a reaction in which heat energy is required is **endothermic.**

The water going through the McNary dam on the Columbia River in Washington is losing some of its potential energy. (Photo Researchers, Inc.)

Where does the energy come from in an exothermic reaction? Where does the energy go in an endothermic reaction? The answer is that the heat energy is obtained from or converted into chemical energy. **Chemical energy** is the energy stored in substances. Chemical energy is a form of potential energy much like the energy stored in water behind a dam. Energy is released as water passes over the dam or through the turbines in the dam. Energy would be required to pump the water from the bottom of the dam back up behind the dam.

Important changes in chemical energy occur both in the photosynthesis that takes place in plants and in the metabolic processes that take place in animals. Green plants store chemical energy in chemicals formed from carbon dioxide and water. The energy needed for the chemical processes that form these compounds is obtained from the radiant energy of light. When the plants are metabolized by animals, this energy is released. No energy has been created or destroyed; the energy has only changed form. You will see many times in this text that the chemical energy stored in various substances determines what reactions they undergo.

1.10 Spontaneity in Chemistry

You hear about artistic performances, athletic feats, or political speeches that are greeted with spontaneous applause. The word *spontaneous* is defined as resulting from a natural impulse or tendency without effort. In natural phenomena it is defined as arising from internal forces or tendencies. In chemistry, a spontaneous process proceeds on its own without any external influence once it starts. In the

Even though quite slow, the rusting of iron is a spontaneous process. (Diana Gongora, Fundamental Photographs)

case of chemical reactions, both the rusting of iron and the combustion of natural gas are spontaneous. Spontaneity is not related to the speed of a reaction.

Photosynthesis, which fortunately does happen, is not spontaneous, because it requires a continued external influence—the sun. If we put a plant in the dark, photosynthesis ceases. Why are reactions such as the combustion of natural gas, which continues after an ignition, spontaneous, while other reactions, such as photosynthesis, are nonspontaneous? One conclusion is that exothermic reactions should be spontaneous because they release chemical energy much like water releases potential energy as it flows downhill. Endothermic reactions require energy similar to the energy required to pump water uphill. Turn off the pump, and no water flows uphill. In general, a process that releases energy is spontaneous, and one that requires energy is nonspontaneous, but not always!

There are spontaneous physical processes that don't release energy. The odor of deodorant or perfume spreads out some distance in the vicinity of a person. These examples of the spontaneous distribution of gases illustrate an important natural phenomenon—the tendency toward disorder. The greater the randomness or disorder of a process, the more likely the process will occur without any external influence. The amount of disorder associated with a material is called its **entropy.** A reaction will tend to be spontaneous if there is more disorder in the products than in the reactants.

The disorder of the maple leaves on a pond in the autumn is the result of a spontaneous process in which entropy increases. (Tony Stone Images)

We have now identified the two factors that control a chemical reaction: the energy change, called the heat of reaction, and the entropy change of the reaction. Clearly, an exothermic process that results in increased disorder will be spontaneous. One example is the combustion of gasoline, which contains large molecules with many carbon–carbon and carbon–hydrogen bonds. Not only is the reaction exothermic but it leads to an increase in disorder: The highly ordered reactant molecules produce many small carbon dioxide and water product molecules. In contrast, photosynthesis is nonspontaneous and it occurs only because radiant energy of the sun "pushes" the reaction uphill. Not only is the reaction endothermic but it results in an increase in order: The many small carbon dioxide and water molecules react to give much larger ordered molecules with many bonds.

The two factors that we have identified don't always operate in the same direction and reinforce each other. What about an exothermic process that leads to increased order or an endothermic process that leads to increased disorder? Is either process spontaneous? The answer is that it depends on the magnitude of the two quantities. We will return to this subject later. For now it is sufficient to state that a combination of the energy contained in a substance, called its *enthalpy,* with its entropy is defined as its *free energy.* Processes that lead to lower free energy are spontaneous regardless of the independent contributions of enthalpy or entropy.

We will qualitatively use free energy to help us understand why physical processes and chemical reactions occur. Changes in the free energy of biochemical reactants are extensively discussed later in the text when we discuss the metabolism of carbohydrates, fats, and proteins.

Summary

Science is a method of repeated, verified observations used to gain knowledge about our universe. Based on observation, the materials of the universe are classified according to characteristics or **properties.**

Science is based on **qualitative** and **quantitative** observations obtained by experiment. **Laws** are summaries of all currently known data. The laws are used to devise models for what may be responsible for a natural phenomenon. **Models** are ideas that can be represented as mathematical equations or pictures. A tentative model used to explain a natural phenomenon is a **hypothesis,** whereas a model that is well established is a **theory.**

Chemistry is the study of matter, its composition, structure, and reactions. Every material or object that we observe consists of matter that has **mass** and occupies space. **Composition** is the identity and amount of the components of a **substance,** and **structure** is the arrangement of the components. Composition and structure of matter at the **submicroscopic** level are described in terms of atoms and molecules. Molecules consist of atoms held together by **bonds.** The compositions of molecules are represented using **molecular formulas.** Two-dimensional representations of atoms and bonds are **molecular structures. Reactions** involve changes in the composition and structure of matter and are represented by **chemical equations.** In an equation, the **reactants** are placed to the left of the reaction arrow, and the **products** are placed to the right.

The three **states** of matter are gas, liquid, and solid. **Gases** do not have definite volume or shape but expand to fill the available space. **Liquids** have definite volume but take the shape of their container. **Solids** have definite volume and shape.

Every pure substance is identified by a set of **physical properties** such as its **boiling point** and **melting point (freezing point). Physical changes,** such as a change of state, do not alter the composition of a substance. **Chemical changes** involve changes in composition in a chemical reaction.

Most matter exists in complex **mixtures** that may be **homogeneous** or **heterogeneous.** Mixtures have variable composition. Pure substances can be separated from mixtures by physical means. **Pure substances** have a definite composition; their components cannot be separated by physical means. The two types of pure substances are **elements** and **compounds.**

Each element is represented by a shorthand notation called a **chemical symbol.** Elements are arranged in a **periodic table** that places similar elements in **groups.** Metals and nonmetals are the two major categories of elements.

Energy can be converted from one form to another but can be neither created nor destroyed. The energy contained in a substance is called **chemical energy.** Chemical reactions may be **exothermic** or **endothermic. Spontaneous reactions** either are exothermic or lead to increased disorder of the atoms in the structures of the products compared with those of the reactants.

Exercises

Methods of Science

1.1 A student states that birds are animals that lay eggs. Comment on the validity of the statement.

1.2 A student states that birds are animals that fly. Comment on the validity of the statement.

1.3 You are visiting a friend at his apartment and decide to turn on the television set while he is out of the room. You push the on button but the screen remains dark. Describe a scientific way to approach this problem.

1.4 You live in an apartment with three other students. One morning, you discover that your calculator is missing. Describe a scientific way to approach this problem.

1.5 Explain the reason for repeating experiments several times before developing a law.

1.6 How many exceptions to the results predicted by a theory are allowed before the theory must be replaced or modified?

1.7 Why is it important that scientific models be published in widely read scientific journals or be presented at scientific meetings?

1.8 It has been said that science progresses most rapidly when scientists can suggest several hypotheses to explain a natural phenomenon. Explain why this statement is valid.

Composition of Molecules

1.9 Propane ("bottled gas") is used in camp stoves and outdoor grills. It consists of molecules containing three carbon atoms and eight hydrogen atoms. Placing carbon before hydrogen, write the molecular formula for propane.

1.10 Hydrazine, used as a fuel in early rockets, consists of molecules containing two nitrogen atoms and four hydrogen atoms. Placing nitrogen before hydrogen, write the molecular formula for hydrazine.

1.11 Ethylene glycol, a substance used in antifreeze, consists of molecules containing two carbon atoms, six hydrogen atoms, and two oxygen atoms. Placing the atoms in the order carbon, hydrogen, and oxygen, write the molecular formula for ethylene glycol.

1.12 Formaldehyde, a substance formerly used to preserve biological specimens, consists of molecules containing one carbon atom, two hydrogen atoms, and one oxygen atom. Placing the atoms in the order carbon, hydrogen, and oxygen, write the molecular formula for formaldehyde.

1.13 When sulfur is burned, sulfur dioxide, represented by SO_2, is formed. What does this symbol mean?

1.14 One of the products of metabolism is represented by CO_2. What does this symbol mean?

1.15 DNA, the molecule that carries the genetic code, contains deoxyribose, whose molecular formula is $C_5H_{10}O_4$. What is the total number of atoms represented by the molecular formula?

1.16 Lysine, one of the essential amino acids required in our diet, has the molecular formula $C_6H_{14}N_2O_2$. What is the total number of atoms represented by the molecular formula?

1.17 A molecule of one of the penicillins has 16 carbon atoms, 18 hydrogen atoms, 2 nitrogen atoms, 5 oxygen atoms, and 1 sulfur atom. Write the molecular formula, keeping the atoms in the order listed.

1.18 The vitamin folic acid, which is essential in the early months of pregnancy, has 19 carbon atoms, 19 hydrogen atoms, 7 nitrogen atoms, and 5 oxygen atoms. Write the molecular formula, keeping the atoms in the order listed.

Molecular Structure

1.19 Hydrogen sulfide, the foul-smelling gas of rotten eggs, consists of molecules made up of two hydrogen atoms and one sulfur atom. The molecule consists of a central sulfur atom bonded to two hydrogen atoms. The two hydrogen atoms are not bonded to each other. Draw the structure of the molecule.

1.20 Phosphine consists of molecules made up of three hydrogen atoms and one phosphorus atom. The molecule consists of a central phosphorus atom bonded to three hydrogen atoms. The three hydrogen atoms are not bonded to each other. Draw the structure of the molecule.

1.21 A hydrazine molecule consists of two nitrogen atoms and four hydrogen atoms. Write a structure for the molecule.

1.22 A hydrogen peroxide molecule consists of two oxygen atoms and two hydrogen atoms. Write a structure for the molecule.

1.23 Examine the following structure for dimethyl ether and compare it with the structure of ethyl alcohol given in Figure 1.4. What can be said about the composition of the two substances? What can be said about the structures of the two substances?

```
    H     H
    |     |
H—C—O—C—H
    |     |
    H     H
```

1.24 Ethyl mercaptan, a strong-smelling compound added to natural gas to make leaks easily detectable, consists of molecules containing two carbon atoms, six hydrogen atoms, and one sulfur atom. Sulfur, like oxygen, can form two bonds. The structure of ethyl mercaptan resembles the structure of ethyl alcohol (Figure 1.4). Draw the structure of ethyl mercaptan.

1.25 Using the color code for atoms (Figure 1.3), determine the molecular formula of each of the following representations. Write a two-dimensional molecular structure for each.

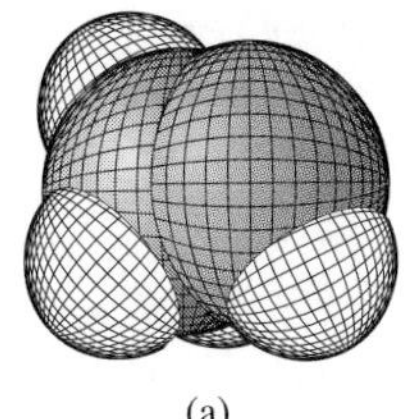
(a)

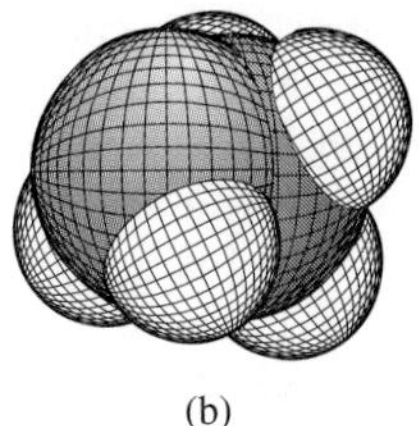
(b)

1.26 Using the color code for atoms (Figure 1.3), determine the molecular formula of each of the following representations. Write a two-dimensional molecular structure for each.

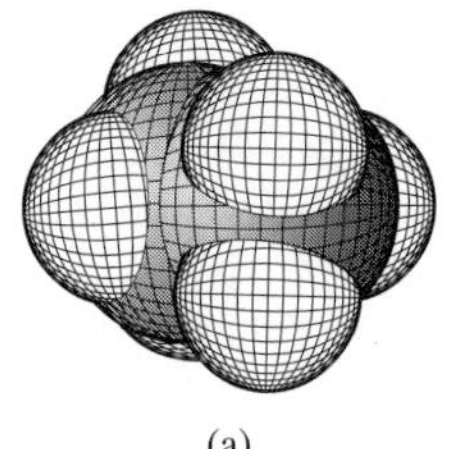
(a)

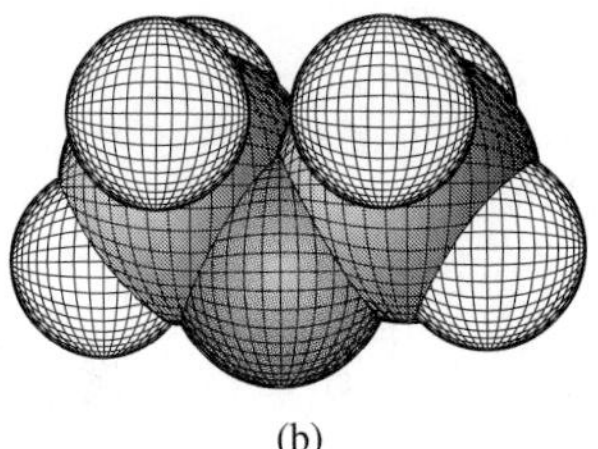
(b)

Chemical Reactions

1.27 Sulfur burns in oxygen to form sulfur dioxide. Write a word equation for the reaction.

1.28 At the high temperatures in an automobile engine, nitrogen combines with oxygen to form nitric oxide. Write a word equation for the reaction.

1.29 Hydrogen peroxide, which is used to cleanse cuts and abrasions, decomposes to give water and oxygen, as given by the following equation. Which substances are reactants? Which are products?

$$\text{hydrogen peroxide} \longrightarrow \text{water} + \text{oxygen}$$

1.30 Water can be decomposed by passing an electric current through it. The reaction is given by the following equation. Which substances are reactants? Which are products?

$$\text{water} \longrightarrow \text{oxygen} + \text{hydrogen}$$

States of Matter

1.31 Nicotine has a melting point of −79 °C and a boiling point of 247 °C. In what state does nicotine exist at 25 °C?

1.32 Methanol, a poisonous alcohol, has a melting point of −93.9 °C and a boiling point of 65.0 °C. In what state does methanol exist at 85 °C?

1.33 Benzene has a melting point of 5.5 °C and a boiling point of 80.1 °C. In what state does benzene exist at 0 °C?

1.34 Isopropyl alcohol, used in water solution as rubbing alcohol, has a melting point of −89.5 °C and a boiling point of 82.4 °C. In what state does isopropyl alcohol exist at −95 °C?

Chemical and Physical Properties

1.35 Indicate whether each process is a physical change or a chemical change.

(a) chopping of wood (b) boiling of water
(c) rusting of iron (d) melting of wax
(e) digesting food (f) distilling moonshine

1.36 Indicate whether each process is a physical change or a chemical change.

(a) photosynthesis (b) burning of gasoline
(c) burning of toast (d) boiling of alcohol
(e) burning of wood (f) melting of iron

Mixtures and Pure Substances

1.37 Some people claim that they avoid foods that have "chemicals" in them. Is this really possible? Explain.

1.38 Is tap water a pure substance or a mixture? How could you support your answer experimentally?

The Periodic Table

1.39 Copper, silver, and gold have been called coinage metals because they are used to manufacture metal coins. Find their chemical symbols and locate them in the periodic table. What relationship exists between these metals based on their location in the periodic table?

1.40 What are the symbols and names of the vertical and horizontal neighbors of tin in the periodic table? Which of these are likely to have chemical properties similar to those of tin?

2

Measurements

The distance jumped by this athlete at a track meet must be measured. (Tony Stone Worldwide)

2.1 What Is a Measurement?
2.2 How Reliable Are Measurements?
2.3 Significant Figures in Measurements
2.4 Expressing Large and Small Numbers
2.5 Using Units in Problem Solving
2.6 The Metric System
2.7 Temperature Units
2.8 Extensive and Intensive Properties

Essays
- *The Scale of Our Universe*
- *Homeostatis, Hyperthermia, and Hypothermia*
- *Calories and Basal Metabolic Rate*

2.1 What Is a Measurement?

Some of the measurements that you make are qualitative, as when you use your hands to describe the size of the fish that got away. However, there are times when you have to use more quantitative measurements. Giving information to a friend who wants to bake a cake like you served should be quantitative, or your friend won't be very happy when the cake isn't as good as yours. On a more serious note, it is critical that the dosage of medicine for a sick child be correctly stated to avoid the extremes of either not curing the patient or killing the patient.

In this chapter we focus on how to make measurements, report them, and use them in calculations. If you stop to think about it, you make measurements every day. You may determine your weight on a scale in the morning. In your car, you may glance at the odometer, which gives the distance that you have traveled, or at the gas gauge, which indicates the volume of gasoline remaining in the gas tank. Looking out the window, you may spot a digital sign with the current temperature.

All measurements that you and scientists make have a common feature. A **measurement** is made by comparison to a scale provided by a standard measuring device. For example, the weight of hamburger in a supermarket is measured on a scale, in pounds. The height of a high jump at a track meet is measured with a tape, in feet. The volume of gasoline you put in your car at a gas station is measured by the pump, in gallons. Your body temperature is measured with a thermometer, in degrees Fahrenheit.

All measurements require both numbers and units. For example, to gain the 10 yards required for a first down in a football game, a team must move the ball a distance that is 10 times the length of a standard called a yard. For every measurement, regardless of whether it is for a scientific experiment or for an everyday event, both the size of the measured quantity and the unit used are stated. If a garage attendant tells you that your car engine is down 2 quarts, the unit is important, because 2 pints or 2 fluid ounces are very different quantities. In this case the unit commonly used is quarts, but in an emergency room the unit of blood volume is pints. A patient might have lost 5 pints of blood but it is hoped not 5 quarts, because that is about all that an average person has.

This sign on the north side of Columbus, Ohio, tells us the distance to Cleveland, Ohio, in English and metric units. (Stock Boston)

Quarts, yards, and pounds are part of the English system of measurement, which is still used in everyday life in the United States. In chemistry, metric units are used to measure matter. The choice of the metric system over the English system by scientists as well as by most of the rest of the world is based on the convenient multiple units and subunits of the metric system. In the metric system the units differ from each other by powers of 10 and are interconverted by moving the decimal point of the number associated with them. We will outline these features of the metric system of measurements in Section 2.6.

Temperature, a measure of the degree of hotness of a material, is given in Celsius degrees or in kelvins in scientific use. However, citizens of the United States are more familiar with the Fahrenheit scale. These three temperature scales are discussed in Section 2.7.

2.2 How Reliable Are Measurements?

Any measurement is uncertain, and the magnitude of that uncertainty is important. In a football game the outcome may depend on how 10 yards is measured. The players rely on how the officials place the football and set the chain on the sidelines as well as whether the chain is fully extended.

Scientists are concerned with the accuracy and precision of measurements. **Precision** means the degree of reproducibility of a measurement. **Accuracy** means the degree to which a measurement represents the true value of what is measured.

The measurement of distance is critical to both teams as well as their fans. (PhotoEdit)

The precision of a measurement is determined by repeated measurements. For example, let's say you determine your temperature three times in succession and the readings are 98.6, 98.5, and 98.7 °F. Your temperature is best represented by the average of these readings, 98.6 °F. The temperature readings differ from this average value by about 0.1 °F, which is the precision of the measurement.

Accuracy depends on the quality of the measuring device. All measuring devices have limited accuracy. For example, some watches have a second hand, whereas others have only a minute hand. You need a second hand to determine your pulse rate accurately. The minute hand, however, is accurate enough for you to get to your chemistry class on time.

If a thermometer is inaccurate, you cannot accurately determine your temperature. If the thermometer always reads 1.0 °F low and reads 98.6 °F, then your temperature is actually 99.6 °F. The measured value differs from the true value by 1.0 °F and is inaccurate by that amount. Note that the accuracy of the measured value cannot be improved by repeated measurements. Repeated measurements can increase only the precision of the measured value.

2.3 Significant Figures in Measurements

If you determine that there are 26 students in a class, there is no uncertainty in the number 26 because it is the result of a direct count; it is an exact number. There also is no uncertainty in the number of feet in 1 yard; it is exactly 3 feet. **Exact numbers** arise from either a direct count or a defined equivalence such as 1 yd = 3 ft.

Scientists use measured numbers, and perfect measurements are not possible. When reporting measured quantities, a scientist must indicate their reliability. The number of digits reported in a measured number that give reliable information is the number of significant figures. The **significant figures** in a measured number include all the certain numbers plus the next digit to the right which is estimated.

What parts of a measurement are certain, and what parts are estimated? Consider the mass of an object determined using a balance. The indicator arrow is between 7 and 8 grams (g) but is closer to 8 g (Figure 2.1a). Divide the space between the marked values into 10 parts. Then estimate that the indicator arrow is $\frac{8}{10}$ of the way toward 8 g. The mass is 7.8 g; the quantity has one certain figure and one estimated figure, or a total of two significant figures. If the object is placed on a more accurate balance (Figure 2.1b), the mass is seen to be 7.8 g. However, the mass can now be estimated to the nearest hundredth of a gram as 7.82 g. In this case there are three significant figures, and the 2 is the estimated figure.

The reliability of any measurement is given by the number of significant figures. It is necessary that the recorded quantity has the number of significant figures that reflects the reliability of the measurement.

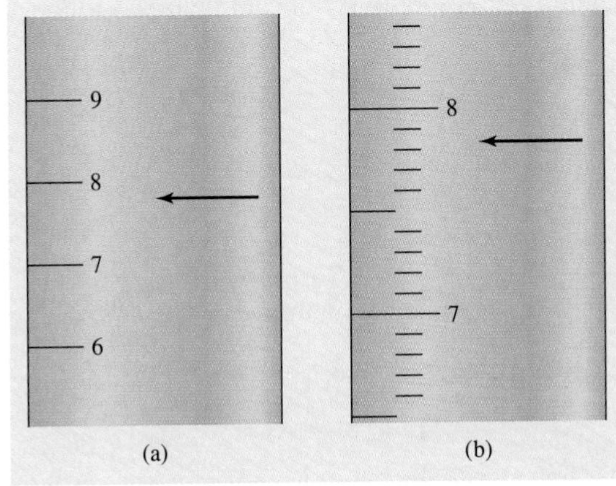

FIGURE 2.1 Significant Figures in Mass Determination

In case (a) the scale of the balance gives the mass in grams only to the nearest gram. The mass is somewhat less than 8 g and is clearly more than 7 g. The mass is estimated as 7.8 g by estimating to the nearest 0.1 g. This value of the mass is known to two significant figures. In case (b) the balance has greater accuracy. The mass is slightly more than 7.8 g and clearly less than 7.9 g. The mass is estimated as 7.82 g, a quantity with three significant figures.

Is Zero a Significant Figure?

With the exception of zero, all digits in a number that is exact or in a number properly expressing the accuracy of a measurement are significant. However, zeros in a quantity may or may not be significant figures. Zeros in measured quantities are divided into three classes:

1. **Leading zeros** are zeros that precede all the nonzero digits.
2. **Captive zeros** are zeros between nonzero digits.
3. **Trailing zeros** are zeros that are to the right of the last nonzero digit.

Suppose that different samples are weighed on a balance that gives the mass accurately to 0.001 g. If the mass of one sample is 0.015 g, the leading zeros before and after the decimal point are not significant figures. The mass of the sample is between 0.014 and 0.016 g, and the number 0.015 has only two significant figures.

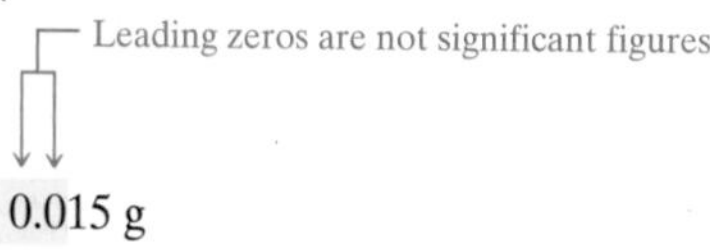

If the mass of a second sample is 2.004 g, the captive zeros "sandwiched" between the 2 and 4 are certain and are significant figures. There are four significant figures in the number, which represents a mass known to be between 2.003 and 2.005 g.

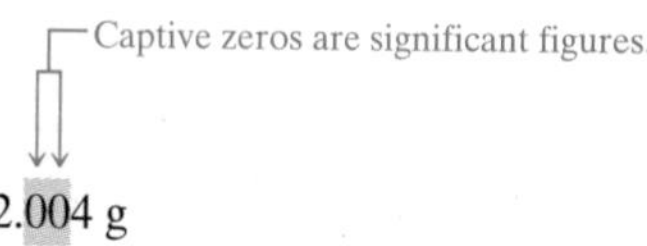

If the mass of a third sample is found to be 5.000 g, its mass is between 4.999 and 5.001 g. Thus, the three trailing zeros following the decimal point are significant figures, and the number of significant figures is four.

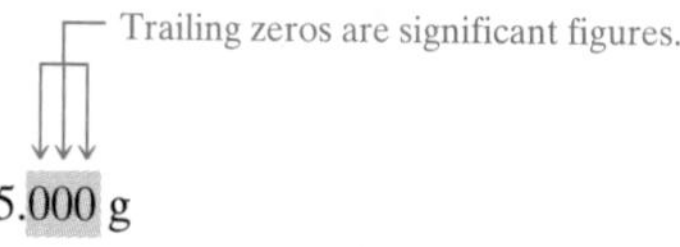

EXAMPLE 2.1

What is the number of significant figures in each of the following numbers?

a. 5041 b. 0.05401 c. 0.5401 d. 0.5410

Solution

There are four significant figures in each case. The zeros that are not significant are underlined.

a. 5041 b. 0.05401 c. 0.5401 d. 0.5410

The captive zeros in a, b, and c are significant figures. The trailing zero in d is a significant figure.

PROBLEM 2.1

What is the number of significant figures in each of the numbers 0.04010, 0.00410, and 0.000401?

Significant Figures in Calculated Quantities

How many significant figures are retained in a calculation using measured quantities? The number of figures retained depends on the number of significant figures in each measured quantity and the type of math operation used. The rules for the use of significant figures in mathematical operations as well as the procedures for rounding off numbers are discussed in Appendix A. Sample calculations are also provided. This appendix will be referred to in later chapters so that you can express calculated quantities with the correct number of significant figures. Simply stated, the rules are

1. When numbers are added or subtracted, the answer must not contain any figures to the right of the decimal that are not common to all the numbers.
2. In multiplication or division of measured numbers, the answer must not contain more significant figures than the least number of significant figures in the measurements.

2.4 Expressing Large and Small Numbers

The national debt is one example of a very large number. (Robert Brenner, PhotoEdit)

You often read about large numbers such as the number of starving people in a country or the size of the national debt, but in everyday life you seldom encounter very small numbers. Both very large and very small numbers are used in chemistry. The large numbers have a long string of trailing zeros that are not significant figures. Similarly, the small numbers have a string of leading zeros that are not significant figures. Scientists express such numbers using powers of 10.

In **scientific notation,** a number is expressed as a product of a coefficient multiplied by a power of 10. The **coefficient** is a number equal to or greater than 1 but less than 10. The power of 10 is an **exponent.**

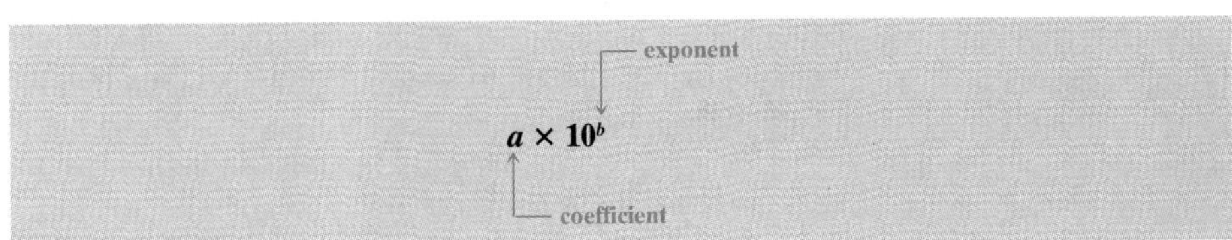

The coefficient retains only the significant figures of the original number; the exponent gives the location of the decimal point. For example, 18.0160 g of water contains 6.02217×10^{23} molecules of water. The mass of a water molecule is 2.99161×10^{-23} g. Each number has six significant figures. The positive exponent for the number of molecules indicates that the quantity is very large; the negative exponent for the mass of one molecule indicates that the quantity is very small.

Appendix A provides some examples for you to work using scientific notation. However, most calculators express small and large numbers using scientific notation, so you will not have to "move" decimal points to write the proper coefficient and exponent.

2.5 Using Units in Problem Solving

Students in chemistry are often concerned about working numerical problems that have unfamiliar units. If you pay attention to units and use conversion factors, you should have fewer difficulties. A **conversion factor** is a multiplier consisting of two or more units used to convert one quantity into a second quantity in different units. The use of conversion factors is called the **factor unit method** because the factor is numerically equivalent to 1. All calculations using conversion factors are summarized by the relationship

information given × conversion factor = information sought

which is worked out by the following steps.

1. Examine the data given, and note the units of all quantities. Determine what is asked for—that is, what units are desired.
2. Write the data given, including the units, to the left of an equal sign and the desired unknown, with its units, to the right of the equal sign.
3. Consider possible conversion factors that relate the units of the given data and the desired unknown. (There are two ways to write a conversion factor—one is the reciprocal of the other.)
4. Choose the conversion factor that eliminates the units that you do not want and gives the desired units. (If the unit to be eliminated is in the numerator, choose the conversion factor with that unit in the denominator. If the unit to be eliminated is in the denominator, choose the conversion factor with that unit in the numerator.)
5. Carry out the arithmetic and determine that the answer is mathematically reasonable.

Let's calculate the length in feet (ft) of a football field that is exactly 100 yd long. You know the answer without using the factor unit method, but the best way to learn the method is to use it with familiar quantities.

Step 1: The information given is 100 yd. The length of the field in feet is the information sought.

Step 2: Write the 100 yd to the left of the times sign and the unknown number of feet to the right of the equal sign.

$$100 \text{ yd} \times \text{factor} = ? \text{ feet}$$

Step 3: We know that 1 yd is equal to 3 ft.

$$1 \text{ yd} = 3 \text{ ft}$$

We can write a factor F_1 that is a mathematical way of saying that 3 ft equals 1 yd.

$$F_1 = \frac{3\text{ ft}}{1\text{ yd}}$$

We also can write a factor F_2 that is a mathematical way of saying that 1 yd equals 3 ft.

$$F_2 = \frac{1\text{ yd}}{3\text{ ft}}$$

Step 4: Since the length in yards is in the numerator and we are interested in feet, F_1 is the proper conversion factor.

$$\cancel{\text{yd}} \times \frac{\text{ft}}{\cancel{\text{yd}}} = \text{ft}$$

Step 5: Multiplication of 100 yd by F_1 gives an answer in feet.

$$100\text{ yd} \times F_1 = 100\text{ }\cancel{\text{yd}} \times \frac{3\text{ ft}}{1\text{ }\cancel{\text{yd}}} = 300\text{ ft}$$

The answer is reasonable because you know that a distance expressed in feet is numerically larger than the same distance expressed in yards. If you use the incorrect conversion factor, the "answer" is 0.333333 ft^2/yd. The number is clearly wrong, and the units are not what you want either.

Two or more conversion factors may be necessary to solve a problem. How many tablespoons (Tbs) are there in 1 quart (qt)? There are exactly 16 Tbs in 1 cup, and there are exactly 4 cups in 1 qt. Thus, we can first convert the volume in quarts into cups and then convert the resulting answer in cups into tablespoons.

$$1\text{ }\cancel{\text{qt}} \times \frac{4\text{ cups}}{1\text{ }\cancel{\text{qt}}} = 4\text{ cups}$$

$$4\text{ }\cancel{\text{cups}} \times \frac{16\text{ Tbs}}{1\text{ }\cancel{\text{cup}}} = 64\text{ Tbs}$$

The conversion can also be done by using one equation with two factors.

$$1\text{ }\cancel{\text{quart}} \times \frac{4\text{ }\cancel{\text{cups}}}{1\text{ }\cancel{\text{quart}}} \times \frac{16\text{ Tbs}}{1\text{ }\cancel{\text{cup}}} = 64\text{ Tbs}$$

Note that there is only one significant figure in 1 qt, but two significant figures are given in the answer. The conversation factors involve defined quantities and are exact numbers. Therefore, in exactly 1 qt there are exactly 64 Tbs.

2.6 The Metric System

Early traders had to contend with the different units of measurement that varied from country to country and in some cases within a country. Eventually this lack of uniformity prompted the French government to appoint a committee of scientists

TABLE 2.1 A Comparison of Metric and English Units

Dimensions	Metric Unit	English Unit
length	1.00 meter	39.4 inches
	2.54 centimeters	1.00 inch
	1 kilometer	0.6 mile
volume	1.00 liter	1.06 quarts
	0.946 liter	1.00 quart
mass	454 grams	1.00 pound
	1.0 kilogram	2.2 pounds

TABLE 2.2 Prefixes Used with Metric and SI Units

Prefix	Symbol	Meaning	Exponential Notation
exa	E	1,000,000,000,000,000,000	10^{18}
peta	P	1,000,000,000,000,000	10^{15}
tera	T	1,000,000,000,000	10^{12}
giga	G	1,000,000,000	10^{9}
mega	M	1,000,000	10^{6}
kilo	k	1,000	10^{3}
hecto	h	100	10^{2}
deka	da	10	10^{1}
		1	10^{0}
deci	d	0.1	10^{-1}
centi	c	0.01	10^{-2}
milli	m	0.001	10^{-3}
micro	μ	0.000 001	10^{-6}
nano	n	0.000 000 001	10^{-9}
pico	p	0.000 000 000 001	10^{-12}
femto	f	0.000 000 000 000 001	10^{-15}
atto	a	0.000 000 000 000 000 001	10^{-18}

TABLE 2.3 The Fundamental SI Units

Physical Quantity	Name of Unit	Abbreviation
mass	kilogram	kg
length	meter	m
time	second	s
temperature	kelvin	K
electric current	ampere	A
amount of substance	mole	mol
luminous intensity	candela	cd

in 1790 to propose a uniform group of measurements. About 10 years later the system that we now call the metric system was adopted by the French. Adoption of the system was delayed not by the slow work of the committee but by the French Revolution, in which even prominent scientists such as Antoine Lavoisier were guillotined. In spite of this less than propitious start, the metric system has swept the world with the notable exception of the United States. Even our neighbors to the north and south use the metric system.

The metric system is simple and convenient to use because all related units are multiples of 10 of the base unit. Conversions are easy, because only the decimal point is moved as the units are changed. The metric equivalents of some English units are given in Table 2.1.

The standard units of the metric system are the second (s) for time, the meter (m) for length, the gram (g) for mass, and the liter (L) for volume. Fractions and multiples of the standard units have prefixes to indicate the size of the unit relative to the standard unit. A list of prefixes appears in Table 2.2. The meanings and the abbreviations for the prefixes kilo, deci, centi, milli, and micro should be memorized. All conversion factors using only metric units are exact. Thus metric unit conversion factors do not affect the number of significant figures in a conversion of units problem.

In 1960 the International Bureau of Weights and Measures proposed the International System of Units, which is a revision and extension of the metric system. These units are called SI after the French name Système International (Table 2.3).

The Scale of Our Universe

As a child you described material qualitatively as small or large. Your comparison or reference was to the size of your body. Early English measurements were also made in terms of human dimensions. The inch was based on the breadth of the thumb, the foot on the length of a human foot, and a fathom on the distance between the fingertips of a man's outstretched arms. Is such an egocentric approach to measurement justified?

The approximate dimensions of a number of objects are given in the table. All distances are given in meters as well as in another metric unit.

The range of known distances is incredible. One of the tiniest objects is the proton contained in the nucleus of the atom. Molecules consisting of atoms are larger, and a human consisting of molecules is larger still. Among the largest objects are the planets, the sun, the Milky Way, and the universe itself.

Where do we fit into the scheme of things on the basis of our dimensions? Examine the table. We are somewhere in the middle—as in the story of Goldilocks, not too big and not too small. Thus, maybe it was not so unreasonable for us to use our human dimensions to describe objects as large and small.

This young boy is looking for small Easter eggs with his big sister near a very big tree. (Robert Ouellette)

Approximate Dimensions of Small and Large Objects

Radius of the universe	3×10^{25} m	3×10^{7} Em
Distance to the sun	1.5×10^{11} m	1.5×10^{-1} Tm
Radius of the sun	7×10^{8} m	7×10^{-1} Gm
Radius of the earth	6.4×10^{6} m	6.4 Mm
Height of Mt. Everest	8.8×10^{3} m	8.8 km
Average human height	1.7 m	
Radius of red blood cell	1×10^{-5} m	1×10^{-2} cm
Radius of a virus	1×10^{-7} m	1×10^{-1} μm
Radius of an atom	1×10^{-10} m	1×10^{-1} nm
Radius of a proton	1×10^{-15} m	1 fm

The SI system has fewer basic units, and all other units are derived from them. However, the SI units are often awkward to use for quantities measured in a chemistry laboratory. For example, a cubic meter is a very large volume. For most subjects covered in this text only the metric system will be used.

The meter (m) is the standard unit of length in both the metric and SI systems, but you will seldom measure length directly in meters in the laboratory. The centimeter, which equals 1/100 m, is more commonly used. The millimeter, which is 1/1000 m or 1/10 cm, may also be used.

EXAMPLE 2.2

The radius of an atom of gold is 0.144 nanometer (nm). Express this radius in centimeters.

Solution

From Table 2.2 you find that the prefix nano means the basic unit times 10^{-9}. Thus, 1 nm = 10^{-9} m, and two conversion factors can be written.

$$F_1 = \frac{1 \text{ nm}}{1 \times 10^{-9} \text{ m}} \qquad F_2 = \frac{1 \times 10^{-9} \text{ m}}{1 \text{ nm}}$$

However, we need the answer in centimeters. A centimeter is 1×10^{-2} m, and again two conversion factors can be written.

$$F_3 = \frac{1 \text{ cm}}{1 \times 10^{-2} \text{ m}} \qquad F_4 = \frac{1 \times 10^{-2} \text{ m}}{1 \text{ cm}}$$

The two factors needed to convert nanometers into centimeters are combined to cancel the units. The factor F_2 converts the given quantity into meters; the factor F_3 converts the resultant meters into centimeters.

$$0.144 \cancel{\text{nm}} \times \frac{1 \times 10^{-9} \cancel{\text{m}}}{1 \cancel{\text{nm}}} \times \frac{1 \text{ cm}}{1 \times 10^{-2} \cancel{\text{m}}} = 1.44 \times 10^{-8} \text{ cm}$$

PROBLEM 2.2

The best light microscope has a resolving power of about 0.2 μm. What is the equivalent resolving power in nanometers?

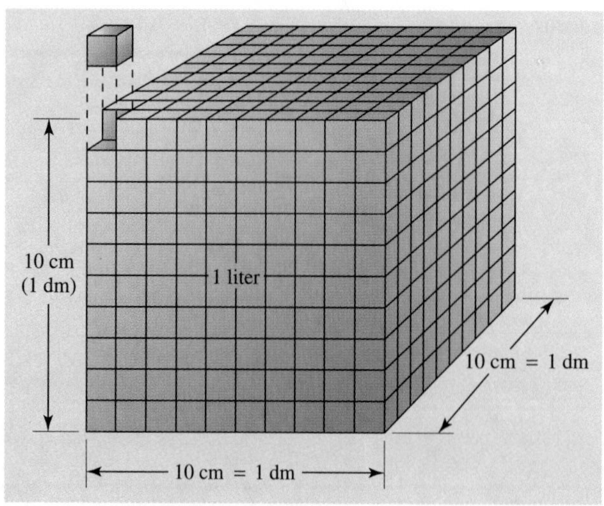

FIGURE 2.2
The Liter and Cubic Centimeters
Each small cube is 1 cm on each side; its volume is 1 cm³. There are 1000 small cubes in the large cube. The liter contains 1000 cm³. Each cubic centimeter is equivalent to 1 milliliter.

FIGURE 2.3 Laboratory Equipment Used to Measure Volume

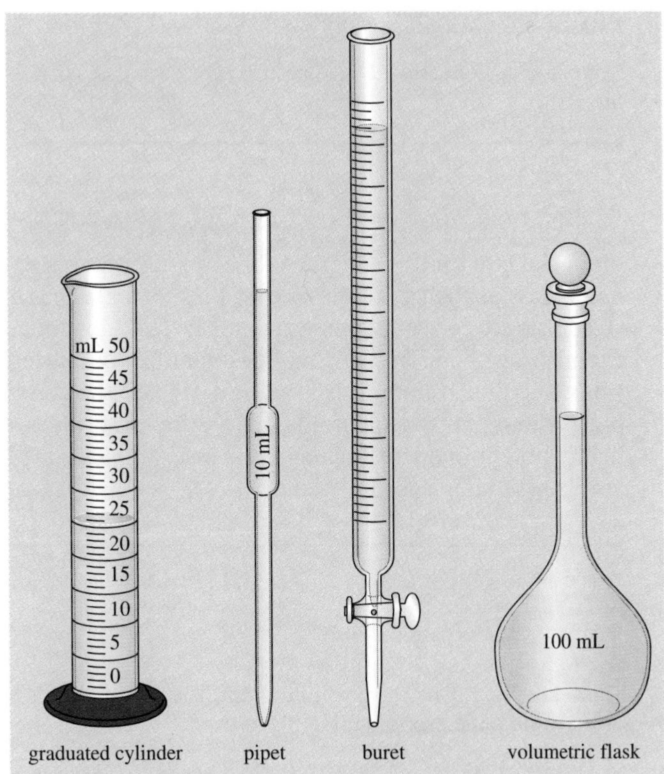

Because the cubic meter (m^3), the SI standard for volume is very large, the liter, the volume unit of the metric system, is preferred by chemists. The liter is the volume of a cube that is 10 cm (1 dm) on each side (Figure 2.2). A liter (L) then has a volume of 1000 cm^3 (1 dm^3). Fractions of a liter are expressed by using the proper prefix. For example, 1/1000 L = 1 milliliter (mL). Note that since a liter is 1000 cm^3, the milliliter and the cubic centimeter are the same volume. These units are used interchangeably.

There are several ways to measure volumes of liquids in the chemistry laboratory (Figure 2.3). The graduated cylinder is used to dispense approximate volumes in much the same way as a measuring cup is used in cooking. A pipet or buret is used to make more accurate measurements in chemistry. The pipet measures a definite amount when filled to its calibration mark. The buret is used to measure out variable amounts of liquids. A volumetric flask filled to its calibration mark contains a specified volume of liquid.

EXAMPLE 2.3

An adult may inhale 1.0×10^4 L of air per day. What is the equivalent volume in milliliters?

Solution

To convert liters into milliliters, the conversion factor must have liters in the denominator and milliliters in the numerator.

$$1.0 \times 10^4 \not{L} \times \frac{1000 \text{ mL}}{1 \not{L}} = 1.0 \times 10^7 \text{ mL}$$

The units of liters cancel, and the unit of milliliters results. Note that the number of milliliters contained in a volume is far greater than the number of liters, as required by the relative sizes of the units.

PROBLEM 2.3

Your stomach releases 2.5 L of gastric juice each day. What is the equivalent volume in milliliters?

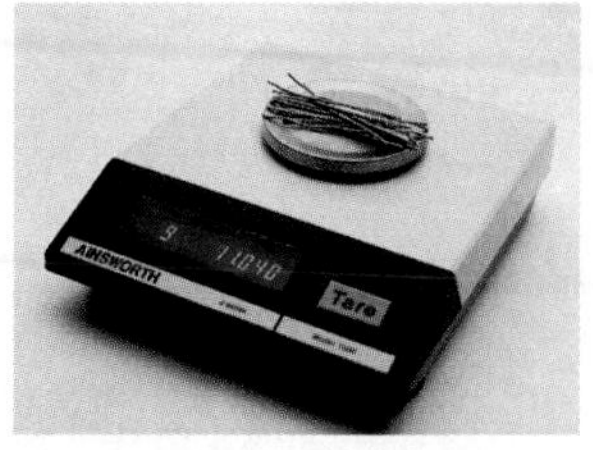

An electronic balance gives the mass of an aluminum sample in grams. (Yoav Levy, Phototake NYC)

A kilogram is the mass of 1 L of water at 4 °C. (The Celsius temperature scale is discussed later in this chapter.) Since 1 L of water is equal to 1000 mL of water and has a mass of 1000 g, it follows that 1 mL of water has a mass of 1 g.

A balance is used to measure mass. The most commonly used unit of mass in chemistry is the gram (g). Smaller quantities of matter are expressed in milligrams (mg). The milligram unit is used to designate the contents of drugs in capsules and pills. Micrograms, nanograms, and even picograms are used to measure some highly potent biological chemicals. For example, some hormone levels in the blood are detected in picogram quantities.

EXAMPLE 2.4

The average mass of a DNA molecule is 1.0×10^{-15} g. What is its mass in nanograms?

Solution

There are 1×10^9 nanograms in a gram. The conversion factor must have the units ng/g in order to yield the proper unit in the answer.

$$1.0 \times 10^{-15}\ \cancel{g} \times \frac{1 \times 10^9\ \text{ng}}{1\ \cancel{g}} = 1.0 \times 10^{-6}\ \text{ng}$$

You should also note that since a nanogram is a smaller unit of mass than a gram, the answer must be numerically larger, and so the exponent will be more positive. Thus, the conversion has been done properly.

PROBLEM 2.4

A blood sample has 65 ng of testosterone in 100 mL. How many grams of testosterone are there in 6.0 L of blood?

2.7 Temperature Units

Temperature is a measure of the degree of "hotness" of a material and does not depend on the quantity of the material. Thus, a large pot of boiling water is as hot as a small pot of boiling water. Water boils at 212°F regardless of the amount of water.

The Celsius scale is used by scientists and by most people throughout the world. The Fahrenheit scale is widely used in the United States, but the Celsius reading is now often given along with the Fahrenheit reading. For example, the digital temperature display outside a bank might tell you that the current temperature is 68 °F (degrees Fahrenheit) or 20°C (degrees Celsius). Although the units are called degrees on both scales, a Celsius degree is not the same as a Fahrenheit degree, and the zero points on the scales also differ.

The temperature of molten iron is too high to be measured by conventional thermometers. (Photo Researchers, Inc.)

On the Celsius scale, the freezing point of water is 0 °C and its boiling point is 100 °C. There are 100 equal intervals or Celsius degrees between these two points. On the Fahrenheit scale, the freezing point of water is 32 °F and its boiling point is 212°F. There are 180 Fahrenheit degrees between the freezing point and boiling point of water. Thus, a 9-degree interval on the Fahrenheit scale corresponds to a 5-degree interval on the Celsius scale. The Celsius and Fahrenheit scales are compared in Figure 2.4.

$$\frac{180 \text{ Fahrenheit degrees}}{100 \text{ Celsius degrees}} = \frac{9 \text{ Fahrenheit degrees}}{5 \text{ Celsius degrees}}$$

The mathematical expressions used to change temperatures from one scale to another are given below. Note the use of the reciprocal factors $\frac{9}{5}$ and $\frac{5}{9}$ in the equations relating °F and °C. These factors are based on the relative sizes of the intervals of the Fahrenheit and Celsius scales.

$$°\text{F} = \left(\frac{9\ °\text{F}}{5\ °\text{C}} \times °\text{C}\right) + 32.0\ °\text{F}$$

$$°\text{C} = \frac{5\ °\text{C}}{9\ °\text{F}} \times (°\text{F} - 32.0\ °\text{F})$$

Scientists also use the Kelvin scale. The units of temperature on this scale are called kelvins, not degrees. No degree sign is used for a value such as 273 K, and it is read 273 kelvin. Both the Celsius and Fahrenheit scales are arbitrarily based on the melting point of ice. The Kelvin scale is based on the lowest possible temperature—that is, absolute zero. No substance can be colder than absolute zero, which is defined as 0 K. Absolute zero is −273.15 °C. Thus, 0 K = −273.15 °C. The intervals on the Kelvin scale are identical to those on the Celsius scale, so that the Kelvin scale is related to the Celsius scale by the expressions

$$\text{K} = °\text{C} + 273.15$$
$$°\text{C} = \text{K} - 273.15$$

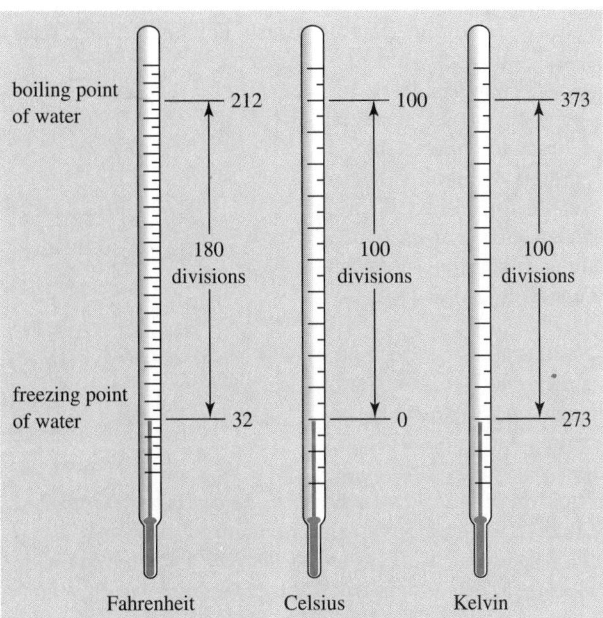

FIGURE 2.4
A Comparison of Temperature Scales
The Fahrenheit and Celsius scales have different sizes of degree intervals. The Celsius and Kelvin scales have the same size intervals but use different reference temperatures to define the scales.

The 0.15 in 273.15 is used for only the most accurate work. Problems in this text will use only 273 to convert Celsius temperature into Kelvin temperature.

EXAMPLE 2.5

Normal body temperature is 98.6 °F. What temperature is considered normal by a doctor who uses the Celsius scale?

Solution

You may substitute directly into either the first or second formula given, but the following is the more efficient.

$$°C = \frac{5\ °C}{9\ °F} \times (98.6\ °F - 32.0\ °F)$$

$$= \frac{5\ °C}{9\ °F} \times (66.6\ °F) = 37.0\ °C$$

Homeostasis, Hyperthermia, and Hypothermia

Your normal body temperature is near 37 °C (98.6 °F) and usually doesn't go below 36 °C or above 40 °C. Body temperature is regulated by a number of control systems. **Homeostasis** is a set of complex processes that the body uses to balance its internal environment while living in an external environment where the temperature can change substantially.

When it is hot, receptors in the skin signal our brain, which sends a message to the blood vessels to expand. Blood flows more readily, and internal heat is transferred to the surface, where it is released into the air. The sweat glands in the skin also respond and evaporate some water, which also causes cooling.

When it is cold, the hormone epinephrine is released, which increases our metabolism to produce heat. The skin also signals the brain to contract blood vessels, which decreases the amount of blood flow. In this way our body heat is conserved for our body core, where our vital organs are located. Some animals can regulate their blood flow, resulting in substantial differences in temperature in their inner core and their extremities.

If you are sick or the air temperature is too high, your body control mechanisms may fail to respond, and your body temperature may increase. **Hyperthermia** is a sustained high body temperature. A temperature of 41.7 °C (107 °F) is usually fatal.

Hypothermia, which is a condition of very low body temperature, is unusual but equally life-threatening. If you fall into near-freezing water, hypothermia occurs in as little as 15 minutes. Hypothermia can also occur when you are exposed to a cold wind without protective clothing or if your clothing is wet.

When the external temperature is lower than body temperature, the rate of metabolism increases to produce heat and maintain body temperature. If too much body heat is lost, the body temperature drops. A drop of 1 °C (about 2 °F) causes shivering. An additional decrease of 1–3 °C (to a body temperature between 91 and 95 °F) may result in amnesia. Muscles don't function when the body temperature drops to between 86 and 90 °F. At lower temperatures, the heart beats irregularly, and heart failure eventually results.

The runners in the bridge-to-bridge race in California must maintain their body temperature. (The Image Works)

PROBLEM 2.5

Industrial diamonds can be produced from graphite under high pressure at 3650 °C. What is this temperature in °F?

EXAMPLE 2.6

Many enzymes that are important to our well-being lose their biological effectiveness if the temperature exceeds 45 °C. What is this temperature in °F?

Solution

You can substitute directly into the following formula and solve for the required degrees Fahrenheit.

$$°F = \left(\frac{9\ °F}{5\ °C} \times 45\ °C\right) + 32.0\ °F$$

$$= 81\ °F + 32.0\ °F = 113\ °F$$

PROBLEM 2.6

Some surgical instruments are sterilized at 120 °C. What is this temperature in °F?

2.8 Extensive and Intensive Properties

Scientists divide properties of matter into two categories: extensive properties and intensive properties. **Extensive properties** depend on the amount of matter; **intensive properties** are independent of the sample size and represent qualities that identify the substance. Both the boiling point and freezing point of a substance are intensive properties. Water boils at 100 °C and melts at 0 °C regardless of whether you have a cup or a gallon of water.

Both the mass and volume of a sample of matter are extensive properties. A 1.00-cm^3 sample of lead has a mass of 11.3 g. Both the 1.00 cm^3 and the 11.3 g are measures of extensive properties. A larger volume of lead would have a larger mass. Thus, the mass and volume do not characterize lead as a substance, but they do describe that particular sample of lead.

Density of Matter

We sometimes say that one substance is "heavier" than another. A metal such as gold is "heavy" and sinks in water. Wood is "light" and floats on water. However, we are comparing the densities, not the weight, of the materials. Density is the mass per unit volume of matter.

$$\textbf{density} = \frac{\textbf{mass}}{\textbf{volume}}$$

Density is an intensive property of a substance because it is the ratio of the mass to the volume and does not depend on the amount of material. A 1.0-mL sample of

Panning for gold involves separating less dense sand and stone from the more dense gold particles. (Photo Researchers, Inc.)

water has a mass of 1.0 g. A 5.0-mL sample of water has a mass of 5.0 g. The ratio of mass to volume is the same. Water has a density of 1.0 g/mL.

$$\text{density} = \frac{\text{mass}}{\text{volume}} = \frac{1.0\text{ g}}{1.0\text{ mL}} = 1.0\text{ g/mL}$$

$$\text{density} = \frac{\text{mass}}{\text{volume}} = \frac{5.0\text{ g}}{5.0\text{ mL}} = 1.0\text{ g/mL}$$

The masses of 1.00-cm^3 samples of lead and maple wood are 11.3 g and 0.490 g, respectively. These values are in accord with our experience that lead is "heavier" than wood. However, what we have really observed is that for the same volume, lead has a greater mass than wood. The densities of lead and maple wood are 11.3 and 0.490 g/cm^3, respectively.

The densities of liquids and solids are expressed in grams per milliliter (g/mL) or grams per cubic centimeter (g/cm^3). The densities of gases are expressed in grams per liter (g/L). A list of the densities of some common substances at 25 °C is given in Table 2.4.

EXAMPLE 2.7

A 1.50-cm^3 sample of aspirin has a mass of 1.74 g. What is the density of aspirin?

Solution

Density is mass per unit volume, and for the units given the density is expressed in grams per cubic centimeter (g/cm^3). The mathematical operation is

$$\frac{\text{mass}}{\text{volume}} = \frac{1.74\text{ g}}{1.50\text{ cm}^3} = 1.16\text{ g/cm}^3 = \text{density}$$

PROBLEM 2.7

Osmium is the densest element known. What is its density in grams per cubic centimeter if 2.72 g has a volume of 0.121 cm^3?

EXAMPLE 2.8

Isopropyl alcohol, used as a solution in rubbing alcohol, has a density of 0.78 g/mL. What is the mass of 35 mL of isopropyl alcohol?

TABLE 2.4 Densities of Some Liquids, Solids, and Gases

Liquid	Density (g/mL)	Solid	Density (g/cm^3)	Gas	Density (g/L)
bromine	3.12	gold	19.3	carbon monoxide	1.14
ether	0.71	lead	11.3	hydrogen	0.08
olive oil	0.92	rock salt	2.2	helium	0.16
turpentine	0.87	sugar	1.59	methane	0.66
water	1.00	uranium	19.0	oxygen	1.31
mercury	13.53	wood, maple	0.49	nitrogen	1.14

Solution
The mass of the sample is obtained by multiplying the volume by the density. In this way the units of volume cancel and the unit of mass remains.

$$35\ \cancel{\text{mL}} \times 0.78\ \frac{\text{g}}{\cancel{\text{mL}}} = 26\ \text{g}$$

PROBLEM 2.8

The density of uranium is 18.96 g/cm^3. What is the mass of 12 cm^3 of uranium?

EXAMPLE 2.9

The density of sulfuric acid is 1.82 g/mL. What volume of sulfuric acid is required to provide 72.8 g of the material?

Solution
The given unit for mass is grams, and the desired unit is volume, which can be calculated in milliliters. If we multiply by the conversion factor mL/g, the unit of grams cancels and the unit of milliliters results. This conversion factor is the reciprocal of the density, so the mathematical operation is the same as dividing by the density.

$$72.8\ \cancel{\text{g}} \times \frac{1\ \text{mL}}{1.82\ \cancel{\text{g}}} = 40.0\ \text{mL}$$

PROBLEM 2.9

Pure silver has a density of 10.5 g/cm^3. What volume of silver will weigh 64 g?

The density of a substance is determined by measuring both its mass and its volume. For a liquid, the mass is measured by adding a volume of a liquid to a graduated cylinder and determining the increase in the mass of the cylinder. In the case of a solid object, the mass is measured with a balance. The volume of the solid object is determined by immersing it in a liquid. The object must be more dense than the liquid so that the object will sink. The object displaces a volume of the liquid that is equal to its own volume (Figure 2.5). The difference between the initial volume of the liquid and the total volume after the solid is immersed is equal to the volume of the object.

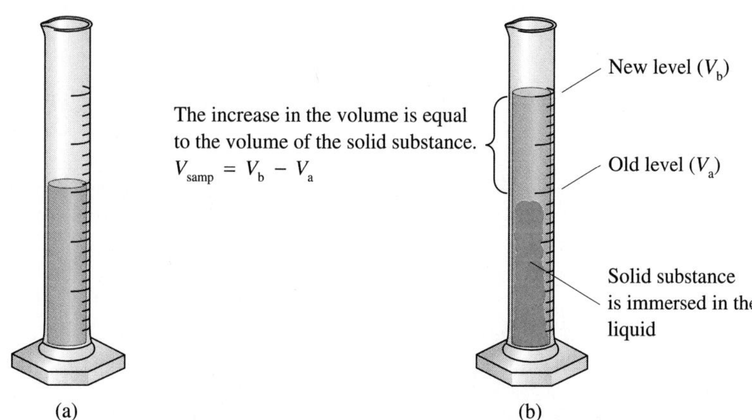

FIGURE 2.5 Determination of the Volume of a Solid
The volume of an object can be determined by the difference between the volumes of (b) and (a), which is the result of the displacement of the liquid by the object.

Specific Gravity

Specific gravity is the ratio of the density of a substance to the density of water at the same temperature.

$$\textbf{specific gravity} = \frac{\textbf{density of substance (g/mL)}}{\textbf{density of water (g/mL)}}$$

Because specific gravity is a ratio of two numbers with the same units, it has no dimensions. Since the density of water is 1 g/mL, the specific gravity of a substance is numerically equal to its density in grams per milliliter. For example, the density of isopropyl alcohol is 0.78 g/mL, so its specific gravity is 0.78.

Energy Is an Extensive Property

As with all other measurements, heat energy is measured by comparison to a standard quantity. The standard quantity of energy is an amount that is required to do something such as increase the temperature of a material. A **calorie** (cal) is the amount of energy required to raise the temperature of 1 g of water 1°C. (The SI unit of energy is the **joule (J);** 1 cal = 4.184 J. The calorie is used extensively in organic chemistry and biochemistry, and it will be used exclusively in this book.)

Temperature and heat energy are easily confused. However, temperature is an intensive property; heat energy is an extensive property. The temperature of a teaspoonful of boiling water, which is 100 °C, is the same as the temperature of a large pot of boiling water. Temperature does not depend on the amount of matter. If a few drops of boiling water are spilled on your skin, a burn might result. However, if a large pot of boiling water were spilled, severe burns would result because the amount of heat energy contained in the water is much larger. The heat energy of matter depends on the amount of matter.

EXAMPLE 2.10

How much energy, in calories, is required to heat 15 g of water from 25.0 °C to 35.0 °C?

Solution

The energy required to heat a sample of matter depends on the mass of the sample. Fifteen times as much energy is required to heat 15 g of water as is required to heat 1 g of water. In addition, 10 times as much heat energy is required to increase the temperature of the sample by 10 °C as to increase it by 1 °C. The heat energy, in calories, required to raise the temperature of 15 g of water by 10 °C is calculated using the definition of the calorie.

$$15 \text{ g} \times (35.0\ ^\circ\text{C} - 25.0\ ^\circ\text{C}) \times 1 \frac{\text{cal}}{\text{g}\ ^\circ\text{C}} = 1.5 \times 10^2 \text{ cal}$$

Note that the number of significant figures in the answer is controlled by the mass of the sample. There are three significant figures in the temperature change. The definition of the calorie is an exact quantity.

PROBLEM 2.10

How much energy, in calories, must be removed to cool 25 g of water from 50.0 °C to 30.0 °C?

Calories and the Basal Metabolic Rate

The kilocalorie (kcal), equivalent to 1000 cal, is the unit used in dietary tables. This unit is the **Calorie (Cal),** that is, with a capital C. Each person's caloric requirement is determined by his or her basal metabolic rate (BMR) and voluntary activities. The caloric requirements for physical activities depend on what muscles are used and for what length of time.

The BMR includes the energy to keep the lungs inhaling and exhaling air, to keep the heart beating, and to maintain the metabolic activities of all types of body cells. The BMR, in Calories per hour, depends on age and physical condition. Your BMR is lowest when you are lying down at a comfortable temperature. A rough estimate of a minimum caloric intake to maintain a patient in a hospital bed is 1.0 Calorie per hour per kilogram of body weight.

The Calories required by a 165-lb patient in a hospital in one day can be calculated in three steps. First, find the mass equivalent of 165 lb.

$$165\ \cancel{\text{lb}} \times \frac{1\text{ kg}}{2.2\ \cancel{\text{lb}}} = 75\text{ kg}$$

Next, calculate the number of Calories required per hour.

$$75\ \cancel{\text{kg}} \times \frac{1.0\text{ Cal}}{\cancel{\text{kg}}\text{ h}} = 75\ \frac{\text{Cal}}{\text{h}}$$

Finally, calculate the number of Calories required per day.

$$75\ \frac{\text{Cal}}{\cancel{\text{h}}} \times 24\ \frac{\cancel{\text{h}}}{\text{day}} = 1800\ \frac{\text{Cal}}{\text{day}} = 1.8 \times 10^{3}\ \frac{\text{Cal}}{\text{day}}$$

Specific Heat

Large bodies of water moderate the surrounding environment because they have high heat capacities. (Pat and Tom Leeson, Photo Researchers)

You may have noticed that it takes a lot of energy to heat water but very little energy to heat a metal. This response of matter to heat energy is an intensive property called *specific heat.* The **specific heat** of a substance is the quantity of heat energy required to change the temperature of 1 g of that substance by 1 °C. From the definition of the calorie, therefore, the specific heat of water is exactly 1 cal/(g deg).

The specific heats of several substances are listed in Table 2.5. Notice that the specific heat of iron is one-tenth that of water. This means that it takes only 0.1 cal to increase the temperature of 1 g of iron by 1 °C. If 1 cal was added to 1 g of iron, the temperature of the iron would increase by 10 °C, but if 1 cal was added to 1 g of water, its temperature would increase by only 1 °C.

The relationship between the heat energy added or removed and the resultant temperature change of a substance is given by the equation

calories = mass of substance × Δt × specific heat

The symbol Δt represents the change in temperature that results from adding or removing heat energy. Any one of the four quantities can be calculated if the other

TABLE 2.5 Specific Heats (cal/(g deg)) of Some Substances at 25°C

Solids	Specific Heat	Liquids	Specific Heat	Gases	Specific Heat
aluminum	0.215	alcohol	0.587	ammonia	0.502
calcium	0.156	bromine	0.113	argon	0.124
copper	0.092	chloroform	0.231	chlorine	0.114
gold	0.031	ether	0.555	oxygen	0.219
iron	0.106	mercury	0.033	methane	0.523
silver	0.057	octane	0.532	nitrogen	0.249
zinc	0.093	water	1.000		

three values are known. For example, the change in temperature that results when 12 cal is added to 75 g of silver, whose specific heat is 0.057 cal/(g deg), is calculated as follows.

$$12 \text{ cal} = 75 \text{ g} \times \Delta t \times 0.057 \frac{\text{cal}}{\text{g °C}}$$

$$\frac{12 \cancel{\text{cal}}}{75 \cancel{\text{g}} \times 0.057 \frac{\cancel{\text{cal}}}{\cancel{\text{g}} \text{°C}}} = \Delta t = 2.8 \text{ °C}$$

If the silver was originally at 25.0 °C, the temperature after adding 12 cal would be 25.0 °C + 2.8 °C = 27.8 °C.

Summary

The emphasis of this chapter is on **measurements** and units of measure. Quantities that arise from a direct count or a defined equivalence are **exact numbers.** Every measurement has some uncertainty in accuracy and precision. **Accuracy** depends on the instrument used to make the measurement. Repetition of measurements improves the **precision** of the measurement.

The certainty of a measurement is indicated by the number of **significant figures. Captive zeros** and **trailing zeros** to the right of the last nonzero digit are significant figures; **leading zeros** are not significant figures. All arithmetic operations on measured quantities must be carried out with, and rounded off to, the proper number of significant figures. **Scientific notation,** used to express large or small numbers, consists of a **coefficient** multiplied by a power of 10 called the **exponent.**

The **metric system** and **SI units** are used for measurements in chemistry. Measurements in a metric unit can be converted into another related metric unit using **conversion factors** that have powers of 10. The **meter, liter,** and **gram** are common units used in chemistry.

Temperatures give the intensity of hotness of matter. U.S. citizens use the **Fahrenheit** temperature scale. Chemists express temperature in either degrees **Celsius** or **kelvins.**

The properties of matter are divided into two categories, **extensive** and **intensive.** Density, specific gravity, and specific heat are examples of intensive properties. The mass of matter per unit volume is its **density.** The **specific gravity** is a related dimensionless quantity. The **specific heat** of a substance is the quantity of heat energy required to change the temperature of 1 g of that substance by 1 °C.

Exercises

Precision and Accuracy

2.1 Consider the data obtained for the length of an object as measured by three students. You know that the length is 14.54 cm. How do you evaluate the work of the three students with regard to precision and accuracy?

	Trial 1	Trial 2	Trial 3	Trial 4
Student 1	14.4 cm	14.6 cm	14.5 cm	14.5 cm
Student 2	14.2 cm	14.1 cm	14.1 cm	14.2 cm
Student 3	14.1 cm	14.8 cm	14.9 cm	14.2 cm

2.2 Consider the data obtained for the mass of an object as measured by three students. You know that the mass is 8.54 g. How do you evaluate the work of the three students with regard to precision and accuracy?

	Trial 1	Trial 2	Trial 3	Trial 4
Student 1	8.4 g	8.6 g	8.5 g	8.5 g
Student 2	8.1 g	8.8 g	8.9 g	8.2 g
Student 3	8.2 g	8.1 g	8.1 g	8.2 g

Significant Figures

2.3 How many significant figures does each of the following numbers have?

(a) 147.89 (b) 0.0375 (c) 2146.8
(d) 0.000408 (e) 21.6489 (f) 0.0000039
(g) 1.230 (h) 24.0500

2.4 How many significant figures does each of the following numbers have?

(a) 147.8 (b) 0.00375 (c) 214.08
(d) 0.00509 (e) 21.6089 (f) 0.0000209
(g) 1.200 (h) 24.0501

2.5 How many significant figures does each of the following numbers have?

(a) 5.02×10^4 (b) 1.256×10^3
(c) 3.15×10^{-4} (d) 1.2×10^{-6}
(e) 2.100×10^{-5} (f) 2.01×10^{-30}

2.6 How many significant figures does each of the following numbers have?

(a) 2.002×10^6 (b) 2.3360×10^2
(c) 4.105×10^{-7} (d) 2.0×10^{-8}
(e) 5.90×10^{-5} (f) 3.05×10^{-23}

Metric Abbreviations

2.7 What is the metric abbreviation for each of the following terms?

(a) milliliter (b) picogram (c) nanometer
(d) decimeter (e) centiliter (f) kilogram

2.8 What is the metric abbreviation for each of the following terms?

(a) milligram (b) picometer (c) nanogram
(d) microliter (e) milligram (f) centimeter

2.9 Write the term represented by each of the following metric abbreviations.

(a) mg (b) pm (c) cL
(d) kg (e) mL (f) km

2.10 Write the term represented by each of the following metric abbreviations.

(a) dg (b) μm (c) pL
(d) cm (e) dL (f) ng

Metric Conversions

2.11 Make the following conversions using conversion factors.

(a) 59 mm into centimeters (b) 153 cm into kilometers
(c) 348 mL into liters (d) 5.328 L into milliliters
(e) 566 cm into decimeters (f) 353 cm into meters

2.12 Make the following conversions using conversion factors.

(a) 0.011 mL into microliters
(b) 243 ng into micrograms
(c) 356.2 pm into nanometers
(d) 456 μg into decigrams
(e) 542 μL into milliliters
(f) 123 μg into nanograms

2.13 The stapes, a bone in the middle ear, is 0.25 cm long. What is the equivalent length in millimeters?

2.14 The femur, a bone in the leg, is about 69 cm long. What is this length in meters?

2.15 One type of floppy disk can hold 800 kilobytes of information. A hard disk holds 400 megabytes. How many floppy disks are required to transfer all information from the hard disk?

2.16 The moon is 3.79×10^5 km from the earth. The speed of light is 3.00×10^8 m/s. How long does it take reflected light from the sun to travel from the moon to earth?

2.17 A frog's egg is about 1.5×10^3 μm in diameter. What is the equivalent diameter in nanometers?

2.18 The best light microscope has a resolving power of about 0.2 μm. What is the equivalent resolving power in nanometers?

2.19 The level of vitamin C in your blood is about 0.2 mg/100 mL of serum. What is the amount of vitamin C in grams per liter?

2.20 An analysis of blood indicates that there is 9.5 mg of calcium in 100 mL of blood. How many grams of calcium would there be in 6.0 L of blood?

2.21 The blood glucose level of a fasting patient drops to 65 mg/dL. What is the number of grams of glucose in 1.0 L of blood?

2.22 The blood urea nitrogen (BUN) of an individual is 1.2 mg/dL. What is this number in grams per milliliter?

2.23 The cholesterol content of the blood of a patient is a high value of 325 mg/dL. What is this number in grams per liter?

2.24 The triglyceride content of the blood of an individual is 121 mg/dL. What is the number of grams of triglycerides in 1.00 mL of this blood?

Temperature

2.25 On a cold day in Vermont, the temperature was −31 °F. What was the equivalent Celsius temperature?

2.26 A doctor tells an American in Paris that his oral temperature is 38.0 °C. What is his temperature in degrees Fahrenheit?

2.27 The coldest recorded temperature at the Russian Antarctic Station was −89.6 °C. What was the equivalent Fahrenheit temperature?

2.28 The highest recorded temperature on earth occurred in Libya. It was 136 °F. What was the equivalent Celsius temperature?

2.29 Heat stroke occurs at a body temperature of 41 °C. What is the equivalent temperature in degrees Fahrenheit?

2.30 In a hypothermia case, the body temperature dropped to 83.7 °F. What was the equivalent temperature in degrees Celsius?

2.31 Helium forms a liquid at −268.6 °C. What is this temperature in kelvins?

2.32 Oxygen solidifies at −218 °C. What is the temperature in kelvins?

Density

2.33 When a 36.9-g sample of iron is placed in a graduated cylinder containing 20.0 mL of water, the water level rises to 24.7 mL. What is the density of iron?

2.34 An empty graduated cylinder weighs 45.8772 g. After 50.0 mL of a liquid is added, the cylinder and its contents weigh 85.7998 g. What is the density of the liquid?

2.35 The density of platinum is 21.45 g/cm^3. What is the mass of 5.9 cm^3 of platinum?

2.36 A 25.0-mL sample of bromine (density 3.12 g/mL) is added to a graduated cylinder. What is the mass of the bromine?

2.37 A patient's urine sample has a density of 1.02 g/mL. How many grams of urine are eliminated on a day in which 1250 mL is excreted?

2.38 Ethylene glycol, which is used as an antifreeze, has a density of 1.1088 g/mL. How many grams of ethylene glycol are in 125 mL?

2.39 The density of ether, a volatile liquid used as an anesthetic, is 0.71 g/mL. What volume would 355 g of ether occupy?

2.40 Ethyl alcohol has a density of 0.789 g/mL. What volume of ethyl alcohol must be poured into a beaker to give 19.8 g of ethyl alcohol?

Specific Gravity

2.41 A syringe contains 10.0 mL of a solution. The mass of the solution is 11.05 g. What is the specific gravity of the solution?

2.42 The specific gravity of turpentine is 0.87. What is the mass of 25.0 mL of turpentine?

2.43 The specific gravity of octane is 0.702. You add 15 mL of octane to a vessel. What is the mass of the octane added?

2.44 The specific gravity of concentrated hydrochloric acid is 1.19. You need 25 g of concentrated hydrochloric acid. What volume should you use?

2.45 A 45.2-g sample of lead is placed in a graduated cylinder containing 20.0 mL of water, and the water level rises to 24.0 mL. What is the specific gravity of lead?

2.46 The specific gravity of alcohol is 0.80. You are told to add 12 g of alcohol to a flask, but you do not have a balance. What can you do?

Energy and Specific Heat

2.47 A quantity of heat equal to 500 cal is added to 25 g of water at 20 °C. What is the final temperature?

2.48 A water bed contains 1000 L of water. How many kilocalories of heat energy are required to heat the water in the bed from 15.0 °C to body temperature (37.0 °C)?

2.49 One ounce of cereal gives 112 Cal of energy on oxidation. How many kilograms of water can be heated from 20 °C to 30 °C by "burning" the cereal?

2.50 A 110-lb woman eats 10 oz of carbohydrates, 2 oz of fat, and 18 oz of protein in one day. The energy released by metabolizing 1 g of either carbohydrate or protein is 4.1 kcal. The energy released by metabolizing 1 g of fat is 9.3 kcal. Calculate the woman's caloric intake in kilocalories.

2.51 The metabolic energy requirement of a laboratory mouse is about 3.8×10^3 cal per day. How many Calories are required to feed the mouse?

2.52 One cup of zucchini contains about 25 Cal. Explain the meaning of this statement.

2.53 It takes 0.092 cal to heat 1.0 g of copper by 1 °C. If 500 cal is required to heat a sample of copper from 15 °C to 90 °C, what is the mass of the copper?

2.54 It takes 0.108 cal to heat 1.0 g of iron by 1.0 °C. How many calories are needed to raise the temperature of 45 g of iron from 14 °C to 47 °C?

The red, blue, and yellow colors in fireworks are the result of electron transitions in salts of strontium, copper, and sodium, respectively. (Michael Dalton, Fundamental Photographs)

3

Atomic Structure and the Periodic Table

3.1 What Is the Smallest Unit of Matter?
3.2 The Atom Is Very Small
3.3 Atoms Contain Subatomic Particles
3.4 Composition of Atoms
3.5 Isotopes
3.6 Mendeleev's Classification of Elements
3.7 The Bohr Atom
3.8 Shells, Subshells, and Orbitals
3.9 Electron Configuration
3.10 Electron-Dot Symbols
3.11 The Periodic Table and Electron Configuration
3.12 Properties of Atoms and the Periodic Table

Essays
- *Isotopes in Biochemistry*
- *Atmospheric Nuclear Tests*

3.1 What Is the Smallest Unit of Matter?

Based on their qualitative observations about the known matter in the world, the ancient Greeks (about 400 B.C.) debated whether matter was infinitely divisible or not. One group of philosophers led by Zeno of Eleas argued that the successive division of any sample of matter could be continued forever. Leucippus and his student Democritus (460–370 B.C.) argued that there is a point at which matter cannot be further divided. At that point the smallest sample of matter is an atom, which means "uncuttable." Democritus suggested that atoms were responsible for processes that we now classify as physical changes and chemical reactions.

Both Plato (427–347 B.C.) and Aristotle (384–322 B.C.) argued against the concept of atoms. Aristotle was held in such high regard that the concept of the atom was ignored for over 2000 years. Although this concept gradually reappeared in the works of Galileo, Francis Bacon, Robert Boyle, and Isaac Newton, it was not until John Dalton (1766–1844) proposed his theory that the idea of atoms took root. Dalton "explained" the quantitative experimental results of two major chemists, Antoine Lavoisier (1743–1794) and Joseph Proust (1754–1826) using the concept of atoms.

The Evidence for Atoms

Antoine Lavoisier proposed the law of conservation of mass, which states that matter can be neither created nor destroyed. His studies of chemical reactions showed that the weight of the products is equal to the weight of the reactants. Dalton's atomic theory "explains" these data because the atoms are postulated to change their arrangement or distribution in a chemical reaction. If they are not destroyed, then their mass is preserved.

Joseph Proust proposed the **law of constant composition,** also known as the law of definite proportions. This law states that any compound always consists of its constituent elements in the same proportion by weight. Thus, water always contains 89% oxygen and 11% hydrogen, no matter where it is found or how it is made by the reaction of hydrogen and oxygen. Again, Dalton's atomic theory "explains" this law, because if water always contains the same atoms and they have characteristic weights, then its composition by weight must be constant.

Dalton's Atomic Theory

With a few modifications to explain additional experimental data, the development of chemistry has been based on the following points of Dalton's atomic theory.

1. Simple substances called elements are composed of indivisible and indestructible particles called atoms.
2. Atoms of the same element are identical but differ from those of any other element. Each type of atom has a mass that is different from that of any other type of atom.
3. Atoms combine only in whole-number combinations to form molecules.
4. Compounds are composed of molecules. All molecules of a compound are identical to each other.
5. In chemical reactions atoms are rearranged to produce different substances. The identity of each atom and its mass are unchanged.

Dalton Was Almost Correct

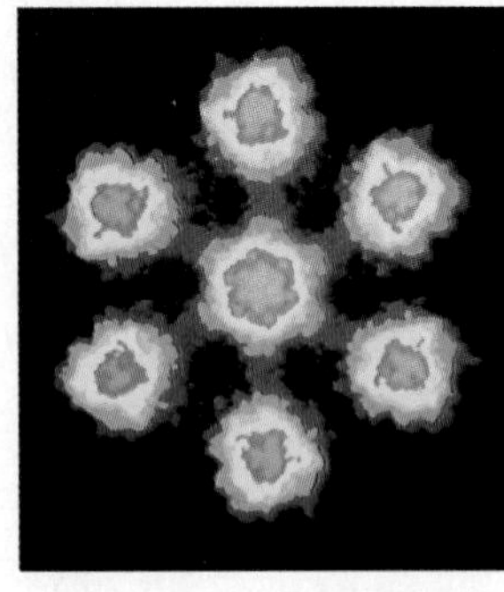

This artificially colored scanning transmission electron micrograph shows uranium atoms. (Photo Researchers)

Dalton's ideas were consistent with the facts known at that time. However, as is the case for most theories, scientists develop concepts in steps, each approaching the truth more closely. Dalton didn't know three facts:

1. Not all elements consist of monatomic species.
2. Atoms are in fact divisible and made up of smaller particles.
3. Some atoms of the same element have different weights.

Some elements, such as helium, do indeed exist as atoms, but other elements, such as hydrogen, chlorine, oxygen, and nitrogen, exist as molecules containing two atoms—diatomic molecules. We now represent these elements as H_2, Cl_2O_2, and N_2, respectively. There are a few other examples of different molecular arrangements for some elements. Phosphorus exists as a molecule with four atoms, that is, as P_4. Sulfur exists as S_8 molecules. These facts don't negate the experimental results of Lavoisier and Proust, nor do they destroy Dalton's atomic concepts. The model just needs to be changed a little. A reinterpretation of the reaction of hydrogen and oxygen to form water molecules is given in Figure 3.1.

We now know that atoms are made up of subatomic particles called electrons, protons, and neutrons. As we will see, with the exception of nuclear reactions

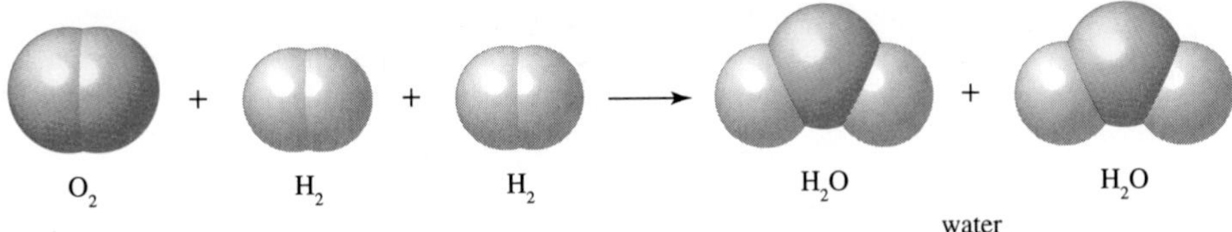

FIGURE 3.1 Formation of Water from Molecular Elements
Water molecules form by reaction of hydrogen and oxygen molecules.

(Chapter 12), the atoms stay fairly intact in ordinary chemical reactions. Electrons are redistributed between atoms in chemical reactions, so no mass is lost.

Different forms of atoms exist because each atom always contains a characteristic number of protons and electrons but may have a different number of neutrons. Thus, these forms of atoms, called *isotopes,* have different masses (Section 3.5).

3.2 The Atom Is Very Small

Hydrogen, the smallest atom, has a mass of 1.673×10^{-24} g. The mass of the uranium atom, which is the heaviest naturally occurring atom, is 3.95×10^{-22} g. As pointed out in the previous chapter, we don't often deal with small numbers, so it is difficult to grasp their significance. However, we do have a good feel for large numbers. To appreciate how small an atom is, let's calculate the number of hydrogen atoms in 1.0 g of hydrogen.

$$1.0\,\cancel{g} \times \frac{1\ \text{atom}}{1.673 \times 10^{-24}\,\cancel{g}} = 6.0 \times 10^{23}\ \text{atoms}$$

To visualize how large this number is, consider having all the people of the world count this number of atoms. If 5 billion people counted one atom a second for 8 h a day, it would take 12 million years to finish the counting!

Since the masses of atoms are so very small, a special unit of mass is used. As in any measurement scale, a unit is selected to give convenient values when any object is compared with the standard. For expressing the mass of atoms that quantity is 1.6605×10^{-24} g, which is called an **atomic mass unit (amu).** Thus, the lightest atom, hydrogen, has a mass of 1.008 amu.

$$1.673 \times 10^{-24}\,\cancel{g} \times \frac{1\ \text{amu}}{1.6605 \times 10^{-24}\,\cancel{g}} = 1.008\ \text{amu}$$

The masses of atoms of elements range between 1 and about 270 amu. Since 1962 the symbol **u** has also been used to represent atomic mass unit. The amu will be used exclusively in this book because it will remind you of the words "atomic mass unit."

EXAMPLE 3.1

The mass of the helium atom is 6.646×10^{-24} g. What is the mass of the helium atom in atomic mass units?

Solution

To convert the quantity in grams into atomic mass units, a conversion factor with the units amu/g is required.

$$6.646 \times 10^{-24}\,\cancel{g} \times \frac{1\ \text{amu}}{1.6605 \times 10^{-24}\,\cancel{g}} = 4.003\ \text{amu}$$

The mass of helium in atomic mass units is four times that of the hydrogen atom in the same units. This must be the case because the helium atom's mass in grams is four times that of the mass of the hydrogen atom in grams.

PROBLEM 3.1

The mass of the argon atom is 6.63×10^{-23} g. What is the mass of the argon atom in atomic mass units?

Now let's consider the volume of an atom to illustrate how small one is. Experiments have shown that atoms are spherical particles. The radii of atoms range from 0.9×10^{-8} to 2.4×10^{-8} cm. These small numbers are conveniently expressed in picometers (pm) to avoid exponential notation ($1 \text{ pm} = 1 \times 10^{-12}$ m). Thus, the range of atomic radii is 90–240 pm. To try to picture the size of 1 pm, consider that the radius of the magnesium atom is 160 pm. If magnesium atoms were arranged in a straight line, it would take 31 million atoms to cover a distance of 1 cm.

3.3 Atoms Contain Subatomic Particles

Dalton's postulate that atoms are indivisible was disproved about a century ago. We now know that atoms consist of electrons, protons, and neutrons, which are collectively called **subatomic particles.** Each has a mass, and electrons and protons have a charge as well (Table 3.1). Neutrons are electrically neutral. Atoms of different elements differ in the number of subatomic particles.

The Electron—The Atom's Negative Charge

If the atom were the size of the Hubert H. Humphrey Metrodome in Minneapolis, the nucleus would be the size of a marble. (Photo Researchers, Inc.)

The electron was discovered by the English physicist J. J. Thomson in 1897. He studied the effect of electricity in a cathode-ray tube, which is similar to a modern television tube. He showed that gaseous atoms can be broken down to yield negatively charged particles called electrons, which were coming off the cathode. The hydrogen atom has one electron, but other atoms have more electrons. The helium atom has two electrons; the oxygen atom has eight electrons.

An **electron** has a mass of 0.0005486 amu, which is only $\frac{1}{1837}$ that of the lightest atom, hydrogen. The small mass of an electron is usually regarded as zero compared with the mass of the atom. The charge of an electron is -1.60×10^{-19} coulomb. Since all charges of subatomic particles and ions are multiples of this quantity, the electron charge is expressed as a relative charge of -1 without a unit.

TABLE 3.1 Properties of Subatomic Particles

Name	Mass (g)	Relative Mass (amu)	Charge (coulombs)	Relative Charge
proton	1.6726×10^{-24}	1.0073	1.60×10^{-19}	+1
neutron	1.6750×10^{-24}	1.0088	0	0
electron	9.110×10^{-28}	0.0005486	-1.60×10^{-19}	−1

The Proton—The Atom's Positive Charge

A **proton** has a mass of 1.6726×10^{-24} g, which equals 1.0073 amu. (For most discussions in this book the mass will be approximated as 1 amu.) The charge of a proton is $+1.60 \times 10^{-19}$ coulomb, which is equal in size to the charge of the electron but of opposite sign. This quantity is expressed as a relative charge of +1 without a unit.

The proton is the unit of positive matter in all atoms, and an atom is chemically characterized by the number of protons that it contains. A hydrogen atom has one proton; the helium atom has two protons; the oxygen atom has eight protons.

Atoms are electrically neutral, that is, they have no charge. Thus, the number of electrons in an atom equals the number of protons. The helium atom has two electrons and two protons; the oxygen atom has eight electrons and eight protons.

The Neutron—A Neutral Subatomic Particle

The masses of atoms cannot be accounted for only by the masses of the protons and electrons that they contain. The mass of a helium atom is 4.00 amu. Based on its two electrons and two protons per atom, the helium atom should weigh only 2.00 amu. The other 2.00 amu of the helium atom is due to two neutrons, subatomic particles whose existence was demonstrated by the English physicist James Chadwick in 1932. The **neutron** has a mass of 1.6750×10^{-24} g. As in the case of the proton, this mass will be approximated as 1 amu.

The Nucleus—The Center of the Atom

In 1911 Ernest Rutherford, an English scientist, showed that most of the mass of the atom as well as its positive charge are concentrated in a small central region called the **nucleus.** Other experiments led to the conclusion that the nucleus is composed of protons and neutrons. The average diameter of the nucleus is 10^{-13} cm. Since the radius of an atom, 10^{-8} cm, is about 10^5 times as large as the nucleus, most of the atom is "empty" space. The electrons occupy this large volume about the nucleus. A close analogy to the atom is the solar system. Although the solar system is large, it is essentially empty space. The planets move in this empty space as they revolve about the sun. We will reexamine this concept later in this chapter.

3.4 Composition of Atoms

The atoms of different elements are characterized by the different number of protons in their nucleus. The number of protons in the nucleus of an atom is called the **atomic number.** Because the atoms of each element have a unique number of protons, we list the atomic number as a physical property of the element. Hydrogen is different from helium because hydrogen atoms contain one proton, and helium atoms contain two protons. The atomic numbers of hydrogen and helium are 1 and 2, respectively.

An atom is electrically neutral, so the number of electrons in an atom must equal the number of protons. Thus, the atomic number is also equal to the number of electrons in a neutral atom.

atomic number = number of protons in atom = number of electrons in atom

The number of protons plus the number of neutrons in the nucleus of an atom is the element's **mass number.** The hydrogen atom has only a single proton, so the mass number of the element hydrogen is 1. The helium atom has two protons and two neutrons, so the mass number of the element helium is 4. The following relationship is used to calculate the number of neutrons in an atom.

number of neutrons = mass number − atomic number

The number of subatomic particles in an atom is shown by adding a preceding superscript and a subscript to the elemental symbol represented by E.

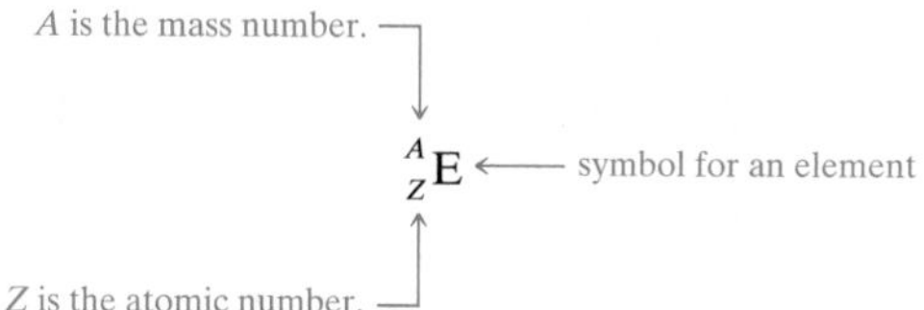

The subscript Z gives the number of protons in the atom. Because the mass number A is equal to the number of protons plus the number of neutrons in the atom, the number of neutrons is equal to $A - Z$.

EXAMPLE 3.2

How many protons, neutrons, and electrons are in an atom of $^{14}_{6}C$?

Solution

The elemental symbol represents the element carbon. There are six protons in the element, as indicated by the subscript. Therefore, carbon also contains six electrons. The number of neutrons given by $A - Z$ is eight.

PROBLEM 3.2

How many neutrons are contained in an atom of $^{238}_{92}U$?

3.5 Isotopes

Dalton had no reason to suspect that atoms in nature actually exist in several different forms called isotopes. **Isotopes** are atoms that have the same number of protons but differ in the number of neutrons they contain. Thus, the isotopes of an element have identical atomic numbers but different mass numbers. Remember that it is the number of protons that determines the identity of the element, not the number of neutrons. Thus, isotopes are just different forms of the same element and have nearly identical chemical properties. A list of the isotopes of some elements is given in Table 3.2.

There are two naturally occurring isotopes of chlorine. Chlorine has an atomic number of 17 and contains 17 protons and 17 electrons. However, one chlorine isotope has 18 neutrons in the nucleus, whereas the other isotope has 20 neutrons. The two isotopes are represented by the symbols $^{35}_{17}Cl$ and $^{37}_{17}Cl$, respectively.

Isotopes have different mass numbers.

$^{35}_{17}Cl$ $^{37}_{17}Cl$

Isotopes have the same atomic number.

EXAMPLE 3.3

What do the symbols $^{235}_{92}U$ and $^{238}_{92}U$ mean? What relationship exists between the atoms represented by these symbols?

Solution

The subscript 92 indicates that uranium contains 92 protons. The symbols represent atoms with the same number of protons, so the atoms are isotopes of each other. The number of neutrons in $^{235}_{92}U$ is $235 - 92 = 143$. In the other isotope, $^{238}_{92}U$, the number of neutrons is $238 - 92 = 146$.

PROBLEM 3.3

The radioisotope $^{131}_{53}I$ is used to treat cancer of the thyroid gland as well as hyperthyroidism. Describe the atomic composition of this isotope.

Atomic Weights

The majority of elements found in nature are mixtures of at least two isotopes. The **atomic weight** of an element is the weighted average of the naturally occurring isotopes of the element expressed in atomic mass units. The atomic weight of each element is listed on the inside cover of this book.

TABLE 3.2 Naturally Occurring Isotopes of Some Common Elements

Element	Isotopes	Natural Abundance (%)
hydrogen	$^{1}_{1}H$	99.985
	$^{2}_{1}H$	0.015
	$^{3}_{1}H$	trace
carbon	$^{12}_{6}C$	98.89
	$^{13}_{6}C$	1.11
	$^{14}_{6}C$	trace
oxygen	$^{16}_{8}O$	99.758
	$^{17}_{8}O$	0.038
	$^{18}_{8}O$	0.204
magnesium	$^{24}_{12}Mg$	78.99
	$^{25}_{12}Mg$	10.00
	$^{26}_{12}Mg$	11.01
chlorine	$^{35}_{17}Cl$	75.77
	$^{37}_{17}Cl$	24.23

Isotopes in Biochemistry

Isotopes made in nuclear laboratories (Chapter 12) are used to probe many chemical processes. Isotopes react chemically in the same fashion. But, because they have different masses they can be distinguished from one another. When incorporated into compounds they can be traced during chemical reactions. This approach has been used to study the many reactions that occur in metabolism, which are described in later chapters. For now, let's see how isotopic labeling of material was used to find out how viruses infect cells.

Until 1944 it was hypothesized that genetic information was carried by chromosomal proteins and that DNA played a lesser role. However, based on his work with a virus that infects the bacterium *Escherichia coli,* Roger Herriot suggested that the hypothesis was incorrect. Viruses consist of DNA surrounded by a protein coat. Herriot suggested that enzymes of the viral protein might be responsible for cutting a hole in an outer membrane through which the DNA then squirts into the cell. Once inside the cell, the viral DNA takes over cell operation and produces copies of itself according to its genetic code.

In 1952, Alfred Hershey and Martha Chase proved Herriot's hypothesis by using radioactive isotopes of phosphorus and sulfur. DNA contains phosphorus but no sulfur; proteins contain sulfur but no phosphorus. Hershey and Chase grew viruses in a medium containing either ^{32}P or ^{35}S, and the succeeding generations of the virus contained the isotope of phosphorus or sulfur. However, the specificity of the isotopes' locations made it possible to monitor each part of the virus as it attacked a cell.

After a period of time, the cells that were infected were mechanically treated to remove any virus that might be adhering to the cell surfaces, and the contents of the cells were examined. Only the phosphorus isotope was found. The proteins within the cells contained none of the sulfur isotope. This study conclusively showed that it is viral DNA that enters the cell and carries with it the instructions for self-replication. This work is one of the landmarks of biochemistry, and it was made possible by the difference of a few neutrons in isotopes.

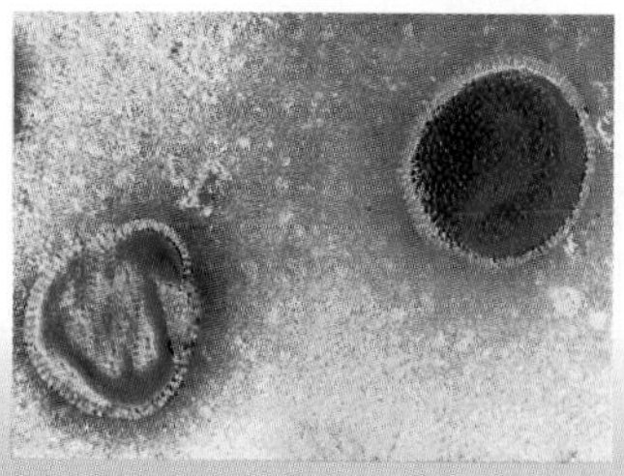

This artificially colored scanning transmission electron micrograph of the influenza virus shows the protein coat as green and the nucleic acid-containing interior as red. (Photo Researchers)

Calculating a weighted average is like calculating your grade in a course. Suppose that two examinations account for 20% and 30% of your grade, and the final examination is 50% of your grade. If you earn 85 on the first examination, 80 on the second examination, and 90 on the final examination, your course grade is 86. This score is obtained by multiplying each grade by the decimal equivalent of the weighting percentage and then summing.

$$
\begin{aligned}
85 \times 0.20 &= 17 \\
80 \times 0.30 &= 24 \\
90 \times 0.50 &= \underline{45} \\
 & 86
\end{aligned}
$$

The atomic weight scale is based on the assignment of 12.0000 amu to the mass of the most common isotope of carbon, $^{12}_{6}C$, also known as carbon-12. Atomic weights can be calculated by multiplying the mass of each isotope by its fractional abundance. For example, a sample of boron is 19.6% atoms of mass 10.013 amu and 80.4% atoms of mass 11.009 amu.

$$
\begin{aligned}
10.013 \text{ amu} \times 0.196 &= 1.96 \text{ amu} \\
11.009 \text{ amu} \times 0.804 &= \underline{8.85 \text{ amu}} \\
\text{atomic weight} &= 10.81 \text{ amu}
\end{aligned}
$$

EXAMPLE 3.4

The two naturally occurring isotopes of copper are $^{63}_{29}Cu$ and $^{65}_{29}Cu$. Their atomic masses are 62.9298 and 64.9278 amu, respectively. The atomic weight of copper is 63.546 amu. Which isotope is the more abundant?

Solution

The atomic weight will be closer to the mass of the more abundant isotope. The value of 63.546 is closer to 62.9298 than to 64.9278. Thus, the more abundant isotope is $^{63}_{29}Cu$.

PROBLEM 3.4

The natural abundances of $^{37}_{17}Cl$ and $^{35}_{17}Cl$ are 24.47% and 75.53%, respectively. The atomic masses of the two isotopes are 36.96590 and 34.96885 amu, respectively. Calculate the atomic weight of chlorine.

3.6 Mendeleev's Classification of Elements

Waves in the ocean are an example of a periodic phenomenon. (Tony Stone Images)

About the middle of the nineteenth century, chemists began to notice similarities among small groups of elements. For example, the metals lithium, sodium, and potassium all react with water to release hydrogen gas and produce caustic solutions. A number of attempts were made to classify and arrange these groups of similar elements in a tabular form. Although many scientists contributed to the idea that there might be a way to organize the elements, it was the Russian chemist Dmitri Mendeleev (1834–1907) who had an insight in 1869 that produced a tabulation that has come to be known as the ***periodic table.*** He found that when the elements are placed in order of increasing atomic weight, those with similar properties occur at periodic intervals. Mendeleev then arranged the 63 known elements in a table to emphasize this periodicity and to group elements with similar properties. He placed the elements in order of increasing atomic weight in a horizontal row until he reached an element with chemical properties similar to those of an earlier element. This element he placed in the next horizontal row beneath the earlier element. The modern version of this periodic table (Figure 3.2) contains almost 50 additional elements discovered since Mendeleev first proposed the table.

Both the atomic number and the atomic weight are given in the rectangular box containing the symbol of the element. The elements are arranged by increasing atomic number in horizontal rows so that elements with similar properties fall into columns. Mendeleev used atomic weights because at that time the atom was thought to be indivisible—subatomic particles had not yet been discovered. However, with very few exceptions, the atomic weights of the elements increase as the atomic number increases. Only by arranging the elements by atomic number does a table result with a logical classification that correlates with the chemical properties of the elements. The modern **periodic law** is stated: The properties of elements are periodic functions of their atomic number.

Organization of the Periodic Table

The horizontal rows in the periodic table are called **periods.** The seven periods are numbered 1 through 7 from the top to the bottom of the table. Note that the periods of the table contain different numbers of elements. The first period contains

transition elements

Period	IA	IIA	IIIB	IVB	VB	VIB	VIIB	VIII	VIII	VIII	IB	IIB	IIIA	IVA	VA	VIA	VIIA	VIIIA
1	1 H 1.0079																	2 He 4.003
2	3 Li 6.941	4 Be 9.012											5 B 10.81	6 C 12.011	7 N 14.007	8 O 15.999	9 F 18.998	10 Ne 20.179
3	11 Na 22.990	12 Mg 24.305											13 Al 26.982	14 Si 28.086	15 P 30.974	16 S 32.06	17 Cl 35.453	18 Ar 39.948
4	19 K 39.098	20 Ca 40.08	21 Sc 44.956	22 Ti 47.88	23 V 50.942	24 Cr 51.996	25 Mn 54.938	26 Fe 55.847	27 Co 58.933	28 Ni 58.69	29 Cu 63.546	30 Zn 65.38	31 Ga 69.72	32 Ge 72.59	33 As 74.922	34 Se 78.96	35 Br 79.904	36 Kr 83.80
5	37 Rb 85.4678	38 Sr 87.62	39 Y 88.906	40 Zr 91.22	41 Nb 92.906	42 Mo 95.94	43 Tc (98)	44 Ru 101.07	45 Rh 102.906	46 Pd 106.42	47 Ag 107.868	48 Cd 112.41	49 In 114.82	50 Sn 118.69	51 Sb 121.75	52 Te 127.60	53 I 126.904	54 Xe 131.29
6	55 Cs 132.905	56 Ba 137.3	57 La* 138.906	72 Hf 178.49	73 Ta 180.948	74 W 183.85	75 Re 186.207	76 Os 190.2	77 Ir 192.22	78 Pt 195.08	79 Au 196.966	80 Hg 200.59	81 Tl 204.383	82 Pb 207.2	83 Bi 208.980	84 Po (209)	85 At (210)	86 Rn (222)
7	87 Fr (223)	88 Ra 226.025	89 Ac** 227.028	104 Rf (261)	105 Ha (262)	106 Sg (263)	107 Ns (262)	108 Hs (265)	109 Mt (266)									

lanthanides *	58 Ce 140.12	59 Pr 140.9077	60 Nd 144.24	61 Pm (145)	62 Sm 150.36	63 Eu 151.96	64 Gd 157.25	65 Tb 158.925	66 Dy 162.50	67 Ho 164.930	68 Er 167.26	69 Tm 168.934	70 Yb 173.04	71 Lu 174.967
actinides **	90 Th 232.0381	91 Pa 231.0359	92 U 238.02	93 Np 237.0482	94 Pu (244)	95 Am (243)	96 Cm (247)	97 Bk (247)	98 Cf (251)	99 Es (252)	100 Fm (257)	101 Md (258)	102 No (259)	103 Lr (260)

FIGURE 3.2
The Modern Periodic Table

only 2 elements, hydrogen and helium. The second and third periods have 8 elements each, whereas the fourth and fifth periods have 18 elements each. Thus, 10 elements in the fourth and fifth periods in the middle of the periodic table have no counterparts in the two earlier periods. Beyond the fifth period, there are even more elements. Elements 58–71 of the sixth period and elements 90–103 of the seventh period have no counterparts in earlier periods. They are placed outside the main table rather than splitting it again.

Each of the eighteen vertical columns in the periodic table contains a **group** of elements; a group is sometimes called a **family.** The **halogens**—fluorine, chlorine, bromine, iodine, and astatine—are in Group VIIA. The **alkali metals** are in Group IA. The elements in Group IIA are **alkaline earth metals.** Group VIIIA at the right side of the periodic table contains the **noble gases,** which are also called **rare gases** or **inert gases.** These elements were unknown at the time Mendeleev proposed his table. The atomic numbers of the noble gases place them after the elements of the halogen family.

Elements in groups designated by Roman numerals I–VIII along with the letter A are called **representative elements.** Elements in Groups IB–VIIB and the three columns designated by VIIIB are **transition metals.** The two rows outside the periodic table are called **inner transition metals.** Elements 58–71 in the first row are **lanthanides,** and elements 90–103 in the second row are **actinides.** (Some periodic tables use Arabic numbers rather than Roman numerals as in 2A rather than IIA. Some tables also number the columns 1 through 18 without the letters A and B.)

EXAMPLE 3.5

Locate gallium (atomic number 31) in the modern periodic table. What are the period number and group number of this element? What elements should resemble gallium?

Solution

Gallium is located in Period 4 and in Group IIIA in the periodic table. The elements aluminum and indium are above and below gallium, respectively. Thus, these two elements should have properties similar to those of gallium.

PROBLEM 3.5

What are the period number and group number of silver (atomic number 47)? What elements should resemble silver?

EXAMPLE 3.6

The melting points of sodium and rubidium are 97.8 °C and 39.0 °C, respectively. Predict the melting point of potassium.

Solution

Potassium is between sodium and rubidium in the Group I column in the periodic table. Therefore, its melting point should be between those of sodium and rubidium. The actual melting point is 63.6 °C, a value that is approximately the average of the other two melting points.

PROBLEM 3.6

The boiling points of fluorine and bromine are − 118.14 °C and 58.78 °C, respectively. Predict the boiling point of chlorine.

Atmospheric Nuclear Tests

The similarities in the chemical properties of elements in the same column of the periodic table can cause some serious biological consequences if an element that is not normally ingested inadvertently enters an organism. Consider strontium, whose chemical properties are similar to those of calcium. Calcium is an important element for humans; our teeth and bones contain calcium phosphate. We require calcium in our diet; one of the common sources is dairy products. Normally we do not ingest much strontium.

Strontium-90 is one of the radioisotopes produced in nuclear explosions. The isotope released in early atmospheric nuclear tests fell to Earth and was incorporated in plants eaten by cows. The cows' milk concentrated the radioactive isotope, which was then consumed by young children and, being similar to calcium, incorporated into the children's teeth and bones. The radioactivity eventually disappears as the isotope decays, but it takes about 28 years for just half the strontium-90 to break down. Exposure to this radiation over such a long time can cause diseases such as leukemia. Because of such health and other environmental concerns, the world's major nuclear powers decided in the early 1960s to ban the testing of atomic weapons in the atmosphere, in outer space, and underwater. Testing continues, but underground, where the nuclear products are contained.

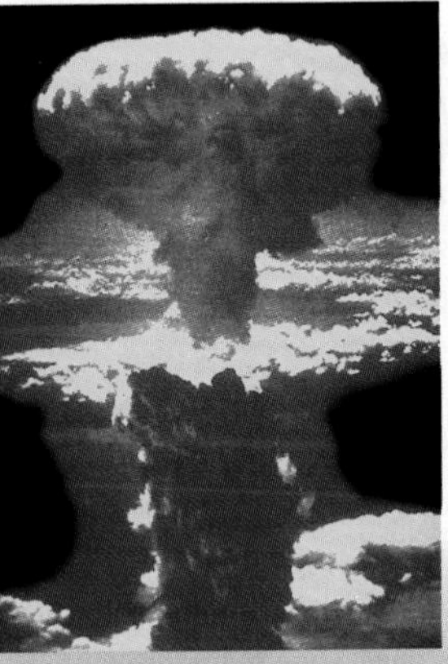

The color of the mushroom cloud of an atomic bomb explosion in 1945 taken at 60,000 feet is computer-enhanced. (Scott Camzine, Photo Researchers)

3.7
The Bohr Atom

The composition of the atom and a simple model for its structure were given in Section 3.3. However, a more detailed picture of the arrangement of the electrons in the atom is needed to understand why and how atoms combine to form bonds. In Chapters 4 and 5 we will see that the arrangement of electrons about the nucleus accounts for how atoms associate to form molecules.

Information about the arrangement of electrons in the atom was first obtained from an analysis of the line spectra of the elements. To understand what is meant by line spectra, let's consider the spectrum obtained when sunlight passes through a prism (Figure 3.3). This is the familiar continuous "rainbow" of all colors from red to violet. Now consider the colored light that results when high voltages are applied to gaseous elements. You have observed this phenomenon in "neon" lights. When this type of light is passed through a prism, a series of narrow colored lines rather than a full rainbow of colors is observed. The number of lines and their colors are characteristic for each element.

These stone steps in France provide one of the examples of quantized quantities in everyday life. You can only stand on the steps, not in the space in between them. (Tony Stone Worldwide)

Niels Bohr (1885–1962), a Danish physicist, suggested that the energies of the electrons are **quantized,** meaning that the energies are limited to certain values and can have no others. The line spectrum of hydrogen reflects what happens to the electron of hydrogen when energy is added. Bohr proposed that electrons of a certain energy travel in a specific orbit about the nucleus much as the planets travel in orbits about the sun. He suggested that there are a limited number of orbits and that the energy of the electrons in the orbits increases with distance from the nucleus. The electron can be in any of these orbits but not between orbits. A precise amount of energy is required for an electron to "jump" from a low-energy orbit near the nucleus to a higher energy orbit farther away (Figure 3.4). When the electron moves to a higher orbit, an "excited atom" forms. The atom loses a precise amount of energy when the electron "falls" back to its lower orbit, and this energy is released as light. Only certain colors of light result, indicating the energy differences between the orbits. Blue light has more energy than red light. Thus, blue light indicates that the electron has "fallen" from a high orbit to a much lower orbit. Red light results from changes between orbits that are closer together.

Bohr's theory of the hydrogen atom was consistent with the facts known at that time. More evidence has since been gathered on more elements, and the Bohr con-

FIGURE 3.3
Spectrum of Sunlight and Line Spectrum of an Element

In sunlight all the colors form a continuous spectrum. In the light emitted from excited atoms of an element, only a few colored lines are observed.

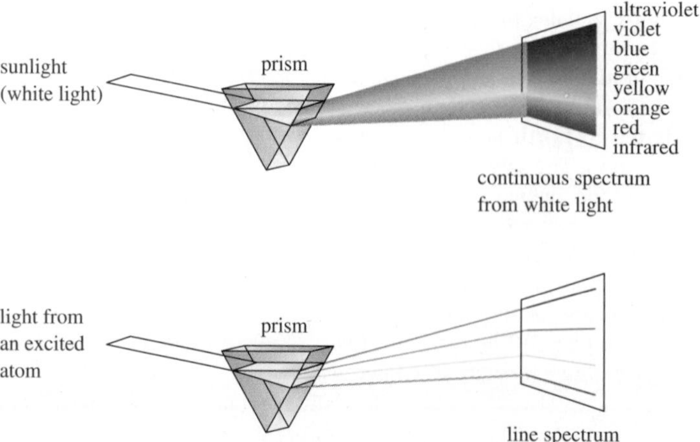

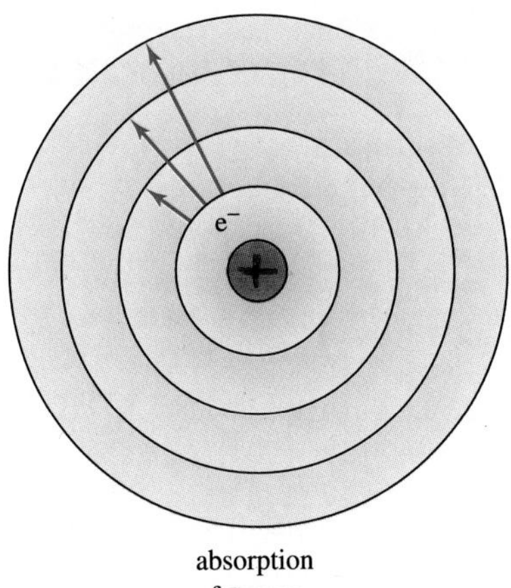

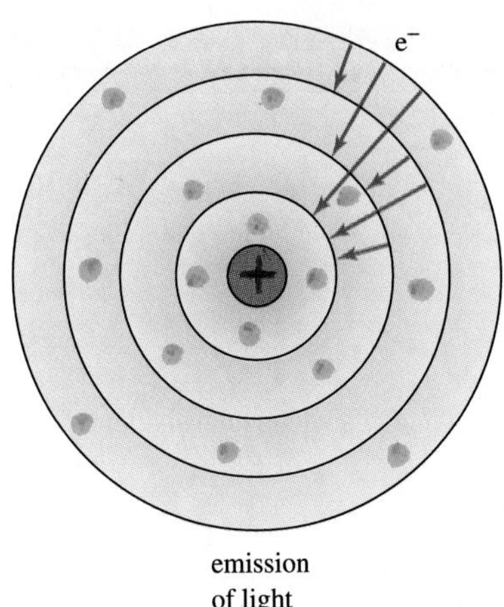

FIGURE 3.4
The Bohr Hydrogen Atom
When energy is added to the atom, the electron moves to higher energy orbits. When the electron returns to a lower orbit, light energy is released. The color of the light is characteristic of the difference in energy between the orbits.

cept has been modified. It is now known that electrons do not exist in planetlike orbits about the nucleus but rather that electrons occupy regions of space about the nucleus called **orbitals.** Although the words *orbit* and *orbital* resemble each other, they have different meanings. An orbit is a *defined path* through space, whereas an orbital is a *volume* of space. The electron is somewhere in a specific volume of space but is not confined to a specific path. There are many orbitals in an atom. They differ in their energy, shape, and location with respect to the nucleus.

3.8 Shells, Subshells, and Orbitals

The modern picture of the electronic arrangement in the atom is based on a theoretical approach called *quantum mechanics,* which involves mathematics that is too advanced to be presented in this book. However, the conclusions of the theory can be summarized simply. The electrons of an atom exist in

1. principal energy levels or shells,
2. subshells that are part of each shell,
3. orbitals that are part of each subshell.

Thus, any electron contained in an orbital is part of a subshell that is part of a shell. Let's now look at the picture of the atom as described by these statements.

Shells

The electrons in the orbitals of an atom are contained in principal energy levels or **shells.** The shells are designated by integers $n = 1, 2, 3, 4$, and so on. The electrons of a shell have a certain average energy that increases in the order $1 < 2 < 3 < 4$, and so on. The distance from the nucleus to a particular shell increases in the same order.

The maximum number of electrons in a shell is $2n^2$, where n is the number of the energy level. For example, the number of electrons that are allowed to exist in energy level 1 is $2 \times 1^2 = 2$. For the second energy level, the maximum number of electrons is $2 \times 2^2 = 8$. The maximum number of electrons in each of the principal energy levels is listed in Table 3.3.

TABLE 3.3 Shells and Subshells

Shell, n	Subshells	Maximum Number of Electrons	Total, $2n^2$
1	*s*	2	2
2	*s*, *p*	2 + 6	8
3	*s*, *p*, *d*	2 + 6 + 10	18
4	*s*, *p*, *d*, *f*	2 + 6 + 10 + 14	32
5[a]	*s*, *p*, *d*, *f*, *g*	2 + 6 + 10 + 14 + 18	50
6[a]	*s*, *p*, *d*, *f*, *g*, *h*	2 + 6 + 10 + 14 + 18 + 22	72
7[a]	*s*, *p*, *d*, *f*, *g*, *h*, *i*	2 + 6 + 10 + 14 + 18 + 22 + 26	98

[a] The *g*, *h*, and *i* subshells do not contain electrons in known elements in their normal state. The theoretical maximum for these energy levels has not been observed because no known elements have the necessary number of electrons to completely fill the shells.

Subshells

Within the energy levels there are sublevels called **subshells;** the higher energy levels contain more subshells. In the atoms of known elements, electrons exist in four types of subshells, which are labeled with the lowercase letters *s, p, d,* and *f.* An "excited atom" may contain electrons in subshells labeled *g, h, i,* and so on; however, we will not consider these in this text.

The energy of a particular type of subshell depends on the energy of the shell. The order of the energy of the subshells within a given shell is $s < p < d < f$.

The number of subshells is equal to the value of n for the shell (Table 3.3). There is only one subshell, in the first energy level. The second energy level has two subshells, an *s* and a *p*. The maximum number of electrons in each of the subshells is 2 in the *s* subshell, 6 in the *p* subshell, 10 in the *d* subshell, and 14 in the *f* subshell. Note that the sum of the numbers of electrons possible within the subshells is equal to $2n^2$, the total number of electrons possible within the shell.

The shells and subshells in an atom are represented by the symbols 2*s*, 4*d*, 5*p*, and so on. The number represents the energy level or shell, and the letter represents the subshell within that energy level.

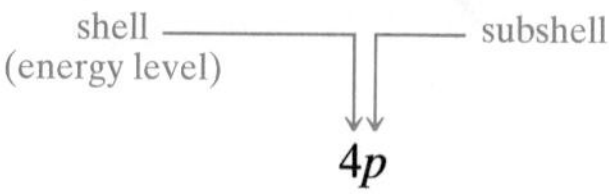

EXAMPLE 3.7

What is the maximum number of electrons that can occupy the 5*p* subshell?

Solution

The indication of the fifth energy level is not required to solve this problem. All *p* subshells, regardless of the energy level, may contain no more than six electrons.

PROBLEM 3.7

What is the maximum number of electrons that can occupy the 7*s* subshell?

Orbitals

An **orbital** is a region about the nucleus in which an electron with a certain energy is most likely to be found. The characteristics of orbitals are as follows:

1. Each type of orbital has a specific shape.
2. All orbitals within a subshell are of the same energy.
3. The orbital volume and average distance from the nucleus increase as the value of n increases.
4. Each orbital can contain a maximum of two electrons, and they must have opposite spins.
5. Electrons move rapidly and "occupy" the entire orbital volume.

The electrons in an orbital can be pictured as a cloud. Clouds in the sky look big but do not contain much matter and are not dense. The electrons within an orbital are moving and occupy a large volume. As a result, the average distribution of matter within the orbital, known as *electron density,* is very low, or cloudlike.

The *s* orbitals are spherical, as shown in Figure 3.5. The 2*s* orbital has a larger volume than the 1*s* orbital, and the 3*s* orbital is larger than the 2*s* orbital. However, each *s* orbital can still contain only two electrons.

There are three *p* orbitals within each *p* subshell, and each orbital can contain two electrons. The three *p* orbitals are identical in shape and are located at right angles to one another (Figure 3.6). Note that each orbital consists of two teardrop shapes. The electrons within a *p* orbital can occupy any of the volume shown as a *p* orbital; they are not confined one to each of the two teardrop shapes. The orbital merely describes a volume of space within which two electrons may exist. There are five *d* orbitals in a *d* subshell and seven *f* orbitals in an *f* subshell. The shapes of *d* and *f* orbitals will not be described in this text.

Electron Spin

Electrons spin or rotate in two possible directions, referred to as clockwise and counterclockwise or as "spin up" and "spin down." These spins are often represented by arrows. An arrow pointing up means that the spin of that electron is the opposite to that of an electron represented by an arrow pointing down.

Electrons repel each other because they have the same negative charge. As a consequence, electrons "spread out" among the orbitals within the subshell. Thus, three electrons within the *p* subshell are distributed one to each of the three or-

FIGURE 3.5
The *s* Orbitals
Electrons are pictured within a volume of space called an orbital. A "cloud" of negative charge surrounds the nucleus which is located at the origin of the intersecting axes. The s *orbital is pictured as a sphere.*

1*s*

2*s*

3*s*

FIGURE 3.6
The Shape of *p* Orbitals
The three orbitals of the p subshell are arranged perpendicular to one another. Each orbital may contain two electrons.

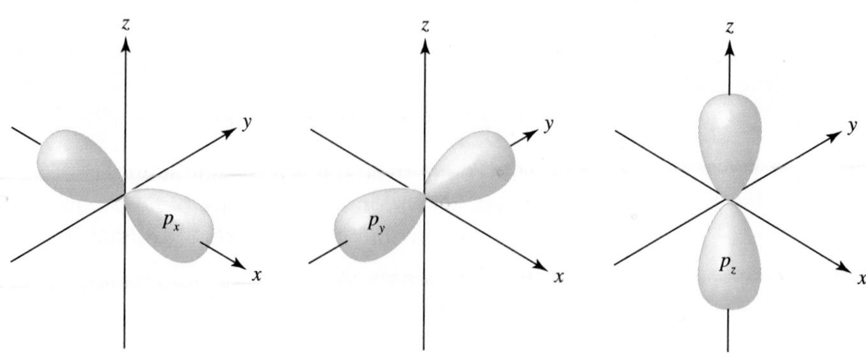

bitals. In addition, the electrons have the same spin. Horizontal lines are used to represent the three orbitals of the *p* subshell.

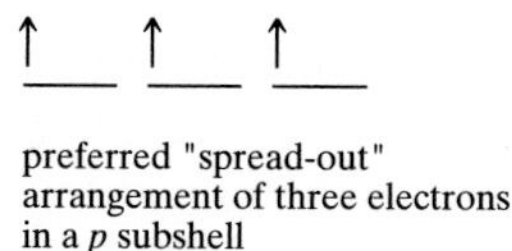

preferred "spread-out" arrangement of three electrons in a *p* subshell

What occurs when there are more electrons than orbitals? Two electrons in the same orbital always have opposite spins. A spinning electron creates a magnetic field; two electrons spinning in opposite directions in an orbital produce opposite magnetic fields. You know that the north pole of one magnet attracts the south pole of another magnet. Thus, electrons with opposite spins magnetically attract each other. Electrons with opposite spins within an orbital are **paired.** A single electron in an orbital or several electrons with the same spin distributed in different orbitals are **unpaired.**

Hund's rule summarizes this arrangement of electrons in orbitals: *An electron will not enter an orbital containing another electron if an empty orbital of the same energy is available.* Remember that when electrons must occupy the same orbital, they must have opposite spins. Thus four electrons in a *p* subshell result in one set of paired electrons and two unpaired electrons.

↑↓ ↑ ↑

preferred "spread-out" arrangement of four electrons

EXAMPLE 3.8

There are seven electrons present in the *d* subshell of an atom. Represent the arrangement of the electrons in this subshell.

Solution

The first five electrons in a *d* subshell must be distributed among the five orbitals—one to each orbital. Then each of the other two electrons must be paired with an electron already in an orbital. The accepted arrangement is

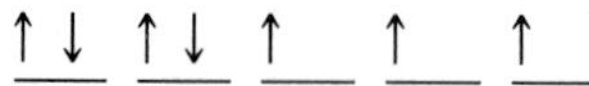

PROBLEM 3.8

If there are nine electrons in a 4*f* subshell, how many unpaired electrons are there?

3.9 Electron Configuration

The picture of the atom developed in the previous section describes where electrons may be placed or "reside" in an atom. Now let's consider all the electrons within an atom and how they are distributed. The arrangement of the electrons within an atom is the **electron configuration.** The rules for predicting the electron configuration are:

1. Electrons are placed in the lowest energy subshell available.
2. Electrons are spread out among the orbitals within the subshell until all orbitals have at least one electron.
3. No more than two electrons may exist in the same orbital, and then only if they have opposite spins.

Each of these rules means that electrons will be located in the lowest energy orbitals first. As more electrons are added, they will occupy the next lowest energy orbital.

The order of increasing energies of subshells is 1s, 2s, 2p, 3s, 3p, 4s, 3d, 4p, 5s, 4d, 5p, 6s, 4f, 5d, 6p, 7s, 5f, 6d as outlined in Figure 3.7. The ordering of shells and subshells may seem complicated because not all of the subshells within a given shell are filled before the next shell starts to fill. For example, the 4s subshell intervenes between the 3p and the 3d subshells. Later in this chapter we will see how the periodic table can provide us with this same order of filling of orbitals.

The convention used to write the electron configuration of atoms gives the energy level, the subshell, and the number of electrons in the subshell.

shell (energy level) → 4; subshell → p

$4p^6$ — number of electrons in the subshell

The description of the entire electron configuration consists of a series of symbols, giving the subshell and the number of electrons contained in that subshell.

The First 20 Elements

The lightest element, hydrogen, contains one electron in the $1s$ subshell, which is the lowest energy subshell. Using the superscript 1 to denote the number of electrons in the subshell, we write the electron configuration of hydrogen as $1s^1$. Helium has two electrons, both occupying the $1s$ subshell. The electron configuration of helium is $1s^2$, and the subshell is filled. The electrons in helium are paired.

H	$1s^1$
He	$1s^2$

Lithium has three electrons. Two electrons fill the $1s$ subshell; the third is in the subshell with the next higher energy, which is the $2s$. The lithium electron configuration is $1s^22s^1$. The electron in the $2s$ orbital is unpaired. Beryllium has the electron configuration $1s^22s^2$. The fourth electron of beryllium completes the $2s$ subshell, which has a capacity of two electrons. The two electrons in the $2s$ subshell are paired.

Li	$1s^22s^1$
Be	$1s^22s^2$

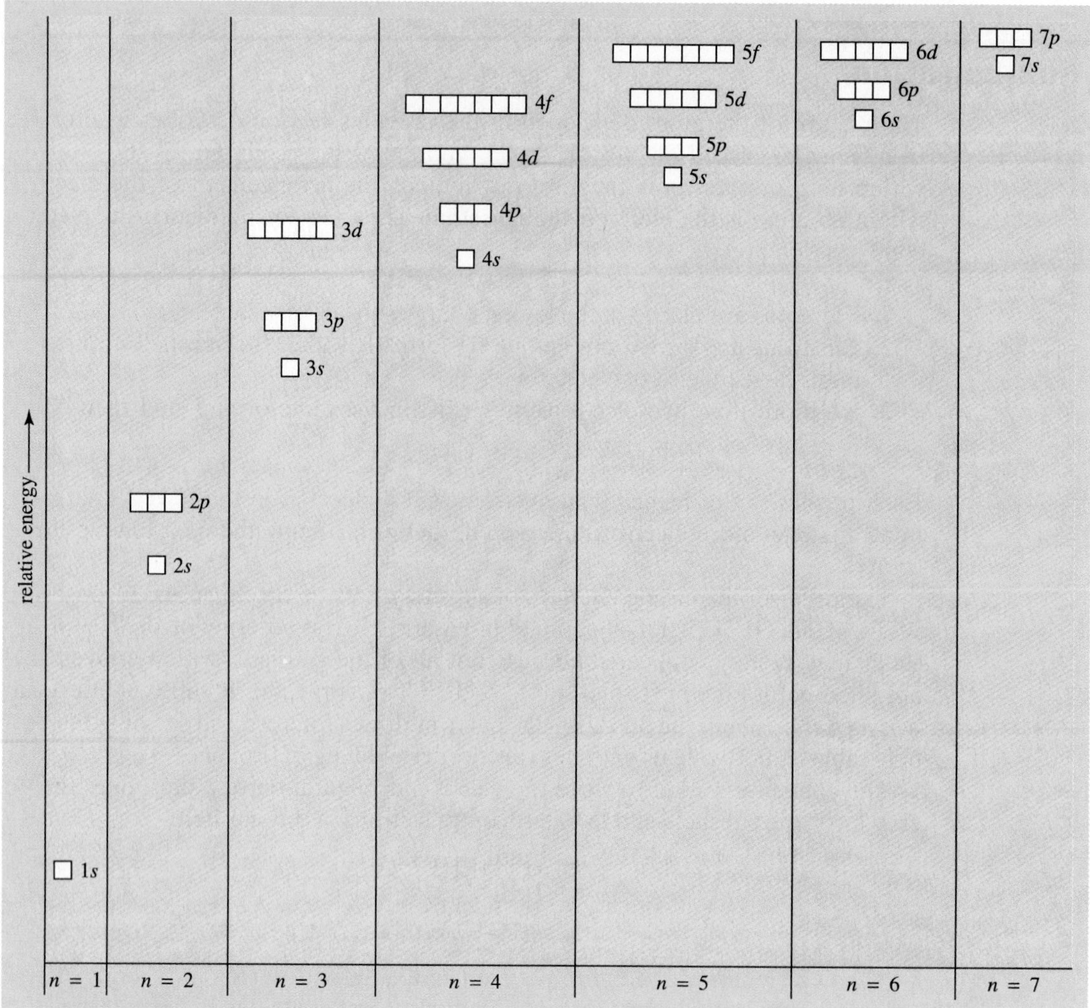

FIGURE 3.7 Subshell Energy Diagram

The boxes represent orbitals within a subshell. Any orbital that is higher on the page is higher in energy than any orbital below it on the page.

In boron, the 2p subshell, the subshell with the next lowest energy, is occupied by a single electron. The electron configuration of boron is $1s^22s^22p^1$. A p subshell can hold a maximum of six electrons, and this subshell is gradually filled as additional electrons are added for the elements carbon, nitrogen, oxygen, fluorine, and neon. The electron configurations for these elements are

B	$1s^22s^22p^1$
C	$1s^22s^22p^2$
N	$1s^22s^22p^3$
O	$1s^22s^22p^4$
F	$1s^22s^22p^5$
Ne	$1s^22s^22p^6$

With neon, the second energy level is complete. This element is a member of a group of elements known as the *noble gases.*

The next eight elements, sodium through argon, have electrons in the third energy level. Continuing to write out the entire electron configuration is time-consuming, so a shorthand method uses [Ne] to represent $1s^22s^22p^6$, and electrons in the third energy level are appended to this symbol. Both the complete electron configuration and the shorthand representation are given for sodium through argon.

	Full Representation	*Shorthand Representation*
Na	$1s^22s^22p^63s^1$	$[Ne]3s^1$
Mg	$1s^22s^22p^63s^2$	$[Ne]3s^2$
Al	$1s^22s^22p^63s^23p^1$	$[Ne]3s^23p^1$
Si	$1s^22s^22p^63s^23p^2$	$[Ne]3s^23p^2$
P	$1s^22s^22p^63s^23p^3$	$[Ne]3s^23p^3$
S	$1s^22s^22p^63s^23p^4$	$[Ne]3s^23p^4$
Cl	$1s^22s^22p^63s^23p^5$	$[Ne]3s^23p^5$
Ar	$1s^22s^22p^63s^23p^6$	$[Ne]3s^23p^6$

With the element argon, the 3*s* and the 3*p* subshells are completely filled, just as the electrons of neon completed the 2*s* and 2*p* subshells. Argon is also a member of the group of noble gases. Thus, chemists use [Ar] as a shorthand to indicate the electron configuration of the 18 electrons in argon and for the first 18 electrons of any other element with more than 18 electrons.

Now we have to be careful. Where are the highest energy electrons located in potassium and calcium? Look at the order of increasing energies of the orbitals (Figure 3.9). Electrons will be located in the 4*s* subshell before any enter the 3*d* subshell. Accordingly, the electron configurations of potassium and calcium are

K	$1s^22s^22p^63s^23p^64s^1$	or	$[Ar]4s^1$
Ca	$1s^22s^22p^63s^23p^64s^2$	or	$[Ar]4s^2$

The Elements Beyond Calcium

The order of the energies of the subshells is complicated beyond the 4s subshell. Nevertheless, if you follow the order outlined in Figure 3.7, you can write the electron configuration of any element. The symbols for the noble gases krypton, xenon, and radon are used as shorthand to write the electron configuration of elements with large numbers of electrons. For each noble gas, the *s* and *p* subshells of the highest energy level have two and six electrons, respectively.

Kr	$1s^22s^22p^63s^23p^64s^23d^{10}4p^6$	or	[Kr]
Xe	$1s^22s^22p^63s^23p^64s^23d^{10}4p^65s^24d^{10}5p^6$	or	[Xe]
Rn	$1s^22s^22p^63s^23p^64s^23d^{10}4p^65s^24d^{10}5p^66s^24f^{14}5d^{10}6p^6$	or	[Rn]

EXAMPLE 3.9

What is the arrangement of electrons in $^{60}_{27}Co$? This isotope, called cobalt-60, is used in cancer therapy.

Solution

There are 27 electrons in this isotope. Write the subshells in order of increasing energy, filling each in order. The first 20 electrons are accounted for by the arrangement

$$1s^22s^22p^63s^23p^64s^2$$

The remaining 7 electrons are placed in an incomplete 3*d* subshell to give

$$1s^2 2s^2 2p^6 3s^2 3p^6 4s^2 3d^7$$

The first 18 electrons are arranged in the same electron configuration as argon. Thus, a shorthand representation is

$$[Ar]4s^2 3d^7$$

PROBLEM 3.9

What is the electron configuration of ${}^{90}_{38}Sr$, an isotope produced in atomic explosions?

Electron Configuration and Electron Pairing

Now let's consider the location of electrons within a subshell in an atom. The 2*p* subshell of the series of elements boron, carbon, nitrogen, oxygen, fluorine, and neon contains from one to six electrons. According to Hund's rule, the electrons are located as shown in the orbital diagrams of Table 3.4. The order of filling the 2*p* subshell dictates the number of pairs of electrons and the number of unpaired electrons. Nitrogen, for example, has three unpaired electrons. The number and type of bonds formed by elements are based on such considerations.

The number of unpaired electrons in any atom can be determined by remembering that the electrons are first distributed one to each orbital in a subshell. Then any additional electrons must be paired.

TABLE 3.4 Hund's Rule and Some Electron Configurations

Element	Atomic Number	1*s*	2*s*	2*p*	Electron Configuration
H	1	↑			$1s^1$
He	2	⇅			$1s^2$
Li	3	⇅	↑		$1s^2 2s^1$
Be	4	⇅	⇅		$1s^2 2s^2$
B	5	⇅	⇅	↑ _ _	$1s^2 2s^2 2p^1$
C	6	⇅	⇅	↑ ↑ _	$1s^2 2s^2 2p^2$
N	7	⇅	⇅	↑ ↑ ↑	$1s^2 2s^2 2p^3$
O	8	⇅	⇅	⇅ ↑ ↑	$1s^2 2s^2 2p^4$
F	9	⇅	⇅	⇅ ⇅ ↑	$1s^2 2s^2 2p^5$
Ne	10	⇅	⇅	⇅ ⇅ ⇅	$1s^2 2s^2 2p^6$

EXAMPLE 3.10

Americium, which is a man-made element used in smoke detectors, has 7 electrons in the 5f subshell. How many unpaired electrons are located in this subshell?

Solution

There are seven electrons in a subshell that can contain up to 14 electrons. There are seven orbitals in the subshell. According to Hund's rule, one electron must first be located in each orbital before any are paired. The final representation of the 5f subshell of americium has seven unpaired electrons.

↑ ↑ ↑ ↑ ↑ ↑ ↑

PROBLEM 3.10

Selenium is a trace mineral required for good health, but in large quantities it is toxic. Write the electron configuration of selenium (atomic number 34) and determine the number of unpaired electrons in the highest energy subshell.

3.10 Electron-Dot Symbols

The highest energy level of an atom containing electrons in *s* and *p* subshells is called the **valence energy level** or **valence shell.** The electrons that occupy these subshells are known as **valence electrons.** All the other electrons, together with the nucleus, are collectively called the **core.** This division is useful because the valence electrons form chemical bonds (Chapters 4 and 5). Because of the importance of valence electrons, we display them in electron-dot symbols as shown for second period elements (Table 3.5). Electron-dot symbols are written according to the following rules:

1. The symbol for the element represents the core.
2. Dots representing valence electrons are placed on either side of, above, and below the symbol.
3. All positions about the symbol are equivalent.
4. A maximum of two electrons is allowed per side, above, and below the symbol, for a total of eight.
5. Valence electrons are indicated by a pair of dots for pairs of electrons and a single dot for an unpaired electron in the valence shell.
6. Hydrogen and helium are exceptions to the maximum of eight electrons rule. They are allowed a maximum of two electrons.

3.11 The Periodic Table and Electron Configuration

Why do the elements as displayed in the periodic table exhibit periodic properties? How are the properties of elements related to the atomic number? We know that the atomic number of an atom is equal to the number of electrons in the atom, and these electrons are systematically located in specific subshells. The periodic table works because it reflects the buildup of the electron configuration of the elements. In this section, the following facts will be established:

TABLE 3.5 Electron-Dot Symbols for Second Period

Li·	Be:	$\dot{\text{B}}$·	$\dot{\text{C}}$· (with dot below)	·N: (with dot below)	:Ö· (with dot below)	:F· (with dots above and below)

1. Rows in the periodic table correspond to principal energy levels of *s* and *p* subshells.
2. Combinations of groups correspond to specific subshells.
3. Elements in any single group have the same number of outer electrons in the same type of orbital.

The relationship between electron configuration and the periodic table is shown in Figure 3.8. In the first period, there are only two elements, hydrogen and helium; these have one and two electrons, respectively, in the 1*s* subshell. The elements lithium (3) and beryllium (4) of the second period have electrons in the 2*s* subshell and are separated from the other six elements in the period. Elements boron (5) through neon (10) involve the addition of six successive electrons to the 2*p* subshell. Thus the two elements adding electrons only in the 2*s* subshell are on the left of the period, whereas the elements having electrons in the 2*p* subshell are on the right of the period. The period number and the energy level number are both labeled 2.

FIGURE 3.8 Relationship Between Subshells and the Periodic Table

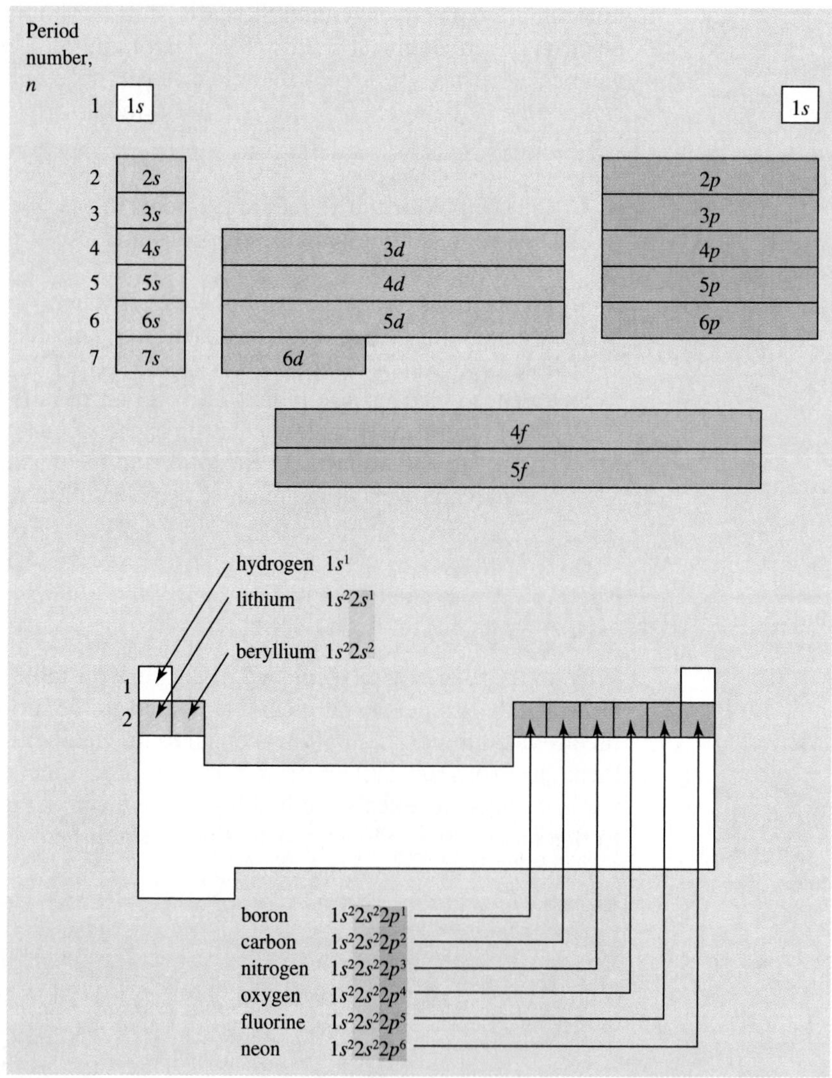

We could continue showing the location of the elements in the periodic table and the related electron configurations of the highest energy electrons. However, the periodic table can be dissected into regions that contain elements with similar electron configurations. The 2 columns on the left have elements with their highest energy electrons in an *s* subshell. The 6 columns on the right have elements with their highest energy electrons in a *p* subshell. The 10 columns in the middle have elements with their highest energy electrons in a *d* subshell. The energy level numbers of the *s* and *p* subshells correspond to the period numbers. The energylevel numbers of *d* subshells are always 1 less than the period number. Outside the general body of the periodic table are located two groups of 14 elements that represent the filling of the 4*f* and 5*f* subshells. The energy-level numbers of *f* subshells are 2 less than the period number.

The electron configuration of an element can be written by reading through the "subshell regions" of Figure 3.8 from left to right and then through the next period until the element is reached. With a little practice you can do this for any element, using an ordinary periodic table without the divisions indicated. Consider iron, for example. The regions of the table that must be passed to reach iron are

$$1s \quad 2s \quad 2p \quad 3s \quad 3p \quad 4s \quad 3d$$

Each of the subshells is filled as electrons are added. Iron is the sixth element in the 3*d* area of the chart, which means that there are six electrons in the 3*d* subshell. The electron configuration for iron is therefore

$$1s^22s^22p^63s^23p^64s^23d^6$$

EXAMPLE 3.12

Using the periodic table, determine the electron configuration of zirconium (atomic number 40).

Solution

In order to reach zirconium by reading through the periodic table, you must pass through the 1*s*, 2*s*, 2*p*, 3*s*, 3*p*, 4*s*, 3*d*, 4*p*, and 5*s* regions. Zirconium is found to be the second element in the 4*d* region. The entire electron configuration is

$$1s^22s^22p^63s^23p^64s^23d^{10}4p^65s^24d^2$$

PROBLEM 3.11

Using the periodic table, determine the electron configuration of tin (atomic number 50).

3.12 Properties of Atoms and the Periodic Table

The periodic table provides a wealth of information. The chemical and physical properties of an element can be predicted from the properties of other elements in the same group. Some of this information is outlined in this section.

Chemical Properties Are Predictable from Groups

In the next two chapters we will learn how the number of valence electrons of an atom determines the number and types of bonds that it forms. The number of valence electrons for a representative element is given directly by the group number.

Since the period number is the same as the energy level being filled, you can tell the valence electron configuration at a glance. Thus, all Group IA elements have an s^1 outer shell configuration. For the top three elements, the configurations are

Li	$1s^22s^1$
Na	$1s^22s^22p^63s^1$
K	$1s^22s^22p^63s^23p^64s^1$

Other elements in the group have similar electron configurations and, as a result, similar chemical properties. Lithium, sodium, and potassium all react with water to form hydrogen gas and caustic solutions. Rubidium and cesium should react similarly. Lithium, sodium, and potassium combine with chlorine to form LiCl, NaCl, and KCl, respectively. Thus RbCl and CsCl should also form from the elements.

Physical Properties

Metals of the representative elements are most often characterized by three or fewer electrons in the valence shell. The transition elements are all metals; they have no more than two electrons in the valence shell, although they have additional electrons in the d and f subshells. Nonmetals are all representative elements, and they generally have more electrons in the valence shell than do metals.

Where do metals end and nonmetals begin in the periodic table? For the representative elements, the metallic properties increase with increasing atomic number within a group, and the nonmetallic properties decrease. Metallic properties within periods increase toward the left, and nonmetallic properties increase toward the right. Thus, within a group, the dividing line between metals and nonmetals occurs at higher atomic numbers for groups on the right than for groups on the left. The stair-step line shown in Figure 3.9 separates the metals and nonmetals. Elements to the right of the line are nonmetals, and elements to the left of the line are metals. The **metalloids** lie immediately to either side of the line. For example, in Group VA the most metallic element is bismuth; antimony and arsenic have borderline properties and are semimetals; phosphorus and nitrogen are nonmetals.

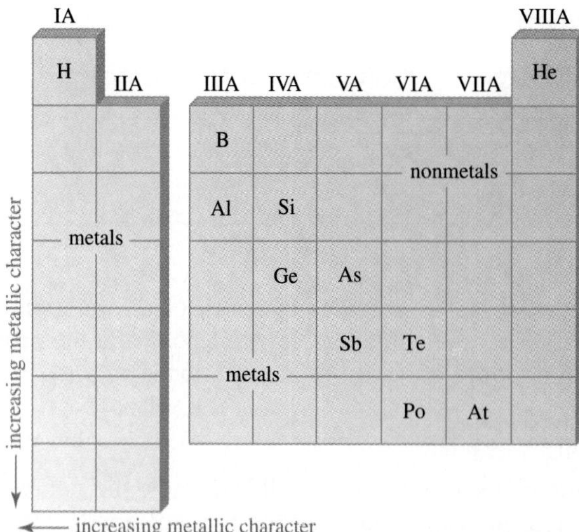

FIGURE 3.9 Metal, Nonmetals, and Metalloids of the Representative Elements

The elements on each side of the stair-step line are metalloids.

Atomic Radius

The size of an atom reflects the number of electrons and their energies, which control the volume of the electron "cloud." The overall shape of the atom is spherical, and the size of an atom is expressed as the radius of a sphere. The radii of atoms are given in nanometers in Figure 3.10 for the representative elements, using a modified version of the periodic table that leaves out the transition elements.

The atomic radii tend to decrease from left to right across a period of the periodic table. Within a period, the nuclear charge increases in the same direction, and the valence electrons are located within the same energy level. The increase in nuclear charge for the elements across the period draws the electrons closer to the nucleus and results in a smaller volume.

From top to bottom in a family of the periodic table, the atomic radii increase. Each successive member has one additional energy level containing electrons. Because the size of a subshell increases with the number of the energy level, the atomic radii increase.

EXAMPLE 3.12

Using the period table, determine the relative sizes of the germanium (atomic number 32) and bromine (atomic number 35) atoms.

Solution
The two elements are in Period 4 of the periodic table. The trend within a period is a decrease in atomic radius in going from left to right. Bromine is in Group VIIA, to the right of germanium in Group IVA. Thus, bromine has a smaller atomic radius than germanium.

PROBLEM 3.12

Using the periodic table, determine the relative sizes of bismuth (atomic number 83) and bromine (atomic number 35) atoms.

FIGURE 3.10 Periodic Trends in Radii (pm) of the Representative Elements

IA	IIA	IIIA	IVA	VA	VIA	VIIA	VIIIA
H 37							He 50
Li 152	Be 111	B 88	C 77	N 70	O 66	F 64	Ne 70
Na 186	Mg 160	Al 143	Si 117	P 110	S 104	Cl 99	Ar 94
K 231	Ca 197	Ga 122	Ge 122	As 121	Se 117	Br 114	Kr 109
Rb 244	Sr 215	In 162	Sn 140	Sb 141	Te 137	I 133	Xe 130
Na 262	Ba 217	Tl 171	Pb 175	Bi 146	Po 150	At 140	Rn 140

Increasing radius (down)

Increasing radius (left)

Ionization Energy

Atoms can lose electrons to form positively charged species called **cations** if sufficient energy is added. The removal of an electron from an atom is called **ionization.** The amount of energy required to remove the electron from a gaseous atom is called the **ionization energy.** Ionization energies are expressed in electron-volts (eV) per atom. The ionization energy of gaseous lithium, Li(g), is 5.4 eV per atom. The (g) represents the phase of the species. The charge of the cation is indicated by a superscript to the right of the elemental symbol.

$$Li(g) \longrightarrow Li^+(g) + 1\,e^-$$

Because the ionization energy is a measure of the energy required to remove an electron from an element, it indicates something about the energy of the electron in the element. Higher energy electrons require less additional energy (lower ionization energy) to remove them from the element; conversely, low-energy electrons are closely bound to the atom and require more energy for ionization. The ionization energies of the representative elements are given in Figure 3.11.

The ionization energies of the representative elements tend to increase from left to right within a period. This trend is due to the increase in the nuclear charge, which causes an increase in the attraction for the electron. The ionization energy for the fluorine atom is 17.4 eV/atom, a value much higher than that for lithium at the start of the period. Metals have lower ionization energies than nonmetals.

Within a group there is a trend to decreasing ionization energies from top to bottom. Thus, the ionization energy of cesium is 3.9 eV/atom, a value less than the 5.4 eV/atom of lithium.

EXAMPLE 3.13

Using the periodic table, determine which element, magnesium or barium, has the higher ionization energy. Which element has the greater tendency to lose electrons?

Solution

The two elements are in the same group of the periodic table. The trend within a group is for ionization energy to decrease in going from top to bottom. Magnesium, which is in the

FIGURE 3.11 Periodic Trends in Ionization Energies (eV/atom) of the Representative Elements

increasing ionization energy ⟶ (left to right); ⟵ increasing ionization energy (bottom to top)

IA	IIA	IIIA	IVA	VA	VIA	VIIA	VIIIA
H 13.6							He 24.6
Li 5.4	Be 9.3	B 8.3	C 11.3	N 14.5	O 13.6	F 17.4	Ne 21.6
Na 5.1	Mg 7.6	Al 6.0	Si 8.2	P 11.0	S 10.4	Cl 13.0	Ar 15.8
K 4.3	Ca 6.1	Ga 6.0	Ge 8.1	As 9.8	Se 9.8	Br 11.8	Kr 14.0
Rb 4.2	Sr 5.7	In 5.8	Sn 7.3	Sb 8.6	Te 9.0	I 10.5	Xe 12.1
Cs 3.9	Ba 5.2	Tl 6.1	Pb 7.4	Bi 7.3	Po 8.4	At	Rn 10.7
Fr	Ra 5.3						

third period, has a higher ionization energy than barium, which is in the sixth period. Barium, which has the lower ionization energy, has the greater tendency to lose electrons.

PROBLEM 3.13

Using the periodic table, determine which element, arsenic or chlorine, has the higher ionization energy. Which element has the greater tendency to lose electrons?

Electronegativity

In Chapters 4 and 5 we will learn that different atoms attract electrons differently in the process of forming bonds. **Electronegativity** is a measure of the ability of an atom to compete for or attract electrons from another atom in a bond. In some cases the atom is sufficiently electronegative to take an electron from another atom and form a negative ion called an **anion.** Fluorine is very electronegative and readily forms an anion.

$$\text{F(g)} + 1\ \text{e}^- \longrightarrow \text{F}^-\text{(g)}$$

The electronegativity values of the representative elements are given in Figure 3.12. They are dimensionless numbers and range from about 1 to 4. The larger the electronegativity value, the greater the tendency to attract electrons. Within a period, electronegativity increases from left to right. Within a group, electronegativity decreases with increasing atomic number. Thus, nonmetals are more electronegative than metals. Metals are often referred to as being electropositive; that is, they have only a small tendency to attract electrons and, in fact, will tend to lose electrons to electronegative elements.

The electronegativity values of atoms determine the types of chemical bonds they form. If the electronegativity values of the atoms are very different, electron transfer will occur, and an ionic bond will result (Chapter 4). If the electronegativity values are similar, electrons are shared in covalent bonds (Chapter 5).

FIGURE 3.12 Periodic Trends in Electronegativities of the Representative Elements

increasing electronegativity ⟶

IA	IIA	IIIA	IVA	VA	VIA	VIIA
H 2.1						
Li 1.0	Be 1.5	B 2.0	C 2.5	N 3.0	O 3.5	F 4.0
Na 0.9	Mg 1.2	Al 1.5	Si 1.8	P 2.1	S 2.5	Cl 3.0
K 0.8	Ca 1.0	Ga 1.6	Ge 1.8	As 2.0	Se 2.4	Br 2.8
Rb 0.8	Sr 1.0	In 1.7	Sn 1.8	Sb 1.9	Te 2.1	I 2.5
Na 0.7	Ba 0.9	Tl 1.8	Pb 1.8	Bi 1.9	Po 2.0	At 2.2
Fr 0.7	Ra 0.9					

⟵ increasing electronegativity

EXAMPLE 3.14

Using the periodic table, decide whether tellurium (Te) is more or less electronegative than bromine (Br).

Solution

Bromine is to the right and above tellurium in the periodic table. Electronegativity tends to increase from left to right within a period and to decrease from top to bottom within a group. Thus, iodine, which is to the right of tellurium, has a greater electronegativity, and bromine, which is above iodine, has a still greater electronegativity. Thus, bromine is more electronegative than tellurium.

PROBLEM 3.14

Using the periodic table, decide whether germanium (Ge) is more or less electronegative than sulfur (S).

Summary

An atom is an electrically neutral particle that is the smallest representative unit of an element; its mass is expressed in **atomic mass units (amu).** The **nucleus** of an atom contains **neutrons** and **protons;** the proton has a positive charge. Surrounding the nucleus are a number of **electrons** equal to the number of protons in the nucleus. An electron is negatively charged.

Atoms are described by their atomic number and their mass number. The **atomic number** is the number of protons in the atom. The **mass number** is the sum of the number of protons and neutrons. Different atoms that have the same number of protons but differ in the number of neutrons are called **isotopes** of the element. The elemental symbol with a subscript to the left of the symbol representing the atomic number and a superscript to the left of the symbol representing the mass number is used to give information about the nuclear composition of the isotope.

A neutral unit consisting of atoms bonded to one another is called a **molecule.** Some elements exist as molecules and are represented by the elemental symbol and a subscript to the right to indicate the number of atoms in the molecule.

The **periodic law** is historically based on the **periodic table,** which provides an ordering of the elements according to their chemical and physical properties. The modern periodic table has been used to establish and describe relationships among the elements, based on the **period** and **group** placement of the elements. Elements within a group in the periodic table constitute a **family;** common names of families include **alkali metals, alkaline earth metals, halogens,** and **noble gases.** Elements in groups IA–VIIIA are **representative elements;** elements in the B groups are **transition elements.** Elements outside the main body of the periodic table—the **lanthanides** and **actinides**—are **inner transition elements.** The classification of elements as metals, semimetals, and nonmetals is shown by their positions in the periodic table.

Our model for the electronic structure of the atom was devised initially from the explanation for line spectra obtained from the elements. The Bohr model of the atom explains the origin of the light emitted from electronically excited hydrogen atoms as the energy released when the electron "falls" back to any of several discrete lower energy orbits from a higher energy orbit.

The modern model of the electronic structure of atoms is more complex. Electrons within an atom have certain energies, but their positions are described in terms of **principal energy levels (shells), subshells,** and **orbitals.**

The maximum number of electrons that can occupy a shell is given by $2n^2$, where n is the number of the shell. Each shell contains one or more subshells designated by *s, p, d,* and *f.* The number of subshells in a particular shell is equal to the value of n. Each subshell can hold a specific maximum number of electrons. These values are 2, 6, 10, and 14 for the *s, p, d,* and *f* subshells, respectively. Each subshell consists of one or more orbitals. For the *s, p, d,* and *f* subshells there are one, three, five, and seven orbitals, respectively. No more than two electrons may occupy any orbital, and the two electrons must have opposite **spins.**

Electron configurations for atoms can be written by following simple rules. First, electrons are located in the lowest energy subshell until the subshell is filled. Second, within a subshell the electrons are distributed according to **Hund's rule**—in different orbitals until each orbital has one electron. When two electrons are located in an orbital, the electrons have opposite spins.

The highest energy level electrons in an atom are in the **valence shell. Electron-dot symbols** are used to show the valence electrons around the **core,** which comprises the nucleus and all electrons other than valence electrons.

The experimental basis of the periodic table prepared by Mendeleev is now understood in terms of a description of the electron configuration of atoms. The groups of the periodic table consist of elements with similar electron configurations. The electron configurations of the elements explain the trends in the **atomic radii, ionization energy,** and **electronegativity** of the elements within the periodic table groupings.

Exercises

Properties of the Atom

3.1 An atom of oxygen has a mass of 2.66×10^{-23} g. How many atoms are in a 16.0-g sample of oxygen?

3.2 An atom of fluorine has a mass of 3.16×10^{-23} g. How many atoms are in a 19.0-g sample of fluorine?

3.3 The atomic radius of arsenic is 1.21×10^{-8} cm. What is its radius in nanometers?

3.4 The atomic radius of indium is 162 pm. What is its radius in centimeters?

3.5 What relationship, if any, exists between the number of electrons and the number of protons in an atom?

3.6 What relationship, if any, exists between the number of electrons and the number of neutrons in an atom?

Elemental Symbols

3.7 What is the atomic number of an element? How is it represented in a symbol?

3.8 What is the mass number of an element? How is it represented in a symbol?

3.9 Indicate the number and type of subatomic particles present in each of the following isotopes.

(a) $^{16}_{8}O$ (b) $^{23}_{11}Na$ (c) $^{27}_{13}Al$
(d) $^{32}_{16}S$ (e) $^{40}_{18}Ar$ (f) $^{40}_{20}Ca$

3.10 Indicate the number and type of subatomic particles present in each of the following isotopes.

(a) $^{19}_{9}F$ (b) $^{31}_{15}P$ (c) $^{39}_{19}K$
(d) $^{79}_{34}Se$ (e) $^{85}_{37}Rb$ (f) $^{96}_{42}Mo$

3.11 Write the symbol for each of the following isotopes, with the indicated numbers of protons and neutrons.

(a) fluorine, 9 protons and 10 neutrons
(b) silicon, 14 protons and 16 neutrons
(c) silicon, 14 protons and 14 neutrons
(d) phosphorus, 15 protons and 16 neutrons

3.12 Write the symbol for each of the following isotopes, with the indicated numbers of protons and neutrons.

(a) cesium, 55 protons and 77 neutrons
(b) selenium, 34 protons and 45 neutrons
(c) cadmium, 48 protons and 64 neutrons
(d) tin, 50 protons and 68 neutrons

Isotopes

3.13 Explain the significance of the symbols used to represent the three isotopes of hydrogen: $^{1}_{1}H, ^{2}_{1}H, ^{3}_{1}H$.

3.14 Explain the significance of the symbols used to represent the three isotopes of oxygen: $^{18}_{8}O, ^{17}_{8}O, ^{16}_{8}O$.

3.15 The radioisotope $^{131}_{53}I$ is used to treat cancer of the thyroid. How does this isotope differ from $^{127}_{53}I$, which is required for normal functioning of the thyroid gland?

3.16 An isotope of chromium whose mass number is 51 is used in compounds to determine the position of the placenta in pregnant women. The atomic number of chromium is 24. What is the symbol for this isotope?

The Periodic Table

3.17 Using the periodic table, indicate the period and group number of each of the following elements. The atomic numbers are given in parentheses.

(a) iron (26) (b) palladium (46)
(c) mercury (80) (d) copper (29)
(e) cadmium (48) (f) osmium (76)

3.18 Identify the element located in the indicated period and group.

(a) 4, VIIB (b) 4, IB (c) 6, VB
(d) 5, IIIB (e) 6, IB (f) 5, VB

3.19 Consider the following elements and select those that are representative elements.

(a) Zn (b) C (c) S
(d) Pd (e) Mn (f) Br

3.20 Consider the following elements and select those that are alkaline earth metals.

(a) Zn (b) Ca (c) Se
(d) Na (e) Sc (f) Ba

Energy Levels

3.21 List the number of electrons contained in each of the following by principal energy levels.

(a) $^{14}_{7}N$ (b) $^{28}_{14}Si$ (c) $^{11}_{5}B$
(d) $^{19}_{9}F$ (e) $^{32}_{16}S$ (f) $^{4}_{2}He$

3.22 List the number of electrons contained in each of the following by principal energy levels.

(a) ${}^{39}_{19}K$ (b) ${}^{24}_{12}Mg$ (c) ${}^{27}_{13}Al$
(d) ${}^{40}_{20}Ca$ (e) ${}^{72}_{32}Ge$ (f) ${}^{70}_{31}Ga$

3.23 How many electrons are contained in the valence shell of each of the following?

(a) ${}^{14}_{7}N$ (b) ${}^{28}_{14}Si$ (c) ${}^{11}_{5}B$
(d) ${}^{19}_{9}F$ (e) ${}^{32}_{16}S$ (f) ${}^{4}_{2}He$

3.24 How many electrons are contained in the valence shell of each of the following?

(a) ${}^{39}_{19}K$ (b) ${}^{24}_{12}Mg$ (c) ${}^{27}_{13}Al$
(d) ${}^{40}_{20}Ca$ (e) ${}^{72}_{32}Ge$ (f) ${}^{70}_{31}Ga$

Subshells

3.25 How many subshells are there in each of the following?

(a) the second energy level
(b) the fourth energy level
(c) the third energy level

3.26 How many subshells are there in each of the following?

(a) the first energy level
(b) the fifth energy level
(c) the sixth energy level

3.27 What subshells are occupied by electrons in the valence level of each of the following atoms?

(a) ${}^{16}_{8}O$ (b) ${}^{36}_{18}Ar$ (c) ${}^{9}_{4}Be$
(d) ${}^{12}_{6}C$ (e) ${}^{31}_{15}P$ (f) ${}^{7}_{3}Li$

3.28 What subshells are occupied by electrons in the valence level of each of the following atoms?

(a) ${}^{39}_{19}K$ (b) ${}^{24}_{12}Mg$ (c) ${}^{27}_{13}Al$
(d) ${}^{40}_{20}Ca$ (e) ${}^{72}_{32}Ge$ (f) ${}^{70}_{31}Ga$

Orbitals

3.29 How many orbitals are in each of the following subshells?

(a) 3*s* (b) 2*p* (c) 4*d*
(d) 6*s* (e) 3*p* (f) 5*d*

3.30 How many orbitals are in each of the following subshells?

(a) 5*s* (b) 4*f* (c) 2*s*
(d) 3*d* (e) 4*p* (f) 4*s*

3.31 How many electrons are allowed to be in each of the following orbitals?

(a) the 3*s* orbital
(b) one of the orbitals in the 3*p* subshell
(c) one of the orbitals in the 4*d* subshell
(d) one of the orbitals in the 5*f* subshell

3.32 How many electrons are allowed to be in each of the following orbitals?

(a) the 6*s* orbital
(b) one of the orbitals in the 6*p* subshell
(c) one of the orbitals in the 5*d* subshell
(d) one of the orbitals in the 4*f* subshell

Electron Configuration

3.33 Write the entire electron configuration for each of the following.

(a) ${}^{16}_{8}O$ (b) ${}^{36}_{18}Ar$ (c) ${}^{9}_{4}Be$
(d) ${}^{12}_{6}C$ (e) ${}^{31}_{15}P$ (f) ${}^{7}_{3}Li$

3.34 Write the entire electron configuration for each of the following.

(a) ${}^{39}_{19}K$ (b) ${}^{24}_{12}Mg$ (c) ${}^{27}_{13}Al$
(d) ${}^{40}_{20}Ca$ (e) ${}^{72}_{32}Ge$ (f) ${}^{70}_{31}Ga$

3.35 Write the entire electron configuration for each of the following.

(a) ${}^{14}_{7}N$ (b) ${}^{28}_{14}Si$ (c) ${}^{11}_{5}B$
(d) ${}^{19}_{9}F$ (e) ${}^{32}_{16}S$ (f) ${}^{4}_{2}He$

3.36 Write the entire electron configuration for each of the following.

(a) ${}^{23}_{11}Na$ (b) ${}^{35}_{17}Cl$ (c) ${}^{13}_{6}C$
(d) ${}^{20}_{10}Ne$ (e) ${}^{85}_{37}Rb$ (f) ${}^{2}_{1}H$

Hund's Rule

3.37 On the basis of the indicated number of electrons in each subshell, determine the number of unpaired electrons.

(a) $3d^6$ (b) $3p^5$ (c) $4s^1$
(d) $2p^3$ (e) $4d^7$ (f) $5f^{10}$

3.38 On the basis of the indicated number of electrons in each subshell, determine the number of unpaired electrons.

(a) $4p^5$ (b) $5d^7$ (c) $5s^2$
(d) $3p^4$ (e) $4d^8$ (f) $4f^9$

3.39 On the basis of the electron configuration, determine the number of unpaired electrons in each of the following elements.

(a) ${}^{14}_{7}N$ (b) ${}^{28}_{14}Si$ (c) ${}^{11}_{5}B$
(d) ${}^{19}_{9}F$ (e) ${}^{32}_{16}S$ (f) ${}^{4}_{2}He$

3.40 On the basis of the electronic configuration, determine the number of unpaired electrons in each of the following elements.

(a) ${}^{16}_{8}O$ (b) ${}^{40}_{18}Ar$ (c) ${}^{9}_{4}Be$
(d) ${}^{12}_{6}C$ (e) ${}^{31}_{15}P$ (f) ${}^{7}_{3}Li$

Electron-Dot Symbols

3.41 Depict each of the following by an electron-dot symbol.

(a) $^{16}_{8}O$ (b) $^{40}_{18}Ar$ (c) $^{9}_{4}Be$
(d) $^{12}_{6}C$ (e) $^{31}_{15}P$ (f) $^{7}_{3}Li$

3.42 Depict each of the following by an electron-dot symbol.

(a) $^{14}_{7}N$ (b) $^{28}_{14}Si$ (c) $^{11}_{5}B$
(d) $^{19}_{9}F$ (e) $^{32}_{16}S$ (f) $^{4}_{2}He$

3.43 Depict each of the following by an electron-dot symbol.

(a) $^{39}_{19}K$ (b) $^{84}_{35}Br$ (c) $^{24}_{12}Mg$
(d) $^{40}_{20}Ca$ (e) $^{72}_{32}Ge$ (f) $^{84}_{36}Kr$

3.44 Depict each of the following by an electron-dot symbol.

(a) $^{131}_{54}Xe$ (b) $^{85}_{37}Rb$ (c) $^{128}_{52}Te$
(d) $^{127}_{52}Te$ (e) $^{119}_{50}Sn$ (f) $^{70}_{31}Ga$

Electron Configuration and The Periodic Table

3.45 In what period and group do the elements with the following electron configurations belong?

(a) $1s^22s^22p^3$ (b) $1s^22s^22p^63s^1$
(c) $1s^22s^22p^63s^23p^5$ (d) $1s^22s^22p^63s^23p^64s^23d^5$

3.46 Identify the groups that have the following general electron configurations.

(a) $ns^2(n-1)d^{10}$ (b) ns^2np^6
(c) $ns^2(n-1)d^5$ (d) $ns^2(n-1)d^8$

3.47 In what period and group is an element with each of the following electron configurations located?

(a) $[Kr]5s^24d^2$ (b) $[Ar]4s^23d^{10}4p^2$
(c) $[Xe]6s^24f^{14}5d^4$ (d) $[Ne]3s^23p^3$

3.48 In what period and group is an element with each of the following electron configurations located?

(a) $[Kr]5s^24d^{10}$ (b) $[Ar]4s^23d^{10}4p^5$
(c) $[Xe]6s^24f^{14}5d^{10}$ (d) $[Ne]3s^23p^4$

3.49 Identify the groups that have the following general electron configurations.

(a) ns^2np^3 (b) $ns^2(n\text{-}1)d^1$
(c) ns^2np^5 (d) ns^2np^2

3.50 Identify the groups that have the following general electron configurations.

(a) $ns^2(n\text{-}1)d^3$ (b) ns^2np^4
(c) ns^2np^1 (d) $ns^2(n\text{-}1)d^2$

3.51 How many electrons are in each of the following?

(a) the *p* subshell of a Group IVA element
(b) the *p* subshell of a Group VA element
(c) the *s* subshell of a Group IA element
(d) the *d* subshell of a Group IVB element

3.52 How many electrons are in each of the following?

(a) the *p* subshell of a Group IIIA element
(b) the *d* subshell of a Group IIIB element
(c) the *p* subshell of a Group VIA element
(d) the *p* subshell of a Group VIIA element

Metals, Nonmetals, and Semimetals

3.53 Consider the following elements, and select those that are metals.

(a) Ca (b) Rb (c) Se
(d) Ag (e) Cl (f) Ni

3.54 Consider the following elements, and select those that are nonmetals.

(a) Zn (b) C (c) S
(d) Pd (e) Mn (f) Br

3.55 Using the periodic table, indicate which member of each pair is the more metallic.

(a) magnesium or silicon (b) germanium or bromine
(c) sulfur or selenium (d) silicon or tin

3.56 Using the periodic table, indicate which member of each pair is the more metallic.

(a) phosphorus or antimony (b) oxygen or tellurium
(c) indium or tin (d) sodium or cesium

Physical Properties

3.57 In each of the following series, estimate the missing value for the indicated property of the element.

(a)	Element	Atomic Radius (nm)
	F	0.064
	Cl	?
	Br	0.114
(b)	Element	Density (g/cm^3)
	Ca	1.54
	Sr	2.60
	Ba	?

3.58 In each of the following series, estimate the missing value for the indicated property of the compound.

(a)	Compound	Boiling Point (°C)
	$SiCl_4$	58
	$GeCl_4$	83
	$SnCl_4$	?
(b)	Compound	Melting Point (°C)
	H_2S	− 86
	H_2Se	?
	H_2Te	− 49

Atomic Radii

3.59 Indicate which member of each of the following pairs of elements has the larger radius.

(a) F or Cl (b) Si or S (c) Li or K
(d) C or F (e) P or Sb (f) Ge or Pb

3.60 Indicate which member of each of the following pairs of elements has the larger radius.

(a) S or F (b) Sr or Cs (c) As or In
(d) C or Ga (e) Ge or Tl (f) Na or Mg

Ionization Energy

3.61 Indicate which member of each of the following pairs of elements has the higher ionization energy.

(a) O or F (b) Li or K (c) Cl or Br
(d) S or Se (e) Ge or Pb (f) Na or Li

3.62 Indicate which member of each of the following pairs of elements has the higher ionization energy.

(a) S or F (b) Sr or Cs (c) As or In
(d) C or Ga (e) Ge or Tl (f) Na or Mg

Electronegativity

3.63 Indicate which member of each of the following pairs of elements has the higher electronegativity.

(a) O or F (b) S or Se (c) Li or K
(d) C or N (e) Cl or Br (f) P or Sb

3.64 Indicate which member of each of the following pairs of elements has the higher electronegativity.

(a) S or F (b) Sr or Cs (c) As or In
(d) C or Ga (e) Ge or Tl (f) Na or Mg

The white cliffs of Dover on the English channel contain ionic compounds such as calcium carbonate. (Paolo Koch)

4 Ionic Compounds

4.1 Bonds, Structure, and Properties
4.2 Types of Bonds
4.3 Cations and Anions
4.4 Ionic Bonds
4.5 Structure of Ionic Compounds
4.6 Names of Ions
4.7 Predicting Formulas of Ionic Compounds
4.8 Names of Ionic Compounds

Essays
- *Ionic Radii Affect Cell Composition and Enzyme Reactions*
- *Calcium Ions and Your Diet*
- *Phosphates in Agriculture*

4.1 Bonds, Structure, and Properties

In Chapter 1 a bond was described as a force that holds atoms together in a molecule. If such forces did not exist, the atoms would fly apart and a molecule couldn't be distinguished from a homogeneous mixture. Furthermore, if atoms weren't bonded in specific ways, then the law of constant proportions could not have been proposed. Now let's move beyond the general concept of a bond and start to examine what a bond really is.

A reporter asks who, what, where, when, and why before writing a story about an event. Chemists ask similar questions about all types of matter in our world.

1. How many atoms of each type are in the substance?
2. Which atoms are bonded to each other and how?
3. What is the shape of the molecule?
4. Why do molecules have characteristic physical and chemical properties?

The answer to the first question is provided by determining the composition of a substance. Questions 2 and 3 deal with structure, which results from bonds, the subject of this and the following chapter. Shapes of molecules are discussed extensively in Chapter 5, where you will learn to predict the shapes of molecules based on their constituent atoms. As you will see in Chapter 8, the types of bonds between atoms in molecules and the structure of molecules govern the forces that exist between the molecules. These forces in turn affect the physical properties of matter. They make sugar brittle and wax soft. They make water a liquid at room temperature and carbon dioxide a gas.

The effects of bonding and structure are central to an understanding of chemical properties, which will be discussed throughout this book. Structure is especially important in biochemical reactions, most of which are catalyzed by enzymes. Each enzyme has a shape designed by nature to use or produce specific compounds. The types of bonds and structures discussed in this and the following chapter are for simple substances only, but the concept of bonding applies equally well to the larger, more complex biological molecules.

4.2 Types of Bonds

Before discussing the models for bonding it is instructive to consider the properties of a few simple compounds that we encounter every day. We can divide compounds into two classes based on their physical properties. One class includes compounds that are brittle crystalline materials and have high melting and boiling points. Table salt (NaCl), for example, is a crystalline solid that melts at 801 °C. Compounds of the other class have very different properties. They tend to be gases, liquids, or low-melting solids. Carbon dioxide, water, and table sugar (sucrose), which melts at 186 °C, are examples.

A clue about how atoms are bonded in compounds is provided by the separation of compounds into groups based on their physical properties. Since two general groups of compounds exist, it has been suggested that two types of bonds exist: ionic bonds and covalent bonds. In this chapter we will study compounds with bonds between oppositely charged ions, or ionic bonds; compounds with electrons shared in covalent bonds are the subject of Chapter 5. In considering both types of bonds we will indicate how the changes in the electronic structure of two or more atoms are responsible for the forces that hold atoms together.

Lewis Octet Rule

In 1916, G. N. Lewis suggested that the filled valence shell electron configuration of the inert gases is responsible for their lack of reactivity. Helium has an electron configuration of $1s^2$. The other inert gases have eight electrons in the *s* and *p* subshells of the highest energy level. Lewis suggested that atoms of other elements react to obtain an electron configuration of eight electrons in their *s* and *p* subshells to electronically resemble the inert gases. This hypothesis is summarized in the **Lewis octet rule:** *Atoms combine and form bonds either by transferring electrons to form ions or by sharing electrons in covalent bonds until each atom is surrounded by eight valence electrons.* Note that the hydrogen atom requires only two electrons.

4.3 Cations and Anions

As we indicated in Section 3.12, electropositive atoms may lose an electron to form a positively charged ion called a **cation.** Electronegative atoms may gain electrons to form a negatively charged ion called an **anion.** All ions are written with the charge as a superscript to the right of the elemental symbol. When the charge is +1 or −1, the number 1 is omitted and only the + or − is retained. For other charges the sign follows the integer. A cation that results from the loss of three electrons from an atom has a +3 charge, which is written as 3+ as a superscript on the elemental symbol, as in Al^{3+}. An anion that results from the gain of two electrons by an atom has a −2 charge, which is written as 2− as a superscript on the elemental symbol, as in S^{2-}.

Formation of Ions and Energy

In Chapter 3 we learned that the energy required to remove an electron from an atom is lowest for elements located to the left and bottom of the periodic table. The atoms of these elements have only one or two electrons in partially filled valence shells. Removing them creates cations whose electron configurations resemble that of an inert gas. Only one electron needs to be removed from the 3*s* subshell of sodium (electron configuration $3s^1$), but two must be removed from the 3*s* subshell of magnesium (electron configuration $3s^2$). Both Na^+ and Mg^{2+} cations have the same electron configuration as neon, but note that energy is required to remove electrons in both cases.

Just how many electrons can be removed from an atom and at what price in terms of energy? Consider the sequential removal of electrons from a metal represented by M to give a series of cations:

$$M \longrightarrow M^+ + 1\,e^-$$
$$M^+ \longrightarrow M^{2+} + 1\,e^-$$
$$M^{2+} \longrightarrow M^{3+} + 1\,e^-$$

Larger amounts of energy are required for each successive ionization, because it is much harder to remove the negatively charged electron from a positively charged species—the cation—than from a neutral atom. In addition, the successive removal of electrons requires much more energy after the electrons of the partially filled valence shell are removed. Removing an electron from the filled energy shell below the valence shell requires a substantially greater amount of energy.

Now let's consider why some atoms gain electrons to form anions. In Chapter 3 we learned that attraction of electrons by atoms of elements is greatest for elements located to the right and top of the periodic table. The atoms of these elements have nearly filled valence shells. Adding electrons creates anions whose electron configurations resemble that of an inert gas. Only one electron needs to be added to the 3*p* subshell of chlorine (electron configuration $3s^23p^5$). Two electrons need to be added to the 3*p* subshell of sulfur (electron configuration $3s^23p^4$). Both Cl^- and S^{2-} anions have the same electron configuration as argon.

Just how many electrons can be added to an atom? First of all, there must be an available subshell. As in the formation of cations, there are energy considerations that limit how many electrons can be added to an available subshell. As electrons are added there are offsetting forces. The negatively charged electrons are always attracted to the positively charged nucleus at any distance. However, extra electrons cause more electron–electron repulsions. Thus, if too many electrons are added, the repulsive forces will outweigh the attractive forces. Consequently, there is a limit to the number of electrons that an atom may accept to form an anion.

EXAMPLE 4.1

Iron (atomic number 26) can lose three electrons to form a cation. How many protons and electrons does the cation have? Write the symbol for the ion.

Solution

The atomic number is equal to the number of protons in an atom, and the number of electrons equals the number of protons in a neutral atom. Thus, iron has 26 electrons and 26 protons. When iron loses 3 electrons, 23 electrons and 26 protons remain. The resulting cation has a +3 charge and is represented by Fe^{3+}.

PROBLEM 4.1

The atomic number of zinc is 30. Zinc can lose two electrons to form a cation, the form in which zinc is found in many vital enzymes. How many protons and electrons does the cation have? Write the symbol for the ion.

Ionic Radii

In Chapter 3 we learned that the atomic radius depends on the energy level containing electrons, and the nuclear charge. Now let's consider the *ionic radius,* which is the radius of an ion as found in an ionic compound. Some examples of ionic radii of Groups IA, IIA, VIA, and VIIA compared with their atomic radii are shown in Figure 4.1. The radius of an anion is larger than the radius of the related atom. Anions have one or more extra electrons without a balancing attraction from the positive charge in the nucleus. Thus, the electrons of anions are more

FIGURE 4.1 Atomic and Ionic Radii of Representative Elements

The radii are given in picometers (10^{-12}m).

Group IA		Group IIA		Group VIA		Group VIIA	
Atoms	Ions	Atoms	Ions	Atoms	Ions	Atoms	Ions
Li 152	Li^+ 60	Be 111	Be^{2+} 31	O 66	O^{2-} 140	F 64	F^- 136
Na 186	Na^+ 95	Mg 160	Mg^{2+} 65	S 104	S^{2-} 184	Cl 99	Cl^- 181
K 227	K^+ 133	Ca 197	Ca^{2+} 88	Se 117	Se^{2-} 198	Br 114	Br^- 195
Rb 248	Rb^+ 148	Sr 215	Sr^{2+} 113	Te 137	Te^{2-} 221	I 133	I^- 216
Cs 265	Cs^+ 169	Ba 217	Ba^{2+} 135				

Ionic Radii Affect Cell Composition and Enzyme Reactions

The size of an ion plays an important role in biological systems. Ions with the same charge, such as Na^+ and K^+, do not behave the same way in cells and are found in quite different places. Sodium ions are concentrated in extracellular fluid, whereas potassium ions are located primarily within cells. The cell membranes (Chapter 22) have imbedded proteins that are responsible for transport of materials across the membrane. They can distinguish between the ions based on ionic radii and keep each ion where it belongs.

Many biological reactions are catalyzed by enzymes (Chapter 24), several of which contain small amounts of a specific ion such as Zn^{2+} that is required for the enzyme to properly catalyze a metabolic reaction. No other ion can serve as a substitute. In fact, replacement of the Zn^{2+} ion, which is so vital to many enzymes, can cause severe disruptions in many functions and can be fatal to an organism. The ionic radius of Cd^{2+} is 92 pm—it is slightly larger than the ionic radius of Zn^{2+}, which is 74 pm. (Note that both cadmium and zinc are in Group IIB.) The Cd^{2+} ion is poisonous because it can replace Zn^{2+} ion in an enzyme, creating a molecule that can no longer catalyze the necessary biological reactions.

spread out in space than those of their corresponding atoms. For example, the chlorine atom has a radius of 99 pm, but the chloride ion, which has one more electron than the chlorine atom—18 electrons attracted by only 17 protons in the nucleus—has a radius of 181 pm. A cation, on the other hand, has a smaller radius than the related atom. There are fewer electrons about the nucleus than there are protons in the nucleus; as a result, the electrons are attracted closer to the nucleus. In the case of the sodium ion, which has "lost" its outer shell, there are 10 electrons attracted by 11 protons in the nucleus. The radius is only 95 pm, compared with the 186 pm radius of the sodium atom.

EXAMPLE 4.2

The atomic radii of zinc and nitrogen atoms are 135 and 70 pm, respectively. Discuss how the sizes of the Zn^{2+} and N^{3-} ions compare with those of the neutral atoms.

Solution

A loss of two electrons from zinc reduces the repulsion between the remaining electrons and they are pulled in closer to the nucleus. Thus, the zinc ion should have an atomic radius of less than 135 pm. The actual radius is 74 pm. The N^{3-} ion has three more electrons than the nitrogen atom. Because there are still only 7 protons in the nucleus, the ionic radius is expected to be larger than that of the nitrogen atom. The actual radius is 171 pm.

PROBLEM 4.2

The atomic radii of calcium and sulfur are 0.197 and 0.104 m, respectively. How would you expect the ionic radii of the Ca^{2+} and S^{2-} ions to compare with those of the neutral atoms?

4.4 Ionic Bonds

Ionic bonds form between two or more atoms by the transfer of one or more electrons between atoms. Electron transfer produces anions and cations that attract each other. In ionic compounds, the electrons lost by one or more atoms to form a cation with a Lewis octet are accepted by the appropriate number of atoms to form an anion with a Lewis octet. Ionic compounds are typically the result of combinations of metals, which are located on the left side of the periodic table, with nonmetals, which are located on the upper right side of the periodic table. Metals are electropositive and tend to lose electrons, whereas nonmetals are electronegative and tend to gain electrons.

Let us examine the ionic bond in sodium chloride. The sodium atom, which has 11 protons and 11 electrons, has a single valence electron in the 3*s* subshell. The chlorine atom, which has 17 protons and 17 electrons, has seven valence electrons represented by $3s^2 3p^5$. An ionic bond forms when the sodium atom, which is electropositive, loses its valence electron to the electronegative chlorine atom. As a result, the sodium ion obtains the same electron configuration as neon ($1s^2 2s^2 2p^6$) and develops a +1 charge, because there are 11 protons in the nucleus but only 10 electrons around the nucleus.

The chlorine atom gains an electron and is converted into a chloride ion that has the same electron configuration as argon ($1s^2 2s^2 2p^6 3s^2 3p^6$). The chloride ion has a −1 charge, because there are 17 protons in the nucleus and 18 electrons around the nucleus.

The formation of sodium chloride from the sodium and chlorine atoms is shown using Lewis structures. **Lewis structures** represent only the valence electrons; the electron pairs are shown as pairs of dots.

$$\underset{[Ne]3s^1}{Na\cdot} + \underset{[Ne]3s^2 3p^5}{:\ddot{\underset{..}{Cl}}\cdot} \longrightarrow \underset{[Ne]}{Na^+} + \underset{[Ar]}{:\ddot{\underset{..}{Cl}}:^-}$$

Note that, by convention, the complete octet is shown for an anion formed from an electronegative element. However, the filled outer shell for a cation that results from loss of electrons from an electropositive element is understood but not shown.

Sodium combines with oxygen to form the ionic compound sodium oxide (Na_2O). In this compound, each of two sodium atoms loses its $3s^1$ valence electron and is converted into a sodium ion with the neon electron configuration and a +1 charge. Oxygen, which has a $1s^2 2s^2 2p^4$ electron configuration, requires two electrons to produce the neon electron configuration. Thus, each of the two sodium atoms transfers its single valence electron to one oxygen atom. The oxide ion produced has two more electrons than protons and has a −2 charge. The formation of sodium oxide from sodium and oxygen atoms is represented with electron-dot symbols as

$$\underset{[Ne]3s^1}{2\,Na\cdot} + \underset{[He]2s^2 2p^4}{:\ddot{\underset{\cdot}{O}}\cdot} \longrightarrow \underset{[Ne]}{2\,Na^+} + \underset{[Ne]}{:\ddot{\underset{..}{O}}:^{2-}}$$

EXAMPLE 4.3

Describe the ionic bond formed between calcium and chlorine in calcium chloride ($CaCl_2$).

Solution

The valence shell electron configurations of calcium and chlorine are $4s^2$ and $3s^2 3p^5$, respectively. Therefore, calcium can lose two electrons to produce the Ca^{2+} cation and achieve the stable electron configuration of argon. Chlorine can gain one electron to achieve the stable electron configuration of argon by forming the Cl^- anion.

$$Ca(4s^2) \longrightarrow Ca^{2+} + 2\,e^-$$
$$1\,e^- + Cl(3s^2 3p^5) \longrightarrow Cl^-(3s^2 3p^6)$$

To balance the electron requirements of each type of atom, each of the two chlorine atoms accepts one electron from a calcium atom to form $CaCl_2$. In terms of electron-dot symbols, the process is

$$\text{Ca}\colon + 2\,\colon\ddot{\underset{\cdot\cdot}{\text{Cl}}}\cdot \longrightarrow \text{Ca}^{2+} + 2\,\colon\ddot{\underset{\cdot\cdot}{\text{Cl}}}\colon^{-}$$

PROBLEM 4.3

Describe the ionic bond formed between magnesium and oxygen in magnesium oxide (MgO).

4.5 Structure of Ionic Compounds

Ionic compounds are solids that consist of ions arranged in a repeating three-dimensional pattern. In a crystal of sodium chloride (Figure 4.2) each sodium ion is surrounded by chloride ions. Each chloride ion is also surrounded by sodium ions. The number of sodium ions equals the number of chloride ions, and the crystal is electrically neutral regardless of its size. The entire crystal is held together by electrostatic forces between the cations and anions.

From the start of our study of chemistry it has been suggested that the molecule is the simplest structural unit in a compound. The ratio of atoms bonded within the molecule is constant. However, ionic compounds do not have molecular units. For example, there is no single unit in a sample of sodium chloride that has an independent existence. We then define a **formula unit,** which is equal to the lowest possible whole number ratio of ions in the compound. Although formulas such as Na_2Cl_2, Na_3Cl_3, and even $Na_{1000}Cl_{1000}$ express the 1:1 ratio of ions in a region of a NaCl crystal, only NaCl is accepted as the simplest formula of sodium chloride.

Energy of Bond Formation

As indicated in Section 1.10, the release of energy is one of the two reasons why chemical processes occur spontaneously. The reaction of sodium and chlorine to produce sodium chloride is exothermic. Let's qualitatively consider why this combination of elements is exothermic, accounting for the energy of hypothetical steps

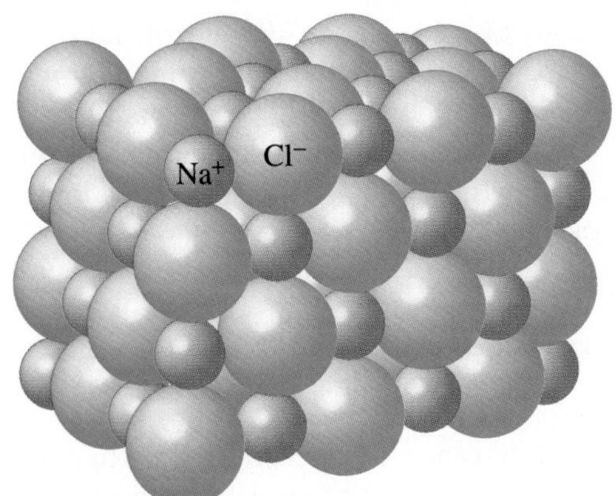

FIGURE 4.2 The Structure of Sodium Chloride

Calcium Ions and Your Diet

Calcium is not contained in many foods, so calcium deficiencies can result from improper diets. Milk and dairy products are the best common sources of calcium, which is present as the Ca^{2+} ion. Other sources include broccoli and sardines. The recommended intake of calcium is 1.0 g per day for adults and 1.4 g per day for pregnant or lactating women. A cup of milk contains about 0.3 g of calcium. Calcium supplements, which are ionic compounds, are also commonly used, especially by postmenopausal women to prevent osteoporosis.

About 99% of the body's calcium is stored in the bones and teeth. The remaining 1% is in the blood. However, calcium deficiency can't be detected by blood analysis because the body maintains the blood level by depleting the bone calcium. Calcium deficiency from improper nutrition causes the slow degeneration of bones. However, even the loss of bone calcium is difficult to detect by X ray until the damage is advanced. Unfortunately, the process of bone degeneration is irreversible. People need adequate calcium in their diet from childhood throughout adulthood. Otherwise, as they age, they may suffer from deformities of the spinal column and frequent fractures.

In addition to building bone tissue, calcium plays a number of other roles in the body. For example, calcium regulates the transport of other ions into and out of cells and the transmission of nerve impulses. It is also required for blood clotting.

that might occur. First, energy is required to remove an electron from the sodium atom. Next, the acceptance of an electron by the chlorine atom releases energy. Finally the two ions are pulled together by electrostatic attraction. Is this last step exothermic or endothermic? We know that it requires energy to pull two oppositely charged particles apart. Thus, for the reverse step, when two ions move closer together, energy is released. In summary, sodium chloride forms because energy is released as a result of the three outlined steps:

1. loss of an electron by sodium	endothermic
2. gain of an electron by chlorine	exothermic
3. attraction of oppositely charged ions	exothermic
formation of sodium chloride	exothermic

You might ask, How do we know that the sum of the reactions is exothermic? All ions that might be formed by transfer of electrons would be attracted to one another and release energy. Thus, the determination of whether two atoms react to form an ionic compound depends on the difference between the energy required to remove an electron from one atom and the energy released by the second atom as it accepts an electron. We won't do such calculations, but you need to remember that combinations of elements on the left of the periodic table with those on the right of the periodic table have the right combination of properties that make ionic bond formation in compounds such as NaCl and $MgCl_2$ possible. Compounds such as CCl_4 are not ionic because too much energy is required to remove four electrons from carbon to give the C^{4+} ion. As will be seen in the next chapter, CCl_4 exists as a covalent molecule.

4.6 Names of Ions

We need to develop some chemical vocabulary to name the cations and anions that make up ionic compounds. For metals that form only one cation, the name is derived by adding the word *ion* to the name of the element (Table 4.1). The term ion

TABLE 4.1 Names of Cations

Group IA	Group IIA	Transition Elements
Li^+ lithium ion		
Na^+ sodium ion	Mg^{2+} magnesium ion	Zn^{2+} zinc ion
K^+ potassium ion	Ca^{2+} calcium ion	Ni^{2+} nickel ion
Rb^+ rubidium ion	Sr^{2+} strontium ion	Ag^+ silver ion
Cs^+ cesium ion	Ba^{2+} barium ion	Cd^{2+} cadmium ion

FIGURE 4.3
Calcium is a metal that reacts with water to produce hydrogen gas. Calcium ions in compounds are different than the metal. (Richard Megna, Fundamental Photographs)

is required to distinguish between the atom and its derived ion. However, in some contexts people use the name of the atom when in fact they should use the name of the ion. For example, bananas are said to contain potassium, and milk is well known as a source of calcium. Neither food contains these metals, which are so reactive that they liberate hydrogen gas from water (Figure 4.3). In these cases the terms potassium and calcium refer to the ions K^+ and Ca^{2+}, respectively.

The majority of the transition metals form several cations. For example, copper forms both a +1 and a +2 ion. Two systems are used to indicate the charge of these related ions. An older system of naming uses a suffix to the name of either the current or the "ancient" name of the metal. The endings *-ous* and *-ic* are used for the ions with the lower and higher charge, respectively. For a few elements the ancient Latin name is used to name the ions. Thus, Cu^+ is the cuprous ion, and Cu^{2+} is the cupric ion (from the Latin *cuprum*). However, the names chromous and chromic ions for Cr^{2+} and Cr^{3+}, respectively, are derived from the current name of the metal. Note that the terms -ous and -ic don't tell you the actual value of the charge of the ion. The -ous suffix represents the +1 ion of copper but the +2 ion for chromium.

The older system of naming ions has been replaced by a system that uses Roman numerals within parentheses to indicate the charge. In modern terms, cuprous and cupric ions are copper(I) and copper(II) ions, respectively. Table 4.2 includes the names in both systems for these cations.

The anions derived from nonmetallic elements are named by adding *-ide* to the root of the element's name followed by the word ion. Thus F^- is the fluoride ion. Table 4.3 lists the names of a number of simple anions. You should memorize the names of these ions and their charges.

TABLE 4.2 Multiple Cations of Transition Metals

Cation	Common Name	Systematic Name
Cu^+	cuprous ion	copper(I) ion
Cu^{2+}	cupric ion	copper(II) ion
Fe^{2+}	ferrous ion	iron(II) ion
Fe^{3+}	ferric ion	iron(III) ion
Cr^{2+}	chromous ion	chromium(II) ion
Cr^{3+}	chromic ion	chromium(III) ion
$Hg_2{}^{2+}$	mercurous ion	mercury(I) ion
Hg^{2+}	mercuric ion	mercury(II) ion

TABLE 4.3 Names of Simple Anions

Anion	Name
F^-	fluoride ion
Cl^-	chloride ion
Br^-	bromide ion
I^-	iodide ion
O^{2-}	oxide ion
S^{2-}	sulfide ion
Se^{2-}	selenide ion
N^{3-}	nitride ion
P^{3-}	phosphide ion

Ions that are composed of several atoms are called **polyatomic ions.** The atoms are covalently bonded to one another. The polyatomic ion may be positive or negative; however, negatively charged polyatomic ions are more common. A list of some common polyatomic ions is given in Table 4.4.

The superscript on the right of the atoms in a polyatomic ion indicates the charge on the ion, whereas the subscripts give the number of atoms of that element in the ion. Thus, the nitrate ion, NO_3^-, consists of one atom of nitrogen and three atoms of oxygen and contains one electron more than the component atoms normally contain. You should learn the symbols for polyatomic ions—both the num-

TABLE 4.4 Names of Polyatomic Ions

Formula	Name
NO_3^-	nitrate ion
NO_2^-	nitrite ion
SO_4^{2-}	sulfate ion
SO_3^{2-}	sulfite ion
HSO_4^-	bisulfate ion
HSO_3^-	bisulfite ion
PO_4^{3-}	phosphate ion
CO_3^{2-}	carbonate ion
HCO_3^-	bicarbonate ion
OH^-	hydroxide ion
CN^-	cyanide ion
MnO_4^-	permanganate ion
ClO_4^-	perchlorate ion
ClO_3^-	chlorate ion
ClO_2^-	chlorite ion
ClO^-	hypochlorite ion
NH_4^+	ammonium ion

ber of atoms in the ion and the charge on the ion. This information is needed to write formulas for ionic compounds.

Polyatomic ions containing oxygen have suffixes *-ate* and *-ite*. The ion with the -ate suffix contains more oxygen atoms than the ion with the -ite suffix. Thus, nitrate and nitrite ions are NO_3^- and NO_2^-, respectively; sulfate and sulfite ions are SO_4^{2-} and SO_3^{2-}, respectively.

Sometimes two elements form more than two polyatomic ions. For example, Cl and O form perchlorate (ClO_4^-), chlorate (ClO_3^-), chlorite (ClO_2^-), and hypochlorite (ClO^-) ions. The prefix *per-* means "over." Thus, perchlorate has one more oxygen atom than chlorate. The prefix *hypo-* means "under." Thus, hypochlorite has one fewer oxygen atom than chlorite.

4.7 Predicting Formulas of Ionic Compounds

Ionic compounds are electrically neutral because the charge of the cations is balanced by the charge of the anions. Sodium chloride has an equal number of Na^+ and Cl^- ions. The single positive charge of the Na^+ ion balances the single negative charge of the Cl^- ion.

Na^+	charge of +1
Cl^-	charge of −1
	net charge 0

As we discussed in Section 4.5, sodium chloride is represented by NaCl even though there is no sodium chloride molecule. The formula indicates only the 1:1 ratio of the ions in the compound.

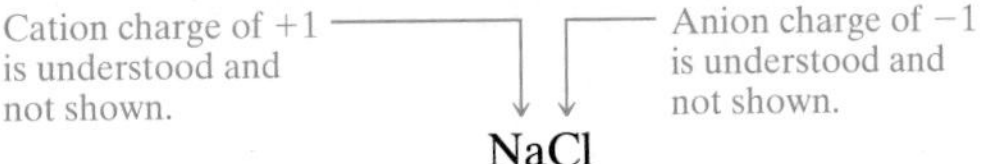

For some combinations of ions, the charge on one cation will not balance the charge on one anion. Therefore, the number of cations will not be equal to the number of anions. The numbers of cations and anions required for electrical neutrality are given by the following relationship:

$$\text{positive charge} + \text{negative charge} = 0$$

$$\text{charge/cation} \times \text{number of cations} + \text{charge/anion} \times \text{number of anions} = 0$$

In calcium chloride the +2 charge of a single Ca^{2+} ion must be balanced by two Cl^- ions. The compound is represented as $CaCl_2$ although there is no $CaCl_2$ molecule. A crystal of $CaCl_2$ contains Ca^{2+} and Cl^- ions in a 1:2 ratio.

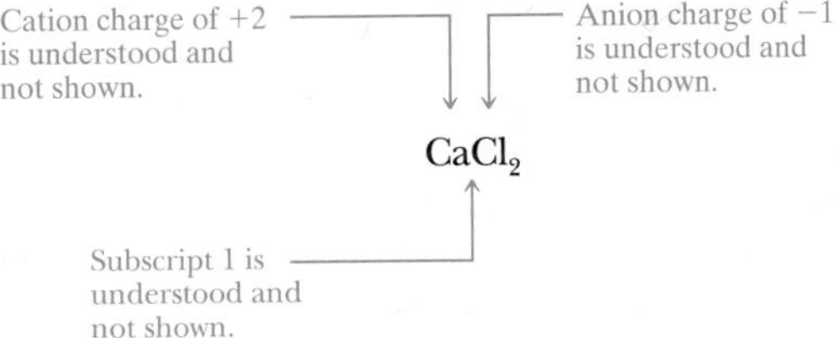

Notice that in writing the formula for an ionic compound, we write the positive ion before the negative ion. Now let's write the formula for a compound that isn't quite so easy. What is the formula for scandium sulfide? Scandium forms a +3 ion;

the sulfide ion has a −2 charge. The charges provide a clue about the subscripts necessary to obtain a neutral compound. Use 3 as the subscript for sulfur and 2 as the subscript for scandium. This "cross-switching" procedure interchanges the numerical values of the charges to serve as subscripts. In mathematical terms we are simply recognizing the fact that 6 is the lowest common multiple of 2 and 3. It takes two Sc^{3+} ions to electrically balance three S^{2-} ions. Thus, the formula is Sc_2S_3, writing the cation first.

Whenever more than one polyatomic ion is required to write the formula unit for an ionic compound, parentheses are used to enclose the polyatomic ion. (Parentheses are never used around monatomic ions.) The subscript to the right of the closing parenthesis indicates the number of polyatomic ions required for electrical neutrality. Thus, $Ca_3(PO_4)_2$ is an ionic compound that consists of three doubly charged calcium ions for every two triply charged phosphate ions.

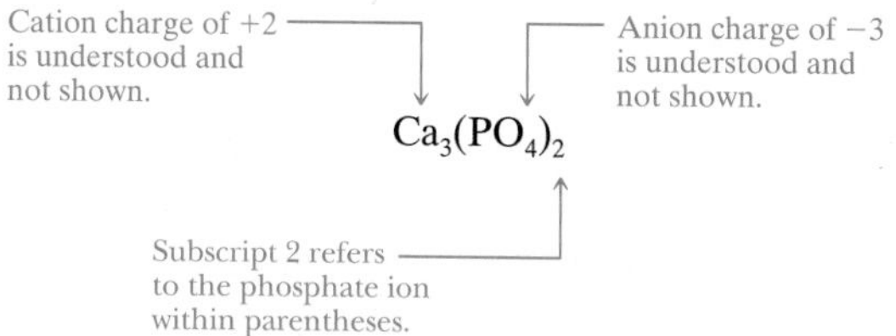

Phosphates in Agriculture

In a different era and at a time with fewer people, farmers could grow sufficient food for their needs without serious damage to the soil. Macronutrients such as nitrogen, in the form of nitrate ions, and phosphorus, in the form of phosphates, could be supplied by "organic" fertilizers when used over a long period of time and in large quantities. However, to feed many more people in societies that are no longer agrarian, farmers have had to increase crop yields per acre by using "chemical" fertilizers.

Phosphate rock, a mineral containing calcium phosphate [$Ca_3(PO_4)_2$], can be pulverized and dusted directly onto soil. However, the process is inefficient because calcium phosphate is not very soluble. The phosphate can be made more soluble by converting it into calcium dihydrogen phosphate [$Ca(H_2PO_4)_2$], or superphosphate, in a reaction with sulfuric acid.

$$\underset{\text{phosphate rock}}{Ca_3(PO_4)_2} + 2\,H_2SO_4 \longrightarrow \underset{\text{superphosphate}}{Ca(H_2PO_4)_2} + 2\,CaSO_4$$

You may have used fertilizers in gardening or for houseplants. These products list their formulations using three numbers, for example, 6-12-12. The percentage of phosphate, calculated in the form of P_2O_5, is given by the middle number. The first and third numbers give the percentages of nitrogen and potassium, respectively. Superphosphate fertilizer has formulations such as 0-45-0.

Phosphate rock contains some cadmium compounds—a matter of some concern because cadmium ions are toxic. Phosphate rock in Russia has as little as 3 mg of cadmium per kilogram of phosphate; U.S. phosphate contains from 20 to 120 mg of cadmium per kilogram of phosphate. In some areas of the world the cadmium content is nearly 300 mg/kg. The cadmium taken up by plants could cause cadmium poisoning in animals or humans. The cadmium can be removed in the manufacturing process but, as in the case of everything else in life, only at a price.

The three numbers on this bag of fertilizer indicate the percentages of nitrogen, phosphorus, and potassium compounds, respectively.

EXAMPLE 4.4

Aluminum forms a +3 ion. Write the formula for aluminum oxide, the material that coats the surface of any aluminum product such as cookware.

Solution

The compound consists of Al^{3+} and O^{2-} ions, and the total charge of the cations and the anions in the formula must be zero. The lowest common multiple of 3 and 2 is 6. Thus, three oxide ions have a total charge of − 6, and two aluminum ions have a total charge of +6. The formula is Al_2O_3.

PROBLEM 4.4

An oxide of chromium has the formula Cr_2O_3. What is the charge of the chromium ion?

EXAMPLE 4.5

Treatment of teeth with fluoride ion produces fluorapatite; the formula is $Ca_5(PO_4)_3F$. Show that this material is electrically neutral.

Solution

Calcium exists in ionic compounds as Ca^{2+}. The phosphate ion has a −3 charge, and the fluoride ion has a −1 charge. The five Ca^{2+} ions have a total charge of $5(+2) = +10$. The three phosphate ions and one fluoride ion have a total charge of $3(-3) + (-1) = -10$. Thus, the compound is electrically neutral.

PROBLEM 4.5

The antacid in Rolaids is $NaAl(OH)_2CO_3$. Show that this substance is electrically neutral.

4.8 Names of Ionic Compounds

In naming an ionic compound, as in writing its formula, the name of the positive ion is given first, then the name of the negative ion. Thus, sodium chloride is the name for NaCl. In the case of elements that form more than one cation, the charge on the ion must be indicated in the name. For example, $FeCl_2$ and $FeCl_3$ are ferrous chloride and ferric chloride, respectively, which tells you that the iron ions in the two compounds are +2 and +3 ions, respectively. The newer names

TABLE 4.5 Names of Ionic Compounds

Formula	Name	Formula	Name	Formula	Name
$LiClO_4$	lithium perchlorate	$CaCl_2$	calcium chloride	$Mg(CN)_2$	magnesium cyanide
LiBr	lithium bromide	CaS	calcium sulfide	$Zn(OH)_2$	zinc hydroxide
$NaNO_3$	sodium nitrate	$CaCO_3$	calcium carbonate	$ZnSO_4$	zinc sulfate
$NaHCO_3$	sodium bicarbonate	$CaSO_4$	calcium sulfate	NH_4Cl	ammonium chloride
KCl	potassium chloride	$Ca_3(PO_4)_2$	calcium phosphate	$(NH_4)_2SO_4$	ammonium sulfate
$KMnO_4$	potassium permanganate	MgO	magnesium oxide	$(NH_4)_3PO_4$	ammonium phosphate
K_3PO_4	potassium phosphate	MgF_2	magnesium fluoride		

Milk of magnesia is a heterogeneous mixture of magnesium hydroxide and water.

TABLE 4.6 Medical Applications of Some Ionic Compounds

Name	Formula	Application
ammoniun carbonate	$(NH_4)_2CO_3$	expectorant, smelling salts
barium sulfate	$BaSO_4$	radiopaque compound for X ray
calcium sulfate	$CaSO_4$	plaster casts
lithium carbonate	Li_2CO_3	antidepressive agent
magnesium hydroxide	$Mg(OH)_2$	antacid (milk of magnesia)
magnesium sulfate	$MgSO_4$	laxative (Epsom salts)
potassium nitrate	KNO_3	diuretic
sodium fluoride	NaF	toothpaste
sodium bicarbonate	$NaHCO_3$	antacid
tin (II) fluoride	SnF_2	external application to teeth
zinc oxide	ZnO	calamine lotion

are iron(II) chloride and iron(III) chloride. A list of ionic compounds and their names is given in Table 4.5. You might want to check them to verify that the formulas are correct and represent electrically neutral compounds. The names of some common ionic compounds that have medical or related applications are listed in Table 4.6.

EXAMPLE 4.6

What is the name for the compound Cu_2O?

Solution

The oxide ion has a −2 charge. Therefore the charge on each of the two copper ions must be +1. The correct name for this ion is cuprous or copper(I) ion. The compound is then named cuprous oxide or copper(I) oxide.

PROBLEM 4.6

What is the name for the compound $(NH_4)_2CO_3$, which is used in smelling salts?

Summary

A loss or gain of electrons by an atom yields an ion. Positive ions are **cations;** negative ions are **anions.** Ions containing two or more atoms are **polyatomic ions.** The charge of an ion is indicated by a superscript to the right of the atomic symbol.

Anions have a larger radius than that of the corresponding atom; cations have a smaller radius compared with the corresponding atom.

The **Lewis octet rule** states that atoms have a tendency to lose, gain, or share valence electrons to form a noble gas electron configuration. In other words, all atoms tend to acquire eight electrons in their valence shell.

Ions held together by electrostatic forces of attraction result in **ionic compounds.** The relative numbers of ions are such that the total charge of the cations equals the total charge of the anions.

Ionic bonds are formed by transfer of electrons between atoms with sufficiently different electronegativity values to form ions. The metal tends to lose enough electrons to achieve the electron configuration of the preceding inert gas; the nonmetal tends to gain electrons to achieve the electron configuration of the next inert gas.

Ionic compounds are named by giving the name of the positive ion followed by the name of the negative ion. The formulas for ionic compounds use subscripts to indicate the simplest ratio of ions contained in the substance.

Exercises

Ions

4.1 How many electrons does each of the following ions formed by the indicated process contain? What is the symbol for the ion?

(a) selenium (atomic number 34) can gain two electrons
(b) phosphorus (atomic number 15) can gain three electrons

4.2 How many electrons does each of the following ions formed by the indicated process contain? What is the symbol for the ion?

(a) manganese (atomic number 25) can lose two electrons
(b) scandium (atomic number 21) can lose three electrons

4.3 Cerium (atomic number 58) can form two ions. One results from the loss of three electrons, and the other from the loss of four electrons. What are the symbols for these ions? How many electrons does each ion contain?

4.4 Thallium (atomic number 81) can form two ions. One results from the loss of one electron, and the other from the loss of three electrons. What are the symbols for these ions? How many electrons does each ion contain?

4.5 Write proper symbols for each of the following ions.

(a) oxide ion (b) sulfide ion (c) iodide ion
(d) bromide ion (e) nitride ion (f) fluoride ion

4.6 Write proper symbols for each of the following ions.

(a) sodium ion (b) calcium ion (c) magnesium ion
(d) potassium ion (e) lithium ion (f) zinc ion

4.7 What is the name of each of the following ions?

(a) S^{2-} (b) I^- (c) O^{2-}
(d) F^- (e) N^{3-} (f) Br^-

4.8 What is the name of each of the following ions?

(a) K^+ (b) Ca^{2+} (c) Zn^{2+}
(d) Mg^{2+} (e) Na^+ (f) Li^+

Ionic Radii

4.9 Which is the larger particle in each of the following pairs? Explain why.

(a) Mg or Mg^{2+} (b) K or K^+
(c) Al or Al^{3+} (d) Br or Br^-
(e) S or S^{2-} (f) N or N^{3-}

4.10 Which is the larger particle in each of the following pairs? Explain why.

(a) O^{2-} or F^- (b) Na^+ or Mg^{2+}
(c) S^{2-} or Cl^- (d) Al^{3+} or Mg^{2+}
(e) Mg^{2+} or Ca^{2+} (f) Br^- or I^-

Polyatomic Ions

4.11 Write the proper symbols for the following ions.

(a) sulfate ion (b) phosphate ion
(c) hydroxide ion (d) ammonium ion
(e) cyanide ion (f) carbonate ion
(g) sulfite ion (h) bicarbonate ion

4.12 Write the proper symbols for the following ions.

(a) perchlorate ion (b) hypochlorite ion
(c) chlorite ion (d) chlorate ion
(e) hypoiodite ion (f) bromate ion
(g) hypobromite ion (h) iodate ion

4.13 Explain what each of the following symbols for ions represents.

(a) WO_4^{2-} (b) HPO_4^{2-} (c) $P_2O_7^{4-}$ (d) VO_2^+

4.14 Explain what each of the following symbols for ions represents.

(a) SeO_3^{2-} (b) $S_2O_8^{2-}$ (c) $H_2BO_3^-$ (d) $P_4O_{12}^{4-}$

Ionic Compounds

4.15 Write the correct formulas of compounds containing the following ions.

(a) Fe^{3+} and Cl^- (b) Na^+ and OH^-
(c) Mg^{2+} and OH^- (d) Cd^{2+} and S^{2-}
(e) Mn^{2+} and F^- (f) Cs^+ and N^{3-}

4.16 Write the correct formulas of compounds containing the following ions.

(a) Al^{3+} and O^{2-} (b) Na^+ and CO_3^{2-}
(c) Mg^{2+} and MnO_4^- (d) Cd^{2+} and PO_4^{3-}
(e) Zn^{2+} and SO_3^{2-} (f) K^+ and S^{2-}

4.17 Write the formula of the ionic compound that results from the reaction of each pair of atoms.

(a) Li and F (b) Mg and Br
(c) Li and O (d) Mg and Se
(e) Al and F (f) Na and S

4.18 Write the formula of the ionic compound that results from the reaction of each pair of atoms.

(a) Ca and Br (b) Li and N
(c) K and S (d) Ba and F
(e) Mg and Cl (f) Cs and P

4.19 The formula of the arsenate ion is AsO_4^{3-}. What other polyatomic ion does it resemble? Lead arsenate was formerly used as an insecticide. What is the formula of this compound?

4.20 The formula of the dihydrogen phosphate ion, an ion eliminated in urine to control the pH of cellular fluids, is

$H_2PO_4^-$. What is the formula of calcium dihydrogen phosphate?

4.21 The formula of the diphosphate ion, an intermediate of metabolic processes, is $P_2O_7^{4-}$. Write the formulas of sodium diphosphate and calcium diphosphate.

4.22 Some calcium supplements are citrate salts. The formula of the citrate ion is $C_6H_5O_7^{3-}$. Write the formula of calcium citrate.

Names of Ionic Compounds

4.23 Name the following compounds.

(a) $Ca(OH)_2$ (b) $LiClO_4$ (c) Na_3PO_4
(d) K_2SO_4 (e) $NaNO_3$ (f) NH_4NO_2

4.24 Name the following compounds.

(a) $Zn(CN)_2$ (b) $Cd(ClO_3)_2$ (c) $Mg_3(PO_4)_2$
(d) $LiNO_3$ (e) $Ba(HSO_3)_2$ (f) $(NH_4)_2SO_4$

4.25 Name the following compounds.

(a) $Ba(OH)_2$ (b) $KClO_2$ (c) Cs_3PO_4
(d) K_2SO_3 (e) $NaNO_2$ (f) NH_4NO_3

4.26 Name the following compounds.

(a) $ZnCO_3$ (b) $Cd(ClO_2)_2$ (c) $Ca_3(PO_4)_2$
(d) KNO_3 (e) $Ba(HSO_4)_2$ (f) $(NH_4)_2SO_3$

4.27 Write the correct formula of each of the following compounds.

(a) lithium fluoride (b) zinc oxide
(c) magnesium carbonate (d) potassium nitrate
(e) sodium cyanide (f) aluminum sulfide

4.28 Write the correct formula of each of the following compounds.

(a) lithium phosphate (b) zinc sulfate
(c) magnesium permanganate (d) sodium bisulfate
(e) calcium hydroxide (f) zinc phosphate

4.29 Write the correct formula of each of the following compounds.

(a) ferric fluoride (b) cuprous oxide
(c) ferrous sulfate (d) cupric cyanide
(e) ferric oxide (f) cupric nitrate

4.30 Write the correct formula of each of the following compounds.

(a) iron(II) chloride (b) copper(I) oxide
(c) lead(II) sulfide (d) mercury(II) bromide
(e) iron(III) sulfide (f) copper(II) sulfide

4.31 Write the formula of each of the following compounds.

(a) calcium hypochlorite, used as a disinfectant in swimming pools
(b) aluminum hydroxide, used in Maalox
(c) copper(II) sulfate, used to kill algae in swimming pools

4.32 Write the formula of each of the following compounds.

(a) titanium(IV) oxide, used as a white paint pigment
(b) silver(I) sulfide, the tarnish on silver
(c) sodium phosphate, used in detergents

This starfish in waters near Fiji is one of the myriad aquatic animals that depend on covalent bonds to sustain life. (Robert Ouellette)

5 Molecular Elements and Compounds

5.1 Covalent Bonds
5.2 Drawing Lewis Structures
5.3 Polar Covalent Bonds
5.4 Coordinate Covalent Bonds
5.5 Resonance Structures
5.6 Molecular Geometry
5.7 Polarity of Molecules
5.8 Orbitals and Molecular Shapes
5.9 Structural Formulas of Organic Compounds
5.10 Structural Models

Essay
- *Coordinate Covalent Bonds and Respiration*

5.1 Covalent Bonds

We learned in the previous chapter that there are two major types of bonds—ionic and covalent. Ionic bonds form between atoms that differ substantially in electronegativity. Fewer than 10% of the known substances are ionic compounds.

We now consider the larger class of compounds, in which covalent bonds hold the atoms of identical or similar electronegativity together in molecules. The simplest molecules are diatomic molecules of the elements such as hydrogen, oxygen, nitrogen, and the halogens. The most complex molecules are biological compounds such as lipids, carbohydrates, proteins, and ribonucleic acids. We will start learning about covalent bonds by first taking little steps and after becoming more proficient, finally taking the larger steps in the latter chapters of this text.

Bonds between the same kinds of atoms or atoms with similar electronegativities cannot form by electron transfer precisely because they are identical or very similar. In other words, those atoms that can lose electrons don't have any tendency to gain electrons, and vice versa. For example, Cl_2 is not an ionic compound consisting of Cl^- and Cl^+ ions. The chloride ion is easily formed because chlorine is electronegative, but the formation of Cl^+ is not feasible because chlorine has a very high ionization energy. The energy required to form Cl^+ is too high to allow formation of an ionic bond. Thus, the bond between chlorine atoms in a chlorine molecule is of a different type: it is a covalent bond, in which a pair of electrons is shared.

Our model for the formation of a covalent bond is most easily seen by considering the simplest molecule, H_2. The atoms are identical and each atom "wants" to have two electrons in its 1*s* subshell to resemble the inert gas helium. As a result, the two hydrogen atoms share their electrons.

$$H\cdot + \cdot H \longrightarrow H:H$$

The electrons shown between the atomic symbols are called **bonding electrons.** They are the "glue" that holds the atoms together in molecular units. In a sense we could say that each hydrogen atom is tugging at the pair of electrons, and neither is willing to give up the struggle, so they stay together. That simple picture reflects the sum of all the attractive and repulsive forces that arise when two protons and two electrons of the two hydrogen atoms are close to each other (Figure 5.1). The two protons of the hydrogen nuclei repel each other, as do the electron clouds of the negatively charged electrons. However, each proton attracts the electron cloud of the other atom. As the atoms move close together there is a point at which the attractive forces between each nucleus and the resulting shared pair of electrons are larger than the repulsive forces. This distance, called the **bond length,** is the distance between the nuclei in the molelcule. The bond length depends on the atoms making up the bond.

Nonbonding Electrons

Now let's consider the bond that forms between atoms with more than a single electron in their valence shell. The two chlorine atoms in the chlorine molecule are joined by a shared pair of electrons. Each chlorine atom, which has seven valence electrons in the third energy level, requires one electron to form an argon-like electron configuration. A bonding electron pair is formed by the sharing of two electrons, one from each chlorine atom. Each chlorine atom has three other pairs of electrons in its valence shell. Pairs of electrons in the valence shell that are not involved in bonding are called **nonbonding electrons,** or **lone-pair electrons,** or **unshared electron pairs.**

$$:\ddot{\underset{..}{Cl}}\cdot + \cdot\ddot{\underset{..}{Cl}}: \longrightarrow :\ddot{\underset{..}{Cl}}-\ddot{\underset{..}{Cl}}: \quad \text{nonbonding electrons}$$

As a result of forming a covalent bond, each chlorine atom is now considered to have a Lewis octet. Each atom has six nonbonding electrons and shares two bonding electrons.

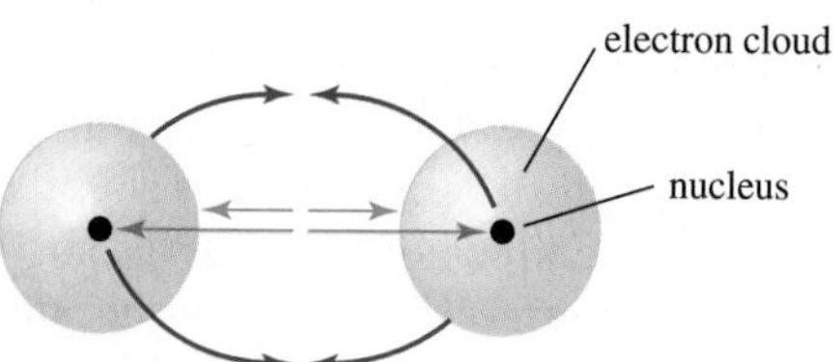

FIGURE 5.1
Bonding Forces in the Hydrogen Molecule
The sum of the attractive forces between the electron of one atom and the proton in the nucleus of the other hydrogen atom is greater than the sum of the electron–electron repulsive forces and the proton–proton repulsive forces.

Molecules such as hydrogen and chlorine are represented by **Lewis structures,** which show the bonding electrons as a line and the nonbonding electrons as pairs of dots located about the core notation for the atoms. The Lewis structures of hydrogen and chlorine molecules are as follows:

$$\text{H—H} \quad :\ddot{\underset{..}{\text{Cl}}}\text{—}\ddot{\underset{..}{\text{Cl}}}:$$

Multiple Covalent Bonds

More than one pair of electrons can be shared between pairs of atoms. If four electrons (two pairs) or six electrons (three pairs) are shared, the bonds are called **double bonds** and **triple bonds,** respectively. The nitrogen molecule consists of two nitrogen atoms bound together by six shared electrons—a triple bond. Each nitrogen atom contributes three electrons to the triple bond.

$$:\dot{\underset{\cdot}{\text{N}}}\cdot + \cdot\dot{\underset{\cdot}{\text{N}}}: \longrightarrow :\text{N}::\text{N}: \quad \text{or} \quad :\text{N}\equiv\text{N}:$$

The carbon atom forms single, double, or triple bonds with other carbon atoms as well as with some other elements (Chapter 13). The carbon atom has four electrons in its valence shell that can be shared to form four covalent bonds. In ethane, ethylene, and acetylene the carbon atoms are bonded by single, double, and triple covalent bonds, respectively. The remaining valence electrons of carbon form single bonds with hydrogen atoms. Each carbon atom has an octet of electrons.

single bond sharing 2 e⁻ — $H_3C\text{—}CH_3$ — ethane

double bond sharing 4 e⁻ — $H_2C{=}CH_2$ — ethylene

triple bond sharing 6 e⁻ — $\text{H—C}\equiv\text{C—H}$ — acetylene

The carbon–carbon bond lengths for single, double, and triple bonds decrease as the number of electrons shared between the same atoms increases; that is, triple bonds are shorter than double bonds, which in turn are shorter than single bonds. The **bond energy,** which is the energy required to break apart the atoms, decreases in the order triple > double > single. This trend makes sense, for one would expect that it would take more energy to break apart atoms that are held together by more electrons.

Covalent Bonds Between Different Atoms

Ethane, ethylene, and acetylene molecules are shown with covalent bonds between carbon and hydrogen as well as between carbon atoms. These carbon–hydrogen bonds result because the difference between the electronegativities of carbon and hydrogen is small, and electron transfer cannot occur. Substances containing carbon, which constitute the largest class of compounds, are known as **organic compounds.** They are the basic units from which biological compounds are formed. We will study organic chemistry, the chemistry of organic compounds, in Chapters 13 through 19, and biochemistry in Chapters 20 through 29.

The carbon atom can form covalent bonds to atoms of a number of other elements with similar electronegativities. Among these are nitrogen, oxygen, sulfur, and chlorine, as shown in the following simple molecules.

```
   H                 H                H                 H
   |    ..           |   ..           |   ..            |  ..
H—C—N—H          H—C—O:           H—C—S:            H—C—Cl:
   |  |              |  |             |  |              |  ..
   H  H              H  H             H  H              H
```

methylamine methanol methyl mercaptan chloromethane

The hydrogen atom and the halogen atoms, such as chlorine, form only one covalent bond to other atoms in stable neutral compounds. However, the carbon, oxygen, and nitrogen atoms can bond to more than one atom. The number of covalent bonds an atom can form is called the **valence** of the atom. The valence of a given atom is the same in most stable, neutral organic compounds. The valences of carbon, nitrogen, oxygen, and sulfur are 4, 3, 2, and 2, respectively. The valences of hydrogen and chlorine are both one.

5.2 Drawing Lewis Structures

Drawing Lewis structures of molecules that contain more than two atoms requires an organized approach. To systematically draw such structures, the following strategy is useful:

1. Determine the sum of the valence electrons of the constituent atoms. For an anion, add to the sum the number of electrons required to produce the charge. For a cation, subtract from the sum the number of electrons required to produce the charge.
2. Based on the known connectivity of the atoms, draw a skeleton structure bonding the necessary atoms with single covalent bonds.
3. For each bond, subtract two electrons from the total number of valence electrons, to give the number of electrons available for nonbonded electron pairs or multiple bonds.
4. Determine the number of electrons necessary to complete the octet about each atom (except for hydrogen, which requires only two electrons). If this number equals the number calculated in step 3, place the electrons around the appropriate atoms.
5. If the number of electrons determined in step 3 is insufficient to provide all atoms with octets, there must be one or more multiple bonds. If the deficiency is 2, there must be a double bond. If the deficiency is 4, there must be either two double bonds or a triple bond.
6. Modify the structure with the appropriate number of multiple bonds. Then place the remaining electrons as nonbonded electron pairs to satisfy the octet rule for each atom.

EXAMPLE 5.1

Carbon tetrachloride, formerly used as a dry-cleaning agent, has the molecular formula CCl_4. Draw its structure given the fact that each chlorine atom is bonded to the central carbon atom.

Solution

Chlorine has seven valence electrons; carbon has four valence electrons. The total number of valence electrons is $(4 \times 7) + 4 = 32$. The arrangement of atoms with single bonds between the carbon atom and the chlorine atoms is

```
         Cl
         |
    Cl—C—Cl
         |
         Cl
```

32 electrons available
−8 electrons used
24 electrons remaining

Eight electrons have been used; 32 − 8 = 24 electrons remain to be placed. The carbon atom has eight electrons that it shares in four bonds with the four chlorine atoms. Each chlorine atom has only two electrons present in one bonded pair. Thus, each chlorine atom requires six additional electrons in the form of three unshared pairs of electrons to form a Lewis octet. This gives 4 × 3 = 12 electron pairs or 24 electrons, the number of electrons available after the bonds are accounted for. The structure is

```
          ..
         :Cl:
          |
    ..    |    ..
   :Cl—C—Cl:
    ..    |    ..
          |
         :Cl:
          ..
```

Electron count
8 electrons in bonds
24 electrons in nonbonded pairs
32 electrons

PROBLEM 5.1

Draw the Lewis structure for NF_3 given the fact that each fluorine atom is bonded to the central nitrogen atom.

EXAMPLE 5.2

Formaldehyde, CH_2O, was formerly used to preserve biological specimens. Draw its Lewis structure given the fact that the hydrogen atoms and oxygen atom are bonded to the carbon atom.

Solution

The number of valence electrons is (2 × 1) + 4 + 6 = 12 electrons for the two hydrogen atoms, one carbon atom, and one oxygen atom. The arrangement of atoms with the necessary bonds is

```
        O
        |
    H—C—H
```

12 electrons available
−6 electrons used
6 electrons remaining

Each of the two hydrogen atoms has the required two electrons present in its covalent bond. The carbon atom has six electrons that it shares in three bonds, and it requires two additional electrons for its Lewis octet. The oxygen atom has only two electrons present in one bonding pair and requires six more electrons for a Lewis octet. The total number of electrons required is 6 + 2 = 8, which exceeds by 2 the number available. Thus, a double bond must be used. The structure is

```
       :O:
        ‖
    H—C—H
```

Electron count
4 electrons in single bonds
4 electrons in a double bond
4 electrons in nonbonded pairs
12 electrons

Note that with two single bonds to hydrogen and a double bond to oxygen, the carbon atom has no unshared pairs of electrons.

PROBLEM 5.2

Draw the Lewis structure for carbon disulfide (CS_2) given the fact that each sulfur atom is bonded to the central carbon atom.

5.3 Polar Covalent Bonds

Two identical atoms share electron pairs equally; two unlike atoms cannot. In a covalent bond between identical atoms, the bonding electrons are symmetrically distributed. The center of the negative charge is between the two atoms, as is the center of the positive charge. Shared electron pairs between unlike atoms are closer to the atom with the greater attraction for electrons, that is, the atom with the higher electronegativity. The center of the negative charge is not the same as the center of the positive charge. One or more pairs of electrons shared unequally between two different atoms form a **polar covalent bond.**

Consider the hydrogen chloride molecule, which consists of two atoms that require one additional electron each to form an inert gas electron configuration. Chlorine is more electronegative than hydrogen, but the chlorine atom's attraction for electrons is not sufficient to completely remove an electron from hydrogen. The bonding electrons in hydrogen chloride are shared but are attracted toward the chlorine atom in a polar covalent bond. Consequently, the chlorine atom obtains a fractional negative charge (δ^-) at the expense of the hydrogen atom, which obtains a fractional positive charge (δ^+). These charges are less than 1 because complete electron transfer does not occur. The molecule is represented by the conventional Lewis structure, even though the shared electron pair is associated to a larger extent with chlorine.

$$^{\delta+}\text{H}-\ddot{\underset{..}{\text{Cl}}}:^{\delta-}$$

The hydrogen chloride (HCl) molecule has electrically distinguishable ends much like the two ends of a magnet. It is a **dipole** (two poles), which means that the centers of positive and negative charge are separated from each other. The dipole is represented by an arrow with a cross at one end. The cross is near the end of the molecule that is positive, and the arrowhead is near the negative end of the molecule.

$$\overset{+\!\!\longrightarrow}{\text{H}-\text{Cl}}$$

There is only a displacement of electrons toward the more electronegative chlorine atom in hydrogen chloride, not a transfer of electrons between atoms, as in sodium chloride. Thus, there are no ions in hydrogen chloride, and the entire molecule is electrically neutral even though there are partial charges on each atom (Figure 5.2). What numerical difference in electronegativity between elements is required to form an ionic bond as opposed to a polar covalent bond? A general rule of thumb is that ionic bonds are formed if the difference in electronegativity is greater than 1.9, but this rule must be used with some caution. If the difference is close to 1.9—say, 1.7—the type of bond cannot be predicted with certainty. Only for a large difference or a small difference in electronegativity can the type of bond be predicted with confidence. Regardless of electronegativity differences, all compounds of hydrogen and nonmetals have covalent bonds, not ionic bonds.

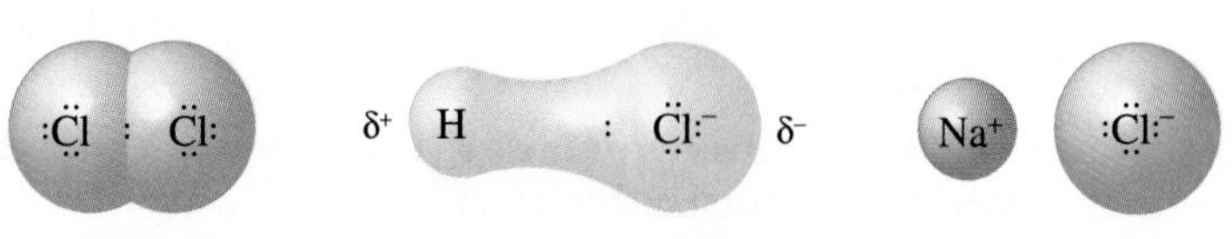

:C̤̈l : C̤̈l: δ+ H : C̤̈l:⁻ δ−   Na+ :C̤̈l:⁻

FIGURE 5.2 Representation of Covalent, Polar Covalent, and Ionic Bonds

The shapes represent the volumes that surround the nuclei and within which electrons are located. In chlorine the volume is symmetrical around the two chlorine nuclei. Note that in HCl the volume around the hydrogen atom is smaller because the electron pair in the bond is more closely associated with the chlorine atom. In sodium chloride electron transfer occurs, and each ion is spherical.

The greater the difference between the electronegativities of the bonded atoms, the more polar the bond. Thus, the direction of the polarity of common bonds is easily predicted. Hydrogen is positive with respect to the common nonmetals.

$$\overset{\longmapsto}{\text{H—N}} \quad \overset{\longmapsto}{\text{H—O}} \quad \overset{\longmapsto}{\text{H—Cl}}$$
(2.1) (3.0) (2.1) (3.5) (2.1) (3.0)

Bonds between carbon and most nonmetals, such as are found in organic molecules, have the positive end of the dipole at the carbon atom.

$$\overset{\longmapsto}{\text{C—N}} \quad \overset{\longmapsto}{\text{C—O}} \quad \overset{\longmapsto}{\text{C—Cl}}$$
(2.5) (3.0) (2.5) (3.5) (2.5) (3.0)

Hydrogen and carbon have similar electronegativity values, and the C—H bond is not normally considered a polar covalent bond. Ethane, ethylene, and acetylene are considered to have only covalent bonds.

EXAMPLE 5.3

Carbon monoxide, CO, is a colorless, odorless and very toxic gas. Draw the Lewis structure for carbon monoxide. What type of bond exists in carbon monoxide?

Solution

The total number of valence electrons for the carbon and oxygen atoms is 4 + 6 = 10 electrons. The arrangement of atoms with a single bond is

C—O

Two electrons are used to form a single bond. The carbon atom and oxygen atom require six additional electrons each for a Lewis octet. The total number of additional electrons required is 6 + 6 = 12, which exceeds by 4 the number available. Thus, a triple bond must be used. The structure is

:C≡O:

Oxygen is more electronegative than carbon. The shared electrons in the triple bond are predicted to be closer to the oxygen atom. The bond is polar covalent.

PROBLEM 5.3

Chloroethane, C_2H_5Cl, is a topical anesthetic that boils at 12 °C. The liquid is sprayed from a pressurized can onto the skin to numb it. Draw its Lewis structure. What types of bonds exist in this molecule? (Refer to the structure of ethane.)

5.4 Coordinate Covalent Bonds

A **coordinate covalent bond** is formed when both electrons of the bonding electron pair shared by two atoms are provided by the same atom. Boron in BF_3 has only six electrons in its valence shell and needs two to complete its octet. A coordinate covalent bond forms in the reaction of ammonia with boron trifluoride.

$$\mathrm{H_3N{:}} + \mathrm{BF_3} \longrightarrow \mathrm{H_3N{-}BF_3}$$

coordinate covalent bond

In ammonia, the nitrogen atom shares three of its five valence electrons with three hydrogen atoms. The remaining two valence-shell electrons are not involved in bonding in the ammonia molecule. However, this unshared pair of electrons forms a coordinate covalent bond with the electron-deficient boron trifluoride.

Consider the structures of hypochlorous acid (HOCl) and chlorous acid ($HClO_2$). The chlorine atom in hypochlorous acid has three unshared pairs of electrons. An oxygen atom has six valence-shell electrons and requires two electrons to complete a Lewis octet. A coordinate covalent bond results when the chlorine atom shares one pair of electrons with the oxygen atom.

coordinate covalent bond

H—Ö—C̈l: H—Ö—C̈l—Ö:

hypochlorous acid chlorous acid

Coordinate Covalent Bonds and Respiration

Red blood cells carry oxygen bonded to iron by a coordinate covalent bond. The iron is contained in a large organic molecule called heme, which in turn is combined with four even larger protein molecules called globin. More details of this structure can be found in Chapter 23.

The interaction of iron with oxygen is represented by a coordinate covalent bond in which both electrons are provided by oxygen.

:Ö=Ö—Fe(heme)

The oxygen is carried to body cells, where it is used for oxidation reactions to provide energy for the organism. This process can be disrupted by any other molecule that forms coordinate covalent bonds to iron. Examples include both carbon monoxide and cyanide ion.

:Ö=Ö:	:O≡C:	:C≡N:⁻
oxygen	carbon monoxide	cyanide

Coordinate covalent bonds from iron to carbon monoxide or cyanide ion are formed in preference to a bond to oxygen. Thus, when a person breathes either carbon monoxide or hydrogen cyanide, the hemoglobin cannot carry oxygen. In a very short time, the body cells can no longer function and the person dies. Hydrogen cyanide is a more lethal poison than carbon monoxide because it also interferes with the electron transport chain (Chapter 25) by reacting with the iron of cytochrome.

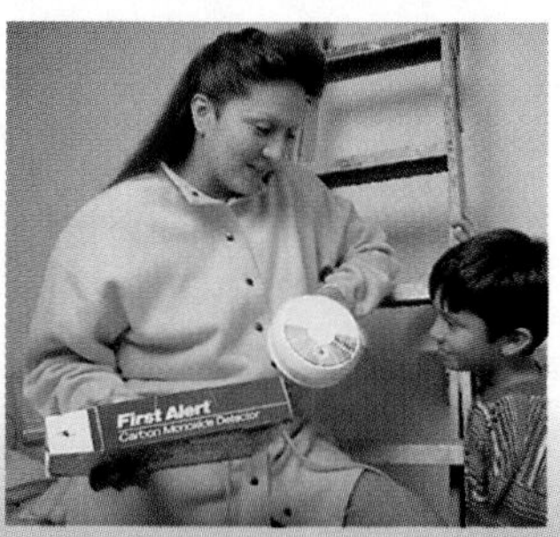

Commercially available carbon monoxide detectors should be placed in areas where the deadly gas may accumulate. (Tony Freeman, PhotoEdit)

EXAMPLE 5.4

Consider the following structures of PCl_3 and $POCl_3$, and describe the bond formed when PCl_3 combines with an oxygen atom.

$$:\ddot{\underset{..}{Cl}}-\ddot{P}-\ddot{\underset{..}{Cl}}: \quad :\ddot{\underset{..}{Cl}}-\overset{:\ddot{O}:}{P}-\ddot{\underset{..}{Cl}}:$$

Solution
The oxygen atom, which has only six valence-shell electrons, requires two electrons from the phosphorus atom to form a bond. The oxygen atom shown in $POCl_3$ is bonded to the phosphorus atom by a coordinate covalent bond.

PROBLEM 5.4

In many biological molecules a carbon atom is bonded to one of the oxygen atoms of phosphoric acid in place of a hydrogen atom. Draw the Lewis structure for phosphoric acid (H_3PO_4). The hydrogen atoms are bonded to oxygen atoms. Describe the phosphorus–oxygen bonds.

5.5 Resonance Structures

A single Lewis structure does not always adequately represent a substance. For example, the structure of ozone (O_3) based on the Lewis octet rule requires a double bond between the central oxygen atom and one of the oxygen atoms but a single bond to the second oxygen atom.

$$:\ddot{O}=\ddot{O}-\ddot{\underset{..}{O}}:$$

A double bond is shorter than a single bond (Section 5.1). However, both oxygen–oxygen bonds in the ozone molecule are 128 pm long. Therefore, the Lewis structure with a single and a double bond does not accurately describe the molecule. The concept of *resonance* is used to describe the bonding in such a molecule.

The **resonance** concept involves depicting two or more structures for a molecule with identical arrangements of atoms but different arrangements of electrons. The real structure of ozone can be better represented as a hybrid of two contributing Lewis structures, neither of which is completely correct. The structures differ only in the arrangement of the electrons.

$$:\ddot{O}=\ddot{O}-\ddot{\underset{..}{O}}: \longleftrightarrow :\ddot{\underset{..}{O}}-\ddot{O}=\ddot{O}:$$

A double-headed arrow between Lewis structures indicates that the actual structure is similar in part to each simple structure but lies somewhere between them. The actual oxygen–oxygen bond is something between a single bond and a double bond. The individual Lewis structures are called contributing structures or **resonance structures.**

Resonance structures are often written using curved arrows to keep track of the electrons. The tail of the arrow is located near the bonded or nonbonded pair of electrons to be "moved" or "pushed," and the arrowhead shows the final destination of the electron pair.

$$:\ddot{O}=\ddot{O}-\ddot{\underset{..}{O}}: \text{ pushing electrons gives } :\ddot{\underset{..}{O}}-\ddot{O}=\ddot{O}:$$

structure 1 structure 2

In resonance structure 1 the nonbonded pair of electrons on the right-hand oxygen atom is moved to form a double bond with the central oxygen atom. One of the bonded pairs of electrons between the central oxygen atom and the oxygen atom on the left is also moved to form a nonbonded pair of electrons on the left oxygen atom. The result is resonance structure 2. This procedure of "pushing" electrons from one position to another is only a bookkeeping formalism. Electrons do not really move this way! The actual ozone molecule has **delocalized** electrons that exist around all three atoms. A single Lewis structure cannot show this phenomenon.

Electrons can be delocalized over many atoms. For example, benzene, C_6H_6, consists of six equivalent carbon atoms contained in a ring in which all carbon–carbon bonds are identical. Each carbon atom is bonded to a hydrogen atom. A single Lewis structure containing alternating single and double bonds can be written to satisfy the Lewis octet requirements.

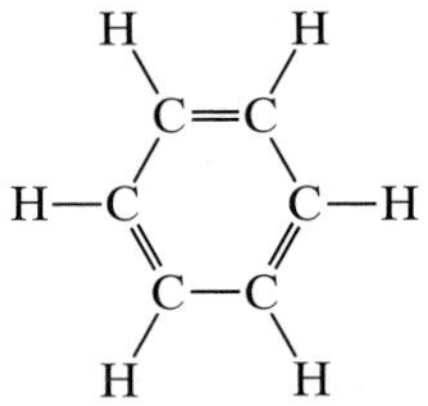

```
     H     H
      \   /
       C=C
      /   \
   H—C     C—H
      \\   //
       C—C
      /   \
     H     H
```

The single and double bonds as shown in the structure should have different bond lengths, but all C—C bonds in benzene are the same length. Like ozone, benzene is represented by two contributing resonance structures separated by a double-headed arrow. The positions of the alternating single and double bonds are interchanged in the two resonance structures.

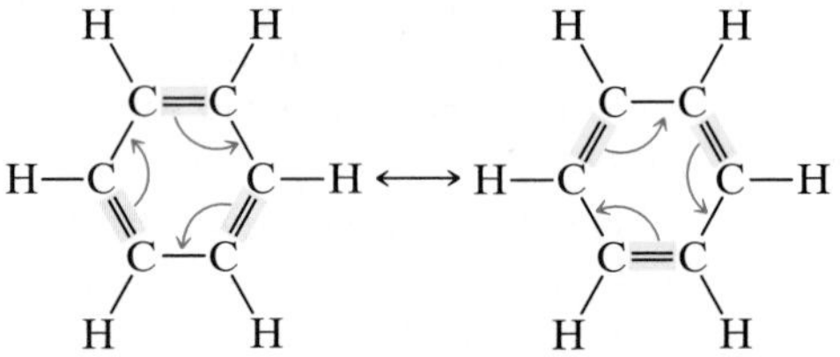

```
     H     H              H     H
      \   /                \   /
       C=C                  C—C
      /   \                //   \\
   H—C     C—H  <——>  H—C     C—H
      \\   //               \   /
       C—C                  C=C
      /   \                /   \
     H     H              H     H
```

Six of the electrons in benzene are delocalized over the six carbon atoms in the ring. This means that there are no true single or double carbon–carbon bonds in benzene. They are of an intermediate type that cannot be represented with a single structure.

EXAMPLE 5.5

Nitrites (NO_2^-) are added as antioxidants in some processed meats. Write resonance structures for the nitrite ion.

Solution

Using the procedure for drawing Lewis structures outlined in Section 5.2, we find that there must be one double bond between the nitrogen atom and one of the oxygen atoms.

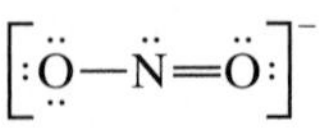

$$\left[:\ddot{\underset{..}{O}}—\ddot{N}=\ddot{O}:\right]^-$$

However, the choice of location of the double bond is arbitrary. The positions of the nitrogen–oxygen single and double bonds can be interchanged as long as the lone-pair electrons are located appropriately on each oxygen atom.

$$\left[:\ddot{O}=\ddot{N}-\ddot{O}:\right]^-$$

Thus, the compound can be represented by two equivalent resonance contributors.

PROBLEM 5.5

Consider the structure of nitromethane, a compound used to increase the power in some specialized race car engines. The length of a nitrogen–oxygen single bond is 136 pm; that of a nitrogen–oxygen double bond is 114 pm. The nitrogen–oxygen bonds in nitromethane are equal and are 122 pm long. Explain the data in terms of the electronic structure of nitromethane.

$$CH_3-N(=\ddot{O}:)-\ddot{O}:$$

5.6 Molecular Geometry

Up to this point, three-dimensional molecules have been represented in two dimensions. However, the electron pairs of atoms, both bonded and nonbonded, extend into space about the atom. The bonded electron pairs attach atoms to each other and create the structure of a molecule. We have to know only the positions of the nuclei to describe the molecular shape (Figure 5.3).

Describing Shapes of Molecules

Computers are used to study the shapes of molecules and to predict how they may interact with other molecules. (Stock Boston)

A molecule of two atoms (diatomic) must necessarily have its nuclei along a line, as in the cases of Cl_2 and HCl. Such molecules are **linear** molecules.

Triatomic molecules, such as H_2O or CO_2, must have all atoms in a plane. The angle formed by the two bonds to the central atom describes the shape of the molecule. A bond angle of 180° results in a linear molecule. Any other angle results in a **bent** (or **angular**) molecule. Carbon dioxide molecules are linear, that is, the O=C=O bond angle is 180°. Water molecules are bent with an H—O—H bond angle of 104.5°.

Molecules that are composed of more than three atoms can be linear or planar, but many other shapes are also possible (Figure 5.3). Formaldehyde (H_2CO) has all four atoms in a plane with 120° bond angles for both the H—C=O and H—C—H bonds. The arrangement of three atoms about a central atom at 120° to each other in a plane is called **trigonal planar.** In contrast, ammonia has a central nitrogen atom bonded to three hydrogen atoms; the hydrogen atoms are in a common plane, but the nitrogen atom is above the plane. The shape of the molecule is described as **trigonal pyramidal.** The H—N—H bond angles in ammonia are all 107°.

Molecules containing four atoms bonded to a central atom exist in a variety of shapes. One such shape is described as **tetrahedral.** Methane (CH_4) is a tetrahedral molecule in which the carbon atom is at the center of the tetrahedron with the four hydrogen atoms at the corners. The H—C—H bond angles are all 109.5°.

Valence-Shell Electron-Pair Repulsion Theory

Do you have to memorize the shapes of all molecules to understand chemistry? Is it possible to make an educated "guess" about molecular shapes? The answer to the latter question is yes. Might it be a matter of assigning the shape based on the

FIGURE 5.3 Shapes of Simple Molecules

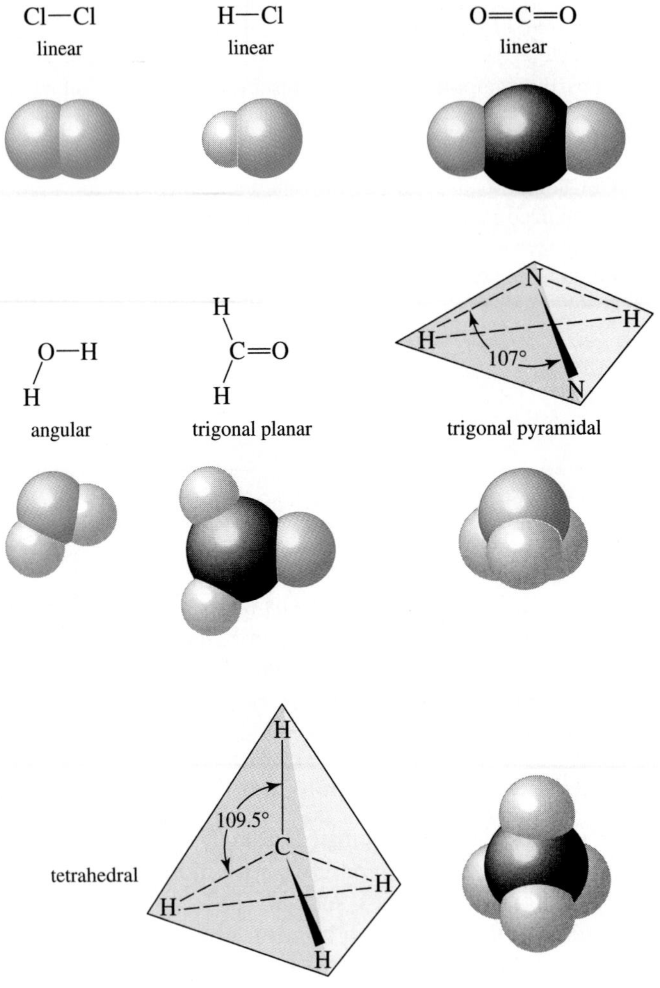

identity of the central atom? The answer to that question is no. Consider carbon in the three molecules shown in Figure 5.3. Carbon dioxide is a linear molecule, formaldehyde (H_2CO) is a trigonal planar molecule, and methane (CH_4) is a tetrahedral molecule.

:Ö=C=Ö: carbon dioxide

formaldehyde (Ö: double-bonded to C, with H and H)

methane (C bonded to four H)

Fortunately the shapes of these molecules and other molecules can be predicted using the **valence-shell electron-pair repulsion (VSEPR) theory.** The name may make the theory sound difficult to understand but it really isn't. The VSEPR (read "vesper") theory is based on the idea that electron pairs located about a central atom repel each other and should be arranged as far apart as possible. Thus, two electron pairs should be arranged at 180° to each other (Figure 5.4); three pairs should be at 120° in a common plane; four electron pairs should have a tetrahedral arrangement, with angles of 109.5°.

FIGURE 5.4 Arrangement of Electron Pairs According to VSEPR Theory

Arrangement of Electron Pairs	Term	Bond Angle
:—A—:	linear	180°
	trigonal planar	120°
	tetrahedral	109.5°

In carbon dioxide, formaldehyde, and methane, all the valence electrons about the central atom are in bonds. Carbon dioxide has two double bonds; each double bond is separated by the maximum distance, and the resulting angle between the bonds is 180°. Formaldehyde has a double bond and two single bonds to the central carbon atom; these bonds correspond to three regions containing electrons, and they are separated by the maximum distance in a trigonal planar arrangement with bond angles of 120°. In methane, there are four bonded electron pairs, and they are best located in a tetrahedral arrangement. Each H—C—H bond angle is predicted to be 109.5°, in agreement with the experimental value.

Now let us consider molecules that have both bonded and nonbonded pairs of electrons in the valence shell of the central atom. Like methane, both water and ammonia have four electron pairs about the central atom. However water and ammonia have shapes described as bent (angular) and trigonal pyramidal, respectively.

bent molecule   trigonal pyramidal molecule

Some of the electron pairs in water and ammonia are bonded to hydrogen atoms, but there are also unshared electron pairs. Methane also has four pairs of electrons about the central atom, but they are all bonding pairs.

The VSEPR theory describes the distribution of electron pairs, including the nonbonded pairs. However, shape is defined by a geometric figure that is described by the positions of the nuclei. Therefore, despite the fact that the four pairs of electrons in both ammonia and water are tetrahedrally arranged, ammonia molecules are pyramidal, and water molecules are bent (Figure 5.5).

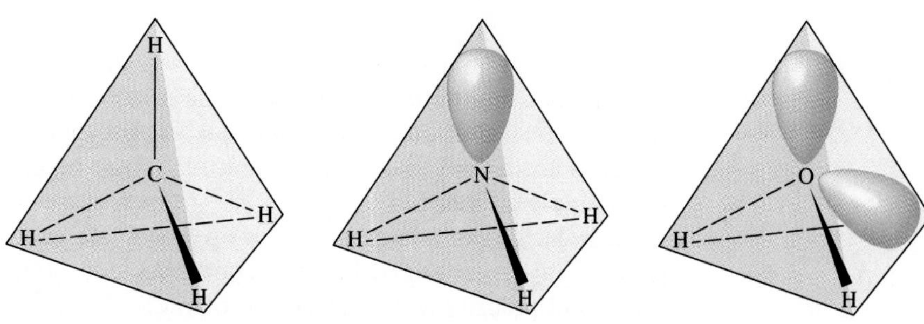

FIGURE 5.5 Location of Electron Pairs in Some Simple Molecules

All electron pairs in methane, ammonia, and water are directed to the corners of a tetrahedron. Note, however, that ammonia is still described as a trigonal pyramidal molecule and water as an angular molecule.

Bond angles are angles between the bonding pairs of electrons. However, nonbonding electron pairs affect the positions of the bonded pairs of electrons. The H—N—H bond angles of 107° are somewhat smaller than the 109.5° tetrahedral bond angles of methane. This smaller angle can be explained by assuming that the nonbonding electron pair is more spread out and pushes the bonding electron pairs closer together.

Water has two bonding pairs and two nonbonding pairs of electrons. The two nonbonding pairs exert an even greater effect on the bonding pairs and reduce the bond angle still further, to 104.5°.

EXAMPLE 5.6

Describe the expected shape of the hydrogen cyanide (HCN) molecule given the fact that there is a single bond between the carbon atom and the hydrogen atom and a triple bond between the carbon atom and the nitrogen atom.

Solution

The electronic arrangement of the molecule based on the rules given in Section 5.2 is

$$\mathrm{H—C{\equiv}N{:}}$$

The central carbon atom has a single bond and a triple bond. The electrons in these two bonds will be separated at a maximum distance when they are at a 180° angle to each other. The molecule is predicted to be linear.

PROBLEM 5.6

Predict the geometry about the C, O, and N atoms of methanol and methylamine.

```
   H                  H
   |  ..              |  ..
H—C—N—H           H—C—O:
   |  |               |  |
   H  H               H  H
 methylamine        methanol
```

5.7 Polarity of Molecules

If a molecule has polar covalent bonds, the molecule itself might be expected to be polar. However, a molecule with polar bonds may in fact be nonpolar. Polarity of molecules depends on both the polarity of the bonds (the bond moments) and the geometry of the molecule. In H—Cl there is only one polar bond, and the molecule is polar.

$$\overset{+\!\!\longrightarrow}{\mathrm{H—Cl}}$$

Chemists can determine whether or not a molecule is polar. Polar molecules tend to line up in an electric field just as magnets are attracted north to south, south to north, and so on. The magnitude of the polarity of a molecule is measured as its **dipole moment.**

A molecule containing two or more polar bonds may be polar or nonpolar depending on its geometry. To understand this idea, consider the arrows used to indicate bond polarity as forces. If two or more forces are exerted on a common ob-

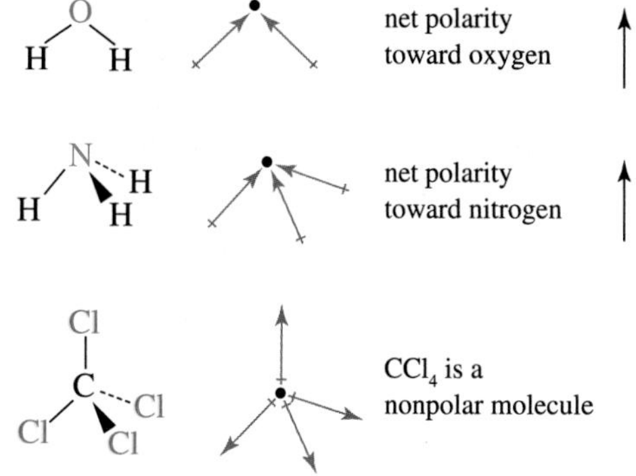

FIGURE 5.6 Polarity and Structure of Molecules

The net dipole moment of molecules is the result of the polarity of individual bonds and the shape of the molecule.

ject, the object may or may not move. For example, if the forces are equal and in opposite directions, they cancel each other, and the object will not move.

⟵+ · +⟶ no net force or movement

+⟶ · ⟵+ no net force or movement

If the two forces are at an angle other than 180°, there is a net force moving the object in a specific direction.

In carbon dioxide (CO_2), the C=O bonds are polar, with the polarity directed from the carbon atom toward the more electronegative oxygen atoms. The two bonds are located along a common line in this linear molecule. As a result, the polarities of the bonds cancel each other, and the molecule has no net dipole moment. Thus, it is possible to have polar bonds in a molecule that is itself nonpolar.

⟵+ +⟶
O=C=O

The water molecule is bent, and the polarities of the bonds do not cancel each other. Think about how an object would move if two people pushed it with equal force but at an angle to each other. The object would move forward on a line bisecting the angle between the directions the people were pushing. Using this physical concept, we expect water to be polar, with the negative end of the net molecular dipole toward the oxygen atom (Figure 5.6).

In ammonia (NH_3), the N—H bonds are polar, with the polarity of the bonds directed from each hydrogen atom toward the nitrogen atom. The net result is a polar molecule with the negative end of the dipole toward the nitrogen atom (Figure 5.6).

In carbon tetrachloride (CCl_4), which is a tetrahedral molecule, the four C—Cl bonds are polar, with the negative end of the dipole toward the chlorine. However, the four equal bonds in a tetrahedral arrangement result in a net nonpolar molecule (Figure 5.6).

5.8 Orbitals and Molecular Shapes

The VSEPR concept is useful to describe molecular geometry. However, let's now get back to the concept of atomic orbitals and consider how the locations of the electrons change when bonds are made. Because carbon is the central atom of organic and biochemistry, let's examine how it forms bonds. The shapes of molecules depend on the location of the electrons in orbitals about the central atom. Two

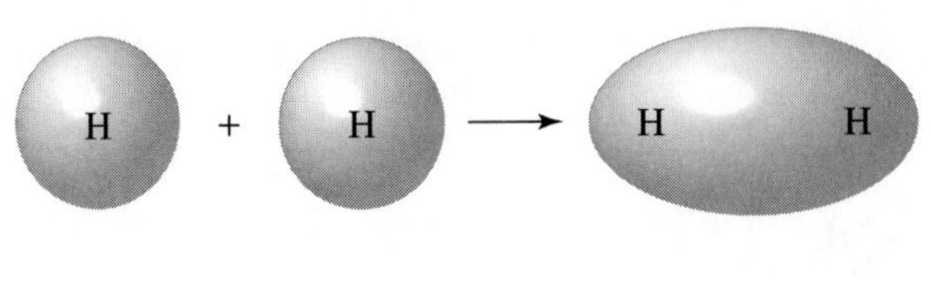

FIGURE 5.7 The Sharing of Electrons in the Sigma Bond of Hydrogen
The region occupied by the electron pair is symmetrical about both hydrogen nuclei. Although the two electrons may be located anywhere within the volume shown, it is most probable that they are between the two nuclei.

electrons are shared in a region of space common to the bonded atoms. Bonding is pictured as an overlap or merging of two orbitals. In hydrogen, the covalent bond results from the overlap of two *s* orbitals to give a **sigma (σ) bond** (Figure 5.7), which is symmetrical around an axis joining the two nuclei. If the hydrogen molecule is rotated about its axis, the appearance of the orbital is unchanged.

Hybridization of Carbon

Many atoms do not use *s* and *p* atomic orbitals in bonding. Instead, bonds are formed from hybrid orbitals resulting from the "mixing" of two or more orbitals in the bonded atoms. The carbon atom of methane (CH_4) uses hybrid orbitals. The electron configuration of an isolated carbon atom ($1s^2 2s^2 2p^2$) has only the two electrons in the 2*p* orbitals available for bonding. To form four tetrahedral bonds in CH_4, the carbon atom must contribute four electrons, and the four electrons must be in orbitals at 109.5° to each other. These features result when four atomic orbitals of carbon are mixed or hybridized to form four new hybrid orbitals (Figure 5.8). The four new orbitals are called sp^3 hybrid orbitals because one *s* and three *p* orbitals are used to create the four hybrid orbitals. One of the four valence electrons of carbon is located in each of the four hybrid orbitals. Each orbital then overlaps a hydrogen 1*s* orbital to form a σ bond.

Now let us consider the bonding electrons in the double bond of ethylene. All six nuclei lie in a plane, and all the bond angles are close to 120°. Each carbon

FIGURE 5.8 Hybridized Atomic Orbitals in Methane

$2p_x$ $2p_y$ $2p_z$

$2s$

isolated C atom

sp^3 sp^3 sp^3 sp^3

hybridized C atom in CH_4

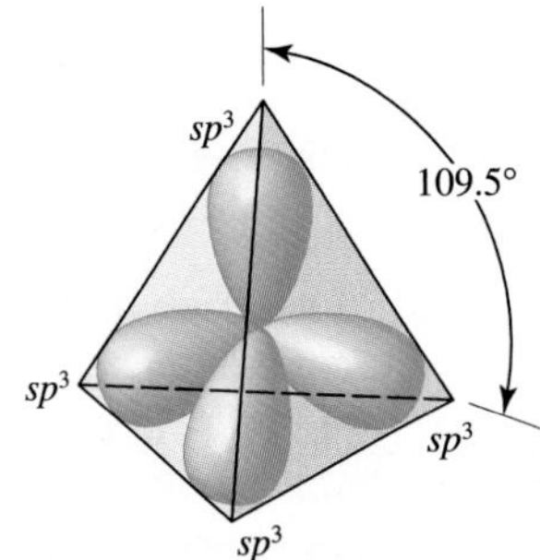

FIGURE 5.9 Hybridization and the Double Bond of Ethylene

$\uparrow$ $\uparrow$ —
$2p_x$ $2p_y$ $2p_z$

$\uparrow\downarrow$
$2s$

isolated C atom

$\uparrow$
$2p_z$

$\uparrow$ $\uparrow$ $\uparrow$
sp^2 sp^2 sp^2

hybridized C atom in C_2H_4

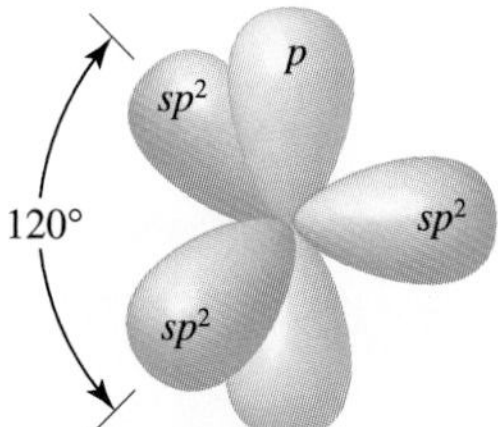

The three sp^2 hybrid orbitals lie in a plane with a 120° angle between them. The remaining *p* orbital is perpendicular to the plane of the sp^2 orbitals.

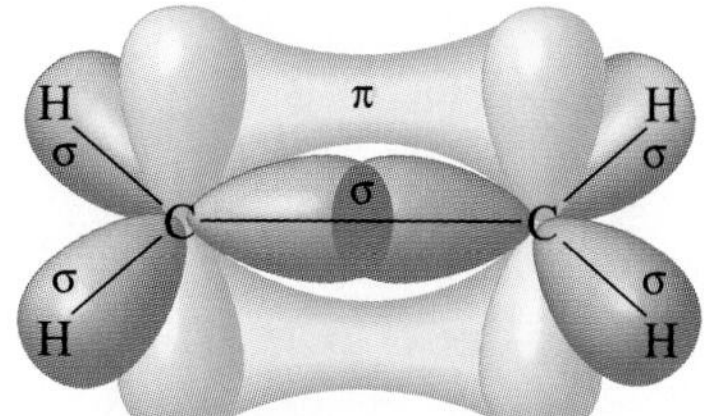

The σ bond between the carbon atoms is formed by overlap of sp^2 orbitals. The σ bonds to hydrogen are formed by overlap of the sp^2 orbitals of carbon with the *s* orbital of individual hydrogen atoms. The π bond is formed by sideways overlap of parallel *p* orbitals.

atom in ethylene is pictured using three sp^2 hybridized orbitals and one *p* orbital. The three sp^2 hybrid orbitals form from a single *s* orbital and two *p* orbitals. The three sp^2 orbitals are separated by 120° in a plane, providing maximum separation of electrons. The four valence electrons are distributed as indicated in Figure 5.9. Two of the sp^2 orbitals, containing one electron each, form σ bonds with hydrogen. The third orbital, with one electron, forms a σ bond with the other carbon atom in ethylene. The second bond of the double bond in ethylene results from a side-by-side overlap of the *p* orbital of each carbon atom. Each *p* orbital is perpendicular to the plane containing the sp^2 orbitals. Each atom contributes one electron in the *p* orbital to form the electron pair for the second bond. The bond formed by sideways overlap of *p* orbitals is a **pi (π) bond.** Note that the electrons in the π bond are not located along an axis between the two atoms but are shared in regions of space both above and below the axis. Although the electron pair occupies two regions of space, it is nevertheless only one bond.

In acetylene the carbon–carbon triple bond consists of one σ bond and two π bonds. The carbon atoms in acetylene form two *sp* hybrid orbitals each with two *p* orbitals remaining (Figure 5.10). The two *sp* hybrid orbitals are at a 180° angle to each other. One *sp* orbital and its unpaired electron form a bond with hydrogen; the other *sp* orbital forms a σ bond with the second carbon atom. The second and third bonds between the carbon atoms result from overlap of *p* orbitals. One set of *p* orbitals overlaps in "front and back" of the molecule to form one π bond. The second set of *p* orbitals overlaps "above and below" the molecule to form the second π bond.

5.9 Structural Formulas of Organic Compounds

The structures of organic molecules are complex, so we will now consider ways to represent them to save both space and time. Consider the following Lewis structures of butane, ethyl alcohol, and ethylene glycol.

FIGURE 5.10 Hybridization and the Triple Bond of Acetylene

$2p_x$ $2p_y$ $2p_z$

$2s$

isolated C atom

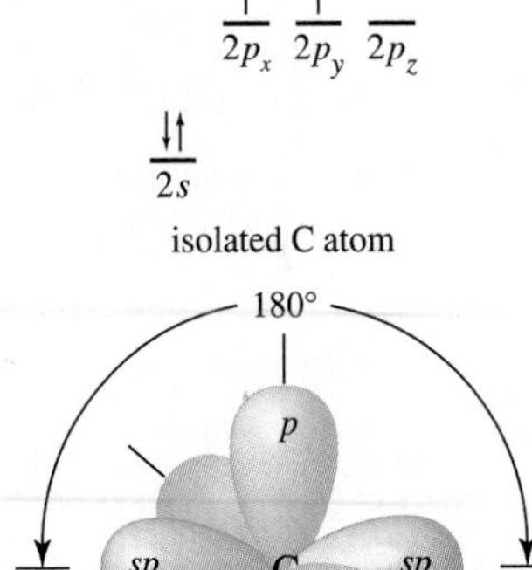

sp hybrid orbitals

The two *sp* hybrid orbitals are at 180° to each other. The two *p* orbitals are mutually perpendicular to each other and to the axis of the *sp* orbitals.

$2p_y$ $2p_z$

sp sp

hybridized C atom in C_2H_2

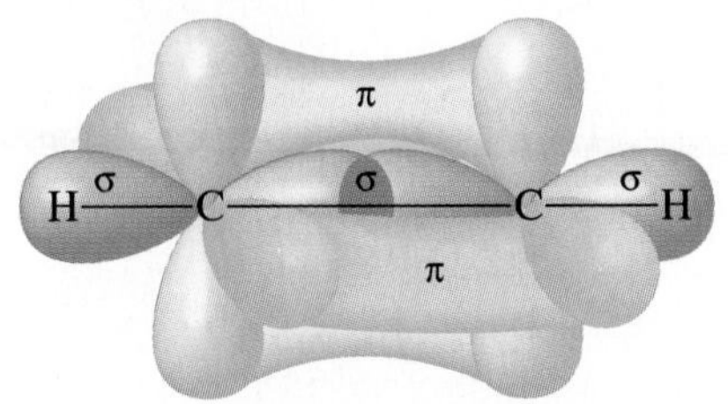

A σ bond is formed by end-to-end overlap of one *sp* orbital from each carbon atom. Two sets of parallel-oriented *p* orbitals form two mutually perpendicular π bonds. The remaining *sp* orbital of each carbon atom forms a σ bond to a hydrogen atom.

```
    H   H   H   H
    |   |   |   |
H—C—C—C—C—H
    |   |   |   |
    H   H   H   H
```

butane (used in camping stoves)

```
    H   H
    |   |
H—C—C—O—H
    |   |
    H   H
```

ethyl alcohol (contained in alcoholic beverages)

```
        H   H
        |   |
H—O—C—C—O—H
        |   |
        H   H
```

ethylene glycol (antifreeze)

Even such relatively simple structures take some time to write. A shorthand method of writing condensed structural formulas has evolved that we will start to learn here. It will be used more extensively in later chapters concentrating on organic chemistry.

Condensed Structural Formulas Save Time

Condensed structural formulas show only specific bonds; other bonds are implied. The degree of condensation depends on which bonds are shown and which are implied. For example, since hydrogen forms only a single bond to carbon, the C—H bond need not be shown, as in the case of the following condensed structure of butane, which shows only C—C bonds.

$$CH_3{-}CH_2{-}CH_2{-}CH_3$$

Since carbon always forms four bonds, the carbon atoms at each end of butane are understood to have three single bonds to hydrogen. Those carbon atoms in the interior of the molecule have two carbon–carbon bonds shown, but the two carbon–hydrogen bonds are understood. Note that by convention the symbol for the hydrogen atom is written to the right of the symbol for the carbon atom.

In a further condensation, the carbon–carbon bonds are also left out.

$$CH_3CH_2CH_2CH_3$$

The carbon atom on the left is understood to be bonded to the three hydrogen atoms and to the carbon atom to its right. The second carbon atom from the left is bonded to the two hydrogen atoms to the right. In addition, that carbon atom is bonded to the two carbon atoms to the immediate right and left, since carbon must have four bonds.

Large structures may have repeated structural subunits that are represented by grouping the subunits within parentheses. The number of times the unit is repeated is given by a subscript following the closing parenthesis. Butane is represented by

$$CH_3(CH_2)_2CH_3$$

The $—CH_2—$ unit occurs twice in butane. Since the units are linked together in a repeating chain, one is placed within the parentheses with a subscript 2.

Two or more identical groups of atoms bonded to a common central atom may also be represented within parentheses with an appropriate subscript in a condensed formula. The groups within parentheses may be placed to the right or left, depending on the way in which the structure of the molecule is drawn.

$$\begin{array}{l} \quad\;\; CH_3 \\ \quad\;\;\, | \\ CH_3—CH—CH_2—CH_2—CH_2—CH_3 \text{ is } (CH_3)_2CHCH_2CH_2CH_2CH_3 \end{array}$$

$$\begin{array}{l} \qquad\qquad\qquad\quad CH_3 \\ \qquad\qquad\qquad\quad\;\, | \\ CH_3—CH_2—CH_2—C—CH_3 \text{ is } CH_3CH_2CH_2C(CH_3)_3 \\ \qquad\qquad\qquad\quad\;\, | \\ \qquad\qquad\qquad\quad CH_3 \end{array}$$

EXAMPLE 5.7

Write three condensed structural formulas for octane.

$$\begin{array}{ccccccccccccccccc} & & H & & H & & H & & H & & H & & H & & H & & H \\ & & | & & | & & | & & | & & | & & | & & | & & | \\ H & — & C & — & C & — & C & — & C & — & C & — & C & — & C & — & C & — H \\ & & | & & | & & | & & | & & | & & | & & | & & | \\ & & H & & H & & H & & H & & H & & H & & H & & H \end{array}$$

Solution

With the C—H bonds understood, we write

$$CH_3—CH_2—CH_2—CH_2—CH_2—CH_2—CH_2—CH_3$$

With both the C—H and C—C bonds understood, we may write

$$CH_3CH_2CH_2CH_2CH_2CH_2CH_2CH_3$$

In the most condensed version, the six $—CH_2—$ units are represented as one CH_2 within parentheses, with a subscript 6 following the closing parenthesis.

$$CH_3(CH_2)_6CH_3$$

PROBLEM 5.7

Write the complete structural formula for $CH_3(CH_2)_2OH$.

Bond-Line Structures Save Even More Time

Condensed structural formulas are convenient but still require time to draw compared to a shorthand method using bond-line structures. Bond-line structures are less cluttered drawings, but you have to recognize many more implied features to understand the structure. The rules for drawing bond-line structures are the following:

1. Carbon atoms are not shown unless they are needed for special emphasis or clarity.
2. All atoms except carbon and hydrogen are shown explicitly.
3. A carbon atom is assumed to be located at the end of each line segment or at the intersection of two lines that are used to depict bonds.
4. Multiple bonds are shown with multiple lines.

To draw a bond-line structure, arrange the carbon atoms in a zigzag manner and then mentally remove both the carbon and hydrogen atoms.

$CH_3{-}CH_2{-}CHBr{-}CH_3$ is CH_3 CH_2 CH(Br) CH_3 is Br

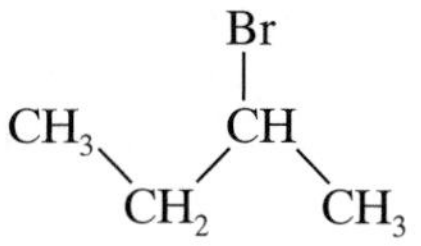

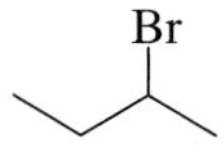

Bond-line formulas are also used to show structures containing rings of carbon atoms. The rings are represented with polygons—an equilateral triangle, square, pentagon, hexagon, and so on.

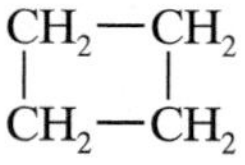

$CH_2{-}CH_2$ / $CH_2{-}CH_2$ is □

Multiple bonds are represented differently. For a double bond between oxygen and carbon, the oxygen atom must be shown, but the carbon atoms of a carbon–carbon double bond are implied by the structure and are not written.

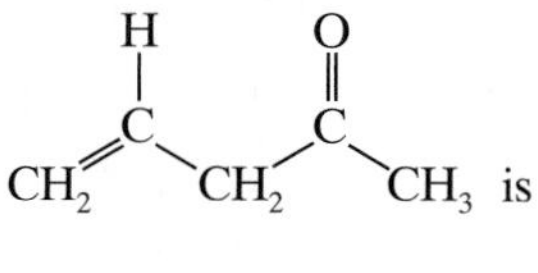

H O C C CH_2 CH_2 CH_3 is O

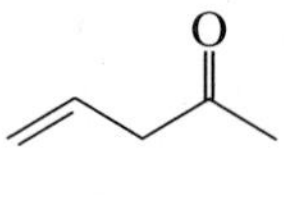

It is important that you know the number of bonds normally formed by each of the common atoms in organic compounds. Carbon, nitrogen, and oxygen form four, three, and two bonds, respectively. The following bond-line diagram indicates how many hydrogen atoms are bonded to each carbon atom based on the number of lines at each intersection.

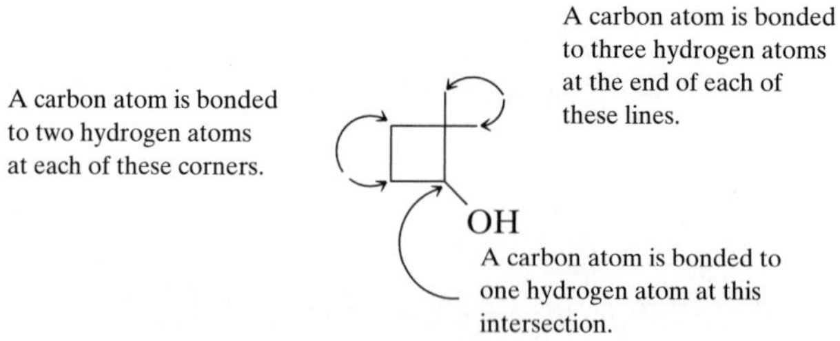

A carbon atom is bonded to two hydrogen atoms at each of these corners.

A carbon atom is bonded to three hydrogen atoms at the end of each of these lines.

OH

A carbon atom is bonded to one hydrogen atom at this intersection.

FIGURE 5.11
Perspective Structures and Molecular Models

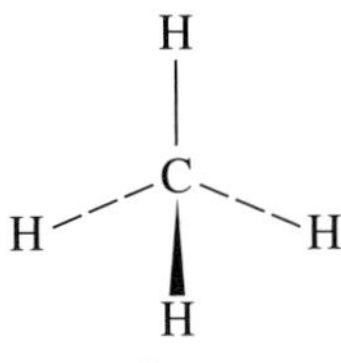

perspective structure

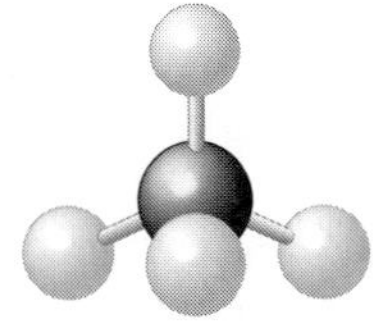
ball-and-stick model

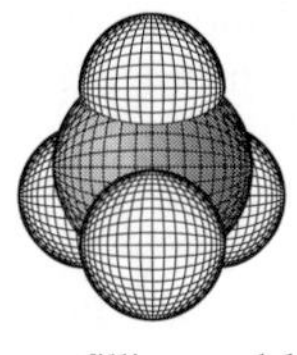
space-filling model

5.10 Structural Models

Because structure is so important to understanding chemical reactions, chemists use models of molecules that can be viewed from a variety of angles. You may find it useful to purchase a molecular model kit to help you understand the structures of organic molecules.

Ball-and-stick and space-filling models are two types of molecular models; each has certain advantages and disadvantages. The ball-and-stick models show the molecular framework and bond angles; the balls represent the nuclei of the atoms, and the sticks represent the bonds (Figure 5.11). The actual volume occupied by the molecule is not shown realistically. Space-filling models do show the entire volume occupied by the electrons surrounding each atom, but as a consequence the carbon skeleton and its bond angles are obscured.

On paper, the three-dimensional shape of molecules is shown by wedges and dashed lines (Figure 5.11). The wedge is viewed as a bond extending out of the plane of the page toward the reader. The dashed lines represent bonds directed behind the plane of the page. The solid line is a bond in the plane of the page. Three-dimensional representations of molecules using wedges and dashed lines are called **perspective structural formulas.**

Summary

Electron pairs shared between atoms are **covalent bonds.** One, two, and three shared pairs of electrons are **single covalent, double covalent,** and **triple covalent bonds,** respectively. A **polar covalent bond** occurs when the electron pairs are shared unequally. A **coordinate covalent bond** occurs when the electron pair is provided by a single atom.

Covalently bonded molecules can usually be represented with **Lewis structures.** The electrons are distributed about the molecular framework according to a set of rules. The total of the **nonbonding electrons (lone-pair electrons** or **unshared electrons**) and **bonding electrons** associated with each atom should be eight.

For some molecules two or more plausible Lewis structures may be drawn. In these cases, according to **resonance theory** the true structure is said to be a **resonance hybrid** represented as a combination of the Lewis structures.

The **valence-shell electron-pair repulsion (VSEPR) theory** can be used to predict the shapes of molecules. The shape of a molecule depends on the geometric distribution of bonded and nonbonded pairs of electrons, which tend to achieve maximum separation from each other. The shape of the molecule is defined in terms of the geometric shape described by the positions of the atoms. Typical molecular shapes are **linear, angular, trigonal planar, trigonal pyramidal,** and **tetrahedral.**

Although a molecule may have polar covalent bonds, the molecule may be polar or nonpolar. The polarity of molecules depends on the polarity of the bonds in the molecule and their arrangement in space. Centers of opposite charge separated from each other by one or more bonds create a **dipole.** The magnitude of the polarity of a molecule is measured as the **dipole moment.**

Organic compounds have chemical and physical properties that are consequences of the carbon–carbon covalent bond. The great variety of organic compounds is the result of stable single and multiple bonds between carbon atoms as well as the incorporation of other atoms into bonds with carbon.

Carbon forms sp^3, sp^2, and sp **hybrid orbitals.** These orbitals are distributed at the maximum distance from each other. Two types of bonds result from orbital overlap. Overlap along a line between the nuclei of the atoms is a **sigma** (σ) bond. Side-to-side overlap of *p* orbitals results in a **pi** (π) bond.

Structural formulas indicate the connectivity of carbon atoms in organic compounds. The same information can be conveyed in abbreviated formulas called **condensed structural formulas** and **bond-line structural formulas.** Three-dimensional representations of molecules using wedges and dashed lines are called **perspective structural formulas.**

Exercises

Types of Covalent Bonds

5.1 Write Lewis structures for the following covalent molecules.

(a) H_2 (b) I_2 (c) F_2
(d) Br_2 (e) Cl_2 (f) N_2

5.2 Write Lewis structures for the following covalent molecules.

(a) H_2S (b) ICl (c) NF_3
(d) H_2O_2 (e) CO_2 (f) N_2H_4

5.3 Indicate which element will be the positive end of the dipole when the following pairs of elements are bonded to each other.

(a) H and Br (b) Br and Cl (c) H and O
(d) O and F (e) O and I (f) Si and F

5.4 Indicate which element will be the positive end of the dipole when the following pairs of elements are bonded to each other.

(a) N and H (b) N and F (c) P and Cl
(d) P and F (e) I and Br (f) C and Cl

Writing Lewis Structures

5.5 Write Lewis structures for the following covalent molecules.

(a) H_2SO_4 (b) PH_3 (c) CF_4
(d) H_2Se (e) CBr_4 (f) $SiCl_4$

5.6 Write Lewis structures for the following covalent molecules.

(a) SiH_4 (b) PCl_3 (c) SbH_3
(d) H_2Te (e) $SnCl_4$ (f) $GeBr_4$

5.7 Write Lewis structures for the following ions.

(a) SH^- (b) PH_4^+ (c) H_3O^+
(d) CN^- (e) SO_4^{2-} (f) NH_4^+

5.8 Write Lewis structures for the following ions.

(a) OCl^- (b) SO_3^{2-} (c) BrO_3^-
(d) IO_3^- (e) ClO_4^- (f) NH_2^-

5.9 The connectivity of atoms in several molecules is given. Using dashes to represent bonds or shared electron pairs and dots for lone-pair electrons, draw Lewis structures for the molecules.

(a) F—C—F (with F above and F below C)

(b) 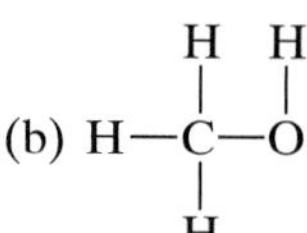

(c) 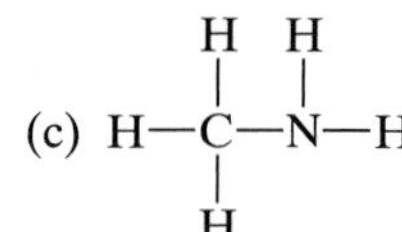

(d) Cl, Cl—C—O (two Cl bonded to C)

(e) 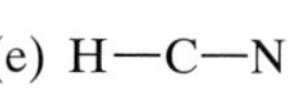

(f) H, H—C—N—O—H (two H bonded to C)

5.10 The connectivity of atoms in several molecules is given. Using dashes to represent bonds or shared electron pairs and dots for lone-pair electrons, draw Lewis structures for the molecules.

(a) Cl—C—Cl (with F above and Br below C)

(b) H—C—S (with H above C)

(c) H—C—P—H (with H above C, H below C, H below P)

(d) H—N—C—N—H (with O above C, H below each N)

(e) H—C—N—H (with H above C)

(f) H—C (with O and O—H bonded to C)

Resonance

5.11 The nitrate ion has identical bonds between each of the three oxygen atoms and the central nitrogen atom. Draw possible resonance structures for this ion.

5.12 The carbonate ion has identical bonds between each of the three oxygen atoms and the central carbon atom. Draw possible resonance structures for this ion.

Molecular Shapes

5.13 What bond angles are present in each of the following compounds? What is the geometry of each molecule?

(a) $SiCl_4$ (b) NF_3 (c) HOCl (d) OF_2

5.14 What bond angles are present in each of the following compounds? What is the geometry of each molecule?

(a) NCl_3 (b) CF_4 (c) HOBr (d) OCl_2

5.15 Predict the shape of each of the following ions according to the VSEPR theory.

(a) SO_4^{2-} (b) NH_4^+ (c) H_3O^+ (d) NH_2^-

5.16 Predict the shape of each of the following ions according to the VSEPR theory.

(a) ClO_2^- (b) ClO_3^- (c) $GeCl_3^-$ (d) PO_4^{3-}

5.17 Based on VSEPR theory, what is the expected value of the indicated bond angle in each of the following compounds?

(a) C—C—N in CH_3—C≡N
(b) C—O—C in CH_3—O—CH_3
(c) C—N—C in CH_3—NH—CH_3
(d) C—C—C in CH_3—C≡C—H

5.18 Based on VSEPR theroy, what is the expected value of the indicated bond angle in each of the following compounds?

```
                    O
                    ‖
(a) C—C—O in CH3—C—H
(b) C—C—C in CH3—CH=CH2
                    O
                    ‖
(c) O—C—O in CH3—C—OH
                  O
                  ‖
(d) O—C—N in H—C—NH2
```

Polarity of Molecules

5.19 Indicate whether you expect the following molecules to be polar or nonpolar, and explain your answers.

(a) H_2S (b) PH_3 (c) CF_4
(d) H_2Se (e) CBr_4 (f) $SiCl_4$

5.20 Indicate whether you expect the following molecules to be polar or nonpolar, and explain your answers.

(a) SiH_4 (b) PCl_3 (c) SbH_3
(d) H_2Te (e) NF_3 (f) $GeBr_4$

Bonding in Carbon

5.21 Describe the four combinations of bonds by which carbon achieves an octet of electrons.

5.22 List a few of the atoms that can form bonds to carbon.

Condensed Structural Formulas

5.23 Write condensed structural formulas in which only the bonds to hydrogen are not shown.

```
         H  H
         |  |
(a) Br—C—C—Br
         |  |
         H  H

       H  H  H  H  H
       |  |  |  |  |
(b) H—C—C—C—C—C—H
       |  |  |  |  |
       H  H  H  H  H

       H  H  H
       |  |  |
(c) H—C—C—C—S—H
       |  |  |
       H  H  H
```

5.24 Write condensed structural formulas in which only the bonds to hydrogen are not shown.

```
       H  H  H  H
       |  |  |  |
(a) H—C—C—C—N—H
       |  |  |
       H  H  H

       H  H  H  H
       |  |  |  |
(b) H—C—C—C—C—O—H
       |  |  |  |
       H  H  H  H

       H  H  H  H  H
       |  |  |  |  |
(c) H—C—C—C—C—C—Cl
       |  |  |  |  |
       H  H  H  H  H
```

5.25 Write a condensed structural formula in which no bonds are shown for each substance in Exercise 5.23.

5.26 Write a condensed structural formula in which no bonds are shown for each substance in Exercise 5.24.

Bond-Line Formulas

5.27 What is the molecular formula for each of the following structures?

(a) the scent marker of the red fox

(b) the compound responsible for the odor of the iris

5.28 What is the molecular formula for each of the following structures?

(a) a compound found in clover and grasses

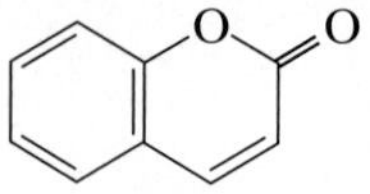

(b) an oil found in citrus fruits

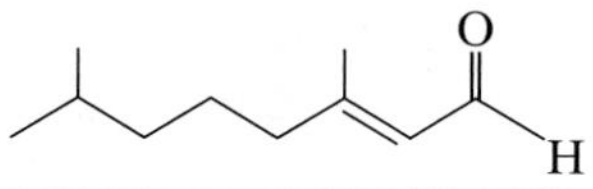

Perspective Structures

5.29 Draw a three-dimensional structure for each of the following.

(a) CH_3Cl (b) CH_2Br_2 (c) CH_3F

5.30 Draw a three-dimensional structure for each of the following.

(a) CHF_3 (b) CH_2ClBr (c) CCl_4

6

Chemical Equations and Reactions

Lightning causes reaction of large quantities of nitrogen with oxygen. This process is called nitrogen fixation. (Tony Stone Images)

6.1 Chemical Equations
6.2 Balancing Equations
6.3 Chemical Reactions
6.4 Combination Reactions
6.5 Decomposition Reactions
6.6 Single Replacement Reactions
6.7 Double Replacement Reactions
6.8 Neutralization Reactions
6.9 Organic Acids and Bases
6.10 Oxidation-Reduction Reactions
6.11 Redox Reactions in Organic Chemistry

Essays
- *The Chemistry of Rockets*
- *Preventing Corrosion, an Unwanted Oxidation*
- *Redox Reactions in Biochemistry*

6.1 Chemical Equations

The law of conservation of mass governs chemical reactions (Section 3.1). At the macroscopic level, this means that matter is neither created nor destroyed and that the total mass of the products is equal to the total mass of the reactants. At the submicroscopic level, we conclude that the number and type of atoms are unchanged in chemical reactions. The atoms arranged in structures in the reactants are merely redistributed into new structures in the products. It makes no difference if the reaction is the simple rusting of iron or one of the complex reactions of DNA.

Equations account for the chemicals in chemical reactions. They indicate the number of atoms involved and how they are distributed in the reactants and products. In chemistry, equations are balanced when the number of atoms of each kind is the same in the products as in the reactants.

In this chapter you will learn how to write balanced equations and how to classify chemical reactions according to the type of process. In the next chapter you will use the balanced chemical reactions equation to express quantitative information about the mass of reactants required to produce a specific mass of products.

Symbols in Equations

Chemical equations are symbolic representations of chemical reactions. The accepted scheme for writing chemical equations is

$$\textbf{reactant}_{(1)} + \textbf{reactant}_{(2)} \longrightarrow \textbf{product}_{(1)} + \textbf{product}_{(2)}$$

The reactants and products may be either elements or compounds. A plus sign is used to separate the reactants on the left as well as the products on the right. It is read as "plus" or "and." The arrow → is like the equal sign in math; it is read as "yields" or "produces."

The balanced equation for the reaction of aluminum with oxygen is

$$4\,Al + 3\,O_2 \longrightarrow 2\,Al_2O_3$$

The numbers before each substance are **coefficients** that indicate the number of units of each substance involved relative to all other substances. This equation reads "four atoms of aluminum and three molecules of oxygen react to yield two formula units of aluminum oxide." The coefficients act as multipliers and apply to all the atoms in the formula that they precede. The coefficient is multiplied by the subscript for each atom to account for all the atoms present.

coefficient × subscript = total number of atoms of element in species

The coefficient 2 for aluminum oxide means that there are $2 \times 2 = 4$ atoms of aluminum as well as $2 \times 3 = 6$ atoms of oxygen on the product side of the equation. There are $4 \times 1 = 4$ atoms of aluminum and $3 \times 2 = 6$ atoms of oxygen on the reactant side of the equation.

The physical states of the reactants and products are denoted by writing (s) for solid, (l) for liquid, or (g) for gas to the right of each chemical symbol. The equation for the rusting of iron with the states indicated is

$$4\,Fe(s) + 3\,O_2(g) \longrightarrow 2\,Fe_2O_3(s)$$

Reactions of substances in water solution are indicated by the symbol (aq), meaning **aqueous.** The reaction of metallic zinc with aqueous hydrochloric acid releases hydrogen gas and forms a solution of zinc chloride. The equation is

$$Zn(s) + 2\,HCl(aq) \longrightarrow ZnCl_2(aq) + H_2(g)$$

The following equation indicates that aqueous solutions of silver nitrate and sodium chloride react to form solid silver chloride, which separates from the solution because it is not soluble. The other product, sodium nitrate, remains in solution.

$$AgNO_3(aq) + NaCl(aq) \longrightarrow AgCl(s) + NaNO_3(aq)$$

A capital Greek delta (Δ) placed above the arrow indicates that the reaction occurs at a high temperature. For example, the conversion of limestone ($CaCO_3$) into lime (CaO) requires heat.

$$CaCO_3(s) \xrightarrow{\Delta} CaO(s) + CO_2(g)$$

The actual temperature required may be written over the arrow, as in the following equation for the industrial preparation of methyl alcohol.

$$2\,H_2(g) + CO(g) \xrightarrow{350\,°C} CH_3OH(g)$$

The symbol for a catalyst may also appear above the arrow. A **catalyst** increases the rate of the reaction but is not consumed. The production of methyl alcohol requires a mixture of zinc and chromium oxides as the catalyst.

$$2\,H_2(g) + CO(g) \xrightarrow{350\,°C,\,ZnO/Cr_2O_3} CH_3OH(g)$$

6.2 Balancing Equations

When balancing an equation, you must obey one very important rule: *Do not change the subscripts in the reactants and products.* The subscripts establish the identity of each substance. Consider the equation for the reaction of hydrogen and oxygen to produce water. Representations of the elements and water are shown to help you to account for the number of atoms of hydrogen and oxygen on each side of the equation.

$$2\,H_2(g) + O_2(g) \longrightarrow 2\,H_2O(l)$$

A balanced equation must represent the actual reactants and products with their correct formulas. Changing subscripts to balance an equation is not an option because those formulas would not represent the actual reactants and products. For example, a balanced equation for the reaction of hydrogen with oxygen can be written by changing the subscript of oxygen in H_2O to 2. The equation is corect "mathematically," but the resulting H_2O_2 is hydrogen peroxide, not water.

$$H_2 + O_2 \longrightarrow H_2O_2$$

hydrogen peroxide

To balance an equation, suitable coefficients must be found to make the number of atoms the same on both sides of the equation. Is there any secret to success? Yes, establish the easiest coefficients first. You'll be most successful with the following procedure.

1. Consider those atoms that appear in only a single substance on each side of the equation.
2. If there are several such atoms, select the substance that has the largest number of atoms of a single element.
3. Balance the number of atoms of that element by placing a proper coefficient in front of the substance containing that element on each side of the equation.
4. Proceed to balance the atoms of other elements by the same process. Check to see if balancing one element in a substance has caused others to become unbalanced. Adjust the coefficient on the other side of the equation to achieve the necessary balance.
5. Next, balance atoms that appear in two or more places on one side of the equation.
6. Finally check and reduce all coefficients to smallest integer ratios.

Methane, the major component of natural gas, burns to give carbon dioxide and water. We start the balancing process by first writing correct formulas for all substances. Remember that oxygen gas is diatomic.

$$CH_4 + O_2 \longrightarrow CO_2 + H_2O \quad \text{(unbalanced)}$$

methane oxygen carbon dioxide water

Oxygen appears in one reactant and two products. Carbon and hydrogen appear in only one substance on each side of the equation. Select hydrogen, which is present in larger numbers of atoms in a single compound. There are four hydrogen atoms in a methane molecule. Therefore place 2 as a coefficient in front of H_2O. The required coefficient before CH_4 is 1, but by convention the 1 is not written.

$$CH_4 + O_2 \longrightarrow CO_2 + 2\,H_2O \quad \text{(unbalanced)}$$

methane

Now balance the carbon atoms. The amount of carbon on the reactant side is controlled by the 1 coefficient on CH_4. Therefore only one CO_2 is required on the product side. Again, by convention the 1 is not written.

Now all coefficients are established except for that of O_2. There are a total of four oxygen atoms in the products—two in the CO_2 molecule and two in the two molecules of H_2O. Balance the O_2 on the reactant side by placing a 2 in front of the symbol for the element. Now the equation is balanced.

$$CH_4 + 2\,O_2 \longrightarrow CO_2 + 2\,H_2O \qquad \text{(balanced)}$$

Next let's balance the equation for the reaction of nitrogen and hydrogen to produce ammonia in the Haber process, which is used to "fix" nitrogen for use in fertilizer. Both nitrogen and hydrogen are diatomic elements.

$$N_2 + H_2 \longrightarrow NH_3 \qquad \text{(unbalanced)}$$

nitrogen hydrogen ammonia

Let's first consider balancing the hydrogen atoms because there are three of them in NH_3. The subscripts in NH_3 and H_2 provide a clue about the coefficients necessary to balance these two substances. Use 3 as the coefficient for H_2 and 2 as the coefficient for NH_3.

$$N_2 + 3\,H_2 \longrightarrow 2\,NH_3$$

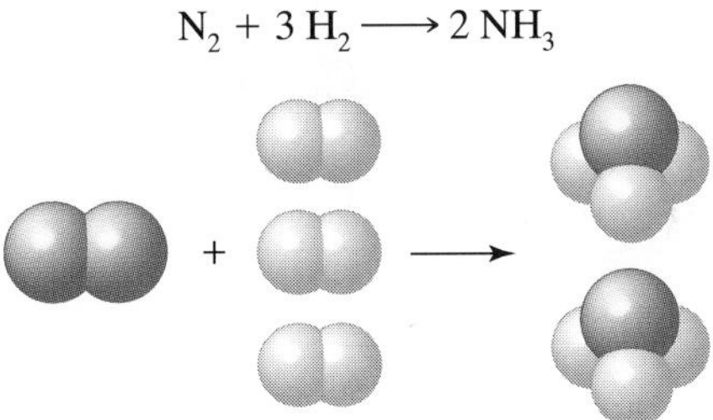

This "cross-switching" procedure interchanges the numerical values of the subscripts to serve as coefficients. In math terms we are simply recognizing the fact that 6 is the lowest common multiple of 2 and 3. The equation is now balanced in terms of hydrogen. There are $2 \times 3 = 6$ hydrogen atoms on the right in the two molecules of NH_3 and $3 \times 2 = 6$ hydrogen atoms on the left in the three molecules of H_2. Now only the coefficient of N_2 needs to be established. There are 2×1 atoms of nitrogen in 2 NH_3, so the coefficient of N_2 is 1 because one molecule of nitrogen contains 1×2 atoms of nitrogen. The equation is already balanced.

Acetylene (C_2H_2) reacts with oxygen in the acetylene torch used for welding. The products are carbon dioxide and water. Let's balance the equation for the reaction.

$$C_2H_2 + O_2 \longrightarrow CO_2 + H_2O \qquad \text{(unbalanced)}$$

Which of the three atoms should we select? Note that the hydrogen atoms are already in balance, and the coefficients for C_2H_2 and H_2O are both established as 1. Oxygen appears in two places on the right-hand side of the equation and can be balanced only after some coefficients are established for other atoms. So, we start the balancing process with carbon. Place the coefficient 2 in front of CO_2 to balance the two carbon atoms of C_2H_2.

$$C_2H_2 + O_2 \longrightarrow 2\,CO_2 + H_2O \qquad \text{(unbalanced)}$$

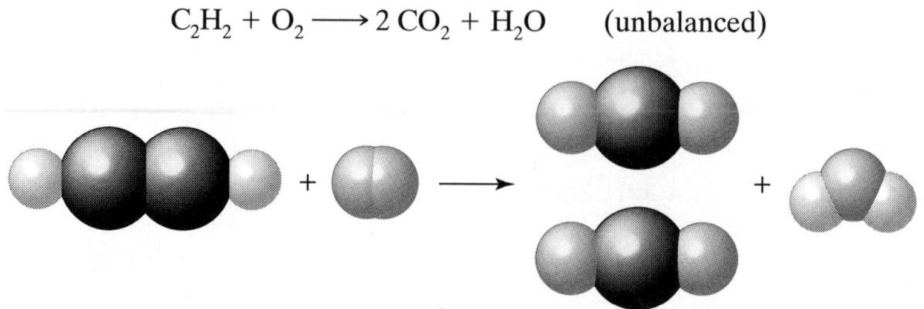

Now balance the oxygen atoms. The oxygen atoms in the products total five, four in the 2 CO_2 and one in the H_2O. In order to provide five oxygen atoms in the form of O_2 molecules, a coefficient of $2\frac{1}{2}$ or $\frac{5}{2}$ is necessary. In $2\frac{1}{2}$ oxygen molecules there are five oxygen atoms.

$$C_2H_2 + 2\tfrac{1}{2}\,O_2 \longrightarrow 2\,CO_2 + H_2O \qquad \text{(balanced)}$$

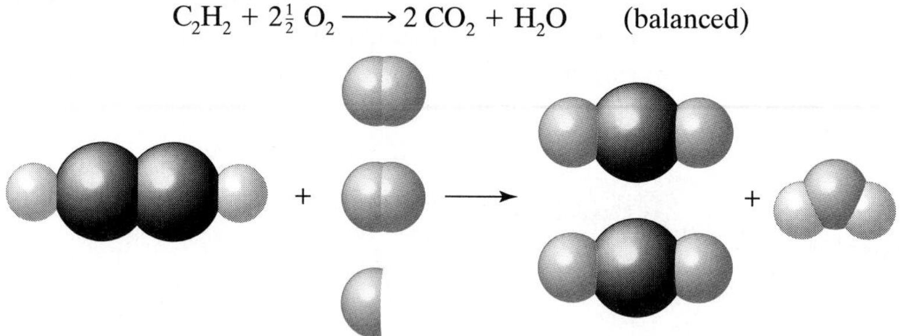

However, by convention, fractional coefficients are not acceptable in balanced equations. All reactants and products must have integer coefficients. The fractional coefficient $2\frac{1}{2}$ is eliminated by multiplying all coefficients in the equation by 2 to obtain an acceptable balanced equation.

$$2\,C_2H_2 + 5\,O_2 \longrightarrow 4\,CO_2 + 2\,H_2O \qquad \text{(balanced)}$$

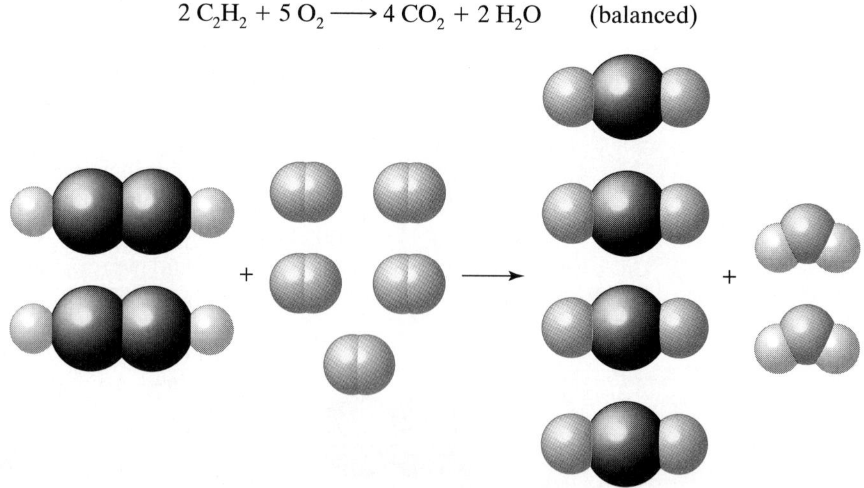

Finally consider the reaction of aluminum hydroxide, $Al(OH)_3$, with sulfuric acid, H_2SO_4, to yield aluminum sulfate and water. In this case we won't use models to represent the reactants and products. Furthermore, we will be a little freer in our approach to show that there may be several ways to achieve the balanced equation.

$$Al(OH)_3 + H_2SO_4 \longrightarrow Al_2(SO_4)_3 + H_2O \quad \text{(unbalanced)}$$

First balance the two atoms of aluminum in $Al_2(SO_4)_3$ on the right side of the equation by placing a coefficient 2 in front of $Al(OH)_3$ on the left side of the equation.

$$2\,Al(OH)_3 + H_2SO_4 \longrightarrow Al_2(SO_4)_3 + H_2O \quad \text{(unbalanced)}$$

Which should be balanced next—the hydrogen, oxygen, or sulfate ions? Balancing the polyatomic ion as a unit is convenient. The three sulfate units in the $Al_2(SO_4)_3$ are balanced by placing the coefficient 3 in front of H_2SO_4.

$$2\,Al(OH)_3 + 3\,H_2SO_4 \longrightarrow Al_2(SO_4)_3 + H_2O \quad \text{(unbalanced)}$$

Now only the oxygen not incorporated in the sulfate ion and the hydrogen remain unbalanced. There are 6 hydrogen atoms contained in the two units of $Al(OH)_3$ and 6 hydrogen atoms in the three units of H_2SO_4, for a total of 12, which must be balanced by a coefficient in front of H_2O. The coefficient needed is 6, since there are two hydrogen atoms in an H_2O molecule.

$$2\,Al(OH)_3 + 3\,H_2SO_4 \longrightarrow Al_2(SO_4)_3 + 6\,H_2O \quad \text{(balanced)}$$

The 6 oxygen atoms in two units of $Al(OH)_3$ are simultaneously balanced by the six units of H_2O. Note that a total of 18 oxygen atoms are incorporated in the two reactants as well as in the two products. There are a total of 12 hydrogen atoms in the two reactants as well as in the two products; therefore, the equation is balanced.

EXAMPLE 6.1

Balance the equation for the combustion of propane (C_3H_8). Propane is used in some camp stoves, and in some rural homes.

$$C_3H_8 + O_2 \longrightarrow CO_2 + H_2O$$

Solution

There are three atoms of carbon in one molecule of propane. Therefore, three molecules of CO_2 must be produced. The coefficient 3 is placed in front of the formula CO_2.

$$C_3H_8 + O_2 \longrightarrow 3\,CO_2 + H_2O$$

There are eight atoms of hydrogen in one molecule of C_3H_8. Therefore, four molecules of H_2O must be produced. Placing the coefficient 4 in front of the formula H_2O yields

$$C_3H_8 + O_2 \longrightarrow 3\,CO_2 + 4\,H_2O$$

The equation is not yet balanced because there are 10 atoms of oxygen on the right of the arrow and only 2 on the left. In order to balance the equation, the coefficient 5 must be placed in front of O_2.

$$C_3H_8 + 5\,O_2 \longrightarrow 3\,CO_2 + 4\,H_2O$$

PROBLEM 6.1

Balance the equation for the combustion of ethanol, which is used in gasohol.

$$C_2H_5OH + O_2 \longrightarrow CO_2 + H_2O$$

EXAMPLE 6.2

Balance the equation for the reaction of magnesium hydroxide and phosphoric acid. Magnesium hydroxide is found in Maalox, and phosphoric acid in soft drinks.

$$Mg(OH)_2 + H_3PO_4 \longrightarrow Mg_3(PO_4)_2 + H_2O$$

Solution

Balance the magnesium atoms in $Mg_3(PO_4)_2$ by placing the coefficient 3 in front of $Mg(OH)_2$.

$$3\,Mg(OH)_2 + H_3PO_4 \longrightarrow Mg_3(PO_4)_2 + H_2O$$

Next, balance the phosphate ions. There are two phosphate ions in the one unit of $Mg_3(PO_4)_2$, whose coefficient has been established. The coefficient 2 is placed before H_3PO_4 to balance the phosphate.

$$3\,Mg(OH)_2 + 2\,H_3PO_4 \longrightarrow Mg_3(PO_4)_2 + H_2O$$

The Chemistry of Rockets

The propulsion for rockets is provided by chemical reactions. Hot product gases emitted from the rocket nozzle propel the rocket upward.

The first stage of the *Saturn V* rocket, used in the Apollo missions to the moon, was powered by the combustion of kerosene with liquid oxygen. Kerosene is a mixture of hydrocarbons that can be represented by $C_{12}H_{26}$. The propulsion is provided by the CO_2 and H_2O emitted from the rocket nozzle.

$$2\,C_{12}H_{26}(l) + 37\,O_2\,(l) \longrightarrow 24\,CO_2(g) + 26\,H_2O(g)$$

The second and third stages of the *Saturn V* used hydrogen and oxygen stored as liquids under pressure in separate chambers. The boiling points of hydrogen and oxygen are −253 °C and −183 °C, respectively. The propulsion is provided by the hot water vapor emitted from the rocket nozzle when the hydrogen and oxygen are allowed to mix and react.

$$2\,H_2(l) + O_2(l) \longrightarrow 2\,H_2O(g)$$

The lunar excursion module of the Apollo missions used the reaction of dimethylhydrazine and dinitrogen tetroxide for propulsion. These compounds were stored as liquids for the trip to the moon.

$$(CH_3)_2NNH_2(l) + 2\,N_2O_4(l) \longrightarrow 3\,N_2(g) + 4\,H_2O(g) + 2\,CO_2(g)$$

Solid propellants are more convenient to handle as rocket fuel. The solid mixture used for the booster rockets for the space shuttle is mixed in the western United States and shipped to Florida. The mixture contains aluminum powder and ammonium perchlorate. The white cloud from the rocket's exhaust that you see when you watch a launch on television is aluminum oxide.

$$3\,Al(s) + 3\,NH_4ClO_4(s) \longrightarrow Al_2O_3(s) + AlCl_3(s) + 3\,NO(g) + 6\,H_2O(g)$$

The white clouds are products of the reaction used to lift the space shuttle which carried a communications satellite into orbit in September 1993. (Photo Researchers, Inc.)

Finally, either the hydrogen or the oxygen may be balanced. There are 12 hydrogen atoms in 3 $Mg(OH)_2$ and 2 H_3PO_4 combined. The coefficient 6 placed in front of H_2O balances the hydrogen.

$$3\,Mg(OH)_2 + 2\,H_3PO_4 \longrightarrow Mg_3(PO_4)_2 + 6\,H_2O$$

Since there are a total of 14 oxygen atoms in both the reactants and products, the equation is balanced.

PROBLEM 6.2

Balance the equation for the reaction of calcium phosphate with sulfuric acid. The reaction is used to produce phosphoric acid, which is used in the production of fertilizers.

$$Ca_3(PO_4)_2 + H_2SO_4 \longrightarrow CaSO_4 + H_3PO_4$$

6.3 Chemical Reactions

Now let's consider how we know when a chemical reaction has occurred. Although there are exceptions, many reactions have one or more characteristics that we can observe. These characteristics are

1. formation of a gas
2. formation of a precipitate
3. generation of a color or a color change
4. evolution or absorption of heat

After two solutions of reactants are mixed, bubbles of a gas may be observed. Examples of this phenomenon are discussed in Section 6.7. The reactants don't have to be in solution for a gas to be released. For example, the compounds in Alka-Seltzer react and release carbon dioxide when the tablets are dropped in water. Heating a solid might also release a gas as a result of decomposition (Section 6.5).

An insoluble solid that deposits from a solution is a **precipitate;** a reaction that forms a precipitate is called a **precipitation reaction.** The formation of a solid from a solution after reactants are mixed is a clear indication that a reaction has occurred.

The reaction of some metals with oxygen gives oxides that have a different color. For example, the rust formed from reaction of iron with oxygen is a red-colored compound whose formula is Fe_2O_3. Many reactions of ionic compounds in solution, especially of transition metals, are accompanied by the development of a color or a change in color.

Changes in temperature when reactants are mixed occur if heat is released or absorbed. These reactions are called exothermic and endothermic, respectively (Section 1.9). We will discuss the role of heat energy in chemical reactions in detail in Chapter 10.

Classification of Chemical Reactions

Millions of chemical reactions are known, and the number of potential reactions among combinations of the 10 million known compounds is astronomically large. What can be done to make sense out of so much information about chemical reactions? First, the reactions are classified to emphasize their similarities and differences. At this point we will consider four simple types of reactions, namely,

1. combination reactions
2. decomposition reactions
3. single-replacement reactions
4. double-replacement reactions

Recognizing a reaction as belonging to one of these classes allows you to predict the products of the reaction. You will learn other ways to classify chemical reactions in organic chemistry and biochemistry later in this text.

6.4 Combination Reactions

A **combination reaction** is the direct union of two or more substances to produce one new substance. The two or more combining materials may be either elements or compounds or a combination of the two. Of course, the product must be a compound. A general equation that summarizes combination reactions with X and Y as reactants follows.

$$X + Y \longrightarrow XY$$

The reactions of metals with the nonmetal oxygen to form metal oxides are combination reactions. Vanadium combines with oxygen to form vanadium(V) oxide.

$$4\,V(s) + 5\,O_2(g) \longrightarrow 2\,V_2O_5(s)$$

Nonmetals also combine with oxygen. For example, sulfur combines with oxygen to yield sulfur dioxide. Sulfur dioxide produced from the burning of sulfur-containing fuels is one of the air pollutants responsible for acid rain.

$$S(s) + O_2(g) \longrightarrow SO_2(g)$$

Carbon combines with oxygen to yield two possible compounds. In a limited amount of oxygen, carbon monoxide is formed. If sufficient oxygen is present, the product is carbon dioxide.

$$2\,C(s) + O_2(g) \longrightarrow 2\,CO(g)$$
$$C(s) + O_2(g) \longrightarrow CO_2(g)$$

Oxygen does not react with nitrogen at ordinary temperatures. However, at high temperatures in an automobile engine, the two elements combine to form nitrogen oxide (nitric oxide). Nitric oxide rapidly combines with oxygen to give nitrogen dioxide, which is a component of smog.

$$N_2(g) + O_2(g) \longrightarrow 2\,NO(g)$$
$$2\,NO(g) + O_2(g) \longrightarrow 2\,NO_2(g)$$

Hydrogen can combine with various elements. The Haber process is an industrial reaction in which ammonia is formed from hydrogen and nitrogen at high temperature. Ammonia is used to produce fertilizer and a variety of other nitrogen compounds.

$$N_2(g) + 3\,H_2(g) \longrightarrow 2\,NH_3\,(g)$$

Compounds can also undergo combination reactions. Metal oxides react with water to produce metal hydroxides, which are called **bases.** Because a base is produced, the oxide from which it is derived is called a **basic oxide.** Bases are discussed briefly in Section 6.8 and in detail in Chapter 11.

$$\underset{}{H_2O(l)} + \underset{\text{basic oxide}}{Na_2O(s)} \longrightarrow \underset{\text{base}}{2\,NaOH(aq)}$$

$$H_2O(l) + \underset{\text{basic oxide}}{MgO(s)} \longrightarrow \underset{\text{base}}{Mg(OH)_2(aq)}$$

Some oxides of nonmetals are called **acidic oxides** because they may combine with water to yield **acids.** These compounds are discussed briefly in Section 6.8 and in detail in Chapter 11.

$$\underset{\text{acidic oxide}}{N_2O_5(g)} + H_2O(l) \longrightarrow \underset{\text{nitric acid}}{2\,HNO_3(aq)}$$

$$\underset{\text{acidic oxide}}{SO_3(g)} + H_2O(l) \longrightarrow \underset{\text{sulfuric acid}}{H_2SO_4(aq)}$$

6.5 Decomposition Reactions

In a **decomposition reaction** a single substance breaks into two or more simpler substances. The reactant is a compound, but the products may be either elements, compounds, or both. Heat is usually required to cause the decomposition of a compound. A general representation of the decomposition reaction follows.

$$XY \longrightarrow X + Y$$

Carbonates, such as calcium carbonate, decompose when heated to yield oxides and carbon dioxide.

$$CaCO_3(s) \xrightarrow{\Delta} CaO(s) + CO_2(g)$$

Calcium carbonate is one of the materials that is deposited in pipes and in equipment when hard water is used. It is the major constituent of coral skeletons and is also produced by the reaction of atmospheric carbon dioxide with calcium ions in the oceans.

Some bicarbonates decompose to yield water, carbon dioxide, and a carbonate salt. Sodium bicarbonate can be used to put out fires because heating it yields carbon dioxide that prevents oxygen from feeding the fire.

$$2\,NaHCO_3(s) \xrightarrow{\Delta} Na_2CO_3(s) + H_2O(g) + CO_2(g)$$

Nitrates may decompose in a variety of ways, as illustrated by the breakdown of sodium nitrate and ammonium nitrate. Ammonium nitrate decomposes with explosive violence at 300 °C.

$$2\ NaNO_3(s) \xrightarrow{\Delta} 2\ NaNO_2(s) + O_2(g)$$

$$NH_4NO_3(s) \xrightarrow{\Delta} N_2O(g) + 2\ H_2O(g)$$

The decomposition of potassium chlorate is a common process used in the undergraduate laboratory to produce oxygen. Manganese dioxide is used as a catalyst.

$$2\ KClO_3(s) \xrightarrow{MnO_2} 2\ KCl(s) + 3\ O_2(g)$$

6.6 Single-Replacement Reactions

In a **single-replacement reaction** one element substitutes for or replaces another element in a compound. Thus, an element and a compound produce another element and a new compound. The elements may be either metals or nonmetals. For example, metal A may replace metal B. In some cases, B may be a hydrogen ion, H^+, of an acid.

$$A + BY \longrightarrow B + AY$$

Zinc replaces the copper in copper(II) sulfate. When a strip of zinc is immersed in a solution of copper(II) sulfate, the zinc disappears, and free copper metal comes out of the solution (Figure 6.1).

$$Zn(s) + CuSO_4(aq) \longrightarrow ZnSO_4(aq) + Cu(s)$$

FIGURE 6.1
A zinc bar is in a solution of a copper(II) salt in the beaker on the left. After a period of time the copper(II) has reacted to coat the zinc bar with copper.
(Yoav Levy, Phototake NYC)

Silver, however, does not replace copper in copper(II) sulfate. The following reaction does not occur.

$$2\ Ag(s) + CuSO_4(aq) \xrightarrow{\text{(doesn't occur)}} Ag_2SO_4(aq) + Cu(s)$$

But the reverse reaction does occur, and in general single-replacement reactions are "one-way" streets.

$$Ag_2SO_4(aq) + Cu(s) \longrightarrow 2\ Ag(s) + CuSO_4(aq)$$

Hydrogen also can be replaced in a single-replacement reaction. Zinc metal reacts with hydrochloric acid and forms hydrogen gas.

$$Zn(s) + 2\ HCl(aq) \longrightarrow ZnCl_2(aq) + H_2(g)$$

Silver, however, does not react with hydrochloric acid. The following potential reaction does not occur.

$$2\ Ag + 2\ HCl \xrightarrow{\text{(doesn't occur)}} 2\ AgCl + H_2$$

These two observations and others provide information about the reactivity of metals in single-replacement reactions. The details of this type of reaction will be presented in Section 6.9.

Single-replacement reactions with nonmetals can be represented as follows, where X and Y represent nonmetals.

$$X + BY \longrightarrow BX + Y$$

The reactivity order for some nonmetals is F > Cl > Br > I. Fluorine will replace chloride, bromide, or iodide ions; chlorine will replace bromide or iodide; bromine will replace only iodide. For example, chlorine gas is used to obtain bromine from the bromide ion in sea water.

$$Cl_2(g) + 2\ NaBr(aq) \longrightarrow 2\ NaCl(aq) + Br_2(g)$$

6.7 Double-Replacement Reactions

In double-replacement reactions two compounds react to exchange atoms or groups of atoms (such as polyatomic ions) to produce two different compounds. A general representation of a double-replacement reaction is

$$AX + BY \longrightarrow AY + BX$$

where any of the letters may represent single atoms or polyatomic ions. In this type of reaction one of the products is usually either a gas or a solid that precipitates from solution.

Gas-Forming Reactions

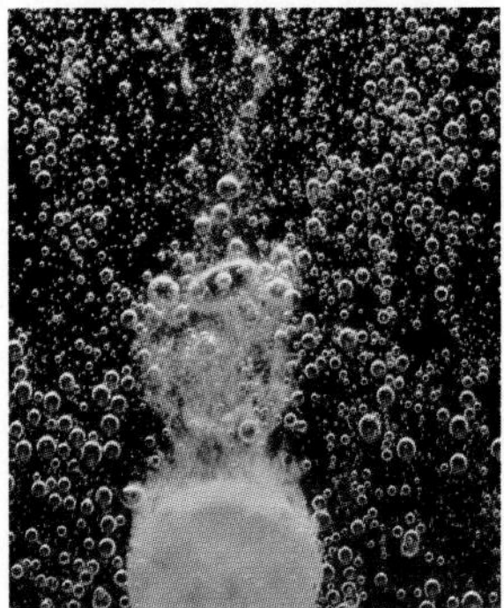

Alka-Seltzer tablets contain sodium bicarbonate and a solid acid that react when placed in water to liberate carbon dioxide. (Paul Silverman, Fundamental Photographs)

A gas forms in some double-replacement reactions and bubbles out of the reaction mixture. Sulfides react with acids to form hydrogen sulfide, a toxic gas that smells like rotten eggs.

$$FeS(s) + 2\ HCl(aq) \longrightarrow FeCl_2(aq) + H_2S(g)$$

When carbonate or bicarbonate compounds react with acid, the double-replacement reaction yields carbonic acid (H_2CO_3). However, carbonic acid is unstable and decomposes to yield carbon dioxide and water.

$$MgCO_3(s) + H_2SO_4(aq) \longrightarrow MgSO_4(aq) + H_2CO_3(aq)$$
$$H_2CO_3(aq) \longrightarrow H_2O(l) + CO_2(g)$$

Magnesium carbonate is an ingredient of several antacids. Antacids react with the hydrochloric acid (HCl) in the stomach to produce carbon dioxide.

Sulfites and bisulfites react with acid to yield sulfurous acid (H_2SO_3), which decomposes to yield sulfur dioxide and water. Sulfur dioxide has a sharp, acrid odor.

$$NaHSO_3(aq) + HCl(aq) \longrightarrow NaCl(aq) + H_2SO_3(aq)$$
$$H_2SO_3(aq) \longrightarrow H_2O(l) + SO_2(g)$$

Precipitation Reactions

A double-replacement reaction may occur when solutions of ionic compounds are mixed. Ionic compounds in water exist as cations and anions. When potassium iodide dissolves in water, an aqueous solution of potassium ions and iodide ions results (Figure 6.2).

FIGURE 6.2 Models of a Precipitation Reaction
When mixed, solutions of silver nitrate and potassium iodide give a precipitate of silver iodide. The solution contains potassium nitrate.

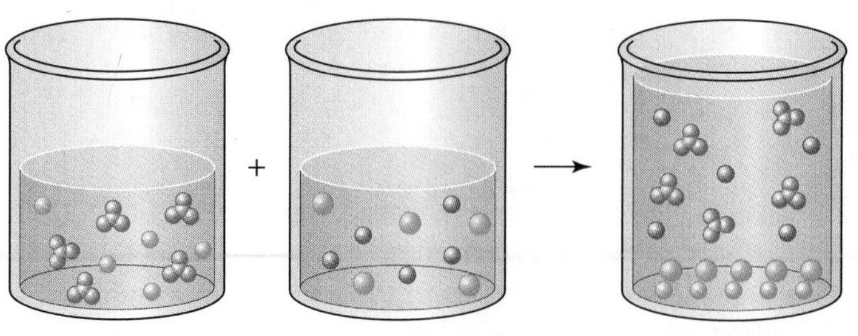

$$KI(s) \xrightarrow{H_2O} K^+(aq) + I^-(aq)$$

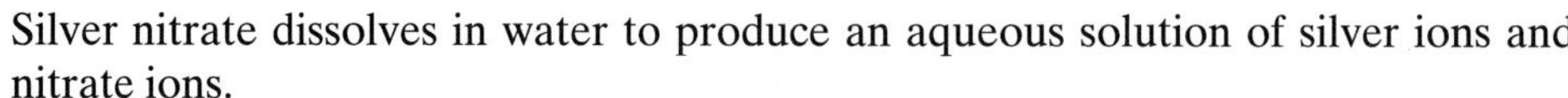

The chromate ions of potassium chromate in the cylinder react with the lead ions of lead nitrate in the beaker to give a precipitate of lead chromate. (Richard Megna, Fundamental Photographs)

Silver nitrate dissolves in water to produce an aqueous solution of silver ions and nitrate ions.

$$AgNO_3(s) \xrightarrow{H_2O} Ag^+(aq) + NO_3^-(aq)$$

When these two solutions are mixed, the cations and anions of both substances are intermingled. The cation of one compound may combine with the anion of the other compound to form a precipitate. In this case, a precipitate of silver iodide forms from the silver and iodide ions. The other product, potassium nitrate, remains in solution.

$$AgNO_3(aq) + KI(aq) \longrightarrow KNO_3(aq) + AgI(s)$$

TABLE 6.1 Solubility Rules of Common Ionic Compounds

1. All ionic compounds of lithium (Li^+), sodium (Na^+), potassium (K^+), rubidium (Rb^+), cesium (Cs^+), and ammonium (NH_4^+) ions are soluble.
2. All nitrates (NO_3^-), chlorates (ClO_3^-), and perchlorates (ClO_4^-) are soluble.
3. The chlorides (Cl^-), bromides (Br^-), and iodides (I^-) of most metals are soluble. The principal exceptions are those of lead (Pb^{2+}), silver (Ag^+), and mercury(I) (Hg_2^{2+}).
4. All sulfates (SO_4^{2-}) are soluble except those of strontium (Sr^{2+}), barium (Ba^{2+}), lead (Pb^{2+}), and mercury(I) (Hg_2^{2+}).
5. All carbonates (CO_3^{2-}), chromates (CrO_4^{2-}), and phosphates (PO_4^{3-}) are insoluble except those of lithium, sodium, potassium, rubidium, cesium, and ammonium.
6. The hydroxides (OH^-) of lithium, sodium, potassium, rubidium, and cesium ions are soluble. The hydroxides of calcium, strontium, and barium are moderately soluble. The rest of the hydroxides are insoluble.
7. The sulfides (S^{2-}) of all metals are insoluble except those of lithium, sodium, potassium, rubidium, cesium, and ammonium.

Net Ionic Equations

The equations for precipitation reactions can be written in an abbreviated form called a **net ionic equation** that shows only the ions that actually react. For the reaction of KI with $AgNO_3$, we write the formula of each species as it actually exists in solution. Since silver iodide is not in solution, it is written as AgI(s).

$$Ag^+(aq) + NO_3^-(aq) + K^+(aq) + I^-(aq) \longrightarrow K^+(aq) + NO_3^-(aq) + AgI(s)$$

The potassium ions and nitrate ions that appear on both sides of the equation do not take part in the reaction. Ions that are present in solution but are not involved in a reaction are called **spectator ions.**

The net ionic equation for a reaction in solution does not include the spectator ions. Only the species that react or are produced are retained in the equation. Ions that appear on both sides of the equation are "canceled." The ions are still present in solution, but they are disregarded. After the spectator potassium and nitrate ions are eliminated, the net ionic equation for the reaction of potassium iodide and silver nitrate is

$$Ag^+(aq) + I^-(aq) \longrightarrow AgI(s)$$

When Will a Precipitation Reaction Occur?

How can you predict if a precipitation reaction will occur? First, consider the two possible combinations of ions that result from exchanging partners. If either of the possible products is insoluble, then a precipitation reaction will occur to form that product. In the reaction of AX with BY, consider AY and BX as possible products. Is either of these compounds insoluble? Consult the general solubility rules (Table 6.1).

EXAMPLE 6.3

An aqueous solution of barium chloride is added to a solution of sodium sulfate, and a precipitate results. Write the net ionic equation for the reaction.

Solution

The possible products are sodium chloride and barium sulfate. According to the solubility rules, sodium chloride is soluble, but barium sulfate is insoluble. Thus, a precipitation reaction will occur to produce solid barium sulfate; sodium chloride will remain in solution. The equation is

$$Na_2SO_4(aq) + BaCl_2(aq) \longrightarrow 2\,NaCl(aq) + BaSO_4(s)$$

According to the rules for expressing ionic equations, the reaction is written as

$$2\,Na^+(aq) + SO_4^{2-}(aq) + Ba^{2+}(aq) + 2\,Cl^-(aq) \longrightarrow 2\,Na^+(aq) + 2\,Cl^-(aq) + BaSO_4(s)$$

The Na^+ and Cl^- ions are not involved in the reaction and are eliminated, giving the net ionic equation.

$$Ba^{2+}(aq) + SO_4^{2-}(aq) \longrightarrow BaSO_4(s)$$

PROBLEM 6.3

An aqueous solution of potassium chromate is added to a solution of lead nitrate. Will a precipitate result? Write the net ionic equation for any reaction that occurs.

6.8 Neutralization Reactions

The terms *acid* and *base* were mentioned in Section 6.4. These substances are so important to all areas of chemistry that Chapter 11 is devoted to the chemistry of acids and bases. We will then see in later chapters that the reactions of acids and bases are a major consideration in explaining both organic and biochemical reactions. Acids and bases have characteristic properties that cancel each other when they react in a **neutralization reaction.**

In a neutralization reaction one of the products is usually water, and the other is an ionic compound, also called a *salt.* The following general equation for a neutralization reaction uses HX and BOH to represent an acid and a base, respectively.

$$HX + BOH \longrightarrow BX + H_2O$$

The general reaction resembles that of a double-replacement reaction, but two of the components are specifically designated. One component of one reactant is the hydrogen ion (H^+), and one component of the second reactant is a hydroxide ion (OH^-). The combination of these two ions to form the covalent water molecule is the driving force for the neutralization reaction.

In Chapter 11 we will establish that acids behave chemically as donors of the hydrogen ion (H^+). The acids that contain this hydrogen ion are written with the hydrogen atom first, followed by a simple or a polyatomic ion. Such acids include hydrochloric acid (HCl), perchloric acid ($HClO_4$), nitric acid (HNO_3), sulfuric acid (H_2SO_4), and phosphoric acid (H_3PO_4). Examples of neutralization reactions of an acid and a base follow.

$$HNO_3(aq) + KOH(aq) \longrightarrow KNO_3(aq) + H_2O(l)$$
$$2\,HCl(aq) + Ca(OH)_2(aq) \longrightarrow CaCl_2(aq) + 2\,H_2O(l)$$

In each balanced equation the number of hydrogen ions provided by the acid equals the number of hydroxide ions provided by the base.

Neutralization reactions also occur for other combinations of reactants, namely,

1. acid and basic oxide
2. acidic oxide and base
3. acidic oxide and basic oxide

Recall from Section 6.4 that metal oxides are basic oxides. In the reaction of calcium oxide with perchloric acid, the calcium ion and the hydrogen ion are interchanged. The products of the reaction are a salt and water.

$$\underset{\text{basic oxide}}{CaO} + \underset{\text{acid}}{2\,HClO_4} \longrightarrow \underset{\text{a salt}}{Ca(ClO_4)_2} + H_2O$$

Recall from Section 6.4 that oxides of nonmetals are acidic oxides. The reaction of carbon dioxide and lithium hydroxide produces a salt and water. This reaction is used to remove carbon dioxide from the air of the space shuttle.

$$\underset{\text{base}}{2\,LiOH} + \underset{\text{acidic oxide}}{CO_2} \longrightarrow \underset{\text{a salt}}{Li_2CO_3} + H_2O$$

In some cases the acidic oxide combines directly with the base, to form a single product.

$$NaOH + SO_2 \longrightarrow NaHSO_3$$

base acidic oxide a salt

Acidic oxides and basic oxides can also combine directly. In such reactions no water is formed.

$$CaO + SO_3 \longrightarrow CaSO_4$$

basic oxide acidic oxide a salt

6.9 Organic Acids and Bases

The examples of acids and bases given in this section are common "inorganic" compounds. In the organic and biochemical sections of this text you will learn about carbon-containing acids and bases. Acetic acid (CH_3CO_2H) is an example of a carboxylic acid. Only one of the four hydrogen atoms is acidic. The acidic hydrogen atom is bonded to an electronegative oxygen atom, as is the hydrogen atom of many inorganic acids such as nitric acid.

$$CH_3-C(=\ddot{O}:)-\ddot{O}-H \qquad :\ddot{O}-N(=\ddot{O}:)-\ddot{O}-H$$

acetic acid nitric acid

Ammonia, like hydroxide ion, is a base because it reacts with the hydrogen ion. The product is the ammonium ion.

$$H-\ddot{N}H_2 + H^+ \longrightarrow [NH_4]^+$$

ammonia ammonium ion

Some organic compounds contain nitrogen and resemble ammonia. These covalent compounds, called amines, have one or more carbon groups bonded to the nitrogen atom. Methylamine ($CH_3—NH_2$) is an example. Amines are organic bases that react like ammonia to give ammonium salts.

$$CH_3-\ddot{N}H_2 + H^+ \longrightarrow [CH_3-NH_3]^+$$

methylamine methylammonium ion

Amino acids are carboxylic acids that also contain an amine group (Chapter 23). These compounds act as both acids and bases. They play an important role in biochemical reactions. Glycine is the simplest amino acid.

$$H-\ddot{N}(H)-CH_2-C(=\ddot{O}:)-\ddot{O}-H$$

glycine

6.10 Oxidation–Reduction Reactions

Some reactions in three of the classes of reactions discussed to this point involve electron transfer reactions, also known as **oxidation–reduction reactions,** or **redox reactions** for short. Each of the following reactions, originally classified in different categories, occurs by electron transfer.

combination	$4\ Al + 3\ O_2 \longrightarrow 2\ Al_2O_3$
single-replacement	$Zn + 2\ AgNO_3 \longrightarrow Zn(NO_3)_2 + 2\ Ag$
decomposition	$2\ NaNO_3 \longrightarrow 2\ NaNO_2 + O_2$

A bookkeeping system using *oxidation numbers* is needed to keep track of the electron transfer that is a common feature of these reactions.

Oxidation Numbers

The **oxidation number** of an atom is a positive or negative integer assigned to describe an element as a free atom, an ion, or part of a polyatomic ion or molecule. To make these assignments we follow the rules listed in Table 6.2.

According to rule 1, the oxidation number of an element, whether it is monatomic, diatomic, or polyatomic, is zero. Thus, each individual atom in the elements H_2, F_2, P_4, and S_8 has an oxidation number of zero.

According to rule 2, the oxidation number is equal to the charge of the cation or the anion. Thus, when an electron is removed from an atom, as from Na to yield Na^+, the oxidation number of the cation is +1. When an electron is gained, as in the case of Cl to yield Cl^-, the oxidation number is −1.

Rule 8 means that in NaCl or in any other ionic compound, the sum of the oxidation numbers must be numerically equal to zero, because the compound is electrically neutral overall.

TABLE 6.2 Assignment of Oxidation Numbers

1. The oxidation number of an element in the elemental state is zero.
2. The oxidation number of a monatomic ion is the same as the charge of the ion.
3. Metals have positive oxidation numbers in compounds.
4. The oxidation number of hydrogen is +1 in most hydrogen compounds.
5. The oxidation number of oxygen is −2 in most oxygen compounds.
6. The oxidation number of fluorine in compounds is −1.
7. The oxidation number of chlorine, bromine, and iodine in compounds is −1 except when combined with oxygen.
8. The algebraic sum of the oxidation numbers of the atoms in a compound is zero.
9. The algebraic sum of the oxidation numbers of the atoms in a polyatomic ion is the same as the charge of the ion.

By convention, the oxidation numbers of atoms in molecular compounds are based on the arbitrary assignment of the electrons to atoms based on their relative tendency to gain electrons. This tendency, known as electronegativity, was discussed in Chapters 3 and 5. Thus, some atoms are viewed as having lost electrons and others as having gained electrons, although none have actually done so. In water the oxidation number of hydrogen is +1 and the oxidation number of oxygen is −2 even though water is not an ionic compound. Note that the sum of the oxidation numbers of atoms in a molecule must equal zero. Abbreviating oxidation number as oxid. no., we write

$$2\,(\text{oxid. no. of hydrogen}) + (\text{oxid. no. of oxygen}) = 0$$
$$2\,(+1) + (-2) = 0$$

Now let's determine the oxidation number of an atom not mentioned in the assignment rules. In carbon tetrachloride (CCl_4) each chlorine atom is assigned an oxidation number of −1. Since the sum of the oxidation numbers of atoms in a molecule must equal zero, the oxidation number of carbon is +4.

$$\text{oxid. no. of carbon} + 4\,(\text{oxid. no. of chlorine}) = 0$$
$$\text{oxid. no. of carbon} + 4\,(-1) = 0$$
$$\text{oxid. no. of carbon} = +4$$

EXAMPLE 6.4

What is the oxidation number of sulfur in SO_2?

Solution

The sum of the oxidation numbers of the atoms in a neutral compound must be zero according to rule 8. Thus, we write

$$\text{oxid. no. of sulfur} + 2\,(\text{oxid. no. of oxygen}) = 0$$

Since the oxidation number of oxygen is −2 according to rule 5, the oxidation number of sulfur can be calculated algebraically.

$$\text{oxid. no. of sulfur} + 2\,(-2) = 0$$
$$\text{oxid. no. of sulfur} - 4 = 0$$
$$\text{oxid. no. of sulfur} = +4$$

PROBLEM 6.4

What is the oxidation number of nitrogen in N_2H_4, a compound used in some rocket fuels?

EXAMPLE 6.5

Solutions of potassium permanganate are used to treat some diseases of tropical fish in home aquariums. What is the oxidation number of manganese in the permanganate ion, MnO_4^-?

Solution

The sum of the oxidation numbers of the atoms in a polyatomic ion is equal to the charge of the ion (rule 9).

$$\text{oxid. no. of manganese} + 4\,(\text{oxid. no. of oxygen}) = -1$$

Since the oxidation number of oxygen is -2 (rule 5), the oxidation number of manganese can be calculated.

$$\text{oxid. no. of manganese} + 4(-2) = -1$$
$$\text{oxid. no. of manganese} = -1 + 8 = +7$$

PROBLEM 6.5

What is the oxidation number of carbon in the oxalate ion, $C_2O_4^{2-}$?

Oxidation Is the Opposite of Reduction

Oxidation is the loss of electrons by an atom or an increase in its oxidation number. **Reduction** is the gain of electrons by an atom or a decrease in its oxidation number. When iron combines with oxygen, the oxidation number of iron increases from 0 to +3; the iron has been oxidized. The oxidation number of oxygen decreases from 0 in the element to −2 in iron(III) oxide; the oxygen has been reduced.

decreases
increases

$$\overset{(0)}{4\,Fe} + \overset{(0)}{3\,O_2} \longrightarrow 2\,\overset{(+3)}{Fe_2}\overset{(-2)}{O_3}$$

When a substance is oxidized, it loses electrons to another substance, which then becomes reduced. For example, the electrons lost by iron in its reaction with oxygen are gained by the oxygen. When any substance is reduced, the electrons it gains are obtained from a substance that is oxidized. This is why reactions in which an electron transfer occurs are called oxidation–reduction reactions. Oxidation cannot occur without reduction, and vice versa.

Oxidizing Agents and Reducing Agents

The relationship between oxidation and reduction is further emphasized by the terms *oxidizing agent* and *reducing agent* (Table 6.3). In an oxidation–reduction reaction, the substance that is reduced is an **oxidizing agent** because, by gaining electrons, it causes oxidation in another substance. The substance that is oxidized is a **reducing agent** because, by losing its electrons, it causes the reduction of another substance. In terms of the changes in oxidation number, an oxidizing agent is a substance whose oxidation number decreases; a reducing agent is a substance whose oxidation number increases.

TABLE 6.3 Oxidation–Reduction Terminology

Term	Electron Change	Oxidation Number Change
oxidation	loss of electrons	increase
reduction	gain of electrons	decrease
oxidizing agent	accepts electrons	decrease
reducing agent	donates electrons	increase

The single-replacement reaction of zinc with copper(II) nitrate is a redox reaction. The oxidation numbers of the elements are given above the elemental symbols. From this information, the substance oxidized, the substance reduced, the oxidizing agent, and the reducing agent are identified.

$$\overset{(0)}{Zn(s)} + \overset{(+2)}{Cu}(NO_3)_2(aq) \longrightarrow \overset{(+2)}{Zn}(NO_3)_2(aq) + \overset{(0)}{Cu(s)}$$

becomes oxidized; is reducing agent — becomes reduced; is oxidizing agent

EXAMPLE 6.6

The black tarnish on silverware results from the reaction of silver with small amounts of hydrogen sulfide, H_2S, in the air. Which element is oxidized, and which one is reduced in the reaction?

$$4\,Ag + 2\,H_2S + O_2 \longrightarrow 2\,Ag_2S + 2\,H_2O$$

Solution

The oxidation numbers of hydrogen and sulfur are +1 and −2, respectively, and are not changed in the reaction. The oxidation number of oxygen in water is −2, but it is 0 in elemental oxygen. Thus, oxygen is reduced. The oxidation number of silver changes from 0 in elemental silver to +1 in Ag_2S. The silver is oxidized.

PROBLEM 6.6

What is the oxidizing agent and what is the reducing agent in the reaction represented by the following equation?

$$4\,NH_3 + 5\,O_2 \longrightarrow 4\,NO + 6\,H_2O$$

What Redox Reactions Will Occur?

The single-replacement reaction of zinc with silver nitrate is a redox reaction that occurs readily. However, silver does not react with zinc nitrate.

$$Zn + 2\,AgNO_3 \xrightarrow{\text{(occurs)}} Zn(NO_3)_2 + 2\,Ag$$

$$Zn(NO_3)_2 + 2\,Ag \xrightarrow[\times]{\text{(doesn't occur)}} Zn + 2\,AgNO_3$$

How can we predict whether zinc will react with silver nitrate? Why doesn't the reverse reaction of silver with zinc nitrate occur? What principle underlies the favored direction of this reaction and other redox reactions?

Oxidation–reduction reactions involve a "battle" for electrons. In the forward reaction the silver ion is the oxidizing agent and gains an electron to form elemental silver.

$$Ag^+ + 1\,e^- \longrightarrow Ag$$

In the reverse reaction the zinc ion would have to be the oxidizing agent and gain electrons to form elemental zinc.

$$Zn^{2+} + 2\,e^- \longrightarrow Zn$$

From this information we conclude that the silver ion has a stronger tendency to gain electrons than does the zinc ion.

The experimentally determined order of attraction for electrons for a few species is given in Table 6.4. Oxidizing agents that have a high tendency to gain electrons are **strong oxidizing agents.** The strength of the oxidizing agents decreases from top to bottom in the list. Fluorine is at the top of the list; it is the strongest oxidizing agent. Species toward the bottom of the list have smaller tendencies to gain electrons and are **weak oxidizing agents.**

Now consider the reverse of the reactions listed in Table 6.4. In these reactions the species on the right would have to release electrons (become oxidized) to form the species on the left. When the species on the right are oxidized, other substances get reduced, so the species on the right are reducing agents. Which species release electrons most easily and are the best reducing agents? To answer this question, consider first the poorest oxidizing agents, that is, the species that have the smallest tendency to gain electrons. The lithium ion, located at the bottom of the table, is the poorest oxidizing agent.

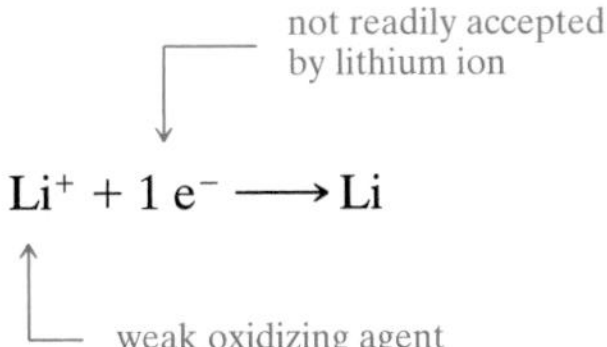

If lithium ion has only a small tendency to gain electrons, what would you expect to be the tendency of lithium metal to lose electrons? Lithium should readily

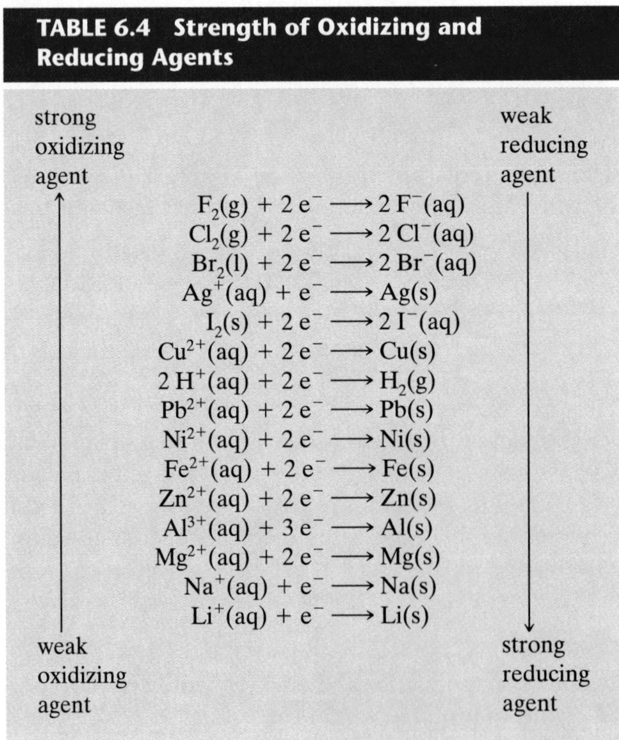
TABLE 6.4 Strength of Oxidizing and Reducing Agents

strong oxidizing agent ↑		weak reducing agent ↓
	$F_2(g) + 2\ e^- \longrightarrow 2\ F^-(aq)$	
	$Cl_2(g) + 2\ e^- \longrightarrow 2\ Cl^-(aq)$	
	$Br_2(l) + 2\ e^- \longrightarrow 2\ Br^-(aq)$	
	$Ag^+(aq) + e^- \longrightarrow Ag(s)$	
	$I_2(s) + 2\ e^- \longrightarrow 2\ I^-(aq)$	
	$Cu^{2+}(aq) + 2\ e^- \longrightarrow Cu(s)$	
	$2\ H^+(aq) + 2\ e^- \longrightarrow H_2(g)$	
	$Pb^{2+}(aq) + 2\ e^- \longrightarrow Pb(s)$	
	$Ni^{2+}(aq) + 2\ e^- \longrightarrow Ni(s)$	
	$Fe^{2+}(aq) + 2\ e^- \longrightarrow Fe(s)$	
	$Zn^{2+}(aq) + 2\ e^- \longrightarrow Zn(s)$	
	$Al^{3+}(aq) + 3\ e^- \longrightarrow Al(s)$	
	$Mg^{2+}(aq) + 2\ e^- \longrightarrow Mg(s)$	
	$Na^+(aq) + e^- \longrightarrow Na(s)$	
	$Li^+(aq) + e^- \longrightarrow Li(s)$	
weak oxidizing agent		strong reducing agent

lose electrons. After all, if Li^+ doesn't want to gain an electron very badly, why should the resulting lithium atom hold on to the electron once it has one?

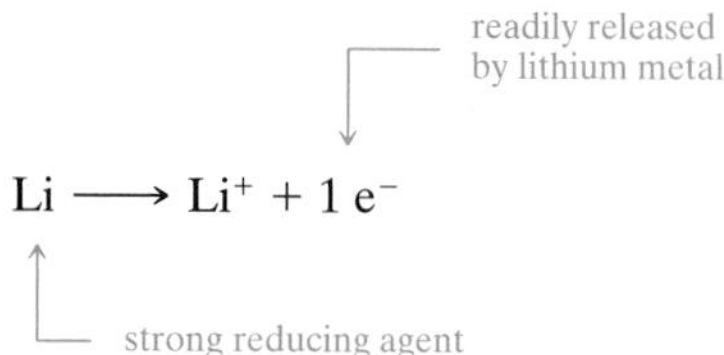

Now we can generalize about the tendency of species to act as reducing agents. A **strong reducing agent** has a strong tendency to release electrons. The best reducing agents are those species on the right side of the equations in Table 6.4 and at the bottom of the list. The species on the right and toward the top of the list are **weak reducing agents.** Thus the trend of the strengths of reducing agents runs opposite that of the strengths of the oxidizing agents. A strong reducing agent will easily lose electrons, and the resultant species, which could conceivably gain an electron, is a weak oxidizing agent.

Preventing Corrosion, an Unwanted Oxidation

The rusting of iron and steel costs the United States about \$10 billion each year. The rust of iron, iron(III) oxide, does not adhere to the metal surface; it flakes off and leaves an iron surface that can be further oxidized. Eventually the iron loses its structural strength because so much iron has rusted.

The first step in the oxidation of iron is loss of electrons to form iron(II).

$$Fe \longrightarrow Fe^{2+} + 2\ e^- \qquad \text{(an oxidation)}$$

The electrons are conducted by the metal to any point where an oxidizing agent is available. Moisture and oxygen combined make an effective oxidizing agent.

$$O_2 + 2\ H_2O + 4\ e^- \longrightarrow 4\ OH^-$$

Further oxidation of Fe^{2+} occurs to produce Fe^{3+}, which reacts with hydroxide ion to form the iron oxide that we call rust.

$$Fe^{2+} \longrightarrow Fe^{3+} + e^- \qquad \text{(an oxidation)}$$

$$2Fe^{3+} + 6\ OH^- \longrightarrow Fe_2O_3 + 3\ H_2O$$

Rusting can be prevented by a protective coating of a corrosion-resistant metal such as chromium, which also gives the metal its "chrome" look. In addition, even some easily oxidized metals can be used to coat iron. Zinc is one such example, which is used to **galvanize** steel. Zinc is more easily oxidized than iron, but the zinc oxide coating does not flake off. Consequently the coating prevents the iron from rusting.

Another method of protecting iron objects that might be oxidized by moisture and oxygen involves placing the iron in contact with a more easily oxidized metal. The electrons for the reduction of oxygen are thus provided by the other metal, which is "sacrificed" to prevent the iron from oxidizing. Both underground steel storage tanks for gasoline and hulls of ships are protected by this method using aluminum or magnesium.

The workers are repairing the corrosion on this fishing vessel in New Bedford, Massachusetts. (Granitasis\The Image Works)

We can predict what redox reaction will occur by using Table 6.4. The stronger oxidizing agent will take electrons from the stronger reducing agent. This combination occurs for the oxidizing agents located on the left and closest to the top of the list and the reducing agents located on the right and closest to the bottom of the list. Reconsider now why zinc reacts with silver nitrate, but the reverse reaction in which silver would react with zinc nitrate does not occur.

$$\underset{\substack{\text{becomes oxidized}\\ \text{is reducing agent}}}{\overset{(0)}{Zn}} + \underset{\substack{\text{becomes reduced}\\ \text{is oxidizing agent}}}{\overset{(+1)}{2\,AgNO_3}} \longrightarrow \overset{(+2)}{Zn(NO_3)_2} + \overset{(0)}{2\,Ag}$$

The oxidizing agent for the written equation is Ag^{+}; in the reverse reaction the oxidizing agent would be Zn^{2+}. The silver ion is located higher in Table 6.4 than the zinc ion. Thus, silver ion is the better oxidizing agent, and the reaction will occur as written. We reach the same conclusion when we examine the reducing agents in the reactions. The reducing agent for the written equation is Zn; in the reverse reaction the reducing agent would be Ag. Both silver and zinc are listed as reducing agents on the right of Table 6.4; zinc is located lower in the table than silver. Thus, zinc is the better reducing agent, and the reaction will occur as written.

6.11 Redox Reactions in Organic Chemistry

In contrast with oxidation–reduction reactions of inorganic compounds, in which the results of electron transfer are easily discernible, oxidation–reduction reactions of organic compounds are less obvious. An organic compound is oxidized if one of its carbon atoms forms a bond to a more electronegative atom, such as oxygen. The carbon atom is oxidized because the electrons in the carbon–oxygen bond are drawn toward the more electronegative oxygen atom. If a carbon atom forms a bond to a less electronegative atom, such as hydrogen, the carbon atom is reduced. The electronegativity of carbon is 2.5 and that of hydrogen is 2.1. Although this difference is small, there is a small net polarity in a carbon–hydrogen bond. The electrons in the carbon–hydrogen bond are drawn slightly toward the carbon atom.

We can determine the change in oxidation state of an organic compound in an oxidation–reduction reaction by counting the number of hydrogen atoms or oxygen atoms gained or lost. The oxidation state of a molecule increases (oxidation) if its hydrogen content decreases or its oxygen content increases. The oxidation state of a molecule decreases (reduction) if its hydrogen content increases or its oxygen content decreases. For example, the conversion of methyl alcohol, CH_3OH, to formaldehyde. CH_2O, is an oxidation because methyl alcohol loses two hydrogen atoms. Further reaction of formaldehyde to produce formic acid occurs with an increase in the oxygen content, so it is also an oxidation process. The symbol [O] represents an unidentified oxidizing agent.

$$\underset{\text{methyl alcohol}}{H{-}\overset{\overset{\displaystyle H}{|}}{\underset{\underset{\displaystyle H}{|}}{C}}{-}O{-}H} \xrightarrow{[O]} \underset{\text{formaldehyde}}{H{-}\overset{\overset{\displaystyle O}{\|}}{C}{-}H} \xrightarrow{[O]} \underset{\text{formic acid}}{H{-}\overset{\overset{\displaystyle O}{\|}}{C}{-}O{-}H}$$

The conversion of a carbon–carbon triple bond into a double bond and finally into a single bond involves reduction because the hydrogen content increases in each step. Hydrogen gas in the presence of a platinum catalyst is a common reducing agent for carbon–carbon double and triple bonds.

$$\text{H—C}\equiv\text{C—H} \xrightarrow[\text{Pt}]{H_2} \text{H}_2\text{C}=\text{CH}_2 \xrightarrow[\text{Pt}]{H_2} \text{H}_3\text{C—CH}_3$$

acetylene ethylene ethane

EXAMPLE 6.7

Ethylene oxide is used to sterilize medical equipment that cannot withstand the steam heat of an autoclave. The gas is produced from ethylene by the following process. Classify the type of reaction. Is an oxidizing or reducing agent required?

Redox Reactions in Biochemistry

Metabolic reactions provide energy in multiple-step reactions in which metabolites are oxidized. Biosynthetic reactions, which build the necessary compounds to maintain organisms, usually are reduction reactions. Thus, organisms require both oxidizing and reducing agents. Furthermore, because there are so many diverse reactions that occur in oxidative degradation and reductive biosynthesis, these oxidizing and reducing agents must have a wide range of reactivity.

Nature has designed chemical reactions that are catalyzed by enzymes—most frequently a specific enzyme for each type of reaction. However, these redox reactions also require a coenzyme that serves as the oxidizing or reducing agent. Nature is efficient in this regard because there are a limited number of these coenzymes. Their structures and details of their reactions are discussed in Chapter 25. Two of these compounds, nicotinamide adenine dinucleotide and flavin adenine dinucleotide, are represented by the shorthand notations NAD^+ and FAD, respectively.

Each of these coenzymes oxidizes biological molecules by removing two hydrogen atoms and an electron from covalent bonds. These reactions, using hydrogen atoms for simplicity, are represented as

$$NAD^+ + 2\,H\cdot \longrightarrow NADH + H^+$$

oxidized form reduced form

$$FAD + 2\,H\cdot \longrightarrow FADH_2$$

oxidized form reduced form

The reduced form of each coenzyme can serve as a reducing agent and be oxidized to regenerate the oxidizing agent by a process discussed in Chapter 25.

NAD^+ and FAD oxidize different classes of compounds. NAD^+ removes hydrogen atoms from a C—H and an O—H bond, whereas FAD removes hydrogen atoms from two C—H bonds. For example, NAD^+ oxidizes malic acid to oxaloacetic acid in one of the steps of the citric acid cycle, which is ultimately responsible for the oxidation of metabolites to carbon dioxide and water.

$$NAD^+ + HO_2C—CH_2—CH(OH)—CO_2H \longrightarrow$$

malic acid

$$NADH + H^+ + HO_2C—CH_2—C(=O)—CO_2H$$

oxaloacetic acid

FAD oxidizes fatty acids such as stearic acid, which is derived from fats—our long-term sources of energy.

$$FAD + CH_3(CH_2)_{14}—CH_2—CH_2—CO_2H \longrightarrow$$

stearic acid

$$FADH_2 + CH_3(CH_2)_{14}—CH=CH—CO_2H$$

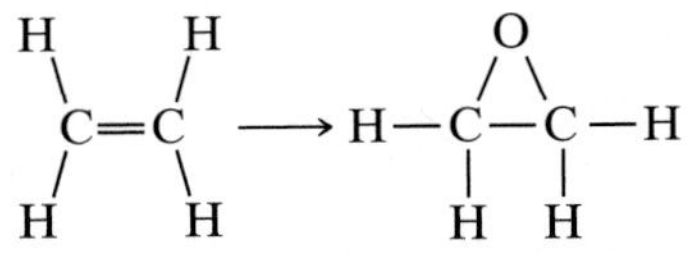

```
H       H                  O
 \     /                  / \
  C = C      ⟶    H — C — C — H
 /     \              |   |
H       H             H   H
```

Solution

The product has one more oxygen atom than the reactant. Thus, the process is an oxidation reaction, and an oxidizing agent is required.

PROBLEM 6.7

Sorbitol, a sugar substitute, is produced in large quantities in chemical industry starting from glucose. What type of chemical reaction is used? What type of reagent is required?

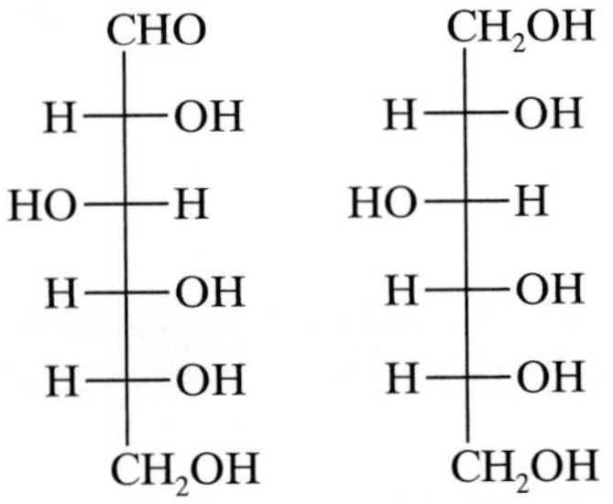

```
       CHO              CH2OH
   H ──┼── OH       H ──┼── OH
  HO ──┼── H       HO ──┼── H
   H ──┼── OH       H ──┼── OH
   H ──┼── OH       H ──┼── OH
      CH2OH            CH2OH
     glucose          sorbitol
```

Summary

A chemical equation expresses the conversion of reactants into products in a chemical reaction. The **coefficients** of a balanced chemical equation provide information about the relative quantities of reactants and products. Additional information that can be conveyed in a chemical equation includes the state of the reactants and products, the conditions necessary to carry out the reaction, and whether a catalyst is required.

Guidelines for balancing equations are useful for establishing the coefficients of a chemical equation. Select an atom that appears only once on each side of the equation and balance it. Continue balancing atoms that appear only once on each side but in unequal numbers. Finally balance atoms that appear in two or more places on one side of the equation. The final coefficients must be the lowest possible whole numbers.

Chemical reactions can be categorized according to the type of process. Simple reactions can be labeled as **combination, decomposition, single-replacement,** or **double-replacement. Acid–base reactions** and **oxidation–reduction** reactions are two additional classes of reactions. Some organic compounds, such as carboxylic acids, are acids; nitrogen-containing compounds that resemble ammonia, known as amines, are bases.

Metal oxides are known as **basic oxides** because they react with water to produce **bases.** Oxides of nonmetals are known as **acidic oxides** because they react with water to produce **acids. Neutralization reactions** may take place between an acid and a base, a basic oxide and an acid, a base and an acidic oxide, or an acidic oxide and a basic oxide.

In an **oxidation–reduction (redox) reaction,** electrons are transferred between reactants. **Oxidation** is the loss of electrons; **reduction** is the gain of electrons. Reactants that become oxidized serve as **reducing agents;** reactants that become reduced serve as **oxidizing agents.** Redox reactions are controlled by the electron-attracting characteristics of the reactants.

Redox reactions in organic compounds are characterized by changes in hydrogen and oxygen content. The loss of hydrogen or gain of oxygen is oxidation; the gain of hydrogen or loss of oxygen is reduction. Biochemical redox reactions require the coenzymes NAD^+ or FAD or their reduced forms, NADH and $FADH_2$.

Exercises

Balancing Equations

6.1 Balance each of the following equations.

(a) $Ba + O_2 \longrightarrow BaO$
(b) $Fe + Cl_2 \longrightarrow FeCl_3$
(c) $H_2 + I_2 \longrightarrow HI$
(d) $HgO \longrightarrow Hg + O_2$
(e) $PbCl_2 + Na_2S \longrightarrow PbS + NaCl$
(f) $Mg + Fe_3O_4 \longrightarrow MgO + Fe$

6.2 Balance each of the following equations.

(a) $Li + Se \longrightarrow Li_2Se$
(b) $H_2O + O_2 \longrightarrow H_2O_2$
(c) $Ca_3P_2 + H_2O \longrightarrow PH_3 + Ca(OH)_2$
(d) $P_4 + Cl_2 \longrightarrow PCl_5$
(e) $CH_4 + Cl_2 \longrightarrow CCl_4 + HCl$
(f) $CO + H_2 \longrightarrow CH_4 + H_2O$

6.3 Balance each of the following equations.

(a) $SiO_2 + HF \longrightarrow H_2SiF_6 + H_2O$
(b) $Li_3N + H_2O \longrightarrow LiOH + NH_3$
(c) $Al_4C_3 + HCl \longrightarrow AlCl_3 + CH_4$
(d) $H_2O_2 + N_2H_4 \longrightarrow N_2 + H_2O$
(e) $PCl_3 + H_2O \longrightarrow H_3PO_3 + HCl$
(f) $Zn + HCl \longrightarrow ZnCl_2 + H_2$

6.4 Balance each of the following equations.

(a) $N_2O \longrightarrow N_2 + O_2$
(b) $Br_2 + NaI \longrightarrow NaBr + I_2$
(c) $ZnO + CO \longrightarrow Zn + CO_2$
(d) $Li + N_2 \longrightarrow Li_3N$
(e) $Cu + Al_2O_3 \longrightarrow CuO + Al$
(f) $ZnS + O_2 \longrightarrow ZnO + SO_2$

6.5 Balance each of the following equations.

(a) $C_2H_6 + O_2 \longrightarrow CO_2 + H_2O$
(b) $C_5H_{12} + O_2 \longrightarrow CO_2 + H_2O$
(c) $C_5H_{10} + O_2 \longrightarrow CO_2 + H_2O$
(d) $C_2H_6O + O_2 \longrightarrow CO_2 + H_2O$
(e) $C_5H_{12}O + O_2 \longrightarrow CO_2 + H_2O$
(f) $C_5H_{10}O + O_2 \longrightarrow CO_2 + H_2O$

6.6 Balance each of the following equations.

(a) $H_2SO_4 + Na_2CO_3 \longrightarrow Na_2SO_4 + H_2O + CO_2$
(b) $H_3PO_4 + KOH \longrightarrow K_3PO_4 + H_2O$
(c) $FeCl_2 + NaOH \longrightarrow Fe(OH)_2 + NaCl$
(d) $Fe(OH)_3 + H_2SO_4 \longrightarrow Fe_2(SO_4)_3 + H_2O$
(e) $H_3PO_4 + Ca(OH)_2 \longrightarrow Ca_3(PO_4)_2 + H_2O$
(f) $NiCO_3 + HCl \longrightarrow NiCl_2 + CO_2 + H_2O$

Combination Reactions

6.7 Write the expected product of each of the following combination reactions, and balance the equation.

(a) $Ba + O_2 \longrightarrow$ (b) $H_2 + I_2 \longrightarrow$
(c) $Li + S \longrightarrow$ (d) $Al + Cl_2 \longrightarrow$
(e) $Ca + Cl_2 \longrightarrow$ (f) $Mg + N_2 \longrightarrow$

6.8 Write the expected product of each of the following combination reactions, and balance the equation.

(a) $Na_2O + H_2O \longrightarrow$ (b) $CaO + H_2O \longrightarrow$
(c) $BaO + CO_2 \longrightarrow$ (d) $MgO + CO_2 \longrightarrow$
(e) $SO_2 + H_2O \longrightarrow$ (f) $Li_2O + H_2O \longrightarrow$

Decomposition Reactions

6.9 Write the expected products of each of the following decomposition reactions, and balance the equation.

(a) $HgO \longrightarrow$ (b) $SrCO_3 \longrightarrow$
(c) $Au_2O_3 \longrightarrow$ (d) $Ag_2O \longrightarrow$

6.10 Write the expected products of each of the following decomposition reactions, and balance the equation.

(a) $N_2O \longrightarrow$ (b) $KClO_3 \longrightarrow$
(c) $MgCO_3 \longrightarrow$ (d) $H_2O \longrightarrow$

Single-Replacement Reactions

6.11 Write the expected products of each of the following single-replacement reactions, and balance the equation.

(a) $KBr + Cl_2 \longrightarrow$ (b) $SnO_2 + H_2 \longrightarrow$
(c) $Mg + Fe_3O_4 \longrightarrow$ (d) $Al_2O_3 + C \longrightarrow$

6.12 Write the expected products of each of the following single-replacement reactions, and balance the equation.

(a) $Sb_4O_6 + C \longrightarrow$ (b) $Br_2 + CaI_2 \longrightarrow$
(c) $TlI_3 + Na \longrightarrow$ (d) $Sb_2S_3 + O_2 \longrightarrow$

6.13 Write the expected products of each of the following single-replacement reactions, and balance the equation.

(a) $Zn + HCl \longrightarrow$ (b) $SiCl_4 + Mg \longrightarrow$
(c) $Br_2 + NaI \longrightarrow$ (d) $ZnO + C \longrightarrow$

6.14 Write the expected products of each of the following single-replacement reactions, and balance the equation.

(a) $Fe + CuCl_2 \longrightarrow$ (b) $NiCl_2 + Zn \longrightarrow$
(c) $F_2 + NaI \longrightarrow$ (d) $Al + HCl \longrightarrow$

Solubility of Ionic Compounds

6.15 State whether each of the following compounds is soluble or insoluble.

(a) Na_2SO_4 (b) $AgBr$ (c) $(NH_4)_2SO_4$
(d) Na_2CO_3 (e) $CaCl_2$ (f) $NaNO_3$

6.16 State whether each of the following compounds is soluble or insoluble.

(a) $BaCO_3$ (b) CdS (c) $(NH_4)_2S$
(d) $KClO_3$ (e) $PbSO_4$ (f) $Fe(OH)_2$

6.17 State whether each of the following compounds is soluble or insoluble.

(a) K_2SO_4 (b) AgI (c) NH_4Cl
(d) Rb_2CO_3 (e) $BaCl_2$ (f) $Sr(NO_3)_2$

6.18 State whether each of the following compounds is soluble or insoluble.

(a) $SrCO_3$ (b) HgS (c) $(NH_4)_2SO_4$
(d) $NaClO_4$ (e) $ZnSO_4$ (f) $Ba(OH)_2$

Double-Replacement Reactions

6.19 Write the expected products of each of the following double-replacement reactions, and balance the equation.

(a) $Pb(NO_3)_2 + HCl \longrightarrow$ (b) $FeCl_2 + NaOH \longrightarrow$
(c) $Cd(NO_3)_2 + H_2S \longrightarrow$ (d) $AgNO_3 + CsCl \longrightarrow$

6.20 Write the expected products of each of the following double-replacement reactions, and balance the equation.

(a) $Pb(NO_3)_2 + Na_2S \longrightarrow$
(b) $NiCO_3 + HCl \longrightarrow$
(c) $(NH_4)_2SO_4 + BaCl_2 \longrightarrow$
(d) $Bi(NO_3)_3 + NaOH \longrightarrow$

6.21 Write the expected products of each of the following double-replacement reactions, and balance the equation.

(a) $MnSO_4 + (NH_4)_2S \longrightarrow$ (b) $CaCO_3 + H_2SO_4 \longrightarrow$
(c) $CdSO_4 + NaOH \longrightarrow$ (d) $ZnCO_3 + H_3PO_4 \longrightarrow$

6.22 Write the expected products of each of the following double-replacement reactions, and balance the equation.

(a) $Bi(NO_3)_3 + H_2S \longrightarrow$
(b) $Ba(NO_3)_2 + KOH \longrightarrow$
(c) $Pb(NO_3)_2 + K_2CrO_4 \longrightarrow$
(d) $AgNO_3 + Na_2S \longrightarrow$

Net Ionic Equations

6.23 Write a net ionic equation for the reaction between each of the following sets of reactants in solution.

(a) $Pb(NO_3)_2 + NaI$ (b) $Cd(NO_3)_2 + Na_2S$
(c) $NH_4Br + AgNO_3$ (d) $Na_2SO_4 + Ba(NO_3)_2$

6.24 Write a net ionic equation for the reaction between each of the following sets of reactants in solution.

(a) $BaCl_2 + Li_2SO_4$ (b) $FeCl_2 + NaOH$
(c) $Li_2CrO_4 + BaCl_2$ (d) $(NH_4)_2S + Pb(NO_3)_2$

6.25 Silver bromide is insoluble in water. Using this fact, write a net ionic equation for the expected reaction when solutions of each of the following pairs of substances are mixed.

(a) $AgNO_3 + NaBr$ (b) $AgNO_3 + HBr$
(c) $AgNO_3 + NH_4Br$

6.26 Lead sulfate is insoluble in water. Using this fact, write a net ionic equation for the expected reaction when solutions of each of the following pairs of substances are mixed.

(a) $Pb(NO_3)_2 + Na_2SO_4$ (b) $Pb(NO_3)_2 + H_2SO_4$
(c) $Pb(NO_3)_2 + (NH_4)_2SO_4$

Neutralization Reactions

6.27 Write the expected products of each of the following neutralization reactions, and balance the equation.

(a) $H_3PO_4 + KOH \longrightarrow$
(b) $Fe(OH)_3 + H_2SO_4 \longrightarrow$
(c) $H_3PO_4 + Ca(OH)_2 \longrightarrow$
(d) $Zn(OH)_2 + HNO_3 \longrightarrow$

6.28 Write the expected products of each of the following neutralization reactions, and balance the equation.

(a) $Fe(OH)_3 + H_3PO_4 \longrightarrow$
(b) $Zn(OH)_2 + H_2SO_4 \longrightarrow$
(c) $Al(OH)_3 + HNO_3 \longrightarrow$
(d) $KOH + H_2SO_4 \longrightarrow$

Oxidation–Reduction Reactions

6.29 Determine which substance is oxidized and which substance is reduced in each of the following reactions.

(a) $2\ Al + 3\ CoCl_2 \longrightarrow 2\ AlCl_3 + 3\ Co$
(b) $Cu + Br_2 \longrightarrow CuBr_2$
(c) $2\ K + Br_2 \longrightarrow 2\ KBr$

6.30 Determine which substance is oxidized and which substance is reduced in each of the following reactions.

(a) $Cl_2 + 2\ NaBr \longrightarrow 2\ NaCl + Br_2$
(b) $WO_3 + 3\ H_2 \longrightarrow W + 3\ H_2O$
(c) $Fe_2O_3 + 2\ Al \longrightarrow Al_2O_3 + 2\ Fe$

6.31 Identify the oxidizing agent and reducing agent in each of the following reactions.

(a) $Sb_2O_3 + 3\ Fe \longrightarrow 2\ Sb + 3\ FeO$
(b) $Sb_2O_5 + 5\ C \longrightarrow 5\ CO + 2\ Sb$
(c) $2\ NaBr + H_2SO_4 \longrightarrow Na_2SO_3 + H_2O + Br_2$

6.32 Identify the oxidizing agent and reducing agent in each of the following reactions.

(a) $NaBrO_3 + F_2 + H_2O \longrightarrow NaBrO_4 + 2\ HF$
(b) $Te + 4\ HNO_3 \longrightarrow TeO_2 + 2\ H_2O + 4\ NO_2$
(c) $As_4 + 6\ Cl_2 \longrightarrow 4\ AsCl_3$

Predicting Redox Reactions

6.33 Predict whether each of the following reactions will occur in the direction written.

(a) $Cl_2 + 2\ Br^- \longrightarrow Br_2 + 2\ Cl^-$
(b) $Pb^{2+} + Zn \longrightarrow Zn^{2+} + Pb$
(c) $Br_2 + 2\ I^- \longrightarrow I_2 + 2\ Br^-$
(d) $Mg^{2+} + Fe \longrightarrow Fe^{2+} + Mg$

6.34 Predict whether each of the following reactions will occur in the direction written.

(a) $Cl_2 + 2\,I^- \longrightarrow I_2 + 2\,Cl^-$
(b) $Br_2 + 2\,F^- \longrightarrow F_2 + 2\,Br^-$
(c) $Cu^{2+} + Fe \longrightarrow Fe^{2+} + Cu$
(d) $2\,Al^{3+} + 3\,Zn \longrightarrow 3\,Zn^{2+} + 2\,Al$

Organic Redox Reactions

6.35 Consider the following reactions for the metabolism of carbamazepine, an anticonvulsant drug. Is oxidation or reduction involved?

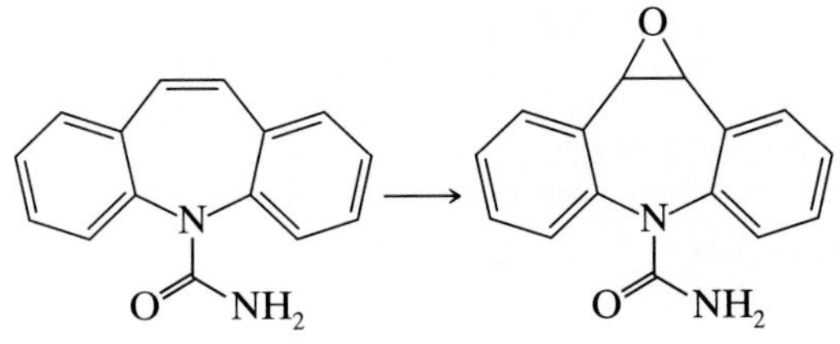

6.36 Consider the following reaction for the metabolism of dantrolene, a muscle relaxant. Is oxidation or reduction involved?

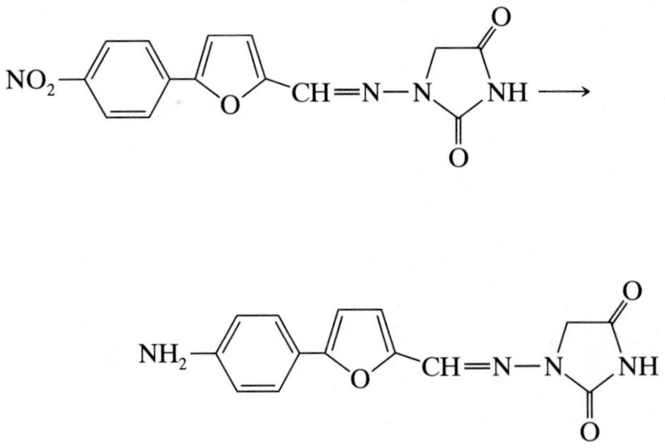

Tanker cars transport large quantities of chemicals to manufacture the many commercial products that a modern society uses. (Charles Thatcher, Tony Stone Worldwide)

7 Composition and Stoichiometry

7.1 Submicroscopic and Macroscopic Views
7.2 Molecular and Formula Weights of Compounds
7.3 Percent Composition
7.4 Avogadro's Number and the Mole
7.5 Calculations Using Moles
7.6 Empirical Formula and Molecular Formula
7.7 Stoichiometry and the Balanced Equation
7.8 Solving Stoichiometry Problems
7.9 Heat of Reaction
7.10 Percent Yield
7.11 Limiting Reagent

Essays
- *Products from the Earth, Sea, and Air*
- *Limiting Reagents in Life Processes*

7.1 Submicroscopic and Macroscopic Views

In Chapters 4 and 5 you learned that on a submicroscopic level compounds are of two types: ionic and molecular. In ionic compounds a collection of cations and anions is held together by the mutual attraction of the positive and negative charges. Molecular compounds have atoms covalently bonded to one another in discrete units.

In this chapter we change our focus to a macroscopic view of the composition of compounds. We will deal with the mass of each element present in a compound. The connection between the submicroscopic and macroscopic levels is the *mole,* a counting unit in chemistry.

After learning how to express composition of compounds with known molecular formulas we will consider the reverse of this process: we will determine the chemical formula given the percent composition.

Combining the concept of the balanced chemical equation established on the submicroscopic level in Chapter 6 and the details of composition developed in this chapter, we will consider the mass relationships between reactants and products.

(Many of the calculations require the use of significant figures and rounding of answers. You may wish to consult Appendix A for information about these calculations.)

7.2 Molecular and Formula Weights of Compounds

We learned in Chapter 3 that the relative weights of atoms are expressed using atomic mass units. Now we extend this concept of mass to collections of ions or atoms. The **molecular weight** of a molecule is the sum of the masses of the atoms indicated in the molecular formula. Methane (CH_4) has a molecular weight of 16.0 amu to the nearest 0.1 amu.

Ionic substances do not exist as molecules but as collections of ions given by a formula unit that is the simplest ratio of ions present in the compound. The **formula weight** of an ionic compound is the sum of the atomic weights of the atoms indicated by a formula unit of the substance.

EXAMPLE 7.1

Glucose ($C_6H_{12}O_6$), often termed blood sugar, is the source of quick energy in metabolic reactions. Use atomic weights to the nearest 0.1 amu and calculate its molecular weight.

Solution

Sum the atomic weights of the constituent atoms, taking into account the number of atoms of each element present in the molecule.

mass of 6 carbon atoms	= 6 × 12.0 amu =	72.0 amu
mass of 12 hydrogen atoms	= 12 × 1.0 amu =	12.0 amu
mass of 6 oxygen atoms	= 6 × 16.0 amu =	96.0 amu
mass of 1 $C_6H_{12}O_6$ molecule	=	180.0 amu

PROBLEM 7.1

What is the molecular weight of saccharin ($C_7H_5NO_3S$), an artificial sweetener?

EXAMPLE 7.2

Barium sulfate is used as a radiopaque substance for X rays of the intestinal tract. The formula unit representing barium sulfate is $BaSO_4$. What is the formula weight of barium sulfate? Use atomic weights to the nearest 0.1 amu.

Solution

Barium sulfate is an ionic compound consisting of barium ions and sulfate ions in a 1:1 ratio. The formula weight is obtained by summing the atomic weights of the atoms.

mass of 1 barium atom	= 1 × 137.3 amu =	137.3 amu
mass of 1 sulfur atom	= 1 × 32.1 amu =	32.1 amu
mass of 4 oxygen atoms	= 4 × 16.0 amu =	64.0 amu
mass of 1 $BaSO_4$ unit	=	233.4 amu

PROBLEM 7.2

What is the formula weight of $NaAl(OH)_2CO_3$, the active ingredient in the antacid Rolaids?

7.3 Percent Composition

Percentage is the ratio of a part to the whole, with the quotient multiplied by 100. The **percent composition** of a given element in a compound is equal to the mass of that element divided by the total mass of all the elements in the compound, with the quotient multiplied by 100. At this point we can calculate the percent composition of a compound using only atomic mass units. We will shortly extend this method to using mass in grams. Let's consider the molecule ethyl alcohol (C_2H_6O), whose molecular weight is 46.0 amu.

mass of 2 carbon atoms	= 2 × 12.0 amu =	24.0 amu
mass of 6 hydrogen atoms	= 6 × 1.0 amu =	6.0 amu
mass of 1 oxygen atom	= 1 × 16.0 amu =	16.0 amu
mass of 1 C_2H_6O molecule	=	46.0 amu

Of this mass, 24.0 amu is due to the two atoms of carbon, 6.0 amu is due to the six atoms of hydrogen, and 16.0 amu is due to the single atom of oxygen. The percent composition is calculated as follows.

$$\%\,\text{C} = \frac{24.0\text{ amu C}}{46.0\text{ amu ethyl alcohol}} \times 100\% = 52.1739\% = 52.2\%\text{ C}$$

$$\%\,\text{H} = \frac{6.0\text{ amu H}}{46.0\text{ amu ethyl alcohol}} \times 100\% = 13.0435\% = 13\%\text{ H}$$

$$\%\,\text{O} = \frac{16.0\text{ amu O}}{46.0\text{ amu ethyl alcohol}} \times 100\% = 34.7826\% = 34.8\%\text{ O}$$

EXAMPLE 7.3

Titanium is used to produce racing bikes. It reacts with oxygen to give TiO_2 which coats the metal. Titanium(IV) oxide is very hard and protects the frame against further oxidation. What is the percent composition of titanium(IV) oxide? Use atomic weights to the nearest 0.1 amu.

Solution

First, determine the formula weight. This process also gives the mass of each component of the compound.

mass of 1 titanium atom	= 1 × 47.9 amu =	47.9 amu
mass of 2 oxygen atoms	= 2 × 16.0 amu =	32.0 amu
mass of 1 TiO_2 formula unit	=	79.9 amu

The percent composition is

$$\%\,\text{Ti} = \frac{47.9\text{ amu Ti}}{79.9\text{ amu TiO}_2} \times 100\% = 59.94993\% = 59.9\%\text{ Ti}$$

$$\%\,\text{O} = \frac{32.0\text{ amu O}}{79.9\text{ amu TiO}_2} \times 100\% = 40.05006\% = 40.1\%\text{ O}$$

PROBLEM 7.3

What is the percent composition of ammonium carbonate, $(NH_4)_2CO_3$, which is used in smelling salts? Use atomic weights to the nearest 0.1 amu.

7.4 Avogadro's Number and the Mole

We can't keep track of individual atoms nor can we weigh them using conventional balances. Thus, we now consider how to deal with quantities of matter that can be seen and weighed, that is, on the macroscopic level. Chemists use a chemical counting system for comparing amounts of substances containing large collections of atoms, molecules, or ions. Three quantities—Avogadro's number, the mole, and the molar mass—provide the connection between the submicroscopic and macroscopic levels of matter.

1. **Avogadro's number** is the number of atoms contained in exactly 12 g of carbon-12. The number is 6.02×10^{23}.
2. A **mole** of any substance contains Avogadro's number of structural units.
3. The **molar mass** of a substance is the mass in grams of 1 mole of that substance.

Avogadro's Number

Amadeo Avogadro was a nineteenth-century Italian scientist who contributed to the development of the concept of atomic weights, but the quantity known as Avogadro's number was not determined by Avogadro. The first determination of this quantity was a calculation of the number of molecules in a mole of a gas by the Austrian chemist Johann Loschmidt in 1865, nine years after Avogadro died. In honor of Avogadro's contributions to chemistry, the quantity 6.02×10^{23}, which connects the submicroscopic and macroscopic levels of matter, is now called Avogadro's number.

Suppose that you need to weigh out the same number of atoms of two different elements. How can you do this if you can't count out the individual atoms? There is a "trick" that chemists use. They use weights as a substitute for counting items such as atoms. Here is how it works. Weigh a collection of atoms in ratios according to their atomic weights, and you will obtain samples containing the same number of atoms. Do you see why? The atomic weights of helium and argon are 4.00 amu and 39.95 amu, respectively, meaning that a helium atom weighs about one-tenth as much as an argon atom. If we have 4.00 g of helium and 39.95 g of argon, the masses of the two samples are in the ratio of approximately 1:10, which is the same ratio as the masses of the individual atoms. Therefore, the two samples contain the same number of atoms. But what is that number?

We know that the number of atoms in exactly 12 g of carbon-12 atoms is Avogadro's number. The 4.00 g of helium is one-third the mass of the 12 g of carbon, but the mass of the individual helium atoms (4.00 amu) is one-third that of the mass of the individual carbon atoms (12.0 amu). Therefore, the same number of atoms must be in each sample, that is, Avogadro's number. We can now generalize that

There is Avogadro's number of atoms of any element in a sample whose mass in grams is numerically equal to its atomic weight.

Now let's do the same thing with molecules. The molecular weights of methane (CH_4) and carbon dioxide (CO_2) are 16.0 and 44.0 amu, respectively. Therefore 16.0 g of methane and 44.0 g of carbon dioxide contain the same number of molecules. The 16.0 g of methane contains Avogadro's number of molecules; the 44.0 g of carbon dioxide also contains Avogadro's number of molecules. We can generalize that

There is Avogadro's number of molecules in any sample whose mass in grams is numerically equal to its molecular weight.

Avogadro's number also relates the number of species in ionic compounds. The formula weight of $CaCO_3$ is 100.1 amu. Thus in 100.1 g of $CaCO_3$ there must be Avogadro's number of $CaCO_3$ formula units. We can generalize that

There is Avogadro's number of formula units of any ionic compound in a sample whose mass in grams is numerically equal to its formula weight.

A summary of how Avogadro's number relates moles of atoms, molecules, and ionic compounds is given in Table 7.1.

The Mole and Molar Mass

There are several terms used to express the number of certain items. A dozen is 12, a gross is 144, and a ream is 500. You know that a dozen eggs is 12 eggs and a dozen oranges is 12 oranges. The number of items in a dozen is a count of the number of items.

Now consider something else that you know by experience. The weight of a dozen items depends on the item. A dozen eggs does not weigh the same as a dozen oranges because the items have different individual masses. Thus, the collective weights of groups of the items, in this case a dozen, must be different.

In chemistry the quantity 6.02×10^{23}, Avogadro's number, is the number of items in a **mole.** This quantity is the "chemist's dozen." The number of items in a mole is large, but the quantity can be handled in the same way as a dozen. The number of items in a mole is a count of the number of items. A mole of any substance has the same number of units as a mole of any other substance, that is, 6.02×10^{23}.

TABLE 7.1 Avogadro's Number and the Mole

Substance	Formula	Number of Particles in a Mole
argon	Ar	6.02×10^{23} Ar atoms
nitrogen	N_2	6.02×10^{23} N_2 molecules
		$2 \times (6.02 \times 10^{23})$ N atoms
carbon dioxide	CO_2	6.02×10^{23} CO_2 molecules
		6.02×10^{23} C atoms
		$2 \times (6.02 \times 10^{23})$ O atoms
zinc fluoride	ZnF_2	6.02×10^{23} ZnF_2 formula units
		6.02×10^{23} Zn^{2+} ions
		$2 \times (6.02 \times 10^{23})$ F^- ions
lithium carbonate	Li_2CO_3	6.02×10^{23} Li_2CO_3 formula units
		$2 \times (6.02 \times 10^{23})$ Li^+ ions
		6.02×10^{23} CO_3^{2-} ions

The **molar mass** is the mass in grams of 1 mole of a substance expressed in units of grams per mole. What is the molar mass of carbon-12? There are Avogadro's number of carbon atoms in exactly 12 g of carbon atoms, and Avogadro's number of atoms is a mole. Thus, the molar mass of carbon is 12 g/mole.

A mole of helium contains 6.02×10^{23} helium atoms. A mole of argon contains 6.02×10^{23} argon atoms. Since the atomic weights of helium and argon are 4.00 and 39.95 amu, respectively, 1 mole of helium weighs 4.00 g, and 1 mole of argon weighs 39.95 g.

For molecular compounds, 1 mole contains 6.02×10^{23} molecules. One mole of carbon dioxide (CO_2) contains 6.02×10^{23} carbon dioxide molecules. One mole of glucose contains 6.02×10^{23} glucose ($C_6H_{12}O_6$) molecules. The molar mass of each of these substances is numerically equal to its individual molecular weight. For carbon dioxide:

One carbon dioxide molecule weighs 44 amu.
One mole of carbon dioxide molecules weighs 44 g.
The molar mass of carbon dioxide is 44 g/mole.

For glucose:

One glucose molecule weighs 180 amu.
One mole of glucose molecules weighs 180 g.
The molar mass of glucose is 180 g/mole.

These relationships for carbon dioxide and glucose are the reason why the mole concept is used. By weighing out a mass of a substance, you are actually counting moles of material as well as individual units of matter. A summary of the number of moles of atoms, molecules, and ions present in a mole of various substances is given in Table 7.2. This information is important in determining how many grams of reactants are needed for a chemical reaction. We will return to this idea in Section 7.8.

TABLE 7.2 Moles and Molar Mass

Substance	Formula	Mass of 1 Mole (g)	Number of Moles of Components
helium	He	4.0	1 mole of He atoms
oxygen	O_2	32.0	1 mole of O_2 molecules 2 moles of O atoms
water	H_2O	18.0	1 mole of H_2O molecules 2 moles of H atoms 1 mole of O atoms
calcium chloride	$CaCl_2$	111.0	1 mole of $CaCl_2$ formula units 1 mole of Ca^{2+} ions 2 moles of Cl^- ions
sodium sulfate	Na_2SO_4	142.0	1 mole of Na_2SO_4 formula units 2 moles of Na^+ ions 1 mole of SO_4^{2-} ions

7.5 Calculations Using Moles

Now let's do some calculations using conversion factors to interconvert grams, moles, and numbers of units of matter. The definition of a mole in terms of Avogadro's number provides two conversion factors that are reciprocals of each other.

$$\frac{6.02 \times 10^{23} \text{ entities}}{1 \text{ mole of entities}} \quad \text{and} \quad \frac{1 \text{ mole of entities}}{6.02 \times 10^{23} \text{ entities}}$$

We use these factors to relate the number of entities on the submicroscopic scale to our counting unit of the macroscopic scale, the mole. If we are considering molecules, then it follows that

$$\textbf{number of molecules} \times \frac{\textbf{1 mole}}{\textbf{6.02} \times \textbf{10}^{23} \textbf{ molecules}} = \textbf{number of moles}$$

$$\textbf{number of moles} \times \frac{\textbf{6.02} \times \textbf{10}^{23} \textbf{ molecules}}{\textbf{1 mole}} = \textbf{number of molecules}$$

The molar mass of a substance is the mass in grams of 1 mole of that substance. We can use the molar mass or its reciprocal to relate the number of moles and the number of grams of a substance. To determine the mass of a given number of moles, we use

$$\textbf{number of moles} \times \textbf{molar mass} = \textbf{mass}$$

$$\textbf{number of moles} \times \frac{\textbf{grams}}{\textbf{mole}} = \textbf{grams}$$

To determine the number of moles in a given mass we use

$$\textbf{number of grams} \times \frac{\textbf{1}}{\textbf{molar mass}} = \textbf{moles}$$

$$\textbf{number of grams} \times \frac{\textbf{mole}}{\textbf{grams}} = \textbf{moles}$$

In summary, the quantities Avogadro's number, the mole, and the molar mass are related as shown in Figure 7.1.

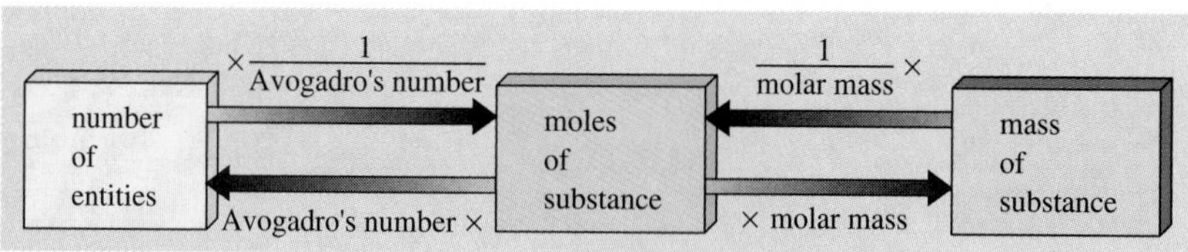

FIGURE 7.1
Mole Calculations
The mass, moles, and number of entities of a substance are interrelated. Each quantity can be obtained using conversion factors based on either Avogadro's number or the molar mass.

Products from the Earth, Sea, and Air

All the calculations in this chapter have been based on the gram, a unit commonly used in the laboratory. However, commercial chemicals are produced in much larger amounts for use in industrial operations on which our society depends. Of the millions of known compounds, only about 10,000 are commercially important. Of these, about 50 are produced in quantities exceeding 1 billion pounds per year in the United States.

The table gives the top 10 commodity chemicals in 1994. With the exception of ethylene and propylene, these chemicals are derived from "inorganic sources." Ethylene and propylene are obtained from "fossil materials," such as petroleum and natural gas, which are carbon-containing (organic) compounds formed from living matter over millions of years.

Both ethylene and propylene are used to produce polymers; over 8 million tons of ethylene went into polyethylene in 1994. Materials with a wide range of properties can be fashioned from polyethylene, including sandwich bags, soft-drink bottles, and milk containers. Polypropylene, a polymer of propylene, is used to make materials such as indoor-outdoor carpeting, artificial turf, and battery cases.

Nitrogen and oxygen are obtained by direct separation from air. Both elements are **cryogens,** from a Greek word meaning "icy cold." Nitrogen, in its liquid form, is used in frozen food preparation and preservation. It is used in the industrial preparation of ammonia, another top 10 commodity chemical. Oxygen is used in steel-making; it is also used in rockets in combination with hydrogen to generate hot reaction gases for propulsion.

Sulfuric acid is used to manufacture fertilizer and in the production of steel. It is produced directly from sulfur obtained from sulfur deposits in the Earth, oxygen from the air, and water. Phosphoric acid, used in the manufacture of baking powder, soft drinks, detergents, and fertilizers, is produced by a series of reactions starting with calcium phosphate obtained from phosphate rocks.

Both sodium hydroxide and chlorine are produced by electrolysis of brine, which is concentrated sea water. Both of these compounds are used to produce a variety of other secondary products.

Ammonia is produced by direct combination of nitrogen and hydrogen at high temperature and pressure. It is used in the production of explosives and fertilizers.

Polyethylene containers are used for liquids such as cider, milk, and water. (Richard Megna, Fundamental Photographs)

TABLE 7A Top 10 Commodity Chemicals Produced in the United States (1994)

Rank	Billions of Pounds per year	Chemical Name	Formula
1	89.2	sulfuric acid	H_2SO_4
2	67.5	nitrogen	N_2
3	49.7	oxygen	O_2
4	48.5	ethylene	C_2H_4
5	38.4	calcium oxide	CaO
6	37.9	ammonia	NH_3
7	28.8	sodium hydroxide	$NaOH$
8	25.8	chlorine	Cl_2
9	25.3	phosphoric acid	H_3PO_4
10	20.5	propylene	C_3H_6

EXAMPLE 7.4

An average adult exhales 1.0×10^3 g of carbon dioxide (CO_2) per day. How many moles of carbon dioxide are exhaled?

Solution

The molar mass of CO_2 is 44.0 g/mole. The number of moles in 1.0×10^3 g is calculated by multiplying by the reciprocal of the molar mass.

$$1.0 \times 10^3 \not{g}\ CO_2 \times \frac{1 \text{ mole } CO_2}{44.0 \not{g}\ CO_2} = 22.727 = 23 \text{ moles } CO_2$$

PROBLEM 7.4

Ferric sulfate, $Fe_2(SO_4)_3$, can be applied to soil that is too alkaline. How many moles are contained in 50.0 lb (22.7 kg) of this substance?

EXAMPLE 7.5

How many molecules of vitamin C are in a 0.500-g tablet of vitamin C ($C_6H_8O_6$)?

Solution
To determine the number of molecules, it is first necessary to determine the number of moles in the tablet, which has a mass of 0.500 g.

$$0.500\ \cancel{g}\ C_6H_8O_6 \times \frac{1\ \text{mole}\ C_6H_8O_6}{176.0\ \cancel{g}\ C_6H_8O_6} = 0.002841\ \text{mole}\ C_6H_8O_6$$

(Note that this quantity is expressed to one place beyond that allowed by significant figures. However, the quantity is obtained as an intermediate quantity. Proper rounding will be done after the final step.)

The number of molecules of $C_6H_8O_6$ in a mole of vitamin C is 6.02×10^{23}. The tablet contains 2.841×10^{-3} mole, so the number of molecules is

$$2.841 \times 10^{-3}\ \cancel{\text{mole}}\ C_6H_8O_6 \times \frac{6.02 \times 10^{23}\ \text{molecules}\ C_6H_8O_6}{1\ \cancel{\text{mole}}\ C_6H_8O_6}$$
$$= 1.7103 \times 10^{21}\ \text{molecules}\ C_6H_8O_6$$
$$= 1.71 \times 10^{21}\ \text{molecules}\ C_6H_8O_6$$

PROBLEM 7.5

Lithium carbonate (Li_2CO_3), is used to treat manic depression. How many lithium ions are in a 0.300-g dose of Li_2CO_3?

7.6 Empirical Formula and Molecular Formula

We have learned how the percent composition can be calculated from the molecular formula. Could this procedure be reversed to determine the molecular formula from the percent composition? Let's see. The percent compositions of ethylene (C_3H_4) and propylene (C_3H_6) are both 85.7% carbon and 14.3% hydrogen. Both compounds have the same ratio of carbon atoms to hydrogen atoms, 1:2. Therefore their percent compositions must be identical. If two compounds with different molecular formulas can have the same percent composition, the percent composition cannot lead to a unique molecular formula. Percent composition data give only the simplest ratio of atoms in a compound, the **empirical formula.** Based on the percent composition of compounds with the molecular formulas C_2H_4 and C_3H_6, only the empirical formula CH_2 can be determined in both cases.

In some cases, the empirical formula and the molecular formula of a compound are the same. From percent composition data the empirical formula of water is H_2O, which means that the simplest ratio of atoms is two hydrogen atoms for every oxygen atom in the molecule. Because each molecule of water actually contains two hydrogen atoms and one oxygen atom, the formula H_2O is also the molecular formula. In contrast, the empirical formula of hydrogen peroxide is HO, but its molecular formula is H_2O_2. The empirical formula reflects the 1:1 ratio of hydrogen atoms to oxygen atoms. The molecular formula indicates that there are two hydrogen atoms and two oxygen atoms per molecule of hydrogen peroxide.

Determining the Empirical Formula

Four arithmetical steps are used to calculate the empirical formula from the mass percent composition. The method is also depicted in Figure 7.2.

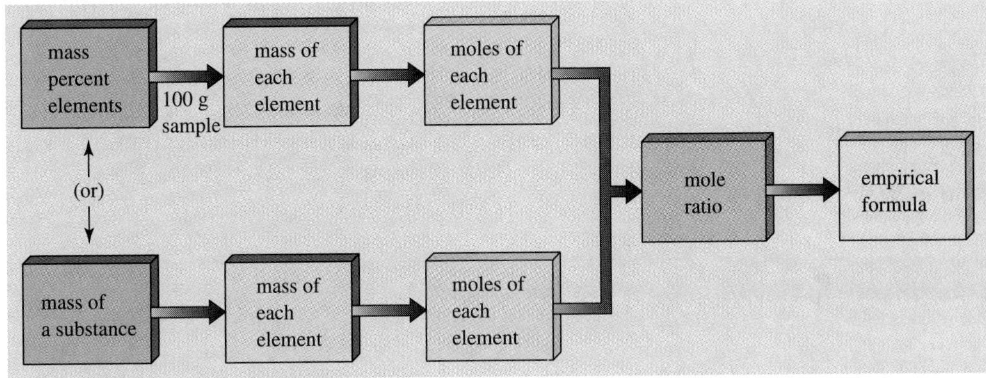

FIGURE 7.2 Determination of an Empirical Formula
The empirical formula can be determined directly from the mass of each element present in a sample or from the mass percent composition.

1. Express the composition of the compound in terms of the mass of each constituent element.
2. Convert the mass of each element into its equivalent in moles.
3. Determine the mole ratio for each element by dividing by the smallest number obtained in step 2.
4. If the mole ratios are not all integers, multiply by the simplest factor required to convert all mole ratios to integers.

If the mass of each element is given for a compound, then you may use these quantities directly to determine the empirical formula. If percent composition is given, convert the percentages to grams for a selected sample. For example, vitamin C is 40.9% carbon, 4.6% hydrogen, and 54.5% oxygen. It is convenient to assume a 100.0-g sample for the calculation, in which case there are 40.9 g of carbon, 4.6 g of hydrogen, and 54.5 g of oxygen in the sample.

Now, convert the mass of each element into its equivalent in moles by multiplying by the reciprocal of the molar mass in units of moles/g. These operations yield the number of moles of each element contained in the 100.0-g sample.

$$40.9 \text{ g C} \times \frac{1 \text{ mole C}}{12.0 \text{ g C}} = 3.41 \text{ moles C}$$

$$4.6 \text{ g H} \times \frac{1 \text{ mole H}}{1.0 \text{ g H}} = 4.6 \text{ moles H}$$

$$54.5 \text{ g O} \times \frac{1 \text{ mole O}}{16.0 \text{ g O}} = 3.41 \text{ moles O}$$

In the third step calculate the mole ratios, which give the relative number of atoms in a molecule or formula unit, by dividing the number of moles of each element by the smallest of these numbers. If all the numbers are close to integers within the accuracy of the calculation, they are the subscripts of the empirical formula. However, in the example considered, the relative numbers of moles are not all integers.

$$\text{mole ratio for C} = \frac{3.41}{3.41} = 1.0$$

$$\text{mole ratio for H} = \frac{4.6}{3.41} = 1.3$$

$$\text{mole ratio for O} = \frac{3.41}{3.41} = 1.0$$

The ratios of the subscripts in a formula must be the same as the mole ratios determined in this calculation. However, the formula $C_1H_{1.3}O_1$ is not acceptable because not all subscripts are integers. Multiplication by 3 yields $C_3H_{3.9}O_3$; within the accuracy of the data, the empirical formula is $C_3H_4O_3$.

EXAMPLE 7.7

Vinegar contains acetic acid. The percent composition is 40.0% carbon, 6.7% hydrogen, and 53.3% oxygen. What is the empirical formula of acetic acid?

Solution

Using a reference 100.0-g sample of acetic acid gives 40.0 g of carbon, 6.7 g of hydrogen, and 53.3 g of oxygen. Convert each of these quantities into moles by using a factor relating moles and mass.

$$40.0\ \not{g}\ C \times \frac{1\ \text{mole C}}{12.0\ \not{g}\ C} = 3.33\ \text{moles C}$$

$$6.7\ \not{g}\ H \times \frac{1\ \text{mole H}}{1.0\ \not{g}\ H} = 6.7\ \text{moles H}$$

$$53.3\ \not{g}\ O \times \frac{1\ \text{mole O}}{16.0\ \not{g}\ O} = 3.33\ \text{moles Oc}$$

Dividing each number of moles by 3.33 yields the numbers 1, 2, and 1. The empirical formula of acetic acid is CH_2O.

PROBLEM 7.7

Cadaverine, a poisonous compound produced when fish rot, is 58.8% C, 13.8% H, and 27.4% N. What is its empirical formula?

Determining the Molecular Formula

A molecular formula is always a whole-number multiple of the empirical formula, but the multiple may be 1, as in the case of water. The empirical formula of ethylene, a gas used in making polyethylene, is CH_2 (Table 7.3). This empirical formula does not necessarily mean that there are only one carbon atom and two hydrogen atoms in one molecule of ethylene. The molecule actually contains two carbon atoms and four hydrogen atoms, so the molecular formula is C_2H_4. In this case the

TABLE 7.3 Empirical and Molecular Formulas

Compound	Molecular Formula	Empirical Formula
acetylene	C_2H_2	CH
benzene	C_6H_6	CH
ethylene	C_2H_4	CH_2
cyclopropane	C_3H_6	CH_2
cyclohexane	C_6H_{12}	CH_2
formaldehyde	CH_2O	CH_2O
acetic acid	$C_2H_4O_2$	CH_2O
glucose	$C_6H_{12}O_6$	CH_2O

subscripts of the empirical formula are multiplied by 2 to obtain the molecular formula. Cyclopropane, a gas used as an anesthetic, also has the empirical formula CH_2, but the molecular formula of cyclopropane is C_3H_6. One molecule of cyclopropane contains three atoms of carbon and six atoms of hydrogen. Therefore, the subscripts of the empirical formula for cyclopropane must be multiplied by 3 to obtain the molecular formula.

Since the molecular formula is always a multiple of the empirical formula, the molecular weight is a multiple of the weight of the empirical formula unit. Thus, after determining the empirical formula of a compound, your next goal is to determine its molecular weight. You will learn how this is done experimentally for gases in Chapter 8 and for liquids and solids in Chapter 9.

The empirical formula of water is H_2O, so the molecular formula is $H_{2n}O_n$, where n may be 1, 2, 3, and so on. Thus, the molecular formula could be H_2O, H_4O_2, H_6O_3, and so on. These various possibilities have molecular weights of 18 amu, 36 amu, 54 amu, and so on. The experimental molecular weight of water is 18 amu, so the molecular formula must be H_2O.

Acetylene and benzene (Table 7.3) both have the empirical formula CH. Their molecular formulas may be any of the possibilities given by C_nH_n. To obtain the molecular formula, we need to know the molecular weight of each substance. The molecular weights of acetylene and benzene are 26 and 78 amu, respectively. Therefore, the molecular formulas are C_2H_2 and C_6H_6, respectively.

EXAMPLE 7.8

The empirical formula of nicotine, a poisonous compound found in tobacco, is C_5H_7N. The molecular weight is 162 amu. What is the molecular formula?

Solution

If the molecular formula were the same as the empirical formula, the molecular weight would be 5×12 amu $+ 7 \times 1$ amu $+ 14$ amu $= 81$ amu. Therefore, the molecular formula must be twice the empirical formula, or $C_{10}H_{14}N_2$.

PROBLEM 7.8

Naphthalene, a compound used as mothballs, is 93.8% C and 6.2% H. Its molecular weight is 128 amu. What is its molecular formula?

7.7 Stoichiometry and the Balanced Equation

Chemical equations tell us how much of each reactant is required on the submicroscopic level. However, chemical reactions in the laboratory or in industry deal with grams, pounds, or even tons of substances. In this section we learn how information about a chemical reaction at the submicroscopic level is translated to the macroscopic level.

The calculation of the amount of reactants required and amount of products obtained in a reaction is called **stoichiometry.** We know that the ratio of the numbers of atoms and molecules in a chemical reaction on a submicroscopic level is given by the coefficients in the chemical equation. The same ratio governs the number of moles of those substances on a macroscopic level. Thus stoichiometric quantities of reactants and products in a reaction can also be calculated on a mole basis using the coefficients of the equation.

The **mole method** is a factor unit method used to solve stoichiometry problems. It consists of the following three steps, which are illustrated in Figure 7.3.

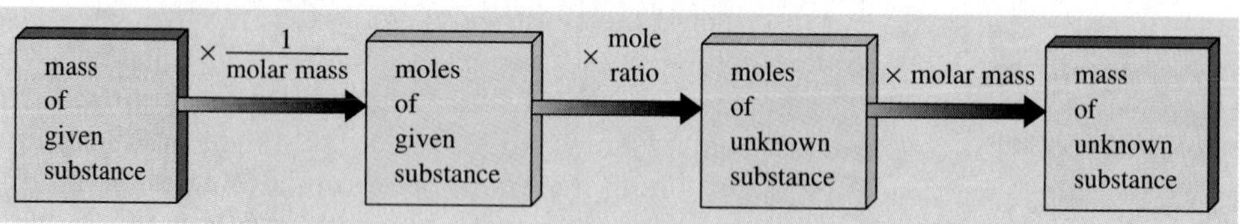

FIGURE 7.3 Stoichiometric Calculations *Stoichiometric calculations use the mole ratio and the molar mass or its reciprocal as conversion factors.*

1. Convert the given quantities into their equivalent mole quantities.
2. Establish the stoichiometric relationship between the given substance and the desired substance from the balanced equation.
3. Convert the moles of the desired substance into the required units.

The number of steps used in working problems by the mole method depends on the units for the given substance and the units required for the desired substance.

The central step of the mole method (step 2) relates the number of moles of one substance to the number of moles of another substance in a mole ratio. The **mole ratio** is a ratio between the numbers of moles of two substances involved in a chemical reaction as given by the balanced equation. Thus, for the reaction of hydrogen and nitrogen to give ammonia, six mole ratios can be written. These ratios contain the coefficients of the corresponding substances in the equation.

$$N_2(g) + 3\,H_2(g) \longrightarrow 2\,NH_3(g)$$

$$\frac{1 \text{ mole } N_2}{3 \text{ moles } H_2} \quad \text{or} \quad \frac{3 \text{ moles } H_2}{1 \text{ mole } N_2}$$

$$\frac{1 \text{ mole } N_2}{2 \text{ moles } NH_3} \quad \text{or} \quad \frac{2 \text{ moles } NH_3}{1 \text{ mole } N_2}$$

$$\frac{3 \text{ moles } H_2}{2 \text{ moles } NH_3} \quad \text{or} \quad \frac{2 \text{ moles } NH_3}{3 \text{ moles } H_2}$$

There are really only three mole ratios that relate different pairs of substances for this reaction. The other three are reciprocals of the same quantities.

Mole ratios are used in stoichiometry problems to convert the known number of moles of the given substance to the number of moles of the desired substance. This relationship is

moles of given substance × mole ratio = moles of desired substance

The mole ratio then must be

$$\textbf{mole ratio} = \frac{\textbf{moles of desired substance in balanced equation}}{\textbf{moles of given substance in balanced equation}}$$

Because of the relationship established between the macroscopic and submicroscopic levels of chemical change, the mole ratio is equal to the ratio of the coefficients in the balanced equation. Thus, we can write

$$\textbf{mole ratio} = \frac{\textbf{coefficient of desired substance in balanced equation}}{\textbf{coefficient of given substance in balanced equation}}$$

Note that the mole ratio is an exact number. Thus, the number of significant figures in an answer depends only on the number of significant figures provided for the given substance.

EXAMPLE 7.9

Consider the following equation and write two mole ratios relating aluminum and oxygen.

$$4\ Al(s) + 3\ O_2(g) \longrightarrow 2\ Al_2O_3(s)$$

Solution

The coefficients for aluminum and oxygen are 4 and 3, respectively. These coefficients give the ratio of the number of moles of each substance required for the reaction. The mole ratios are

$$R_1 = \frac{4 \text{ moles Al}}{3 \text{ moles } O_2} \qquad R_2 = \frac{3 \text{ moles } O_2}{4 \text{ moles Al}}$$

The ratio R_1 is used to convert a given number of moles of O_2 into the number of moles of Al required for the reaction. The ratio R_2 is used to convert a given number of moles of Al into the number of moles of O_2 required for the reaction.

PROBLEM 7.9

Consider the reaction used to produce tungsten for incandescent light bulbs. Write two mole ratios relating tungsten (VI) oxide and hydrogen.

$$WO_3(s) + 3\ H_2(g) \longrightarrow W(s) + 3\ H_2O(g)$$

7.8 Solving Stoichiometry Problems

First let's do some stoichiometry problems using only moles of reactants and products. The steps and math in problems of this type are easy. However, the method is essential to solution of the more complex problems using the masses of reactants and products.

In step 2 of the mole method, the mole ratio is needed to calculate the number of moles of one substance that react with or are produced from a given number of moles of a second substance. How many moles of oxygen are required to react with 0.20 mole of iron? The coefficients in the balanced equation tell us that 3 moles of oxygen are required for 4 moles of iron.

$$4\ Fe + 3\ O_2 \longrightarrow 2\ Fe_2O_3$$

Using the mole ratio method, we calculate the number of moles of oxygen required.

$$0.20 \text{ mole Fe} \times \frac{3 \text{ moles } O_2}{4 \text{ moles Fe}} = 0.15 \text{ mole } O_2$$

EXAMPLE 7.10

Ammonia is produced by direct combination of nitrogen and hydrogen. How many moles of ammonia can be produced from 9 moles of hydrogen according to the following equation?

$$3\ H_2 + N_2 \longrightarrow 2\ NH_3$$

Solution

The coefficients of the equation indicate that 3 moles of hydrogen yield 2 moles of ammonia. The mole ratio necessary for the solution of the problem must cancel the unit moles of hydrogen and leave moles of ammonia.

$$9 \text{ moles } H_2 \times \text{mole ratio} = ? \text{ moles } NH_3$$

$$9 \text{ moles } H_2 \times \frac{2 \text{ moles } NH_3}{3 \text{ moles } H_2} = 6 \text{ moles } NH_3$$

You could have solved this problem quickly by using "ratio and proportion" methods learned in other courses. However, the factor unit method employing mole ratios is helpful in solving more difficult problems, so it is worth practicing.

PROBLEM 7.10

Oxygen can be generated by the decomposition of potassium chlorate. How many moles of potassium chlorate are required to produce 0.15 mole of oxygen according to the following equation?

$$2\ KClO_3 \longrightarrow 2\ KCl + 3\ O_2$$

Now let's solve mass–mass stoichiometry problems. The quantity of the known substance is given in mass units, and the desired unknown substance must be found in mass units. All steps of the mole method must be followed.

EXAMPLE 7.11

How many grams of oxygen are required to metabolically convert 45 g of glucose ($C_6H_{12}O_6$) into carbon dioxide and water?

$$C_6H_{12}O_6 + 6\ O_2 \longrightarrow 6\ CO_2 + 6\ H_2O$$

Solution

First, the number of moles of glucose is calculated as required by step 1 of the mole method. The molecular weight of $C_6H_{12}O_6$ is 180 amu. Thus, 1 mole of glucose has a mass of 180 g.

$$45 \text{ g } C_6H_{12}O_6 \times \frac{1 \text{ mole } C_6H_{12}O_6}{180 \text{ g } C_6H_{12}O_6} = 0.25 \text{ mole } C_6H_{12}O_6$$

Next, the number of moles of oxygen required is calculated from the mole ratio obtained from the coefficients of the balanced equation. This is step 2 of the mole method.

$$0.25 \text{ mole } C_6H_{12}O_6 \times \frac{6 \text{ moles } O_2}{1 \text{ mole } C_6H_{12}O_6} = 1.5 \text{ moles } O_2$$

Finally, in step 3 the mass of oxygen required is calculated using the molar mass of oxygen to convert moles into grams.

$$1.5 \text{ moles } O_2 \times \frac{32.0 \text{ g } O_2}{1 \text{ mole } O_2} = 48 \text{ g } O_2$$

PROBLEM 7.11

How many grams of ethyl alcohol (C_2H_5OH) can be produced by fermentation of 171 g of cane sugar ($C_{12}H_{22}O_{11}$)?

$$C_{12}H_{22}O_{11} + H_2O \longrightarrow 4\, C_2H_5OH + 4\, CO_2$$

7.9 Heat of Reaction

When the chemical energy stored in the reactants is greater than that stored in the products, energy is released by the reaction, and it is termed *exothermic.* The thermal energy associated with a chemical reaction is represented by ΔH, which is called the enthalpy change. The symbol Δ stands for "change in." For exothermic reactions, the quantity of energy released is expressed as ΔH and is assigned a negative value. When the total chemical energy stored in the reactants is less than that stored in the products, addition of energy is required by the reaction and it is termed *endothermic.* The ΔH for such a reaction is assigned a positive value.

The combustion of propane is exothermic. The reaction can be represented with energy written on the product side of the equation. However, the reaction is also represented by placing the enthalpy change to the right of the equation regardless of whether the reaction is exothermic or endothermic.

$$C_3H_8(g) + 5\, O_2(g) \longrightarrow 3\, CO_2(g) + 4\, H_2O(l) + 530 \text{ kcal}$$
$$C_3H_8(g) + 5\, O_2(g) \longrightarrow 3\, CO_2(g) + 4\, H_2O(l) \qquad \Delta H = -530 \text{ kcal}$$

The reaction of nitrogen and oxygen to give nitric oxide, a process that occurs to some extent in an automobile engine, is endothermic. The equation can be represented with the required energy written on the reactant side of the equation or the ΔH with a positive sign to the right side of the equation.

$$N_2(g) + O_2(g) + 43 \text{ kcal} \longrightarrow 2\, NO(g)$$
$$N_2(g) + O_2(g) \longrightarrow 2\, NO(g) \qquad \Delta H = +43 \text{ kcal}$$

The ΔH listed for an accompanying reaction represents the amount of heat released or required for reactants in mole quantities corresponding to the coefficients of the balanced equation. Thus, for the combustion of propane the 530 kcal released is for 1 mole of propane. The amount of energy released for any quantity of propane is calculated in much the same way as the amount of product formed is determined using the stoichiometry of the equation. If 10 moles of propane are burned, the quantity of energy released is 10 times 530 kcal, or 5300 kcal.

EXAMPLE 7.12

The following equation represents the photosynthesis reaction in plants. Is the reaction exothermic or endothermic? What is ΔH for the reaction?

$$6\,CO_2(g) + 6\,H_2O(g) + 678\text{ kcal} \longrightarrow C_6H_{12}O_6(aq) + 6\,O_2(g)$$

Solution

Energy is required for the reaction because it is written on the reactant side of the equation. The reaction is endothermic, and $\Delta H = +678$ kcal.

PROBLEM 7.12

How much heat energy is required to produce 4.5 g of nitric oxide by the reaction of nitrogen and oxygen?

$$N_2(g) + O_2(g) + 43\text{ kcal} \longrightarrow 2\,NO(g)$$

7.10 Percent Yield

The yield of a reaction as calculated in Section 7.8 is the **theoretical yield,** which is the maximum amount of product that could be obtained. However, there are several reasons why the amount of product actually obtained for a given amount of reactant may be a smaller quantity. Some of the product may be lost in the isolation or purification process. In addition there may be other reactions that occur in competition with the desired reaction. For example, when carbon and oxygen react to form carbon dioxide, some carbon monoxide also forms.

$$C + O_2 \longrightarrow CO_2 \quad \text{(desired reaction)}$$
$$2\,C + O_2 \longrightarrow 2\,CO \quad \text{(side reaction)}$$

If complete combustion to give carbon dioxide is the desired reaction, then the formation of carbon monoxide is the side reaction. To the extent that some of the carbon is converted to undesired material, the theoretical yield of carbon dioxide will not be obtained.

The amount of desired product obtained in a chemical reaction is the **actual yield.** The **percent yield** is the actual yield divided by the theoretical yield multiplied by 100.

$$\%\text{ yield} = \frac{\text{actual yield}}{\text{theoretical yield}} \times 100\%$$

Industrial processes used to produce large quantities of products such as nylon, aspirin, and ammonia are selected to give the highest possible percent yield under the most economic conditions. Reactants not converted to product, and products formed in side reactions, raise the cost to the consumer. Furthermore, the separation of a desired product from leftover reactants or undesired products is often costly as well.

If the reactants are gases, the unused material can be separated from products that are liquid or solid. For example, the industrial production of urea, a compound widely used in the production of certain plastics as well as barbiturates, is manufactured from carbon dioxide and ammonia.

$$2\,NH_3 + CO_2 \longrightarrow \underset{\text{urea}}{NH_2-\overset{\overset{\displaystyle O}{\|}}{C}-NH_2} + H_2O$$

Under the best conditions of pressure and temperature, a 60% yield is obtained. However, in this case the leftover reactants can be separated easily and can be recycled in subsequent reactions to produce more urea. Unfortunately, such reclamation is not always possible, and the wasted materials must be placed in landfills or incinerated. Both methods are expensive and can cause some pollution.

7.11 Limiting Reagent

Using a balanced equation we can calculate the quantities of each substance required and the amount of product that can be produced. However, there are circumstances when stoichiometric quantities are not used. How does this affect the amount of product formed? The reactant that is in short supply is called the **limiting reagent,** because the amount of product that can be formed is limited by the amount of this reactant. The other reactants are excess reagents and cannot be further converted to product after the limiting reagent has run out.

The concept of a limiting reagent may make sense to you if you consider the number of hot dogs and the number of hot dog buns sold in typical packages. If you buy one package of each, you may have 10 hot dogs and 8 hot dog buns. The number of hot dogs that can be served with buns is limited to 8 because the hot dog buns are limited.

In chemical equations the situation is a little more complex because the number of moles of each reactant required may not be a 1:1 ratio. The limiting reagent is the substance that has the lowest mole-to-coefficient ratio, which is calculated by dividing the number of moles of each reactant available by the coefficient in the balanced equation. Consider the reaction of nitrogen with hydrogen to give ammonia.

$$N_2(g) + 3\,H_2(g) \longrightarrow 2\,NH_3(g)$$

If 25 moles of nitrogen and 45 moles of hydrogen are available, we calculate the mole-to-coefficient ratio as follows

$$\frac{25 \text{ moles } N_2}{1 \text{ mole } N_2} = 25 \qquad \frac{45 \text{ moles } H_2}{3 \text{ moles } H_2} = 15$$

Although there are more moles of hydrogen than nitrogen, hydrogen is the limiting reagent. Based on the balanced equation only 30 moles of ammonia can be produced from these reactants.

You can confirm that hydrogen is the limiting reagent by calculating the amount of nitrogen required to react with the 45 moles of hydrogen.

$$45 \text{ moles } H_2 \times \frac{1 \text{ mole } N_2}{3 \text{ moles } H_2} = 15 \text{ moles } N_2$$

Since 15 moles of N_2 is required, and 25 moles is available, this calculation confirms that nitrogen is the excess reagent. An alternative way of checking that hydrogen is the limiting reagent is to assume that nitrogen is the limiting reagent and calculate the number of moles of hydrogen required.

$$25 \text{ moles } N_2 \times \frac{3 \text{ moles } H_2}{1 \text{ mole } N_2} = 75 \text{ moles } H_2$$

Since 75 moles is required, and only 45 moles of H_2 is available, there is insufficient hydrogen, and the assumption that nitrogen is the limiting reagent was incorrect.

The amounts of reactants used in industrial processes are often governed by economic considerations. In order to convert an expensive substance into product, the quantity of a second, less expensive reactant may be used in excess. For example, ethylene glycol, which is used in antifreeze, is produced by the reaction of ethylene oxide with water.

$$\underset{\text{ethylene oxide}}{\overset{\;O\;}{CH_2{-}CH_2}} + H_2O \longrightarrow \underset{\text{ethylene glycol}}{HO{-}CH_2{-}CH_2{-}OH}$$

Water is the less expensive reactant. It is used in excess to make sure that the more expensive ethylene oxide is more completely converted into product.

Limiting Reagents in Life Processes

Stoichiometric ratios for important biochemical reactions are not always available to cells at the time of need. For example, under conditions of extreme exertion, the supply of oxygen required to oxidize energy-containing compounds may be insufficient. The resulting condition, called oxygen debt and characterized by aching muscles, results from formation of products other than carbon dioxide and water. These products can subsequently be oxidized when activity is diminished and sufficient oxygen is moved to the cells.

There are four major elements contained in living tissue—carbon, hydrogen, oxygen, and nitrogen—as well as smaller amounts of sulfur and phosphorus. Under certain conditions, phosphorus is a limiting reagent in life processes. Phosphorus is not available to living tissue from any gaseous molecules. In addition, the major forms of phosphorus in minerals are insoluble in water. Thus phosphorus in living organisms, present as phosphate units bonded to carbon structures, must be largely recycled between various life forms. Microorganisms in bodies of water may feed on dead material and obtain phosphorus. Going up the food chain, larger organisms feed on smaller organisms, and the phosphorus is conserved.

The amount of phosphorus-containing compounds in a lake determines what life forms can exist. Only so many species can share the available phosphorus. The addition of phosphates from detergents that may flow into a water supply may actually lead to a decrease in life forms in spite of the increased amount of phosphorus. The first life form that grows in large amounts is algae. Under normal circumstances algae are a food source for other life forms. However, if algae grow rapidly, they may cover the surface of a lake and cut off light to other photosynthetic organisms. In addition, the rapid growth of algae depletes the supply of dissolved oxygen that is required by organisms including fish. The death of a range of life forms in a body of water is called *eutrophication.*

The algae on this eutrophied pond is the result of excessive growth caused by phosphates. (Daniel Brady, Stock Boston)

Summary

Atomic weight is used to express the mass of atoms. The **molecular weight** of a molecule is the sum of the atomic weights of its component atoms. The **formula weight** of an ionic compound is the sum of the atomic weights of the atoms indicated by the formula unit of the substance.

The **mole** is a count of the number of units of a substance present in matter. These units may be atoms, molecules, or ions. The number of units in a mole of a substance is called **Avogadro's number** (6.02×10^{23}). The **molar mass** of a substance is equal to the mass of a mole of the substance expressed in grams.

The composition of a compound is given by a **molecular formula** or **formula unit** in terms of the number of constituent atoms or ions. The formula of a compound and the atomic weights of the elements are used to calculate the **percent composition** of the elements.

The **empirical formula,** or simplest ratio of atoms in a compound, can be calculated from the percent composition. The molecular formula is a multiple of the empirical formula.

Stoichiometry is the mathematical relationship among the quantities of reactants and products of a reaction as indicated by a chemical equation. A variety of calculations may be done, but all are based on the mole method and use a mole ratio obtained from the coefficients in the balanced equation. The **mole ratio** is a conversion factor used to relate the stoichiometric amounts of the reactants and products in a chemical reaction.

The **heat of reaction** is expressed as ΔH. The quantity is negative for exothermic reactions and positive for endothermic reactions. The specific amount of heat energy required or released is calculated using the stoichiometry of the reaction.

The **percent yield** of a product is calculated by dividing the **actual yield** by the **theoretical yield** and multiplying by 100%.

The **limiting reagent** controls the amount of product that can be formed in a chemical reaction.

Exercises

Molecular Weights

7.1 Calculate the molecular weight of each of the following substances.

(a) CO (b) CO_2 (c) SO_2 (d) SO_3

7.2 Calculate the molecular weight of each of the following substances.

(a) CCl_4 (b) SiF_4 (c) $GeBr_4$ (d) PBr_3

7.3 Calculate the molecular weight of rotenone ($C_{23}H_{22}O_6$), a garden pesticide.

7.4 Calculate the molecular weight of aspartame ($C_{14}H_{18}N_2O_5$), the sweetener sold as NutraSweet.

Formula Weights

7.5 Calculate the formula weight of each of the following compounds.

(a) Na_2O (b) K_2S (c) Li_2Se (d) Sb_2O_3

7.6 Calculate the formula weight of each of the following compounds.

(a) $TiCl_4$ (b) $ScBr_3$ (c) $SnCl_4$ (d) $Ni(OH)_2$

7.7 Calculate the formula weight of each of the following compounds.

(a) sodium sulfate (b) potassium carbonate
(c) lithium sulfite (d) calcium nitrate

7.8 Calculate the formula weight of each of the following compounds.

(a) sodium carbonate (b) lithium sulfate
(c) potassium phosphate (d) iron(II) sulfate

Percent Composition

7.9 Calculate the percent composition of each of the following substances.

(a) CCl_4 (b) SiF_4 (c) $GeBr_4$ (d) PBr_3

7.10 Calculate the percent composition of each of the following substances.

(a) Na_2SO_4 (b) K_2CO_3
(c) Li_2SO_3 (d) $Ca(NO_3)_2$

7.11 Calculate the percent composition of each of the following substances.

(a) sodium carbonate (b) lithium sulfate
(c) potassium phosphate (d) iron(II) sulfate

7.12 Calculate the percent composition of each of the following substances.

(a) sodium oxide (b) lithium sulfide
(c) potassium iodide (d) iron(II) chloride

7.13 Calculate the percent composition of each of the following vitamins.

(a) $C_6H_8O_6$, vitamin C (b) $C_{20}H_{30}O$, vitamin A
(c) $C_{63}H_{88}N_{14}O_{14}PCo$, vitamin B_{12}

7.14 Calculate the percent composition of each of the following minerals.

(a) Na_3AlF_6, cryolite (b) $CaWO_4$, scheelite
(c) $Be_3Al_2Si_6O_{18}$, beryl

The Mole

7.15 Calculate the number of moles of atoms in each of the following samples.

(a) 46.0 g of sodium (b) 16.0 g of sulfur
(c) 2.0 g of mercury (d) 24.3 g of magnesium

7.16 Calculate the number of moles of atoms in each of the following samples.

(a) 78.2 g of potassium (b) 0.60 g of carbon
(c) 1.97 g of gold (d) 59.3 g of tin

7.17 Calculate the number of moles of molecules in each of the following samples.

(a) 180 g of H_2O (b) 0.44 g of CO_2
(c) 92 g of NO_2 (d) 3.00 g of NO

7.18 Calculate the number of moles of molecules in each of the following samples.

(a) 6.8 g of C_3O_2 (b) 0.76 g of N_2O_3
(c) 0.108 g of N_2O_5 (d) 2.2 g of P_4O_6

7.19 Calculate the number of moles of formula units in each of the following samples.

(a) 0.585 g of NaCl (b) 5.95 g of KBr
(c) 25.9 g of LiF (d) 28.2 g of MgS

7.20 Calculate the number of moles of formula units in each of the following samples.

(a) 12.4 g of Na_2O (b) 0.274 g of K_2S
(c) 0.186 g of Li_2Se (d) 0.623 g of MgF_2

7.21 The recommended daily allowance of vitamin C ($C_6H_8O_6$) in the adult human diet is 45 mg. How many moles of vitamin C is this?

7.22 The minimum daily requirement of the amino acid lysine ($C_6H_{14}N_2O_2$) is 0.80 g. How many moles is this?

7.23 Myocrisin, a drug used to treat rheumatoid arthritis, has the molecular formula $Na_2C_4H_3O_4SAu$. How many moles of gold atoms are contained in an injection of 50 mg of the compound?

7.24 A solution of potassium permanganate ($KMnO_4$) is used to treat ick, a disease of tropical fish maintained in aquariums. How many moles of permanganate ions are in 8 mg, a quantity commonly used?

Avogadro's Number

7.25 Calculate the number of atoms present in each of the following samples.

(a) 20.0 g of mercury (b) 120 g of carbon
(c) 53.2 g of palladium (d) 0.635 g of copper

7.26 Calculate the number of atoms present in each of the following samples.

(a) 46.0 g of sodium (b) 16.0 g of sulfur
(c) 2.0 g of mercury (d) 24.3 g of magnesium

7.27 Calculate the number of molecules present in each of the following samples.

(a) 2.8 g of CO (b) 90 g of H_2O
(c) 32 g of O_2 (d) 0.40 g of CH_4

7.28 Calculate the number of molecules present in each of the following samples.

(a) 34.0 g of C_3O_2 (b) 19.0 g of N_2O_3
(c) 22.0 g of P_4O_6 (d) 14.2 g of P_4O_{10}

7.29 Calculate the number of formula units present in each of the following samples.

(a) 1.24 g of Na_2O (b) 27.6 g of K_2S
(c) 6.23 g of MgF_2 (d) 55.5 g of $CaCl_2$

7.30 Calculate the number of formula units present in each of the following samples.

(a) 1.64 g of Na_3PO_4 (b) 7.90 g of $KMnO_4$
(c) 11.0 g of Li_2SO_4 (d) 8.57 g of $Ba(OH)_2$

7.31 Calculate the number of indicated atoms present in each of the following.

(a) oxygen atoms in 0.1 mole of SO_3
(b) carbon atoms in 0.02 mole of C_3O_2
(c) oxygen atoms in 0.2 mole of N_2O_5

7.32 Calculate the number of indicated atoms present in each of the following.

(a) chlorine atoms in 2.0 moles of CCl_4
(b) fluorine atoms in 0.020 mole of SF_6
(c) fluorine atoms in 0.030 mole of IF_7

7.33 A cup (236 mL) of water may contain 3×10^{-6} g of mercury. How many atoms of mercury are present?

7.34 Grain treated with EDB ($C_2H_4Br_2$) may contain 2.5×10^{-6} g of EDB per kilogram of grain. How many molecules of EDB are present in this quantity?

Empirical Formula

7.35 The molecular formula of ethylene glycol, used in antifreeze, is $C_2H_6O_2$. What is the empirical formula?

7.36 Hydroquinone, a compound used in developing photographs, has the molecular formula $C_6H_6O_2$. What is the empirical formula?

7.37 The compound in the light-emitting diode used in some electronic calculators contains 69.2% Ga and 30.8% P. What is the empirical formula of the compound?

7.38 A sample of a freon propellant in an aerosol can contains 0.423 g of C, 2.50 g of Cl, and 1.34 g of F. What is the empirical formula of this substance?

7.39 Given the following percent compositions, determine the empirical formula of each compound.

(a) 85.7% C and 14.3% H

(b) 80.0% C and 20.0% H

(c) 92.3% C and 7.7% H

7.40 Given the following percent compositions, determine the empirical formula of each compound.

(a) 38.7% C, 9.67% H, and 51.6% O
(b) 40.0% C, 6.66% H, and 53.3% O
(c) 54.6% C, 9.09% H, and 36.3% O

Molecular Formulas

7.41 The composition of adipic acid, used in the manufacture of nylon, is 49.3% C, 6.9% H, and 43.8% O by mass. The molecular weight is 146 amu. What is the molecular formula?

7.42 A compound used in detergents is 22.5% Na, 30.4% P, and 47.1% O. The molecular weight is 612 amu. What is the molecular formula?

7.43 The stimulant caffeine has a molecular weight of 194.2 amu, and the percent composition by mass is 49.5% C, 5.2% H, 28.8% N, and 16.5% O. What is the molecular formula of caffeine?

7.44 Fructose, a sugar found in honey and fruit juice, is 40.0% C, 6.7% H, and 53.3% O. It has a molecular weight of 180 amu. What is its molecular formula?

The Mole Ratio

7.45 Balance the following equation and write the mole ratio necessary to convert the number of moles of MnO_2 available into the number of moles of Al required for the reaction.

$$MnO_2 + Al \longrightarrow Al_2O_3 + Mn$$

7.46 Balance the following equation and write the mole ratio necessary to convert the number of moles of Na_2O available into the number of moles of P_4O_{10} required for the reaction.

$$Na_2O + P_4O_{10} \longrightarrow Na_3PO_4$$

7.47 Balance the following equation and write the mole ratio necessary to convert the number of moles of $BaCl_2$ available into the number of moles of AgCl produced.

$$BaCl_2 + AgNO_3 \longrightarrow AgCl + Ba(NO_3)_2$$

7.48 Balance the following equation and write the mole ratio necessary to convert the number of moles of $C_4H_{10}O_2$ available into the number of moles of O_2 required for the reaction.

$$C_4H_{10}O_2 + O_2 \longrightarrow CO_2 + H_2O$$

7.49 Balance the following equation and write the mole ratio necessary to convert the number of moles of Bi_2S_3 available into the number of moles of SO_2 produced.

$$Bi_2S_3 + O_2 \longrightarrow Bi_2O_3 + SO_2$$

7.50 Balance the following equation and write the mole ratio necessary to convert the number of moles of Ca_3P_2 available into the number of moles of PH_3 produced.

$$Ca_3P_2 + H_2O \longrightarrow Ca(OH)_2 + PH_3$$

Mole–Mole Stoichiometry

7.51 Ammonium nitrate (NH_4NO_3) may decompose with explosive violence to produce N_2 gas. How many moles of NH_4NO_3 are required to produce 2.4 moles of N_2?

$$2\,NH_4NO_3 \longrightarrow 2\,N_2 + O_2 + 4\,H_2O$$

7.52 Nitrogen oxide (NO), which is produced during lightning flashes in the atmosphere, reacts rapidly with O_2 to yield NO_2. How many moles of NO_2 result from the reaction of 8 moles of NO?

$$2\,NO + O_2 \longrightarrow 2\,NO_2$$

7.53 Nitrogen dioxide (NO_2) reacts with water to form HNO_3. Calculate the number of moles of HNO_3 that can be formed from 0.06 mole of NO_2.

$$3\,NO_2 + H_2O \longrightarrow 2\,HNO_3 + NO$$

7.54 Sulfur dioxide (SO_2) reacts with water to form H_2SO_3. How many moles of SO_2 are required to form 0.05 mole of H_2SO_3?

$$SO_2 + H_2O \longrightarrow H_2SO_3$$

7.55 How many moles of oxygen are required to react completely with 5.0 moles of sulfur dioxide to yield sulfur trioxide?

$$2\,SO_2 + O_2 \longrightarrow 2\,SO_3$$

7.56 How many moles of carbon dioxide are produced in the combustion of 0.50 mole of ethane?

$$2\,C_2H_6 + 7\,O_2 \longrightarrow 4\,CO_2 + 6\,H_2O$$

Mass–Mass Stoichiometry

7.57 Solutions of sodium hypochlorite (NaClO) are sold as laundry bleach. How many grams of chlorine (Cl_2) are needed to react with 60.0 g of sodium hydroxide (NaOH) to produce NaClO?

$$2\,NaOH + Cl_2 \longrightarrow NaCl + NaClO + H_2O$$

7.58 White phosphorus (P_4) is used in military incendiary devices because it ignites spontaneously in air. How many grams of phosphorus (P_4) will react with 25.0 g of oxygen (O_2)?

$$P_4 + 5\,O_2 \longrightarrow P_4O_{10}$$

7.59 Freshly exposed aluminum surfaces react with oxygen to form a tough oxide coating that protects the metal from further corrosion. How many grams of oxygen are required to react with 8.09 g of aluminum?

$$4\,Al + 3\,O_2 \longrightarrow 2\,Al_2O_3$$

7.60 Silicon carbide, used as an abrasive for sandpaper, is prepared as indicated in the following equation. How much SiO_2 (in grams) is required to produce 738 g of silicon carbide?

$$SiO_2 + 3\,C \longrightarrow SiC + 2\,CO$$

7.61 Ethanol can be prepared by the fermentation of glucose. How many grams of ethanol can be made from 1.0 kg of glucose?

$$C_6H_{12}O_6 \longrightarrow 2\,C_2H_5OH + 2\,CO_2$$

7.62 Tungsten (W) metal, used to make incandescent bulb filaments, is produced by the following reaction. How many grams of W can be obtained from 3.67 g of WO_3?

$$WO_3 + 3\,H_2 \longrightarrow W + 3\,H_2O$$

Heats of Reaction

7.63 Using the following equation, determine the amount of heat energy released by the combustion of 11 g of propane in a camp stove.

$$C_3H_8(g) + 5\,O_2(g) \longrightarrow 3\,CO_2(g) + 4\,H_2O \quad \Delta H = -530 \text{ kcal}$$

7.64 Using the following equation, determine the amount of heat energy released by the combustion of 52 g of acetylene in a welder's torch.

$$2\,C_2H_2(g) + 5\,O_2(g) \longrightarrow 4\,CO_2(g) + 2\,H_2O \quad \Delta H = -302 \text{ kcal}$$

Percent Yield

7.65 A student reacts 4.16 g of $BaCl_2$ with excess Na_2SO_4 and obtains 4.06 g of $BaSO_4$. What is the percent yield?

$$BaCl_2(aq) + Na_2SO_4(aq) \longrightarrow BaSO_4(s) + 2\,NaCl(aq)$$

7.66 A student reacts 4.25 g of $AgNO_3$ with excess NaCl and obtains 3.58 g of AgCl. What is the percent yield?

$$AgNO_3(aq) + NaCl(aq) \longrightarrow AgCl(s) + NaNO_3(aq)$$

Limiting Reagents

7.67 How many grams of water will be produced from the reaction of 1.5 moles of hydrogen and 1.5 moles of oxygen?

$$2\,H_2 + O_2 \longrightarrow 2\,H_2O$$

7.68 Zinc and sulfur combine to give zinc sulfide, a substance used in phosphors that coat the inner surface of television tubes. How many grams of ZnS can be formed when 12.0 g of Zn reacts with 6.50 g of S?

7.69 How many grams of silver chloride can be produced from 16.99 g of silver nitrate and 2.92 g of sodium chloride?

$$AgNO_3(aq) + NaCl(aq) \longrightarrow AgCl(s) + NaNO_3(aq)$$

7.70 How many grams of carbon dioxide will be produced by reaction of 1.6 g of oxygen and 1.1 g of propane?

$$C_3H_8(g) + 5\,O_2(g) \longrightarrow 3\,CO_2(g) + 4\,H_2O$$

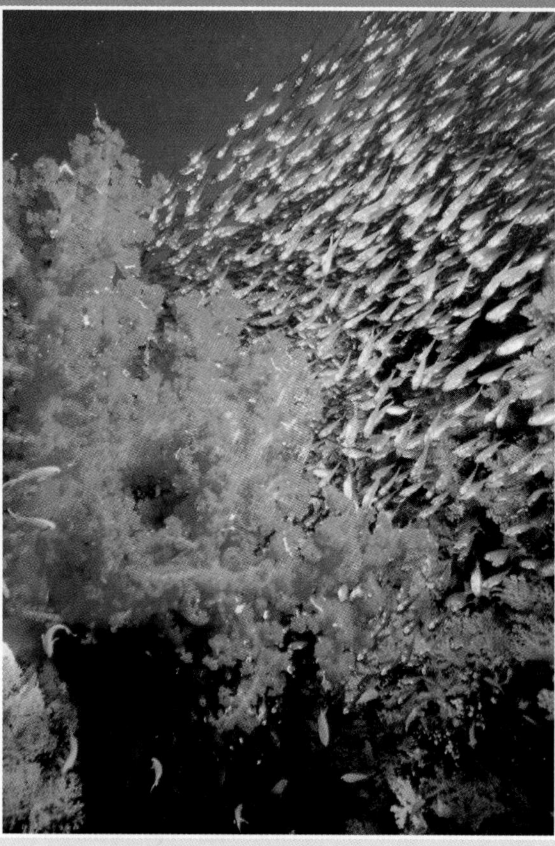

The variety and organization of life forms amidst disorder is seen in this school of sweeper fish near soft coral in the Red Sea near Egypt. (Freddy Storheil, Tony Stone Worldwide)

8

Gases, Liquids, and Solids

8.1 The Organization of Matter
8.2 Pressure
8.3 The Gas Laws
8.4 Avogadro's Hypothesis and the Ideal Gas Law
8.5 Dalton's Law of Partial Pressures
8.6 Liquids
8.7 Solids
8.8 Le Châtelier's Principle and Changes of State
8.9 Types of Intermolecular Forces

Essays
- *Breathing and Boyle's Law*
- *Partial Pressures in Medicine*
- *Water—An Unusual Compound*

8.1 The Organization of Matter

Matter is held together by a variety of forces. Electrons are attracted to the protons in the nuclei of atoms, and atoms are bound to other atoms by the bonding forces that result from attractions between nuclei and bonding electrons.

Attractive forces between molecules are called **intermolecular forces;** these forces are much weaker than bonding forces, but they determine whether a substance is a gas, liquid, or solid. Understanding the forces that hold matter together in the various states allows chemists to design new compounds that might have desirable physical properties. In Section 8.9 the types of intermolecular forces are presented and discussed in terms of the bonding and structure of molecules.

Organization Versus Disorganization

Molecules are in constant motion and have kinetic energies that increase with temperature. As a result, molecules tend to separate from each other at higher temperatures. However, the intermolecular attractive forces provide a balancing force. If the kinetic energy of the molecules is greater than the energy of the intermolecular attractive forces, the gaseous state results. At a lower temperature, the kinetic energy is sufficiently reduced that the energy of the attractive forces is more nearly

FIGURE 8.1 Representation of States of Matter for Halogens

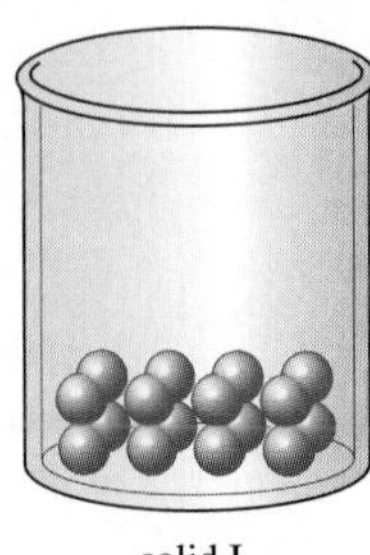

solid I_2

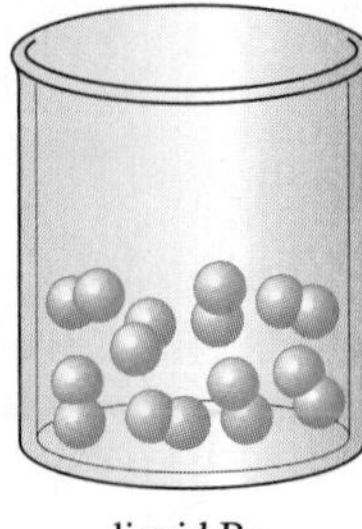

liquid Br_2

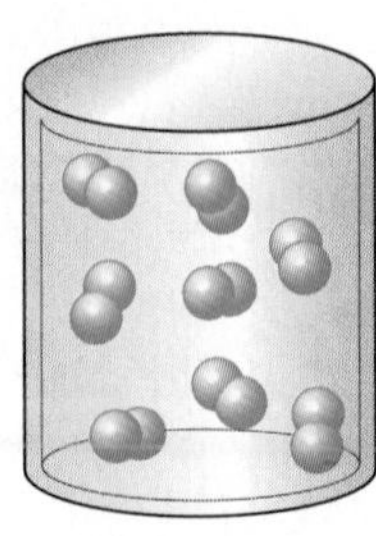

gaseous Cl_2

equal to the kinetic energy, and the liquid state results. Finally, at a still lower temperature, the kinetic energy is so low that the intermolecular attractive forces are sufficient to hold the molecules in place, and the solid state results.

At a common temperature, the state of different substances indicates the magnitude of their respective attractive forces. For example, at 20 °C iodine is a solid, bromine is a liquid, and chlorine is a gas. All of these elements consist of diatomic molecules (Figure 8.1). Iodine molecules are regularly arranged in a solid, indicating that they have the strongest intermolecular forces of these three elements. Bromine molecules have weaker intermolecular forces, and can move with greater freedom in the liquid phase. Chlorine molecules have still weaker intermolecular forces, so weak that they stay apart from one another in the gas phase.

Kinetic Molecular Theory

The effect of moving gas molecules is seen in the devastation caused by Hurricane Andrew in Homestead, Florida. (Chris Brown, Stock Boston)

The **kinetic molecular theory** describes the properties of moving molecules. Although the theory is usually applied to the gaseous state, it also extends to the liquid and solid states. The kinetic molecular theory is summarized as follows:

1. Atoms or molecules in the gaseous state are widely separated from one another. The space occupied by the atoms or molecules is very small compared with the space between them.
2. The atoms or molecules move rapidly and randomly in straight lines. The direction is maintained until collision occurs with a wall of the container or with another atom or molecule.
3. There are no attractive forces between molecules or atoms of a gas.
4. Collisions of molecules or atoms don't change the total energy of the gas, although a transfer of energy between the molecules or atoms occurs.
5. Individual atoms or molecules of a gas move at different speeds (Figure 8.2). However, the average velocity for a specific gas is constant at a given temperature. As the temperature increases, the average velocity increases, and therefore the average kinetic energy also increases. The average kinetic energy is directly proportional to the temperature on the Kelvin scale. At the same temperature, all gases have the same average kinetic energy.

A hypothetical gas that conforms to all the assumptions of the kinetic theory is an **ideal gas.** However, if there were no attractive forces between the atoms or molecules of an ideal gas, it could not be condensed to form a liquid. Real gases do have intermolecular attractive forces and can be condensed to liquids at high pressures and low temperatures. However, at low pressure and high temperature, real gases behave approximately as ideal gases.

FIGURE 8.2 Distribution of Velocities of Molecules and Temperature

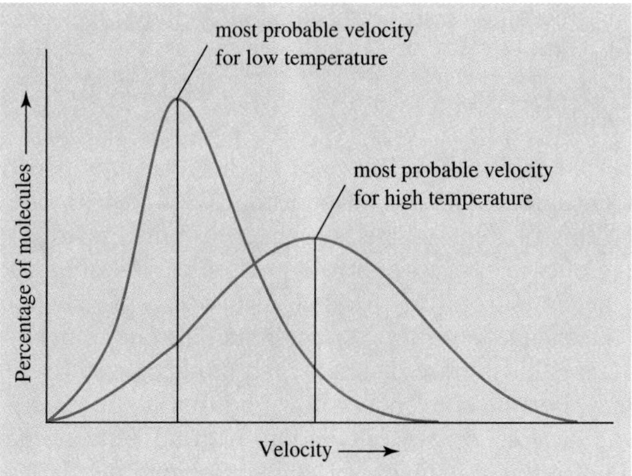

Equilibrium Between States

A **static equilibrium** exists when two or more opposing forces are in balance with one another and no further change occurs in the system. A **dynamic equilibrium** exists when there is a continuation of opposing processes at the submicroscopic level even though, at the macroscopic level, the system appears to be static because there is no net observable change.

Consider your finances as an example of a dynamic equilibrium. If the deposits to and withdrawals from your checking account are controlled to give a constant balance, you are in financial equilibrium. Your checking account balance remains unchanged, but the same dollars are not always there. Your finances are dynamic.

Condensation, the conversion of a gas into a liquid, occurs when a gas is cooled. **Vaporization** is the conversion of a liquid into a gas. The gas and liquid exist in equilibrium at the boiling point (Figure 8.3). At this temperature the average kinetic energy balances the energy of the average forces of attraction between molecules. Below the boiling point, the molecules tend to "stick together." Above the boiling point the molecules exist as a gas because their kinetic energy exceeds the energy of the attractive forces.

When a liquid is cooled, the attractive forces between neighboring particles exert a greater influence. At the freezing point (or melting point), the liquid exists in equilibrium with the solid (Figure 8.3). In the solid, the motion of the particles is restricted to vibration about fixed positions. The conversion of a liquid into a solid is called **freezing.** Conversion of a solid into a liquid is **melting.** Below the freezing

FIGURE 8.3 Equilibria Between States of Matter

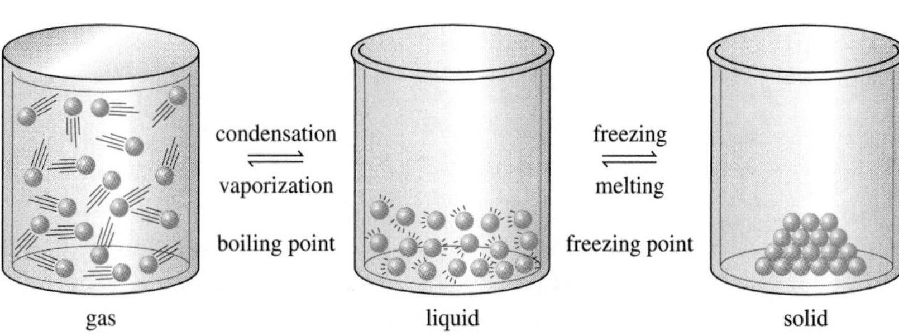

point, the solid exists because the attractive forces exceed the kinetic energy of the molecules.

Entropy and Changes of State

The concept that an increase in disorder may contribute to the spontaneity of chemical reactions was introduced in Section 1.10. The disorder in a system such as a sample of a compound is given by its entropy, which is designated by S. Entropy is an elusive concept, but the greater the disorder, the larger the value of S. The change in entropy, ΔS, for a physical process or chemical reaction, gives a measure of the change in disorder from the initial state to the final state.

Gases have more disorder than liquids, which in turn have more disorder than solids. For the conversion of a solid to a liquid or a liquid to a gas the ΔS is positive. We won't do any calculations with ΔS, but the change in entropy for vaporization of a liquid is much larger than for melting the corresponding solid to form a liquid. This is because the liquid and solid occupy about the same volume and have molecules in close proximity to one another. When the gaseous state forms, the molecules move far apart, so the increase in the disorder is greater.

The tendency for a physical process or chemical reaction to occur depends on both the enthalpy change of the process or reaction, ΔH, and the entropy change, ΔS. For example, the increase in entropy for the vaporization of a liquid is favorable, but it is necessary to add heat energy to overcome the attractive forces between molecules. At a temperature below the boiling point the energy available is insufficient to allow the molecules to escape to the gas phase. Only at the boiling point has sufficient energy been added to allow the molecules to escape into the more random gaseous phase.

8.2 Pressure

This blue spotted stingray has a flat shape, but not because of the high pressure of its environment. (Klaus Hilgert\Okapia, Photo Researchers)

Pressure is the force per unit area on an object. We live at the bottom of an ocean of air and are under pressure from the air much as a diver is under pressure from water in the ocean. However, we have lived under air pressure all our lives and we don't feel it. The pressure results from the weight of air extending from the outer edge of the atmosphere down to the surface of the Earth. Normally the weight of a column of air above an area of 1 in^2 is 14.7 lb at sea level. We say that the pressure is 14.7 lb/in^2 or 14.7 psi. At higher altitude the pressure is less because the column of air from that point to the edge of the atmosphere weighs less (Figure 8.4).

Pressure under water increases substantially with depth because water is much denser than air. When a scuba diver descends to a depth of 120 ft, the pressure due to the mass of water above the diver is about 59 lb/in^2. The diver experiences a total pressure of about 74 lb/in^2 because the total pressure is equal to that of the water plus the pressure of the atmosphere.

Measurement of Pressure

The **barometer,** a device to measure atmospheric pressure, was invented by Evangelista Torricelli in the seventeenth century. It is a long tube closed at one end, filled with mercury, and inverted in a vessel of mercury. Part of the mercury

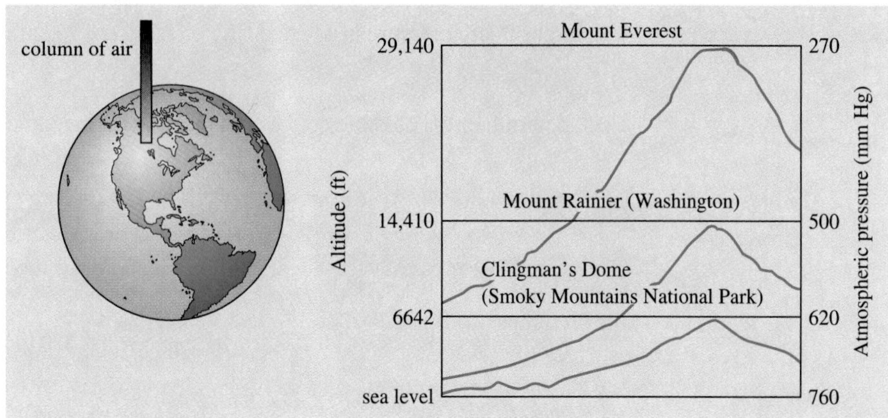

FIGURE 8.4 Atmospheric Pressure and Altitude

A column of air from the outer atmosphere to a given point on Earth contains gases that exert a force equal to 14.7 lb on each square inch. This pressure is 1 atm or 760 mm Hg. At a higher altitude the pressure is less because the mass of the column of air from the outer atmosphere to that point is less. Examples of pressure on three mountains are given on the right.

runs out of the tube when it is inverted, leaving a vacuum in the closed end. A column of mercury approximately 76 cm (760 mm) high remains in the tube (Figure 8.5).

The column of mercury remains because the atmosphere exerts pressure on the surface of the mercury outside the tube. This pressure, transmitted through the mercury to the base of the column, supports the mercury in the tube. There is pressure at the base of the mercury column due to the mass of the mercury in the column. The mercury falls until the pressure exerted by its mass equals the atmospheric pressure.

Pressure is usually expressed in units of force per unit area. In the barometer, the downward force exerted by the mercury column is proportional to the mass of the liquid supported in the column, which in turn is proportional to the height of the column. Therefore, atmospheric pressure can be expressed by the height of the mercury in the barometric tube. The terms centimeters of mercury (cm Hg) and millimeters of mercury (mm Hg) indicate the pressure or force per unit area exerted by the mass of mercury contained in a stated length of a mercury column.

Atmospheric pressure depends on altitude and the local weather, which fluctuates from day to day. At 0 °C at sea level, the average atmospheric pressure is 760 mm Hg. This pressure is called a standard **atmosphere** (1 atm). The unit of pressure 1 mm Hg is also called a **torr.** Therefore, standard pressure (1 atm) is 760 torr.

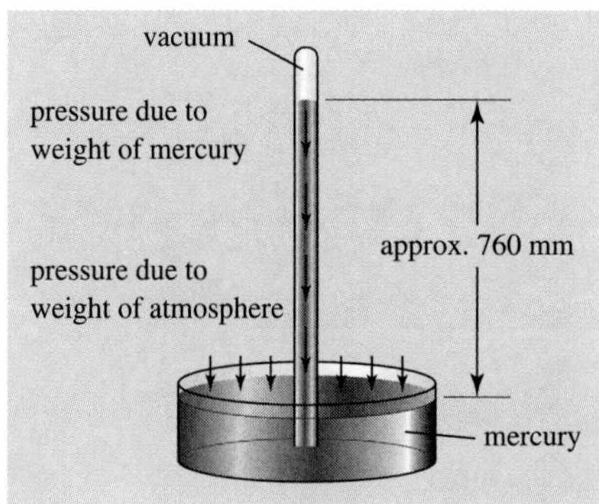

FIGURE 8.5 The Barometer

The pressure due to the atmosphere and the pressure due to the column of mercury are equal.

EXAMPLE 8.1

The atmospheric pressure in Mexico City one day is 0.750 atm. What is the pressure in torr?

Solution

First, determine two possible conversion factors relating atmosphere and torr.

$$\frac{1 \text{ atm}}{760 \text{ torr}} \quad \text{and} \quad \frac{760 \text{ torr}}{1 \text{ atm}}$$

The second factor cancels the given unit of atm and gives the required value in torr.

$$0.750 \cancel{\text{atm}} \times \frac{760 \text{ torr}}{1 \cancel{\text{atm}}} = 570 \text{ torr}$$

PROBLEM 8.1

The pressure on the surface of Venus is about 1.0×10^2 atm. What is the pressure in cm Hg?

8.3 The Gas Laws

The physical properties of gases depend on pressure, volume, and temperature. Each of these variables affects the others. For example, if the pressure on a gas sample is changed, a change in the volume of the sample occurs. In this section we will calculate the consequence of changing any of the physical conditions of the gas.

Boyle's Law

In 1662, Robert Boyle studied the relationship between volume and pressure of gases. He measured "the spring of air," the pressure with which a gas sample pushes back when it is compressed. He found a quantitative relationship now

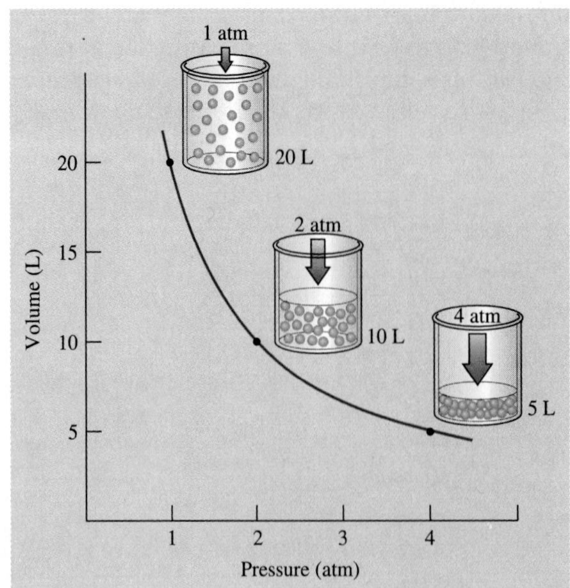

FIGURE 8.6 An Illustration of Boyle's Law

As pressure increases, volume decreases. As the external pressure is increased, the number of molecules per unit volume increases. Therefore, the molecules collide more often with the walls of the container, and the internal pressure is increased.

known as **Boyle's law:** *At constant temperature, the volume of a sample of gas varies inversely with the pressure exerted on it.* Thus, if the pressure on a given volume of gas is doubled, the volume is decreased by half. A mathematical expression of Boyle's law is

$$V = k \times \frac{1}{P} \quad \text{or} \quad PV = k$$

where V stands for volume and P for pressure, and k is a proportionality constant which depends on the temperature and quantity of gas. Thus, the product of the volume and pressure of a gas is constant only if the quantity of the gas sample and the temperature are unchanged. An example of data illustrating Boyle's law is given in Figure 8.6. Boyle's law is also expressed by the following equation, in which the subscripts 1 and 2 refer to the two sets of experimental conditions.

$$P_2 \times V_2 = k = P_1 \times V_1$$

For a change in pressure, we calculate the new volume from the rearranged equation

$$V_2 = V_1 \times \frac{P_1}{P_2} \quad \text{or} \quad V_2 = V_1 \times P_{\text{factor}}$$

Similarly, if a new pressure must be calculated to produce a desired volume change, then

$$P_2 = P_1 \times \frac{V_1}{V_2} \quad \text{or} \quad P_2 = P_1 \times V_{\text{factor}}$$

In either calculation, we determine the necessary factor by recalling the inverse proportionality of Boyle's law. Thus, if the pressure is increased, the volume must decrease, and only a pressure factor less than unity will give the correct answer.

EXAMPLE 8.2

An oxygen cylinder for medical use contains 35.5 L at 25 °C and 149 atm. What volume in liters will this amount of gas occupy at 1.00 atm at the same temperature?

Solution

Since the temperature for the sample of the gas is unchanged, Boyle's law applies. First, arrange the information to help you analyze the problem.

$P_1 = 149$ atm	$V_1 = 35.5$ L
$P_2 = 1.00$ atm	$V_2 = ?$ L
Pressure decreases.	Volume must increase.

Direct substitution into the Boyle's law equation provides the required volume. However, you can also solve the problem by noting that the pressure has decreased from 149 to 1.00 atm and that the volume must therefore increase by a pressure factor containing these two numbers.

$$V_2 = V_1 \times \frac{P_1}{P_2} \quad \text{or} \quad V_2 = V_1 \times P_{\text{factor}}, \quad \text{where} \quad P_{\text{factor}} > 1$$

$$V_2 = 35.5 \text{ L} \times \frac{149 \text{ atm}}{1.00 \text{ atm}}$$

$$V_2 = 5.29 \times 10^3 \text{ L}$$

Breathing and Boyle's Law

Our thoracic (chest) cavity is sealed from outside air. The two lungs in this cavity are inflated with outside air when we inhale, and they are deflated when we exhale. The intrapleural space within the thoracic cavity is equal to the volume of the cavity minus the volume occupied by the lungs.

The diaphragm is at the bottom of the thoracic cavity. When it moves down and the rib cage expands, there is a temporary expansion in the volume of the thoracic cavity. As a result, the volume of the intrapleural space increases and the pressure decreases. However, as soon as the pressure starts to drop, air moves into the lungs. The lungs expand, the intrapleural space is restored to its original volume, and the intrapleural pressure returns to its original value.

When the diaphragm relaxes and the rib cage contracts, the thoracic cavity decreases in volume. As a result, the intrapleural pressure should increase. However, as soon as the pressure starts to increase, the same pressure is exerted on the lungs, which then decrease in size. This decrease in volume would increase the pressure in the lungs, but air is forced out or exhaled. Artificial respiration simulates the sequence of pressure changes experienced by the lungs in normal breathing. Expanding and contracting the thoracic cavity forces air to enter or leave the lungs.

The Heimlich maneuver, an emergency procedure used to remove food lodged in the trachea, also works on the principle of Boyle's law. You wrap your arms below the rib cage of the choking individual. A sharp upward bear hug raises the diaphragm and increases the pressure on the lungs, which forces expulsion of the lodged material.

The Mechanism of Breathing

The volume of the thoracic cavity is changed as the diaphragm and rib-cage positions are changed. Since the volume of the intrapleural space cannot change, the volume of the lungs must change by expanding and contracting in order to maintain the pressure.

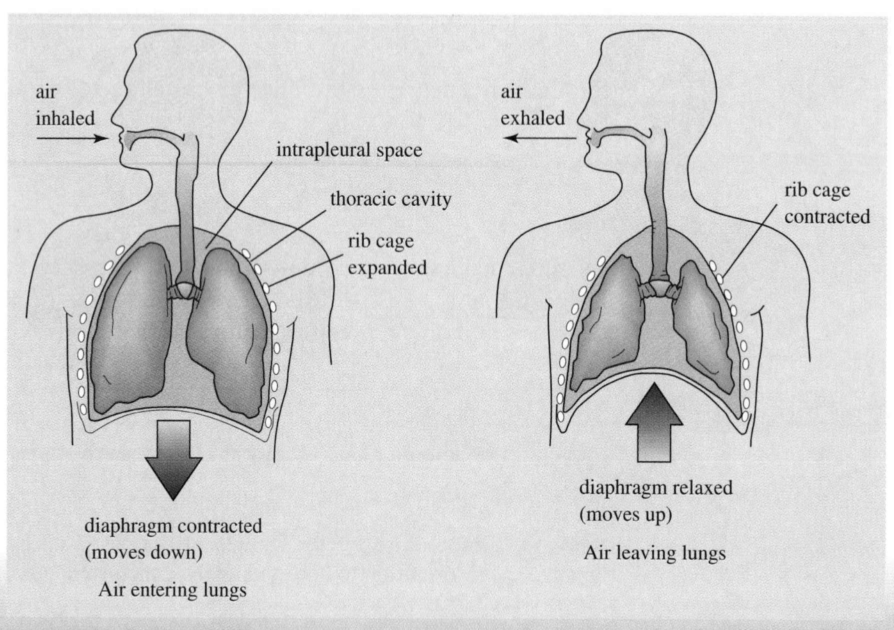

PROBLEM 8.2

The volume of one cylinder in an automobile is 0.45 L when the oxygen and fuel mixture is pulled into the cylinder at 1.0 atm. When the piston moves upward it compresses the volume to 0.075 L. Assuming that the temperature is constant, what pressure results when the gas is fully compressed?

Charles's Law

In 1787, the French physicist J. A. C. Charles observed that gases expand when heated under constant pressure and contract when cooled. From the data of

FIGURE 8.7
An Illustration of Charles's Law

As temperature increases, volume increases. The kinetic energy of the molecules causes them to move faster. As a result, the number of impacts increases as does the force of each impact, causing the volume to increase.

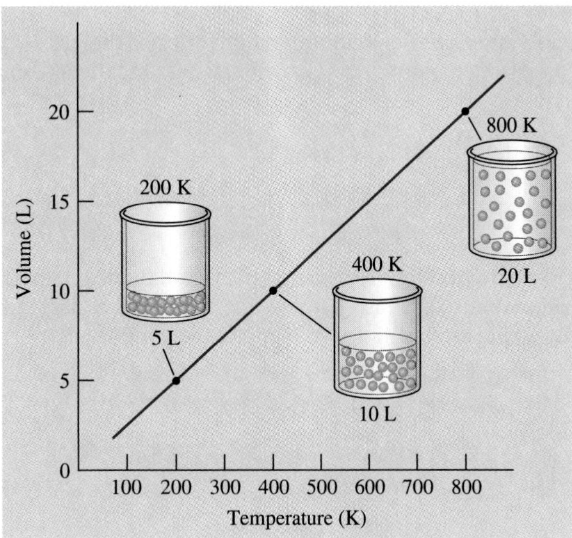

Charles and other scientists, we have **Charles's law:** *At constant pressure the volume of a fixed mass of a gas is directly proportional to the absolute (Kelvin) temperature.* An example of such data is given in Figure 8.7.

The mathematical expression of Charles's law is

$$V = kT \quad \text{or} \quad \frac{V}{T} = k$$

where k is a proportionality constant. The constant remains unchanged at constant pressure for a fixed sample of gas. Thus, for constant pressure, the following relationship holds under any two sets of experimental conditions.

$$\frac{V_2}{T_2} = k = \frac{V_1}{T_1}$$

For a change in temperature, we calculate the new volume by using a temperature factor.

$$V_2 = V_1 \times \frac{T_2}{T_1} \quad \text{or} \quad V_2 = V_1 \times T_{\text{factor}}$$

Alternatively, we calculate the temperature required to change the volume by using a volume factor. Remember, with either formula, the temperature must be in kelvins.

$$T_2 = T_1 \times \frac{V_2}{V_1} \quad \text{or} \quad T_2 = T_1 \times V_{\text{factor}}$$

EXAMPLE 8.3

A child's balloon is filled with helium to give a volume of 5.0 L in an air-conditioned store at 21 °C. The child takes the balloon outdoors, where the air temperature is 38 °C. What happens to the balloon?

Solution

Since the atmospheric pressure and quantity of the gas remain constant, Charles's law applies to solving the problem. Arrange the data in tabular form and calculate the temperature in kelvins.

$T_1 = 21 + 273 = 294\text{ K}$	$V_1 = 5.0\text{ L}$
$T_2 = 38 + 273 = 311\text{ K}$	$V_2 = ?\text{ L}$
Temperature increases.	Volume must increase.

Direct substitution into the Charles's law equation will give the unknown volume. Note that the volume must increase because the temperature increases. Therefore, multiply the volume by a temperature factor to increase the volume.

$$V_2 = V_1 \times \frac{T_2}{T_1} \quad \text{or} \quad V_2 = V_1 \times T_{\text{factor}}, \quad \text{where} \quad T_{\text{factor}} > 1$$

$$V_2 = 5.0\text{ L} \times \frac{311\cancel{\text{K}}}{294\cancel{\text{K}}}$$

$$V_2 = 5.3\text{ L}$$

PROBLEM 8.3

A 250-mL sample of hydrogen gas at 127 °C is cooled to 150 K. What is the volume of the sample, assuming that the pressure is constant?

Gay-Lussac's Law

In 1802, Joseph Gay-Lussac observed that the pressure of a gas increases when the gas is heated under constant volume and decreases when it is cooled. From the data of Gay-Lussac and other scientists, we have **Gay-Lussac's law:** *At constant volume the pressure of a fixed mass of a gas is directly proportional to the absolute (Kelvin) temperature.* The mathematical expression of Gay-Lussac's law is

$$P = kT \quad \text{or} \quad \frac{P}{T} = k$$

where k is a proportionality constant. For a fixed volume and mass of a gas, the following relationship is obeyed under any two sets of experimental conditions when only the temperature and pressure change.

$$\frac{P_2}{T_2} = k = \frac{P_1}{T_1}$$

For a change in temperature, we calculate the pressure using a temperature factor.

$$P_2 = P_1 \times \frac{T_2}{T_1} \quad \text{or} \quad P_2 = P_1 \times T_{\text{factor}}$$

Alternatively, to calculate the temperature required to change the pressure, we use a pressure factor. Remember, the temperature must be in kelvins.

$$T_2 = T_1 \times \frac{P_2}{P_1} \quad \text{or} \quad T_2 = T_1 \times P_{\text{factor}}$$

Heating a gas in a closed container may cause an explosion if the container cannot withstand the pressure. For this reason, spray cans of deodorants, paints, and so on have a warning label indicating that the can should not be heated.

EXAMPLE 8.4

What will happen to the pressure of a "used" spray can containing only the propellant at a pressure of 1.1 atm at 23 °C if it is thrown into a fire at 475 °C?

Solution

State the conditions for Gay-Lussac's law and arrange the data as follows, after converting Celsius degrees to kelvins.

$T_1 = 296$ K	$P_1 = 1.1$ atm
$T_2 = 748$ K	$P_2 = ?$ atm
Temperature increases.	Pressure must increase.

The pressure increases by a temperature factor. Therefore,

$$P_2 = P_1 \times \frac{T_2}{T_1} \quad \text{or} \quad P_2 = P_1 \times T_{\text{factor}}, \quad \text{where} \quad T_{\text{factor}} > 1$$

$$P_2 = 1.1 \text{ atm} \times \frac{748 \cancel{\text{K}}}{296 \cancel{\text{K}}}$$

$$P_2 = 2.8 \text{ atm}$$

If the can cannot withstand this pressure, it will explode.

PROBLEM 8.4

An autoclave is used to sterilize medical instruments. Steam is produced at 100 °C and 1.0 atm. If the steam is heated to increase the pressure to 1.15 atm, what is the temperature? Note that the volume of the autoclave is unchanged.

Combined Gas Law

Boyle's, Charles's, and Gay-Lussac's laws can be combined into one mathematical expression:

$$\frac{PV}{T} = k$$

Therefore, for a fixed sample of a gas, we write

$$\frac{P_2V_2}{T_2} = k = \frac{P_1V_1}{T_1}$$

This equation need not be memorized because any variable can be changed by the factors of the other two variables.

$$P_2 = P_1 \times V_{\text{factor}} \times T_{\text{factor}}$$
$$V_2 = V_1 \times P_{\text{factor}} \times T_{\text{factor}}$$
$$T_2 = T_1 \times P_{\text{factor}} \times V_{\text{factor}}$$

To solve problems in which two variables change, consider each variable separately. Thus, for a change in both pressure and temperature, determine first what effect the pressure change has on the volume. If the pressure increases, the volume decreases. After determining the pressure factor, consider the temperature factor. If the temperature decreases, the volume decreases. Multiplying by both the pressure factor and the temperature factor gives the final volume.

EXAMPLE 8.5

A weather balloon is filled with 2.0×10^4 L of helium at 20 °C and 730 torr. What volume will the balloon occupy at an altitude where the pressure is 69 torr and the temperature is −45 °C?

Solution

Convert the Celsius temperature to kelvins, and arrange the data as follows:

$$\begin{array}{ll} V_1 = 2.0 \times 10^4 \text{ L} & V_2 = ?\text{ L} \\ T_1 = 293 \text{ K} & T_2 = 228 \text{ K} \\ P_1 = 730 \text{ torr} & P_2 = 69 \text{ torr} \end{array}$$

Note the effect of both pressure and temperature changes on the volume. Because the temperature decreases, the volume must decrease by a factor of $\frac{228}{293}$. Because the pressure decreases, the volume must increase by a factor of $\frac{730}{69}$.

$$V_2 = V_1 \times T_{\text{factor}} \times P_{\text{factor}}$$

$$V_2 = V_1 \times \frac{T_2}{T_1} \times \frac{P_1}{P_2}$$

$$V_2 = 2.0 \times 10^4 \text{ L} \times \frac{228 \text{ K}}{293 \text{ K}} \times \frac{730 \text{ torr}}{69 \text{ torr}}$$

$$V_2 = 1.64 \times 10^5 \text{ L}$$

PROBLEM 8.5

A scuba diver releases a 25-mL bubble at the bottom of a cold lake, where the total pressure and temperature are 5.6 atm and 10 °C, respectively. Calculate the volume of the bubble when it reaches the surface, where the temperature is 23 °C and the pressure is 1.0 atm.

8.4 Avogadro's Hypothesis and the Ideal Gas Law

Avogadro's hypothesis, first proposed in 1811, states that equal volumes of gases under the same conditions of temperature and pressure contain the same number of submicroscopic particles. For example, equal volumes of hydrogen and oxygen at 1 atm pressure and 0 °C (273 K), called **standard temperature and pressure** or **STP,** contain the same number of molecules. A mathematical statement of Avogadro's hypothesis is given by the following expression, where n represents the number of moles of gases identified by subscripts 1 and 2:

$$\frac{V_2}{V_1} = \frac{n_2}{n_1}$$

Avogadro's hypothesis can be used to determine the relative masses of atoms and molecules. Because equal volumes of gases under the same conditions of temperature and pressure contain the same number of submicroscopic particles, the masses of the gas particles must be in the same ratio as the gas densities. The densities of the monatomic gases helium and argon at standard temperature and pressure are 0.179 and 1.79 g/L, respectively, indicating that one argon atom is 10 times as massive as one helium atom.

The molecular weight of an unknown gaseous compound can be determined by using the mole concept and Avogadro's hypothesis. The volume occupied by 1 mole (32 g) of oxygen molecules at standard temperature and pressure is 22.4 L. Since equal volumes of gases contain the same numbers of particles, 22.4 L must be the volume occupied by 1 mole of any gaseous substance at standard temperature and pressure. Thus 22.4 L at STP is the **molar volume** of any gas.

Avogadro's hypothesis can be combined with the gas laws into a more general expression known as the **ideal gas law:**

$$\frac{PV}{T} = k = nR \quad \text{or} \quad PV = nRT$$

where n is the number of moles of a gas, and R is the **universal gas constant.** To three significant figures, the value for R is 0.0821, L atm mole^{-1} K^{-1}.

EXAMPLE 8.6

A mass of 1.34 g of a gas occupies 2.00 L at 91°C and 0.500 atm. What is the molecular weight of the gas?

Solution

The number of moles in this sample is

$$n = \frac{PV}{RT} = \frac{0.500 \text{ atm} \times 2.00 \text{ L}}{0.0821 \text{ L atm mole}^{-1} \text{K}^{-1} \times 364 \text{ K}} = 0.0335 \text{ mole}$$

The 0.0335 mole of the gas has a mass of 1.34 g. Thus, the mass of 1 mole is

$$\frac{1.34 \text{ g}}{0.0335 \text{ mole}} = 40.0 \text{ g/mole}$$

PROBLEM 8.6

A gas used in a car air conditioner has a density of 7.3 g/L at 1.5 atm and 27 °C. What is the molecular weight of the gas?

8.5 Dalton's Law of Partial Pressures

John Dalton studied the properties of gaseous mixtures. Dalton's **law of partial pressures** states that the sum of the partial pressures of all the gases in a gas mixture is equal to the total pressure. Dalton's law expressed mathematically is

$$P_{\text{total}} = P_1 + P_2 + P_3 + \cdots$$

where P_1, P_2, P_3, and so forth, are the partial pressures of the gases in the mixture. The **partial pressure** of a gas is equal to the pressure it would exert if it were the only gas present under the experimental conditions. For a specific volume at a given temperature, the partial pressure of a specific gas is related to the number of moles of that gas in the mixture.

$$P_1 = n_1 \times \frac{RT}{V}$$

Thus, the total pressure of a gaseous mixture is related to the total number of moles of all gases. Furthermore, we see that the ratio of the partial pressures of two gases is equal to the ratio of the number of moles of those gases, which in turn can be related to the volume percent composition of the mixture.

$$\frac{P_2}{P_1} = \frac{n_2}{n_1}$$

Partial Pressures of Atmospheric Gases

Air consists mostly of nitrogen and oxygen, with smaller amounts of water vapor and carbon dioxide. The total pressure exerted by air is the sum of the partial pressures of the individual gases. At 1 atm (760 torr), the partial pressures of nitrogen, oxygen, and water vapor in air of moderate humidity are approximately 595, 159, and 6 torr, respectively. The 159 torr pressure of oxygen is 20.9% of the total atmospheric pressure. At higher altitudes, the total pressure of air decreases, but the percent composition remains the same. Thus, the partial pressures of all the components of air decrease. For example, at an altitude of 1 mile, the total pressure is 630

Partial Pressures in Medicine

A high partial pressure of oxygen can be detrimental to the body. For example, newborn infants who are placed in oxygen atmospheres for a long period of time may suffer damage to the retinal tissue and become either partially or totally blind. On the other hand, if too little oxygen is available to body tissues, even for brief periods, coma and/or irreversible tissue damage may occur. Brain tissue is especially sensitive to oxygen levels.

Respiratory therapists administer gases to treat a number of diseases. For example, if a patient has difficulty breathing, a respirator is used to increase the rate of breathing and thus to increase the volume of oxygen reaching the patient's lungs. Increasing the partial pressure of oxygen above its normal value in air may also be a prescribed treatment. The extra oxygen can be provided by means of an oxygen tent or a face mask. The proper partial pressure of water vapor must be maintained in the gas mixture. We exchange water vapor from our membranes as we breathe. If the air we inhale is too dry, the membranes tend to dry out.

A patient may be placed in a hyperbaric chamber that contains air or additional oxygen at higher than atmospheric pressure. The resulting larger partial pressure of oxygen is useful in treating gangrene and in enhancing the effect of radiation in cancer therapy. Gangrene is caused by bacteria whose growth is inhibited by oxygen. In the case of cancer therapy, the cancerous cells, which grow more rapidly than normal cells, are destroyed more effectively by X rays under high partial pressures of oxygen. The effect of the X rays on normal cells is proportionately less.

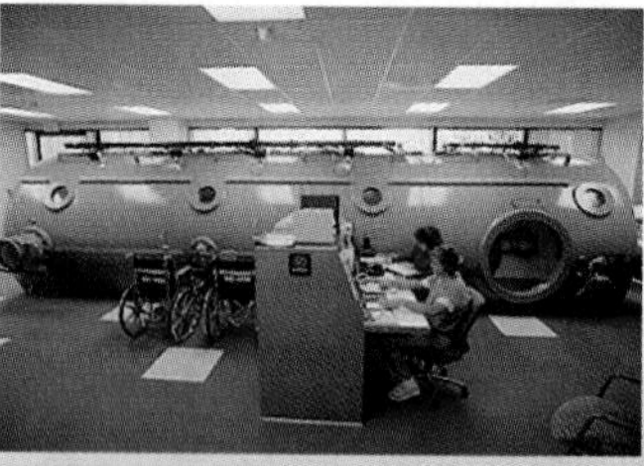

Some hyperbaric chambers must be large enough for patients and medical personnel. (Gregory B. Dimijian, Photo Researchers, Inc)

torr, and the partial pressures of nitrogen, oxygen, and water vapor are 494, 132, and 4 torr, respectively.

Our bodies function efficiently if the partial pressure of oxygen is about 160 torr. Thus, an individual accustomed to living at sea level will not operate as efficiently and will tire more easily at higher altitudes. At the higher altitude the oxygen component of air is still about 21%, but the partial pressure is less than 160 torr. As a result of the lower partial pressure of oxygen, the red blood cells absorb a smaller amount of oxygen. However, if the person stays at the higher altitude for some time, the body adapts by producing more red blood cells.

EXAMPLE 8.7

The compressed air in a scuba diver's tank is approximately 21% oxygen, as it is in air at atmospheric pressure. At a depth of 100 feet the diver's regulator delivers air at 4 atm because that is the pressure on the diver's body at that depth. What is the partial pressure of oxygen in atmospheres?

Solution

The partial pressure of oxygen in air at atmospheric pressure is proportional to the percentage of oxygen—that is, 0.21 atm. When air is compressed, the percentage of oxygen is still 21%. Thus, the partial pressure of oxygen when the total air pressure is 4 atm is approximately 0.8 atm.

PROBLEM 8.7

Air that is exhaled (expired) differs from normal air. The amount of oxygen in it is 15% by volume. What is the partial pressure of oxygen in expired air if the atmospheric pressure is 740 torr?

8.6 Liquids

Molecules are much closer to one another in the liquid state than in the gaseous state. This state results because the attractive forces are sufficiently strong to restrict the tendency of molecules to move apart. The properties of liquids depend on these forces, which in turn reflect the molecular structure. In this section we examine several physical properties of liquids that can be used to propose models for intermolecular attractive forces, which are outlined in Section 8.9.

Compressibility

Compressibility is a measure of the decrease in volume of a state of matter that results from an increase in pressure. In contrast with gases, liquids are essentially incompressible. If the pressure on a 1000-mL sample of a gas at 20 °C is increased from 1 atm to 100 atm, the volume is reduced to 10 mL. When the pressure on a 1000-mL sample of water at 20 °C is increased from 1 atm to 100 atm, the volume is reduced to only 997.6 mL because the molecules are already so close that very little free space remains. It is for this reason that a liquid is used to transmit pressure applied by a foot on the brake pedal through the brake lines to the brakes. If liquids were compressible, the force applied on the brake pedal would just compress the liquid with no effect on the brakes.

Viscosity

The effect of viscosity of honey is seen when it is poured. (Diane Schiumo, Fundamental Photographs)

Each liquid has a resistance to flow, a phenomenon known as **viscosity.** For example, the antifreeze that you pour into your car radiator flows more slowly than the gasoline that you pour into a lawn mower. This difference is explained by a model in which intermolecular forces tend to hold molecules together and make them more resistant to flow. Polar molecules have strong intermolecular attractive forces and tend to flow more slowly than nonpolar molecules having the same general molecular shape. Consider the viscosity at 20 °C of the following three organic liquids with similar molecular weights and similar molecular shapes. Although the unit of viscosity is a rather unusual Newton-second · meter^{-2} (N s m^{-2}) we need recognize only that an increase in the number tells us that the viscosity increases.

	$CH_3CH_2CH_2CH_2CH_3$	$CH_3CH_2OCH_2CH_3$	$CH_3CH_2CH_2CH_2OH$
	pentane	diethyl ether	1-butanol
molecular weight	72 amu	74 amu	72 amu
viscosity	0.240 N · s · m^{-2}	0.233 N · s · m^{-2}	2.948 N · s · m^{-2}

We can conclude from this information that the oxygen atom bonded to two carbon atoms in diethyl ether doesn't significantly affect the viscosity of the compound, because the viscosities of pentane and diethyl ether are similar. However, the —OH group of 1-butanol greatly increases the viscosity. In Section 8.9 we will learn how this group is responsible for strong intermolecular attractive forces.

The shape of molecules also affects viscosity. Complex molecules don't "slide" past each other as well as small symmetrical molecules do because they entangle one another.

Surface Tension

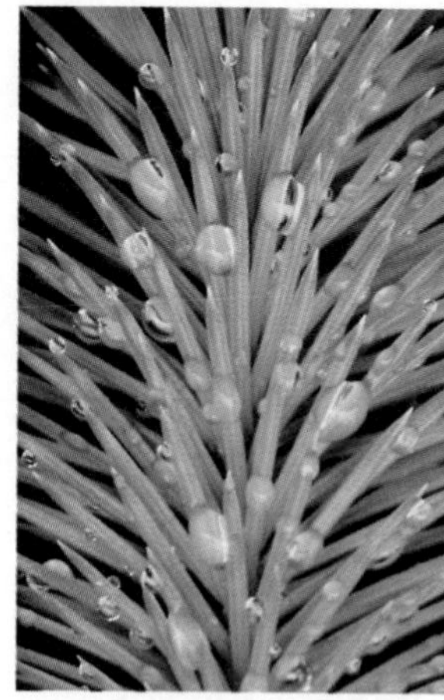

The surface tension of water holds water molecules together in raindrops on this fir branch. (Charles Krebs, Tony Stone Images)

The **surface tension** of a liquid is a property of a liquid that prevents it from spreading out on a surface. We are familiar with this phenomenon in water. Water beads up and tends not to spread on a newly waxed car or other clean surfaces. You may also have seen how a needle can "float" if it is carefully placed on the surface of water. However, when pushed below the surface, the needle sinks. In both cases water behaves as if it has a film on the surface that resists penetration by other materials.

Molecules in the interior of a liquid are attracted to other molecules in all directions (Figure 8.8). Forces between molecules on the surface are not as balanced as the forces between molecules in the interior of the liquid. As a result, there is an uneven distribution of forces on the surface molecules that makes the liquid "pull together" and assume a shape with the smallest possible surface area.

High surface tension indicates that the molecules in a liquid have strong intermolecular attractive forces. Surface tension is expressed in units of J m^{-2}. Again, as in the case of viscosity, we need consider only the changes in the value to compare surface tensions. Water has a very high surface tension, 0.0729 J m^{-2} at 20°C, indicative of very strong intermolecular attractive forces. The surface tensions of the following three compounds at 20°C convey the same trend in intermolecular attractive forces suggested by viscosity measurements. Pentane and diethyl ether have similar intermolecular attractive forces. The molecules of 1-butanol have stronger intermolecular attractive forces.

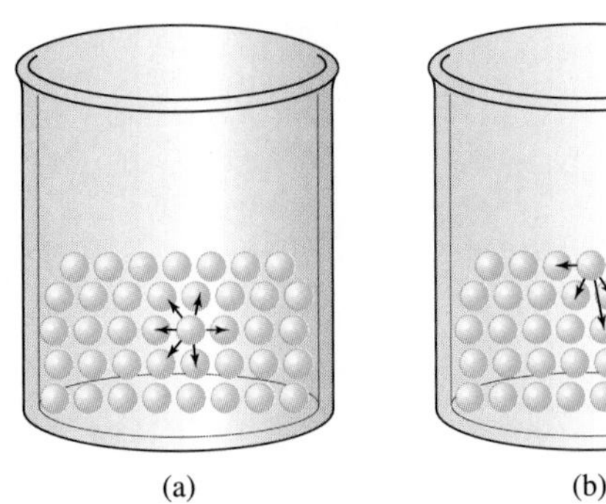

FIGURE 8.8 Model for Surface Tension

(a) Molecules in the interior of the liquid experience intermolecular attractive forces in all directions. (b) Molecules at the surface are pulled into the liquid by intermolecular forces because there are no balancing forces between the surface and the air.

	$CH_3CH_2CH_2CH_2CH_3$	$CH_3CH_2OCH_2CH_3$	$CH_3CH_2CH_2CH_2OH$
	pentane	diethyl ether	1-butanol
molecular weight	72 amu	74 amu	72 amu
surface tension	0.016 J m^{-2}	0.017 J m^{-2}	0.0246 J m^{-2}

At higher temperature, the surface tension decreases because the molecules have higher kinetic energy, which counteracts the attractive forces. The surface tension of water decreases from 0.073 at 20 ° C to 0.068 J m^{-2} at 50 ° C.

Evaporation

Evaporation is a phenomenon by which a liquid is slowly converted into a gas. In a liquid, individual molecules travel at different speeds. Those with high velocity have enough kinetic energy to pull away from the attractive forces of other molecules and enter the gas phase. As the high kinetic energy molecules depart, the average kinetic energy of the molecules remaining in the liquid phase decreases. As a result, the temperature of the liquid decreases.

If you have stood in a breeze immediately after emerging from a swimming pool, you know about the cooling effect of evaporation. The water molecules that leave the surface of your skin most readily are the most energetic particles, and the remaining liquid is cooler. If the rate of evaporation is slow, the temperature of a liquid stays constant because the liquid gains heat from the surroundings. For example, a glass of water in a room slowly evaporates without any noticeable cooling. As the most energetic molecules leave the liquid phase, heat is transferred from the surroundings to the liquid, and the temperature is maintained.

Evaporation of liquids is slower at lower temperatures because the average kinetic energy of the molecules is smaller. This is the reason why it is more difficult to dry clothes outdoors on a cool day than on a hot day.

At the same temperature, different liquids evaporate at different rates. For example, gasoline evaporates faster than lubricating oil. Because the average kinetic energies of the particles in two different liquids at the same temperature are identical, this implies that the escaping tendency of the molecules depends on the attractive forces between neighboring molecules. If the attractive forces between neighboring molecules are large, the ability of the particles to escape is smaller.

The molecules of a liquid in a closed container still leave the liquid phase and enter the gas phase. As the number of molecules in the gas phase increases, they are more likely to collide with the liquid surface and return to the liquid phase. Eventually the rates at which the particles leave and return to the liquid phase become equal (Figure 8.9). The system is in equilibrium, and no net change is ob-

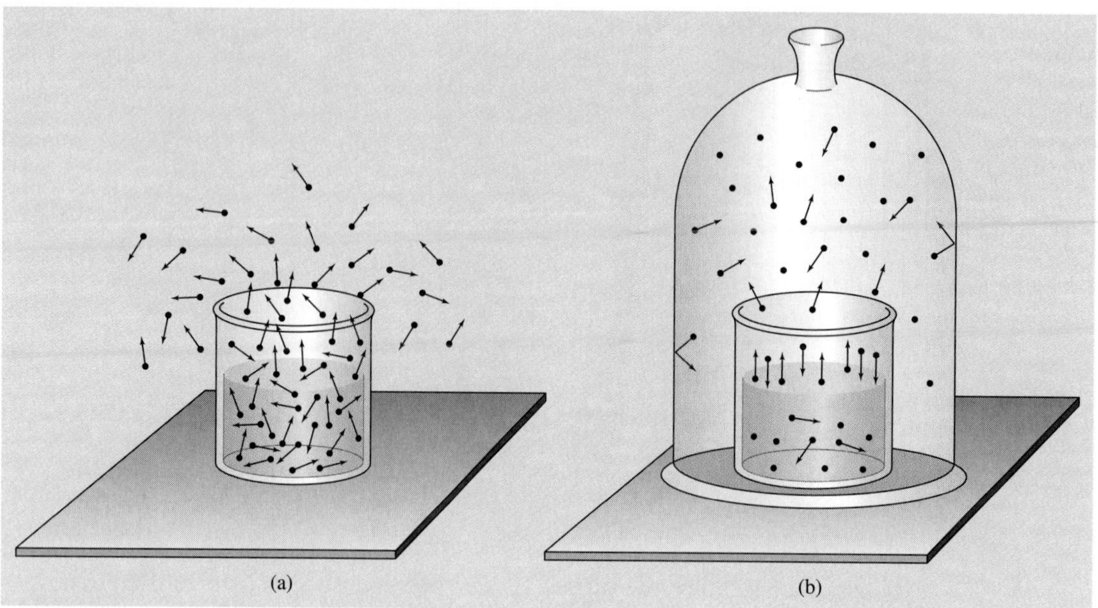

FIGURE 8.9
Evaporation of a Liquid and Vapor Pressure
(a) The molecules that enter the gas phase tend to move away from the container and not return to the liquid phase. As a consequence, all the liquid will eventually evaporate. (b) Molecules of liquid leave the liquid phase and enter the gas phase, and some molecules leave the gas phase and return to the liquid phase. The pressure exerted by the gaseous molecules in equilibrium with the liquid is the vapor pressure.

served on a macroscopic level, although evaporation and condensation are still occurring at the submicroscopic level.

$$\text{liquid} \underset{\text{condensation}}{\overset{\text{vaporization}}{\rightleftharpoons}} \text{gas}$$

Vapor Pressure

Like any other gas, "vapor" from a liquid exerts a pressure. The pressure of a gas in equilibrium with its liquid phase is the **vapor pressure** of the liquid. The vapor pressure is a physical property that indicates the tendency of molecules in the liquid phase to escape. The vapor pressure increases as the temperature increases.

Although the average kinetic energies of various liquids at the same temperature are equal, all liquids do not have the same vapor pressure at the same temperature. Figure 8.10 shows a graph of the vapor pressures of diethyl ether ($C_4H_{10}O$), ethyl alcohol (C_2H_6O), and water. At any temperature, the order of vapor pressures is diethyl ether > ethyl alcohol > water. The molecular weights of diethyl ether, ethyl alcohol, and water are 74, 46, and 18 amu, respectively. Clearly, the observed vapor pressure does not depend on molecular weight. If it did, the order of vapor pressures would be water > ethyl alcohol > diethyl ether, the reverse of what it actually is.

The observed order of vapor pressures depends on the attractive forces between particles which affect the ease with which molecules can escape from the liquid phase. Water molecules attract one another very strongly, and their vapor pressure is low even though water has a low molecular weight. We will return to discuss this

FIGURE 8.10 The Relationship between Vapor Pressure of Liquids and Temperature

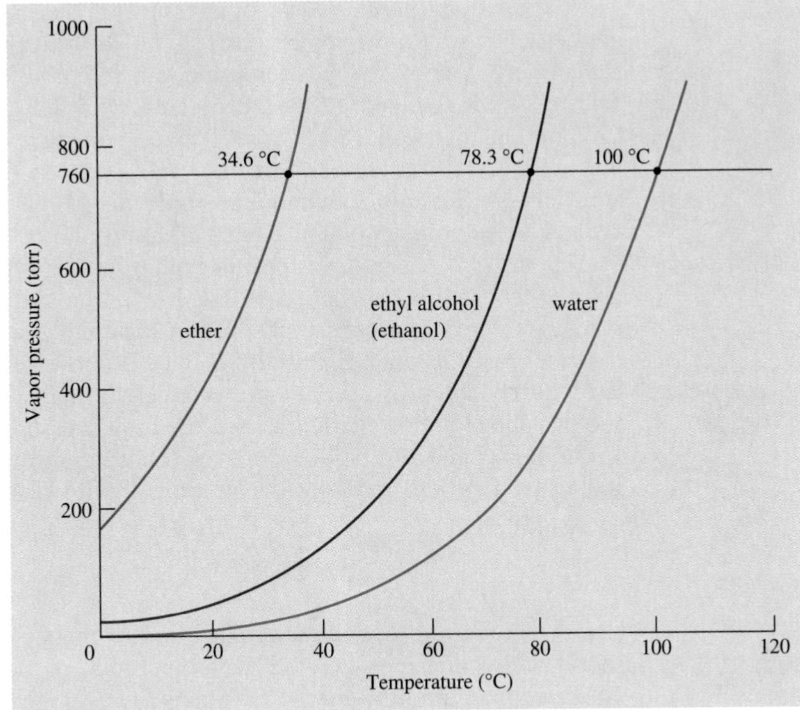

very common but unusual liquid in later sections. Ethyl alcohol also has strong attractive forces between molecules, and it has a lower vapor pressure than diethyl ether, which has a higher molecular weight but weaker attractive forces.

EXAMPLE 8.8

The vapor pressures of methyl alcohol (CH_3OH) and methyl iodide (CH_3I) at 25 °C are 120 and 400 mm Hg, respectively. Interpret these data in terms of the intermolecular forces that are present in each compound.

Solution

The molecular weights of methyl alcohol and methyl iodide are 32 and 142 amu, respectively. Although we might expect the lower molecular weight compound to have the higher vapor pressure, and the higher molecular weight compound to have the lower vapor pressure, the opposite is true. Methyl alcohol must have much stronger intermolecular forces than methyl iodide.

PROBLEM 8.8

The vapor pressures of $GeCl_4$ and $GeBr_4$ are 760 and 40 mm Hg, respectively, at 84 °C. What conclusions can you draw about the intermolecular forces in these two compounds?

Boiling Point

When a liquid is heated, at a certain temperature it undergoes a pronounced transformation. Bubbles form throughout the liquid, rise rapidly to the surface, burst, and release vapor in large quantities. This process is called **boiling,** and the temperature at which it occurs is called the **boiling point** of the liquid.

This hiker in the Grand Tetons will wait a longer time at a high altitude to cook food in boiling water. (C. Seghers, Photo Researchers)

At the boiling point, the vapor pressure of the liquid equals the atmospheric pressure. For water, the temperature at which the vapor pressure equals 760 mm Hg is 100 °C. Water, or any other liquid, can boil at other temperatures if the external pressure is changed. The vapor pressure of water at 80 °C is 355 mm Hg. If we lived on a planet with a normal atmospheric pressure of 355 mm Hg, water would boil at 80 °C. At this temperature, the vapor pressure of the liquid would equal the atmospheric pressure. To avoid ambiguity, the standard or **normal boiling point** is defined as the boiling point at 1 standard atmosphere.

Anyone who has cooked food in boiling water at high altitudes knows that it takes longer than at low altitudes. The higher the altitude, the lower is the boiling point of water because the atmospheric pressure is less. It is necessary to use cooking times twice as long at altitudes of 7000 ft above sea level. Food can be cooked faster at the high pressures in a pressure cooker. When the pressure gauge is set for 5 lb/in.^{2} above atmospheric pressure, the total pressure inside the pressure cooker is 19.7 lb/in.^{2} and the boiling point of the water in the pressure cooker is about 108 °C. Under these conditions, food cooks about twice as fast as at 100 °C.

Heat of Vaporization

Energy is required to maintain boiling and to change a liquid into a gas. The heat energy added does not increase the temperature of the liquid at the boiling point. The continued input of energy maintains the average kinetic energy of the liquid. The heat energy required to transform 1 g of a substance at its boiling point from a liquid into a gas is its **heat of vaporization.** The **molar heat of vaporization** (in calories per mole) is the product of the heat of vaporization and the molar mass. This quantity is used to compare the intermolecular attractive forces of molecules. The high heat of vaporization of water (Table 8.1) reflects the strong attractive forces among water molecules.

EXAMPLE 8.9

The heats of vaporization of ethyl alcohol and diethyl ether are 204 and 84 cal/g, respectively, and the respective molecular weights are 46 and 74 amu. Calculate and compare the heats of vaporization of 1 mole of each liquid. What do these values indicate about the intermolecular forces in these two compounds?

Solution

The heat of vaporization per mole is obtained by multiplying the heat of vaporization per gram by the molecular weight.

$$\frac{204\text{ cal}}{1\,\cancel{g}} \times \frac{46\,\cancel{g}}{1\text{ mole}} = 9.4 \times 10^3\text{ cal/mole} \qquad \text{(ethyl alcohol)}$$

$$\frac{84\text{ cal}}{1\,\cancel{g}} \times \frac{74\,\cancel{g}}{1\text{ mole}} = 6.2 \times 10^3\text{ cal/mole} \qquad \text{(diethyl ether)}$$

On a mole basis, the heat of vaporization of ethyl alcohol is larger than that of diethyl ether. Thus, the intermolecular forces must be larger for ethyl alcohol than for diethyl ether.

PROBLEM 8.9

The heats of vaporization of ethyl mercaptan ($CH_3—CH_2—SH$) and dimethyl sulfide ($CH_3—S—CH_3$) are 6728 and 6742 cal/mole, respectively. What do these values indicate about the strengths of the intermolecular forces in these two compounds?

TABLE 8.1 Normal Boiling Points and Heats of Vaporization of Liquids

Substance	Boiling Point (K)	Molar Mass	Heat of Vaporization (cal/g)	Molar Heat of Vaporization (kcal/mole)
argon	87	39.9	39.1	1.56
krypton	121	83.8	25.8	2.16
xenon	166	131.3	23.0	3.02
methane	112	16.0	122	1.95
ammonia	240	17.0	329	5.60
water	373	18.0	540	9.73

8.7 Solids

When heat energy is added to a solid, the temperature increases until the solid starts to melt. The temperature at which added heat energy melts the solid without raising the temperature of the solid or liquid is called the **melting point.** At the melting point, the solid and liquid states exist in equilibrium. Particles from the solid state, which are in ordered arrangements, escape and enter the more random and mobile liquid state, while particles from the liquid may be deposited on the surface of the solid.

$$\text{solid} \underset{\text{freezing}}{\overset{\text{melting}}{\rightleftharpoons}} \text{liquid}$$

This figure skater depends on the pressure of her blade to glide gracefully over the ice. (David Ulmer, Stock Boston)

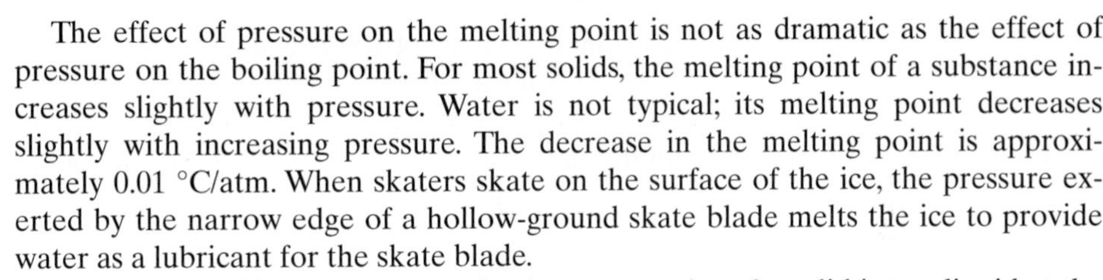

The effect of pressure on the melting point is not as dramatic as the effect of pressure on the boiling point. For most solids, the melting point of a substance increases slightly with pressure. Water is not typical; its melting point decreases slightly with increasing pressure. The decrease in the melting point is approximately 0.01 °C/atm. When skaters skate on the surface of the ice, the pressure exerted by the narrow edge of a hollow-ground skate blade melts the ice to provide water as a lubricant for the skate blade.

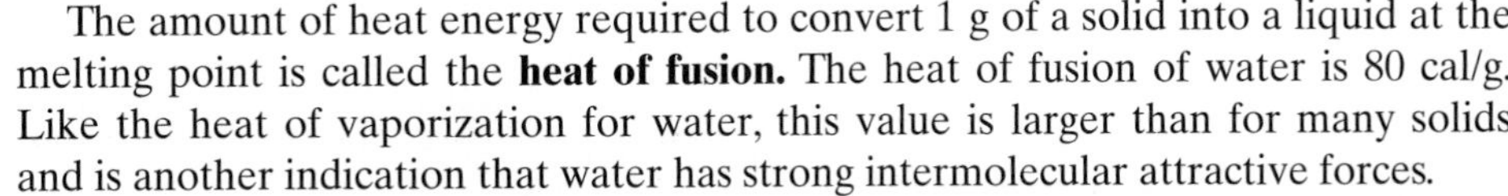

The amount of heat energy required to convert 1 g of a solid into a liquid at the melting point is called the **heat of fusion.** The heat of fusion of water is 80 cal/g. Like the heat of vaporization for water, this value is larger than for many solids and is another indication that water has strong intermolecular attractive forces.

For substances of similar molecular weight, those with the higher melting points have the stronger intermolecular forces. However, other factors also affect the melting point of a solid. Foremost among these is the packing—the geometric arrangement of one particle with respect to its neighbors. The shapes of molecules affect how closely they can approach each other in the solid state. Substances with molecules that can pack closely together will have high melting points.

8.8 Le Châtelier's Principle and Changes of State

In 1888 the French chemist Henri Le Châtelier proposed a simple generalization to explain how equilibrium systems are affected by changes in conditions. A summary of his theory is known as **Le Châtelier's principle:** *If an external stress is applied to* *a system in equilibrium, the system readjusts to achieve equilibrium by reducing the stress applied to it.* You have experienced this effect. Consider how you react to temperature changes. If the temperature of your surroundings drops, your body responds by shivering to produce heat. If the temperature increases, your body cools itself by perspiring and driving off high-energy water molecules.

Le Châtelier's principle explains the effect of pressure on the boiling point of a liquid and the melting point of a solid. Exerting pressure on matter makes it occupy a smaller volume. When a liquid is boiled to form a gas, there is a large increase in volume. Applying pressure causes the equilibrium system of a liquid and its vapor at its normal boiling point to shift toward the liquid state, which has a lower volume. Therefore, to boil a liquid at higher pressures, higher temperatures are necessary.

$$\text{liquid} \underset{\text{increased pressure}}{\overset{\text{decreased pressure}}{\rightleftharpoons}} \text{gas}$$

Most substances are about 10% more dense as solids than as liquids. Applying pressure causes an equilibrium system of liquid and solid to shift toward the solid state because the solid occupies the smaller volume. Therefore, to melt most solids under pressure, higher temperatures are necessary. An increase in pressure increases the melting point. Usually only small increases in melting points are observed, owing to the small difference in volume between solids and liquids.

$$\underset{\text{less dense}}{\text{liquid}} \underset{\text{decreased pressure}}{\overset{\text{increased pressure}}{\rightleftharpoons}} \underset{\text{more dense}}{\text{solid}}$$

The abnormal behavior of water is due to the lower density of ice compared with that of water. Solid water occupies a larger volume than liquid water. Thus, the volume increases as water becomes ice, and an increase in pressure shifts the ice–water equilibrium toward water. In other words, an increase in pressure decreases the melting point of water.

$$\underset{\text{smaller volume}}{\text{water}} \underset{\text{increased pressure}}{\overset{\text{decreased pressure}}{\rightleftharpoons}} \underset{\text{larger volume}}{\text{ice}}$$

8.9 Types of Intermolecular Forces

Now that we have considered some experimental observations on the behavior of the states of matter, we can interpret this information using models for intermolecular forces. The energy of these forces is usually less than 10 kcal/mole. Intermolecular forces in general are called **van der Waals forces,** after the Dutch scientist Johannes van der Waals (1837–1923) who studied the effect of high pressure on the gaseous state. It is convenient to subdivide van der Waals forces into three categories: London forces, dipole–dipole forces, and hydrogen bonding forces. Regardless of the type, all three forces are the result of attraction of sites of opposite charge in neighboring atoms or molecules.

London Forces

Although nonpolar substances such as argon have no permanent charged sites, they can be liquefied by decreasing the temperature. On average, the electrons in a nonpolar molecule or atom are distributed uniformly around the nucleus. However, at some instant the electrons may be closer to one side of an atom. The result is a **temporary dipole** (Figure 8.11). A temporary dipole affects the electron distribution of nearby atoms. The ease with which an electron cloud can be distorted by nearby charges is called **polarizability.** The temporary dipole polarizes neighboring atoms, forming an **induced dipole.** The attractive forces between a temporary dipole and an induced dipole are called **London forces.**

In the case of molecules, the strength of London forces depends on the number of electrons and the distance from the nuclei of the constituent atoms as well as on the size and shape of the molecule. The polarizability of the molecule as a whole increases with the number of electrons and their distance from the nucleus. Consider the boiling points of the nonpolar molecules fluorine, chlorine, and bromine, which are −188, −35, and 59°C, respectively—they increase with increasing molecular weight. The London forces are stronger for the liquid with larger molecules because the electrons of the larger atoms are farther from the nuclei and are more polarizable. These forces hold the larger molecules together more strongly, and therefore the temperature required to boil the larger molecules is higher.

The shape of molecules can also affect the boiling point. Consider the boiling points of pentane and 2,2-dimethylpropane, two organic compounds with the same composition but different structures.

$$CH_3{-}CH_2{-}CH_2{-}CH_2{-}CH_3$$

pentane
(bp 36 °C)

$$CH_3{-}C(CH_3)_2{-}CH_3$$

2, 2-dimethylpropane
(bp 10 °C)

Pentane molecules have an elongated, cylindrical shape, but 2,2-dimethylpropane is a spherically shaped molecule (Figure 8.12). The more spread-out shape of the pentane molecule allows more contact between neighboring molecules. The more compact shape of the spherical 2,2-dimethylpropane molecule provides only limited points of contact with neighboring molecules. The weaker London forces in

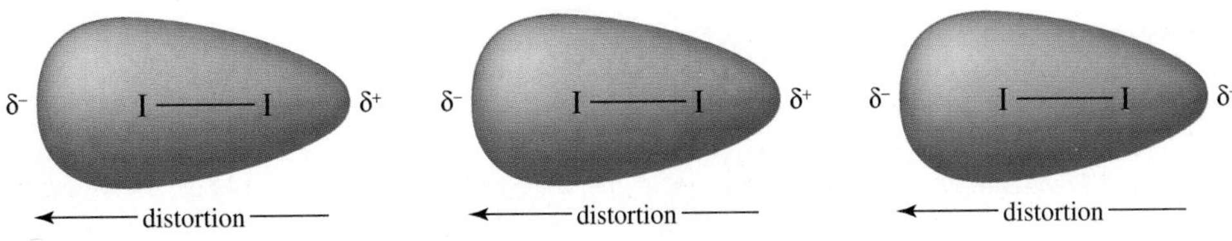

FIGURE 8.11 London Attractive Forces
The electron distribution in an atom or molecule becomes distorted, and a temporary dipole results. A dipole is induced in adjacent atoms or molecules by the redistribution of electrons. There is a net attraction between the temporary dipoles. Molecules containing atoms with large atomic radii such as iodine are easily polarized.

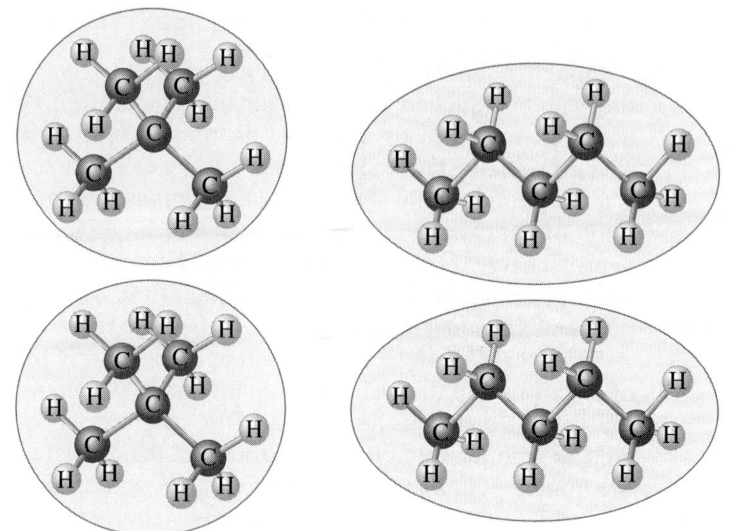

FIGURE 8.12 Shapes of Molecules and Intermolecular Attractive Forces *The shape of 2.2-dimethylpropane is nearly spherical. Neighboring molecules cannot approach each other closely, and the points of contact are limited. Pentane is an extended molecule with a cylindrical shape, so neighboring molecules can approach each other side-by-side. As a consequence the intermolecular attractive forces are stronger.*

liquid 2,2-dimethylpropane are the reason for its lower boiling point compared with pentane.

Dipole-Dipole Forces

Intermolecular attractive forces between polar molecules are easier to visualize than London forces. A polar molecule has a negative end and a positive end, that is, it has a dipole (Section 5.7). As a consequence, the positive end of one molecule attracts the negative end of another molecule, and vice versa. This interaction between dipoles of neighboring molecules is a **dipole–dipole force.**

The molecular weight of ICl (162 amu) is close to the molecular weight of Br_2 (160 amu); yet the boiling point of ICl is 97 °C, which is 38 °C higher than that of Br_2. In Br_2 the bonding pair of electrons is shared equally in a nonpolar covalent bond between the two atoms. In ICl, the bonding pair of electrons is more strongly attracted toward chlorine, the more electronegative element, and the bond is polar covalent. As a result, the ICl molecules are strongly attracted to one another by intermolecular forces between dipoles (Figure 8.13). To boil ICl, it is necessary to

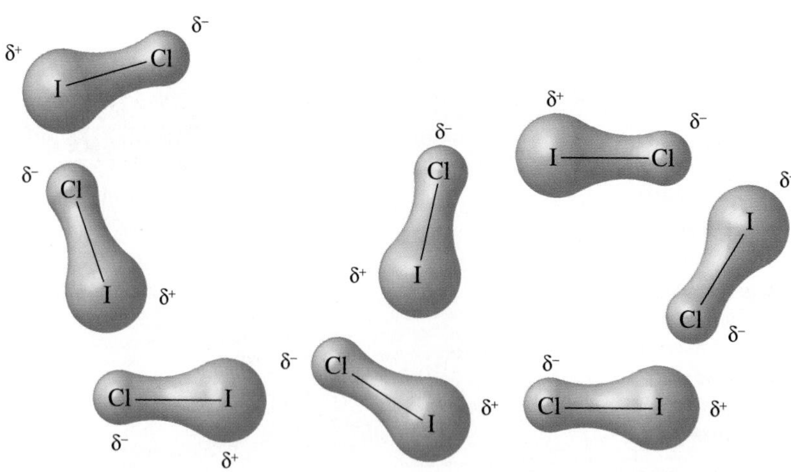

FIGURE 8.13 Intermolecular Attractive Forces *The iodine atom is attracted to the chlorine atom of a neighboring ICl molecule. The dipole is the result of a difference in the electronegativities of the two atoms.*

heat the molecules to a higher temperature than is needed for a nonpolar molecule of comparable molecular weight such as Br_2.

Dipole–dipole forces are responsible for many of the physical properties of organic compounds containing electronegative atoms. For example, the boiling points of 2-methylpropane and acetone are substantially different.

$$\underset{\substack{\text{2-methylpropane} \\ \text{(bp } -12\ ^\circ\text{C)}}}{CH_3{-}\overset{\overset{\displaystyle CH_3}{|}}{C}H{-}CH_3} \qquad \underset{\substack{\text{acetone} \\ \text{(bp 56 }^\circ\text{C)}}}{CH_3{-}\overset{\overset{\displaystyle O}{\|}}{C}{-}CH_3}$$

These two substances have the same molecular weight, and the molecules have similar shapes (Figure 8.14). Thus the London forces in each substance are about the same. However, acetone has a polar carbon–oxygen double bond, and the molecule is polar. The dipole–dipole attraction between positive and negative sites of neighboring molecules is responsible for the higher boiling point of acetone.

EXAMPLE 8.10

The boiling points of liquid krypton and liquid argon are −152.9 and −185.8 °C, respectively. Suggest a reason for this difference, and predict the boiling point of xenon.

Solution

The atomic numbers of krypton and argon are 36 and 18, respectively. The electrons in krypton are in higher energy levels and farther from the nucleus than are the electrons in argon. As a result, the electrons in krypton are more polarizable, and stronger London forces result. One would predict the boiling point of xenon to be still higher. It is −108.1 °C.

PROBLEM 8.10

The boiling points of nitrogen (N_2) and carbon monoxide (CO) are 77 and 83 K, respectively. Suggest a reason for this difference in boiling points.

Hydrogen Bonding Forces

Compounds that contain a hydrogen atom bonded to a fluorine, oxygen, or nitrogen atom have unusually high intermolecular forces. The compounds NH_3, H_2O, and HF all have much higher boiling points than CH_4.

	CH_4	NH_3	H_2O	HF
boiling point	− 161 °C	− 33 °C	100 °C	− 19.5 °C

These three molecules have two structural features in common. (1) Each has at least one hydrogen atom joined by a polar covalent bond to a very electronegative atom. (2) Each has at least one unshared pair of electrons on the electronegative

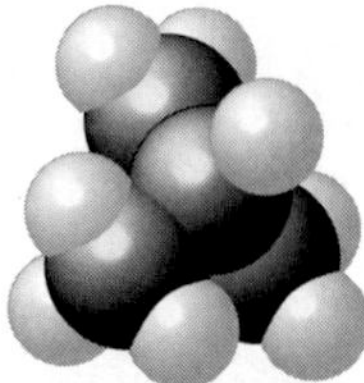

2-methylpropane

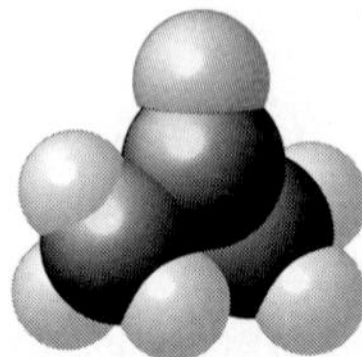

acetone

FIGURE 8.14 Molecular Models of 2-Methylpropane and Acetone

The shapes of the two molecules are similar. However, acetone has a polar carbon–oxygen bond.

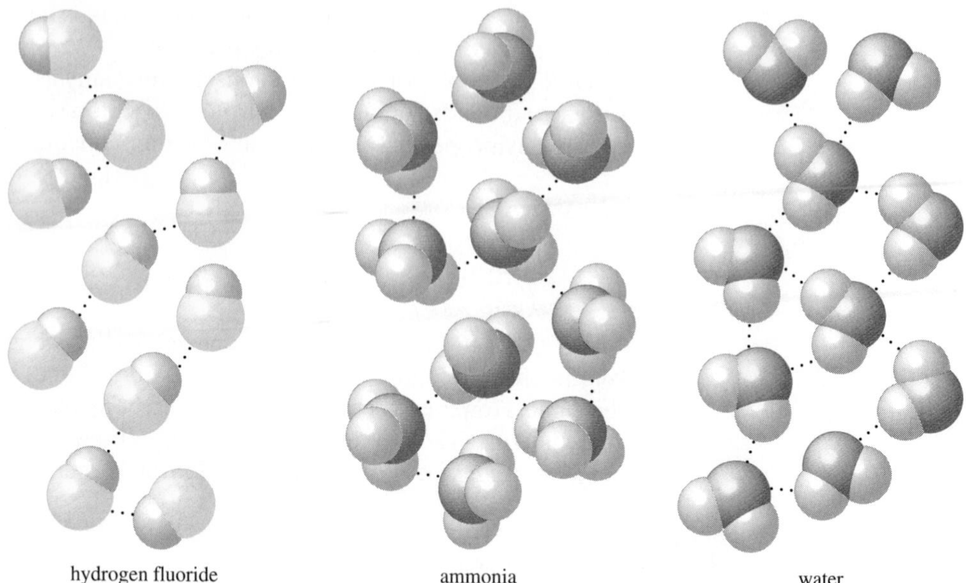

FIGURE 8.15 Molecules That Form Hydrogen Bonds
Hydrogen fluoride is limited in forming hydrogen bonds by its single hydrogen atom, in spite of its three nonbonding electron pairs. Ammonia is limited in forming hydrogen bonds by its single electron pair, in spite of its three hydrogen atoms. Only water has an equal number of hydrogen atoms and nonbonding electron pairs.

atom (Figure 8.15). The very small hydrogen atom has a partial positive charge because the bonding electrons are attracted toward the electronegative element to which it is bonded. An intermolecular attraction called a **hydrogen bond** exists when a hydrogen atom of one molecule is attracted to a nonbonding electron pair of a highly electronegative atom of a neighboring molecule.

$$\overset{\delta+}{\mathrm{H}}-\overset{\delta-}{\ddot{\underset{..}{\mathrm{F}}}}\!:\cdots\overset{\delta+}{\mathrm{H}}-\overset{\delta-}{\ddot{\underset{..}{\mathrm{F}}}}\!:$$

hydrogen bond

The carbon–hydrogen bond of methane is essentially nonpolar. In addition, there are no nonbonding electron pairs on carbon. As a consequence, methane does not form intermolecular hydrogen bonds. Thus, CH_4 has a substantially lower boiling point than NH_3, H_2O, or HF.

Hydrogen bonds are quite strong, on the order of 3–10 kcal/mole. The positive charge on the hydrogen atom decreases as the electronegativity difference between the hydrogen atom and the bonded atoms decreases. Thus, you might predict that the order of boiling points should be $HF > H_2O > NH_3$. However, that order is incorrect—both HF and NH_3 boil at a lower temperature than H_2O. The reason for the apparent anomaly is that both the HF molecule and the NH_3 molecule can form only one hydrogen bond per molecule (Figure 8.15). The fluorine atom has three unshared electron pairs but only one hydrogen atom. The ammonia (NH_3) molecule has three hydrogen atoms but only one unshared electron pair. However, water has two hydrogen atoms and two unshared electron pairs per molecule, so the extent of aggregation of H_2O molecules is greater than that of HF or NH_3 molecules. Water has the proper balance of hydrogen atoms and electron pairs to form the maximum number of hydrogen bonds. As a consequence it has a much higher boiling point.

Hydrogen bonding affects the physical properties of organic molecules as well. For example, the boiling points of ethanol and propane are 78 and − 42 °C respectively. The molecular weights and shapes of the two compounds are similar. However, ethanol is a polar molecule and is expected to have a higher boiling point.

$$CH_3-CH_2-CH_3 \qquad CH_3-CH_2-OH$$

propane (bp −42°C) ethanol (bp −78.5 °C)

The difference in the boiling points of propane and ethanol is due to more than the polarity of the carbon–oxygen bond in ethanol. Consider the boiling points of acetone and isopropyl alcohol. The polarity of the carbon–oxygen double bond of acetone is larger than the polarity of the carbon–oxygen single bond of isopropyl alcohol. Yet, isopropyl alcohol has the higher boiling point.

$$CH_3-\overset{\overset{\displaystyle O}{\|}}{C}-CH_3 \qquad CH_3-\overset{\overset{\displaystyle OH}{|}}{CH}-CH_3$$

acetone (bp 56 °C) isopropyl alcohol (bp 82 °C)

Isopropyl alcohol has an —OH group that can form hydrogen bonds via its hydrogen atom and the unshared pairs of electrons on the oxygen atoms of neighboring molecules. Acetone molecules, although polar, cannot form hydrogen bonds with one another because no hydrogen atom is attached to oxygen. The hydrogen bonding in isopropyl alcohol is responsible for the higher boiling point. As a result, we can also conclude that ethanol has a substantially higher boiling point than propane because of hydrogen bonding in ethanol (Figure 8.16).

Many molecules of biological importance contain O—H or N—H groups that can form hydrogen bonds to water or with each other within the molecule. Both proteins and DNA form hydrogen bonds that affect their molecular shapes.

FIGURE 8.16 Molecular Models and Hydrogen Bonding
Ethanol can form hydrogen bonds between the oxygen atom of one molecule and the hydrogen atom of another molecule. The resulting attractive forces are responsible for the relatively high boiling point of ethanol.

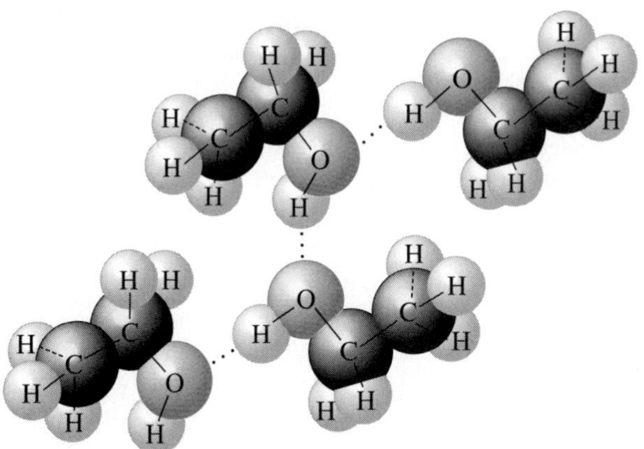

EXAMPLE 8.11

The boiling points of ethyl alcohol and dimethyl ether are 78.5 and − 23 °C, respectively. The compounds have the same molecular formula, C_2H_6O, but different structures. Explain the large difference between the boiling points of the two compounds.

$$H-\underset{\underset{\displaystyle H}{|}}{\overset{\overset{\displaystyle H}{|}}{C}}-\underset{\underset{\displaystyle H}{|}}{\overset{\overset{\displaystyle H}{|}}{C}}-O-H \qquad H-\underset{\underset{\displaystyle H}{|}}{\overset{\overset{\displaystyle H}{|}}{C}}-O-\underset{\underset{\displaystyle H}{|}}{\overset{\overset{\displaystyle H}{|}}{C}}-H$$

ethyl alcohol dimethyl ether

Solution

Ethyl alcohol has an—OH group as part of its structure. This group can form hydrogen bonds via its hydrogen atom and the unshared pairs of electrons on the oxygen atoms of neighboring molecules. Dimethyl ether molecules cannot form hydrogen bonds with one another because there is no hydrogen atom attached to oxygen. The hydrogen bonding in ethyl alcohol is responsible for the higher boiling point.

PROBLEM 8.11

The boiling points of ethyl mercaptan and dimethyl sulfide are 35 and 37 °C, respectively. The compounds have the same molecular formula, C_2H_6S, but different structures. Why is there very little difference between the boiling points of the two compounds?

```
   H  H              H     H
   |  |              |     |
H—C—C—S—H        H—C—S—C—H
   |  |              |     |
   H  H              H     H
```

ethyl mercaptan   dimethyl sulfide

Water—An Unusual Compound

At 100°C the density of water is 0.958 g/mL. The density increases as the temperature decreases until at 4 °C the density of water is 1.0000 g/mL, which is its maximum density. Below 4°C, as the graph shows, the density decreases until the density of 0.99987 g/mL is reached at 0°C.

Ice at 0°C has a density of 0.917 g/mL and is less dense than liquid water at 0°C. For that reason ice floats in water, displacing its own weight. The volume of ice that remains floating above the surface of the water is approximately 8%. Since 92% of a mass of ice is below the surface, the visible part of an iceberg is indeed only "the tip of the iceberg."

The facts that ice is less dense than water and that the maximum density of liquid water occurs at 4°C rather than at 0°C have profound ecological significance. As the temperature of the water near the surface of a lake is lowered in cold weather, the cooler and denser water sinks toward the bottom of the lake. The warmer water in the lake rises to the top because it is less dense. This circulation of water continues until the overall water temperature is 4°C. Further cooling results

The density of water is a maximum at 4 °C. The density decreases from 4 °C to 0 °C.

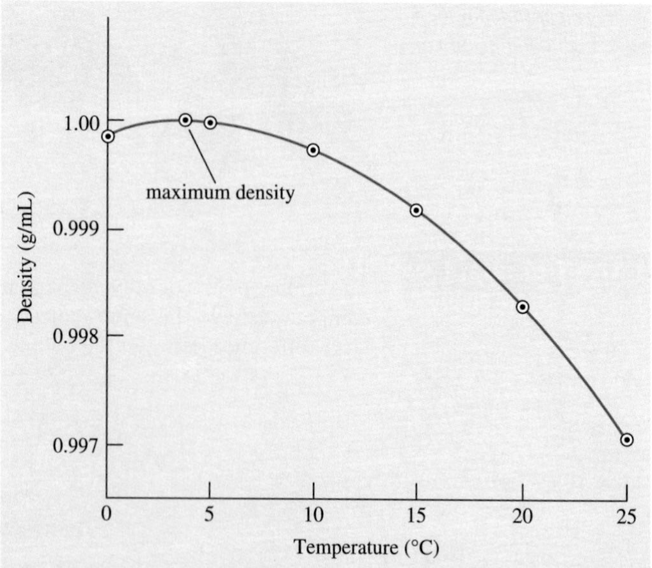

The danger of icebergs lies beneath the surface of the ocean. (Delphine Star, Tony Stone Images)

in a lower density, so the colder water remains at the surface and no longer sinks to the bottom of the lake. Thus, the warmest water is 4°C at the lake bottom.

The water at the surface eventually freezes. Since ice is less dense than water, the ice remains on the surface. If it were more dense, it would sink to the bottom. A continued cold spell would then cause the lake to freeze from the bottom to the top, and living organisms in the lake would not survive. The lower density of ice results in a sheet of ice that protects the water beneath it from further decreases in temperature. The ice thickness slowly increases, but the water temperature beneath the ice is maintained comfortably enough for aquatic life to survive.

The change in the density of water at temperatures below 4°C and the lower density of ice compared with that of liquid water are due to the hydrogen bonding in water. Most liquid materials become more dense as the temperature decreases and become still more dense when the solid state is formed. For liquids the decrease in temperature lowers the kinetic energy and allows more of the matter to become aggregated in "clumps" of molecules. In the solid state the molecules get as close as possible and leave a minimum of free space.

In water the aggregation has some unique structural requirements as a result of hydrogen bonding. As long as the "clumps" are small, as is the case at higher temperatures, water behaves like other liquids and becomes more dense as the temperature decreases. However, at 4°C a pattern of molecules emerges as the number of hydrogen bonds is increased. Then, as freezing occurs, the pattern of hydrogen-bonded molecules crystallizes. This pattern is an open cagelike network as shown. "Holes" in the cages make ice less dense than water.

Structure of Ice and the Hydrogen Bond

The three-dimensional structure shows each covalent oxygen–hydrogen bond as a solid line. The hydrogen bonds are shown by colored dotted lines. Each oxygen atom in the structure has two covalent bonds and two hydrogen bonds. Each hydrogen atom is covalently bonded to one oxygen atom and hydrogen-bonded to another oxygen atom. The empty space created by the structure accounts for the lower density of ice.

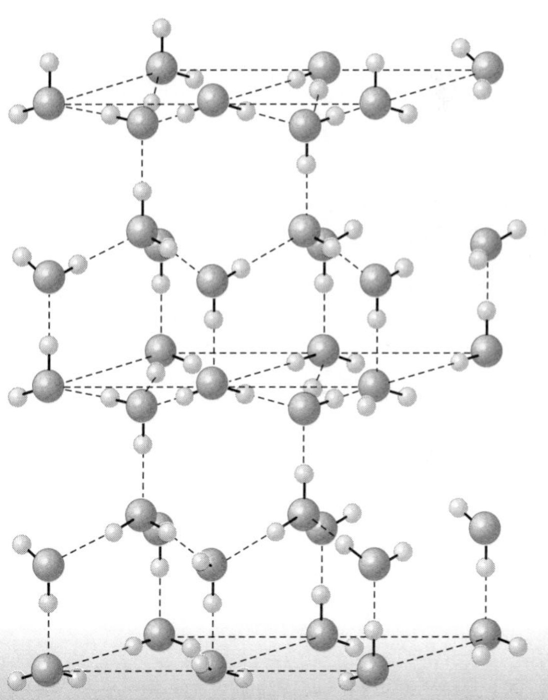

Summary

Pressure is force per unit area. Gas pressure, which is measured using a **barometer,** can be expressed in units of **atmospheres, centimeters of mercury (cm Hg),** and **millimeters of mercury (mm Hg)** or **torr.**

The model for understanding the physical properties of matter is the **kinetic molecular theory.** As applied to gases the theory describes an **ideal gas. Real gases** behave according to the theory only at high temperatures and low pressures.

A dynamic **equilibrium** exists between the states of matter at the **melting point** and **boiling point.** Changes in state reflect the **intermolecular forces** of the molecules and their kinetic energy. When the temperature of a gas is decreased, **condensation** occurs to form a liquid; when the temperature of a liquid is increased, **vaporization** occurs to form a gas. When the temperature of a liquid is decreased, **freezing** occurs to form a solid; when the temperature of a solid is increased, **melting** occurs to form a liquid.

Relationships among a pair of the four gas variables (P, V, T, and n), the other two remaining constant, are defined by **Boyle's law,** which relates pressure and volume; **Charles's law,** which relates volume and temperature; **Gay-Lussac's law,** which relates pressure and temperature; and **Avogadro's hypothesis,** which relates volume and the amount of the gas. The individual gas laws can be combined into one comprehensive gas law. By using a constant called the **universal gas law constant,** a general expression known as the **ideal gas law** is obtained. The condition of 0 °C and 1 atm is **standard temperature and pressure (STP).** The volume of a mole of a gas at STP is the **molar volume,** 22.4 L/mole.

Dalton's **law of partial pressures** applies to the properties of mixtures of gases, each of which exerts a **partial pressure.** Any gas in a mixture can be treated using any of the gas laws.

Liquids are incompressible. The **viscosity** and **surface tension** of liquids depend on the strength of intermolecular attractive forces. A liquid has a characteristic **vapor pressure** that depends on the temperature. When the vapor pressure equals the atmospheric pressure, boiling occurs. The **normal boiling point** of a liquid is the temperature at which the vapor pressure equals 1 atm. The energy required to transform 1 g of a liquid to a gas at its boiling point is the **heat of vaporization** of the liquid.

Pressure exerts little effect on the melting point. The energy required to transform 1 g of a solid into a liquid is the **heat of fusion** of the substance.

Intermolecular forces are of three types: dipole–dipole forces, London forces, and hydrogen bonds. **Dipole–dipole forces,** the attraction of the permanent dipoles of molecules, are the consequence of the unequal distribution of electrons in the bonds between atoms of different electronegativities. **London forces,** the attraction between a **temporary dipole** and an **induced dipole,** are the consequence of the **polarizability** of electrons in atoms and molecules. **Hydrogen bonds** are the attraction between a bonded hydrogen atom and an electron pair on a highly electonegative atom of a neighboring molecule.

Le Châtelier's principle provides an explanation for the physical changes of the states of matter. Systems at equilibrium change to a new equilibrium condition to reduce the stress caused by an external change. Under pressure, a gas is converted to a liquid, which has a smaller volume. The equilibrium between the liquid and the solid states is less sensitive to pressure because there is only a small difference in the volume of the substance in the two states. Water is an anomalous liquid; it has a larger volume in the solid state than in the liquid state as a result of its ability to form networks of hydrogen bonds.

Exercises

Pressure

8.1 Why is the pressure at the top of Denali (formerly Mt. McKinley) 350 torr?

8.2 Why is the pressure in a mine a mile underground greater than that on the surface of the Earth?

8.3 Convert each of the following pressures into the indicated unit.

(a) 380 mm Hg into atm
(b) 0.500 atm into cm Hg
(c) 0.100 atm into torr
(d) 190 torr into cm Hg
(e) 13.5 lb/in.^{2} into atm
(f) 9.8 lb/in.^{2} into cm Hg

8.4 Convert each of the following pressures into the indicated unit.

(a) 0.750 atm into mm Hg
(b) 38 cm Hg into atm
(c) 520 torr into atm
(d) 19.0 cm Hg into torr
(e) 0.90 atm into psi
(f) 57 cm Hg into psi

8.5 The pressure outside an airplane at 35,000 ft is 222 mm Hg. What is the pressure in atmospheres?

8.6 The pressure at the top of Mount Everest is 4.9 psi. What is the pressure in atmospheres?

Boyle's Law

8.7 A 50.0-L cylinder containing oxygen gas at 175 atm is used for respiration therapy. What volume will the oxygen occupy when gradually released to a patient at 1.00 atm if the temperature is unchanged?

8.8 A scuba tank has a volume of 13.2 L. What volume of air in liters at 720 torr is required to fill the tank to a pressure of 150 atm, assuming no change in temperature?

8.9 Scuba divers are told not to hold their breath and to breathe continuously. Assume that a diver holds his breath at a depth of 66 feet in the ocean, where the total pressure is 3 atm. The diver's lungs have a volume of 5 L. What volume of air would be released by the diver after returning to the surface, assuming that his lungs are inelastic and don't burst? (Incidentally, both of these assumptions are incorrect).

8.10 What pressure in atmospheres is required to compress 1.0×10^3 L of methane gas at 1.00 atm into a 45-L cylinder?

Charles's Law

8.11 On a cold day, a person takes in 425 mL of air at − 10°C. Assuming that the pressure in the lungs is the same as that of the atmosphere, what is the volume of this air in the lungs at 37°C?

8.12 Charle's law can be demonstrated by drawing air into a large syringe that can be sealed. What will the volume be if 25 mL of air at 25°C is drawn into the syringe and then it is placed in boiling water at 100°C and the plunger moves freely?

8.13 A 300-mL sample of helium at 127°C is cooled to 150 K. What is its volume if the pressure is kept constant?

8.14 A sample of argon gas in a 150-cm^3 cylinder exerts a pressure of 820 mm Hg at 25°C. At what temperature will the volume decrease to 125 cm^3 if the pressure is kept constant?

Gay-Lussac's Law

8.15 The pressure in a gas storage tank is 7.0 atm on a summer day when the temperature is 35°C. The tank is left undisturbed, and the pressure drops to 6.0 atm on a winter day. What is the temperature?

8.16 The pressure in an automobile tire is 44.7 psi at 27°C at the start of a trip. At the end of the trip the pressure is 47.7 psi. What is the temperature of the tire?

8.17 A storage container of a gas is filled at 4.0 atm at 20°C. What will be its pressure in a warehouse on a hot day when the temperature is 35°C?

8.18 Scuba divers are told not to leave their air tanks in the trunk of a car on a hot day. If the air tank is filled to a pressure of 2900 psi at 20°C, what will be the pressure in the tank in a trunk at 50°C?

Combined Gas Law

8.19 The pressure in a natural gas well with a volume of 1.0×10^9 L is 1.2 atm at 42°C. If the gas is transferred to tanks in which the pressure cannot exceed 5.0 atm at 20°C, what will be the total volume of the required tanks?

8.20 The pressure at the top of the troposphere is about 0.20 atm, and the temperature may be − 50°C. What will be the volume of a weather balloon at that point if its volume was 100.0 L at 1.00 atm at ground level, where the temperature is 20°C?

8.21 A 5.0-mL bubble of swamp gas is released from the bottom of a swamp where the temperature is 10°C and the pressure is 3.5 atm. What will be the volume of the bubble at the surface at 1 atm and 25°C?

8.22 The temperature and pressure in the stratosphere, where the ozone layer occurs, are 250 K and 1.0×10^{-3} atm, respectively. What will be the volume of a liter of stratospheric gas when the conditions are changed to 20°C and 1.0 atm?

Avogadro's Hypothesis and Molar Volume

8.23 Calculate the volume of each of the following samples at STP, and determine how many moles of gas are present in the sample.

(a) 1.12 L at 2.00 atm and 0°C
(b) 2.99 L at 1.00 atm and 91°C
(c) 560 mL at 5.00 atm and 182°C

8.24 Calculate the volume of each of the following samples at STP and determine how many molecules of gas are present in the sample.

(a) 2.24 L of H_2 at 2.00 atm and 0°C
(b) 5.98 L of O_2 at 1.00 atm and 91°C
(c) 280 mL of CO_2 at 5.00 atm and 182°C

8.25 How many molecules of sulfur dioxide (SO_2) are contained in a 1-mole sample of the gas at STP? How many molecules are in 1 L at STP?

8.26 How many molecules of carbon dioxide (CO_2) are contained in a 1-mole sample of the gas at STP? How many molecules are in 1 L at STP?

Ideal Gas Law

8.27 A hydrocarbon sample weighing 0.185 g occupies 110 cm^3 at 25 °C and 74 cm Hg. What is the molecular weight of the gas?

8.28 The mass of a 51-mL sample of a diatomic gas at 10 °C and 725 torr is 58 mg. What is the atomic weight of the element of the diatomic gas?

8.29 The atmosphere of Venus is mostly carbon dioxide. The pressure is 100 atm and the temperature is 460 °C. How many moles of carbon dioxide are in 1.0 L under these conditions?

8.30 The volume of a neon sign is 1250 mL. It is filled at 25 °C to a pressure of 1.8 torr. How many moles of neon are in the sign?

8.31 The ozone molecules in the stratosphere absorb ultraviolet radiation from the sun. The pressure of ozone (O_3) is 1.5×10^{-7} atm, and the temperature is − 20 °C. How many molecules are in 1.0 L of ozone under these conditions?

8.32 A pressure of 1.0×10^{-3} torr is obtained by using a vacuum pump. Calculate the number of molecules in 1.0 mL of gas at this pressure at 20 °C.

Dalton's Law of Partial Pressures

8.33 The partial pressure of nitrogen on Denali is 288 torr on a day when the atmospheric pressure is 0.480 atm. What is the partial pressure of oxygen?

8.34 The pressure on Mount Everest is 265 mm Hg. What is the partial pressure of oxygen, assuming that the composition of the air is 20% oxygen by volume?

8.35 A scuba diver at a depth of 60 ft in the ocean breathes air at 3 atm. The air is about 20% oxygen and 80% nitrogen by volume. What is the partial pressure of oxygen?

8.36 The excessive absorption of nitrogen in the blood of divers can cause nitrogen narcosis, otherwise known as "rapture of the deep." This phenomenon can occur at 130 ft, where the pressure is 5 atm. The diver receives air from her regulator at the existing external pressure. What is the partial pressure of nitrogen that the diver breathes?

8.37 Oxygen toxicity occurs if a partial pressure of oxygen of about 1 atm is breathed for too long. At what total pressure of compressed air delivered to a diver can toxicity occur? Air is 20% oxygen.

8.38 To avoid nitrogen narcosis and oxygen toxicity, special gas mixtures are breathed by researchers and navy personnel at great depths in the ocean. For example, a 2% oxygen and 98% helium mixture may be selected for work at 300 ft, where the total pressure is 10 atm. What is the partial pressure of oxygen at this depth?

Liquids

8.39 The vapor pressures of $SbCl_3$ and $SbBr_3$ are 100 and 10 mm Hg, respectively, at 145 °C. Which substance should evaporate faster at 145 °C?

8.40 The vapor pressure of chloroform is 400 mm Hg at 43°C, and the vapor pressure of carbon tetrachloride is 400 mm Hg at 58 °C. Which substance will evaporate faster at 43 °C?

8.41 The vapor pressure of $SnCl_4$ is 400 mm Hg at 92 °C. The vapor pressure of SnI_4 is 400 mm Hg at 315°C. Which substance should have the higher boiling point?

8.42 The vapor pressure of methanol is 400 mm Hg at 50 °C, and it is 760 mm Hg at 65 °C. What can one conclude about the atmospheric pressure on a day that methanol is boiling at 60 °C?

8.43 The boiling points of $SiCl_4$ and SiF_4 at 1 atm are 56.8 and − 94.8 °C, respectively. What do these values indicate about intermolecular forces in these two compounds?

8.44 The boiling point of CS_2 is 46.5 °C. What can be said about the intermolecular forces in CS_2 compared with those in CO_2, which is a gas at 46.5 °C?

8.45 The heats of vaporization of CCl_4 and CBr_4 are 53.7 and 32.8 cal/g, respectively. Calculate the molar heats of vaporization. Which compound has the larger intermolecular forces?

8.46 The heats of vaporization of CH_3OH and CH_3SH are 294 and 132 cal/g, respectively. Calculate the molar heats of vaporization. Which compound has the larger intermolecular forces?

Solids

8.47 A figure skater skates on only a small portion of the edge of a blade and may exert 300 atm of pressure on the ice. What is the melting point of the ice under this pressure?

8.48 A speed-skating blade is long to provide stability and control. Explain why it is made very thin in order to provide sufficient pressure to melt the ice.

8.49 Mercury thermometers cannot be used below − 39 °C. Explain why.

8.50 If the density of a solid is less than that of the corresponding liquid, what will be the effect of increased pressure on the melting point?

Le Châtelier's Principle

8.51 The melting point of CH_4 is − 182.5 °C at 1 atm. Assuming that solid methane is more dense than liquid methane, how will the melting point of methane be affected by increased pressure?

8.52 The melting point of bismuth is 271.3 °C. The density of liquid bismuth is greater than that of solid bismuth at the melting point. How will the melting point be affected by increased pressure?

Types of Intermolecular Forces

8.53 The surface tensions of ethyl alcohol, ethylene glycol, and glycerol are 0.024, 0.0477, and 0.0634 J m^{-2}, respectively. Explain this trend based on intermolecular forces.

$CH_3—CH_2—OH$
ethyl alcohol

$HO—CH_2—CH_2—OH$
ethylene glycol

$HO—CH_2—CH(OH)—CH_2—OH$
glycerol

8.54 The surface tensions of 2-methylbutane, isobutylamine, and isobutyl alcohol are 0.0137, 0.0179, and 0.023 J m^{-2}, respectively. Explain this trend based on intermolecular forces.

$CH_3—CH(CH_3)—CH_2—CH_3$
2-methylbutane

$CH_3—CH(CH_3)—CH_2—NH_2$
isobutylamine

$CH_3—CH(CH_3)—CH_2—OH$
isobutyl alcohol

8.55 The viscosities of heptane and dipropyl ether are 0.448 and 0.461 $N \cdot s \cdot m^{-2}$, respectively. Explain the similarity of these values based on intermolecular forces.

$CH_3CH_2CH_2CH_2CH_2CH_2CH_3$

heptane

$CH_3CH_2CH_2OCH_2CH_2CH_3$

dipropyl ether

8.56 The viscosities of acetone and isopropyl alcohol are 0.316 and 1.77 $N \cdot s \cdot m^{-2}$, respectively. Explain the difference between these values based on intermolecular forces.

$CH_3-C(=O)-CH_3$ $CH_3-CH(OH)-CH_3$

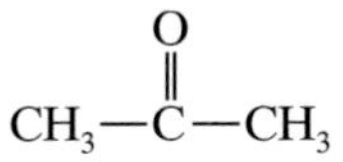

acetone 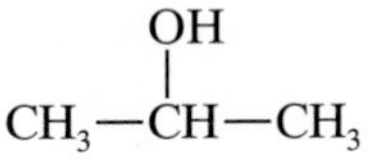   isopropyl alcohol

8.57 The compounds CF_4 and CCl_4 are tetrahedral molecules. Neither is polar. Explain why the boiling points of CF_4 and CCl_4 are − 129 and 76.8 °C, respectively.

8.58 The boiling point of CS_2 is 46.5 °C, whereas CO_2 is a gas at room temperature. What types of intermolecular forces are responsible for this difference?

8.59 The boiling points of ethyl alcohol and ethylene glycol (used in antifreeze) are 78.5 and 190 °C, respectively. Explain the high boiling point of ethylene glycol.

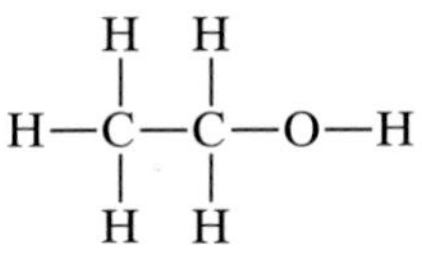

```
    H  H                H  H
    |  |                |  |
 H—C—C—O—H       H—O—C—C—O—H
    |  |                |  |
    H  H                H  H
```

ethyl alcohol 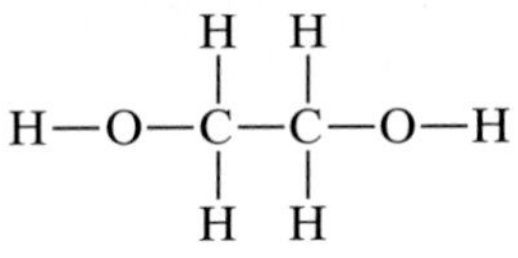   ethylene glycol

8.60 Which of the following can form intermolecular hydrogen bonds in the pure liquid?

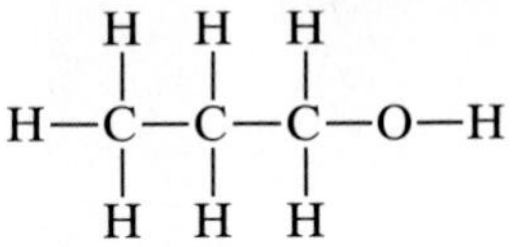

```
    H  H  H               H  H     H
    |  |  |               |  |     |
 H—C—C—C—O—H       H—C—C—O—C—H
    |  |  |               |  |     |
    H  H  H               H  H     H
```

propanol 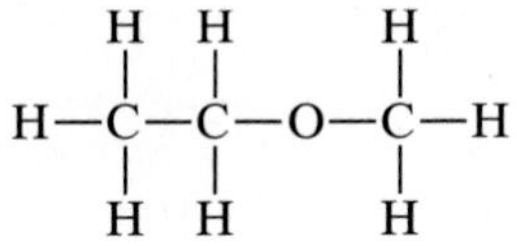   ethyl methyl ether

8.61 The boiling points of propylamine and trimethylamine are 49 °C and 3 °C, respectively. Explain why these values differ for two compounds that have the same composition.

$CH_3-CH_2-CH_2-NH_2$ $CH_3-N(CH_3)-CH_3$

propylamine   trimethylamine

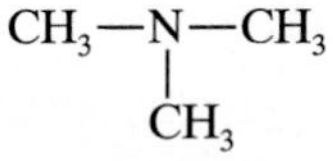

8.62 Arrange the following compounds in order of increasing boiling point.

$CH_3-CH_2-O-CH_2-CH_2-OH$

I

$CH_3-O-CH_2-CH_2-O-CH_3$

II

$HO-CH_2-CH_2-CH_2-CH_2-OH$

III

Solutions

These falls in Yosemite National Park are one of the many examples of solutions that move dissolved materials around our planet. (Robert J. Ouellette)

9.1 The Nature of Solutions
9.2 Types of Solvents and Solutes
9.3 Concentration—The Amount of Solute in Solution
9.4 Percent Concentration
9.5 Molarity
9.6 Dilution of Solutions
9.7 Electrolytes and Nonelectrolytes
9.8 Electrolytes and Body Fluids
9.9 The Formation of a Solution
9.10 Structure and Solubility
9.11 Factors that Affect Solubility
9.12 Colligative Properties
9.13 Osmosis
9.14 Osmosis and Cells

Essays
- *Gout, Kidney Stones, and Solubility*
- *Hot Packs and Cold Packs*
- *Solubility of Vitamins*
- *Gas Solubility and the Bends*
- *Design of Medicines*

9.1 The Nature of Solutions

Up to this point we have focused on the properties of pure substances. However, most of the matter that you encounter in your daily life exists as mixtures. In this chapter we will examine the properties of homogeneous mixtures which are usually called solutions. A solution is uniform all the way down to the submicroscopic level.

Although solutions may contain any number of components, we will concentrate on two-component mixtures. The substance present in the largest quantity in a solution is called the **solvent;** the substance dissolved in the solvent is the **solute.** Among the questions that we will consider about solutions are the following:

1. How does a solute dissolve in a solvent?
2. How does molecular structure affect the amount of solute that can dissolve in a solvent to form a solution?
3. What effect do temperature and pressure have on the amount of solute that can dissolve in a solvent?
4. What effect does the amount of solute dissolved have on the physical properties of the solvent?

Solutions—Around Us and in Us

Solutions abound in our environment and in our bodies. The largest volume of solution in our environment is the ocean, which contains about 1×10^{21} L of water. About 70 elements, as various ions, have been found dissolved in sea water, although many are present only in very small amounts. For example, there are about 3×10^{-6} g of uranium and about 6×10^{-9} g of gold per liter of sea water. The major dissolved constituents in this solution are listed in Table 9.1.

The average person has about 40 L of water as the major component of body fluids (Chapter 30). Blood plasma, which contains our blood cells (red cells and others), and spinal fluid are both solutions. The major cation in blood plasma is Na^+; very small amounts of K^+, Ca^{2+}, and Mg^{2+} are also present. The major anion is Cl^-; some HCO_3^- is also present. Both intracellular (*intra* = within) fluids and extracellular (*extra* = outside) fluids are solutions. The major cation in intracellular fluid is K^+; Mg^{2+} and Na^+ occur in smaller amounts. The major intracellular anions are HPO_4^{2-} and $H_2PO_4^-$ accompanied by smaller amounts of HCO_3^- and SO_4^{2-}.

9.2 Types of Solvents and Solutes

The solvent and solute of a solution may be a gas, liquid, or solid (Figure 9.1). We will briefly discuss each type of solvent before we consider water solutions in detail. A solution of a solute in water is an **aqueous solution.**

All gases mix with each other in all proportions. In gaseous solutions, the molecules are far apart (Figure 9.1) and move independently. Clean dry air is a gaseous solution that consists of approximately 78% nitrogen molecules, 21% oxygen molecules, and 1% argon atoms. Neither liquids nor solids can be solutes in air. Fog, which is small water droplets in air, and smoke, which contains minute particles of solids in air, are colloids, not solutions. Each contains large particles distributed in the air, and these mixtures are heterogeneous.

Gases, liquids, and solids may dissolve in liquid solvents. In liquid solutions the molecules of solute and solvent are closer together than in gaseous solutions (Figure 9.1). Oxygen dissolved in water—a solution of a gas in a liquid—maintains aquatic life. Carbon dioxide in carbonated beverages is another example of a solution of a gas in a liquid. Ethyl alcohol and acetic acid are two common examples of liquids that dissolve in water. Ethyl alcohol is a component of all alcoholic beverages; acetic acid is present in vinegar. Both salt and sugar are examples of solids that dissolve in water.

Solid solutions are less common, but many metals dissolve in one another to form solid solutions called **alloys** (Figure 9.1). Brass (zinc and copper) and sterling

TABLE 9.1 Concentration of Dissolved Ions in Sea Water

Cation	g/L	Anion	g/L
Na^+	10.8	Cl^-	19.3
Mg^{2+}	1.3	SO_4^{2-}	2.7
Ca^{2+}	0.4	HCO_3^-	0.11
K^+	0.4	Br^-	0.067
Sr^{2+}	0.008	F^-	0.001

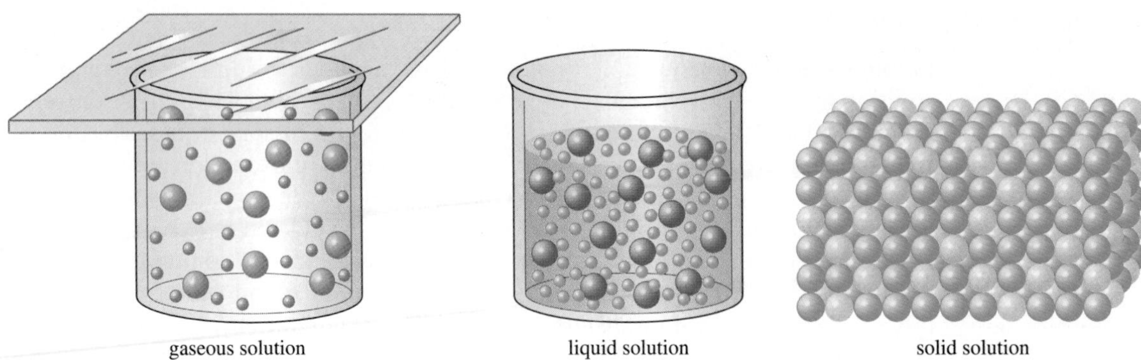

FIGURE 9.1
Solutions and the States of Matter

silver (silver and copper) are alloys. The alloy used to make hip and knee replacements is 67% cobalt, 27% chromium, and 6% molybdenum.

Liquid mercury dissolves in metals to form **amalgams** such as the silver amalgam used in dental work. Hydrogen gas dissolved in metals such as palladium is used to reduce organic compounds known as oils to fats (Section 15.4).

9.3 Concentration—The Amount of Solute in Solution

Concentration is a term that indicates the amount of solute in a given amount of solution or solvent. A **concentrated solution** contains a larger amount of solute per given amount of solvent or solution. A **dilute solution** contains a smaller amount of solute per given amount of solvent or solution. "Concentrated" and "dilute" are relative terms.

Now let's consider how much solute can dissolve in a solvent. When a few milligrams of sodium chloride are added to 10 mL of water, the solid readily dissolves. As more solid sodium chloride is added, a point is reached where no additional salt dissolves. At this point the solution is **saturated.** As long as more solute will dissolve, the solution is **unsaturated.** Thus, the term saturated means "full," whereas unsaturated means that the solution could hold more. The **solubility** of a solute is the concentration of a saturated solution.

Sodium chloride is quite soluble in water: 36 g is required to saturate 100 mL of water at room temperature. On the other hand, only 9.0×10^{-5} g of silver chloride is required to saturate 100 mL of water. The term *saturated* should not be confused with *concentrated,* nor should *unsaturated* be confused with *dilute.* A solution may be "full" yet contain very little material. Solutions of 9.0×10^{-5} g of sodium chloride and 9.0×10^{-5} g of silver chloride in 100 mL of water are both very dilute solutions, but only the silver chloride solution is saturated.

Quantitative expressions of the amount of solute in a solvent or solution are required in chemical laboratories, industries, hospitals, and pharmacies. Units used to express concentration include percent concentration and molarity. Each method has been chosen for convenience under a particular set of circumstances. Percent concentration is used in some clinical situations; molarity is used in chemistry.

Gout, Kidney Stones, and Solubility

Although gout is not a common problem today, you may have heard that the nobility and kings of England, including King Henry VIII, suffered from this ailment. If not, you may know someone who has suffered from kidney stones. Both of these unpleasant conditions are the result of limited solubility and the precipitation of substances in places that the body does not tolerate. Uric acid is such a culprit.

Uric acid is a product of the metabolic breakdown of nucleic acids. Under normal conditions, about 0.5 g of uric acid is excreted daily in urine. The solubility of uric acid is about 7 g/L at body temperature. If too much uric acid is produced, its solubility may be exceeded, causing it to precipitate. If uric acid precipitates in the soft tissue around the joints, the crystals can cause the painful inflammation known as gout. In addition, continued persistent precipitation can cause degradation of bone. There may be a genetic disposition toward gout, but it usually can be controlled by dietary modification that avoids food rich in nucleoproteins which contain nucleic acids.

Kidney stones are caused by precipitation of uric acid and some phosphate and carbonate salts. These stones cause excruciating pain when "passed" through the ureter. Large stones can cause blockage, and still larger stones may have to be treated medically.

9.4 Percent Concentration

The concentration of a solution may be expressed as the percent of the solute in the solution. A 3% solution means 3 parts of solute out of a total of 100 parts of solution. There are, of course, 97 parts of solvent present to make the total 100 parts. Percent concentrations can be expressed in units of mass or volume in ratios such as weight/weight, weight/volume, and volume/volume. Parts per million is a related concentration unit that is used to express low concentrations of solute.

Weight/Weight Percent

The weight/weight percent (w/w %) is defined by the following equation. The weights are usually given in grams.

$$\text{w/w \%} = \frac{\text{weight of solute}}{\text{weight of solute} + \text{weight of solvent}} \times 100\%$$

Weight/Volume Percent

In clinical situations, weight/volume percent (w/v %), which compares the weight of the solute to the total volume of the solution, is used.

$$\text{w/v \%} = \frac{\text{mass of solute}}{\text{volume of solution}} \times 100\%$$

For example, the normal saline solution used for intravenous therapy is 0.9 w/v % aqueous sodium chloride. The solution contains 0.9 g of NaCl in 100 mL of solution. To prepare a liter (1000 mL) of normal saline, 9 g of sodium chloride is weighed out, and enough water is then added to make 1000 mL of solution.

Volume/Volume Percent

Volume/volume percent (v/v %) can be used with liquid solutes. The equation is

$$\textbf{v/v \%} = \frac{\textbf{volume of solute}}{\textbf{total volume of solution}} \times \textbf{100\%}$$

The volume/volume percent concentration states directly the number of milliliters of solute per 100 mL of solution. This method is used to express the concentration of alcohol in beverages. A wine that is 12 v/v % alcohol contains 12 mL of alcohol in 100 mL of solution. Proof, a term that is also used to express the alcohol content in alcoholic beverages, is equal to twice the volume/volume percent of the alcohol. An alcoholic beverage that is 40% alcohol is 80 proof. Since pure alcohol is 100%, it is 200 proof.

EXAMPLE 9.1

A 2.00-mL sample of blood plasma contains 6.80 mg of sodium ions. Calculate the concentration of sodium ions in milligrams per 100 mL.

Solution
First, calculate the weight/volume ratio in the units given.

$$\frac{6.80 \text{ mg sodium}}{2.00 \text{ mL plasma}} = 3.40 \text{ mg/mL}$$

Since there are 3.40 mg of sodium ions in 1 mL of plasma, in a 100-mL volume there would be

$$3.40 \text{ mg/}\cancel{\text{mL}} \times 100 \cancel{\text{mL}} = 340 \text{ mg}$$

Therefore, the concentration is 340 mg/100 mL. The equivalent milligrams per deciliter (mg/dL) is now the preferred unit in clinics.

You may have to look carefully but all wines are labeled with the percent alcohol by volume. (Peter Arnold, Inc)

PROBLEM 9.1

How many grams of tartaric acid ($C_4H_6O_6$) are present in 750 mL of a wine that is 3.0 w/v % tartaric acid?

EXAMPLE 9.2

A solution is prepared by dissolving 200 mL of ethyl alcohol in water to produce 1.000 L of solution. What are the volume/volume percent concentration and proof of the solution?

Solution
Both volumes must be expressed in the same units

$$\frac{200 \text{ mL of ethyl alcohol}}{1000 \text{ mL of solution}} \times 100\% = 20.0\% \text{ alcohol}$$

The proof is twice the concentration, or 40 proof.

PROBLEM 9.2

An intoxicated driver has 0.2 v/v % alcohol (C_2H_5OH) in his blood. How much alcohol is contained in this individual's total 6 L of blood?

Parts per Million

Some solutes are present in such small amounts that decimal percent concentrations like 0.0001% would have to be used. For such dilute solutions, parts per million (ppm) is a convenient concentration unit. One **ppm** is one part of solute per million parts of solution. Like percent concentration, the ppm unit may refer to weight/weight, weight/volume, or volume/volume. Consider a pollutant in a water source at a concentration of 0.0005 g per 100 mL of solution. The weight/volume percent is 0.0005%. However, given in parts per million, the concentration is

$$\frac{0.0005 \text{ g}}{100 \text{ mL}} \times \frac{10^4}{10^4} = \frac{5 \text{ g}}{1{,}000{,}000 \text{ mL}} = 5 \text{ ppm}$$

EXAMPLE 9.3

The U.S. Food and Drug Administration (FDA) has set the human tolerance for mercury in fish at 0.5 ppm. A 5.0-g sample of fish contains 20 μg of mercury. Calculate the concentration of mercury in parts per million, and determine if the FDA tolerance is exceeded.

Solution

First, convert 20 μg into grams so that the masses of mercury and fish are in the same units. The answer (in ppm) will be on a weight/weight basis.

$$20 \cancel{\mu g} \times \frac{1 \text{ g}}{10^6 \cancel{\mu g}} = 20 \times 10^{-6} \text{ g} = 2.0 \times 10^{-5} \text{ g}$$

Now, determine the ratio of grams of mercury to the grams of fish. Multiplication by 10^6 gives the number of grams of mercury per million grams of fish, or parts per million.

$$\frac{2.0 \times 10^{-5} \text{ g Hg}}{5.0 \text{ g fish}} \times 10^6 = 4.0 \text{ ppm}$$

The concentration exceeds the tolerance limit of the FDA.

PROBLEM 9.3

The DDT concentration in the mud of one of the Great Lakes is 0.015 ppm. How many grams of DDT are in 1.0 kg of mud?

9.5 Molarity

A concentration unit known as **molarity,** which is the number of moles of solute per liter of solution, is used in chemistry. The abbreviation for molarity is M.

$$\textbf{molarity} = \frac{\textbf{moles of solute}}{\textbf{liters of solution}} = \textbf{M}$$

A 1.00 M NaCl solution is prepared by dissolving 58.5 g (1.00 mole) of NaCl in sufficient water to give 1.00 L of solution. The molarity of any solution is calculated by dividing the number of moles of solute by the number of liters of solution. A solution of 5.85 g of NaCl in 500 mL of solution is 0.200 M.

$$5.85 \text{ g NaCl} \times \frac{1 \text{ mole NaCl}}{58.5 \text{ g NaCl}} = 0.100 \text{ mole NaCl}$$

$$500 \text{ mL} \times \frac{1 \text{ L}}{1000 \text{ mL}} = 0.500 \text{ L}$$

$$\frac{0.100 \text{ mole NaCl}}{0.500 \text{ L}} = 0.200 \text{ M NaCl}$$

Molarity gives the number of moles contained in a given volume of solution or the volume of solution required to provide a desired number of moles. The number of moles of solute contained in a volume of solution is given by the equation

liters of solution × molarity = moles of solute

$$\textbf{liters} \times \frac{\textbf{moles}}{\textbf{liter}} = \textbf{moles}$$

The volume of solution required to provide a stated number of moles of solute is

$$\textbf{moles of solute} \times \frac{\textbf{1}}{\textbf{molarity}} = \textbf{volume}$$

$$\textbf{moles} \times \frac{\textbf{liters}}{\textbf{mole}} = \textbf{liters}$$

Volumes are measured with volumetric laboratory glassware such as a buret or pipet. Thus, the required amount of a solute can be provided from a calculated volume of solution of known molarity.

EXAMPLE 9.4

Calculate the volume of 0.10 M glucose solution required to provide 45 g of glucose ($C_6H_{12}O_6$).

Solution

First, calculate the number of moles of glucose contained in 45 g.

$$45 \text{ g} \times \frac{1 \text{ mole}}{180 \text{ g}} = 0.25 \text{ mole}$$

Now, calculate the volume that contains 0.25 mole of glucose using the reciprocal of the molarity.

$$0.25 \text{ mole} \times \frac{1 \text{ L}}{0.10 \text{ mole}} = 2.5 \text{ L}$$

PROBLEM 9.4

The maximum permissible amount of Cd^{2+} in drinking water is 0.010 mg/L. What is the maximum permissible molarity of Cd^{2+}?

9.6 Dilution of Solutions

Sometimes a solution of a desired concentration must be prepared by diluting a more concentrated solution. Solvent is added to increase the volume and, as a result, to decrease the concentration of the solute. However, the number of moles of solute is unchanged. Because the product of the volume and molarity gives the number of moles, we can write

$$\mathbf{V_{initial} \times M_{initial} = moles = V_{final} \times M_{final}}$$

EXAMPLE 9.5

If 1.00 L of $AgNO_3$ solution is made by adding 900 mL of water to 100 mL of 2.0 M $AgNO_3$, what is the molarity of the dilute solution?

Solution

The number of moles of $AgNO_3$ is obtained by multiplying the volume by the molarity of the initial solution.

$$0.100\,\cancel{L} \times \frac{2.0 \text{ mole}}{1\,\cancel{L}} = 0.20 \text{ mole } AgNO_3$$

The volume of the diluted solution is 1.00 L, but the number of moles of $AgNO_3$ in the solution is the same as in the concentrated solution. Thus, the concentration is calculated by dividing the number of moles by the final volume.

$$\frac{0.20 \text{ mole}}{1.00 \text{ L}} = 0.20 \text{ M } AgNO_3$$

PROBLEM 9.5

How many milliliters of water must be added to 50 mL of a 0.40 M NaCl solution to obtain a 0.050 M NaCl solution?

9.7 Electrolytes and Nonelectrolytes

Pure water is a nonconductor, as illustrated by the apparatus shown in Figure 9.2. Two electrodes are immersed in water. If water could conduct electricity between the electrodes, the light bulb would glow. The bulb does not glow because pure water is not a conductor. Some aqueous solutions conduct electricity, whereas others do not. Thus, these solutes differ in some way when dissolved in aqueous solutions.

Any substance that forms a solution in water that conducts electricity is an **electrolyte.** When a solution of sodium chloride is used in the apparatus in Figure 9.2, the bulb glows brightly. The explanation for this phenomenon was provided in 1884 by the Swedish scientist Svante Arrhenius, who suggested that substances whose aqueous solutions conduct electricity contain ions in solution. The cations in the solution migrate to the negative electrode, called the **cathode,** while the anions migrate to the positive electrode, called the **anode.** As a result of the movement of ions, charge is conducted through the solution.

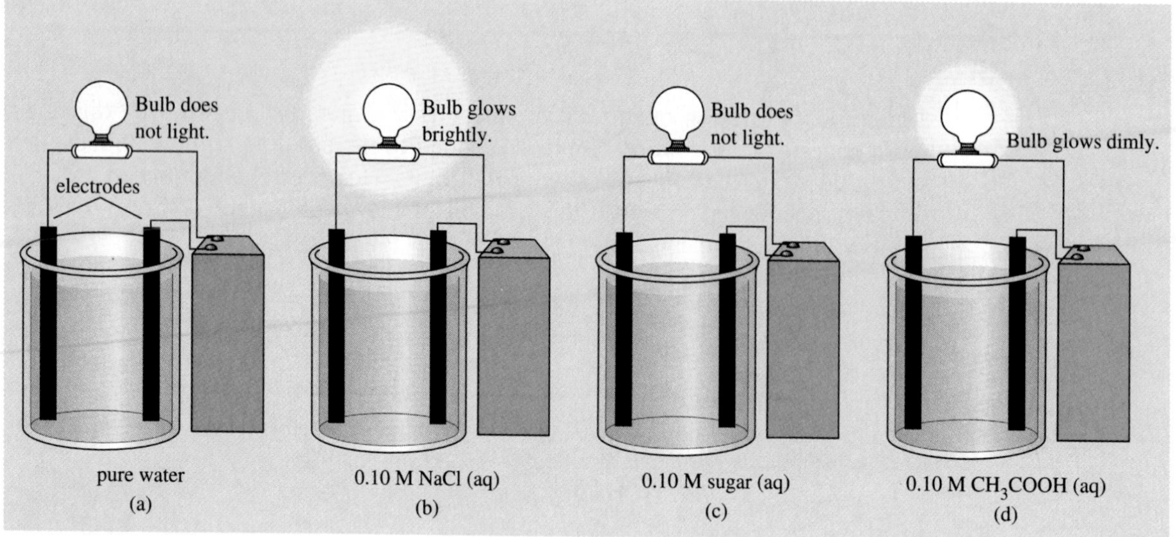

FIGURE 9.2
Electrolytes
(a) Water does not conduct electricity. (b) A solution of sodium chloride does conduct electricity, and the bulb glows brightly. (c) A sugar solution does not conduct electricity. (d) Acetic acid is a weak electrolyte and does not efficiently conduct electricity.

When ionic substances dissolve in water, ions are separated from the crystal structure by the water molecules. The separation of ions from a solid ionic compound such as NaCl is called **dissociation.** Some covalent substances produce ions when they are dissolved in water by a process called **ionization.** For example, the gaseous covalent molecule hydrogen chloride yields ions when dissolved in water.

$$HCl(g) + H_2O(l) \longrightarrow H_3O^+(aq) + Cl^-(aq)$$

Any substance that does not conduct a current when dissolved in water is a **nonelectrolyte.** Ordinary cane sugar, which is sucrose ($C_{12}H_{22}O_{11}$), is a nonelectrolyte. Sucrose does not contain ions in the solid state nor does it produce ions in solution. Ethyl alcohol (C_2H_5OH) is another common nonelectrolyte.

Electrolytes are divided into two classes called strong electrolytes and weak electrolytes. **Strong electrolytes** dissociate or ionize completely, causing the light bulb in the device shown in Figure 9.2 to glow brightly. The degree of brightness is related to the number of ions present in solution. All ionic substances that are soluble in water are strong electrolytes. Some covalent compounds, such as HCl, are also strong electrolytes, indicating that they ionize completely in water. Other covalent compounds, such as acetic acid, do not ionize completely in water; they are **weak electrolytes,** and the bulb in the device glows only dimly.

9.8 Electrolytes and Body Fluids

The concentrations of ions in body fluids are responsible for acid–base balance, control of the nervous system, and muscle action. In this section we examine the method of expressing the concentrations of body electrolytes.

Molarity is a suitable concentration unit for solutions containing a single solute, such as 0.2 M NaCl or 0.01 M K_2SO_4. There are no sodium chloride or potassium sulfate units in these solutions because both are strong electrolytes. But we do know that there is a specific stoichiometric ratio between the anions and cations identified by the name of the compound. Now consider a solution that contains both sodium chloride and potassium sulfate. It contains sodium ions, potassium ions, chloride ions, and sulfate ions. So, we really can't label such a solution based on substances that don't exist. Blood and a variety of body fluids are mixtures of solutes. To express the concentrations of the electrolytes in such solutions, the unit *equivalent* is used.

An **equivalent** (equiv) of an ion is the amount in grams that provides 1 mole of charge. For ions with a +1 or −1 charge, the weight of an equivalent is the same as the weight of a mole. Thus the **equivalent weight** of the sodium ion is 23 g; the equivalent weight of the nitrate ion is 62 g. However, for ions such as the calcium ion or the sulfate ion, the equivalent weight is equal to the formula weight divided by the charge (neglecting the sign).

$$\textbf{equivalent weight} = \frac{\textbf{molar mass of ion}}{\textbf{charge of ion}}$$

For the sulfate ion, the molar mass is 96 g, and the equivalent weight is 48 g. Thus, 1 mole of sulfate ions counts as 2 equiv.

The number of equivalents in a sample can be calculated from the number of moles.

$$\textbf{number of equivalents} = \textbf{number of moles} \times \textbf{charge on ion}$$

Concentrations of ions in body fluids such as blood are quite low, so it is more convenient to use the *milliequivalent* for the amount of ion contained in a solvent. One **milliequivalent** (mequiv) is $\frac{1}{1000}$ of an equivalent. Thus, the concentration of potassium ion in blood, which might be 0.004 equiv/L, is expressed as 4 mequiv/L.

All solutions, including body fluids, are electrically neutral. That is, the total concentration of positive charges is equal to the total concentration of negative charges. A 0.1 M solution of Na_2SO_4 is 0.2 M in sodium ions and 0.1 M in sulfate ions. But in terms of equivalents, the concentration of sodium ions is 0.2 equiv/L, and the concentration of sulfate ions is 0.2 equiv/L.

$$\text{For } Na^+\text{: } 0.2 \text{ mole/L} \times 1 \text{ equiv/mole} = 0.2 \text{ equiv/L}$$
$$\text{For } SO_4^{2-}\text{: } 0.1 \text{ mole/L} \times 2 \text{ equiv/mole} = 0.2 \text{ equiv/L}$$

The average concentrations of the major ions contained in blood are listed in Table 9.2. The total concentration of all cations is about 168 mequiv/L, which is larger than the total concentration of the anions, which is about 140 mequiv/L. However, a number of organic anions that are not listed account for this difference. They are derived from carboxylic acids and proteins, and each is present in low concentration.

TABLE 9.2 Average Concentration of Major Electrolytes in Blood Plasma

Cation	mequiv/L	Anion	mequiv/L
Na^+	140	Cl^-	102
K^+	4.3	HCO_3^-	27
Ca^{2+}	4.8	SO_4^{2-}	2.0
Mg^{2+}	1.5	HPO_4^{2-}	3.7

EXAMPLE 9.6

The concentration of SO_4^{2-} in a blood plasma sample is 2.0 mequiv/L. How many milligrams of sulfate ion are in 1.0 L of blood plasma?

Solution

First, determine the equivalent weight of the sulfate ion. The molar mass of the sulfate ion is 96 g/mole. Dividing the molar mass by 2, the charge of the ion, gives 48 g/equiv as the equivalent weight of the sulfate ion. The mass of a milliequivalent is $\frac{1}{1000}$ that of an equivalent, or 0.048 g.

$$\frac{48\text{ g}}{1\text{ equiv}} \times \frac{1\text{ equiv}}{1000\text{ mequiv}} = 0.048\text{ g/mequiv}$$

Now, determine the mass of sulfate ion in a 2 mequiv/L solution.

$$\frac{0.048\text{ g}}{1\text{ mequiv}} \times \frac{2.0\text{ mequiv}}{1\text{ L}} = 0.096\text{ g/L}$$

Finally, express the mass in milligrams.

$$\frac{0.096\text{ g}}{1\text{ L}} \times \frac{1000\text{ mg}}{1\text{ g}} = 96\text{ mg/L}$$

PROBLEM 9.6

The concentration of calcium ions in a blood sample is 5.0 mequiv/L. How many milligrams of calcium are in 100 mL of blood?

9.9 The Formation of a Solution

In an ionic solid such as sodium chloride, the sodium and chloride ions are attracted to each other. What happens when the solid dissolves in water? Water is a polar molecule and is attracted to ions. A tug of war begins. The ions on the surface of the sodium chloride crystals can be pulled away by the polar water molecules. As the ions separate from the crystal, the polar water molecules surround them, that is, they *solvate* them (Figure 9.3). The partially negatively charged oxygen atom of water is oriented toward the sodium ion, whereas the partially positively charged hydrogen atoms of water are oriented toward the chloride ion. New surfaces are exposed to the water, and the crystal continues to dissolve.

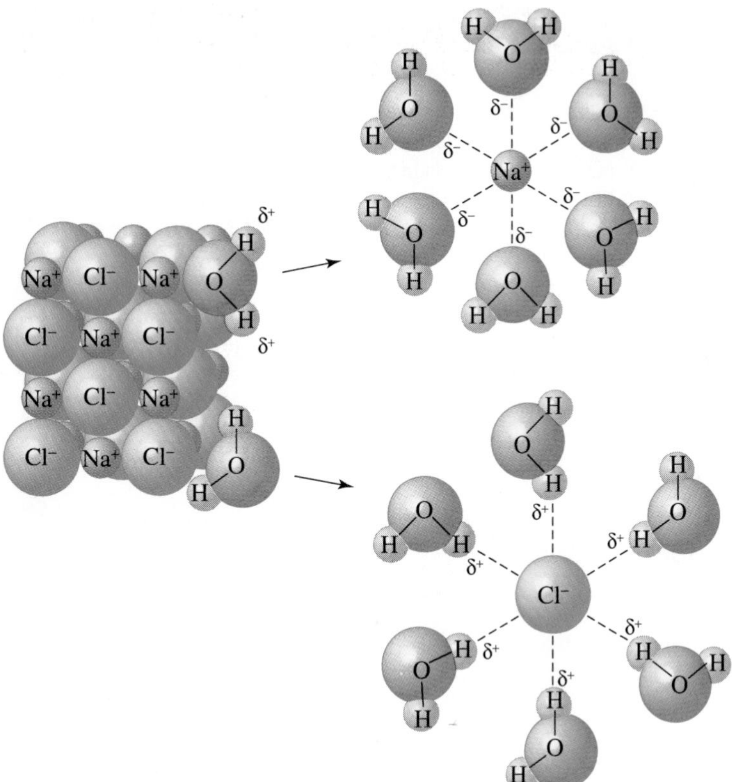

FIGURE 9.3 Dissolution of an Ionic Compound in Water
The attractive forces between ions and the polar water molecules are sufficient in soluble compounds to overcome the attractive forces within the crystal. The cations are approached by the negative end of the dipole of the water molecule and enter the solution surrounded by water molecules. The anions are approached by the positive end of the dipole of the water molecule and enter the solution surrounded by water molecules.

The dissolution process is reversible. The ions in solution are still attracted to the ions in the crystal and they may return to the solid state and crystallize. The balance in this tug of war depends on the identity of the solid, that is, on how strongly the ions are attracted to one another and how many ions the solvent can hold. If there is enough solvent and the attraction of the ions in the crystal is low, dissolution continues until the solid completely dissolves.

What happens when a saturated solution is finally formed? Although no further change is apparent in a saturated solution in contact with undissolved solid, the dissolution and crystallization processes continue. Some of the solid still dissolves in the solvent, while some of the dissolved solute crystallizes. There is a state of equilibrium much like the liquid–vapor and liquid–solid equilibria studied in Chapter 8. A **saturated solution** is one in which the rate of dissolution of the solute equals the rate of crystallization of the solute.

$$\textbf{solvent + solute} \underset{\text{crystallization}}{\overset{\text{dissolution}}{\rightleftharpoons}} \textbf{solution}$$

Heats of Solution and Entropy

Now let's consider the energy change that occurs when a solute dissolves in water. Some salts such as calcium chloride dissolve in an exothermic process. However, ammonium chloride dissolves in an endothermic process. Why are there such dif-

Hot Packs and Cold Packs

Commercially produced hot and cold packs take advantage of the exothermic and endothermic behavior, respectively, of the dissolution of various solids in water. Athletic trainers use instant hot or cold packs to provide first aid to injured athletes when hot water or ice might not be available. These packs are plastic bags containing a pouch of water and a solid salt. The pouch of water can be ruptured within the bag, and the released water dissolves the salt. Depending on which salt it contains, the plastic bag becomes hot or cold and can be applied to the athlete's injury.

Calcium chloride is used in hot packs; ammonium nitrate is used in cold packs. Dissolution of the solids produces solutions of ions.

$$CaCl_2(s) \longrightarrow Ca^{2+}(aq) + 2\,Cl^-(aq) \qquad \Delta H^\circ = -348 \text{ kcal/mole}$$

$$NH_4NO_3(s) \longrightarrow NH_4^+(aq) + NO_3^-(aq) \qquad \Delta H^\circ = 121 \text{ kcal/mole}$$

Calcium chloride releases 348 kcal/mole; ammonium nitrate absorbs 121 kcal/mole. The packs are designed to release or absorb heat in sufficient amounts to produce suitable temperatures. A hot pack can heat to 60 °C; a cold pack cools to 0 °C.

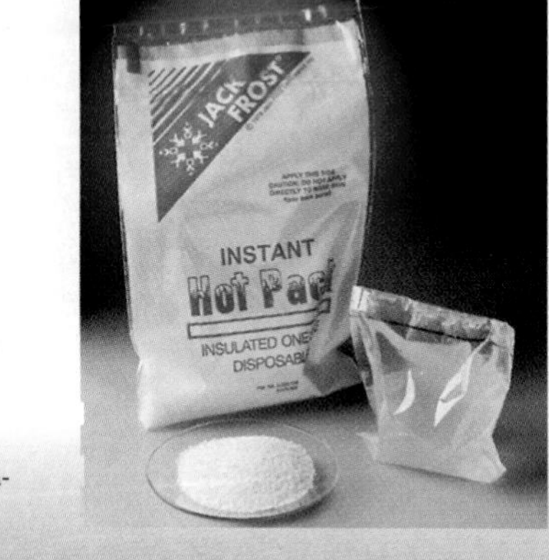

The two components contained within the hot pack are shown. (Richard Megna, Fundamental Photographs)

ferences? Energy is required to separate the ions from the crystals, but energy is released by the solvation of the ions. The net result may be exothermic or endothermic depending on the magnitude of these two energy terms.

One of the criteria known to determine spontaneity is the heat energy released or absorbed. Thus, we accept the fact that calcium chloride dissolves in water, because the process is exothermic. But how do we explain the fact that ammonium chloride dissolves even though the process is endothermic? Recall that it is the change of free energy that determines the spontaneity of a process, and two terms

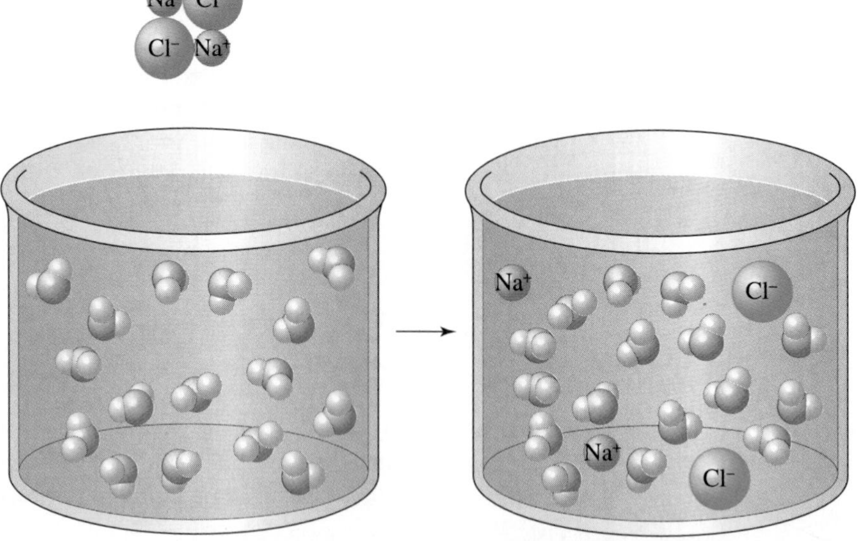

FIGURE 9.4 Increase in Entropy and Solutions

The molecules and ions are more randomly distributed in the solution than in the separate solute and solvent.

contribute to free energy—the change in enthalpy and the change in entropy. Thus, salts that dissolve in an endothermic process do so because the change in entropy is sufficiently favorable to counterbalance the unfavorable enthalpy change. Ionic solids are highly ordered. When the ions dissolve, the solvated ions are distributed throughout the solvent, and the entropy increases (Figure 9.4).

9.10 Structure and Solubility

The generalization that "like dissolves like" suggests that molecules of solute and molecules of solvent have similar intermolecular attractive forces. Water, which is a polar solvent, is a good solvent for polar solutes, ionic compounds, and substances that produce ions in water. Carbon tetrachloride (CCl_4), a nonpolar substance, does not dissolve ionic compounds. However, fats and waxes readily dissolve in this nonpolar solvent because they are nonpolar substances.

Liquids that dissolve in each other in all proportions are said to be **miscible.** Liquids that do not dissolve in each other are immiscible. Immiscible liquids form separate layers in a container. Ethyl alcohol is miscible with water, but carbon tetrachloride is immiscible with water. The miscibility of ethyl alcohol with water can be rationalized based on its structure.

$$\begin{array}{ccccccc} & & H & & H & & \\ & & | & & | & & \\ H & - & C & - & C & - & \ddot{O}: \\ & & | & & | & & | \\ & & H & & H & & H \end{array}$$

Ethyl alcohol is polar and the structural feature —OH resembles water. The nonbonded electron pairs on the oxygen atom in ethyl alcohol form hydrogen bonds with water, thus making it soluble (Figure 9.5).

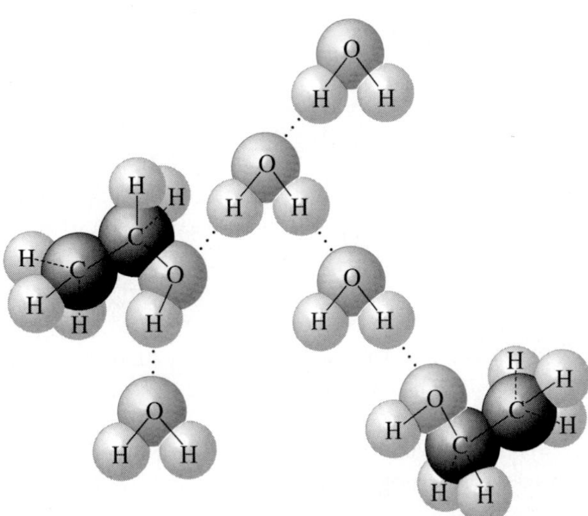

FIGURE 9.5
Solutions of Ethyl Alcohol and Water
The —OH group of ethyl alcohol forms hydrogen bonds to water molecules and fits into the solvent because the carbon-containing part of the molecule is not large.

Solubility of Vitamins

Vitamin C is one of the water-soluble vitamins, whereas vitamin A is a fat-soluble vitamin. The vitamin C molecule is small and has many —OH groups that can form hydrogen bonds to water. The vitamin A molecule, with the exception of one —OH group, is nonpolar. It is not "like" water and has a very low solubility in water, but vitamin A does dissolve in fatty tissue because it structurally resembles the carbon compounds in fats.

Because of its water solubility, vitamin C is not stored in the body and should be taken in as part of your daily diet. Unneeded vitamin C is eliminated from the body.

Fat-soluble vitamins are stored by the body for future use. If excessive quantities of fat-soluble vitamins are consumed in vitamin supplements, illness can result. The condition is known as *hypervitaminosis.*

vitamin A

vitamin C

9.11 Factors that Affect Solubility

Solubility depends on the experimental conditions as well as on structure. These conditions are temperature and, for gases, pressure. For that reason, it is common to report solubilities of solutes at one or more specific temperatures.

Effect of Pressure on Solubility

The carbon dioxide in Canada Dry is released once the cap is removed. (Charles D. Winters, Photo Researchers, Inc.)

You learned in Chapter 8 that pressure changes cause large changes in the volume of gases but not in the volume of liquids and solids. Because the volume of a gas changes as it dissolves in a solvent, the solubility of gases in liquids depends on pressure. The solubility of liquid and solid solutes is not affected by pressure.

The solubility of a gas, C_g, is directly proportional to the partial pressure of that gas, P_g. This relationship is known as **Henry's law.**

$$C_g = k_g \times P_g$$

All carbonated beverages are bottled under a pressure of carbon dioxide. When the bottle is opened, the partial pressure of carbon dioxide above the liquid decreases because there is little carbon dioxide in the atmosphere. As a result, the solubility of carbon dioxide decreases, and the solution effervesces as the carbon dioxide bubbles off. Table 9.3 lists the solubilities for some common gases in water. The values are for a pressure of 1 atm of the specific gas. Thus, the solubility of oxygen is 4.0×10^{-2} g/L at 25 °C under a pressure of 1 atm of oxygen. Since the partial pressure of oxygen in air is 0.2 atm, the solubility of oxygen is only 8×10^{-3} g/L. This amount of oxygen in our blood would be insufficient to support human life; however, hemoglobin molecules in our blood cells chemically bind oxygen molecules, which greatly increases the amount of oxygen available to the tissues.

TABLE 9.3 Solubility of Gases in 1 L of Water at 25 °C and 1 Atm

Gas	Mass (g)	Moles
He	0.0006	0.00015
Ar	0.052	0.0013
N_2	0.017	0.00062
O_2	0.03	0.0012
CO_2	1.5	0.032
CO	0.025	0.00093

Effect of Temperature on Solubility

The solubility of solids can either decrease or increase with increasing temperature. For example, the solubility of potassium chloride (KCl) increases from 28 g per 100 g of water at 0 °C to 57 g per 100 g of water at 100 °C. Similarly, the solubility of sodium chloride (NaCl) increases from 36 g to 39 g per 100 g of water over

Gas Solubility and the Bends

The condition called the bends, which scuba divers may experience, is due to the effect of pressure on the solubility of a gas. Divers breathe compressed air that is 80% nitrogen. At atmospheric pressure the partial pressure of nitrogen is 0.8 atm. However, at 100 ft, where the pressure is 4 atm, the diver breathes compressed air at 4 atm, which is provided by a regulator. The partial pressure of nitrogen is 0.8×4 atm = 3.2 atm. As a consequence, larger quantities of nitrogen, which is not used by the body, dissolve in fluids and tissue under pressure. Depending on the length of the dive, the amount of nitrogen that dissolves may exceed that which the blood can hold at atmospheric pressure. Upon the diver's return to the surface, the nitrogen in blood comes out of solution and forms small bubbles that can block capillaries and cut off blood flow. Nitrogen released by tissues accumulates in the joints and prevents the affected individual from straightening his or her arms or legs. Attempts to do so cause pain, and the individual assumes a bent position to gain some relief. Other more serious conditions may result, such as embolisms, nerve damage, and stroke.

To avoid excessively high blood concentrations of nitrogen and the possibilty of getting the bends, a diver must not exceed the allowable times listed in dive tables. Those times depend on depth, because the pressure depends on depth. For example, the limit at 60 ft is 55 min, but at 100 ft it is only 20 min. If such times are exceeded, the diver must ascend to designated depths and remain there for a specified time to allow some of the nitrogen to be eliminated. These stops on ascent are called *decompression stops.* Recreational divers should avoid diving under conditions that require decompression stops.

There are individual variations in response of blood and body tissues to elimination of nitrogen, and the process is slow. As a result, a second dive is limited by how long the diver has been at the surface. Even after an hour, the allowable time for a second dive is significantly reduced. After a 55-min dive at 60 ft followed by a 1-hr stay at the surface, the limit on the second dive to 60 ft is 30 min. The reduced time is calculated based on the residual nitrogen in the blood from the first dive.

Artificial mixtures for breathing, consisting of helium and oxygen, are used by research divers and military personnel for lengthy stays in the ocean at great depths. Helium is less soluble than nitrogen in water and in blood. When the pressure on the diver is decreased, there is less dissolved gas to form bubbles, and the bends can be prevented more easily.

This diver is checking her dive table to avoid getting the bends. (Robert J. Ouellette)

FIGURE 9.6
Effect of Temperature on the Solubility of Ionic Compounds

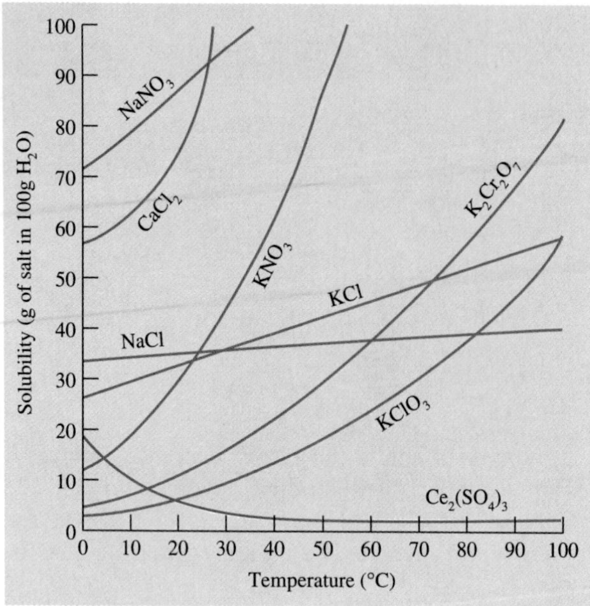

the same temperature range. However the solubilities of some salts, such as cerium sulfate ($Ce_2(SO_4)_3$), decrease with increasing temperature (Figure 9.6).

For solutes that dissolve exothermically, the solubility decreases at higher temperature. For endothermic processes the solubility increases at higher temperature. These effects of temperature on solubility can be explained by Le Châtelier's principle. Consider the equation for the dissolution of a solute that occurs exothermically.

$$\text{solute} + \text{solvent} \rightleftharpoons \text{solution} + \text{heat}$$

The equilibrium responds to increased temperature by proceeding in the direction to use up the added energy. That direction is from right to left, which results in a decreased solubility.

For an endothermic dissolution process the heat energy is written on the left side of the equation.

$$\text{solute} + \text{solvent} + \text{heat} \rightleftharpoons \text{solution}$$

This equilibrium also responds to increased temperature by proceeding in the direction to use up the added energy. However, in this case that direction is from left to right, which results in an increased solubility.

EXAMPLE 9.7

The heat of solution of lithium chloride is −8.8 kcal/mole. Is lithium chloride more soluble or less soluble at 50 °C than at 25 °C?

Solution

For this exothermic process we can write

$$\text{LiCl(s)} + H_2O \rightleftharpoons \text{LiCl(aq)} + \text{heat}$$

Design of Medicines

There is more to a pill or capsule of medicine than its color, shape, and size. Often the active ingredient is only a small component. Most medicines have a binder, one or more fillers, and often a coating. Binders hold the pill together and prevent its crumbling while in the container. Fillers make a pill the right size, because the dosage of many medicines is so low that a pill of it alone would be too small to handle. Finally, the pill may be coated with a compound to determine where it will dissolve in the digestive tract. The film around capsules serves the same role.

To be effective, a medicine must survive in the digestive tract until it reaches a point where it may effectively be absorbed. Thus, to survive the acidic medium in the stomach some pills are coated with acid-resistant material. This material dissolves at a later time in the basic medium of the intestines and releases the medicine there.

Timed-release medicines allow the gradual release of medication over a desired period of time. The first commercial success of this process was Contac decongestant in 1961. This medicine, as well as many more examples today, consists of beads coated with water-soluble polymers of varying thicknesses. The beads with the thinner coat of polymer dissolve quickly. Those with thicker coats dissolve more slowly. These thicknesses are chosen to provide release of the medicine over a desired time period.

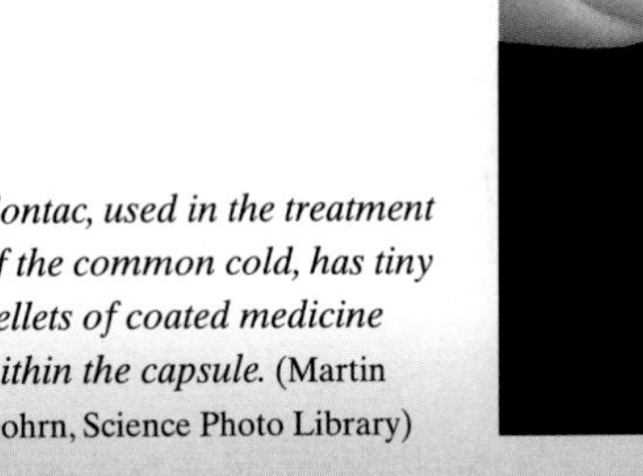

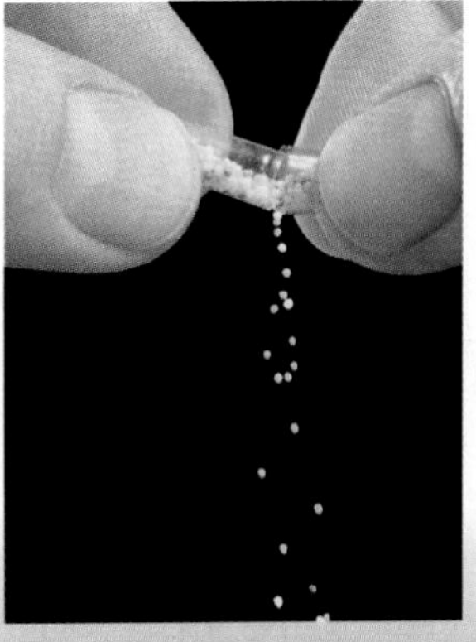

Contac, used in the treatment of the common cold, has tiny pellets of coated medicine within the capsule. (Martin Dohrn, Science Photo Library)

According to Le Châtelier's principle, increasing the temperature should shift the equilibrium position toward the left. Therefore, LiCl is less soluble at the higher temperature.

PROBLEM 9.7

The solubility of KBr increases with increasing temperature. Is the ΔH for the solution process positive or negative?

Gases are less soluble in water at higher temperatures and dissolved air forms bubbles that escape from the liquid. Many industrial plants and power plants use water for cooling, and the hot water may be passed into lakes and streams. As these waterways warm up, the solubility of oxygen decreases. This gas is less available to the fish. The water is said to be *thermally polluted.* Fish cannot easily adapt to large temperature changes. An increase of 10 °C in the temperature approximately doubles their metabolic rate. The increased metabolism requires more oxygen at a time when the amount of oxygen dissolved in the water is less.

9.12 Colligative Properties

The properties of a solution are different from those of the pure solvent. Those properties that depend on the number of dissolved particles and not on their chemical identity are called **colligative properties.** These include vapor pressure, boiling point, freezing point, and osmotic pressure.

How Does a Solute Affect the Properties of a Solvent?

When a solute is added to a solvent, the solute molecules occupy some positions formerly occupied by solvent molecules. In a sense the solvent is diluted, and its molecules are more separated from each other. For example, solute molecules occupy a portion of the surface of a solution. In other words, there are fewer solvent molecules on the surface in a solution than in the pure solvent.

Consider a 1.0 M solution of sugar in water. Sugar is nonvolatile—it cannot escape into the gas phase. However, because the sugar molecules occupy positions on the surface of the solution, there is a decreased chance that water molecules will escape. As a result, the vapor pressure of the solution is lower than that of pure water. The difference in vapor pressure is proportional to the concentration of the sugar molecules in the solution.

The difference between the vapor pressure of a 1.0 M sodium chloride solution and that of pure water is not the same as the difference for a 1.0 M sugar solution. Experiments show that the difference for the salt solution is twice that for the sugar solution. Remember that colligative properties do not depend on the identity of the solute but do depend on the number of particles. When NaCl dissolves in water, two ions form but a covalent compound such as sugar exists as molecules. You should therefore expect the decrease in the vapor pressure of water on the addition of any substance to depend on the total concentration of all the species formed when the solute dissolves.

Boiling Points of Solutions

Because the vapor pressure of a solution of a nonvolatile solute is lower than that of the pure solvent, a higher temperature is required to increase the vapor pressure of the solution to atmospheric pressure. Thus, the boiling point of the solution is higher than that of the solvent; the increase in the boiling point of the solvent is directly proportional to the concentration of solute.

The boiling point of water increases by 0.52 °C per mole of solute particles per kilogram of water. If the 1 mole of solute is sugar, the boiling point of the aqueous solution is 100.52 °C. What if the 1 mole of solute is NaCl? The solution contains 2 moles of particles, 1 mole of Na^+ and 1 mole of Cl^-. Thus, the boiling point is increased by 2×0.52 °C, so it is 101.04 °C.

Freezing Points of Solutions

Cities in northern states use salt to melt ice on roads and sidewalks in the winter. Why does the ice melt? Consider a mixture of water and ice at equilibrium at 0 °C. If a solute is added, how does it affect the release of water molecules in ice? How does it affect the return of water molecules in the liquid to the solid?

The tendency of water molecules to escape from the solid state to the liquid state is not affected by the addition of solute because only water molecules are in the ice. However, the tendency of the water molecules of the solution to enter the solid phase is decreased by the presence of the dissolved solute. You could picture this decreased tendency to return to the solid as due to an obstruction by the solute. It is less likely that the water molecules can become oriented as required in the solid state. Thus, if solute is added to an equilibrium mixture of solid and liquid, the conversion of solid into liquid is unaffected, but the conversion of liquid into solid is impeded. As a consequence, there will be more solid melting than there is liquid forming solid. In order to equalize the relative tendencies of solvent to go between the two phases, the temperature must be lowered. At some lower temperature, the equilibrium between liquid and solid phases will be reestablished, and this temperature is the freezing point of the solution.

When salt is spread on snow and ice, the tendency of the water molecules to freeze and form ice is decreased. An equilibrium results between ice and salt solution at a temperature controlled by the concentration of the salt in the solution. There is a practical limit on the amount of ice that can be melted. The resulting aqueous solution becomes more dilute as the ice melts.

Antifreeze in car radiators decreases the freezing point of water. The antifreeze consists of ethylene glycol ($C_2H_6O_2$), which is very soluble in water. The addition of antifreeze prevents the water from freezing at temperatures above a temperature which is determined by the amount of antifreeze added. The antifreeze also increases the boiling point of the solution and makes it less likely that the radiator fluid will boil over in the summer.

9.13 Osmosis

You may know students who think that they can learn by "osmosis." They think that knowledge will seep into the brain without doing any work. Water molecules can seep across cell membranes but knowledge cannot. The transfer of water molecules through a semipermeable membrane is called **osmosis.** A **semipermeable membrane** is a membrane through which water molecules can pass but other molecules cannot. The phenomenon of osmosis and the related osmotic pressure is a colligative property. Osmosis is important in organisms, as it adjusts the water balance between intracellular fluid and extracellular fluid. Certain nutrients can also cross cell walls, but not by osmosis. Special transport processes (Chapter 22) are necessary to move molecules other than water across cell membranes.

How Does Osmosis Occur?

Consider a U-shaped tube containing more water on the right side of a semipermeable membrane than on the left side, as in Figure 9.7. The water pressures on the two sides are unequal, and water molecules tend to pass from right to left to equalize the pressure and the water levels. Even at this point, the transfer of water continues; water also passes through the membrane from left to right, and an equilibrium is maintained. The water molecules were going from left to right all along but at a slower rate than from right to left owing to the pressure on the right side.

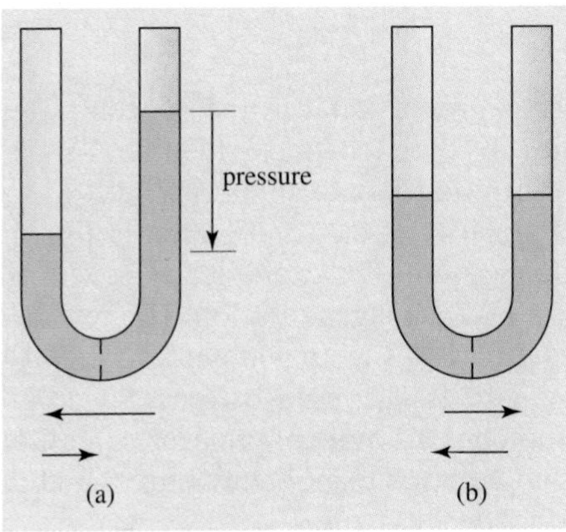

FIGURE 9.7
Osmosis
(a) The larger amounts of water on the right side creates a pressure that pushes water from the right to the left across the semipermeable membrane in the center of the tube. (b) The water continues to pass in both directions across the membrane, and the system is at equilibrium.

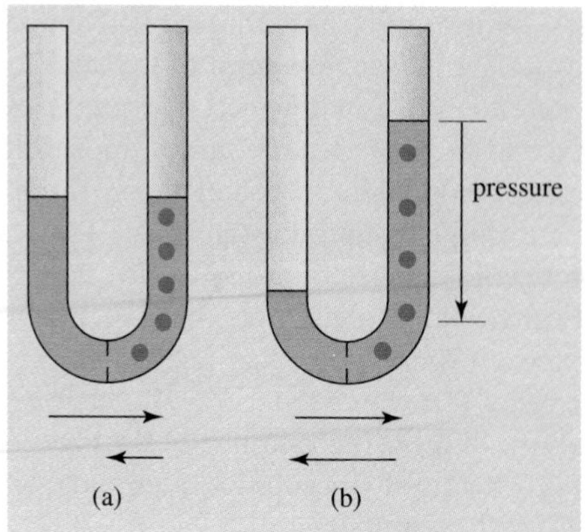

FIGURE 9.8
Osmotic Pressure of Solutions
(a) With columns of liquid initially at the same height, water molecules pass from left to right to dilute the sugar solution on the right. (b) Eventually, pressure due to the column of solution on the right prevents the continued net transfer of water from left to right.

Now consider a sugar solution separated from pure water by a semipermeable membrane (Figure 9.8). Initially the heights of the two columns of a liquid are equal. With time, the volume of the solution on the right increases while the volume of pure water on the left decreases. The water remains pure; no sugar molecules pass through the membrane. From the first experiment, we know that water molecules can cross the membrane, and that they do so in both directions. The same process must occur in this experiment as well. Why don't the levels remain the same?

The tendency of water molecules to pass from the solution side to the pure water side of the membrane is diminished by the solute molecules. Since there is no such restriction on the passage of water molecules from the pure water side to the solution side of the membrane, a net transfer of water molecules occurs and dilution of the sugar solution results.

Eventually, no further net transfer of water molecules occurs, and the level of the solution remains constant at some height above that of the water. The difference in the levels of the solution and the water is related to the net tendency of water to dilute the solution. However, there is a "back pressure" due to the height of the column of liquid on the solution side of the membrane. In effect the column of liquid opposes the tendency of the water molecules to push across the membrane. The difference in pressure caused by the difference in column heights is the **osmotic pressure.** If pressure is applied on the solution side of the tube, the net flow of water molecules can be reversed, and the height of the two columns of liquid can be made equal. The pressure required to maintain equal levels of the water and the solution is equal to the osmotic pressure.

Osmotic Pressure

Osmotic pressure is proportional to the concentration of the solute particles and does not depend on their chemical identity—it is a colligative property. The greater the concentration of the solution, the higher the osmotic pressure. However, it is not the molarity of the solution that is important but the total molarity of all solute particles present. For example, a 0.1 M NaCl solution has the same osmotic pressure as a 0.2 M sugar solution. The reason for this phenomenon is that a 0.1 M NaCl solution is 0.1 M in sodium ions and 0.1 M in chloride ions. Thus, in 1 L of solution there is 0.2 mole of ions. Sugar is a covalent molecule and a 0.2 M sugar solution has 0.2 mole of molecules per liter.

The relationship between osmotic pressure Π and the molarity of the solute particles resembles the ideal gas law,

$$\Pi V = nRT$$

where n is the number of moles of all solute particles, and V is the volume of the solution. The equation can be rearranged to obtain the quotient n/V.

$$\Pi = \frac{n}{V} \times RT$$

The quotient n/V is the molarity of all solute particles. For a molecular substance, the quotient is equal to the molarity of the solution. For an ionic compound it is equal to the sum of the molarities of all ions. (Physiologists use the term **osmolarity** to represent total molarity.)

EXAMPLE 9.8

What is the osmotic pressure of a physiological saline solution that is 0.15 M NaCl at 37 °C?

Solution

The total molarity of the solution is twice the calculated molarity, since 1 mole of NaCl yields 2 moles of ions in solution. Remember that temperature must be in kelvins.

$$\frac{n}{V} = 2 \times (0.15 \text{ mole/L}) = 0.30 \text{ mole/L}$$

$$\Pi = \frac{n}{V} \times RT$$

$$\Pi = 0.30 \frac{\text{mole}}{\text{L}} \times 0.0821 \frac{\text{L atm}}{\text{mole K}} \times 310 \text{ K} = 7.6 \text{ atm}$$

PROBLEM 9.8

The osmotic pressure of a solution containing 5.0 g of insulin per liter of solution is 16.3 mm Hg at 25 °C. Calculate the approximate molecular weight of insulin. Insulin is a nonelectrolyte.

These old sequoias in Sequoia National Park move water from the soil to the leaves by osmosis. (Marc Muench, Tony Stone Images)

Water transfers between aqueous solutions of different molarities separated by a semipermeable membrane. When the osmotic pressures of the solutions are unequal, the tendency of water to cross the membrane is not the same on both sides of the membrane. Water will flow from less concentrated to more concentrated solutions.

Osmotic pressure differences account for the movement of water from soil into plants and trees. Cells in root systems have high concentrations of solutes. Water tends to move from the less concentrated solutions of ground water into the roots and up through the plant or tree. When water evaporates from leaf cells (transpiration), the local concentration of solute increases and water tends to move to the leaves from the stems and into the stems from the roots. Transpiration is one reason water continues to move from the ground into the plant.

Reverse Osmosis

If a pressure greater than the osmotic pressure is applied to a solution, water flows from the solution to the other side of the membrane. The osmotic pressure of sea water is about 25 atm, so pressure greater than 25 atm is needed to obtain pure water from sea water. As reverse osmosis occurs, the saline solution becomes more concentrated and its osmotic pressure increases. Thus, a reverse osmosis purification of water often requires pressures up to 100 atm.

We find water palatable if dissolved salts are below 500 ppm. A large reverse osmosis plant in Yuma, Arizona, produces 100 million gallons per day using water from the Colorado river. The dissolved salts are decreased from over 3000 ppm to about 280 ppm. On Sanibel Island in Florida a smaller reverse osmosis plant produces about 3.5 million gallons of water per day to provide water needed to replace well water that has become increasingly saline. Reverse osmosis plants provide much of the water in Middle Eastern countries such as Saudi Arabia. One of the fears of the Gulf War was the possibililty that the oil released into the gulf by Iraq might reach the water intake pipes of the reverse osmosis plants. Fortunately this potential disaster was averted.

9.14 Osmosis and Cells

Both animals and plants have membranes through which water passes, and if the concentrations of the solutes in the water solutions are not balanced, water transport occurs that may damage cells (Figure 9.9). Water moves from solutions of low osmotic pressure to solutions of high osmotic pressure. If the osmotic pressure of a solution surrounding red blood cells is less than that within the cells, water passes into the cells and the cell walls may burst, a process called **hemolysis.** If the osmotic pressure of the surrounding solution is greater than that within the red blood cells, water leaves the cells and causes them to shrivel, a process called **crenation.**

The intravenous administration of dextrose and the ionic constituents of body fluids must be carefully controlled. A solution whose osmotic pressure is equal to that within cells is **isotonic.** A solution with an osmotic pressure less than that within a cell is **hypotonic;** if the solution's osmotic pressure is greater than that within the cell, the solution is **hypertonic.** The osmotic pressure of the fluid in a red blood cell is 7.7 atm at a body temperature of 98.6 °F. This pressure can be balanced by low concentrations of solute in the surrounding fluid. A 0.9% sodium chloride solution has an osmotic pressure equal to that of a red blood cell.

It is sometimes necessary to intravenously administer controlled concentrations of solutions to change the water balance in a patient. A hypertonic solution will transfer water from tissues into the blood, from which the kidneys will remove the water. A hypotonic solution may be used to transfer water out of the blood into the surrounding tissues and thus decrease the blood pressure.

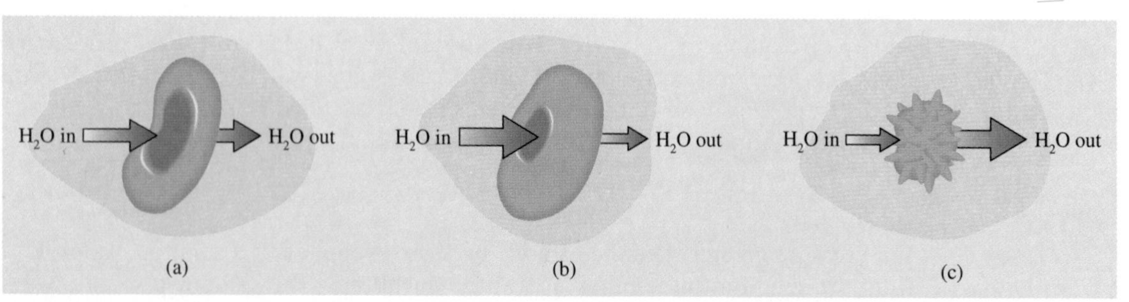

FIGURE 9.9 Effect of Osmotic Pressure On Red Blood Cells
(a) In an isotonic solution the cell shape does not change because water enters and leaves at the same rate. (b) In a hypotonic solution the rate of water passing into the cell exceeds the rate of loss of water. As a result, the cell distends and may burst. (c) In a hypertonic solution the rate of water leaving the cell exceeds the rate of water entering the cell. This causes the cell to shrink.

Summary

In describing a **solution** (homogeneous mixture), we indicate the relative amounts of **solute** and **solvent.** Solutions in water as solvent are **aqueous solutions. Alloys** are solid solutions of two or more metals.

Solubility is the amount of solute that can dissolve in a solvent under specified conditions. Solubility of a solute in a solvent depends on temperature, pressure, and the structures of the solvent and the solute. Solutions in which no additional solute will dissolve are **saturated;** if additional solute can dissolve, the solution is **unsaturated.** Liquids that dissolve in each other in all proportions are **miscible.**

Concentrations can be qualitatively described as **dilute** or **concentrated.** A variety of quantitative expressions for concentration are used. Percent concentration units include weight/weight, weight/volume, and volume/volume. Parts per million **(ppm)** may be used for very dilute solutions. **Molarity** is a unit that provides direct information about the number of moles of solute in a given volume of solution.

Solutes in aqueous solution may be electrolytes or nonelectrolytes. **Electrolytes** are substances that are ionic or are converted into ions when dissolved in water. The ions may result from **dissociation** of an ionic compound or **ionization** of a covalent compound. **Strong electrolytes** ionize completely in water; **weak electrolytes** do not ionize completely in water. **Nonelectrolytes** are substances that exist as molecules in solution.

An **equivalent** of an electrolyte is the amount in grams that gives 1 mole of charge. One **milliequivalent** is $\frac{1}{1000}$ of an equivalent.

Properties of a solution that depend on the number of solute particles in solution and not on their identity are called **colligative properties.** These properties include lowering the vapor pressure, raising the boiling point, decreasing the freezing point, and increasing the **osmotic pressure** of a solvent.

Osmosis is the net flow of water across a **semipermeable membrane.** Transfer of material across membranes is an important biological process. A solution whose osmotic pressure is equal to that within cells is **isotonic.** A solution with an osmotic pressure less than that within a cell is **hypotonic;** one with an osmotic pressure greater than that within the cell is **hypertonic.** If the osmotic pressure of a solution surrounding red blood cells is less than that within the cells, the cell walls may burst, a process called **hemolysis.** If the osmotic pressure of the surrounding solution is greater than that within the red blood cells, the cells will shrivel, a process called **crenation.**

Exercises

Saturated Solutions

9.1 The solubility of sodium chloride in water is 36 g/100 mL at 25 °C. A solution contains 6 g of NaCl in 20 mL at 25 °C. Is the solution saturated?

9.2 The solubility of sodium nitrate in water is 81 g/100 mL at 25 °C. A solution contains 8.1 g of $NaNO_3$ in 10 mL at 25 °C. Is the solution saturated?

Percent Concentrations

9.3. Calculate the w/v % concentration of each of the following.

(a) 10.0 g of NaH_2PO_4 in 1.00 L of solution
(b) 16.0 g of $NaNO_3$ in 400 mL of solution
(c) 1.50 g of NaCl in 50 mL of solution

9.4 Calculate the volume of the given w/v % solution required to provide the indicated mass of solute.

(a) a 0.900 w/v % solution; 18.0 g of NaCl
(b) a 5.0 w/v % solution; 30 g of $C_6H_{12}O_6$ (glucose)
(c) a 1.0 w/v % solution; 60.0 g of Na_2CO_3

9.5 Calculate the v/v % concentration of each of the following.

(a) 100 mL of ethyl alcohol in 500 mL of solution
(b) 3 mL of acetic acid in 60 mL of vinegar solution
(c) 0.003 mL of alcohol in 1 mL of blood

9.6 Calculate the number of grams of solute needed to make each of the following w/w % solutions.

(a) 1.0 kg of 5.00 w/w % NH_4Cl
(b) 500 g of 3.00 w/w % H_2SO_4
(c) 100 g of 1 w/w % $C_6H_{12}O_6$

Parts per Million

9.7 You breathe 1×10^4 L of air per day. The concentration of sulfur dioxide (SO_2), a pollutant in the air, is 0.1 ppm. What volume of SO_2 molecules do you breathe?

9.8 Dizziness and headaches occur when incomplete combustion of fossil fuels causes the carbon monoxide concentration in air to reach 100 ppm. Express this concentration in v/v %.

9.9 A gaseous pollutant is present at a concentration of 0.2 ppm. What volume of pollutant is contained in 10 L of air?

9.10 The Salton Sea in California has 1.9 ppm of lithium ion. How many grams of lithium ions are contained in 100 L of water?

9.11 A preparation of aspirin (acetylsalicylic acid) contains 0.0002 g of salicylic acid impurity per 1.0 kg of aspirin. Express this amount in ppm.

9.12 A fluoridated water source contains 35 mg of NaF per 2.0×10^4 L of solution. Express this amount in ppm on a weight/weight basis.

Molarity

9.13 Calculate the molarity of each of the following solutions.

(a) 4.0 g of NaOH in 500 mL of solution
(b) 5.6 g of KOH in 200 mL of solution
(c) 410.0 g of H_2SO_4 in 1.00 L of solution

9.14 Calculate the molarity of each of the following solutions.

(a) 0.0058 g of $Mg(OH)_2$ in 2000 mL of solution
(b) 2.0 g of NaOH in 2000 mL of solution
(c) 20.7 g of H_2SO_4 in 400 mL of solution

9.15 How many grams of solute are there in each of the following solutions?

(a) 250 mL of 1.00 M glucose ($C_6H_{12}O_6$)
(b) 250 mL of 1.00 M sucrose ($C_{12}H_{22}O_{11}$)
(c) 1.50 L of 0.250 M NaOH

9.16 How many grams of solute are there in each of the following solutions?

(a) 125 mL of 1.00 M ethyl alcohol (C_2H_6O)
(b) 250 mL of 0.500 M glycerin ($C_3H_8O_3$)
(c) 50 mL of 0.125 M KOH

9.17 What volume of the given concentration of solution will provide the required number of moles of solute?

(a) 0.20 mole of NaOH from 0.50 M NaOH
(b) 1.00 mole of H_2SO_4 from 0.100 M H_2SO_4
(c) 0.010 mole of NaCl from 0.100 M NaCl

9.18 What volume of the given concentration of solution will provide the required number of moles of solute?

(a) 0.10 mole of KOH from 0.25 M KOH
(b) 0.50 mole of H_2SO_4 from 0.200 M H_2SO_4
(c) 0.250 mole of NaCl from 0.05 M NaCl

9.19 A 2.00-mL sample of blood contains 0.490 mg of calcium ions. Calculate the molarity of the calcium ions.

9.20 A tincture of iodine contains 0.1 g of I_2 in 10.0 mL of ethyl alcohol. What is the molarity of iodine?

9.21 Rubbing alcohol contains 60 g of isopropyl alcohol (C_3H_8O) in 100 mL of solution. What is the molarity of the isopropyl alcohol solution?

9.22 A water sample has 0.1 mg of Cd^{2+} in 10 mL of solution. What is the molarity of the Cd^{2+} solution?

9.23 An antiseptic solution is 0.85 M hydrogen peroxide (H_2O_2). How many moles of H_2O_2 are in 225 mL of the solution?

9.24 Household bleach is 0.65 M NaOCl. How many moles of NaOCl are in 2.0 L?

9.25 The concentration of glucose ($C_6H_{12}O_6$) in spinal fluid is 75 mg/100 mL. What is the molarity?

9.26 An individual has 0.15 g of cholesterol ($C_{27}H_{46}O$) in 100 mL of blood serum. What is the molarity?

Dilution of Solutions

9.27 What will be the molarity of the solution that results from diluting to the indicated final volume?

(a) 10.0 mL of 12 M HCl to 210.0 mL of solution
(b) 25.0 mL of 16 M HNO_3 to 275.0 mL of solution
(c) 100.0 mL of 18 M H_2SO_4 to 500.0 mL of solution

9.28 What volume of the indicated concentrated reagent is required to prepare the indicated dilute solution?

(a) 15 M NH_3 to give 100.0 mL of 1.0 M NH_3
(b) 18 M H_2SO_4 to give 1.00 L of 0.10 M H_2SO_4
(c) 16 M HNO_3 to give 250.0 mL of 1.0 M HNO_3

9.29 What must the final volume be to dilute the indicated sample to give the desired final concentration of solution?

(a) 100.0 mL of 0.50 M NaOH to give 0.20 M NaOH
(b) 50.0 mL of 1.0 M NH_3 to give 0.020 M NH_3
(c) 25.0 mL of 0.20 M NaCl to give 0.010 M NaCl

9.30 What must the final volume be to dilute the indicated sample to give the desired final concentration of solution?

(a) 25.0 mL of 2.0 M NH_3 to give 0.010 M NH_3
(b) 50.0 mL of 1.0 M HCl to give 0.025 M HCl
(c) 100.0 mL of 1.5 M NaCl to give 0.15 M NaCl

Electrolytes

9.31 Hydrogen bromide (HBr) is a covalent molecule, but an aqueous solution of HBr conducts electricity. Write an equation that explains this observation.

9.32 Ethylene glycol, used as an antifreeze, is a nonelectrolyte. What does this tell us about the structure of ethylene glycol in water?

Equivalents

9.33 Ringer's solution, used in treating burns, contains 0.866 g NaCl, 0.030 g KCl, and 0.033 g $CaCl_2$ per 100 mL. Calculate the concentration of chloride ion in equiv/L.

9.34 A solution used by the World Health Organization to rehydrate victims of cholera contains 3.5 g NaCl/L, 1.5 g KCl/L, 2.5 g $NaHCO_3$/L, and 20 g/L glucose. Calculate the concentration of sodium ion in equiv/L.

Solubility

9.35 Why do bubbles escape from a bottle of soft drink after the cap is removed?

9.36 Why does an open bottle of soft drink go "flat" at room temperature faster than one stored in a refrigerator?

9.37 The heat of solution of $CaCl_2$ is -19.4 kcal/mole. Does the concentration of calcium chloride increase or decrease with increasing temperature?

9.38 The solubility of NH_4ClO_4 increases with increasing temperature. Is the solution process exothermic or endothermic?

9.39 Ammonia (NH_3) is very soluble in water. Explain why.

9.40 Some ionic solids are insoluble in water. What does this fact indicate?

9.41 Glycerol has the following structure. Predict its solubility in water and in carbon tetrachloride (CCl_4).

```
     H   H   H
     |   |   |
 H—C—C—C—H
     |   |   |
    OH  OH  OH
```

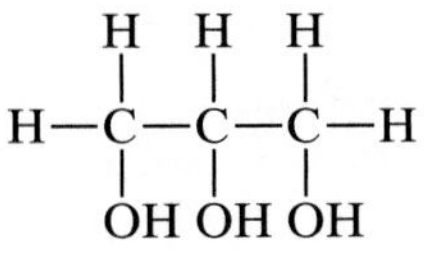

9.42 Mothballs have the following structure. Predict their solubility in water and in carbon tetrachloride (CCl_4).

Cl—⟨benzene ring⟩—Cl

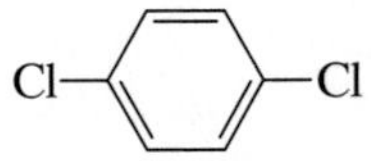

9.43 Which compound is more soluble in water, hexane or 1-pentanol?

$CH_3(CH_2)_4CH_3$	$CH_3(CH_2)_3CH_2OH$
hexane	1-pentanol

9.44 Methyl alcohol (CH_3OH) is not soluble in octane, but 1-hexanol is soluble. Explain why.

$CH_3(CH_2)_6CH_3$	$CH_3(CH_2)_4CH_2OH$
octane	1-hexanol

9.45 Consider the structure of biotin, a vitamin, and predict whether it is a water-soluble or a fat-soluble compound.

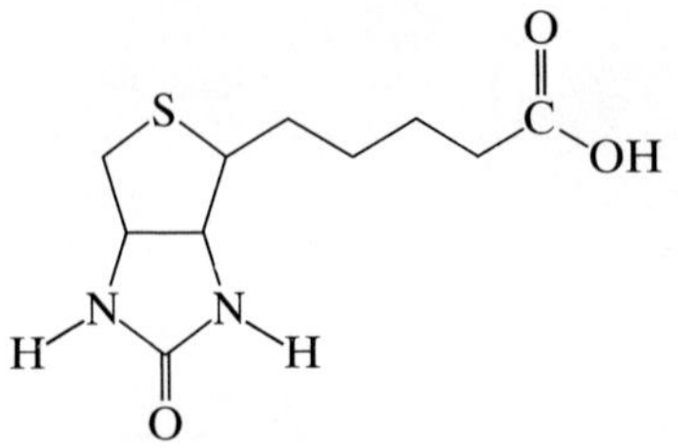

9.46 Based on the molecular formula for vitamin K ($C_{31}H_{40}O_2$), predict whether it is a water-soluble or a fat-soluble compound.

Colligative Properties

9.47 Celery that is kept in a refrigerator for a long time can become limp. Its crispness can be restored by putting it in water. Explain why.

9.48 Salt spread on driveways to melt ice may be injurious to nearby plants and shrubs. Explain why.

9.49 Amylose, a soluble starch, has a molecular weight of approximately 30,000 amu. What is the osmotic pressure of a solution of 2.0 g of amylose in 1 L of water at 25 °C?

9.50 The osmotic pressure of a solution of 10 g of a milk protein in 1 L of water is 5 mm Hg at 25 °C. What is the molecular weight of the protein?

9.51 The osmotic pressure of 0.100 g of hemoglobin in 10.0 mL of solution is 2.87 torr at 25 °C. Calculate the molecular weight of hemoglobin.

9.52 Dextran is used as a blood plasma substitute. The osmotic pressure of a solution containing 0.69 g of dextran in 125 mL of water at 25 °C is 1.49 torr. What is the molecular weight of dextran?

Automobile races depend on spontaneous reactions that also occur at a fast rate. (J. DiMaggio/J.Kalish, Peter Arnold, Inc.)

10

Reaction Rates and Equilibrium

10.1 Reactions—An Overview
10.2 Energy and Chemical Reactions
10.3 Spontaneity and Chemical Reactions
10.4 Free Energy
10.5 Kinetics
10.6 Activation Energy
10.7 Chemical Equilibrium
10.8 Chemical Equilibrium and Le Châtelier's Principle

Essays

- ***Explosions and Entropy***
- ***Free Energy and Biological Processes***
- ***The Effect of Temperature on Organisms***
- ***Catalysts in Living Systems***
- ***Catalytic Converters***
- ***Production of Ammonia for Fertilizer Products***

10.1 Reactions—An Overview

Every chemical reaction is a unique event in which atoms change from their original arrangement in reactants to form a new arrangement in products. Certain bonds in the reactants are broken, and new bonds are formed in the products.

Chemical reactions occur at a variety of rates. The formation of oil in the Earth over centuries is a slow process, whereas gasoline burns at explosive speed in a car engine. Reaction rates change with temperature. Milk spoils readily at room temperature but will last for a week or more in a refrigerator. The state of matter and the concentration of reactants affect the rate of reaction. For example, chunks of coal in a furnace burn slowly over a period of hours, but coal dust in the air of a mine may explode. In this chapter we will examine theories that are based on each of these facts. Our goals are to answer each of the following questions.

1. Why do different substances react at different rates?
2. Why does the rate of reaction increase with temperature?
3. How is the rate of reaction affected by concentration of reactants?

The answers to these questions are provided by chemical **kinetics,** the study of the speed or rate of chemical reactions. Chemists use such studies to learn how to pre-

The number of people on a floor and whether or not equilibrium is established depend on the rates of entry and exit from this escalator. (Bob Daemmrich, Stock Boston)

dict and control the speed of chemical reactions. The kinetics of chemical reactions are important in industry, where the desired chemical product must be formed rapidly and in preference to other products. A reaction may be too slow to be economically practical, or a reaction may be so fast that it is difficult to control.

In Section 7.10 we learned that the yield of a chemical reaction is less than 100% for at least two reasons: (1) There are usually losses due to mechanical factors associated with the separation and purification of the product. (2) There are also competing reactions that convert some of the reactants to undesired products. In this chapter we consider a third reason why the yield of product may be less than 100%. Many reactions do not reach "completion"—they appear to stop, and some of the reactants remain mixed with the products. Chemical reactions are reversible and a condition is reached where there is no net macroscopic change, although submicroscopic processes continue. This balance between reactants and products is a chemical **equilibrium.**

It is important to determine how to convert reactants to products as completely as possible. An understanding of equilibrium processes and competing reactions is necessary in both industrial and biological chemistry. In industry, reactants must be efficiently converted into products. Not only is the inefficient conversion of chemicals costly, but the unwanted reactants must be removed to purify the product. In biological chemistry, equilibrium also has practical consequences. The equilibrium of any biochemical process may be affected by disease, improper diet, or the introduction of foreign substances, so we need to understand how to restore the normal equilibrium state.

In this chapter we will distinguish between kinetics and equilibrium. The composition of a reaction mixture at equilibrium reflects the net direction of chemical change; kinetics deals with the time required to make that change. A rapid reaction may not give a 100% yield of product; a substantial amount of reactant may remain at equilibrium. In contrast, a very slow reaction may give a 100% yield of product.

10.2 Energy and Chemical Reactions

Molecules in a mixture of reactants are in constant motion and collide with one another. Without collisions the atoms could not be redistributed to form new combinations in products. However, chemical reactions require more than a gentle meeting of molecules. Collisions must occur with sufficient energy to break bonds in the reactant molecules (Figure 10.1).

Why Are Reactions Exothermic or Endothermic?

All substances contain stored chemical energy in their bonds. The cleavage of bonds requires energy; the formation of bonds liberates energy. When reactants are converted to products, the stored chemical energies are not the same, because the number and types of bonds are altered.

Heat energy is liberated in an exothermic reaction because more energy is released as bonds are formed than is required by the bonds that are broken. If 50 kcal/mole is required to break the A—B bond (Figure 10.1), and 65 kcal/mole is released when the A—C bond forms, the reaction is exothermic and liberates 15 kcal/mole. Energy is consumed in an endothermic process because the energy re-

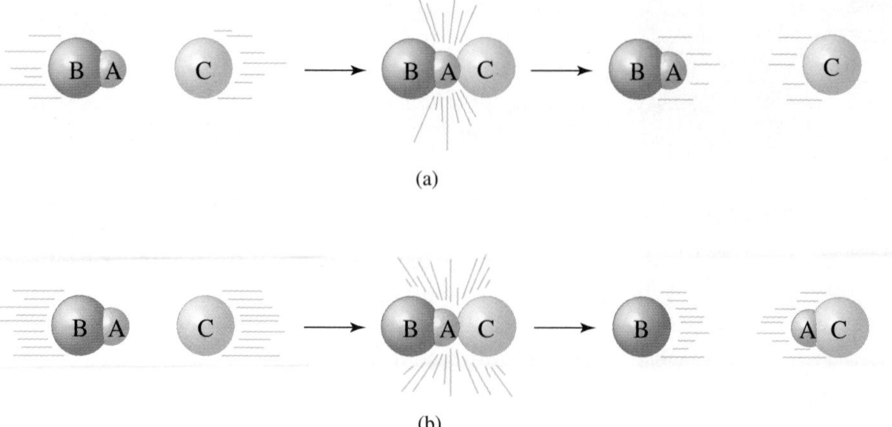

FIGURE 10.1 Molecular Collisions and Reactions
Collisions in a single replacement reaction: (a) The slow-moving molecule AB collides with C and bounces off without reaction. (b) The fast-moving AB molecule collides with enough energy to break the bond between A and B and form a bond between A and C.

quired to break some bonds is greater than the energy released as other bonds form. If 60 kcal/mole is required to break the A—B bond (Figure 10.1), and 55 kcal/mole is released when the A—C bond forms, the reaction is endothermic and requires 5 kcal/mole.

Heat of Reaction

The combustion of methane (natural gas) is exothermic, releasing 211 kcal per mole of methane. The energy released in a chemical reaction can be written on the right side of the chemical equation.

$$CH_4 + 2\,O_2 \longrightarrow CO_2 + 2\,H_2O + 211 \text{ kcal}$$

The energy difference between the products and the reactants is called the **enthalpy change, $\Delta H°$**. We recall that $\Delta H°$ is given a negative sign in an exothermic process, that is, when energy is released. Thus, for the combustion of methane, $\Delta H° = -211$ kcal/mole (Figure 10.2).

$$CH_4 + 2\,O_2 \longrightarrow CO_2 + 2\,H_2O \qquad \Delta H° = -211 \text{ kcal}$$

Light energy is required in some reactions, such as in photosynthesis. Plants produce carbohydrates from carbon dioxide and water, and light energy equivalent to 686 kcal/mole is required for the reaction. The energy required for the reaction is written on the left-hand side as if it were a reactant.

$$686 \text{ kcal} + 6\,CO_2 + 6\,H_2O \longrightarrow C_6H_{12}O_6 + 6\,O_2$$

The enthalpy change for endothermic processes is positive and it is written to the right of the equation. For the photosynthesis process, $\Delta H° = +686$ kcal/mole of glucose formed (Figure 10.2).

$$6\,CO_2 + 6\,H_2O \longrightarrow C_6H_{12}O_6 + 6\,O_2 \qquad \Delta H° = +686 \text{ kcal}$$

FIGURE 10.2 Exothermic and Endothermic Reactions
The difference in enthalpy between reactants and products is determined by the heat energy released or required by the reaction. The combustion of methane (a) is exothermic; the photosynthesis reaction (b) is endothermic.

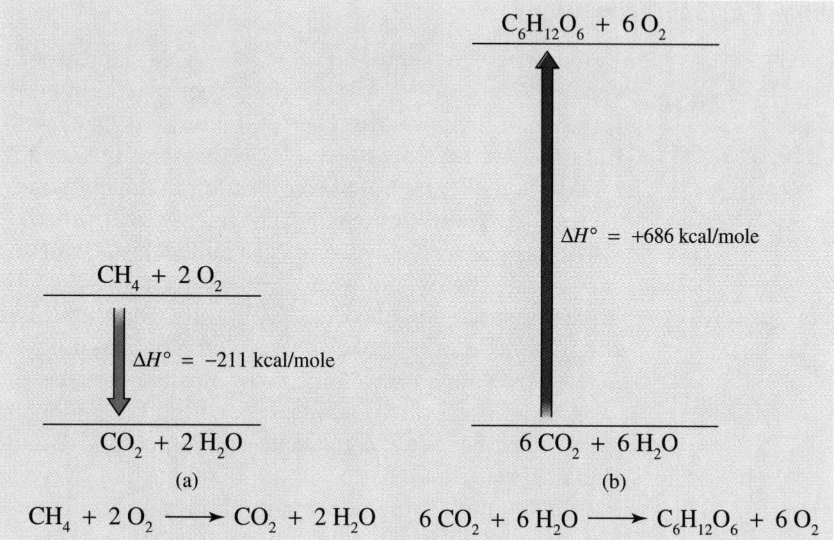

10.3 Spontaneity and Chemical Reactions

You have seen many processes that happen all by themselves—that is, they are **spontaneous.** No one has to make them happen. For example, rocks naturally roll downhill in a landslide. An apple falls from the tree to the ground. Why are these processes spontaneous? Why don't rocks roll uphill, or why doesn't an apple jump from the ground back to the tree? The answer is that such **nonspontaneous** processes would require an input of energy from an outside source.

Most spontaneous chemical reactions are exothermic, as in the case of the combustion of methane. The reaction does require a lighted match to start it; however, the reaction is still considered to be spontaneous because the energy released in combustion is enough to keep the reaction going once it starts. A nonspontaneous reaction will not occur unless energy is constantly added to keep the reaction going. Photosynthesis is a nonspontaneous reaction. Take away the light source and photosynthesis stops.

FIGURE 10.3 A Nonspontaneous Process
This uphill "landslide" is not a spontaneous process. The rocks will roll uphill only if they receive energy from another source.

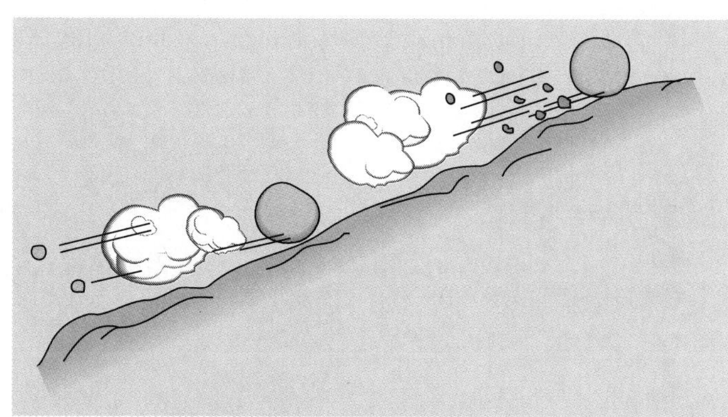

Entropy and Chemical Reactions

You have seen spontaneous processes occur in which there is no net energy change. A hot object and a cold object placed in contact reach some common intermediate temperature. The total energy is conserved, but energy flows from the hot object to the cold object. Heat does not flow spontaneously in the opposite direction to make the hot object hotter and the cold object colder.

The expansion of a gas is another physical process that is spontaneous and involves no energy change. Also, gas molecules do not congregate to give a high pressure in one region and a vacuum in another region. They mix spontaneously; they do not separate spontaneously. The nitrogen molecules in the air of a room won't congregate in the space where you are standing while the oxygen molecules needed for your survival hide away in some corner.

Now let's return to chemical reactions. We understand how an exothermic reaction (negative $\Delta H°$) can occur spontaneously. But there are also endothermic chemical reactions that occur spontaneously. The reaction of solid barium hydroxide octahydrate and solid ammonium chloride is endothermic and spontaneous. This reaction is driven by a feature of nature called entropy, which is related to the disorder of matter.

$$Ba(OH)_2 \cdot 8H_2O + 2\,NH_4Cl \longrightarrow BaCl_2 + 2\,NH_3 + 10\,H_2O$$

This pile of cards has a high entropy. (Kristen Borchmann, Fundamental Photographs)

Matter tends to exist in random or disordered arrangements and it takes special effort to counter this natural tendency. Consider taking a pack of cards and throwing them across the room. The cards spread about and become disordered. They do not become ordered in a neat pile arranged by suit and in increasing card value. To form a perfectly arranged deck of cards, you have to intervene and expend some energy.

The degree of randomness or disorder is called the **entropy** of a system. By definition, the entropy change, $\Delta S°$, is positive for increasing disorder (Figure 10.4). The ordered arrangement of particles changes when a solid melts and the particles are released to form the more random liquid state. An even greater change occurs when a liquid is converted into a gas. In each case the entropy of the material increases. When two gases are mixed, the molecules of both gases are scattered among one another, and a more random arrangement results. The entropy of the gas mixture is greater than the entropy of the individual gases.

Zero entropy is a state of perfect order and no motion which corresponds to a perfect crystal at absolute zero. As the crystal is heated, motion occurs and entropy increases. As components of the solid are converted to liquid, entropy continues to increase.

The entropy change of a chemical reaction depends on the way in which the atoms are bonded. For reactions in which atoms or molecules with simple struc-

FIGURE 10.4 Entropy Changes in Changes of States

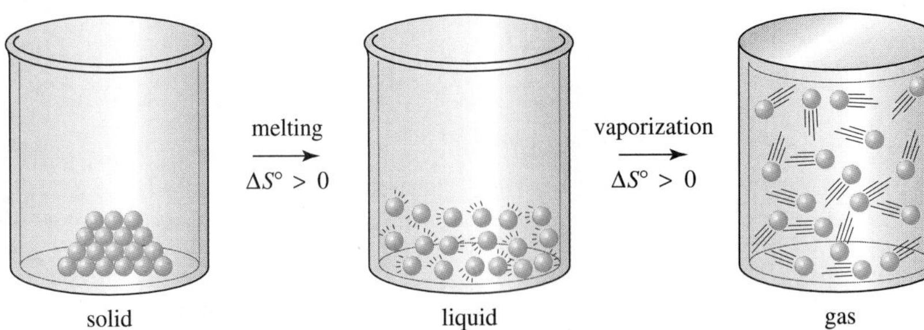

Explosions and Entropy

Explosives are used to quarry granite and to blow rock and soil away to form roads. The military also uses explosives. Regardless of the type of explosive, all explosions are strongly exothermic reactions that occur rapidly. The rapid release of heat energy causes some of the destruction, but the entropy change for the reaction is more important. Consider the reaction for the decomposition of nitroglycerin, the explosive in dynamite.

$$2\ C_3H_6N_3O_6(s) \longrightarrow 6\ CO(g) + 3\ N_2(g) + 6\ H_2O(g)$$

Two molecules of nitroglycerin produce 15 molecules of products. This redistribution of atoms into many molecules results in a large increase in entropy. In addition, the products are all gases, including the water, because the reaction generates a high temperature. The change of state of the substances during the reaction greatly increases entropy. Thus, explosive reactions are spontaneous not only because they are exothermic but also because they result in a large increase in entropy.

The destruction caused by an explosive is due to the gases from the reaction. They are released in a short time period and exert great pressure against everything in the vicinity. If the explosive is confined, the resultant high pressure has to be released by rearranging the confined area—clearly an increase in entropy.

The demolition of a high-rise is often accomplished with explosives. (David Hoffman, Tony Stone Images)

tures assemble to give a product with a complex structure, the entropy change is negative. The number of product molecules formed relative to the number of reactant molecules also contributes to the entropy change of a reaction. If the number of moles of products exceeds the number of moles of the reactants, there is an increase in disorder, and the entropy change is positive.

10.4 Free Energy

We now know that there are two quantities that determine whether a reaction is spontaneous or nonspontaneous. Reactions tend to occur in a direction that releases energy (negative ΔH°) or increases the disorder (positive ΔS°). Some reactions are spontaneous for both reasons. However, there are spontaneous reactions that create order (decrease the entropy), as in the growth of animals. The increase in order (negative ΔS°) must be offset by the exothermic metabolic reactions (negative ΔH°). Similarly spontaneous chemical reactions that are endothermic occur only because of a large positive entropy change.

The relationship between changes in enthalpy and changes in entropy is given by the following expression, in which ΔG° symbolizes the change in the **free energy** of a system at constant pressure:

free energy change ↘ heat of reaction ↓ entropy change ↙

$$\Delta G^\circ = \Delta H^\circ - T\,\Delta S^\circ$$

The free energy change is a measure of the tendency of a reaction to proceed spontaneously. In a spontaneous reaction, the free energy of the products is lower than the free energy of the reactants and ΔG° is negative. A negative enthalpy change contributes to making ΔG° negative, as does a positive entropy change.

We know that reactions with a decrease in enthalpy (negative ΔH°) and an increase in entropy (positive ΔS°) are spontaneous. Similarly a reaction with an increase in enthalpy (positive ΔH°) and a decrease in entropy (negative ΔS°) cannot occur spontaneously, because ΔG° would be positive.

Now let's consider some situations that are not as clear cut. A reaction with a negative ΔS° can occur, but only if the reaction is sufficiently exothermic. A positive entropy change might make a reaction favorable in spite of a positive enthalpy change. A summary of the contributions of ΔH° and ΔS° to ΔG° and the resultant prediction of the spontaneity of a chemical reaction is given in Table 10.1.

The contribution of the entropy change to the free energy change depends on temperature. At low temperatures the product $T\Delta S^\circ$ will be less important than the ΔH°. Thus, an exothermic reaction can counter an unfavorable decrease in entropy at some sufficiently low temperature. At high temperatures the product $T\Delta S^\circ$ can become very important and overwhelm an unfavorable ΔH° term. Thus, an endothermic reaction can occur along with a favorable increase in entropy at some sufficiently high temperature.

The decomposition of dinitrogen tetroxide to give nitrogen dioxide illustrates how temperature affects the spontaneity of a reaction. The decomposition is not spontaneous at 25 °C.

$$N_2O_4 \rightleftharpoons 2\,NO_2 \qquad \Delta H^\circ = 13.7 \text{ kcal/mole}$$

The ΔH° for the reaction is positive; however, the ΔS° is positive because 1 mole of reactant gives 2 moles of product. Let's see what effect temperature has on the reaction. Using the experimental value for ΔS°, which is 42 cal mole^{-1} deg^{-1}, we calculate a positive ΔG° at 25 °C and the reaction is not spontaneous.

$$\begin{aligned}\Delta G^\circ &= \Delta H^\circ - T\,\Delta S^\circ \\ &= 13{,}700 \text{ cal/mole} - 298 \text{ K } [42 \text{ cal/(mole}\cdot\text{deg)}] \\ &= 13{,}700 \text{ cal/mole} - 12{,}600 \text{ cal/mole} \\ &= 1100 \text{ cal/mole}\end{aligned}$$

TABLE 10.1 Contribution of ΔH° and ΔS° to ΔG°

ΔH°	ΔS°	$\Delta G^\circ = \Delta H^\circ - T\,\Delta S^\circ$	Result
negative	positive	negative at all temperatures	Reaction will be spontaneous.
positive	negative	positive at all temperatures	Reaction will not be spontaneous.
negative	negative	negative if the temperature is sufficiently low	Reaction might be spontaneous at sufficiently low temperature.
positive	positive	negative if the temperature is sufficiently high	Reaction might be spontaneous at sufficiently high temperature.

Free Energy and Biological Processes

Plants increase their free energy but only by obtaining energy from sunlight. Photosynthesis produces biological molecules by processes in which ΔH° is positive. In addition, ΔS° is negative because the biological molecules are more highly organized than carbon dioxide and water, the two reactants.

Animals obtain chemical energy contained in the biological molecules of plants by metabolic processes. This energy is used by cells in a series of reactions to form the biological molecules necessary to maintain the cell. This energy is not used immediately but rather is stored in energy-rich molecules that move throughout the organisms and transport the energy to sites where it is required. Adenosine triphosphate (ATP) is the most important transport molecule. It is formed when adenosine diphosphate (ADP) links with phosphoric acid in a nonspontaneous reaction that occurs only because free energy is provided from the metabolism of plant biological molecules.

$$\text{ADP} + H_3PO_4 \rightleftarrows \text{ATP} + H_2O \qquad \Delta G^\circ = 7.3 \text{ kcal/mole}$$

When required, the energy stored in ATP is provided by the reverse of the reaction by which it is formed. The reaction is spontaneous.

$$\text{ATP} + H_2O \rightleftarrows \text{ADP} + H_3PO_4 \qquad \Delta G^\circ = -7.3 \text{ kcal/mole}$$

This energy can be transferred to a nonspontaneous reaction such as the conversion of the amino acid glutamic acid into another amino acid, glutamine. The ΔG° for this reaction is +3.3 kcal/mole.

The conversion of glutamic acid into glutamine can occur only because ATP provides free energy by its hydrolysis. The net effect of the two reactions is obtained by summing them. Thus, although the conversion of glutamic acid into glutamine is not spontaneous, the conversion can occur in a cell because ATP is simultaneously converted into ADP to provide the necessary energy.

This runner in a high school cross country final in Texas uses stored ATP for energy. (The Image Works)

$$\text{glutamic acid} + NH_3 \longrightarrow \text{glutamine} + H_2O \qquad \Delta G^\circ = +3.3 \text{ kcal/mole}$$

(structures: glutamic acid, HO–C(=O)–CH₂–CH₂–CH(NH₂)–C(=O)–OH; glutamine, NH₂–C(=O)–CH₂–CH₂–CH(NH₂)–C(=O)–OH)

glutamic acid | glutamine

$$\begin{array}{lr} \text{glutamic acid} + NH_3 \longrightarrow \text{glutamine} + H_2O & \Delta G^\circ = +3.3 \text{ kcal/mole} \\ \text{ATP} + H_2O \longrightarrow \text{ADP} + H_3PO_4 & \Delta G^\circ = -7.3 \text{ kcal/mole} \\ \hline \text{glutamic acid} + NH_3 + \text{ATP} \longrightarrow \text{glutamine} + H_3PO_4 + \text{ADP} & \Delta G^\circ = -4.0 \text{ kcal/mole} \end{array}$$

At 75 °C we calculate a negative free energy change and the reaction is spontaneous. This difference reflects the greater contribution of ΔS° at the higher temperature.

$$\begin{aligned} \Delta G^\circ &= \Delta H^\circ - T\Delta S^\circ \\ &= 13{,}700 \text{ cal/mole} - 348 \text{ K } [42 \text{ cal/(mole·deg)}] \\ &= 13{,}700 \text{ cal/mole} - 14{,}700 \text{ cal/mole} \\ &= -1000 \text{ cal/mole} \end{aligned}$$

You won't do calculations of this type in this book. The example for the decomposition of N_2O_4 is provided only to illustrate the roles of enthalpy, entropy, and temperature in determining the spontaneity of a reaction. We will return to a discussion of free energy changes in Chapter 25, where we give an overview of metabolic reactions. Living systems cannot significantly change their temperature. Thus, we will consider the free energy change for metabolic reactions only at 37 °C.

10.5 Kinetics

You have just learned that spontaneous reactions proceed naturally without continued help from us. However, you know that the detonation of dynamite requires that someone light a fuse. The reason that some reactions have to be initiated will be discussed in Section 10.6. For now, spontaneity means only that given the opportunity, the reaction will occur.

Reactions occur at a variety of rates. The detonation of dynamite doesn't occur in slow motion over a period of hours, days, or years. The rusting of iron is also a spontaneous reaction but requires weeks and years under most conditions—you can't see the rust form before your eyes. Thus, there is no connection between the spontaneity of a reaction and the speed with which it occurs.

Kinetics is the study of the rates of chemical reactions, that is, of how quickly reactants are converted into products. Factors that affect reaction rates include the nature of the reactants, temperature, concentration of reactants, and the presence of substances called catalysts. We will first review the factors that control the rates of reaction and then propose a model in Section 10.6 to account for the facts.

Reactants and Reaction Rates

The identity and concentrations of the reactants control the reaction rate. For example, in the reaction of CO with NO_2, an oxygen atom from NO_2 is transferred to CO. Bonding occurs between carbon and oxygen, and a bond between nitrogen and oxygen is broken.

The shape of the reacting molecules and their orientation at collision is also important. If the molecules don't collide in just the right way, the atoms will not be aligned to form bonds. The carbon atom of CO must collide with the oxygen atom of NO_2 to break and form bonds (Figure 10.5). If the carbon atom of CO collides

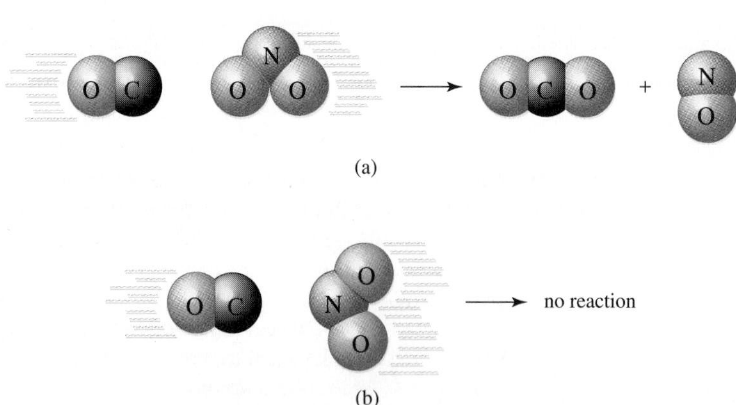

FIGURE 10.5 Molecular Orientation in Chemical Reactions *The orientation of CO and NO_2 is correct for an effective collision leading to a reaction in (a). The ineffective collision shown in (b) will not lead to a reaction.*

with the nitrogen atom of NO_2, no reaction occurs. Furthermore the orientation of the atoms at collision must be quite precise. Just as in basketball, close doesn't count. You have to do more than just hit the rim to score points.

Molecules with similar structures may react at different rates. The colorless gas NO (nitric oxide) produced in an internal combustion engine reacts very quickly with oxygen at room temperature to form the reddish brown gas NO_2 (nitrogen dioxide) that is responsible in part for the smog in some metropolitan areas.

$$2\,NO(g) + O_2(g) \longrightarrow 2\,NO_2(g)$$

In contrast, the colorless compound CO (carbon monoxide), produced from the incomplete combustion of gasoline in a car engine, reacts very slowly with oxygen at room temperature.

$$2\,CO(g) + O_2(g) \longrightarrow 2\,CO_2(g)$$

As a consequence, the CO level may build up in the atmosphere. If the reaction of CO with oxygen were as rapid as that of NO, carbon monoxide poisoning would not occur.

Part of the difference in the reactivity of NO and CO with O_2 is due to differences in the bonding in these two molecules. The carbon atom in CO has an octet of electrons, but the nitrogen atom in nitric oxide does not. The reactivity of NO is the result of a deficiency of electrons and the unpaired electron located on the nitrogen atom.

$$:C{\equiv}O: \qquad \cdot\ddot{N}{=}\ddot{O}:$$

Concentration and Reaction Rates

The burning of wood involves a heterogeneous mixture in which oxygen gas reacts with a solid. The reactants must come into contact with each other at the surface of the solid to react. You know that wood shavings burn faster than a solid log in a fireplace. When wood is chopped into fine kindling, more surface area is created and more of the solid comes into contact with the oxygen of air.

For reactions in homogeneous mixtures, the reactants must also come into contact. As the concentration of reactants, is increased, the rate of reaction increases because there is a higher probability that the reactant molecules will collide. In the liquid phase, reactant concentrations are increased by adding reactant. In a gas the reaction rate is increased either by increasing the amount of reactants in a constant volume or by decreasing the volume containing the reactants (increasing the pressure).

The effect of pressure as well as temperature on a reaction rate is seen in the reaction of nitrogen and oxygen to give nitrogen oxides in the cylinder of an automobile engine. Nitrogen and oxygen do not react to any noticeable degree in an ordinary flame at atmospheric pressure. In the cylinder of the engine the hot gases are compressed, so the molecules are closer to each other than in an open fire. Thus, reaction is more likely at high pressures. To decrease the amount of nitrogen oxides produced in automobiles, the compression ratio of automobile engines has been decreased. This change has also decreased the power of engines.

EXAMPLE 10.1

A pile of flour will burn slowly if heated with a flame. However, flour distributed as dust in the air forms an explosive mixture that can blow a flour mill building apart. Why?

Solution

The flour dust consists of much smaller particles. The millions of tiny specks of flour have more surface area than the pile of flour. Furthermore, each of the flour dust particles is surrounded by oxygen molecules in the air. A larger proportion of the flour molecules are in contact with oxygen molecules, and the chances of a collision to give a reaction are increased.

PROBLEM 10.1

Airline passengers on international flights are advised to extinguish all smoking materials if oxygen masks are deployed in an emergency. Explain why.

Temperature and Reaction Rates

Chemical reactions occur when food is heated. You can decrease the time required to cook food in an oven by using a hotter oven. Of course, you may also end up with a flat cake or a tough roast, but that's another matter.

Each reaction has a different sensitivity to temperature changes. A general rule of thumb is that on the average a 10 °C rise in temperature doubles the reaction rate.

$$\text{rate increase} = 2^n$$

where n = number of 10 °C increments. How much faster will a cooking reaction proceed at 100 °C than at 70 °C based on this general rule? The 30° change equals three 10° increments, so the reaction will be $2^3 = 8$ times faster.

Decreasing the temperature causes a decrease in the rates of all metabolic reactions. The body temperature of some patients is lowered in some surgical procedures. Without oxygen from blood circulation, the brain tissue would be damaged in a few minutes at normal body temperature. However, if the body temperature is lowered from 37 °C to 20 °C, the brain can be deprived of oxygen for about an

The Effect of Temperature on Organisms

Chemical reactions in living organisms are affected by changes in temperature. Increasing the temperature of the organism results in an increase in its metabolic rate. In humans with a fever, the metabolic rate increases by 5% for each degree Fahrenheit above normal body temperature. As a consequence, food intake must increase or a weight loss will result.

Lowering the temperature slows down the reaction rates of the body. Animals that hibernate survive because their body temperatures are so low that their need for food is minimal. A woodchuck's normal heart rate decreases from about 75 beats/min to about 5 beats/min during hibernation. The rates of metabolic reactions of stored body fat are decreased, and the woodchuck can survive the hibernation period.

Studies on some animals have shown that those with slower metabolic rates live longer. For example, the average life span of a fruit fly at 30 °C is 25 days but at 20 °C it is 100 days because its metabolic rate is decreased at the lower temperature. The same effect is not observed in warmblooded animals. Cold temperatures actually produce shorter life spans. Warmblooded animals increase their metabolic rate in the cold in order to stay warm.

hour and the tissue does not deteriorate. This procedure is used in some types of heart surgery when circulation must be interrupted for a prolonged period. After surgery is complete, circulation is resumed, and the patient's body temperature is gradually returned to normal.

EXAMPLE 10.2

Milk spoils rapidly at room temperature but remains fresh for a week or more in a refrigerator. Explain why.

Solution

The spoiling of milk involves chemical reactions, and these reactions occur faster at room temperature than at the temperature in the refrigerator. However, given sufficient time, even milk stored in a refrigerator will spoil. It is for this reason that expiration dates are stamped on milk and milk products such as cottage cheese.

PROBLEM 10.2

Unripened fruit is left at room temperature rather than placed in a refrigerator. Explain why.

Catalysts and Reaction Rates

A **catalyst** is a substance that increases a reaction rate but undergoes no permanent change. The catalyst is said to catalyze the reaction, and its effect is known as catalysis. A catalyst is required only in small amounts; it is present in the same amount

Catalysts in Living Systems

Enzymes are natural catalysts that make reactions in an organism occur fast enough for its survival. Every species requires many enzymes because each enzyme is usually highly specific in its catalysis of one reaction. In our body, glucose is metabolized efficiently at 37 °C.

$$C_6H_{12}O_6 + 6\,O_2 \longrightarrow 6\,H_2O + 6\,CO_2$$

The conversion occurs in several steps, each catalyzed by a specific enzyme. Outside the body, the same combustion occurs rapidly only above 600 °C. At body temperature, the chemical conversion without enzymes would require months, a rate that would be too slow to provide the energy necessary to support life.

Nitrogen fixation in plants is another important process catalyzed by enzymes. The nitrogen gas in the atmosphere is the principal natural source of ammonia and other nitrogen compounds essential for protein formation. Nitrogen gas is quite unreactive, although it is converted into nitrogen compounds by lightning discharges. Under less dramatic conditions, the bacteria in the roots of leguminous plants such as alfalfa, beans, clover, and peas readily convert nitrogen gas into ammonia. These bacteria contain enzymes that catalyze the conversion under mild conditions and provide the largest source of nitrogen compounds for organisms.

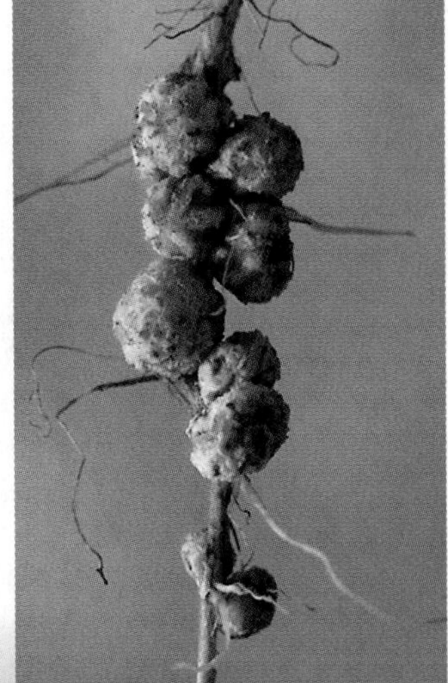

Nodules in bean roots contain nitrogen fixing bacteria. (Photo Researchers)

before and after the reaction, even though it must interact with the reactant during the reaction. Consider the decomposition reaction of hydrogen peroxide.

$$2\,H_2O_2 \longrightarrow 2\,H_2O + O_2$$

Although hydrogen peroxide is unstable, it can be stored in opaque plastic bottles in your medicine cabinet. A platinum catalyst can increase the rate of decomposition by a factor of 10^4 at room temperature. An enzyme in our cells accelerates the reaction rate by about 10^8 times the uncatalyzed rate. You see evidence of this rapid rate when you use hydrogen peroxide to clean a wound. The bubbling that immediately results is from the oxygen gas released.

The success of industrial chemical processes depends on finding specific catalysts. A specific catalyst accelerates one reaction in preference to another possible competing reaction. For example, the reaction of carbon monoxide and hydrogen can produce either methane or methyl alcohol depending on the catalyst.

$$CO + 3\,H_2 \xrightarrow{\text{Ni}} CH_4 + H_2O$$

$$CO + 2\,H_2 \xrightarrow{\text{ZnO/Cr}_2\text{O}_3} CH_3OH$$

Catalysts are economically desirable for industrial processes because they help make reactions go faster at lower temperatures. Hence, the energy costs are lower.

10.6 Activation Energy

Not every collision between reactant molecules generates products. In most collisions the molecules simply bounce off each other. Collisions between molecules that result in a chemical reaction are called *effective collisions.* The minimum energy of an effective collision in a chemical reaction is called the **activation energy,** E_a. Molecules colliding with less than the activation energy rebound without reaction. This is analogous to the collision of two cars at 5 mph while stopping for a traffic light; the bumpers may withstand the collision without any change. However, at higher speeds, damage will occur. The damage in a collision also depends on the construction of the car. Similarly, each reaction has a characteristic activation energy that depends on the number and type of bonds in the reactants.

How does temperature affect the rate of a reaction? Recall from Chapter 8 that the kinetic energy of molecules is directly proportional to the absolute temperature. As the kinetic energy increases, so does the number of collisions. However, more important is the increase in the number of molecules that have energy equal to or greater than the activation energy.

Before going into more detail about what happens in a reaction, let's consider the process of moving a rock as shown in Figure 10.6. The rock would have a lower potential energy at the lower level, but it does not spontaneously move there because the rock must first be moved to the top of the hill. It takes a certain amount of energy to push the rock up the hill, and the potential energy of the rock is increased by that amount. When the rock rolls down the other side of the hill, its potential energy decreases. The difference between the energy required to move the rock to the top of the hill and the energy released as the rock rolls to the bottom is equal to the difference between its potential energy in the original position and its potential energy in the final position.

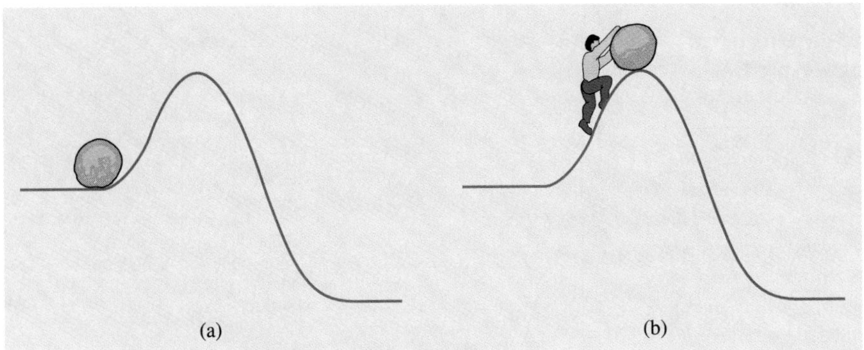

FIGURE 10.6
A Physical Analogy for Activation Energy
The potential energy of the rock in (a) is higher on the left side of the hill than it would be on the right side. However, energy must first be expended to get the rock to the top of the hill (b), where it is in position to roll spontaneously to the lower level.

Like the movement of the rock, a spontaneous reaction still requires energy to get the molecules into position to react. In the process of reacting, the atoms change their arrangement as bonds are distorted and eventually broken and new bonds are formed. Thus, the energy of the reacting system changes as the configuration of the atoms in the reactants is altered to give the configuration of the atoms in the products. All such atomic redistributions initially have a higher energy than that of the reactants. This increase in energy is the result of repulsions between atoms in the reactants as they are moved close together. Eventually a structure of maximum energy results—the **activated complex** (Figure 10.7). Subsequent changes in structure lead to products or back to the original reactants by a reversal of the changes that took place up to that point.

The vertical axis of the **reaction coordinate diagram** is the potential energy of the reacting system. The horizontal axis represents the progress of the reaction from the beginning (reactants) at the left to the end (products) at the right.

In the combustion of methane a flame can be used to increase the temperature so that more methane molecules have the necessary activation energy to react with oxygen. The potential energy of the reactant molecules must be increased by an amount at least equal to their activation energy to boost them to the top of the en-

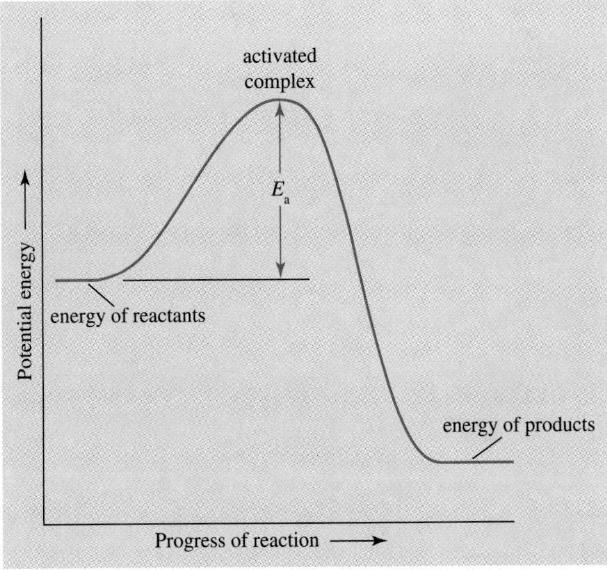

FIGURE 10.7
Potential Energy and Progress of a Reaction
This reaction coordinate diagram shows that the potential energy of the activated complex is greater than the energy of either the reactants or the products.

FIGURE 10.8 Activation Energy of an Exothermic Reaction
After the highest energy point is reached by adding energy, the activation energy and energy of reaction are released. Since the activation energy was needed to cause the reactants to reach the high point of potential energy, the net energy released is equal to the heat of the reaction.

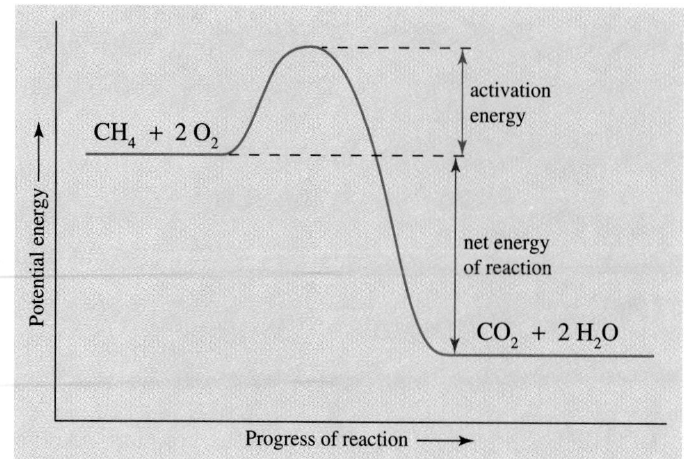

ergy hump shown in Figure 10.8. The energy released is equal to the activation energy originally added plus the energy released in the exothermic reaction.

Exothermic and Endothermic Reactions

In the combustion of methane the net release of energy is 211 kcal/mole. The energy released in this exothermic reaction is enough to continue to increase the kinetic energy of the remaining reactant molecules, to give them energy at least equal to the activation energy. The reaction is then self-sustaining, and the methane continues to burn.

Endothermic reactions also require an activation energy for the molecules to react (Figure 10.9). As the molecules proceed from the point of highest energy to the

FIGURE 10.9 Activation Energy of an Endothermic Reaction
After the highest energy point is reached by adding the activation energy, some energy is released. The energy released is less than the activation energy. Thus, a net energy input is required for the reaction.

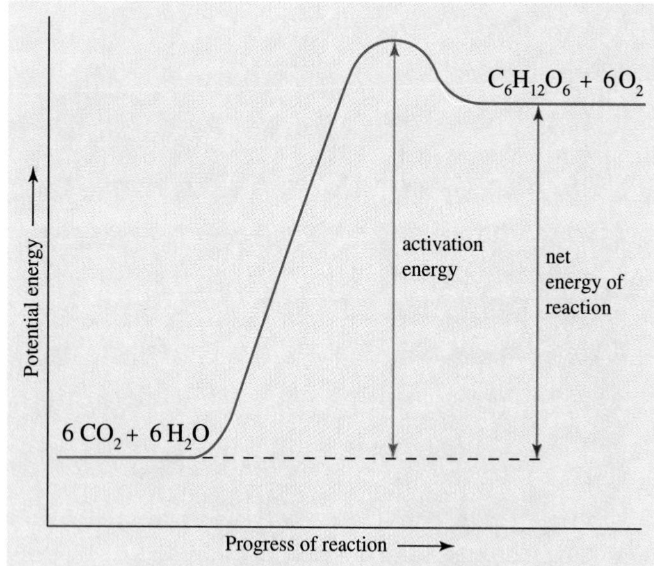

product molecules, some energy is released, but it is less than the activation energy. Thus, energy must be continually added to keep the reaction going.

Catalysts and the Reaction Path

Why do catalysts increase the rate of a reaction? A catalyst often provides a different path for the progress of the reaction and makes more of the collisions effective collisions. The path still starts at the reactants and ends at the products, but the catalyzed-reaction path has a different activation energy (Figure 10.10). For example, the activation energy for the decomposition of hydrogen peroxide (Section 10.5) is 18 kcal/mole. For the platinum-catalyzed reaction the activation energy is 13 kcal/mole, and for the enzyme-catalyzed reaction it is only 7 kcal/mole.

How can a catalyst provide a different path for a reaction? Consider the hypothetical reaction of A and B.

$$A + B \longrightarrow A{-}B$$

If few molecules have the necessary activation energy at a specific temperature, the reaction is slow. However, in the presence of a catalyst, such as an enzyme, represented by E, the following reactions may occur.

$$\text{Step 1} \quad A + E \longrightarrow A{-}E$$
$$\text{Step 2} \quad A{-}E + B \longrightarrow A{-}B + E$$

If the activation energy of each step is low, a larger fraction of molecules will be able to react faster via this enzyme-catalyzed pathway than they could without the enzyme at the same temperature.

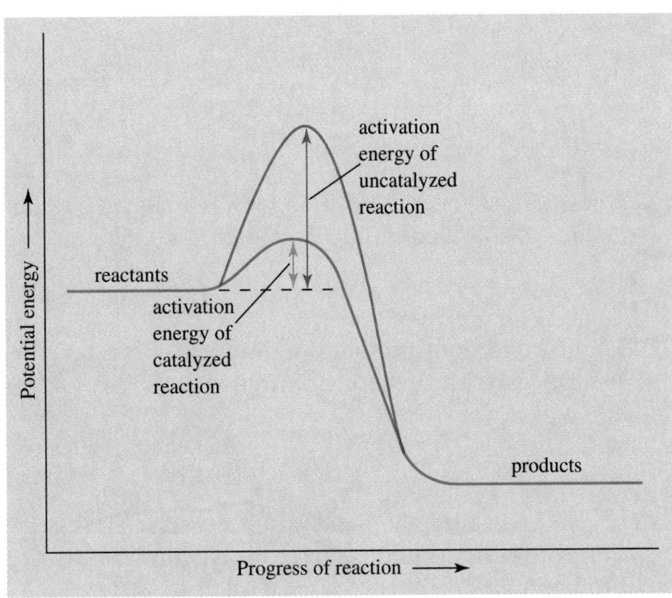

FIGURE 10.10 Activation Energy and the Effect of Catalysts *The catalyst provides an alternative pathway for a reaction. The activation energy for the process is lower than for the uncatalyzed reaction. This lower energy requirement results in a faster reaction. The catalyzed reaction may involve a different number of steps.*

Catalytic Converters

The internal combustion engine that contributes so much to a modern society is also responsible for one of our major problems—air pollution. As a result of incomplete combustion, both carbon monoxide and unburned gasoline are released to the atmosphere. At the high temperature and pressure of the reaction, nitrogen and oxygen combine to give NO, another air pollutant. These substances can be removed before they are released to the air by passing the combustion products through a catalytic converter.

Catalytic converters for automobiles contain mixtures of the transition metals platinum, palladium, rhodium, and nickel. Components of the catalytic converter speed the complete oxidation of both carbon monoxide and gasoline (represented here as C_8H_{18}).

$$2\,CO(g) + O_2(g) \longrightarrow 2\,CO_2(g)$$
$$2\,C_8H_{18}(g) + 25\,O_2(g) \longrightarrow 16\,CO_2(g) + 18\,H_2O(g)$$

The decomposition of nitric oxide to form nitrogen and oxygen at atmospheric pressure is spontaneous; however, the rate of the reaction is slow. The catalytic converter accelerates the reaction sufficiently to rid the exhaust of the nitric oxide.

$$2\,NO(g) \longrightarrow N_2(g) + O_2(g)$$

10.7 Chemical Equilibrium

In Chapters 8 and 9 we discussed equilibrium in physical processes. For example, water vapor can exist in equilibrium with liquid water. At equilibrium the rates at which molecules leave and return to the liquid are equal. Similarly, in osmosis, water molecules cross the semipermeable membrane in both directions until an equilibrium is achieved based on concentration and pressure. In physical equilibria opposing processes occur constantly, although there is no net macroscopic change.

As a reaction occurs, product molecules form that may then collide with each other and re-form the original reactants. Thus, two opposing reactions occur, and when the rate of product formation equals the rate of reactant formation, an equilibrium is established. Because catalysts increase the rate of both forward and reverse reactions, they do not affect the position of the equilibrium.

$$\text{reactants} \underset{\text{reverse reaction}}{\overset{\text{forward reaction}}{\rightleftharpoons}} \text{products}$$

The Equilibrium Constant

Consider the general equation for a reaction at equilibrium, where A and B are reactants, X and Y are products, and *m, n, p,* and *q* are coefficients.

$$m\,A + n\,B \rightleftharpoons p\,X + q\,Y$$

For this reaction the following ratio is a constant at a specific temperature. The brackets indicate molar concentrations

$$\frac{[X]^p[Y]^q}{[A]^m[B]^n} = K$$

where K is called the **equilibrium constant** and is calculated from the concentrations of reactants and products at equilibrium. By this convention, reactions that proceed to the right have large K values.

The expression of the equilibrium constant for any reaction is written by inspection of the balanced equation. For example, hydrogen and iodine combine to form hydrogen iodide.

$$H_2 + I_2 \rightleftharpoons 2\,HI$$

According to the generalized form, the equilibrium constant expression is

$$K = \frac{[HI]^2}{[H_2]\,[I_2]}$$

The concentration of hydrogen iodide is squared because the coefficient of hydrogen iodide in the balanced equation is 2.

EXAMPLE 10.3

Ammonia is produced commercially by the reaction of nitrogen and hydrogen gases at high pressure and temperature in the presence of a catalyst. Write an equilibrium constant expression for the reaction.

$$N_2 + 3\,H_2 \rightleftharpoons 2\,NH_3$$

Solution

First, write the concentration of ammonia, the product, in the numerator, and square it because the coefficient of NH_3 in the equation is 2.

$$[NH_3]^2$$

Next, place the concentration of nitrogen, a reactant, in the denominator. The power of 1 is assumed.

$$\frac{[NH_3]^2}{[N_2]}$$

Finally, place the concentration of hydrogen, the other reactant, in the denominator, and cube it because the coefficient of H_2 in the equation is 3.

$$\frac{[NH_3]^2}{[N_2]\,[H_2]^3} = K$$

PROBLEM 10.3

In the commercial production of sulfuric acid, sulfur dioxide reacts with oxygen to produce sulfur trioxide. Write the equilibrium constant expression.

$$2\,SO_2 + O_2 \rightleftharpoons 2\,SO_3$$

Equilibrium Calculations

The equilibrium constant expression gives the relationship between the concentrations of reactants and products at equilibrium. If the concentrations of all reactants and products are known at equilibrium, the equilibrium constant can be calculated by substituting the concentrations into the equilibrium expression. If the equilibrium constant is known, the equilibrium concentration of one substance can be calculated provided the concentrations of the other substances are known.

EXAMPLE 10.4

Consider the following reaction at equilibrium, where the concentrations are 5.6×10^{-3} M CO, 5.6×10^{-3} M H_2O, 4.4×10^{-3} M CO_2, and 4.4×10^{-3} M H_2 at 1000 °C. What is the equilibrium constant?

$$CO(g) + H_2O(g) \rightleftharpoons CO_2(g) + H_2(g)$$

Solution

First, write the equilibrium constant expression.

$$\frac{[CO_2]\,[H_2]}{[CO]\,[H_2O]} = K$$

Next, substitute the equilibrium concentrations into the expression, and solve the equation.

$$\frac{(4.4 \times 10^{-3}\text{ M})\,(4.4 \times 10^{-3}\text{ M})}{(5.6 \times 10^{-3}\text{ M})\,(5.6 \times 10^{-3}\text{ M})} = K = 0.62$$

PROBLEM 10.4

Consider the following equilibrium reaction in which the equilibrium concentrations of H_2 and I_2 are both 0.25 M. What is the equilibrium concentration of HI if the equilibrium constant is 57?

$$H_2(g) + I_2(g) \rightleftharpoons 2\,HI(g)$$

Equilibrium and Completeness of Reactions

For an equilibrium in which the product of the concentrations of the materials raised to the appropriate powers in the numerator is larger than the product of the concentrations of the materials raised to the appropriate powers in the denominator, the equilibrium constant is large. Conversely, for a small equilibrium constant the numerical value of the numerator is smaller than the numerical value of the denominator.

Consider the combination reaction of ethylene and hydrogen bromide to form bromoethane.

$$CH_2{=}CH_2 + HBr \rightleftharpoons CH_3CH_2Br$$

$$\frac{[CH_3CH_2Br]}{[CH_2{=}CH_2]\,[HBr]} = K = 10^8$$

For all practical purposes very little of the reactants remain, at equilibrium, since the equilibrium constant is very large. Reactions with equilibrium constants greater than 10^3 can be considered "complete" because the amount of reactant left is about 0.1%.

Now let's consider the reaction of acetic acid and ethyl alcohol to produce ethyl acetate, and the related equilibrium constant expression.

$$CH_3-\overset{\overset{\displaystyle :\ddot{O}}{\|}}{C}-\ddot{\underset{..}{O}}-H + CH_3CH_2-OH \rightleftharpoons CH_3-\overset{\overset{\displaystyle :\ddot{O}}{\|}}{C}-\ddot{\underset{..}{O}}-CH_2CH_3 + H_2O$$

acetic acid — ethyl alcohol — ethyl acetate

$$\frac{[CH_3CO_2CH_2CH_3]\,[H_2O]}{[CH_3CH_2OH]\,[CH_3CO_2H]} = K = 4.0$$

The equilibrium constant for this reaction is close to 1. There will be a significant amount of the reactants still present at equilibrium. Thus, the yield of product is less than 100% based on the balanced equation.

What Affects Equilibrium Constants?

Catalysts do not change the equilibrium constant. The catalyst affects both the forward and reverse reactions and there is no difference in the amount of products and reactants at equilibrium between the catalyzed and uncatalyzed reaction.

Equilibrium constants do depend on temperature. For reactions that are exothermic as written in the forward direction, the equilibrium constant decreases with increasing temperature. For reactions that are endothermic in the forward direction as written, the equilibrium constant increases with increasing temperature (Table 10.2).

10.8 Chemical Equilibrium and Le Châtelier's Principle

Le Châtelier's principle applies to chemical as well as physical equilibria even though the physical concept of stress is not as easily visualized in chemical reactions. Experimentally it has been shown that a change in the conditions of a chemical equilibrium usually causes a change in concentrations of reactants and products to establish a new equilibrium. Le Châtelier's principle provides some handy rules to predict the directions of these changes. The three "stresses" that we will consider are changes in concentration, pressure, and temperature. The first two changes do not affect the value of the equilibrium constant; changes in concentration and pressure affect only the concentrations of the substances at equilibrium. Temperature changes alter the value of the equilibrium constant as well as the concentrations of the substances at equilibrium.

TABLE 10.2 Effect of Temperature on the Equilibrium Constant

Exothermic Reaction		**Endothermic Reaction**	
$CO + H_2O \rightleftharpoons CO_2 + H_2$	$\Delta H° = -9.9$ kcal	$N_2 + O_2 \rightleftharpoons 2\,NO$	$\Delta H° = 43.2$ kcal
***T*(°C)** (increase *T* ↓)	***K*** (decrease *K* ↓)	***T*(°C)** (increase *T* ↓)	***K*** (increase *K* ↓)
600	3.5	1600	2.5×10^{-4}
800	1.23	1800	7.3×10^{-4}
1000	0.60	2000	1.9×10^{-3}
1200	0.35	2200	4.0×10^{-3}

Concentration Changes

If some reactant is added to a reaction mixture to increase its concentration, the concentrations of other reactants as well as products change in a way that maintains the equilibrium constant. Some of the added reactant reacts, and the concentration of the products increases. In short, the stress imposed on the system is relieved by the reaction of some of the added material. If a product is removed from a chemical reaction at equilibrium, more reactants react to replace part of the product removed. Regardless of the condition imposed on the chemical equilibrium, the concentrations are ultimately changed so that the same value of the equilibrium constant results.

Consider the equilibrium that exists between glucose in the blood and glycogen stored in the liver.

$$n \text{ glucose} \rightleftharpoons n\ H_2O + \text{glycogen}$$

Glycogen is a large molecule containing many covalently bonded glucose molecules. These bonds can be broken with water to give free glucose molecules. At equilibrium the body has 65–100 mg of glucose per 100 mL of blood. When a person does work, the blood glucose level decreases. The system responds by producing more glucose from stored glycogen in the liver. Conversely, when excess glucose is ingested, the liver converts it into glycogen.

EXAMPLE 10.5

Consider the following reaction used in the commercial production of diethyl ether from ethyl alcohol. What change in the position of equilibrium would result by removing H_2O?

$$2\ CH_3CH_2OH\ (l) \rightleftharpoons CH_3CH_2OCH_2CH_3\ (l) + H_2O\ (l)$$

Solution

The removal of H_2O, one of the products, will cause the reaction to replace some of the water. The reaction will proceed to the right, and more of the ethyl alcohol will be converted into the desired diethyl ether product.

PROBLEM 10.5

Consider the following reaction at equilibrium, which is used to produce carbon disulfide, a solvent used in manufacturing cellophane. What change in the position of the equilibrium would result if H_2 were added?

$$CH_4(g) + 2\ H_2S(g) \rightleftharpoons CS_2(g) + 4\ H_2(g)$$

Pressure Changes

If the pressure of a gas in a reaction is increased, the system responds by processes that tend to decrease the increased pressure. In the hemoglobin–oxyhemoglobin reaction, the amounts of the oxygenated forms of hemoglobin are controlled by the pressure of oxygen.

$$\begin{aligned} Hb + O_2 &\rightleftharpoons HbO_2 \\ HbO_2 + O_2 &\rightleftharpoons Hb(O_2)_2 \\ Hb(O_2)_2 + O_2 &\rightleftharpoons Hb(O_2)_3 \\ Hb(O_2)_3 + O_2 &\rightleftharpoons Hb(O_2)_4 \end{aligned}$$

Both humans and animals in Nepal develop higher than normal concentrations of hemoglobin. (Burbank, The Image Works)

High pressures of oxygen, such as those in a hyperbaric chamber, convert the hemoglobin (Hb) into oxygenated hemoglobin. The increased amount of oxyhemoglobin transports more oxygen to the cells. Conversely, at lower oxygen concentrations at high altitudes, the amount of oxyhemoglobin is decreased. Oxygen deprivation, or *hypoxia,* occurs because insufficient oxygen is available for the cells in all tissues, and the body cannot function as efficiently. This is why you tire more easily at high altitudes. The percent saturation of hemoglobin carrying four oxygen molecules is shown in Figure 10.11 in relation to the partial pressure of oxygen. Air contains about 21% oxygen; thus, the partial pressure of oxygen at sea level is about 159 mm Hg. However as air mixes with water vapor and carbon dioxide in the alveoli of the lungs, the partial pressure decreases to about 100 mm Hg. At this partial pressure about 97% of the hemoglobin is oxygenated. At higher altitudes the partial pressure of oxygen is smaller, and the percent saturation of hemoglobin is less. Individuals who live at high altitudes adapt by producing a higher concentration of hemoglobin in their blood.

Reactions respond to changes in pressure only if there is a difference in the moles of gaseous reactants compared with moles of gaseous products. The volume of liquids and solids is not affected very much by pressure changes. Thus, it is only gaseous reactants and products that are considered in deciding which way the equilibrium will shift when the pressure is changed.

EXAMPLE 10.6

Consider the following reactions. Will either reaction proceed more to the right if the pressure is increased?

$$2\,CO(g) + O_2(g) \rightleftharpoons 2\,CO_2(g) \qquad (1)$$
$$CO_2(g) + C(s) \rightleftharpoons 2\,CO(g) \qquad (2)$$

Solution

In reaction 1 there are 2 moles of gaseous product and 3 moles of gaseous reactants. An increase in pressure causes the reaction to proceed to the side with the smaller number of gaseous molecules, that is, toward the right.

In reaction 2 there are 2 moles of gaseous product but only 1 mole of gaseous reactant. (Note that carbon is a solid.) Under increased pressure the reaction proceeds toward the side with the smaller number of gaseous molecules, which is the reactant, or left side.

FIGURE 10.11 Partial Pressure and Saturation of Hemoglobin

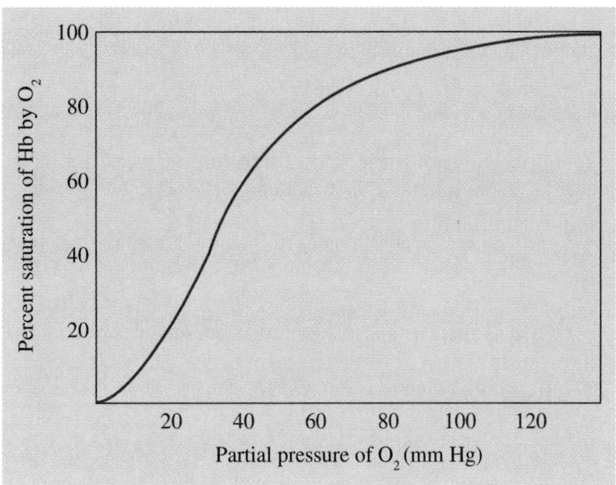

PROBLEM 10.6

Petroleum refining involves "cracking" large molecules into smaller molecules at high temperatures. How would the following reaction for the cracking of decane into octane and ethylene be affected by increasing the pressure?

$$\underset{\text{decane}}{CH_3(CH_2)_8CH_3\ (g)} \rightleftharpoons \underset{\text{octane}}{CH_3(CH_2)_6CH_3\ (g)} + \underset{\text{ethylene}}{CH_2{=}CH_2\ (g)}$$

Temperature Changes

Changes in temperature affect the value of the equilibrium constant. We can understand the change by considering heat as a reactant or product. For a reaction that is exothermic, heat can be treated as a product.

$$A \rightleftharpoons B + \text{heat}$$

Recall from our discussion of ΔH° that an exothermic reaction has a negative enthalpy change. If this system is heated, the added heat energy must be used, so the reverse reaction is favored. Some of B is consumed, and the value of the equilibrium constant decreases. If the system is cooled, it reacts to produce heat; the amount of B increases, as does the equilibrium constant.

For a reaction that is endothermic in the forward direction, heat can be treated as a reactant.

$$A + \text{heat} \rightleftharpoons B$$

This system responds to an increase in temperature by using up the added heat energy; the forward reaction is favored, and more B is produced. Thus, the value of the equilibrium constant increases. If the system is cooled, it will react to produce heat energy, and the amount of B will decrease, as will the equilibrium constant.

EXAMPLE 10.7

Ethyl acetate, a solvent used in a variety of commercial applications such as fingernail-polish remover, is manufactured using ethyl alcohol and acetic acid. How does the equilibrium constant for the formation of ethyl acetate change with increasing temperature?

$$CH_3{-}\overset{\overset{\displaystyle :\ddot{O}}{\|}}{C}{-}\ddot{\underset{..}{O}}{-}H + CH_3CH_2{-}OH \longrightarrow \underset{\text{ethyl acetate}}{CH_3{-}\overset{\overset{\displaystyle :\ddot{O}}{\|}}{C}{-}\ddot{\underset{..}{O}}{-}CH_2CH_3} + H_2O$$

$$\Delta H^\circ = -0.7 \text{ kcal}$$

Solution

The reaction is exothermic and generates heat energy as it proceeds to the right. Thus, increasing the temperature will decrease the extent of the reaction to the right and will decrease the equilibrium constant.

PROBLEM 10.7

The reaction of nitrogen and oxygen in an automobile engine to produce nitric oxide, an air pollutant, occurs at a high temperature. How does the equilibrium constant change with increasing temperature?

$$N_2(g) + O_2(g) \rightleftharpoons 2\,NO(g) \qquad \Delta H^\circ = 44 \text{ kcal}$$

Production of Ammonia for Fertilizer Products

The principles governing a chemical equilibrium for a reaction as well as those for controlling the rate of the same reaction are nicely illustrated in the commercial process for the production of ammonia. The annual production of this simple substance is 35 billion tons at a price of about 7 cents a pound. There are few commodities that can be produced at such a low cost. Ammonia is used directly as a fertilizer and to produce ammonium sulfate and ammonium hydrogen sulfate, which are also used as fertilizers.

The Haber process, developed in 1912 by the German chemist Fritz Haber, is still used today to produce ammonia. The process directly combines nitrogen and hydrogen.

$$N_2(g) + 3\,H_2(g) \rightleftharpoons 2\,NH_3(g) \quad \Delta H^\circ = -22 \text{ kcal/mole}$$

Because 4 moles of reactants are converted to 2 moles of product, the ΔS° for the reaction is negative. Nevertheless, the ΔG° for the reaction is negative, so the reaction is spontaneous at 25 °C with an equilibrium constant of about 4×10^8. Unfortunately, the rate of the reaction is so slow that centuries would be required to form even a trace of product.

As with any other reaction, increasing the temperature leads to an increase in the reaction rate. However, as we have learned in our study of Le Châtelier's principle, the equilibrium is shifted to the left at higher temperatures if ΔH° is negative. In addition, the $T\,\Delta S^\circ$ term also contributes to making the reaction nonspontaneous at a higher temperature. The equilibrium constant decreases to 0.2 at 450 °C.

The reaction is carried out at high pressures to increase the yield of NH_3. Le Châtelier's principle predicts that increasing the pressure will cause the reaction to go more to the right because converting 4 moles of gaseous reactants to 2 moles of gaseous product will reduce the pressure.

The reaction is also made feasible by using a catalyst that consists of Fe_3O_4 mixed with KOH, SiO_2, and Al_2O_3. This catalyst not only increases the rate of the reaction but allows it to run at a somewhat lower temperature, where the equilibrium constant is more favorable.

The final technique that makes the Haber process feasible is the condensation of ammonia from the mixture of products and reactants. Liquid ammonia is formed by cooling the equilibrium mixture of gases. Thus, it is separated from the reactant gases, which then are passed over the catalyst to react again. This technique is an application of Le Châtelier's principle. Since we have removed a product, the reaction tends to go forward to restore the equilibrium concentration of that product.

Summary

A **spontaneous** reaction is one that has a natural tendency to occur. The **free energy** change of all spontaneous reactions is negative due to the contribution of either a negative **enthalpy** change or a positive **entropy** change or both. Reactions with a negative ΔH° are **exothermic;** reactions with a positive ΔH° are **endothermic.**

Reaction rates are determined by the identity of the reactants, concentration, temperature, and catalysts. Increasing the temperature of a reaction increases the reaction rate by increasing the fraction of molecules that have the necessary **activation energy** for reaction. **Catalysts** cause an increase in reaction rate by providing a reaction pathway with a lower activation energy.

When the rates of conversion of reactants to products and of products to reactants are equal, a chemical equilibrium results. The **equilibrium constant** is a measure of the relative concentrations of reactants and products at equilibrium. Altering the concentration of one or more substances in the reaction causes a shift in the equilibrium that changes the concentrations of all reactants and products in a way that maintains the equilibrium constant. The equilibrium constant is affected by temperature changes but not by changes in concentration or pressure or by the presence of a catalyst.

Exercises

Enthalpy, Entropy, and Free Energy

10.1 What does a change in enthalpy for a reaction indicate about the heat evolved from, or required for, the reaction?

10.2 What does the change in entropy indicate about a chemical reaction?

10.3 What is meant by the term spontaneous in reference to chemical reactions?

10.4 A reaction has a negative $\Delta S°$. What is meant by this statement?

10.5 What predictions can be made about reactions with the following combinations of $\Delta H°$ and $\Delta S°$?

(a) positive $\Delta H°$ and negative $\Delta S°$
(b) negative $\Delta H°$ and negative $\Delta S°$

10.6 What predictions can be made about reactions with the following combinations of $\Delta H°$ and $\Delta S°$?

(a) positive $\Delta H°$ and positive $\Delta S°$
(b) negative $\Delta H°$ and positive $\Delta S°$

10.7 Explain why the energy required in the photosynthesis of glucose is equal to the energy released in the metabolism of glucose.

10.8 Are the products in an exothermic reaction at a higher or lower potential energy than the reactants?

Reaction Rates

10.9 Why might an industry choose to carry out a reaction at a high temperature in spite of the high cost of supplying the heat?

10.10 Substances burn more rapidly in pure oxygen than in air. Explain why. This phenomenon was responsible for the tragedy during a ground test of an Apollo spacecraft in which three American astronauts died.

10.11 Explain why persons with elevated temperatures for a prolonged period must increase their food intake to avoid losing weight.

10.12 Individuals who "drown" in very cold water and have stopped breathing sometimes can be revived even after a prolonged period of time underwater. Explain why.

10.13 Ripened tomatoes are stored in the refrigerator, whereas unripened tomatoes are left at room temperature. Explain why.

10.14 Certain antibiotic drugs must be stored under refrigeration. Suggest a reason for this requirement.

Activation Energy

10.15 Reaction A has $\Delta H° = -20$ kcal/mole and $E_a = 25$ kcal/mole. Reaction B has $\Delta H° = -15$ kcal/mole and $E_a = 20$ kcal/mole. Which reaction occurs faster at the same temperature?

10.16 Reaction A has $\Delta H° = -15$ kcal/mole and $E_a = 35$ kcal/mole. Reaction B has $\Delta H° = -20$ kcal/mole and $E_a = 20$ kcal/mole. Which reaction releases more heat energy?

10.17 Comment on the following statement: Reaction rates increase with increasing temperature because the activation energy becomes smaller.

10.18 Comment on the following statement: Reaction rates are increased by a catalyst because the catalyst increases the kinetic energy of the reactants.

Equilibrium Constants

10.19 Methanol (CH_3OH) is manufactured by a reaction of hydrogen gas and carbon monoxide using a catalyst of ZnO and Cr_2O_3. Write a balanced equation for the reaction, and write the equilibrium constant expression.

10.20 Acetone, an industrial solvent, is produced from isopropyl alcohol by a decomposition reaction. Write a balanced equation for the reaction, and write the equilibrium constant expression.

$CH_3-C(=O)-CH_3$ (acetone) $CH_3-CH(OH)-CH_3$ (isopropyl alcohol)

10.21 Write the equilibrium constant expression for each of the following reactions.

(a) $3\,O_2(g) \rightleftharpoons 2\,O_3(g)$
(b) $N_2(g) + 3\,H_2(g) \rightleftharpoons 2\,NH_3(g)$
(c) $CH_4(g) + Cl_2(g) \rightleftharpoons CH_3Cl(g) + HCl(g)$

10.22 Write the equilibrium constant expression for each of the following reactions.

(a) $CO(g) + 3\,H_2(g) \rightleftharpoons CH_4(g) + H_2O(g)$
(b) $H_2(g) + Br_2(g) \rightleftharpoons 2\,HBr(g)$
(c) $2\,H_2O(g) \rightleftharpoons 2\,H_2(g) + O_2(g)$

10.23 Write the equilibrium constant expression for each of the following reactions.

(a) $CS_2(g) + 4\,H_2(g) \rightleftharpoons CH_4(g) + 2\,H_2S(g)$
(b) $2\,Cl_2(g) + 2\,H_2O(g) \rightleftharpoons 4\,HCl(g) + O_2(g)$
(c) $N_2O_4(g) \rightleftharpoons 2\,NO_2(g)$

10.24 Write the equilibrium constant expression for each of the following reactions.

(a) $COCl_2(g) \rightleftharpoons CO(g) + Cl_2(g)$
(b) $NO(g) + Br_2(g) \rightleftharpoons NOBr_2(g)$
(c) $4\,NH_3(g) + 5\,O_2(g) \rightleftharpoons 4\,NO(g) + 6\,H_2O(g)$

Equilibrium Calculations

10.25 If $[H_2] = 0.0343$ M, $[I_2] = 0.326$ M, and [HI] = 0.826 M for the following reaction at equilibrium at 400 °C, calculate the equilibrium constant.

$$I_2(g) + H_2(g) \rightleftharpoons 2\ HI(g)$$

10.26 If $[N_2O_4] = 0.208$ M and $[NO_2] = 3.11 \times 10^{-2}$ M for the following reaction at equilibrium at 25 °C, calculate the equilibrium constant.

$$N_2O_4(g) \rightleftharpoons 2\ NO_2(g)$$

10.27 The equilibrium constant for the following reaction is 4.66×10^{-3} at 25 °C. At equilibrium $[N_2O_4] = 0.208$ M. What is the concentration of NO_2?

$$N_2O_4(g) \rightleftharpoons 2\ NO_2(g)$$

10.28 The equilibrium constant for the following reaction is 54.8 at 425 °C. If $[H_2] = 0.0378$ M and [HI] = 0.728 M at equilibrium, what is the concentration of I_2?

$$I_2(g) + H_2(g) \rightleftharpoons 2\ HI(g)$$

Effect of Concentration on Equilibrium

10.29 Consider the following reaction, and indicate the effect of each of the changes given.

$$CO(g) + 3\ H_2(g) \rightleftharpoons CH_4(g) + H_2O(g)$$

(a) addition of CO (b) addition of CH_4
(c) removal of H_2 (d) removal of H_2O

10.30 Consider the following reaction and indicate the effect of each of the changes given.

$$CS_2(g) + 4\ H_2(g) \rightleftharpoons CH_4(g) + 2\ H_2S(g)$$

(a) addition of CS_2 (b) addition of CH_4
(c) removal of H_2 (d) removal of H_2S

10.31 Consider each of the following reactions at equilibrium and determine whether the indicated change will cause the reaction to proceed to the right.

(a) $CO(g) + 3\ H_2(g) \rightleftharpoons CH_4(g) + H_2O(g)$ (removal of H_2O)
(b) $CO_2(g) + C(s) \rightleftharpoons 2\ CO(g)$ (addition of CO)
(c) $N_2(g) + 3\ H_2(g) \rightleftharpoons 2\ NH_3(g)$ (addition of NH_3)

10.32 Consider each of the following reactions at equilibrium, and determine whether the indicated change will cause the reaction to proceed to the right.

(a) $2\ CO_2(g) \rightleftharpoons 2\ CO(g) + O_2(g)$ (addition of CO)
(b) $CO(g) + 2\ H_2(g) \rightleftharpoons CH_3OH(g)$ (addition of CO)
(c) $CO(g) + 3\ H_2(g) \rightleftharpoons CH_4(g) + H_2O(g)$ (removal of CO)

10.33 Consider the following reaction and indicate the effect of each of the changes given.

$$I_2(g) + H_2(g) \rightleftharpoons 2\ HI(g)$$

(a) addition of HI (b) addition of H_2
(c) removal of I_2 (d) removal of HI

10.34 Consider the following reaction. What effect would each of the listed changes have on the reaction?

$$O_3(g) + NO(g) \rightleftharpoons O_2(g) + NO_2(g)$$

(a) addition of O_3 (b) addition of O_2
(c) removal of NO (d) removal of NO_2

Effect of Pressure on Equilibrium

10.35 What will be the effect of increased pressure on the following reaction, which can occur in a fuel cell?

$$CH_4(g) + 2\ O_2(g) \rightleftharpoons CO_2(g) + 2\ H_2O(l)$$

10.36 What will be the effect of increased pressure on the following reaction, which is one of the steps in the manufacture of sulfuric acid?

$$O_2(g) + 2\ SO_2(g) \rightleftharpoons 2\ SO_3(g)$$

10.37 Consider the following reactions. In which case(s) will the reaction proceed more to the right if the pressure is increased?

(a) $2\ CO(g) + O_2(g) \rightleftharpoons 2\ CO_2(g)$
(b) $2\ NO(g) \rightleftharpoons N_2(g) + O_2(g)$
(c) $N_2O_4(g) \rightleftharpoons 2\ NO_2(g)$

10.38 Consider the following reactions. In which case(s) will the reaction proceed more to the right if the pressure is increased?

(a) $Ni(s) + 4\ CO(g) \rightleftharpoons Ni(CO)_4(g)$
(b) $N_2(g) + 3\ H_2(g) \rightleftharpoons 2\ NH_3(g)$
(c) $I_2(g) + H_2(g) \rightleftharpoons 2\ HI(g)$

10.39 Consider the following reactions. In which case(s) will the reaction proceed more to the right if the pressure is decreased?

(a) $4\ HCl(g) + O_2(g) \rightleftharpoons 2\ Cl_2(g) + 2\ H_2O(g)$
(b) $CS_2(g) + 4\ H_2(g) \rightleftharpoons CH_4(g) + 2\ H_2S(g)$
(c) $N_2(g) + O_2(g) \rightleftharpoons 2\ NO(g)$

10.40 Consider the following reactions. In which case(s) will the reaction proceed more to the right if the pressure is decreased?

(a) $CO(g) + 3\ H_2(g) \rightleftharpoons CH_4(g) + H_2O(g)$
(b) $COCl_2(g) \rightleftharpoons CO(g) + Cl_2(g)$
(c) $CO(g) + 2\ H_2(g) \rightleftharpoons CH_3OH(g)$

Effect of Temperature on Equilibrium

10.41 Consider the following reactions. In which case(s) will product formation be favored by a temperature increase?

(a) $CO(g) + 3\ H_2(g) \rightleftharpoons CH_4(g) + H_2O(g)$ $\Delta H° = -49.4$ kcal
(b) $CO_2(g) + C(s) \rightleftharpoons 2\ CO(g)$ $\Delta H° = 41.2$ kcal
(c) $H_2(g) + I_2(g) \rightleftharpoons 2\ HI(g)$ $\Delta H° = -2.2$ kcal

10.42 Consider the following reactions. In which case(s) will product formation be favored by a temperature increase?

(a) $3\,O_2(g) \rightleftharpoons 2\,O_3(g)$ $\Delta H° = 68.2$ kcal
(b) $2\,H_2O(g) \rightleftharpoons 2\,H_2(g) + O_2(g)$ $\Delta H° = 116$ kcal
(c) $N_2(g) + 3\,H_2(g) \rightleftharpoons 2\,NH_3(g)$ $\Delta H° = -21.9$ kcal

Effect of Pressure and Temperature on Equilibrium

10.43 What conditions of temperature and pressure would most favor the reaction of isopropyl alcohol to give acetone if both reactant and products are gases under the reaction conditions?

$$CH_3-\underset{|}{\overset{:\ddot{O}H}{}}CH-CH_3 \longrightarrow CH_3-\overset{:\ddot{O}}{\overset{\|}{C}}-CH_3 + H_2 \qquad \Delta H° = 13.6 \text{ kcal}$$

10.44 What conditions of temperature and pressure would most favor the reaction of carbon monoxide and hydrogen to give methanol?

$$CO(g) + 2\,H_2(g) \longrightarrow CH_3OH(g) \qquad \Delta H° = -21.9 \text{ kcal}$$

10.45 For which of the following reactions is product formation favored by low pressure and high temperature?

(a) $CO(g) + 3\,H_2(g) \rightleftharpoons CH_4(g) + H_2O(g)$ $\Delta H° = -49.5$ kcal
(b) $CO_2(g) + C(s) \rightleftharpoons 2\,CO(g)$ $\Delta H° = 41.2$ kcal
(c) $H_2(g) + I_2(g) \rightleftharpoons 2\,HI(g)$ $\Delta H° = -2.2$ kcal

10.46 For which of the following reactions is product formation favored by high pressure and low temperature?

(a) $N_2(g) + O_2(g) \rightleftharpoons 2\,NO(g)$ $\Delta H° = 43.3$ kcal
(b) $N_2(g) + 2\,O_2(g) \rightleftharpoons 2\,NO_2(g)$ $\Delta H° = 16.1$ kcal
(c) $N_2(g) + 3\,H_2(g) \rightleftharpoons 2\,NH_3(g)$ $\Delta H° = -21.9$ kcal

Acids occur naturally in citrus fruits, which contain citric acid.
(Seth Joel, Photo Researchers)

11 Acid–Base Equilibria

11.1 Acids and Bases
11.2 Properties of Acids and Bases
11.3 Reactions of Acids and Bases
11.4 Conjugate Acids and Conjugate Bases
11.5 Self-Ionization of Water
11.6 Strengths of Acids and Bases
11.7 The pH Scale
11.8 Hydrolysis of Salts
11.9 Buffers
11.10 Buffers in the Body
11.11 Titration

Essays
- *Acids in the Environment*
- *Antacids*
- *Acids and Bases in Agriculture*

11.1 Acids and Bases

Acids and bases play an important role in the maintenance of life processes and in commercial products that we use. Many acids have a sour taste; the word acid comes from the Latin *acidus,* meaning "sour." Dilute solutions of citric acid in lemons, and acetic acid in vinegar, are examples of acids that taste sour. Bases taste bitter. Antacids and the quinine in tonic water are examples of bases that taste bitter. Acids react with bases to neutralize each other—which means that their individual chemical properties cancel each other.

Acids and bases change the tint of the colored material in some vegetables and plants. The color of the juice of red cabbage changes from red in acid solution to blue green in basic solution. The color of the petals of a red rose changes to light pink if it is placed in a mild acid solution. The colors of flowers such as the orchid vary with the acidity of the soil in which they are grown. Some naturally occurring pigments can be used to test whether a substance in solution is acidic or basic. For example, litmus paper consists of strips of paper impregnated with a plant pigment. The pigment changes from blue to red when a drop of acid is placed on the strip.

Litmus paper changes from red to blue with base. Substances that change color and indicate whether a substance is an acid or a base are called **indicators** (Section 11.7).

Acids and Bases in Life Processes

Acids and bases are involved in most of the reactions in the body. Gastric acid, a solution of hydrochloric acid, is required for the digestion that occurs in the stomach. The metabolism of food produces organic acids, which are eventually converted into carbon dioxide and water. The concentrations of these acids are controlled by body mechanisms because slight changes in chemical equilibria disrupt physiological reactions. Hemoglobin–oxygen binding in respiration is affected by acid concentration. The activity of enzymes such as the digestive enzymes pepsin and trypsin is critically dependent on the acidity. Pepsin is most active in highly acidic solutions. Trypsin is most active in slightly basic solution (Chapter 25).

Household Acids and Bases

Many items in the kitchen and medicine cabinet are acids or bases. Citric acid in lemon juice, acetic acid in vinegar, and acetylsalicylic acid in aspirin tablets are acids. Baking soda, household ammonia, and antacids are bases.

Acids in the Environment

Acids and bases occur naturally in the environment, but the amount of these substances is also affected by industrial processes. Rain is slightly acidic but becomes substantially more acidic as a result of dissolved gases from the combustion of fossil fuels containing high concentrations of sulfur. The phenomenon of *acid rain* was recognized in 1872 by Robert Smith, who discovered that the rain in England became much more acidic in the period now known as the Industrial Revolution. The sulfur dioxide produced in combustion processes combined with other components of the air to give sulfuric acid (H_2SO_4).

Acid rain adversely affects the growth of forests by changing the natural acidity of soil, which damages the root structure of trees, as well as by altering the mineral balance of the soil. The effect of acid on the growth of trees can be seen in the Black Forest of Germany as well as in the forests of the northeastern United States.

Acid rain has also changed the characteristics of small lakes in the Adirondack Mountains of New York. As a consequence, whole populations of fish have been destroyed, and the lakes cannot be restocked as long as the acidity is too high.

An understanding of the effects of acids and bases on soil is important in agriculture. Each crop is most efficiently grown in soils having a characteristic acidity. Soils that are too acidic are called *sour soils.* The term sour is related to the property of acids, which are described as sour to the taste. Of course, agriculturalists don't taste soil to determine its acidity—they use indicators or a pH meter. Both methods are described in Section 11.7. Sour soil can be made less acidic by adding limestone ($CaCO_3$), which is a base. The neutralization reaction of an acid and a base such as limestone is also a subject of later sections.

This researcher is examining the effect of acid rain on damaged trees. (Will McIntyre, Photo Researchers)

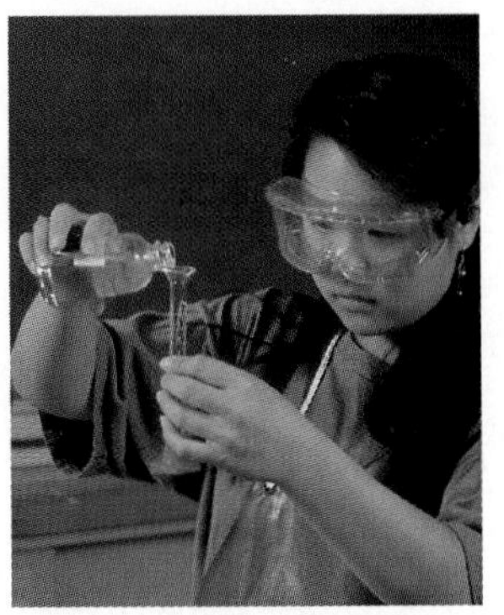

Students must wear goggles in the laboratory to protect their eyes. (Russell D. Curtis, Photo Researchers, Inc.)

Some household acids and bases are extremely dangerous. Both battery acid (sulfuric acid) and muriatic acid (hydrochloric acid), used to clean mortar from bricks, are very corrosive. If these acids are accidentally spilled on skin, they react with the proteins in the skin and cause chemical burns. Contact of these acids with the eyes causes severe eye damage because corneal tissues are more sensitive than skin. Removal of acid by immediate washing with large quantities of water is vital to prevent eye or skin damage.

The bases in oven cleaners, drain cleaners, and paint strippers are also dangerous. They react with fats and oils to produce water-soluble compounds. They also will react with fats and oils in cell membranes to form products that feel slippery or soapy as your skin dissolves. Bases are even more damaging to your eyes than acids. Strongly basic commercial products splattered in the eyes cause severe damage to the cornea because they react at a rapid rate to cause ulceration and scar tissue that is not transparent, sometimes necessitating a corneal transplant.

11.2 Properties of Acids and Bases

Acids and bases have "opposite" properties. In a sense they are antagonists that react to cancel each others' individual properties. What process occurs when these chemical opposites react and neutralize each other?

The Danish chemist Johannes Brønsted and the English chemist Thomas Lowry in 1923 provided a useful definition of acids and bases: An **acid** is a substance that can donate a hydrogen ion (H^+). A **base** is a substance that can accept a hydrogen ion. Note that the hydrogen ion is actually a proton; these two terms are often used interchangeably. Thus, we often refer to acids as proton donors.

Acidic Solutions Contain Hydronium Ions

When gaseous hydrogen chloride dissolves in water, 100% of the HCl molecules react to produce ions in an **ionization reaction.** The polar covalent bond is broken, and the more electronegative chlorine atom retains the bonding pair of electrons. The resulting solution is called hydrochloric acid.

$$H{-}\ddot{\underset{..}{Cl}}{:}(g) \xrightarrow{H_2O} H^+(aq) + {:}\ddot{\underset{..}{Cl}}{:}^-(aq)$$

The hydrogen ion (proton) does not actually exist in water. It reacts with water and bonds to the lone-pair electrons of the oxygen atom forming the hydronium ion. However, the terms hydrogen ion and hydronium ion are often used interchangeably to describe aqueous acid solutions.

$$H^+ + {:}\underset{\substack{|\\H}}{\ddot{O}}{-}H \longrightarrow H{-}\underset{\substack{|\\H}}{\ddot{O}^+}{-}H$$

hydronium ion

Strong and Weak Acids

Acids differ in their tendency to transfer a proton to water. **Strong acids** are acids that ionize completely in water—virtually none of the original acid molecules remain. Hydrochloric acid, nitric acid (HNO_3), and perchloric acid ($HClO_4$) are strong acids. (Other strong acids include HBr and HI. However, HF is a weak acid.) Each of the strong acids completely donates its proton to water.

$$HCl + H_2O \longrightarrow H_3O^+ + Cl^-$$
$$HNO_3 + H_2O \longrightarrow H_3O^+ + NO_3^-$$
$$HClO_4 + H_2O \longrightarrow H_3O^+ + ClO_4^-$$

The ionization of an acid transfers a hydrogen ion from the acid molecule to a water molecule. The reaction of the acid HCl with water is shown in Figure 11.1. The polar covalent bond of HCl is broken and the bonding pair of electrons is retained by the chlorine atom as the chloride ion. The hydrogen ion forms a bond with one of the nonbonding pairs of electrons on the oxygen atom of water.

The properties of acids are due to the hydronium ion formed when each donates a proton to water. The resulting hydronium ion is an acid because it has a bonded hydrogen atom that it can donate to bases. Note that water acts as a base in each of these reactions because it accepts a hydrogen ion (proton). We will return to this chemical property of water in Section 11.4.

Acids that are only partially ionized in water, such as hydrocyanic acid, (HCN) are **weak acids.** An equilibrium exists between HCN and the hydronium ion and cyanide ion.

$$HCN + H_2O \rightleftharpoons H_3O^+ + CN^-$$

Most organic acids are weak. The most common organic acids are carboxylic acids. They structurally resemble acetic acid. The hydrogen atoms bonded to the carbon atoms do not ionize. The only hydrogen atom that does ionize to some degree is the one bonded to the oxygen atom of the $—CO_2H$ group.

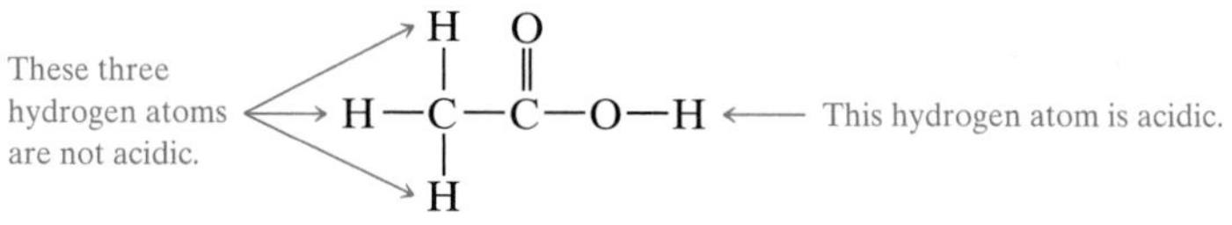

These three hydrogen atoms are not acidic.

H—C(H)(H)—C(=O)—O—H ← This hydrogen atom is acidic.

acetic acid

Any solution of a weak acid, represented by HA, contains both HA molecules and A^- ions. We will return to discuss the quantitative aspects of the acidity of weak acids in Section 11.6.

Polyprotic Acids Can Transfer More Than One Hydrogen Ion

Acids like hydrochloric acid, nitric acid, and perchloric acid that can transfer only one hydrogen ion (proton) per molecule are **monoprotic.** Acids that transfer more

FIGURE 11.1
Transfer of a Proton from an Acid to Water

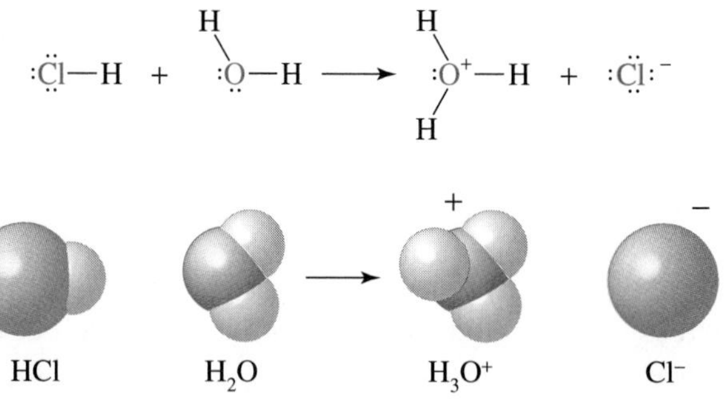

$$:\ddot{Cl}—H + H—\ddot{O}—H \longrightarrow H_2\dot{O}^+—H + :\ddot{Cl}:^-$$

HCl H$_2$O → H_3O^+ Cl^-

than one hydrogen ion are **polyprotic. Diprotic acids** like sulfuric acid can transfer two hydrogen ions to water.

$$H_2SO_4 + H_2O \longrightarrow H_3O^+ + HSO_4^-$$
$$HSO_4^- + H_2O \rightleftharpoons H_3O^+ + SO_4^{2-}$$

Transfer of one hydrogen ion yields the hydrogen sulfate (or bisulfate) ion, HSO_4^-; transfer of the second hydrogen ion yields the sulfate ion, SO_4^{2-} (Figure 11.2). The first transfer is complete; sulfuric acid is a strong acid. The second transfer occurs to a lesser extent, and a substantial amount of HSO_4^- exists at equilibrium. In general, the loss of a second positively charged H^+ does not occur as easily because it is more difficult to release the cation from an ion that has a negative charge.

Carbonic acid is a weak diprotic acid that is important in respiration and in the acid–base balance of the blood. As we would expect, the extent of transfer of the second proton to water is even smaller than that of the first.

$$H_2CO_3 + H_2O \rightleftharpoons H_3O^+ + HCO_3^-$$
$$HCO_3^- + H_2O \rightleftharpoons H_3O^+ + CO_3^{2-}$$

Triprotic acids can transfer three hydrogen ions to water or a base. Phosphoric acid is a weak triprotic acid. The $H_2PO_4^-$ and HPO_4^{2-} ions play an important role in maintaining the acid–base balance in cells.

$$H_3PO_4 + H_2O \rightleftharpoons H_3O^+ + H_2PO_4^-$$
$$H_2PO_4^- + H_2O \rightleftharpoons H_3O^+ + HPO_4^{2-}$$
$$HPO_4^{2-} + H_2O \rightleftharpoons H_3O^+ + PO_4^{3-}$$

Basic Solutions Contain the Hydroxide Ion

The properties of bases are due to the hydroxide ion, OH^-. Solutions of hydroxide ion are called basic or alkaline. The hydroxide ion exists in ionic compounds such

FIGURE 11.2 Reaction of a Diprotic Acid with Water

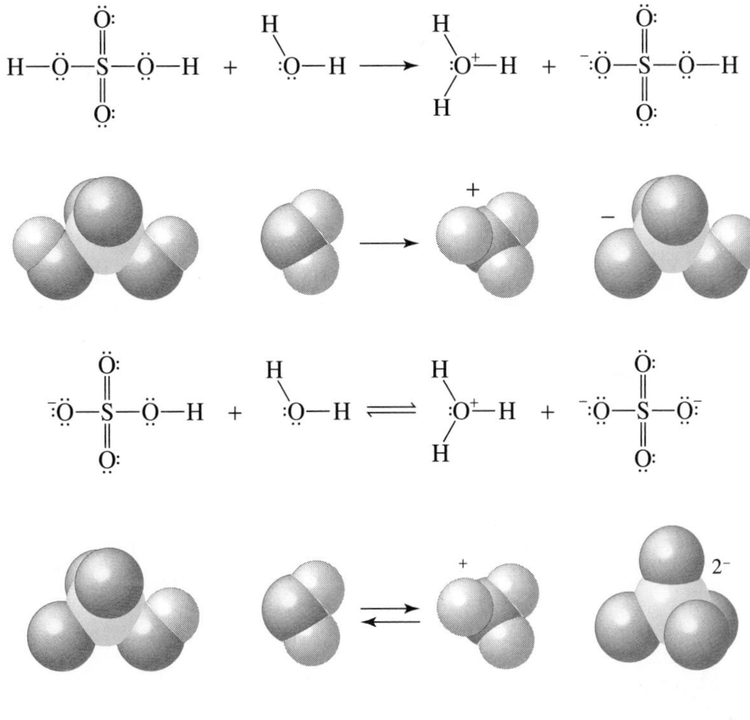

FIGURE 11.3 Transfer of a Proton from Water to a Base

$$H{-}\ddot{N}H_2 + H{-}\ddot{\underset{..}{O}}{-}H \rightleftharpoons H{-}NH_2{-}H^{+} + {}^{-}{:}\ddot{\underset{..}{O}}{-}H$$

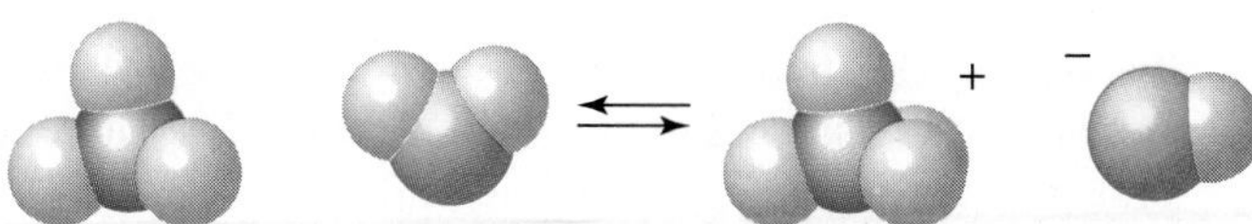

as solid NaOH and KOH. When added to water, these solids dissolve and form a solution containing the ions. These metal hydroxides are classified as **strong bases.**

$$NaOH(s) \xrightarrow{H_2O} Na^+(aq) + OH^-(aq)$$

Some bases do not contain the hydroxide ion but produce that ion by reacting with water. Ammonia accepts a hydrogen ion from water and forms small amounts of ammonium and hydroxide ions (Figure 11.3).

$$\underset{\text{ammonia}}{NH_3(aq)} + H_2O \rightleftharpoons \underset{\text{ammonium ion}}{NH_4(aq)^+} + OH^-(aq)$$

Ammonia is classified as a **weak base** because only a small percentage of the ammonia molecules react by accepting a proton from water. Ammonia solutions are sometimes called ammonium hydroxide. However, the major component of the solution is dissolved ammonia with small amounts of ammonium and hydroxide ions.

11.3 Reactions of Acids and Bases

An acid reacts with a base in a **neutralization reaction.** The products of the reaction are water and a salt.

$$\underset{\text{an acid}}{HClO_4(aq)} + \underset{\text{a base}}{NaOH(aq)} \longrightarrow \underset{\text{water}}{H_2O(l)} + \underset{\text{a salt}}{NaClO_4(aq)}$$

The cation of the salt comes from the base, and the anion from the acid. Since the cation and anion come from compounds that are fully ionized and were originally present in aqueous solutions of the base and the acid, they are spectator ions. The net ionic equation is

$$H_3O^+(aq) + OH^-(aq) \longrightarrow 2\,H_2O(l)$$

This equation is the same for the neutralization of all strong acids with strong bases. In the process a hydrogen ion is transferred from the bonding pair of electrons of one oxygen atom to the nonbonding electron pair of another oxygen atom (Figure 11.4).

Bases other than the hydroxide ion also react with acids. For example, ammonia accepts a hydrogen ion from the hydronium ion.

$$\underset{\text{a base}}{NH_3} + \underset{\text{an acid}}{H_3O^+} \longrightarrow NH_4^+ + H_2O$$

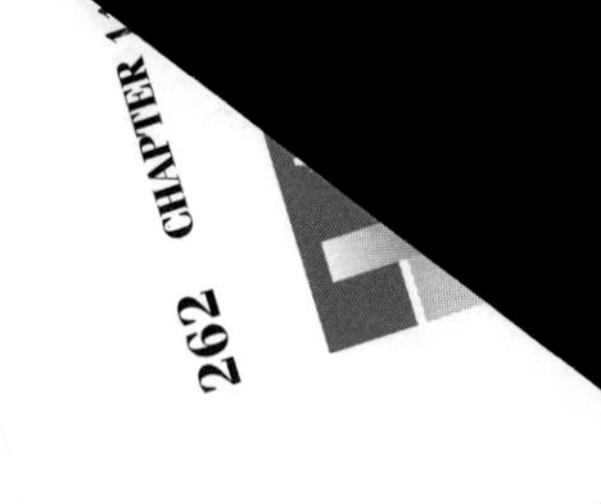

FIGURE 11.4
Neutralization of an Acid and a Base

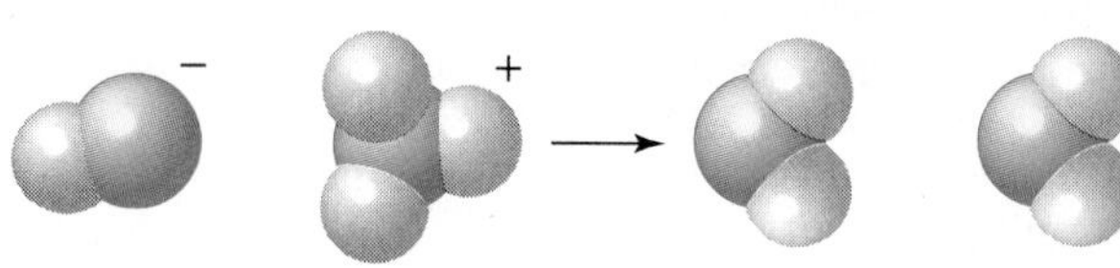

The carbonate ion is also a base. If the carbonate salt is soluble, as is the case for sodium carbonate, the carbonate ion reacts with the hydronium ion. The carbonic acid formed is unstable and decomposes to water and carbon dioxide, which bubbles out of the solution.

$$\underset{\text{an acid}}{2\,H_3O^+(aq)} + \underset{\text{a base}}{CO_3^{2-}(aq)} \rightleftharpoons H_2CO_3(aq) + 2\,H_2O(l)$$

$$H_2CO_3(aq) \rightleftharpoons CO_2(g) + H_2O(l)$$

Insoluble carbonate salts such as calcium carbonate dissolve by reacting with strong acids.

$$CaCO_3(s) + 2\,H_3O^+(aq) \longrightarrow Ca^{2+}(aq) + CO_2(g) + 3\,H_2O(l)$$

The shells of shellfish are made of calcium carbonate. One shellfish, the conch, produces hydrochloric acid, which it uses to dissolve the shells of other animals such as clams and sea urchins. It attacks a shellfish and releases hydrochloric acid to dissolve a little hole in the shell. The conch then sucks out the insides of its victim. You can see the results of the chemical reaction in the remaining shell you find at the beach. A small round hole of approximately 1 cm tells you what happened to the shellfish.

EXAMPLE 11.1

Based on the following structure for amphetamine, predict its acid–base properties.

$$C_6H_5-CH_2-\underset{\displaystyle CH_3}{\underset{|}{CH}}-NH_2$$

Solution

This organic compound contains a nitrogen atom bonded to two hydrogen atoms and a group of carbon atoms. It structurally resembles ammonia in the vicinity of the nitrogen atom. Ammonia has three bonds to hydrogen atoms. Because ammonia is a base, we expect amphetamine to be one as well.

PROBLEM 11.1

Based on the following structure for ibuprofen, the active ingredient in a number of commercial analgesics, predict its acid–base properties.

$$CH_3-\underset{\displaystyle CH_3}{\underset{|}{CH}}-CH_2-C_6H_4-\underset{\displaystyle CH_3}{\underset{|}{CH}}-\overset{\displaystyle O}{\overset{\|}{C}}-OH$$

Antacids

The walls of the stomach contain cells that secrete hydrochloric acid, which suppresses the growth of bacteria and aids in the digestion of food. Although the gastric juice in the stomach is normally about 0.1 M HCl, the inner lining of the stomach is not harmed. The concentration of gastric juice varies during the day and reaches a maximum about an hour after eating. However, some medical conditions result in secretion of excess acid, which may react with the stomach lining and cause painful ulcers. Antacids are bases that neutralize some of the excess stomach acid. They may be taken for mild or temporary discomfort. Longer term therapy requires drugs such as Tagamet or Zantac, which affect the body's control mechanism for the release of HCl.

Rolaids tablets contain dihydroxyaluminum sodium carbonate, $NaAl(OH)_2CO_3$. Both the hydroxide ion and the carbonate ion neutralize some stomach acid, but continued use is not recommended for individuals on a low-sodium diet. Steady use of AlkaSeltzer, which contains sodium bicarbonate ($NaHCO_3$), must also be avoided for the same reason—two tablets contain 1 g of sodium. The base is the bicarbonate ion; the carbon dioxide formed by the reaction with stomach acid may cause a belch.

Both Tums and Pepto-Bismol contain calcium carbonate. These antacids may cause constipation, and excessive use can cause kidney stones. Milk of magnesia contains a suspension of insoluble magnesium hydroxide, $Mg(OH)_2$. Large amounts of this antacid act as a laxative.

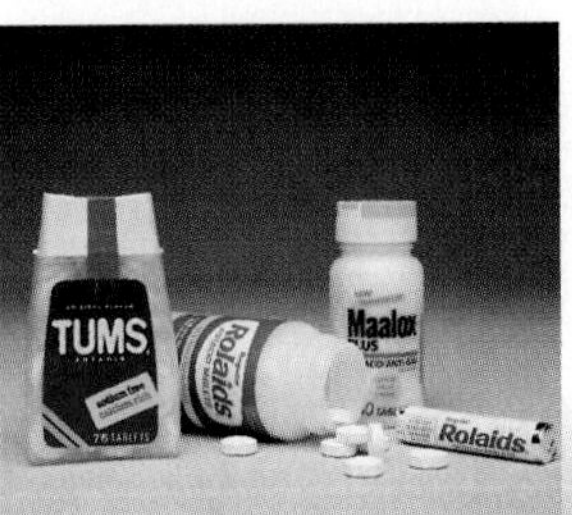

Antacids are common products that can be purchased without a prescription. (Robert Mathena, Fundamental Photographs)

11.4 Conjugate Acids and Conjugate Bases

When an acid gives up its hydrogen ion, the resulting anion is a base. It can accept a hydrogen ion in the reverse reaction or from some other acid. The Brønsted–Lowry terms are extended to cover the relationship between the original acid and the product base. The species remaining after an acid transfers a hydrogen ion is its **conjugate base.**

acid $\rightleftharpoons$ conjugate base + hydrogen ion

After HCl transfers its hydrogen ion to water, the chloride ion is formed, which is the conjugate base of HCl.

$$\underset{\text{acid}}{HCl} + H_2O \longrightarrow H_3O^+ + \underset{\text{conjugate base}}{Cl^-}$$

A similar relationship exists between a base and the product of its reaction with an acid. A base reacts by accepting a hydrogen ion and is transformed into a substance that now has a hydrogen ion and can act as an acid. The species produced by the reaction of a base with a hydrogen ion is its **conjugate acid.**

base + hydrogen ion $\rightleftharpoons$ conjugate acid

When ammonia accepts a hydrogen ion from the hydronium ion, the ammonium ion that forms is the conjugate acid of ammonia.

$$NH_3 + H_3O^+ \longrightarrow H_2O + NH_4^+$$

base conjugate acid

Amphoteric Substances

Acids lose hydrogen ions; bases gain hydrogen ions. Some substances can do both. You have seen one such compound—water. Substances that can either lose or gain a hydrogen ion are **amphoteric substances.** Water acts as an acid when, for example, it donates a hydrogen ion to ammonia; it acts as a base in its reaction with hydrogen chloride.

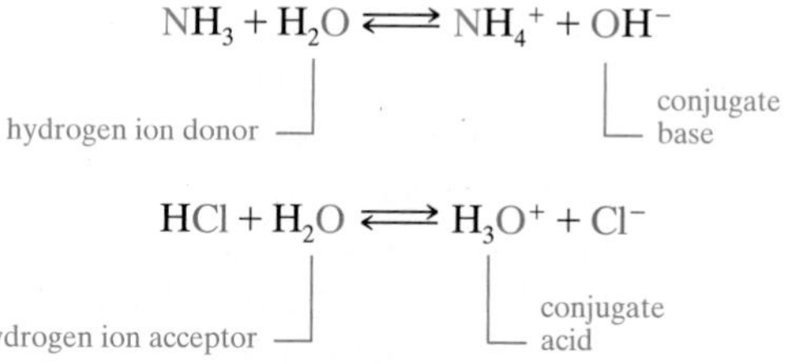

EXAMPLE 11.2

The amphoteric dihydrogen phosphate ion, $H_2PO_4^-$, which is excreted in urine, is involved in the control of the acidity of the fluids of the body. What are the conjugate acid and the conjugate base of this amphoteric ion?

Solution

When a substance acts as an acid, it donates a hydrogen ion to a base such as water. The substance related to the acid is the conjugate base. The conjugate base of $H_2PO_4^-$ is HPO_4^{2-}.

$$H_2PO_4^- + H_2O \longrightarrow HPO_4^{2-} + H_3O^+$$

an acid conjugate base

When a substance acts as a base, it accepts a hydrogen ion from an acid such as water. The substance formed from the base is the conjugate acid. The conjugate acid of $H_2PO_4^-$ is H_3PO_4.

$$H_2PO_4^- + H_2O \longrightarrow H_3PO_4 + OH^-$$

a base conjugate acid

PROBLEM 11.2

Amino acids are amphoteric. Write the structural formula of the conjugate base and conjugate acid of glycine, the simplest amino acid.

```
    H  O
    |  ‖
H—C—C—OH
    |
    NH2
```

11.5 Self-Ionization of Water

You have learned that water can act both as an acid in the presence of a base and as a base in the presence of an acid. However, pure water can act as an acid and a base with itself. Water self-ionizes at 25 °C to produce hydronium ions and hydroxide ions (Figure 11.5).

FIGURE 11.5 Self-Ionization of Water

$$H_2O + H_2O \longrightarrow OH^- + H_3O^+$$

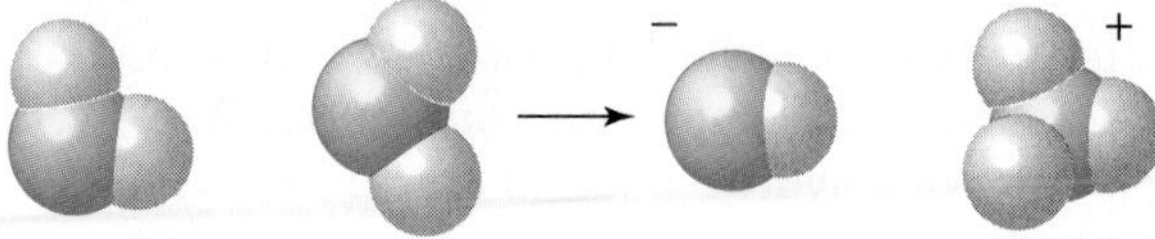

Transfer of a hydrogen ion occurs in a small fraction of the collisions of water molecules. One water molecule behaves as an acid, and another behaves as a base in the reaction. The resulting hydronium ion and hydroxide ion may also react to form two water molecules in a neutralization reaction.

At equilibrium the concentration of H_3O^+ in pure water is 1×10^{-7} M. Note from the equation that for every hydronium ion formed, one hydroxide ion forms. Thus, the concentration of OH^- is also 1×10^{-7} M.

$$2\,H_2O(l) \rightleftharpoons H_3O^+(aq) + OH^-(aq)$$

The concentrations of hydronium and hydroxide ions are very low compared with the concentration of water molecules. In 1 L of water (1000 g) there are 55.5 moles of water molecules. In that liter there are only 1×10^{-7} mole of hydronium ions and 1×10^{-7} mole of hydroxide ions. Thus, for each pair of ions formed, there are 555 million molecules of water.

Pure water is considered neither acidic nor basic because the acidic hydronium ion and the basic hydroxide ion are present in equal amounts. Pure water or any aqueous solution whose $[OH^-] = [H_3O^+]$ is **neutral.**

The equilibrium constant expression for the self-ionization of water is written as follows:

$$K = \frac{[H_3O^+][OH^-]}{[H_2O]^2}$$

The concentration of water is so large that it remains virtually constant at 55.5 M. Thus, it is included in an alternate equilibrium constant, the **ion product constant** of water, K_w. The value of K_w is 1.0×10^{-14}, since $[H_3O^+]$ and $[OH^-]$ are both 1×10^{-7} M at 25 °C.

$$K_w = K[H_2O]^2 = [H_3O^+][OH^-] = (1.0 \times 10^{-7})(1.0 \times 10^{-7}) = 1.0 \times 10^{-14}$$

What Is Going On in Acidic Solutions?

What happens to the concentrations of hydronium ion and hydroxide ion when an acid is added to water? The hydronium ion concentration becomes larger than it was in pure water because hydrogen ions are transferred to water molecules. For example, in 0.010 M HCl, the concentration of the H_3O^+ formed is 1.0×10^{-2} M, which is 100,000 times its concentration in pure water. An **acidic solution** has a higher concentration of hydronium ion than does pure water.

How is the concentration of hydroxide ion in the acid solution affected by the higher concentration of hydronium ion? Le Châtelier's principle gives the answer.

As hydronium ions are added, the hydroxide ions tend to resist this change by reacting with some of them, and the OH^- concentration decreases.

A large number of hydronium ions are formed from the added acid.

$$2\,H_2O \rightleftharpoons H_3O^+ + OH^-$$

Some of the hydroxide ions must react.

For a 1.0×10^{-2} M H_3O^+ solution, the new hydroxide ion concentration is calculated from the K_w expression. The equilibrium expression requires that the product of the hydronium ion concentration and the hydroxide ion concentration be a constant. The hydronium ion concentration as a result of the added acid is 1.0×10^{-2} M.

$$K_w = [H_3O^+][OH^-] = 1.0 \times 10^{-14}$$
$$[1.0 \times 10^{-2}\ M][OH^-] = 1.0 \times 10^{-14}$$
$$[OH^-] = \frac{1.0 \times 10^{-14}}{1.0 \times 10^{-2}}$$
$$[OH^-] = 1.0 \times 10^{-12}\ M$$

Thus, the hydroxide ion concentration is reduced from 1.0×10^{-7} M in pure water to 1.0×10^{-12} M in a 0.010 M HCl solution. In any acidic solution the concentration of hydroxide ions is less than in pure water.

What Is Going On in Basic Solutions?

Now let's consider a basic solution. In a 0.0010 M sodium hydroxide solution, the OH^- concentration from this base is 1.0×10^{-3} M, 10,000 times the concentration of hydroxide ion in pure water. A **basic solution** has a higher concentration of hydroxide ion than pure water.

How is the concentration of hydronium ion affected by the increased concentration of hydroxide ion? Le Châtelier's principle again gives the answer. As hydroxide ions are added, some of the hydronium ions react with them, and the H_3O^+ concentration is decreased.

A large number of hydroxide ions come from the added base.

$$2\,H_2O \rightleftharpoons H_3O^+ + OH^-$$

Some of the hydronium ions must react.

A basic solution then has a hydronium ion concentration that is lower than that of pure water. For a 1.0×10^{-3} M OH^- solution, the hydronium ion concentration is calculated from the K_w expression.

$$[H_3O^+][OH^-] = 1.0 \times 10^{-14}$$
$$[H_3O^+][1.0 \times 10^{-3}] = 1.0 \times 10^{-14}$$
$$[H_3O^+] = \frac{1.0 \times 10^{-14}}{1.0 \times 10^{-3}} = 1.0 \times 10^{-11}\ M$$

EXAMPLE 11.3

The hydronium ion concentration of a human blood sample is 4.5×10^{-8} M. What is the hydroxide ion concentration?

Solution

Human blood is slightly basic, since the hydronium ion concentration is less than 1.0×10^{-7} M. The hydroxide ion concentration must be greater than 1.0×10^{-7} M. Substitution into the K_w expression gives the expected answer.

$$K_w = [H_3O^+][OH^-] = 1.0 \times 10^{-14}$$
$$(4.5 \times 10^{-8}\ M)[OH^-] = 1.0 \times 10^{-14}$$
$$[OH^-] = 2.2 \times 10^{-7}\ M$$

PROBLEM 11.3

Morphine is used medically as a pain reliever. A 0.0050 M solution of morphine has $[OH^-] = 8.8 \times 10^{-5}$ M. What is the H_3O^+ concentration of the solution? Is the solution acidic or basic?

11.6 Strengths of Acids and Bases

The description of acids and bases as *strong* and *weak* refers to the degree of ionization, not to the concentrations of the acids and bases. A strong acid such as HCl is strong even in a dilute solution because it is completely ionized. Similarly, a weak acid such as acetic acid (CH_3CO_2H) is weak even in a concentrated solution because most of the compound is present in the molecular form.

When hydrogen chloride dissolves in water, virtually no covalent HCl remains.

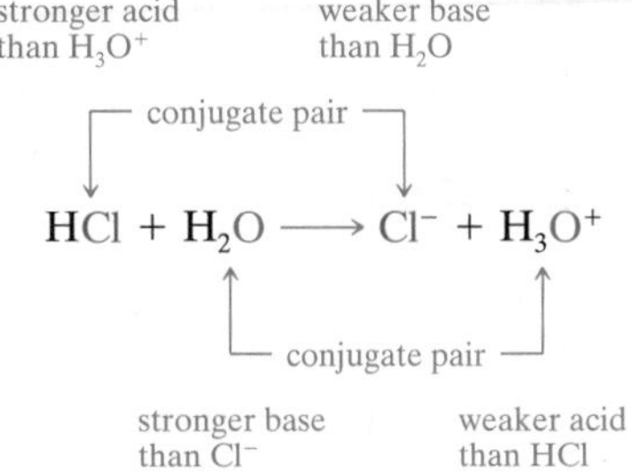

We conclude that HCl has a stronger tendency to donate protons than does H_3O^+ in the reverse reaction; therefore, HCl is a stronger acid than H_3O^+. The position of the equilibrium also reflects the abilities of H_2O and Cl^- to accept protons. The water molecule is a stronger base than Cl^-.

Acetic acid does not ionize completely in water, and few ions are produced. Acetic acid is a weaker acid than H_3O^+, and the acetate ion ($CH_3CO_2^-$) is a stronger base than H_2O. Note that the equilibrium lies to the side of the equation containing the weaker acid and weaker base.

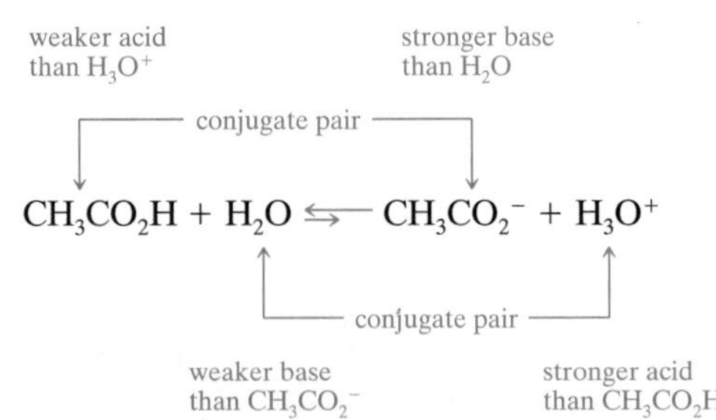

The equilibria between acids and bases and their conjugate bases and acids can be viewed as a battle for hydrogen ions. A strong acid, with its larger tendency to lose hydrogen ions, is conjugate to a weak base, which has a low affinity for hydrogen ions. The more willing the acid is to lose a hydrogen ion, the less willing its conjugate base is to take it back. Strong bases attract hydrogen ions strongly and are conjugate to weak acids, which do not readily lose hydrogen ions. The stronger a base, the weaker its conjugate acid. The more willing the base is to accept a hydrogen ion, the less willing its conjugate acid is to give it back.

A list showing the relationship between common acid–base pairs is given in Table 11.1. Any acid at the top of the list can donate its proton to the conjugate base of any acid below it.

A strong base such as the hydroxide ion removes hydrogen ions from even weak acids such as acetic acid. The equilibrium lies far to the right, favoring the weaker base and weaker acid.

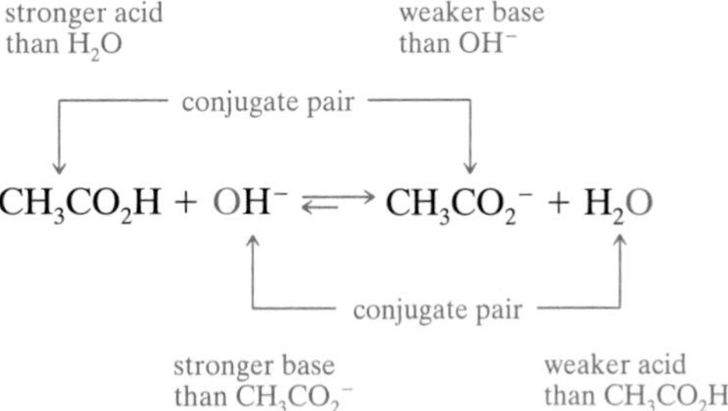

Only a small fraction of the molecules of a weak base accept hydrogen ions at equilibrium. When ammonia dissolves in water, a low concentration of hydroxide ion forms as a result of the transfer of a hydrogen ion from water to ammonia. The equilibrium lies far to the left, favoring the weaker acid and weaker base.

TABLE 11.1 Brønsted–Lowry Conjugate Acid–Base Pairs

	Acid	Base	
strongest ↑	$HClO_4$	ClO_4^-	weakest ↓
	H_2SO_4	HSO_4^-	
	HCl	Cl^-	
	HNO_3	NO_3^-	
	H_3O^+	H_2O	
	HSO_4^-	SO_4^{2-}	
	H_3PO_4	$H_2PO_4^-$	
	HF	F^-	
	CH_3CO_2H	$CH_3CO_2^-$	
	H_2CO_3	HCO_3^-	
	H_2S	HS^-	
	$H_2PO_4^-$	HPO_4^{2-}	
	NH_4^+	NH_3	
	HCO_3^-	CO_3^{2-}	
	HPO_4^{2-}	PO_4^{3-}	
	H_2O	OH^-	
	HS^-	S^{2-}	
weakest	OH^-	O^{2-}	strongest

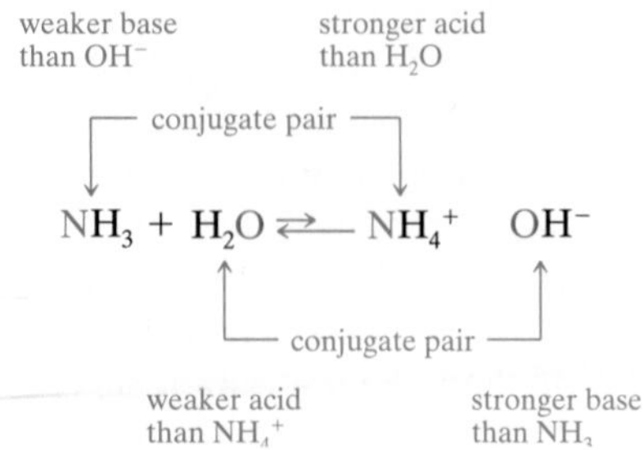

Acid Ionization Constant

The strengths of acids are measured by their tendencies to transfer protons to a base, usually water. For an acid with the general formula HA, the equilibrium constant for ionization is obtained from the equation for ionization.

$$HA + H_2O \rightleftharpoons H_3O^+ + A^-$$

$$K = \frac{[H_3O^+][A^-]}{[HA][H_2O]}$$

The concentration of water is 55.5 M at 25 °C. This concentration is so large compared with the concentrations of the ions formed in the equilibrium that its value changes very little when the acid HA is added. Therefore, the $[H_2O]$ is included in another equilibrium constant, the **acid ionization constant, K_a.**

$$K_a = K[H_2O] = \frac{[H_3O^+][A^-]}{[HA]}$$

The acid ionization constants of some acids and the percent ionization of 1 M solutions are given in Table 11.2. The larger the value of K_a, the larger the percent ionization at the same concentration.

EXAMPLE 11.4

Lactic acid is a monoprotic organic acid produced in metabolic reactions. A 1.0 M solution of lactic acid is about 1% ionized. Calculate the K_a for lactic acid.

TABLE 11.2 Acidity of Acids

Acid	K_a	Percent Ionization of 1.0 M Solution
HSO_4^-	1.3×10^{-2}	11
H_2SO_3	1.2×10^{-2}	11
H_3PO_4	7.5×10^{-3}	8.3
HNO_2	7.2×10^{-4}	2.7
HF	6.6×10^{-4}	2.5
CH_3CO_2H	1.8×10^{-5}	0.42
H_2CO_3	4.3×10^{-7}	0.065
$H_2PO_4^-$	6.2×10^{-8}	0.025
H_2S	5.7×10^{-8}	0.024
HCN	4.4×10^{-10}	0.0020
HCO_3^-	5.6×10^{-11}	0.00075
HPO_4^{2-}	2.2×10^{-13}	0.000047
HS^-	1.3×10^{-13}	0.000036

Solution

The concentration of hydronium ion will be one-hundredth that of the lactic acid. Thus, the concentration of hydronium ion is

$$[H_3O^+] = (0.01)\,(1.0\ M) = 0.01\ M$$

Ionization of a monoprotic acid produces hydronium ions and the conjugate base of the acid in a 1:1 ratio. If the acid is represented as HL, then the concentration of the conjugate base L^- is 0.01 M. The concentration of lactic acid, which is slightly diminished by ionization, is 0.99 M. The acid ionization constant is then calculated.

$$\frac{[H_3O^+]\,[L^-]}{[HL]} = \frac{[0.01]\,[0.01]}{[0.99]} = 1.010101 \times 10^{-4} = 1 \times 10^{-4}$$

PROBLEM 11.4

Formic acid (HCO_2H) is an irritating monoprotic organic acid produced by red ants and used in their sting. A 0.50 M solution of formic acid is about 2% ionized. Calculate the K_a for formic acid.

Base Ionization Constants

To compare bases, we use their tendency to accept protons from water. For an electrically neutral base with the general formula B, the base ionization reaction and the related equilibrium constant expression are

$$B + H_2O \rightleftharpoons BH^+ + OH^-$$

$$K = \frac{[BH^+]\,[OH^-]}{[B]\,[H_2O]}$$

If the base is NH_3, the conjugate base BH^+ is NH_4^+. (Some bases are negatively charged. In these cases the conjugate acid would be electrically neutral and represented by BH.)

The concentration of water is large compared with the concentrations of the ions formed in the equilibrium for the reaction of a base with water. Therefore, as in the case of the acid ionization constant, the $[H_2O]$ is included in the **base ionization constant, K_b.**

$$K_b = K[H_2O] = \frac{[BH^+]\,[OH^-]}{[B]}$$

The base ionization constants of some bases are given in Table 11.3. The larger the value of K_b, the stronger is the base.

11.7 The pH Scale

The concentration of hydronium ion in solution is an important concern in many fields in addition to chemistry. It is awkward to describe the acidity of a solution in moles per liter. Because those units involve negative powers of 10, many people must take a few seconds to recognize the meaning of concentrations such as 3×10^{-4} M and 4×10^{-3} M. An easier method of expressing acidity uses values be-

TABLE 11.3 Base Ionization Constants

Name of Base	Formula	K_b
sulfide ion	S^{2-}	8.3×10^{-2}
phosphate ion	PO_4^{3-}	2.1×10^{-2}
methyl amine	CH_3NH_2	3.7×10^{-4}
carbonate ion	CO_3^{2-}	1.8×10^{-4}
ammonia	NH_3	1.8×10^{-5}
hypochlorite ion	ClO^-	3.3×10^{-7}
bisulfide ion	HS^-	1.8×10^{-7}
hydrogen phosphate ion	HPO_4^{2-}	1.6×10^{-7}
sulfite ion	SO_3^{2-}	1.6×10^{-7}
bicarbonate ion	HCO_3^-	2.4×10^{-8}
acetate ion	$CH_3CO_2^-$	5.6×10^{-10}
formate ion	HCO_2^-	5.9×10^{-11}
nitrite ion	NO_2^-	2.2×10^{-11}
cyanide ion	CN^-	2.0×10^{-11}
fluoride ion	F^-	1.4×10^{-11}
dihydrogen phosphate ion	$H_2PO_4^-$	1.3×10^{-12}
bisulfite ion	HSO_3^-	7.7×10^{-13}

tween 0 and 14 on a pH scale. The pH scale is used for many common applications, as in the health sciences and agriculture. On this scale, a pH of 7 corresponds to a neutral solution. If the pH is less than 7, the solution is acidic; a pH greater than 7 corresponds to a basic (alkaline) solution. On this scale, the larger the pH, the more alkaline is the solution; the smaller the pH, the more acidic is the solution.

pH is defined as the negative logarithm of the molarity of the hydronium ion. This expression indicates the acidity of a solution more concisely than exponential notation.

$$pH = -\log[H_3O^+]$$

For a hydronium ion concentration of 1.0×10^{-7} M, the pH is 7.0. This is the pH of a neutral solution.

$$pH = -\log[H_3O^+] = -\log(1.0 \times 10^{-7}) = -(-7.0) = 7.0$$

The advantage of the pH scale is immediately apparent. It is easier to say "the pH is 7" than to say "the hydronium ion concentration is 1.0×10^{-7} M."

At higher hydronium ion concentrations—that is, in acidic solutions—the pH is smaller. In a 0.010 M HCl solution, the hydronium ion concentration is 1.0×10^{-2} M, and the pH is 2.0.

Some personal products indicate that they are pH balanced. (Peter Arnold, Inc.)

$$pH = -\log(1.0 \times 10^{-2}) = -(-2.0) = 2.0$$

In basic solutions, the hydronium ion concentration is lower than in pure water. For a 0.0010 M NaOH solution, the hydroxide ion concentration is 1.0×10^{-3} M. Recall from Section 11.5 that the hydronium ion concentration, as calculated from the K_w expression, is 1.0×10^{-11} M. The pH is thus 11.0.

$$pH = -\log(1.0 \times 10^{-11}) = -(-11.0) = 11.0$$

s shown in Table 11.4. For any hydronium ion concentration of pH is x.

ine the relationship between pH and the hydronium ion concen- 11.4. A pH change from 4 to 3 may look small. However, this change of 1 pH unit corresponds to a change in the hydronium ion concentration by a factor of 10. Thus, a change of a few pH units corresponds to a large change in hydronium ion concentration. A pH change of three units corresponds to change in the hydronium ion concentration by a factor of 10^3, or 1000.

Dealing with pH for solutions with hydronium ion concentrations written as 1.0×10^{-x} M is fairly simple. The pH is obtained by changing the sign of the exponent. Now let us calculate the pH of a blood sample with a hydronium ion concentration equal to 4.5×10^{-8} M. A hand-held calculator can be used to calculate pH. On most calculators you first enter 4.5, press the exp key, and then enter the exponent. Then you press the log key to obtain the logarithm of the number. Rounded to the hundredth place, the logarithm of 4.5×10^{-8} is -7.35. Since pH is the negative logarithm, the pH $= -(-7.35) = 7.35$. Don't rely completely on the displayed answer on a calculator. You may have entered the numbers incorrectly. You should have a feel for the correctness of the answer. For a lower hydronium ion concentration of 1.0×10^{-8} M, the pH is 8.00. For a higher hydronium ion concentration of 1.0×10^{-7} M, the pH is 7.00. Thus, for 4.5×10^{-8} M, the pH must be between 7.00 and 8.00.

Note that the pH calculated for blood appears to have three significant figures, whereas the hydronium ion concentration is expressed with only two significant figures. However, the first integer in the pH is related to the power of the exponent in the hydronium ion concentration. Only the number of integers to the right of the decimal point correspond to the number of significant figures in 4.5, which is two.

The pH of blood is slightly greater than 7, as expected from the hydronium ion concentration, which is less than that of pure water. The pH values of several of our body fluids are given in Table 11.5. Except for gastric juices, in which the main acid is hydrochloric acid, the majority of body fluids have pH values near 7. No body fluid is very basic.

TABLE 11.4 The pH Scale

$[H_3O^+]$	pH	$[OH^-]$	
10^0	0	10^{-14}	acidic solutions
10^{-1}	1	10^{-13}	
10^{-2}	2	10^{-12}	
10^{-3}	3	10^{-11}	
10^{-4}	4	10^{-10}	
10^{-5}	5	10^{-9}	
10^{-6}	6	10^{-8}	
10^{-7}	7	10^{-7}	neutral solution
10^{-8}	8	10^{-6}	basic solutions
10^{-9}	9	10^{-5}	
10^{-10}	10	10^{-4}	
10^{-11}	11	10^{-3}	
10^{-12}	12	10^{-2}	
10^{-13}	13	10^{-1}	
10^{-14}	14	10^0	

TABLE 11.5 pH Values of Body Fluids

blood	7.35–7.45
gastric juices	1.6–1.8
bile	7.8–8.6
urine	5.5–7.0
saliva	6.2–7.4
interstitial fluid	7.4
muscle intracellular fluid	6.1
liver intracellular fluid	6.9
pancreatic juice	7.8–8.0

EXAMPLE 11.5

The hydronium ion concentration of a sample of urine is 2.0×10^{-6} M. What is the pH?

Solution

The sample is more acidic than a solution with a hydronium ion concentration of 1×10^{-6} M. Therefore, the pH must be less than 6. Determine the logarithm of 2.0×10^{-6} on your calculator. The display reads -5.69897, and the calculated pH is 5.69897. However, there are only two significant figures in the quantity 2.0×10^{-6}. Thus, only two integers to the right of the decimal point can be kept in the calculated pH. The pH is rounded to 5.70. The value of the pH is somewhat less than 6, as expected.

PROBLEM 11.5

A 0.0050 M morphine solution has $[OH^-] = 8.8 \times 10^{-5}$ M. What is the pH of the solution?

Measuring pH

To determine the degree of ionization of a weak acid or the concentration of a solution of a strong acid, chemists measure the H_3O^+ concentration or the pH of the solution. Hospital technicians measure the pH of blood or other body fluids. An agricultural specialist measures the acidity or pH of a soil sample. All use a pH meter, which gives the pH either as a needle reading on a scale or a digital display. The instrument measures the voltage across a special glass membrane that passes only H^+ when immersed in the solution. Such meters can measure pH to the nearest 0.01 pH unit.

An older method of determining pH uses chemical indicators. Indicators are available as solutions or impregnated in strips of paper called *pH test paper.* Chemical indicators are weak acids, represented by HInd, that exist as the acid and its conjugate base, Ind^-, in water.

$$\text{HInd} + H_2O \rightleftharpoons H_3O^+ + \text{Ind}^-$$

$$K_a = \frac{[H_3O^+][\text{Ind}^-]}{[\text{HInd}]}$$

Each member of the conjugate pair of an indicator is a different color, so an indicator changes color as the concentrations of the two species change. In acidic solution the indicator exists as HInd, and the color is that of the acid form of the indicator.

In basic solution the hydroxide ion removes a proton to give the conjugate base, Ind^-, which is a different color.

The ratio of the concentrations of the conjugate base and the acid form depends on the K_a of the indicator and the hydronium ion concentration.

$$\frac{[Ind^-]}{[HInd]} = \frac{K_a}{[H_3O^+]}$$

When the ratio of the concentrations of the two colored species is about 0.1, the color is that of the acid form of the indicator. When the ratio of the concentrations is 10, the color is that of the conjugate base of the indicator. Therefore, the range of sensitivity of the indicators is about a 100-fold range of $[H_3O^+]$, or two pH units. Phenolphthalein is colorless in solutions of pH less than 8.3, where it exists in its acidic form; it is pink in solutions of pH greater than 10.0, where it exists in its basic form. The colors of a number of other indicators are listed in Table 11.6 along with the approximate pH at which the color change occurs.

11.8 Hydrolysis of Salts

We might expect that when a salt is added to water the resulting solution would be neutral. That is the result for salts such as sodium chloride. However there are salts that give basic solutions as well as salts that give acidic solutions.

When sodium acetate is added to water, a basic solution ($pH > 7$) results. Therefore in the sodium acetate solution the hydroxide ion concentration is greater than that of pure water. A solution of ammonium chloride is acidic ($pH < 7$). The hydronium ion concentration is greater than that of pure water. What is responsible for these differences? How do the ions in the two salts affect the amount of hydronium ions or hydroxide ions in solution? In both instances an ion of the salt reacts with water to produce hydronium or hydroxide ions. The reaction of ions with water is called **hydrolysis.**

TABLE 11.6 Colors of Indicators

Indicator	Color in the More Acidic Range	pH Range	Color in the More Basic Range
methyl violet	yellow	0–2	violet
thymol blue	pink	1.2–2.8	yellow
bromophenol blue	yellow	3.0–4.7	violet
methyl orange	pink	3.1–4.4	yellow
bromocresol green	yellow	4.0–5.6	blue
bromocresol purple	yellow	5.2–6.8	purple
litmus	red	4.7–8.2	blue
phenolphthalein	colorless	8.3–10.0	pink
thymolphthalein	colorless	9.3–10.5	blue
alizarin yellow G	colorless	10.1–12.1	yellow
trinitrobenzene	colorless	12.0–14.3	orange

Why Some Salt Solutions Are Basic or Acidic

Sodium acetate is completely ionized in solution and contains Na^+ and $CH_3CO_2^-$ ions. The Na^+ ion does not react with water, but the $CH_3CO_2^-$ ions do. Because acetic acid is a weak acid, the acetate ion is a moderately strong base. When acetate ions dissolve in water, hydrogen ions are removed from some water molecules.

$$CH_3CO_2^- + H_2O \rightleftharpoons CH_3CO_2H + OH^-$$

Although the equilibrium lies to the left, enough OH^- ions are produced to increase the pH to about 9. For this reason sodium acetate is called a **basic salt.**

Acidic solutions result from hydrolysis of cations. They are less common than the basic solutions that result from the hydrolysis of anions. Ammonium chloride gives an acidic solution and is called an **acidic salt.** When ammonium chloride dissolves in water, a solution of NH_4^+ and Cl^- ions results. The Cl^- ions do not react with water, but the NH_4^+ ions do. The NH_4^+ ions can donate hydrogen ions to water to yield some hydronium ions and ammonia. Although the equilibrium lies to the left, the hydronium ions produced make the solution slightly acidic and the pH is less than 7.

$$NH_4^+ + H_2O \rightleftharpoons NH_3 + H_3O^+$$

Group IA cations such as Na^+ and Group IIA ions such as Ca^{2+} do not affect the acidity of solutions. However, solutions of many transition-metal ions, especially those with a +3 charge, are acidic. These ions bind strongly to water molecules, giving a complex ion. The complex ion can react as an acid and donate a proton to water. The pH values of 0.1 M solutions of some metal nitrates at 25 °C are listed in Table 11.7.

$$Fe(H_2O)_6^{3+}(aq) + H_2O \rightleftharpoons Fe(H_2O)_5(OH)^{2+}(aq) + H_3O^+(aq)$$

How Can We Predict the Properties of Salts?

All salts can be viewed as being formed from reaction of an acid and a base. We can predict whether a salt is neutral, basic, or acidic by considering the strengths of the acid and base. There are four combinations of acids and bases to consider.

1. *Salt of a strong acid and a strong base.* In the case of the neutral salt NaCl, the two ions are derived from the strong base NaOH and the strong acid HCl. Neither ion reacts with water, and the resulting solution is neutral.

TABLE 11.7 pH of 0.1 M Solutions of Metal Ions

Metal Ion	pH
Ca^{2+}	7.0
Mg^{2+}	6.2
Zn^{2+}	5.0
Cu^{2+}	4.5
Al^{3+}	3.0
Cr^{3+}	2.5
Fe^{3+}	1.8

2. *Salt of a strong base and a weak acid.* In the case of the basic salt sodium acetate, the Na^+ ion is derived from the strong base NaOH, but the $CH_3CO_2^-$ ion is derived from acetic acid, which is a weak acid. The $CH_3CO_2^-$ ion is sufficiently basic to react with water and produce hydroxide ions.
3. *Salt of a weak base and a strong acid.* In the case of the acidic salt ammonium chloride, the NH_4^+ ion is the conjugate acid of ammonia, which is a weak base. The Cl^- ion, which is the conjugate base of the strong acid HCl, is a very weak base and does not react with water. The hydronium ions formed by reaction of the NH_4^+ ion with water make the solution slightly acidic.
4. *Salt of a weak base and a weak acid.* Salts such as ammonium acetate fit into this category. Both ions can react with water. The NH_4^+ ion tends to generate hydronium ions; the $CH_3CO_2^-$ ion tends to produce hydroxide ions. As a result the two effects tend to largely cancel each other. The resulting solution may be slightly acidic or basic, but we will not further consider this case.

Calculations of pH of Salt Solutions

We will restrict the calculation of the pH of salt solutions to basic salts such as sodium acetate. The equilibrium equation and equilibrium constant expression for the hydrolysis reaction are

$$CH_3CO_2^- + H_2O \rightleftharpoons CH_3CO_2H + OH^-$$
$$K = \frac{[CH_3CO_2H][OH^-]}{[CH_3CO_2^-]} = K_b$$

This expression is the same as for the base ionization constant presented in Section 11.6, because in this hydrolysis reaction the anion reacts as a base.

The base ionization constant for the acetate anion is related to the acid ionization constant of acetic acid, the conjugate acid of the acetate ion.

$$K_b = \frac{K_w}{K_a} = \frac{[H_3O^+][OH^-]}{\left(\frac{[H_3O^+][CH_3CO_2^-]}{[CH_3CO_2H]}\right)} = \frac{[CH_3CO_2H][OH^-]}{[CH_3CO_2^-]}$$

This is a general relationship. For any anion B^-, the equilibrium constant for hydrolysis is K_b, which is equal to K_w/K_a. If the K_b is not tabulated, it may be calculated from the acid ionization constant of the related acid HB. The weaker the acid, the smaller is the acid dissociation constant. As a consequence, the base ionization constant of the conjugate base is larger, and the hydroxide ion concentration of the solution of the conjugate base is larger.

EXAMPLE 11.6

Determine the hydroxide ion concentration of a 0.10 M solution of sodium acetate in water. The K_a of acetic acid is 1.8×10^{-5}.

Solution

First, calculate the base ionization constant of the acetate ion.

$$K_b = \frac{K_w}{K_a} = \frac{1.0 \times 10^{-14}}{1.8 \times 10^{-5}} = 5.6 \times 10^{-10}$$

The concentration of hydroxide ion produced by the hydrolysis of acetate ion is equal to the concentration of the acetic acid produced.

Acids and Bases in Agriculture

An understanding of acid–base chemistry and the pH scale is important in agriculture and sometimes even in growing green grass. Each of the many types of soil has its own characteristic acidity. Sandy soil is close to neutral, but a variety of other soils are not. Clay soil is usually acidic; soil containing limestone is basic because it contains $CaCO_3$, which contains the basic CO_3^{2-} ion. Soils in humid areas are acidic, whereas soils in arid areas tend to be basic or neutral.

Some crops flourish under acidic conditions and others require more basic conditions. Plants differ in how they take up nutrients in their roots, and the form of these nutrients depends on the pH. Thus, the successful cultivation of plants depends on the proper soil pH.

Soil is often tested to determine whether acidic or basic fertilizers are required for a particular crop. If the soil is too acidic, it can be "limed" by adding calcium carbonate. The acid of the soil reacts with the carbonate ion.

$$2\,H^+ + CaCO_3 \longrightarrow Ca^{2+} + H_2O + CO_2$$

If the soil is too basic, organic matter such as leaves may be plowed into it. These materials decompose to produce acids. Chemicals such as aluminum sulfate, $Al_2(SO_4)_3$, or ferric sulfate, $Fe_2(SO_4)_3$, may also be added. Solutions of Al^{3+} or Fe^{3+} are acidic as a result of hydrolysis.

$$CH_3CO_2^- + H_2O \rightleftharpoons CH_3CO_2H + OH^-$$

The concentration of the acetate ion is still essentially 0.10 M because the base ionization constant is small. The concentration of the hydroxide ion may be represented by x. Then, since the concentration of acetic acid is equal to that of the hydroxide ion, its concentration is also x. Substituting into the equilibrium constant expression, we have

$$K_b = \frac{[CH_3CO_2H]\,[OH^-]}{[CH_3CO_2^-]} = \frac{(x)\,(x)}{0.10} = 5.6 \times 10^{-10}$$

$$x^2 = (5.6 \times 10^{-10})\,(0.10)$$

$$x = 7.483315 \times 10^{-6} = 7.5 \times 10^{-6}\text{ M}$$

PROBLEM 11.6

Which solution is more basic, a 0.1 M solution of sodium cyanide or a 0.1 M solution of sodium acetate?

11.9 Buffers

Consider what happens to the pH of water when a little acid is added. Let's say 0.1 mL of concentrated (12 M) hydrochloric acid is added to 100 mL of water. The hydronium ion concentration jumps from 1×10^{-7} M to 1×10^{-2} M. The pH changes from 7 to 2—a "shocking" change for the addition of such a small amount of acid. In an organism a change of even a few tenths of a pH unit would shock the organism and it would die. How do life forms regulate their hydronium ion concentration to avoid large changes in pH? We will deal with that subject later. First, let's see how a chemist prevents substantial changes in pH in solutions.

The term *to buffer* means "to prevent changes" or "to lessen the shock of changes." In chemistry a **buffer** is a solution that prevents a drastic pH change when either H_3O^+ or OH^- is added to it. In other words, a buffer maintains a reasonably steady acidity level even when a strong acid or strong base is added.

A chemical buffer is prepared by dissolving both a weak acid and a salt of its conjugate base—for example, acetic acid and sodium acetate—in water. (A buffer can also be made from a weak base and a salt of its conjugate acid.) The buffer reacts to neutralize any strong acid or base that is added to the solution. Hydronium ions react with the acetate ions; hydroxide ions react with the acetic acid.

$$H_3O^+ + CH_3CO_2^- \longrightarrow H_2O + CH_3CO_2H$$
$$OH^- + CH_3CO_2H \longrightarrow H_2O + CH_3CO_2^-$$

Thus, either H_3O^+ or OH^-, which could change the pH, is neutralized by one of the components of the buffer. In everyday terms, the weak acid in the buffer "gobbles up" added base, and the conjugate base in the buffer "gobbles up" added acid.

A variety of hydronium ion concentrations result from various ratios of a weak acid and a salt of its conjugate base. A relationship that gives the H_3O^+ concentration in terms of the acid ionization constant and the two components of the buffer can be derived from the acid ionization constant.

$$K_a = \frac{[H_3O^+][A^-]}{[HA]}$$

$$[H_3O^+] = K_a \times \frac{[HA]}{[A^-]}$$

Buffers can be prepared from any ratio of concentrations of a weak acid and its conjugate base. A 1:1 ratio of acid to conjugate base is the most efficient for handling either a base or an acid. If the buffer contains more acid than conjugate base, it is less efficient in handling added acid. Conversely, a buffer with more conjugate base than acid cannot efficiently counteract the addition of a base.

EXAMPLE 11.7

Determine the hydronium ion concentration of a buffer consisting of 0.10 M sodium acetate and 0.20 M acetic acid in water. The K_a of acetic acid is 1.8×10^{-5}.

Solution

The concentration of the hydronium ion in equilibrium with both acetic acid and acetate ion is controlled by the equilibrium

$$CH_3CO_2H + H_2O \rightleftharpoons CH_3CO_2^- + H_3O^+$$

The concentration of the acetate ion is essentially 0.10 M because the base ionization constant is small. Similarly, the concentration of acetic acid is essentially 0.20 M because the ionization of acetic acid is small. Using x as the concentration of the hydronium ion and substituting into the equilibrium constant expression, we have

$$K_a = \frac{[CH_3CO_2^-][H_3O^+]}{[CH_3CO_2H]} = \frac{(0.10)(x)}{0.20} = 1.8 \times 10^{-5}$$

$$x = \frac{(1.8 \times 10^{-5})(0.20)}{0.10} = 3.6 \times 10^{-5}\text{ M} = [H_3O^+]$$

PROBLEM 11.7

The first ionization constant of H_3PO_4 is 7.5×10^{-3}. What is the pH of a solution containing 0.10 M H_3PO_4 and 0.10 M NaH_2PO_4?

11.10 Buffers in the Body

Normal metabolic reactions produce a variety of acids. There are two ways to decrease the concentration of these acids. One is by breathing and expelling the carbon dioxide that is formed from carbonic acid. The second is by excreting an acid in urine. Both mechanisms depend on buffers. The carbonate buffer H_2CO_3/HCO_3^- is present in the bloodstream, whereas the phosphate buffer $H_2PO_4^-/HPO_4^{2-}$ is involved in kidney functions.

Blood has a normal pH range of 7.35–7.45, which corresponds to hydronium ion concentrations of 4.5×10^{-8} to 3.6×10^{-8} M. Any acids formed in the blood react with bicarbonate ions to give carbonic acid, which then decomposes to give carbon dioxide. The carbon dioxide is removed from the blood by the lungs and exhaled.

$$HCO_3^-(aq) + H_3O^+(aq) \rightleftharpoons H_2CO_3(aq) + H_2O(l)$$
$$H_2CO_3(aq) \rightleftharpoons H_2O(l) + CO_2(g)$$

Although the reaction is less common, the H_2CO_3/HCO_3^- buffer can prevent an increase in base concentration. The carbonic acid can react to neutralize a base.

$$H_2CO_3(aq) + OH^-(aq) \rightleftharpoons HCO_3^-(aq) + H_2O(l)$$

However, the ability of the buffer to handle a base is much more limited than its ability to neutralize acid because the $[H_2CO_3]/[HCO_3^-]$ ratio is about 1:20.

The $H_2PO_4^-/HPO_4^{2-}$ buffer is important within cells. Many reactions in cells involve complex compounds with covalently bonded phosphate groups. The pH in cellular fluids is maintained by the reaction of a base with $H_2PO_4^-$ and the reaction of an acid with HPO_4^{2-}.

$$H_2PO_4^- + OH^- \rightleftharpoons HPO_4^{2-} + H_2O$$
$$HPO_4^{2-} + H_3O^+ \rightleftharpoons H_2PO_4^- + H_2O$$

However, acids are the more common product of metabolism, so HPO_4^{2-} is the more important component of the buffer. The normal $[H_2PO_4^-]/[HPO_4^{2-}]$ ratio in the cell is about 1:4, which allows the buffer to neutralize acid more efficiently than base. The $H_2PO_4^-$ formed from the reaction of HPO_4^{2-} with acid is then eliminated by excretion in the urine.

Acidosis and Alkalosis

When the blood pH is lower than 7.35 for a period of time, the condition is known as **acidosis.** When the blood pH is higher than 7.45 for a period of time, the condition is known as **alkalosis.** The medical conditions leading to these states are listed in Table 11.8. The consequence of each physical condition can be explained using acid–base equilibria.

The condition of acidosis results from (1) excessive production of acid, (2) decreased concentration of HCO_3^-, or (3) increased concentration of H_2CO_3. Note that according to Le Châtelier's principle, either increasing $[H_2CO_3]$ or decreasing $[HCO_3^-]$ causes an increase in $[H_3O^+]$.

$$H_2CO_3(aq) + H_2O(l) \rightleftharpoons HCO_3^-(aq) + H_3O^+(aq)$$

Increased acid production from metabolic reactions, called **metabolic acidosis,** occurs both in diabetes mellitus and in some low-carbohydrate, high-protein diets.

TABLE 11.8 Causes of Acidosis and Alkalosis

Condition	Causes
respiratory acidosis	increase in CO_2 partial pressure due to: deterioration of pulmonary function, as in asthma, emphysema, pneumonia cardiac insufficiency, cardiac arrest hypoventilation
metabolic acidosis	increased formation of metabolic acids due to: diabetes, fasting or starvation, excessive exercise decreased excretion of acid in urine due to: renal failure, chemical poisoning decreased bicarbonate concentration due to: diarrhea
respiratory alkalosis	excessive loss of CO_2 due to: high fever, nervousness, hyperventilation
metabolic alkalosis	increased bicarbonate concentration due to: diuretics, overdose of antacid, vomiting

Lack of insulin prevents glucose metabolism, and glucose is excreted in the urine. The body metabolizes more of its stored fats, a process that yields large amounts of acetoacetic acid and 3-hydroxybutyric acid, known as ketone bodies. The continued formation and excretion of these acids increase the hydronium ion concentration and affect the acid–base balance. Coma and death can result if the diabetic individual suffers from acidosis.

The specific condition of lowered blood pH as a result of increased carbonic acid is called **respiratory acidosis.** Emphysema, pneumonia, poliomyelitis, anesthesia, and heart failure are some of the conditions that decrease the respiration rate and lead to an increased concentration of H_2CO_3.

Temporary acidosis results from holding your breath for a period of time. As your body continues to produce acid, the blood pH decreases. Only by exhalation of carbon dioxide can the pH be returned to normal. The nervous system forces you to breathe, so you cannot voluntarily hold your breath for too long.

Although less common than acidosis, alkalosis results from the ingestion of bases such as antacids or from the loss of stomach acid through severe vomiting. In such cases, the kidneys excrete alkaline urine containing Na_2HPO_4.

Excessive loss of carbonic acid results in a condition known as **respiratory alkalosis.** This condition results from hyperventilation, which may occur in extreme fevers or severe hysteria. Hyperventilation can increase the pH to 7.6 in a few minutes. The breakdown of H_2CO_3 and expulsion as CO_2 result in a shift of the H_2CO_3/HCO_3^- equilibrium to decrease $[H_3O^+]$, that is, to increase the pH. To compensate, the body lowers the respiration rate. In some cases of hysteria, fainting occurs and the respiration rate is lowered.

11.11 Titration

An acid–base **titration** is a procedure used to determine the total concentration or amount of an acid (or base) by neutralizing it with a known amount of a strong base (or strong acid). The strong base (or acid) reacts with both the ionized and nonionized portions of the acid (or base). The point of neutralization is known as the **endpoint** or **neutralization point.** The endpoint is detected by using an indicator.

In a titration, a measured volume or mass of an acid (or base), whose concentration is the unknown, is placed in a flask, and a small amount of an indicator is added. A solution of a base (or acid) of known concentration is then added dropwise from a buret. When the indicator changes color, the addition is terminated and the volume of solution added from the buret is recorded (Figure 11.6).

Normality—A Concentration Unit

You learned in Section 9.8 that it is useful to think of electrolytes in terms of the number of equivalents of an ion. Recall that an equivalent of an ion is the amount that provides 1 mole of charge. Because some acids have more than one available proton and some bases have more than one hydroxide ion per formula unit, it is useful to extend the concept of equivalents to calculations involving acids and bases.

An **equivalent** (1 equiv) of an acid is the amount of the acid containing 1 mole of protons. An equivalent of a base is the amount that will react with 1 mole of protons. The **equivalent weight** of an acid is the mass that contains 1 mole of protons; it is calculated by dividing the molar mass of the acid by the number of moles of ionizable hydrogen atoms contained in 1 mole of the acid. Similarly, the equivalent weight of a base is calculated by dividing the molar mass of the base by the number of moles of hydrogen ions that will react with 1 mole (or formula unit) of the base. The equivalent weights of several acids and bases are listed in Table 11.9.

Because the concept of equivalents is useful in determining how much acid is required to neutralize a given amount of base, a concentration unit using equivalents has been devised. **Normality** (N) is the number of equivalents of acid or base per liter of solution.

$$\mathrm{N} = \frac{\textbf{equivalents}}{\textbf{liters of solution}}$$

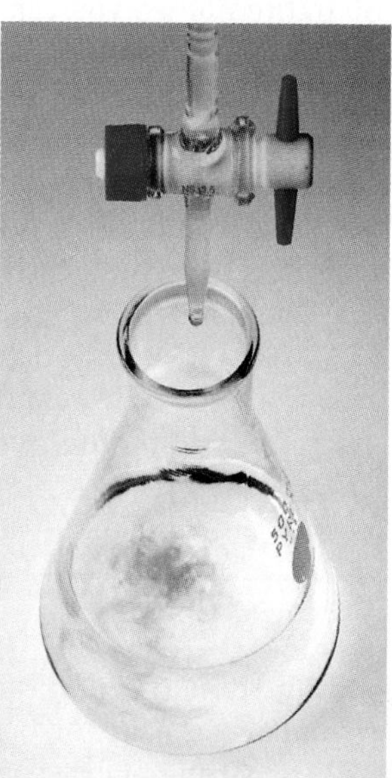

FIGURE 11.6 Titration
(a) At the pH of the acid solution the phenolphthalein indicator is colorless. (b) Near the endpoint some pink color of the indicator develops. (c) When neutralization is complete the last drop makes the solution basic and a pink color develops. (Richard Megna, Fundamental Photographs)

TABLE 11.9 Equivalent Weights of Acids and Bases

Compound	Molar Mass (g/mole)	Equivalent Weight (g/equiv)
HCl	36.5	$\frac{36.5}{1} = 36.5$
H_2SO_4	98.0	$\frac{98.0}{2} = 49.0$
H_3PO_4	98.0	$\frac{98.0}{3} = 32.7$
NaOH	40.0	$\frac{40.0}{1} = 40.0$
$Ca(OH)_2$	74.0	$\frac{74.0}{2} = 37.0$
$Al(OH)_3$	78.0	$\frac{78.0}{3} = 26.0$

The advantage of the normality unit is that 1 equiv of any acid reacts with 1 equiv of any base. If molarity is used, a mole ratio factor derived from the balanced equation must always be used to determine the quantity of acid or base required for neutralization. For example, whereas 1 L of a 1 M solution of HCl will neutralize 1 L of a 1 M solution of NaOH, a 1 M solution of H_2SO_4 contains twice as many hydrogen ions (protons) and will require 2 L of a 1 M NaOH solution. However, 1 L of a 1 N solution of any acid will neutralize 1 L of a 1 N solution of any base.

EXAMPLE 11.8

What is the normality of a solution of H_3PO_4 obtained by dissolving 49.0 g of the acid in sufficient water to produce 3.00 L of solution?

Solution

First, calculate the equivalent weight of H_3PO_4 to determine the number of equivalents present in 49.0 g.

$$\text{molar mass } H_3PO_4 = 98.0 \text{ g}$$

There are 3 moles of available hydrogen ions per mole of the acid. Divide the molar mass by 3 to obtain the equivalent weight.

$$\text{equiv wt } H_3PO_4 = \frac{98 \text{ g/}\cancel{\text{mole}}}{3 \text{ equiv/}\cancel{\text{mole}}} = 32.66 \text{ g/equiv}$$

$$49.0 \cancel{\text{g}} \times \frac{1 \text{ equiv}}{32.66 \cancel{\text{g}}} = 1.50 \text{ equiv}$$

Calculate the normality by dividing the number of equivalents by the volume.

$$\frac{1.50 \text{ equiv}}{3.00 \text{ L}} = 0.500 \text{ N}$$

PROBLEM 11.8

Calculate the number of grams of H_2SO_4 required to make 100 mL of a 0.20 N H_2SO_4 solution.

Volumetric Analysis

Volumetric analysis is an experimental technique used to determine the concentration of an unknown sample by the measurement of volumes. Titration of an acid sample such as gastric juice followed by the calculation of its normality is an example of volumetric analysis. In such a titration the number of equivalents of base added from the buret is calculated by multiplying the volume of base used to reach the endpoint by its normality.

$$\frac{\text{equivalents of base}}{\text{liter}} \times \text{liters} = \text{equivalents of base}$$

This number of equivalents of base is equal to the number of equivalents of the acid of unknown concentration because 1 equiv of a base is required to neutralize 1 equiv of an acid.

$$\text{equivalents of base} = \text{equivalents of acid}$$

In a laboratory, small volumes of solutions are frequently used. Therefore, the units of equivalents and liters are not convenient. One-thousandth of a liter is a milliliter, and one-thousandth of an equivalent is a milliequivalent (mequiv). Thus, the ratio of the number of milliequivalents to the number of milliliters is also equal to the normality.

$$\frac{\text{milliequivalents}}{\text{milliliter}} = \frac{\text{mequiv}}{\text{mL}} = \text{N}$$

The number of milliequivalents of base or acid added from the buret is calculated by multiplying the volume used by its normality.

$$\text{normality} \times \text{milliliters} = \text{milliequivalents}$$

$$\frac{\text{milliequivalents}}{\text{milliliter}} \times \text{milliliters} = \text{milliequivalents}$$

This number of milliequivalents is equal to the number of milliequivalents of the unknown because 1 mequiv of an acid is required to neutralize 1 mequiv of a base.

$$\text{milliequivalents of acid} = \text{milliequivalents of base}$$

EXAMPLE 11.9

A 5.00-mL sample of household ammonia is titrated with 48.0 mL of a 0.200 N solution of hydrochloric acid to a methyl red endpoint. What is the molarity of the household ammonia?

Solution

The acid solution contains 0.200 mequiv/mL of hydrochloric acid. Thus, the number of milliequivalents of acid used is

$$\frac{0.200 \text{ mequiv}}{\cancel{\text{mL}}} \times 48.0 \cancel{\text{mL}} = 9.60 \text{ mequiv}$$

At the endpoint the number of milliequivalents of ammonia is also equal to 9.60 mequiv. This quantity is contained in 5.00 mL of the ammonia solution, so the normality must be

$$\frac{9.60 \text{ mequiv}}{5.00 \text{ mL}} = 1.92 \text{ N}$$

Because ammonia reacts with only 1 equiv of acid per mole, its equivalent weight is the same as its molecular weight. Therefore, the molarity is the same as the normality, or 1.92 M.

PROBLEM 11.9

A 0.0420-g sample of sodium bicarbonate (baking soda) is dissolved in water in a flask, and 20.0 mL of a hydrochloric acid solution is required to neutralize it. What is the normality of the HCl solution?

$$NaHCO_3 + HCl \longrightarrow NaCl + CO_2 + H_2O$$

Summary

The Brønsted–Lowry concept of acid–base reactions focuses on proton transfer from an acid to a base: **acids** are hydrogen ion (proton) donors; **bases** are proton acceptors. Substances that can both donate and accept protons are **amphoteric.** Loss of a hydrogen ion from an acid produces its **conjugate base;** acceptance of a hydrogen ion by a base produces its **conjugate acid.** Acids that can donate one, two, or three hydrogen ions are **monoprotic, diprotic,** and **triprotic,** respectively.

The tendency of acids and bases to lose and gain protons, respectively, determines the direction of an acid–base reaction. The tendency of an acid to lose hydrogen ions, **ionization,** is given by the **acid ionization constant. Strong acids** transfer their protons completely to water to form **hydronium ions; weak acids** exist in equilibrium with their conjugate bases and the hydronium ion in water. The tendency of a base to gain hydrogen ions is given by the **base ionization constant. Strong bases** ionize completely to produce **hydroxide ions. Weak bases** produce small quantities of hydroxide ions.

Water undergoes **self-ionization** to produce small amounts of hydronium and hydroxide ions in equal concentration; water is **neutral.** The addition of acid or base to water affects the concentrations of both hydronium ions and hydroxide ions in a manner predicted by Le Châtelier's principle. **Acidic** solutions have a concentration of hydronium ion in excess of that in pure water. **Basic** solutions have a concentration of hydroxide ion in excess of that in pure water.

The concentration of hydronium ions in water may be expressed in **pH** units. The pH of a solution can be measured using chemical **indicators.**

The **hydrolysis** of some salts results in solutions with a pH other than 7. **Basic salts** contain as the anion the conjugate base of a weak acid. **Acidic salts** contain as the cation the conjugate acid of a weak base. The equilibrium constant for hydrolysis of the salt of a weak acid is equal to K_b, which may be calculated by dividing K_w by the K_a of the conjugate acid of the anion.

Buffers have two components, a weak acid and a salt of its conjugate base or a weak base and a salt of its conjugate acid, that will react with added amounts of either a base or an acid. The effectiveness of a buffer against an added acid or base depends on the ratio of the concentrations of the two components of the buffer. Buffers control the pH of fluids in living organisms.

Concentrations of acids or bases may be expressed in **normality** units. This concentration indicates the number of **equivalents** of an acid or base in a liter of solution. The **equivalent weight** of an acid or base is equal to the molecular weight divided by the number of protons that can be donated or accepted by the substance.

Titration is a procedure in which an acid and a base reach a point of **neutralization** at the **endpoint.** An indicator is used in a titration to identify the endpoint.

Exercises

Brønsted–Lowry Theory

11.1 Can a substance act as both an acid and a base? Explain, giving examples.

11.2 What is meant by the term *amphoteric?*

11.3 Write equations for the stepwise ionization of the diprotic acid H_2SO_3.

11.4 Write equations for the stepwise ionization of the diprotic acid $H_2C_2O_4$.

11.5 Describe the relationship between an acid and its conjugate base.

11.6 Describe the relationship between a base and its conjugate acid.

11.7 What is the conjugate acid of each of the following?

(a) ClO_4^- (b) Cl^- (c) NO_3^-
(d) CO_3^{2-} (e) $CH_3CO_2^-$ (f) HSO_4^-

11.8 What is the conjugate base of each of the following?

(a) HSO_4^- (b) H_3PO_4 (c) $H_2PO_4^-$
(d) HPO_4^{2-} (e) CH_3CO_2H (f) HBr

11.9 Veronal ($HC_8H_{11}N_2O_3$), a barbiturate drug, is an acid. What is its conjugate base?

11.10 Niacin ($HC_6H_4NO_2$), also known as nicotinic acid, is a vitamin. What is its conjugate base?

11.11 Pyruvic acid, an acid produced in metabolic reactions, has the following structure. What is the molecular formula of the conjugate base? What is the charge of this ion?

```
     H  O  O
     |  ‖  ‖
  H—C—C—C—O—H
     |
     H
```

11.12 Methanol is a weak base. From its structure, write the formula for its conjugate acid. What is the charge of the ion?

```
     H
     |
  H—C—O—H
     |
     H
```

11.13 Codeine ($C_{18}H_{21}NO_3$) is a cough suppressant that is a base. What is the formula of the conjugate acid that results from the gain of a proton?

11.14 Aspartic acid is a diprotic amino acid with the molecular formula $C_4H_7NO_4$. What is the molecular formula and charge of the ion that results from the loss of two protons?

Acid–Base Strength

11.15 Ammonia is very soluble in water. Why is concentrated aqueous ammonia, a 27% solution of ammonia, still considered weakly basic?

11.16 A 0.1 M solution of HCl is a solution of a strong acid. Why is a solution of this low concentration considered a *strong* acid?

11.17 Classify each of the following as a weak or strong acid.

(a) CH_3CO_2H (b) HCl (c) HNO_3
(d) HCN (e) H_2SO_4 (f) H_2O

11.18 Classify each of the following as a weak or strong acid.

(a) HF (b) HNO_2 (c) H_2SO_3
(d) HBr (e) $HClO_4$ (f) H_3PO_4

Acid Ionization Constant

11.19 Write an acid ionization constant expression for each ionization step of the triprotic acid H_3PO_4.

11.20 Write an acid ionization constant expression for each ionization step of the diprotic acid H_2SO_3.

11.21 What is the ionization constant of an acid if the hydronium ion concentration of a 0.100 M solution is 7.66×10^{-4} M?

11.22 What is the ionization constant of an acid if the hydronium ion concentration of a 0.400 M solution is 1.40×10^{-4} M?

Self-Ionization of Water

11.23 What are the hydronium ion and hydroxide ion concentrations in each of the following solutions?

(a) 0.1 M HCl (b) 1×10^{-2} M HNO_3
(c) 1×10^{-3} M KOH (d) 0.05 M H_2SO_4
(e) 1×10^{-4} M NaOH (f) 0.005 M $Mg(OH)_2$

11.24 What are the hydronium ion and hydroxide ion concentrations in each of the following solutions?

(a) 0.02 M HBr (b) 2×10^{-2} M HNO_3
(c) 3×10^{-4} M LiOH (d) 0.04 M H_2SO_4
(e) 5×10^{-4} M KOH (f) 0.002 M $Sr(OH)_2$

pH Values

11.25 Calculate the pH of each of the following solutions.

(a) 1×10^{-3} M HCl (b) 2×10^{-3} M HNO_3
(c) 1×10^{-2} M NaOH (d) 3×10^{-3} M NaOH

11.26 Calculate the pH of each of the following solutions.

(a) 1×10^{-2} M HBr (b) 2×10^{-4} M $HClO_4$
(c) 1×10^{-1} M KOH (d) 4×10^{-2} M LiOH

11.27 Calculate the pH of each of the following items.

(a) a soft drink with $[H_3O^+] = 2 \times 10^{-4}$ M
(b) milk with $[H_3O^+] = 2 \times 10^{-7}$ M

(c) ammonia with $[H_3O^+] = 2 \times 10^{-12}$ M
(d) gastric juice with $[H_3O^+] = 0.1$ M

11.28 Calculate the pH of each of the following items.

(a) vinegar with $[H_3O^+] = 7.5 \times 10^{-3}$ M
(b) orange juice with $[H_3O^+] = 2.9 \times 10^{-4}$ M
(c) lye with $[OH^-] = 0.05$ M
(d) arterial blood with $[OH^-] = 2.5 \times 10^{-7}$ M

Hydrolysis of Salts

11.29 Given the following K_a values, determine which conjugate base is the strongest base.

HF (6.8×10^{-4}) HNO_2 (7.2×10^{-4})
HCNO (2.2×10^{-4})

11.30 Given the following K_a values, determine which conjugate base is the strongest base.

H_2SO_3 (1.2×10^{-2}) HNO_2 (7.2×10^{-4})
HCN (4.4×10^{-10})

11.31 What is the K_b of the OI^- ion? The acid ionization constant of HOI is 2.0×10^{-11}.

11.32 What is the K_b of the CN^- ion? The acid ionization constant of HCN is 4.4×10^{-10}.

11.33 What is the hydroxide ion concentration in 0.810 M OBr^-? The acid ionization constant of HOBr is 2.0×10^{-9}.

11.34 What is the hydroxide ion concentration in 0.750 M OCl^-? The acid ionization constant of HOCl is 3.0×10^{-8}.

Indicators

11.35 Some natural fruit juices change color when a base is added. Explain why.

11.36 How does an indicator work?

11.37 What will be the color of methyl orange at pH 2? What will its color be at pH 7?

11.38 What will be the color of phenolphthalein at pH 7? What will its color be at pH 11?

Buffers

11.39 How does the acid/conjugate base ratio affect the pH of a buffer?

11.40 How does the acid/conjugate base ratio affect the effectiveness of a buffer toward acid or base?

11.41 Explain why a solution containing HCl and NaCl cannot serve as a buffer.

11.42 Which of the following combinations of an acid and its conjugate base cannot serve as a buffer?

(a) $HClO_4$ and $KClO_4$ (b) H_2SO_4 and $NaHSO_4$
(c) NH_4Cl and NH_3

11.43 What is the hydronium ion concentration of a solution that contains 0.260 M acid and 0.640 M conjugate base if the acid ionization constant is 2.06×10^{-6}?

11.44 What is the hydronium ion concentration of a solution that contains 0.700 M acid and 0.440 M conjugate base if the acid ionization constant is 3.59×10^{-8}?

11.45 What is the pH of a solution that contains 0.170 M acid and 0.500 M conjugate base if the acid ionization constant is 2.87×10^{-9}?

11.46 What is the pH of a solution that contains 0.810 M acid and 0.650 M conjugate base if the acid ionization constant is 4.92×10^{-7}?

11.47 Describe the effects that hyperventilation and hypoventilation have on blood pH.

11.48 What two processes are used by the body to eliminate the acids formed in metabolic reactions?

Normality

11.49 How many milliequivalents of acid are present in each of the following solutions?

(a) 50 mL of 0.010 N HCl
(b) 5.0 mL of 0.10 N HNO_3
(c) 1.0 L of 1.0 N HCl

11.50 How many milliequivalents of acid are present in each of the following solutions?

(a) 75 mL of 0.020 N HBr
(b) 25 mL of 0.30 N $HClO_4$
(c) 30 mL of 0.25 N HI

11.51 What is the normality of each solution?

(a) 4.0 g of NaOH in 500 mL of solution
(b) 3.65 g of HCl in 1.0 L of solution
(c) 4.9 g of H_2SO_4 in 100 mL of solution

11.52 What is the normality of each solution?

(a) 0.56 g of KOH in 250 mL of solution
(b) 4.05 g of HBr in 500 mL of solution
(c) 3.27 g of H_3PO_4 in 100 mL of solution

11.53 Lactic acid ($C_3H_6O_3$) is a monoprotic acid. How many grams of lactic acid are required to produce 250 mL of a 0.010 N solution?

11.54 Oxalic acid ($C_2H_2O_4$) is a diprotic acid. How many grams of oxalic acid are required to produce 500 mL of a 0.020 N solution?

Titration

11.55 What quantity of 0.10 N NaOH is required to titrate 25.0 mL of a 0.060 N HCl solution?

11.56 A 50.0-mL sample of $Ca(OH)_2$ is titrated with 36.0 mL of a 0.05 N HCl solution to an endpoint. What is the molarity of the $Ca(OH)_2$ solution?

11.57 A 0.1060-g sample of Na_2CO_3 is neutralized with 48.0 mL of $HClO_4$ solution. What is the normality of the $HClO_4$ solution?

11.58 A 0.100-g sample of NaOH is titrated to an endpoint with 12.5 mL of HCl. What is the normality of the HCl?

11.59 What mass (g) of calcium carbonate ($CaCO_3$) is required to react with the sulfuric acid in 38.5 mL of a 0.973 M solution of H_2SO_4?

11.60 A 10.0-mL sample of sulfuric acid solution from an automobile battery reacts exactly with 35.0 mL of a 1.87 M solution of sodium hydroxide (NaOH). What is the molar concentration of the battery acid?

Commercially available radon test kits are available to detect this radioactive gas in the home. (Richard Megna, Fundamental Photographs)

12

Nuclear Chemistry

12.1 Chemistry of the Nucleus
12.2 Radioactivity and Nuclear Reactions
12.3 Balancing Nuclear Equations
12.4 Half-Lives
12.5 Nuclear Transmutations
12.6 Nuclear Fission Reactions
12.7 Nuclear Fusion
12.8 Detection of Radiation
12.9 Biological Effects of Radiation
12.10 Radioisotopes in Diagnosis
12.11 Radioisotopes in Therapy

Essays
- *Dating Old Rocks and Fossils*
- *Chernobyl*
- *Positron Emission Tomography*

12.1 Chemistry of the Nucleus

Up to this point chemistry has been presented only in terms of the changes in the number and types of bonds between atoms in compounds. The same atoms are still there and are merely redistributed in different structures. Now we will examine reactions that occur in the nuclei of atoms that convert them into different atoms. This chapter is devoted to **nuclear chemistry,** which deals with the reactions of atomic nuclei.

Recall from Chapter 3 that isotopes of an element have the same number of protons but different numbers of neutrons. In the general symbol ${}^{A}_{Z}E$, Z is the atomic number or number of protons, and A is the mass number or the sum of the numbers of protons and neutrons. Thus, ${}^{24}_{11}Na$ represents an isotope of sodium that contains 11 protons and 13 neutrons.

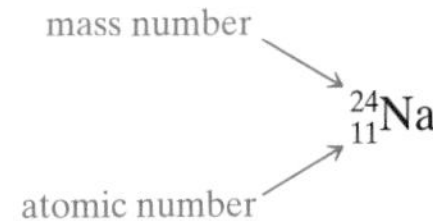

In this chapter we will also use the symbol Na-24 and sodium-24. The number following the elemental symbol or element name is the mass number of the isotope.

Isotopes of elements are also called nuclides, and the protons (p) and neutrons (n) are collectively called nucleons. The nuclear properties of each nuclide of an element differ substantially, but the chemical properties are nearly indistinguishable. Some nuclei are stable indefinitely, whereas others undergo spontaneous reactions that change the composition of the nucleus and produce either another isotope or, more commonly, another element.

Nuclear Decay Reactions

Isotopes that undergo nuclear decay and produce radiation are often called **radioisotopes.** The reactant nucleus or **parent nucleus** decays to form another nucleus called the **daughter nucleus.** In this process high-energy particles collectively known as *radiation* are formed.

parent nucleus ⟶ daughter nucleus + radiation

The daughter nucleus may further decay. Thus, a series of decay processes may occur before a stable nucleus is ultimately formed.

Both carbon-12 and carbon-13 are naturally occurring stable nuclides. However, carbon-14 decays slowly to give nitrogen-14 and a high-energy electron called a beta particle. The reaction is written as follows:

$$^{14}_{6}C \longrightarrow ^{14}_{7}N + ^{0}_{-1}e$$

The superscript 0 in the symbol $^{0}_{-1}e$ indicates that on the atomic mass scale the mass of an electron is essentially zero. The subscript -1 that occupies the position of the atomic number in the symbol for an atom is not an atomic number but indicates that the charge is -1. Of course, in atomic symbols the atomic number is equal to the positive charge of the nucleus.

The energy of ordinary chemical reactions, which involves changes in the electron configurations of atoms, is usually on the order of 10–50 kcal/mole. In comparison the energy of nuclear reactions is extremely high. The decay of carbon-14 releases about 3.6×10^6 kcal/mole. A nuclear fission reaction of uranium-235 releases about 4×10^9 kcal/mole. When suitably arranged, a sufficient sample of the uranium-235 isotope can react rapidly and cause an atomic explosion.

Uses of Radioisotopes

Radioisotopes were used in the atomic bombs that devastated two Japanese cities and ended World War II. These bombs were small compared with the hydrogen bombs of today. In recent years the major powers have attempted to limit their arms and prevent the spread of nuclear devices to smaller countries.

Use of radioisotopes for the generation of power is controversial in spite of the fact that the world has a limited supply of fossil fuels. Nuclear energy is currently being used as an alternative energy source for the generation of electricity in many countries (Table 12.1). However, there is concern among Americans about the radioactivity hazards and environmental side effects of nuclear power plants, and no new plants have been constructed in many years. In contrast the nuclear power industry in France is well supported as a matter of public policy. France has continued to construct plants safely and with none of the delays experienced in the United States.

The use of radioisotopes in medicine is one major benefit of nuclear chemistry. Those radioisotopes which do not occur in nature have been produced by nuclear reactions. They are used in the diagnosis of abnormalities in body functions as well

TABLE 12.1 Nuclear Share of Electricity in Industrialized Countries

Country	% Nuclear
France	65
Sweden	42
West Germany	31
Japan	23
United Kingdom	19
United States	16
Canada	13

as in therapy. For example, iodine-123 is used to determine the thyroid gland's activity. Another radioisotope, iodine-131, is used in therapy to combat certain thyroid cancers. In this chapter we will learn how different isotopes decay and how these processes can be used in medicine.

12.2 Radioactivity and Nuclear Reactions

Radioactivity, the emission of radiation from certain nuclei, was discovered a century ago. In 1897 the English physicist Ernest Rutherford found three common types of radiation, which are designated by the first three letters of the Greek alphabet—alpha (α), beta (β), and gamma (γ). A radioactive isotope does not emit all three types of radiation in a single process. However, gamma radiation may be produced with either alpha or beta radiation.

Alpha Radiation

Alpha radiation consists of a stream of **alpha particles,** which are helium nuclei, consisting of two protons and two neutrons each. An alpha particle is represented as ${}^{4}_{2}He^{2+}$. However, the +2 charge is ordinarily omitted because nuclear equations deal only with the contents of the nucleus. Furthermore, the energy associated with the gain or loss of electrons from the valence shell is significantly smaller than the energy of radioactive processes that produce radiation.

The emission of alpha particles is common for radioactive isotopes of high atomic number, such as uranium-238, which yields thorium-234.

$$ {}^{238}_{92}U \longrightarrow {}^{234}_{90}Th + {}^{4}_{2}He $$

Alpha particles emitted from the nucleus of an atom have little penetrating power and can be stopped by this sheet of paper (Figure 12.1). A 0.05-mm layer of dead cells on the surface of skin will stop alpha particles. However, an intense dose of alpha radiation will cause skin burns. If radioactive dust particles get inside the body, the alpha particles emitted will affect cells and cause biological damage.

Beta Radiation

Beta radiation consists of a stream of high-energy electrons called **beta particles.** Although the electrons do not exist as such in the nucleus, they are produced in a process in which a neutron is converted into a proton and a beta particle.

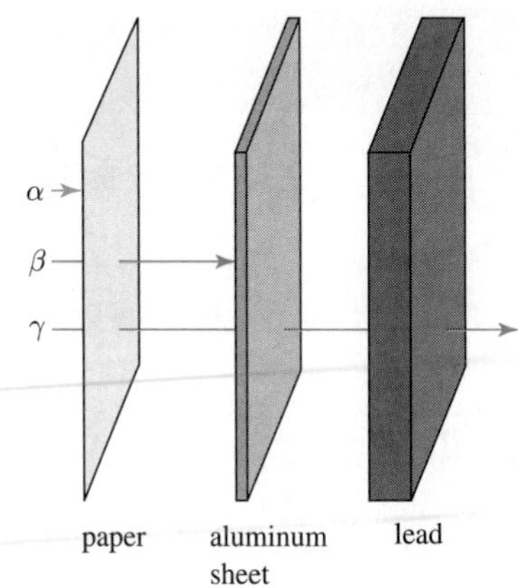

FIGURE 12.1 Illustration of Penetrating Power of Radiation

neutron ⟶ proton + electron

$$^{1}_{0}n \longrightarrow {}^{1}_{1}H + {}^{0}_{-1}e$$

The product nucleus has the same mass number as the reactant nucleus because the total count of protons and neutrons hasn't changed—a neutron is lost but a proton is gained. However, the atomic number of the product is one larger than that of the reactant because of the proton created in the nucleus. The decay of the beta emitter phosphorus-32, which is used in leukemia therapy, gives sulfur-32.

$$^{32}_{15}P \longrightarrow {}^{32}_{16}S + {}^{0}_{-1}e$$

Beta particles have greater penetrating power than alpha particles but they are stopped by thin sheets of metal (Figure 12.1). Beta particles can penetrate about 4 mm into living tissue. Radiation burns result, but the vital internal organs are not affected unless a radioactive source is inhaled or ingested.

Gamma Rays

Gamma radiation does not consist of particles but is a high-energy radiation similar to X-rays. Gamma rays emitted by a nucleus have high penetrating power. A 5-cm sheet of lead will not stop all gamma rays (Figure 12.1).

Positron Emission and K Electron Capture

Two additional types of radioactive decay processes have been discovered since Rutherford's original work, namely, *positron emission* and K *electron capture.* In both processes the mass numbers of the reactant and product are identical but the atomic number of the product is one less than that of the reactant. In each case the net effect is conversion of a proton into a neutron.

In **positron emission** processes, a proton in the nucleus is converted into a neutron while ejecting a positron, ${}^{0}_{1}e$ or β^{+}. A positron has the same mass as an electron but has the opposite charge. It may be regarded as a positive electron. The conversion of potassium-40 to argon-40, a process used to establish the age of rocks, occurs by positron emission.

$$^{40}_{19}K \longrightarrow {}^{40}_{18}Ar + {}^{0}_{1}e$$

In **K electron capture** processes, a proton in the nucleus captures an inner shell electron and is converted into a neutron.

proton + electron ⟶ neutron

Although the net change of forming a neutron in place of a proton is the same as for positron emission, the process is very different. Mercury-197 reacts by electron capture to form gold-197. The process is written using an electron as a reactant, but remember that the electron was already part of the mercury-197 atom and did not come from another source such as a reducing agent.

$$^{197}_{80}\text{Hg} + {}^{0}_{-1}\text{e} \longrightarrow {}^{197}_{79}\text{Au}$$

12.3 Balancing Nuclear Equations

We can write equations for nuclear reactions much as we do for ordinary chemical reactions. Both mass and nuclear charge are conserved. That means that the sum of the mass numbers of the reactants equals the sum of the mass numbers of the products. Thus the sums of the superscripts included with the elemental symbol, the mass numbers, are the same on both sides of the equation. The total charge of the nuclear reactants and products are also equal. Thus the sums of the subscripts of the elemental symbols, the atomic numbers or charge, are the same on both sides of the equation.

Consider the decay of uranium-238 which produces alpha particles and thorium-234, which is always found with uranium in natural ore deposits. There are two fewer protons in thorium than in uranium. The isotope of thorium formed also has two fewer neutrons than uranium, because the two neutrons are lost to the alpha particle.

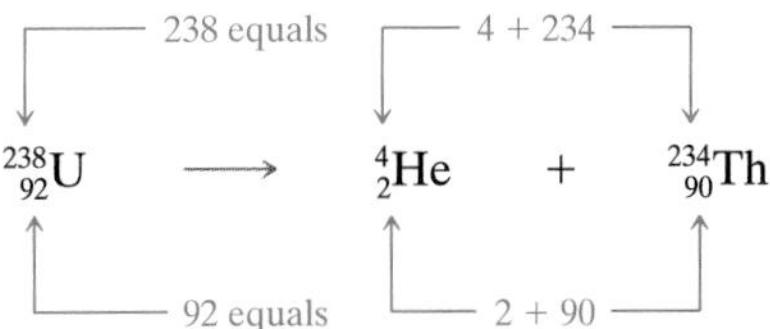

The sum of the mass numbers of the products (4 + 234) equals the mass number of the uranium reactant, 238. In addition, the sum of the atomic numbers of the products (2 + 90) equals the atomic number of the uranium reactant, 92.

Thorium-234 is radioactive and produces protactinium-234 and a beta particle. As the result of conversion of a neutron in the nucleus into a proton and an electron, the atomic number is increased by 1. However, the product has the same mass number as the reactant.

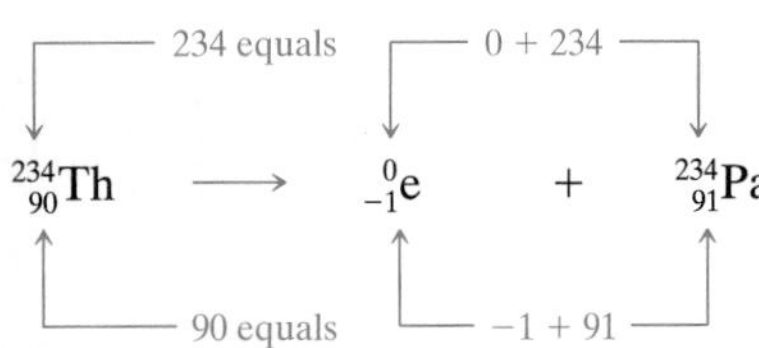

In electron capture processes it is necessary to write the symbol for the electron separately even though it is part of the reacting nuclide. The reaction of ${}^{26}_{13}Al$ to give ${}^{26}_{12}Mg$ occurs by electron capture.

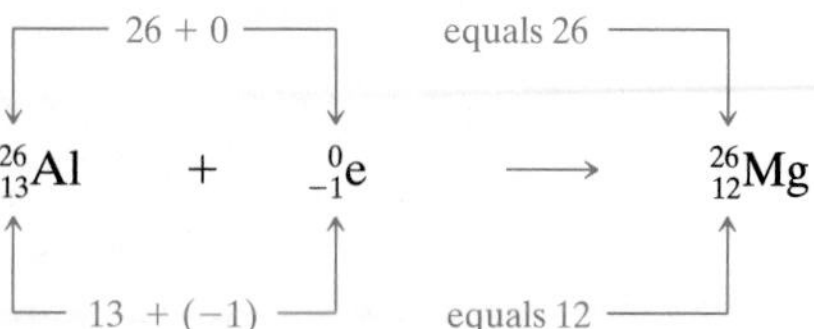

EXAMPLE 12.1

The isotope polonium-212 undergoes alpha decay to yield one alpha particle per atom and a single element. What is the element?

Solution

First, write the symbol for polonium-212 to the left of an arrow. Its atomic number is 84. Place ${}^{4}_{2}He$ to the right of the arrow along with ${}^{A}_{Z}E$ to represent the unknown element.

$${}^{212}_{84}Po \longrightarrow {}^{4}_{2}He + {}^{A}_{Z}E$$

The sum of the atomic numbers of the products must equal that of polonium.

$$84 = 2 + Z$$
$$82 = Z$$

Similarly, the sum of the mass numbers of the products must equal the mass number of polonium.

$$212 = 4 + A$$
$$208 = A$$

The unknown element has an atomic number of 82 and a mass number of 208. In the list of elements we find that lead has an atomic number of 82; therefore, the element is lead-208. The balanced equation is

$${}^{212}_{84}Po \longrightarrow {}^{4}_{2}He + {}^{208}_{82}Pb$$

PROBLEM 12.1

The carbon-14 isotope is unstable and emits one beta particle per atom. What element is produced?

12.4 Half-Lives

The rate of a radioactive decay reaction does not depend on external conditions that affect ordinary chemical reactions such as temperature and pressure. The rate does not depend on whether the nuclide is in its elemental form or in a compound. However, the rate is very much a characteristic of the type of nuclide. Some nuclides last for millions of years, whereas others decay in seconds.

TABLE 12.2 Half-Lives of Elements in Uranium-238 Series

Isotope	Radiation	Half-Life
$^{238}_{92}U$	alpha	4.5 billion years
$^{234}_{90}Th$	beta	24.1 days
$^{234}_{91}Pa$	beta	1.2 min
$^{234}_{92}U$	alpha	250,000 years
$^{230}_{90}Th$	alpha	80,000 years
$^{226}_{88}Ra$	alpha	1620 years
$^{222}_{86}Rn$	alpha	3.8 days
$^{218}_{84}Po$	alpha, beta	3.05 min
$^{214}_{82}Pb$	beta	26.8 min
$^{214}_{83}Bi$	alpha, beta	19.7 min
$^{210}_{81}Tl$	beta	1.3 min
$^{210}_{82}Pb$	beta	22 years
$^{210}_{83}Bi$	beta	5.0 days
$^{210}_{84}Po$	alpha	138 days

All radioactive decay reactions, regardless of the type of process, have one characteristic in common. In a period of time called the **half-life,** the amount of nuclide that decays is 50%. Each subsequent period of time equal to the half-life results in the decay of half of the nuclide that remains after the preceding half-life. The half-life is different for each nuclide (Table 12.2).

The graph in Figure 12.2 illustrates the decay pattern of radioactive materials. The strontium-90 isotope, one of many produced in an atomic bomb explosion, has a half-life of 28.1 years. Atmospheric nuclear tests in the 1950s produced significant quantities of strontium-90. As shown in Figure 12.2, for 80 mg of the isotope deposited somewhere on Earth, the decay process leaves 40 mg after 28.1 years. After another 28.1 years 20 mg remains. Thus, in the year 2006, 25% of the isotope from a 1950 test will still remain.

EXAMPLE 12.2

Sodium-24 is used in medicine to determine the effectiveness of blood circulation. Its half-life is 15 h. What percentage of the radioisotope will remain after 60 h?

Solution

After each half-life, the amount remaining will be half of the amount remaining after the previous half-life. The table shows this behavior. After 60h, 6.25% of the radioisotope will remain.

Time (h)	Amount (%)
0	100
15	50
30	25
45	12.5
60	6.25

FIGURE 12.2
The Half-Life of Strontium-90

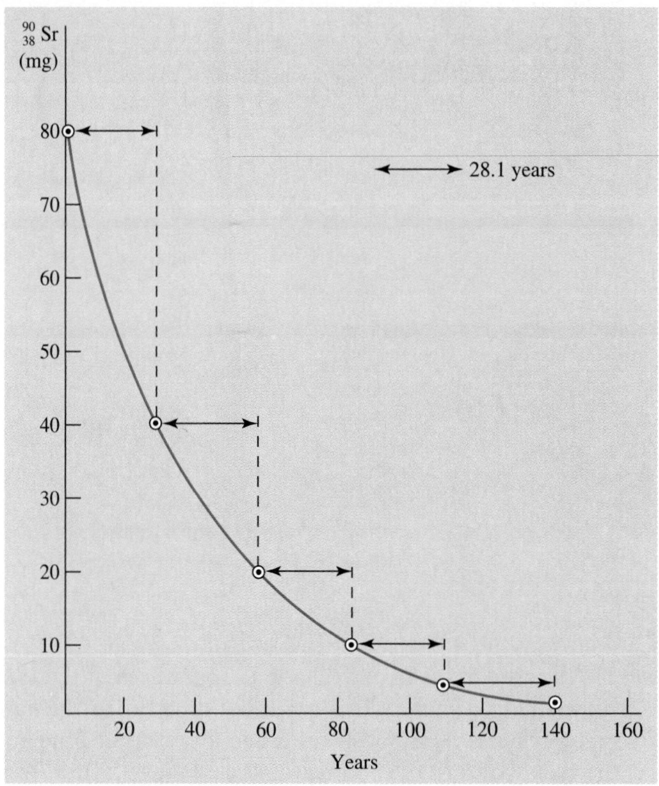

PROBLEM 12.2

The half-life of iodine-131 is 8 days. If 16 mg of the radioisotope is administered to a patient to treat cancer of the thyroid, how much will remain in the body after 24 days?

Atomic Dating

A sample containing a radioisotope can be dated—that is, its age can be determined. Uranium-238 decomposes to lead-206 in a series of steps in which 50% of the uranium decays in 4.5 billion years. After one half-life, a sample that originally contained 1.00 g of uranium-238 would contain 0.50 g of uranium and 0.43 g of lead. This amount of lead is calculated from the atomic weights based on the 0.50 g of uranium that decomposed.

$$\frac{206 \text{ g/mole Pb}}{238 \text{ g/mole U}} \times 0.50 \text{ g U} = 0.43 \text{ g Pb}$$

A rock sample that contains uranium-238 and lead-206 in the ratio 0.50/0.43 is 4.5 billion years old. If the ratio is larger or smaller, the sample is younger or older. Analysis of rocks from Earth, the moon, and meteorites all show that our solar system is about 4.5 billion years old.

Carbon-14 Dating

One of the best-known applications of radioactive decay is the carbon-14 method of dating artifacts. The age of materials made of plant or animal matter can be es-

Dating Old Rocks and Fossils

Potassium-40 dating provides a very convenient way to date igneous rocks such as granite and basalt. Potassium compounds are commonly found in such samples. The decay product, argon-40, is formed by positron emission.

$$^{40}_{19}K \longrightarrow {}^{40}_{18}Ar + {}^{0}_{1}e$$

Because argon is a gas, any of it trapped in a rock escapes when the rock melts, as in a lava flow. When the lava solidifies, potassium-40 remains without any argon-40. With time, argon-40 is slowly formed and is trapped in the new rock. Thus, a geologist can date a rock by determining the relative amounts of argon-40 and potassium-40. The half-life of potassium-40 is 1.25 billion years, which means that a rock that has equal proportions of potassium-40 and argon-40 is 1.25 billion years old. If a fossil is found within or directly below the lava flow, it can safely be concluded that it is at least 1.25 billion years old.

The age of rocks dated by the potassium-40 method is a minimum value—they could be older. If some of the gas has escaped from the rock, the ratio of argon-40 to potassium-40 will be less than the value that would have been determined if the gas had remained in the rock. As a result we will incorrectly conclude that the rock is younger than it actually is.

Lava is so hot that no gases remain. After cooling, subsequent decay of potassium yields argon, which is trapped in the solid. (Photo Researchers)

tablished based on their $^{14}_{6}C$ content. Carbon dioxide in the atmosphere consists of carbon-12 with trace amounts of carbon-14, which is radioactive and decays. However, the concentration of carbon-14 does not decrease because it is constantly formed in the atmosphere from the action of cosmic rays on nitrogen ($^{14}_{7}N$).

All plants absorb carbon dioxide from the atmosphere. As long as the plant is living, the amount of carbon-14 incorporated into the molecules it produces and uses is a constant fraction of the amount of carbon present. When the plant dies, the carbon compounds in its cells no longer exchange carbon atoms with the carbon dioxide of the atmosphere. The amount of carbon-14 in these plants, therefore, diminishes with time as the carbon-14 decays. Since the half-life of carbon-14 is 5730 years, the age of an object can be estimated by determining relative amounts of ^{14}C and ^{12}C present in the sample.

Carbon-14 dating has been immensely useful in archaeology. The Dead Sea Scrolls, found in the early 1950s in caves south of Jericho, were thought to have been hidden by Essene Jews about 68 B.C. Because the leather and linen of the scrolls were once living materials, the material in the scrolls could be dated. They were confirmed to be about 2000 years old.

The carbon-14 dating technique cannot be used accurately for objects that are more than 50,000 years old. After many half-lives have elapsed, the amount of carbon-14 remaining is too small for accurate measurement.

12.5 Nuclear Transmutations

The conversion of one element or isotope into another is called **transmutation.** There are about 300 naturally occurring nuclides, and about 3000 have been made by reactions in which a nuclear projectile collides with another nuclide. To produce

An accelerator in Geneva, Switzerland is 100 meters underground and is about 2 kilometers in circumference as indicated by the outlined circle. (Photo Researchers)

the necessary radioisotopes for medical and other applications, nuclear scientists must make them using specialized instruments. Unlike ordinary chemical reactions, changes in the composition of the nucleus require extremely high energy. Nuclei and subatomic particles must be slammed together with great force to cause nuclear change. The speed of selected charged "projectiles" is increased in particle accelerators; then they are focused to collide with a target of a selected isotope. The collision results in one or more nuclear products and may release radiation.

The first high-energy projectiles used were electrons ($_{-1}^{0}e$), neutrons ($_{0}^{1}n$), protons ($_{1}^{1}H$), deuterons ($_{1}^{2}H$), and alpha particles ($_{2}^{4}He$). However, more powerful particle accelerators can now impart sufficient energy to nuclei of light atoms to produce projectiles carrying more neutrons and protons to the target nucleus.

Rutherford first observed a transmutation reaction in the reaction of nitrogen with alpha particles. He showed that protons were produced and concluded that oxygen-17 must have formed because mass must be conserved in the process.

$$_{7}^{14}N + _{2}^{4}He \longrightarrow _{1}^{1}H + _{8}^{17}O$$

The neutron is another "light" projectile that can be used for the transmutation of atoms. Reaction of nitrogen-14 with high-energy neutrons produces carbon-14.

$$_{7}^{14}N + _{0}^{1}n \longrightarrow _{6}^{14}C + _{1}^{1}H$$

Carbon-14 is a beta emitter; it is used as a radioactive "tag" in studying organic reactions. The carbon-14 atoms are incorporated in an organic molecule, which then is used in either an organic or a biochemical reaction. As the compound reacts to form products the location of the carbon-14 tag is followed among the products formed. Photosynthesis was studied by this method.

Nuclear transmutation reactions have produced elements of higher atomic number than exist in nature. Before the nuclear age, uranium was the element with the highest atomic number. Elements with atomic numbers higher than uranium are **transuranium elements.** Some of the isotopes of transuranium elements, with the transmutation reactions used to prepare them, are listed in Table 12.3.

TABLE 12.3 Transmutation Reactions to Form Transuranium Elements

Element	Atomic Number	Reaction
neptunium, Np	93	$_{92}^{238}U + _{0}^{1}n \longrightarrow _{93}^{239}Np + _{-1}^{0}e$
plutonium, Pu	94	$_{92}^{238}U + _{1}^{2}H \longrightarrow _{93}^{238}Np + 2\,_{0}^{1}n$
		$_{93}^{238}Np \longrightarrow _{94}^{238}Pu + _{-1}^{0}e$
americium, Am	95	$_{94}^{239}Pu + _{0}^{1}n \longrightarrow _{95}^{240}Am + _{-1}^{0}e$
curium, Cm	96	$_{94}^{239}Pu + _{2}^{4}He \longrightarrow _{96}^{242}Cm + _{0}^{1}n$
berkelium, Bk	97	$_{95}^{241}Am + _{2}^{4}He \longrightarrow _{97}^{243}Bk + 2\,_{0}^{1}n$
californium, Cf	98	$_{96}^{242}Cm + _{2}^{4}He \longrightarrow _{98}^{245}Cf + _{0}^{1}n$
einsteinium, Es	99	$_{92}^{238}U + 15\,_{0}^{1}n \longrightarrow _{99}^{253}Es + 7\,_{-1}^{0}e$
fermium, Fm	100	$_{92}^{238}U + 17\,_{0}^{1}n \longrightarrow _{100}^{255}Fm + 8\,_{-1}^{0}e$
mendelevium, Md	101	$_{99}^{253}Es + _{2}^{4}He \longrightarrow _{101}^{256}Md + _{0}^{1}n$
nobelium, No	102	$_{96}^{246}Cm + _{6}^{12}C \longrightarrow _{102}^{254}No + 4\,_{0}^{1}n$
lawrencium, Lr	103	$_{98}^{252}Cf + _{5}^{10}B \longrightarrow _{103}^{257}Lr + 5\,_{0}^{1}n$

Note that both nobelium and lawrencium are produced using light atoms as projectiles.

12.6 Nuclear Fission Reactions

In the previous section we examined nuclear transmutations in which the majority of the nucleons remained in a single atom. However, in processes called **nuclear fission** the nucleons of a "heavy" atom are split between two lighter atoms. Evidence for nuclear fission was discovered in 1938 by two German scientists, Otto Hahn and Fritz Strassman. They found small amounts of barium-141 when uranium-235 was bombarded with neutrons, which indicated that the uranium nucleus was fragmented. As a result of further work by Lise Meitner and Otto Frisch it was concluded that the uranium nucleus captured a neutron to form an unstable uranium-236, which then fragmented into barium, krypton, and three neutrons. This process liberates 4×10^9 kcal/mole.

$$^{235}_{92}\text{U} + {}^{1}_{0}\text{n} \longrightarrow {}^{236}_{92}\text{U} \longrightarrow {}^{92}_{36}\text{Kr} + {}^{141}_{56}\text{Ba} + 3\,{}^{1}_{0}\text{n}$$

The fission of a heavy nucleus does not occur exactly the same way each time. Krypton and barium were selected in the previous equation to illustrate the fission reaction. Among other pairs of elements produced are ${}^{89}_{37}\text{Rb}$ and ${}^{144}_{55}\text{Cs}$ in a reaction that also generates three neutrons.

$$^{235}_{92}\text{U} + {}^{1}_{0}\text{n} \longrightarrow {}^{236}_{92}\text{U} \longrightarrow {}^{89}_{37}\text{Rb} + {}^{144}_{55}\text{Cs} + 3\,{}^{1}_{0}\text{n}$$

The Atomic Bomb

There is an important consequence to the fact that more neutrons are produced from the fission of uranium-235 than are used to initiate the process. Continued reaction of neutrons can result in a self-sustaining chain reaction that produces ever more energy. This occurs only if the neutrons are not allowed to escape from the sample. A nuclear fission chain reaction producing three neutrons per atom is illustrated in Figure 12.3.

It takes a certain minimum amount of fissionable material, called the **critical mass,** to sustain a nuclear chain reaction. If the amount is less than the critical mass, too many neutrons escape and a self-sustaining chain reaction cannot occur. Only if a critical mass of material is properly shaped can a sufficient number of neutrons encounter nuclei and as a result maintain a chain reaction. For uranium-235 the critical mass is about 56 kg. Two specifically shaped quantities of uranium-235, each less than the critical mass, are kept apart in an atomic bomb. The nuclear explosion occurs when the two shapes are "shoved" together within the bomb.

Nuclear Fission in Power Plants

Nuclear fission can be controlled in a nuclear reactor to produce electric power. The reaction is slower than that which occurs in a nuclear bomb for two reasons. First, the uranium in the core of nuclear reactors is of lower purity than weapons-grade uranium, and the quantities used are less than the critical mass. Second, the rate of the chain reaction is controlled by a moderator and control rods. Water is the **moderator;** it slows down the neutrons released by the uranium, which is sus-

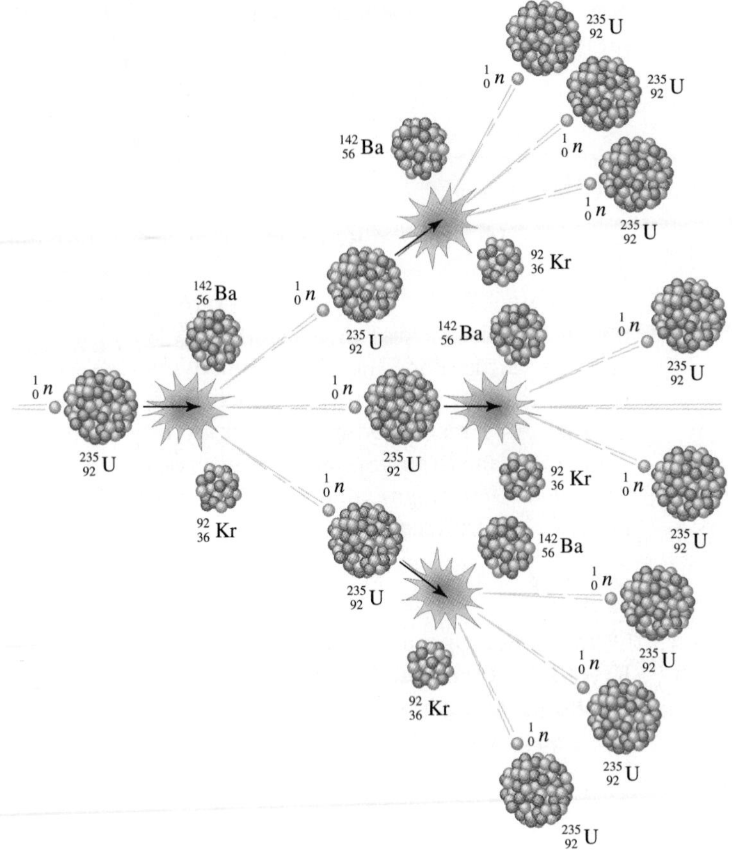

FIGURE 12.3
Nuclear Chain Reaction
Each nuclear fission releases more neutrons than are used as reactants. The rate of the reaction accelerates as more and more nuclei undergo fission.

pended in water. **Control rods** made of cadmium and boron absorb some of the neutrons generated in the nuclear reaction.

$$^{1}_{0}n + {}^{113}_{48}Cd \longrightarrow {}^{114}_{48}Cd$$
$$^{1}_{0}n + {}^{10}_{5}B \longrightarrow {}^{11}_{5}B$$

The control rods are positioned within the uranium fuel (Figure 12.4). When the reaction needs to be slowed, the control rods are lowered to absorb neutrons. To speed up the reaction, the rods are raised and the neutrons continue the chain reaction of the nuclear fuel.

Controlled nuclear fission is used as a source of energy in industrial societies. Consumption of the limited fossil-derived fuels (oil, gas, and coal), will eventually exhaust this energy source. In addition, there are concerns about the effect of the carbon dioxide produced by combustion of fossil fuels. The release of carbon dioxide into the atmosphere may cause global warming, the climate may change, and the sea level may rise as the polar ice caps melt.

Nuclear fission is an alternative source of energy. However, the availability of uranium-235 is also limited. This isotope accounts for only 0.71% of naturally occurring uranium. With the estimated world reserves and anticipated growth of nuclear power, the supply of uranium-235 may be exhausted in two to three decades. Another important consideration is the disposal of the spent material. The waste products are themselves radioactive and have long half-lives. The technological problem of separation and disposal of these waste products is complicated by the political problem of where to store the material.

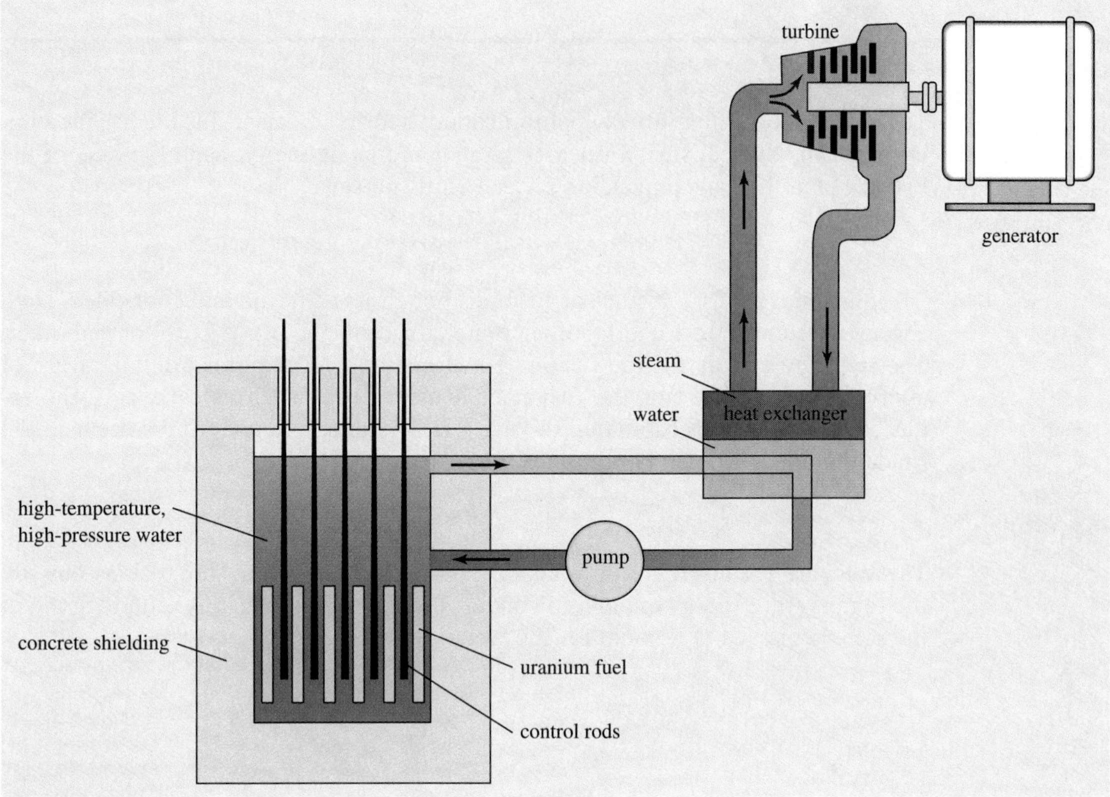

FIGURE 12.4
Generation of Electricity in a Nuclear Power Plant
The hot radioactive water is pumped to a heat exchanger that converts nonradioactive water in a second system into nonradioactive steam. The steam drives a turbine to produce electricity.

Chernobyl

Although an atomic explosion at a power plant is not possible, the operation of these plants is not without danger. The main danger is loss of control of the fission process, which can cause overheating of the core. The resulting uncontrolled release of heat can cause *meltdown,* in which the fuel as well as components of the nuclear reactor literally melt. If the meltdown was contained within the specially designed areas of the power plant, no radioactivity would be released. However there is a possible danger that an ordinary chemical explosion could result from the heat released. This explosion could blow radioactive material from within the plant into the surrounding environment.

In spite of numerous fail-safe mechanisms built into nuclear power plants, accidents have occurred. Most have been minor. However, in the case of Chernobyl in the Ukraine, human errors caused a tragedy of immense proportions that is predicted will eventually cause thousands of deaths due to cancer. Several of the safety systems used to cool the reactor were turned off while tests were performed on the reactor. The control rods were then removed, and as a result, the nuclear fuel started to react at a rate that was beyond control. Because the necessary cooling mechanisms had been turned off there was no way to prevent a meltdown and an explosion resulted. The large quantities of radioactive material released made areas miles from the plant uninhabitable. In addition, radioisotopes were carried by the atmosphere hundreds of miles away into Europe.

12.7 Nuclear Fusion

The reaction of comparatively light nuclides to form a heavier nuclide is called **nuclear fusion.** Nuclear fusion releases large amounts of energy, as in the case of the fusion of four hydrogen nuclei to give a helium nucleus.

$$4\,{}^{1}_{1}\text{H} \longrightarrow {}^{4}_{2}\text{He} + 2\,{}^{0}_{1}\text{e} \qquad \Delta H^{\circ} = -6 \times 10^{8}\ \text{kcal}$$

Unlike nuclear fission, nuclear fusion is not easily achieved and controlled. High energy is required to force together the charged nuclei, which repel one another. The requisite energy for the fusion of hydrogen nuclei requires a temperature of approximately 10^7 K. For the fusion of heavier nuclei such as carbon-12 the required temperature is estimated to be 10^9 K. These high temperatures occur in the sun and in the hydrogen bomb.

Fusion in the Sun

The stars are fueled by fusion reactions. In a relatively young star such as our sun the temperature is high enough to convert hydrogen isotopes into helium. Some of the steps proposed to account for the fusion reaction are

$$\begin{aligned}
{}^{1}_{1}\text{H} + {}^{1}_{1}\text{H} &\longrightarrow {}^{2}_{1}\text{H} + {}^{0}_{1}\text{e} \\
{}^{1}_{1}\text{H} + {}^{2}_{1}\text{H} &\longrightarrow {}^{3}_{2}\text{He} \\
{}^{3}_{2}\text{He} + {}^{3}_{2}\text{He} &\longrightarrow {}^{4}_{2}\text{He} + 2\,{}^{1}_{1}\text{H} \\
{}^{3}_{2}\text{He} + {}^{1}_{1}\text{H} &\longrightarrow {}^{4}_{2}\text{He} + {}^{0}_{1}\text{e}
\end{aligned}$$

As hydrogen in a star diminishes, new fusion reactions occur using helium as the reactant. The fusion of helium-4 can give beryllium-8, which in turn can react with helium-4 to give carbon-12.

$$\begin{aligned}
{}^{4}_{2}\text{He} + {}^{4}_{2}\text{He} &\longrightarrow {}^{8}_{4}\text{Be} \\
{}^{8}_{4}\text{Be} + {}^{4}_{2}\text{He} &\longrightarrow {}^{12}_{6}\text{C}
\end{aligned}$$

The Hydrogen Bomb

On Earth, nuclear fusion has been achieved by using atomic bomb "triggers" to provide the energy to initiate fusion in the hydrogen bomb. The reactants are lithium and deuterium in the salt lithium deuteride (LiD). The lithium absorbs a neutron produced in the fission reaction and the tritium (${}^{3}_{1}\text{H}$) produced reacts with the deuterium to produce helium and energy.

$$\begin{aligned}
{}^{6}_{3}\text{Li} + {}^{1}_{0}\text{n} &\longrightarrow {}^{3}_{1}\text{H} + {}^{4}_{2}\text{He} \\
{}^{3}_{1}\text{H} + {}^{2}_{1}\text{H} &\longrightarrow {}^{4}_{2}\text{He} + {}^{1}_{0}\text{n}
\end{aligned}$$

The reactions occur because the fission reaction pushes the nuclei of light elements together with enough force to initiate the fusion process.

Fusion Power

Nuclear fusion has considerable appeal as a source of power. Deuterium is available from water in the oceans. About 1 out of every 6000 hydrogen atoms is deuterium, and of course, there is a lot of sea water. The amount of energy that would be released from the fusion of even a small amount of the deuterium in the ocean is incredibly large. A single cubic meter of sea water could provide the energy equivalent of 1.4×10^{12} barrels of oil. That amount exceeds the known oil reserves on the planet. Controlled nuclear fusion would for practical purposes eliminate any concern about future sources of energy.

The goal of harnessing the energy of fusion reactions in a nuclear plant has not been achieved. The United States has spent several billion dollars in research to develop fusion reactors. Although progress is being made, even optimistic scientists estimate that fusion power will not become available until after the year 2010.

Three technical problems have prevented the fusion of atoms in a controlled reaction. First, extremely high temperatures are required. Second, the reactants must be confined, and no known material could contain the reactants at the temperature required for fusion—they would vaporize. Third, the energy released must be convertible into a usable form such as electricity. The necessary energy to initiate fusion has been achieved using powerful lasers. The reactants have been contained with a strong magnetic field. However, until recently the amount of energy used in these fusion experiments has exceeded that released by the small amount of fusion that occurred. In these experiments the time of fusion is extremely short. As yet, the problem of converting the energy released into usable form remains unsolved.

12.8 Detection of Radiation

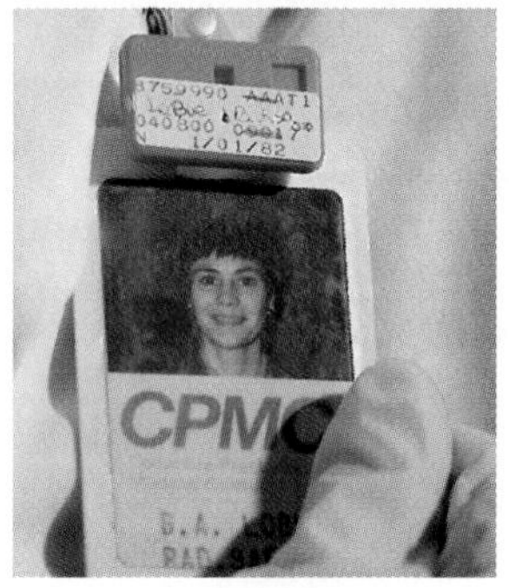

FIGURE 12.5
The Film Badge
A photographic negative contained within this workers identification tag is processed and replaced on a regular basis.
(Photo Researchers)

The detection of radiation depends on both the amount and the source of radiation. Radiation causes atoms and molecules to lose electrons and form ions—hence the term **ionizing radiation.** Radiation detectors include the Geiger counter, the film badge, and the scintillation counter.

A **Geiger counter** is a metal tube containing a gas at low pressure. The gas chamber is connected to a battery. In the absence of radiation, there is no current flow through the gas chamber. Exposure to radiation produces ions and electrons as the gas ionizes, and a current flows through the chamber. The higher the intensity of the radiation, the stronger the current flow, which is indicated by a swing of the needle on the Geiger counter or by an audible clicking sound.

A **film badge** measures the amount of radiation that workers near radioactive sources receive. The film badge contains photographic film in a holder sealed from light (Figure 12.5). Radiation passes through the holder, and the film is "exposed." Higher radiation exposure causes the developed film to be darker. Film badges are periodically developed to check the amount of radiation exposure of the wearer.

Scintillation counters are used to detect radiation in medical tests. A scintillation counter contains crystals of chemicals that produce flashes of light when hit by radiation. An electronic device called a **photomultiplier** records and amplifies the light. Scintillation counters are used to measure the location and extent of accumulation of injected diagnostic radioisotopes in a patient's body.

Units of Radiation

Radiation is measured quantitatively by two different methods. One method measures the intensity of the radiation emitted from the radioactive source. The other method measures the radiation absorbed by matter such as air or tissue.

The **curie** (Ci), a unit of radiation intensity, is equal to 3.7×10^{10} nuclear disintegrations per second. This unit equals the number of disintegrations that 1 g of radium undergoes in 1 s. The microcurie, 37,000 disintegrations per second, is a common unit used to report radiation. We are exposed to less than 10^{-4} μCi from our natural environment.

The **roentgen** (R) is a chemical measure of exposure to gamma radiation or X-rays. One roentgen is the dose of radiation that produces 2.1×10^9 ions in 1 cm^3 of dry air at 0 °C and 1 atm pressure. Geiger counters, which operate by detecting ionized air, are calibrated to read in milliroentgens per hour (mR/h).

The **rad,** short for **radiation absorbed dose,** is a measure of the energy absorbed by tissue. One rad is equal to 100 ergs (2.4×10^{-6} cal) of energy per gram of irradiated tissue. Although the roentgen and the rad are defined differently, they are numerically similar. One roentgen of gamma radiation delivers 0.97 rad in muscle tissue. In bone, 1 R delivers 0.92 rad. A dose of about 500 rads of gamma radiation is lethal for a human.

Even the dosage actually absorbed does not necessarily indicate the biological damage to an organ. In other words, a 10-rad dose of one type of radiation may result in different biological consequences than a 10-rad dose of another type of radiation. The **relative biological effectiveness (RBE)** accounts for differences in biological damage caused by radiation. The standard for RBE is gamma radiation from cobalt-60. The RBE of a radiation source is the ratio of the dose of gamma radiation from cobalt-60 to the dose of the radiation in question required to cause the same biological effect. For example, 1 rad of alpha particles may have the same effect as 10 rads of gamma radiation from cobalt-60. The RBE of alpha particles is then 10. Highly charged and heavy particles such as alpha particles cause more ionization in matter than lighter and singly charged beta particles. The RBE of beta particles is about 1.

The **rem,** short for roentgen equivalent for man, is defined as the product of rads and RBE. For alpha particles, 1 rad equals 10 rem, whereas for beta particles 1 rad equals 1 rem.

$$\text{rem} = \text{rad} \times \text{RBE}$$

The radiation exposure of workers in nuclear chemistry laboratories is monitored in millirem (mrem).

12.9 Biological Effects of Radiation

The energy of ionizing radiation causes chemical reactions when it interacts with matter. Consider the effect of radiation on the covalent bond of a hypothetical molecule A : B.

$$\text{A:B} \longrightarrow \text{A}\cdot\text{B}^+ + e^-$$
$$\text{A}\cdot\text{B}^+ \longrightarrow \text{A}^+ + \text{B}\cdot$$

After an electron is "knocked out" of the covalent bond, the cation does not have the necessary bonding electrons to remain together. Decomposition of the cation produces another cation and a species called a **radical,** which has an unpaired electron. The subsequent reactions of cations and radicals with other molecules can seriously disrupt the normal functions of cells.

We are exposed to radiation every day, mostly from natural sources such as cosmic rays and radon gas, and less frequently from technological sources such as X-rays. The ionizing radiation of X-rays is a more hazardous form of radiation, but low dosages of X-rays are necessary for certain medical and dental diagnoses.

TABLE 12.4 Effects of Short-Term Radiation Doses

Dose (rem)	Probable Effect
<25	no detectable short-term effect
25–100	decrease in white blood cell count, which lowers resistance to infection
100–200	reduction in number of blood cells, fatigue, nausea
200–300	nausea and vomiting on first day of exposure; fever, diarrhea, loss of hair by the third week
300–500	nausea, vomiting, and diarrhea in a few hours; probability of death about 50%
>500	vomiting, severe changes in blood and gastrointestinal system; death within 2 months in essentially 100% of the cases

Overexposure to radiation may result from a single large dose or many cumulative small doses. It is estimated that in one year the average American receives 200 mrem of radiation, half of which is from natural environmental background radiation and cannot be decreased. The second major radiation source is X rays used in medical diagnosis. The exact effect of radiation can be stated only in terms of the probability of biological consequences. A listing is provided in Table 12.4.

Effects of Radiation

Somatic effects of radiation are the consequences to the actual person exposed. **Genetic effects** are those passed on to subsequent generations. In either case, the effect of radiation on a cell is one of breaking apart molecules to produce reactive intermediates that react with neighboring molecules. All such reactions are foreign to the cell and cause serious damage. Cells are able to repair themselves up to a point, and low radiation exposure might not cause permanent biological damage. However, no exposure to radiation is ever 100% risk free.

Somatic effects may not be seen immediately. Cancers that are the result of radioactive exposure may appear years later. Slightly more than 10% of leukemias are thought to originate from natural radiation in our environment. Similarly, 10% of lung cancers are due to radiation, and most of these are from radon gas. However, smoking is the major contributor to lung cancer—not radiation. There is evidence that smoking in combination with exposure to radiation has a disastrous synergistic effect: Smokers who are exposed to radon gas are far more likely to develop lung cancer.

Tissues that reproduce at a rapid rate, such as bone marrow, lymphatic system tissue, and embryonic tissue, are the most sensitive to radiation damage. Since bone marrow is the site of red cell formation, one of the early signs of radiation overexposure is a drop in the red cell count.

The nucleus of every cell contains DNA (deoxyribonucleic acid) that copies itself and passes on genetic information for the formation of new cells (Chapter 29). Radiation can breakdown the DNA molecule or alter it. Because DNA produces RNA (ribonucleic acid) that in turn directs the synthesis of enzymes and all the other proteins in the body, exposure to radiation can have serious health consequences.

The most serious effect of radiation is genetic. The mutation of a species will occur if radiation affects the DNA of the nuclei of germ cells of the testes that produce sperm or the ovaries that produce eggs. Although there is no effect on the exposed individual, the altered DNA is passed on to the next generation, which may suffer debilitating diseases. Thus, individuals of child-bearing age should minimize their exposure to radiation of all types.

Radiation Safety

Shielding material provides some protection for individuals who are exposed to radiation in their occupation. Such precautions are required for technicians who use X-ray machines. Both the thickness and composition of the shielding material are important in decreasing the radiation exposure. As indicated earlier, alpha particles and beta particles are easily stopped. Gamma rays require concrete or lead shielding. The thickness required depends on the energy of the gamma radiation. For example, half of the radiation of technetium-99 is stopped by 0.2 mm of lead, whereas iodine-131 requires 6 mm for the same protection.

The level of exposure to radiation is related to the distance between the individual and the radiation source. The dosage decreases as the square of the distance. Thus at 2 ft the dosage is $(\frac{1}{2})^2$ or $\frac{1}{4}$ the dosage at 1 ft. Every incremental increase in the distance from the radiation source provides increased protection. At 4 ft the dosage is $(\frac{1}{4})^2 = \frac{1}{16}$ the dosage at 1 ft. This relationship can be stated as an inverse square law where I_a and I_b are the intensities at distance d_a and d_b respectively.

$$\frac{I_a}{I_b} = \frac{d_b^2}{d_a^2}$$

12.10 Radioisotopes in Diagnosis

Physicians use radioisotopes to locate and determine internal disorders without surgery. Radioisotopes that can accumulate at the site of the disorder are adminis-

TABLE 12.5 Radioisotope Half-Lives and Uses in Diagnosis

Isotope	Half-Life	Part of Body	Use in Diagnosis
barium-131	11.6 days	bone	detection of bone tumors
chromium-51	27.8 days	blood	determination of blood volume and red blood cell lifetime
copper-64	12.8 h	liver	diagnosis of Wilson's disease
gold-198	64.8 h	kidney	assessment of kidney activity
iodine-123	60 days	thyroid	assessment of iodine uptake by thyroid
iron-59	45 days	blood	evaluation of iron metabolism in blood
krypton-79	34.5 h	blood	evaluation of cardiovascular system
mercury-197	65 h	spleen	evaluation of spleen function
		brain	brain scans
selenium-75	120 days	pancreas	determination of size and shape of pancreas
technetium-99	6.0 h	brain	detection of brain tumors, hemorrhages, or blood clots

tered to the patient, and a radiation-detecting instrument is used to determine the location of the radioisotope and its concentration.

Since any radiation entails some health risks, a radiologist must minimize the exposure of the patient. There are five criteria for the selection of a radioisotope for diagnosis.

1. It must be effective for diagnosis at a low concentration and yet be reliably detectable.
2. It must have a short half-life to produce a large number of detectable nuclear disintegrations in a short time.
3. It should be readily eliminated by the body after the diagnosis.
4. It should be selectively absorbed in the part of the body where diagnosis is desired.
5. Its radiation must have sufficient penetrating power to reach the instruments placed outside the body. Thus, only radioisotopes that emit gamma rays are appropriate for diagnosis (Table 12.5).

Positron Emission Tomography

New technologies are constantly being developed to improve the accuracy of diagnosis and to diminish the side effects to the patient. Positron emission tomography (PET) is an imaging method using positron emitters such as carbon-11, nitrogen-13, and oxygen-15.

A positron, or "positive electron", is a type of "antimatter" that can exist only for very short times in the presence of matter. When a positron encounters an electron, they are both annihilated and gamma rays are formed.

$${}^{0}_{1}e + {}^{0}_{-1}e \longrightarrow \text{gamma rays}$$

The key to PET is that gamma rays are emitted at exactly a 180° angle to each other so that the source can be precisely determined by "drawing lines" back to a point where the "lines" cross.

A small number of positrons generates a large amount of energy, so this diagnostic technique requires only small amounts of positron emitters. Another advantage of the method is the short half-life of the positron emitters, which decreases the residence time inside the body.

The positron emitters all contain fewer neutrons than the naturally occurring isotopes. An isotope such as carbon-11 is produced in a cyclotron shortly before use because its half-life is short. The products of positron emitters are safe because they are normally nonradioactive isotopes. Let's see why. The mass number of the product is unchanged but the atomic number is decreased by 1.

$${}^{11}_{6}C \longrightarrow {}^{0}_{1}e + {}^{11}_{5}B$$

One of the medically useful applications of the PET method involves making glucose with carbon-11 in place of one of the six carbon atoms. Glucose can cross the blood–brain barrier—that is, it is required by the brain, so it is allowed into the brain cells. Scanning the brain to image the positrons emitted from the labeled glucose gives a picture of where glucose is being metabolized. Abnormal metabolism occurs in manic depressive and schizophrenic individuals. Neuropsychiatric disorders such as Parkinson's disease are being studied with other compounds containing carbon-11.

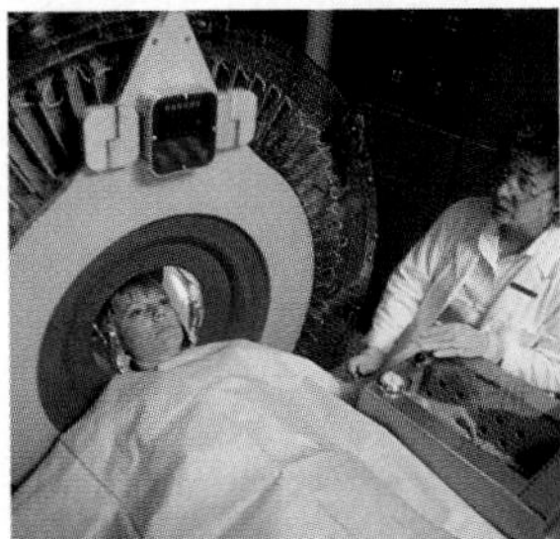
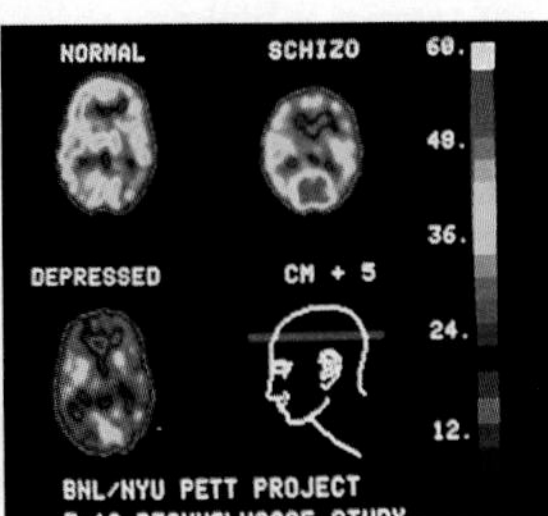

The PET scan instrument at Brookhaven Hospital in Maryland gives scans of the brain. Scans of normal, schizophrenic and depressed individuals are shown. (Photo Researchers)

Iodine-123 is used to diagnose thyroid activity. A patient drinks water containing the radioisotope in the form of sodium iodide. If the thyroid is functioning normally, about one-sixth of the radioisotope accumulates in the thyroid gland within 24 h. A radiation detector placed at the neck measures the concentration of iodine in the thyroid. If a hypothyroid condition exists, the amount accumulated is less than normal. If a greater than average amount accumulates, then a hyperthyroid condition exists. A thyroid scan shows the distribution of the iodine-123 in the thyroid as a radiation "picture." If any part of the thyroid is missing from the "picture," isotopic iodine was not absorbed and a hypothyroid condition exists. The absence of radiation in an area is called a "cold spot."

Body fluids are also studied with radioisotopes. The radioisotope sodium-24 is used to determine the effectiveness of blood circulation. A dose of sodium-24 is injected as $^{24}NaCl$ directly into the blood, and its progress is monitored. If it takes a longer time than average to reach a certain part of the body, then impaired circulation is indicated.

12.11 Radioisotopes in Therapy

The objectives of radiation therapy are different from those of diagnostic procedures. The radioisotope is placed in the proper part of the body to selectively destroy diseased cells or tissue. There is no need to monitor the radioisotope by an external detector. Thus, gamma ray emitters are not needed. This means that alpha or beta emitters can be used to destroy tissue or cells in a localized area.

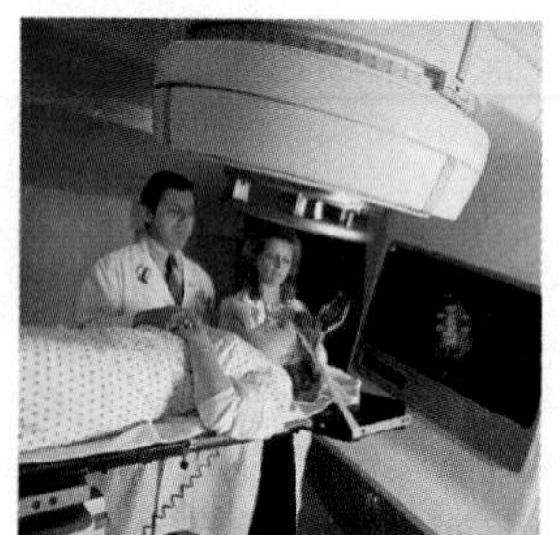

Physicians prepare a patient for radiation treatment of cancer. (Erik Hildebrandt, (Photo Researchers)

Cells in the process of dividing are the most susceptible to radiation damage. Thus, cancer, which is the rapid, uncontrolled division of abnormal cells, is affected by radiation. It is this principle that is applied in cancer therapy. Choosing a radioisotope that will do maximum damage to the cancerous cells while producing minimum damage to healthy cells is the goal of the radiologist.

Salts of the radioisotope $^{226}_{88}Ra$ were among the earliest used in cancer therapy; however, in addition to alpha particles, the radioactive decay of radium results in the emission of highly penetrating gamma rays.

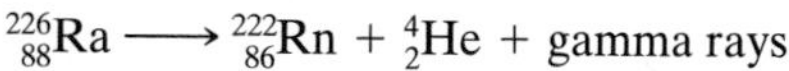

$$^{226}_{88}Ra \longrightarrow ^{222}_{86}Rn + ^{4}_{2}He + \text{gamma rays}$$

Cobalt-60, which is used for high intensity treatment of cancers has a 5.2-year half-life. This isotope is about 200 times as active as radium. It emits beta particles and gamma rays. The cobalt-60 is contained in a lead cylinder with an opening that releases radiation. The beam of radiation is focused on the small area of the body where the tumor is located. Although the radiation is carefully directed at the site of the cancer cells, some normal cells are always destroyed as the radiation passes through the body. The patient suffers some effects of radiation sickness while undergoing treatment.

Radiation therapy also utilizes radiation sources placed inside the body. For example, small hollow gold needles containing a radium salt are sometimes inserted into the tumor in the body. This therapeutic method has several disadvantages. Radium is very expensive. Also it has a half-life of 1620 years, so the implant has to be removed at some later date to prevent continued radiation exposure.

Yttrium-90 has advantages over radium-226 in the localized treatment of tumors by the implant procedure. It is a beta emitter with a half-life of 64 h. Since no gamma rays are produced, healthy tissue a few centimeters away is unaffected. In addition, because of its short half-life, the radioisotope decays rapidly.

$$^{90}_{39}Y \longrightarrow ^{90}_{40}Zr + ^{0}_{-1}e$$

Some radioisotopes are injected into the body or ingested. They are chosen because they concentrate at specific locations in the body. For example, iodine tends to concentrate in the thyroid gland. An overactive thyroid gland may be treated by orally administering a solution of sodium iodide containing some iodine-131. The radioisotope concentrates in the thyroid and destroys some cells. Iodine-131 has a half-life of 8.1 days, which allows it to be used with relative safety.

Phosphorus-32 has been used in phosphate ions to treat some forms of leukemia. The phosphate ion is transported to many parts of the body and is eventually eliminated in the urine. However, phosphate ions are also incorporated in bones, where they affect the production of white blood cells in the bone marrow.

Summary

Many naturally occurring isotopes have existed since the Earth was formed. **Radioisotopes** undergo radioactive transformations that produce a **daughter nucleus** and generate radiation. Radiation from natural sources includes **alpha particles, beta particles,** and **gamma radiation.**

Radioactive isotopes are made by **nuclear transmutation** by bombarding a **target nucleus** with **projectiles** of subatomic particles. This method is used to produce both isotopes for medical applications and new elements. **Transuranium elements,** elements with atomic numbers greater than that of uranium, have been produced by transmutation reactions.

A radioisotope decays at a rate that is given by its **half-life.** One of the applications of half-life data is the use of carbon-14 in dating ancient materials made from plant and animal sources.

Two types of nuclear reactions, **fusion** and **fission,** are alternative fuel sources for the future. Fission processes are currently being used but are limited by the availability of uranium-235 and the problem of waste disposal. Fusion has not been achieved in a self-sustaining reaction, but research continues to develop this promising source of energy.

Radiation detectors include **Geiger counters, film badges,** and **scintillation counters.** The unit that measures the number of radioactive disintegrations is the **curie. Roentgens, rads,** and **rems** are units of dosage. The use of radioisotopes requires shielding, but the effect of radiation also depends on the distance between the worker and the radiation source.

Radiation from radioactive substances causes surrounding material to ionize to produce cations and **radicals** that undergo abnormal reactions. The effect of **ionizing radiation** on living tissue can be minor, or it can be severe enough to damage DNA and cause mutation of the species or the death of the organism.

Radioisotopes are used in medicine for both diagnosis and therapy. The choice of radioisotope is dictated by the purpose as well as the target organ.

Exercises

Balancing Nuclear Equations

12.1 In 1919 Ernest Rutherford proved the existence of a nuclear particle with the following reaction. What is the nuclear particle?

$$^{14}_{7}N + ^{4}_{2}He \longrightarrow ^{17}_{8}O + ?$$

12.2 In 1932 James Chadwick proved the existence of a nuclear particle with the following reaction. What is the nuclear particle?

$$^{9}_{4}Be + ^{4}_{2}He \longrightarrow ^{12}_{6}C + ?$$

12.3 Write a balanced nuclear equation for beta emission from each of the following.

(a) fluorine-21 (b) silicon-31 (c) magnesium-28

12.4 Write a balanced nuclear equation for beta emission from each of the following.

(a) calcium-40 (b) potassium-40 (c) strontium-89

12.5 Write a balanced nuclear equation for alpha emission from each of the following.

(a) polonium-212 (b) curium-240 (c) einsteinium-252

12.6 Write a balanced nuclear equation for alpha emission from each of the following.

(a) thorium-229 (b) boron-11 (c) berkelium-245

12.7 Supply the correct symbol for the product of each of the following reactions.

(a) $^{243}_{96}Cm \longrightarrow ^{4}_{2}He + ?$ (b) $^{222}_{86}Rn \longrightarrow ^{4}_{2}He + ?$

(c) $^{234}_{94}Pu \longrightarrow ^{4}_{2}He + ?$ (d) $^{245}_{97}Bk \longrightarrow ^{4}_{2}He + ?$

12.8 Supply the correct symbol required to balance each of the following equations.

(a) $^{140}_{56}Ba \longrightarrow ? + ^{140}_{57}La$ (b) $^{245}_{96}Cm \longrightarrow ? + ^{241}_{94}Pu$
(c) $^{30}_{13}Al \longrightarrow ? + ^{30}_{14}Si$ (d) $^{11}_{5}B \longrightarrow ? + ^{7}_{3}Li$

12.9 What element is produced in each of the following transformations?

(a) $^{20}_{8}O \longrightarrow ? + ^{0}_{-1}e$ (b) $^{234}_{92}U \longrightarrow ? + ^{4}_{2}He$
(c) $^{230}_{90}Th \longrightarrow ? + ^{4}_{2}He$ (d) $^{27}_{12}Mg \longrightarrow ? + ^{0}_{-1}e$

12.10 What element is produced in each of the following transformations?

(a) $^{218}_{84}Po \longrightarrow ? + ^{4}_{2}He$ (b) $^{186}_{73}Ta \longrightarrow ? + ^{0}_{-1}e$
(c) $^{82}_{35}Br \longrightarrow ? + ^{0}_{-1}e$ (d) $^{210}_{84}Po \longrightarrow ? + ^{4}_{2}He$

12.11 Write the symbol for the product formed by positron emission from each of the following light elements.

(a) fluorine-18 (b) carbon-11 (c) oxygen-15

12.12 Write the symbol for the product formed by positron emission from each of the following heavy elements.

(a) iron-52 (b) copper-64 (c) polonium-207

12.13 Write the symbol for the product formed by K electron capture from each of the following elements.

(a) aluminum-26 (b) calcium-41 (c) beryllium-7

12.14 Write the symbol for the product formed by K electron capture from each of the following elements.

(a) copper-64 (b) iron-55 (c) argon-37

Half-lives

12.15 A wooden object has 12.5% of the usual abundance of carbon-14 remaining. How old is the wood?

12.16 A wooden object found in a cave in Greece has 6.2% of the normal abundance of carbon-14. How old is the wood?

12.17 The half-life of gold-198 is 2.7 days. The isotope is used to diagnose liver functions. How much of a 8.4-mg sample will remain after 8.1 days?

12.18 The half-life of $^{32}_{15}P$, which is used in the form of Na_2HPO_4 to treat myeloid leukemia, is 14 days. How many grams of the isotope in a sample containing 10.0-mg of the isotope will remain after 56 days?

12.19 The half-life of radon-224 is 3.7 days. How many days are required for 10.00-g sample to decay to 0.62 g of this isotope?

12.20 The half-life of gallium-67, which is used for medical diagnosis of certain cancers, is 78 h. If 160 μg is given to a patient, how much will remain after 13 days?

Transmutation of Elements

12.21 Nickel-58 produces an element and an alpha particle when bombarded with a proton. What is the elemental symbol of the product?

12.22 The bombardment of $^{7}_{3}Li$ by an alpha particle produces a neutron and an element. What is the element?

12.23 Complete the following equations.

(a) $^{23}_{11}Na + ? \longrightarrow ^{24}_{11}Na + ^{1}_{1}H$
(b) $^{10}_{5}B + ? \longrightarrow ^{13}_{7}N + ^{1}_{0}n$
(c) $^{59}_{27}Co + ? \longrightarrow ^{56}_{25}Mn + ^{4}_{2}He$

12.24 Complete the following equations.

(a) $^{209}_{83}Bi + ? \longrightarrow ^{210}_{84}Po + ^{1}_{0}n$
(b) $^{14}_{7}N + ? \longrightarrow ^{17}_{8}O + ^{1}_{1}H$
(c) $^{27}_{13}Al + ? \longrightarrow ^{25}_{12}Mg + ^{4}_{2}He$

12.25 Complete the following equations.

(a) $^{235}_{92}U + ^{2}_{1}H \longrightarrow ? + ^{0}_{-1}e$
(b) $^{27}_{13}Al + ^{4}_{2}He \longrightarrow ? + ^{1}_{0}n$
(c) $^{130}_{52}Te + ^{1}_{0}n \longrightarrow ? + ^{0}_{-1}e$

12.26 Complete the following equations.

(a) $^{239}_{94}Pu + ^{4}_{2}He \longrightarrow ? + ^{0}_{1}n$
(b) $^{45}_{21}Sc + ^{1}_{0}n \longrightarrow ? + ^{1}_{1}H$
(c) $^{27}_{13}Al + ^{1}_{0}n \longrightarrow ? + ^{4}_{2}He$

12.27 Complete the following equations.

(a) $^{27}_{13}Al + ^{4}_{2}He \longrightarrow ^{30}_{15}P + ?$
(b) $^{15}_{7}N + ^{1}_{1}H \longrightarrow ^{12}_{6}C + ?$
(c) $^{242}_{96}Cm + ^{4}_{2}He \longrightarrow ^{245}_{98}Cf + ?$

12.28 Complete the following equations.

(a) $^{7}_{3}Li + ^{2}_{1}H \longrightarrow ^{8}_{4}Be + ?$
(b) $^{75}_{33}As + ^{2}_{1}H \longrightarrow ^{76}_{33}As + ?$
(c) $^{10}_{5}B + ^{1}_{0}n \longrightarrow ^{7}_{3}Li + ?$

12.29 Complete the following equations.

(a) $^{253}_{99}Es + ^{4}_{2}He \longrightarrow ? + ^{1}_{0}n$
(b) $^{249}_{98}Cf + ^{18}_{8}O \longrightarrow ? + 4\,^{1}_{0}n$
(c) $^{238}_{92}U + ^{14}_{7}N \longrightarrow ? + 5\,^{1}_{0}n$
(d) $^{238}_{92}U + ^{16}_{8}O \longrightarrow ? + 5\,^{1}_{0}n$

12.30 Complete the following equations.

(a) $^{238}_{92}U + ^{14}_{7}N \longrightarrow ? + 6\,^{1}_{0}n$
(b) $^{252}_{98}Cf + ^{10}_{5}B \longrightarrow ? + 5\,^{1}_{0}n$
(c) $^{238}_{92}U + 15\,^{1}_{0}n \longrightarrow ? + 7\,^{0}_{-1}e$
(d) $^{245}_{96}Cm + ^{13}_{6}C \longrightarrow ? + 5\,^{1}_{0}n$

Radiation

12.31 Explain the term *ionizing radiation.*

12.32 Why is ionizing radiation harmful to living cells?

12.33 Explain the operation of the Geiger counter.

12.34 How are film badges processed to detect radiation received by workers in nuclear laboratories?

12.35 The becquerel (Bq) is a unit of radiation intensity. The equivalence is 1 Bq = 1 distintegration/s. How many Bq are in 1 Ci (Curie)?

12.36 The gray (Gy) is a unit of radiation absorbed dose equal to the absorption of 1 J of energy per kilogram of tissue. (1 J = 10^7 ergs) Show that 1 Gy = 100 rad.

12.37 A radiologist measures 128 mrem at a distance of 2 m from a source. At what distance will the radiation measure 2 mrem?

12.38 A radiologist measures 25 mrem radiation at 10 m. What will the radiation be at 1 m?

12.39 Which is more dangerous, a radioisotope with a short half-life or one with a long half-life, if they emit the same type of radiation? Explain.

12.40 A person receives 5 rem of radiation per year for 40 years and does not show any biological effect. However, a person who receives 200 rem in one dose may die. Explain the difference.

Radiation in Diagnosis and Therapy

12.41 Explain why radioisotopes chosen for use in diagnosis usually emit gamma radiation.

12.42 Explain why radioisotopes chosen for use in diagnosis should have a short half-life.

12.43 How do isotopes used in therapy differ from isotopes used in diagnosis?

12.44 Explain why radioisotopes are used in cancer therapy even though they can cause cancer.

Even in the desert, organic material such as wild flowers exist within a terrain that is largely inorganic. (S. Maimone FPG International)

13

The Organization of Organic Chemistry

13.1 Organic Compounds
13.2 Bonding and Molecular Shapes
13.3 Structural Formulas and Conformations
13.4 Functional Groups
13.5 Isomers
13.6 Organic Reactions and Mechanisms
13.7 Acid–Base Reactions in Organic Chemistry
13.8 Oxidation–Reduction Reactions
13.9 Classification of Organic Reactions

Essays
- *Organic Compounds in Daily Life*
- *Chemical Communications in the Insect World*

13.1 Organic Compounds

Compounds have traditionally been divided into two classes called organic and inorganic compounds. **Organic compounds** contain carbon in combination with a limited number of other elements such as hydrogen, oxygen, and nitrogen. There are about 10 million organic compounds. **Inorganic compounds** contain all other elements, but account for fewer than 1 million compounds. Some inorganic compounds contain carbon, as in Na_2CO_3.

Carbon makes up less than 0.03% of the Earth's crust, the "inorganic" part of our world. Carbon compounds are concentrated in life forms. The cellulose in trees contains carbon, as do the proteins in animals. The term organic was originally selected because it was thought that carbon compounds could be derived only from living matter. Early attempts to make these compounds in the laboratory failed, so it was thought that a "vital force" was necessary to produce organic compounds. However, organic compounds were eventually synthesized in the laboratory just like compounds of any other elements. Nevertheless "organic" is still appropriate to describe the chemistry of carbon compounds, because carbon is the critically important element for all life forms.

The Unique Carbon Atom

What is so special about the carbon atom and why are there so many organic compounds? The first reason is that carbon can form very stable covalent bonds with other carbon atoms as well as with hydrogen atoms. Carbon is in the middle of the second period of the periodic table and has an intermediate electronegativity value. With its four valence-shell electrons, a carbon atom forms four covalent bonds to other atoms. The atoms bonded to one carbon atom contribute a total of four electrons, giving carbon an octet of electrons. As a result, a carbon atom forms covalent bonds with other carbon atoms, linking them together in chains and rings.

$$CH_3{-}CH_2{-}CH_2{-}CH_3$$

a chain of carbon atoms

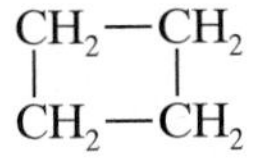

$$\begin{array}{c} CH_2{-}CH_2 \\ | \quad\quad | \\ CH_2{-}CH_2 \end{array}$$

a ring of carbon atoms

Many organic compounds contain only hydrogen atoms bonded to the carbon framework. No other element can make so many large and diverse networks of atoms combined only with hydrogen.

The second reason that there are so many organic compounds is that carbon forms stable covalent bonds with atoms such as nitrogen, oxygen, sulfur, and the halogens. Examples of simple organic compounds containing these atoms follow.

$$NH_2{-}CH_2{-}CH_2{-}CH_2{-}CH_2{-}NH_2$$

putrescine
(odor of decaying tissue)

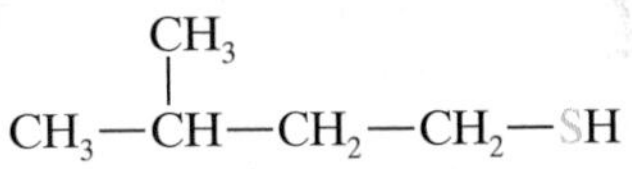

$$\begin{array}{c} \quad CH_3 \\ \quad | \\ CH_3{-}CH{-}CH_2{-}CH_2{-}SH \end{array}$$

3-methyl-1-butanethiol
(odor of skunk)

$$CH_3{-}CH_2{-}O{-}CH_2{-}CH_3$$

diethyl ether
(anesthetic)

$$CH_3{-}CH_2{-}Cl$$

chloroethane
(spray for bruises)

A third reason for the variety of organic compounds is that carbon forms multiple bonds to other carbon atoms, as in ethylene and acetylene. In addition, carbon atoms form multiple bonds to oxygen and nitrogen. There is a double bond to oxygen in acetone and a triple bond to nitrogen in acetonitrile.

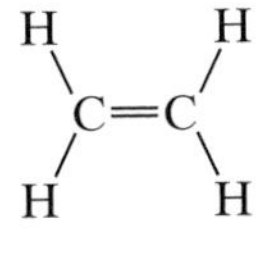

$$\begin{array}{ccc} H & & H \\ & \diagdown\ \ \diagup & \\ & C{=}C & \\ & \diagup\ \ \diagdown & \\ H & & H \end{array}$$

ethylene
(used to produce polyethylene)

$$H{-}C{\equiv}C{-}H$$

acetylene
(used in welding torches)

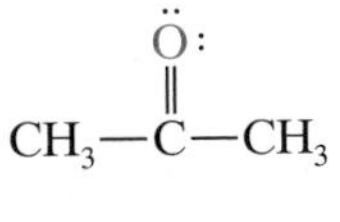

$$\begin{array}{c} \ddot{O}: \\ \| \\ CH_3{-}C{-}CH_3 \end{array}$$

acetone
(a polar solvent)

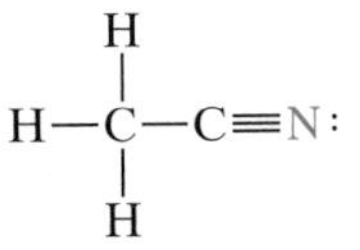

$$\begin{array}{c} H \\ | \\ H{-}C{-}C{\equiv}N: \\ | \\ H \end{array}$$

acetonitrile
(a polar solvent)

The fourth reason there are many carbon compounds is the variety of skeletons that form with the same number of carbon atoms. Compounds that have the same number and kinds of atoms arranged in different structures are **isomers.** Ethanol and dimethyl ether have the molecular formula, C_2H_6O, but different structures. They are isomers of one another.

Organic Compounds in Daily Life

Chemists can make compounds in the laboratory that are identical to those in nature. There is no difference between vitamin C made in nature and the same compound made in the laboratory. In addition, they can make carbon compounds that have never been produced in living systems. Foremost among these are medicines. Among them are antiseptics and disinfectants that prevent the growth of microorganisms outside the body. Drugs to lower blood pressure and to lower cholesterol help control heart disease. Chemotherapeutic agents are used to treat some forms of cancer. These and many other compounds, from the common aspirin to birth control pills, all contain carbon.

Most personal consumer products are also organic compounds. These include the creams and lotions that you use as skin moisturizers as well as the sun blocks you apply to prevent ultraviolet radiation damage to skin. Soaps, detergents, shampoos, deodorants, and perfumes all contain carbon compounds.

Your car also depends on a number of organic compounds; foremost among these are gasoline and rubber tires. Plastics reduce the weight of automobiles and increase gas mileage. Antifreeze, brake fluid, greases, and lubricants are also carbon-containing compounds produced in the laboratory from petroleum sources.

Your life has been irreversibly changed by the synthesis of large organic molecules called polymers. In the vernacular these compounds are called plastics. The list of common polymeric products includes floor tiles, rugs, latex paints, and building materials. Both phonograph records and compact disks are made of plastics. Audio and video tapes are also made of various types of plastics. Clothing is produced from polyesters (Dacron) and polyamides (nylon). In the future we will undoubtedly use even more polymers.

With the exception of the metal used to frame the car, the remainder of the interior of the car is plastic in various forms. (J. Greenberg, The Image Works)

```
   H  H                 H        H
   |  |                 |        |
H—C—C—O—H           H—C—O—C—H
   |  |                 |        |
   H  H                 H        H
   ethanol              dimethyl ether
```

13.2 Bonding and Molecular Shapes

The chemical and physical properties of organic molecules depend on the types of bonds between carbon atoms as well as those between the carbon atoms and other atoms and the resulting molecular shape. In Section 5.8 we learned that the carbon atom can assume a number of hybridized states that control the number and type of bonds that it forms. The carbon atoms in most organic compounds are found in one of three bonding arrangements:

1. four single bonds
2. two single bonds and one double bond
3. one single bond and one triple bond

Compounds containing a carbon atom with two double bonds are less common and will not be discussed in this text.

Compounds with Four Single Bonds

Methane is the simplest example of a compound whose carbon atom has four single bonds. The H—C—H bond angles in this compound are 109.5°. The three-dimensional shape of methane is shown in a **perspective structural formula.** Only two of the C—H bonds can be represented in the plane of the page. The other two bonds are above and below the plane of the page and are represented by wedge-shaped and dashed lines, respectively.

H
109.5° bond angle
H---C—H
H

The tetrahedral shape of CH_4 is represented using sp^3 hybrid orbitals for the carbon atom. One of the four valence electrons of carbon is located in each of the

FIGURE 13.1 Hybridization of Carbon

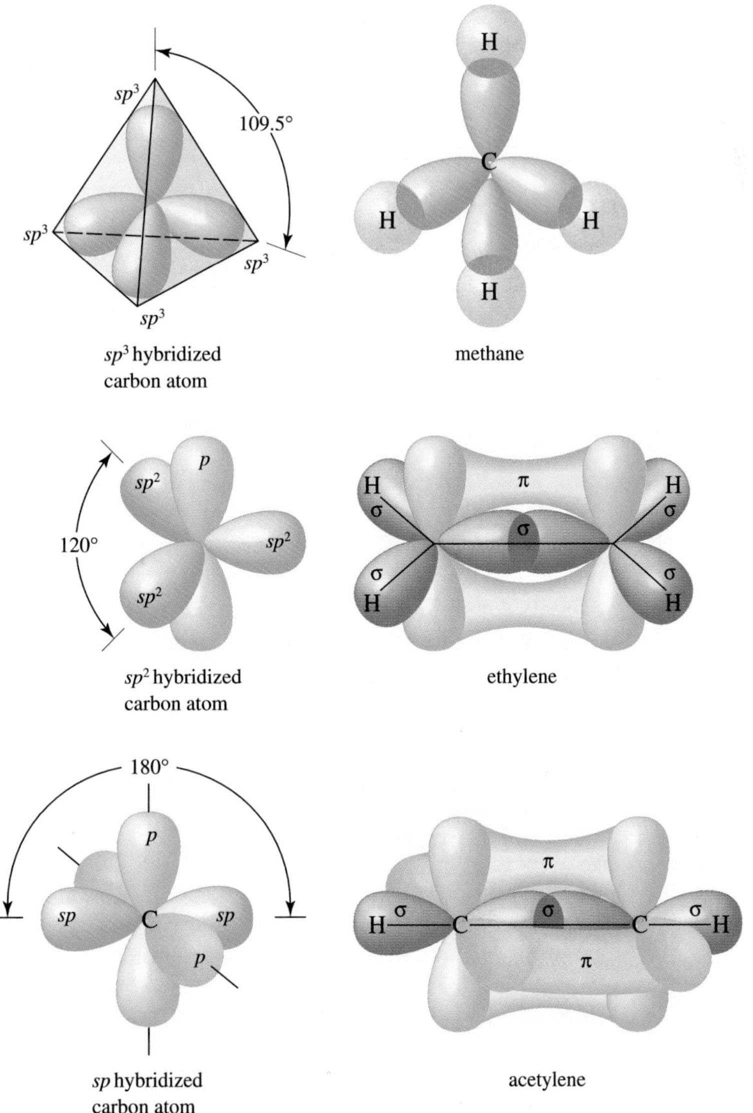

four hybrid orbitals. Each orbital then overlaps a hydrogen 1*s* orbital to form a σ bond (Figure 13.1).

Compounds with One Double Bond and Two Single Bonds

Ethylene is the simplest compound with a carbon–carbon double bond. The other two bonds of each carbon atom are to hydrogen atoms. The molecule is planar—all atoms lie in a common plane. Each bond angle is approximately 120°.

bond angles

122°, 116°

$H_2C{=}CH_2$

ethylene

Each carbon atom in ethylene is pictured using three sp^2 hybridized orbitals and one *p* orbital (Figure 13.1). The three sp^2 orbitals are separated by 120° in a plane. Two of the sp^2 orbitals, containing one electron each, form σ bonds with hydrogen. The third sp^2 orbital, with one electron, forms a σ bond with the other carbon atom in ethylene. The second bond of the double bond in ethylene results from a side-by-side overlap of the *p* orbital of each carbon atom to give a **pi (π) bond.**

Compounds with One Triple Bond and One Single Bond

Acetylene is the simplest compound with carbon atoms bonded by a triple bond. The molecule is linear—all atoms are arranged along a straight line. Each bond angle is 180°.

bond angle

180°

$H{-}C{\equiv}C{-}H$

acetylene

In acetylene the carbon–carbon triple bond consists of one σ bond and two π bonds. The carbon atoms in acetylene form two *sp* hybrid orbitals each, with two *p* orbitals remaining (Figure 13.1). One *sp* orbital and its unpaired electron form a bond with hydrogen; the other *sp* orbital forms a σ bond with the second carbon atom. The second and third bonds between the carbon atoms result from side-by-side overlap of *p* orbitals to form two π bonds.

13.3 Structural Formulas and Conformations

In Section 5.9 you learned how structural formulas and bond-line structures are used to represent molecular structure. Condensed structural formulas, which are often used, show only specific bonds; other bonds are implied. The degree of condensation depends on which bonds are shown and which are implied. For example, the structural formula of butane can be represented many ways.

$H{-}CH_2{-}CH_2{-}CH_2{-}CH_2{-}H$ (full structural formula, all C–H bonds shown) $CH_3{-}CH_2{-}CH_2{-}CH_3$ $CH_3CH_2CH_2CH_3$

FIGURE 13.2
Ball-and Stick Model of Butane

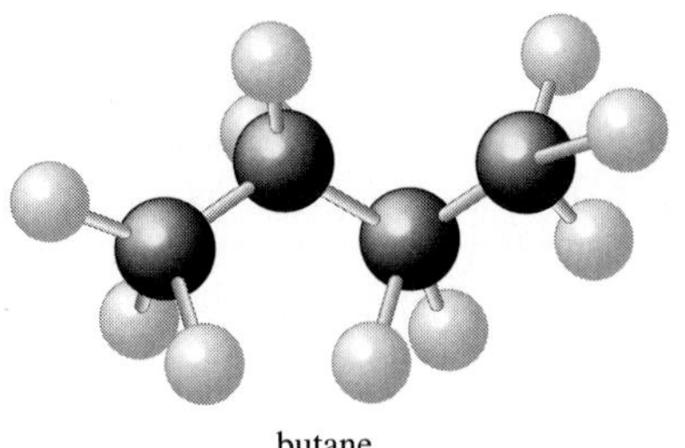

Bond-line structures are even more abbreviated and you have to recognize many more implied features. For butane, the bond-line structure consists of only three zig-zag lines.

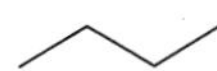

If you construct a ball-and-stick model of butane you might obtain the representation shown in Figure 13.2. The shape of butane can be represented as a perspective structural formula. The carbon–carbon bonds as well as two carbon–hydrogen bonds are all written in the plane of the page. The remaining carbon–hydrogen bonds are above and below the plane of the page. The dashed bonds are below the plane of the page and the wedge-shaped bonds have their wide end above the plane of the page.

perspective structural formula

You may write two structural formulas or construct two molecular models that appear to be different but in fact represent the same molecule. Consider the models of ethane shown in Figure 13.3. In both examples, the two carbon atoms are bonded to each other, and each carbon atom is bonded to three hydrogen atoms. The two representations differ in the position of the hydrogen atoms of one carbon

FIGURE 13.3
Conformations of Ethane
Rotation of the methyl group on the right by 60° for the space-filling, the ball-and-stick model, or the perspective structural formula converts a staggered conformation into an eclipsed conformation.

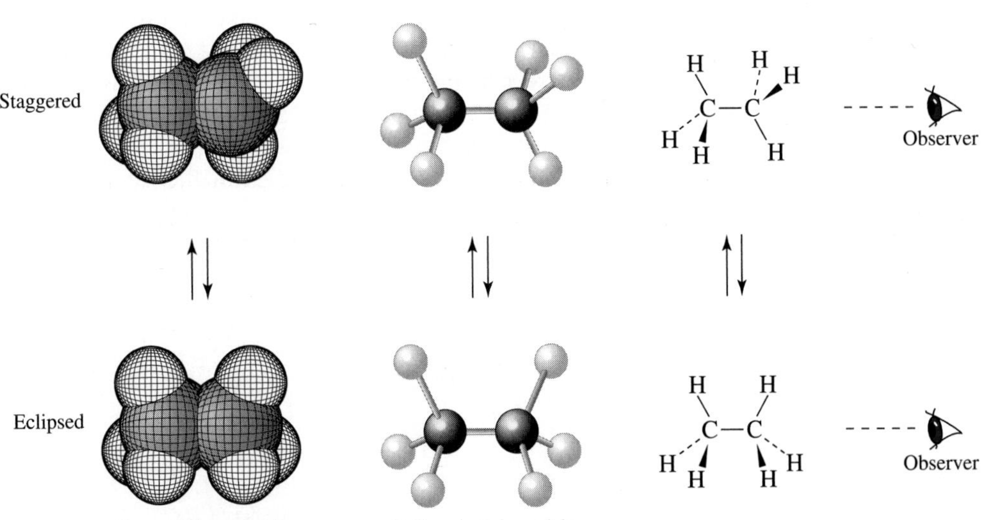

atom relative to the hydrogen atoms of the other carbon atom. Which of the two forms represents ethane? The answer is that both do to some extent.

When an sp^3 hybrid orbital of one carbon atom overlaps an sp^3 hybrid orbital of another carbon atom, it does so along the axis between the two nuclei. Each carbon atom and its three bonded hydrogen atoms can rotate about the carbon–carbon σ bond. The two carbon atoms are still bonded, but the spatial positions of the hydrogen atoms are altered.

Atoms in molecules such as ethane are constantly rotating. However, the connectivity of the carbon atoms and of the carbon–hydrogen bonds are unchanged. The motion can be compared to the twisting and turning of your body while dancing. You may look different, but the parts of your body are still connected to the normal places. Only the orientation of your limbs changes. A molecule can change its orientation or **conformation** by rotation about single bonds.

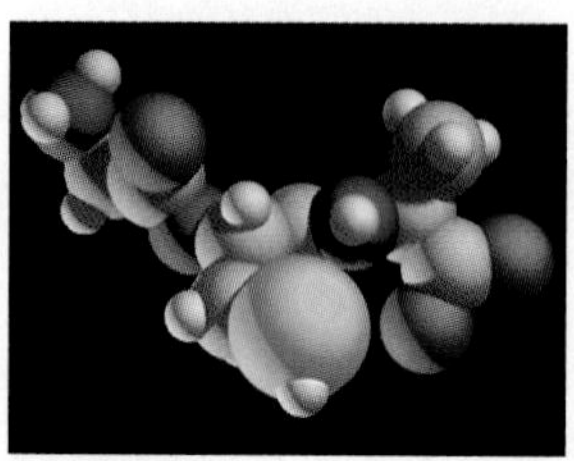

Computer generated pictures of molecules such as this peptide give chemists a view of the molecular world. (Ken Eward/ Biografx, Photo Researchers, Inc.)

Ethane can exist in a number of conformations, but the staggered conformation has the lowest energy because the hydrogen atoms are as far apart as possible. The eclipsed conformation has the highest energy because the hydrogen atoms are closest to each other. Each C—H bond on one carbon atom eclipses a C—H bond on another carbon atom, much as the Earth can eclipse the moon. The energy of the eclipsed conformation is 3 kcal/mole higher than that of the staggered conformation. This small difference in energy means that the rotation about the C—C bond is essentially unrestricted. Different conformations of ethane are not different compounds; ethane is a mixture of conformations that cannot be separated from each other.

When we study enzymes and other complex biochemical molecules, you will find that rotation about carbon–carbon bonds gives flexibility to these molecules. Thus, conformations are important in explaining both the chemical and physical properties of biological molecules.

13.4 Functional Groups

The sheer numbers of organic compounds can make the study of organic compounds and their related physical and chemical properties a daunting objective. Fortunately, organic chemists have found ways to handle the immense number of facts by organizing organic compounds into families of compounds based on functional groups. Atoms or groups of bonded atoms responsible for similar physical and chemical properties in a family of compounds are **functional groups.** Thus, the study of 10 million compounds is organized into more manageable groups of compounds whose reactivity is predictable.

Some functional groups are a part of the molecular framework. These include the carbon–carbon double bond in compounds such as ethylene and the carbon–carbon triple bond in compounds such as acetylene. Double or triple

```
                                        H       H
                                         \     /
  H       H                                C=C
   \     /                                /   \
    C=C          H—C≡C—H           H—C       C—H
   /     \                                \\   //
  H       H                                C—C
                                          /   \
                                         H     H

  ethylene        acetylene              benzene
```

bonds are also present in molecules that have more complex structures. Although benzene is represented as a series of alternating carbon–carbon single and double bonds, it reacts differently from ethylene.

Chemical reactions in organic compounds occur at functional groups while the rest of the structure remains unchanged. Once you learn the properties and reactions of one functional group, you will know the properties and reactions of thousands of compounds in that class. For example, the carbon–carbon double bond in ethylene reacts with one mole of hydrogen gas in the presence of a platinum catalyst to give ethane.

$$H_2C{=}CH_2 + H_2 \xrightarrow{Pt} H{-}CH_2{-}CH_2{-}H$$

ethylene ethane

All **alkenes**—compounds containing a carbon–carbon double bond—readily undergo a similar reaction. **Alkynes**—compounds containing a carbon–carbon triple bond—react with two moles of hydrogen gas. The double bonds of benzene are far less reactive (Chapter 15).

Functional groups can contain a variety of elements. The most common elements are oxygen and nitrogen, although sulfur or the halogens may also be present. **Alcohols** and **ethers** are two classes of compounds that contain carbon–oxygen single bonds. The —OH unit in alcohols is the **hydroxyl group.**

$-\overset{|}{\underset{|}{C}}-\ddot{O}-H$ as in methyl alcohol $H-\overset{H}{\underset{H}{C}}-\ddot{O}-H$

(an alcohol)

$-\overset{|}{\underset{|}{C}}-\ddot{O}-\overset{|}{\underset{|}{C}}-$ as in dimethyl ether $H-\overset{H}{\underset{H}{C}}-\ddot{O}-\overset{H}{\underset{H}{C}}-H$

(an ether)

Aldehydes and ketones contain double bonds to oxygen. The unit C=O is called the **carbonyl group.** The carbon atom of the carbonyl group is called the **carbonyl carbon atom,** and the oxygen atom is called the **carbonyl oxygen atom.** Note that an **aldehyde** has at least one hydrogen atom bonded to the carbonyl carbon atom. In **ketones** the carbonyl carbon atom is bonded to two other carbon atoms.

$-\overset{:\ddot{O}}{\overset{\|}{C}}-H$ as in acetaldehyde $H-\overset{H}{\underset{H}{C}}-\overset{:\ddot{O}}{\overset{\|}{C}}-H$

(an aldehyde)

$-\overset{|}{\underset{|}{C}}-\overset{:\ddot{O}}{\overset{\|}{C}}-\overset{|}{\underset{|}{C}}-$ as in acetone $H-\overset{H}{\underset{H}{C}}-\overset{:\ddot{O}}{\overset{\|}{C}}-\overset{H}{\underset{H}{C}}-H$

(a ketone)

Carboxylic acids and esters contain both single and double bonds from a carbon atom to oxygen atoms. In a **carboxylic acid** the carbonyl group is bonded to a hydroxyl group and to either a hydrogen or a carbon atom. In an **ester** the carbonyl group is bonded to an OR group, where R contains one or more carbon atoms, and to either a hydrogen or a carbon atom.

```
 :Ö                                H :Ö
  ‖                                |  ‖
—C—Ö—H      as in acetic acid    H—C—C—Ö—H
                                   |
                                   H
```

(a carboxylic acid)

(a carboxylic acid)

```
 :Ö                                     H :Ö     H
  ‖     |                               |  ‖     |
—C—Ö—C—      as in methyl acetate     H—C—C—Ö—C—H
        |                               |        |
                                        H        H
```

A nitrogen atom can form single, double, or triple bonds to a carbon atom. Compounds with carbon–nitrogen single bonds are **amines.** The remaining two bonds to nitrogen can be to hydrogen or to carbon atoms.

```
  |                                    H
  |  ..                                |  ..
—C—N—      as in methylamine         H—C—N—H
  |  |                                 |  |
                                       H  H
```

(an amine)

Compounds with carbon–nitrogen double and triple bonds are **imines** and **nitriles,** respectively.

```
     :N—H                              H :N—H
  |   ‖                                |  ‖
—C—C—H      as in ethylimine         H—C—C—H
  |                                    |
                                       H
```

(an imine)

```
  |                                    H
  |                                    |
—C—C≡N:      as in acetonitrile      H—C—C≡N:
  |                                    |
                                       H
```

(a nitrile)

Amides have a carbonyl carbon atom linked by a single bond to a nitrogen atom and to either a hydrogen or a carbon atom. The remaining two bonds to the nitrogen atom may be to either hydrogen or carbon atoms.

```
 :Ö                                  H :Ö
  ‖  ..                              |  ‖  ..
—C—N—      as in acetamide         H—C—C—N—H
     |                               |     |
                                     H     H
```

(an amide)

Sulfur forms single bonds to carbon in two classes of compounds. **Thiols** (also called mercaptans) and **thioethers** (also called sulfides) structurally resemble alcohols and ethers, but contain sulfur in place of oxygen.

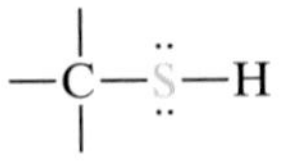

$$-\overset{|}{\underset{|}{C}}-\ddot{\underset{..}{S}}-H \quad \text{as in methanethiol} \quad H-\overset{H}{\overset{|}{\underset{\underset{H}{|}}{C}}}-\ddot{\underset{..}{S}}-H$$

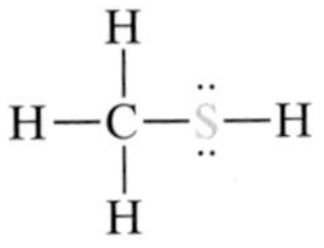

(a thiol)

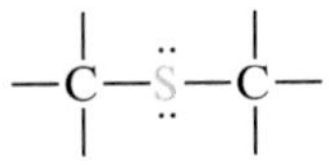

$$-\overset{|}{\underset{|}{C}}-\ddot{\underset{..}{S}}-\overset{|}{\underset{|}{C}}- \quad \text{as in dimethyl sulfide} \quad H-\overset{H}{\overset{|}{\underset{\underset{H}{|}}{C}}}-\ddot{\underset{..}{S}}-\overset{H}{\overset{|}{\underset{\underset{H}{|}}{C}}}-H$$

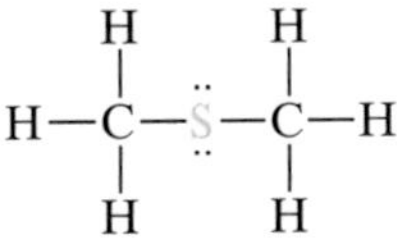

(a thioether)

The halogens form single bonds to carbon. Chlorine and bromine are the more common halogens in organic compounds.

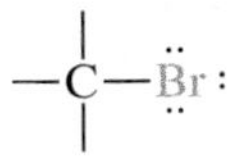

$$-\overset{|}{\underset{|}{C}}-\ddot{\underset{..}{Br}}: \quad \text{as in bromoethane} \quad H-\overset{H}{\overset{|}{\underset{\underset{H}{|}}{C}}}-\overset{H}{\overset{|}{\underset{\underset{H}{|}}{C}}}-\ddot{\underset{..}{Br}}:$$

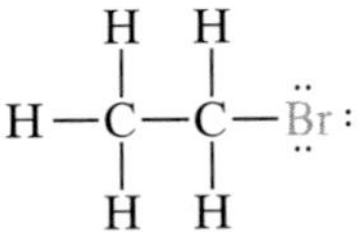

(a haloalkane)

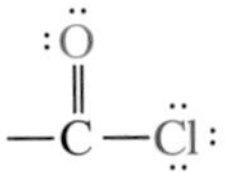

$$-\overset{:\ddot{O}}{\overset{\|}{C}}-\ddot{\underset{..}{Cl}}: \quad \text{as in ethanoyl chloride} \quad H-\overset{H}{\overset{|}{\underset{\underset{H}{|}}{C}}}-\overset{:\ddot{O}}{\overset{\|}{C}}-\ddot{\underset{..}{Cl}}:$$

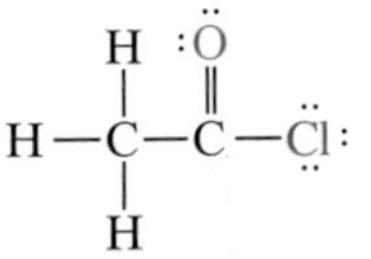

(an acyl halide)

EXAMPLE 13.1

Identify the functional groups in cysteine, one of the components of proteins.

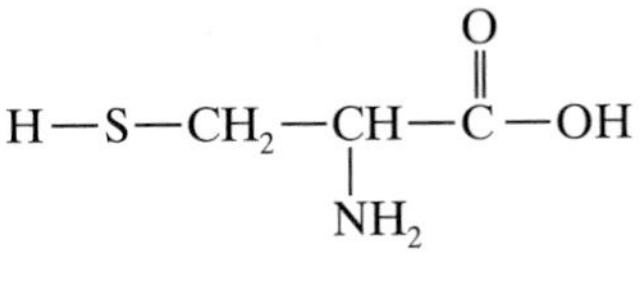

$$H-S-CH_2-\underset{\underset{NH_2}{|}}{CH}-\overset{O}{\overset{\|}{C}}-OH$$

Solution

There are three functional groups. The HS— group bonded to the leftmost carbon atom is characteristic of thiols. The $-NH_2$ group bonded to the middle carbon atom makes cysteine an amine. Finally, the carbonyl carbon atom on the right is also bonded to an —OH group, which means that cysteine is also a carboxylic acid.

PROBLEM 13.1

Phenacetin, a drug once used in pills for headaches, has the following structure. Identify its functional groups.

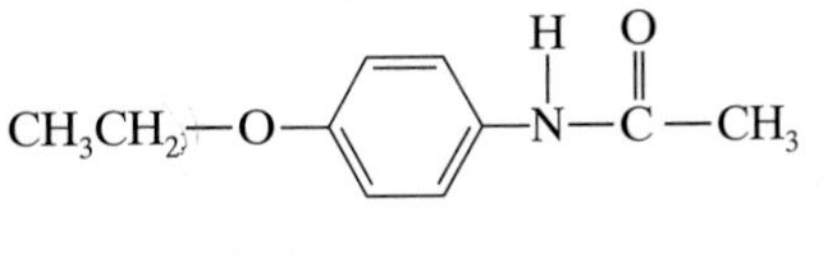

$$CH_3CH_2-O-C_6H_4-\overset{H}{\overset{|}{N}}-\overset{O}{\overset{\|}{C}}-CH_3$$

Chemical Communications in the Insect World

The scope of organic chemistry is rapidly changing, and the field contributes to many areas. For example, we cannot understand modern biology without a foundation in organic chemistry and indirectly without an understanding of functional groups. Organic chemistry underlies all life forms. As an example, we will consider the structure and functional groups of some pheromones. Pheromones (from the Greek *pherein,* "to transfer" + *hormon,* "to excite") are compounds (occasionally mixtures of compounds) that insects and other animals use to communicate. The major identified species that use pheromones are insects, but it is thought that even higher animals such as mammals may emit pheromones.

Pheromones are used to mark trails, warn of dangers, cause aggregation of species, defend against aggressors, and attract members of the opposite sex. The whip scorpion ejects a defensive spray to ward off predators. Some species of ants warn other ants of danger by an alarm pheromone. Bark beetles responsible for Dutch elm disease emit an aggregation pheromone that results in the gathering of a large number of beetles, who carry a fungus that kills the tree. The sex attractants, usually emitted by the female of the species, attract members of the opposite sex. Pheromones are signals that the female is ready to mate. They also aid the male in locating the female.

All moths that have been studied have sex attractants that are species specific. The compounds usually are derived from long chains of carbon atoms. However, the functional groups in the pheromones vary considerably. Two examples are the sex attractants of the gypsy moth and the grape berry moth, whose structures are shown below. The oxygen atom in the three-membered ring of the sex attractant of the gypsy moth is in an ether linkage. The oxygen atoms in the sex attractant of the grape berry moth are part of an ester functional group. Note that this compound also contains a double bond.

When the structures of sex attractants were determined, some scientists predicted that it might be possible to bait traps with the compound and, by removing one sex, break the reproductive cycle. Unfortunately, this "ideal" way to control insects and eliminate the use of pesticides has not proved effective for most species. However, there is some evidence that pheromone traps cause some confusion and may inhibit mating. In addition, pheromone traps serve as an early warning of possible infestation of that species. The ultimate goal of replacing pesticides with pheromones may yet be possible.

The gypsy moth sends out an aggregation pheromone which brings together many moths, leading to the destruction of trees. (Michael Gadomski, Photo Researchers, Inc.)

$$CH_3CH_2CH_2CH_2CH_2CH_2CH_2CH_2CH_2CH_2CH_2CH\overset{O}{—}CHCH_2CH_2CH_2CH_2CH(CH_3)_2$$

sex attractant of gypsy moth

$$CH_3CH_2—CH=CH—CH_2CH_2CH_2CH_2CH_2CH_2CH_2CH_2—O—\overset{\overset{O}{\|}}{C}—CH_3$$

sex attractant of grape berry moth

13.5 Isomers

Compounds that have the same molecular formula but differ in structure—that is, in the sequence of the bonded atoms—are **isomers.** As you examine the structures of organic compounds in detail, you will learn how differences in the structures of isomers affect the physical and chemical properties of compounds.

FIGURE 13.4 Molecular Models of Isomeric Butanes

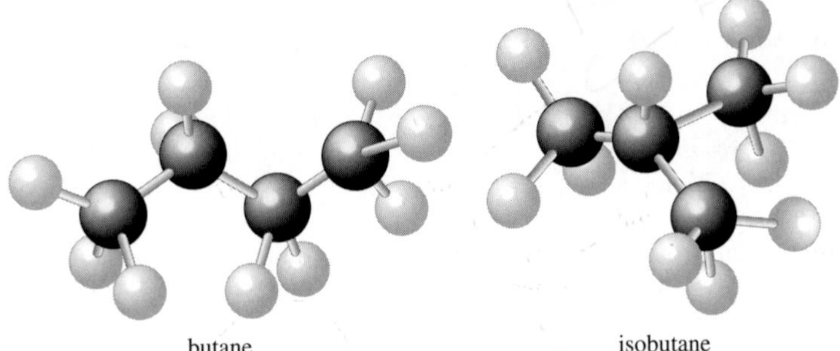

Consider the structural differences in the two isomers of C_4H_{10}. Butane has an uninterrupted chain of carbon atoms (Figure 13.4), but isobutane has only three carbon atoms connected in sequence and a fourth carbon atom appended to the chain. Isomers that differ in their carbon skeleton are **skeletal isomers.** The number of skeletal isomers for a given molecular formula increases exponentially as the number of carbon atoms increases (Table 13.1). We will return to this subject and learn how to name isomers in Section 14.3.

Isomers that have different functional groups are **functional group isomers.** The molecular formula for ethyl alcohol is C_2H_6O; the molecular formula for dimethyl ether is also C_2H_6O. The composition of the two compounds is identical, but their functional groups differ. The atomic sequence is C—C—O in ethyl alcohol, and the oxygen atom is present as an alcohol (Figure 13.5). The C—O—C sequence in the isomer corresponds to that of an ether.

```
   H  H              H       H
   |  |              |       |
H—C—C—O—H         H—C—O—C—H
   |  |              |       |
   H  H              H       H
```

ethyl alcohol   dimethyl ether

TABLE 13.1 Number of Isomers

Molecular Formula	Number of Isomers
CH_4	1
C_2H_6	1
C_3H_8	1
C_4H_{10}	2
C_5H_{12}	3
C_6H_{14}	5
C_7H_{16}	9
C_8H_{18}	18
C_9H_{20}	35
$C_{10}H_{22}$	75
$C_{20}H_{42}$	336,319
$C_{30}H_{62}$	4,111,846,763
$C_{40}H_{82}$	62,491,178,805,831

FIGURE 13.5 Molecular Models of Functional Group Isomers

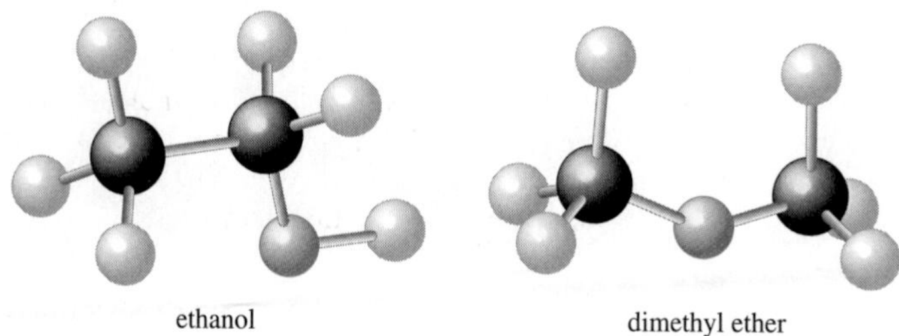

ethanol dimethyl ether

Positional isomers are compounds that have the same functional group in different locations. The isomeric alcohols 1-propanol and 2-propanol differ in the location of the hydroxyl group.

OH group on the end of the chain

OH group in the middle of the chain

```
                                    OH
                                    |
CH3—CH2—CH2—OH              CH3—CH—CH3
  1-propanol                   2-propanol
```

$CH_3—CH_2—CH_2—OH$ (1-propanol) $CH_3—CH(OH)—CH_3$ (2-propanol)

You may sometimes think that two structures represent isomers when in fact they represent the same compound written in different ways. It is important that you learn to recognize isomers and distinguish them from different representations of the same compound. For example, 1,2-dichloroethane can be written in a variety of ways, but the bonding sequence is still Cl—C—C—Cl.

```
   H  H               H  Cl               H  Cl
   |  |               |  |                |  |
Cl—C—C—Cl   or   Cl—C—C—H    or    H—C—C—H
   |  |               |  |                |  |
   H  H               H  H                Cl H
```

1,2-dichloroethane (CH_2ClCH_2Cl)

1,1-Dichloroethane, which is an isomer of 1,2-dichloroethane has a different placement of chlorine atoms.

```
   Cl H
   |  |
Cl—C—C—H
   |  |
   H  H
```

1,1-dichloroethane ($CHCl_2CH_3$)

In 1,1-dichloroethane, the two chlorine atoms are bonded to the same carbon atom. In 1,2-dichloroethane, the two chlorine atoms are bonded to different carbon atoms. The different condensed structural formulas, $CHCl_2CH_3$ and CH_2ClCH_2Cl, also convey this information about these isomers.

EXAMPLE 13.2

Consider the following structural formulas. Do they represent isomers?

```
   H  H              H  H
   |  |              |  |
Br—C—C—Cl        Cl—C—C—Br
   |  |              |  |
   H  H              H  H
```

Solution

The atomic compositions given in these structural formulas are identical. The molecular formula is C_2H_4BrCl. The condensed structural formula of the first structure is $BrCH_2CH_2Cl$, and that of the second is $ClCH_2CH_2Br$. However, the two structural formulas do not represent isomers. The order of bonds between atoms is Br—C—C—Cl in both cases. The second structural formula is merely the first structural formula written in reverse.

PROBLEM 13.2

Consider the following structural formulas. Do they represent isomers?

```
     H  Br              H  H
     |  |               |  |
  H—C—C—Cl         Cl—C—C—Br
     |  |               |  |
     H  H               H  H
```

13.6 Organic Reactions and Mechanisms

The description of how a reaction occurs is the **reaction mechanism.** It details the steps in a reaction and shows the order in which bonds are broken in the reactant and formed in the product.

Concerted and Multistep Reactions

Some organic reactions occur in a single step, forming and breaking bonds simultaneously. Such processes are called **concerted reactions.** The reaction mechanism resembles that of an ordinary balanced chemical equation.

$$\text{A(reactant)} \longrightarrow \text{B(product)}$$

Many reactions occur in a series of steps. For example the conversion of reactant A into product B may occur in two steps involving an intermediate, M.

$$\text{A(reactant)} \longrightarrow \text{M(intermediate)}$$
$$\text{M(intermediate)} \longrightarrow \text{B(product)}$$

In a multistep sequence the order of bond formation and bond cleavage is part of the reaction mechanism. Note that the events in the second step cannot occur until certain events have occurred in the first step.

The rates of the individual steps in a multistep mechanism are also important. The overall rate of conversion of reactant into intermediate into product can occur no faster than the slowest individual step, which is the **rate-determining step.**

Types of Bond Cleavage and Bond Formation

There are two types of bond cleavage. When a bond breaks so that one electron remains with each of the two fragments, the process is **homolytic cleavage.** The fragments that contain an unpaired electron are **radicals.** In a second alternative, called **heterolytic cleavage,** two electrons remain with one fragment, giving it a negative charge. The second fragment is electron deficient and thus has a positive charge.

$$X:Y \longrightarrow X\cdot + \cdot Y$$ homolytic cleavage—one electron remains with each fragment

$X:Y \longrightarrow X^+ + Y:^-$ heterolytic cleavage—two electrons stay with one of the fragments

The two ways to form two-electron covalent bonds from fragments are the reverse of the two cleavage reactions. Bond formation from fragments that contain one electron each is a **homogenic** process. Formation of a bond using two electrons from one fragment and no electrons from the second fragment is a **heterogenic** process.

$X\cdot + \cdot Y \longrightarrow X:Y$ homogenic process—one electron provided from each fragment

$X^+ + :Y^- \longrightarrow X:Y$ heterogenic process—two electrons donated from one of the fragments

Homolysis of a bond to carbon produces a **carbon radical,** which tends to react in a homogenic process to obtain another electron. The electron is acquired along with another atom or group of atoms that becomes bonded to the carbon atom.

$$CH_3{-}Y \xrightarrow[\text{cleavage}]{\text{homolytic}} \underset{\text{methyl radical}}{CH_3\cdot} + \cdot Y$$

$$CH_3\cdot + Cl{-}Cl \xrightarrow[\text{process}]{\text{homogenic}} CH_3{-}Cl + Cl\cdot$$

Processes that involve either heterolytic bond cleavage or heterogenic bond formation are common reactions in organic chemistry. In many cases these processes will be compared with acid-base reactions, which you have already studied in detail. Heterolysis of a bond to carbon can produce two different carbon species. If the electrons of the bond remain with the carbon atom, a **carbanion** results. If the electrons of the bond are retained by the atom that was bonded to the carbon atom, a **carbocation** results.

$$CH_3{-}Y \xrightarrow{\text{heterolysis}} \underset{\text{methyl carbanion}}{CH_3:^-} + Y^+$$

$$CH_3{-}Y \xrightarrow{\text{heterolysis}} \underset{\text{methyl carbocation}}{CH_3^+} + :Y^-$$

A carbocation has a sextet of electrons and is electron-deficient. Carbocations behave as **electrophiles** (electron lovers) and seek negatively charged centers to neutralize the positive charge and obtain an octet of electrons.

A carbanion has an octet of electrons and so is not electron-deficient. Because it has an unshared electron pair, a carbanion reacts as a base. Carbanions behave as **nucleophiles** (nucleus lovers) and seek positively charged centers to neutralize the negative charge.

13.7 Acid–Base Reactions in Organic Chemistry

The definitions and principles of acid–base reactions were studied in Chapter 11. However, the emphasis at that time was on the proton as it "moved" from one pair of bonding electrons to another pair of electrons. Organic chemists prefer to focus on the electrons involved in the transfer. The electron pair of an atom is "moved" toward the proton. This movement of an electron pair is shown by a curved arrow formalism. Consider acetic acid, which has a carboxyl group that can donate a proton to a base such as water. A curved arrow "moves" the electrons from the oxygen atom of water toward the hydrogen atom of acetic acid. The bonding pair of electrons in the O—H bond of acetic acid is "moved" to become a nonbonding electron pair position of oxygen in the acetate ion.

$$CH_3-\overset{\overset{O}{\|}}{C}-\ddot{\underset{..}{O}}-H + :\underset{..}{\overset{\overset{H}{|}}{O}}-H \longrightarrow CH_3-\overset{\overset{O}{\|}}{C}-\ddot{\underset{..}{O}}:^- + H-\overset{\overset{H}{|+}}{\underset{..}{O}}-H$$

acetic acid　　　　acetate ion

Amines are organic bases whose chemistry is similar to the reactions of ammonia. Thus, methylamine behaves as a base because the nonbonding electron pair of the nitrogen atom can accept a proton from an acid such as the hydronium ion.

$$CH_3-\underset{\underset{H}{|}}{\overset{\overset{H}{|}}{N}}: + H-\underset{\underset{H}{|}}{\ddot{O}^+}-H \longrightarrow CH_3-\underset{\underset{H}{|}}{\overset{\overset{H}{|}}{N^+}}-H + :\underset{\underset{H}{|}}{\ddot{O}}-H$$

methylamine　　　　methylammonium ion

Lewis Acids and Bases

There are many organic chemical reactions that do not involve a transfer of protons but that nevertheless are also regarded as acid–base reactions. The Lewis concept of acids and bases focuses on the electron pair. A **Lewis acid** is a substance that accepts an electron pair; a **Lewis base** is a substance that donates an electron pair. Thus HCl, which is an acid in the Brønsted–Lowry sense, is also a Lewis acid—it contains a proton that can accept an electron pair. Similarly, ammonia is a Lewis base as well as a Brønsted–Lowry base, because it can donate an electron pair. However, the Lewis classification of acids and bases is more extensive because it is not restricted to a transfer of protons.

In many organic reactions the Lewis acid and Lewis base combine with each other to give a single product. The conjugate base of HCN is the cyanide ion, CN^-. In the reaction of the cyanide ion with the carbonyl group of ethanal (acetaldehyde), the cyanide ion is a Lewis base and ethanal is a Lewis acid.

$$CH_3-\overset{\overset{:\ddot{O}}{\|}}{C}-H + ^-:C\equiv N: \longrightarrow CH_3-\underset{\underset{CN}{|}}{\overset{\overset{:\ddot{O}:^-}{|}}{C}}-H$$

ethanal

EXAMPLE 13.3

Consider the reaction of a proton with ethylene to give a carbocation intermediate. Classify the reactants using Lewis acid–base nomenclature.

$$H_2C=CH_2 + H^+ \longrightarrow {}^+\underset{\underset{H}{|}}{\overset{\overset{H}{|}}{C}}-\underset{\underset{H}{|}}{\overset{\overset{H}{|}}{C}}-H$$

Solution

The proton accepts an electron pair from the π bond of ethylene to form a carbon–hydrogen bond. Thus, the proton acts as a Lewis acid, and ethylene acts as the Lewis base.

PROBLEM 13.3

Consider the reaction of the bromide ion, Br^-, with the carbocation intermediate in Example 13.3. Classify the reactants using Lewis acid–base nomenclature.

$$:\ddot{\underset{..}{Br}}:^- + {}^{+}\underset{H}{\overset{H}{C}}-\underset{H}{\overset{H}{C}}-H \longrightarrow :\ddot{\underset{..}{Br}}-\underset{H}{\overset{H}{C}}-\underset{H}{\overset{H}{C}}-H$$

13.8 Oxidation–Reduction Reactions

In Section 6.11 we learned that the oxidation of an organic compound is characterized by either a decrease in its hydrogen content or an increase in its oxygen content. Conversely, any organic molecule that either gains in hydrogen content or decreases in oxygen content is reduced. Consider the reaction of the indicated alcohol with potassium dichromate, which yields a carboxylic acid. An oxidation has occurred because there is a decrease in the hydrogen content as well as an increase in the oxygen content.

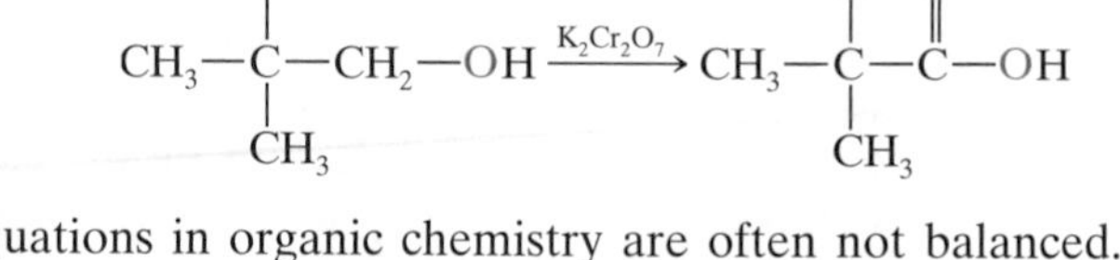

The oil fires set by Iraq at the conclusion of the war in Kuwait generated combustion products that polluted the atmosphere. (Wesley Bocxe, Photo Researchers, Inc.)

Chemical equations in organic chemistry are often not balanced. The focus is on the conversion of one functional group into another. For that reason, reactants such as potassium dichromate are simply placed above the reaction arrow. Note that catalysts are also placed above the reaction arrow. You will learn the difference between a catalyst and a reactant as you study specific reactions. In the preceding reaction, which converts an alcohol into an acid, an oxidation has occurred, so the substance above the arrow must be a reactant which is an oxidizing agent.

Some reactions in which there is a change in both the hydrogen and oxygen content are not oxidation–reduction reactions. For example, the conversion of ethylene to ethanol is not an oxidation–reduction reaction. There is an increase of two hydrogen atoms, a reduction, but there is also an increase of one oxygen atom, an oxidation. Thus, there is no net oxidation or reduction.

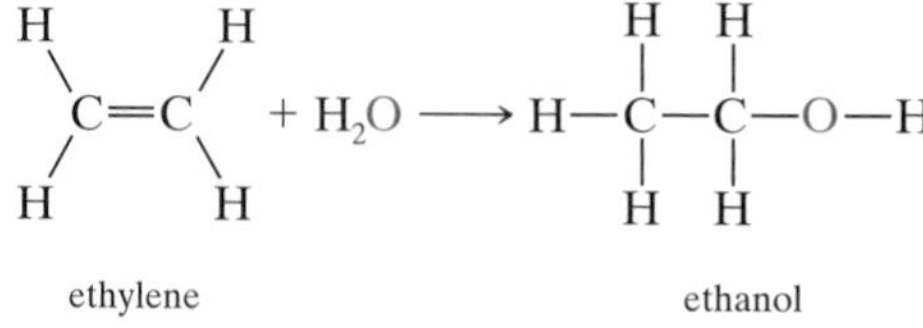

EXAMPLE 13.4

One of the steps in the metabolism of alcohol in the body involves reaction with a coenzyme NAD^+ to give the indicated product. What type of process occurs? How is the coenzyme NAD^+ classified?

$$CH_3-\underset{H}{\overset{OH}{C}}-H \xrightarrow{NAD^+} CH_3-\overset{O}{\overset{\|}{C}}-H$$

Solution

The reaction results in a decrease in the hydrogen content of the molecule. An oxidation has occurred. The NAD^+ must have served as an oxidizing agent.

PROBLEM 13.4

Disulfiram, a drug used to treat alcoholism, is split in two by the following metabolic reaction. What type of process occurs? What type of reagent is required?

$$(CH_3CH_2)_2N{-}\overset{S}{\overset{\|}{C}}{-}S{-}S{-}\overset{S}{\overset{\|}{C}}{-}N(CH_2CH_3)_2 \longrightarrow 2\ (CH_3CH_2)_2N{-}\overset{S}{\overset{\|}{C}}{-}S{-}H$$

13.9 Classification of Organic Reactions

In the preceding two sections we reviewed two classes of reactions presented in earlier chapters. Now we will look at other common examples of organic reactions. These reactions will be discussed in greater detail in subsequent chapters.

Addition reactions occur when two reactants combine to give a single product. An example of an addition reaction is the reaction of ethylene with HBr to form bromoethane. The hydrogen and bromine atoms are added to adjacent atoms, a common characteristic of addition reactions.

These two atoms are added to adjacent atoms.

$$H_2C{=}CH_2 + H{-}Br \longrightarrow H{-}\underset{H}{\overset{H}{\underset{|}{\overset{|}{C}}}}{-}\underset{H}{\overset{Br}{\underset{|}{\overset{|}{C}}}}{-}H$$

Elimination reactions occur when a compound is split into two compounds. Most elimination reactions form a product with a double bond containing the majority of the atoms in the reactant, and a second smaller molecule such as H_2O. The atoms eliminated to form the smaller molecule are usually located on adjacent carbon atoms in the reactant. For example, 2-propanol reacts with concentrated sulfuric acid to produce propene; water is eliminated in this reaction.

These groups of atoms are eliminated.

$$H{-}\underset{H}{\overset{H}{\underset{|}{\overset{|}{C}}}}{-}\underset{H}{\overset{OH}{\underset{|}{\overset{|}{C}}}}{-}\underset{H}{\overset{H}{\underset{|}{\overset{|}{C}}}}{-}H \xrightarrow{H_2SO_4} H{-}\underset{H}{\overset{H}{\underset{|}{\overset{|}{C}}}}{-}\underset{H}{\underset{|}{C}}{=}\underset{H}{\underset{|}{C}}{-}H + H_2O$$

A single bond is converted into a double bond.

In **substitution reactions** one atom or group of atoms displaces a second atom or group of atoms. The generalized reaction is

$$A{-}X + Y \longrightarrow A{-}Y + X$$

In the preceding reaction Y substitutes for X. An example of a substitution reaction is the conversion of bromomethane into methanol.

The hydroxide ion is the reactant.

The hydroxyl group is covalently bonded in the alcohol product.

$$CH_3{-}Br + OH^- \longrightarrow CH_3{-}OH + Br^-$$

This covalently bonded bromine atom is replaced.

A bromide ion is formed.

In **hydrolysis reactions** (from the Greek *hydro,* "water" + *lysis,* "splitting") water splits a large reactant molecule into two smaller product molecules. The generalized reaction is

$$A{-}B + H_2O \longrightarrow A{-}H + HO{-}B$$

One product molecule contains a hydrogen atom derived from water. The other product contains an —OH group derived from water. The hydrolysis of an amide to produce a carboxylic acid and an amine is an example of this process.

This N—C bond bond is cleaved.

The —OH group is bonded to the carbon atom.

$$CH_3{-}\overset{H}{\overset{|}{N}}{-}\overset{O}{\overset{\|}{C}}{-}CH_3 + H_2O \longrightarrow CH_3{-}\overset{H}{\overset{|}{N}}{-}H + HO{-}\overset{O}{\overset{\|}{C}}{-}CH_3$$

This N—H bond is formed.

an amide | an amine | a carboxylic acid

In **condensation reactions** two reactants combine to form one larger product and simultaneously form a second, smaller product. When the second product is water, the reaction is the reverse of a hydrolysis reaction. The general reaction is

$$A{-}H + H{-}O{-}B \longrightarrow A{-}B + H_2O$$

The formation of an ester from an alcohol and a carboxylic acid is an example of this process.

This C—O bond bond is cleaved.

This C—O bond is formed.

$$CH_3{-}\overset{O}{\overset{\|}{C}}{-}O{-}H + H{-}O{-}CH_3 \longrightarrow CH_3{-}\overset{O}{\overset{\|}{C}}{-}O{-}CH_3 + H_2O$$

a carboxylic acid | an alcohol | an ester

Rearrangement reactions result from the reorganization of bonds within a single reactant to give an isomeric product. This type of reaction will not be encountered as frequently in this text as other reactions. One example is a rearrangement in which the location of a double bond changes to give an isomer.

$$CH_3{-}\overset{Br}{\overset{|}{C}H}{-}CH{=}CH_2 \longrightarrow CH_3{-}CH{=}CH{-}CH_2{-}Br$$

Note that it is necessary to relocate the bromine atom to form the product.

EXAMPLE 13.5

Chloroform is metabolized via an intermediate to phosgene, a compound that causes liver damage. What type of reactions are involved in the formation and decomposition of the intermediate?

```
       H              O—H            O
       |              |              ‖
  Cl—C—Cl ⟶ Cl—C—Cl ⟶ Cl—C—Cl
       |              |
       Cl             Cl
  chloroform                      phosgene
```

Solution

In the first step an —OH group replaces a hydrogen atom. The reaction can be classified as a substitution reaction. In addition, because the product has an oxygen atom and there is no change in the number of hydrogen atoms, the reaction is also an oxidation reaction. In the second step a hydrogen atom from the —OH group and a chlorine atom are removed from adjacent atoms. This process is typical of an elimination reaction.

PROBLEM 13.5

The sedative–hypnotic chloral hydrate is metabolized as follows. What type of reaction occurs in each step?

```
      Cl OH              Cl O              Cl H
      |  |               |  ‖              |  |
  Cl—C—C—OH ⟶ Cl—C—C—H ⟶ Cl—C—C—OH
      |  |               |                 |  |
      Cl H               Cl                Cl H
```

Summary

The chemical and physical properties of **organic compounds** are consequences of the carbon–carbon covalent bond. The great variety of organic compounds results from the ability of carbon atoms to form stable single and multiple bonds with other carbon atoms as well as with atoms of other elements.

Carbon forms ***sp*^{3}, *sp*^{2}, and *sp* hybrid orbitals.** These orbitals are distributed at the maximum distance from each other and give tetrahedral, trigonal planar, and linear structures, respectively.

Structural formulas indicate the connectivity of carbon atoms in organic compounds in two dimensions. Three-dimensional representations of molecules using wedges and dashed lines are called **perspective structural formulas.**

Groupings of atoms called **functional groups** are used to classify organic compounds. These functional groups confer common chemical reactivities on a class of compounds.

Isomers have identical molecular formulas but different structures. Two or more models, or written structural representations, for the same organic compound in which atoms are differently oriented in space are called **conformations. Conformations** can be interconverted by rotation about carbon–carbon single bonds.

The description of how a reaction occurs is the **reaction mechanism.** It details the various stages in a transformation and shows the order in which bonds are broken in the reactant and formed in the product.

Reactions that occur in a single step are **concerted reactions.** The slowest step in a multistep reaction is the **rate-determining step.** Bond cleavage may be **homolytic,** giving a carbon **radical,** or **heterolytic,** giving either a **carbanion** or a **carbocation.** Bond formation may be **homogenic** or **heterogenic** depending on the reactants. Carbocations behave as **electrophiles;** carbanions behave as **nucleophiles.**

Acid–base reactions in organic chemistry are viewed as the movement of electrons toward an electron pair acceptor such as a proton. A **Lewis acid** is an electron pair acceptor. A **Lewis base** is an electron pair donor.

Organic oxidation–reduction reactions involve changes in the hydrogen or oxygen content of a molecule. Oxidation is the loss of hydrogen or the gain of oxygen. Reduction is the gain of hydrogen or the loss of oxygen.

Organic reactions can be classified based on common structural changes. These include **addition reactions, elimination reactions, substitution reactions, hydrolysis reactions, condensation reactions,** and **rearrangement reactions.**

Exercises

Isomers

13.1 State whether each of the following pairs are isomers.

(a) CH_3OH and CH_3CH_2OH
(b) $CH_3C{\equiv}CH$ and $CH_3CH{=}CH_2$
(c) $CH_3CH_2CH_2OH$ and $CH_3CH_2OCH_3$

13.2 State whether each of the following pairs are isomers.

(a) $CH_3CHClCH_3$ and $CH_3CH_2CH_2Cl$
(b) $CH_2ClCH_2CH_2Cl$ and $CH_3CH_2CHCl_2$
(c) $CH_3CHClCH_2Cl$ and $CHCl_2CH_2CH_3$

13.3 There are two isomers for each of the following molecular formulas. Draw their structural formulas.

(a) $C_2H_4Br_2$ (b) C_2H_6O (c) C_2H_4BrCl

13.4 There are three isomers for each of the following molecular formulas. Draw their structural formulas.

(a) $C_2H_3Br_2Cl$ (b) C_3H_8O (c) C_3H_8S

13.5 Consider the following structural formulas for two compounds used as general anesthetics. Do they represent isomers? How do they differ?

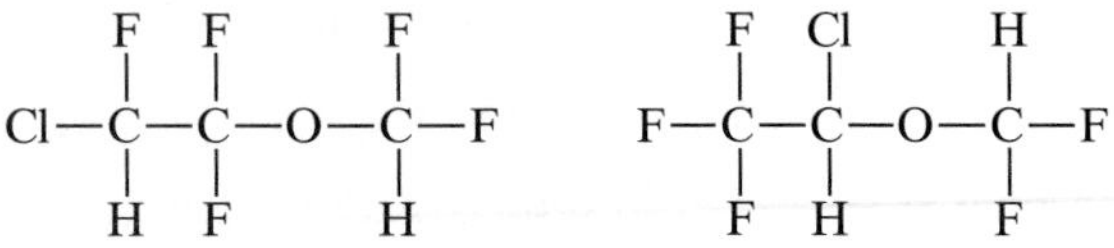

```
     F  F     F              F  Cl     H
     |  |     |              |  |      |
 Cl—C—C—O—C—F           F—C—C—O—C—F
     |  |     |              |  |      |
     H  F     H              F  H      F

        I                       II
```

13.6 Compare the following structural formulas of two intermediates in the metabolism of glucose. Do they represent isomers? How do they differ?

```
      H            H     O
      |             \  //
  H—C—OH             C
      |              |
     C=O         H—C—OH
      |              |
  H—C—OH         H—C—OH
      |              |
      H              H
```

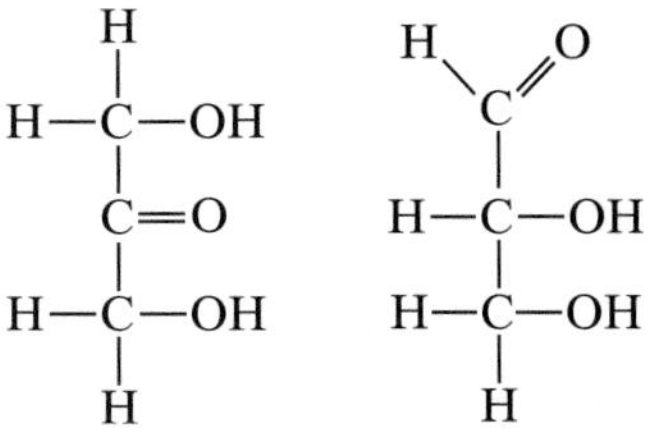

Representations of Molecules

13.7 Indicate whether the following pairs of structures are isomers or different representations of the same molecule.

```
       Br H                  H  H
       |  |                  |  |
(a) H—C—C—Br     and    Br—C—C—Br
       |  |                  |  |
       H  H                  H  H

(b) CH3—CH2      and    CH3—CH2—CH2—Cl
         |
        CH2—Cl

(c) CH3—CH—Cl    and    CH3—CH2—CH2—Cl
         |
        CH3
```

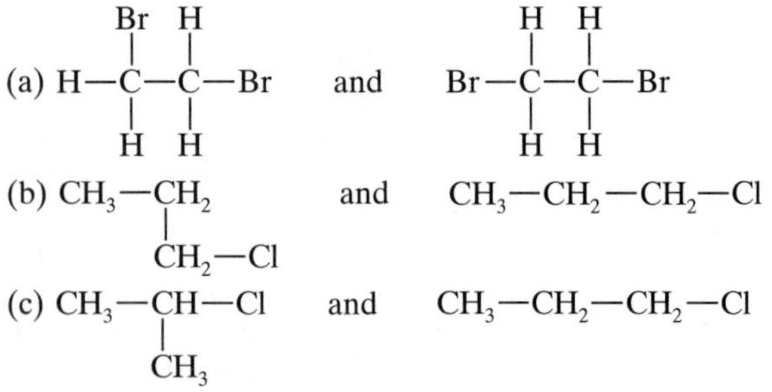

13.8 Indicate whether the following pairs of structures are isomers or different representations of the same molecule.

```
       H  Cl                 H  H
       |  |                  |  |
(a) H—C—C—Br     and    Cl—C—C—Br
       |  |                  |  |
       H  H                  H  H

(b) CH3—CH2      and    CH3—CH—CH3
         |                   |
        CH2—Cl               Cl

(c) CH3CHCH2Cl   and    CH3CH2CH2CH2Cl
       |
      CH3
```

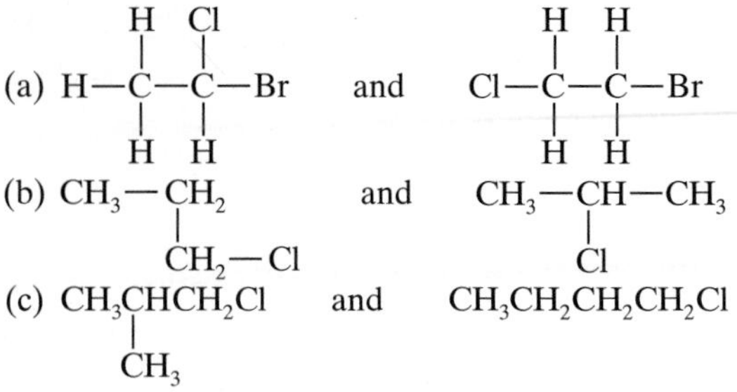

Functional Groups

13.9 Name the class of compounds to which each of the following belongs.

```
            O
            ||
(a) CH3—C—O—CH2—CH3
```
(b) $CH_3{-}CH_2{-}O{-}CH_2{-}CH_3$
(c) $CH_3{-}CH_2{-}CH{=}CH_2$

13.10 Name the class of compounds to which each of the following belongs.

```
                  O
                  ||
(a) CH3—CH2—C—H
```
(b) $CH_3{-}C{\equiv}C{-}CH_3$
(c) $CH_3{-}CH_2{-}CH_2{-}OH$

13.11 Identify the functional groups in each of the following structures.

(a) a compound responsible for odor of the iris

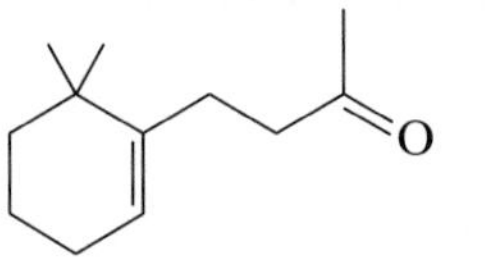

(b) a compound found in clover and grasses

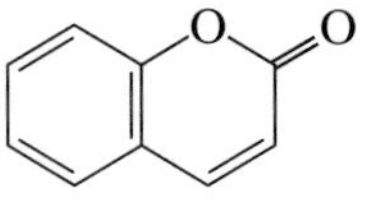

(c) an oil found in citrus fruits

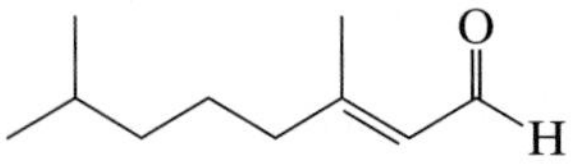

13.12 Identify the functional groups in each of the following structures.

(a) a compound used to produce a type of nylon

(b) a compound used in perfumes

(c) the active ingredient in some insect repellents

Organic Reactions and Mechanisms

13.13 Identify the processes of bond cleavage and bond formation for each of the following reactions.

(a) $CH_4 + \cdot\ddot{Cl}: \longrightarrow H_3C\cdot + H-\ddot{Cl}:$

(b) $H_3C\cdot + :\ddot{Cl}-\ddot{Cl}: \longrightarrow H_3C-\ddot{Cl}: + \cdot\ddot{Cl}:$

13.14 Identify the processes of bond cleavage and bond formation for each of the following reactions.

(a) $(CH_3)_3C^+ + HO^- \longrightarrow (CH_3)_3C-OH$

(b) $(CH_3)_3C-Cl \longrightarrow (CH_3)_3C^+ + Cl^-$

Acids and Bases

13.15 Write the structure of the conjugate acid for each of the following species. Fill in the nonbonding electrons where needed.

(a) $H_2C{=}O$ (b) $CH_3-NH-CH_3$ (c) $CH_3-C{\equiv}N$

13.16 Write the structure of the conjugate base for each of the following species. Fill in the nonbonding electrons where needed.

(a) CH_3-OH (b) CH_3-NH_2 (c) CH_3-SH

13.17 Using Lewis acid–base terminology, describe the reactants in the following reaction. Fill in the nonbonding electrons where needed.

$$CH_3CH_2OCH_2CH_3 + BF_3 \longrightarrow CH_3CH_2O(BF_3)CH_2CH_3$$

13.18 Using Lewis acid–base terminology, describe the reactants in the following reaction. Fill in the nonbonding electrons where needed.

$$(CH_3)_3C^+ + H-O-H \longrightarrow (CH_3)_3C-\overset{+}{O}(H)-H$$

Oxidation–Reduction Reactions

13.19 Identify the following transformations given by unbalanced equations as oxidation, reduction, or neither.

(a) $CH_3-C{\equiv}N \longrightarrow CH_3-CH_2-NH_2$

(b) $2\ CH_3-SH \longrightarrow CH_3-S-S-CH_3$

(c) $CH_3-S-CH_3 \longrightarrow CH_3-S({=}O)-CH_3$

13.20 None of the following unbalanced reactions involves oxidation or reduction although they may appear to be redox reactions. Explain why.

(a) $CH_3-CH{=}CH_2 \longrightarrow CH_3-CH(OH)-CH_3$

(b) $CH_3-C{\equiv}CH \longrightarrow CH_3-C({=}O)-CH_3$

(c) $CH_3-C({=}O)-CH_3 \longrightarrow CH_3-C({=}NH)-CH_3$

13.21 Consider the following reaction for the metabolism of dantrolene, a muscle relaxant. What type of reaction occurs?

13.22 Consider the following reaction for the metabolism of tolmetin, an anti-inflammatory drug. What type of reaction occurs?

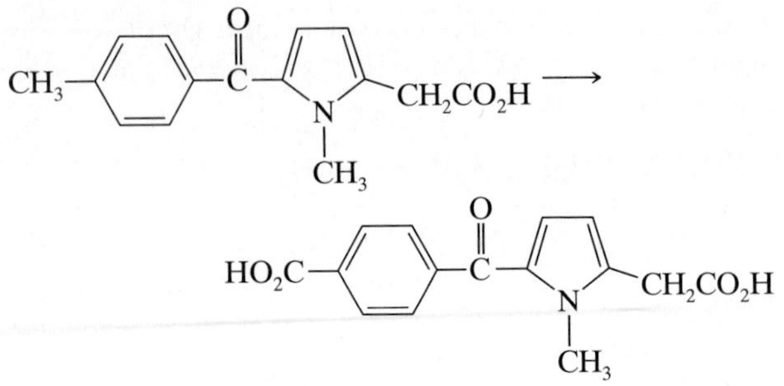

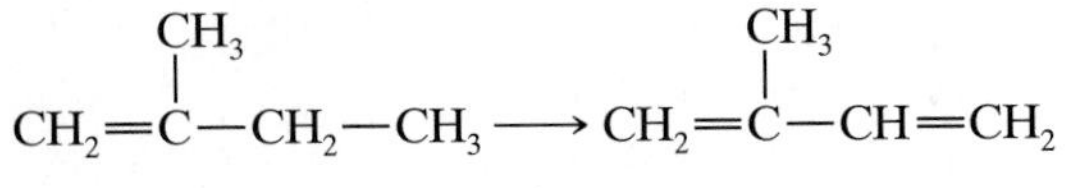

$$CH_3-C_6H_4-C(=O)-(\text{N-methylpyrrol-2-yl})-CH_2CO_2H \longrightarrow HO_2C-C_6H_4-C(=O)-(\text{N-methylpyrrol-2-yl})-CH_2CO_2H$$

Types of Reactions

13.23 The following reaction is used in industry to prepare the monomer chlorotrifluoroethylene, which is used to produce a polymer. What type of reaction occurs? What is the byproduct of the reaction?

$$Cl-CClF-CF_2-Cl \xrightarrow{Zn} ClFC=CF_2$$

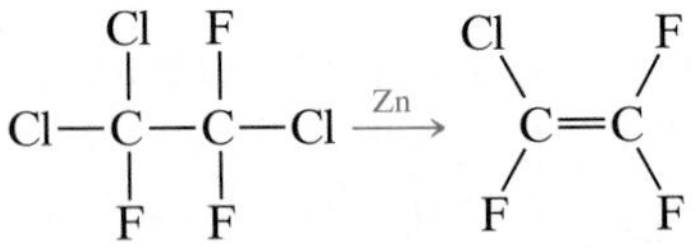

13.24 The following reaction is used in industry to prepare isoprene, which is converted into polymers used for the manufacture of rubber products. What type of reaction occurs? What is the byproduct of the reaction?

$$CH_2=C(CH_3)-CH_2-CH_3 \longrightarrow CH_2=C(CH_3)-CH=CH_2$$

isoprene

13.25 The following series of reactions is used as an industrial synthesis of epichlorohydrin, a compound used in the production of epoxy resins. What type of reaction occurs in each step? What are the byproducts of each step?

$$CH_2=CH-CH_3 \xrightarrow{Cl_2} CH_2=CH-CH_2-Cl \xrightarrow[H_2O]{Cl_2} HO-CH_2-CH(Cl)-CH_2-Cl \xrightarrow{Ca(OH)_2} \overset{O}{CH_2-CH}-CH_2-Cl$$

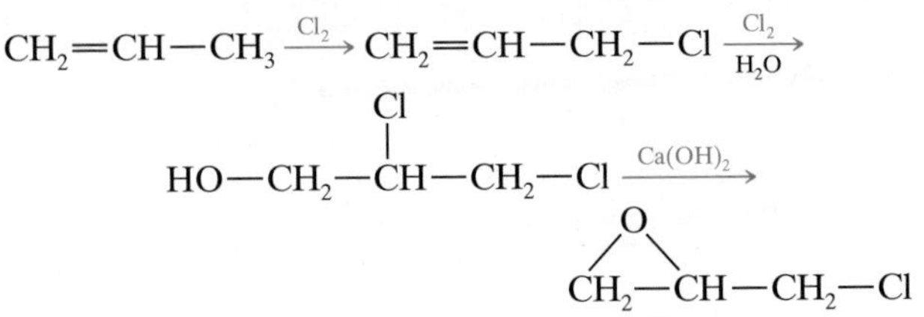

epichlorohydrin

13.26 The following series of reactions is used as an industrial synthesis of vinyl chloride, a compound used in the production of polyvinyl chloride (PVC). What type of reaction occurs in each step? What are the byproducts of each step?

$$CH_2=CH_2 \xrightarrow{Cl_2} Cl-CH_2CH_2-Cl \xrightarrow{heat} CH_2=CH-Cl$$

The separation of crude oil into fractions based on boiling points is a vital process to maintain our industrialized society. (Peter Arnold, Inc.)

14 Alkanes and Cycloalkanes

14.1 Classes of Hydrocarbons
14.2 Alkanes
14.3 Nomenclature of Alkanes
14.4 Cycloalkanes
14.5 Physical Properties of Saturated Hydrocarbons
14.6 Oxidation of Alkanes and Cycloalkanes
14.7 Halogenation of Saturated Hydrocarbons

Essays
- *Physiological Properties of Alkanes*
- *Octane Numbers*
- *Freon, Radicals, and the Ozone Layer*

14.1 Classes of Hydrocarbons

Hydrocarbons are compounds that contain only hydrogen and carbon. They occur in natural gas, petroleum, and coal, which are collectively known as fossil fuels. Hydrocarbons fall into two broad classes based on the types of bonds between carbon atoms. A hydrocarbon that has only carbon–carbon single bonds is **saturated.** Hydrocarbons that contain carbon–carbon multiple bonds are **unsaturated.** Saturated hydrocarbons are of two types: alkanes and cycloalkanes. **Alkanes** have carbon atoms bonded in chains; **cycloalkanes** have carbon atoms bonded to form a ring.

$$CH_3{-}CH_2{-}CH_2{-}CH_3 \qquad \begin{array}{ccc} CH_2 & - & CH_2 \\ | & & | \\ CH_2 & - & CH_2 \end{array}$$

butane (an alkane) — cyclobutane (a cycloalkane)

Compounds that have a chain of carbon atoms, some of which are attached to functional groups, are called **acyclic** compounds, meaning *not cyclic.* Compounds that contain rings of carbon atoms, and that may also contain functional groups, are **carbocyclic** compounds, commonly called cyclic compounds. Some cyclic compounds contain at least one atom in the ring that is not a carbon atom; those atoms are called **heteroatoms.** Cyclic compounds containing one or more heteroatoms are called **heterocyclic** compounds.

an acyclic compound — 2-heptanone (in oil of cloves)

a cyclic compound — carvone (in spearmint oil)

a heterocyclic compound — nicotinic acid (niacin, a B vitamin)

14.2 Alkanes

Hydrocarbons with all carbon atoms in a continuous chain are **normal alkanes.** Their structures are drawn with the carbon chain in a horizontal line.

$$CH_3{-}CH_2{-}CH_2{-}CH_2{-}CH_2{-}CH_2{-}CH_2{-}CH_3$$

octane (a normal alkane)

The names and condensed structural formulas of 20 normal alkanes are given in Table 14.1. The first four compounds have common names. The names of the

TABLE 14.1 Names of Normal Alkanes

Number of Carbon Atoms	Name	Molecular Formula
1	methane	CH_4
2	ethane	C_2H_6
3	propane	C_3H_8
4	butane	C_4H_{10}
5	pentane	C_5H_{12}
6	hexane	C_6H_{14}
7	heptane	C_7H_{16}
8	octane	C_8H_{18}
9	nonane	C_9H_{20}
10	decane	$C_{10}H_{22}$
11	undecane	$C_{11}H_{24}$
12	dodecane	$C_{12}H_{26}$
13	tridecane	$C_{13}H_{28}$
14	tetradecane	$C_{14}H_{30}$
15	pentadecane	$C_{15}H_{32}$
16	hexadecane	$C_{16}H_{34}$
17	heptadecane	$C_{17}H_{36}$
18	octadecane	$C_{18}H_{38}$
19	nonadecane	$C_{19}H_{40}$
20	eicosane	$C_{20}H_{42}$

other compounds are derived from Greek numbers indicating the number of carbon atoms. Each name has the suffix *-ane,* which identifies the compound as an alkane.

Hydrocarbons that have carbon atoms bonded to more than two other carbon atoms are called **branched alkanes.** The carbon atom bonded to three or four other carbon atoms is the branching point. The carbon atom attached to the chain of carbon atoms at the branching point is part of an **alkyl group.** The branched alkane isobutane has three carbon atoms in the chain and one branch, a $-CH_3$ group.

$$CH_3-CH_2-CH_2-CH_3$$

butane

$$CH_3-\underset{\displaystyle CH_3}{\underset{|}{CH}}-CH_3$$

isobutane

Both normal and branched alkanes have the general molecular formula C_nH_{2n+2}. For example, the molecular formula of hexane is C_6H_{14}.

$$\begin{array}{ccccccccccccc} & & H & & H & & H & & H & & H & & H & \\ & & | & & | & & | & & | & & | & & | & \\ H & - & C & - & C & - & C & - & C & - & C & - & C & - H \\ & & | & & | & & | & & | & & | & & | & \\ & & H & & H & & H & & H & & H & & H & \end{array}$$

C_nH_{2n}

$$H_1 + C_6H_{12} + H_1 = C_6H_{14}$$

Each carbon atom in this normal alkane, where $n = 6$, has at least two hydrogen atoms bonded to it, which accounts for the $2n$ in the general formula. Each of the two terminal carbon atoms has another hydrogen atom bonded to it, which accounts for the +2 in the subscript on hydrogen in the general formula.

The molecular formulas of the series of alkanes differ from one another in the number of $-CH_2-$ units. A series of compounds whose members differ from adjacent members by a repeating unit is called a **homologous series.**

EXAMPLE 14.1

One of the components of the wax of a cabbage leaf is a normal alkane containing 29 carbon atoms. What is the molecular formula of the compound?

Solution

The value of n in the general formula for alkanes is 29. There must be $(2 \times 29) + 2$ hydrogen atoms. The molecular formula is $C_{29}H_{60}$.

PROBLEM 14.1

Hectane is a normal alkane with 100 carbon atoms. What is its molecular formula?

Classification of Carbon Atoms

Hydrocarbon structures are classified according to the number of carbon atoms directly bonded to a specific carbon atom. This classification is used to describe the reactivity of functional groups attached at those carbon atoms.

FIGURE 14.1 Classification of Carbon Atoms

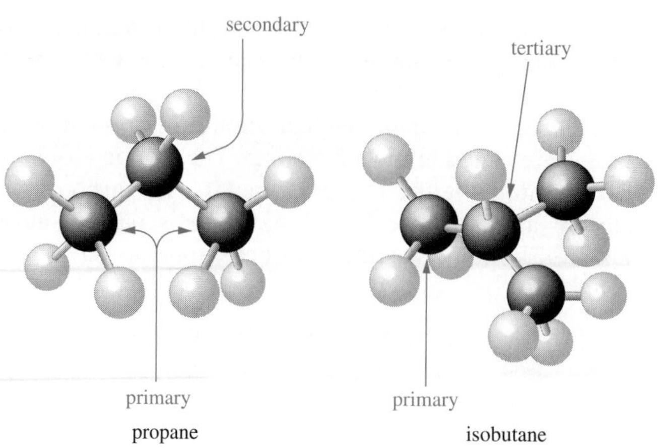

Propane can be transported to fill propane tanks near homes in rural areas. (The Image Bank)

A carbon atom bonded to only one other carbon atom is called a **primary carbon atom.** A primary carbon atom is designated by the symbol 1°. The carbon atom at each end of a carbon chain is primary. For example, ethane and propane each have two primary carbon atoms. In contrast, the middle carbon atom in propane is not primary because it is bonded to two other carbon atoms (Figure 14.1).

A carbon atom that is bonded to two other carbon atoms is a **secondary carbon atom,** designated by the symbol 2°. For example, the middle carbon atom of propane is secondary. A **tertiary carbon atom** is bonded to three other carbon atoms and is designated by 3°. For example, in the structure of isobutane, one of the four carbon atoms is tertiary; the other three are primary (Figure 14.1). A **quaternary carbon atom** is bonded to four other carbon atoms.

EXAMPLE 14.2

The following compound is a sex attractant released by the female tiger moth. Classify the carbon atoms in this compound as primary, secondary, or tertiary.

$$\begin{array}{l} \quad\;\; CH_3 \\ \quad\;\;\; | \\ CH_3CHCH_2CH_2CH_2CH_2CH_2CH_2CH_2CH_2CH_2CH_2CH_2CH_2CH_2CH_2CH_3 \end{array}$$

Solution

The two terminal carbon atoms and the carbon atom in the branching $-CH_3$ group are primary carbon atoms because each is bonded to only one other carbon atom. The second carbon atom from the left is bonded to two atoms in the chain as well as to the branching $-CH_3$ group and so is tertiary. The remaining 14 carbon atoms are bonded to two carbon atoms each and are secondary.

PROBLEM 14.2

Pentaerythritol tetranitrate is used to reduce the frequency and severity of angina attacks. Classify the carbon atoms in this compound.

$$\begin{array}{c} CH_2-O-NO_2 \\ | \\ O_2N-O-CH_2-C-CH_2-O-NO_2 \\ | \\ CH_2-O-NO_2 \end{array}$$

14.3 Nomenclature of Alkanes

Alkanes are named by the rules of the International Union of Pure and Applied Chemistry (IUPAC). When these rules are followed, a unique name describes each compound. The IUPAC name consists of three parts: prefix, parent, and suffix.

prefix—parent—suffix

The **parent** is the longest carbon chain in a molecule. A parent alkane has the ending *-ane.* Other suffixes identify functional groups such as a hydroxyl group. Some functional groups are identified in the prefix. For example, the prefixes chloro and bromo identify chlorine and bromine, respectively.

The prefix also indicates the identity and location of any branching alkyl groups and some functional groups on the parent chain. An alkane that has "lost" one hydrogen atom is called an **alkyl** group. Alkyl groups are named by replacing the *-ane* ending of an alkane with *-yl.* The parent name of CH_4 is methane. Thus, CH_3— is a methyl group. The parent name of C_2H_6 is ethane, so CH_3CH_2— is an ethyl group.

$$\begin{array}{c}\mathrm{H}\\ |\\ \mathrm{H{-}C{-}H}\\ |\\ \mathrm{H}\end{array} \quad \text{removing H gives} \quad \begin{array}{c}\mathrm{H}\\ |\\ \mathrm{H{-}C{-}}\\ |\\ \mathrm{H}\end{array} \quad \text{or} \quad CH_3{-}$$

methane — methyl group

$$\begin{array}{c}\mathrm{H\;\;H}\\ |\;\;\;|\\ \mathrm{H{-}C{-}C{-}H}\\ |\;\;\;|\\ \mathrm{H\;\;H}\end{array} \quad \text{removing H gives} \quad \begin{array}{c}\mathrm{H\;\;H}\\ |\;\;\;|\\ \mathrm{H{-}C{-}C{-}}\\ |\;\;\;|\\ \mathrm{H\;\;H}\end{array} \quad \text{or} \quad CH_3CH_2{-}$$

ethane — ethyl group

The general shorthand representation of an alkyl group is R—, which stands for the "rest" or "remainder" of the molecule.

The names of alkanes specify the length of the carbon chain and the location and identity of alkyl groups attached to it. The IUPAC rules for naming alkanes are as follows.

1. The longest chain of carbon atoms is the parent. Its name is listed in Table 14.1. The longest chain is not always immediately apparent.

$$\begin{array}{l}\qquad\qquad\quad CH_2{-}CH_3\\ \qquad\qquad\quad |\\ CH_3{-}CH_2{-}CH{-}CH_3\end{array}$$

There are five carbon atoms in the longest carbon chain, not four.

If two possible chains have the same number of carbon atoms, the parent is the one with the larger number of branch points.

$$\begin{array}{l}\qquad\qquad\quad CH_2{-}CH_3\\ \qquad\qquad\quad |\\ CH_3{-}CH{-}CH{-}CH_2{-}CH_2{-}CH_3\\ \qquad\;\; |\\ \qquad\;\; CH_3\end{array}$$

The compound should be considered a six-carbon parent chain with two branches, a methyl group and an ethyl group.

$$\begin{array}{l}\qquad\qquad\quad CH_2{-}CH_3\\ \qquad\qquad\quad |\\ CH_3{-}CH{-}CH{-}CH_2{-}CH_2{-}CH_3\\ \qquad\;\; |\\ \qquad\;\; CH_3\end{array}$$

A six-carbon parent chain with only a single three-carbon alkyl group is not a correct choice.

2. Number the carbon atoms in the longest chain starting from the end of the chain nearer the first branch.

$$CH_2{-}CH_3$$
$$CH_3{-}CH{-}CH_2{-}CH{-}CH_2{-}CH_3$$
$$CH_3$$

(Carbons numbered 1–6 from left to right; CH_3 branch on C-2, $CH_2{-}CH_3$ branch on C-4.)

This substituted hexane has a methyl group at C-2 and an ethyl group at C-4, not an ethyl group at C-3 and a methyl group at C-5.

If the first branch occurs at an equal distance from each end of the chain, number from the end that is nearer the second branch.

$$CH_3{-}CH_2 \qquad CH_3$$
$$CH_3{-}CH{-}CH_2{-}CH_2{-}CH{-}CH_2{-}CH{-}CH_3$$
$$CH_3$$

(Carbons numbered 8, 7, 6, 5, 4, 3, 2, 1 from left to right; CH_3 branch on C-7, $CH_3{-}CH_2$ branch on C-4, CH_3 branch on C-2.)

The ethyl group is closest to the right side of the molecule.

3. Each branch or substituent has a number that indicates its location on the parent chain. When two substituents are located on the same carbon atom, each must be assigned the same number.

$$CH_3{-}CH_2 \qquad CH_3$$
$$CH_3{-}CH_2{-}CH{-}CH_2{-}C{-}CH_2{-}CH{-}CH_3$$
$$CH_3 \qquad CH_3$$

(Carbons numbered 8, 7, 6, 5, 4, 3, 2, 1 from left to right; CH_3 on C-6, $CH_3{-}CH_2$ and CH_3 on C-4, CH_3 on C-2.)

This octane has methyl groups on the C-2, C-4, and C-6 atoms and an ethyl group on the C-4 atom.

4. The number for the position of each alkyl group is placed immediately before the name of the group and is joined to the name by a hyphen. Alkyl groups as well as halogen atoms are listed in alphabetical order.

$$CH_2{-}CH_3$$
$$CH_3{-}CH{-}CH_2{-}CH{-}CH_2{-}CH_3$$
$$CH_3$$

(Carbons numbered 1–6 from left to right; CH_3 on C-2, $CH_2{-}CH_3$ on C-4.)

This is 4-ethyl-2-methylhexane, not 2-methyl-4-ethylhexane.

Two or more groups of the same type are indicated by the prefixes di-, tri-, and so forth. The numbers that indicate the locations of the branches are separated by commas.

$$CH_3$$
$$CH_3{-}CH{-}CH_2{-}CH{-}CH_2{-}CH_3$$
$$CH_3$$

(Carbons numbered 1–6 from left to right; CH_3 on C-2 and on C-4.)

This is 2,4-dimethylhexane.

5. The prefixes di-, tri-, tetra, and so forth, do not alter the alphabetical ordering of the alkyl groups.

$$CH_2{-}CH_3$$
$$CH_3{-}CH_2{-}CH{-}CH_2{-}C{-}CH_2{-}CH_2{-}CH_3$$
$$CH_3 \qquad CH_3$$

(Carbons numbered 1–8 from left to right; CH_3 on C-3, $CH_2{-}CH_3$ and CH_3 on C-5.)

This is 5-ethyl-3,5-dimethyloctane, not 3,5-dimethyl-5-ethyloctane.

EXAMPLE 14.3

Name the following compound, which is produced by the *Spirogyra* alga.

$$CH_3 \qquad CH_3 \qquad CH_3 \qquad CH_3$$
$$CH_3CHCH_2CH_2CH_2CHCH_2CH_2CH_2CHCH_2CH_2CH_2CHCH_2CH_3$$

(Each CH_3 is attached to one of the CH carbons in the chain below it.)

Solution

The longest chain has 16 carbon atoms and is named as a substituted hexadecane. The chain is numbered from left to right to locate the four methyl groups at positions 2, 6, 10, and 14. The compound is 2,6,10,14-tetramethylhexadecane.

PROBLEM 14.3

Name the following compound.

$$\begin{array}{rcll} & CH_3 & & \\ & | & & \\ CH_3-CH_2- & C & - & CH-CH_3 \\ & | & & | \\ & CH_3 & & CH_2-CH_3 \end{array}$$

Names of Alkyl Groups

There is only one alkyl group each derived from methane and ethane. However, for a longer chain of carbon atoms, there are isomeric alkyl groups depending on which carbon atom "loses" a hydrogen atom. For example, propane has two primary carbon atoms and a secondary carbon atom. If a primary carbon atom loses a hydrogen atom, a primary alkyl group, propyl, is produced. Propyl and other primary alkyl groups derived from normal alkanes are **normal alkyl groups.** If the 2° carbon atom of propane loses a hydrogen atom, a secondary alkyl group known as the *isopropyl* group is formed.

$$CH_3-CH_2-CH_2- \qquad\qquad \begin{array}{rc} & CH_3 \\ & | \\ CH_3- & CH- \end{array}$$

propyl   isopropyl

Next, let's look at the alkyl groups that can be derived from the two isomers of butane, C_4H_{10}. These alkyl groups all have the formula C_4H_9-. Two alkyl groups are derived from butane and two from isobutane. If a primary carbon atom of butane loses a hydrogen atom, a *butyl* group results; if a secondary carbon atom of butane loses a hydrogen atom, a secondary alkyl group called the *sec-butyl* group forms.

$$\overset{}{CH_3-CH_2-CH_2-\overset{1^\circ}{CH_2}-} \qquad\qquad \begin{array}{rc} & CH_3 \\ & | \\ CH_3-CH_2- & \overset{2^\circ}{CH}- \end{array}$$

butyl   *sec*-butyl

Removal of a hydrogen atom from a primary carbon atom of isobutane gives a primary alkyl group called the *isobutyl* group. Removal of a hydrogen atom from the tertiary carbon atom gives a tertiary alkyl group called the *tert-butyl* (*t*-butyl) group. Thus, there are four isomeric C_4H_9- alkyl groups.

$$\begin{array}{rcl} & CH_3 & \\ & | & \\ CH_3- & CH & -\overset{1^\circ}{CH_2}- \end{array} \qquad\qquad \begin{array}{rc} & CH_3 \\ & | \\ CH_3- & \overset{3^\circ}{C}- \\ & | \\ & CH_3 \end{array}$$

isobutyl   *tert*-butyl

EXAMPLE 14.4

Identify the alkyl group on the left of the benzene ring in ibuprofen, an analgesic present in Nuprin, Advil, and Motrin.

$$CH_3-\underset{|}{\overset{CH_3}{CH}}-CH_2-C_6H_4-\overset{CH_3}{\underset{}{CH}}-C(=O)OH$$

Solution

There are four carbon atoms in the alkyl group, which is derived from isobutane—not butane. The benzene ring is bonded to the terminal carbon atom—not the internal carbon atom. This group is the isobutyl group.

$$CH_3-\overset{CH_3}{\overset{|}{CH}}-CH_2-\text{benzene ring}$$

PROBLEM 14.4

The food preservative BHT has the following structure. Identify the alkyl groups bonded to the benzene ring.

OH

$(CH_3)_3C$ — benzene ring — $C(CH_3)_3$

CH_3

14.4 Cycloalkanes

Cycloalkanes contain only carbon–carbon single bonds and are saturated hydrocarbons. However, they have the general formula C_nH_{2n}, with two fewer hydrogen atoms than alkanes because an additional carbon–carbon bond is needed to form the ring. Cycloalkanes are drawn as simple polygons (Section 5.9) in which the sides represent the carbon–carbon bonds. It is understood that each corner of the polygon is a carbon atom attached to two hydrogen atoms.

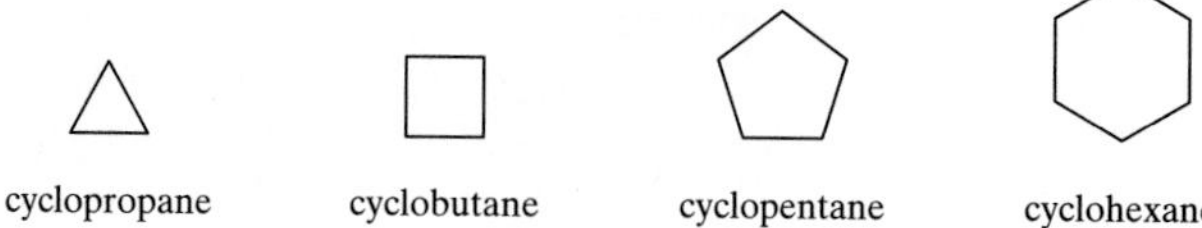

cyclopropane cyclobutane cyclopentane cyclohexane

Multiple rings in a molecule can share one or more common atoms. **Spirocyclic compounds** share one carbon atom between two rings. These compounds are relatively rare in nature. **Fused ring compounds** share two atoms and the bond between them. These compounds are prevalent in steroids, which contain four fused rings. **Bridged ring compounds** share two nonadjacent carbon atoms, which are called the **bridgehead carbon** atoms. These compounds are less prevalent in nature than fused ring compounds.

bridgehead carbon atoms

spirocyclic compound fused ring compound bridged ring compound

EXAMPLE 14.5

Adamantane has a carbon skeleton that is also found as part of the structure of diamond. Amantadine, which contains an amino group (—NH_2) bonded to the adamantane structure, is useful in the prevention of infection by influenza A viruses. What are the molecular formulas of adamantane and amantadine?

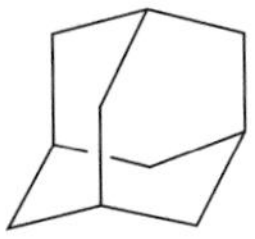

NH_2

adamantane 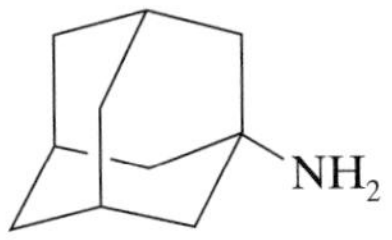 amantadine

Solution

Adamantane has 10 carbon atoms. Four of these carbon atoms are tertiary; they have bonds to three other carbon atoms and one bond to a hydrogen atom. The remaining six carbon atoms are secondary; they have bonds to two other carbon atoms and two bonds to hydrogen atoms. The total number of hydrogen atoms is $4(1) + 6(2) = 16$. The molecular formula of adamantane is $C_{10}H_{16}$.

Amantadine has an amino group (—NH_2) in place of a hydrogen atom at one of the tertiary carbon atoms. Thus, the molecular formula of amantadine differs from adamantane by one nitrogen and one hydrogen atom. The molecular formula is $C_{10}H_{17}N$.

PROBLEM 14.5

Determine the molecular formula of menthol based on the following bond-line structure.

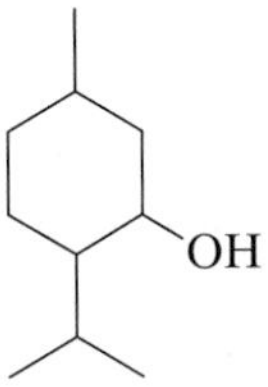

OH

Geometric Isomerism

In Chapter 13 we learned that isomers exist with different carbon skeletons, different functional groups, and different locations of functional groups. Now let us consider a different type of isomerism. Compounds that have the same sequential arrangement of atoms but different spatial arrangements are **geometric isomers.**

Consider cyclopropane, whose three carbon atoms are in a single plane. Any group attached to the ring may be held "above" or "below" the plane of the ring. If two methyl groups on adjacent carbon atoms are on the same side of the plane of the ring, the substance is called a **cis** isomer; it is *cis*-1,2-dimethylcyclopropane. If the two methyl groups are attached on the opposite sides of the plane of the ring, the compound is the **trans** isomer. These are geometric isomers. (Note that when used as prefixes in names of compounds, *cis*- and *trans*- are italicized.) In the following structures the cyclopropane ring is viewed as perpendicular to the plane of the page, and the CH_2 is pointed away from the viewer.

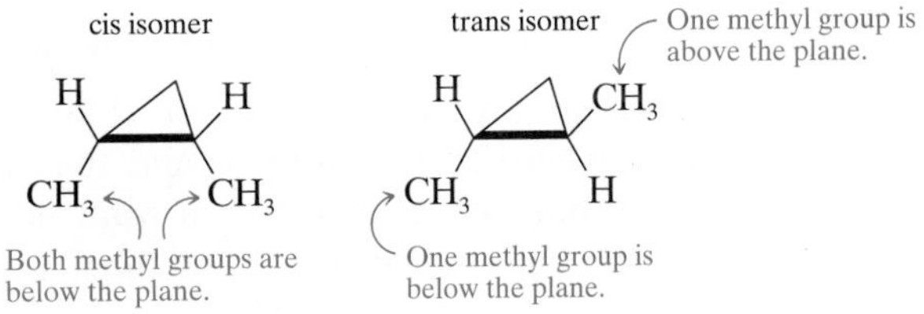

cis isomer

H H

CH_3 CH_3

Both methyl groups are below the plane.

trans isomer

One methyl group is above the plane.

H CH_3

CH_3 H

One methyl group is below the plane.

In the following structures the cyclopropane ring is viewed in the plane of the page. Wedge-shaped lines denote bonds above the plane of the ring and dashed lines show bonds below the plane of the ring.

cis isomer trans isomer

Note that cis and trans compounds are not two conformations of the same molecule but are isomeric substances that have different physical properties. It is impossible to convert one isomer into the other without breaking a bond. Rotation cannot occur about the double bond.

EXAMPLE 14.6

Disparlure, the sex attractant of the female gypsy moth, has the following structure. Are geometric isomers possible for this structure?

$$(CH_3)_2CHCH_2CH_2CH_2CH_2CH{-}CHCH_2CH_2CH_2CH_2CH_2CH_2CH_2CH_2CH_2CH_3$$

(the two CH carbons are joined through O in a three-membered ring)

Solution

The three-membered heterocyclic ring contains two carbon atoms and one oxygen atom. Each carbon atom has a hydrogen atom and a large alkyl group bonded to it. These alkyl groups could be located cis or trans with respect to the plane of the ring. The cis isomer is the biologically active compound.

PROBLEM 14.6

Brevicomin, the sex attractant of a species of pine beetle, has the following structure. Write the structure of a geometric isomer of brevicomin.

Nomenclature of Cycloalkanes

Cycloalkanes are named according to the IUPAC system by using the prefix *cyclo-*. When only one position contains a functional group or alkyl group, only one compound is possible, as in the case of ethylcyclopentane and isopropylcyclobutane.

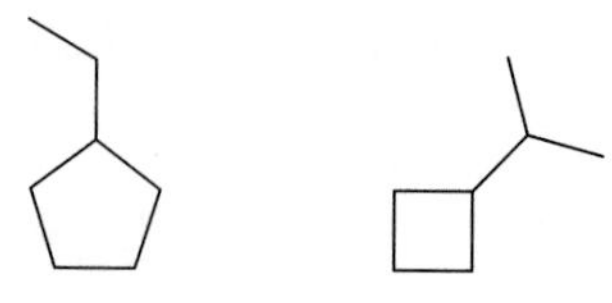

ethylcyclopentane isopropylcyclobutane

When more than one group is attached to the ring, the ring is numbered. One substituent is at position 1, and the ring is numbered in a direction to give the lower number to the position with the next substituent attached to the ring, as in 1,1,4-trichlorocyclodecane and *trans*-1-bromo-3-ethylcyclohexane.

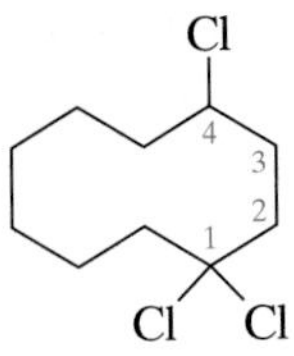

Cl
4 3 2 1
Cl Cl

1,1,4-trichlorocyclodecane

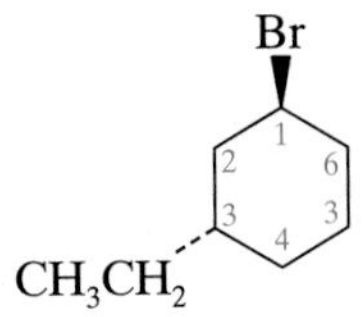

Br
1 2 6 3 3 4
CH_3CH_2

trans-1-bromo-3-ethylcyclohexane

EXAMPLE 14.7

What is the name of the following compound?

Cl Cl

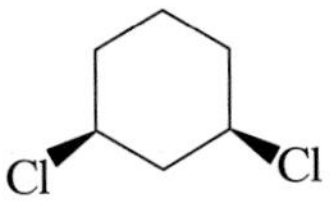

Solution

The ring must be numbered starting from one carbon atom with a chlorine atom and counting toward the other carbon atom with a chlorine atom in the direction that gives the lower number. Starting with the carbon atom at the "4 o'clock" position and numbering clockwise gives the number 3 to the atom at the "8 o'clock" position.

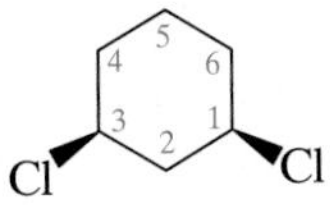

4 5 6 3 2 1
Cl Cl

The two chlorine atoms are on the same side of the plane of the ring, so the correct name is *cis*-1,3-dichlorocyclohexane.

PROBLEM 14.7

What is the name of each of the following compounds?

(a) 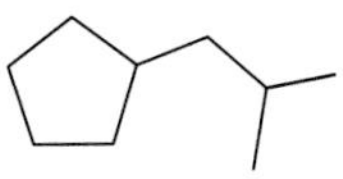 (b) 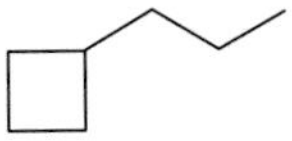 (c) 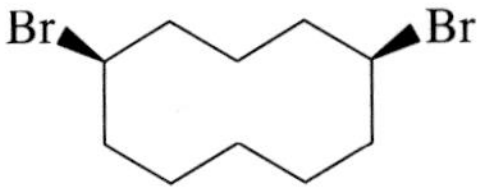 Br Br

Cycloalkanes in Nature

Cycloalkanes containing five- and six-membered rings are common in nature. Multiple fused rings occur in steroids such as cholesterol. Note that there are several sites where the geometry of the groups at the fusion points is indicated.

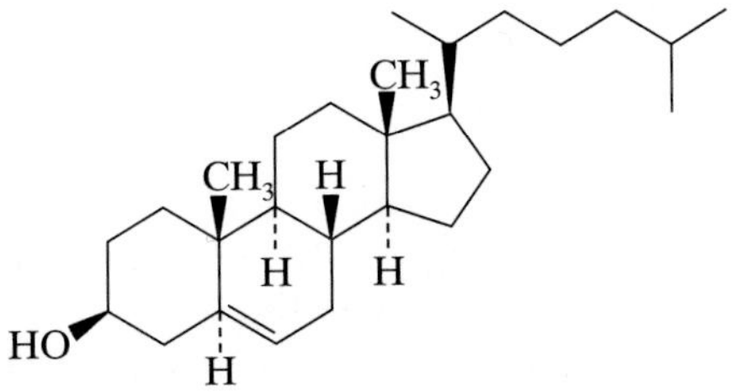

CH_3 CH_3 H H H HO H

cholesterol

The physiological activity of steroids depends not only on the types of functional groups bonded to the rings and their location but also on their geometric relationship to the ring. Steroids will be considered in Chapter 22.

14.5 Physical Properties of Saturated Hydrocarbons

Alkanes have densities between 0.6 and 0.8 g/cm^3, so they are less dense than water. Thus gasoline, which is largely a mixture of alkanes, is less dense than water and will float on water. Pure alkanes are colorless, tasteless, and nearly odorless. However, gasoline does have an odor and some color because dyes are added to gasoline by refiners to indicate its source and composition. Gasoline also contains aromatic compounds (Chapter 15), which have characteristic odors.

Alkanes contain only carbon–carbon and carbon–hydrogen bonds. Because carbon and hydrogen have similar electronegativity values, the C—H bonds are nonpolar. Thus, only London forces govern the physical properties of alkanes such as solubility and boiling point.

Alkanes are not soluble in water, a polar substance. The two substances do not meet the usual criterion of solubility: "like dissolves like." Water molecules are too strongly attracted to each other by hydrogen bonds to allow nonpolar alkanes to slip in between them and dissolve.

The boiling points of the normal alkanes increase with increasing molecular weight (Table 14.2). As the molecular weight increases, London forces increase because more atoms are present to increase the surface area of the molecules. Simply put, there are more points of contact between neighboring molecules, and the London forces are larger.

Physiological Properties of Alkanes

Our bodies may be affected by alkanes in the atmosphere or in liquid petroleum-based products. Under most circumstances, methane is the only alkane in the atmosphere. It constitutes only 0.00002% of "clean" air and is a natural product of biological processes such as decay of vegetable matter in swamps and the flatulence of animals such as cows. Higher local concentrations of alkanes occur in gas leaks, around gasoline pumps, and when we use products such as paint thinner.

Low concentrations of alkanes in the air are not physiologically dangerous. For example, breathing low concentrations of methane does not disturb any physiological processes. However, higher concentrations are dangerous because they decrease the partial pressure of oxygen and consequently may cause asphyxia.

Liquid alkanes, such as mixtures in gasoline, can cause physiological damage depending on how they contact the body. Alkanes dissolve nonpolar organic materials such as fats and oils. Thus, gasoline vapors cause severe damage to lung tissue because they dissolve the fatty material in cell membranes.

Body oils maintain the "moisture" of human skin. Contact of the skin with low molecular weight alkanes removes these skin oils and can cause soreness and blisters. For this reason you should minimize contact with hydrocarbon solvents such as paint thinner.

Higher molecular weight alkanes are emollients (skin softeners). Mineral oil, a mixture of alkanes, can replace natural skin oils that have been washed away. Petroleum jelly is a soft semi-solid that, when applied to the skin, provides a protective film. It is used on baby's skin to protect against diaper rash.

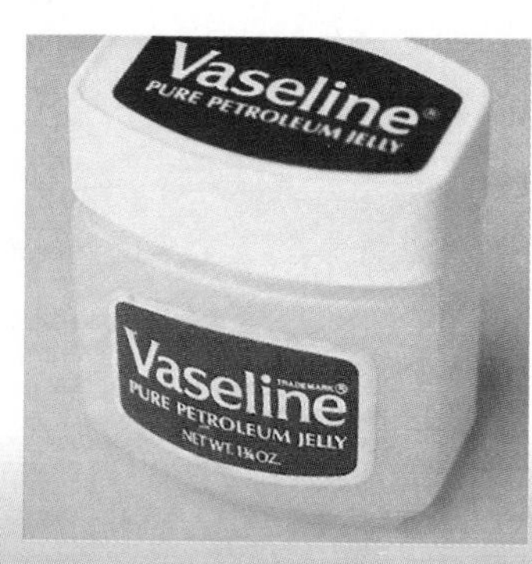

Petroleum jelly is a mixture of high molecular weight alkanes. (Rainbow)

TABLE 14.2 Boiling Points of Alkanes

Name	Boiling Point (°C)	Name	Boiling Point (°C)
methane	−162	hexane	69
ethane	−89	3-methylpentane	63
propane	−42	2-methylpentane	60
butane	−1	2,3-dimethylbutane	58
2-methylpropane	−12	2,2-dimethylbutane	50
pentane	36	heptane	98
2-methylbutane	28	octane	126
2,2-dimethylpropane	10	nonane	151
		decane	174

TABLE 14.3 Physical Properties of Cycloalkanes

Alkane	Boiling Point (°C)	Density (g/mL)
cyclopropane	−32.7	
cyclobutane	12.0	
cyclopentane	49.3	0.7547
cyclohexane	80.7	0.7786
cycloheptane	118.5	0.8098
cyclooctane	148.5	0.8349

Normal alkanes have efficient contact between chains (Section 8.9). A branched alkane is more spherical and has a smaller surface area than a normal alkane. In general, London forces are lower for more spherical isomers. The order of boiling points of the isomeric C_5H_{12} compounds illustrates this phenomenon.

$CH_3—CH_2—CH_2—CH_2—CH_3$
pentane (bp 36 °C)

$CH_3—CH(CH_3)—CH_2—CH_3$
isopentane (bp 28 °C)

$CH_3—C(CH_3)_2—CH_3$
neopentane (bp 10 °C)

The physical properties of a series of cycloalkanes of increasing molecular weight are similar to those of a series of alkanes. The densities increase, as do the boiling points (Table 14.3). The boiling points of the cycloalkanes are higher than those of the alkanes containing the same number of carbon atoms.

14.6 Oxidation of Alkanes and Cycloalkanes

The carbon–carbon and carbon–hydrogen bonds of alkanes and cycloalkanes are not very reactive. Alkanes are also called **paraffins** from the Latin *parum affinis,* "little activity". The carbon–carbon bonds are σ bonds, and the bonding electrons are tightly held between the carbon atoms. The carbon–hydrogen bonds are located about the carbon skeleton and are more susceptible to reaction but usually do so only under extreme conditions. One such process is oxidation.

Octane Numbers

In an automobile engine the fuel and air are drawn into the cylinder on its downward stroke, and the piston compresses the mixture on the upward stroke. Ideally, the mixture ignites at the top of the stroke. The resulting explosion drives the piston downward. Normal alkanes are not suitable fuels in an automobile engine because they ignite prematurely and uncontrollably. A knocking or pinging sound results, which indicates that a force is resisting the upward motion of the piston. Branched hydrocarbons burn smoothly and are more efficient fuels.

The burning efficiency of gasoline is rated by an octane number scale (see the table). An octane number of 100 is assigned to 2,2,4-trimethylpentane, an excellent fuel. Heptane, a poor fuel, has an octane number of zero. Gasoline with the same burning characteristics as a 90% mixture of 2,2,4-trimethylpentane and 10% heptane is rated at 90 octane. Hydrocarbons that burn more efficiently than 2,2,4-trimethylpentane have octane numbers greater than 100. Hydrocarbons that burn less efficiently than heptane have negative octane numbers. Octane numbers decrease with increasing molecular weight. In isomeric compounds, increased branching increases the octane number (see the table).

All gasoline pumps display the octane number of the gasoline. (Fundamental Photographs)

Octane Numbers of Alkanes

Formula	Compound	Octane Number
C_4H_{10}	butane	94
C_5H_{12}	pentane	62
	2-methylbutane	94
C_6H_{14}	hexane	25
	2-methylpentane	73
	2,2-dimethylbutane	92
C_7H_{16}	heptane	0
	2-methylhexane	42
	2,3-dimethylpentane	90
C_8H_{18}	octane	−19
	2-methylheptane	22
	2,3-dimethylhexane	71
	2,2,4-trimethylpentane	100

Methane, the major component of natural gas, yields 212 kcal/mole when burned. Although the reaction is spontaneous, a spark or flame is required to provide the activation energy for the reaction. Thus, natural gas can accumulate from a gas leak and not explode. But the gas–oxygen mixture is very dangerous.

$$CH_4 + 2\,O_2 \longrightarrow CO_2 + 2\,H_2O \qquad \Delta H^\circ = -212 \text{ kcal}$$

As in the case of natural gas leaks, vapors of any alkane can form explosive mixtures with air. Gasoline can safely be stored in closed containers such as gas tanks and pipelines. However, a spark or flame in the vicinity can cause the hydrocarbon vapors to explode. For this reason, you should not smoke while filling your automobile gas tank, lawn mower, or outboard motor with gasoline.

14.7 Halogenation of Saturated Hydrocarbons

Alkanes react with halogens at high temperature or in the presence of light to give substitution products which have halogen atoms in place of hydrogen atoms. For example, methane reacts with chlorine when heated to a high temperature or when exposed to ultraviolet light.

$$CH_4 + Cl_2 \longrightarrow CH_3Cl + HCl$$

The reaction is difficult to control and the product can continue to react with additional chlorine to produce several substitution products.

$$CH_3Cl + Cl_2 \longrightarrow \underset{\substack{\text{dichloromethane} \\ \text{(methylene chloride)}}}{CH_2Cl_2} + HCl$$

$$CH_2Cl_2 + Cl_2 \longrightarrow \underset{\substack{\text{trichloromethane} \\ \text{(chloroform)}}}{CHCl_3} + HCl$$

$$CHCl_3 + Cl_2 \longrightarrow \underset{\substack{\text{tetrachloromethane} \\ \text{(carbon tetrachloride)}}}{CCl_4} + HCl$$

Chlorination of higher molecular weight alkanes yields many monosubstituted products. The chlorine atom is not selective in its substitution for hydrogen atoms. Thus, reaction with pentane yields 1-chloropentane, 2-chloropentane, and 3-chloropentane. A large number of polysubstituted products is also possible.

Halogenated hydrocarbons are used for many industrial purposes. Unfortunately, many of these compounds can cause liver damage and cancer. In the past, chloroform was used as an anesthetic, and carbon tetrachloride was used as a dry-cleaning solvent. They are no longer used for these purposes.

Mechanism of Chlorination of Alkanes

The mechanism of chlorination involves homolytic bond cleavage and homogenic bond formation. In the first step the chlorine molecule absorbs either heat energy or light energy and the bond breaks to give two chlorine radicals. These species are electron-deficient and highly reactive. The chlorination reaction can occur only after this initial homolytic bond cleavage, which has a large activation energy. This step that starts the reaction is the **initiation step.**

$$\text{Step 1} \quad :\ddot{\underset{..}{Cl}}-\ddot{\underset{..}{Cl}}: \longrightarrow :\ddot{\underset{..}{Cl}}\cdot + \cdot\ddot{\underset{..}{Cl}}:$$

Two steps, collectively known as **propagation steps**, then occur. These steps involve radicals as reactants and products and have low activation energies.

$$\text{Step 2} \quad H-\overset{\displaystyle H}{\underset{\displaystyle H}{\overset{|}{\underset{|}{C}}}}-H + \cdot\ddot{\underset{..}{Cl}}: \longrightarrow \underset{\text{methyl radical}}{H-\overset{\displaystyle H}{\underset{\displaystyle H}{\overset{|}{\underset{|}{C}}}}\cdot} + H-\ddot{\underset{..}{Cl}}:$$

$$\text{Step 3} \quad \underset{\text{methyl radical}}{H-\overset{\displaystyle H}{\underset{\displaystyle H}{\overset{|}{\underset{|}{C}}}}\cdot} + :\ddot{\underset{..}{Cl}}-\ddot{\underset{..}{Cl}}: \longrightarrow H-\overset{\displaystyle H}{\underset{\displaystyle H}{\overset{|}{\underset{|}{C}}}}-\ddot{\underset{..}{Cl}}: + \cdot\ddot{\underset{..}{Cl}}:$$

In step 2 a C—H bond is broken and an H—Cl bond is produced; in step 3 a Cl—Cl bond is broken and a C—Cl bond is formed. Furthermore, in each step a radical reacts and a radical is produced. One radical generates another in this chain propagation sequence. Therefore, only a small number of chlorine atoms are required to initiate the process, which continues as long as there are reactants.

EXAMPLE 14.8

How many mono-, di-, and trichlorinated compounds result from the chlorination of ethane?

Solution

The two carbon atoms in ethane are equivalent, and only one monochlorinated compound, CH_3CH_2Cl, can result. The two carbon atoms in this product are not equivalent. Substitution by a second chlorine atom results in two isomers.

Freon, Radicals, and the Ozone Layer

Halogenated alkanes are less flammable than alkanes. Extensively halogenated compounds such as carbon tetrachloride will not burn at all. At one time carbon tetrachloride was used in fire extinguishers to provide an inert atmosphere to prevent oxygen from reaching the flames. Reduced combustibility makes haloalkanes useful for many purposes. For example, hydrocarbons have been used as refrigerants and as aerosol propellants, but the danger of combustion is a serious drawback. Thus, certain halogenated alkanes are now used. Freon-12 is used in air conditioners. It was also used until recently as an aerosol propellant.

$$\begin{array}{ccc} & F & \\ & | & \\ F- & C & -Cl \\ & | & \\ & Cl & \end{array}$$

Unfortunately, although inert on Earth, Freon-12 decomposes in the stratosphere to produce radicals. Ultraviolet radiation ruptures the carbon–chlorine bond.

$$CF_2Cl_2 \xrightarrow{\text{UV light}} :\ddot{Cl}\cdot + \cdot CF_2Cl$$

This process is partially responsible for the destruction of the ozone layer in the stratosphere. Ozone in the stratosphere protects us from solar ultraviolet radiation, which splits ozone molecules into molecular oxygen and atomic oxygen. These products then recombine and release heat energy.

$$:\ddot{O}=\ddot{O}-\ddot{O}: \xrightarrow{\text{UV light}} :\ddot{O}=\ddot{O}: + \cdot\ddot{O}\cdot$$

$$:\ddot{O}=\ddot{O}: + \cdot\ddot{O}\cdot \longrightarrow :\ddot{O}=\ddot{O}-\ddot{O}: + \text{heat energy}$$

As a result of the combination of the two reactions, the Earth is protected from extensive doses of ultraviolet radiation that would be harmful to life and would increase the incidence of skin cancer.

The chlorine radical from Freon-12 reacts with ozone in the stratosphere, producing ClO, which reacts with atomic oxygen.

$$:\ddot{O}=\ddot{O}-\ddot{O}: + \cdot\ddot{Cl}: \longrightarrow :\ddot{O}=\ddot{O}: + \cdot\ddot{O}-\ddot{Cl}:$$

$$\cdot\ddot{O}\cdot + \cdot\ddot{O}-\ddot{Cl}: \longrightarrow :\ddot{O}=\ddot{O}: + \cdot\ddot{Cl}:$$

Note that a chlorine radical reacts in the first equation and is a product in the second equation. Thus, the chlorine is a catalyst for the destruction of ozone. The net reaction of these two steps is the destruction of an ozone molecule.

$$:\ddot{O}=\ddot{O}-\ddot{O}: + \cdot\ddot{O}\cdot \longrightarrow 2\,:\ddot{O}=\ddot{O}:$$

These reactions remove ozone and atomic oxygen from the atmosphere, and the protection provided by the ozone layer is diminished. The destruction of the ozone layer is pronounced in the Antarctic; a similar effect has recently been detected in the Arctic region. Destruction of the ozone layer is now occurring at midlatitudes as well, where more of the Earth's inhabitants live. As the amount of radiation reaching the Earth's surface increases, the incidence of skin cancer may increase.

The summit of Mt. Shinn in Antarctica at 16,000 ft is exposed to ultraviolet radiation as a result of a depleted ozone layer. (Peter Arnold Inc.)

CH_3CHCl_2 $ClCH_2CH_2Cl$

1,1-dichloroethane 1,2-dichloroethane

In subsequent reactions of these products, three chlorine atoms may be located on a single carbon atom. Another product has two chlorine atoms on the same carbon atom and one on the other.

CH_3CCl_3 $ClCH_2CHCl_2$

1,1,1,-trichloroethane 1,1,2-trichloroethane

PROBLEM 14.8

How many mono- and dichlorinated compounds result from chlorination of cyclobutane?

Summary

Saturated hydrocarbons contain only carbon–carbon single bonds and carbon–hydrogen bonds. Each carbon atom of a saturated hydrocarbon has four sp^3 hybrid orbitals forming four single covalent bonds with other atoms. The atoms bonded to each carbon atom form a tetrahedron.

There are two classes of saturated hydrocarbons: **alkanes** and **cycloalkanes.** The general molecular formula for an alkane is C_nH_{2n+2}. A series of compounds, such as alkanes, whose members differ from adjacent members by a repeating unit is called a homologous series. Each ring of a cycloalkane has two fewer hydrogen atoms than an alkane with the same number of carbon atoms. The general formula for a cycloalkane is C_nH_{2n}.

Normal alkanes are hydrocarbons with a continuous chain of carbon atoms. **Branched alkanes** have carbon atoms bonded to more than two other carbon atoms.

A carbon atom is classified according to the number of other carbon atoms that are attached to it. A **primary carbon atom** is bonded to only one other carbon atom. A **secondary carbon atom** is bonded to two other carbon atoms. A **tertiary carbon atom** is bonded to three other carbon atoms. A **quaternary carbon atom** is bonded to four other carbon atoms.

The rules of the International Union of Pure and Applied Chemistry (IUPAC) are used to name organic compounds. An alkane is named using the longest continuous chain of carbon atoms as the parent compound. The partial structure that remains after one hydrogen atom is removed from an alkane is an **alkyl group.** The name of the alkyl group is derived from the alkane containing the same number of carbon atoms. Both the position and the identity of the alkyl group are prefixed to the parent name for a branched alkane.

In **geometric** or **cis–trans isomers,** atoms are bonded to each other in the same sequence but are oriented differently in space. Cis–trans isomers cannot be interconverted by rotation about the carbon–carbon bonds. A substituted cycloalkane that has two substituents on the same side of a ring is called the cis isomer, whereas a cycloalkane with the two substituents on opposite sides of the ring is the trans isomer.

Saturated hydrocarbons are not soluble in water and have a lower density than water. Saturated hydrocarbons have only weak London forces between molecules, and the compounds have low boiling points. Boiling points of alkanes increase with increasing molecular weight and decrease with increasing branching of isomeric substances.

Saturated hydrocarbons are quite unreactive. The burning efficiency of gasoline is rated on an **octane number** scale. The carbon–hydrogen bond undergoes a **substitution reaction** in which a hydrogen atom is replaced by a halogen atom. Halogen-substituted hydrocarbons are less flammable than hydrocarbons. The mechanism of chlorination involves free radicals in a series of **chain-propagating steps.**

Exercises

Molecular Formulas

14.1 Beeswax contains approximately 10% hentriacontane, which is a normal alkane with 31 carbon atoms. What is the molecular formula for hentriacontane? Write a completely condensed formula for hentriacontane.

14.2 Currently the largest known normal alkane has 390 carbon atoms. What is the molecular formula for this alkane? Write a completely condensed formula for it.

Structural Formulas

14.3 Redraw each of the following so that the longest continuous chain is written horizontally.

```
(a) CH3—CH2
        |
        CH2—CH3
(b) CH2—CH2—CH—CH2—CH3
    |       |
    CH3     CH2—CH3
(c) CH3—CH—CH2—CH3
        |
    CH3—CH2
```

14.4 Redraw each of the following so that the longest continuous chain is written horizontally.

```
(a) CH3—CH—CH2
        |   |
        CH3 CH3
(b) CH3—CH—CH2—CH2
        |      |
    CH3—CH2    CH3
(c) CH3—CH—CH2—CH3
        |
    CH3—CH—CH2—CH3
```

14.5 Which of the following represent the same compound?

```
CH3—CH—CH—CH2—CH3
    |  |
    CH3 CH2—CH2—CH3
        (I)

CH3—CH—CH—CH2—CH2
    |  |         |
    CH3 CH2—CH3  CH3
        (II)

CH3—CH—CH2—CH3
    |
CH3—CH—CH2—CH2—CH3
        (III)

CH3—CH—CH—CH3
    |  |
CH3—CH2 CH2—CH2—CH3
        (IV)
```

14.6 Which of the following represent the same compound?

```
CH3—CH—CH—CH2—CH3
    |  |
    CH3 CH3
        (I)

CH3—CH—CH2—CH2
    |       |
CH3—CH2     CH3
        (II)

CH3—CH—CH2—CH3
    |
CH3—CH—CH3
        (III)

CH3—CH—CH2
    |  |
CH3—CH2 CH2—CH3
        (IV)
```

Alkyl Groups

14.7 What is the common name for each of the following alkyl groups?

```
(a) CH3—CH2              (b) CH3—CH—CH2—CH3
        |                        |
        CH2—
(c) CH3—CH—CH2—
        |
        CH3
```

14.8 What is the common name for each of the following alkyl groups?

```
(a) CH3—CH—              (b) CH3—CH2—CH2—CH2—
        |
        CH3

        |
(c) CH3—C—CH3
        |
        CH3
```

Classification of Carbon Atoms

14.9 Classify each carbon atom in the following compounds.

```
(a) CH3—CH2—CH2—CH2—CH3
(b) CH3—CH2—CH—CH2—CH3
            |
            CH3
        CH3
        |
(c) CH3—C—CH2—CH3
        |
        CH3
```

14.10 Classify each carbon atom in the following compounds.

```
(a) CH3—CH—CH2—CH—CH3
        |      |
        CH3    CH3
(b) CH3—CH2—CH2—CH3
        CH3 CH3
        |   |
(c) CH3—C———C—CH3
        |   |
        CH3 CH3
```

14.11 Determine the number of primary, secondary, tertiary, and quaternary carbon atoms in each of the following compounds.

$$\text{(a)}\ CH_3-\overset{\displaystyle CH_3}{\overset{|}{\underset{\underset{\displaystyle CH_3}{|}}{C}}}-CH_3$$

$$\text{(b)}\ CH_3-\underset{\underset{\displaystyle CH_3}{|}}{CH}-CH_2-CH_3$$

$$\text{(c)}\ CH_3-\underset{\underset{\displaystyle CH_3}{|}}{CH}-CH_2-CH_2-CH_3$$

14.12 Determine the number of primary, secondary, tertiary, and quaternary carbon atoms in each of the following compounds.

$$\text{(a)}\ CH_3-\overset{\displaystyle CH_3}{\overset{|}{\underset{\underset{\displaystyle CH_3}{|}}{C}}}-CH_2-\overset{\displaystyle CH_3}{\overset{|}{\underset{\underset{\displaystyle CH_3}{|}}{C}}}-CH_3$$

$$\text{(b)}\ CH_3-\underset{\underset{\displaystyle CH_3}{|}}{CH}-CH_2-\underset{\underset{\displaystyle CH_3}{|}}{CH}-CH_3$$

$$\text{(c)}\ CH_3-CH_2-\underset{\underset{\displaystyle CH_3}{|}}{CH}-CH_2-CH_3$$

Nomenclature of Alkanes

14.13 Give the IUPAC name for each of the following compounds.

$$\text{(a)}\ CH_3-\underset{\underset{\displaystyle CH_2-CH_3}{|}}{CH}-CH_3$$

$$\text{(b)}\ \underset{\underset{\displaystyle CH_3}{|}}{CH_2}-CH_2-\underset{\underset{\displaystyle CH_3}{|}}{CH}-CH_2-CH_3$$

$$\text{(c)}\ CH_3-\underset{\underset{\displaystyle CH_3}{|}}{CH}-CH_2-\underset{\underset{\displaystyle CH_3}{|}}{CH_2}$$

$$\text{(d)}\ CH_3-CH_2-\underset{\underset{\displaystyle CH_2-CH_2-CH_2}{|}}{CH}-CH_3\quad \overset{\displaystyle CH_3}{|}$$

14.14 Give the IUPAC name for each of the following compounds.

$$\text{(a)}\ CH_3-\underset{\underset{\displaystyle CH_3}{|}}{CH}-\underset{\underset{\displaystyle CH_2-CH_2-CH_3}{|}}{CH}-CH_3$$

$$\text{(b)}\ CH_3-\underset{\underset{\displaystyle CH_3}{|}}{CH}-\underset{\underset{\displaystyle CH_2-CH-CH_3}{|}}{CH_2}\quad \overset{\displaystyle CH_3}{|}$$

$$\text{(c)}\ CH_3-\underset{\underset{\displaystyle CH_3-CH-CH_2-CH_2-CH_3}{|}}{CH}-CH_2-CH_3$$

$$\text{(d)}\ CH_3-\underset{\underset{\displaystyle CH_3-CH_2}{|}}{CH}-CH_2-\underset{\underset{\displaystyle CH_2-CH_3}{|}}{CH}-CH_3$$

14.15 Write the structural formula for each of the following compounds.

(a) 3-methylpentane (b) 3,4-dimethylhexane
(c) 2,2,3-trimethylpentane (d) 4-ethylheptane

14.16 Write the structural formula for each of the following compounds.

(a) 2-methylpentane (b) 3-ethylhexane
(c) 2,2,4-trimethylhexane (d) 2,4-dimethylheptane

Isomers

14.17 There are nine isomeric C_7H_{16} compounds. Which isomers have a single methyl group as a branch and are named as methyl substituted hexanes?

14.18 There are nine isomeric C_7H_{16} compounds. Which isomers have two methyl groups as branches and are named as dimethyl-substituted pentanes?

Cycloalkanes

14.19 Write bond-line structures for each of the following compounds.

(a) chlorocyclopropane (b) 1,1-dimethylcyclobutane
(c) cyclooctane

14.20 Write bond-line structures for each of the following compounds.

(a) bromocyclopentane (b) 1,1-dichlorocyclopropane
(c) cyclopentane

14.21 Name each of the following compounds.

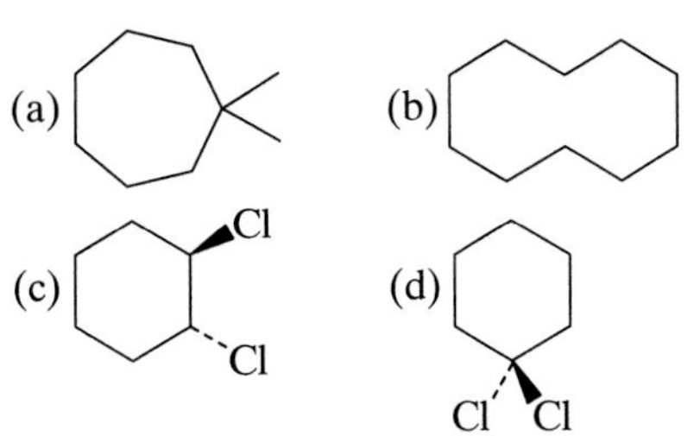

14.22 Name each of the following compounds.

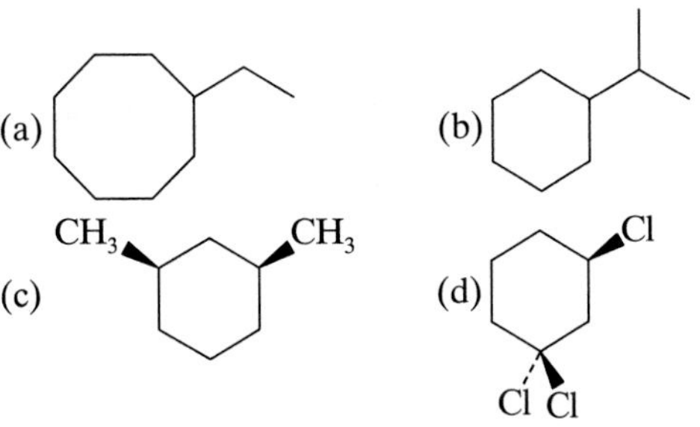

14.23 What is the molecular formula for each of the following compounds?

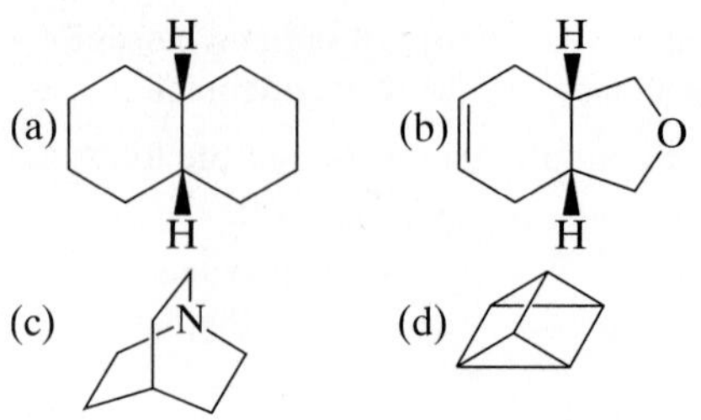

14.24 What is the molecular formula for each of the following compounds?

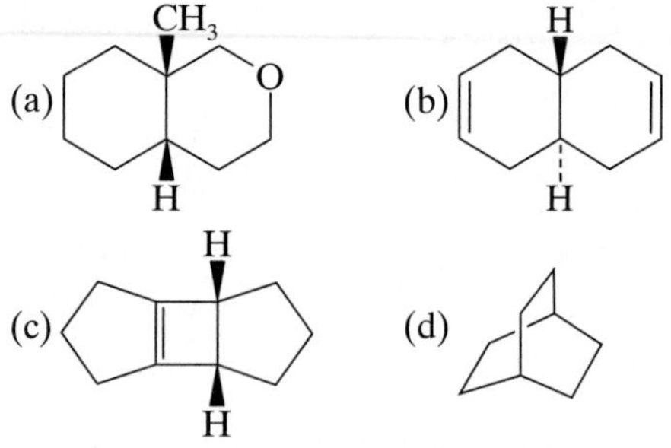

Properties of Hydrocarbons

14.25 Cyclopropane is an anesthetic, but it cannot be used in operations in which electrocauterization of tissue is done. Why?

14.26 Which compound should have the higher octane number, cyclohexane or methylcyclopentane?

14.27 Which of the isomeric C_8H_{18} compounds has the highest boiling point? Which has the lowest boiling point?

14.28 The boiling point of methylcyclopentane is lower than the boiling point of cyclohexane. Suggest a reason why.

Halogenation of Hydrocarbons

14.29 How many products can result from the substitution of a chlorine atom for one hydrogen atom in each of the following compounds?

(a) propane (b) butane
(c) methylpropane (d) pentane

14.30 How many products can result from the substitution of a chlorine atom for one hydrogen atom in each of the following compounds?

(a) 2-methylbutane (b) 2,2-dimethylbutane
(c) 2,3-dimethylbutane (d) 3-methylpentane

14.31 Halothane, an anesthetic, has the formula C_2HF_3ClBr. Draw structural formulas for the four possible isomers with this molecular formula.

14.32 A saturated refrigerant has the molecular formula C_4F_8. Draw structural formulas for two possible isomers of this compound.

15 Unsaturated Hydrocarbons

Compact disks are made from plastics produced from the polymerization of unsaturated hydrocarbons. (Rainbow)

15.1 Types of Unsaturated Hydrocarbons
15.2 Geometric Isomerism
15.3 Nomenclature of Alkenes
15.4 Oxidation and Reduction of Alkenes
15.5 Addition Reactions of Alkenes
15.6 Polymerization of Alkenes
15.7 Nomenclature of Alkynes
15.8 Reactions of Alkynes
15.9 Aromatic Hydrocarbons
15.10 Nomenclature of Aromatic Hydrocarbons
15.11 Polycyclic Aromatic Hydrocarbons
15.12 Heterocyclic Aromatic Compounds

Essays
- *Physiologically Active Alkynes*
- *Geometric Isomers and the Sex Life of Moths*
- *Carcinogenic Aromatic Hydrocarbons*

15.1 Types of Unsaturated Hydrocarbons

Organic compounds with carbon–carbon multiple bonds have fewer hydrogen atoms than structurally related alkanes or cycloalkanes. For this reason these compounds are said to be unsaturated. There are three main classes of unsaturated compounds based on the type of carbon–carbon multiple bonds.

1. **Alkenes** are compounds that contain a carbon–carbon double bond.
2. **Alkynes** are compounds that contain a carbon–carbon triple bond.
3. **Aromatic** hydrocarbons are compounds that contain a benzene ring or structural units like a benzene ring.

All classes of unsaturated compounds have pi (π) bonds. The π bond is the result of a side-by-side overlap of p orbitals on adjacent carbon atoms (see Section 5.8). Both the structure and the chemical reactivity of unsaturated molecules are a consequence of the π bond.

FIGURE 15.1 Molecular Models of Unsaturated Hydrocarbons

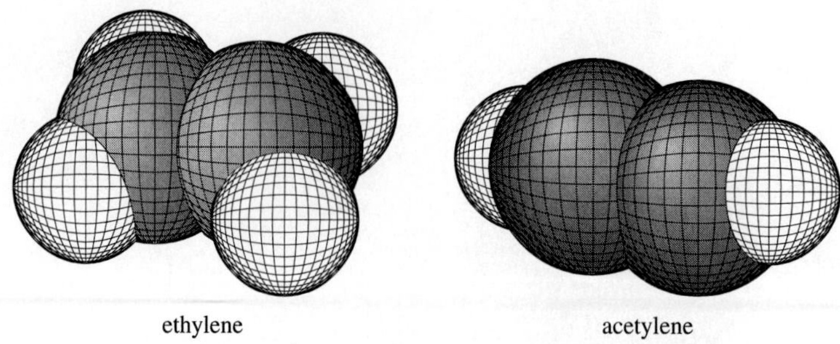

ethylene acetylene

Alkenes

The simplest alkene, C_2H_4, is called ethylene; its IUPAC name is ethene. A space-filling model of ethylene is shown in Figure 15.1. All six atoms, two carbon atoms and four hydrogen atoms, are in the same plane. In perspective structural formulas the plane may be either in the plane of the page or perpendicular to it. If the plane is perpendicular to the printed page, the carbon–hydrogen bonds project in front of and behind the page. Wedge-shaped lines represent the bonds in front of the page, and dashed lines those behind the page.

```
H       H            H.         .H
 \     /               `C=C´
  C=C                H◄     ►H
 /     \
H       H
```

in plane of page perpendicular to plane of page

The double bond decreases the number of hydrogen atoms in a molecule by two compared with the number in the corresponding alkane, so the general formula for an alkene is C_nH_{2n}. Note that this is also the general formula for cycloalkanes, which are thus isomeric with alkenes. Each additional double bond reduces the number of hydrogen atoms by another two.

Polyenes

Some alkenes, called **polyenes,** contain two or more carbon–carbon double bonds. Alkenes with two, three, and four double bonds are called dienes, trienes, and tetraenes, respectively. Multiple double bonds are present in polyunsaturated oils. These substances are esters of unsaturated carboxylic acids such as linolenic acid. A tetraene unit occurs in arachidonic acid, a precursor to physiologically active molecules called prostaglandins (Chapter 22).

$$CH_3CH_2CH{=}CHCH_2CH{=}CHCH_2CH{=}CH(CH_2)_7CO_2H$$

linolenic acid

$$CH_3(CH_2)_4CH{=}CHCH_2CH{=}CHCH_2CH{=}CHCH_2CH{=}CH(CH_2)_3CO_2H$$

arachidonic acid

β-Carotene, found in carrots, and vitamin A, which is derived from β-carotene, are polyunsaturated. β-Carotene is converted to vitamin A in mammals by an enzyme-catalyzed reaction that oxidizes β-carotene into two molecules of vitamin A.

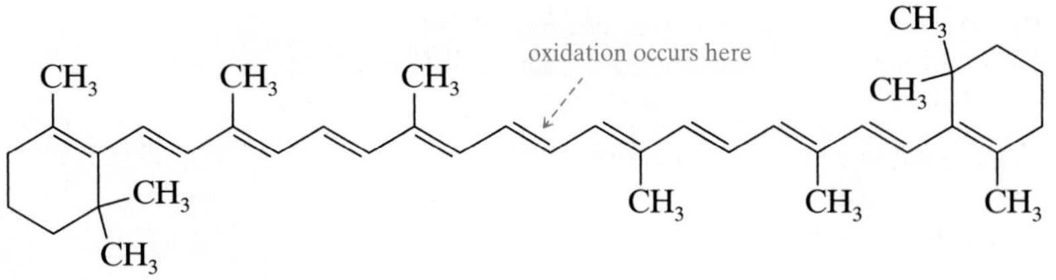

oxidation occurs here

β-carotene

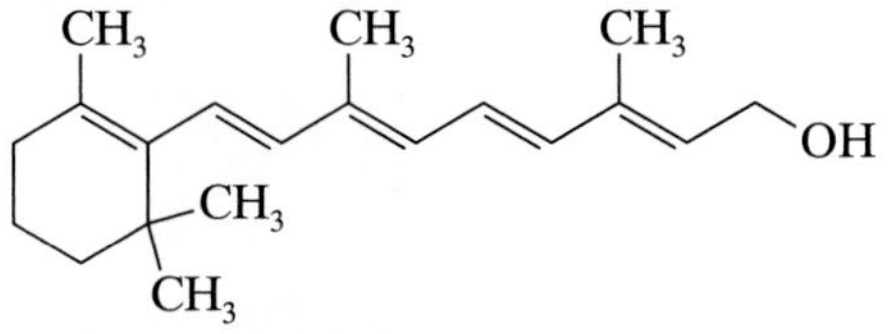

vitamin A

EXAMPLE 15.1

What is the molecular formula for a hydrocarbon with six carbon atoms that contains a ring and a double bond?

Solution

A ring decreases the number of hydrogen atoms by two compared with an alkane, to give C_nH_{2n}. For the double bond another two hydrogen atoms must be subtracted to give C_nH_{2n-2}. For six carbon atoms the molecular formula must be C_6H_{10}.

PROBLEM 15.1

What is the molecular formula for farnesene, a compound found in the coating of apples?

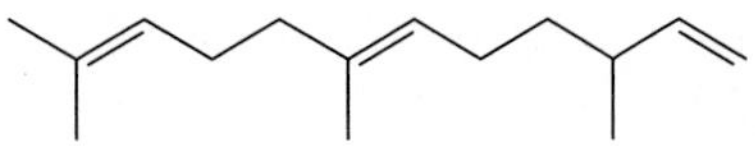

Alkynes

The simplest alkyne, C_2H_2, is commonly called acetylene. The IUPAC name for C_2H_2 is ethyne. Unfortunately, the common name ends in -ene, which suggests that the compound contains a double bond. Such confusion is one reason IUPAC names are so important for clear communication in chemistry.

A space-filling model of ethyne is shown in Figure 15.1. Each H—C≡C bond angle is 180°. The two triple-bonded carbon atoms and the two atoms directly attached to them all lie in a straight line in all alkynes.

The triple bond in an alkyne decreases the number of hydrogen atoms in the molecule by four compared with the corresponding alkane. As a result, the general molecular formula for alkynes is C_nH_{2n-2}.

Physical Properties of Alkenes and Alkynes

The physical properties of the homologous series of alkenes (C_nH_{2n}) and alkynes (C_nH_{2n-2}) are similar to those of the homologous series of alkanes (C_nH_{2n+2}). The molecules of both classes of unsaturated hydrocarbons are nonpolar and have low boiling points. The members of the series containing fewer than five carbon atoms are gases at room temperature. As in the case of alkanes, the boiling points in-

Physiologically Active Alkynes

Although less common than polyenes, alkynes containing several triple bonds are found in natural products. One of the more interesting examples is the triyne ichthyothereol—secreted from the skin of a species of frog that lives in the Lower Amazon basin. It is a defensive venom that acts on the nervous system and is a mucous-membrane irritant that wards off predators. The Indians of the area coat their arrowheads with the secretion. When the arrow pierces the skin of the prey, the compound causes convulsions.

$$CH_3{-}C{\equiv}C{-}C{\equiv}C{-}C{\equiv}C{-}CH{=}CH{-}(C_6H_{10}OH)$$

ichthyothereol

The skin of this frog is coated with a secretion that can cause paralysis in predators. (Photo Researchers, Inc.)

Natural products with several triple bonds are also found in plants. Cicutoxin is a poisonous compound contained in the water extract of hemlock, which was used in the execution of Socrates.

$$HOCH_2CH_2CH_2{-}C{\equiv}C{-}C{\equiv}C{-}CH{=}CH{-}CH{=}CH{-}CH{=}CH{-}\underset{\displaystyle OH}{\underset{|}{CH}}{-}CH_2CH_2CH_3$$

cicutoxin

TABLE 15.1 Boiling Points of Alkanes, Alkenes, and Alkynes

Alkanes	Boiling Point (°C)	Alkenes	Boiling Point (°C)	Alkynes	Boiling Point (°C)
pentane	36	1-pentene	30	1-pentyne	40
hexane	69	1-hexene	63	1-hexyne	71
heptane	98	1-heptene	94	1-heptyne	100
octane	126	1-octene	121	1-octyne	125

crease with an increase in the number of carbon atoms in the molecule, because the London forces increase (Table 15.1).

15.2 Geometric Isomerism

As a result of free rotation about carbon–carbon single bonds, alkanes can exist in many conformations. Thus, the structure of 1,2-dichloroethane can be drawn in several ways.

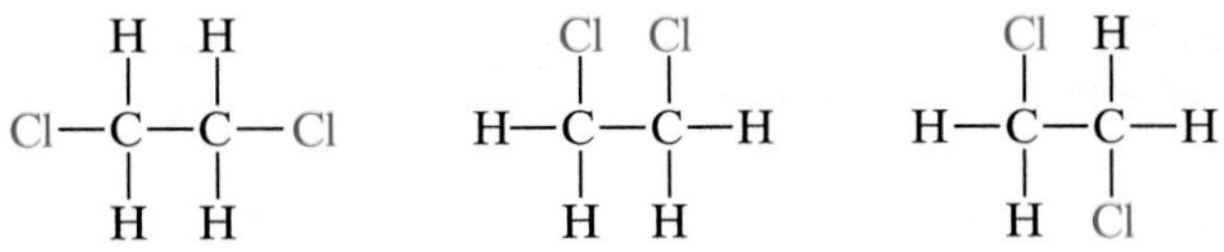

```
    H  H              Cl Cl              Cl H
    |  |              |  |               |  |
Cl—C—C—Cl          H—C—C—H            H—C—C—H
    |  |              |  |               |  |
    H  H              H  H               H  Cl
```

Free rotation cannot occur about the carbon–carbon double bond of alkenes. To maintain the π bond of an alkene, the two p orbitals must remain side by side. If the two carbon atoms were rotated about the bond axis, the σ bond would remain but the π bond would be broken. About 60 kcal/mole is required to break the π bond, so the bond is not easily broken.

As a consequence of the restricted rotation, the groups bonded to the carbon atoms of the double bond can exist in different geometric arrangements. These isomers, which have the same connectivity of atoms, are called **geometric isomers** or cis–trans isomers.

Consider the general alkene formula CXY=CXY. When we draw a structural formula we find that two representations result.

```
Y     Y              Y     X
 \   /                \   /
  C=C        and       C=C
 /   \                /   \
X     X              X     Y
```

These two structures represent different molecules. The structure on the left has two X groups on the same "side" of the molecule; this is the cis isomer. The structure on the right is the trans isomer; the Xs are on opposite "sides" of the molecule.

There are two geometric isomers of 1,2-dichloroethylene. Like all geometric isomers, they have different physical properties. For example, the boiling points of *cis*- and *trans*-1,2-dichloroethylene are 60° and 47 °C, respectively.

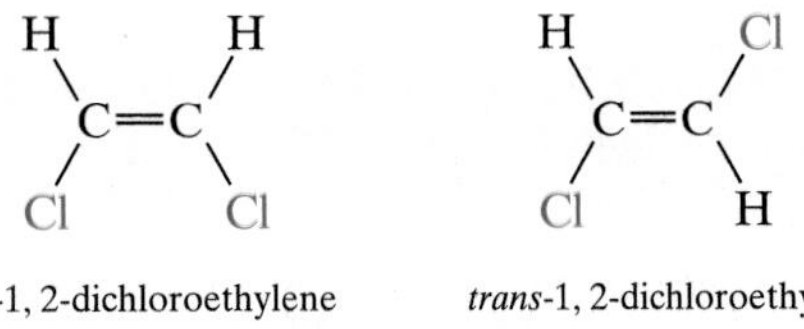

```
H     H              H     Cl
 \   /                \   /
  C=C                  C=C
 /   \                /   \
Cl    Cl             Cl    H
```

cis-1, 2-dichloroethylene *trans*-1, 2-dichloroethylene

Only an alkene that has two different atoms or groups of atoms attached to each of the double-bonded carbon atoms can exist as cis and trans isomers. Each of the two unsaturated carbon atoms in 1,2-dichloroethylene has a chlorine atom and a hydrogen atom attached to it. These groups are different.

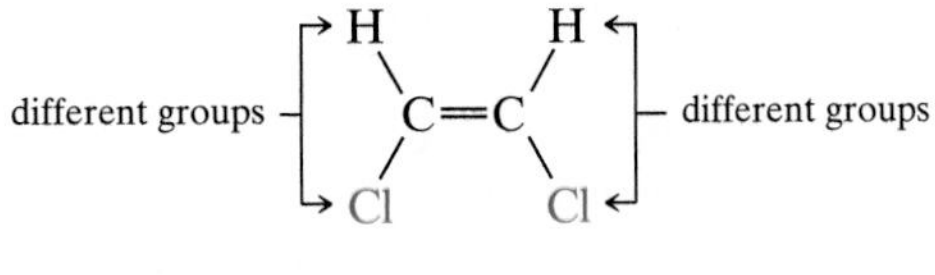

```
                  ┌→H     H←┐
                  │  \   /  │
different groups ─┤   C=C   ├─ different groups
                  │  /   \  │
                  └→Cl    Cl←┘
```

If one of the unsaturated carbon atoms is attached to two identical groups, cis–trans isomerism is not possible. Neither chloroethylene nor 1,1-dichloroethylene can exist as a cis–trans pair of geometric isomers. There is only one chloroethylene and one 1,1-dichloroethylene.

```
                  ┌→H     H←┐                                     ┌→Cl    H←┐
                  │  \   /  │                                     │  \   /  │
different groups ─┤   C=C   ├─ identical groups  identical groups ─┤   C=C   ├─ identical groups
                  │  /   \  │                                     │  /   \  │
                  └→Cl    H←┘                                     └→Cl    H←┘
```

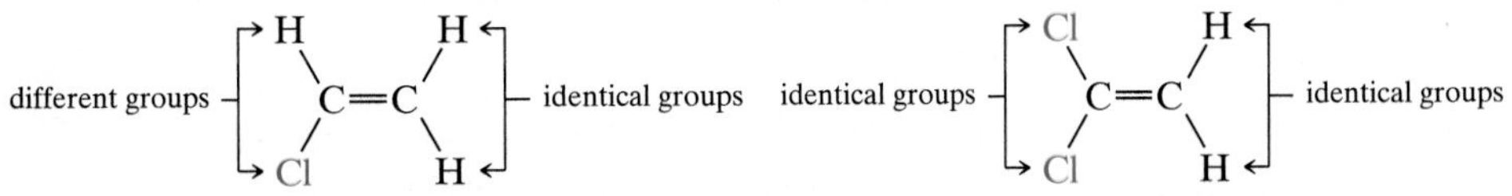

chloroethylene 1,1-dichloroethylene

EXAMPLE 15.2

Is cis–trans isomerism possible around either of the double bonds of geraniol, a natural oil?

```
         CH3                   CH3
         |                     |
CH3—C=CH—CH2—CH2—C=CH—CH2—OH
```

Solution

The double bond on the left of the structure has two methyl groups bonded to one of the double-bonded carbon atoms. Thus, cis–trans isomers are not possible about this double bond. The double bond on the right of the structure has two different groups bonded to each of the two double-bonded carbon atoms. One carbon atom has a hydrogen atom and a —CH_2OH group. The other carbon atom has a methyl group and the remainder of the carbon chain. Thus, cis–trans isomers are possible about this double bond.

PROBLEM 15.2

Indicate the geometry around each of the double bonds in the following compound, a trail pheromone of termites. (The chain is numbered starting from the carbon atom with the hydroxyl group.)

```
                    H     H
                     \   /
          H           C=C          H
           \         /   \        /
            C=C          CH2—C
           /   \              \\
  CH3CH2CH2     H              C—H
                              /
                       HOCH2CH2
```

Geometric Isomers and the Sex Life of Moths

Pheromones are compounds released by some animals, such as insects, that transmit messages within the species but not to other species. One of the uses of these compounds is to communicate between the male and female for the purpose of mating. The female usually emits the pheromones, and the male responds by seeking the female. The amount required is as little as 1 pg (1 pg = 10^{-12} g).

The female silkworm moth secretes bombykol, a diene that has a trans–cis arrangement of the double bonds. Numbering the carbon chain from the carbon atom bearing the hydroxyl group, we find a trans double bond at C-10 and a cis double bond at C-12.

```
               H       (CH2)9OH
                \     /
  CH3(CH2)2      C=C
          \     /   \
           C=C       H
          /   \
         H     H
```

bombykol

The other three geometric isomers (cis–cis, cis–trans, and trans–trans), which have been synthesized in the laboratory, do not attract the male silkworm moth. This lack of biological activity suggests that recognition of shape is an important feature of the cell-surface receptor molecules that interact with the pheromone.

The silkworm moth communicates with the opposite sex by using chemical methods. (Tony Stone Images)

15.3 Nomenclature of Alkenes

The IUPAC rules for naming alkenes are similar to those for alkanes, but the position of the double bond in the chain and the geometric arrangement of substituents around the double bond must be indicated. A few common names are allowed by IUPAC, including vinyl, allyl, and isopropenyl.

$CH_2{=}CH{-}$	$CH_2{=}CHCH_2{-}$	$CH_2{=}C(CH_3){-}$
vinyl	allyl	isopropenyl

The IUPAC rules follow.

1. The longest chain that contains the double bond is the parent.

$$\overset{1}{C}H_3{-}\overset{2}{C}H_2{-}\overset{3}{C}(CH_3){=}\overset{4}{C}(CH_3){-}\overset{5}{C}H_2{-}\overset{6}{C}H(CH_3){-}\overset{7}{C}H_2{-}\overset{8}{C}H_3$$

There are eight carbon atoms in this chain, so it is an octene.

2. The parent chain is given the same stem name as an alkane, but the suffix -ene replaces -ane. The parent name of the preceding structure is octene.
3. The carbon atoms are numbered consecutively from the end nearer the double bond. The number of the first carbon atom with the double bond is used as a prefix to the parent name and is separated from it by a hyphen.

$$\overset{1}{C}H_3{-}\overset{2}{C}H_2{-}\overset{3}{C}(CH_3){=}\overset{4}{C}(CH_3){-}\overset{5}{C}H_2{-}\overset{6}{C}H_2{-}\overset{7}{C}H_3$$

This is a substituted 3-heptene, not a substituted 4-heptene.

4. Alkyl groups and other substituents are named, and their positions on the chain are identified, according to the numbering established by rule 3. Names and numbers are prefixed to the parent name.

$$\overset{5}{C}H_3{-}\overset{4}{C}H_2{-}\overset{3}{C}(CH_3){=}\overset{2}{C}(CH_3){-}\overset{1}{C}H_3$$

This is 2,3-dimethyl-2-pentene, not 3,4-dimethyl-3-pentene.

5. If the compound is a cis or trans isomer with respect to the parent chain the appropriate prefix followed by a hyphen is placed in front of the name.

$$\overset{1}{C}H_3{-}\overset{2}{C}H_2{-}\overset{3}{C}(CH_3){=}\overset{4}{C}H{-}\overset{5}{C}H_2{-}\overset{6}{C}H_3$$

This is *trans*-3-methyl-3-hexene.

6. If the compound contains more than one double bond, the location of each double bond is specified by a number. A prefix to -ene indicates the number of double bonds.

$$\overset{1}{CH_2}=\overset{2}{CH}-\overset{3}{CH}=\overset{4}{CH}-\overset{5}{CH}=\overset{6}{CH}-\overset{7}{CH_3}$$

1,3,5-heptatriene

7. The ring of a cycloalkene is numbered to give the double-bonded carbon atoms the numbers 1 and 2. The direction of numbering is chosen so that the first substituent on the ring receives the lower number. The position of the double bond is not given because it is known to be between the C-1 and C-2 atoms.

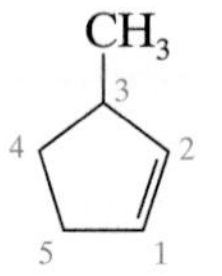

3-methylcyclopentene 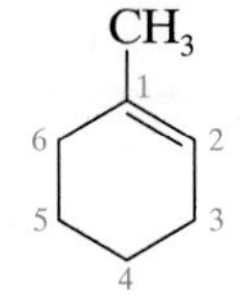 1-methylcyclohexene

EXAMPLE 15.3

Name the following compound.

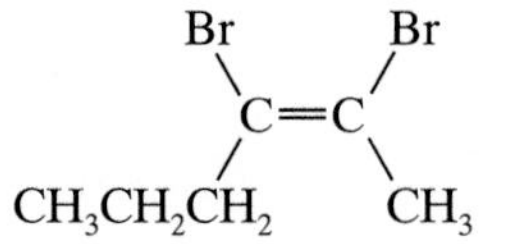

Br Br / C=C / $CH_3CH_2CH_2$ CH_3

Solution

There are six carbon atoms in the longest chain. It is numbered from right to left so that the double bond is at the carbon atom in position 2. The parent is then 2-hexene. The bromine atoms are at positions 2 and 3.

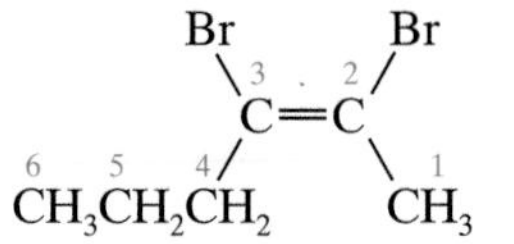

Since there are two different groups of atoms on each unsaturated carbon atom, cis and trans isomers are possible. This molecule is *cis*-2,3-dibromo-2-hexene.

PROBLEM 15.3

Draw the structure of 3-methyl-1,4-cyclohexadiene and 2-methyl-1,3-cyclohexadiene.

15.4 Oxidation and Reduction of Alkenes

Alkenes can be oxidized selectively at the π bond by oxidizing agents that do not affect the single bonds of the molecule. For example, potassium permanganate oxidizes an alkene to a dialcohol, or diol (two OH groups).

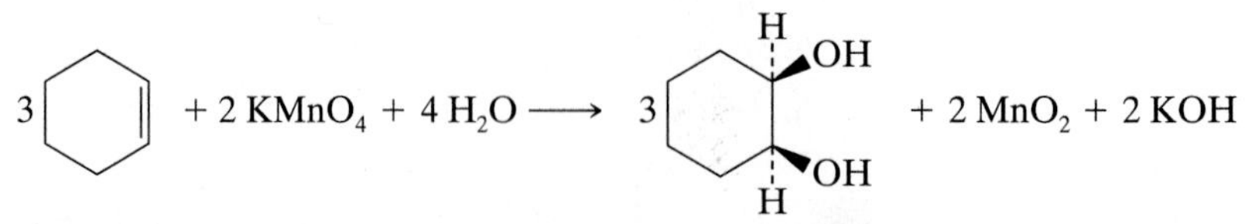

$$3 \text{ cyclohexene} + 2\,KMnO_4 + 4\,H_2O \longrightarrow 3 \text{ cis-1,2-cyclohexanediol} + 2\,MnO_2 + 2\,KOH$$

cyclohexene *cis*-1,2-cyclohexanediol

Potassium permanganate is purple in aqueous solution. Manganese dioxide (MnO_2) is a brown solid that precipitates from solution. Since there is a color change, oxidation with potassium permanganate can be used to test visually for the presence of a double bond. Alkanes and cycloalkanes are not oxidized by $KMnO_4$, so the purple color remains.

Alkenes and cycloalkenes combine with hydrogen gas in an addition reaction to give saturated compounds. In this process the alkene is reduced. The reaction is also called **hydrogenation.** The hydrogenation of an alkene is an exothermic process, but the reaction has a high activation energy, so it occurs only slowly even at high temperatures. However, the hydrogenation of an alkene, such as 1-octene to give octane, occurs rapidly at room temperature in the presence of certain transition metal catalysts. One such catalyst is palladium dispersed on carbon (Pd/C).

The double bond → is changed into a single bond

$$\underset{\text{1-octene}}{CH_3(CH_2)_5CH{=}CH_2} + H_2 \xrightarrow{Pd/C} \underset{\text{octane}}{CH_3(CH_2)_5CH_2{-}CH_3}$$

The melting point of oils and fats depends on the degree of saturation. (Fundamental Photographs)

The hydrogenation of oils derived from plants to produce more saturated products called fats is an important industrial process. Fats are esters of long-chain saturated carboxylic acids. Oils are structurally related esters, but the acids have one or more double bonds. The difference between Crisco oil and its companion solid Crisco is just the degree of saturation of the carboxylic acids in the esters.

Alkenes can be hydrogenated using finely dispersed nickel known as *Raney nickel.* Hydrogenation reactions using Raney nickel require higher temperatures or pressures than those required for palladium or platinum catalysts. However, nickel is less expensive than palladium or platinum, so it is used in industrial processes, such as the conversion of triolein, an oil, into tristearin, a fat.

$$\begin{array}{l} CH_2{-}O{-}C({=}O){-}(CH_2)_7CH{=}CH(CH_2)_7CH_3 \\ | \\ CH{-}O{-}C({=}O){-}(CH_2)_7CH{=}CH(CH_2)_7CH_3 \\ | \\ CH_2{-}O{-}C({=}O){-}(CH_2)_7CH{=}CH(CH_2)_7CH_3 \\ \quad\quad\text{triolein} \end{array} \xrightarrow[Ni]{3\,H_2} \begin{array}{l} CH_2{-}O{-}C({=}O){-}(CH_2)_{16}CH_3 \\ | \\ CH{-}O{-}C({=}O){-}(CH_2)_{16}CH_3 \\ | \\ CH_2{-}O{-}C({=}O){-}(CH_2)_{16}CH_3 \\ \quad\quad\text{tristearin} \end{array}$$

EXAMPLE 15.4

How many moles of hydrogen gas will react with one mole of cembrene, which is present in pine oil? What is the molecular formula for the product?

cembrene

Solution

There are four double bonds in the compound, which will react with four moles of hydrogen gas. The product will be a cycloalkane. There are 14 carbon atoms in the ring, three methyl groups, and an isopropyl group, for a total 20 carbon atoms. The general formula for a cycloalkane is C_nH_{2n}, so the molecular formula of the product is $C_{20}H_{40}$.

PROBLEM 15.4

Write the structure obtained by hydrogenation of ipsdienol, a pheromone of the Norwegian spruce beetle.

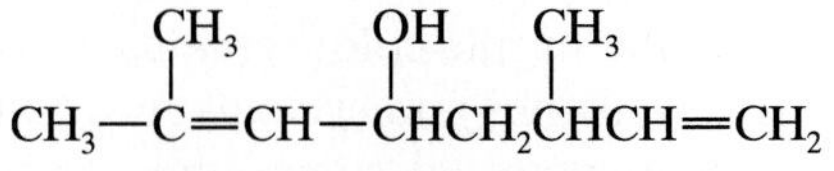

$$CH_3-\overset{\displaystyle CH_3}{\overset{|}{C}}=CH-\overset{\displaystyle OH}{\overset{|}{C}}HCH_2\overset{\displaystyle CH_3}{\overset{|}{C}}HCH=CH_2$$

2,6-dimethyl-2,7-octadien-4-ol
(ipsdienol)

15.5 Addition Reactions of Alkenes

The reaction of ethylene with Br_2 forms 1,2-dibromoethane in an addition reaction. The atoms that add to ethylene are placed one each on adjacent carbon atoms of the double bond. This is a characteristic of addition reactions of alkenes.

These two atoms — add to adjacent atoms.

$$\begin{matrix} H & & H \\ & C=C & \\ H & & H \end{matrix} + Br-Br \longrightarrow H-\underset{H}{\overset{Br}{C}}-\underset{H}{\overset{Br}{C}}-H$$

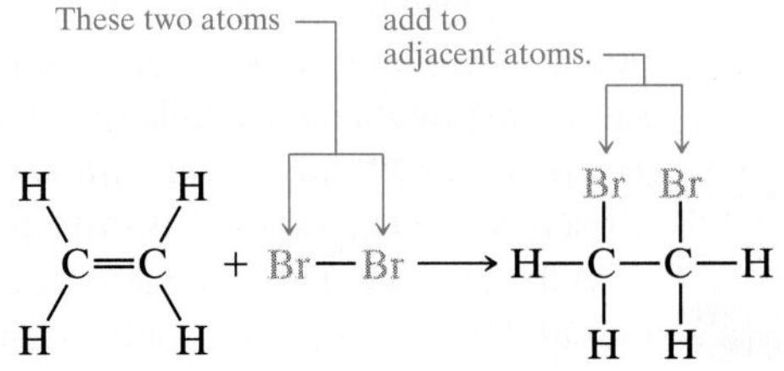

1,2-dibromoethane

Evidence for the addition of bromine to an alkene is easily seen. Bromine is reddish; reaction leads to a colorless organic compound. The disappearance of the reddish color of bromine is, then, a useful test to determine if a compound is unsaturated. Drops of Br_2 dissolved in CCl_4 are added to a compound. If the bromine color rapidly disappears, the compound is unsaturated.

$$\underset{\text{colorless}}{CH_3(CH_2)_3CH=CH_2} + \underset{\text{red}}{Br_2} \longrightarrow \underset{\text{colorless}}{CH_3(CH_2)_3CHBr-CH_2Br}$$

Symmetrical and Unsymmetrical Reagents

Some examples of addition reactions of ethylene (C_2H_4) with some common reagents are

$$CH_2=CH_2 + Br_2 \longrightarrow BrCH_2CH_2Br$$
$$CH_2=CH_2 + HCl \longrightarrow CH_3CH_2Cl$$
$$CH_2=CH_2 + H_2O \xrightarrow{H^+} CH_3CH_2OH$$

These reagents and others that add to alkenes are classified as symmetrical or unsymmetrical. **Symmetrical reagents** consist of two identical groups which will add to the carbon atoms of the double bond. Bromine is a symmetrical reagent. **Unsymmetrical reagents** consist of two different groups which bond to the carbon atoms of the double bond. Both HCl and H_2O are unsymmetrical reagents.

There is only one possible product of the addition of a symmetrical reagent to an alkene. For example, bromine reacts with propene to yield one compound. It makes no difference which bromine atom bonds to which carbon atom.

$$CH_3-CH{=}CH_2 + Br_2 \longrightarrow CH_3-\underset{}{\overset{Br}{\overset{|}{C}}}H-\overset{Br}{\overset{|}{C}}H_2$$

propene 1,2-dibromopropane

Two products could result from the addition of HCl to an unsymmetrical alkene, but only one is formed. Addition of HCl to propene could yield either 1-chloropropane or 2-chloropropane. However, only the latter is formed. The X written through one reaction arrow indicates that a reaction does not occur.

$$CH_3CH{=}CH_2 + HCl \not\longrightarrow CH_3\overset{H}{\overset{|}{C}}H-\overset{Cl}{\overset{|}{C}}H_2 \quad \text{(not formed)}$$

$$CH_3CH{=}CH_2 + HCl \longrightarrow CH_3\underset{Cl}{\underset{|}{C}}H-\underset{H}{\underset{|}{C}}H_2$$

Markovnikov's Rule

The Russian chemist Vladimir Markovnikov observed that unsymmetrical reagents add to unsymmetrical double bonds in a specific way. **Markovnikov's rule** states that a molecule of the general formula HX adds to a double bond so that the hydrogen atom bonds to the unsaturated carbon atom that has the greater number of directly bonded hydrogen atoms.

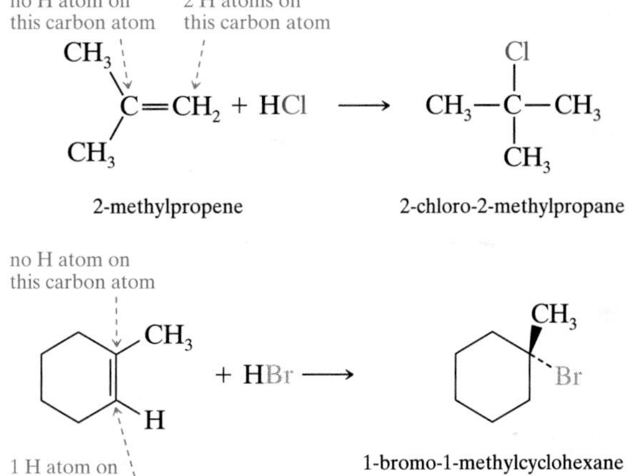

EXAMPLE 15.5

Predict what product will be formed when HCl adds to 2-methyl-2-butene.

Solution

One of the unsaturated carbon atoms has one attached hydrogen atom, and the other has no attached hydrogen atoms. The predicted product is 2-chloro-2-methylbutane.

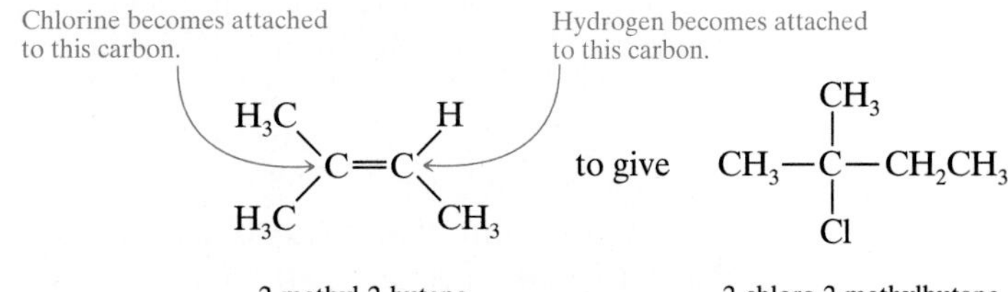

PROBLEM 15.5

Predict the product formed when HBr adds to 1-ethylcyclobutene.

Mechanism of Addition Reactions

Markovnikov's rule is the result of experimental observation and has predictive value. However, it doesn't tell us why the reagents react the way they do. An understanding of the reaction is provided by the mechanism that has now been established. The π electrons in an alkene serve as a Lewis base (Section 13.6) and can be donated to an electrophile (electron-loving species). Consider the reaction of propene with HBr. The first step is written using a curved arrow to show the movement of two electrons in the π bond to form a σ bond to hydrogen.

$$CH_3CH{=}CH_2 + H{-}\ddot{Br}: \longrightarrow CH_3\overset{+}{C}HCH_3 + :\ddot{Br}:^-$$

isopropyl carbocation

You can view the movement of the π electrons in this step like the movement of a swinging gate. One end of the gate stays attached, and the other end is free. The electron pair stays as part of a bond to one of the two carbon atoms.

In the second step of the addition reaction, the carbocation acts as a Lewis acid and accepts an electron pair from the bromide ion, which acts as a nucleophile (nucleus-loving species).

$$CH_3\overset{+}{C}HCH_3 + :\ddot{Br}:^- \longrightarrow CH_3CHBrCH_3$$

In the first step, the hydrogen atom attaches to one of the two possible carbon atoms of the original π bond; this placement accounts for the product predicted by Markovnikov's rule. Why is the isopropyl carbocation formed? If the hydrogen atom had bonded to the interior carbon atom, the electrons would have had to "swing" in the other direction and a propyl carbocation would result.

$$CH_3CH{=}CH_2 + H{-}\ddot{Br}: \longrightarrow CH_3CH_2\overset{+}{C}H_2 + :\ddot{Br}:^-$$

propyl carbocation

Both the isopropyl carbocation and the propyl carbocation are unstable intermediates. However, the isopropyl carbocation is more stable and is formed preferentially. Alkyl groups attached to a positively charged carbon atom help stabilize the charge because the electrons in the carbon–carbon bonds can be polarized toward the positive center. The isopropyl carbocation has the charge on a secondary carbon atom, one that has two alkyl groups attached. The propyl carbocation has the charge on a primary carbon atom, one that has only one alkyl group attached.

By the same reasoning, it follows that a tertiary carbocation is the most stable carbocation because it has three alkyl groups attached to the positive carbon atom. The order of stability of carbocations is

$$\underset{\text{methyl}}{CH_3^+} < \underset{\text{primary}}{R{-}CH_2^+} < \underset{\text{secondary}}{R{-}\overset{\displaystyle R}{\overset{|}{C}}H^+} < \underset{\text{tertiary}}{R{-}\overset{\displaystyle R}{\overset{|}{\underset{\underset{\displaystyle R}{|}}{C^+}}}}$$

$$\xrightarrow{\text{increasing stability}}$$

This order of stability "explains" Markovnikov's rule. Addition of the electrophile occurs to give the most stable carbocation, which controls the product formed.

15.6 Polymerization of Alkenes

Polymers are high molecular weight molecules made up of as many as thousands of low molecular weight repeating units called **monomers** (meaning one part). The process of joining monomers to produce polymers is called **polymerization.**

Alkenes can be polymerized by a multiple addition reaction. For example, with the appropriate catalyst, ethylene molecules successively add to neighboring molecules, forming carbon–carbon bonds. Polyethylene is the product.

$$H_2C{=}CH_2 + H_2C{=}CH_2 + H_2C{=}CH_2 + \longrightarrow \cdots{-}CH_2{-}CH_2{-}CH_2{-}CH_2{-}CH_2{-}CH_2{-}\cdots$$

polyethylene

When $CH_2{=}CHCl$ (vinyl chloride) is polymerized, the resultant polymer is polyvinyl chloride, or PVC, as it is commonly known.

$$n\ H_2C{=}CHCl \longrightarrow \cdots{-}CH_2{-}CHCl{-}CH_2{-}CHCl{-}CH_2{-}CHCl{-}\cdots$$

Artificial turf is a polymeric product that can withstand the forces of nature as well as football players. (Photo Researchers, Inc.)

An exact formula for a polymer cannot be written because the size of the molecule may vary depending on how it is formed. There is no single "polyethylene" or "polyvinyl chloride" molecule; these compounds are really mixtures of compounds with a range of molecular weights. For this reason polymers are represented by placing the repeating units of the monomer within a set of parentheses. For ethylene, the unit is $({-}CH_2CH_2{-})$. To show that a large number of units are present, the subscript n is used. For the polymerization of ethylene, we write

$$n\ CH_2{=}CH_2 \longrightarrow {+}CH_2{-}CH_2{+}_n$$

The properties of polymers depend on the monomer used and the molecular weight of the product. A list of some useful polymers is given in Table 15.2. Vinyl chloride has been linked to a form of liver cancer, so chemical workers must take special precautions when producing it. However, its polymer, polyvinyl chloride, is not carcinogenic and is widely used in plastic products.

TABLE 15.2 Uses and Structures of Polymers

Monomer	Polymer	Uses
propylene $CH_2{=}CHCH_3$	polypropylene $-CH_2CH(CH_3)CH_2CH(CH_3)-$	carpet fibers, heart valves, bottles
vinyl chloride $CH_2{=}CHCl$	polyvinyl chloride (PVC) $-CH_2CH(Cl)CH_2CH(Cl)CH_2CH(Cl)-$	floor covering, records, garden hoses
dichloroethylene $CH_2{=}CCl_2$	polydichloroethylene $-CH_2CCl_2CH_2CCl_2CH_2CCl_2-$	plastic food wrap
tetrafluoroethylene $CF_2{=}CF_2$	polytetrafluoroethylene $-CF_2CF_2CF_2CF_2CF_2CF_2-$	Teflon, bearings
acrylonitrile $CH_2{=}CHCN$	polyacrylonitrile $-CH_2CH(CN)CH_2CH(CN)CH_2CH(CN)-$	Orlon, Acrilan
styrene $CH_2{=}CHC_6H_5$	polystyrene $-CH_2CH(C_6H_5)-CH_2CH(C_6H_5)-CH_2CH(C_6H_5)-$	toys, styrofoam
methyl methacrylate $CH_2{=}C(CH_3)-C(=O)OCH_3$	polymethyl methacrylate $-CH_2-C(CH_3)(COOCH_3)-CH_2-C(CH_3)(COOCH_3)-CH_2-C(CH_3)(COOCH_3)-$	Plexiglas, Lucite

15.7 Nomenclature of Alkynes

The rules for naming alkynes by the IUPAC system are analogous to those for alkanes and alkenes.

1. The longest chain containing the triple bond is the parent.
2. The parent is given the same stem name as an alkane, but the suffix -ane is replaced with -yne.
3. The carbon atoms are numbered consecutively from the end of the chain nearer the triple bond. The number of the first carbon atom with the triple bond is used as a prefix separated by a hyphen from the parent name.

$$\overset{6}{C}H_3-\overset{5}{C}H_2-\overset{4}{C}H_2-\overset{3}{C}\equiv\overset{2}{C}-\overset{1}{C}H_3$$

This compound is numbered from right to left. It is 2-hexyne.

4. Alkyl groups are named, and their positions on the chain, are identified by the numbering established by rule 3.

$$\underset{1}{CH_3}-\underset{2}{C}\equiv\underset{3}{C}-\underset{4}{\overset{\overset{\displaystyle CH_3}{|}}{CH}}-\underset{5}{CH_3}$$

This compound is 4-methyl-2 pentyne

5. Compounds with multiple triple bonds are diynes, triynes, and so on. Compounds with both double and triple bonds are called enynes, not ynenes. The numbering of compounds with both double and triple bonds is started from the end nearer the first multiple bond, regardless of type. When a choice is possible, double bonds are assigned lower numbers than triple bonds.

$$CH_2{=}CH-CH_2-C\equiv C-H$$

This is 1-penten-4-yne, not 4-penten-1-yne.

EXAMPLE 15.6

Why is 2-bromo-4-hexyne an incorrect name for the following compound? What is the correct IUPAC name?

$$CH_3-\underset{\underset{\displaystyle Br}{|}}{CH}-CH_2-C\equiv C-CH_3$$

Solution

The name is incorrect because the numbering of the carbon chain should be based on the position of bromine rather than the triple bond. Starting from the rightmost carbon atom places the triple bond at the C-2 atom. The bromine atom is on the C-5 atom, and the correct name is 5-bromo-2-hexyne.

PROBLEM 15.6

1,3,11-Tridecatriene-5,7,9-triyne is a compound found in safflowers and used as a chemical defense against nematode infestations. Write the structure of the compound.

15.8 Reactions of Alkynes

The oxidation of acetylene in the oxyacetylene torch generates sufficient heat to melt metals. (Photo Researchers, Inc.)

The triple bond is made up of one σ bond and two π bonds, and its reactivity is similar to that of a double bond in an alkene. Thus, triple bonds undergo addition reactions just like double bonds. However, they can add two molecules of a reagent. For example, 5-decyne reacts with hydrogen to give decane.

$$CH_3(CH_2)_3C\equiv C(CH_2)_3CH_3 + 2\,H_2 \xrightarrow{Pt} CH_3(CH_2)_3-CH_2-CH_2-(CH_2)_3CH_3$$

The catalytic hydrogenation reaction can be stopped after adding one mole of hydrogen gas to form a cis alkene by using a particular palladium catalyst, called the **Lindlar catalyst**. It consists of palladium coated on calcium carbonate that contains a small amount of lead acetate.

$$\underset{\text{5-decyne}}{CH_3(CH_2)_3C\equiv C(CH_2)_3CH_3} + H_2 \xrightarrow{\text{Lindlar catalyst}} \underset{\textit{cis}\text{-5-decene}}{\overset{CH_3(CH_2)_2CH_2 \qquad CH_2(CH_2)_2CH_3}{\underset{H \qquad\qquad H}{C{=}C}}}$$

Reduction of an alkyne with sodium metal in liquid ammonia gives a trans alkene. The reaction occurs by a different mechanism than for the catalytic hydrogenation reaction.

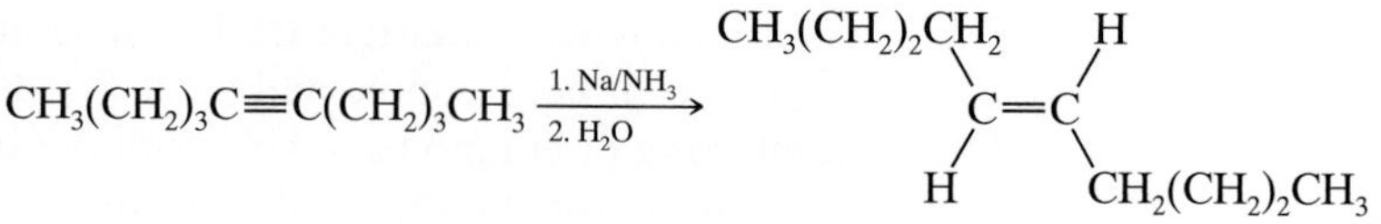

$$CH_3(CH_2)_3C{\equiv}C(CH_2)_3CH_3 \xrightarrow[2.\ H_2O]{1.\ Na/NH_3} \begin{array}{ccc} CH_3(CH_2)_2CH_2 & & H \\ & C{=}C & \\ H & & CH_2(CH_2)_2CH_3 \end{array}$$

5-decyne — *trans*-5-decene

EXAMPLE 15.7

The IUPAC name of muscalure, the sex hormone of the housefly, is *cis*-9-tricosene. How can this compound be produced from a structurally related alkyne? Name that alkyne.

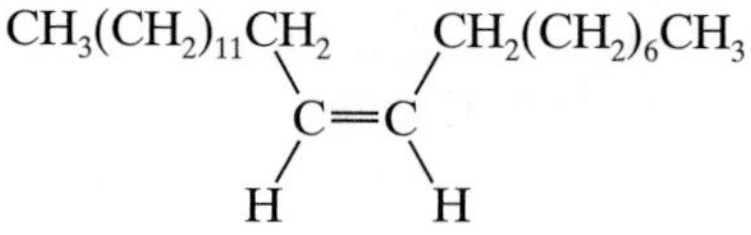

$$\begin{array}{ccc} CH_3(CH_2)_{11}CH_2 & & CH_2(CH_2)_6CH_3 \\ & C{=}C & \\ H & & H \end{array}$$

muscalure

Solution

The alkene has two alkyl groups in a cis arrangement about a carbon–carbon double bond between C-9 and C-10. Reduction of a 23 carbon atom alkyne with a carbon–carbon triple bond between C-9 and C-10 using the Lindlar catalyst will give the alkene.

$$CH_3(CH_2)_{11}CH_2{-}C{\equiv}C{-}CH_2(CH_2)_6CH_3$$

PROBLEM 15.7

One of the intermediate compounds in the synthesis of the spruce budworm sex pheromone is *trans*-11-tetradecen-1-ol. How can this compound be produced from an alkyne?

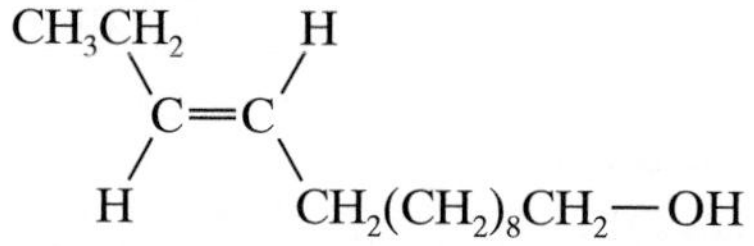

$$\begin{array}{ccc} CH_3CH_2 & & H \\ & C{=}C & \\ H & & CH_2(CH_2)_8CH_2{-}OH \end{array}$$

trans-11-tetradecen-1-ol

Hydrogen bromide adds to alkynes giving the product predicted by Markovnikov's rule.

$$CH_3(CH_2)_3C{\equiv}CH + HBr \longrightarrow CH_3(CH_2)_3CBr{=}CH_2$$

1-hexyne — 2-bromo-1-hexene

The alkenyl bromide reacts with a second mole of HBr in a slower second step to give a product predicted by Markovnikov's rule.

$$\begin{array}{ccc} Br & & H \\ & C{=}C & \\ CH_3(CH_2)_2CH_2 & & H \end{array} + HBr \longrightarrow CH_3(CH_2)_3{-}\underset{Br}{\overset{Br}{C}}{-}\underset{H}{\overset{H}{C}}{-}H$$

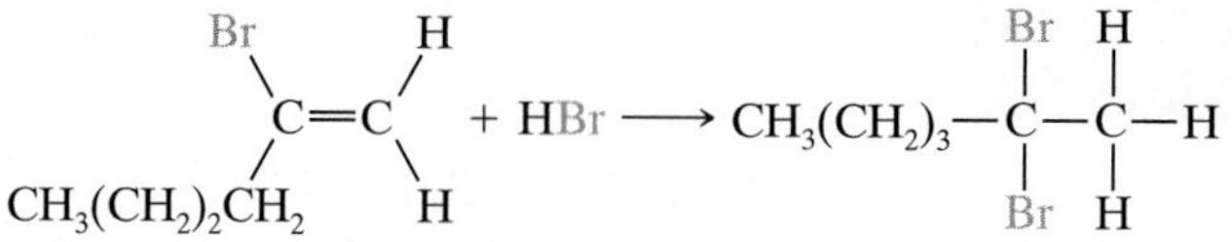

2-bromo-1-hexene — 2,2-dibromohexane

Alkynes react with chlorine or bromine to produce tetrahaloalkanes, which contain two halogen atoms on each of the original carbon atoms of the triple bond. Hence, the reaction consumes two moles of the halogen.

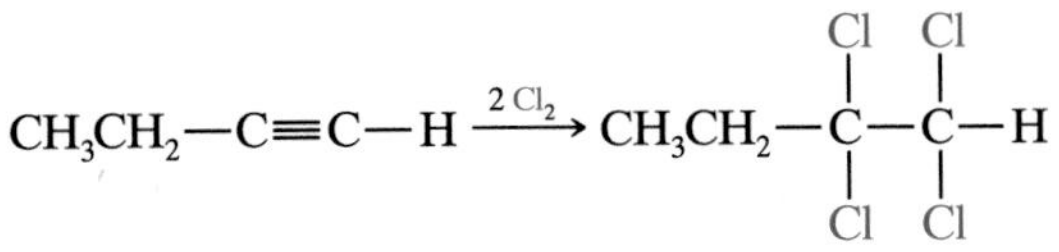

$$CH_3CH_2{-}C{\equiv}C{-}H \xrightarrow{2\,Cl_2} CH_3CH_2{-}CCl_2{-}CCl_2{-}H$$

1-butyne 1,1,2,2-tetrachlorobutane

If only one mole of the halogen is used, the reaction product has the halogen atoms on the opposite sides of the double bond.

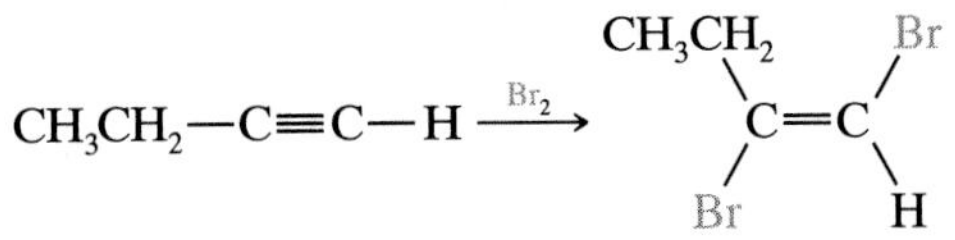

$$CH_3CH_2{-}C{\equiv}C{-}H \xrightarrow{Br_2} (CH_3CH_2)(Br)C{=}C(Br)(H)$$

15.9 Aromatic Hydrocarbons

Early chemists used the term *aromatic* to describe substances with an aroma. Many of these fragrant compounds, it turned out, contain a benzene ring bonded to one or more substituents. Oil of sassafras, oil of wintergreen, and vanillin, for example, all contain a benzene ring.

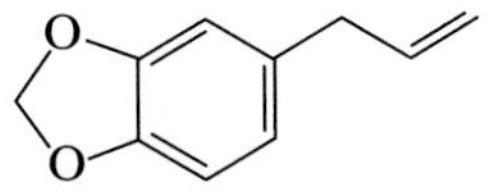

safrole (oil of sassafras) | methyl salicylate (oil of wintergreen) | vanillin (vanilla)

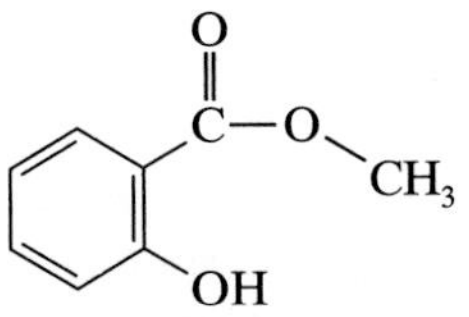

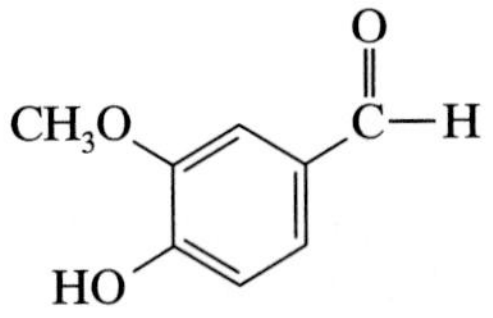

Not all compounds with a benzene ring are fragrant. For example, some aromatic compounds are solids with little or no odor. Solid aromatic compounds include the pain relievers or analgesics, aspirin, ibuprofen, and acetaminophen.

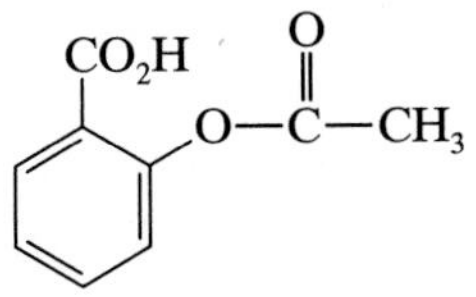

aspirin 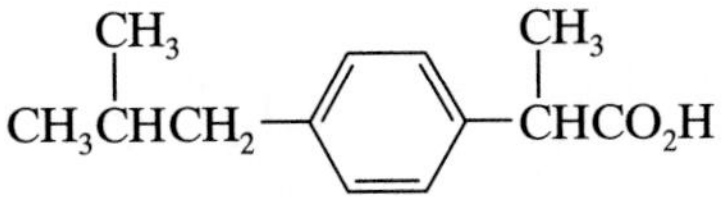 | ibuprofen 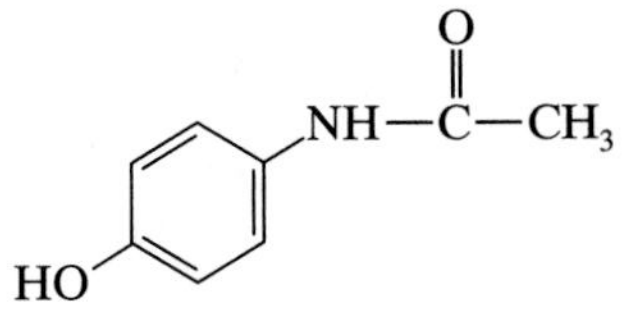 | acetaminophen

Substituted benzene compounds belong to a class of compounds called **arenes.** The common structural feature of arenes is a monocyclic or polycyclic system of π electrons that results in a special stability called **aromaticity.**

Benzene, often represented by a hexagon containing three double bonds, is unsaturated. However, benzene does not react with most reactants that attack π bonds to form addition products. That is, it does not behave like the "1,3,5-cyclohexatriene" depicted by its Lewis structure. Benzene does not undergo addition reactions with HBr, and does not react with the powerful oxidizing agent potassium permanganate. Benzene does react with bromine but gives a substitution product in which a bromine atom replaces a hydrogen atom. The reaction requires iron(III) bromide as a catalyst. Only one compound, C_6H_5Br, forms.

$$C_6H_6 + Br_2 \xrightarrow{FeBr_3} C_6H_5Br + HBr$$

The low reactivity of benzene contradicts what we know about unsaturated compounds. In 1865 a German chemist, F. August Kekulé, suggested that benzene has a single ring of six carbon atoms linked by alternating single and double bonds, now called a *Kekulé structure.* He further proposed that the single bonds rapidly become double, as the adjacent double bonds become single.

H H
H C H H C H
C C C C
or
C C C C
H C H H C H
H H

Kekulé proposed that the oscillation of single and double bonds makes benzene resist addition reactions. He also reasoned that the oscillation of single and double bonds around the ring makes all six carbon atoms equivalent.

Modern measurements show benzene as a planar molecule in which all carbon–carbon bonds are equivalent. The carbon–carbon bond length, 140 pm, lies between that of a single bond, 154 pm, and that of a double bond, 133 pm. The C—C—C bond angles of the ring are all 120°. Each carbon atom in benzene is sp^2 hybridized. The carbon atoms link by σ bonds in which the carbon atom shares one electron in each of its σ bonds. Two σ bonds link to adjacent carbon atoms. The third links to a hydrogen atom. Each sp^2-hybridized carbon atom has an electron in a $2p$ orbital perpendicular to the plane of the benzene ring (Figure 15.2). The six $2p$ orbitals of benzene overlap to share electrons in a six-π-electron system that extends over the entire ring. These electrons are located both above and below the plane of the ring. The delocalization of the electrons over all the carbon atoms of benzene accounts for the unique chemical stability of benzene.

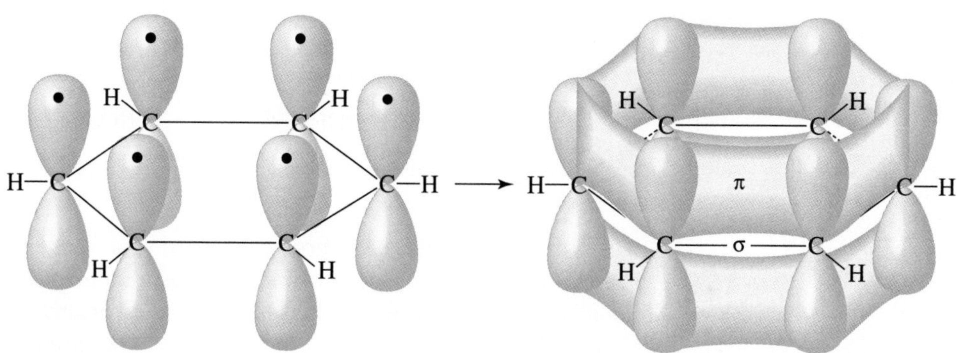

FIGURE 15.2
Bonding in the Benzene Ring

The lines between carbon atoms represent the σ bonds of the benzene ring. In addition, each carbon atom has one p *orbital that contributes one electron to the π system. Overlap of the six* p *orbitals that are maintained mutually parallel results in a delocalized system that distributes the electrons over the entire carbon framework.*

Two resonance structures depict the structure of benzene. They are the Kekulé structures, which differ only in the position shown for double bonds. Because the two resonance structures are otherwise identical, they contribute equally to the resonance hybrid of benzene. Kekulé structures are not "real," but the benzene molecule can be viewed as a resonance hybrid of these two structures. We indicate the relationship between the contributing structures by a single double-headed arrow.

The structure of benzene is usually represented in chemical equations as one of the two possible Kekulé structures. Each corner of the hexagon represents a carbon atom with one attached hydrogen atom, often not shown.

15.10 Nomenclature of Aromatic Hydrocarbons

Like the other classes of compounds we have discussed, benzene compounds have common, nonsystematic names and IUPAC names. Common names often stem from the sources of the compounds, and they have been used for so long that they have become accepted by IUPAC. One example is toluene, which used to be obtained from the South American gum tree *Toluifera balsamum.* Other examples are shown in Figure 15.3.

FIGURE 15.3 Nomenclature of Aromatic Hydrocarbons

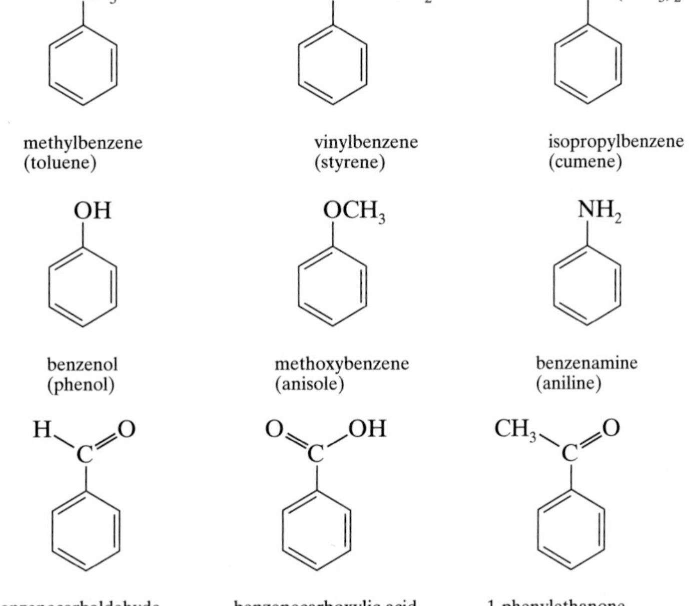

A polymer of styrene has excellent insulating properties and is used in the construction of houses. (The Image Bank)

The IUPAC system of naming substituted aromatic hydrocarbons uses the names of the substituents as prefixes to benzene. Examples include

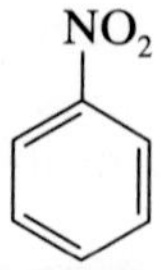

NO_2 CH_2CH_3 Br

nitrobenzene 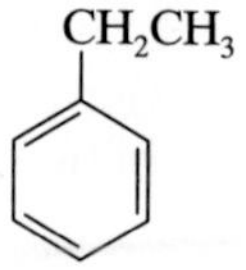 ethylbenzene 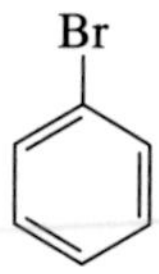 bromobenzene

Structures often have the substituent shown at the "12 o'clock" position. However, all six positions on the benzene ring are equivalent, and you should be able to recognize a compound, such as bromobenzene, no matter where the bromine atom is written.

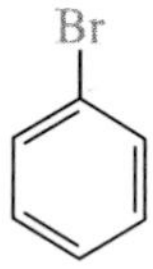

Br is the same as Br is the same as Br

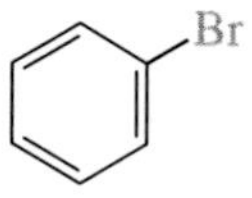

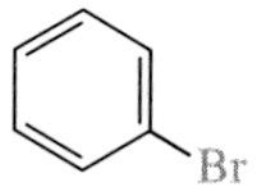

Disubstituted compounds result from replacement of two hydrogen atoms of the benzene ring by other groups. Two substituents on a benzene ring can give three different isomers. These isomers are designated *ortho, meta,* and *para.* These terms are abbreviated with the prefixes *o*-, *m*-, and *p*-, respectively.

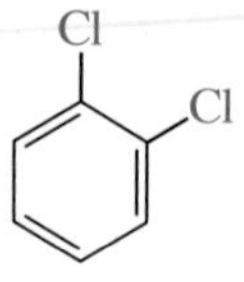

Cl Cl Cl Cl Cl Cl

o-dichlorobenzene
1,2-dichlorobenzene

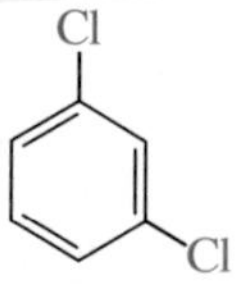

m-dichlorobenzene
1,3-dichlorobenzene

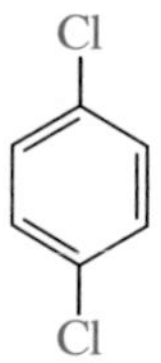

p-dichlorobenzene
1,4-dichlorobenzene

The ortho isomer has two groups on adjacent carbon atoms—that is, in a 1,2 relationship. In the meta and para isomers, the two groups are in a 1,3 and a 1,4 relationship, respectively. The IUPAC name of a disubstituted aromatic compound is obtained by numbering the benzene ring to give the lowest possible numbers to the carbon atoms bearing the substituents. When three or more substituents are present, the ring carbon atoms must be numbered.

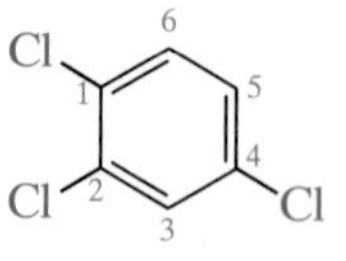

1,2,4-trichlorobenzene 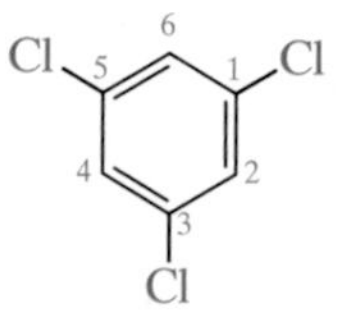 1,3,5-trichlorobenzene

Many derivatives of benzene are named with the common name of the monosubstituted aromatic compound as the parent. The position of the substituent of the "parent" is designated 1, but the number is not used in the name. The remaining substituents are prefixed in alphabetical order to the parent name, along with numbers indicating their locations.

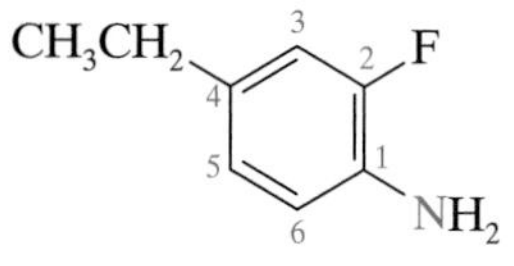

4-ethyl-2-fluoroaniline

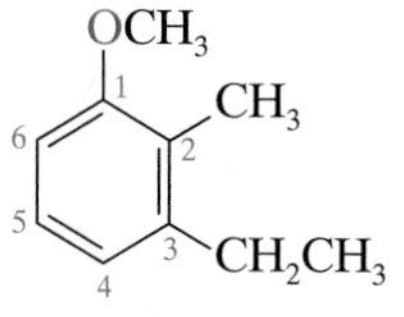

3-ethyl-2-methylanisole

An aromatic ring residue attached to a larger parent structure is called an **aryl group.** It is indicated by Ar (not to be confused with argon), just as the symbol R is used for an alkyl group. Two groups whose names unfortunately do not make much sense are the phenyl (fen′-nil) and benzyl groups. We might reasonably expect that the aryl group derived from benzene (C_6H_5—) would be named benzyl. Alas, it is a phenyl group. A benzyl group, derived from toluene, has the formula $C_6H_5CH_2$—.

C_6H_5— or

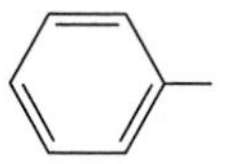

phenyl group

$C_6H_5CH_2$— or

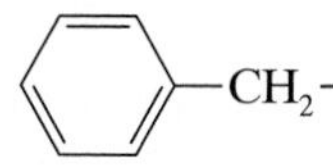

benzyl group

If alkyl groups containing fewer than six carbon atoms are bonded to a benzene ring, the compound is named an alkyl-substituted benzene. For more complex molecules, the term phenyl designates the aryl group on the parent chain of carbon atoms, as in 3-phenylheptane.

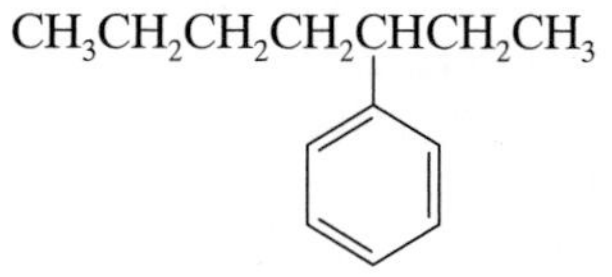

$CH_3CH_2CH_2CH_2CHCH_2CH_3$

3-phenylheptane

EXAMPLE 15.8

Indicate how a name could be derived for the following trisubstituted compound known as BHT, used as an antioxidant in some food products.

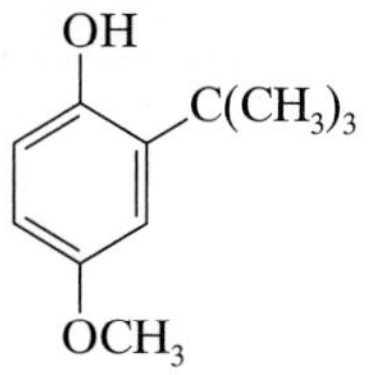

Solution

Either the —OH or the —OCH_3 group could provide the parent name, which would be phenol or anisole, respectively. Let's assume that the compound is a substituted anisole. When we assign the number 1 to the carbon atom bearing the —OCH_3 group at the "six o'clock" position, we must number the ring in a counterclockwise direction. A *tert*-butyl group is located at the 3 position, and a hydroxyl group is at the 4 position, so the compound is 3-*tert*-butyl-4-hydroxyanisole.

PROBLEM 15.8

What is the name of the following compound?

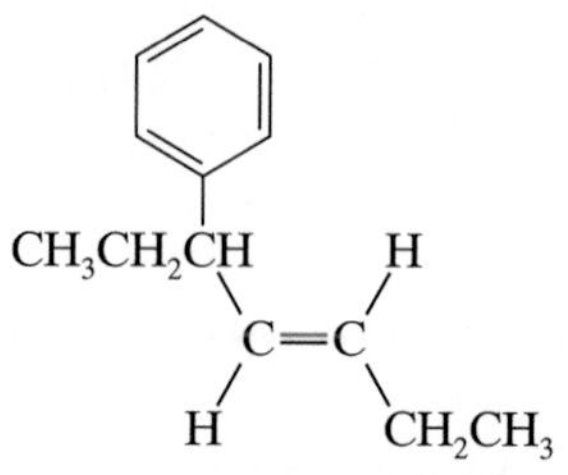

CH_3CH_2CH(C$_6H_5$)–CH=CH–CH_2CH_3 (H and H trans across the C=C)

Carcinogenic Aromatic Hydrocarbons

Fused polycyclic aromatic hydrocarbons containing four or more rings that are not colinear (that is, there is an "angle" in the series of rings) are carcinogenic. Thus, these compounds resemble phenanthrene, not anthracene. Three of the most potent carcinogens are 1,2-benzanthracene, 1,2,5,6-dibenzanthracene, and 3,4-benzpyrene. The angular area is bold faced in the structures shown here.

When applied to the skin of mice, small amounts of these angular fused-ring aromatic hydrocarbons cause cancer in about a month. These compounds are present in effluent from coal-burning power plants and in automobile exhaust. In addition, they are found in tobacco smoke and in meat cooked over charcoal. The incidence of lung cancer among smokers and inhabitants of large urban areas may be partially the result of breathing these airborne compounds over a period of time.

Although the reason was not understood at the time, it was common for chimney sweeps in England as long as 200 years ago to develop cancer. Today the reason is understood. While they worked, the chimney sweeps became covered with chimney soot and inhaled the dust that contained angular fused aromatic hydrocarbons.

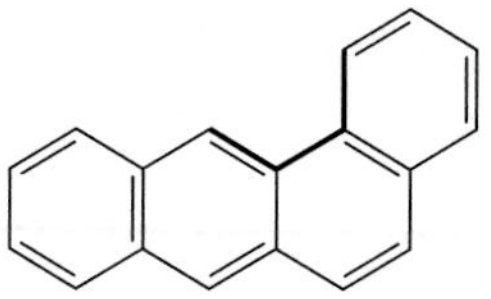

1,2-benzathracene

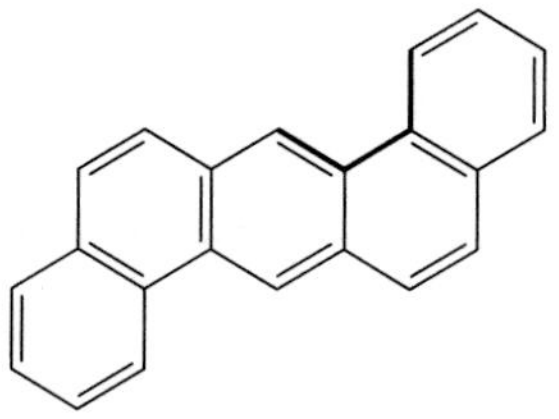

1,2,5,6-dibenzanthracene

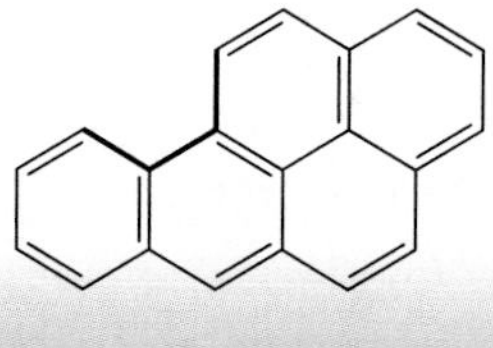

3,4-benzpyrene

15.11 Polycyclic Aromatic Hydrocarbons

Some aromatic compounds contain two or more rings "fused" together. **Fusion** of carbon rings means that two carbon atoms are common to two rings. Compounds of this type are called **polycyclic aromatic hydrocarbons.** Several examples of these compounds follow. Note that all the carbon atoms have a bond to a hydrogen atom, except those that are at the points of fusion. The molecules are planar—all atoms in the rings and those directly attached to the rings are in a plane.

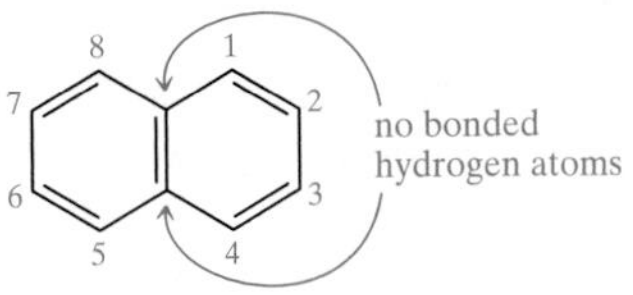

naphthalene

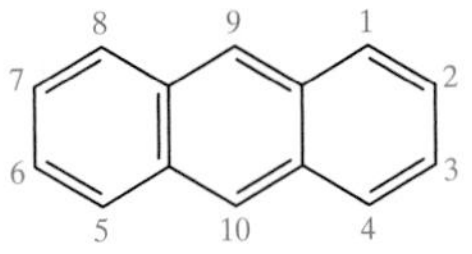

no bonded hydrogen atoms

anthracene

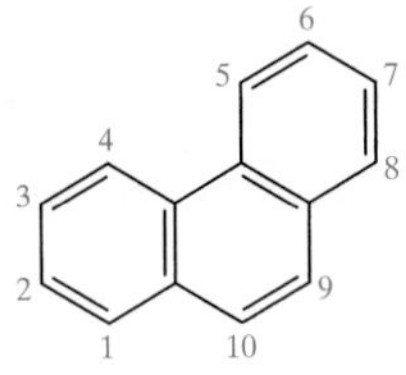

phenanthrene

15.12 Heterocyclic Aromatic Compounds

Heterocyclic compounds that have the π electron systems of arenes are called **heterocyclic aromatic compounds.** Nitrogen and oxygen are commonly encountered in naturally occurring heterocyclic compounds. Sulfur-containing compounds also exist. The structures of a few heterocyclic aromatic compounds are shown.

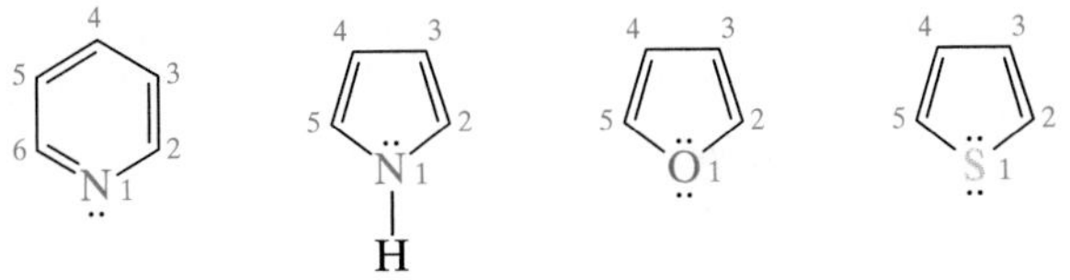

pyridine　pyrrole　furan　thiophene

Many naturally occurring, biologically important compounds—such as vitamins B_1, B_3, and B_6— have one or more aromatic heterocyclic rings of five or six atoms. Vitamin B_1 has two rings with hetero atoms.

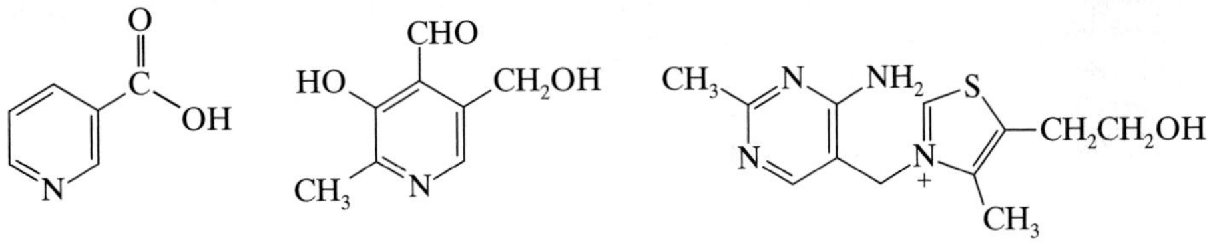

niacin (B_3)　pyridoxine (B_6)　thiamine (B_1)

Heterocyclic rings are present in many pharmaceutical compounds. For example, Tagamet (generic name cimetidine), an antiulcer drug, contains a heterocyclic aromatic ring with two nitrogen atoms. Methotrexate, a chemotherapeutic agent used to treat some kinds of cancer, contains four nitrogen atoms in a ring resembling naphthalene.

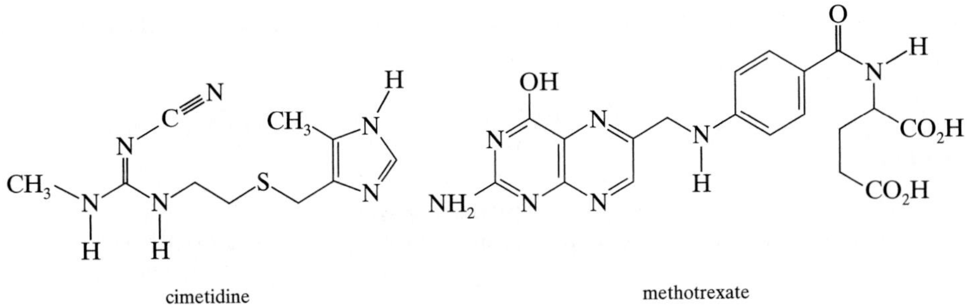

cimetidine　methotrexate

Summary

The **alkenes** are one class of unsaturated hydrocarbons. They contain a carbon–carbon double bond that consists of a σ bond and a π bond. The double-bonded carbon atoms are sp^2 hybridized. **Alkynes** are a second class of unsaturated hydrocarbons. They contain a carbon–carbon triple bond that consists of a σ bond and two π bonds. The triple-bonded carbon atoms are *sp* hybridized.

Arrangement of atoms bonded to the double-bonded carbon atoms in alkenes is spatially restricted. Two **geometric** isomers are possible for suitably substituted alkenes; these are called **cis** and **trans** isomers.

The IUPAC system for naming alkenes uses the -ene suffix and designates the position of the double bond by a number. The numbering of substituents is dictated by the numbers assigned to double-bonded carbon atoms.

The characteristic reaction of alkenes is **addition** to the double bond. The product of the reaction can be predicted from **Markovnikov's rule.** Bromine is a **symmetrical reagent;** water and hydrogen halides such as HCl and HBr are **unsymmetrical reagents.**

Polymers are high molecular weight molecules produced by the **polymerization** of thousands of repeating **monomer** units.

The IUPAC system of naming alkynes uses the -yne suffix. The atoms in a compound are numbered so as to assign the lowest numbers to the triple-bonded carbon atoms.

The reactions of alkynes are additions across the triple bond. The reactions resemble those of the alkenes, but 2 moles of reagent may add to 1 mole of the alkyne.

Aromatic hydrocarbons or **arenes** are a third type of unsaturated hydrocarbon. These compounds have a special stability as a result of **delocalization** of π electrons about a ring of carbon atoms. Each carbon atom in the ring is sp^2 hybridized. The aromatic group C_6H_5— derived from benzene is called a **phenyl group.** The general name for aromatic groups that are to be treated as substituents is **aryl. Polycyclic aromatic hydrocarbons** have **fused** carbon rings, meaning that two carbon atoms are common to two rings. **Heterocyclic aromatic hydrocarbons** have one or more heteroatoms in place of carbon in aromatic rings.

Exercises

Molecular Formulas

15.1 Write the molecular formula for each of the following compounds.

(a) 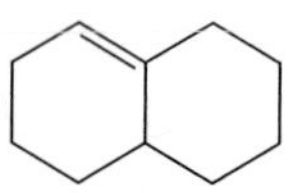 (b)

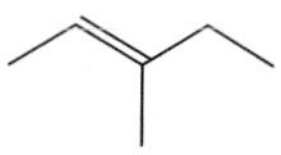

(c) 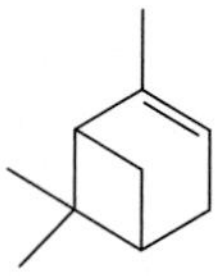 (d)

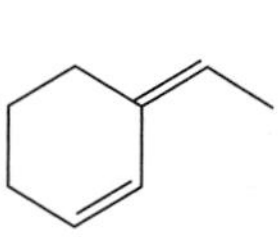

15.2 Write the molecular formula for each of the following compounds.

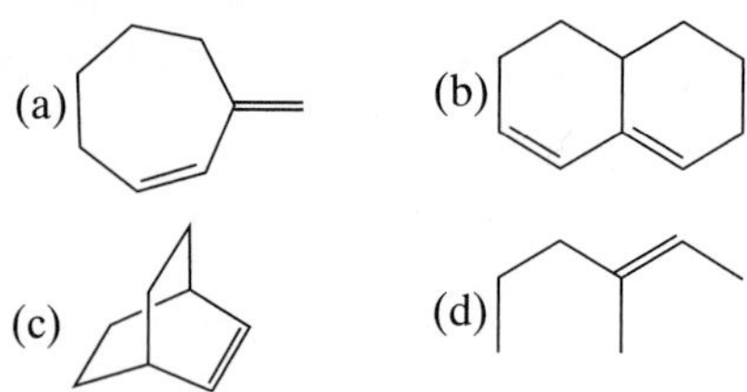

Isomers

15.3 Draw structural formulas for four isomeric alkenes with the molecular formula C_4H_8, and name each compound.

15.4 Draw structural formulas for two isomeric cycloalkanes with the molecular formula C_4H_8, and name each compound.

15.5 Draw structural formulas for six isomeric alkenes with the molecular formula C_5H_{10}, and name each compound.

15.6 Draw structural formulas for two isomeric alkynes with the molecular formula C_4H_6, and name each compound.

15.7 Draw structural formulas for the four isomeric alkenes with the molecular formula C_3H_5Cl, and name each compound.

15.8 Draw structural formulas for the three isomeric alkynes with the molecular formula C_5H_8, and name each compound.

Geometric Isomers

15.9 Which of the following molecules can exist as cis and trans isomers?

(a) 1-hexene (b) 3-heptene
(c) 4-methyl-2-pentene (d) 2-methyl-2-butene

15.10 Which of the following molecules can exist as cis and trans isomers?

(a) 3-methyl-1-hexene (b) 3-ethyl-3-heptene
(c) 2-methyl-2-pentene (d) 3-methyl-2-pentene

15.11 The sex attractant of the European vine moth has the following structural formula. What is the geometry about each of the double bonds?

H H H H $(CH_2)_6-O-\overset{O}{\overset{\|}{C}}-CH_3$

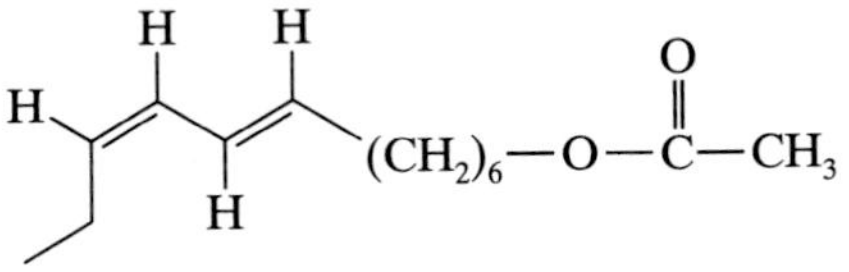

15.12 The sex attractant of the pink bollworm moth has the following structural formula. What is the geometry about each of the double bonds?

H H H H $(CH_2)_6-O-\overset{O}{\overset{\|}{C}}-CH_3$

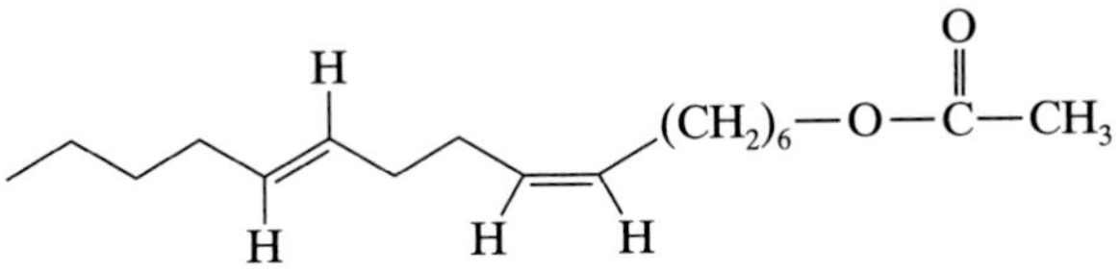

Nomenclature of Alkenes

15.13 Name each of the following compounds.

(a) 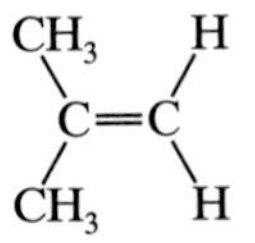 CH_3, CH_3 / C=C / H, H (b) CH_3, CH_3 / C=C / CH_3, CH_3

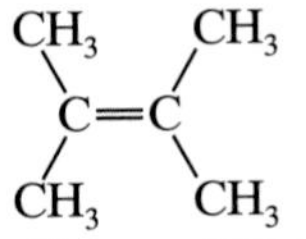

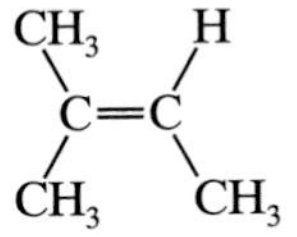

(c) CH_3, CH_3 / C=C / H, CH_3 (d) 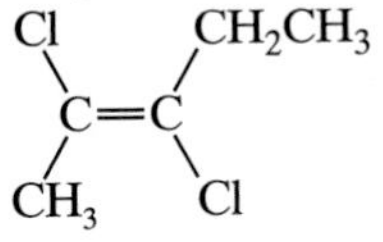 Cl, CH_3 / C=C / CH_2CH_3, Cl

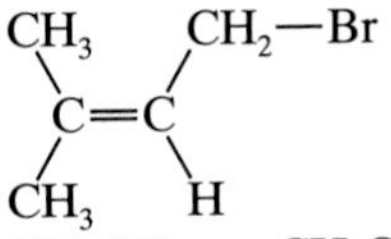

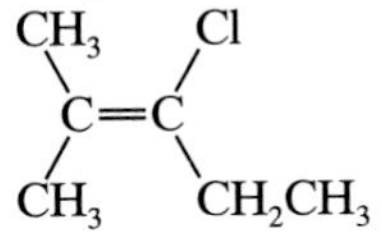

15.14 Name each of the following compounds.

(a) 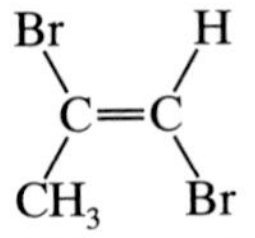 Br, CH_3 / C=C / H, Br (b) CH_3, CH_3 / C=C / CH_2-Br, H

(c) CH_3, CH_3 / C=C / Cl, CH_2CH_3 (d) 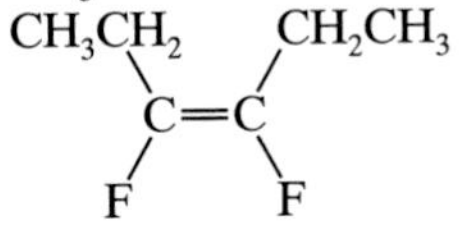 CH_3CH_2, F / C=C / CH_2CH_3, F

15.15 Name each of the following compounds.

(a) 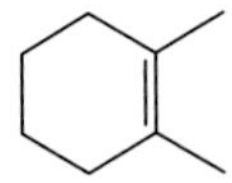 (b)

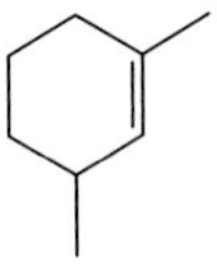

(c) 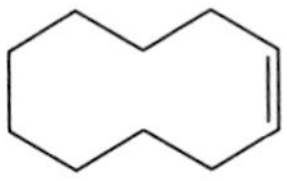 (d)

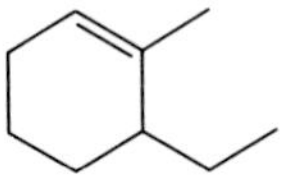

15.16 Name each of the following compounds.

(a) (b)

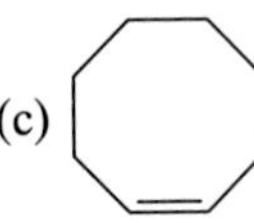

(c)

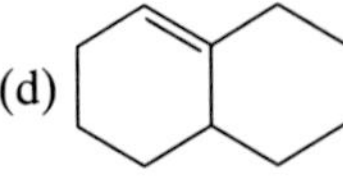

(d)

15.17 Draw a structural formula for each of the following compounds.

(a) 2-methyl-2-pentene
(b) 1-hexene
(c) *cis*-2-methyl-3-hexene
(d) *trans*-5-methyl-2-hexene

15.18 Draw a structural formula for each of the following compounds.

(a) *trans*-1-chloropropene
(b) *cis*-2,3-dichloro-2-butene
(c) 3-chloropropene
(d) 4-chloro-2,4-dimethyl-2-hexene

15.19 Draw a structural formula for each of the following compounds.

(a) cyclohexene
(b) 1-methylcyclopentene
(c) 1,2-dibromocyclohexene
(d) 4,4-dimethylcyclohexene

15.20 Draw a structural formula for each of the following compounds.

(a) cyclopentene
(b) 3-methylcyclohexene
(c) 1,3-dibromocyclopentene
(d) 3,3-dichlorocyclopentene

Chemical Reactions of Alkenes

15.21 Describe the observation that is made when 1-hexene reacts with bromine. How could bromine be used to distinguish between 1-hexene and cyclohexane?

15.22 Describe the observation that is made when *cis*-2-pentene reacts with potassium permanganate. How could this reagent be used to distinguish between *cis*-2-pentene and cyclopentane?

15.23 Write the structure of the product of the reaction of bromine with 2-methyl-1-butene.

15.24 Write the structure of the product of the reaction of HBr with 2-methyl-2-butene.

15.25 Write the structure of the product of the reaction of HBr with each of the following compounds.

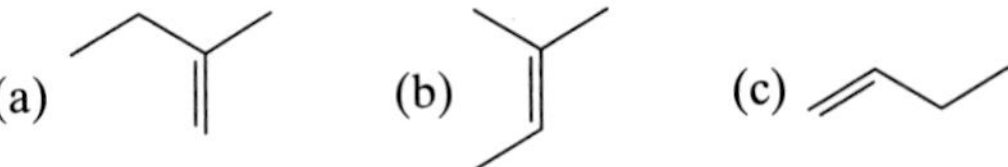

(a) (b) (c)

15.26 Write the product of the reaction of HBr with each of the following compounds.

(a) 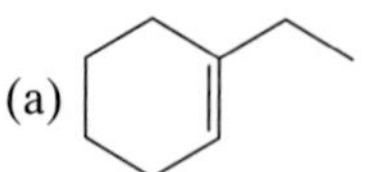 (b) 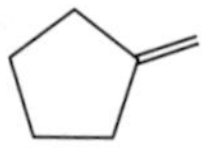 (c)

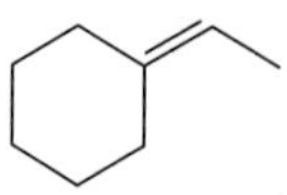

Polymers

15.27 Draw a representation of the polymer produced from $CH_2{=}CHCN$.

15.28 Draw a representation of the polymer produced from $CH_2{=}CCl_2$.

Nomenclature of Alkynes

15.29 Name each of the following compounds

(a) $CH_3CH_2CH_2C{\equiv}CH$
(b) $(CH_3)_3CC{\equiv}CCH_2CH_3$
(c) $CH_3{-}C{\equiv}C{-}CH(CH_2{-}CH_3){-}CH_3$

15.30 Name each of the following compounds.

(a) $CH_3CHBrCHBrC{\equiv}CCH_3$
(b) $Cl(CH_2)_2C{\equiv}C(CH_2)_3CH_3$
(c) $CH_3{-}CH(CH_2{-}CH_3){-}CH_2{-}C{\equiv}C{-}CH(Cl){-}CH_3$

15.31 Write the structural formula for each of the following compounds.

(a) 2-hexyne (b) 3-methyl-1-pentyne
(c) 5-ethyl-3-octyne

15.32 Write the structural formula for each of the following compounds.

(a) 3-heptyne (b) 4-methyl-1-pentyne
(c) 5-methyl-3-heptyne

Chemical Reactions of Alkynes

15.33 Complete the following equations.

(a) $CH_3C{\equiv}CH + 2\ H_2 \longrightarrow$
(b) $CH_3C{\equiv}CH + 2\ Br_2 \longrightarrow$
(c) $CH_3CH_2C{\equiv}CH + HBr \longrightarrow$
(d) $CH_3CH_2C{\equiv}CH + 2\ HBr \longrightarrow$

15.34 Complete the following equations.

(a) $CH_3C{\equiv}CCH_3 + 2\ H_2 \longrightarrow$
(b) $CH_3C{\equiv}CCH_3 + 2\ Br_2 \longrightarrow$
(c) $CH_3C{\equiv}CCH_3 + HBr \longrightarrow$
(d) $CH_3C{\equiv}CCH_3 + 2\ HBr \longrightarrow$

15.35 How many moles of hydrogen gas will react with each of the following compounds?

(a) $CH_3{-}CH{=}CH{-}C{\equiv}CH$
(b) $HC{\equiv}C{-}C{\equiv}CH$
(c) $CH_2{=}CH{-}C{\equiv}C{-}CH{=}CH_2$

15.36 How many moles of bromine will react with each of the following compounds?

(a) $CH_3{-}CH{=}CH{-}C{\equiv}C{-}CH_3$
(b) $HC{\equiv}C{-}C{\equiv}C{-}CH{=}CH_2$
(c) $CH_2{=}CH{-}C{\equiv}C{-}CH{=}CH_2$

Aromatic Compounds

15.37 Identify each of the following as an *ortho-*, *meta-*, or *para-*substituted compound.

(a) methylparaben, a food preservative

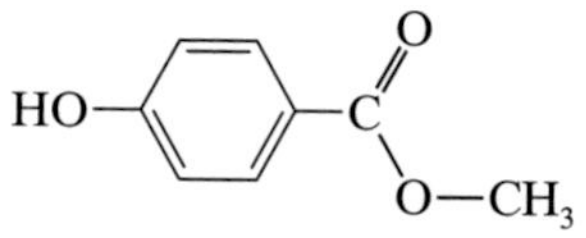

HO–C₆H₄–C(=O)–O–CH₃

(b) crotamiton, used for topical treatment of scabies

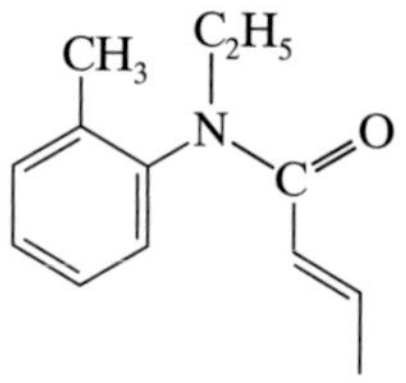

(c) diethyltoluamide, an insect repellent

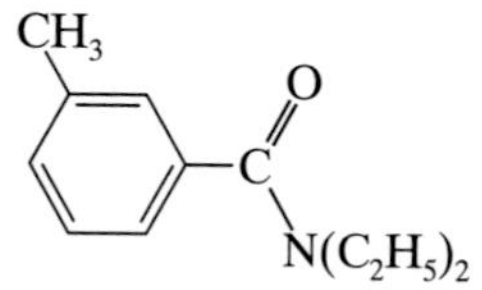

15.38 Identify each of the following as an *ortho-*, *meta-*, or *para-*substituted compound.

(a) resorcinol monoacetate, a germicide

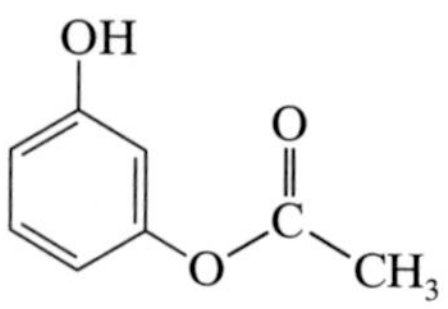

(b) halazone, used to disinfect water

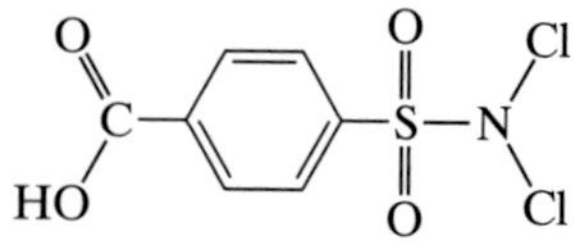

(c) salicylamide, an analgesic

NH_2 C O OH

15.39 Name each of the following compounds.

(a) CH_2CH_3 (b) CH_3 CH CH_3

(c) C_2H_5 C_2H_5

(d) CH_3 CH_3 CH_3

15.40 Name each of the following compounds.

(a) Cl Br (b) OH Br

(c) NH_2 Br (d) CH_3 Cl

15.41 Name the following compound, which is an antiseptic agent used to treat athlete's foot.

OH CH_3 CH_3 Cl

15.42 Name the following compound, which is used to make a local anesthetic.

CH_3 NH_2 CH_3

15.43 Write the formula of each of the following compounds.

(a) 3-phenylheptane
(b) 1,1-diphenylethane
(c) 1-chloro-2-phenylpropane

15.44 Write the formula of each of the following compounds.

(a) 1,2-diphenylethane
(b) 1,3-diphenylpropane
(c) 2-chloro-1-phenylpropane

15.45 There are three isomeric dichlorobenzenes. One compound has no dipole moment. Which one?

15.46 There are three isomeric trichlorobenzenes. One compound has no dipole moment. Which one?

Polycyclic Aromatic Hydrocarbons

15.47 What is the molecular formula of each of the following compounds?

(a)

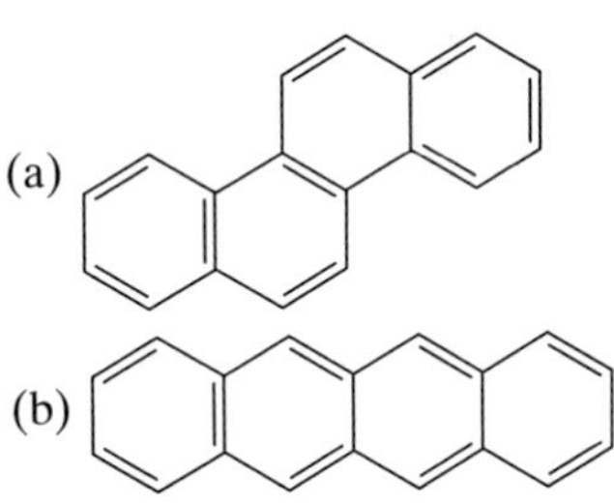

(b)

15.48 What is the molecular formula of each of the following compounds?

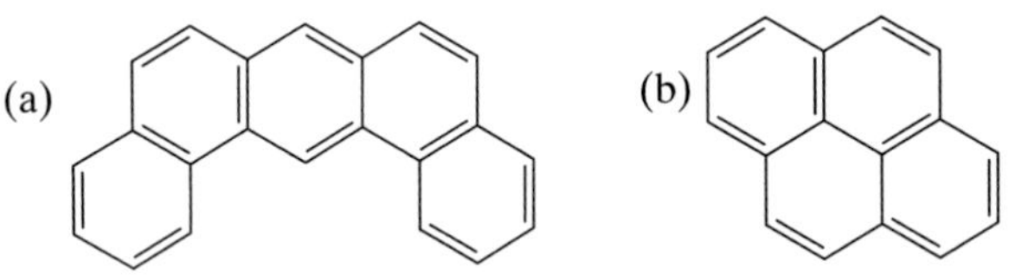

15.49 Two isomeric bromonaphthalenes exist. Draw their structures.

15.50 Three isomeric bromoanthracenes exist. Draw their structures.

Heterocyclic Aromatic Compounds

15.51 There are three isomeric diazines ($C_4N_2H_4$) that resemble benzene but have two nitrogen atoms in place of carbon in the ring. Draw their structures. Which of the isomers should have no dipole moment?

15.52 There are three isomeric triazines ($C_3N_3H_3$) that resemble benzene but have three nitrogen atoms in place of carbon in the ring. Draw their structures. Which of the isomers should have no dipole moment?

15.53 Identify the heterocyclic ring structure contained in the following compound, which has been investigated as possible male contraceptive.

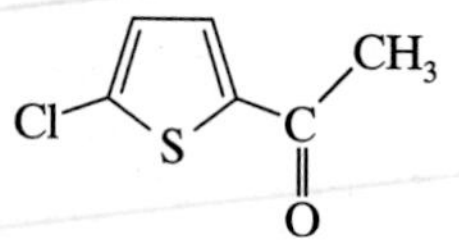

Cl, S, C, CH_3, O

15.54 Identify the aromatic heterocyclic ring structure contained in tolmetin, a drug used to lower blood sugar levels.

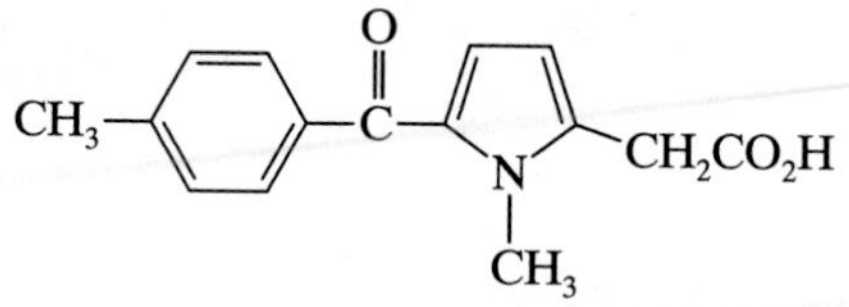

CH_3, O, C, N, CH_2CO_2H, CH_3

Grapes have been used to make wine since ancient times by many societies. (The Image Bank)

16

Alcohols, Phenols, and Ethers

16.1 Compounds of Oxygen
16.2 Common Alcohols and Phenols
16.3 Nomenclature and Classification of Alcohols
16.4 Hydrogen Bonding in Alcohols
16.5 Acid–Base Reactions
16.6 Substitution Reactions
16.7 Dehydration of Alcohols
16.8 Oxidation of Alcohols
16.9 Ethers
16.10 Sulfur Compounds

Essays
- *Fetal Alcohol Syndrome*
- *Phosphate and Pyrophosphate Esters*
- *Ethanol—An Antidote for Methanol Poisoning*
- *Polyethers in the Laboratory and in Medicine*

16.1 Compounds of Oxygen

Carbohydrates, fats, proteins, and nucleic acids all contain oxygen. To understand the chemistry of these molecules we first consider simpler organic compounds with oxygen-containing functional groups. Families of organic compounds that contain oxygen include alcohols, phenols, ethers, aldehydes, ketones, acids, esters, and amides. Alcohols are the main focus of this chapter; phenols and ethers will be discussed briefly.

Alcohols are compounds that contain the hydroxyl (—OH) functional group bonded to an sp^3-hybridized carbon atom. They can be viewed as the organic analogs of water in which one hydrogen atom is replaced by an alkyl group. Compounds with a hydroxyl group bonded to an aromatic ring are **phenols.** The distinction between an alcohol and a phenol is illustrated by the following isomeric compounds.

This hydroxyl group is bonded to an sp^2-hybridized carbon atom. The compound is a phenol.

This hydroxyl group is bonded to an sp^3-hybridized carbon atom. The compound is an alcohol.

HO ... OH

There are differences in the reactions of the carbon–oxygen bonds of alcohols and phenols because the oxygen atom is bonded to a differently hybridized carbon atom. However, there are similarities in the chemical properties of the oxygen–hydrogen bonds of alcohols and phenols. For example, both types of compounds react with alkali metals the same way water does to produce hydrogen gas.

$$2\ Na + 2\ ROH \longrightarrow H_2 + 2\ RO^- + 2\ Na^+$$
$$2\ Na + 2\ ArOH \longrightarrow H_2 + 2\ ArO^- + 2\ Na^+$$
$$2\ Na + 2\ H_2O \longrightarrow H_2 + 2\ HO^- + 2\ Na^+$$

Ethers contain two groups, which may be alkyl or aryl groups, bonded to oxygen. Like alcohols, ethers can be viewed as organic analogs of water. However, ethers are different from alcohols because they do not have an —OH group, and they have different chemical and physical properties. The simplest ether, dimethyl ether, has two methyl groups bonded to the ether oxygen atom.

$$H_3C{-}O{-}CH_3$$

dimethyl ether

The oxygen atoms of both alcohols and ethers are sp^3-hybridized. As a result both the C—O—H bond angle of alcohols and the C—O—C bond angle of ethers are close to the tetrahedral bond angle (Figure 16.1). Using VSEPR theory,

FIGURE 16.1 Structure of Methanol and Dimethyl Ether

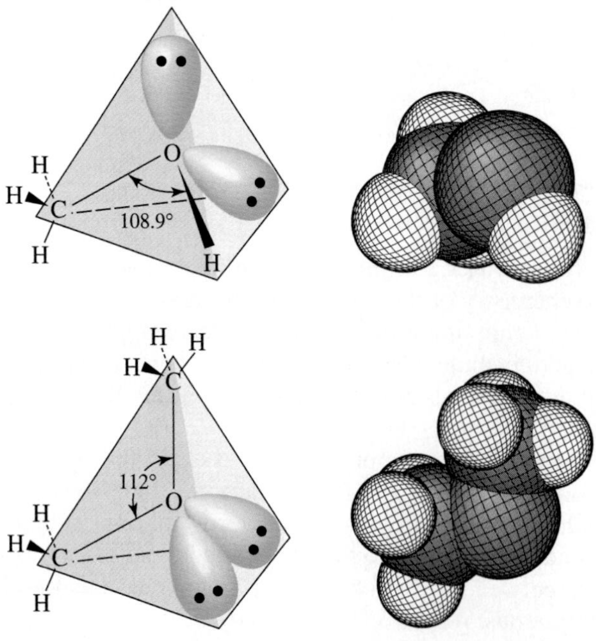

we show the two lone pairs of electrons of the oxygen atom directed to two corners of a tetrahedron. Space-filling models of both methanol and dimethyl ether are also shown in Figure 16.1.

16.2 Common Alcohols and Phenols

Methanol

Methanol (CH_3OH) is a toxic substance. Ingestion of as little as 15 mL of pure methanol can cause blindness; 30 mL will cause death. Temporary blindness, permanent blindness, or death can result from consumption of the methanol present as an impurity in illegally produced ethanol. Methanol is used in car windshield washer fluids and has been used as a fuel in racing cars for many years.

Ethanol

Ethanol (CH_3CH_2OH), the substance popularly known as alcohol, is a depressant, not a stimulant, as some people believe. It is rapidly absorbed into the blood. The effects of various blood alcohol levels are given in Table 16.1. Only about 1 oz of pure alcohol per hour is removed from the human body by oxidation in the liver. Even in moderate amounts, alcohol causes drowsiness and depresses brain functions. As a consequence, activities that require judgment or coordination, like driving a car, are impaired.

Fermentation of almost any substance containing sugar produces ethanol, but only in concentrations up to about 14%. The rate of the reaction, catalyzed by yeast cells, is inhibited as the alcohol concentration increases. The taste and color of wine are controlled by the variety of grape used and the way the grape juice and skins are treated in the crushing process. Although there are differences in quality and price between common and vintage wines, the major components—water and ethanol—are the same and constitute 99% of all wines. The price difference has to do with the remaining 1% of the compounds in the wine.

Ethanol is used as an industrial solvent. Tax-free permits are issued for this use. Pure ethanol costs about $1/gallon, but the federal tax on pure ethanol is about $20/gallon! To ensure the proper use of untaxed alcohol, substances are added in small amounts to render such alcohol unfit for drinking. Alcohol containing adulterants is called **denatured alcohol.**

Polyhydroxy Alcohols

Ethylene glycol and glycerol are polyhydroxy alcohols that are widely used in commercial products. Ethylene glycol contains two hydroxyl groups, one replacing a

TABLE 16.1 Effect of Blood Alcohol Level

% by Volume	Effect
0.05–0.15	loss of coordination
0.15–0.20	severe intoxication
0.20–0.40	loss of consciousness
0.50	death

Fetal Alcohol Syndrome

Adults can metabolize alcohol in several steps. The alcohol is first oxidized in the liver to acetaldehyde and then further oxidized to acetic acid. After release from the liver the acetic acid is converted to acetyl CoA, which can be oxidized to give carbon dioxide and water or can be used to synthesize fat (Section 28.5).

$$CH_3CH_2OH \xrightarrow[\text{dehydrogenase}]{\text{alcohol}} CH_3{-}\overset{\overset{\large O}{\|}}{C}{-}H \xrightarrow[\text{dehydrogenase}]{\text{acetaldehyde}}$$

$$CH_3{-}\overset{\overset{\large O}{\|}}{C}{-}OH \xrightarrow[\text{synthetase}]{\text{acetyl CoA}} CH_3{-}\overset{\overset{\large O}{\|}}{C}{-}SCoA$$

Small amounts of ethyl alcohol are produced by microbial fermentation in the large intestine. This small quantity as well as ingested alcohol can be oxidized by adults. Drinking excessive amounts of alcohol increases the fat content of the body and can cause cirrhosis of the liver.

The liver of a fetus does not contain the enzymes necessary to metabolize alcohol. Alcohol ingested by a pregnant woman efficiently crosses the placental membrane and causes changes in the fetus that are observed as deformities in the newborn infant. The group of problems experienced by young children is collectively called **fetal alcohol syndrome.** They include mental retardation, poor growth, and facial deformities. These facial characteristics include a flattened face with a shorter nasal bridge, indistinct infranasal depression (tissue folds between the nose and lip), thin upper lip, and short eyelid tissues.

Exactly how much alcohol is required to cause varying degrees of deformities is unknown. However, there is considerable evidence that even "social drinking" during pregnancy can cause fetal alcohol effects. These include substandard birth weight, slow growth and development of the newborn, some learning disabilities, and some behavioral problems. Considering the severity of the consequences for the child, it is recommended that pregnant women abstain from alcohol. There is no "cure" for fetal alcohol syndrome—it is a lifelong affliction that is totally preventable.

Ethylene glycol is produced in large quantities by chemical industry and sold in small containers. (PhotoEdit)

hydrogen atom on each of the adjacent carbon atoms of ethane. Glycerol, also known as glycerin, contains three hydroxyl groups, one replacing a hydrogen atom on each of the three carbon atoms of propane.

$$\underset{\text{ethylene glycol}}{\underset{OH}{\underset{|}{CH_2}}{-}\underset{OH}{\underset{|}{CH_2}}} \qquad \underset{\text{glycerol}}{\underset{OH}{\underset{|}{CH_2}}{-}\underset{OH}{\underset{|}{CH}}{-}\underset{OH}{\underset{|}{CH_2}}}$$

Ethylene glycol is used in automobile antifreeze. However, one of the largest commercial uses of ethylene glycol is in the production of Dacron fibers and Mylar film (Chapter 18), the latter of which is used in tapes for recorders and computers.

The moisture-retaining properties of glycerol make it useful in skin lotions, inks, and pharmaceuticals. Glycerol is present as the backbone of several important biological compounds. These include phospholipids, which are present in cell membranes, and fats and oils, which are used as a source of energy in metabolic reactions (Chapter 22).

Phenols

Phenols are germicides. The English surgeon Joseph Lister used phenol itself as a hospital antiseptic in the late nineteenth century. Phenol can cause severe burns to skin, and it is always used as a dilute solution. A 2% solution can be used to decontaminate medical instruments. Alkyl groups on the aromatic ring increase the biological activity of phenols. *p*-Cresol and its ortho and meta isomers are used in Lysol, the commercial disinfectant. Thymol is used by dentists to sterilize a tooth prior to filling it.

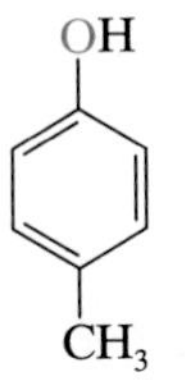

OH CH$_3$ OH CH(CH$_3$)$_2$ CH$_3$

p-cresol 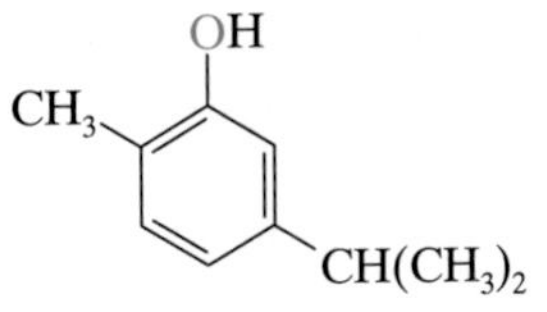 thymol

The efficiency of germicides is measured in terms of the phenol coefficient. A 1% solution of a germicide that is as effective as a 10% solution of phenol has a phenol coefficient of 10. Mouthwash solutions usually contain phenolic substances. The antiseptic hexylresorcinol is used in throat lozenges. Hexachlorophene (phenol coefficient = 120) has been used in some toothpastes, deodorants, and soaps.

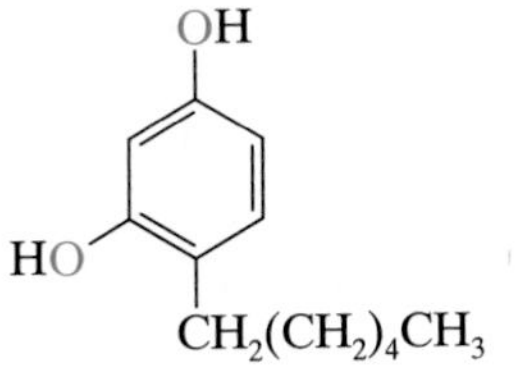

OH HO CH$_2$(CH$_2$)$_4$CH$_3$ Cl OH HO Cl CH$_2$ Cl Cl Cl Cl

hexylresorcinol 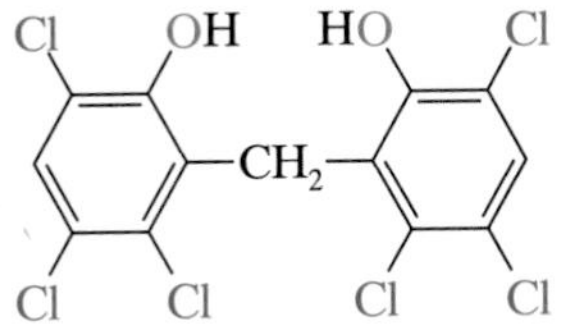 hexachlorophene

16.3 Nomenclature and Classification of Alcohols

Alcohols that contain one to four carbon atoms have common names consisting of the name of the alkyl group followed by the term "alcohol." For example, CH_3CH_2OH is ethyl alcohol, and $CH_3CH(OH)CH_3$ is isopropyl alcohol. Other common names are benzyl alcohol and allyl alcohol.

OH CH$_2$OH

allyl alcohol 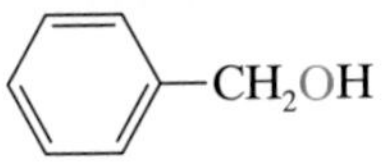 benzyl alcohol

The IUPAC system of naming alcohols is as follows.

1. The longest chain of carbon atoms that includes the hydroxyl group is designated as the parent chain.

$$CH_3{-}CH_2{-}\underset{\substack{|\\ CH_2{-}OH}}{CH}{-}CH_2{-}CH_3$$

The longest chain that contains the hydroxyl group has four carbon atoms, although the longest chain has five carbon atoms.

2. The parent is named by substituting the suffix -ol for the final "-e" of the corresponding alkane.

$$CH_3{-}\overset{\substack{CH_3\\ |}}{CH}{-}CH_2{-}CH_2{-}OH$$

The parent alkane is butane. This is a substituted butanol. A methyl branch is attached to the butanol chain.

3. The position of the hydroxyl group is indicated using the number of the carbon atom to which it is attached. The chain is numbered so that the carbon atom bearing the hydroxyl group has the lower number.

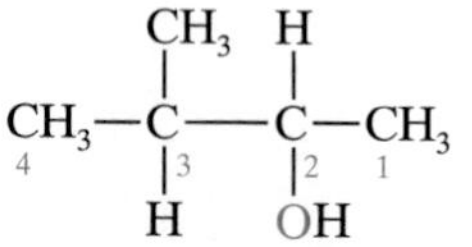

$$\begin{array}{c} \quad CH_3 \quad H \\ CH_3-\underset{\substack{|3\\H}}{\overset{|}{C}}-\underset{\substack{|2\\OH}}{\overset{|}{C}}-CH_3 \end{array}$$

This is 3-methyl-2-butanol, not 2-methyl-3-butanol.

4. When the hydroxyl group is attached to a ring, the ring is numbered starting with the carbon atom bearing the hydroxyl group. Numbering is continued in the direction that gives the lowest numbers to carbon atoms with substituents such as alkyl groups. The number 1 is not used in the name to indicate the position of the hydroxyl group.

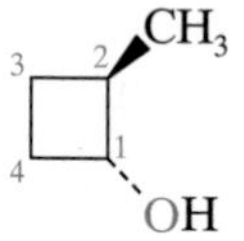

trans-2-methvlcvclobutanol 3.3-dimethvlcvclohexanol

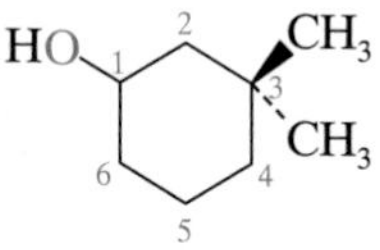

5. Alcohols that contain two or more hydroxyl groups are called diols, triols, and so on. The terminal -e is retained in the name of the parent alkane, and the suffix -diol or -triol is added. The positions of the hydroxyl groups in the parent chain are indicated by numbers.

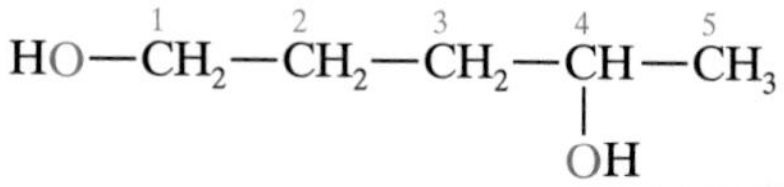

$$HO-CH_2-CH_2-CH_2-\underset{\substack{|\\OH}}{CH}-CH_3$$

1,4-pentanediol

6. When an alcohol contains a double or triple bond, the hydroxyl group takes precedence in the numbering of the carbon chain. The number that indicates the position of the multiple bond is placed in front of the name of the alkene (or alkyne). Delete the final e and append a number that indicates the position of the hydroxyl group along with the suffix -ol.

$$CH_2{=}CH-CH_2-CH_2-CH_2-OH$$

4-penten-1-ol

$$CH_3-C{\equiv}C-CH_2-\underset{\substack{|\\OH}}{CH}-CH_2-OH$$

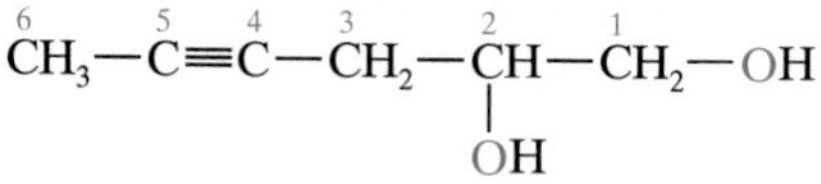

4-hexyn-1, 2-diol

EXAMPLE 16.1

Assign the IUPAC name to citronellol, a compound found in geranium oil used in perfumes.

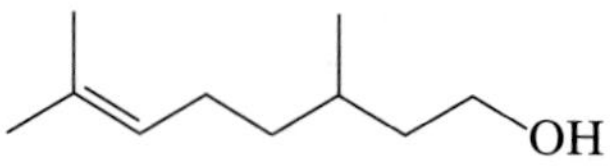

OH

Solution

The longest carbon chain that contains the hydroxyl group has eight carbon atoms. The hydroxyl group is on the carbon atom located on the right side of the chain. This carbon atom is C-1. Thus, the methyl groups are at the C-3 and C-7 atoms. A double bond is located at the C-6 atom. The name is 3,7-dimethyl-6-octen-1-ol.

PROBLEM 16.1

Assign the IUPAC name to the sex attractant of the Mediterranean fruit fly.

H

OH

H

Classification of Alcohols

Although all alcohols have some reactions in common, there are differences in reactivity depending on the location of the hydroxyl group. Alcohols are classified by the number of alkyl groups bonded to the carbon atom bearing the hydroxyl group.

1° carbon atom — $CH_3CH_2CH_2CH_2OH$ — a primary alcohol

2° carbon atom — $CH_3CH_2CH(OH)CH_3$ — a secondary alcohol

3° carbon atom — $CH_3—C(OH)(CH_3)—CH_3$ — a tertiary alcohol

EXAMPLE 16.2

Classify the carbon atoms bearing hydroxyl groups in the antibiotic chloramphenicol.

$O_2N—C_6H_4—CH(OH)—CH(CH_2OH)—NH—C(=O)—CHCl—Cl$

Solution

First, locate the oxygen atoms in the structure. There are five, but only two are found in hydroxyl groups. Two oxygen atoms are bonded to a nitrogen atom in a group bonded to the benzene ring. One oxygen atom has a double bond to a carbon atom in an amide group. The hydroxyl group in the middle of the structure is bonded to a carbon atom that has two hydrogen atoms and a carbon atom bonded to it; this is a primary alcohol. The hydroxyl group near the benzene ring is bonded to a carbon atom with two other carbon atoms bonded to it; one carbon atom is part of a substituted alkyl group, the other carbon atom is part of the benzene ring. This alcohol is secondary.

PROBLEM 16.2

Ethychlorvynol is a sedative–hypnotic. Classify this compound as a primary, secondary, or tertiary alcohol.

$H—C≡C—C(OH)(CH_2CH_3)—CH=CHCl$

16.4 Hydrogen Bonding in Alcohols

Alcohols have polar C—O and O—H bonds, which makes alcohols more polar than alkanes. In addition, the O—H group can participate in hydrogen bonds, which strongly affects the physical properties of alcohols. Alcohols, like water, form very strong hydrogen bonds.

Boiling Points of Alcohols

Alcohols boil at higher temperatures than alkanes of approximately the same molecular weight. Propane boils at −42 °C, and ethanol boils at 78 °C. These two compounds have approximately the same London forces, but ethanol has larger dipole–dipole forces of attraction. However, the large difference in boiling points is due mainly to hydrogen bonding between alcohol molecules in the liquid state. More energy is needed to separate the molecules because of the attraction between the hydroxyl hydrogen atom in one molecule and the oxygen atom of a neighboring molecule.

hydrogen bond

A comparison of the boiling points of the 1-alkanols and alkanes of approximately the same molecular weight is given in Figure 16.2. Note the large difference in boiling points for compounds at lower molecular weights. As the molecular weights in-

FIGURE 16.2 Boiling Points of Alcohols and Alkanes

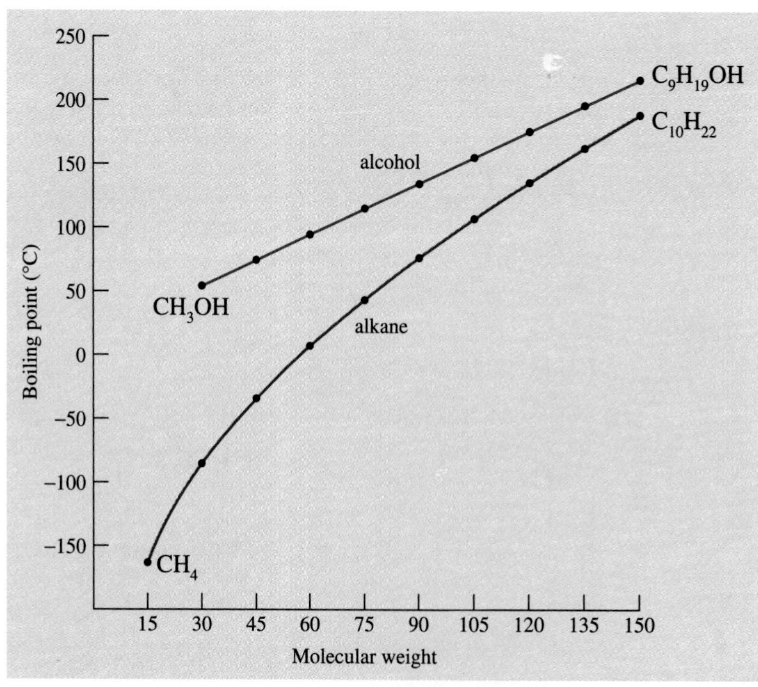

TABLE 16.2 Boiling Points and Solubilities of Alcohols

Name	Formula	Boiling Point (°C)	Solubility (g/100 mL in water)
methanol	CH_3OH	65	miscible
ethanol	CH_3CH_2OH	78	miscible
1-propanol	$CH_3CH_2CH_2OH$	97	miscible
1-butanol	$CH_3CH_2CH_2CH_2OH$	117	7.9
1-pentanol	$CH_3CH_2CH_2CH_2CH_2OH$	137	2.7
1-hexanol	$CH_3(CH_2)_4CH_2OH$	158	0.59
1-heptanol	$CH_3(CH_2)_5CH_2OH$	176	0.09
1-octanol	$CH_3(CH_2)_6CH_2OH$	194	insoluble
1-nonanol	$CH_3(CH_2)_7CH_2OH$	213	insoluble
1-decanol	$CH_3(CH_2)_8CH_2OH$	229	insoluble

crease, the two curves approach each other. At high molecular weights the hydroxyl portion of the molecule is less important, and the alcohol behaves more like an alkane.

Solubility of Alcohols in Water

Propylene glycol is used to de-ice airplanes in inclement weather. (Rainbow)

Table 16.2 lists the solubilities in water of some alcohols with normal alkyl groups. The lower molecular weight alcohols are completely soluble in water. You should expect this behavior based on the general rule that "like dissolves like." However, with increasing size of the alkyl group, alcohols more closely resemble alkanes, and the hydroxyl group is less important to the physical properties of the alcohols. Water still can form hydrogen bonds to the hydroxyl group. However, the long chain interferes too much with the other water molecules and prevents them from hydrogen bonding to each other. Disrupting the extensive hydrogen bonding of water requires too much energy. As a result, the solubility of alcohols decreases with increasing size of the alkyl group.

16.5 Acid–Base Reactions

Alcohols, like water, can act as acids or bases. Alcohols are weak acids, as evidenced by the small K_a of ethanol. The conjugate base of an alcohol is an **alkoxide ion.**

$$CH_3CH_2OH + H_2O \rightleftharpoons \underset{\text{ethoxide ion}}{CH_3CH_2O^-} + H_3O^+ \qquad K_a = 1 \times 10^{-17}$$

Alkoxides are slightly stronger bases than the hydroxide ion. They are used as bases in organic solvents because they are more soluble than hydroxide salts.

Alcohols react with strong acids to give conjugate acids in a reaction analogous to that of water with a strong acid to give the hydronium ion. The conjugate acids of alcohols are intermediates in a number of reactions catalyzed by strong acids.

$$CH_3CH_2OH + HA \rightleftharpoons CH_3CH_2OH_2^+ + A^-$$

Phenols Are Stronger Acids Than Alcohols

Phenols are stronger acids than alcohols, but they are still quite weak acids.

$$R{-}O{-}H + H_2O \rightleftharpoons R{-}O^- + H_3O^+ \qquad K_a = 1 \times 10^{-17}$$
$$Ar{-}O{-}H + H_2O \rightleftharpoons Ar{-}O^- + H_3O^+ \qquad K_a = 1 \times 10^{-10}$$

This increased acidity is due to the greater stability of the phenoxide ion compared with the alkoxide ion. In the alkoxide ion, the negative charge is localized at the oxygen atom. In the phenoxide ion, the negative charge is delocalized by resonance over the benzene ring. Thus, the loss of a proton from phenol gives a more stabilized anion.

Phenols react with hydroxide ions to produce the phenoxide ion. Thus, phenols dissolve in basic solution, whereas high molecular weight alcohols are insoluble.

16.6 Substitution Reactions

The hydroxyl group of an alcohol can be substituted for or replaced by a halogen. Either HBr or PBr_3 may be used to convert an alcohol into a bromoalkane. In these substitution reactions the carbon–oxygen bond is broken, and the halogen replaces the hydroxyl group.

$$CH_3CH_2CH_2CH_2OH + HBr \longrightarrow CH_3CH_2CH_2CH_2Br + H_2O$$
$$3\ CH_3CH_2CH_2OH + PBr_3 \longrightarrow 3\ CH_3CH_2CH_2Br + H_3PO_3$$

The reaction with HCl to form a chloroalkane requires $ZnCl_2$ as a catalyst.

$$CH_3CH_2CH_2OH + HCl \xrightarrow{ZnCl_2} CH_3CH_2CH_2Cl + H_2O$$

The rate for the reaction with HCl, in the presence of $ZnCl_2$ as catalyst, depends on the type of alcohol. The reactivity decreases in the order tertiary > secondary > primary alcohols. When this reaction is used for classification purposes it is called the **Lucas test.** Chloroalkanes are insoluble in water, so they are easily detected in the test. If an alcohol is added to Lucas reagent and turbidity occurs immediately, the alcohol is probably tertiary. The turbidity results from the insoluble halide formed in the reaction. If the turbidity develops within 5–10 min, a secondary structure is indicated. Primary alcohols may never cause turbidity at room temperature.

EXAMPLE 16.3

How could the following isomeric compounds be distinguished?

$C_6H_{11}CH_2CH_2OH$ (I) $\qquad$ $C_6H_{11}CHOHCH_3$ (II) $\qquad$ 1-ethylcyclohexanol: OH, CH_2CH_3 (III)

Solution

First, classify each compound. Compounds (I), (II), and (III) are primary, secondary, and tertiary alcohols, respectively. Thus, the rate of reaction with HCl and $ZnCl_2$ will be (III) > (II) > (I). Compound (III) should react readily at room temperature. Compound (II) should react more slowly than (III). The reaction with (I) should be extremely slow or not occur at all at room temperature.

PROBLEM 16.3

How could 3-methyl-2-butanol and 2-methyl-2-butanol be distinguished?

Mechanism of the Nucleophilic Substitution Reaction

The reaction of a nucleophile at a carbon atom to substitute for (replace) another group is a common reaction in organic chemistry. The symbol Nu is used to represent the nucleophile. L designates the "leaving group" that is displaced from the carbon center. The leaving group is usually an electronegative atom or a group that

Phosphate and Pyrophosphate Esters

Many of the reactions that occur in cells require displacement of a hydroxyl group by a nucleophile. Although the hydroxyl group of an alcohol is not a good leaving group, it can be converted into a phosphate or pyrophosphate ester. Both phosphoric and pyrophosphoric acids are stronger acids than water. Thus, the related conjugate bases—phosphate and pyrophosphate—are weaker bases than the hydroxide ion. Both conjugate bases are then excellent leaving groups. Let's focus on the pyrophosphates and pyrophosphoric acid.

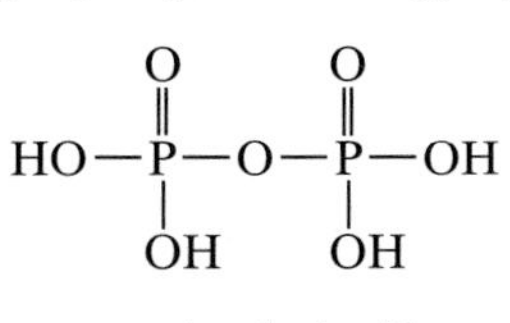

$$\mathrm{HO{-}P(=O)(OH){-}O{-}P(=O)(OH){-}OH}$$

pyrophosphoric acid

$$\mathrm{RO{-}P(=O)(OH){-}O{-}P(=O)(OH){-}OH}$$

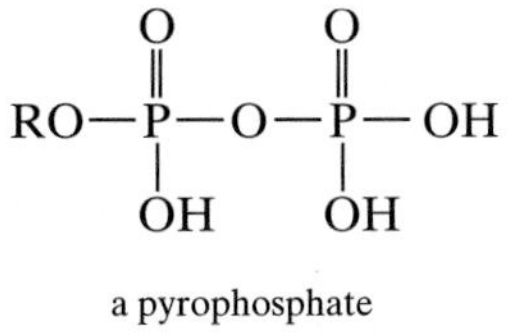

a pyrophosphate

The concentration of pyrophosphate in cells is very low, and the conversion of alcohols to pyrophosphates almost never occurs by itself. So how are alcohols converted to pyrophosphates in cells? The answer is that the source of pyrophosphate is not the pyrophosphate ion or phosphoric acid but a molecule called adenosine 5′-triphosphate, ATP, which contains a pyrophosphate group. The adenosine of the triphosphate is represented by A in the simplified structure.

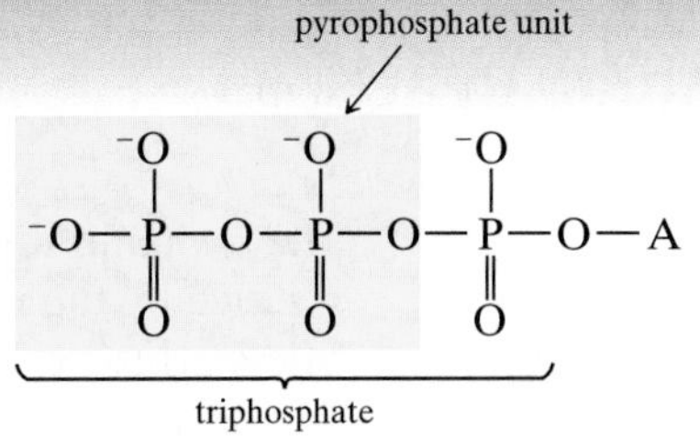

pyrophosphate unit

$$\mathrm{{}^{-}O{-}P(=O)(O^-){-}O{-}P(=O)(O^-){-}O{-}P(=O)(O^-){-}O{-}A}$$

triphosphate

adenosine 5′-triphosphate (ATP)

Phosphate and pyrophosphate esters of alcohols form when a phosphorus-containing group transfers from ATP to oxygen. The reaction occurs by nucleophilic attack on phosphorus by the hydroxyl group's oxygen atom. The net enzyme-catalyzed reaction for formation of a pyrophosphate ester is shown using A to represent the adenosine unit.

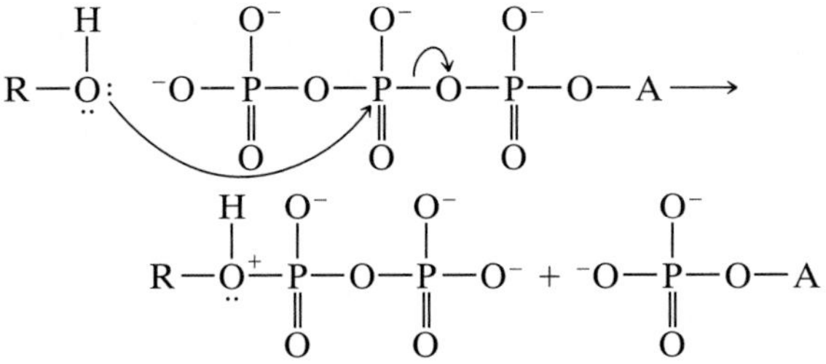

$$\mathrm{R{-}\ddot{O}H} + \mathrm{{}^{-}O{-}P(=O)(O^-){-}O{-}P(=O)(O^-){-}O{-}P(=O)(O^-){-}O{-}A} \longrightarrow$$

$$\mathrm{R{-}\overset{+}{O}H{-}P(=O)(O^-){-}O{-}P(=O)(O^-){-}O^-} + \mathrm{{}^{-}O{-}P(=O)(O^-){-}O{-}A}$$

(conjugate acid of a pyrophosphate)

Pyrophosphate derivatives of alcohols participate in the biosynthesis of many molecules, including cholesterol and steroid hormones.

can exist as a stable anion. R represents the remainder or the rest of the molecule. The curved arrow shows the movement of electrons in the reaction.

$$\text{Nu}:^- + \text{R—L} \longrightarrow \text{Nu—R} + :\text{L}^-$$

Note that the nucleophile has an unshared pair of electrons that bonds to the carbon residue in a heterogenic process. The leaving group departs with an electron pair, which is a heterolytic process.

The leaving group in a nucleophilic substitution reaction may also be electrically neutral as in the case of the reaction of alcohols with HBr. The mechanism involves two steps. First, the alcohol is protonated in an acid–base reaction.

$$\text{R—CH}_2\text{—}\ddot{\text{O}}\text{—H} + \text{H—}\ddot{\text{Br}}: \longrightarrow \text{R—CH}_2\text{—}\overset{+}{\text{O}}\text{H}_2 + :\ddot{\text{Br}}:^-$$

Water, a neutral leaving group, can be displaced more easily than hydroxide, a negatively charged leaving group. In general, as the basicity of leaving groups decreases, the better they are as leaving groups. Water is substantially less basic than hydroxide ion and is therefore a much better leaving group than hydroxide ion.

In the second step the bromide ion is a nucleophile and displaces water.

$$:\ddot{\text{Br}}:^- + \text{R—CH}_2\text{—}\overset{+}{\text{O}}\text{H}_2 \longrightarrow \text{R—CH}_2\text{—}\ddot{\text{Br}}: + :\ddot{\text{O}}\text{—H}_2$$

The bromide ion approaches the carbon atom, which has been made somewhat positive by the electronegative oxygen atom of the water leaving group. As a pair of electrons of the bromide ion begins to bond to the carbon atom, the bonding pair to the leaving group is weakened. The formation of the bond to the nucleophile provides the energy required to break the bond to the leaving group. The entire process is **concerted**—that is, both bond breaking and bond formation occur simultaneously.

16.7 Dehydration of Alcohols

The elimination of water from an alcohol is a **dehydration reaction.** An acid catalyst such as sulfuric acid or phosphoric acid is used. The reaction is illustrated by the formation of ethylene from ethyl alcohol.

These atoms are eliminated to form water.

$$\text{H—CH}_2\text{—CH}_2\text{—OH} \xrightarrow{H_2SO_4} \text{H}_2\text{C=CH}_2 + H_2O$$

The carbon–oxygen bond and a carbon–hydrogen bond of adjacent carbon atoms break in this process. Alcohols such as 2-butanol have two adjacent carbon atoms that can release a hydrogen atom to form water.

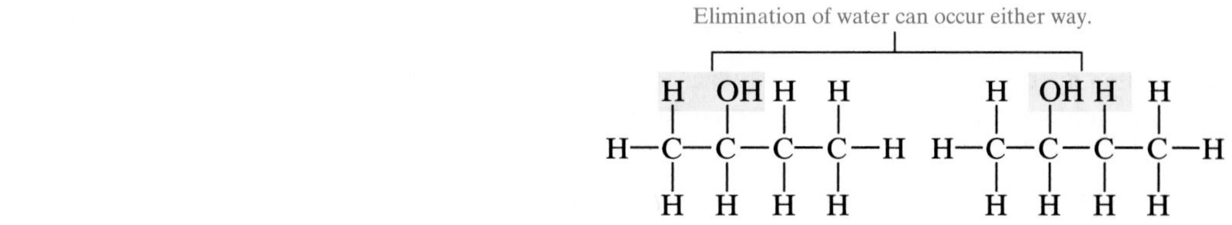

Elimination of water can occur either way.

Thus, the dehydration produces a mixture of products. The isomer that contains the greatest number of alkyl groups attached to the double bond predominates in the mixture.

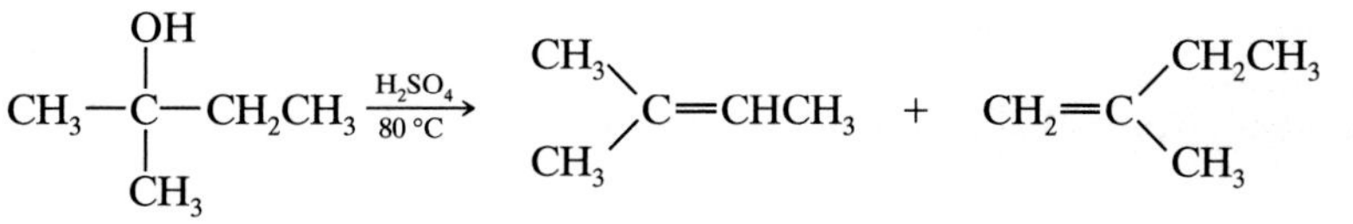

2-methyl-2-butanol

2-methyl-2-butene (90%) (major product; trisubstituted)

2-methyl-1-butene (10%) (minor product; disubstituted)

Dehydration reactions (and the reverse hydration reactions) occur in metabolic reactions. These reactions are catalyzed by enzymes and they are rapid even though high concentrations of acids are not present and the reaction temperature is only 37 °C. One example is the dehydration of citric acid, catalyzed by the enzyme aconitase, to give *cis*-aconitic acid.

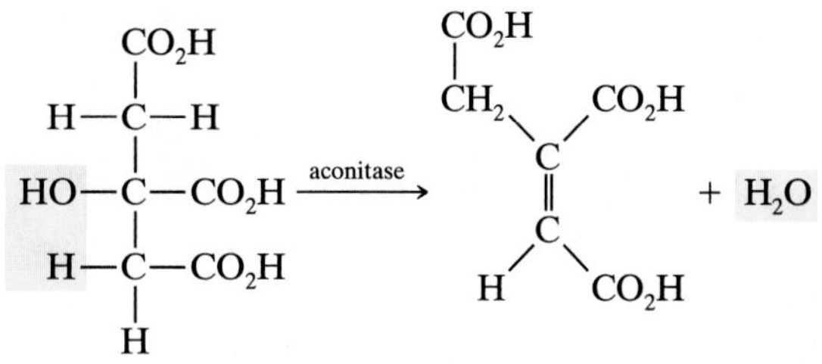

citric acid

cis-aconitic acid

EXAMPLE 16.4

What is produced by the dehydration of 1-methylcyclohexanol?

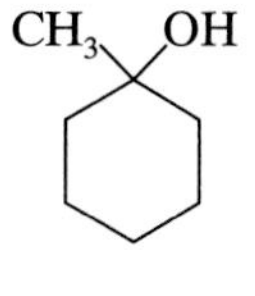

Solution

This tertiary alcohol has three adjacent carbon atoms that may lose a hydrogen atom in a dehydration reaction. However, the loss of a hydrogen atom from either C-2 or C-6 will result in the same product. Thus, only two isomers result.

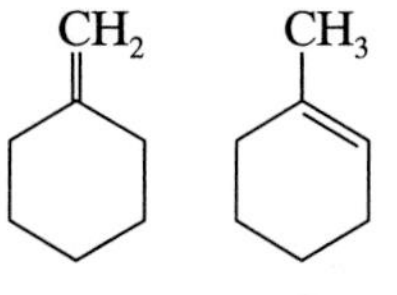

The second structure, with the double bond within the six-membered ring, predominates. This compound has three carbon atoms attached to the two double-bonded carbon atoms.

PROBLEM 16.4

One of the steps in the metabolism of glucose involves the enzymatic dehydration of 2-phosphoglyceric acid. Write the structure of the product.

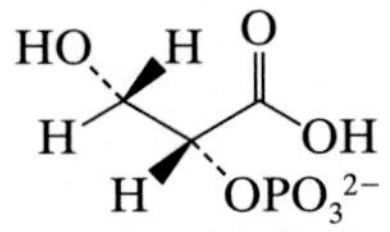

HO, H, O, H, OH, H, OPO_3^{2-}

Mechanism of the Dehydration Reaction

The first step in the mechanism of the dehydration reaction is the protonation of the electron pair of the oxygen atom in an acid–base reaction. The acid is represented as HA.

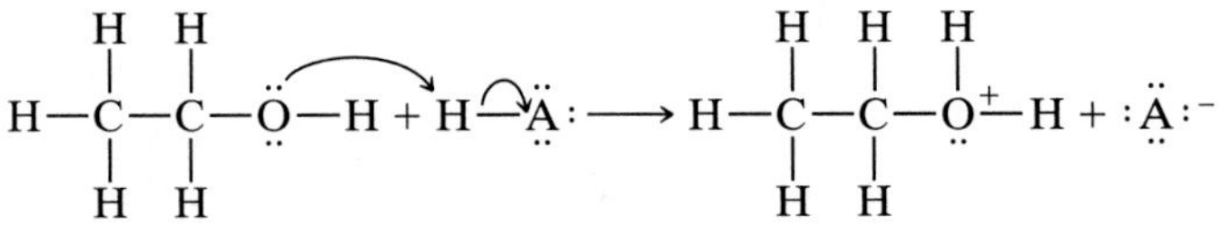

$$H-CH_2-CH_2-\ddot{O}-H + H-\ddot{A}: \longrightarrow H-CH_2-CH_2-\overset{+}{O}(H)-H + :\ddot{A}:^-$$

Then a proton is removed from the neighboring carbon atom by the conjugate base of the acid. The electron pair in the C—H bond moves to form a carbon–carbon bond, while the electron pair of the C—O bond leaves as a nonbonding pair of electrons on the oxygen atom of water. The entire process is concerted. Note that the acid HA is a catalyst and is regenerated at the conclusion of the series of steps.

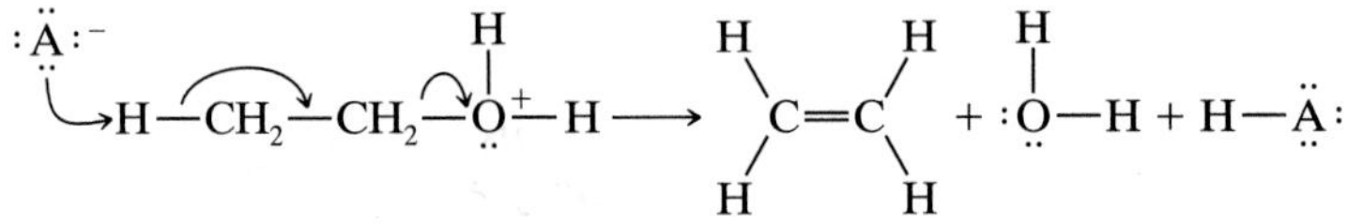

$$:\ddot{A}:^- \; H-CH_2-CH_2-\overset{+}{O}(H)-H \longrightarrow H_2C{=}CH_2 + :\ddot{O}(H)-H + H-\ddot{A}:$$

16.8 Oxidation of Alcohols

Alcohols can be oxidized selectively using a variety of oxidizing agents. The net effect of the oxidation of an alcohol is the removal of two hydrogen atoms, one from the —OH group and the other from the carbon atom bearing the —OH group. The generalized reaction, where [O] represents an oxidizing agent, is thus

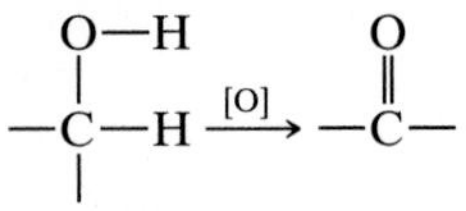

$$-\underset{|}{\overset{O-H}{\overset{|}{C}}}-H \xrightarrow{[O]} -\overset{O}{\overset{\|}{C}}-$$

The type of carbonyl compound formed depends on the class of alcohol as well as the oxidizing agent used. Primary alcohols are first oxidized to aldehydes and then are usually rapidly further oxidized to carboxylic acids unless special reagents are used. The second oxidation step corresponds to the gain of one oxygen atom.

$$\text{R}-\underset{\text{H}}{\underset{|}{\overset{\text{O}-\text{H}}{\overset{|}{\text{C}}}}}-\text{H} \xrightarrow{[\text{O}]} \text{R}-\overset{\text{O}}{\overset{\|}{\text{C}}}-\text{H} \xrightarrow{[\text{O}]} \text{R}-\overset{\text{O}}{\overset{\|}{\text{C}}}-\text{OH}$$

a primary alcohol — an aldehyde — a carboxylic acid

Secondary alcohols are oxidized to form ketones, which cannot be further oxidized because there is no hydrogen atom on the oxygen-bearing carbon atom of the ketone. Tertiary alcohols are not oxidized because the carbon atom bearing the —OH group has no hydrogen atom.

$$\text{R}-\underset{\text{R}}{\underset{|}{\overset{\text{O}-\text{H}}{\overset{|}{\text{C}}}}}-\text{H} \xrightarrow{[\text{O}]} \text{R}-\overset{\text{O}}{\overset{\|}{\text{C}}}-\text{R} \xrightarrow{[\text{O}]} \text{no further oxidation}$$

a secondary alcohol — a ketone

$$\text{R}-\underset{\text{R}}{\underset{|}{\overset{\text{O}-\text{H}}{\overset{|}{\text{C}}}}}-\text{R} \xrightarrow{[\text{O}]} \text{no oxidation products}$$

a tertiary alcohol

Potassium dichromate, $K_2Cr_2O_7$, in an aqueous acidic solution oxidizes primary alcohols to carboxylic acids and secondary alcohols to ketones. Solutions of dichromate ion ($Cr_2O_7^{2-}$) are yellow orange, whereas solutions of chromium(III) ion (Cr^{3+}), the reduced form of the reagent, are green. Thus, the oxidation of an alcohol by potassium dichromate produces a color change. This color change is used to indicate the blood alcohol concentration in the Breathalyzer test used to test individuals suspected of driving while intoxicated.

In the laboratory, the Jones reagent is used to oxidize alcohols. The **Jones reagent** consists of chromium trioxide (CrO_3) in a solution of aqueous acetone and sulfuric acid. Primary alcohols are oxidized to aldehydes, which then are immediately oxidized to carboxylic acids. This reagent also converts secondary alcohols to ketones. Oxidation by the Jones reagent occurs rapidly at room temperature.

$$\text{(skeletal)}-CH_2OH \xrightarrow[\text{acetone}/H_2SO_4]{CrO_3} \text{(skeletal)}-\overset{\text{O}}{\overset{\|}{\text{C}}}-\text{OH}$$

$$\text{cyclooctanol (OH)} \xrightarrow[\text{acetone}/H_2SO_4]{CrO_3} \text{cyclooctanone (=O)}$$

Alcohols can also be oxidized with an oxidizing agent known as pyridinium chlorochromate **(PCC).** The PCC reagent, which is made by dissolving CrO_3 in HCl and then adding pyridine, is soluble in methylene chloride (CH_2Cl_2). PCC also oxidizes secondary alcohols to ketones. However, the principal advantage of PCC is that primary alcohols are converted to aldehydes without continued oxidation to carboxylic acids.

$$\text{(skeletal)}-CH_2OH \xrightarrow[CH_2Cl_2]{\text{PCC}} \text{(skeletal)}-\overset{\text{O}}{\overset{\|}{\text{C}}}-\text{H}$$

Ethanol—An Antidote for Methanol Poisoning

Although ethanol is the least toxic of alcohols, it is still a poisonous substance that the body must oxidize to prevent high blood alcohol levels, which can "poison" the brain. The liver enzyme alcohol dehydrogenase (ADH) catalyzes the oxidation of alcohols. ADH requires a coenzyme, nicotinamide adenine dinucleotide (NAD^+), as an oxidizing agent. The coenzyme can exist in an oxidized form, NAD^+, and a reduced form, NADH. NAD^+-dependent liver ADH catalyzes the oxidation of ethanol to ethanal (acetaldehyde) and its subsequent oxidation to ethanoic acid (acetic acid), which is nontoxic.

$$CH_3-CH_2-OH + NAD^+ \xrightarrow{ADH} CH_3-C(=O)-H + NADH + H^+$$

ethanal (acetaldehyde)

$$CH_3-C(=O)-H + NAD^+ + H_2O \xrightarrow{ADH} CH_3-C(=O)-OH + NADH + H^+$$

ethanoic acid (acetic acid)

The oxidation products of some other alcohols are toxic. In the case of methanol, oxidation catalyzed by ADH gives methanal (formaldehyde) and then methanoic acid (formic acid).

$$H-CH_2-OH + NAD^+ \xrightarrow{ADH} H-C(=O)-H + NADH + H^+$$

methanal (formaldehyde)

$$H-C(=O)-H + NAD^+ + H_2O \xrightarrow{ADH} H-C(=O)-OH + NADH + H^+$$

methanoic acid (formic acid)

Formaldehyde travels in the blood throughout the body and reacts with proteins, destroying their biological function. Methanol causes blindness because formaldehyde destroys a protein involved in the vision process. Formaldehyde reacts with an amine functional group of the amino acid lysine in the protein, called rhodopsin. Formaldehyde also reacts with amino groups in other proteins, including many enzymes, and loss of function of these biological catalysts causes death.

Ethylene glycol is also toxic. This sweet-tasting substance is the primary component of antifreeze. Dogs sometimes ingest antifreeze left in open containers. Oxidation occurs to give oxalic acid, which crystallizes as a calcium salt in the kidneys. This process can lead to kidney failure and death.

$$H-CH(OH)-CH(OH)-H \longrightarrow HO-C(=O)-C(=O)-OH$$

ethylene glycol — oxalic acid

Physicians treat methanol or ethylene glycol poisoning with intravenous injections of ethanol before substantial oxidation has occurred. ADH binds more tightly to ethanol than to methanol or ethylene glycol, and the rate of oxidation of ethanol is about six times faster than that of methanol. The ethanol concentration can be kept higher because it is directly injected. As a result, neither methanol nor ethylene glycol is competitively oxidized to toxic products, and the kidneys can slowly excrete them.

EXAMPLE 16.5

Which of the isomeric $C_4H_{10}O$ alcohols will react with potassium dichromate to produce a ketone (C_4H_8O)?

Solution

Two of the $C_4H_{10}O$ alcohols are primary, one is secondary, and one is tertiary. Only 2-butanol (*sec*-butyl alcohol) can give a ketone.

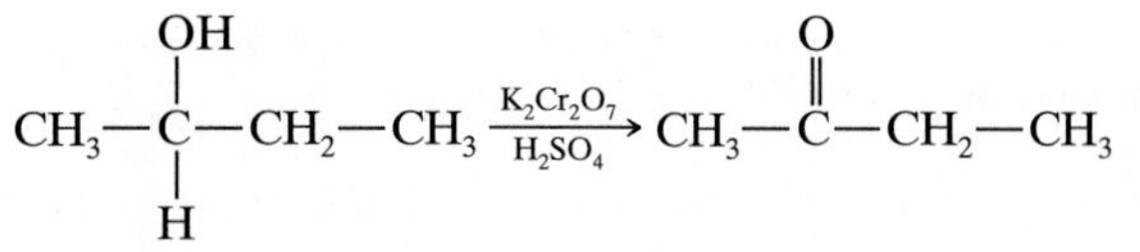

$$CH_3-\underset{H}{\overset{OH}{C}}-CH_2-CH_3 \xrightarrow[H_2SO_4]{K_2Cr_2O_7} CH_3-\overset{O}{\overset{\|}{C}}-CH_2-CH_3$$

PROBLEM 16.5

What are the products of the reaction of each of following isomeric alcohols with Jones reagent? How do the products differ when PCC is used as the oxidizing agent?

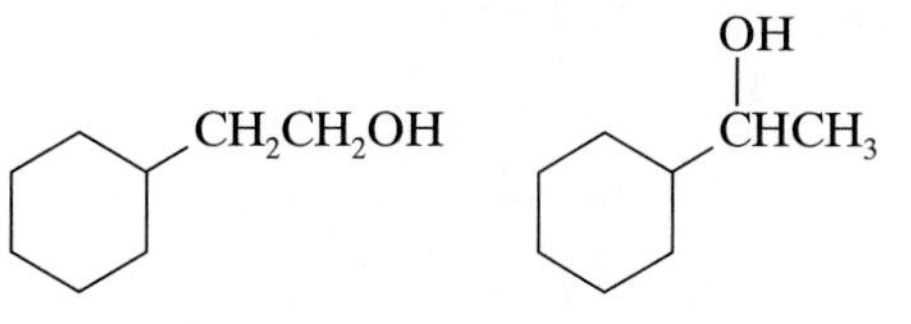

(I: cyclohexane–CH_2CH_2OH; II: cyclohexane–$\overset{OH}{CHCH_3}$)

I II

Oxidation of alcohols occurs as one of the steps in many sequences of biological reactions. For example, in the oxidation of fats a β-hydroxy acid is oxidized to a β-keto acid.

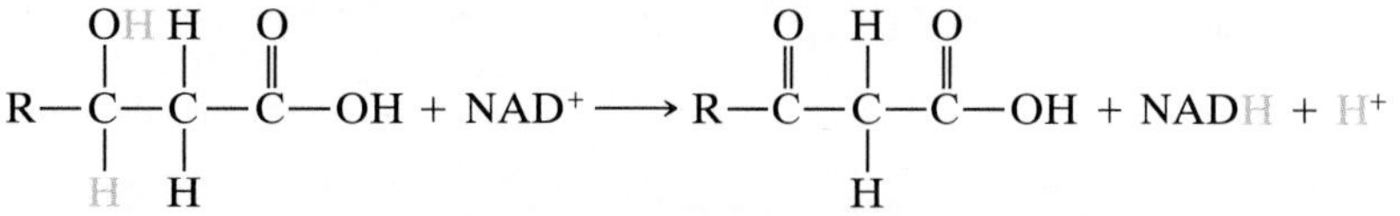

$$R-\underset{H}{\overset{OH}{C}}-\underset{H}{\overset{H}{C}}-\overset{O}{\overset{\|}{C}}-OH + NAD^+ \longrightarrow R-\overset{O}{\overset{\|}{C}}-\underset{H}{\overset{H}{C}}-\overset{O}{\overset{\|}{C}}-OH + NADH + H^+$$

The β-position of the acid is the second one down the carbon chain after the carbon atom of the acid. The oxidizing agent is a coenzyme, nicotinamide adenine dinucleotide (NAD^+). One of the steps in the citric acid cycle is the oxidation of the secondary alcohol of malic acid. The oxidizing agent is again NAD^+.

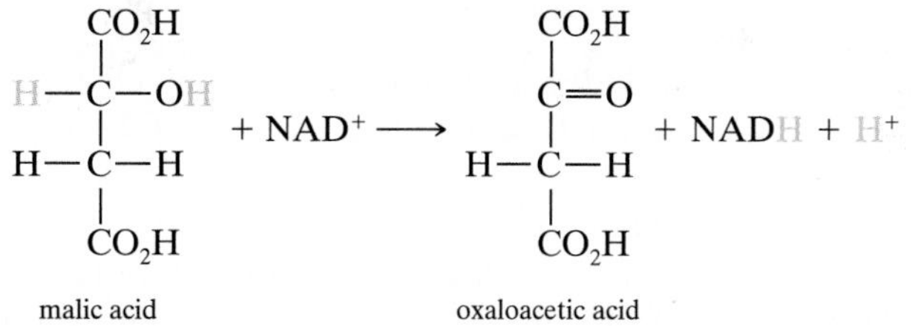

$$HO_2C-CH(OH)-CH_2-CO_2H + NAD^+ \longrightarrow HO_2C-C(=O)-CH_2-CO_2H + NADH + H^+$$

malic acid oxaloacetic acid

16.9 Ethers

Ethers contain two alkyl or aryl groups bonded to an oxygen atom. The two groups are identical in a **symmetrical ether** and different in an **unsymmetrical ether.**

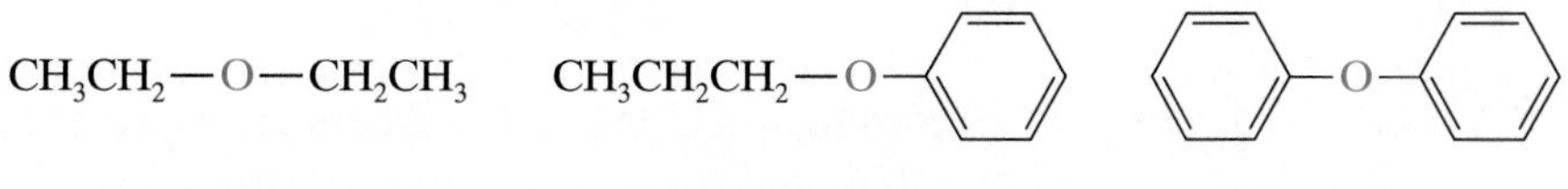

$CH_3CH_2-O-CH_2CH_3$ $CH_3CH_2CH_2-O-C_6H_5$ $C_6H_5-O-C_6H_5$

diethyl ether (symmetrical ether) phenyl propyl ether (unsymmetrical ether) diphenyl ether (symmetrical ether)

Figure 16.3 shows examples of naturally occurring, physiologically active cyclic ethers. Tetrahydrocannabinol (THC), the principal active ingredient in marijuana,

FIGURE 16.3 Structures of Naturally Occurring Ethers

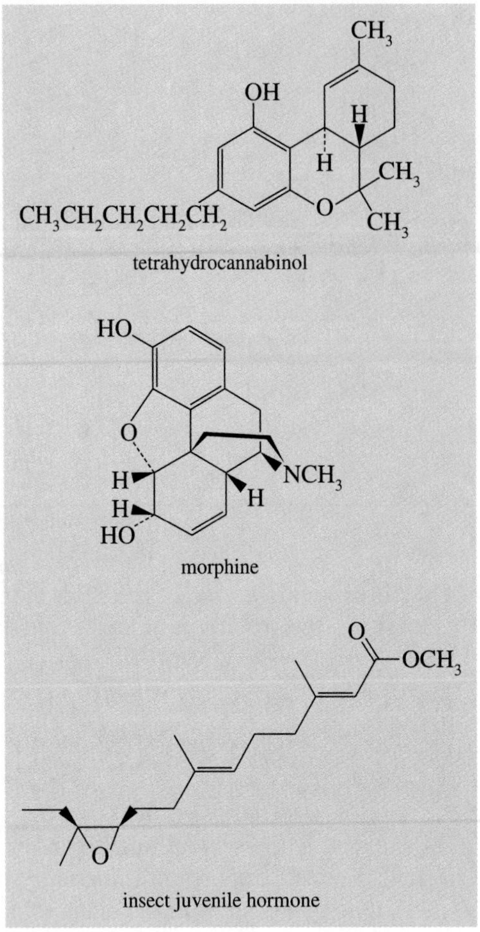

tetrahydrocannabinol

morphine

insect juvenile hormone

includes a six-membered ring ether. Morphine contains a five-membered ring ether. Three-membered cyclic ethers (epoxides) are rare in nature. The juvenile hormone of some insects contains a cis-substituted epoxide. This hormone controls the rate of maturation of insects.

Simple ethers are commonly called *alkyl alkyl* ethers. The name consists of a list of the alkyl (or aryl) groups in alphabetical order followed by the name *ether.* For example, an ether with a butyl group and a methyl group is named butyl methyl ether. Ethers with two equivalent groups are named by using the prefix di- in conjunction with the name of the alkyl group. For example, an ether with two isopropyl groups is called diisopropyl ether.

$$CH_3{-}CH_2{-}CH_2{-}CH_2{-}O{-}CH_3$$

butyl methyl ether

$$CH_3{-}\underset{}{\overset{\overset{\displaystyle CH_3}{|}}{C}H}{-}O{-}\overset{\overset{\displaystyle CH_3}{|}}{C}H{-}CH_3$$

diisopropyl ether

According to IUPAC rules, ethers are named *alkoxyalkanes,* where the smaller alkyl group and the oxygen atom constitute an **alkoxy group.** An alkoxy group is treated as a substituent on the larger, parent alkane chain. For example, a five-carbon chain with an $-OCH_3$ group at the C-2 atom is 2-methoxypentane.

The larger group is the parent chain

$CH_3CH_2CH_2CHCH_3$ with O—CH_3 on C-2 (the smaller group is the substituent.)

2-methoxypentane

Because ethers have two polar C—O bonds arranged at an angle, they are polar compounds, as are alcohols. However, there is no O—H bond in ethers; hence, no hydrogen bonding exists between ether molecules. As a consequence, the boiling points of ethers are substantially lower than the boiling points of alcohols. The boiling points of ethers are very close to the boiling points of alkanes of similar molec-

Polyethers in the Laboratory and in Medicine

Cations represented as M^{n+} are solvated by several water molecules in aqueous solution. In 1967 Charles J. Pederson of the Du Pont Company suggested that cyclic polyethers might similarly solvate cations and increase the solubility of ionic compounds in nonpolar organic solvents. These compounds, called **crown ethers,** are named *x*-crown-*y* where *x* is the total number of atoms in the ring, and *y* is the number of oxygen atoms. The 18-crown-6 ether can coordinate to a potassium ion in a cavity within the ring.

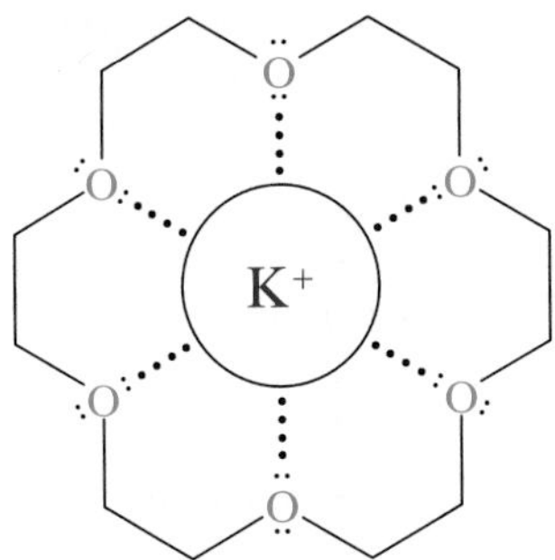

potassium ion solvated by 18-crown-6

The solubility of ions in crown ethers depends on the match between the size of the cavity and the ionic radius of the ion. The internal cavity of 18-crown-6 spans between 260 and 310 pm, close to the ionic radius of the potassium ion, about 270 pm. Thus, not only does the potassium ion fit within the cavity, but all the oxygen atoms of the crown ether lie close enough to the potassium ion to effectively complex it. Because of this coordination, the solubilities of inorganic salts such as KCN or $KMnO_4$ are increased.

Several cyclic and acyclic polyethers act as antibiotics by transporting ions across biological membranes. These ethers are called **ionophores** (ion carriers). They function by disrupting the electrolyte balance between the interior and exterior of cells that is necessary for normal maintenance of the cell.

The cyclic ether antibiotic Nonactin (see figure) selectively transports potassium out of bacterial cells. This compound, which contains four five-membered ring ethers linked by ester units, binds potassium about 10 times better than it binds sodium. Because cells must maintain a higher internal concentration of potassium ions than of sodium ions, the selective removal of potassium ions kills bacteria.

Structure of Antibiotic Polyethers

nonactin

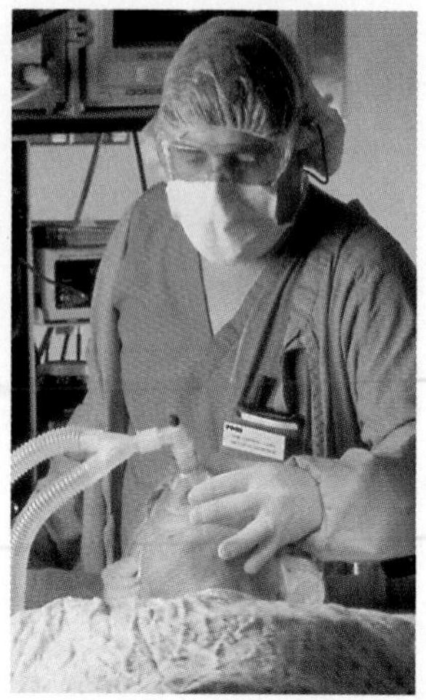

Diethyl ether, although widely used at one time, has now been replaced by other anesthetics. (Photo Researchers, Inc.)

ular weight. The boiling points of diethyl ether and pentane are 35 °C and 36 °C, respectively.

The slight polarity of ethers makes them more water soluble than alkanes. The solubility of diethyl ether is about 10 g per 100 mL of water. As the molecular weight increases, ethers more closely resemble alkanes in their solubilities because the functional group contributes less to the overall structure of the compound.

Diethyl ether is most familiar as a general anesthetic (Figure 16.4). It is administered as a vapor and acts as a depressant on the central nervous system, causing unconsciousness. Ethers such as ethyl vinyl ether, divinyl ether, and methyl propyl ether have also been used as anesthetics. All of these ethers are potentially explosive when mixed with oxygen. The lower flammability of halogenated compounds has led to the development of anesthetics that contain chlorine and fluorine, such as enflurane and isoflurane.

$$\text{F}-\underset{\text{Cl}}{\overset{\text{H}}{\text{C}}}-\underset{\text{F}}{\overset{\text{F}}{\text{C}}}-\text{O}-\underset{\text{F}}{\overset{\text{F}}{\text{C}}}-\text{H} \qquad \text{F}-\underset{\text{F}}{\overset{\text{F}}{\text{C}}}-\underset{\text{Cl}}{\overset{\text{H}}{\text{C}}}-\text{O}-\underset{\text{F}}{\overset{\text{F}}{\text{C}}}-\text{H}$$

enflurane isoflurane

16.10 Sulfur Compounds

Sulfur is in the same group of the periodic table as oxygen and forms compounds structurally similar to alcohols. The —SH group is called the **sulfhydryl group.** Compounds containing the —SH group are called **mercaptans** or **thiols.** The nomenclature of these compounds resembles that of alcohols with -thiol used in place of -ol. The -e of the alkane name is retained.

$$CH_3-CH_2-\overset{\text{SH}}{\overset{|}{CH}}-CH_2-CH_3$$

3-pentanethiol

Thiols can be considered as derivatives of hydrogen sulfide in which one hydrogen atom is replaced by an alkyl or aryl group. Although there are some similarities between alcohols and thiols, there are pronounced differences as well. Thiols have lower boiling points than the corresponding alcohols because sulfur does not form hydrogen bonds. (Only nitrogen, oxygen, and fluorine form hydrogen bonds.)

One of the distinguishing properties of thiols is their strong, disagreeable odor. The odor of skunk is due to a thiol! Thiols can be detected by the human nose at the level of parts per billion in air. For this reason, small amounts of thiols are added to natural gas so that leaks can easily be detected.

The strong odor of skunk is due to thiols contained in its defensive spray. (Photo Researchers, Inc.)

Although they are weak acids, thiols are stronger acids than alcohols.

$$R-S-H + H_2O \longrightarrow R-S^- + H_3O^+ \qquad K_a = 10^{-8}$$
$$R-O-H + H_2O \longrightarrow R-O^- + H_3O^+ \qquad K_a = 10^{-17}$$

The sulfhydryl group is acidic enough to react with hydroxide ions to form salts.

$$R-S-H + NaOH \longrightarrow R-S^- + Na^+ + H_2O$$

Thiols are easily oxidized but yield disulfides rather than the structural analogs of aldehydes and ketones. In the following equation the symbol [O] represents an unspecified oxidizing agent that removes the hydrogen atoms.

$$2\,R{-}S{-}H \xrightarrow{[O]} \underset{\text{a disulfide}}{R{-}S{-}S{-}R}$$

A biological reaction of great importance involves the oxidation of the —SH group of the amino acid cysteine (Chapter 23).

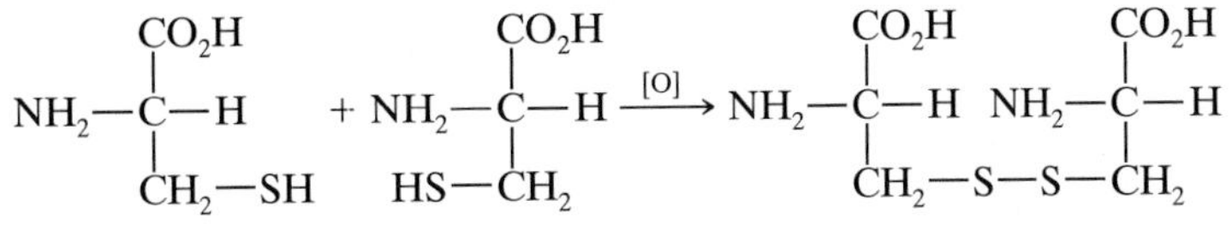

$$\underset{\text{cysteine}}{NH_2{-}\underset{CH_2-SH}{\overset{CO_2H}{C}}{-}H \;+\; NH_2{-}\underset{HS-CH_2}{\overset{CO_2H}{C}}{-}H} \xrightarrow{[O]} \underset{\text{cystine}}{NH_2{-}\overset{CO_2H}{\underset{CH_2}{C}}{-}H \quad NH_2{-}\overset{CO_2H}{\underset{CH_2}{C}}{-}H}$$

(in cystine the two CH_2 groups are joined: $CH_2{-}S{-}S{-}CH_2$)

Frequently the activity of an enzyme is due to free sulfhydryl groups. The oxidation of cysteine contained in enzymatic proteins alters the action of the molecule and deactivates the enzyme.

Salts of lead and mercury are toxic because they react with the sulfhydryl group of enzymes (E—S—H).

$$2\,E{-}S{-}H + Hg^{2+} \longrightarrow E{-}S{-}Hg{-}S{-}E + 2\,H^+$$

The relationship of disulfide bonds to the structure and activity of enzymes will be presented in Chapter 24.

Summary

In **alcohols** a **hydroxyl group** is bonded to a saturated carbon atom. There are three classes of alcohols: **primary, secondary,** and **tertiary.** Alcohols have higher boiling points than alkanes of similar molecular weight because of hydrogen bonding between hydroxyl groups. Lower molecular weight alcohols are soluble in water, but as the size of the alkyl group increases, the solubility decreases.

Phenols have the general formula Ar—OH, where Ar represents an aryl group derived from an aromatic compound. Phenols are effective germicides. Their effectiveness is designated by a **phenol coefficient.**

Alcohols are weaker acids than water. Phenols are stronger acids than alcohols. Three other important reactions of alcohols are substitution, dehydration, and oxidation. The **Lucas test,** which can be used to classify alcohols, is a substitution reaction. **Dehydration** of alcohols can form a mixture of alkenes. Oxidation of primary alcohols produces **aldehydes** first and then carboxylic acids. Secondary alcohols are oxidized to **ketones,** whereas tertiary alcohols are not oxidized at all.

In **ethers** either alkyl or aryl groups replace both hydrogen atoms of water. Ethers are named by indicating the **alkoxy group** attached to a larger alkane chain. Ethers are slightly polar compounds.

Thiols or **mercaptans** have a **sulfhydryl group** and have the general formula R—SH. These compounds are more acidic than alcohols. Oxidation of thiols forms **disulfides,** represented by R—S—S—R. The most distinctive physical property of thiols is their foul odor.

Exercises

Nomenclature of Alcohols

16.1 Write the structural formula for each of the following.

(a) 2-methyl-2-pentanol
(b) 2-methyl-1-butanol
(c) 2,3-dimethyl-1-butanol

16.2 Write the structural formula for each of the following.

(a) 2-methyl-3-pentanol
(b) 3-ethyl-3-pentanol
(c) 4-methyl-2-pentanol

16.3 Write the structural formula for each of the following.

(a) cyclopentanol
(b) 1-methylcyclohexanol
(c) *trans*-2-methylcyclohexanol
(d) *cis*-3-ethylcyclopentanol

16.4 Write the structural formula for each of the following.

(a) 1,2-hexanediol (b) 1,3-propanediol
(c) 1,2,4-butanetriol (d) 1,2,3,4,5,6-hexanehexol

16.5 Name each of the following compounds.

(a) OH
(b) OH
(c) OH

16.6 Name each of the following compounds.

(a) CH_3 HO
(b) $CH_2CH_2CH_2CH_2OH$
(c) H OH H Br

Classification of Alcohols

16.7 Classify each of the following vitamins as alcohols.

(a) vitamin B_6

H O C HO CH_2OH CH_3 N

(b) vitamin B_1

CH_3 N NH_2 S N N + CH_2CH_2OH CH_3

16.8 Classify each of the following steroids as alcohols.

(a) hydrocortisone, an anti-inflammatory drug

CH_2OH C=O CH_3 HO OH H_3C H H H O

(b) norethindrone, an oral contraceptive

CH_3 OH C≡CH H H H H O

Physical Properties of Alcohols

16.9 1,2-Hexanediol is very soluble in water. Why?

16.10 Explain why ethylene glycol boils at 198 °C whereas 1-propanol boils at 97 °C.

16.11 Explain why 1-butanol is less soluble in water than 1-propanol.

16.12 Suggest a reason why 2-methyl-1-propanol is more soluble in water than 1-butanol.

Substitution Reactions

16.13 Describe how the following three compounds could be distinguished by using the Lucas reagent.

$CH_3-CH_2-C(CH_3)_2-OH$ (I)

$CH_3-CH_2-CH(CH_3)-CH_2-OH$ (II)

$CH_3-CH(OH)-CH(CH_3)-CH_3$ (III)

16.14 Order the following compounds according to their reactivity with the Lucas reagent.

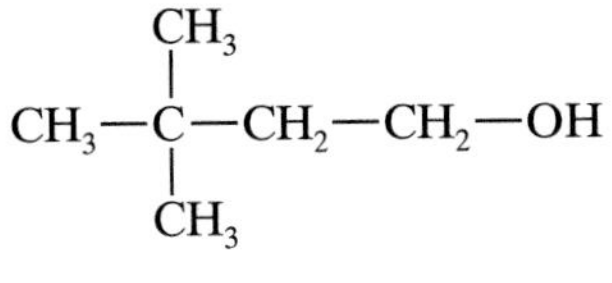

$$CH_3-\overset{CH_3}{\underset{CH_3}{C}}-CH_2-CH_2-OH$$

(I)

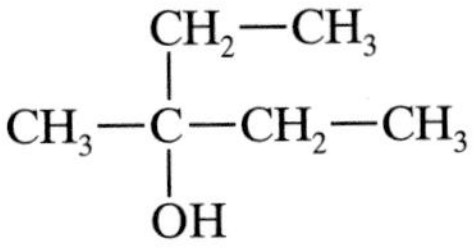

$$CH_3-\overset{CH_2-CH_3}{\underset{OH}{C}}-CH_2-CH_3$$

(II)

$$CH_3-\overset{H}{\underset{CH_3}{C}}-CH_2-\underset{OH}{CH}-CH_3$$

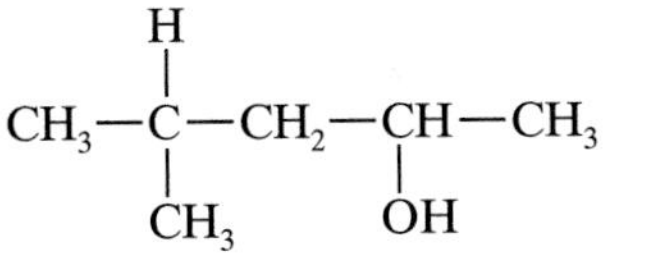

(III)

16.15 Write the structure of the product of reaction of each of the following with PBr_3.

(a) 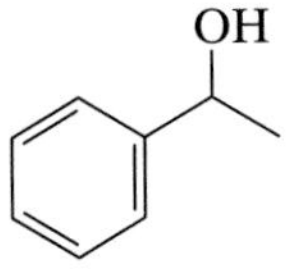 OH (b) 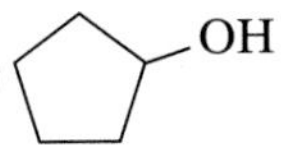 OH

(c) 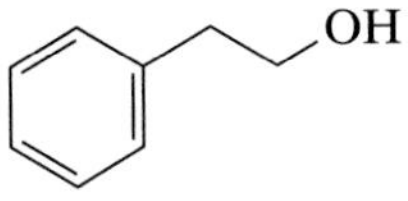 OH

16.16 Write the structure of the product of reaction of each of the following with HBr.

(a) 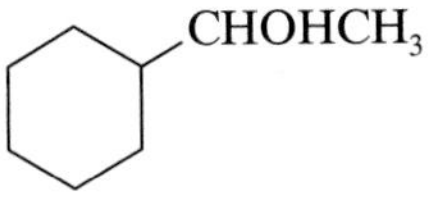 CHOHCH$_3$ (b) 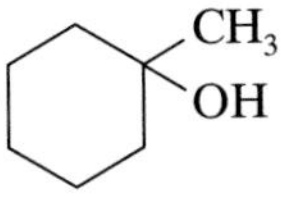 CH$_3$ OH

(c) 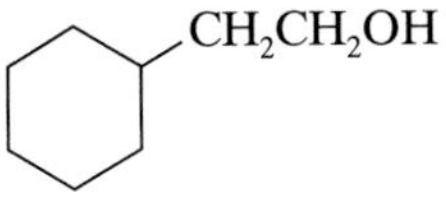 CH_2CH_2OH

Dehydration of Alcohols

16.17 Draw the structure of the dehydration products when each of the following compounds reacts with sulfuric acid.

(a) $CH_3-CH_2-CH_2-OH$

(b) 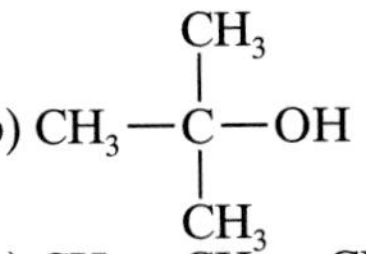 $CH_3-\overset{CH_3}{\underset{CH_3}{C}}-OH$

(c) $CH_3-CH_2-CH_2-CH_2-CH_2-OH$

(d) 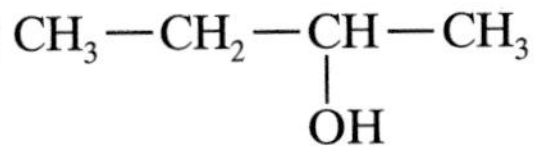 $CH_3-CH_2-\underset{OH}{CH}-CH_3$

16.18 Draw the structure of the dehydration products when each of the following compounds reacts with sulfuric acid.

(a) 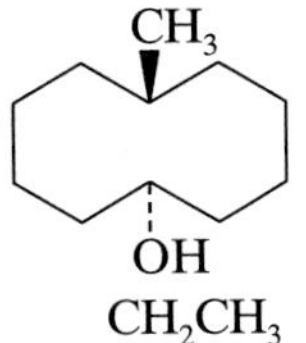 CH$_3$ OH (b) 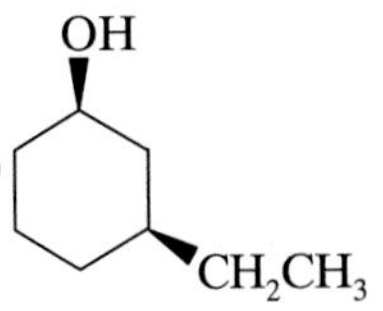 OH CH$_2$CH$_3$

(c) 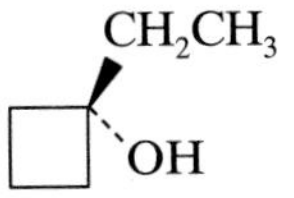 CH$_2$CH$_3$ OH

Oxidation of Alcohols

16.19 Both 1-octanol and 2-octanol react with aqueous potassium dichromate. The product of the reaction of 2-octanol is insoluble in aqueous base, but the product of the reaction of 1-octanol is soluble. Write the structure of each product and explain the difference in solubility.

16.20 1,2-Propanediol is oxidized in the liver to pyruvic acid ($C_3H_4O_3$), which can be metabolized by the body. Draw the structure of pyruvic acid.

16.21 Draw the structure of the product of each of the following reactions.

(a) 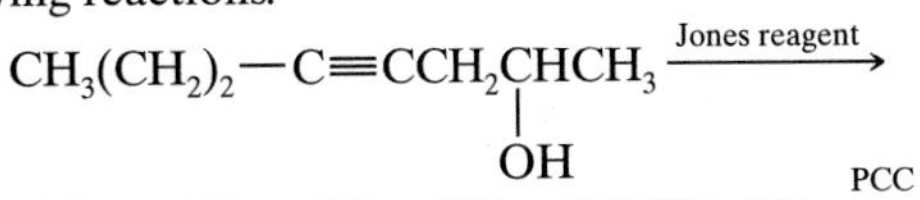 $CH_3(CH_2)_2-C{\equiv}CCH_2\underset{OH}{CH}CH_3 \xrightarrow{\text{Jones reagent}}$

(b) $CH_2{=}CH-CH{=}CH-CH_2CH_2OH \xrightarrow{\text{PCC}}$

(c) 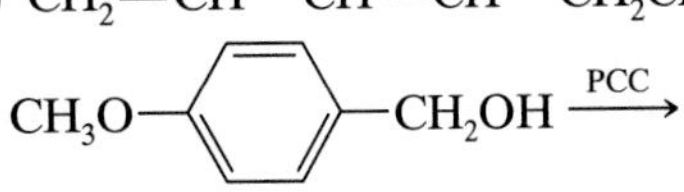 $CH_3O-C_6H_4-CH_2OH \xrightarrow{\text{PCC}}$

(d) 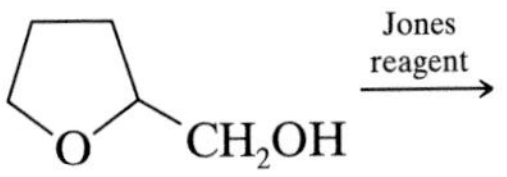 O CH$_2$OH $\xrightarrow{\text{Jones reagent}}$

16.22 Write the product formed by oxidation with PCC of each of the compounds in Exercise 16.16.

16.23 Write the product formed by oxidation with PCC of the sex attractant of the Mediterranean fruit fly.

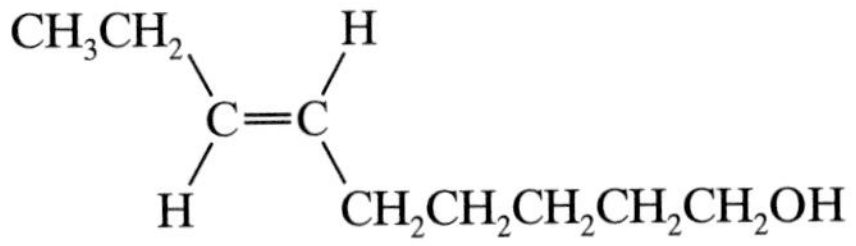

CH$_3$CH$_2$(H)C=C(H)CH$_2$CH$_2$CH$_2$CH$_2$CH$_2$OH

16.24 Write the product formed by oxidation with PCC of the following mosquito repellent.

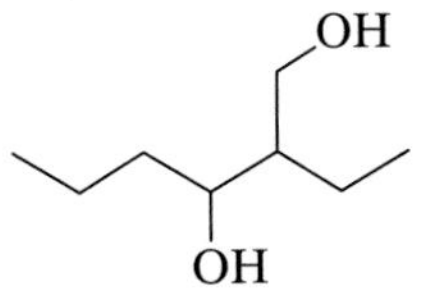

OH OH

Phenols

16.25 Suggest a way to distinguish cyclohexanol and phenol based on differences in solubility.

16.26 Which of the following compounds is a phenol?

(a) 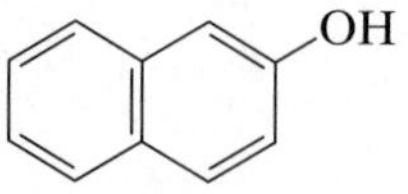 [structure with OH]

(b) 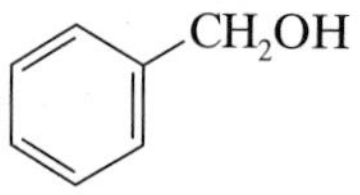 [structure with CH_2OH]

(c) 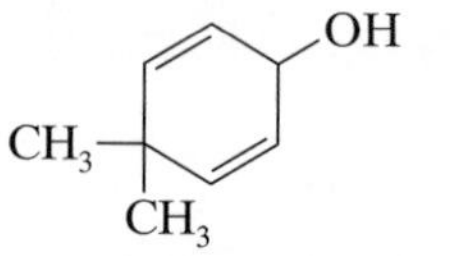 [structure with OH, CH_3, CH_3]

16.27 What is the IUPAC name for chlorophene, an effective germicide?

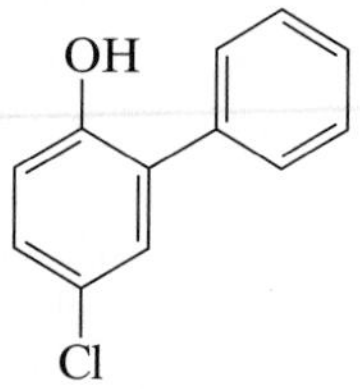

[structure: OH, Cl, phenyl]

16.28 4-Chloro-3,5-dimethylphenol is used in topical preparations for athlete's foot. Draw its structure.

Ethers

16.29 Draw the three isomeric ethers with the molecular formula $C_4H_{10}O$.

16.30 Draw the four isomeric methyl ethers with the molecular formula $C_5H_{12}O$.

16.31 Give the common name for each of the following ethers.

(a) 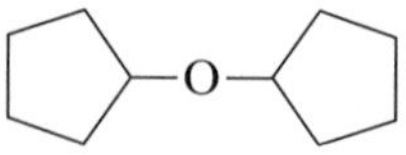 [cyclopentyl]—O—[cyclopentyl]

(b) 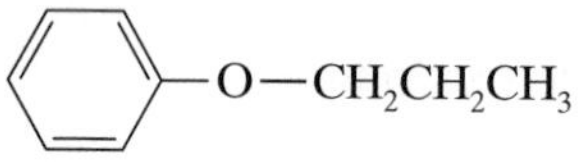 [benzene ring]—O—$CH_2CH_2CH_3$

(c) 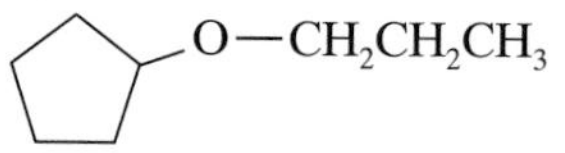 [cyclopentyl]—O—$CH_2CH_2CH_3$

16.32 Give the common name for each of the following ethers.

(a) 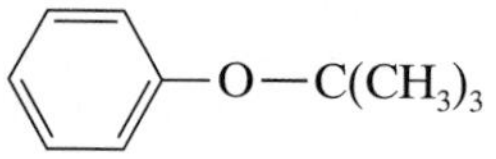 [benzene ring]—O—$C(CH_3)_3$

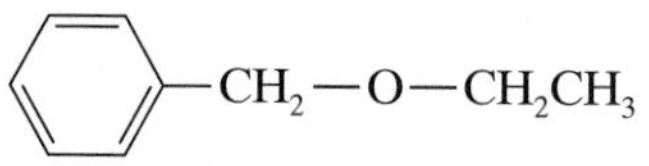

(b) [benzene ring]—CH_2—O—CH_2CH_3

(c) [cyclohexyl]—O—$CH(CH_3)_2$

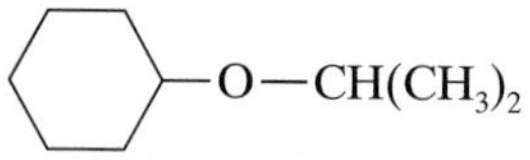

16.33 Assign the IUPAC name to each of the following compounds.

(a) $CH_3CH_2CH_2CHCH_3$ with OCH_3 on C-2 (the CH)

(b) $CH_3CHCH_2CHCH_3$ with CH_3 and OCH_3 substituents

(c) $CH_3CH_2CH_2CHCH_2CH_3$ with OCH_2CH_3 substituent

16.34 Assign the IUPAC name to each of the following compounds.

(a) $CH_3CH_2CH_2CHCH_2OCH_3$ with OCH_3 substituent

(b) $CH_3CHCH_2CHCH_3$ with CH_3O and OCH_3 substituents

(c) $CH_3CH_2CHCH_2CHCH_3$ with CH_3CH_2O and OCH_2CH_3 substituents

Sulfur Compounds

16.35 There are two isomeric compounds C_3H_8S with an —SH group. Draw the structures of the compounds.

16.36 There are four isomeric compounds $C_4H_{10}S$ with an —SH group. Draw the structures of the compounds.

16.37 Addition of sodium hydroxide to an aqueous solution of $CH_3CH_2CH_2SH$ eliminates much of the odor. Explain why.

16.38 Thiols have distinctly lower boiling points than alcohols. Explain why.

16.39 Draw the structure of each of the following compounds.

(a) 1-propanethiol
(b) 2-methyl-3-pentanethiol
(c) cyclopentanethiol

16.40 Draw the structure of each of the following compounds.

(a) 2-propanethiol
(b) 2-methyl-1-propanethiol
(c) cyclobutanethiol

17

Aldehydes and Ketones

17.1 *The Carbonyl Group*
17.2 *Naturally Occurring Aldehydes and Ketones*
17.3 *Nomenclature of Aldehydes and Ketones*
17.4 *Physical Properties of Aldehydes and Ketones*
17.5 *Oxidation of Carbonyl Compounds*
17.6 *Reduction of Carbonyl Compounds*
17.7 *Addition Reactions of Carbonyl Compounds*
17.8 *Addition of Oxygen Compounds*
17.9 *Addition of Nitrogen Compounds*
17.10 *Reactivity of the α-Carbon Atom*
17.11 *The Aldol Condensation*

Essays
- *Monitoring Blood Sugar and Diabetes*
- *Biological Oxidation with Enzymes*
- *Isomerism, Addition Reactions, and Vision*

Carbonyl compounds are responsible for the characteristic odor of the iris. (The Image Works)

17.1 The Carbonyl Group

A **carbonyl group** consists of a double bond linking a **carbonyl carbon atom** and a **carbonyl oxygen atom.** The carbonyl oxygen atom shares two of its six valence electrons with the carbonyl carbon atom. Its remaining four valence electrons remain as two sets of lone-pair electrons. The carbonyl carbon atom shares two of its four valence electrons with the carbonyl oxygen atom, and its remaining two electrons form two single bonds to other atoms. Formaldehyde, CH_2O, shown below, is the simplest compound with a carbonyl group. Note that the lone pairs on the oxygen atom of the carbonyl group are not always shown in structural formulas.

nonbonding electrons are shown — $H_2C{=}\ddot{O}{:}$

nonbonding electrons are not shown — $H_2C{=}O$

representations of formaldehyde

The carbonyl carbon atom, which is sp^2 hybridized, contributes one electron to each of the three hybrid orbitals, forming three σ bonds. Formaldehyde has two σ bonds to hydrogen atoms and one σ bond to the carbonyl oxygen atom. These coplanar bonds lie at approximately 120° to each other. The fourth electron of the carbonyl carbon atom occupies a $2p$ orbital perpendicular to the plane of the three sp^2 hybrid orbitals. The carbonyl oxygen atom, also sp^2 hybridized, contributes one of its six valence electrons to the sp^2 hybrid orbital that forms a σ bond with the carbonyl carbon atom. Four valence electrons remain as two sets of nonbonding electron pairs in the other two sp^2 hybrid orbitals. They lie at approximately 120° to each other and to the carbon–oxygen bond (Figure 17.1). The last valence electron occupies a $2p$ orbital perpendicular to the plane of the sp^2 hybrid orbitals. The $2p$ orbitals of the carbon and oxygen atoms overlap to form a π bond.

Because oxygen is more electronegative than carbon, the oxygen atom attracts the electrons in the carbon–oxygen double bond, making the carbonyl bond polar. The carbonyl group is resonance-stabilized, as shown by the charged contributing structure (2).

FIGURE 17.1
Bonding and Structure in Carbonyl Compounds

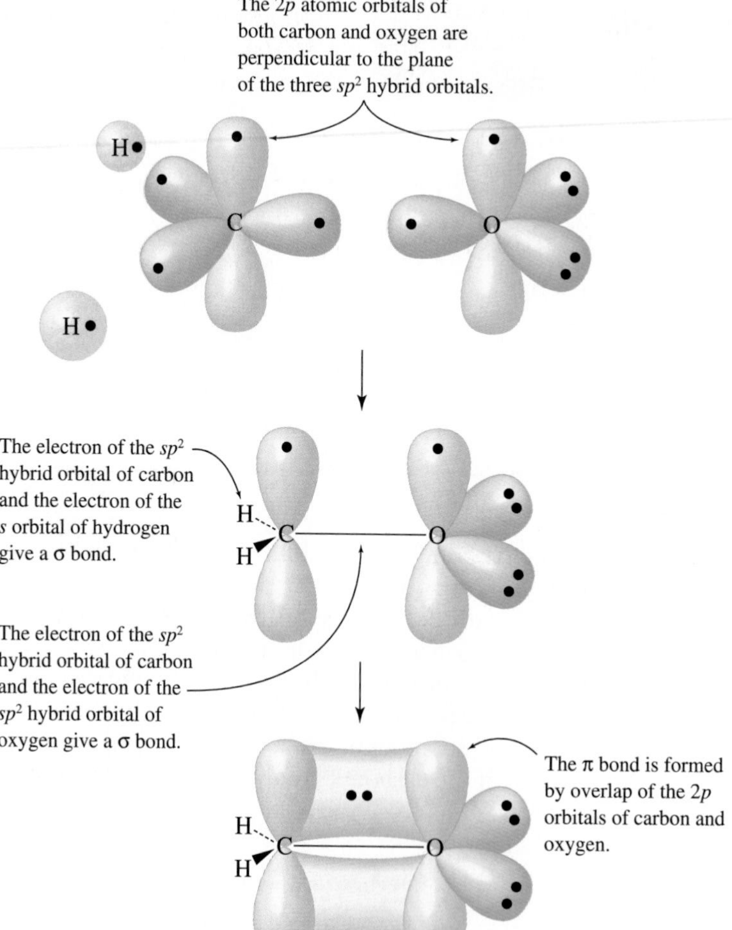

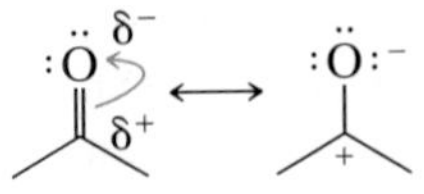

(1) ⟷ (2)

Contributing structure (1) is more important because each atom has a Lewis octet. However, the dipolar structure (2) is often used to account for the chemical reactions of aldehydes and ketones. The carbonyl carbon atom reacts with nucleophiles, and the carbonyl oxygen atom reacts with electrophiles.

Carbonyl Compounds

When a carbonyl carbon atom is bonded to at least one hydrogen atom, the compound is an **aldehyde.** The simplest aldehyde, formaldehyde, has a carbonyl carbon atom bonded to two hydrogen atoms. The carbonyl group is bonded to one hydrogen atom and either an alkyl group (R) or an aromatic group (Ar) in other aldehydes. Although the bond angles around the carbonyl carbon atom are approximately 120°, structures are often written with a linear arrangement of carbon atoms.

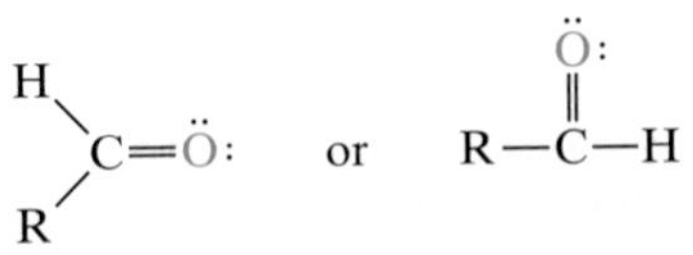

$$\begin{matrix} H \\ & \diagdown \\ & & C{=}\ddot{O}: \\ & \diagup \\ R \end{matrix} \quad \text{or} \quad R{-}\overset{\overset{\Large\ddot{O}:}{\|}}{C}{-}H$$

general structural formula of aldehydes

When a carbonyl carbon atom is bonded to two other carbon atoms, the compound is a **ketone.** The bonded groups may be any combination of alkyl or aromatic groups. A ketone has 120° bond angles at the carbonyl carbon atom, but structures are often written with a linear arrangement of carbon atoms.

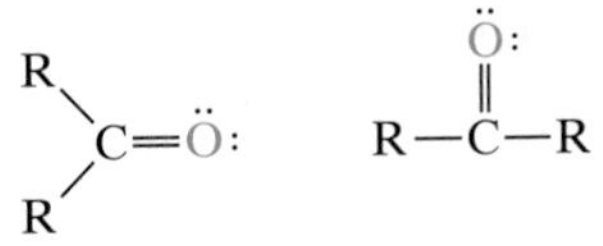

$$\begin{matrix} R \\ & \diagdown \\ & & C{=}\ddot{O}: \\ & \diagup \\ R \end{matrix} \qquad R{-}\overset{\overset{\Large\ddot{O}:}{\|}}{C}{-}R$$

general structural formulas of ketones

An aldehyde can be written with the condensed formula RCHO or ArCHO, where the symbol CHO indicates that both hydrogen and oxygen atoms are bonded to the carbonyl carbon atom. A ketone has the condensed formula RCOR. In this condensed formula the symbol CO represents the carbonyl group, and the two R groups flanking the CO group are bonded to the carbonyl carbon atom.

17.2 Naturally Occurring Aldehydes and Ketones

The carbonyl group is the most common functional group in oxygen-containing organic compounds isolated from biological sources. One of two suffixes in common names may indicate the presence of a carbonyl group in a molecule. If the carbonyl compound is an aldehyde, we find the suffix *-al.* If the carbonyl compound is a ketone, we use the suffix *-one.* For example, retinal is an aldehyde required for vision. The first part of the name indicates that this compound is present in the retina, and the suffix tells us that it is an aldehyde. Noethindrone, a steroid hormone, has a ketone group.

retinal

norethindrone

17.3 Nomenclature of Aldehydes and Ketones

Aldehydes and ketones with low molecular weights are often referred to by their common names. These names are derived from the related common names of carboxylic acids (Chapter 18).

$H-C(=O)-H$ formaldehyde

$CH_3-C(=O)-H$ acetaldehyde

$CH_3-C(=O)-CH_3$ acetone

The common names of some aromatic aldehydes and ketones include the following:

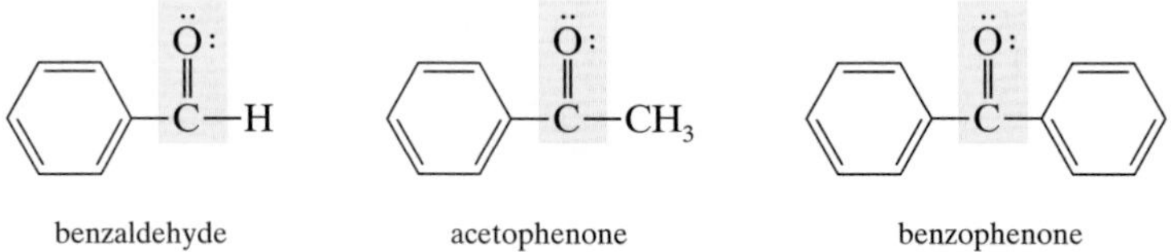

benzaldehyde acetophenone benzophenone

IUPAC Names of Aldehydes

The IUPAC rules for naming aldehydes are similar to those outlined for alcohols.

1. The longest carbon chain that contains the carbonyl carbon atom is named as the parent chain. The final *-e* of the parent hydrocarbon is replaced by the ending *-al.*
2. The parent chain is numbered to make the carbonyl carbon atom C-1. The number 1 is not used to indicate the position of the carbonyl carbon atom, because it is understood to be located at the end of the chain. The name of each substituent and the number of the carbon atom to which it is attached are added to the parent name as a prefix.

$\overset{4}{CH_3}-\overset{3}{CH}(CH_3)-\overset{2}{CH}(CH_3)-\overset{1}{C}(=O)-H$

This is 2,3-dimethylbutanal, not 2,3-dimethyl-1-butanal.

3. The aldehyde functional group has a higher priority than alkyl, halogen, hydroxyl, and alkoxy groups. If any of these groups are present, their names and positions are indicated as prefixes to the name of the parent aldehyde.

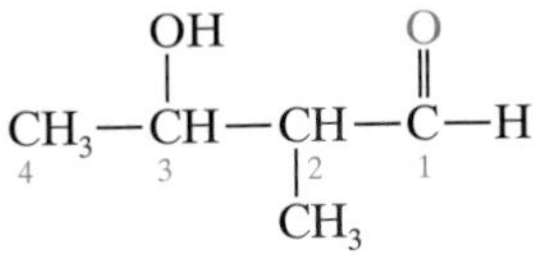

$$\underset{4}{CH_3}-\underset{3}{\overset{OH}{\overset{|}{CH}}}-\underset{\underset{CH_3}{|2}}{CH}-\underset{1}{\overset{O}{\overset{\|}{C}}}-H$$

3-hydroxy-2-methylbutanal

4. The aldehyde functional group has a higher priority than double or triple bonds. When the parent chain contains a double or triple bond, the final -e of the name of the parent alkene or alkyne is replaced with the suffix -al. The position of the multiple bond is indicated with a prefix.

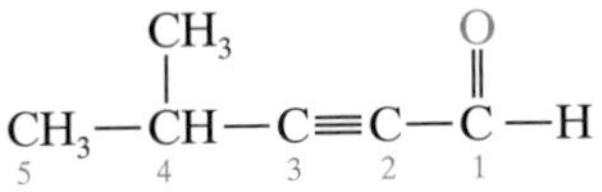

$$\underset{5}{CH_3}-\underset{4}{\overset{CH_3}{\overset{|}{CH}}}-\underset{3}{C}\equiv\underset{2}{C}-\underset{1}{\overset{O}{\overset{\|}{C}}}-H$$

4-methyl-2-pentynal

5. If an aldehyde or ketone contains other groups with a higher priority, such as carboxylic acids, the carbonyl group is given the prefix *oxo*-. A number is used to indicate the position of the oxo group. The priority order is carboxylic acid > aldehyde > ketone.

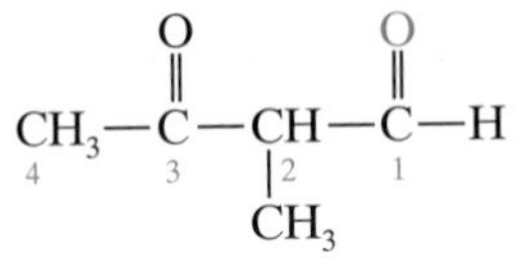

$$\underset{4}{CH_3}-\underset{3}{\overset{O}{\overset{\|}{C}}}-\underset{\underset{CH_3}{|2}}{CH}-\underset{1}{\overset{O}{\overset{\|}{C}}}-H$$

2-methyl-3-oxobutanal

6. If an aldehyde group is attached to a ring, the suffix *-carbaldehyde* is used.

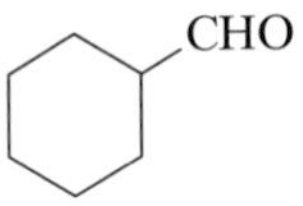

CHO (on cyclohexane ring)

cyclohexanecarbaldehyde

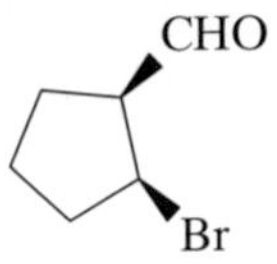

CHO, Br (on cyclopentane ring)

cis-2-bromocyclopentanecarbaldehyde

IUPAC Names of Ketones

The IUPAC rules for naming ketones are similar to those used for aldehydes. However, because the carbonyl group in a ketone is not on a terminal carbon atom, a number is used to indicate its position.

1. The longest carbon chain that contains the carbonyl carbon atom is named as the parent chain. The final *-e* of the parent hydrocarbon is replaced by the ending *-one*.
2. The carbon chain is numbered so that the carbonyl carbon atom has the lowest possible number. This number is used as a prefix to the parent name. The identity and location of substituents are added as a prefix to the parent name.

$$\underset{5}{CH_3}-\underset{4}{\overset{CH_3}{\overset{|}{CH}}}-\underset{3}{CH_2}-\underset{2}{\overset{O}{\overset{\|}{C}}}-\underset{1}{CH_3}$$

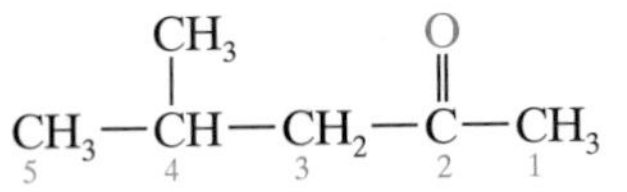

This is 4-methyl-2-pentanone, not 2-methyl-4-pentanone.

3. Cyclic ketones are named as cycloalkanones. The carbonyl carbon atom receives the number 1. The ring is numbered in the direction that gives the lower number to the first substituent encountered.

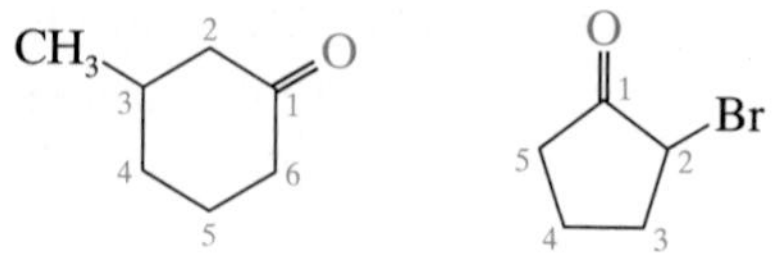

3-methylcyclohexanone 2-bromocyclopentanone

4. Halogen, hydroxyl, alkoxy groups, and multiple bonds have lower priorities than the ketone group. These substituted ketones are named in the same way as described for aldehydes.

EXAMPLE 17.1

Give the IUPAC name for the following alarm pheromone in some species of ants.

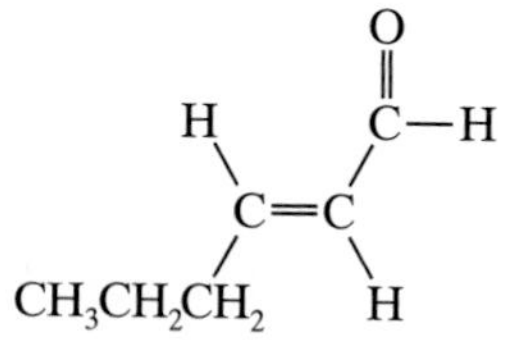

$$\underset{CH_3CH_2CH_2}{\overset{H}{}}\;C{=}C\;\overset{C(=O)-H}{\underset{H}{}}$$

Solution

The aldehyde carbon atom on the right of the structure is assigned the number 1. The double bond is therefore located at the C-2 atom, and the name, disregarding stereochemistry, is 2-hexenal. The groups bonded to the unsaturated carbon atoms—the CHO and propyl groups—are in a trans arrangement. The IUPAC name is *trans*-2-hexenal.

PROBLEM 17.1

The IUPAC name for Capillin, used against skin fungi, is 1-phenyl-2,4-hexadiyn-1-one. Draw its structure.

17.4 Physical Properties of Aldehydes and Ketones

Since oxygen is more electronegative than carbon, the shared electrons are pulled toward the oxygen atom, and the carbonyl group is polar. The polarity of the group is shown by use of the symbols $\mapsto$ or $\delta+$ and $\delta-$.

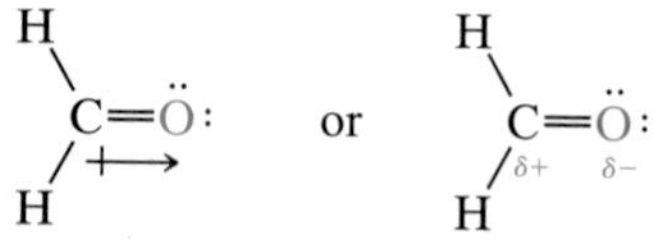

$$H_2C{=}\ddot{O}: \;(\longmapsto) \quad \text{or} \quad H_2\overset{\delta+}{C}{=}\overset{\delta-}{\ddot{O}}:$$

The physical properties of aldehydes and ketones are distinctly different from those of the alkanes or alcohols of similar molecular weight (Table 17.1). They have higher boiling points than the alkanes because there are dipole–dipole attractive forces between neighboring polar molecules. However, these attractive forces are weaker than in alcohols because aldehydes and ketones lack a positive hydrogen atom bonded to an electronegative atom as in the —OH group of alcohols. Therefore, alcohols have higher boiling points than aldehydes and ketones of similar molecular weight (Figure 17.2). As the molecular weights of the carbonyl compounds increase, their dipole–dipole attractive forces become less important. As a result, aldehydes and ketones become more like hydrocarbons with increasing

TABLE 17.1 Boiling Points of Alkanes, Alcohols, and Carbonyl Compounds

Compound	Structure	Molecular Weight (amu)	Boiling Point (°C)
ethane	CH_3CH_3	30	−89
methanol	CH_3OH	32	64.6
methanal	HCHO	30	−21
propane	$CH_3CH_2CH_3$	44	−42
ethanol	CH_3CH_2OH	46	78.3
ethanal	CH_3CHO	44	20
butane	$CH_3CH_2CH_2CH_3$	58	−1
1-propanol	$CH_3CH_2CH_2OH$	60	97.1
propanal	CH_3CH_2CHO	58	48.8
methylpropane	$CH_3CH(CH_3)_2$	58	−12
2-propanol	$CH_3CH(OH)CH_3$	60	82.5
propanone	CH_3COCH_3	58	56.1

chain lengths. The boiling point differences become smaller, although the order of boiling points is still alcohol > carbonyl compound > alkane.

Aldehydes and ketones form hydrogen bonds with water. As a consequence, the lower molecular weight compounds formaldehyde, acetaldehyde, and acetone are soluble in water in all proportions.

The lone-pair electrons of the carbonyl oxygen atom serve as a hydrogen bond acceptor.

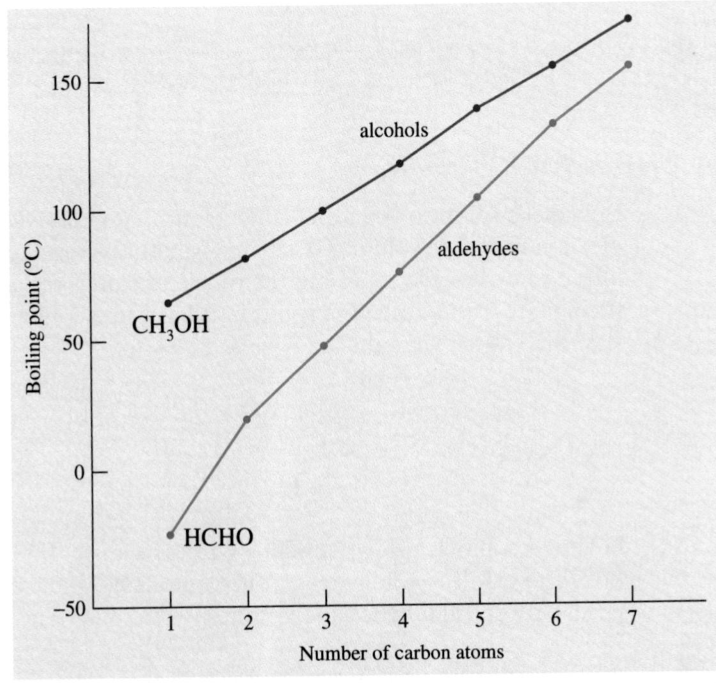

FIGURE 17.2 Boiling Points of Aldehydes and Alcohols

The boiling points of both aldehydes and alcohols increase with the number of carbon atoms. However, alcohols have higher boiling points than aldehydes.

For the same reason as was discussed for the solubility of alcohols (Chapter 16), the solubility of carbonyl compounds in water decreases as the chain length increases, and they become more like hydrocarbons.

17.5 Oxidation of Carbonyl Compounds

In Chapter 16 you learned that primary alcohols are oxidized to aldehydes, which are then easily oxidized to acids. Under the same conditions, secondary alcohols are oxidized to ketones, which are not oxidized further. The difference in the reactivities of aldehydes and ketones toward oxidizing agents is used to distinguish between these two classes of compounds.

Tollens' Reagent Gives a Silver Mirror

This child can see herself because the silver coating on the back of the mirror reflects light. (PhotoEdit)

Tollens' reagent is a mild oxidizing agent that reacts with all aldehydes but not with ketones. The reagent is a basic solution of a silver ammonia complex ion. When an aldehyde is oxidized by $Ag(NH_3)_2{}^+$, metallic silver is deposited as a mirror on the wall of a test tube.

$$R{-}\overset{\overset{\displaystyle O}{\|}}{C}{-}H + 2\,Ag(NH_3)_2{}^+ + 3\,OH^- \longrightarrow R{-}\overset{\overset{\displaystyle O}{\|}}{C}{-}O^- + 2\,Ag(s) + 4\,NH_3 + 2\,H_2O$$

The addition of Tollens' reagent to each of the following isomeric carbonyl compounds gives a clear result that distinguishes between the two isomers.

$$C_5H_9{-}\underset{}{\overset{\overset{\displaystyle CH_3}{|}}{C}H}{-}CHO \xrightarrow{Ag(NH_3)_2{}^+} C_5H_9{-}\overset{\overset{\displaystyle CH_3}{|}}{C}H{-}CO_2{}^- + Ag(s)$$

$$C_5H_9{-}\overset{\overset{\displaystyle O}{\|}}{C}{-}CH_2CH_3 \xrightarrow{Ag(NH_3)_2{}^+} \text{no oxidation product}$$

Benedict's Solution Gives a Red Precipitate

Benedict's solution is an oxidizing agent that contains cupric ion (Cu^{2+}) as a complex ion in a basic solution. The Cu^{2+} will not oxidize ketones, but aldehydes are oxidized to carboxylic acids and the blue Cu^{2+} ion is reduced to Cu^+, which forms the precipitate Cu_2O. Since the solution is basic, the conjugate base of a carboxylic acid is formed rather than the acid itself.

$$R{-}\overset{\overset{\displaystyle O}{\|}}{C}{-}H + \underset{\text{blue solution}}{2\,Cu^{2+}} + 5\,OH^- \longrightarrow R{-}\overset{\overset{\displaystyle O}{\|}}{C}{-}O^- + \underset{\text{red precipitate}}{Cu_2O(s)} + 3\,H_2O$$

Fehling's solution, which contains Cu^{2+} as a different complex ion in a basic solution, also oxidizes aldehydes but not ketones. Either reagent can be used to distinguish between aldehydes and ketones.

EXAMPLE 17.2

Describe how the following isomeric carbonyl-containing compounds of molecular formula C_8H_8O could be distinguished from each other using Fehling's solution.

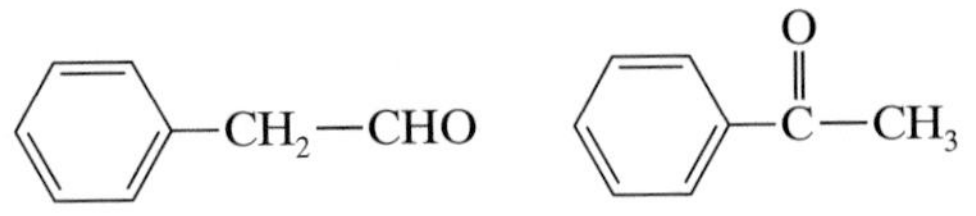

$$C_6H_5{-}CH_2{-}CHO \qquad C_6H_5{-}\overset{O}{\overset{\|}{C}}{-}CH_3$$

I II

Solution

Compound I is an aldehyde; compound II is a ketone. Fehling's solution will react only with compound I, as evidenced by formation of a red precipitate of Cu_2O.

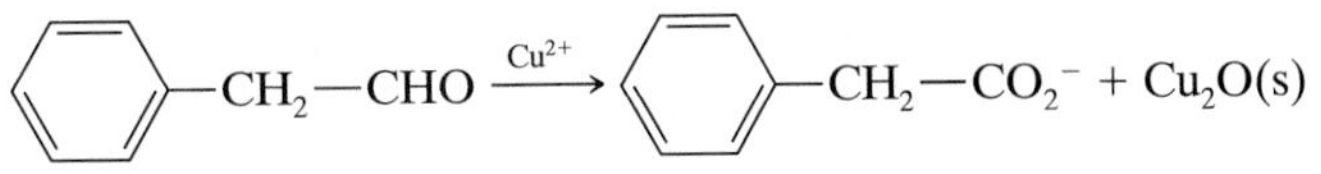

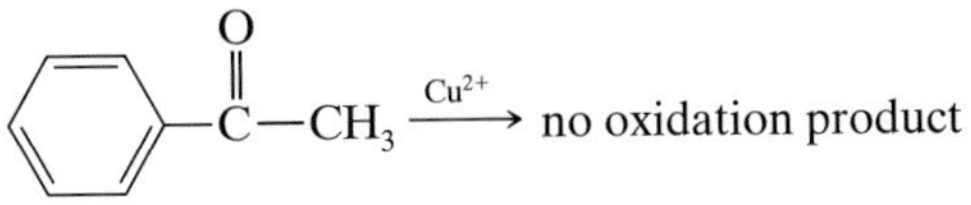

$$C_6H_5{-}CH_2{-}CHO \xrightarrow{Cu^{2+}} C_6H_5{-}CH_2{-}CO_2^- + Cu_2O(s)$$

$$C_6H_5{-}\overset{O}{\overset{\|}{C}}{-}CH_3 \xrightarrow{Cu^{2+}} \text{no oxidation product}$$

PROBLEM 17.2

Four of the seven isomeric carbonyl-containing compounds with the molecular formula $C_5H_{10}O$ react with Tollens' reagent. Draw their structures.

Monitoring Blood Sugar and Diabetes

The disease Type I diabetes mellitus results from a lack of insulin, a polypeptide hormone that regulates glucose transport into cells. Loss of insulin-producing cells in the pancreas causes a sudden onset of the disease in childhood. Without enough insulin, body cells cannot take up glucose, and its concentration in the blood rises. The kidneys, which normally reabsorb glucose into the blood, cannot handle the overload and pass it into the urine. A person who suffers from Type I diabetes used to determine the timing of insulin injections required to maintain the proper blood glucose concentration by performing the Benedict's test on a urine sample.

Glucose has an aldehyde group that reacts with Benedict's solution. Benedict's solution contains blue Cu^{2+}. If the solution turns a greenish yellow color, the sample contains about 0.5% glucose. If the sample solution turns red (the color of Cu_2O), the concentration of glucose in the original sample is greater than 2%.

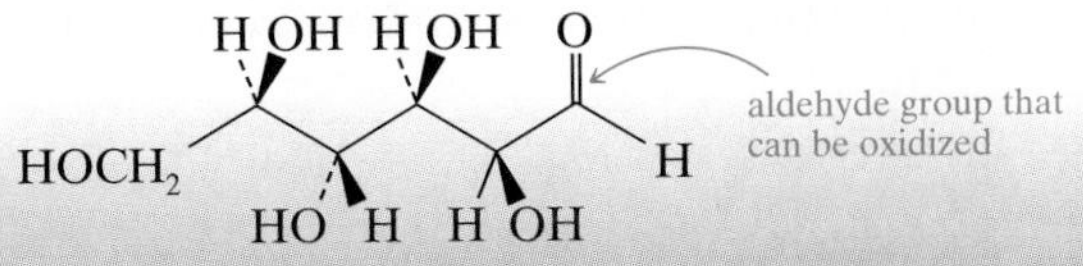

H OH H OH O
HOCH$_2$ H
HO H H OH
aldehyde group that can be oxidized

Modern methods allow much closer monitoring of the level of glucose in the blood by use of glucose oxidase, an enzyme that specifically oxidizes glucose. Hydrogen peroxide is produced in addition to gluconic acid, the oxidized form of glucose. A second enzyme, peroxidase, catalyzes an oxidation–reduction reaction in which a colorless dye is oxidized. A colored oxidation product results.

$$\text{glucose} + O_2 + H_2O \xrightarrow{\text{glucose oxidase}} \text{gluconic acid} + H_2O_2$$

$$H_2O_2 + \underset{\text{(colorless)}}{\text{dye}} \xrightarrow{\text{peroxidase}} \underset{\text{(colored)}}{\text{oxidized dye}} + H_2O$$

Both enzymes are contained on a paper strip. The individual pricks a finger to release a blood drop that is smeared on the strip. The intensity of the resulting colored dye is measured electronically using a device the size of a calculator.

17.6 Reduction of Carbonyl Compounds

Both aldehydes and ketones can be reduced to alcohols. The reaction with hydrogen gas, called *hydrogenation,* is catalyzed by nickel, palladium, or platinum.

$$CH_3CH_2CH_2CHO + H_2 \xrightarrow[100\text{ atm}]{Pt} CH_3CH_2CH_2CH_2OH$$

butanal (an aldehyde) → 1-butanol (a primary alcohol)

$$CH_3\overset{\overset{\displaystyle O}{\|}}{C}CH_2CH_3 + H_2 \xrightarrow[100\text{ atm}]{Pt} CH_3\overset{\overset{\displaystyle OH}{|}}{C}HCH_2CH_3$$

butanone (a ketone) → 2-butanol (a secondary alcohol)

Recall that carbon–carbon double bonds also are reduced by hydrogen (Chapter 15). However, the reduction of carbon–carbon double bonds is far easier, occurring at 1 atm in comparison with the 50–100 atm H_2 required for the reduction of carbonyl compounds. In any compound containing both carbon–carbon double bonds and carbonyl groups, both are hydrogenated at a high pressure of H_2.

$$CH_2{=}CH{-}CH_2{-}\overset{\overset{\displaystyle O}{\|}}{C}{-}CH_3 + H_2 \xrightarrow[100\text{ atm}]{Pt} CH_3{-}CH_2{-}CH_2{-}\overset{\overset{\displaystyle OH}{|}}{C}H{-}CH_3$$

The order of reactivity of the π bonds can be reversed by using sodium borohydride ($NaBH_4$) in ethanol as solvent. Aqueous acid is then added to obtain the final alcohol product. The carbonyl group is reduced by sodium borohydride without affecting a carbon–carbon double bond. Aldehydes and ketones are converted into primary and secondary alcohols, respectively.

$$CH_2{=}CH{-}CH_2{-}CH_2{-}\overset{\overset{\displaystyle O}{\|}}{C}{-}CH_3 \xrightarrow[CH_3CH_2OH]{NaBH_4} CH_2{=}CH{-}CH_2{-}CH_2{-}\overset{\overset{\displaystyle OH}{|}}{C}H{-}CH_3$$

The two hydrogen atoms added to the carbonyl group in the reduction come from different sources. A hydride ion is transferred from the borohydride reagent to the carbonyl carbon atom. The hydrogen atom bonded to the oxygen atom is transferred from the aqueous acid. For an aldehyde the reaction is

$$R{-}\overset{\overset{\displaystyle :\ddot{O}}{\|}}{C}{-}H \xrightarrow{NaBH_4} R{-}\underset{\underset{\displaystyle H}{|}}{\overset{\overset{\displaystyle :\ddot{O}:^-}{|}}{C}}{-}H \xrightarrow{H_3O^+} R{-}\underset{\underset{\displaystyle H}{|}}{\overset{\overset{\displaystyle :\ddot{O}{-}H}{|}}{C}}{-}H$$

(H on carbon: from $NaBH_4$; H on oxygen: from H_3O^+)

These reactions of aldehydes and ketones occur because the carbonyl group is polarized, making the carbon atom partially positive and the oxygen atom partially negative. The negative hydride ion transferred from the borohydride reagent is a nucleophile that attacks the partially positive carbon atom. The second reaction is an acid–base reaction of an alkoxide ion.

$$H{-}\underset{\underset{\displaystyle H}{|}}{\overset{\overset{\displaystyle H}{|}}{B}}{-}H \quad R_2C{=}\ddot{O}: \longrightarrow H{-}\underset{\underset{\displaystyle R}{|}}{\overset{\overset{\displaystyle R}{|}}{C}}{-}\ddot{O}:^-$$

Biological Oxidation with Enzymes

The oxidation of an alcohol to a carbonyl group is a common metabolic reaction. The reverse reaction, reduction of a carbonyl compound to an alcohol, is also common. In fact, the two related reactions often occur by the same mechanism, with the direction of the reaction controlled by needs of the cell. The two reactions transfer a hydride ion either to or from carbon atoms bonded to the oxygen atom in the substrate from or to a carbon atom of a coenzyme.

We recall that the oxidation of ethanol in a human liver cell to yield ethanal (acetaldehyde) is catalyzed by alcohol dehydrogenase (ADH), involving the reduction of nicotinamide adenine dinucleotide, NAD^+ (Section 16.9).

$$CH_3CH_2OH + NAD^+ \xrightarrow{ADH} CH_3CHO + NADH + H^+$$

The oxidized and reduced forms of nicotinamide adenine dinucleotide are drawn using a shorthand in which the R group attached to the heterocyclic ring represents a part of the molecule not directly involved in its reactions. The oxidation or reduction of the two forms of the coenzyme occurs within the six-membered pyridine ring.

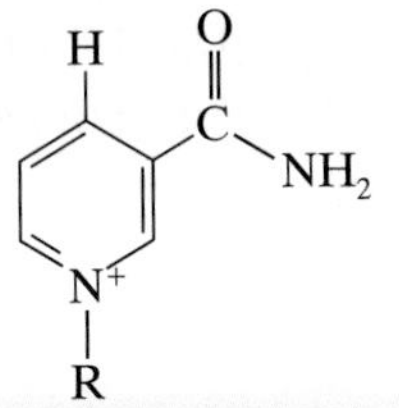

NAD^+ (oxidized form)

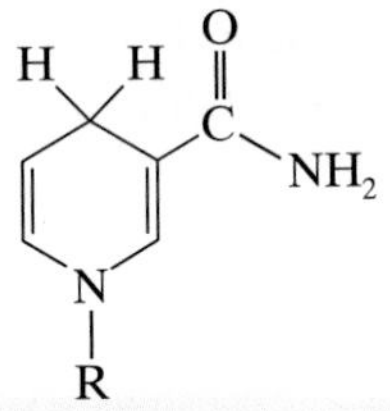

NADH (reduced form)

When ethanol is oxidized by NAD^+, a hydride ion is transferred from the C-1 atom of ethanol to the C-4 atom of the heterocyclic ring. In the reverse reaction, acetaldehyde is reduced by transfer of a hydride ion from the C-4 atom of NADH to the carbonyl carbon atom.

alcohol dehydrogenase

$$CH_3-\overset{O}{\overset{\|}{C}}-H + \ldots + H^+$$

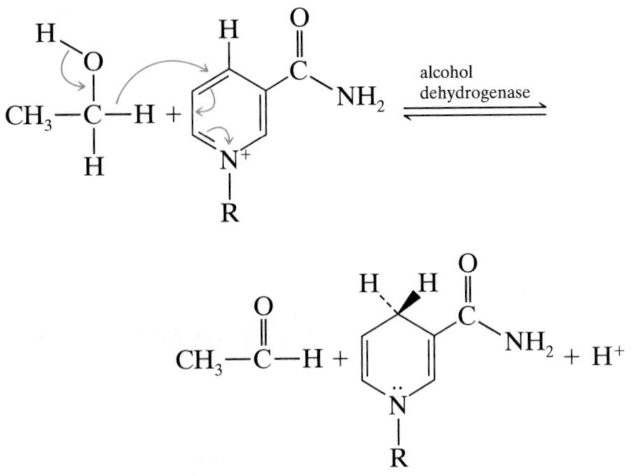

Thus, NADH in the cell serves a role similar to that of $NaBH_4$ in the laboratory. Both transfer a hydride ion to the carbonyl carbon atom.

EXAMPLE 17.3

Write the structure of the product of the reaction of the following compound with $NaBH_4$.

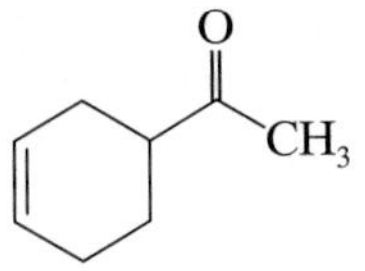

Solution

Sodium borohydride reduces ketones to secondary alcohols. The resulting structure is obtained by converting the carbon–oxygen double bond into a carbon–oxygen single bond and adding hydrogen atoms to both the carbon and oxygen atoms.

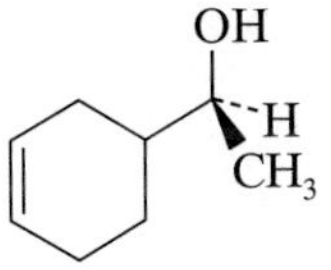

PROBLEM 17.3

What is the product of the reaction of the compound in Example 17.3 with hydrogen gas at 100 atm pressure in the presence of a platinum catalyst?

17.7 Addition Reactions of Carbonyl Compounds

Addition reactions are characteristic reactions of π bonds. In alkenes the order of addition is determined by the stability of the carbocation formed. The polarity of the carbonyl group determines the mode of addition of a reagent to the π bond of carbonyl compounds. The electrophilic part of the reagent attacks the electronegative oxygen atom; the nucleophilic part attacks the electropositive carbon atom.

nucleophile attacks here ⟶ $R_2C{=}\ddot{O}\!:$ ⟵ electrophile attacks here

Mechanism of Carbonyl Addition Reactions

Many of the reagents that add to carbonyl compounds can be represented by HNu. The electrophile is H^+; the nucleophile is Nu^-. The mechanism of these reactions is written in two steps. In the first step a proton bonds to the oxygen atom to produce a carbocation. After this acid–base reaction, a second acid–base reaction occurs. The carbocation acts as an acid and reacts with Nu^-, which is a nucleophile.

$$R{-}C({=}\ddot{O}{:}){-}H + H{-}Nu \rightleftharpoons R{-}C^{+}(\ddot{O}{-}H){-}H + {:}Nu^-$$

$$R{-}C^{+}(\ddot{O}{-}H){-}H + {:}Nu^- \rightleftharpoons R{-}C(\ddot{O}{-}H)(Nu){-}H$$

The sequence of the steps may be different depending on the reagent HNu. The conjugate base Nu^- may react as a nucleophile and attack the carbonyl carbon atom, which is somewhat positive owing to the polarity of the carbonyl group. The carbonyl carbon atom is the Lewis acid, and the Nu^- is the Lewis base in this step. Next, the oxygen atom is protonated in a second acid–base reaction.

$$R{-}C({=}\ddot{O}{:}){-}H + {:}Nu^- \rightleftharpoons R{-}C({:}\ddot{O}{:}^-)(Nu){-}H$$

$$R{-}C({:}\ddot{O}{:}^-)(Nu){-}H + H{-}Nu \rightleftharpoons R{-}C(\ddot{O}{-}H)(Nu){-}H + {:}Nu^-$$

17.8 Addition of Oxygen Compounds

The HNu groups that add to a carbonyl group may contain oxygen or nitrogen bonded to hydrogen. The HO group of both water and alcohols add to aldehydes and ketones. The reaction of nitrogen compounds with carbonyl compounds will be presented in Section 17.9.

Addition of Water Gives Unstable Hydrates

Aldehydes and ketones add water to form hydrates. The proton of water bonds to the oxygen atom of the carbonyl group; the hydroxide ion adds to the carbon atom.

$$\mathrm{H_2C{=}O} + \mathrm{H{-}O{-}H} \rightleftharpoons \mathrm{H{-}C(H)(OH){-}O{-}H}$$

Formalin, formerly used to preserve biological specimens, is a 37% by weight solution of formaldehyde in water. Formaldehyde is more than 99% hydrated, whereas other aldehydes are substantially less hydrated. Ketones are hydrated to a small extent, usually less than 1%. The hydrates of aldehydes and ketones cannot be isolated. They exist only in solution, where the large amount of water forces the equilibrium to the right.

Addition of Alcohols Gives Hemiacetals and Hemiketals

Alcohols add to carbonyl compounds in much the same way that water does. The hydrogen atom of the alcohol adds to the carbonyl oxygen atom, and the —OR′ portion (alkoxy group) adds to the carbon atom. The product is a **hemiacetal** or **hemiketal** that has both an —OH group and an —OR′ group attached to the same carbon atom. The hemiacetal has a hydrogen atom and an alkyl group attached to the carbon atom, whereas the hemiketal has two alkyl groups attached.

$$\mathrm{RC(=O)H} + \mathrm{H{-}O{-}R'} \underset{}{\overset{H^+}{\rightleftharpoons}} \mathrm{R{-}C(H)(OH){-}O{-}R'}$$

a hemiacetal

$$\mathrm{RC(=O)R} + \mathrm{H{-}O{-}R'} \underset{}{\overset{H^+}{\rightleftharpoons}} \mathrm{R{-}C(R)(OH){-}O{-}R'}$$

a hemiketal

Hemiacetals and hemiketals generally are unstable compounds. The equilibrium for both of the preceding reactions lies to the left. However, when both the carbonyl group and the alcohol are part of the same molecule, they form a stable cyclic product. The cyclization is favorable because the two functional groups are close to each other. Carbohydrates (Chapter 21), which contain both carbonyl and hydroxyl groups, exist to only a small extent as open-chain molecules. They exist largely as cyclic hemiacetals or hemiketals.

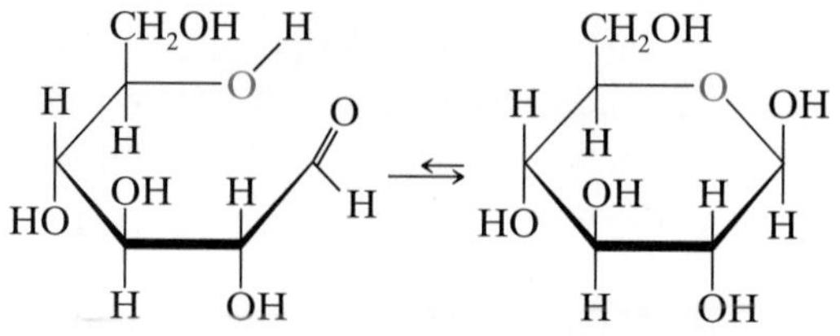

glucose ⇆ cyclic hemiacetal form of glucose

Formation of Acetals and Ketals

In acidic solution a substitution reaction of —OH by —OR′ in a hemiacetal or hemiketal occurs to produce an acetal or ketal.

$$\text{R—C(OH)(H)—OR}' + \text{H—O—R}' \xrightleftharpoons{H^+} \text{R—C(H)(OR}')\text{—OR}' + \text{H—OH}$$

a hemiacetal → an acetal (acetal bonds; original carbonyl carbon atom)

$$\text{R—C(OH)(R)—OR}' + \text{H—O—R}' \xrightleftharpoons{H^+} \text{R—C(R)(OR}')\text{—OR}' + \text{H—OH}$$

a hemiketal → a ketal (ketal bonds; original carbonyl carbon atom)

Note that both **acetals** and **ketals** have two alkoxy groups (—OR′) attached to the same carbon atom. The acetal has a hydrogen atom and an alkyl group attached to the carbon atom, whereas the ketal has two alkyl groups attached.

The reaction is reversible in acid solution, but formation of the acetal or ketal is increased by removing the water formed in the reaction or by increasing the amount of the alcohol.

$$\text{hemiacetal + alcohol} \xrightleftharpoons{H^+} \text{acetal + water}$$

adding alcohol "pushes" reaction to the right

removing water "pulls" reaction to the right

In total, 2 moles of alcohol react per mole of carbonyl compound.

$$CH_3CH_2CH_2C(=O)\text{—H} + 2\ CH_3OH \xrightleftharpoons{H^+} CH_3CH_2CH_2\text{—C(OCH}_3)_2\text{—H} + H_2O$$

The reverse reaction is favored when water is added. Acetals and ketals react with water in a hydrolysis reaction catalyzed by acid to give the carbonyl compound and the alcohol. Acetals and ketals do not react in neutral or basic solution.

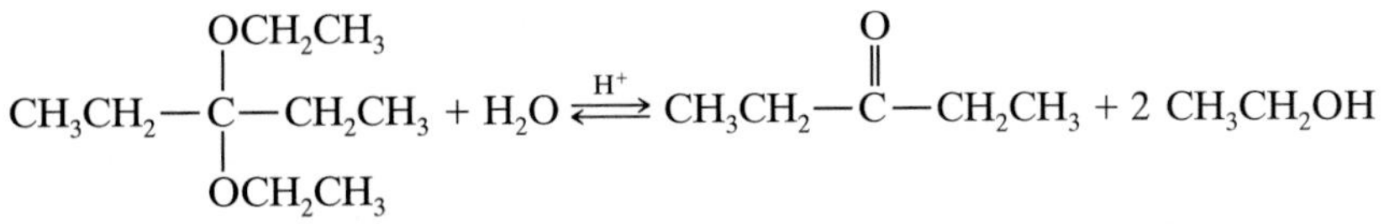

$$CH_3CH_2-\underset{\underset{OCH_2CH_3}{|}}{\overset{\overset{OCH_2CH_3}{|}}{C}}-CH_2CH_3 + H_2O \overset{H^+}{\rightleftharpoons} CH_3CH_2-\overset{\overset{O}{\|}}{C}-CH_2CH_3 + 2\ CH_3CH_2OH$$

Cyclic hemiacetals or hemiketals react with alcohols to produce cyclic acetals or ketals. In these compounds one oxygen atom was originally part of a hydroxyl group of the hydroxy aldehyde compound. This oxygen atom is in the ring in the final product. The second oxygen atom is derived from the alcohol represented here by ROH.

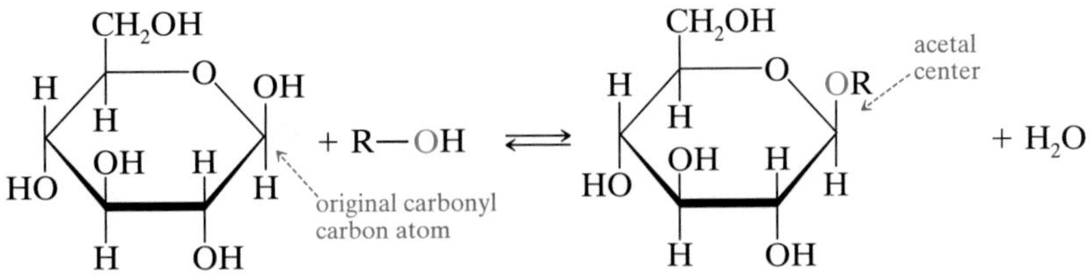

CH_2OH ... + R—OH ⇌ ... + H_2O

original carbonyl carbon atom

acetal center

a cyclic hemiacetal

a cyclic acetal

You will see examples of these compounds in the study of carbohydrates in which the alcohol, ROH, is a complex structure such as molecule of glucose (Chapter 21).

EXAMPLE 17.4

Identify the class to which each of the following compounds belongs.

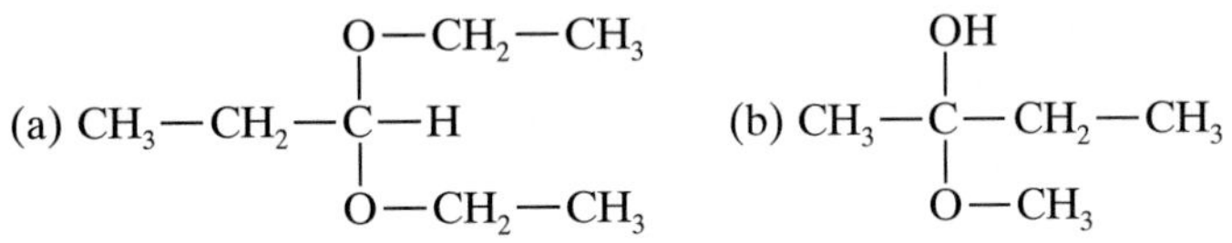

$$\text{(a) } CH_3-CH_2-\underset{\underset{O-CH_2-CH_3}{|}}{\overset{\overset{O-CH_2-CH_3}{|}}{C}}-H \qquad \text{(b) } CH_3-\underset{\underset{O-CH_3}{|}}{\overset{\overset{OH}{|}}{C}}-CH_2-CH_3$$

Solution

The carbon atom in (a) has two $-OCH_2CH_3$ groups bonded to it and is also bonded to hydrogen. The compound is an acetal that can be formed from propanal and ethanol. In (b) the carbon atom with one $-OCH_3$ group and an $-OH$ group bonded to it has two bonds to alkyl groups. The compound is a hemiketal that can be formed from 2-butanone and methanol.

PROBLEM 17.4

Brevicomin, the sex attractant of a species of pine beetle, has the following structure. Identify the functional group class to which it belongs.

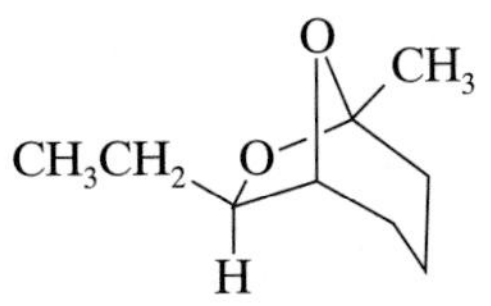

O, CH_3, CH_3CH_2, O, H

Mechanism of Acetal and Ketal Formation

The mechanism of acetal and ketal formation from a hemiacetal or hemiketal involves four steps. First, the acid protonates the oxygen atom of the hydroxyl group. In the second step, water leaves and a carbocation forms. In the third step, the carbocation combines with the alcohol. Finally, a proton is removed. Note that the acid is a catalyst—it starts the reaction and is regenerated in the last step.

$$R'-\ddot{O}-C(R)_2-\ddot{O}-H + H^+ \xrightleftharpoons{\text{step 1}} R'-\ddot{O}-C(R)_2-\overset{+}{\ddot{O}}(H)-H$$

$$R'-\ddot{O}-C(R)_2-\overset{+}{\ddot{O}}(H)-H \xrightleftharpoons{\text{step 2}} R'-\ddot{O}-C^+(R)_2 + :\ddot{O}(H)-H$$

$$R'-\ddot{O}-C^+(R)_2 + :\ddot{O}(H)-R' \xrightleftharpoons{\text{step 3}} R'-\ddot{O}-C(R)_2-\overset{+}{\ddot{O}}(H)-R'$$

$$H^+$$

$$R'-\ddot{O}-C(R)_2-\overset{+}{\ddot{O}}(H)-R' \xrightleftharpoons{\text{step 4}} R'-\ddot{O}-C(R)_2-\ddot{O}-R' + H^+$$

17.9 Addition of Nitrogen Compounds

The nitrogen–hydrogen bond of amines (Chapter 19) can add to a carbonyl group. A hydrogen atom bonds to the oxygen atom of the carbonyl group and the nitrogen atom bonds to the carbonyl carbon atom. This first step resembles the addition of an alcohol to a carbonyl compound. However, the addition product then loses a molecule of water in an elimination reaction to give an **imine.**

$$C_6H_5-C(H)=O + NH_2CH_3 \xrightleftharpoons{H^+} C_6H_5-C(H)(OH)-N(H)CH_3 \xrightleftharpoons{H^+} C_6H_5-C(H)=NCH_3 + H_2O$$

an imine

The net result of reacting a compound of the type G—NH_2 with an aldehyde or ketone is the replacement of the carbonyl oxygen atom with G—N=.

Imines can be isolated by removing the water from the equilibrium mixture. They react with water in a reverse of the two preceding reactions to produce a carbonyl compound and a nitrogen compound. The interconversion of carbonyl compounds and nitrogen derivatives is the basis for biochemical reactions such as transamination (Chapter 28), which interconverts keto acids and amino acids.

Isomerism, Addition Reactions, and Vision

We were told as children to eat carrots because our eyesight would be better. This homely injunction is true because carrots contain β-carotene, which mammals require for vision. β-Carotene (Section 15.1) is a pigment largely responsible for the color of carrots. Persons who do not have adequate β-carotene in their diet—which is available in egg yolk, liver, and various fruits and vegetables in addition to carrots—suffer from *night blindness.* Mammals have a liver enzyme system that splits β-carotene in half to give two molecules of an aldehyde named *retinal.* Retinal has a series of alternating single and double bonds. Geometric isomers can exist around each of the double bonds. The all-trans compound and the isomer with a cis orientation around the C-11 and C-12 atoms play an important role in vision.

CHO

retinal

CHO

cis-11-retinal

All animals, regardless of their reputed visual acuity, rely on the same chemical phenomena. (Peter Arnold, Inc.)

cis-11-Retinal undergoes an addition reaction with a protein in the retina called *opsin* to form a substance called *rhodopsin.* The aldehyde group of *cis*-11-retinal reacts with a specific amino group in the protein to form an imine. The shape of the imine adduct of *cis*-11-retinal allows it to "fit" into the protein.

CHO + opsin—NH_2 ⟶

cis-11-retinal

H—C=N^+(H)—opsin

rhodopsin

Rhodopsin absorbs visible light, acting in the retina as a visual receptor. Light strikes rhodopsin, isomerizing the cis double bond at C-11 to a trans double bond, a process called *photoisomerization.* The resulting all-trans isomer no longer fits into the opsin, the imine spontaneously hydrolyzes, and the all-trans retinal is released from opsin. The reaction generates a nerve impulse, which travels to the brain. Millions of nerve impulses combine to form a visual image.

If *cis*-11-retinal cannot bind opsin to give rhodopsin, vision is impaired. We recall from earlier discussions that formaldehyde, produced by the oxidation of methyl alcohol, can cause blindness. Blindness occurs because formaldehyde reacts faster than does *cis*-11-retinal with the reactive amine group of opsin. If no rhodopsin forms, then no "light-induced" messages reach the brain.

17.10 Reactivity of the α-Carbon Atom

Up to this point all the reactions of the carbonyl compounds have been transformations of the functional group itself. In fact, all the organic reactions in previous chapters have likewise only modified the functional group. Now for the first time we consider a reaction in which carbon–carbon bonds are made, and the size and framework of the organic molecule changes. The process of synthesizing large molecules from small ones by making carbon–carbon bonds is an important aspect of organic chemistry. In addition, many biochemical reactions also involve carbon–carbon bond formation. One such reaction is the aldol condensation in which two molecules of an aldehyde or ketone are joined. This important reaction is considered in Section 17.11. However, we first have to learn some chemistry of the α-carbon atom to understand this process.

α-Hydrogen Atoms Are Acidic

Part of the chemistry of carbonyl compounds is due to reactivity of the **α-carbon atom,** the carbon atom directly attached to the carbonyl carbon atom.

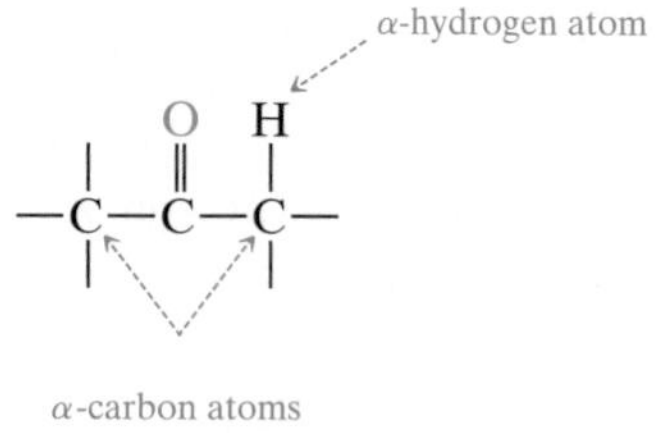

α-hydrogen atom

$$-\underset{|}{\overset{|}{C}}-\overset{O}{\overset{\|}{C}}-\underset{|}{\overset{\overset{H}{|}}{C}}-$$

α-carbon atoms

The carbon atom of the carbonyl group has a partial positive charge and pulls electrons in neighboring bonds to itself. As a result, the electrons bonding the **α-hydrogen atom** to the α-carbon atom are to some degree pulled away from the hydrogen atom. Thus, the α-hydrogen atom is more acidic than a hydrogen atom in a hydrocarbon C—H bond. The acid dissociation constant of ethane is about 10^{-50}, whereas that of acetone is about 10^{-20}.

$$CH_3-CH_3 + H_2O \rightleftharpoons CH_3-CH_2^- + H_3O^+ \qquad K_a = 10^{-50}$$

$$CH_3-\overset{O}{\overset{\|}{C}}-CH_3 + H_2O \rightleftharpoons CH_3-\overset{O}{\overset{\|}{C}}-CH_2^- + H_3O^+ \qquad K_a = 10^{-20}$$

The conjugate base of a carbonyl compound is stabilized by resonance, which places the negative charge on the oxygen atom. It is this resonance stabilization of the product that increases the acidity of carbonyl compounds.

$$CH_3-\overset{:\ddot{O}}{\overset{\|}{C}}-\ddot{C}H_2^- \longleftrightarrow CH_3-\overset{:\ddot{O}:^-}{\overset{|}{C}}=CH_2$$

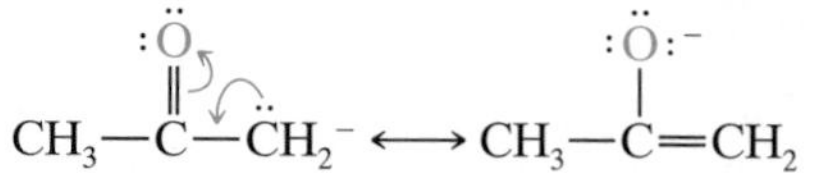

resonance structures of conjugate base

Keto–Enol Tautomers

Aldehydes and ketones exist as an equilibrium mixture of compounds called the **keto** and the **enol** forms.

$$CH_3-\overset{O}{\overset{\|}{C}}-CH_3 \rightleftharpoons CH_3-\overset{OH}{\overset{|}{C}}=CH_2$$

acetone (keto form) | propen-2-ol (enol form)

A hydrogen atom bonded to an α-carbon atom is removed and transferred to the carbonyl oxygen atom. The carbon–oxygen double bond becomes a single bond, and the carbon–carbon single bond becomes a double bond. The isomeric keto and enol forms are **tautomers**—they differ in the location of a proton and a double bond. For most aldehydes and ketones, the keto form predominates; less than 1% of the mixture is the isomeric enol form. Nevertheless, the enol form is often responsible for the reactivity of carbonyl compounds.

Tautomerism is important in the chemistry of carbohydrates (Chapter 21) and in the metabolism of these compounds (Chapter 26). For example, dihydroxyacetone phosphate is isomerized in an enzyme-catalyzed reaction to form glyceraldehyde 3-phosphate.

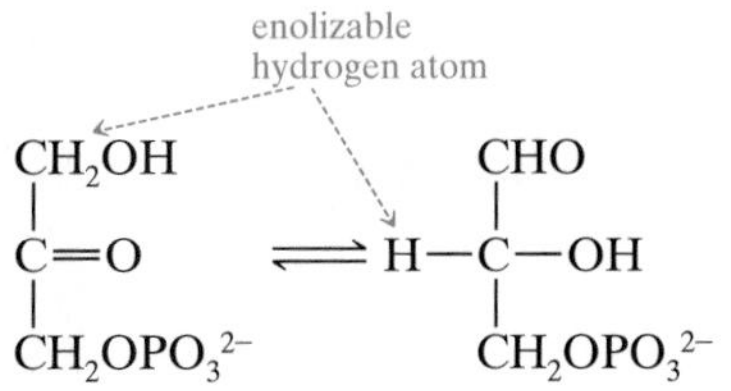

enolizable hydrogen atom

$$\begin{array}{c} CH_2OH \\ | \\ C{=}O \\ | \\ CH_2OPO_3^{2-} \end{array} \rightleftharpoons \begin{array}{c} CHO \\ | \\ H{-}C{-}OH \\ | \\ CH_2OPO_3^{2-} \end{array}$$

The transfer of a hydrogen atom from the α-carbon atom of dihydroxyacetone phosphate to the carbonyl group yields an enediol intermediate.

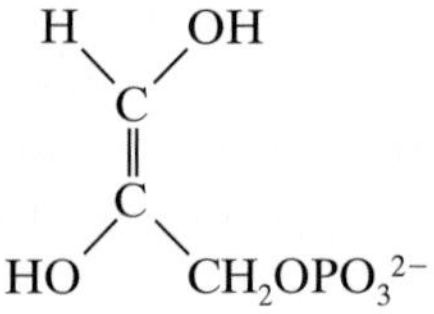

$$(H)(OH)C{=}C(OH)(CH_2OPO_3^{2-})$$

an enediol intermediate

The transfer of a hydrogen atom from the α-carbon atom of glyceraldehyde 3-phosphate produces the same enediol intermediate. Isomerization occurs because the enediol intermediate is in equilibrium with the two compounds.

17.11 The Aldol Condensation

Base-catalyzed reaction between 2 moles of a carbonyl compound is a second consequence of the acidity of an α-hydrogen atom.

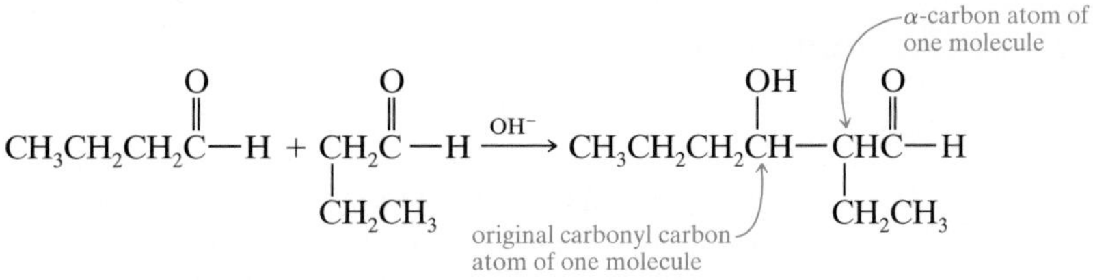

$$CH_3CH_2CH_2\overset{O}{\overset{\|}{C}}{-}H + \underset{CH_2CH_3}{\underset{|}{CH}}\overset{O}{\overset{\|}{C}}{-}H \xrightarrow{OH^-} CH_3CH_2CH_2\overset{OH}{\overset{|}{C}H}{-}\underset{CH_2CH_3}{\underset{|}{C}H}\overset{O}{\overset{\|}{C}}{-}H$$

α-carbon atom of one molecule

original carbonyl carbon atom of one molecule

an aldol product

The acidic α-hydrogen atom adds to the oxygen atom of the carbonyl group of another molecule. The α-carbon atom adds to the carbon atom of the carbonyl group. The product is called an **aldol** because it is both an aldehyde and an alcohol.

The sequence of steps in the mechanism for the aldol condensation is shown for acetaldehyde, but the same reactions occur for any other aldehyde with α-hydrogen atoms.

Step 1 : Formation of an enolate anion.

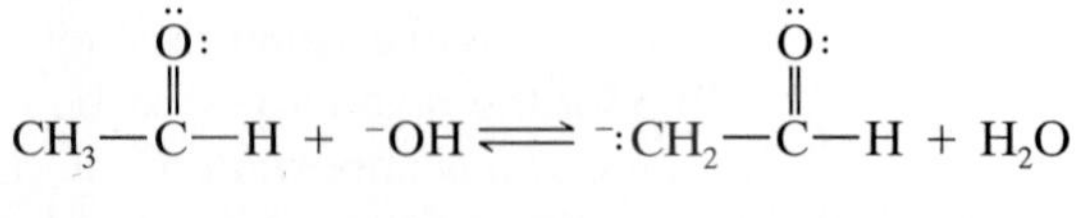

$$CH_3-\overset{\overset{\large\ddot{O}:}{\|}}{C}-H + {}^{-}OH \rightleftharpoons {}^{-}:CH_2-\overset{\overset{\large\ddot{O}:}{\|}}{C}-H + H_2O$$

Step 2: Nucleophilic attack of the enolate anion on the carbonyl carbon atom.

$$CH_3-\overset{\overset{\large\ddot{O}:}{\|}}{C}-H \;\; {}^{-}:CH_2-\overset{\overset{\large\ddot{O}:}{\|}}{C}-H \rightleftharpoons CH_3-\overset{\overset{:\ddot{O}:^{-}}{|}}{C}H-CH_2-\overset{\overset{\large\ddot{O}:}{\|}}{C}-H$$

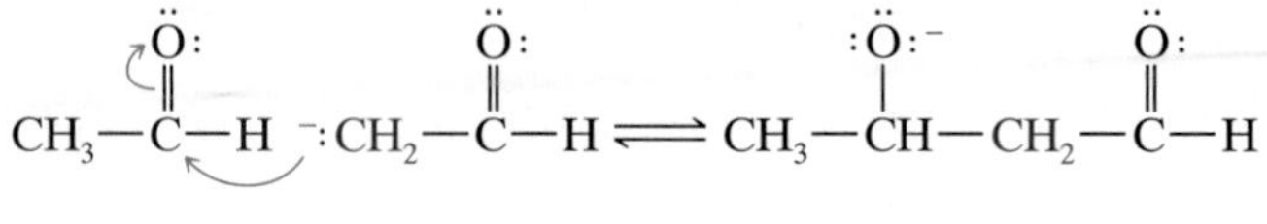

Step 3: Acid–base reaction to give an aldol and regenerate the hydroxide ion.

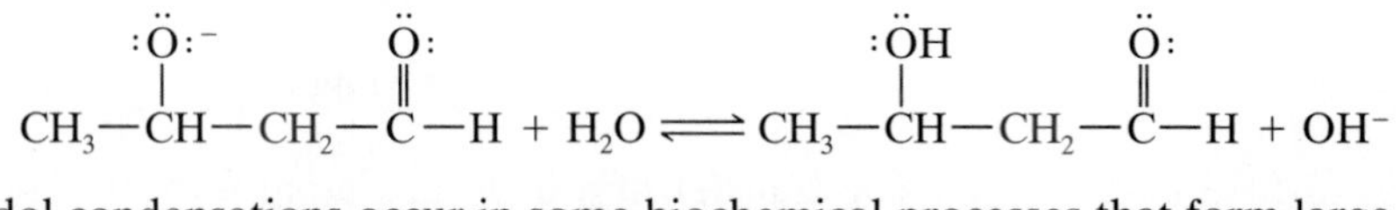

$$CH_3-\overset{\overset{:\ddot{O}:^{-}}{|}}{C}H-CH_2-\overset{\overset{\large\ddot{O}:}{\|}}{C}-H + H_2O \rightleftharpoons CH_3-\overset{\overset{:\ddot{O}H}{|}}{C}H-CH_2-\overset{\overset{\large\ddot{O}:}{\|}}{C}-H + OH^{-}$$

Aldol condensations occur in some biochemical processes that form large molecules. One step in the formation of the sugar fructose is an aldol reaction. The α-hydrogen atom of dihydroxyacetone phosphate adds to the carbonyl oxygen atom of glyceraldehyde 3-phosphate. The α-carbon atom of the ketone adds to the carbonyl carbon atom of the aldehyde in a "mixed" aldol condensation.

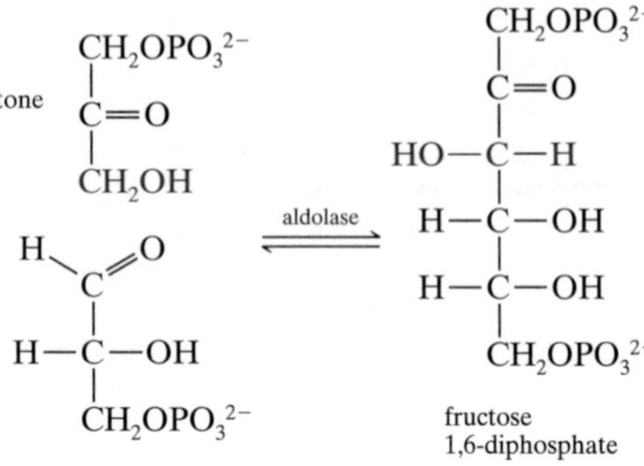

dihydroxy acetone phosphate: $CH_2OPO_3^{2-}$—C=O—CH_2OH

glyceraldehyde 3-phosphate: H—C(=O)—(H—C—OH)—$CH_2OPO_3^{2-}$

$\xrightleftharpoons{\text{aldolase}}$

fructose 1,6-diphosphate: $CH_2OPO_3^{2-}$—C=O—(HO—C—H)—(H—C—OH)—(H—C—OH)—$CH_2OPO_3^{2-}$

The reverse of this reaction, known as a retro-aldol reaction, occurs in the metabolism of carbohydrates (Chapter 26).

EXAMPLE 17.5

Draw the product of the aldol condensation of propanal.

Solution

First, draw the structural formula of propanal with the carbonyl group arranged toward the right. Next, draw a second structural formula of propanal with the *a*-carbon atom near the carbonyl carbon atom of the first structure. Place the *a*-hydrogen atom of the second structure so that it is close to the carbonyl oxygen atom of the first molecule.

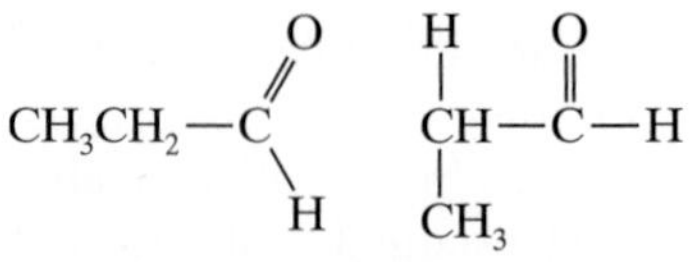

$$CH_3CH_2-\overset{O}{\overset{/\!/}{C}}\underset{H}{\diagdown} \qquad \overset{\overset{H}{|}}{C}H\underset{\underset{CH_3}{|}}{}-\overset{\overset{O}{\|}}{C}-H$$

Transfer a proton from the *a*-carbon atom of the structure on the right to the carbonyl oxygen atom of the structure on the left. Then form a bond between the *a*-carbon atom of the structure on the right and the carbonyl carbon atom of the structure on the left. The aldol product results.

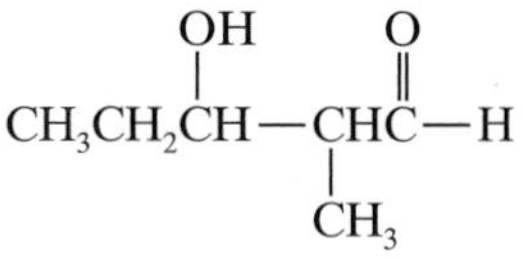

$$CH_3CH_2CH(OH){-}CH(CH_3)C(=O){-}H$$

(OH on the third carbon; O double-bonded to the carbonyl carbon; CH_3 on the alpha carbon)

PROBLEM 17.5

What aldehyde is required to produce the following aldol?

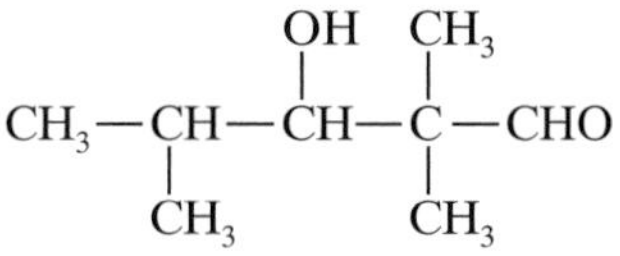

$$CH_3{-}CH(CH_3){-}CH(OH){-}C(CH_3)_2{-}CHO$$

(OH and CH_3 above; CH_3 and CH_3 below)

Summary

Aldehydes and ketones are compounds that have a **carbonyl group. Aldehydes** have one bond from the carbonyl carbon atom to a hydrogen atom and the other bond to an alkyl or aryl group. **Ketones** have both bonds from the carbonyl group to alkyl or aryl groups.

The carbonyl group is polar because of the difference in the electronegativities of the carbon atom and the oxygen atom. As a result of this polarity, carbonyl compounds have higher boiling points than hydrocarbons of similar molecular weight. Since carbonyl compounds cannot form hydrogen bonds to themselves, these compounds have lower boiling points than the structurally related alcohols. Lower molecular weight aldehydes and ketones are soluble in water.

In the IUPAC system of nomenclature, aldehydes are designated by dropping the terminal -e of the alkane and adding -al. Ketones are named by replacing the -e of the alkane with -one.

Aldehydes are oxidized by **Benedict's, Fehling's,** and **Tollens'** reagents, whereas ketones are not oxidized by these reagents. Both aldehydes and ketones can be reduced to alcohols.

Aldehydes and ketones undergo addition reactions with water, alcohols, and nitrogen compounds. Addition of water gives an unstable hydrate. Addition of an alcohol to an aldehyde or ketone gives a **hemiacetal** or **hemiketal,** respectively. Subsequent reactions yield an **acetal** or **ketal.** Addition of nitrogen compounds produces **imines.**

Carbonyl compounds exist in equilibrium with a small amount of an **enol** form, which is a **tautomer** of the **keto** form. The **α-hydrogen atom** bonded to the **α-carbon atom** of aldehydes and ketones is acidic. The conjugate base is stabilized by resonance. Carbonyl compounds undergo an **aldol condensation** reaction in which the α-carbon atom of one molecule bonds to the carbonyl carbon atom of another molecule. These self-addition reactions are important in building a large molecule from two smaller ones in living systems.

Exercises

Nomenclature

17.1 Write the structural formula of each of the following compounds.

(a) 2-methylbutanal (b) 3-ethylpentanal
(c) 2-bromopentanal (d) 3,4-dimethyloctanal

17.2 Write the structure of each of the following compounds.

(a) 3-bromo-2-pentanone
(b) 4-methyl-2-pentanone
(c) 2,4-dimethyl-3-pentanone
(d) 3,4-dimethyl-2-pentanone

17.3 Give the IUPAC name of each of the following compounds.

(a) $CH_3CH_2CH_2CHO$

(b) 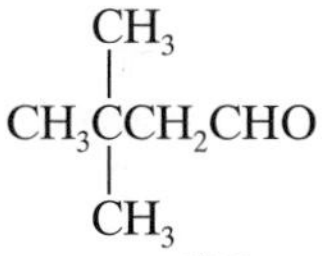 $CH_3C(CH_3)_2CH_2CHO$ (one CH_3 above and one below the second carbon)

(c) 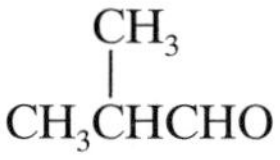 $CH_3CH(CH_3)CHO$ (CH_3 above the second carbon)

(d) 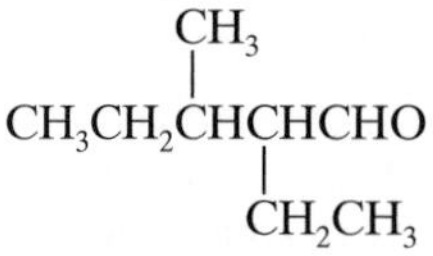 $CH_3CH_2CH(CH_3)CH(CH_2CH_3)CHO$ (CH_3 above the third carbon; CH_2CH_3 below the fourth carbon)

17.4 Give the IUPAC name of each of the following compounds.

(a) 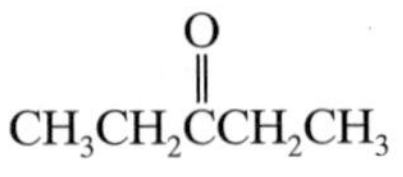 $CH_3CH_2C(=O)CH_2CH_3$

(b) 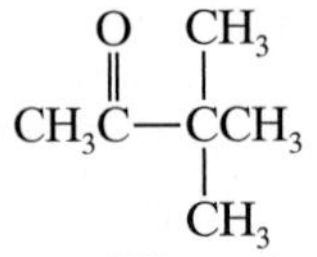 $CH_3C(=O)-C(CH_3)_2CH_3$

(c) $CH_3CH(CH_3)C(=O)CH_2CH_3$

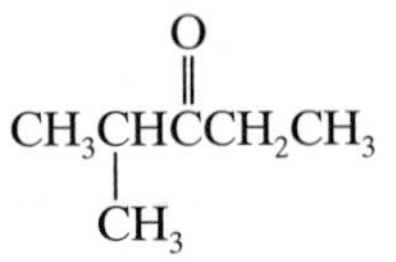

(d) 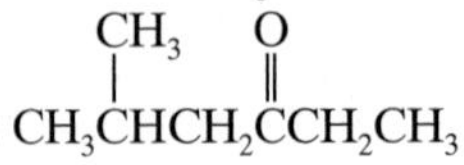 $CH_3CH(CH_3)CH_2C(=O)CH_2CH_3$

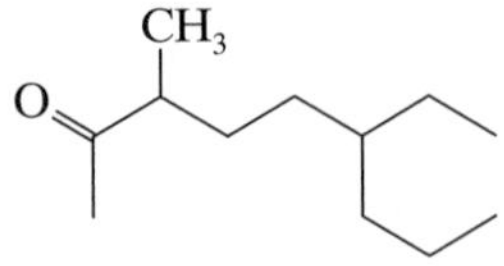

17.5 Give the IUPAC name of each of the following compounds.

(a) 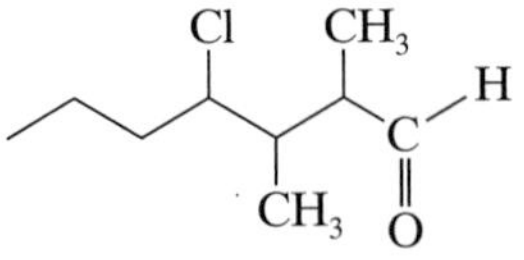 Cl, CH_3, CH_3, C=O, H

(b) O, CH_3

(c) CH_3, C=O, H, CH_3

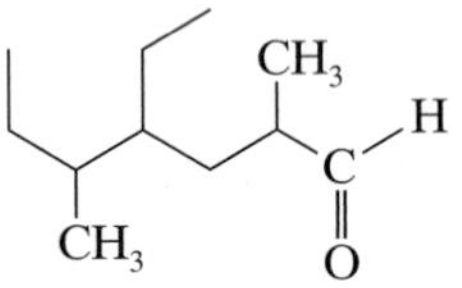

(d) 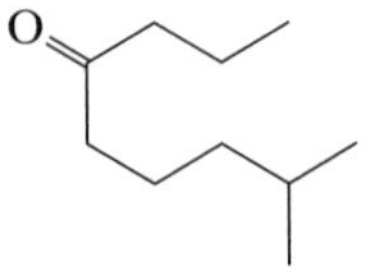 O

17.6 Give the IUPAC name of each of the following compounds.

(a) 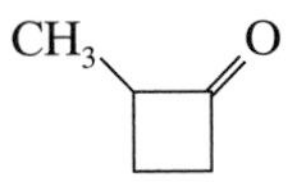 CH_3, O

(b) 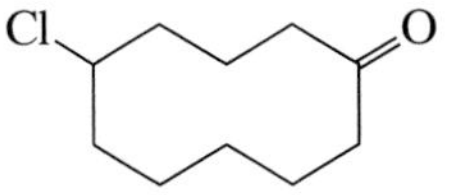 Cl, O

(c) 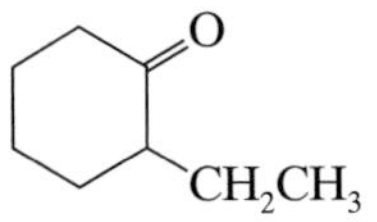 O, CH_2CH_3

(d) 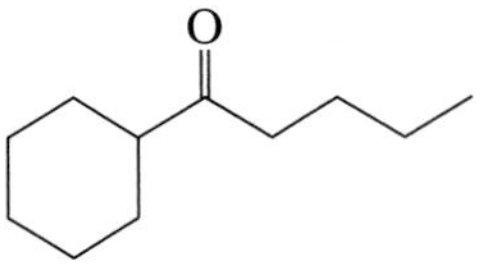 O

17.7 Draw the structural formula of each of the following carbonyl compounds. Their common names are given within parentheses.

(a) 2,2-dimethylpropanal (pivaldehyde)
(b) 2-hydroxy-1,2-diphenyl-1-ethanone (benzoin)
(c) 2-propenal (acrolein)
(d) 4-methyl-3-penten-2-one (mesityl oxide)

17.8 Draw the structural formula of each of the following carbonyl compounds. The common name of each compound is given within parentheses.

(a) 3,3-dimethyl-2-butanone (pinacolone)
(b) 4-hydroxy-4-methyl-2-pentanone (diacetone alcohol)
(c) *trans*-2-butenal (crotonaldehyde)
(d) 1,3-diphenyl-2-buten-1-one (dypnone)

Physical Properties

17.9 Why can't aldehydes and ketones form intermolecular hydrogen bonds with themselves?

17.10 Arrange hexanal, 1-hexanol, and heptane in order of increasing boiling points.

17.11 Aldehydes in general are less soluble in water than the corresponding alcohols of similar molecular weight. Explain why.

17.12 Which compound has the higher boiling point, cyclohexanol or cyclohexanone?

Oxidation and Reduction

17.13 What is observed when an aldehyde reacts with Benedict's solution? What is observed when an aldehyde reacts with Tollens' reagent?

17.14 What class of compounds results from the reduction of ketones with sodium borohydride? What class of compounds results from the reduction of aldehydes by hydrogen gas at high pressure in the presence of a platinum catalyst?

17.15 What is the product when each of the following reacts with hydrogen gas at high pressure in the presence of a platinum catalyst?

(a) acetone (b) 3-pentanone (c) hexanal

17.16 What is the product when each of the following compounds reacts with sodium borohydride?

(a) cyclohexanone
(b) 4,4-dimethylcyclohexanone
(c) 3,3-dimethylpentanal

17.17 Draw the structure of the product of each of the following reactions.

(a) 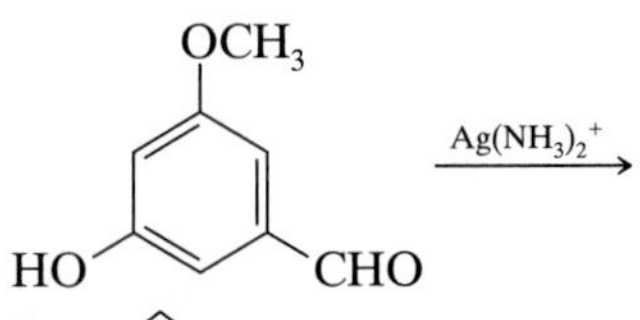 OCH_3, HO, CHO $\xrightarrow{Ag(NH_3)_2^+}$

(b) 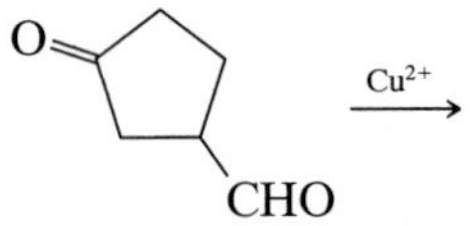 O, CHO $\xrightarrow{Cu^{2+}}$

(c) 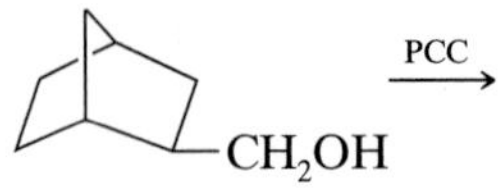 CH_2OH $\xrightarrow{PCC}$

17.18 Draw the structure of the product of each of the following reactions.

(a) 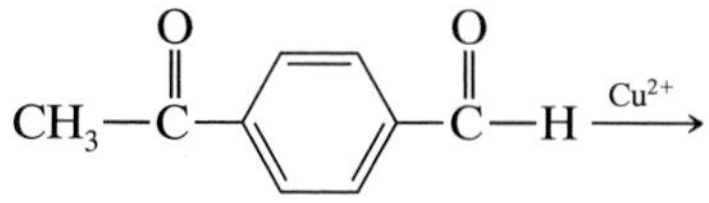 $CH_3-\overset{O}{\overset{\|}{C}}-C_6H_4-\overset{O}{\overset{\|}{C}}-H \xrightarrow{Cu^{2+}}$

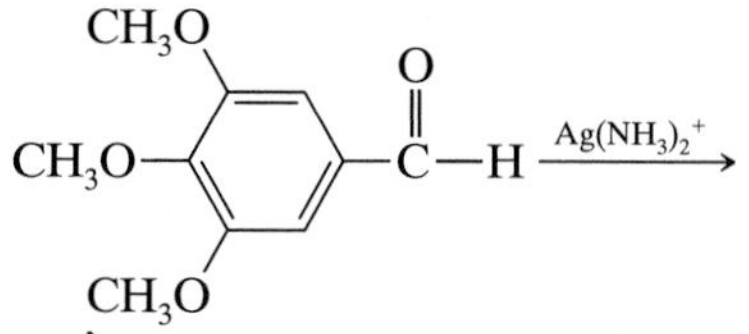

(b) $(CH_3O)_3C_6H_2-\overset{O}{\overset{\|}{C}}-H \xrightarrow{Ag(NH_3)_2^+}$

(c) bicyclic $-\overset{O}{\overset{\|}{C}}-H \xrightarrow[H_2SO_4]{CrO_3}$

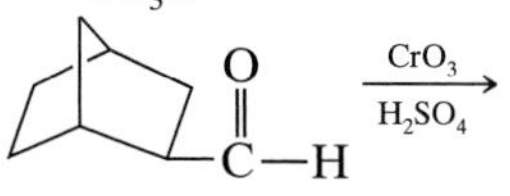

Addition Reactions

17.19 Formaldehyde has been used to disinfect surgical instruments because of its reaction with the nucleophilic centers of contaminating compounds. Why is this compound so effective compared with other carbonyl compounds?

17.20 Glutaraldehyde (5-oxopentanal) is used as a sterilizing solution for instruments that cannot be heated in an autoclave. Explain its action in sterilizing objects.

17.21 Identify each of the following as a hemiacetal, hemiketal, acetal, or ketal.

(a) $CH_3CH_2CH(OCH_3)_2$
(b) $CH_3CH_2C(OCH_3)_2CH_3$
(c) $CH_3CH(OCH_2CH_3)OCH_2CH_3$ (d) $CH_3CH_2CH(OCH_3)OH$

17.22 Identify each of the following as a hemiacetal, hemiketal, acetal, or ketal.

(a) $CH_3CH_2C(OCH_3)_2CH_3$ (b) $(CH_3O)_2CHCH_2CH(OCH_3)_2$

(c) cyclohexane with OCH_3, OCH_3 (d) 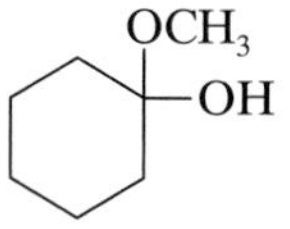 cyclohexane with OCH_3, OH

17.23 Identify each of the following as a hemiacetal, hemiketal, acetal, or ketal.

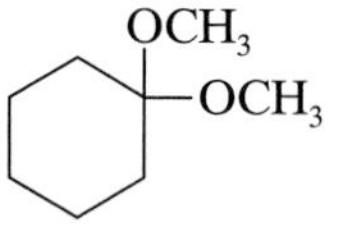

(a) 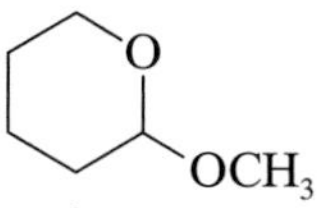 ring O, OCH_3 (b) 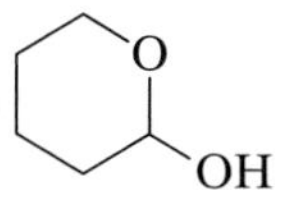 ring O, OH (c) 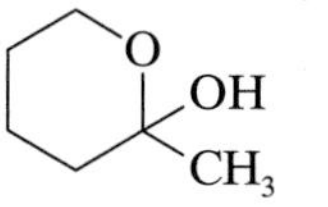 ring O, OH, CH_3 (d) 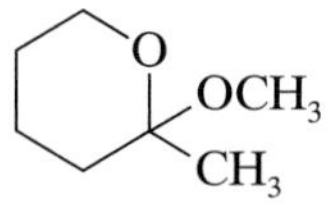 ring O, OCH_3, CH_3

17.24 Identify each of the following as a hemiacetal, hemiketal, acetal, or ketal.

(a) 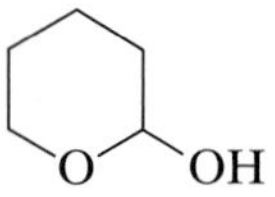 ring O, OH (b) 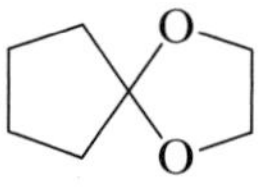 spiro O, O (c) 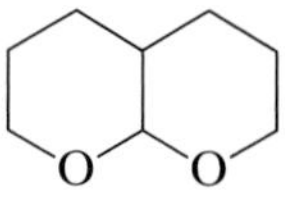 bicyclic O, O (d) 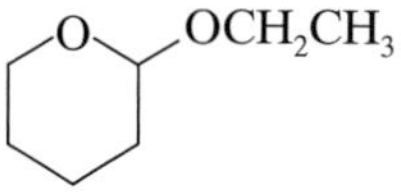 ring O, OCH_2CH_3

17.25 Identify the functional groups in talaromycin A, a substance found in the fungus that grows in poultry litter.

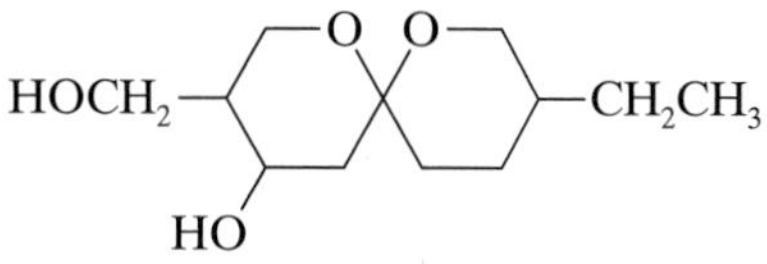

$HOCH_2$, HO, O O, CH_2CH_3

17.26 Identify the functional groups in daunosamine, a component of Adriamycin, used in cancer chemotherapy.

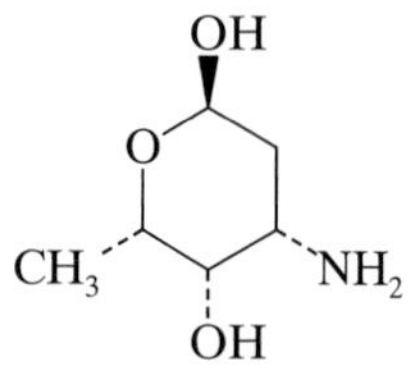

OH, O, CH_3, NH_2, OH

Addition of Nitrogen Compounds

17.27 Write the structure of the product for each of the following combinations of reactants.

(a) ethanal and methylamine (CH_3NH_2)
(b) acetone and ethylamine ($CH_3CH_2NH_2$)
(c) 3-pentanone and hydroxylamine (NH_2OH)

17.28 Identify the reactants required to form each of the following structures.

(a) 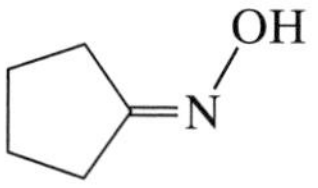 cyclopentyl=N–OH (b) $C_6H_5CH=N-NH-C_6H_5$

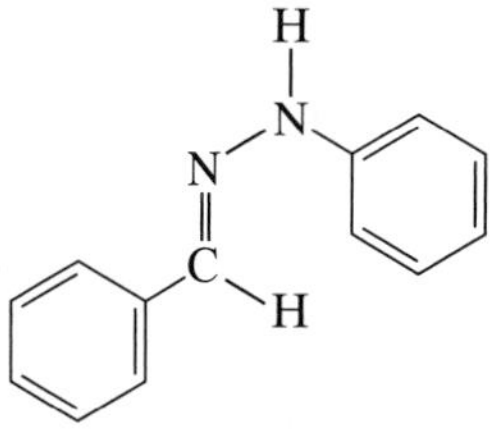

(c) cyclohexyl–$C(CH_3)$=N–NH–C(=O)–NH_2

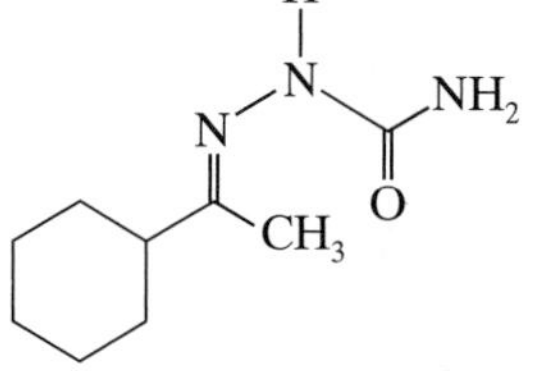

17.29 Reaction of cyclohexanone with hydroxylamine (NH_2-OH) yields a single product. However, cyclopentanecarbaldehyde yields two isomeric products. Consider the hybridization of the nitrogen atom and explain why.

17.30 Draw the structure of the product of reaction of hydrazine (NH_2NH_2) with two moles of benzaldehyde.

Chemistry of the α-Carbon Atom

17.31 Formaldehyde cannot exist as an enol. Why?

17.32 2,2-Dimethylpropanal cannot exist as an enol. Why?

17.33 Write the enol form of each of the following compounds.

(a) acetone (b) butanal
(c) cyclohexanone (d) acetaldehyde

17.34 There are three possible enols of 2-butanone. Explain why and write the structures.

17.35 Draw the keto structure of each of the following enols.

(a) 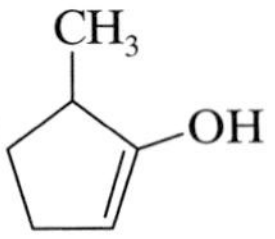 CH_3 OH (b) 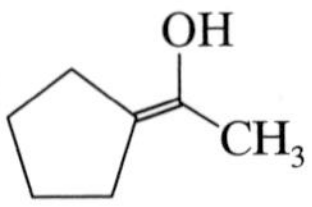 OH CH_3

17.36 Draw the keto structure of each of the following compounds.

(a) 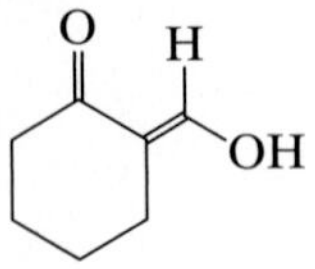 O H OH (b) 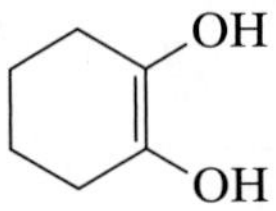 OH OH

The Aldol Condensation

17.37 Draw the structure of the aldol product formed from 2-methylpropanal reacting with base.

17.38 Draw the structure of the aldol product formed from phenylethanal reacting with base.

Windsurfers and sailors alike rely on the strength of sails made of polyesters. (The Image Works)

18

Carboxylic Acids and Esters

18.1 The Carboxyl and Acyl Groups
18.2 Nomenclature of Carboxylic Acids
18.3 Physical Properties of Carboxylic Acids
18.4 Acidity of Carboxylic Acids
18.5 Salts of Carboxylic Acids
18.6 Decarboxylation of Carboxylic Acids
18.7 Esters
18.8 Hydrolysis and Saponification of Esters
18.9 Nucleophilic Acyl Substitution
18.10 Esters and Anhydrides of Phosphoric Acid
18.11 The Claisen Condensation

Essays
- *Polyfunctional Carboxylic Acids in Biochemistry*
- *Carboxylate Salts as Food Additives*
- *Polyesters*
- *Soaps and Detergents*
- *Thioesters Are Nature's Active Acyl Compounds*

18.1 The Carboxyl and Acyl Groups

In this chapter we consider the structure, properties, and reactions of carboxylic acids. These compounds have a **carboxyl group** bonded to a hydrocarbon unit, which may be saturated, unsaturated, aromatic, or heterocyclic.

$$-\overset{\overset{\Large\ddot{O}:}{\|}}{C}-\ddot{\underset{\cdot\cdot}{O}}-H \qquad R-\overset{\overset{\Large\ddot{O}:}{\|}}{C}-\ddot{\underset{\cdot\cdot}{O}}-H \qquad Ar-\overset{\overset{\Large\ddot{O}:}{\|}}{C}-\ddot{\underset{\cdot\cdot}{O}}-H$$

carboxyl group — carboxylic acids

The carbonyl carbon atom is sp^2 hybridized, and three of its valence electrons form three σ bonds at 120° angles to one another (Figure 18.1). One of the σ bonds is to a hydrogen atom or a carbon atom of an alkyl, aromatic, or heterocyclic group. The other two σ bonds are to oxygen atoms: one to the hydroxyl oxygen atom and the other to the carbonyl oxygen atom. The carbonyl carbon atom also has one electron in a $2p$ orbital forming a π bond with an electron in a $2p$ orbital of the carbonyl oxygen atom.

FIGURE 18.1
Bonding in Carboxylic Acids

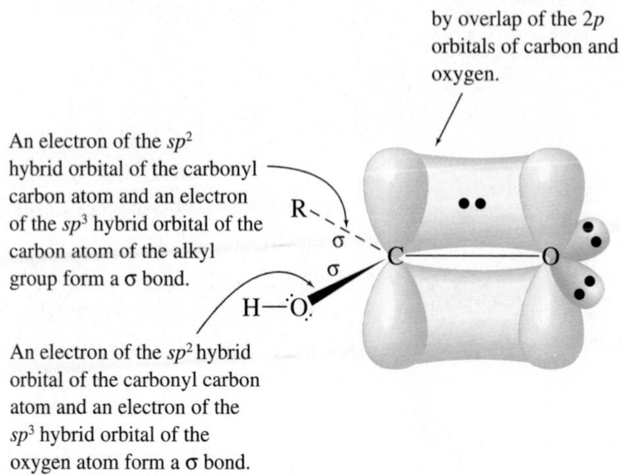

Although the bond angles at the carbonyl carbon atom are all approximately 120°, the carboxyl group is often represented with only vertical and horizontal lines. To save space, two condensed representations of the carboxyl group are commonly used. Unless required to account for the mechanism of a reaction, the nonbonding electrons are not shown.

$$R-\overset{\overset{\displaystyle O}{\|}}{C}-OH \quad \text{or} \quad R-COOH \quad \text{or} \quad R-CO_2H$$

representations of a carboxylic acid

The Acyl Group and Acid Derivatives

The RCO unit contained in a carboxylic acid is called an **acyl group.** Several families of acid derivatives have oxygen-containing groups or electronegative atoms such as nitrogen or halogens in place of the OH group of a carboxylic acid. This group, represented by L, can be replaced in a nucleophilic acyl substitution reaction (Section 18.9). The reactivity of each class of acid derivatives in these reactions varies over a wide range.

$$R-\overset{\overset{\displaystyle O}{\|}}{C}-L + Nu\!: \longrightarrow R-\overset{\overset{\displaystyle O}{\|}}{C}-Nu + L\!:$$

If an alkoxy (—OR) or phenoxy (—OAr) group is bonded to the acyl group, the derivative is an **ester.** Esters are formed from a carboxylic acid and an alcohol in a condensation reaction (Section 18.7). We will see that esters are moderately reactive toward water in hydrolysis reactions and produce a carboxylic acid and an alcohol (Section 18.8).

$$R-\overset{\overset{\displaystyle O}{\|}}{C}-O-R \qquad R-\overset{\overset{\displaystyle O}{\|}}{C}-O-Ar \qquad Ar-\overset{\overset{\displaystyle O}{\|}}{C}-O-Ar$$

esters

If the substituent is linked to the acyl group through a nitrogen atom, the compound is called an **amide.** The classification of amides depends on the number of carbon groups, including the acyl group, bonded to the nitrogen atom. These compounds are much less reactive than esters in hydrolysis reactions. The amide functional group is responsible for the structural stability of proteins. Amides are discussed in Chapter 19.

$R{-}C(=O){-}NH_2$ — a primary amide

$R{-}C(=O){-}NHR$ — a secondary amide

$R{-}C(=O){-}NR_2$ — a tertiary amide

Polyfunctional Carboxylic Acids in Biochemistry

Carboxylic acids and their derivatives are among the more important functional groups in biomolecules. Both lipids and proteins are carboxylic acid derivatives. In addition, carboxylic acids are involved in metabolic reactions that release stored energy of fats (Chapter 27). The citric acid cycle, a central pathway for the generation of energy by conversion of acyl groups into carbon dioxide and water, is a series of reactions in which all reactants and products are carboxylic acids.

Most of the carboxylic acids in biochemical reactions are polyfunctional; they contain other functional groups in addition to the carboxylic acid. Dicarboxylic acids are compounds with two carboxyl groups. Succinic acid and glutaric acid are dicarboxylic acids that are intermediates in the citric acid cycle (Chapter 25).

$HO{-}C(=O){-}C(=O){-}OH$ — oxalic acid

$HO{-}C(=O)CH_2C(=O){-}OH$ — malonic acid

$HO{-}C(=O)CH_2CH_2C(=O){-}OH$ — succinic acid

$HO{-}C(=O)CH_2CH_2CH_2C(=O){-}OH$ — glutaric acid

$HO{-}C(=O)CH_2CH_2CH_2CH_2C(=O){-}OH$ — adipic acid

Unsaturated acids of importance include the two geometric isomers maleic acid and fumaric acid. Fumaric acid is an intermediate in the citric acid cycle.

$(HO_2C)(H)C{=}C(H)(CO_2H)$ (cis) — maleic acid

$(HO_2C)(H)C{=}C(CO_2H)(H)$ (trans) — fumaric acid

Hydroxyl groups are common in naturally occurring compounds. Citric acid and malic acid are intermediates in the citric acid cycle.

$CO_2H{-}CH(OH){-}CH_3$ — lactic acid

$CO_2H{-}CH(OH){-}CH_2{-}CO_2H$ — malic acid

$CO_2H{-}CH_2{-}C(OH)(CO_2H){-}CH_2{-}CO_2H$ — citric acid

Keto acids that occur in the citric acid cycle include oxaloacetic acid and α-ketoglutaric acid. Pyruvic acid is a central metabolic intermediate (Chapter 26).

$CO_2H{-}C(=O){-}CH_3$ — pyruvic acid

$CO_2H{-}C(=O){-}CH_2{-}CO_2H$ — oxaloacetic acid

$CO_2H{-}C(=O){-}CH_2{-}CH_2{-}CO_2H$ — α-ketoglutaric acid

When the substituent attached to an acyl group is a chlorine atom, the derivative is called an **acid chloride.** These highly reactive compounds do not occur in nature but are used in the laboratory synthesis of esters and amides. When two acyl groups are bonded to a common oxygen atom, the compound is an **acid anhydride.** These compounds are also used in the laboratory synthesis of esters and amides.

$R-C(=O)-Cl$ $\quad$ $R-C(=O)-O-C(=O)-R$

an acid chloride 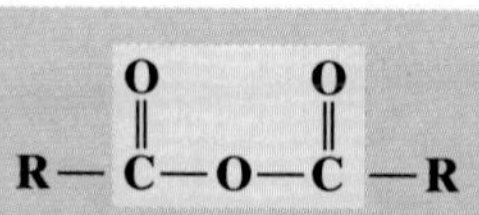 $\quad$ an acid anhydride

When a substituent is linked to an acyl group through a sulfur atom, the derivative is called a **thioester.** Thioesters are less reactive than acid chlorides and acid anhydrides, but they are sufficiently reactive to participate in many biochemical acyl transfer reactions.

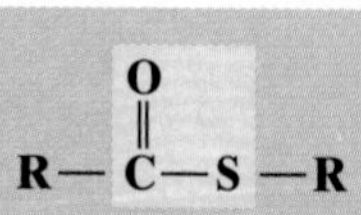

$R-C(=O)-S-R$

a thioester

Esters, amides, anhydrides, and thioesters may make up part of a cyclic structure. Cyclic esters are called **lactones.** Cyclic amides are called **lactams.** Cyclic acyl derivatives behave chemically like acyclic compounds.

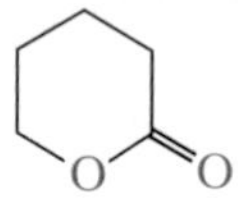

a lactone (a cyclic ester) $\quad$ a lactam (a cyclic amide)

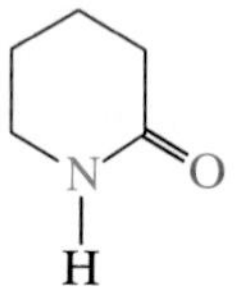

EXAMPLE 18.1

Identify the oxygen-containing group in the sex pheromone of the female Japanese beetle.

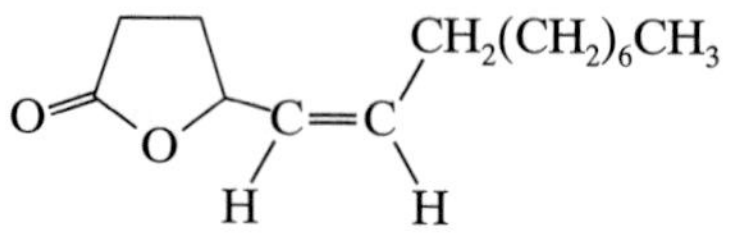

$CH_2(CH_2)_6CH_3$

Solution

The carbonyl carbon atom that is part of the five-membered ring is bonded to an oxygen atom. That oxygen atom is bonded to a carbon atom that is also part of the ring. The functional group is an ester, but because both the carbonyl carbon atom and the oxygen atom are part of the same ring, the compound is a lactone.

PROBLEM 18.1

Identify all the functional groups of indomethacin, an anti-inflammatory drug.

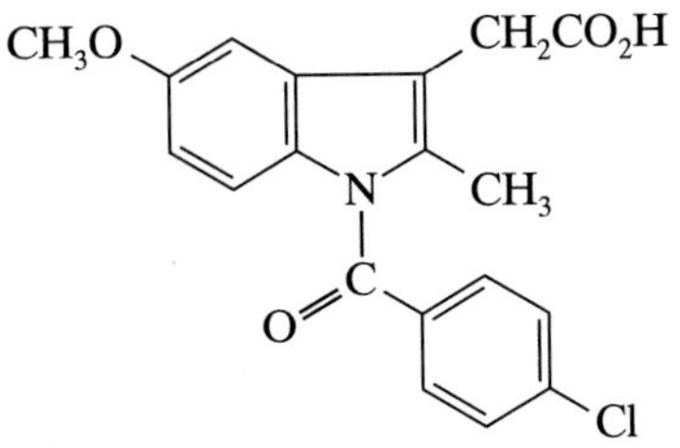

CH_3O CH_2CO_2H N CH_3 C O Cl

18.2 Nomenclature of Carboxylic Acids

Carboxylic acids were among the first organic substances isolated from nature and their common names are still used. Both the common and the IUPAC names of carboxylic acids are given in Table 18.1.

Common Names

In common names, the positions of groups attached to the parent chain are designated alpha (α), beta (β), gamma (γ), delta (δ), and so forth. The COOH group itself is not designated by a Greek letter.

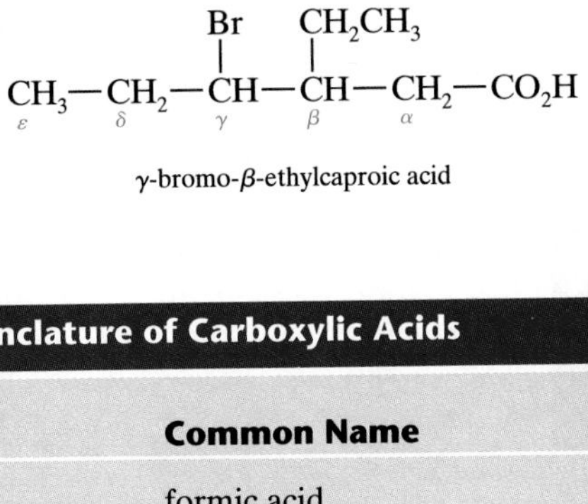

$$\underset{\varepsilon}{CH_3}-\underset{\delta}{CH_2}-\underset{\gamma}{\overset{Br}{\overset{|}{CH}}}-\underset{\beta}{\overset{CH_2CH_3}{\overset{|}{CH}}}-\underset{\alpha}{CH_2}-CO_2H$$

γ-bromo-β-ethylcaproic acid

TABLE 18.1 Nomenclature of Carboxylic Acids

Formula	Common Name	IUPAC Name
HCO_2H	formic acid	methanoic acid
CH_3CO_2H	acetic acid	ethanoic acid
$CH_3CH_2CO_2H$	propionic acid	propanoic acid
$CH_3(CH_2)_2CO_2H$	butyric acid	butanoic acid
$CH_3(CH_2)_3CO_2H$	valeric acid	pentanoic acid
$CH_3(CH_2)_4CO_2H$	caproic acid	hexanoic acid
$CH_3(CH_2)_6CO_2H$	caprylic acid	octanoic acid
$CH_3(CH_2)_8CO_2H$	capric acid	decanoic acid
$CH_3(CH_2)_{10}CO_2H$	lauric acid	dodecanoic acid
$CH_3(CH_2)_{12}CO_2H$	myristic acid	tetradecanoic acid
$CH_3(CH_2)_{14}CO_2H$	palmitic acid	hexadecanoic acid
$CH_3(CH_2)_{16}CO_2H$	stearic acid	octadecanoic acid

Aromatic acids containing one or more carboxyl groups are often referred to by their common names.

CO_2H / OH (ortho)	CO_2H / CO_2H (ortho)	CO_2H / CO_2H (para)
salicylic acid	phthalic acid	terephthalic acid

IUPAC Names

Carboxylic acids are named by rules similar to those outlined for aldehydes.

1. The longest carbon chain containing the carboxyl carbon atom is named as the parent chain. The final *-e* of the parent hydrocarbon is replaced by the ending *-oic acid.*
2. The parent chain is numbered by assigning the number 1 to the carbonyl carbon atom. The number 1 is not used in the name because the carbonyl carbon atom is understood to be located at the end of the chain. The names and locations of any substituents are added as prefixes to the parent name.

$$\overset{5}{CH_3}-\overset{4}{CH}(Br)-\overset{3}{CH}(CH_3)-\overset{2}{CH_2}-\overset{1}{C}(=O)-OH$$
4-bromo-3-methylpentanoic acid

3. The carboxylic acid group has a higher priority than double or triple bonds. A carboxylic acid that contains a double or triple bond is named by replacing the final *-e* of the name of the parent alkene or alkyne name with the suffix *-oic acid.* The position of the multiple bond is indicated with a prefix.

$$\overset{5}{CH_3}-\overset{4}{CH}(CH_3)-\overset{3}{C}\equiv\overset{2}{C}-\overset{1}{C}(=O)-OH$$
4-methyl-2-pentynoic acid

4. The carboxylic acid functional group has a higher priority than aldehyde, ketone, halogen, hydroxyl, and alkoxy groups. The names and locations of these substituents are indicated with prefixes to the name of the parent carboxylic acid.

$$\overset{4}{CH_3}-\overset{3}{CH}(OH)-\overset{2}{CH}(CH_3)-\overset{1}{C}(=O)-OH$$
3-hydroxy-2-methylbutanoic acid

5. If a carboxylic acid contains an aldehyde or ketone, the carbonyl group is indicated with the prefix *oxo-*. The priority order is carboxylic acid > aldehyde > ketone.

$$\overset{4}{CH_3}-\overset{3}{C}(=O)-\overset{2}{CH}(CH_3)-\overset{1}{C}(=O)-OH$$
2-methyl-3-oxobutanoic acid

6. Compounds that have a $-CO_2H$ group bonded to a cycloalkane ring are named as derivatives of the cycloalkane, and the suffix *carboxylic acid* is

Rhubarb has significant quantities of oxalic acid whose IUPAC name is ethanedioic acid. (Photo Researchers, Inc.)

added. The carbon atom to which the carboxyl group is bonded is assigned the number 1, but this number is not included in the name.

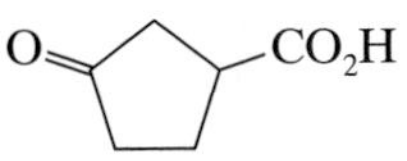

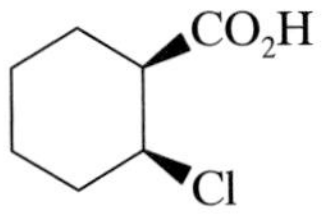

3-oxocyclopentanecarboxylic acid

cis-2-chlorocyclohexanecarboxylic acid

7. Dicarboxylic acids are named by adding the suffix *-dioic acid* to the name of the parent alkane that includes the carbon atoms of both carboxyl groups. The chain is numbered starting with the carboxyl carbon atom closest to the first substituent.

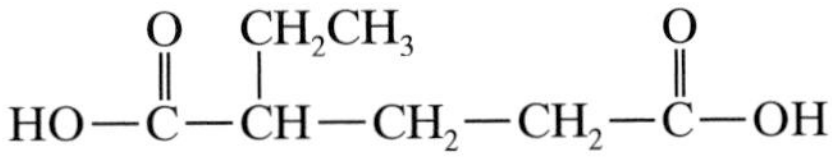

$$HO-\overset{O}{\overset{\|}{C}}-\overset{CH_2CH_3}{\overset{|}{C}H}-CH_2-CH_2-\overset{O}{\overset{\|}{C}}-OH$$

2-ethylpentanedioic acid

EXAMPLE 18.2

The structure of oleic acid, an unsaturated carboxylic acid present as an ester in vegetable oils, is shown below. What is the IUPAC name of oleic acid?

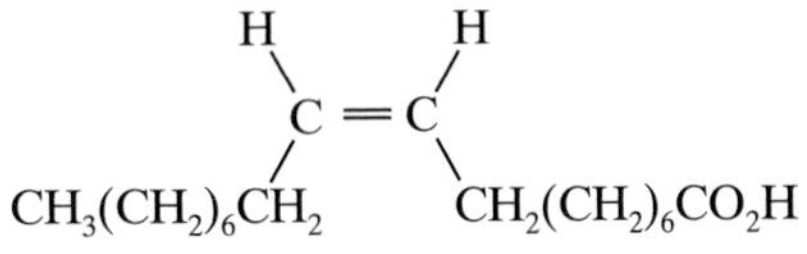

$CH_3(CH_2)_6CH_2(H)C{=}C(H)CH_2(CH_2)_6CO_2H$

Solution

First, determine the length of the chain that contains the CO_2H group; it contains 18 carbon atoms. The double bond is located at the C-9 atom in the chain, numbering from the carboxyl group on the right. Thus, the compound is a 9-octadecenoic acid. The configuration about the double bond is cis. Thus, the complete name is *cis*-9-octadecenoic acid.

PROBLEM 18.2

Mevalonic acid is required to form isopentenyl pyrophosphate, an intermediate in terpene synthesis. It has the following structure. Assign its IUPAC name.

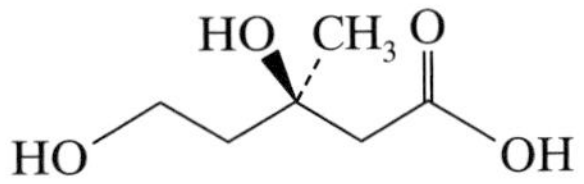

18.3 Physical Properties of Carboxylic Acids

Liquid carboxylic acids have unpleasant odors. For example, butanoic acid is responsible for the odor of rancid butter and aged cheese. Caproic, caprylic, and capric acids have the smell of goats. (The Latin word for "goat," *caper*, is the source of the common names of these acids.) The odors of these compounds depend on poorly understood physiological responses. However, physical properties such as boiling points, melting points, and solubility are easily related to structure and the types of intermolecular forces in these compounds.

Boiling Point

Low molecular weight carboxylic acids are liquids at room temperature. Those with higher molecular weights are waxlike solids. The boiling points of carboxylic acids are higher than other organic compounds with the same molecular weight. Carboxylic acids have high boiling points because they form hydrogen-bonded dimers. The hydroxyl group of one molecule acts as a hydrogen bond donor to the carbonyl oxygen atom of the second molecule. The hydroxyl group of the second molecule acts as a hydrogen bond donor to the carbonyl oxygen atom of the first molecule.

```
       Ö:···H—Ö:
      //        \
 R—C            C—R
      \         //
      :O—H···:O
```

hydrogen-bonded dimer of a carboxylic acid

The two hydrogen bonds stabilize the dimer so that it remains even in the gas phase. As a result, the boiling points of carboxylic acids are higher than those of substances of comparable molecular weights because hydrogen bond dimerization doubles their "effective" molecular weight. The boiling points of some representative carboxylic acids are given in Table 18.2.

Melting Point

Unsaturation lowers the melting points of carboxylic acids, especially if the configuration around the double bond is cis. The "bend" in the unsaturated carboxylic acid hinders efficient packing of the molecules, and the London forces are weaker than those in saturated carboxylic acids. The melting points of unsaturated acids are discussed in Chapter 22.

The melting points of dicarboxylic acids are very high because they can form twice as many hydrogen bonds as carboxylic acids.

$HO-C(=O)-CH_2-C(=O)-OH$

malonic acid
mp = 136 °C

$HO-C(=O)-CH_2-CH_2-C(=O)-OH$

succinic acid
mp = 185 °C

TABLE 18.2 Physical Properties of Carboxylic Acids

IUPAC Name	Melting Point (°C)	Boiling Point (°C)	Solubility (g/100 g H_2O, at 20°C)
methanoic acid		101	miscible
ethanoic acid		118	miscible
propanoic acid		141	miscible
butanoic acid		164	miscible
pentanoic acid		186	4.97
hexanoic acid		205	0.96
octanoic acid		239	0.068
decanoic acid	32	270	0.015
dodecanoic acid	44	299	0.0055

Solubility

Carboxylic acids with low molecular weights dissolve in water because the carboxyl group forms several hydrogen bonds with water. A carboxylic acid serves both as a hydrogen bond donor through its hydroxyl hydrogen atom and as a hydrogen bond acceptor through the lone-pair electrons of both oxygen atoms. The solubility of carboxylic acids, like that of alcohols, decreases with increasing chain length because the hydrocarbon chain dominates the physical properties of the acid.

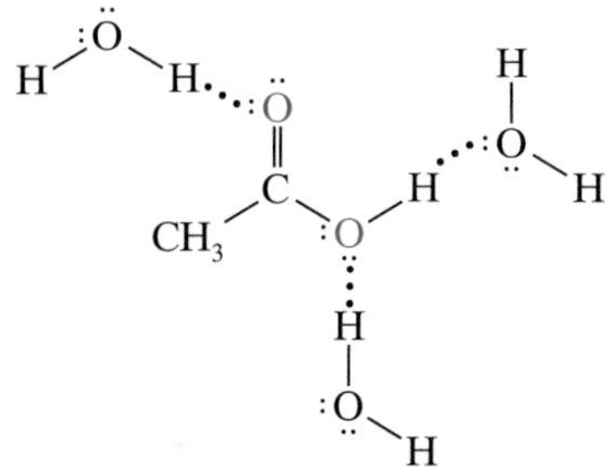

18.4 Acidity of Carboxylic Acids

The ionization of an acid HA, as reflected in the acid dissociation constant, K_a, depends on both the strength of the H—A bond and on the stability of the conjugate base A^- in the solvent. Although acetic acid and other carboxylic acids are weak acids, they are much more acidic than alcohols or phenols. For example, the K_a of acetic acid is about 10^{11} times that of ethanol.

$$CH_3CH_2OH + H_2O \rightleftharpoons CH_3CH_2O^- + H_3O^+ \qquad K_a = 1 \times 10^{-16}$$
$$CH_3CO_2H + H_2O \rightleftharpoons CH_3CO_2^- + H_3O^+ \qquad K_a = 1.8 \times 10^{-5}$$

Resonance Stabilization of the Carboxylate Ion

Acetic acid is much more acidic than ethanol because the resulting carboxylate anion is resonance stabilized. Dispersal of charge in the acetate ion between the two oxygen atoms is shown in two equivalent resonance forms. In the ethoxide ion ($CH_3CH_2O^-$), the negative charge is localized on one oxygen atom.

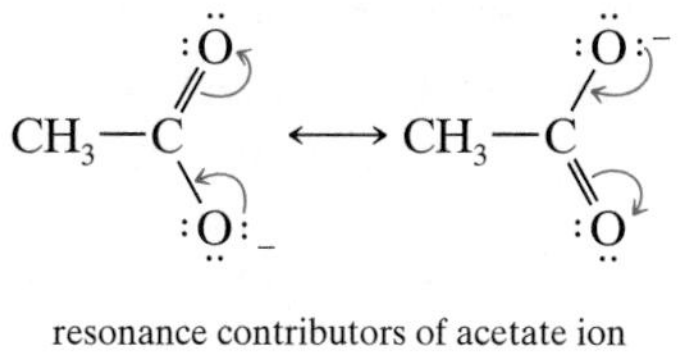

resonance contributors of acetate ion

Inductive Effect on Acidity

The acidity of carboxylic acids is affected by an **inductive effect,** which is the ability of an atom to polarize nearby bonds. The carbonyl group polarizes the H—O bond by attracting electrons through the σ bonds. The withdrawal of electron density from the H—O bond weakens it, increasing the acidity of the hydrogen atom. An electronegative group attached to the α-carbon atom of a carboxylic acid fur-

TABLE 18.3 Effect of Substituents on Acidity of Carboxylic Acids

Name of Acid	Formula	K_a	pK_a
acetic acid	CH_3CO_2H	1.8×10^{-5}	4.7
butyric acid	$CH_3CH_2CH_2CO_2H$	1.3×10^{-5}	4.9
isobutyric acid	$(CH_3)_2CHCO_2H$	1.6×10^{-5}	4.8
fluoroacetic acid	FCH_2CO_2H	2.5×10^{-3}	2.6
chloroacetic acid	$ClCH_2CO_2H$	1.4×10^{-3}	2.9
dichloroacetic acid	Cl_2CHCO_2H	5.0×10^{-2}	1.3
trichloroacetic acid	Cl_3CCO_2H	2.5×10^{-1}	0.6

ther increases its acidity. For example, halogen atoms pull electron density away from the carbon skeleton, and indirectly from the O—H bond. As a consequence, the proton is more easily removed and K_a increases (Table 18.3).

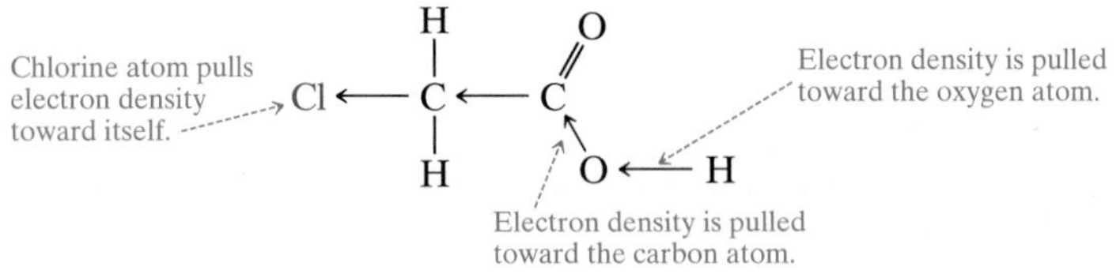

Chlorine atom pulls electron density toward itself.

Cl ← CH_2 ← C(=O) ← O ← H

Electron density is pulled toward the oxygen atom.

Electron density is pulled toward the carbon atom.

As the distance between the halogen atom and the carboxyl group increases, the inductive effect falls off dramatically. For β- and γ-substituted acids, the K_a values approach that of an unsubstituted carboxylic acid.

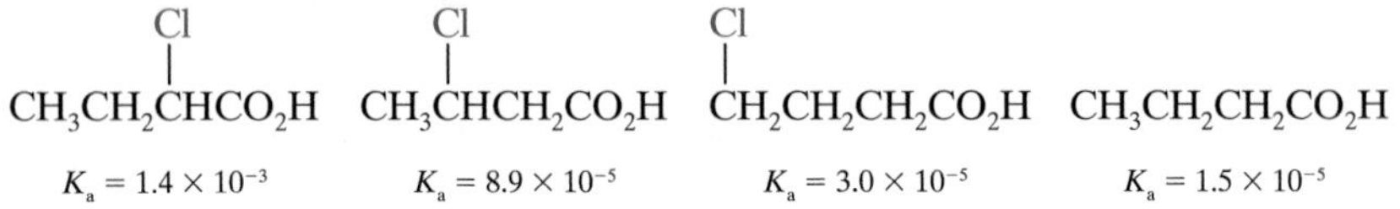

$CH_3CH_2CHClCO_2H$	$CH_3CHClCH_2CO_2H$	$ClCH_2CH_2CH_2CO_2H$	$CH_3CH_2CH_2CO_2H$
$K_a = 1.4 \times 10^{-3}$	$K_a = 8.9 \times 10^{-5}$	$K_a = 3.0 \times 10^{-5}$	$K_a = 1.5 \times 10^{-5}$

pK_a Scale

The pK_a scale expresses the acid strength of compounds without using exponential notation. It is defined as follows:

$$\mathrm{p}K_a = -\log K_a$$

The definition is analogous to that of pH; the stronger the acid, the smaller is its pK_a. Understanding the relationship between the pK_a of an acid and the pH of a solution is important for understanding many biological processes. The pK_a values of biologically important acids will be discussed later in this text.

EXAMPLE 18.3

Pyruvic acid is a key metabolic intermediate in oxidative processes that provide energy for the growth and maintenance of cells. Its pK_a is 2.5, indicating that it is about 100 times more acidic than propanoic acid (pK_a = 4.7). Explain why.

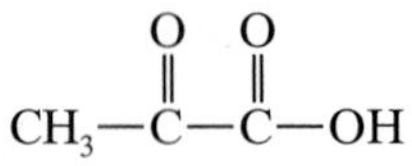

$CH_3—C(=O)—C(=O)—OH$

pyruvic acid

Solution

The α-carbon atom is a carbonyl carbon atom, which is a polar group. The electrons in the carbonyl group are drawn toward the more electronegative oxygen atom. This electron withdrawal decreases the electron density of the O—H bond and increases its acidity.

PROBLEM 18.3

The pK_a values for the dissociation of the first of the two carboxyl groups of malic acid and oxaloacetic acid are 3.41 and 1.70, respectively. Which compound is the stronger acid? Which of the two carboxyl groups in each compound dissociates to a greater degree?

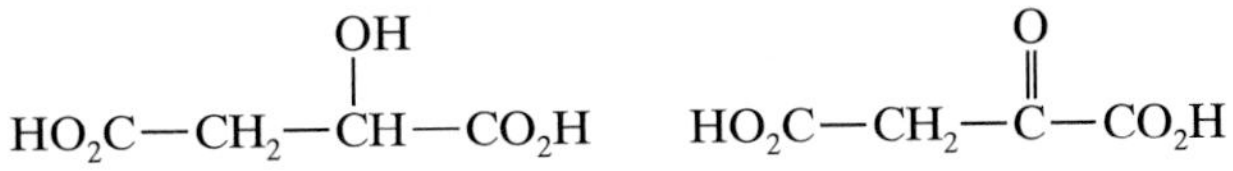

$HO_2C{-}CH_2{-}CH(OH){-}CO_2H$ — malic acid

$HO_2C{-}CH_2{-}C(=O){-}CO_2H$ — oxaloacetic acid

18.5 Salts of Carboxylic Acids

Carboxylic acids react with strong bases to produce carboxylate salts. The name of a salt is derived from the acid name by changing the *-ic* ending to *-ate* and preceding this name with the name of the metal ion.

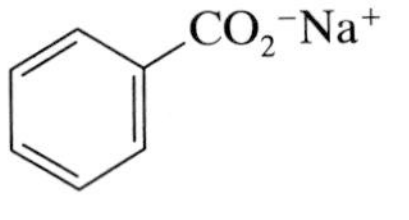

$C_6H_5CO_2^-Na^+$ — sodium benzoate

$[CH_3CH_2CO_2^-]_2Ca^{2+}$ — calcium propanoate

Because carboxylate salts are ionic compounds they are more soluble in water than carboxylic acids. For example, the solubility of benzoic acid at 25 °C is 3.5 g/L, but the solubility of sodium benzoate is 550 g/L. In the laboratory, carboxylic acids are separated from nonpolar organic compounds by adding a solution of sodium hydroxide. For example, decanol is not soluble in water and does not react with sodium hydroxide. However, decanoic acid reacts with sodium hydroxide and thus dissolves in the basic solution.

$$\underset{\text{(insoluble in water)}}{CH_3(CH_2)_8CH_2OH} + OH^- \longrightarrow \text{no reaction}$$

$$\underset{\text{(insoluble in water)}}{CH_3(CH_2)_8CO_2H} + OH^- \longrightarrow \underset{\text{(soluble in water)}}{CH_3(CH_2)_8CO_2^-} + H_2O$$

The decanol is separated from the basic solution. Then HCl is added to neutralize the solution, and the insoluble decanoic acid separates from the solution.

$$\underset{\text{(soluble in water)}}{CH_3(CH_2)_8CO_2^-} + H_3O^+ \longrightarrow \underset{\text{(insoluble in water)}}{CH_3(CH_2)_8CO_2H} + H_2O$$

This procedure is very useful for isolating acids from mixtures in nature. It is also used to purify acids produced by chemical synthesis in the laboratory.

Carboxylate Salts as Food Additives

The labels of ingredients on the packages of processed foods often list chemicals whose names end in *-ate,* indicating that a carboxylate salt has been added. The purposes of these additives are manifold but perhaps the most important is as preservatives. The sodium and calcium salts of propanoic acid are used in baked goods to prevent the growth of bacteria and in cheese to inhibit the formation of mold. Sodium benzoate is added to acidic foods to prevent the growth of microorganisms. Potassium sorbate is an inhibitor of mold and fungus growth.

$$CH_3CH{=}CH{-}CH{=}CHCO_2^-\ K^+$$

potassium sorbate

Other salts are added for quite different reasons. The trisodium salt of citric acid is added to buffer the citric acid in some fruit products. Monosodium glutamate is added to enhance the meaty flavor of products.

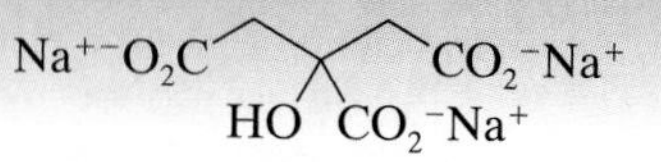

$Na^{+}\ {}^{-}O_2C$ … $CO_2^-Na^+$; HO $CO_2^-Na^+$

sodium citrate

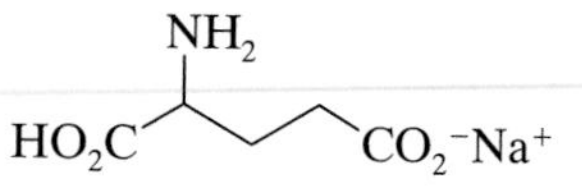

NH_2; HO_2C … $CO_2^-Na^+$

monosodium glutamate

All food additives must be approved by the FDA. This agency requires that the additives be effective for the stated purpose on the label. However, it also considers the possible side effects that may result from the ingestion of the additive. Every additive must be metabolized, so it is important that the resulting metabolites not damage organs or interfere with other biological processes.

18.6 Decarboxylation of Carboxylic Acids

The loss of a carboxyl group and its replacement by a hydrogen atom is called **decarboxylation.**

$$R{-}\overset{O}{\overset{\|}{C}}{-}OH \longrightarrow R{-}H + CO_2$$

Decarboxylation does not occur readily for carboxylic acids without other functional groups. However, β-keto acids such as acetoacetic acid, decarboxylate at relatively low temperatures.

$$CH_3{-}\overset{O}{\overset{\|}{C}}{-}CH_2{-}CO_2H \longrightarrow CH_3{-}\overset{O}{\overset{\|}{C}}{-}CH_2{-}H + CO_2$$

acetoacetic acid

The decarboxylation occurs by way of a cyclic mechanism in which a proton is transferred from the carboxyl group to the carbonyl oxygen atom. Subsequent rapid tautomerization of the enol product gives acetone.

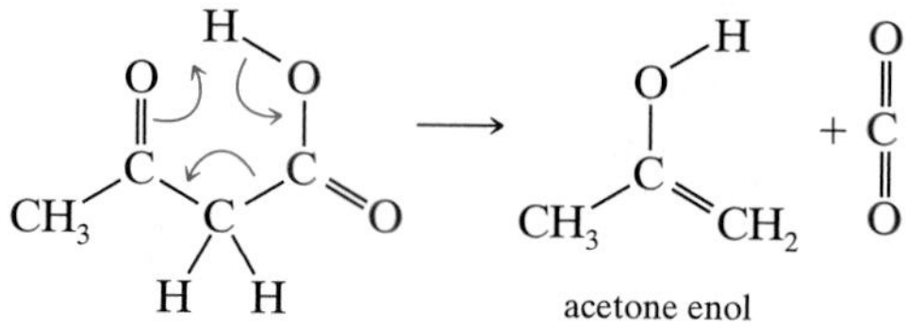

$$CH_3{-}C({=}O{\cdots}H){-}CH_2{-}C({=}O){-}O \longrightarrow CH_3{-}C(OH){=}CH_2 + O{=}C{=}O$$

acetone enol

The decarboxylation of β-keto acids is an important biological process. The conversion of isocitrate to α-ketoglutarate in the citric acid cycle is one example. This reaction is catalyzed by the enzyme isocitrate dehydrogenase, and NAD^+ is the oxidizing agent. In the first step, the hydroxyl group of isocitrate is oxidized to a carbonyl group. The oxalosuccinate product, a β-ketoacid, subsequently loses carbon dioxide.

$$\underset{\text{isocitrate}}{\begin{array}{c} CO_2^- \\ | \\ HO-C-H \\ | \\ H-C-CO_2^- \\ | \\ H-C-H \\ | \\ CO_2^- \end{array}} + NAD^+ \xrightarrow[\text{dehydrogenase}]{\text{isocitrate}} \underset{\text{oxalosuccinate}}{\begin{array}{c} CO_2^- \\ | \\ C=O \\ | \\ H-C-CO_2^- \\ | \\ H-C-H \\ | \\ CO_2^- \end{array}} + NADH \longrightarrow \underset{\text{glutarate}}{\begin{array}{c} CO_2^- \\ | \\ C=O \\ | \\ H-C-H \\ | \\ H-C-H \\ | \\ CO_2^- \end{array}} + CO_2$$

18.7 Esters

An acid reacts with an alcohol in a condensation reaction to give an ester and water. In the following general reaction H—O—A represents any acid.

Water comes from here

part of the acid   part of the alcohol

$$\underset{\text{an acid}}{A-O-H} + \underset{\text{an alcohol}}{H-O-R} \longrightarrow \underset{\text{an ester}}{A-O-R} + H_2O$$

Esters of Inorganic Acids

Esters of nitric acid are called nitrate esters. Many of these esters, such as glyceryl trinitrate (nitroglycerin), are powerful explosives. Glyceryl trinitrate is also a smooth-muscle relaxant and vasodilator used to lower blood pressure and to treat angina pectoris, a heart disorder.

$$\underset{\text{nitric acid}}{\begin{array}{c} O_2N-O-H \\ O_2N-O-H \\ O_2N-O-H \end{array}} + \underset{\text{glycerol}}{\begin{array}{l} H-O-CH_2 \\ \quad\quad\quad | \\ H-O-CH \\ \quad\quad\quad | \\ H-O-CH_2 \end{array}} \longrightarrow \underset{\text{glyceryl trinitrate}}{\begin{array}{l} O_2N-O-CH_2 \\ \quad\quad\quad\quad | \\ O_2N-O-CH \\ \quad\quad\quad\quad | \\ O_2N-O-CH_2 \end{array}} + 3\ H_2O$$

Esters of sulfuric acid are called sulfate esters. The sulfate ester of 1 mole of an alcohol and 1 mole of sulfuric acid has an acidic proton, which can react with sodium hydroxide in a neutralization reaction. The conjugate bases of sulfate esters of high molecular weight alcohols such as lauryl alcohol are detergents.

$$\underset{\text{lauryl alcohol}}{CH_3(CH_2)_{10}CH_2OH} + HO-\overset{\overset{\ddot{O}:}{\|}}{\underset{\underset{\ddot{O}:}{\|}}{S}}-OH \longrightarrow \underset{\text{lauryl sulfate}}{CH_3(CH_2)_{10}CH_2-O-\overset{\overset{\ddot{O}:}{\|}}{\underset{\underset{\ddot{O}:}{\|}}{S}}-OH} + H_2O$$

$$CH_3(CH_2)_{10}CH_2{-}O{-}\overset{\overset{\displaystyle O}{\|}}{\underset{\underset{\displaystyle O}{\|}}{S}}{-}OH + NaOH \longrightarrow CH_3(CH_2)_{10}CH_2{-}O{-}\overset{\overset{\displaystyle O}{\|}}{\underset{\underset{\displaystyle O}{\|}}{S}}{-}O^-Na^+ + H_2O$$

sodium lauryl sulfate

Esters of Carboxylic Acids

Carboxylic acids react with alcohols in an acid-catalyzed reaction to form esters.

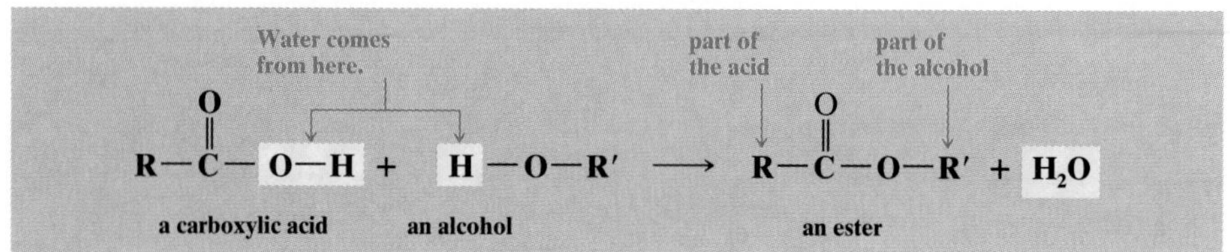

Substantial amounts of both the reactants and the ester are present at equilibrium. The yield of an ester can be increased by using an excess of the alcohol or by distilling the water out of the reaction mixture. Removal of the water shifts the position of the equilibrium as predicted by Le Châtelier's principle.

Adding alcohol "pushes" the reaction to the right.

Removal of water "pulls" the reaction to the right.

$$CH_3(CH_2)_2\overset{\overset{\displaystyle O}{\|}}{C}{-}OH + CH_3(CH_2)_3OH \overset{H^+}{\rightleftharpoons} CH_3(CH_2)_2\overset{\overset{\displaystyle O}{\|}}{C}{-}O(CH_2)_3CH_3 + H_2O$$

butanoic acid bp = 164 °C; 1-butanol bp = 118 °C; butyl butanoate bp = 167 °C

Nomenclature of Esters

Esters are named like the salts of acids with the name of the alkyl group of the alcohol replacing the name of the metal ion. In the following examples the alcohol (or phenol) portion of each molecule is on the right side of the structure. However, regardless of how the structure is drawn, the name of the alcohol is written first in the name of the ester.

This part is from formic acid. This part is from ethanol.

$$H{-}\overset{\overset{\displaystyle O}{\|}}{C}{-}O{-}CH_2CH_3$$

ethyl methanoate (ethyl formate)

This part is from acetic acid. This part is from phenol.

$$CH_3{-}\overset{\overset{\displaystyle O}{\|}}{C}{-}O{-}C_6H_5$$

phenyl ethanoate (phenyl acetate)

This part is from benzoic acid. This part is from methanol.

$$C_6H_5{-}\overset{\overset{\displaystyle O}{\|}}{C}{-}O{-}CH_3$$

methyl benzoate (methyl benzoate)

Several esters of salicylic acid are commercially produced in large quantities. Salicylic acid acts as an antipyretic (fever reducer) and an analgesic (pain reliever). However, it irritates the stomach lining, so the acetate ester of the phenolic group—called acetylsalicylic acid, better known as aspirin—is preferred as a medicine. Approximately 12,000 tons of aspirin are produced annually in the United States. The methyl ester of the carboxylic acid is methyl salicylate (oil of wintergreen) and is used in liniments such as Ben Gay.

acetylsalicyclic acid   methyl salicylate

Properties of Esters

These fruits all have odors characteristic of naturally occurring esters. (Photo Researchers, Inc.)

Esters are polar molecules but have lower boiling points than carboxylic acids and alcohols of similar molecular weight. This lower boiling point reflects the absence of hydrogen bonds between ester molecules.

$$CH_3CH_2CH_2CH_2CH_2OH \qquad CH_3CH_2CH_2\overset{\overset{\displaystyle O}{\|}}{C}-OH \qquad CH_3\overset{\overset{\displaystyle O}{\|}}{C}-O-CH_2CH_3$$

1-pentanol, bp = 138 °C; butanoic acid, bp = 164 °C; ethyl ethanoate, bp = 77 °C

Esters can form hydrogen bonds with water molecules via their oxygen atoms and thus have some degree of water solubility. However, since esters do not have a hydrogen atom to bond to an oxygen atom of water, they are less soluble than carboxylic acids. The solubility and boiling points of esters are listed in Table 18.4.

The odors of esters are different from those of the corresponding acids. Esters have pleasant fruity smells. Many of the odors of fruits are due to esters. For example, ethyl ethanoate is found in pineapples, 3-methylbutyl ethanoate in apples and bananas, 3-methylbutyl 3-methylbutanoate in apples, and octyl acetate in oranges.

TABLE 18.4 Physical Properties of Esters

Common Name	Formula	Boiling Point (°C)	Solubility (g/100 g H_2O, at 20°C)
methyl formate	HCO_2CH_3	32	miscible
methyl acetate	$CH_3CO_2CH_3$	57	24.4
methyl propionate	$CH_3CH_2CO_2CH_3$	80	1.8
methyl butyrate	$CH_3(CH_2)_2CO_2CH_3$	102	0.5
methyl valerate	$CH_3(CH_2)_3CO_2CH_3$	126	0.2
methyl caproate	$CH_3(CH_2)_4CO_2CH_3$	151	0.06
methyl caprylate	$CH_3(CH_2)_6CO_2CH_3$	208	0.007
ethyl acetate	$CH_3CO_2CH_2CH_3$	77	7.4
propyl acetate	$CH_3CO_2CH_2CH_2CH_3$	102	1.9
butyl acetate	$CH_3CO_2CH_2CH_2CH_2CH_3$	125	0.9

Polyesters

Many commercial products called **condensation polymers** are made by reacting monomers to give large molecules and some small molecule, such as water, as a byproduct. This polymerization process differs from addition polymerization, in which the entire monomer remains in the polymer.

In condensation polymerization reactions, each monomer has two functional groups (each is said to be difunctional). An example of condensation polymerization is the reaction of terephthalic acid and ethylene glycol. One step in the reaction sequence is shown. Although the product looks like a simple ester, it is also both an acid and an alcohol. Its acid end can react with more ethylene glycol, and its alcohol end can react with more terephthalic acid. Each reaction produces a larger molecule that continues to react to form a high molecular weight polyester. The polymer is industrially processed into a fiber called Dacron. Dacron is used in woven and knitted fabrics in combination with cotton or wool. Dacron fabric is physiologically inert, so it can be used in the form of a mesh to replace diseased sections of blood vessels.

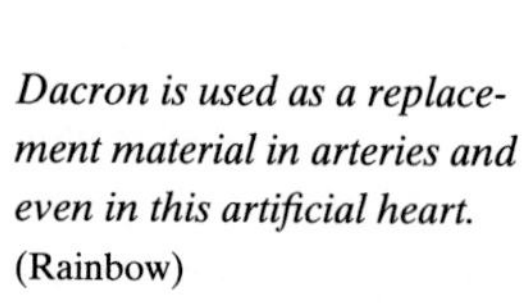

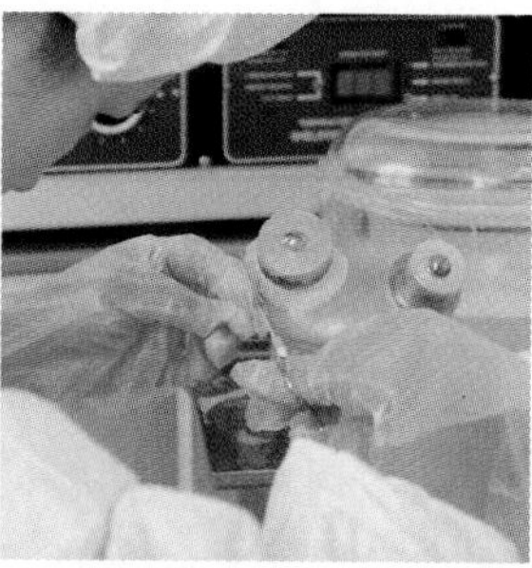

Dacron is used as a replacement material in arteries and even in this artificial heart. (Rainbow)

$$HO-\overset{O}{\overset{\|}{C}}-C_6H_4-\overset{O}{\overset{\|}{C}}-OH + HOCH_2CH_2OH \longrightarrow HO-\overset{O}{\overset{\|}{C}}-C_6H_4-\overset{O}{\overset{\|}{C}}-OCH_2CH_2OH + H_2O$$

terephthalic acid; ethylene glycol

This end (HO—C) can react with ethylene glycol. This end (—OH) can react with terephthalic acid.

$$HOCH_2CH_2O\left[\overset{O}{\overset{\|}{C}}-C_6H_4-\overset{O}{\overset{\|}{C}}-OCH_2CH_2O\right]_n\overset{O}{\overset{\|}{C}}-C_6H_4-\overset{O}{\overset{\|}{C}}-OH$$

EXAMPLE 18.4

Assign the IUPAC name to clofibrate, a drug used to lower the concentration of blood triglycerides and cholesterol.

$$Cl-C_6H_4-O-C(CH_3)_2-\overset{O}{\overset{\|}{C}}-O-CH_2CH_3$$

clofibrate

Solution

First, identify the alcohol portion of the ester; it is located at the right of the molecule. The alcohol portion contains two carbon atoms and the compound is an ethyl ester.

$$Cl-C_6H_4-O-C(CH_3)_2-\overset{O}{\overset{\|}{C}}-OH \qquad HO-CH_2CH_3$$

acyl portion of ester; ethanol

The acyl portion is a substituted propanoic acid with a methyl group and an aryl-containing group at the C-2 atom. Imagine removing the aryl-containing group from the acid and adding a hydrogen atom to its oxygen atom. The resulting compound is *p*-chlorophenol. The original group is *p*-chlorophenoxy.

$$\text{Cl}-\text{C}_6\text{H}_4-\text{OH} \qquad \text{H}-\overset{\text{CH}_3}{\underset{\text{CH}_3}{\text{C}}}-\overset{\text{O}}{\overset{\|}{\text{C}}}-\text{O}-\text{H}$$

p-chlorophenol 2-methylpropanoic acid

The name of the acid is 2-(*p*-chlorophenoxy)-2-methylpropanoic acid. Change the -ic ending of the acid to -ate. Write the name of the alkyl group of the alcohol as a separate word before the modified acid name. The ester is ethyl 2-(*p*-chlorophenoxy)-2-methylpropanoate.

PROBLEM 18.4

Isobutyl methanoate has the odor of raspberries. 3-Methylbutyl ethanoate has the odor of bananas. Draw the structural formula of each ester.

18.8 Hydrolysis and Saponification of Esters

Ester hydrolysis involves breaking the ester bond to produce an acid and an alcohol. Ester hydrolysis, then, is the reverse of the esterification reaction.

The C—O bond is cleaved.

The —OH group is bonded to the carbonyl carbon atom.

$$CH_3-\overset{O}{\overset{\|}{C}}-O-CH_2CH_3 + H_2O \xrightarrow{H^+} CH_3-\overset{O}{\overset{\|}{C}}-OH \quad H-O-CH_2CH_3$$

This O—H bond is formed.

Ester hydrolysis is catalyzed by acid and is favored by an excess of water.

$$R-\overset{O}{\overset{\|}{C}}-O-R' + H_2O \overset{H^+}{\rightleftharpoons} R-\overset{O}{\overset{\|}{C}}-O-H + R'O-H$$

Adding water "pushes" the reaction to the right.

Saponification is the hydrolysis of an ester by a strong base to give an alcohol and the conjugate base of a carboxylic acid. For example, in the saponification of ethyl acetate, the acetate ion is formed, in contrast with the hydrolysis reaction in which acetic acid is produced.

$$CH_3-\overset{O}{\overset{\|}{C}}-O-CH_2CH_3 + HO^- \longrightarrow CH_3-\overset{O}{\overset{\|}{C}}-O^- + CH_3CH_2-OH$$

There is an important difference between hydrolysis and the related saponification reaction. Hydrolysis is *catalyzed* by an acid. In the saponification reaction hydroxide is a *reagent,* not a catalyst, because equal numbers of moles of hydroxide and ester are required. The hydroxide is consumed in the reaction, and the position of equilibrium lies overwhelmingly to the right.

Soaps and Detergents

Soaps are salts of long-chain acids called *fatty acids.* The best soaps are made from saturated acids with 14–18 carbon atoms. Soaps fabricated as bars are usually sodium salts, whereas the potassium salts, which are softer, are used in shaving creams.

$$CH_3(CH_2)_{16}CO_2^- \ Na^+$$

sodium stearate
(a soap)

Soaps were originally produced from animal fats, triesters of glycerol and carboxylic acids containing 12–18 carbon atoms.

$$\begin{array}{l} CH_2-O-C(=O)-(CH_2)_{16}CH_2 \\ | \\ CH-O-C(=O)-(CH_2)_{16}CH_3 \\ | \\ CH_2-O-C(=O)-(CH_2)_{16}CH_3 \end{array}$$

tristearin
(a fat)

The detergent industry depends on chemistry to make substitutes for soap. (PhotoEdit)

The carboxylate salts of fatty acids have long, nonpolar hydrocarbon chains. Therefore, they do not form solutions of individual ions but exist as spherical aggregations of ions called **micelles.** In a micelle of carboxylate salts, the nonpolar hydrocarbon chains point toward the interior of the sphere, and the polar carboxylate "heads" lie on the surface of the sphere (see figure). This arrangement encloses the maximum amount of "hydrocarbon" material for the smallest surface area. As a consequence, the hydrogen-bonded structure of water is disrupted to the smallest extent possible.

The nonpolar hydrocarbon chain of a fatty acid repels water and is called **hydrophobic** ("water fearing"). In contrast, the polar "head" of the carboxylate group forms hydrogen bonds to water and is called **hydrophilic** ("water loving"). London forces between hydrocarbon chains hold the micelle together. The tendency of nonpolar solutes to aggregate in aqueous solution is called the **hydrophobic effect.** The micelle surface, which may contain as many as 100 carboxylate groups, has a large number of negative charges. As a result, individual micelles repel each other and remain suspended in water.

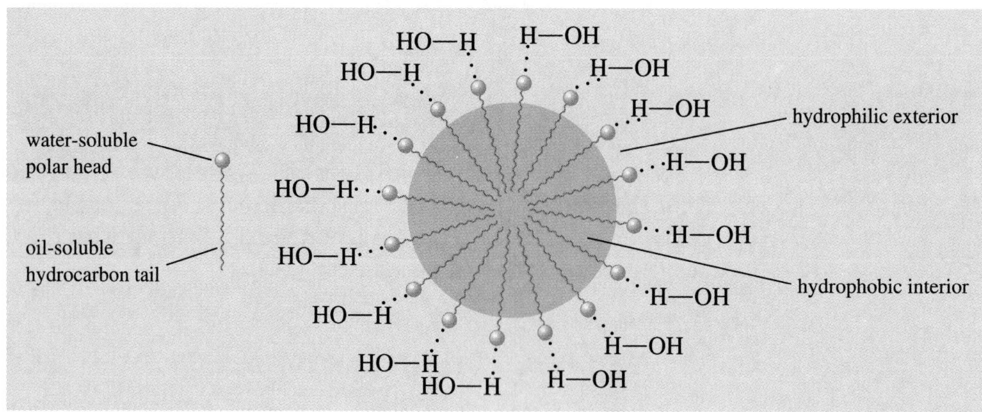

Micelle of a Soap

The long hydrocarbon chains of carboxylates are nonpolar and associate within the micelle. The polar carboxylate heads are located on the exterior and are hydrogen-bonded to water.

An example of a hydrophobic substance is grease, which does not dissolve in water because it is nonpolar. However, grease will dissolve in the hydrocarbon region of the micelle. This process accounts for the cleansing action of a soap.

Micelles also interact with the ions in hard water, which contains relatively high concentrations of Ca^{2+} and Mg^{2+} ions. These ions react with the carboxylate ions of soaps and form precipitates that reduce its cleansing power. For this reason, **detergents**—salts of organic sulfate esters—work better than soaps in hard water. Detergents have long hydrophobic tails and hydrophilic heads and form micelles. However, they do not form precipitates with Ca^{2+} and Mg^{2+} ions.

$$CH_3(CH_2)_{11}{-}O{-}SO_3^-Na^+$$

sodium dodecyl sulfate

$$CH_3(CH_2)_9CH(CH_3){-}C_6H_4{-}SO_3^-Na^+$$

sodium-(2-dodecyl) benzene sulfonate

EXAMPLE 18.5

The antibiotic chloramphenicol tastes bitter. Its palatability for children is improved by using a suspension of the palmitate ester. Once orally administered, enzymes in the intestine hydrolyze the ester. Given the structure of the ester, write the structure of the antibiotic.

$$NO_2{-}C_6H_4{-}CH(OH){-}CH(NHCCHCl_2)CH_2{-}O{-}\overset{O}{\overset{\|}{C}}{-}(CH_2)_{14}CH_3 \quad (\text{NHCCHCl}_2 \text{ with } C{=}O)$$

chloramphenicol palmitate

Solution

First, locate the ester functional group by examining the carbonyl carbon atoms. The carbonyl group at the bottom of the structure is bonded to a nitrogen atom. This is an amide. The carbonyl carbon atom toward the right of the structure is bonded to an oxygen atom. This is the ester functional group. The carbon chain to the right of the carbonyl group is part of the acid. A total of 16 carbon atoms are contained in the carboxylic acid. Chloramphenicol is bonded in the ester through its primary hydroxyl group.

$$NO_2{-}C_6H_4{-}CH(OH){-}CH(NHCCHCl_2)CH_2{-}OH \qquad HO{-}\overset{O}{\overset{\|}{C}}{-}(CH_2)_{14}CH_3$$

chloramphenicol

hexadecanoic acid (palmitic acid)

PROBLEM 18.5

Saponification of an extract of apples followed by acidification gives 3-methyl-1-butanol and 3-methylbutanoic acid. Draw the structure and give the name of the compound contained in the extract.

18.9

Nucleophilic Acyl Substitution

In Chapter 16 you learned that a nucleophile (Nu) can displace a leaving group (L) bonded to an alkyl carbon atom.

$$Nu:^- + R{-}L \longrightarrow Nu{-}R + :L^-$$

Carboxylic acids and acyl derivatives react with nucleophiles to form new acyl derivatives, but there are two steps in the mechanism for nucleophilic acyl substitution.

Step 1: $Nu:^- + R{-}C({=}\ddot{O}:){-}L \longrightarrow R{-}C(L)(Nu){-}\ddot{O}:^-$

Step 2: $R{-}C(L)(Nu){-}\ddot{O}:^- \longrightarrow R{-}C(={\ddot{O}}){-}Nu + :L^-$

The order of reactivity of acyl derivatives is acid chloride > acid anhydride > ester = acid > amide. This order means that it is possible to convert a more reactive acyl derivative into a less reactive acyl derivative.

Acid chlorides and acid anhydrides do not exist in nature. They are prepared in the laboratory by methods that will not be discussed in this text. They are used to prepare other acyl compounds.

$R{-}C(=O){-}Cl$ acid chloride

$R{-}C(=O){-}O{-}C(=O){-}R$ acid anhydride

Acid chlorides are very reactive and react with nucleophiles including water vapor in air. Reaction of an acid chloride with an alcohol gives a quantitative yield of an ester. Amines are converted into amides by reaction with acid chlorides.

$CH_3{-}C(=O){-}Cl \xrightarrow{H{-}O{-}H} CH_3{-}C(=O){-}O{-}H + HCl$

$CH_3{-}C(=O){-}Cl \xrightarrow{H{-}O{-}R} CH_3{-}C(=O){-}O{-}R + HCl$

$CH_3{-}C(=O){-}Cl \xrightarrow{H{-}NH{-}R} CH_3{-}C(=O){-}NH{-}R + HCl$

Acid anhydrides also react readily with nucleophiles. The most common acid anhydride is acetic anhydride; over 1 million tons are produced each year and used to convert various alcohols into acetates. Water hydrolyzes acid anhydrides to acids.

Thioesters Are Nature's Active Acyl Compounds

The interconversion of acyl compounds in cells occurs by the transfer of acyl groups from one molecule to another. The acyl group of the donor or acceptor molecule is often in the form of a thioester. Thioesters are more reactive than esters, and an alkoxy group from an alcohol easily replaces the thiol group.

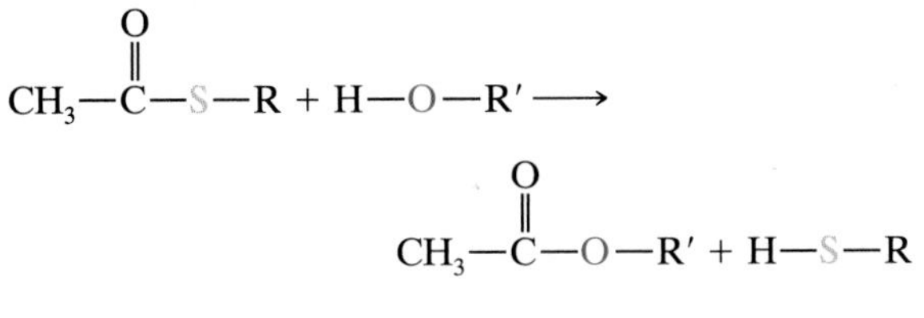

$$CH_3{-}\overset{\overset{\displaystyle O}{\|}}{C}{-}S{-}R + H{-}O{-}R' \longrightarrow CH_3{-}\overset{\overset{\displaystyle O}{\|}}{C}{-}O{-}R' + H{-}S{-}R$$

The most important thioester is acetyl coenzyme A (acetyl CoA), formed from the thiol group of coenzyme A, a complex thiol abbreviated as CoA—SH. The thiol group of CoA—SH is bonded to an acyl group in acyl—S—CoA derivatives.

When an acetyl group is linked to CoA—SH, the addition compound, or adduct, is acetyl CoA. This important metabolite results from the degradation of long-chain carboxylic acids contained in fats. Acetyl CoA also results from metabolic degradation of many amino acids and carbohydrates.

Acetyl CoA is a donor of the acetyl group in the biosynthesis of carboxylic acids. It also reacts with nucleophiles in biological reactions to give new acyl compounds. For example, acetyl CoA provides the acetyl group in the biosynthesis of the neurotransmitter acetylcholine. Choline contains a hydroxyl group that is acetylated by acetyl CoA to make acetylcholine.

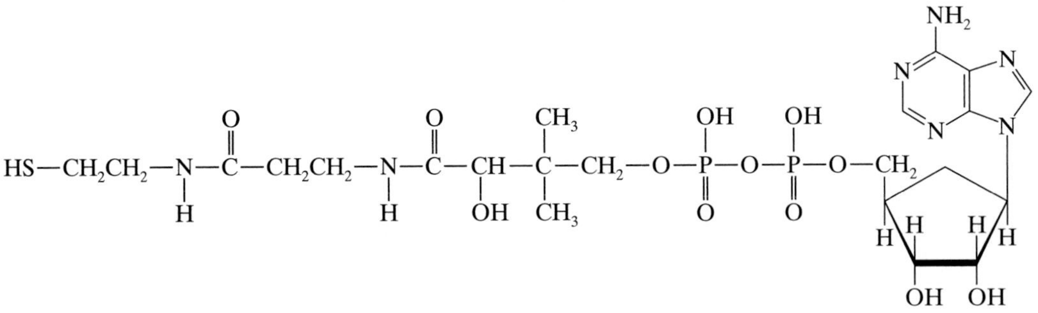

$$HS{-}CH_2CH_2{-}\underset{\underset{\displaystyle H}{|}}{N}{-}\overset{\overset{\displaystyle O}{\|}}{C}{-}CH_2CH_2{-}\underset{\underset{\displaystyle H}{|}}{N}{-}\overset{\overset{\displaystyle O}{\|}}{C}{-}\underset{\underset{\displaystyle OH}{|}}{CH}{-}\overset{\overset{\displaystyle CH_3}{|}}{\underset{\underset{\displaystyle CH_3}{|}}{C}}{-}CH_2{-}O{-}\overset{\overset{\displaystyle OH}{|}}{\underset{\underset{\displaystyle O}{\|}}{P}}{-}O{-}\overset{\overset{\displaystyle OH}{|}}{\underset{\underset{\displaystyle O}{\|}}{P}}{-}O{-}CH_2{-}\text{(ribose, H, H, H, H, OH, OH; adenine with } NH_2\text{)}$$

coenzyme A

$$CH_3{-}\overset{\overset{\displaystyle O}{\|}}{C}{-}S{-}CoA + H{-}Nu \longrightarrow CH_3{-}\overset{\overset{\displaystyle O}{\|}}{C}{-}Nu + CoA{-}SH$$

$$CH_3{-}\overset{\overset{\displaystyle O}{\|}}{C}{-}SCoA + CH_3{-}\overset{\overset{\displaystyle CH_3}{|}}{\underset{\underset{\displaystyle CH_3}{|}}{N^+}}{-}CH_2CH_2OH \longrightarrow CH_3{-}\overset{\overset{\displaystyle O}{\|}}{C}{-}OCH_2CH_2{-}\overset{\overset{\displaystyle CH_3}{|}}{\underset{\underset{\displaystyle CH_3}{|}}{N^+}}{-}CH_3 + CoA{-}SH$$

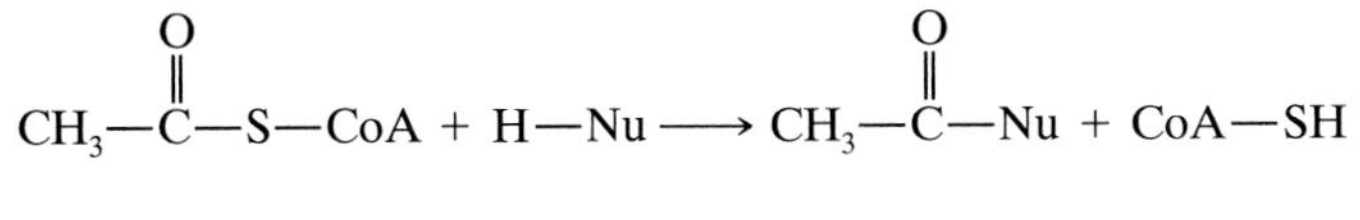

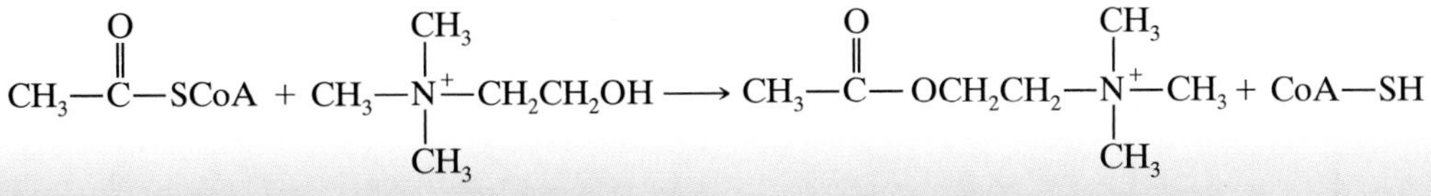

choline acetylcholine

Alcohols react to give esters, and amines give amides. The byproduct in each case is 1 mole of acetic acid.

$$CH_3{-}\overset{\overset{\displaystyle O}{\|}}{C}{-}O{-}\overset{\overset{\displaystyle O}{\|}}{C}{-}CH_3 \xrightarrow{H{-}O{-}H} CH_3{-}\overset{\overset{\displaystyle O}{\|}}{C}{-}O{-}H + H{-}O{-}\overset{\overset{\displaystyle O}{\|}}{C}{-}CH_3$$

$$CH_3{-}\overset{\overset{\displaystyle O}{\|}}{C}{-}O{-}\overset{\overset{\displaystyle O}{\|}}{C}{-}CH_3 \xrightarrow{H{-}O{-}R} CH_3{-}\overset{\overset{\displaystyle O}{\|}}{C}{-}O{-}R + H{-}O{-}\overset{\overset{\displaystyle O}{\|}}{C}{-}CH_3$$

$$CH_3{-}\overset{\overset{\displaystyle O}{\|}}{C}{-}O{-}\overset{\overset{\displaystyle O}{\|}}{C}{-}CH_3 \xrightarrow{H{-}NH{-}R} CH_3{-}\overset{\overset{\displaystyle O}{\|}}{C}{-}NH{-}R + H{-}O{-}\overset{\overset{\displaystyle O}{\|}}{C}{-}CH_3$$

18.10 Esters and Anhydrides of Phosphoric Acid

Phosphoric acid, pyrophosphoric acid, and triphosphoric acid occur in cells as esters. Just as alcohols react with carboxylic acids, they also react with phosphoric acid, pyrophosphoric acid, and triphosphoric acid.

$$R{-}OH + HO{-}\underset{\underset{\displaystyle O}{\|}}{\overset{\overset{\displaystyle OH}{|}}{P}}{-}OH \longrightarrow RO{-}\underset{\underset{\displaystyle O}{\|}}{\overset{\overset{\displaystyle OH}{|}}{P}}{-}OH + H_2O$$

phosphoric acid — alkyl phosphate (ester bond: RO—P)

$$R{-}OH + HO{-}\underset{\underset{\displaystyle O}{\|}}{\overset{\overset{\displaystyle OH}{|}}{P}}{-}O{-}\underset{\underset{\displaystyle O}{\|}}{\overset{\overset{\displaystyle OH}{|}}{P}}{-}OH \longrightarrow RO{-}\underset{\underset{\displaystyle O}{\|}}{\overset{\overset{\displaystyle OH}{|}}{P}}{-}O{-}\underset{\underset{\displaystyle O}{\|}}{\overset{\overset{\displaystyle OH}{|}}{P}}{-}OH + H_2O$$

pyrophosphoric acid — alkyl diphosphate (ester bond: RO—P)

$$R{-}OH + HO{-}\underset{\underset{\displaystyle O}{\|}}{\overset{\overset{\displaystyle OH}{|}}{P}}{-}O{-}\underset{\underset{\displaystyle O}{\|}}{\overset{\overset{\displaystyle OH}{|}}{P}}{-}O{-}\underset{\underset{\displaystyle O}{\|}}{\overset{\overset{\displaystyle OH}{|}}{P}}{-}OH \longrightarrow RO{-}\underset{\underset{\displaystyle O}{\|}}{\overset{\overset{\displaystyle OH}{|}}{P}}{-}O{-}\underset{\underset{\displaystyle O}{\|}}{\overset{\overset{\displaystyle OH}{|}}{P}}{-}O{-}\underset{\underset{\displaystyle O}{\|}}{\overset{\overset{\displaystyle OH}{|}}{P}}{-}OH$$

triphosphoric acid — alkyl triphosphate (ester bond: RO—P)

Note that these esters are also acids. The protons in the —OH groups of these esters exist as ions at physiological pH. That's why phosphoric acid derivatives are soluble in the aqueous environment of living systems.

The esters of pyrophosphoric and triphosphoric acids are also acid anhydrides. The oxygen atom between the phosphorus atoms constitutes the anhydride bond. These compounds are analogs of acid anhydrides of carboxylic acids.

$$R{-}O{-}P(=O)(OH){-}O{-}P(=O)(OH){-}OH \qquad R{-}O{-}P(=O)(OH){-}O{-}P(=O)(OH){-}O{-}P(=O)(OH){-}OH$$

anhydride bond — anhydride bonds

The anhydride bonds of diphosphates and triphosphates are hydrolyzed in many biological reactions. Adenosine triphosphate (ATP), a triphosphate ester of adenosine, stores some of the energy released in the degradation of carbohydrates, fats, and amino acids (Figure 18.2).

The phosphorus atoms in ATP are very susceptible to attack by nucleophiles. The leaving group is a phosphate or diphosphate derivative, depending on which phosphorus atom the nucleophile attacks. For example, attack at the terminal phosphorus atom releases a diphosphate as the leaving group. To simplify the equation, the diphosphate portion of the molecules is represented by —ADP.

$$Nu{:}^- + {}^-O{-}P(=\ddot{O})(O^-){-}ADP \rightleftharpoons Nu\cdots P(\ddot{O}{:}^-)(O^-)(O^-)\cdots ADP \rightleftharpoons Nu{-}P(=\ddot{O})(O^-){-}O^- + ADP$$

Either of the P—O—P bonds of adenosine triphosphate can be hydrolyzed. Nucleophilic attack by water at a terminal phosphorus atom displaces adenosine 5′-diphosphate (ADP). Attack at the internal phosphorus atom displaces adenosine 5′-monophosphate (AMP). The ΔG°_{rxn} for both hydrolysis reactions is negative, as indicated by the following equations using the ionic forms of all species as they exist in biological systems.

$$ATP^{4-} + H_2O \longrightarrow ADP^{3-} + HPO_4^{2-} + H^+ \qquad \Delta G^{\circ} = -8.2\ \text{kcal/mole}$$
$$ATP^{4-} + H_2O \longrightarrow AMP^{2-} + HP_2O_7^{3-} + H^+ \qquad \Delta G^{\circ} = -8.4\ \text{kcal/mole}$$

Phosphate and diphosphate esters of alcohols form when a phosphorus-containing group transfers from ATP to the oxygen atom of an alcohol. The reaction occurs by nucleophilic attack on phosphorus by the hydroxyl group's oxygen atom. The net enzyme-catalyzed reaction for formation of a diphosphate group is shown. For simplicity, charges have been left out, and —OPP is used to represent the covalently bonded diphosphate group.

$$ROH + ATP \longrightarrow R{-}OPP + AMP$$

FIGURE 18.2
Structure of ATP

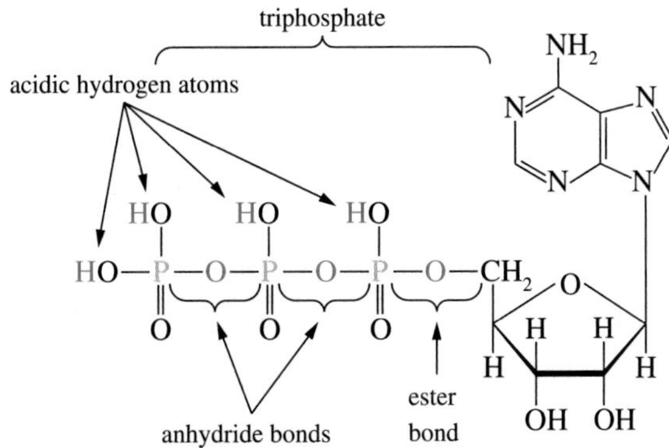

18.11
The Claisen Condensation

Two molecules of an ester react in the presence of an alkoxide base to produce a β-keto ester in a reaction called the **Claisen condensation.**

$$2\ CH_3-\overset{O}{\overset{\|}{C}}-OCH_2CH_3 \xrightarrow[CH_3CH_2OH]{CH_3CH_2ONa} CH_3-\overset{O}{\overset{\|}{C}}-CH_2-\overset{O}{\overset{\|}{C}}-OCH_2CH_3 + CH_3CH_2OH$$

ethyl acetate ethyl acetoacetate

Some of the steps in the mechanism of the Claisen condensation resemble those of the aldol condensation (Section 17.10). Both mechanisms have several reversible steps. In the first step sodium ethoxide reacts with an ester to produce an ester enolate.

$$CH_3CH_2\ddot{\underset{..}{O}}:^- + H-CH_2-\overset{O}{\overset{\|}{C}}-OCH_2CH_3 \rightleftharpoons {}^-:CH_2-\overset{O}{\overset{\|}{C}}-OCH_2CH_3 + CH_3CH_2\ddot{\underset{..}{O}}-H$$

The ester enolate reacts with a molecule of ester to form a carbon–carbon bond.

$$CH_3-\overset{\ddot{O}:}{\overset{\|}{C}}-OCH_2CH_3 + {}^-:CH_2-\overset{\ddot{O}:}{\overset{\|}{C}}-OCH_2CH_3 \rightleftharpoons CH_3-\overset{:\ddot{O}:^-}{\overset{|}{\underset{OCH_2CH_3}{\underset{|}{C}}}}-CH_2-\overset{\ddot{O}:}{\overset{\|}{C}}-OCH_2CH_3$$

conjugate base of a hemiacetal

The addition product is the conjugate base of a hemiacetal, which loses an alkoxide to give a β-keto ester.

$$CH_3-\overset{:\ddot{O}:^-}{\overset{|}{\underset{:\ddot{\underset{..}{O}}CH_2CH_3}{\underset{|}{C}}}}-CH_2-\overset{\ddot{O}:}{\overset{\|}{C}}-OCH_2CH_3 \rightleftharpoons CH_3-\overset{\ddot{O}:}{\overset{\|}{C}}-CH_2-\overset{\ddot{O}:}{\overset{\|}{C}}-OCH_2CH_3 + CH_3CH_2\ddot{\underset{..}{O}}:^-$$

Condensation Reactions in Biochemistry

An aldol condensation of acetyl coenzyme A occurs in the first step of the citric acid cycle. The acetyl group is transferred to oxaloacetic acid to form a compound that is hydrolyzed to citric acid.

$$HO_2CCH_2-\overset{O}{\overset{\|}{C}}-CO_2H + CH_3-\overset{O}{\overset{\|}{C}}-S-CoA \longrightarrow HO_2CCH_2-\overset{OH}{\overset{|}{\underset{CO_2H}{\underset{|}{C}}}}-CH_2-\overset{O}{\overset{\|}{C}}-S-CoA$$

oxaloacetic acid acetyl coenzyme A

$$HO_2CCH_2\overset{OH}{\overset{|}{\underset{CO_2H}{\underset{|}{C}}}}-CH_2-\overset{O}{\overset{\|}{C}}-SCoA + H_2O \longrightarrow HO_2CCH_2\overset{OH}{\overset{|}{\underset{CO_2H}{\underset{|}{C}}}}-CH_2-\overset{O}{\overset{\|}{C}}-OH + CoA-SH$$

citric acid

In normal metabolism, the majority of acetyl CoA reacts with oxaloacetic acid to form citric acid. However, in illnesses such as diabetes, the metabolism of fats predominates over the metabolism of carbohydrates (Chapter 27). As a result there is an insufficient amount of oxaloacetic acid and the acetyl CoA reacts with itself in a Claisen condensation.

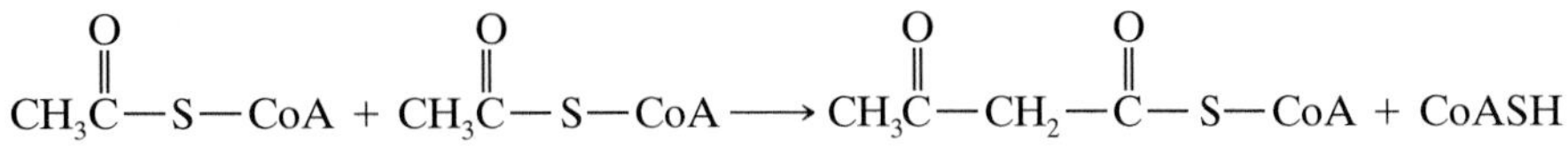

$$CH_3\overset{O}{\overset{\|}{C}}{-}S{-}CoA + CH_3\overset{O}{\overset{\|}{C}}{-}S{-}CoA \longrightarrow CH_3\overset{O}{\overset{\|}{C}}{-}CH_2{-}\overset{O}{\overset{\|}{C}}{-}S{-}CoA + CoASH$$

Hydrolysis of the resulting thioester yields acetoacetic acid (3-ketobutanoic acid). Reduction of this compound yields 3-hydroxybutanoic acid; decarboxylation yields acetone. These three compounds are collectively called ketone bodies (Section 27.4). Detection of these compounds in the urine is indicative of diabetes.

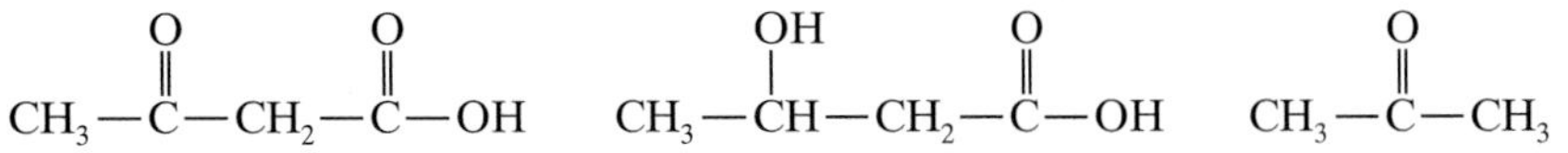

$$CH_3{-}\overset{O}{\overset{\|}{C}}{-}CH_2{-}\overset{O}{\overset{\|}{C}}{-}OH \qquad CH_3{-}\overset{OH}{\overset{|}{C}H}{-}CH_2{-}\overset{O}{\overset{\|}{C}}{-}OH \qquad CH_3{-}\overset{O}{\overset{\|}{C}}{-}CH_3$$

3-ketobutanoic acid   3-hydroxybutanoic acid   acetone

Summary

The structural characteristic of **carboxylic acids** is the **carboxyl group,** CO_2H. Carboxylic acids are represented as RCO_2H or RCOOH. The R group may contain other structural features such as double bonds, hydroxyl groups, or carbonyl groups.

The carboxyl group is very polar and forms hydrogen bonds which accounts for the high boiling points of carboxylic acids compared with those of ethers, alcohols, aldehydes, and ketones. Since carboxylic acids can hydrogen bond to water, the low molecular weight compounds are very soluble in water. Carboxylic acids react with strong bases to form carboxylate salts. Carboxylic acids are only slightly ionized in water. The equilibrium constant is represented by the symbol K_a and is called the acid dissociation constant. The pK_a of an acid is equal to $-\log K_a$. Weak acids have small K_a values and large pK_a values.

Esters are formed by the reaction of an acid with an alcohol and have the general formula RCO_2R' or RCOOR′. Esters are polar compounds, but they cannot form hydrogen bonds to each other. Consequently, their boiling points are lower than those of acids of similar molecular weight.

The **hydrolysis** of an ester occurs in the presence of a strong acid catalyst to give an equilibrium mixture containing an alcohol and a carboxylic acid. The reaction of an ester with a strong base results in **saponification** to yield an alcohol and a carboxylate salt.

An ester can undergo the **Claisen condensation** the reaction to give a β-keto ester.

Acid anhydrides of phosphoric acid are important in the chemistry of the cell. ATP, a triphosphate ester of adenosine, contains anhydride bonds that store chemical energy.

Exercises

Nomenclature of Acids

18.1 Give the IUPAC name of each of the following acids.

(a) $CH_3\underset{CH_2CH_3}{\underset{|}{C}H}CH_2CO_2H$ (b) $CH_3\underset{CH_3}{\underset{|}{C}H}CH_2\underset{CH_3}{\underset{|}{C}H}CO_2H$

(c) $CH_3\underset{CH_2CH_2CH_3}{\underset{|}{C}H}CH_2CO_2H$ (d) $CH_3\overset{CH_3}{\overset{|}{\underset{CH_3}{\underset{|}{C}}}}CH_2\underset{CH_2-CH_3}{\underset{|}{C}H}CH_2CO_2H$

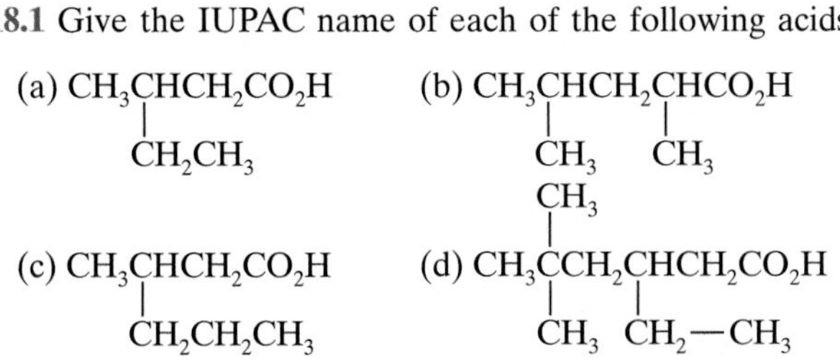

18.2 Give the IUPAC name of each of the following acids.

(a) $CH_3\underset{Cl}{\underset{|}{C}H}CH_2CO_2H$ (b) $Br\underset{CH_3}{\underset{|}{C}H}CO_2H$

(c) $CH_3\overset{CH_3}{\overset{|}{C}H}\underset{Br}{\underset{|}{C}H}CH_2CO_2H$ (d) $CH_3\overset{CH_3}{\overset{|}{\underset{CH_3}{\underset{|}{C}}}}CH_2\overset{Cl}{\overset{|}{C}H}CO_2H$

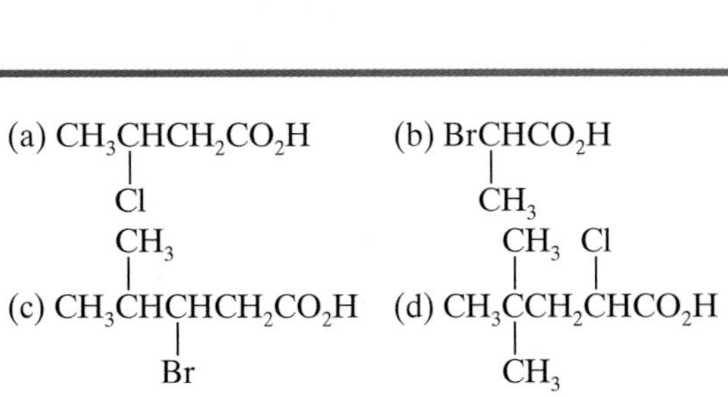

18.3 The IUPAC name of ibuprofen, the analgesic in Motrin, Advil, and Nuprin, is 2-(4-isobutylphenyl)propanoic acid. Write the structure.

18.4 10-Undecenoic acid is the antifungal agent contained in Desenex and Cruex. Write the structure.

Properties of Acids

18.5 Why is 1-butanol less soluble in water than butanoic acid?

18.6 Adipic acid is much more soluble in water than hexanoic acid. Why?

18.7 The boiling point of decanoic acid is higher than the boiling point of nonanoic acid. Explain why.

18.8 The boiling points of 2,2-dimethylpropanoic acid and pentanoic acid are 164 and 186 °C, respectively. Explain why.

18.9 Why is butyric acid a stronger acid than 1-butanol?

18.10 Which is the stronger acid, chloroacetic acid or bromoacetic acid?

18.11 The K_a values of formic acid and acetic acid are 1.8×10^{-4} and 1.8×10^{-5}, respectively. Which compound is the stronger acid?

18.12 The pK_a values of acetic acid and benzoic acid are 4.74 and 4.19, respectively. Which acid is stronger?

18.13 The K_a of methoxyacetic acid is 2.7×10^{-4}. Explain why this value is larger than the K_a of acetic acid.

18.14 Estimate the K_a values of each of the two carboxyl groups in 3-chlorohexanedioic acid.

18.15 The pK_a of amoxicillin, a synthetic penicillin, is 2.4. Recalling the inductive effect of electronegative atoms, explain why it is more acidic than acetic acid.

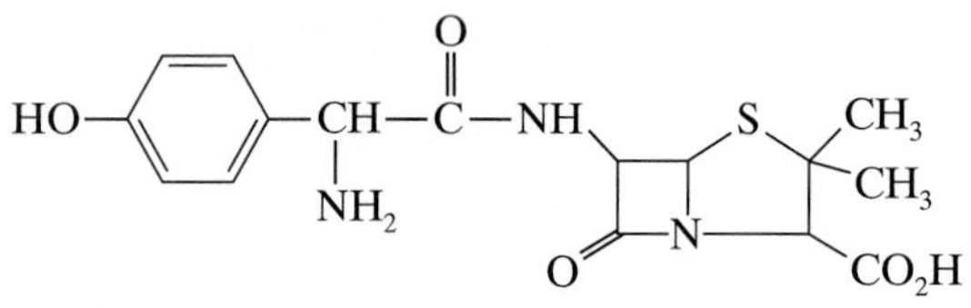

HO—C₆H₄—CH(NH₂)—C(=O)—NH— [β-lactam/thiazolidine ring: S, N, O, CH_3, CH_3, CO_2H]

amoxicillin

18.16 The pK_a of indomethacin, an anti-inflammatory analgesic used to treat rheumatoid arthritis, is 4.5 and is of comparable acidity to acetic acid. Explain why.

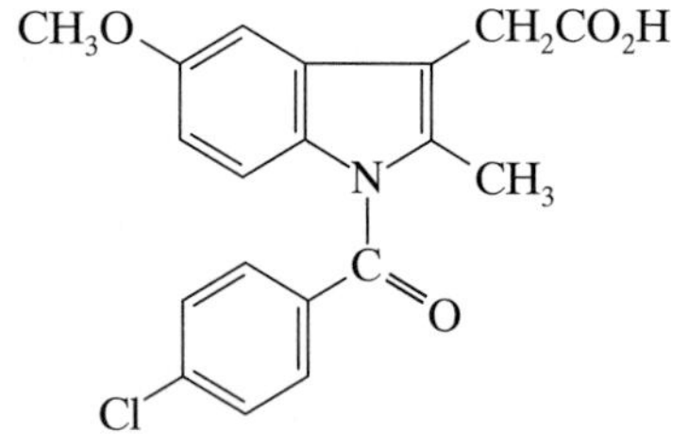

[Structure: indole ring with CH_3O, CH_2CO_2H, CH_3; N—C(=O)—C₆H₄—Cl]

indomethacin

Salts of Carboxylic Acids

18.17 Write an equation for the reaction of lactic acid with sodium hydroxide.

18.18 Write an equation for the reaction of adipic acid with potassium hydroxide.

18.19 Give the common and IUPAC names of each of the following.

(a) $CH_3CH_2CH_2CO_2Na$ (b) $CH_3(CH_2)_{16}CO_2K$
(c) $(CH_3CH_2CO_2)_2Ca$

18.20 Give the common and IUPAC names of each of the following.

(a) $CH_3CH_2CH_2CH_2CO_2Li$ (b) $CH_3(CH_2)_{14}CO_2Rb$
(c) $(CH_3CH_2CH_2CO_2)_2Mg$

18.21 Write a balanced equation for the reaction of each compound in Exercise 18.19 with hydrochloric acid.

18.22 Write a balanced equation for the reaction of each compound in Exercise 18.20 with hydrobromic acid.

Molecular Formulas

18.23 What is the general molecular formula for a carboxylic acid?

18.24 What is the general molecular formula for a dicarboxylic acid?

18.25 How many isomeric acids have the molecular formula $C_4H_8O_2$?

18.26 How many isomeric esters have the molecular formula $C_4H_8O_2$?

Nomenclature of Esters

18.27 Draw the structure of each of the following.

(a) octyl acetate (b) *tert*-butyl formate
(c) ethyl butyrate (d) propyl valerate

18.28 Draw the structure of each of the following.

(a) methyl pyruvate (b) dimethyl glutarate
(c) ethyl lactate (d) diethyl fumarate

18.29 Give the IUPAC name of each of the following.

(a) $CH_3CH_2{-}O{-}C(=O){-}H$

(b) $CH_3{-}O{-}C(=O){-}CH_2CH_2CH_3$

(c) $CH_3(CH_2)_7{-}O{-}C(=O){-}CH_3$

18.30 Give the IUPAC name of each of the following.

(a) $CH_3CH_2{-}O{-}C(=O){-}CH_2CH_2CH_3$

(b) $CH_3(CH_2)_4{-}O{-}C(=O){-}CH_2CH_2CH_3$

(c) $CH_3(CH_2)_4{-}O{-}C(=O){-}CH_3$

18.31 Give the common name of each of the following.

(a) $CH_3CH_2CH_2CO_2CH_2CH_2CH_2CH_3$
(b) $CH_3CO_2CH_2CH_2CH_2CH_2CH_3$
(c) $CH_3CO_2CH_2(CH_2)_6CH_3$

18.32 Give the common name of each of the following.

(a) $CH_3CH_2CH_2CH_2CO_2CH_2CH_2CH_3$
(b) $HCO_2CH_2CH_2CH_2CH_3$
(c) $CH_3CH_2CO_2CH_2(CH_2)_4CH_3$

Properties of Esters

18.33 Explain why carboxylic acids are more soluble in water than esters of the same molecular weight.

18.34 Explain why esters have lower boiling points than acids of the same molecular weight.

18.35 The boiling points of methyl pentanoate and butyl ethanoate are 126 and 125°C, respectively. Explain why the values are similar.

18.36 The boiling points of methyl pentanoate and methyl 2,2-dimethylpropanoate are 126 and 102°C, respectively. Explain why these values differ.

Reactions of Esters

18.37 Write the products of hydrolysis of each of the following esters.

(a) 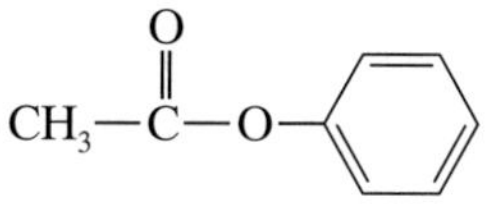

(b) 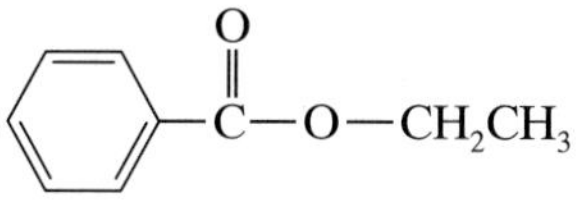

(c) $H{-}C({=}O){-}O{-}CH_2CH_3$

18.38 Write the products of hydrolysis of each of the following esters.

(a) $CH_3{-}C({=}O){-}O{-}CH_2CH_2CH_3$
(b) $CH_3CH_2CH_2CH_2{-}C({=}O){-}O{-}CH_2CH_2CH_2CH_3$
(c) $CH_3CH_2{-}C({=}O){-}O{-}CH_2CH_3$

18.39 Write the products of saponification of each compound in Exercise 18.37 with sodium hydroxide.

18.40 Write the products of saponification of each compound in Exercise 18.38 with potassium hydroxide.

18.41 Draw the structure of the ester that can be formed from each acid and alcohol.

(a) $CH_3CH_2CH_2CO_2H$ and $HOCH_2CH_2CH_2CH_3$
(b) CH_3CO_2H and $HOCH_2CH_2CH_2CH_2CH_3$
(c) CH_3CO_2H and $HOCH_2(CH_2)_6CH_3$

18.42 Draw the structure of the ester that can be formed from each acid and alcohol.

(a) $CH_3CH_2CH_2CH_2CO_2H$ and $HOCH_2CH_2CH_3$
(b) HCO_2H and $HOCH_2CH_2CH_2CH_3$
(c) $CH_3CH_2CO_2H$ and $HOCH_2(CH_2)_4CH_3$

18.43 What alcohol and acid are required to form each of the following esters?

(a) $CH_3CH_2{-}O{-}C({=}O){-}H$
(b) $CH_3CH_2CH_2CH_2CH_2{-}O{-}C({=}O){-}CH_3$
(c) $CH_3{-}C({=}O){-}O{-}CH_2CH_3$

18.44 What alcohol and acid are required to form each of the following esters?

(a) $CH_3CH_2CH_2CH_2{-}O{-}C({=}O){-}CH_2CH_2CH_3$
(b) $CH_3{-}O{-}C({=}O){-}CH_2CH_2CH_3$
(c) $CH_3CH_2{-}O{-}C({=}O){-}CH_2CH_2CH_3$

18.45 The common name of the vasodilator cyclandelate is 3,5,5-trimethylcyclohexyl mandelate. Write the structure of the acid component and assign its name based on the name of the ester.

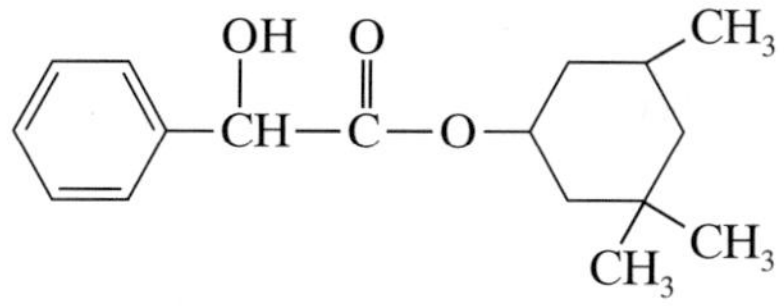

18.46 Hydrolysis of the ester group of diloxanide furanoate is required for the drug to be effective against intestinal amebiasis. Write the structure of the acid component of the drug and assign its name based on the name of the ester.

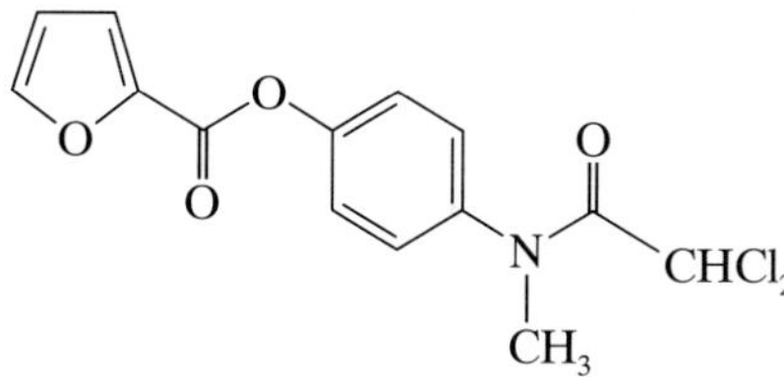

Polyesters

18.47 List some dicarboxylic acids that can be used to make polyesters.

18.48 Could glycerol be used to produce polyesters?

18.49 What monomers are needed to form the following polyester?

$$-O(CH_2)_4O\overset{O}{\overset{\|}{C}}(CH_2)_3\overset{O}{\overset{\|}{C}}O(CH_2)_4O\overset{O}{\overset{\|}{C}}(CH_2)_3\overset{O}{\overset{\|}{C}}O-$$

18.50 What monomers are needed to form the following polyester?

$$-O\overset{O}{\overset{\|}{C}}(CH_2)_4\overset{O}{\overset{\|}{C}}O(CH_2)_3O\overset{O}{\overset{\|}{C}}(CH_2)_4\overset{O}{\overset{\|}{C}}O(CH_2)_3O-$$

Nucleophilic Acyl Substitution

18.51 Which of the following reactions will occur?

(a) $CH_3\overset{O}{\overset{\|}{C}}-Cl + CH_3OH \longrightarrow CH_3\overset{O}{\overset{\|}{C}}-OCH_3 + HCl$

(b) $CH_3\overset{O}{\overset{\|}{C}}-NH_2 + CH_3OH \longrightarrow CH_3\overset{O}{\overset{\|}{C}}-OCH_3 + NH_3$

(c) $CH_3\overset{O}{\overset{\|}{C}}-OCH_3 + CH_3NH_2 \longrightarrow CH_3\overset{O}{\overset{\|}{C}}-NHCH_3 + CH_3OH$

18.52 Which of the following reactions will occur?

(a) $CH_3\overset{O}{\overset{\|}{C}}-O-\overset{O}{\overset{\|}{C}}CH_3 + NH_3 \longrightarrow CH_3\overset{O}{\overset{\|}{C}}-NH_2 + HO-\overset{O}{\overset{\|}{C}}CH_3$

(b) $CH_3\overset{O}{\overset{\|}{C}}-O-\overset{O}{\overset{\|}{C}}CH_3 + HCl \longrightarrow CH_3\overset{O}{\overset{\|}{C}}-OH + Cl-\overset{O}{\overset{\|}{C}}CH_3$

(c) $CH_3\overset{O}{\overset{\|}{C}}-O-\overset{O}{\overset{\|}{C}}CH_3 + CH_3OH \longrightarrow CH_3\overset{O}{\overset{\|}{C}}-OCH_3 + HO-\overset{O}{\overset{\|}{C}}CH_3$

Thioesters

18.53 What are the products of the following reaction?

$$CH_3CH_2-\overset{O}{\overset{\|}{C}}-S-CH_2CH_3 + CH_3OH \longrightarrow$$

18.54 What are the products of the following reaction?

$$CH_3CH_2-\overset{O}{\overset{\|}{C}}-S-CH_3 + CH_3CH_2OH \longrightarrow$$

Phosphoric Acid Esters and Anhydrides

18.55 Draw the structure of the ester formed from reaction of 1 mole of methanol with 1 mole of phosphoric acid. How many acidic protons are there in the product?

18.56 Draw the structure of the ester formed from reaction of 1 mole of ethanol with 1 mole of pyrophosphoric acid. How many acidic protons are there in the product?

18.57 Determine the number of anhydride and ester bonds in the following compound. How many acidic protons are present?

$$CH_3CH_2-O-\underset{OH}{\underset{|}{\overset{O}{\overset{\|}{P}}}}-O-\underset{OH}{\underset{|}{\overset{O}{\overset{\|}{P}}}}-O-CH_2CH_3$$

18.58 Determine the number of anhydride and ester bonds in the following compound. How many acidic protons are present?

$$CH_3CH_2-O-\underset{OH}{\underset{|}{\overset{O}{\overset{\|}{P}}}}-O-\underset{OH}{\underset{|}{\overset{O}{\overset{\|}{P}}}}-O-\underset{OH}{\underset{|}{\overset{O}{\overset{\|}{P}}}}-O-CH_2CH_3$$

Claisen Condensation

18.59 What ester is required to form each of the following by a Claisen condensation?

(a) $CH_3-\overset{O}{\overset{\|}{C}}-CH_2-CO_2CH_3$

(b) $CH_3CH_2\overset{O}{\overset{\|}{C}}\underset{CH_3}{\underset{|}{C}}HCO_2CH_2CH_3$

18.60 What ester is required to form each of the following by a Claisen condensation?

(a) $C_5H_9-CH_2-\overset{O}{\overset{\|}{C}}-\underset{C_5H_9}{\underset{|}{C}}H-CO_2CH_3$ (cyclopentyl rings)

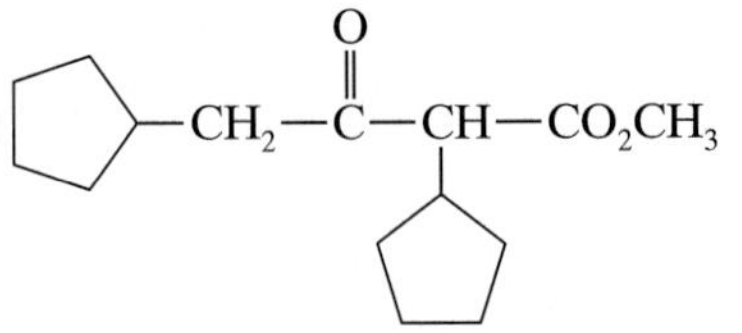

(b) $CH_3-\overset{O}{\overset{\|}{C}}-CH_2-CO_2CH_2CH_3$

18.61 Write the product of the Claisen condensation of methyl acetate.

18.62 Write the product of the Claisen condensation of ethyl propanoate.

Amines and Amides

Parachutes use nylon, a polyamide, which has high tensile strength. (Peter Arnold, Inc.)

19.1 Organic Nitrogen Compounds
19.2 Bonding and Structure of Amines
19.3 Classification and Nomenclature of Amines
19.4 Physical Properties of Amines
19.5 Basicity of Amines
19.6 Quaternary Ammonium Salts
19.7 Reactions of Amines with Nitrous Acid
19.8 Amides
19.9 Formation of Amides
19.10 Hydrolysis of Amides

Essays
- *Heterocyclic Compounds From the Ocean*
- *Amphetamines*
- *Muscle Relaxants*
- *Nitrates and Cancer*
- *Polyamides*

19.1 Organic Nitrogen Compounds

Nitrogen is the fourth most common element in living systems, after carbon, hydrogen, and oxygen. Organic compounds containing nitrogen are widely distributed in plants and animals. Nitrogen is found in amino acids and proteins (Chapter 23) as well as in nucleotides and polynucleotides (Chapter 29). Some nitrogen-containing compounds are important industrial products, including polymers such as nylon, many dyes, explosives, and pharmaceutical agents.

A nitrogen atom has five valence electrons and forms a total of three covalent bonds to carbon or hydrogen atoms. It can form single, double, or triple bonds. In this chapter we will focus on amines and amides which contain carbon-nitrogen single bonds.

$R{-}\ddot{N}H_2$ (an amine) $\quad R{-}C(={:}\ddot{O}{:}){-}{:}NH_2$ (an amide) $\quad R_2C{=}N{:}{-}R$ (an imine) $\quad R{-}C{\equiv}N{:}$ (a nitrile)

Many amines are physiologically active, affecting the brain, spinal cord, and nervous system. These compounds include the neurotransmitters epinephrine, serotonin, and dopamine. Epinephrine, commonly called adrenaline, stimulates the conversion of glycogen into glucose. Serotonin is a hormone that causes sleep. Serotonin deficiency is responsible for some forms of mental depression. Parkinson's disease is accompanied by a low concentration of dopamine.

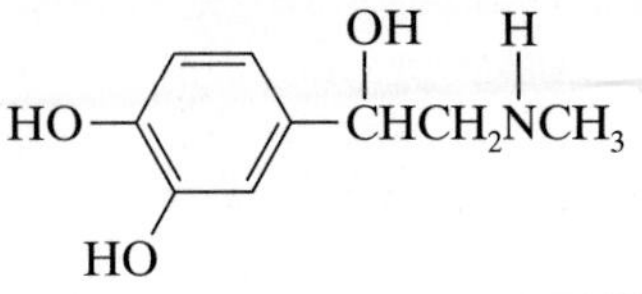

OH H
HO— —$CHCH_2NCH_3$
HO

$CH_2CH_2NH_2$
N
H

HO— —$CH_2CH_2NH_2$
HO

epinephrine 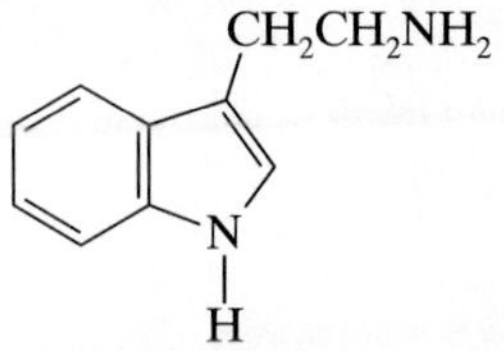 serotonin 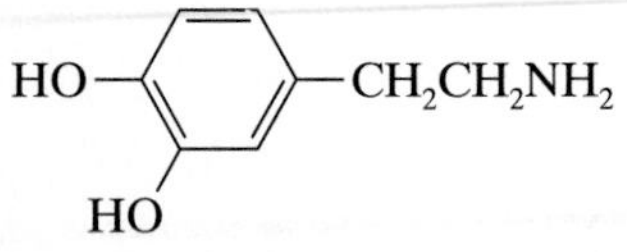 dopamine

Heterocyclic compounds containing two or more nitrogen atoms are required for the transmission of genetic information. DNA (deoxyribonucleic acid) and RNA (ribonucleic acid) contain substituted pyrimidine and purine rings.

pyrimidine 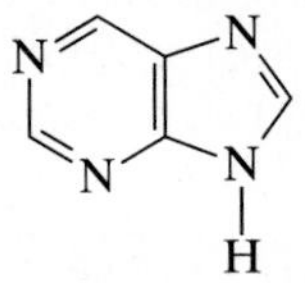 purine

Proteins, the most important and versatile biological compounds, consist of nitrogen-containing molecules called α-amino acids. The amine functional group of one α-amino acid reacts with the carboxyl group of another α-amino acid to form an amide bond. The chemistry of these compounds will be presented in Chapter 23.

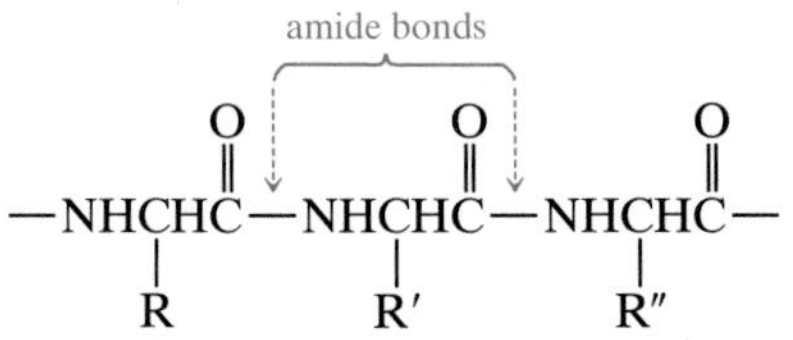

amide bonds

—NHCHC—NHCHC—NHCHC—
R R′ R″

EXAMPLE 19.1

Identify the functional groups in Valium.

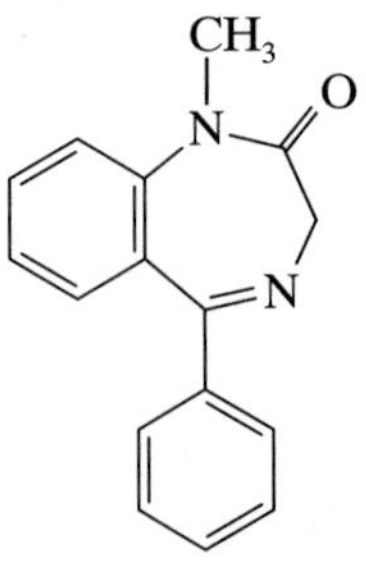

CH_3
N O
N

Solution

The nitrogen atom located at the top of this structure is bonded to a carbonyl carbon atom and is part of an amide group. The other nitrogen atom is bonded only to two carbon atoms—one by a double bond. This functional group is an imine.

PROBLEM 19.1

Classify the functional groups in Mepivacaine, a local anesthetic.

19.2 Bonding and Structure of Amines

In the simplest amine, methylamine (CH_3NH_2), a methyl group replaces one hydrogen atom of ammonia (Figure 19.1). The C—N—H and H—N—H bond angles are approximately 112° and 106°, respectively, so methylamine has a pyramidal shape around the nitrogen atom. In methylamine and other amines the nitrogen atom has five valence electrons in four sp^3 hybrid orbitals. As expected from VSEPR theory, these orbitals point to the corners of a tetrahedron. Three are half-filled and form three covalent bonds. The fourth orbital contains a pair of nonbonding electrons that plays an important role in the chemical properties of amines.

19.3 Classification and Nomenclature of Amines

Just as we can regard alcohols and ethers as organic derivatives of water, we can regard amines as organic derivatives of ammonia. However, amines are not classified like alcohols. The classification of alcohols is based on the number of groups attached to the carbon atom bearing the hydroxyl group. Amines are classified by the number of alkyl (or aryl) groups attached to the nitrogen atom.

$H—\ddot{N}(H)—H$	$R—\ddot{N}(H)—H$	$R—\ddot{N}(R)—H$	$R—\ddot{N}(R)—R$
ammonia	a primary (1°) amine	a secondary (2°) amine	a tertiary (3°) amine

For example, *tert*-butylamine has a *tert*-butyl group attached to an $—NH_2$ group. However, the amine is primary because only one alkyl group is bonded to the ni-

FIGURE 19.1 Structure of Methylamine

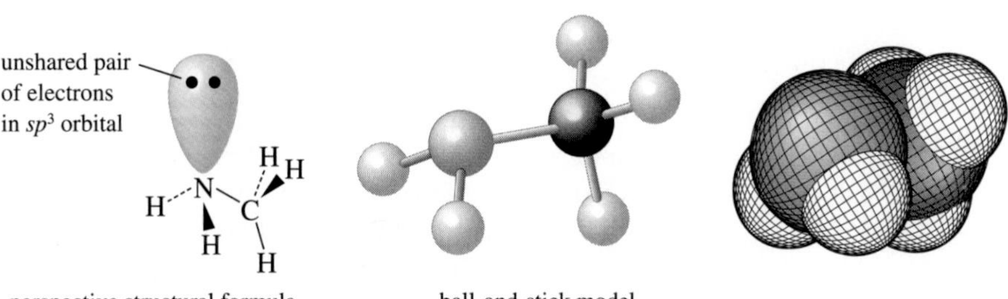

trogen atom. In contrast, *tert*-butyl alcohol is a tertiary alcohol because the carbon atom bonded to the —OH group is bonded to three alkyl groups. Trimethylamine is a tertiary amine because the nitrogen atom is bonded to three alkyl groups.

$$CH_3-\overset{\displaystyle CH_3}{\underset{\displaystyle CH_3}{C}}-\ddot{O}H \qquad CH_3-\overset{\displaystyle CH_3}{\underset{\displaystyle CH_3}{C}}-\ddot{N}H_2 \qquad CH_3-\overset{\displaystyle CH_3}{\underset{\displaystyle CH_3}{N}}:$$

tert-butyl alcohol (a tertiary alcohol) | *tert*-butylamine (a primary amine) | trimethylamine (a tertiary amine)

Compounds that have one or more atoms other than carbon in a ring are heterocyclic compounds. For example, pyrrolidine and piperidine are five- and six-membered heterocyclic compounds that are secondary amines. Pyridine is an aromatic amine.

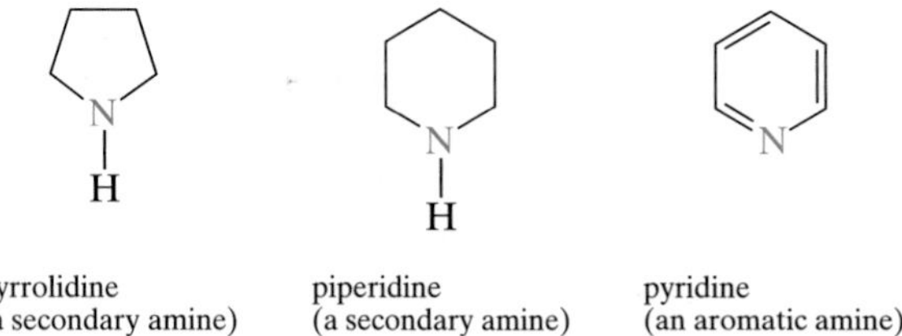

pyrrolidine (a secondary amine) | piperidine (a secondary amine) | pyridine (an aromatic amine)

Common Names of Amines

In common nomenclature, amines are described as *alkylamines.* The common name of a primary amine results from naming the alkyl group bonded to the amino group ($-NH_2$) and adding the suffix *-amine.* The entire name is written as one word. The common name for a secondary or tertiary amine is obtained by listing the alkyl groups alphabetically. When two or more identical alkyl groups are present, the prefixes *di-* and *tri-* are used.

$$C_6H_{11}-NH_2 \qquad CH_3CH_2-NH-CH_3 \qquad CH_3CH_2-NH-CH_2CH_3$$

cyclohexylamine | ethylmethylamine | diethylamine

For more complex primary amines, the amino group is treated as a substituent. The nitrogen-containing substituent in complex secondary and tertiary amines is named as an *N*-alkylamino (—NHR) or *N,N*-dialkylamino (—NRR′) group. The *N*- indicates that the alkyl group is bonded to the nitrogen atom and not to the parent chain. The largest or most complicated group is used as the parent molecule.

$$NH_2-CH_2-CH_2-CH_2-CO_2H \qquad CH_3-CH_2-CH_2-\overset{\displaystyle N(CH_3)_2}{CH}-CH_2-CO_2H$$

γ-aminobutyric acid | β-(N,N-dimethylamino)caproic acid

As we already know, amino-substituted benzene compounds are anilines. They are numbered starting at the carbon atom bearing the amino group if the other groups bonded to the ring have a lower priority for citation than the amino group. The direction of the numbering is based on the location of the second group, using the concept of the first point of difference established in our study of alkanes. Substituents are listed in alphabetical order.

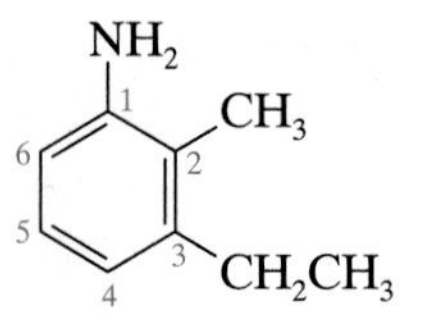

3-ethyl-2-methylaniline 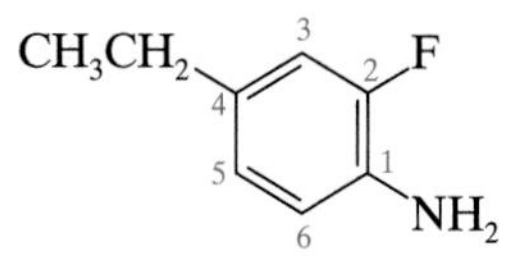 4-ethyl-2-fluoroaniline

However, the priority for citation of an amino group is low. It ranks below all acyl derivatives (acids, acid derivatives, aldehydes, ketones) and even the hydroxyl group. Thus, a benzene compound containing both a carboxylic acid group and an amino group is an amino-substituted benzoic acid, not a carboxylic acid–substituted aniline.

IUPAC Names

The IUPAC names of amines are based on the same system used for alcohols. The longest continuous chain to which the amino group is attached is the parent alkane. The *-e* ending of the alkane is dropped and *-amine* is appended. Substituents on the carbon chain, including the amino group, are designated by number. The prefix *N*- is used for each substituent on the nitrogen atom.

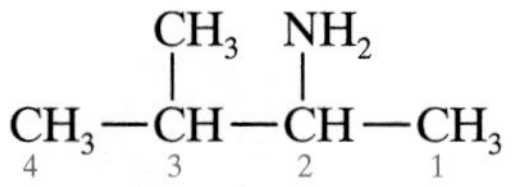

$$\underset{4}{CH_3}-\underset{3}{\overset{CH_3}{\overset{|}{CH}}}-\underset{2}{\overset{NH_2}{\overset{|}{CH}}}-\underset{1}{CH_3}$$

3-methyl-2-butanamine

$$\underset{1}{CH_3}-\underset{2}{\overset{CH_3}{\overset{|}{CH}}}-\underset{3}{\overset{NH-CH_2-CH_3}{\overset{|}{CH}}}-\underset{4}{CH_2}-\underset{5}{CH_3}$$

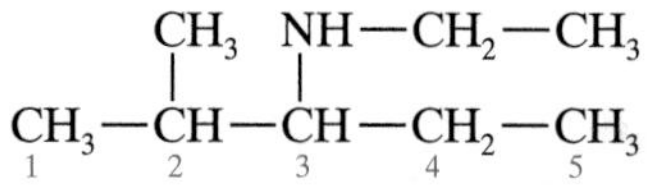

N-ethyl-2-methyl-3-pentanamine

The term *areneamine* is used for aromatic amines. Thus, aniline is benzeneamine. However, this systematic name is rarely used. Saturated three-, four-, five-, and six-membered heterocyclic rings containing one nitrogen atom are named aziridine, azetidine, pyrrolidine, and piperidine, respectively. The rings are numbered from the heteroatom.

Amines in which the nitrogen atom is part of an aromatic ring are called **heterocyclic aromatic amines.** In these compounds the positions of substituents are established by assigning the number 1 to a nitrogen atom, and numbering in the direction that provides the lowest possible numbers to other heteroatoms.

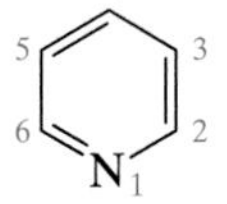

pyridine 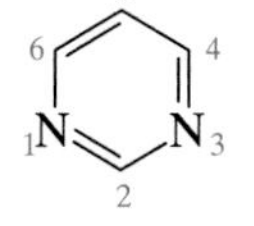 pyrimidine 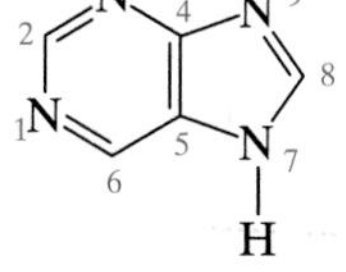 purine 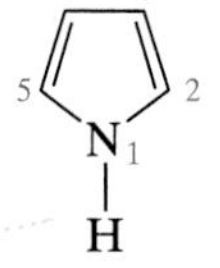 pyrrole 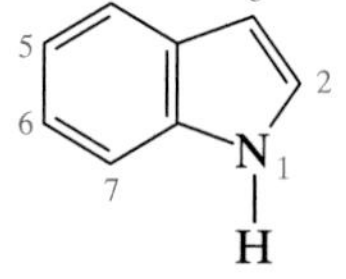 indole

EXAMPLE 19.2

Classify the amine in Demerol, a synthetic narcotic analgesic.

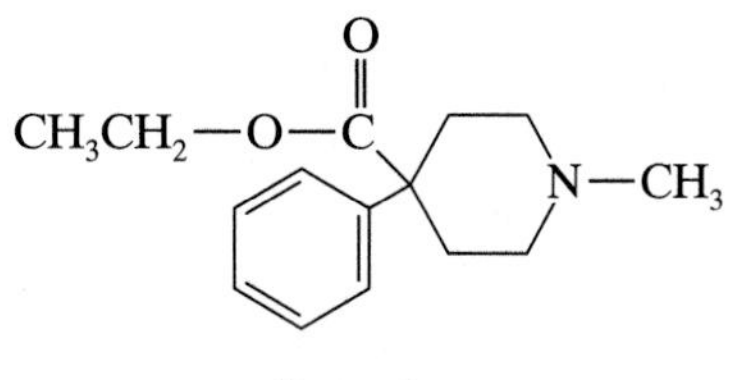

Demerol

Heterocyclic Compounds From the Ocean

Heterocyclic compounds of nitrogen, oxygen, and sulfur occur widely in plants and terrestrial animals and may constitute more than half of all of the known compounds. A high percentage of drugs are heterocyclic compounds that were developed using naturally occurring compounds found in terrestrial plants. Only recently has the chemistry of the organisms of the ocean been examined in the search for new compounds that might be useful in the design of drugs. Given the diversity of species in the ocean there may be many useful heterocyclic compounds to be discovered.

As in terrestrial plants and animals, most of the heterocyclic compounds in the ocean have five- and six-membered rings. In contrast with compounds obtained from terrestrial sources, a high percentage of heterocyclic compounds from ocean organisms are often halogenated. For example, a Bahamian sponge produces several chlorinated indoles or indole-derived compounds such as batzelline.

[Structure: batzelline — H, N, Cl, SCH_3, O, O, N, CH_3]

batzelline

Although less common, three-membered heterocyclic compounds have also been found in marine products. A species of sponge found in Fiji produces a cyclopropene ring containing nitrogen. The origin of this imine is not clear, but it may be derived from sphingosine, a component of cell membranes (Chapter 22).

[Structure: $CH_3(CH_2)_{11}$–CH=CH– azirine ring (N), H, CO_2CH_3]

dysidazirine

[Structure: $CH_3(CH_2)_{11}$–CH=CH–C(H)(OH)–C(H)(NH_2)–CH_2OH]

sphingosine

The many varieties of sponges in the ocean may provide many new medicinal compounds. (Robert Ouellette)

Solution

There are three carbon atoms bonded to nitrogen. Two of the carbon atoms are in the heterocyclic ring. The third carbon atom bonded to the nitrogen atom is a methyl group. The compound is a tertiary amine.

PROBLEM 19.2

Classify the amine in the illicit drug methamphetamine (speed).

[Structure: phenyl–CH_2–CH(CH_3)–N(H)–CH_3]

EXAMPLE 19.3

Name the following compound produced by the marine acorn worm.

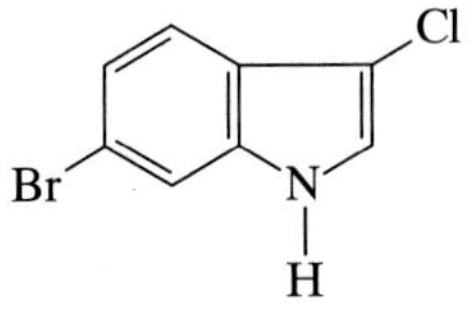

Solution

The compound has an indole ring. A chlorine atom is located at position 3 and a bromine atom at position 6. Listing the groups alphabetically, the name is 6-bromo-3-chloroindole.

PROBLEM 19.3

2-(3,4,5-Trimethoxyphenyl)ethanamine is the systematic name of mescaline, a hallucinogen. Write its structure.

19.4 Physical Properties of Amines

Amines with low molecular weights are gases at room temperature, but amines with higher molecular weights are liquids or solids (Table 19.1). Amines have higher boiling points than alkanes of similar molecular weight but lower boiling points than alcohols.

$$CH_3{-}CH_2{-}CH_3 \qquad CH_3{-}CH_2{-}NH_2 \qquad CH_3{-}CH_2{-}OH$$

$$\text{bp} = -42\ °\text{C} \qquad \text{bp} = 17\ °\text{C} \qquad \text{bp} = 78\ °\text{C}$$

Amines have higher boiling points than hydrocarbons of comparable molecular weight because the C—N bond is more polar than a C—C bond. Also, primary and secondary amines can form intermolecular hydrogen bonds because they can serve as both hydrogen bond donors and acceptors.

TABLE 19.1 Boiling Points of Some Amines

Name	Structure	Boiling Point (°C)
ammonia	NH_3	−33
methylamine	CH_3NH_2	−6
ethylamine	$C_2H_5NH_2$	17
propylamine	$C_3H_7NH_2$	49
butylamine	$C_4H_9NH_2$	77
tert-butylamine	t-$C_4H_9NH_2$	44
dimethylamine	$(CH_3)_2NH$	7
trimethylamine	$(CH_3)_3N$	3
aniline	$C_6H_5NH_2$	184

R—N̈(R)—H···:N(H)(R)—R

a hydrogen bond

Tertiary amines have no hydrogen atoms bonded to the nitrogen atom and therefore are not hydrogen bond donors. Thus, these amines cannot form intermolecular hydrogen bonds. As a consequence, they have lower boiling points than primary and secondary amines of comparable molecular weight (Table 19.1).

Amines have lower boiling points than alcohols because nitrogen is less electronegative than oxygen. As a result the N—H bond is less polar than the O—H bond, and the N—H···N hydrogen bond in amines is weaker than the O—H···O hydrogen bond in alcohols.

less polar bond — R—N̈(R)—H···:N(H)(R)—R — weaker hydrogen bond

more polar bond — R—Ö—H···:Ö(H)—R — stronger hydrogen bond

Solubility in Water

Amines with five or fewer carbon atoms are miscible with water. The solubility of amines decreases with increasing molecular weight because the functional group is a less significant part of the structure. Primary and secondary amines function as both hydrogen bond donors and acceptors, and they form hydrogen bonds with water. Even tertiary amines are soluble in water because the nonbonding electron pair of the nitrogen atom is a hydrogen bond acceptor of a hydrogen atom of water.

:Ö(H)—H···:N(R)(R)—R

a hydrogen bond

The solubilities of toluene (0.05 g/100 mL) and aniline (3.5 g/100 mL) illustrate the effect of hydrogen bonding on the solubilities of arylamines. Aniline forms hydrogen bonds with water; toluene does not.

Odor and Toxicity of Amines

Amines with low molecular weights have sharp penetrating odors similar to that of ammonia. Amines with higher molecular weights smell like decaying fish. Two compounds responsible for the odor of decaying animal tissue are appropriately named putrescine and cadaverine.

$NH_2CH_2CH_2CH_2CH_2NH_2$ — putrescine

$NH_2CH_2CH_2CH_2CH_2CH_2NH_2$ — cadaverine

Amphetamines

There are recurring structures in physiologically active compounds that contain nitrogen. One such structure is 2-phenylethanamine, which is required for binding at certain receptor sites that stimulate the central nervous system. The adrenal gland produces the hormones epinephrine and norepinephrine, which contain a 2-phenylethanamine unit. Epinephrine, also known as adrenaline, makes glucose available to tissues under conditions of excitement. Norepinephrine maintains the muscle tone of blood vessels and hence controls blood pressure.

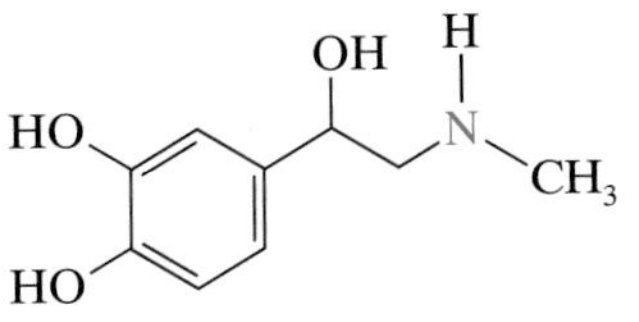

epinephrine

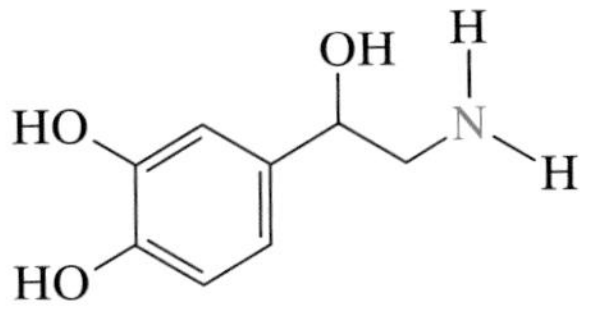

norepinephrine

Amphetamines are synthetic drugs known as "uppers" that mimic the action of epinephrine and norepinephrine. Amphetamine (also known as Benzedrine) has a structure similar to epinephrine's. It is a moderate appetite suppressant and stimulates the cortex of the brain, which effectively counters fatigue. For that reason this illegal drug has been used by long-distance truck drivers and students pulling "all-nighters" before exams. The structurally related illegal drugs methamphetamine and methoxyamphetamine are "speed" and STP, respectively.

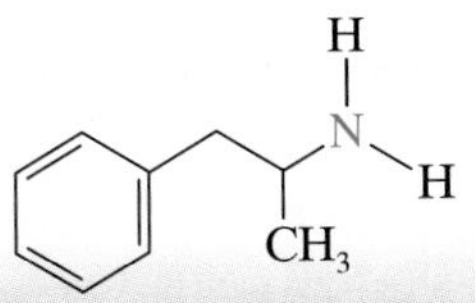

amphetamine

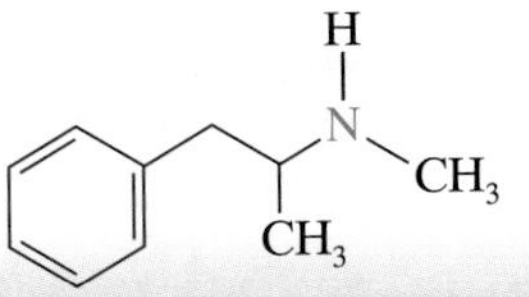

methamphetamine

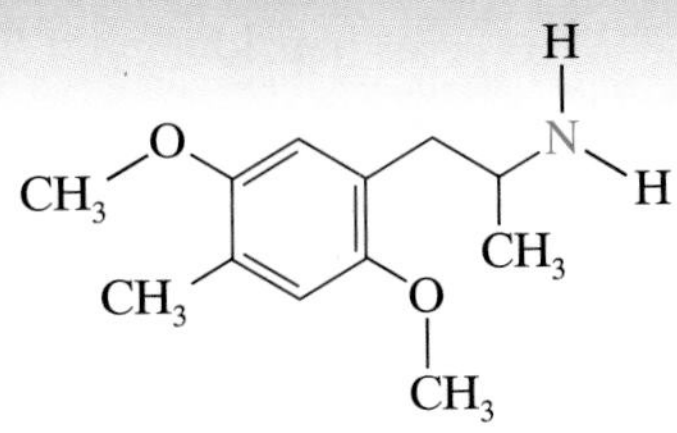

methoxyamphetamine

Abuse of these drugs produces severe physiological reactions. In addition, once the drug wears off, the user tends to "crash" into a state of physical and mental exhaustion.

Phenylpropanolamine is an over-the-counter amphetamine that has been used as an appetite suppressant by those on crash diets. Its use is not recommended by physicians, and the drug can cause health hazards for individuals suffering from hypertension.

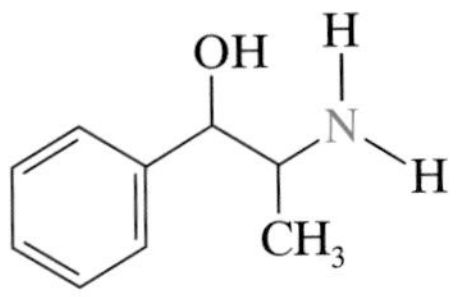

phenylpropanolamine

Although 2-phenylethanamine derivatives generally act as stimulants, methylphenidate (Ritalin) is a structurally related compound that seems to have a calming effect. This drug has been used in the treatment of hyperactivity, a condition largely restricted to young boys. The use of this drug is controversial.

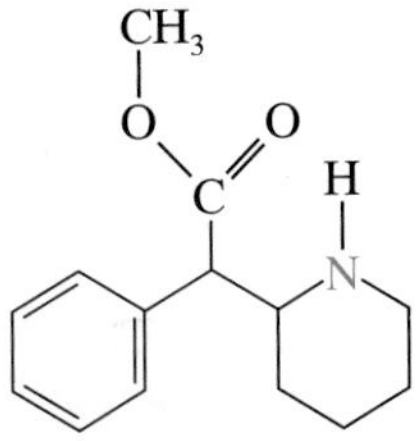

methylphenidate (Ritalin)

Because amines are physiologically active, the ingestion of an amine not normally used by a living organism can cause poisoning and death. In addition, the skin absorbs arylamines, so care should be used in handling them. Some arylamines such as benzidine and β-naphthylamine are carcinogenic.

NH_2 NH_2 NH_2

benzidine β-naphthylamine

19.5 Basicity of Amines

Like ammonia, amines are weak bases that have an unshared pair of electrons on the nitrogen atom that can accept a proton. The position of the following equilibrium for methylamine is to the left, as indicated by the small **base ionization constant, K_b**.

$$CH_3NH_2 + H_2O \rightleftharpoons CH_3NH_3^+ + OH^-$$

$$K_b = \frac{[OH^-][CH_3NH_3^+]}{[CH_3NH_2]} = 4.4 \times 10^{-4}$$

The basicity of amines is expressed by pK_b values much like the pK_a values used to express the acidity for acids.

$$pK_b = -\log K_b$$

For methylamine, with $K_b = 4.6 \times 10^{-4}$, the pK_b is 3.34. The base ionization constants and pK_b values of several amines are given in Table 19.2. Weak bases have small K_b values and large pK_b values. Alkyl-substituted amines are slightly stronger bases than ammonia. In contrast, aryl-substituted amines are much weaker bases than ammonia because the unshared pair of electrons of nitrogen is delocalized over the π orbital system of the benzene ring. As a result, the unshared electron pair of the nitrogen atom of aniline is less available for bonding with a proton than is the electron pair of the nitrogen atom of an amine such as cyclohexylamine.

TABLE 19.2 Basicity of Amines and Acidity of Ammonium Salts

	K_b	K_a	pK_b	pK_a
ammonia	1.8×10^{-5}	5.5×10^{-10}	4.74	9.26
methylamine	4.6×10^{-4}	2.2×10^{-11}	3.34	10.66
ethylamine	4.8×10^{-4}	2.1×10^{-11}	3.20	10.80
dimethylamine	4.7×10^{-4}	2.1×10^{-11}	3.27	10.73
diethylamine	3.1×10^{-4}	3.2×10^{-11}	3.51	10.49
triethylamine	1.0×10^{-3}	1.0×10^{-11}	3.00	11.00
cyclohexylamine	4.6×10^{-4}	2.2×10^{-11}	3.34	10.66
aniline	4.3×10^{-10}	2.3×10^{-5}	9.37	4.63

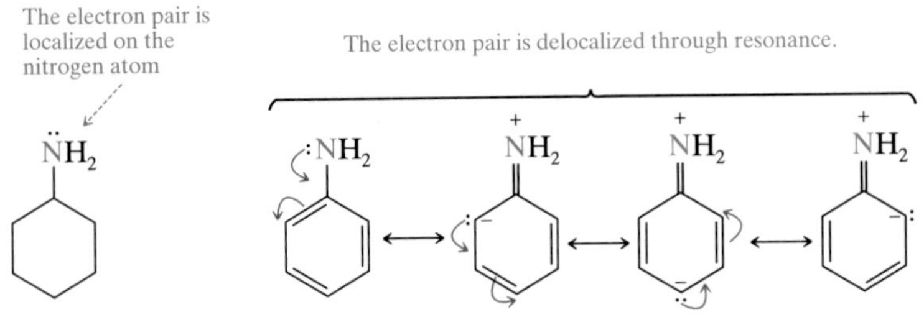

The electron pair is localized on the nitrogen atom

The electron pair is delocalized through resonance.

cyclohexylamine

contributing resonance forms of aniline

Heterocyclic Amines

The basicity of heterocyclic amines varies over a wide range and reflects both the hybridization of the nitrogen orbital containing the lone pair electrons and the effects of delocalization. Pyridine is a substantially weaker base than alkylamines. The electron pair of pyridine occupies an sp^2-hybridized orbital and lies closer to the nitrogen nucleus than the electron pair in the sp^3-hybridized orbital of alkylamines. As a result, pyridine is a weaker base (larger pK_b) than an alkylamine.

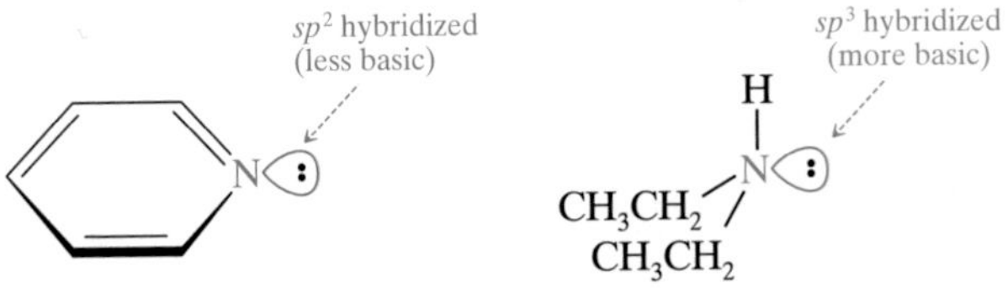

sp^2 hybridized (less basic)

sp^3 hybridized (more basic)

pyridine
$pK_b = 8.75$

diethylamine
$pK_b = 3.51$

Pyrrole is an exceedingly weak base. The pair of electrons on the nitrogen atom interacts with the four electrons of the two carbon–carbon double bonds to give an aromatic six-π-electron system similar to that of benzene's. Thus, the electron pair is not available for protonation because it is required to maintain the sextet of electrons in the aromatic ring. Pyrrolidine has a pK_a similar to acyclic amines.

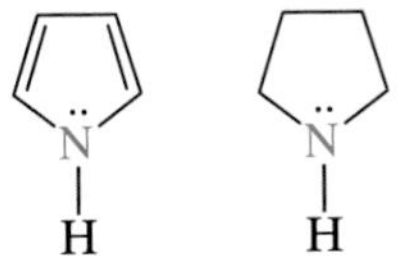

pyrrole
$pK_b = 15$

pyrrolidine
$pK_b = 2.73$

Imidazole is an important aromatic ring found in many biological molecules. One of its nitrogen atoms resembles that of pyrrole and is not basic. The second nitrogen atom, which is structurally similar to the nitrogen atom of pyridine, acts as a base. However, imidazole is about 100 times more basic than pyridine. The increased basicity results from resonance stabilization of the positive charge.

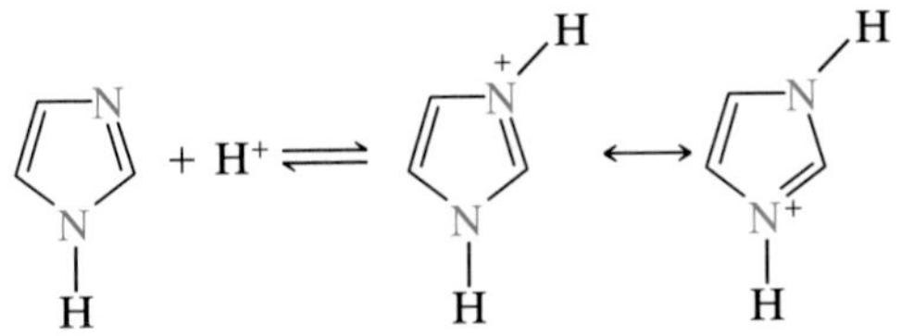

imidazole
$pK_b = 7.05$

resonance-stabilized conjugate acid

Ammonium Salts Are Acids

The strength of a base is related to the value of the ionization constant K_a of its conjugate acid. For methylamine the reaction is that of the methylammonium ion.

$$CH_3NH_3^+ + H_2O \rightleftharpoons CH_3NH_2 + H_3O^+$$

$$K_a = \frac{[CH_3NH_2][H_3O^+]}{[CH_3NH_3^+]} = 2.2 \times 10^{-11}$$

The K_b value of methylamine and the K_a value of the methylammonium ion illustrate the relationship between a base and its conjugate acid. If a base is relatively strong, its conjugate acid is relatively weak. Therefore, the K_b value for methylamine is relatively large (small pK_b), and the K_a value for the methylammonium ion is relatively small (large pK_a).

If this ammonium ion is a weaker acid (small K_a),

then this amine is a stronger base (large K_b)

$$RNH_3^+ + H_2O \rightleftharpoons RNH_2 + H_3O^+$$

Often the pK_a values of the conjugate acids of amine bases are listed rather than the pK_b values of the amines. The sum of the pK_a and pK_b values is 14 (Table 19.2).

EXAMPLE 19.4

Estimate the pK_b of pheneinamine, an antihistamine.

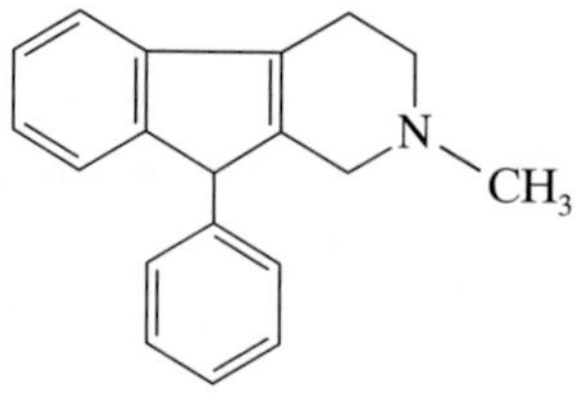

N
CH_3

Solution

The compound is a tertiary amine, and the nitrogen atom is bonded only to alkyl groups. Thus, its pK_b should be similar to that of a simpler tertiary amine like triethylamine, which is 3.0 from Table 19.2.

PROBLEM 19.4

Which of the two nitrogen atoms of chlorpromazine (Thorazine), an antipsychotic drug, is the more basic?

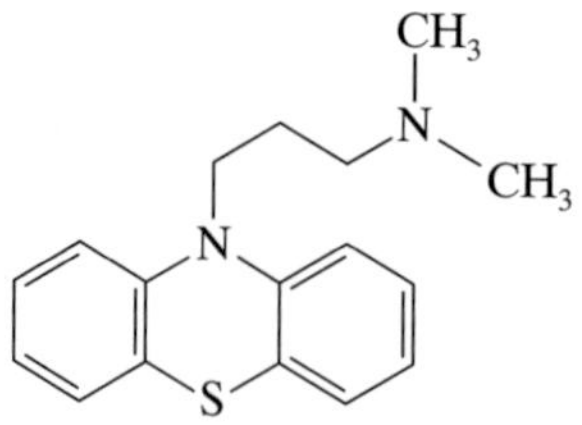

CH_3
N
CH_3
N
S

Amines Dissolve in Acid Solution

When an amine is added to a solution of a strong acid, such as hydrochloric acid, the amine nitrogen atom is protonated to produce an ammonium salt.

$$CH_3NH_2 + HCl \longrightarrow CH_3NH_3^+ + Cl^-$$

The ammonium salts are soluble in water if the hydrocarbon portion of the amine is small. Drugs containing an amino group are often prepared as ammonium salts to improve their solubility in body fluids. For example, the solubility of procaine (Novocain) is only 0.5 g/100mL, but the solubility of its ammonium salt is 100 g/100mL.

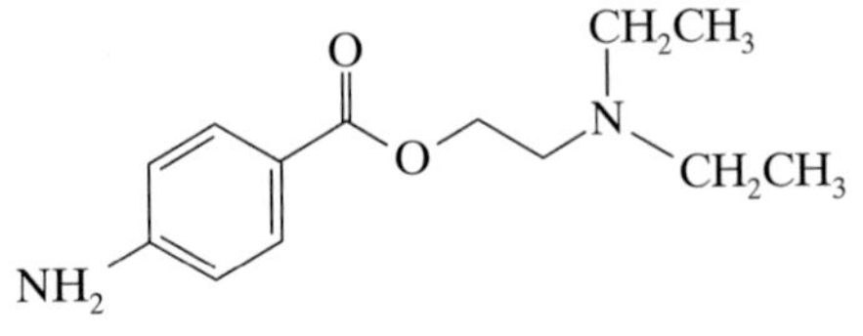

procaine

The ammonium salts have higher melting points than the amines and have virtually no odor. For example, ephedrine melts at 79°C and has a fishy odor. Its hydrochloride salt, used in cold and allergy medications, melts at 217°C and has no odor.

Amines can be separated from other substances by converting them to ammonium salts. Consider, for example, the separation of octane from 1-octanamine. Both compounds are insoluble in water. Addition of HCl converts the 1-octanamine into its ammonium salt, whereas octane is not affected.

$$\underset{\text{(insoluble in water)}}{CH_3(CH_2)_6CH_2NH_2} + HCl \longrightarrow \underset{\text{(soluble in water)}}{CH_3(CH_2)_6CH_2NH_3^+ + Cl^-}$$

$$\underset{\text{(insoluble in water)}}{CH_3(CH_2)_6CH_3} + HCl \longrightarrow \text{no reaction}$$

The octane is separated from the aqueous acid solution. Then the acid solution is neutralized with sodium hydroxide to form the free amine, which is insoluble and can be separated from the aqueous solution.

$$\underset{\text{(soluble in water)}}{CH_3(CH_2)_6CH_2NH_3^+} + OH^- \longrightarrow \underset{\text{(insoluble in water)}}{CH_3(CH_2)_6CH_2NH_2} + H_2O$$

19.6 Quaternary Ammonium Salts

Quaternary ammonium salts have four alkyl or aryl groups bonded to nitrogen. Some quaternary ammonium salts containing a long carbon chain are **invert soaps.** Invert soaps differ from soaps and detergents in that the polar end of the ion has a positive rather than negative charge.

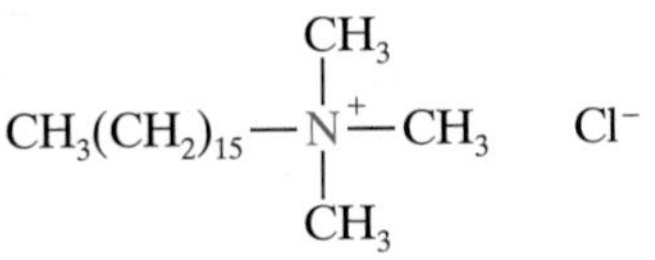

$$CH_3(CH_2)_{15}-\overset{+}{N}(CH_3)_3 \quad Cl^-$$

cetyltrimethylammonium chloride

$$CH_3(CH_2)_{14}CH_2-\overset{+}{N}(CH_3)_2-CH_2-C_6H_5 \quad Cl^-$$

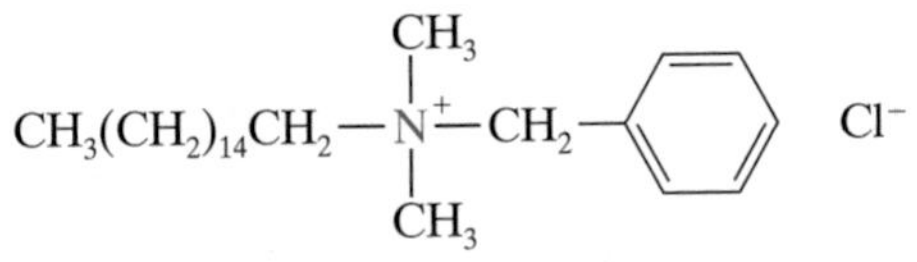

benzalkonium chloride

Invert soaps such as cetyltrimethylammonium chloride are used in hospitals for their bactericidal properties. They are active against bacteria, fungi, and protozoa, but they are not effective against spore-forming microorganisms. One type of in-

Muscle Relaxants

Choline is a quaternary ammonium ion and a component of lipids found in cell membranes (Section 22.7). Acetylcholine, an acetyl ester of its primary hydroxyl group, is a neurotransmitter. When a nerve impulse is transmitted, the nerve cell secretes acetylcholine, which diffuses across the synapse between nerve and muscle cells and bonds to a receptor protein in the muscle cell membrane, triggering contraction. The acetylcholine must be removed to return the muscle to the relaxed state. The enzyme acetylcholinesterase catalyzes the hydrolysis of the ester in acetylcholine, and the muscle returns to its original state.

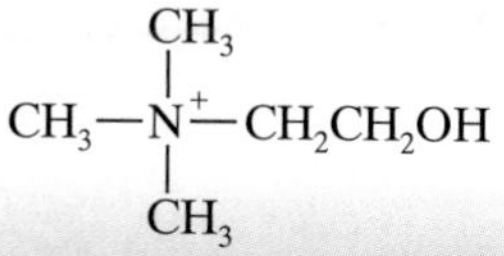

$$CH_3-\overset{CH_3}{\underset{CH_3}{\overset{|}{\underset{|}{N^+}}}}-CH_2CH_2OH \qquad CH_3-\overset{CH_3}{\underset{CH_3}{\overset{|}{\underset{|}{N^+}}}}-CH_2CH_2-O-\overset{O}{\overset{\|}{C}}-CH_3$$

choline 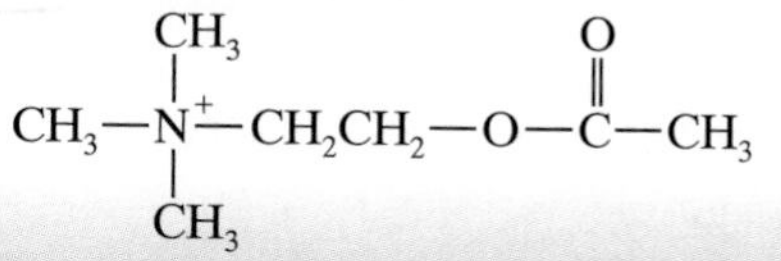 — acetylcholine

vert soap is the family of benzalkonium chlorides. The alkyl groups of these compounds contain from 8 to 16 carbon atoms. These compounds are effective at concentrations of 0.1% to 0.005%. The more complex benzethonium chloride is also an effective antiseptic.

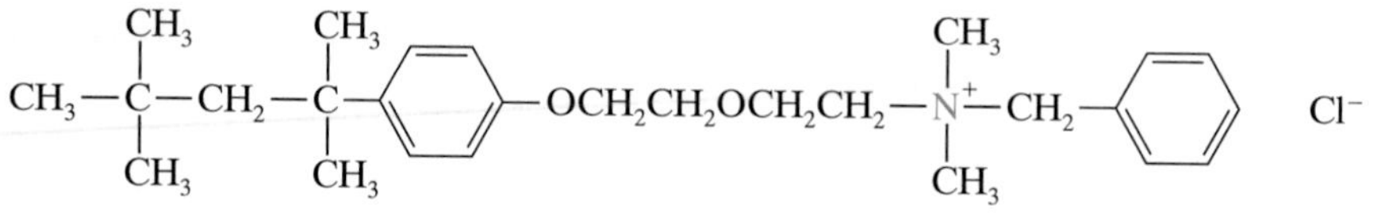

$$CH_3-\overset{CH_3}{\underset{CH_3}{\overset{|}{\underset{|}{C}}}}-CH_2-\overset{CH_3}{\underset{CH_3}{\overset{|}{\underset{|}{C}}}}-C_6H_4-OCH_2CH_2OCH_2CH_2-\overset{CH_3}{\underset{CH_3}{\overset{|}{\underset{|}{N^+}}}}-CH_2-C_6H_5 \quad Cl^-$$

benzethonium chloride

19.7 Reactions of Amines with Nitrous Acid

Nitrous acid (HNO_2) is an unstable compound produced by the reaction of a nitrite salt and a strong acid.

$$H_3O^+ + NO_2^- \longrightarrow H_2O + HNO_2$$

An amine can be classified as primary, secondary, or tertiary by its reaction with nitrous acid. Primary amines produce unstable diazonium salts which decompose to produce nitrogen gas and a mixture of organic products. The evolution of nitrogen gas is a visual confirmation that the amine is primary.

$$R-NH_2 + HNO_2 + H_3O^+ \longrightarrow R-N{\equiv}N^+ + 3\,H_2O$$
$$R-N{\equiv}N^+ \longrightarrow N_2 + \text{organic products}$$

Secondary amines react with HNO_2 to form yellow oily compounds called *N-nitrosoamines.* The yellow oil separates and floats on top of the acid solution.

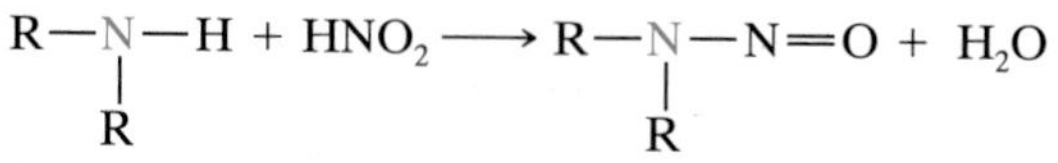

$$R-\underset{R}{\underset{|}{N}}-H + HNO_2 \longrightarrow R-\underset{R}{\underset{|}{N}}-N{=}O + H_2O$$

an *N*-nitrosodialkylamine

Tertiary amines do not react with nitrous acid. As a result, the three classes of amines can be distinguished from one another by using HNO_2. However, this test is no longer done because newer methods using spectroscopy provide more information on the structure of amines.

EXAMPLE 19.5

How can the following isomeric compounds be distinguished using nitrous acid?

(I) N–CH_2CH_3 (II) $NHCH_3$ (III) CH_2NH_2

Solution

Compound III is a primary amine that will give off nitrogen gas when treated with HNO_2. Compound II is a secondary amine and will form a yellow oil with the same reagent. Compound I is a tertiary amine. It will simply dissolve in the HCl solution used to produce HNO_2.

PROBLEM 19.5

There are four isomeric compounds of molecular formula C_3H_9N. Which compounds can be identified on the basis of their reactivity with nitrous acid?

Nitrites and Cancer

Studies of the effects of nitrosoamines on animals indicate that they cause liver cancers. It has been suggested that nitrites added to bacon, hot dogs, and sandwich meats to retard spoilage may cause stomach cancer. In the stomach, the gastric acid (HCl) can react with nitrite ions to form nitrous acid, which can react with the amines present in food to produce nitrosoamines.

The amount of nitrites in the body also depends on environmental factors over which we may have little control. These include water supplies and even some "natural" foods. For example, fresh spinach contains about 5 mg of nitrites per kilogram, and the level can increase to several hundred milligrams per kilogram when spinach is stored in the refrigerator.

There is considerable controversy about the use of nitrites in foods and their possible effect on the human body. It is clear that processed meats must contain preservatives to have a reasonable shelf life. However, how much can we tolerate? The amount of nitrites used in laboratory tests with rats is many factors of 10 higher than that used in commercial food products. Thus, the amount of nitrites ingested by most individuals is small, whereas we are exposed to many other cancer-producing chemicals of more serious and proven concern, such as cigarette smoke. Thus, as in many other aspects of life, there are trade-offs that we have to make. In 1976 the FDA limited the amount of nitrites allowed to preserve meats to a maximum of 125 ppm. However, the rule has been relaxed. At this point the carcinogenic potential of nitrites in meats eaten by humans is considered to be quite low.

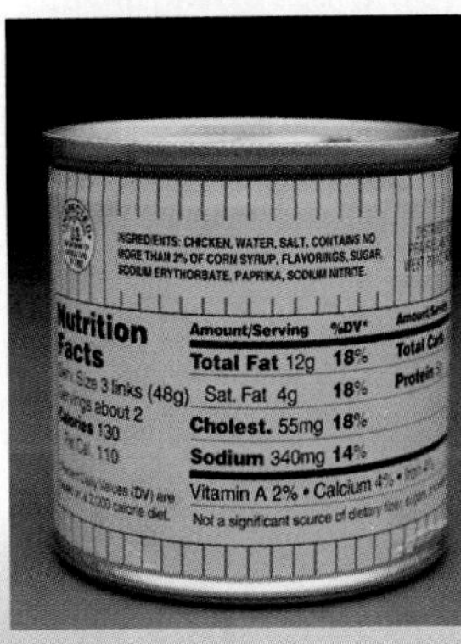

The list of ingredients of processed meats indicates that sodium nitrite is added. (Fundamental Photographs)

19.8 Amides

Amides have an amino group or substituted amino group bonded to a carbonyl carbon atom. The other two bonds of the nitrogen atom may be to hydrogen atoms, alkyl groups, or aryl groups. The amides are classified based on the total number of alkyl or aryl groups bonded to the nitrogen atom, including the acyl group.

$$\underset{\text{a primary amide}}{R{-}\overset{\overset{\displaystyle O}{\|}}{C}{-}NH_2} \qquad \underset{\text{a secondary amide}}{R{-}\overset{\overset{\displaystyle O}{\|}}{C}{-}NHR} \qquad \underset{\text{a tertiary amide}}{R{-}\overset{\overset{\displaystyle O}{\|}}{C}{-}NR_2}$$

The structures of amides resemble those of other carbonyl compounds; the three atoms bonded to carbon are in the same plane. However, there is an important difference in the bonding between the carbonyl carbon atom and the nitrogen atom. The nitrogen atom has an unshared pair of electrons that is delocalized with the π electrons of the carbonyl group (Figure 19.2). Two resonance structures represent formamide.

$$H{-}C({=}\ddot{O}{:}){-}\ddot{N}H{-}H \longleftrightarrow H{-}C({-}{:}\ddot{O}{:}^{-}){=}\overset{+}{N}H{-}H$$

contributing resonance forms of an amide

The bond between carbon and nitrogen is therefore intermediate between a single bond and a double bond. As a consequence, rotation about the carbon–nitrogen bond is somewhat restricted. In Chapter 23 we will study the chemistry of proteins, which contain amide bonds. The restricted rotation about the carbon–nitrogen bond affects the conformation of these compounds.

FIGURE 19.2
Bonding in Formamide

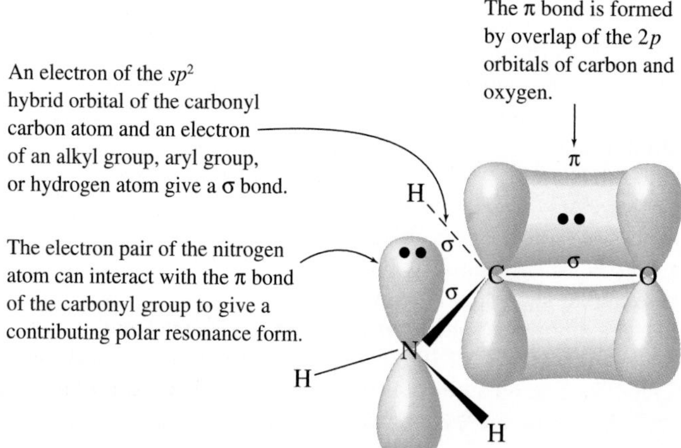

Nomenclature of Amides

The common names of amides are formed by dropping the suffix *-ic* of the related acid and adding the suffix *-amide*. When there is a substituent on the nitrogen atom, the prefix *N*- followed by the name of the group bonded to nitrogen is attached to the name. Substituents on the acyl group are designated by α, β, γ, and so on, as in the case for the common names of acids. In the IUPAC system the final *-e* of the alkane is replaced by *-amide*. The substituents on nitrogen are indicated by the same method as in the common system. However, numbers are used for substituents on the parent chain.

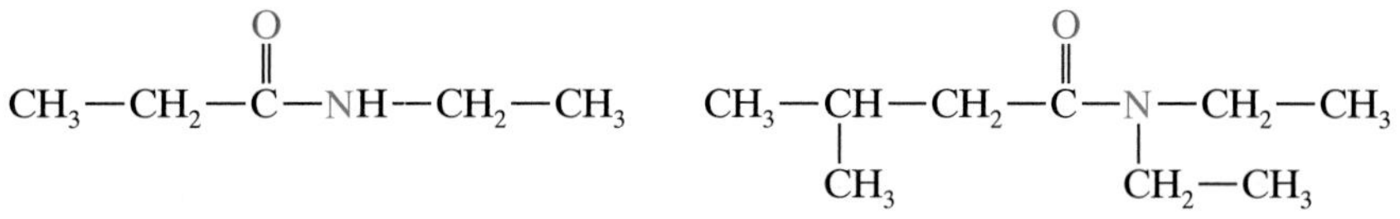

$$CH_3-CH_2-\overset{\displaystyle O}{\overset{\|}{C}}-NH-CH_2-CH_3$$

N-ethylpropanamide
(*N*-ethylpropionamide)

$$CH_3-\underset{\displaystyle CH_3}{\underset{|}{CH}}-CH_2-\overset{\displaystyle O}{\overset{\|}{C}}-\underset{\displaystyle CH_2-CH_3}{\underset{|}{N}}-CH_2-CH_3$$

N,*N*-diethyl-3-methylbutanamide
(*N*,*N*-diethyl-β-methylbutyramide)

Properties of Amides

In contrast with amines, amides are not basic. This difference in basicity is due to the carbonyl group, which draws electrons away from the nitrogen atom. The resulting delocalization of electrons decreases their availability to accept a proton.

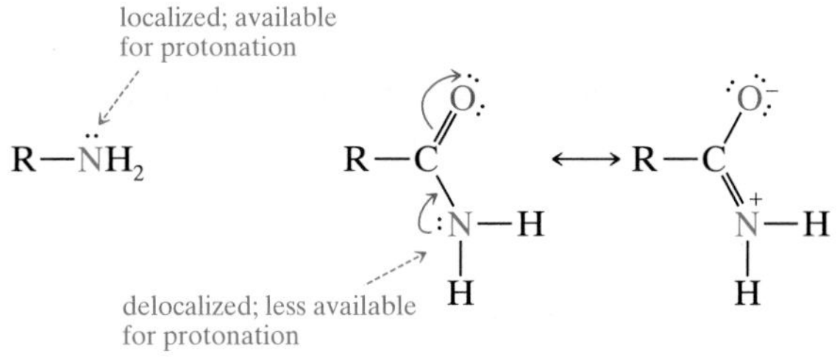

$$R-\ddot{N}H_2 \qquad R-C(=\ddot{O}:)-\ddot{N}H-H \longleftrightarrow R-C(-\ddot{O}:^-)=\overset{+}{N}H-H$$

localized; available for protonation

delocalized; less available for protonation

With the exception of formamide, the unsubstituted amides are solids at room temperature. There is strong intermolecular hydrogen bonding between the amide hydrogen atom and the carbonyl oxygen atom, which is responsible for the high melting and boiling points of primary amides. Substitution of the hydrogen atoms on the nitrogen atom by alkyl or aryl groups reduces the number of possible intermolecular hydrogen bonds and lowers the melting and boiling points. Tertiary amides cannot form intermolecular hydrogen bonds.

$$CH_3-\overset{\displaystyle :\ddot{O}}{\overset{\|}{C}}-\underset{\displaystyle H}{\underset{|}{\ddot{N}}}-H \;\cdots\; :\ddot{O}=C(CH_3)-\ddot{N}H-H$$

hydrogen bond

$$CH_3-\overset{\displaystyle :\ddot{O}}{\overset{\|}{C}}-\underset{\displaystyle H}{\underset{|}{\ddot{N}}}-CH_3 \;\cdots\; :\ddot{O}=C(CH_3)-\ddot{N}H-CH_3$$

hydrogen bond

$$CH_3-\overset{\displaystyle :\ddot{O}}{\overset{\|}{C}}-\underset{\displaystyle CH_3}{\underset{|}{\ddot{N}}}-CH_3 \qquad :\ddot{O}=C(CH_3)-\ddot{N}(CH_3)-CH_3$$

no hydrogen bond possible

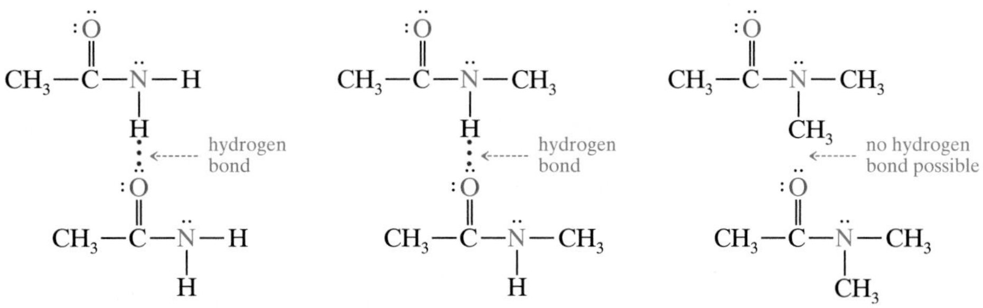

Low molecular weight amides are soluble in water. The solubility is due to hydrogen bonds between the amide and water. Even tertiary amides of low molecular are water soluble because the carbonyl oxygen atom can form hydrogen bonds to the hydrogen atoms of water.

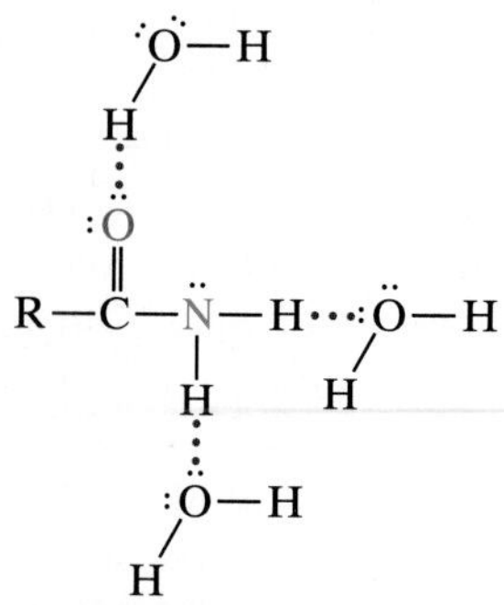

$$\begin{array}{c} H-\ddot{O}-H \cdots :O \\ \| \\ R-C-\ddot{N}-H\cdots:\ddot{O}-H \\ | \quad\quad\quad | \\ H \quad\quad\quad H \\ \vdots \\ :\ddot{O}-H \\ | \\ H \end{array}$$

19.9 Formation of Amides

When heated to a high temperature, carboxylic acids react with ammonia, a primary amine, or a secondary amine to form an amide and water. The reaction occurs by loss of water from the ammonium salt.

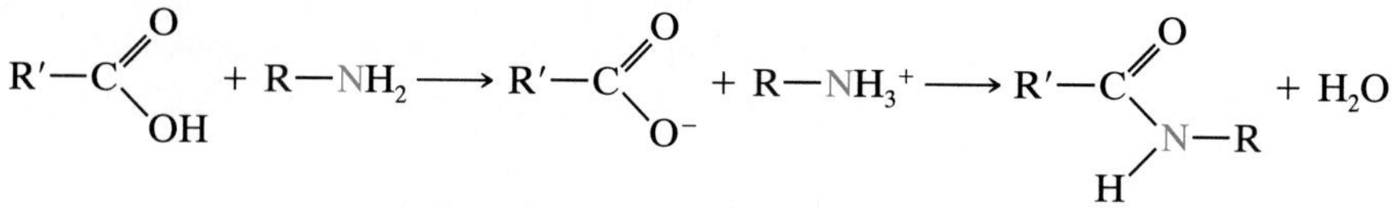

$$R'-C(=O)-OH + R-NH_2 \longrightarrow R'-C(=O)-O^- + R-NH_3^+ \longrightarrow R'-C(=O)-N(H)-R + H_2O$$

Amides can be more easily prepared using the more reactive acid chloride. A base such as pyridine is used to neutralize the HCl formed.

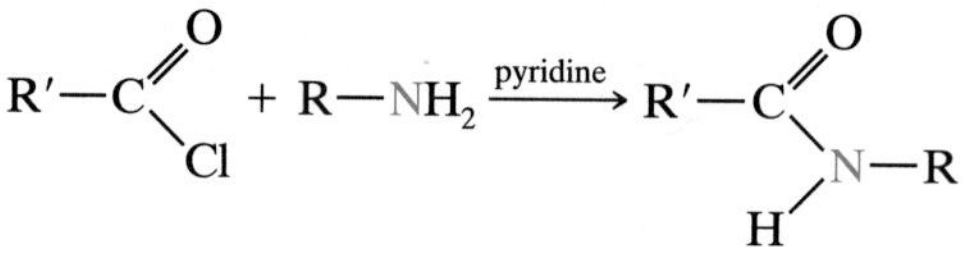

$$R'-C(=O)-Cl + R-NH_2 \xrightarrow{\text{pyridine}} R'-C(=O)-N(H)-R$$

EXAMPLE 19.6

Flecainide, an antiarrhythmic drug used to treat irregular heartbeats, is an amide. What amine and what carboxylic acid are required to form the drug? What other product might be formed?

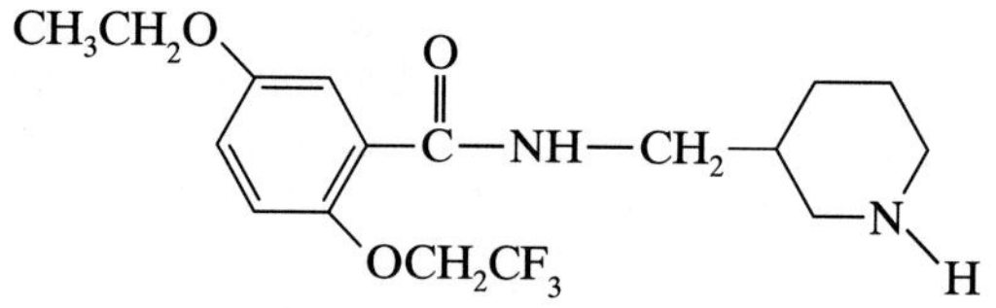

$$CH_3CH_2O-C_6H_3(OCH_2CF_3)-C(=O)-NH-CH_2-(C_5H_9NH)$$

Solution

Mentally separate the amide into two components by breaking the bond between the nitrogen atom and the carbonyl carbon atom. Place a hydrogen atom on the nitrogen atom. Place a hydroxyl group on the carbonyl carbon atom.

$$CH_3CH_2O-C_6H_3(OCH_3CF_3)-C(=O)-OH \qquad NH_2-CH_2-(C_5H_9NH)$$

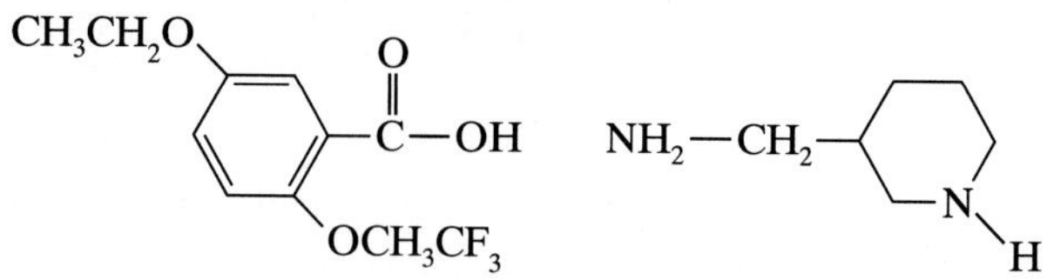

Note that the "amine" is actually a diamine: one is a primary amine, the other a secondary amine. Thus, the diamine could react at either nitrogen atom to form two isomeric amides. The structure of the second isomer is

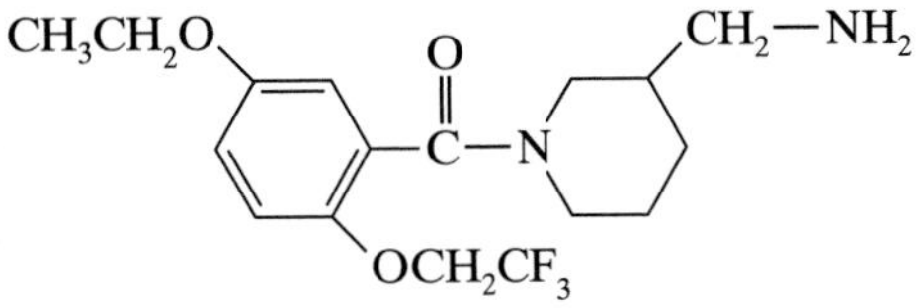

CH_3CH_2O $CH_2—NH_2$ OCH_2CF_3

The primary amine is actually more reactive because it is in a more open structural environment. Primary amines are more reactive than secondary amines.

Polyamides

Many polymers are polyamides. Perhaps the most famous of these is nylon. Synthetic polyamides, such as nylon, are produced from diamines and dicarboxylic acids. One type of nylon is made from adipic acid and hexamethylenediamine (1,6-diaminohexane).

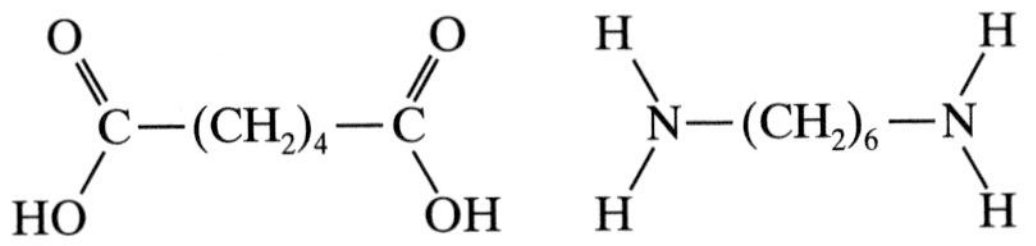

adipic acid

hexamethylenediamine

The product of the first condensation reaction is an amide that also contains a free amino group and a free carboxylic acid group.

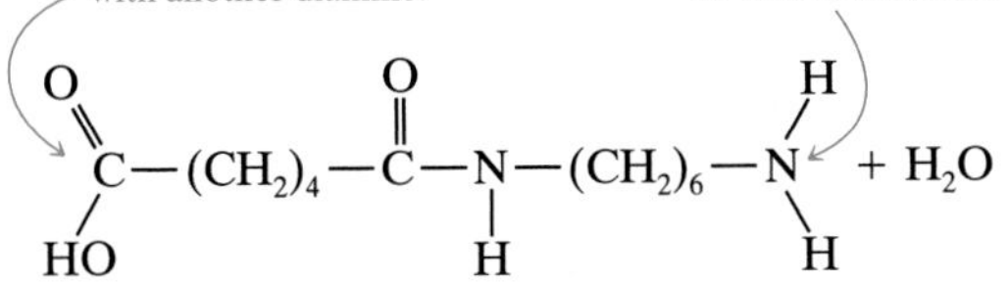

This end can react with another diamine.

This end can react with another diacid.

$+ H_2O$

an amide

The amine end of this molecule reacts with another molecule of adipic acid to produce another amide linkage. The carboxylic acid end of the new molecule then reacts with another molecule of hexamethylenediamine. This sequence of reactions occurs again and again to produce a polyamide.

The nylon formed from adipic acid and hexamethylenediamine is called nylon 66. The "66" refers to the six-carbon diacid and six-carbon diamine reactants.

Polyamides containing aromatic rings (aramides) have many special properties. The presence of aromatic rings in the polymer produces a stiff and tough fiber.

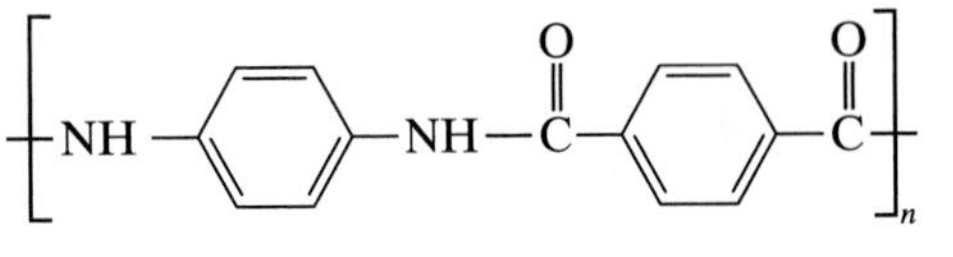

an aramide

One commercially important aramide is Kevlar, a polyamide made from *p*-phenylenediamine and terephthaloyl chloride.

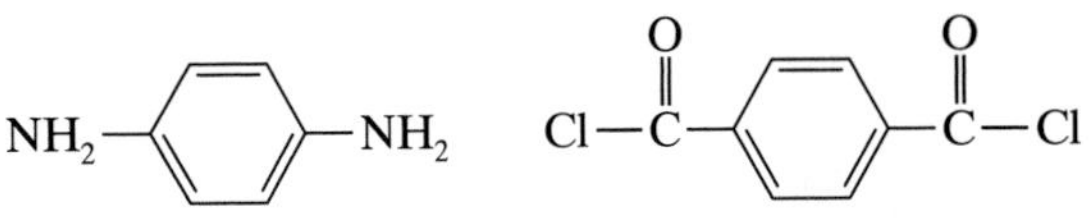

p-phenylenediamine

terephthaloyl chloride

It is used instead of steel in bullet-resistant vests. These vests are so light and flexible that they can be worn inconspicuously under normal clothing.

An aramide called Nomex has a structure resembling that of Kevlar. The monomers in Nomex are meta rather than para isomers. Nomex is used in flame-resistant clothing for fire fighters and race-car drivers. It is so strong that it can be used in flame-resistant building materials.

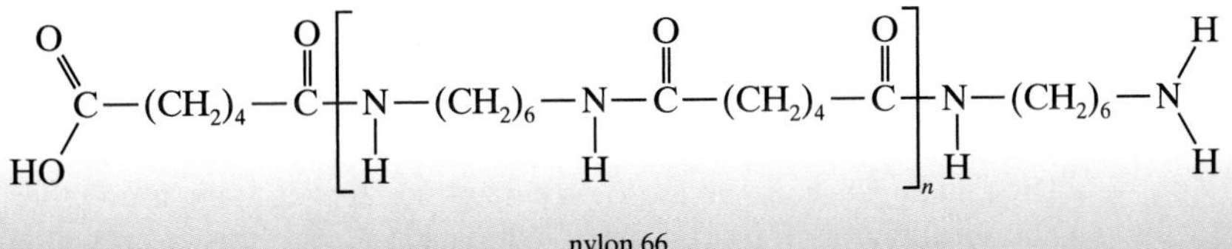

nylon 66

PROBLEM 19.6

What reactants are required to synthesize diethyltoluamide (DEET), a compound used as a mosquito repellent?

$$CH_3-C_6H_4-\overset{O}{\overset{\|}{C}}-N(CH_2CH_3)_2$$

19.10 Hydrolysis of Amides

The hydrolysis of an amide breaks the carbon–nitrogen bond to form an acid and either ammonia or an amine. A similar hydrolysis was presented for esters in Chapter 18. There are, however, important differences. The hydrolysis of esters occurs much more easily than for amides. Amides are hydrolyzed only by heating for hours using a strong base or strong acid. The salt of the carboxylic acid is formed, and a mole of base is required per mole of amide under basic conditions.

$$C_6H_5-\overset{O}{\overset{\|}{C}}-N(H)CH_2CH_3 + OH^- \longrightarrow C_6H_5-\overset{O}{\overset{\|}{C}}-O^- + CH_3CH_2-NH_2$$

The ammonium salt of the amine is formed, and a mole of acid is required per mole of amide under acidic conditions.

$$C_6H_5-\overset{O}{\overset{\|}{C}}-N(H)CH_2CH_3 + H_3O^+ \longrightarrow C_6H_5-\overset{O}{\overset{\|}{C}}-OH + CH_3CH_2-NH_3^+$$

The resistance of amides toward hydrolysis means that proteins will not hydrolyze at physiological pH and at body temperature. However, in the presence of specific enzymes, the hydrolysis of amides is rapid (Chapter 23).

EXAMPLE 19.7

What are the products of the hydrolysis of phenacetin by a base? Phenacetin was formerly used in APC tablets, an analgesic consisting of aspirin, phenacetin, and caffeine.

$$CH_3-\overset{O}{\overset{\|}{C}}-\underset{H}{\underset{|}{N}}-C_6H_4-OCH_2CH_3$$

Solution

The functional group on the right side of the benzene ring is an ether, which is very stable. The functional group on the left is an amide. Hydrolysis of an amide breaks the bond between the nitrogen atom and the carbonyl group. The acid fragment will be acetic acid. The amine fragment will be an aniline containing an ether substituent.

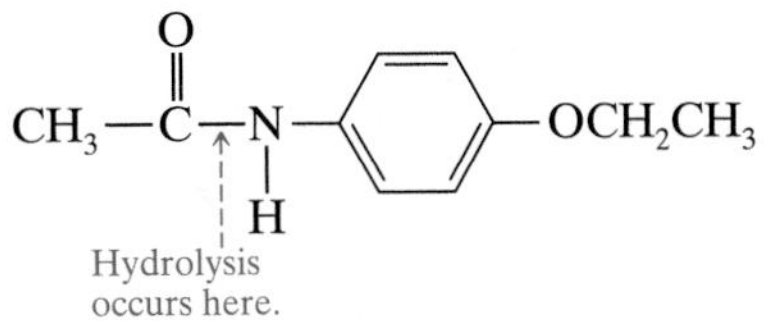

$CH_3-\overset{O}{\overset{\|}{C}}-\underset{H}{\underset{|}{N}}-C_6H_4-OCH_2CH_3$

Hydrolysis occurs here.

Because a base is used in the hydrolysis, the acid product will exist as the acetate ion. The amine is *p*-ethoxyaniline.

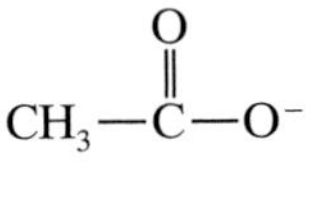

$CH_3-\overset{O}{\overset{\|}{C}}-O^-$ $\quad$ $H-\underset{H}{\underset{|}{N}}-C_6H_4-OCH_2CH_3$

acetate ion 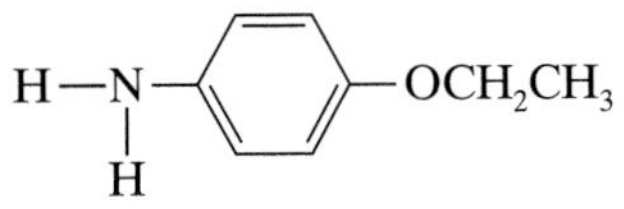 $\quad$ *p*-ethoxyaniline

PROBLEM 19.7

What are the products of the hydrolysis of Mepivicaine, a local anesthetic, under acidic conditions?

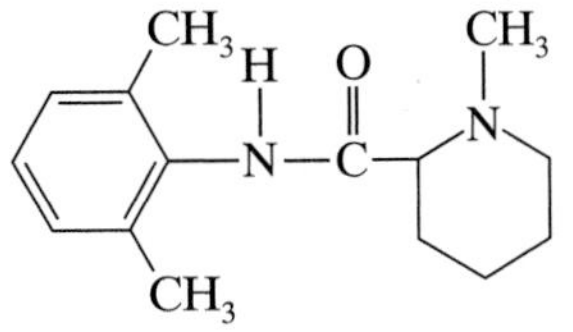

CH_3 $\quad$ H $\quad$ O $\quad$ CH_3

N—C — N

CH_3

Summary

Amines can be viewed as derivatives of ammonia in which one, two, or three hydrogen atoms are replaced by hydrocarbon groups. Amines are classified as **primary, secondary,** or **tertiary** depending on the number of hydrocarbon groups replacing hydrogen atoms. If the nitrogen atom of an amine is one of the atoms of a ring, the compound is called a **heterocyclic amine.**

The common names of amines use the names of the alkyl groups bonded to the nitrogen atom followed by the suffix -amine. The entire name is written as one word. In naming more complex amines, the $—NH_2$ is called the **amino group.** It is treated as a substituent, and its position on a carbon chain or ring must be indicated. Complex secondary and tertiary amines are named using *N*-alkylamino (—NHR) or *N,N*-dialkylamino ($—NR_2$) groups as substituents. The *N*- indicates that the alkyl group is bonded to nitrogen and not to the parent molecule. The largest or most complicated group is used as the parent molecule.

In the IUPAC system the -e ending of the alkane is changed to -amine. Substituents on the carbon chain are designated by number. The prefix *N*- is used for each substituent on the nitrogen atom.

Primary and secondary amines form intermolecular hydrogen bonds, but amines form weaker hydrogen bonds than alcohols. The primary and secondary amines have higher boiling points than the nonpolar hydrocarbons but lower boiling points than alcohols. Tertiary amines cannot form intermolecular hydrogen bonds with each other and have low boiling points.

Amines containing a small number of carbon atoms are soluble in water. These compounds have a fishy odor. Amines are weak bases; they have small **base ionization constants** (K_b) and large **pK_b** values. The reaction of a strong acid with an amine forms an ammonium salt, which is much more soluble in water than the amine. **Quaternary ammonium salts** are ammonium salts that have four alkyl or aryl groups bonded to nitrogen. Some quaternary ammonium salts containing a long carbon chain are **invert soaps.**

The reactions of amines with nitrous acid give distinctly different products. Primary amines form nitrogen gas; secondary amines form oily ***N*-nitroso compounds;** tertiary amines dissolve in the acid solution with no apparent reaction.

An **amide** is derived from a carboxylic acid by replacing the hydroxyl group with an $—NH_2$, —NHR, or $—NR_2$ group. The resulting carbon–nitrogen bond has double-bond character, and the restricted rotation is an important factor in proteins, which consist of amino acids bonded via amide bonds.

Unsubstituted amides have high melting points and boiling points owing to intermolecular hydrogen bonds. Substitution of hydrocarbon groups for hydrogen atoms on nitrogen reduces the number of intermolecular hydrogen bonds and lowers the melting point and boiling point.

Amide bonds may be hydrolyzed by acid or base, but this hydrolysis requires more vigorous conditions than does the hydrolysis of esters.

Exercises

Functional Groups

19.1 Identify all functional groups in acebutolol, an antiarrhythmia drug.

19.2 Identify all functional groups in haloperidol, an antipsychotic drug.

19.3 Classify the nitrogen-containing functional groups in labetalol, an antihypertensive drug.

19.4 Classify the nitrogen-containing functional groups in encainide, an antiarrhythmic drug.

Classification of Amines

19.5 Classify the amine site in methadone, the drug used to treat heroin addicts.

$CH_3{-}CH_2{-}C(=O){-}C(C_6H_5)_2{-}CH_2{-}CH(CH_3){-}N(CH_3){-}CH_3$

19.6 Classify the amine site in the amino acid tyrosine.

19.7 Classify each of the following amines.

(a) $CH_3{-}NH{-}CH_2CH_3$

(b) $CH_3CH_2{-}N(CH_2CH_3){-}CH_2CH_3$

(c) $CH_3CH_2{-}NH{-}CH(CH_3)CH_3$

19.8 Classify each of the following amines.

(a) $CH_3CH_2{-}N(CH_3){-}CH_2CH_2{-}NH_2$

(b) $NH_2{-}CH_2CH_2{-}NH{-}CH_3$

(c) $CH_3{-}NH{-}CH_2CH_2{-}N(CH_3){-}CH_3$

19.9 Classify each of the following amines.

(a) (b) (c) (d)

19.10 Classify each of the following amines.

(a) (b) (c) (d)

Nomenclature

19.11 Give the IUPAC name of each of the following compounds.

(a) $CH_3CH_2CH(NH_2)CH_2CH_2CH_3$

(b) $CH_3CH_2CH_2CH_2{-}N(CH_3){-}CH_3$

(c) $CH_3CH(CH_3)CH_2CH(NH_2)CH_3$

19.12 Give the IUPAC name of each of the following compounds.

(a) $CH_3CH(NH_2)CH_2CH_3$

(b) $CH_3CH(NHCH_3)CH_2CH_2OH$

(c) $CH_3-N(CH_3)-CH(CH_3)CH_2CH_3$

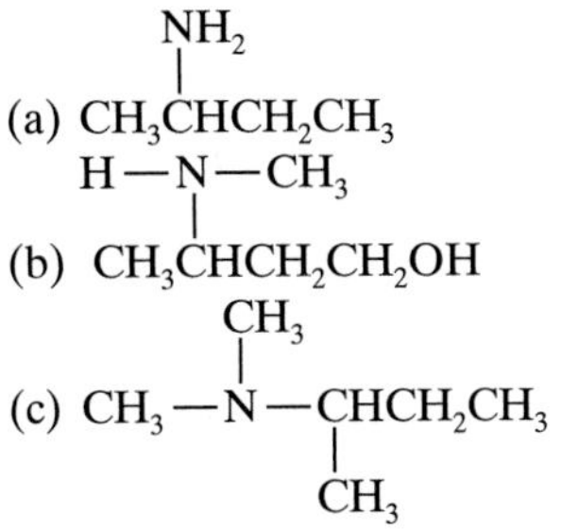

19.13 Give the IUPAC name of each of the following compounds.

(a) 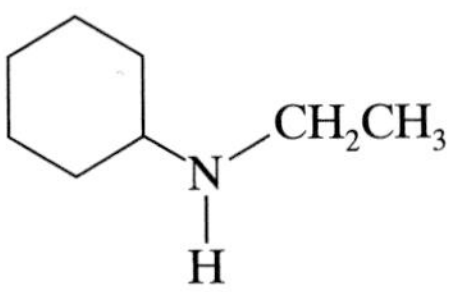 cyclohexyl–NH–CH_2CH_3

(b) 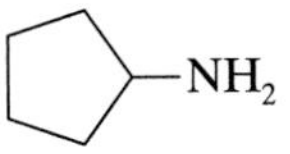 cyclopentyl–NH_2

(c) 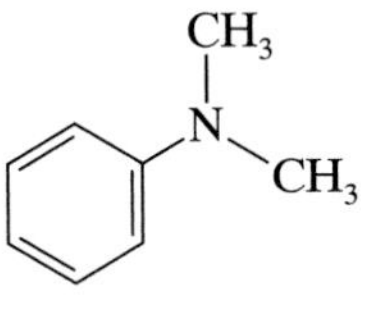 phenyl–$N(CH_3)$–CH_3

(d) 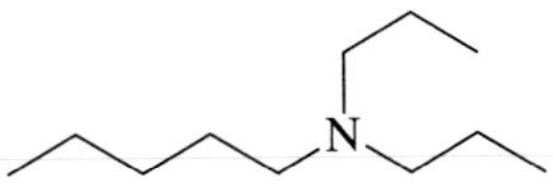 [line-angle structure: N with three chains]

19.14 Give the IUPAC name of each of the following compounds.

(a) 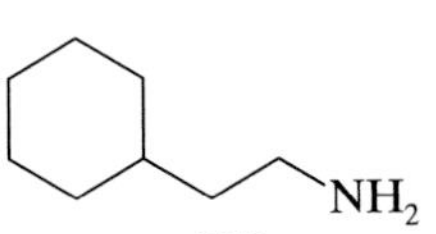 cyclohexyl–CH_2CH_2–NH_2

(b) 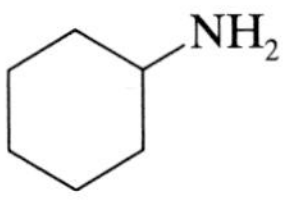 cyclohexyl–NH_2

(c) 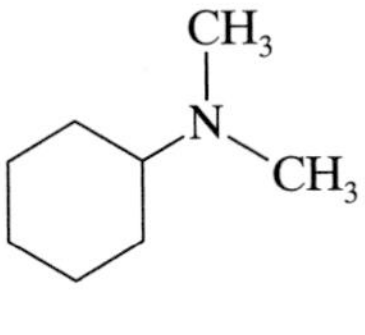 cyclohexyl–$N(CH_3)$–CH_3

(d) 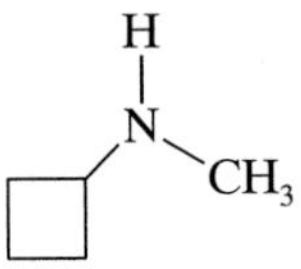 cyclobutyl–NH–CH_3

19.15 Write the structure for *trans*-2-phenylcyclopropylamine, an antidepressant drug.

19.16 Write the structure for *trans*-4-(aminomethyl)cyclohexanecarboxylic acid, a drug that aids blood clotting.

19.17 Write the structure of each of the following compounds

(a) 2-ethylpyrrole (b) 3-bromopyridine
(c) 2,5-dimethylpyrimidine (d) 2,6,8-trimethylpurine

19.18 Name each of the following compounds.

(a) 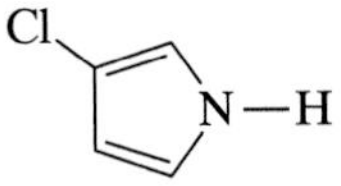 Cl-substituted pyrrole, N—H

(b) 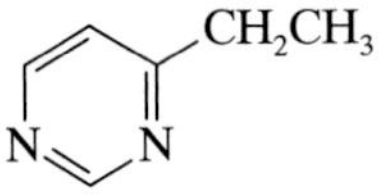 pyrimidine ring with CH_2CH_3

(c) 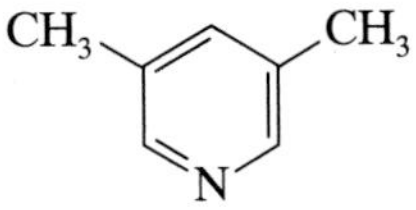 pyridine ring with CH_3 and CH_3

(d) 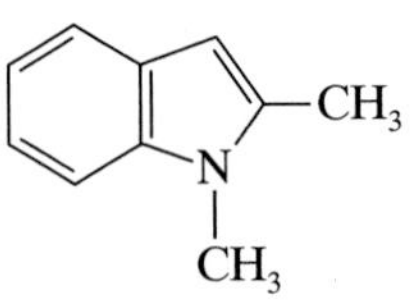 indole ring with CH_3 and N–CH_3

Isomers of Amines

19.19 How many isomers are possible with the molecular formula C_2H_7N?

19.20 How many isomers are possible with the molecular formula C_3H_9N?

19.21 How many isomers are possible for primary amines with the molecular formula $C_4H_{11}N$?

19.22 How many isomers are possible for tertiary amines with the molecular formula $C_5H_{13}N$?

Properties of Amines

19.23 The boiling points of propylamine and trimethylamine are 49 and 3.5 °C, respectively. Explain this difference.

19.24 The boiling point of 1,2-diaminoethane is 116 °C. Explain why this compound boils at a much higher temperature than propylamine (49 °C).

19.25 Give a likely reason why lemon juice, which contains citric acid, is put on fish dishes.

19.26 The tertiary amine *N,N*-dimethylaniline is not soluble in water but does dissolve in dilute HCl. Explain why.

Basicity of Amines

19.27 Estimate the K_b of each of the following.

(a) 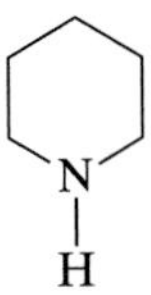 piperidine ring, N–H

(b) 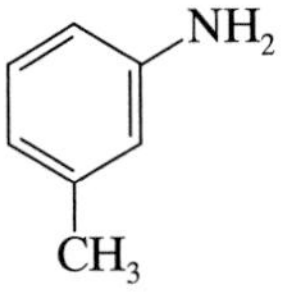 benzene ring with NH_2 and CH_3

(c) 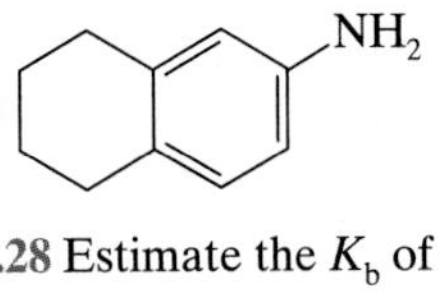 tetrahydronaphthalene ring with NH_2

(d) 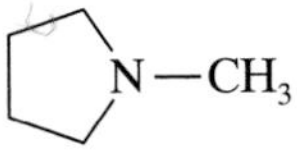 pyrrolidine ring, N—CH_3

19.28 Estimate the K_b of each of the following.

(a) 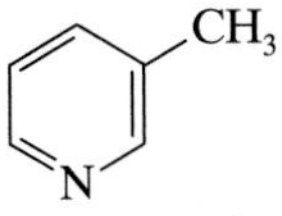 pyridine ring with CH_3

(b) 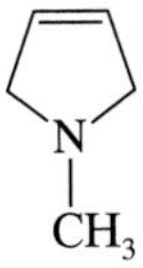 dihydropyrrole ring, N–CH_3

(c) NH_2 (d) N—CH_3

19.29 The pK_b values for cyclohexylamine and triethylamine are 3.34 and 2.99, respectively. Which compound is the stronger base?

19.30 The K_b values for dimethylamine and diethylamine are 4.7×10^{-4} and 3.1×10^{-4}, respectively. Which compound is the stronger base?

Reaction of Amines with Nitrous Acid

19.31 A compound $C_4H_{11}N$ reacts with nitrous acid to yield a yellow oil. What structures are possible for $C_4H_{11}N$?

19.32 A compound $C_4H_{11}N$ reacts with nitrous acid to yield nitrogen gas. What structures are possible for $C_4H_{11}N$?

19.33 What reaction will occur when each compound in Exercise 19.11 reacts with nitrous acid?

19.34 What reaction will occur when each compound in Exercise 19.12 reacts with nitrous acid?

Nomenclature of Amides

19.35 Give the IUPAC name of each of the following compounds.

(a) $CH_3CH_2C(=O)-N(CH_3)-CH_3$

(b) $H-C(=O)-N(CH_3)-CH_2CH_3$

(c) $CH_3CH(CH_3)CH_2C(=O)-N(H)-CH_3$

(d) $C_6H_5-C(=O)-N(H)-CH_3$

19.36 Give the IUPAC name of each of the following compounds.

(a) $CH_3CH_2C(=O)-NH_2$ (b) $CH_3CH_2C(=O)-NHCH_2CH_3$

(c) $CH_3CH_2CH(CH_3)C(=O)-NH_2$

(d) $CH_3CH_2CH_2C(=O)-N(CH_3)_2$

Properties of Amides

19.37 At what sites can acetamide form hydrogen bonds to water?

19.38 Explain why *N,N*-dimethylformamide is soluble in water.

19.39 The boiling points of methyl ethanoate and propanamide are 57°C and 213°C, respectively. Explain why these two compounds with similar molecular weights have such different boiling points.

19.40 The boiling points of *N,N*-dimethylformamide and propanamide are 153°C and 213°C, respectively. Explain why these two isomeric compounds have such different boiling points.

Hydrolysis of Amides

19.41 Write the products of each of the following reactions.

(a) $C_6H_5-C(=O)-N(H)-CH_2CH_3 \xrightarrow{OH^-}$

(b) $CH_3CH_2-C(=O)-N(CH_3)_2 \xrightarrow{H_3O^+}$

(c) $H_2N-C(=O)-C_6H_4-C(=O)-NH_2 \xrightarrow{H_3O^+}$

19.42 Write the products of each of the following reactions.

(a) $CH_3CH_2-C(=O)-NHCH_3 \xrightarrow{H_3O^+}$

(b) $C_6H_5-C(=O)-N(H)-CH_3 \xrightarrow{OH^-}$

(c) $C_6H_5-C(=O)-NH_2 \xrightarrow{H_3O^+}$

Synthesis of Amides

19.43 Write the structure of the amine and carboxylic acid required to form the amides contained in each of the following compounds.

(a) acetaminophen, an analgesic

HO—C_6H_4—N(H)—C(=O)—CH_3

(b) nubucaine, a local anesthetic

O=C(NH—$CH_2CH_2N(CH_2CH_3)_2$) attached to naphthalene ring bearing $OCH_2CH_2CH_2CH_3$

19.44 Write the structure of the amine and carboxylic acid required to form the amides contained in each of the following compounds.

(a) DEET, an insect repellent

CH_3—C_6H_4—C(=O)—N(CH_2CH_3)(CH_2CH_3)

(b) crotamiton, used to treat scabies

(CH_3)$_2C_6H_3$—NH—C(=O)— piperidine ring, N—$CH_2CH_2CH_2CH_3$

Polyamides

19.45 Nylon is resistant to dilute acids or bases, but polyesters are damaged by acids or bases. Explain this difference.

19.46 Draw a representation of the condensation polymer formed by adipic acid and 1,3-diaminopropane.

All life forms from the very small to the large are based on common chemical patterns. (The Image Bank)

20 The Organization of Biochemistry

20.1 What Is Biochemistry?
20.2 Composition and Structure of Biochemicals
20.3 Energy and Biochemical Reactions
20.4 Types of Cells
20.5 The Eukaryotic Cell
20.6 Stereoisomers
20.7 Molecular Chirality
20.8 Optical Activity
20.9 Fischer Projection Formulas
20.10 Multiple Chiral Centers
20.11 How Are Stereoisomers Formed?

Essay
- *Chirality and the Senses*

20.1 What Is Biochemistry?

Biochemistry is the study of the composition, structure, and reactions of substances in living systems—in other words, the study of the molecular basis of life. In your everyday life you might see only a few of the astonishing number of plants and animals. There are over a million species of living organisms, and not all are found in your local environment. Some are found only in rain forests and others only in tropical waters. However, there are many more species than you think even on your college campus when you include insects and earthworms as well as microscopic organisms.

In spite of the diversity of life, from bacteria in the saliva of your mouth to whales of the ocean, there are repeated patterns of both molecular structure and chemical reactivity. The same classes of compounds create the structures of all organisms. Several classes of biomolecules are presented in this chapter and are discussed in detail in subsequent chapters.

In this chapter we will examine the cell—the smallest unit of life—and its subcellular structures called *organelles.* These cells are organized to form tissues. A variety of tissues constitute organs, which in turn function as cooperative units in the organism. All cells contain four classes of biomolecules—carbohydrates, proteins, lipids, and nucleic acids. Each class of biomolecules contains functional groups whose chemistry you have studied in earlier chapters. Thus, many of the reactions of biomolecules will be of functional groups with which you are already familiar.

Living systems and the ways in which they differ from inanimate matter have long fascinated chemists. The transmission of nerve impulses or the mechanism of vision may appear to be very different from the rusting of iron. However, reactions in living matter obey the same chemical laws as those that occur in any other type of matter. Many series of reactions occur in life processes, but each reaction can be understood based on the chemistry of organic compounds. In essence, biochemical reactions are just examples of organic reactions that occur in cells. One important feature of biochemical reactions is catalysis by enzymes, which is discussed in Chapter 24.

Living organisms extract and use energy from food to maintain their structure, to grow, and to reproduce. The method of using energy is a distinctive characteristic of living organisms. Energy is extracted, stored, and used in the form of chemical energy contained in adenosine triphosphate (ATP). This process is discussed in Chapter 25. Chapters 26–28 examine specific metabolic reactions of carbohydrates, lipids, and proteins.

Much of the excitement of biochemistry in the 1960s came from the discovery of the structure of deoxyribonucleic acid (DNA) and the evaluation of the genetic code. All forms of life, from the simplest bacterium to the highest life forms, store and transmit genetic information in the form of DNA. Chapter 29 is devoted to this subject.

20.2 Composition and Structure of Biochemicals

The human body contains only 10 elements in amounts greater than 0.1% (Table 20.1). Carbon, hydrogen, and oxygen, which constitute 93% of the body, are in covalent molecules, whereas sodium, potassium, calcium, and chlorine are present as ions in "inorganic" substances.

Abundance alone does not indicate the importance of the elements to life. Iron is a part of hemoglobin, which is responsible for carrying oxygen in the blood. Iodine is vital to the proper functioning of the thyroid gland. Cobalt is contained in vitamin B_{12}. Zinc, copper, and manganese are present in small amounts in some enzymes. In each of these compounds there are thousands of carbon, hydrogen, oxygen, and nitrogen atoms for every metal atom. Nevertheless, without the metal the compounds could not serve their biological function.

TABLE 20.1 Composition of the Human Body

Element	Percent	Element	Percent
oxygen	65	sulfur	0.2
carbon	18	sodium	0.1
hydrogen	10	chlorine	0.1
nitrogen	3	magnesium	0.05
calcium	2	iron	trace
phosphorus	1.1	iodine	trace
potassium	0.3	bromine	trace

TABLE 20.2 Composition of an *E. coli* Cell

Component	Percentage by Weight	Number of Types of Components
water	70	1
proteins	15	3000
carbohydrates	3	50
lipids	2	40
nucleic acids		
DNA	1	1
RNA	6	1000
simple organic compounds	2	500
inorganic ions	1	12

Water is the most abundant compound in living organisms. Human tissue is about 75% water; the *E. coli* cell is about 70% water (Table 20.2). After water the four most abundant classes of compounds in life forms are proteins, carbohydrates, lipids, and nucleic acids.

The Modular Nature of Biomolecules

Large biomolecules are not assembled one atom at a time but are built up of smaller molecules or fragments. Life processes that produce proteins, carbohydrates, lipids, and nucleic acids assemble modular subunits containing many atoms in a way analogous to the assembly of monomers to form polymers. In fact, the manufacture of polymers such as polyamides and polyesters is a result of our imitating nature.

The most common way that biological molecules are assembled is through condensation reactions. Subunits are joined one to another with the concurrent loss of water. Biological molecules often react to form their constituent subunits by hydrolysis reactions.

Carbohydrates

Carbohydrates are polyhydroxy aldehydes or ketones or compounds that can be hydrolyzed to produce polyhydroxy aldehydes or ketones. The simplest subunit of a carbohydrate is a **monosaccharide**—it cannot be hydrolyzed to form simpler molecules. Glyceraldehyde is the simplest monosaccharide. More complex monosaccharides have four to seven carbon atoms, usually with a hydroxyl group located on each additional carbon atom. Relatively few monosaccharides account for the majority of the molecules in this class of biomolecules.

$$\begin{array}{c} CHO \\ | \\ H-C-OH \\ | \\ CH_2OH \end{array}$$

glyceraldehyde

You already know that alcohols can be oxidized and can form esters and that aldehydes can be oxidized or reduced. Thus, based on the information in Chapters 17 and 18, you are prepared for the study of monosaccharides presented in Chapter 21. For example, you know that aldehydes and ketones form acetals and ketals. It is this type of linkage that holds monosaccharides together in more complex carbohydrates termed *disaccharides, trisaccharides,* and even *polysaccharides.*

Hydrolysis of acetal or ketal bonds gives the component alcohol and carbonyl compound.

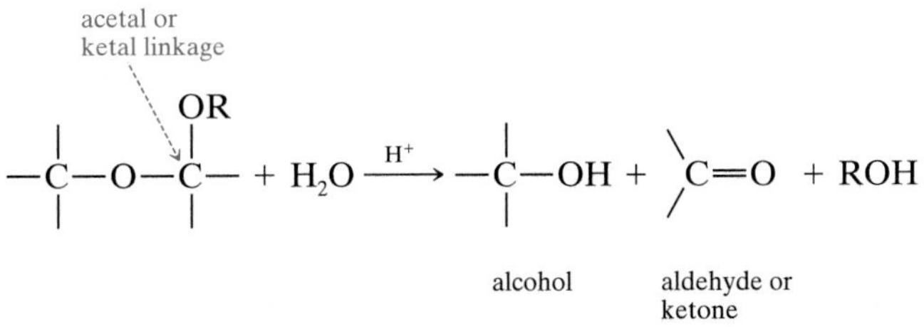

acetal or ketal linkage

$$-\overset{|}{\underset{|}{C}}-O-\overset{OR}{\overset{|}{\underset{|}{C}}}- + H_2O \xrightarrow{H^+} -\overset{|}{\underset{|}{C}}-OH + \;\rangle C{=}O + ROH$$

alcohol — aldehyde or ketone

We will learn in Chapter 21 that the alcohol functional group is actually part of another monosaccharide unit that also has a carbonyl functional group. In addition, the aldehyde or ketone also has alcohol functional groups. Thus, polysaccharides are polymers containing acetal or ketal linkages that hold many monosaccharide units together.

Proteins

Proteins are polyamides of amino acids, which are molecules that contain both a carboxylic acid group and an amino group. Alanine is one of the amino acids found in proteins.

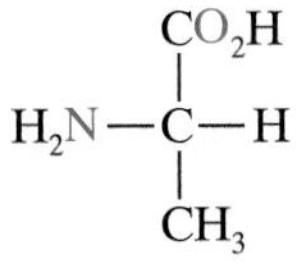

$$H_2N-\overset{CO_2H}{\overset{|}{\underset{\underset{CH_3}{|}}{C}}}-H$$

alanine

There are about 20 amino acids that are commonly found in the proteins of virtually all animals. Some proteins are enzymes that catalyze almost all of the reactions inside cells. Others serve as structural material as found in skin, hair, and horns. Additional roles of proteins are discussed in Chapter 23.

The larger number of types of subunits of proteins as compared with carbohydrates gives proteins considerably more structural diversity as well as function. There are almost as many amino acids as there are letters in the alphabet. We know that thousands of words can be formed from the alphabet, so we might anticipate that a lot of proteins are possible.

Even though they are difficult to hydrolyze, all proteins can be broken down into smaller molecules and eventually into their component amino acids by strong acid.

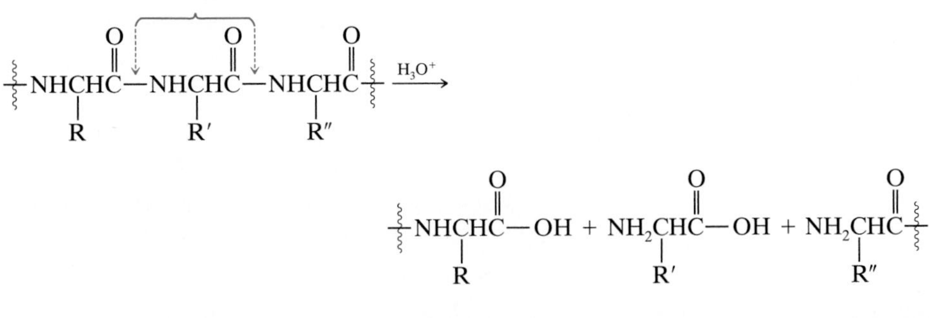

amide bonds

$$\sim NH\underset{\underset{R}{|}}{C}H\overset{O}{\overset{\|}{C}}-NH\underset{\underset{R'}{|}}{C}H\overset{O}{\overset{\|}{C}}-NH\underset{\underset{R''}{|}}{C}H\overset{O}{\overset{\|}{C}}\sim \xrightarrow{H_3O^+}$$

$$\sim NH\underset{\underset{R}{|}}{C}H\overset{O}{\overset{\|}{C}}-OH + NH_2\underset{\underset{R'}{|}}{C}H\overset{O}{\overset{\|}{C}}-OH + NH_2\underset{\underset{R''}{|}}{C}H\overset{O}{\overset{\|}{C}}\sim$$

This reaction occurs more readily when catalyzed by enzymes, which are required for digestion of protein-containing foods.

Lipids

The three classes of compounds in the diet—proteins, carbohydrates, and fats—are also important classes of compounds in biochemistry. Although *proteins* and *carbohydrates* are familiar words, the term *lipids* is less well known. **Lipids** are a class of compounds that includes fats and oils.

Fats are esters of glycerol and long-chain fatty acids such as stearic acid. The chemistry of these substances is based on the reactions given in Chapter 18. Fats are hydrolyzed much more easily than proteins.

$$\begin{array}{l} CH_2{-}O{-}\overset{O}{\overset{\|}{C}}{-}(CH_2)_{16}CH_3 \\ | \\ CH{-}O{-}\overset{O}{\overset{\|}{C}}{-}(CH_2)_{16}CH_3 \\ | \\ CH_2{-}O{-}\overset{O}{\overset{\|}{C}}{-}(CH_2)_{16}CH_3 \end{array} + 3\,H_2O \xrightarrow{H^+} \begin{array}{l} CH_2{-}OH \\ | \\ CH{-}OH \\ | \\ CH_2{-}OH \end{array} + 3\,HO{-}\overset{O}{\overset{\|}{C}}{-}(CH_2)_{16}CH_3$$

a fat — glycerol — stearic acid

Phosphoric acid also forms esters with glycerol as part of the structure of phospholipids (Chapter 22). The phosphate group is also bonded to a small, often nitrogen-containing alcohol.

$$\begin{array}{l} CH_2{-}O{-}\overset{O}{\overset{\|}{C}}{-}(CH_2)_{16}CH_3 \\ | \\ CH{-}O{-}\overset{O}{\overset{\|}{C}}{-}(CH_2)_{16}CH_3 \\ | \\ CH_2{-}O{-}\overset{O}{\overset{\|}{P}}{-}O{-}CH_2CH_2NH_2 \\ \qquad\quad\;\; | \\ \qquad\quad\; OH \end{array} + 4\,H_2O \xrightarrow{H^+}$$

a phospholipid

$$\begin{array}{l} CH_2{-}OH \\ | \\ CH{-}OH \\ | \\ CH_2{-}OH \end{array} + 2\,HO{-}\overset{O}{\overset{\|}{C}}{-}(CH_2)_{16}CH_3 + H_3PO_4 + HO{-}CH_2CH_2NH_2$$

Phospholipids are contained in cell membranes, but they are not joined together by covalent bonds. Rather, they associate by London attractive forces to give a flexible structure that allows diffusion and transport of materials across the membrane.

Nucleic Acids

Nucleic acids are the central biomolecules that spell out all the directions for control, growth, and duplication of cells. **Nucleic acids** are polymers of nucleotides, which in turn consist of a five-carbon carbohydrate, a nitrogen-containing base, and a phosphate group (Chapter 29). The carbohydrate is ribose in ribonucleic acids

(RNA) and deoxyribose in deoxyribonucleic acids (DNA). Adenine, guanine, cytosine, and thymine are the bases in DNA. Adenine, guanine, cytosine, and uracil are the bases in RNA.

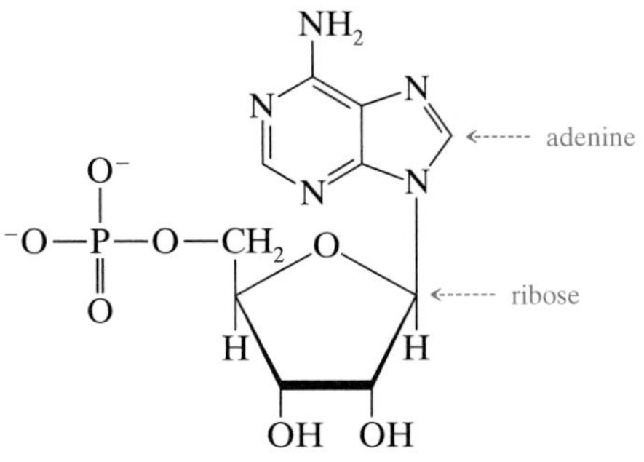

a nucleotide

20.3 Energy and Biochemical Reactions

Biochemists use the general term **metabolism** to describe all the reactions that occur in a living organism. The molecules that are absorbed by the body may be either built into more complex molecules or broken down into simpler molecules. The intermediate compounds of metabolic reactions are **metabolites.**

Anabolic reactions (anabolism) are metabolic reactions that build (synthesize) more complex molecules and require energy. **Catabolic reactions** (catabolism) are reactions that break down (degrade) larger molecules into smaller molecules and release energy. Both anabolic and catabolic reactions constantly occur in living organisms to maintain cellular functions.

Most anabolic and catabolic reactions are catalyzed by enzymes that make reactions go fast enough to support the cell. Many enzymes are entirely protein, whereas others consist of protein and another substance called a *coenzyme.*

A compound contains a certain amount of stored free energy (Section 10.4) in its bonds. When a reaction occurs, the product may contain more or less stored free energy than the reactants. If the product has lower free energy, the ΔG° for the reaction is negative; the reaction is **exergonic.** If the product has a higher free energy, the ΔG° for the reaction is positive; the reaction is **endergonic.**

The energy balance of living organisms is provided by adenosine triphosphate (ATP) and its related compound adenosine diphosphate (ADP). Energy released in catabolic reactions is stored to provide energy for anabolic reactions by the conversion of ADP into ATP (Chapter 25).

The energy released by the catabolic reaction is saved in ATP, much as money earned is deposited in a bank until needed. The energy required for an anabolic reaction at some other site in a cell is supplied by the conversion of ATP into ADP as the organism draws on its bank account of energy.

Energy is constantly being produced and used in living systems. Endergonic reactions occur only because exergonic reactions also occur to provide the energy to make the unfavorable process possible. For example, the ATP formed in the metabolism of glucose may be used to synthesize proteins from amino acids in another part of the cell (Figure 20.1). The overall combination of reactions is exergonic, and the remaining energy is released as heat.

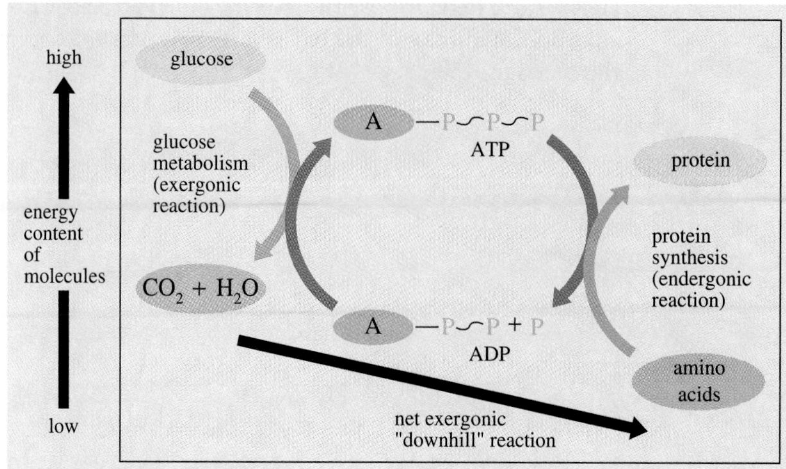

FIGURE 20.1 Reactions and ATP in Cells
The exergonic reactions of glucose metabolism supply ATP used to drive endergonic reactions such as synthesis of proteins. More energy is released by the exergonic reaction than is used by the endergonic reaction. As a result the extra energy is released as heat. (Taken from Biology by Audesirk—page 78)

EXAMPLE 20.1

The hydrolysis of phosphoenolpyruvate has $\Delta G^\circ = -14.8$ kcal/mole. What is the ΔG° for the reverse of this reaction, in which phosphoenolpyruvate is formed?

$$\mathrm{CH_2{=}C(CO_2H){-}O{-}P(=O)(OH){-}OH + H_2O \longrightarrow CH_3{-}C(=O){-}CO_2H + H_3PO_4} \qquad \Delta G^\circ = -14.8 \text{ kcal/mole}$$

Solution

Since energy can be neither created nor destroyed in chemical reactions, the energy released by a reaction must be gained in the reverse reaction. Thus, the formation of phosphoenolpyruvate requires energy, and $\Delta G^\circ = +14.8$ kcal/mole.

PROBLEM 20.1

The free energy content of the hydrolysis products of creatine phosphate is lower than that of the reactants by 10.3 kcal/mole. What is ΔG° for the reaction?

20.4 Types of Cells

The chemical reactions of an organism occur within cells, which are the basic units of life. Cells are small—ranging from 1 to 100 μm. Some organisms such as bacteria consist of one cell, but most life forms consist of aggregates of cells that act cooperatively. All cells have at least a plasma membrane, genetic material, and cytoplasm. Cells can be divided into two classes: eukaryotic and prokaryotic. Eukaryotic comes from the Greek *eu* for "true" and *karyon* for "nucleus." **Eukaryotes** are cells that contain a membrane-bound nucleus and other membrane-bound components of the cell called *organelles.* Fungi, plants, and animals are eukaryotic organisms. The most widely studied eukaryotic cell is the liver parenchymal cell because of its availability and the ease with which it can be separated into cell components.

Prokaryotes do not have either a membrane-bound nucleus or other internal membrane-bound organelles. The term prokaryote (from the Greek *pro,* "before") indicates that biologists feel that these cells existed before the more complex eukaryotes. The bacterium *Escherichia coli (E. coli)* is the best-studied example of a prokaryotic cell. It is about 0.5 μm in diameter and 1.5 μm long.

The Prokaryotic Cell

Most prokaryotic cells are less than 5 μm in length. They have a cell wall that provides protection against osmotic pressure changes. Without a cell wall, which also provides mechanical strength, water from the medium in which the cell is located could enter the cell, causing it to swell and burst. Inside the cell wall is the **cell membrane,** which is a few molecules thick and consists of lipids and proteins. The cell membrane is selectively permeable and is responsible for controlling the passage of nutrients into the cell and waste products out of the cell.

The part of the cell enclosed within the cell membrane is the **cytoplasm,** which consists of the cytosol and various suspended units, such as **ribosomes,** the factories for protein production, and the nucleoid or **nuclear region,** which controls the growth and reproduction of the cell. Both the ribosomes and nuclear region of prokaryotes are structurally different from the ribosomes and nucleus of eukaryotes. These differences are exploited in the treatment of bacterial infections. Some antibiotics function by disrupting the protein-producing functions of the ribosomes of prokaryotic bacteria. These agents are designed not to affect the ribosomes of eukaryotic host organisms.

The Eukaryotic Cell

Eukaryotic cells are larger than prokaryotic cells. In addition they are structurally more complex. A eukaryotic cell has a cell membrane but lacks a cell wall, because the composition of the body fluids within the organism and within the cell is controlled to prevent large differences in osmotic pressure. A eukaryotic cell contains many **organelles** (Figure 20.2). These organelles provide for the separation of the reactions that simultaneously occur within a single cell. Thus, both catabolic and anabolic reactions can take place within a cell. For example, in one part of the cell, glucose may be converted into glycogen while in another part of the cell glucose may be oxidized to carbon dioxide and water. In addition to the organelles, there is a network of protein fibers called the *cytoskeleton,* which organizes the cytoplasm by attaching the various organelles. The relationship of the organelles to the cell, the macromolecules, and simpler biochemicals is shown in Figure 20.3.

20.5 The Eukaryotic Cell

Membranes separate cells from their environment as well as organelles from one another within the cell. They are responsible for the organization and proper operation of the cell. Membranes have selective permeability that allows certain molecular and ionic species to pass through them while denying passage to others.

Although membranes differ widely in both structure and function, they have some common features. Membranes are usually only a few molecules thick, or between 6 and 10 nm thick. The membranes consist mainly of lipids and proteins in ratios varying from 1:4 to 4:1. Carbohydrates are bonded to some lipids to form

FIGURE 20.2 Components of the Eukaryotic Cell

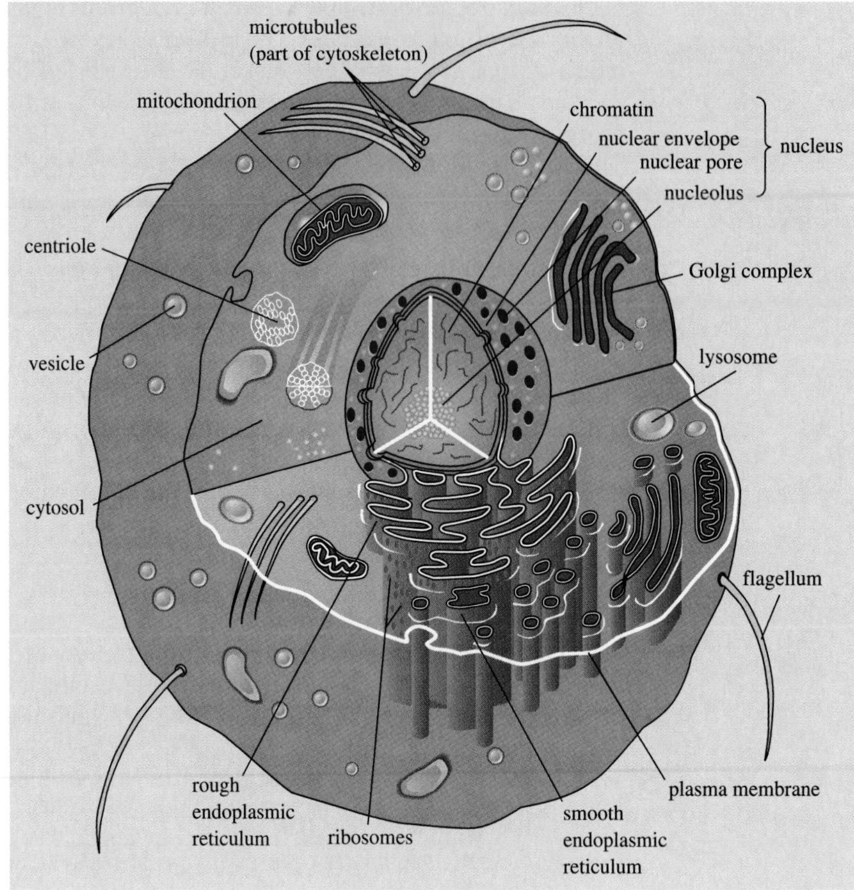

glycolipids and to some proteins to form glycoproteins. The proteins and lipids are held together in the membrane by London forces, and the structure is flexible.

Whether particular molecules pass through a membrane depends on the components of the membrane. In a sense the control is governed by the rule of solubilities that "like dissolves like." Lipids allow the passage of some nonpolar molecules and

FIGURE 20.3 Molecular Organization of the Components of a Cell

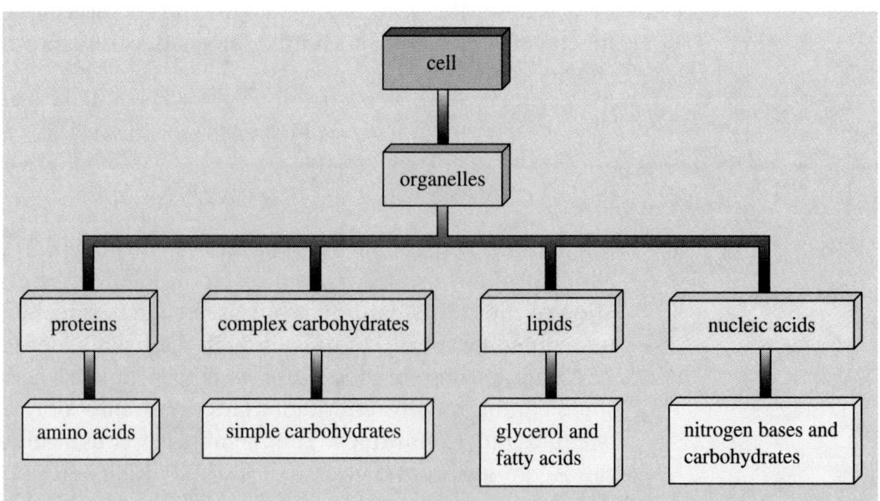

restrict that of polar molecules. Specific proteins with polar sites are responsible for the passage and transport of polar molecules and ions (Chapter 22).

The Nucleus

The nucleus is about 5 μm in diameter and is surrounded by a double membrane called the **nuclear envelope.** This structure has pores that allow the passage of water, certain ions, and ATP but not large molecules. The nucleus also contains a granular-looking material called **chromatin** and a second region called the **nucleolus.**

Chromatin contains more than 95% of the cell's DNA and is a control center that is responsible for directing both the chemical reactions in the cell and the reproduction of the cell. DNA molecules are associated with proteins in large strands called chromosomes, which occur in pairs. The DNA of eukaryotes is much larger than the DNA contained in the cytoplasm of prokaryotic cells.

The nucleolus is the region where the synthesis of ribosomes occurs. Ribosomes consist of ribosomal RNA and ribosomal proteins. The RNA is synthesized under the direction of DNA (Chapter 29); the ribosomal proteins are synthesized in the cytoplasm and pass through the nuclear envelope to the nucleolus. Once assembled, the ribosomes pass through the nuclear envelope to the cytoplasm.

Mitochondria

A mitochondrion (plural: *mitochondria*) is oval, about 2 μm long and 0.5 μm in diameter. A cell may contain several hundred to a thousand mitochondria depending on its function. Although some metabolism of food sources occurs in the cytosol, these reactions are anaerobic (without oxygen). The mitochondria are the only region where aerobic (with oxygen) catabolic reactions occur. Approximately 95% of the ATP generated in a cell is produced by aerobic reactions. The energy released by the oxidation of food is stored in ATP in the mitochondria. For this reason the mitochondria are often called the power packs of the cell.

Endoplasmic Reticulum

The endoplasmic reticulum is a continuous network of membrane-bound channels throughout the eukaryotic cell. The fluid within the endoplasmic reticulum is different from the fluid outside it. For example, this fluid in liver cells contains enzymes that catalyze the formation of lipids.

Near the nucleus the endoplasmic reticulum merges with the membrane of the nucleus. The **smooth endoplasmic reticulum** is a region without ribosomes. Enzymes are imbedded in the smooth endoplasmic reticulum at sites of lipid synthesis, including the phospholipids of the endoplasmic reticulum itself. The region near the nucleus, which is coated with ribosomes, is the **rough endoplasmic reticulum.** The ribosomes are the sites of protein synthesis (Chapter 29). Proteins synthesized at the ribosomes accumulate in pockets within the rough endoplasmic reticulum near the Golgi complex. The material in these pockets buds off in membrane-bound sacs called *vesicles* and migrates to the Golgi complex.

Golgi Complex

The **Golgi complex** or Golgi apparatus is a stack of flattened membrane sacs that is closely associated with the endoplasmic reticulum. This organelle receives the proteins formed at the endoplasmic reticulum on one side of its membrane via transfer vesicles. In the Golgi complex the temporarily stored proteins are chemically altered; some are converted into glycoproteins and lipoproteins. Proteins are released from the other side of the membrane via secretory vesicles.

Lysosomes

Some of the proteins synthesized in the endoplasmic reticulum are digestive enzymes. In the Golgi complex these enzymes are packaged as digestive organelles called *lysosomes.* A lysosome contains granules of enzymes and is surrounded by a single membrane of protein and lipids. These enzymes can hydrolytically break down other molecules, including food particles and microorganisms.

The membrane surrounding the lysosome prevents the contents from contacting other "inappropriate" parts of the cell. Without this protection, the cell itself would be digested. However, when a cell dies, the lysosome ruptures, and its enzymes hydrolyze the components of the dead cell. How these cells are identified for destruction is not well understood.

Food particles are enclosed within **food vacuoles,** which are recognized as such by lysosomes. Fusion of the two units allows the contents to mix and the lysosomes to digest the food, giving amino acids from proteins, monosaccharides from carbohydrates, and fatty acids from lipids.

20.6 Stereoisomers

In Chapters 14 and 15 we considered the structures of geometric isomers of cycloalkanes and alkenes. Geometric isomerism is only one type of the general phenomenon known as *stereoisomerism.* **Stereoisomers** have the same connectivity—the same sequence of bonded atoms—but different configurations. The three-dimensional arrangement of the atoms in a molecule in space, as in geometric isomers, determines its **configuration.**

The configuration of a molecule plays a major role in its biological function. Two isomeric molecules that differ in configuration often have entirely different biological properties. For example, bombykol, the sex attractant of the male silkworm moth, has a trans–cis arrangement around the double bonds at the C-10 and C-12 positions. This geometric isomer is 10^9 to 10^{13} times more potent at attracting male moths than the other three possible geometric isomers. Disparlure, the sex attractant of the female gypsy moth, is active only if the large alkyl groups bonded to the three-membered ring are cis.

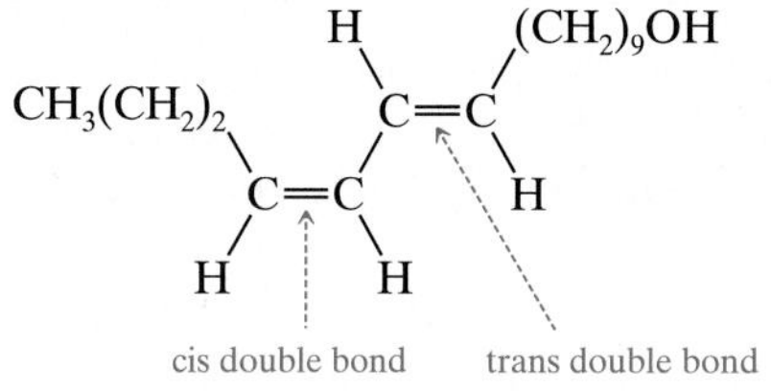

bombykol

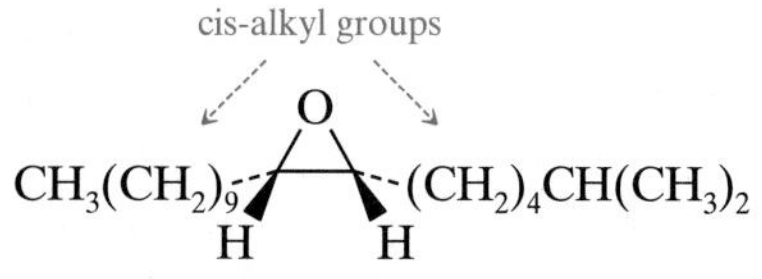

disparlure

Mirror Images and Chirality

Another type of stereoisomerism, based on mirror-image relationships between molecules, is not as easily visualized as geometric isomerism. However, the consequences of this type of isomerism are even more vital to life processes.

We live in a three-dimensional world that has important personal consequences. When you look into a mirror, you see someone who does not actually exist, your mirror image. Every object has a mirror image, but this reflected image need not be identical with the actual object.

Let's consider the mirror image of a common three-dimensional object. A simple wooden chair looks exactly like its mirror image (Figure 20.4). Similarly, the

FIGURE 20.4 Objects and Their Mirror Images

(a) The simple chair and its mirror image are identical. The mirror image can be superimposed on the original chair. (b) The sidearm chair has a mirror image that is different. The chair is a right-handed object; the mirror image is a left-handed object. The mirror image cannot be superimposed on the original chair.

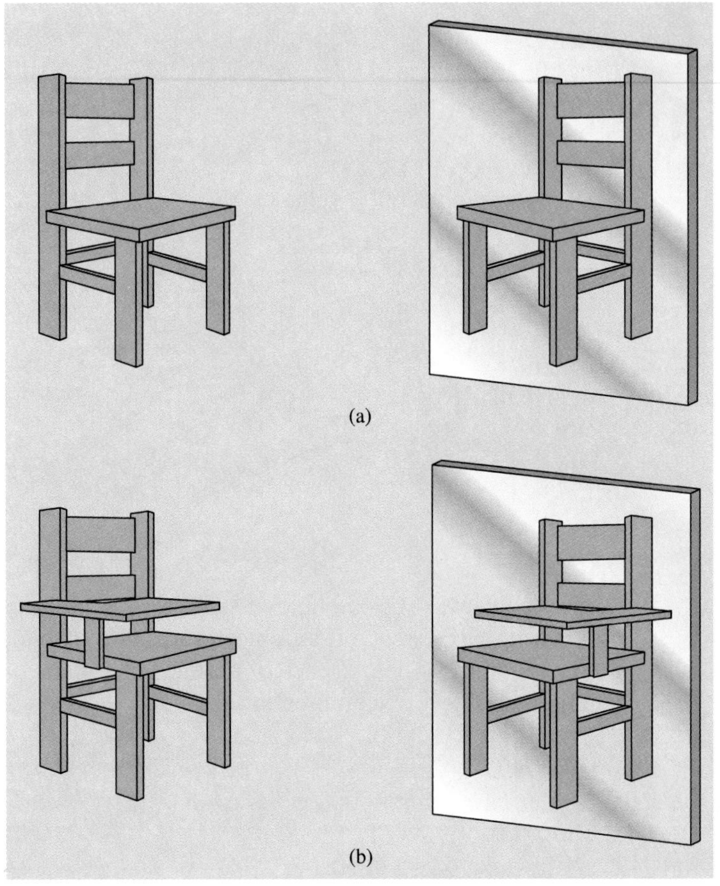

mirror image of an item such as a cup is identical with the object itself. When an object and its mirror image exactly match, we say that they are **superimposable.** Superimposable objects can be "placed" on each other so that each feature of one object precisely coincides in space with an equivalent feature in the mirror image.

Now let's consider objects that cannot be superimposed on their mirror images. One example is the sidearm chair found in many classrooms. When a chair with a "right-handed arm" is reflected in a mirror, it becomes a chair with a "left-handed arm" (Figure 20.4). We can convince ourselves of this by imagining sitting in the chair or its mirror image.

The chirality of life forms is seen in many animals in the ocean, as in the gills of the Christmas tree worm which lives within corals. (Robert J. Ouellette)

Now consider the nonsuperimposability of hands, which are also related as mirror images. The mirror image of a left hand looks like a right hand. But, when we try to superimpose our hands, we find that it can't be done. Therefore, our hands are related as nonsuperimposable mirror images.

An object that is not superimposable on its mirror image is called **chiral** (from the Greek *chiron,* "hand"). Objects such as gloves and shoes have a "handedness" and are chiral. An object that can be superimposed on its mirror image, such as a cup or a hammer, is **achiral.** We can determine whether an object is chiral or achiral without trying to superimpose its mirror image. One way is to determine if it has a plane of symmetry. A **plane of symmetry** bisects an object so that one half is the mirror image of the other half. For example, a cup has a plane of symmetry that divides it so that one half of it is the mirror image of the other half (Figure 20.5).

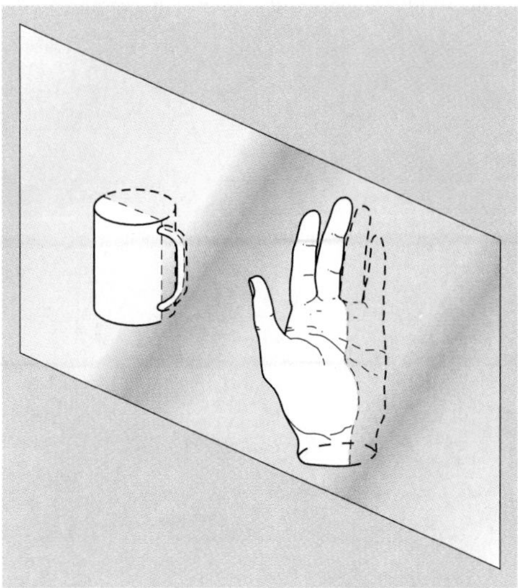

FIGURE 20.5
Plane of Symmetry
Any object with a plane of symmetry is achiral. The half on one side of the plane is the mirror image of the half on the other side of the plane. The cup shown can be divided into two equal halves that are mirror images of each other. It is achiral. Any object that does not have a plane of symmetry is chiral. The plane shown does not split a hand into two equal halves. A hand is chiral.

Any object that has a plane of symmetry is superimposable on its mirror image and is achiral. However, if an object, such as your hand, has no such plane of symmetry, it is chiral. The presence or absence of a plane of symmetry is one way to tell whether an object is achiral or chiral.

20.7 Molecular Chirality

The concept of chirality can be extended from macroscopic objects to molecules. Most molecules produced in living organisms are chiral. A molecule is chiral if it contains a single tetrahedral carbon atom attached to four non-equivalent atoms or groups of atoms. Such a carbon atom is a **stereogenic center.** A stereogenic center is also called a **chiral center,** and the carbon atom is called a **chiral carbon atom.**

The four atoms or groups of atoms at a stereogenic center can be arranged in two ways to correspond to two stereoisomers. Let's consider the stereoisomers of bromochlorofluoromethane. One stereoisomer of this molecule and its mirror image are illustrated in Figure 20.6. The two structures cannot be superimposed. Therefore, bromochlorofluoromethane is chiral.

Molecules, like macroscopic achiral objects, can contain a plane of symmetry. Dichloromethane has two planes of symmetry; bromochloromethane has one plane of symmetry (Figure 20.7). Each molecule can be superimposed on its mirror image and is therefore achiral. In contrast, bromochlorofluoromethane does not have a plane of symmetry and is chiral.

Enantiomers Are Mirror Image Isomers

Two stereoisomers related as nonsuperimposable mirror images are called **enantiomers** (from the Greek *enantios,* "opposite" + *meros,* "part"). We can tell that a substance is chiral and predict that two enantiomers exist by identifying the substituents on each carbon atom. A carbon atom with four non-equivalent substituents is stereogenic, and a molecule with one stereogenic center is chiral. This molecule can exist as either of a pair of enantiomers. For example, 2-bromobutane is chiral because the C-2 atom is attached to four non-equivalent groups (CH_3,

CH_3CH_2, H, and Br). In contrast, 2-bromopropane does not contain any carbon atom attached to four non-equivalent groups. The C-2 atom has two methyl groups attached to it.

stereogenic center

$CH_3CH_2-\underset{\displaystyle H}{\overset{\displaystyle Br}{C}}-CH_3$

2-bromobutane
(a chiral molecule)

not a stereogenic center

$CH_3-\underset{\displaystyle H}{\overset{\displaystyle Br}{C}}-CH_3$

2-bromopropane
(an achiral molecule)

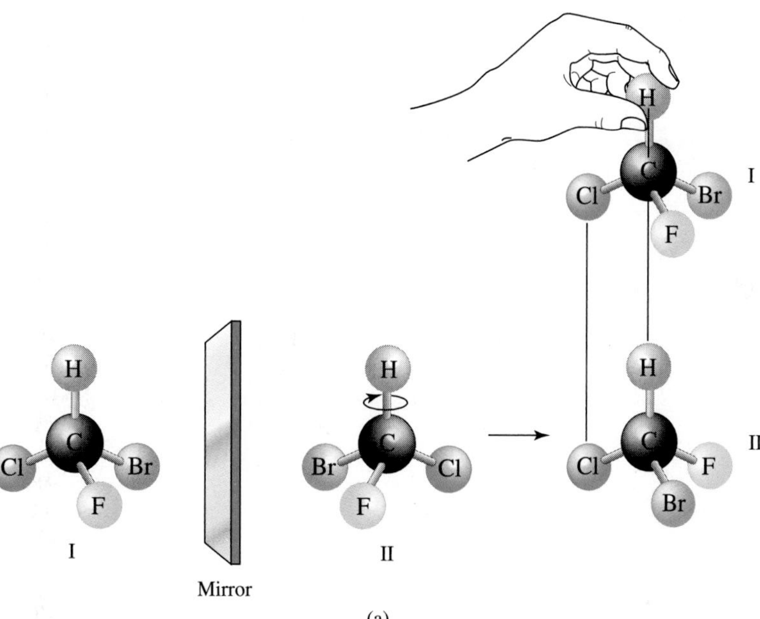

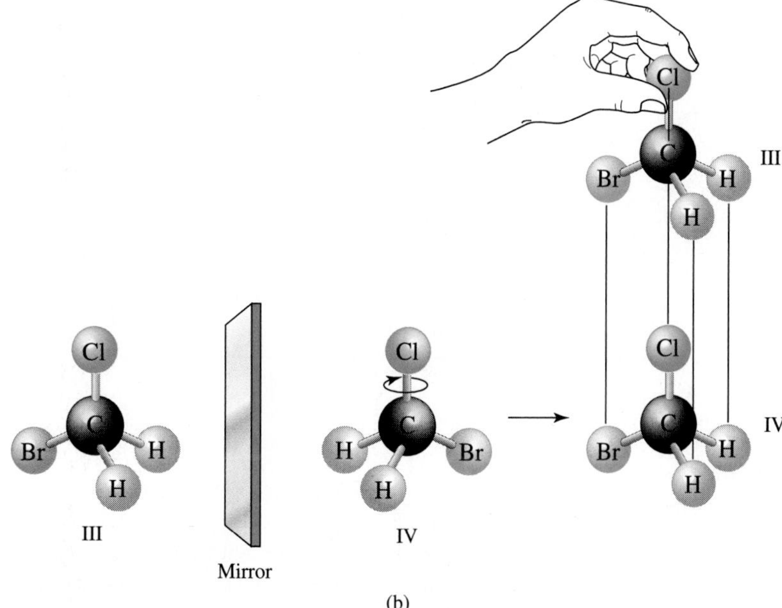

FIGURE 20.6 Criteria for Chirality in Molecules
(a) The two molecular models representing bromochlorofluoromethane cannot be superimposed. The spheres of structure I held in the hand do not line up with those of structure II, which has been rotated to line up the chlorine atoms. (b) The mirror image of chlorobromomethane can be superimposed on the original structure. When structure IV is rotated, all spheres line up with those of structure III.

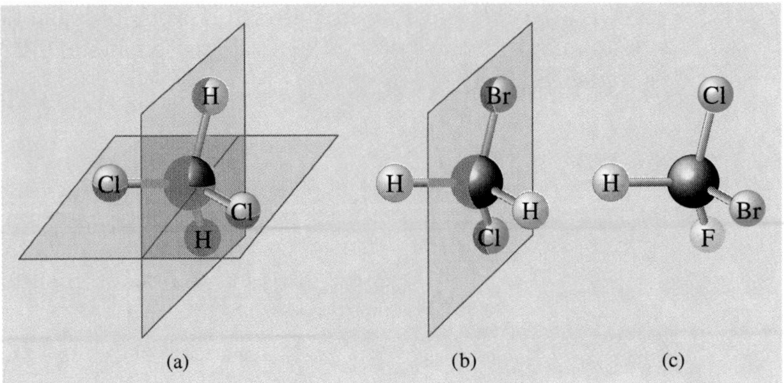

FIGURE 20.7 Planes of Symmetry for Molecules
(a) Dichloromethane is bisected by two planes of symmetry. For each plane, one part of the structure mirrors the part of the molecule lying on the other side of the plane. (b) Bromochloromethane has one plane of symmetry. (c) Bromochlorofluoromethane has no plane of symmetry.

EXAMPLE 20.2

Consider phenytoin, a compound with anticonvulsant activity. Is the molecule chiral or achiral? Determine your answer by identifying the number of non-equivalent groups bonded to tetrahedral carbon atoms and by determining whether or not the molecule has a plane of symmetry.

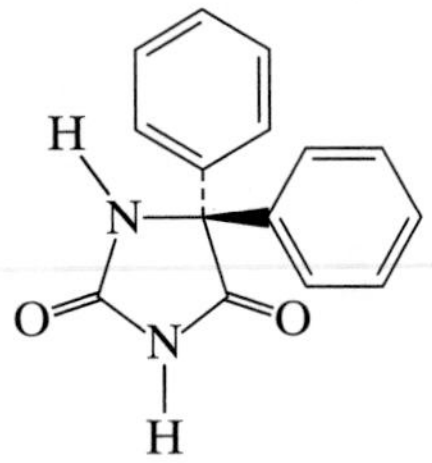

Solution

Phenytoin has only one tetrahedral carbon atom in the entire molecule! That carbon atom is bonded to a nitrogen atom, a carbonyl group, and two benzene rings. Because the tetrahedral carbon atom is attached to two identical benzene rings, the molecule is achiral.

Phenytoin has a plane of symmetry that lies in the plane of the page. One of the benzene rings of phenytoin is above and the other below the symmetry plane. Note that the other atoms of phenytoin are bisected by the plane.

PROBLEM 20.2

Consider the following structural formula for nicotine. Is the molecule chiral?

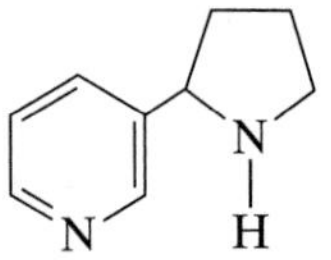

Properties of Enantiomers

We can regard hands as analogous to the enantiomers of a chiral molecule. Let's consider the interaction of hands with a symmetrical object such as a pair of tweezers. The tweezers are symmetrical. They can be used equally well with either hand because there is no preferred way to pick up or manipulate a pair of tweezers. However, even if blindfolded, you could easily use your hands to distinguish right- and left-handed gloves. Your hands are "a chiral environment," and in this environment, mirror image gloves do not interact with hands in the same way. The right glove will fit only the right hand. We can distinguish chiral objects only because we are chiral.

Pairs of enantiomers have the same density, melting point, and boiling point. They also have the same chemical properties in an achiral environment. However, enantiomers can be distinguished in a chiral environment. This difference is important in many processes in living cells. Only one of a pair of enantiomers fits into a specific site in a biological molecule such as an enzyme catalyst, because the site on the enzyme that binds the enantiomer is chiral. The binding of one specific enantiomer is called **stereospecificity.**

20.8 Optical Activity

Although enantiomers have identical chemical properties in achiral environments, they differ in one physical property: Enantiomers behave differently toward plane-polarized light. This difference is used to distinguish a chiral molecule from its enantiomer in the laboratory.

Plane-Polarized Light

Light consists of waves oscillating in an infinite number of planes at right angles to the direction of propagation of the light. When a beam of "ordinary" light passes through a **polarizing filter,** it is converted to **plane-polarized light** vibrating in a single plane. We are familiar with this phenomenon in everyday life: Plane-polarized light can be produced by Polaroid sunglasses, which reduce glare by acting as a polarizing filter.

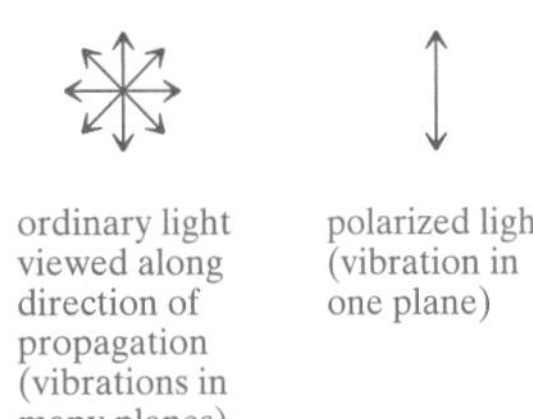

The Polarimeter

Plane-polarized light interacts with chiral molecules, and this interaction can be measured with an instrument called a **polarimeter** (Figure 20.8). In a polarimeter, light with a single frequency of vibration—that is, monochromatic light—passes through a polarizing filter. The polarized light then traverses a tube containing a solution of the compound to be examined. While passing through the solution the polarized light is unaffected by achiral molecules. But an interaction with a chiral compound causes the plane of polarized light to rotate. After the plane-polarized light leaves the sample tube, it passes through a second polarizing filter called an analyzer. The analyzer must be rotated either clockwise or counterclockwise to match the plane of polarization of the light and allow it to pass. An angle α can be

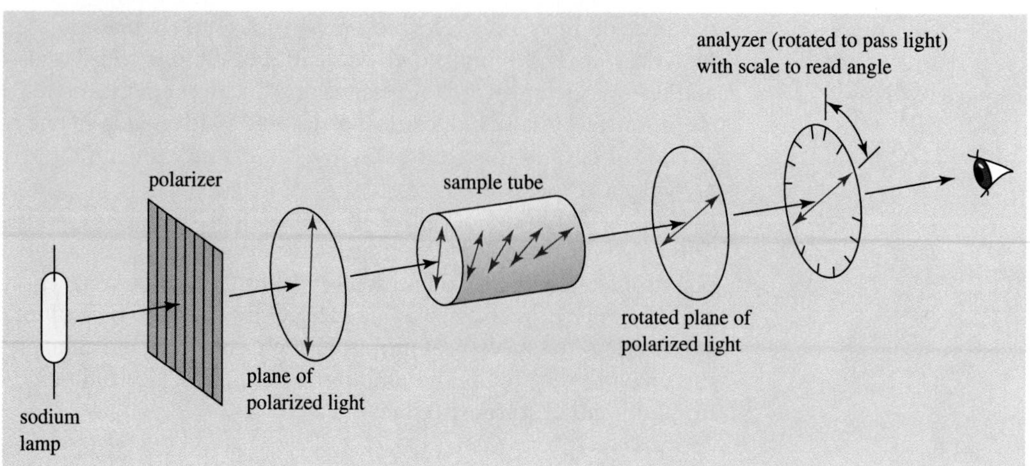

FIGURE 20.8 Representation of a Polarimeter

read off the analyzer and is called the *observed rotation.* It equals the angle by which the light has been rotated by the chiral compound. Because chiral molecules rotate plane-polarized light, they are **optically active.** Achiral molecules do not rotate plane-polarized light, so they are **optically inactive.**

Specific Rotation

The amount of rotation observed in a polarimeter depends on the structure of the substance and the number of molecules encountered by the light. The optical activity of a pure chiral substance is reported as its **specific rotation,** $[\alpha]_D$. It is the number of degrees of rotation of a solution at a concentration of 1 g/mL in a tube 1 dm long and is expressed without a degree sign. The standard conditions selected for these experiments are 25 °C and the yellow light (D line, 589 nm) of the sodium vapor lamp. The symbols *l* and *c* refer to length and concentration, respectively.

$$[\alpha]_D = \frac{\alpha}{l \times c}$$

If a chiral substance rotates plane-polarized light to the right—that is, in a positive (+) or clockwise direction—the substance is **dextrorotatory** (from the Latin *dextro,* "right"). If a chiral substance rotates plane-polarized light to the left—in a negative (−) or counterclockwise direction—the substance is **levorotatory** (from the Latin *laevus,* "left"). The two enantiomers of a chiral substance—called dextrorotatory and levorotatory isomers—rotate polarized light the same number of degrees but in opposite directions. Therefore, they are sometimes called **optical isomers.**

We often refer to an enantiomer by prefixing the sign of the optical rotation at 589 nm to the name of the compound. For example, one of the enantiomers of 2-iodobutane has $[\alpha]_D = -15.15$. It is called (−)-2-iodobutane. The other enantiomer, (+)-2-iodobutane, has $[\alpha]_D = +15.15$.

TABLE 20.3 Specific Rotations of Common Compounds

Compound	$[\alpha]_D$
azidothymidine (AZT)	+99
cefotaxine (a cephalosporin)	+55
cholesterol	−31.5
cocaine	−16
codeine	−136
epinephrine (adrenaline)	−5.0
heroin	−107
levodopa	−13.1
monosodium glutamate (MSG)	−25.5
morphine	−132
oxacillin (a penicillin)	+201
progesterone (female sex hormone)	+172
sucrose (table sugar)	+66.5
testosterone (male sex hormone)	+109

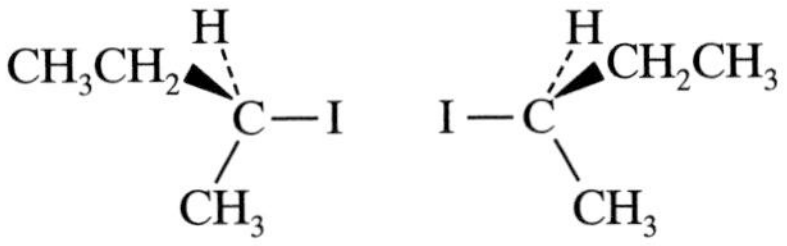

(+)-2-iodobutane (−)-2-iodobutane

The (+) isomer is sometimes called the **d form,** and the (−) isomer is called the **l form.** The specific rotations of some common substances are listed in Table 20.3.

20.9 Fischer Projection Formulas

Drawing molecules is often difficult, especially for compounds that contain several chiral carbon atoms. The structural formulas of chiral substances can be conveniently drawn, however, using a convention proposed by the German chemist Emil Fischer more than a century ago. The configurations of chiral substances are indicated by comparing them with the configuration of a reference compound called glyceraldehyde.

Glyceraldehyde contains a carbon atom bonded to four different groups, so it exists as two enantiomeric forms (Figure 20.9). The enantiomers of glyceraldehyde are written according to the projection method proposed by Fischer. The carbonyl group (—CHO), the hydroxymethyl group (—CH_2OH), and the chiral carbon atom are arranged vertically, with the most oxidized group (—CHO) at the "top." The chiral carbon atom is placed in the plane of the paper. Because this carbon atom is tetrahedral, the CHO group and the CH_2OH group extend behind the plane of the page, and the hydrogen atom and the hydroxyl group extend up and

FIGURE 20.9 Projection Formula of Enantiomers of Glyceraldehyde

The Fischer projection of the two enantiomers of glyceraldehyde has crossed lines at a point where the chiral carbon atom would be. However, the carbon atom is not usually shown. The vertical lines are assumed to project away from the viewer. The horizontal lines project out toward the viewer.

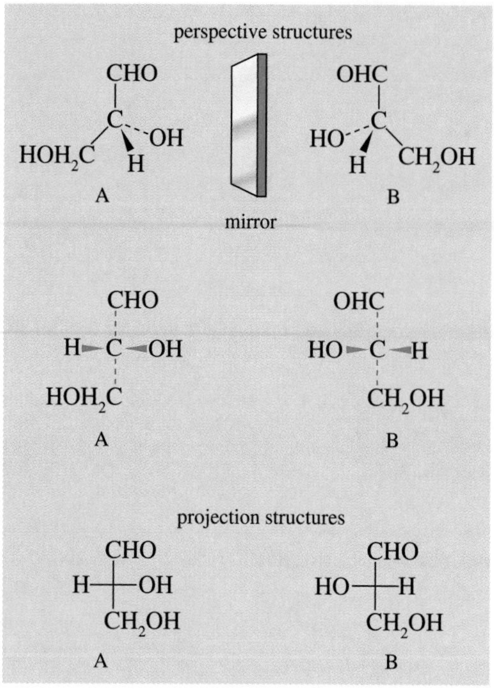

out of the plane. When these four groups are projected onto a plane, the projection is called the **Fischer projection formula.** The chiral carbon atom is not shown in this convention. It is located at the point where the bond lines cross. The vertical lines are assumed to project away from the viewer. The horizontal lines project out toward the viewer. In most examples in this text we will retain the carbon atoms for clarity.

A Fischer projection formula is a two-dimensional representation. It might appear that if we lifted one formula out of the plane and rotated it 180° around the carbon backbone, we would obtain the formula of the enantiomer. However, if this were done for molecule A in Figure 20.9 the carbonyl group and the hydroxymethyl group, originally behind the plane, would be in front of the plane. These groups would not occupy identical positions with respect to the carbonyl group and hydroxymethyl group of molecule B, which are behind the plane. Therefore, to avoid the error of apparently achieving a two-dimensional equivalence of nonequivalent three-dimensional molecules, it is important not to mentally lift a two-dimensional representation out of the plane of the paper and flip it over.

Fischer projection formulas can be drawn to depict any pair of enantiomers. These formulas imply that we "know" the configuration at the chiral carbon atom. However, the true configuration could not be determined by early chemists because there was no way to "see" the arrangement of the atoms in space. Therefore, Fischer arbitrarily assigned a configuration to one member of the enantiomeric pair of glyceraldehydes. The dextrorotatory enantiomer of glyceraldehyde, which rotates plane-polarized light in a clockwise direction (+ 13.5), was assigned to the Fischer projection with the hydroxyl group on the right side. Fischer called the compound D-glyceraldehyde. The mirror image compound, (−)-glyceraldehyde, corresponds to the structure in which the hydroxyl group is on the left. It rotates plane-polarized light in a counterclockwise direction. Fischer called the compound L-glyceraldehyde.

$$\begin{array}{ccc} & \text{CHO} & \\ & | & \\ \text{H}- & \text{C} & -\text{OH} \\ & | & \\ & \text{CH}_2\text{OH} & \end{array} \qquad \begin{array}{ccc} & \text{CHO} & \\ & | & \\ \text{HO}- & \text{C} & -\text{H} \\ & | & \\ & \text{CH}_2\text{OH} & \end{array}$$

D-(+)-glyceraldehyde $[\alpha]_D = +13.5$ L-(−)-glyceraldehyde $[\alpha]_D = -13.5$

The designations D and L refer to configuration only; they do not indicate the direction of rotation. The direction of rotation of light depends on both configuration and the identity of the atoms bonded to the chiral carbon atom (a subject that will not be further discussed).

The configurations of other chiral compounds are determined by chemically relating them to either D- or L-glyceraldehyde. If D-(+)-glyceraldehyde is converted into lactic acid by converting the CH_2OH group into CH_3 and the CHO group into CO_2H, the resulting lactic acid must still have the D configuration. The lactic acid produced rotates light in a counterclockwise direction; it is levorotatory and is designated D-(−)-lactic acid. This one transformation illustrates the fact that the D and L notations refer to configuration of the chiral carbon, not the direction of rotation of plane-polarized light.

$$\begin{array}{ccc} & \text{CHO} & \\ & | & \\ \text{H}- & \text{C} & -\text{OH} \\ & | & \\ & \text{CH}_2\text{OH} & \end{array} \xrightarrow{\text{several steps}} \begin{array}{ccc} & \text{CO}_2\text{H} & \\ & | & \\ \text{H}- & \text{C} & -\text{OH} \\ & | & \\ & \text{CH}_3 & \end{array}$$

D-(+)-glyceraldehyde $[\alpha]_D = +13.5$ D-(−)-lactic acid $[\alpha]_D = -13.8$

In 1950 the actual configuration of an optically active substance was finally determined by X-ray analysis. The arrangement of the atoms corresponded to those assigned to the configuration of glyceraldehyde. Therefore, the original assignment was correct. When we know the exact position of atoms in space, we know the molecule's **absolute configuration.**

EXAMPLE 20.3

What is the configuration of alanine shown in the following Fischer projection? What is the sign of its optical rotation?

$$\begin{array}{ccc} & \text{CO}_2\text{H} & \\ & | & \\ \text{NH}_2- & \text{C} & -\text{H} \\ & | & \\ & \text{CH}_3 & \end{array}$$

Solution

The carbon chain is written vertically in the Fischer projection, and the amino group is on the left. Thus, the compound resembles the configuration for L-glyceraldehyde and is designated L-alanine. The sign of rotation is not related to configuration in any predictable way and therefore cannot be determined from the information given.

PROBLEM 20.3

The molecular formula of valine is $(CH_3)_2CHCH(NH_2)CO_2H$. The L isomer is found in most proteins. However, the D isomer is found in the antibiotic fungisporin produced by some *Penicillium* species. Draw the projection formula of the D isomer.

20.10 Multiple Chiral Centers

So far we have considered only molecules with a single stereogenic center. However, many compounds contain several stereogenic centers. For example, the antibiotic erythromycin contains 18 chiral centers!

How is the number of stereoisomers of a molecule related to the number of stereogenic centers? What relationships exist among these isomers, and how are their optical rotations related? The answers to these questions depend on the relationship among the groups at each stereogenic center. Are the centers equivalent or nonequivalent? If the chiral carbon atoms are not bonded to identical sets of substituents, the stereogenic centers are **nonequivalent.** In contrast, if the stereogenic centers are bonded to identical sets of substituents, the centers are **equivalent.**

Nonequivalent Stereogenic Centers

The number of optically active isomers for a molecule containing n nonequivalent chiral carbon atoms is 2^n. Consider 2, 3, 4-trihydroxylbutanal.

$$\overset{4}{CH_2}(OH)-\overset{3}{CH}(OH)-\overset{2}{CH}(OH)-\overset{1}{CHO}$$

The C-2 and C-3 atoms are chiral and nonequivalent. Thus there are $2^2 = 4$ possible stereoisomers (Figure 20.10). Both chiral centers contribute to the total rotation of the molecule, but their contributions are not identical because the bonds from each atom are not identical.

Structures I and II are mirror images, are not superimposable on each other, and are enantiomers. This can be verified in two dimensions by imagining a mirror

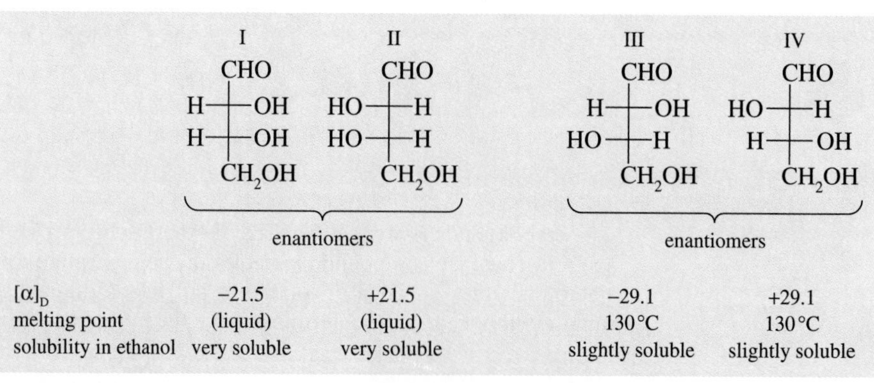

FIGURE 20.10 Enantiomers and Diastereomers
There are four stereoisomers of a compound containing two nonequivalent chiral centers. There are two sets of enantiomers. Any combination of stereoisomers that are not enantiomers are diastereomers.

placed between I and II. Structures III and IV are also nonsuperimposable mirror images and are enantiomers.

Now consider structures I and III. These stereoisomers do not have a mirror-image relationship. Stereoisomers that are not enantiomers are called **diastereomers.** Although the physical properties of enantiomers are the same, diastereomers have different physical properties.

EXAMPLE 20.4

L-Threonine, an amino acid, has the following condensed molecular formula. Write the Fischer projections of the possible stereoisomers.

$$CH_3CH(OH)CH(NH_2)CO_2H$$

Solution

Both C-2 and C-3 are chiral. Thus, four structures are possible. The Fischer projections are written by placing the carboxyl group at the top of the vertical chain. The amino and hydroxyl groups may be on the right or left side of the projection formulas.

```
     CO2H              CO2H               CO2H              CO2H
      |                 |                  |                 |
H — C — NH2     NH2 — C — H        H — C — NH2     NH2 — C — H
      |                 |                  |                 |
H — C — OH       HO — C — H         HO — C — H       H — C — OH
      |                 |                  |                 |
     CH3               CH3                CH3               CH3
```

The actual structure of L-threonine in proteins is given by the Fischer projection at the right.

PROBLEM 20.4

Isoleucine has the following condensed structural formula. Determine the number of chiral centers in the molecule, and write projection formulas for all the stereoisomers. Identify all pairs of enantiomers.

$$CH_3CH_2CH(CH_3)CH(NH_2)CO_2H$$

Equivalent Stereogenic Centers

Let's consider tartaric acids as example of compounds with equivalently substituted chiral centers. The eclipsed conformations of the tartaric acids are shown in Figure 20.11. In each structure the C-2 and C-3 atoms are connected to four different groups. The number of stereoisomers of tartaric acids and their optical properties differ from those of the stereoisomers of the 2,3,4-trihydroxybutanals. Only three stereoisomers exist. Of these, one is optically inactive! Look at structures III and IV. Although the structures are drawn as "mirror images," they are superimposable and, in fact, are identical. To show that this is so, rotate one structure 180° in the plane of the paper; the resulting structure superimposes on the original structure. Thus, the two structures represent the same molecule. It is achiral and cannot be optically active.

The stereoisomeric tartaric acid that is optically inactive has a plane of symmetry. We recall from Section 20.5 that a structure with a plane of symmetry is achiral and that it is superimposable on its mirror image. In the case of the tartaric acid, the plane of symmetry is between the C-2 and C-3 atoms, so the top half of the molecule is the mirror image of the bottom half.

FIGURE 20.11 Tartaric Acids—Optically Active and Meso Compounds
There are only three stereoisomers of a compound with two equivalent chiral centers. Two compounds are enantiomers. The third compound has a plane of symmetry. It is optically inactive and is called a meso compound.

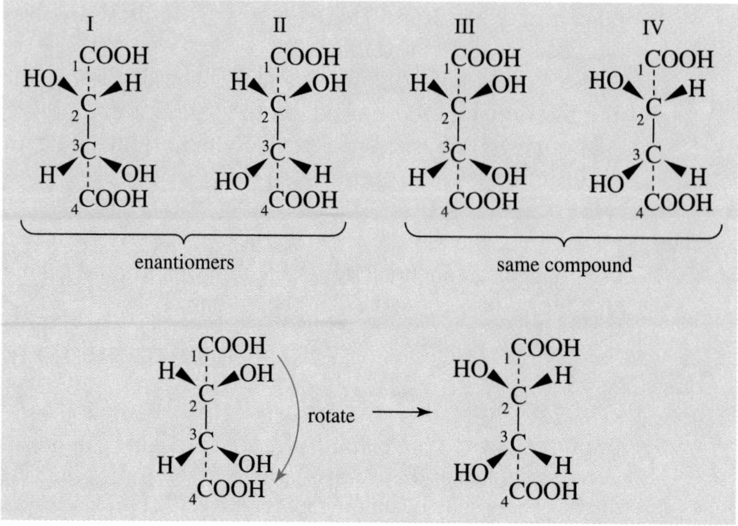

Chirality and the Senses

Senses are sensitive to the configuration of molecules. Both the sense of taste and the sense of smell result from changes induced in each sensory receptor when it binds a specific molecule. The binding causes a conformational change that triggers a sequence of events culminating in transmission of a nerve impulse to the brain by sensory neurons. The brain interprets the input from sensory neurons as the "odor" of, say, spearmint.

Enantiomeric and diastereomeric molecules interact differently with sensory receptors in living systems. Differences in biological response to diastereomeric compounds are perhaps easier to understand. After all, diastereomeric compounds have different physical properties. For example, mannose, a carbohydrate, exists in two diastereomeric forms that differ in configuration at one stereogenic center. The α form tastes sweet, but the β form tastes bitter.

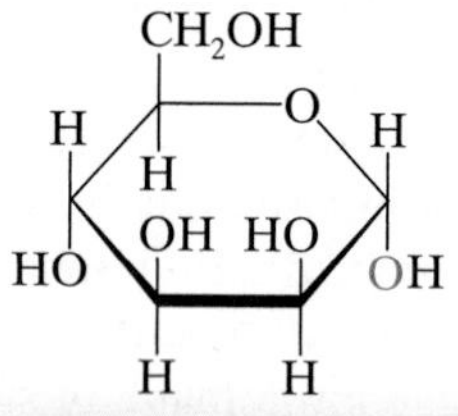

α-D-mannose (sweet)

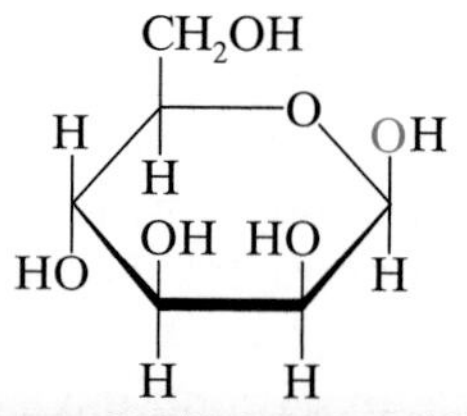

β-D-mannose (bitter)

Sensory receptors also readily distinguish enantiomers. The specificity of response is similar to the relationship between our hands and how they fit into gloves. Because sensory receptors are chiral, they interact stereospecifically with only one of a pair of enantiomers. The two enantiomeric forms of carvone have very different odors. (+)-Carvone is present in spearmint oil, imparting the familiar smell. In contrast, its enantiomer, (−)-carvone, is present in caraway seed. It has the odor associated with rye bread.

(+)-carvone (spearmint)

(−)-carvone (caraway)

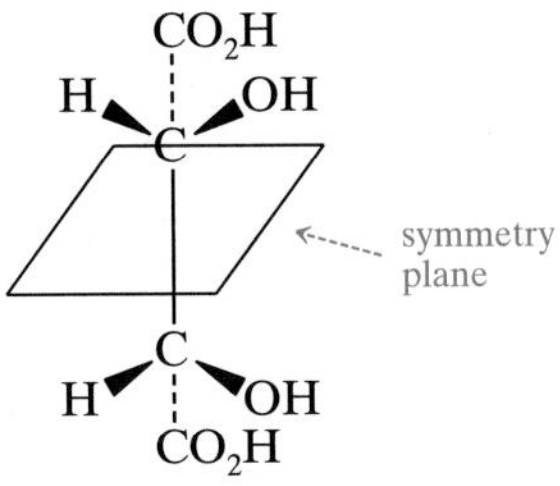

CO_2H
H OH
C
symmetry plane
C
H OH
CO_2H

Stereoisomers of compounds, such as tartaric acid, which have two or more chiral centers but are nevertheless achiral, are called **meso compounds** (from the Greek *meso,* "middle"). Meso compounds are not optically active.

EXAMPLE 20.5

Determine the number of chiral centers in vitamin K_1. How many stereoisomers are possible?

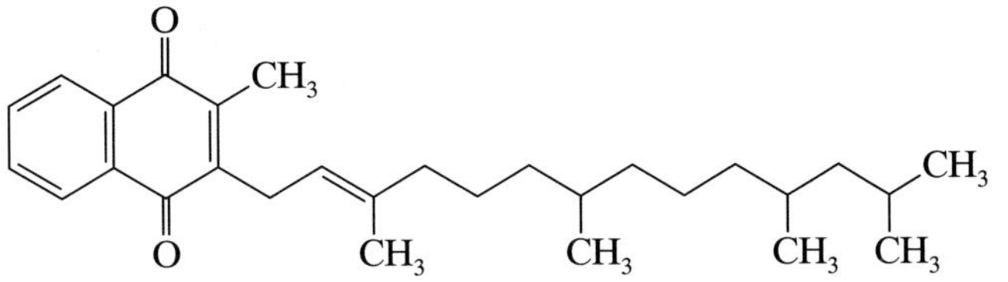

vitamin K_1

Solution

The carbon atoms in the two rings are not chiral because they are not tetrahedral. The long alkyl chain contains eight methylene units, none of which is chiral because a carbon atom in a methylene group is bonded to two hydrogen atoms. The tertiary carbon atom near the end of the alkyl chain has two methyl groups and also is not chiral.

Next, consider the positions in the middle of the alkyl chain that have methyl group branches. The methyl group on the left is bonded to a double-bonded carbon atom, which does not have four groups bonded to it. Therefore, it is not chiral. The next two methyl groups are located on chiral centers. Because there are two chiral carbon atoms, $2^2 = 4$ stereoisomers are possible.

PROBLEM 20.5

Write the Fischer projection formulas of the stereoisomeric 2,3-dibromobutanes. What relationships should exist between the optical activities of these isomers?

20.11 How Are Stereoisomers Formed?

Chemical reactions in living systems occur in the presence of enzymes, hormones, carbohydrates, lipids, proteins, and nucleic acids that have multiple chiral centers. Only one enantiomer will react in each case because it physically and chemically fits with the highly chiral biological molecules in the organism. Similarly, when a compound that contains a chiral center is produced in a living system, only one enantiomer results. For example, fumaric acid is hydrated to give only the L isomer of malic acid. Additional examples of this common phenomenon will be given in subsequent chapters.

$$\text{HO}_2\text{C(H)C=C(H)CO}_2\text{H} + \text{H}_2\text{O} \xrightarrow{\text{fumarase}} \text{HO—C(H)(CO}_2\text{H)—CH}_2\text{—CO}_2\text{H}$$

fumaric acid — L-malic acid

Any laboratory synthesis using chiral reactants will give chiral products provided that the bonds at the chiral center are not affected by the reagents. Esterification of D-lactic acid gives D-methyl lactate.

$$\text{H—C(CO}_2\text{H)(CH}_3\text{)—OH} \xrightarrow{\text{CH}_3\text{OH}} \text{H—C(CO}_2\text{CH}_3\text{)(CH}_3\text{)—OH} + \text{H}_2\text{O}$$

D-(−)-lactic acid $[\alpha]_D = -3.8$ — D-(+)-methyl lactate $[\alpha]_D = +7.5$

However, formation of a product with a chiral center starting from an achiral reactant gives a 50:50 mixture of enantiomers called a **racemic mixture.** Reduction of pyruvic acid using achiral reagents gives a racemic mixture of lactic acids.

$$\text{CO}_2\text{H—C(=O)—CH}_3 \xrightarrow{\text{NaBH}_4} \text{H—C(CO}_2\text{H)(CH}_3\text{)—OH} + \text{HO—C(CO}_2\text{H)(CH}_3\text{)—H}$$

pyruvic acid — racemic mixture of lactic acids

The individual lactic acid molecules are optically active, but any rotation of plane-polarized light by the D form is canceled by the opposite rotation of the L form.

The reduction of pyruvic acid in biological systems is quite different. Pyruvic acid is produced in muscles under anaerobic conditions and causes muscle cramps. The cramping is the result of lactic acid formed by reduction of pyruvic acid by NADH. Only L-lactic acid is produced because the biological reducing agent is chiral.

$$\text{CO}_2\text{H—C(=O)—CH}_3 \xrightarrow{\text{NADH}} \text{HO—C(CO}_2\text{H)(CH}_3\text{)—H}$$

pyruvic acid — L-lactic acid

Summary

Biochemistry is the study of the composition, structure, and reactions of substances in living systems. The cell is the functional unit of biological activity. The two basic types of cells are the **eukaryotes,** which contain nuclei, and the **prokaryotes,** which do not. The principal components of eukaryotes are the **cell membrane** and the **organelles:** the **nucleus, mitochondria, endoplasmic reticulum, Golgi complex,** and **lysosomes.**

Although cells have great structural complexity, the compounds that make up the cell are of four types: **carbohydrates, lipids, proteins,** and **nucleic acids.** These classes are responsible for the structure of all organisms.

Metabolism is a series of chemical reactions to obtain energy from food and to construct cellular material. **Catabolic reactions** degrade larger molecules into smaller molecules. **Anabolic reactions** build complex molecules from smaller molecules. The amount of energy released in catabolic reactions and required in anabolic reactions is symbolized by the change in free energy. If $\Delta G^\circ < 0$, the reaction is **exergonic.** If $\Delta G^\circ > 0$, the reaction is **endergonic.**

Stereoisomers have the same structure but different configurations. **Configuration,** the spatial arrangement of atoms in a molecule, is important to an understanding of biomolecules. A **stereogenic** center is an atom that bears four different substituents. **Chiral molecules** lack a plane of symmetry and can exist in enantiomeric forms. **Enantiomers** rotate the plane of **polarized light** in opposite directions. Compounds that rotate the plane of polarized light in a clockwise direction are **dextrorotatory** and are denoted with the prefix (+), whereas those that rotate the plane of polarized light in a counterclockwise direction are **levorotatory** and are denoted with the prefix (−). An equimolar mixture of enantiomers is called a **racemic mixture.**

Glyceraldehyde is used as a standard to define the configuration of optically active compounds. Compounds with n multiple chiral carbon atoms can exist as 2^n stereoisomers. Stereoisomers that are not enantiomers are **diastereomers.** A **meso** compound has chiral centers but is nevertheless achiral.

Exercises

Composition of Biochemicals

20.1 What are the three most abundant elements in the human body?

20.2 List three elements that exist in covalent biochemicals in less than 10% abundance.

20.3 List some inorganic ions that occur in significant amounts in the body.

20.4 What are the major classes of biochemicals?

Energy and Biochemistry

20.5 What is the sign of $\Delta G°$ for an anabolic reaction?

20.6 A reaction has a negative $\Delta G°$. Is the reaction exergonic or endergonic?

20.7 The hydrolysis of creatine phosphate is exergonic by 10.3 kcal/mole. What is the sign of $\Delta G°$ for the reaction?

20.8 What central compound provides the energy for anabolic reactions?

The Cell

20.9 What is the function of the mitochondria?

20.10 What is the function of the nucleus?

20.11 What two principal types of biochemicals are found in cell membranes?

20.12 What types of forces hold the components together in a cell membrane?

Chirality

20.13 Which of the following molecules have chiral centers?

(a) $CH_3CH_2CHBrCH_3$ (b) $CH_3CH_2CBr_2CH_3$
(c) $CH_3CH(OH)CO_2CH_3$ (d) $HOCH_2CH_2CO_2CH_3$

20.14 Which of the following molecules have chiral centers?

(a) $CH_3CH_2CHBrCH_2CH_3$
(b) $CH_3CH_2CH_2CHBrCH_3$
(c) $CH_3C(=O)CO_2H$
(d) $CH_3CH(OH)CH_2CO_2H$

20.15 Which of the following compounds has a plane of symmetry?

(a) *cis*-1,2-dibromocyclobutane
(b) *trans*-1,2-dibromocyclobutane

20.16 Which of the following compounds has a plane of symmetry?

(a) *cis*-1,3-dibromocyclobutane
(b) *trans*-1,3-dibromocyclobutane

20.17 How many chiral centers does each of the following barbiturates have?

phenobarbital

H, O, CH_2CH_3, N, O, N, O, H

secobarbital

CH_3, O, $CHCH_2CH_2CH_3$, H, N, $CH_2CH{=}CH_2$, O, N, O, H

20.18 How many chiral centers does each of the following drugs have?

(a) phenylbutazone, used to treat gout

C_6H_5, C_6H_5, N—N, O, O, $CH_2CH_2CH_2CH_3$

(b) ibuprofen, an analgesic

$CH_3CH(CH_3)CH_2-C_6H_4-CH(CH_3)CO_2H$

(c) chloramphenicol, an antibiotic

$NO_2-C_6H_4-CH(OH)-CH(CH_2OH)-NH-C(=O)-CHCl_2$

20.19 How many chiral carbon atoms does Dianabol, a synthetic anabolic steroid have?

20.20 Determine the number of chiral centers in the male sex hormone testosterone.

Optical Activity

20.21 Lactic acid in the blood has a specific rotation of +2.6. A sample of lactic acid obtained from sour milk has a specific rotation of −2.6. What is the difference in these two compounds?

20.22 Optically pure β-citronellol from citronella oil has a specific rotation of +5.3. An enantiomer of optically pure β-citronellol is obtained from geranium oil. What is its specific rotation?

20.23 D-(−)-Lactic acid is converted into a methyl ester when it reacts with methanol. What is the configuration of the ester? Can you predict its sign of rotation?

20.24 The amino acid L-threonine has a specific rotation of +28.3. What is the specific rotation of D-threonine?

20.25 Consider the following four projection formulas. Determine the two missing specific rotations.

CHO; H—OH; H—OH; CH_2OH — $[\alpha]_D = -14.8$

CHO; HO—H; H—OH; CH_2OH — $[\alpha]_D = +19.6$

CHO; H—OH; HO—H; CH_2OH

CHO; HO—H; HO—H; CH_2OH

20.26 What relationships among the specific rotations would exist for the following four compounds?

I: CH_2OH; =O; H—OH; H—OH; CH_2OH

II: CH_2OH; =O; HO—H; HO—H; CH_2OH

III: CH_2OH; =O; H—OH; HO—H; CH_2OH

IV: CH_2OH; =O; HO—H; H—OH; CH_2OH

Projection Formulas

20.27 Convert the following into Fischer projection formulas.

(a) C with CO_2H, H, Br, CH_3

(b) C with CO_2H, NH_2, H, CH_3

20.28 Draw three-dimensional structures for each of the Fischer projection formulas.

(a) CHO; H—C—OH; CH_2; CH_3

(b) CO_2CH_3; Br—C—H; C_6H_5

20.29 Draw the projection formula for L-serine from the following three-dimensional structure.

CO_2H — C (with $HOCH_2$, H, NH_2)

20.30 The projection formula for L-allothreonine is given below. Draw the projection formula for D-allothreonine.

CO_2H
NH_2—H
HO—H
CH_3

Diastereomers

20.31 How many optically active isomers are there of $CH_3CH(OH)CH(OH)CH(OH)CO_2H$?

20.32 Examine the four structures listed in Example 20.4 (Section 20.10). What stereochemical relationships exist among them?

20.33 Ribose is optically active, but ribitol, its reduction product, is optically inactive. Why?

ribose	ribitol
CHO	CH_2OH
H—OH	H—OH
H—OH	H—OH
H—OH	H—OH
CH_2OH	CH_2OH

20.34 Which of the following carbohydrate derivatives are meso compounds?

(a)
CH_2OH
H—OH
=O
H—OH
CH_2OH

(b)
CH_2OH
H—OH
HO—H
HO—H
H—OH
CH_2OH

(c)
CO_2H
H—OH
H—OH
CH_2OH

20.35 There are four isomeric 2,3-dichloropentanes but only three isomeric 2,4-dichloropentanes. Explain why.

20.36 Which of the following structures represents a meso compound?

(a)
CH_2Cl
Br—H
H—Cl
Br—H
CH_2Cl

(b)
CH_3
H—OH
H—OH
HO—H
HO—H
CH_2OH

(c)
CH_2OH
H—OCH_3
H—OCH_3
CH_2OH

20.37 Consider the structure of 5-hydroxylysine, and determine the number of stereoisomers possible.

$NH_2CH_2CH_2CH(OH)CH_2CH_2CH(NH_2)C(=O)OH$

20.38 Consider the structure of pantothenic acid (vitamin B_3), and determine the number of stereoisomers possible.

$HOCH_2—C(CH_3)_2—CH(OH)—C(=O)—NH—CH_2—CH_2—CO_2H$

Stereoisomers in Biochemistry

20.39 Why do biochemical reactions produce one member of a pair of possible enantiomeric compounds?

20.40 D-Glucose is a sugar that the body can metabolize. Suggest what would happen if one were to eat L-glucose.

20.41 The isomer of hydroxycitronellal shown has the odor of lily of the valley. Its mirror image has a minty odor. Explain why.

(Structure: H, CH_3, CHO, CH_3, CH_3, OH)

20.42 L-Dopa is used to treat Parkinson's disease, a disease of the central nervous system. D-Dopa has no effect. Explain why.

(Structure: HO_2C, H, NH_2, OH, OH)

Pasta, in all of its diverse shapes, is an excellent source of carbohydrates. (The Image Bank)

21 Carbohydrates

21.1 Carbohydrates and Life
21.2 Classification of Carbohydrates
21.3 Configuration of Monosaccharides
21.4 Metabolic Reactions of Glucose
21.5 Cyclic Forms of Monosaccharides
21.6 Glycosides
21.7 Oxidation of Monosaccharides
21.8 Reduction of Monosaccharides
21.9 Disaccharides
21.10 Polysaccharides

Essays
- *Lactose Intolerance*
- *Carbohydrates and Human Blood Groups*

21.1 Carbohydrates and Life

Carbohydrates are polyhydroxy aldehydes or ketones or compounds that can be hydrolyzed to form them. Sugars, starches, and cellulose found in the roots, stems, and leaves of all plants are carbohydrates, which makes carbohydrates the most abundant class of biomolecules on Earth. These compounds, formed by photosynthesis in plants, are produced from carbon dioxide of the atmosphere and water.

$$n\,CO_2 + n\,H_2O \xrightarrow{\text{light}} (CH_2O)_n + n\,O_2$$

Starch and cellulose are the two major carbohydrates produced in plants. Starch forms their nutritional reservoir, and cellulose serves a structural role. Phytoplankton, which are the major photosynthetic source of carbohydrates and oxygen in the oceans, convert about 40 trillion kg of carbon dioxide each year.

In photosynthesis solar energy is captured and changed into chemical energy. Carbon dioxide is converted into partially reduced organic compounds that store free energy in carbon–hydrogen bonds. These compounds are important sources of energy for animals who eat the partially reduced carbon compounds and obtain the stored energy by oxidative metabolic reactions. However, not all ingested carbohydrates are immediately oxidized. The body has a storage form for carbohydrates called *glycogen,* which provides energy to drive metabolic processes for a considerable time after the last meal.

$$(CH_2O)_n + n\,O_2 \longrightarrow n\,CO_2 + n\,H_2O$$

Carbohydrates also provide the carbon atoms for intermediates in biochemical reactions. However, carbohydrates are not absolutely required in the diet because the body's metabolic reactions can produce them from other dietary compounds (Chapter 26). Only 75 g of carbohydrates per day is required for the normal operation of the citric acid cycle (Chapter 25).

Carbohydrates serve a variety of roles other than as a biochemical fuel. They are attached to cell membranes and they play a vital role in cellular recognition and in communication between cells. Carbohydrates are also part of the structural backbone of the nucleic acids RNA and DNA.

21.2 Classification of Carbohydrates

Carbohydrates are classified into two groups—simple and complex. **Simple carbohydrates** are monosaccharides like glucose and fructose that cannot be hydrolyzed into smaller molecules. **Complex carbohydrates** are composed of two or more monosaccharides. **Oligosaccharides** are small polymers containing from 2 to 20 linked monosaccharides. The number of monosaccharides is indicated by terms such as disaccharide and trisaccharide, referring to two and three monosaccharide units, respectively. The disaccharide lactose contains two monosaccharides, glucose and galactose.

Polysaccharides have more than 20 monosaccharide units—usually thousands of monosaccharides linked together. Polysaccharides containing only one type of monosaccharide are **homopolysaccharides.** Starch and cellulose in plants and glycogen in animals are homopolysaccharides; they yield only glucose when hydrolyzed. **Heteropolysaccharides** are polysaccharides that contain more than one type of monosaccharide. Hyaluronic acid, found in the vitreous humor of the eye and the synovial fluid of joints, is a heteropolysaccharide.

Heteropolysaccharides make up the shell of a lobster. (The Image Works)

The monosaccharides in complex carbohydrates are linked by acetal or ketal bonds called **glycosidic linkages** in carbohydrate chemistry. These bonds are between the aldehyde or ketone site of one monosaccharide and a hydroxyl group of another monosaccharide. Hydrolysis of this glycosidic linkage yields the component monosaccharides.

Monosaccharides are classified by the number of carbon atoms and the type of carbonyl group. An *-ose* suffix indicates that a compound is a carbohydrate. An *aldo-* or *keto-* prefix indicates that the compound is an aldehyde or ketone, respectively. The number of carbon atoms is given by the prefixes *tri-*, *tetr-*, *pent-*, and *hex-*. Aldoses are numbered from the carbonyl carbon atom, whereas ketoses are numbered from the end of the carbon chain closest to the carbonyl carbon atom.

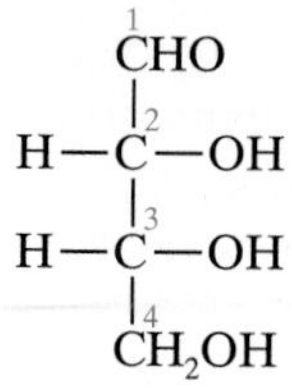

$$\begin{array}{c} {}^{1}\text{CHO} \\ | \\ \text{H}-{}^{2}\text{C}-\text{OH} \\ | \\ \text{H}-{}^{3}\text{C}-\text{OH} \\ | \\ {}^{4}\text{CH}_2\text{OH} \end{array} \qquad \begin{array}{c} {}^{1}\text{CH}_2\text{OH} \\ | \\ {}^{2}\text{C}=\text{O} \\ | \\ \text{H}-{}^{3}\text{C}-\text{OH} \\ | \\ \text{H}-{}^{4}\text{C}-\text{OH} \\ | \\ {}^{5}\text{CH}_2\text{OH} \end{array}$$

an aldotetrose 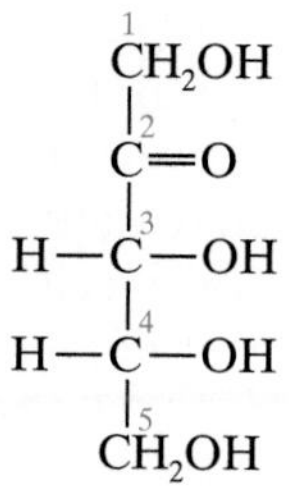 a ketopentose

21.3 Configuration of Monosaccharides

The simplest aldose, glyceraldehyde, occurs naturally only in the D form. Its hydroxyl group is on the right at the chiral carbon atom. Its enantiomer, L-glyceraldehyde, which can be synthesized in the laboratory, has the hydroxyl group on the left at the chiral carbon atom.

$$\begin{array}{c} \text{CHO} \\ | \\ \text{H}-\text{C}-\text{OH} \\ | \\ \text{CH}_2\text{OH} \end{array} \qquad \begin{array}{c} \text{CHO} \\ | \\ \text{HO}-\text{C}-\text{H} \\ | \\ \text{CH}_2\text{OH} \end{array}$$

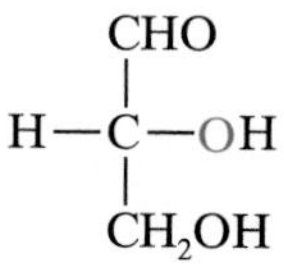

D-glyceraldehyde 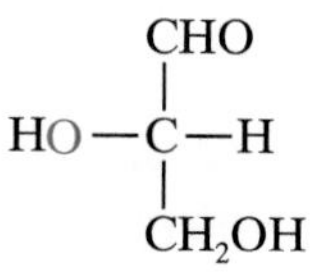 L-glyceraldehyde

Monosaccharides with multiple chiral centers are represented by a Fischer projection formula using eclipsed conformations at all carbon–carbon bonds with the carbon backbone continually pushed back behind the plane of the page. The attached hydrogen atoms and hydroxyl groups point out from the backbone of the carbon chain.

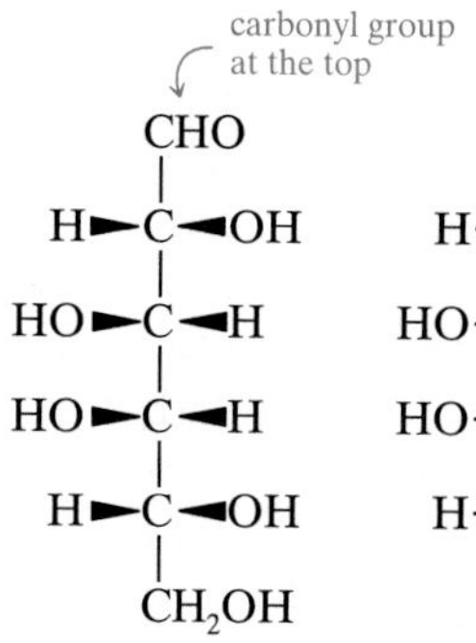

carbonyl group at the top

carbonyl group near the top

$$\begin{array}{c} \text{CHO} \\ | \\ \text{H}\blacktriangleright\text{C}\blacktriangleleft\text{OH} \\ | \\ \text{HO}\blacktriangleright\text{C}\blacktriangleleft\text{H} \\ | \\ \text{HO}\blacktriangleright\text{C}\blacktriangleleft\text{H} \\ | \\ \text{H}\blacktriangleright\text{C}\blacktriangleleft\text{OH} \\ | \\ \text{CH}_2\text{OH} \end{array} \quad \begin{array}{c} \text{CHO} \\ \text{H}-\!\!+\!\!-\text{OH} \\ \text{HO}-\!\!+\!\!-\text{H} \\ \text{HO}-\!\!+\!\!-\text{H} \\ \text{H}-\!\!+\!\!-\text{OH} \\ \text{CH}_2\text{OH} \end{array} \quad \begin{array}{c} \text{CH}_2\text{OH} \\ | \\ \text{C}=\text{O} \\ | \\ \text{HO}\blacktriangleright\text{C}\blacktriangleleft\text{H} \\ | \\ \text{H}\blacktriangleright\text{C}\blacktriangleleft\text{OH} \\ | \\ \text{H}\blacktriangleright\text{C}\blacktriangleleft\text{OH} \\ | \\ \text{CH}_2\text{OH} \end{array} \quad \begin{array}{c} \text{CH}_2\text{OH} \\ =\text{O} \\ \text{HO}-\!\!+\!\!-\text{H} \\ \text{H}-\!\!+\!\!-\text{OH} \\ \text{H}-\!\!+\!\!-\text{OH} \\ \text{CH}_2\text{OH} \end{array}$$

galactose 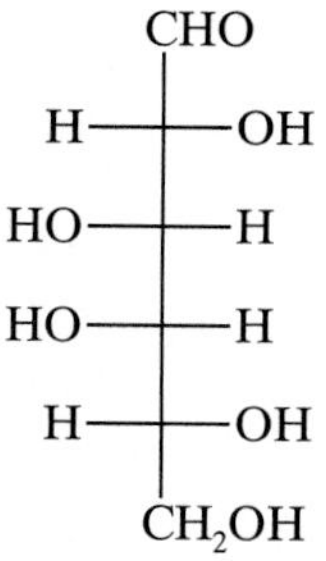 Fischer projection 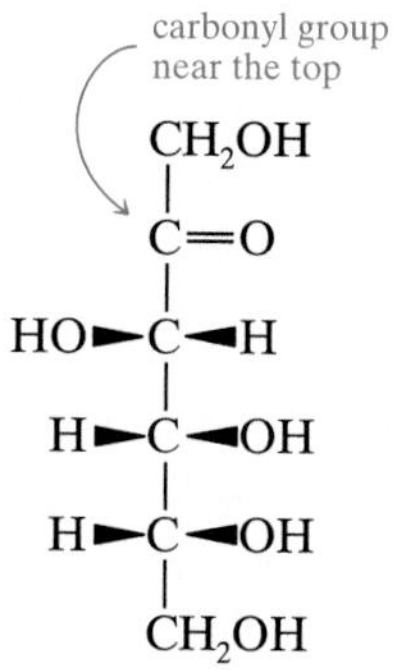 fructose 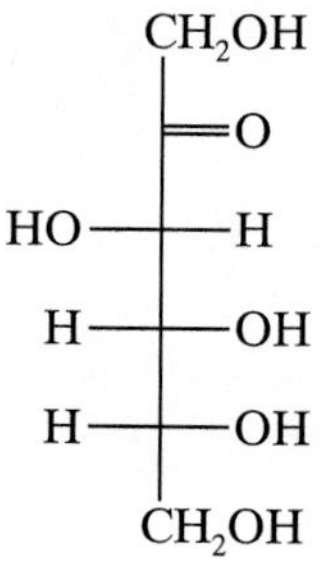 Fischer projection

Because of the way monosaccharides are synthesized in nature, most monosaccharides have the same configuration as D-glyceraldehyde at the chiral carbon atom farthest from the carbonyl group. Each can be derived by "inserting" chiral H—C—OH groups between the carbonyl carbon atom and the terminal —CH_2OH group. Thus, the hydroxyl groups on C-4 of D-ribose and on C-5 of D-glucose and D-fructose are on the right side in the projection formulas.

D-glyceraldehyde	D-ribose	D-glucose	D-fructose
		CHO	CH_2OH
	CHO	H—OH	=O
	H—OH	HO—H	HO—H
CHO	H—OH	H—OH	H—OH
H—OH	H—OH	H—OH	H—OH
CH_2OH	CH_2OH	CH_2OH	CH_2OH

There are few L monosaccharides in nature. They have the hydroxyl group on the left at the chiral center farthest from the carbonyl group in the projection formula. In addition, the configuration of all centers is reversed compared with the D monosaccharides. D and L monosaccharides are enantiomers and are mirror images of each other.

mirror plane

CHO		CHO
HO—C—H		H—C—OH
H—C—OH		HO—C—H
CH_2OH		CH_2OH
determines D configuration		determines L configuration

An aldotetrose has two nonequivalent chiral centers, making possible $2^2 = 4$ stereoisomers or two D,L pairs of enantiomers called erythrose and threose. The D compounds are shown in Figure 21.1. D-Erythrose and L-erythrose are enantiomers, as are D-threose and L-threose. Other combinations, such as D-erythrose and D-threose, are diastereomers. Recall that enantiomers possess the same physical properties; diastereomers do not.

Aldopentoses have three chiral centers and a total of eight stereoisomers, which consist of four pairs of enantiomers. There are four compounds of the D configuration and four of the L configuration. The four compounds of the D configuration shown in Figure 21.1 are diastereomers and have different properties. D-Ribose, which is part of RNA (ribonucleic acid), is produced in the pentose phosphate pathway (Chapter 26) from D-glucose.

Aldohexoses have four chiral centers and a total of 16 stereoisomers comprising eight pairs of enantiomers. The eight diastereomeric compounds of the D configuration are given in Figure 21.1. D-Glucose and D-galactose are the most widely found in nature; D-mannose and D-talose occur in smaller amounts.

Ketoses

The simplest ketose, dihydroxyacetone, does not have any chiral carbon atoms. This ketose is produced in the metabolism of glucose (Chapter 26). Longer chain ketones are related to dihydroxyacetone much like aldoses are related to glyceraldehyde (Figure 21.2). Ketotetroses and ketopentoses are named by inserting *ul* in the name of the structurally related aldose. However, the ketohexoses have names unrelated to the aldohexoses. The ketopentoses ribulose and xylulose are both intermediates in the pentose phosphate pathway (Chapter 26). Fructose is produced by isomerization of glucose in the glycolysis of glucose (Chapter 26).

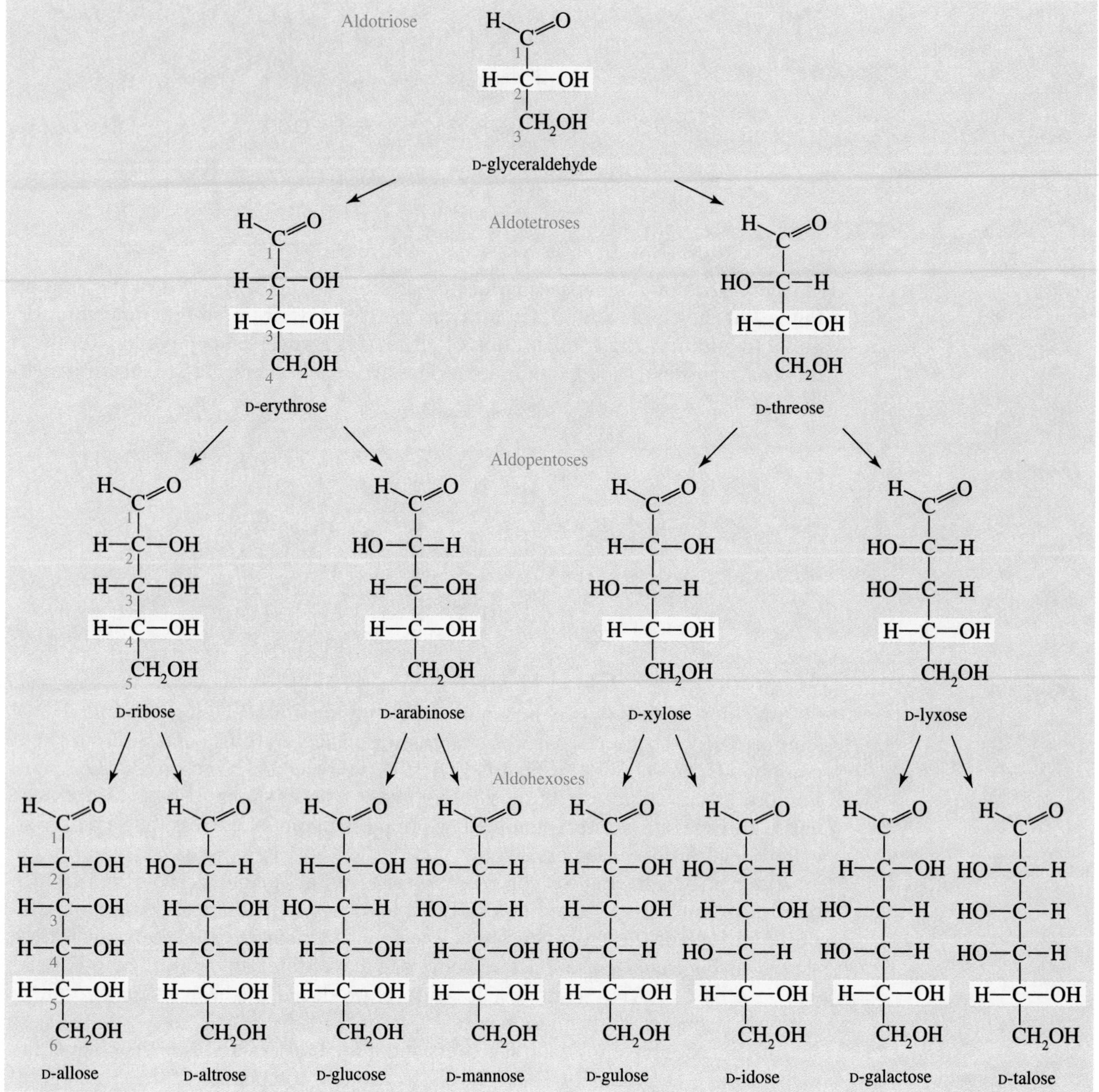

FIGURE 21.1 Structures of the D-Aldoses

EXAMPLE 21.1

L-Arabinose is part of the antiviral drug vidarabine used to control DNA viruses. What is the structure of L-arabinose?

Solution

L-Arabinose, is the enantiomer of D-arabinose—its mirror image. Therefore, the isomer can be drawn by reflecting the planar projection formula in an imagined mirror perpendicular to the plane of the page and parallel to the carbon chain. Each hydroxyl group that is on the right in D-arabinose is on the left in L-arabinose.

FIGURE 21.2 Structures of the D-2-Ketoses

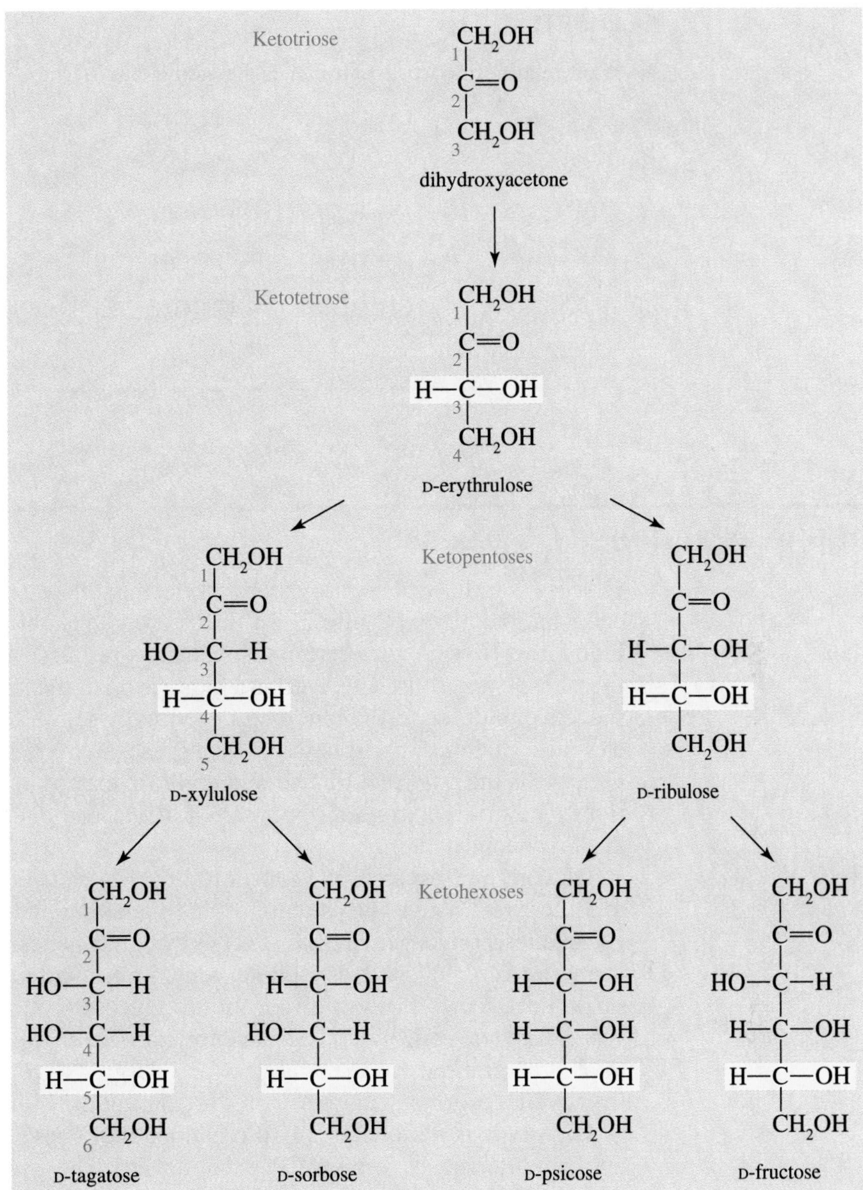

mirror plane

determines D configuration

determines L configuration

D-arabinose

L-arabinose

PROBLEM 21.1

What relationship exists between each possible pair of these compounds?

$$
\begin{array}{ccc}
\begin{array}{c}
CH_2OH \\
| \\
C{=}O \\
| \\
HO{-}C{-}H \\
| \\
H{-}C{-}OH \\
| \\
CH_2OH
\end{array}
&
\begin{array}{c}
CH_2OH \\
| \\
C{=}O \\
| \\
H{-}C{-}OH \\
| \\
H{-}C{-}OH \\
| \\
CH_2OH
\end{array}
&
\begin{array}{c}
CH_2OH \\
| \\
C{=}O \\
| \\
H{-}C{-}OH \\
| \\
HO{-}C{-}H \\
| \\
CH_2OH
\end{array}
\\
\text{(I)} & \text{(II)} & \text{(III)}
\end{array}
$$

21.4 Metabolic Reactions of Glucose

Glucose is rapidly phosphorylated, producing glucose 6-phosphate when it enters a cell from the blood (Figure 21.3). To release glucose from a cell and return it to the blood, the enzyme glucose 6-phosphatase is required to catalyze the hydrolysis of the phosphate group. The liver and kidneys have this enzyme and therefore are exporters of glucose to the blood and eventually to other organs. The brain and muscles lack glucose 6-phosphatase. Once glucose enters these cells, it is metabolized. Glucose is the only fuel for the human brain except under severe starvation conditions. The brain consumes about 120 g of glucose daily, which provides about 400 kcal of chemical energy.

Glucose 6-phosphate plays a central role in cells of organs such as the muscles and the liver. When glucose 6-phosphate is abundant, or if the ATP level is high, **glycogenesis** (glycogen creation) occurs. In glycogenesis, the glucose 6-phosphate is isomerized to glucose 1-phosphate, which is converted to glycogen. The reverse of this process, conversion of glycogen into glucose 6-phosphate, is called **glycogenolysis** (glycogen hydrolysis). Glycogenolysis occurs when glucose 6-phosphate and ATP are required by the muscles for energy. Glycogenolysis also occurs in the liver, which exports glucose to other organs via the blood.

Glycolysis is a sequence of 10 reactions that converts glucose into pyruvic acid and simultaneously forms ATP (Chapter 26). The pyruvic acid produced may be converted into the acetyl group of acetyl coenzyme A (acetyl CoA). Acetyl CoA

FIGURE 21.3 Metabolic Pathways of Glucose

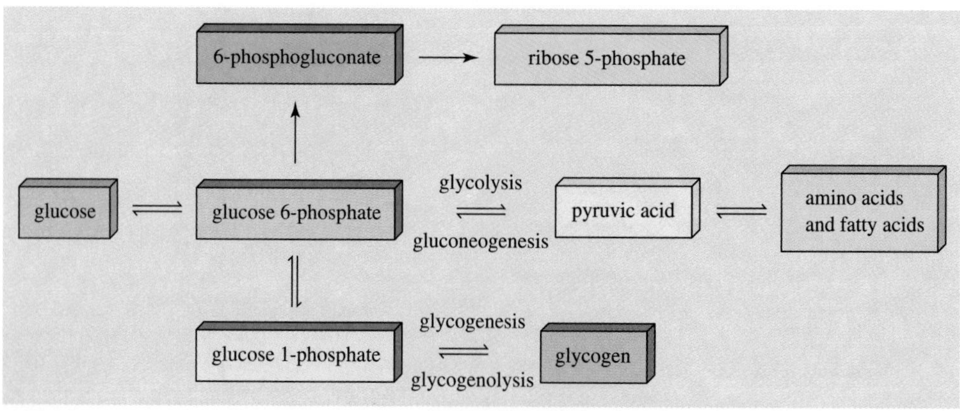

fuels the citric acid cycle, which releases more stored chemical energy and forms more ATP. Thus, glycolysis occurs when ATP is required by the organism or when pyruvic acid is needed as a source of carbon compounds in anabolic processes.

Glucose, and ultimately glycogen, can be synthesized from pyruvic acid. Conversion of pyruvic acid into glucose is called **gluconeogenesis** (Chapter 26). Compounds such as glycerol from fats, lactic acid, and alanine can yield pyruvic acid for gluconeogenesis.

CO_2H	CH_2OH	CO_2H	CO_2H
C=O	CHOH	CHOH	$CHNH_2$
CH_3	CH_2OH	CH_3	CH_3
pyruvic acid	glycerol	lactic acid	alanine

The major site of gluconeogenesis is the liver, but it also occurs in the kidneys. Very little gluconeogenesis occurs in the brain, skeletal muscle, or heart muscle. Thus, gluconeogenesis in the liver is an important process for producing glucose to maintain the glucose level in the blood for use by other organs.

The conversion of glucose 6-phosphate by the cell into ribulose 5-phosphate occurs in the **pentose phosphate pathway.** This process generates nicotinamide dinucleotide phosphate ($NADP^+$) in its reduced form, NADPH. The NADPH formed then reduces biochemicals in anabolic reactions. Ribulose 5-phosphate is converted into ribose 5-phosphate, which is required for the synthesis of ribonucleic acids (Chapter 29).

21.5 Cyclic Forms of Monosaccharides

Recall that aldehydes and ketones react reversibly with alcohols to form hemiacetals and hemiketals, respectively (Section 17.7). When the hydroxyl group and the carbonyl group are part of the same molecule, a cyclic hemiacetal is formed. Cyclic hemiacetals containing five or six atoms in the ring are favored because of the proximity of the two functional groups.

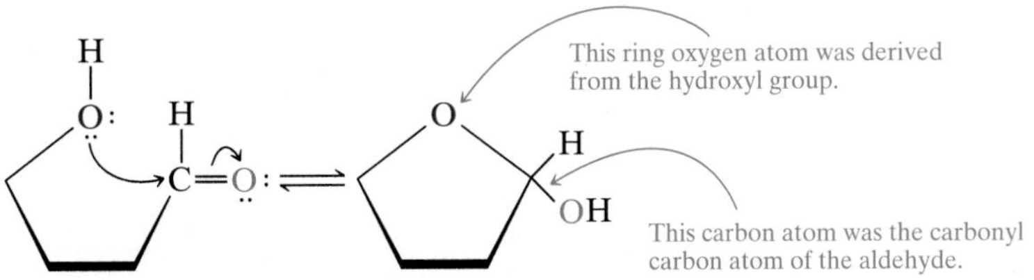

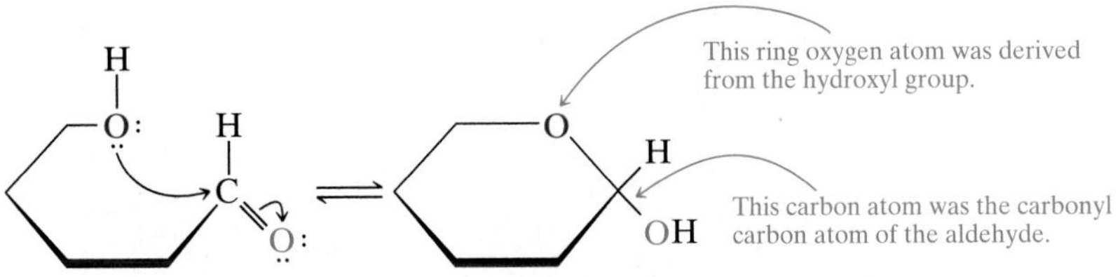

This ring oxygen atom was derived from the hydroxyl group.

This carbon atom was the carbonyl carbon atom of the aldehyde.

This ring oxygen atom was derived from the hydroxyl group.

This carbon atom was the carbonyl carbon atom of the aldehyde.

Five-membered cyclic hemiacetals and hemiketals of carbohydrates are called **furanoses;** six-membered cyclic hemiacetals and hemiketals are called **pyranoses.** Both glucose and fructose react via the hydroxyl group on the C-5 atom. In the case of glucose the product is a six-membered ring and is a pyranose. For fructose the cyclic hemiketal is a five-membered ring and is a furanose. These structures can be represented using the Haworth projection convention.

Haworth Projection Formulas

A Haworth projection formula represents a cyclic hemiacetal or hemiketal as a planar structure viewed edge-on. Bond lines representing atoms toward the viewer appear as heavy wedges, bond lines away from the viewer appear as unaccentuated lines. The carbon atoms are arranged clockwise with the C-1 atom of the aldohexose or aldopentose on the right. For hemiketals the C-2 atom is placed on the right.

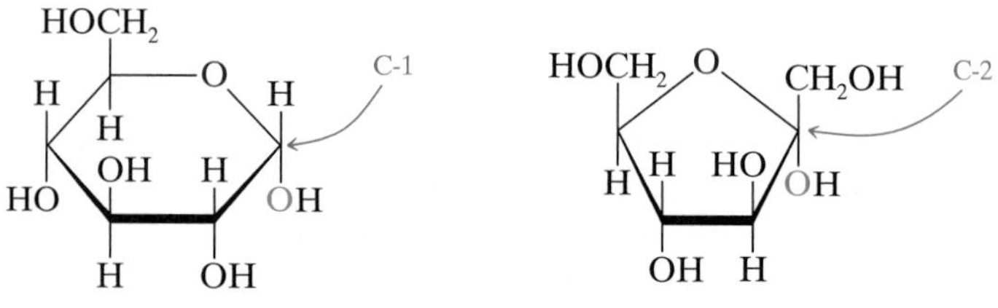

Haworth projection of a pyranose

Haworth projection of a furanose

Let's see how the Fischer projection formula of D-glucose can be converted into a hemiacetal written as a Haworth projection formula. The open-chain form of D-glucose looks like a "C" arranged vertically on the page (Figure 21.4). (Recall that the carbon chain in a Fischer projection formula projects away from the reader.) Now, tilt this curved chain to the right so that it is horizontal. Groups on the right in the Fischer projection project downward, whereas groups on the left project upward.

In this arrangement the C-5 OH group is not near enough to the carbonyl carbon atom to form a ring. To bring the C-5 OH group nearer the carbonyl carbon atom, rotate that part of the structure about the bond between the C-4 and C-5 atoms. The CH_2OH group is now above the plane of the curved carbon chain, and

FIGURE 21.4 Conversion of Fischer Projection into Haworth Projection

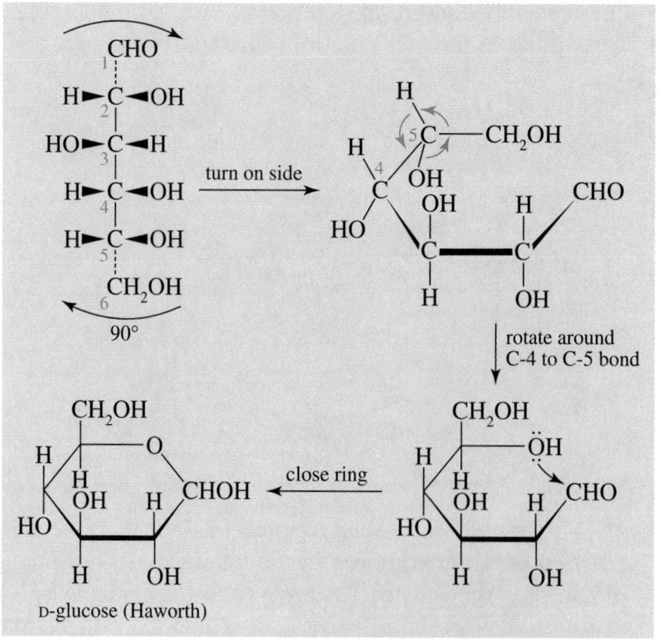

turn on side

rotate around C-4 to C-5 bond

close ring

D-glucose (Haworth)

the C-5 hydrogen atom is below the plane. The oxygen atom of the C-5 OH group in the plane adds to the carbonyl carbon atom. A six-membered ring containing five carbon atoms and one oxygen atoms results. All carbohydrates with a D configuration have the $—CH_2OH$ group located above the ring in a Haworth projection.

When glucose forms a cyclic hemiacetal, four different groups are attached to the C-1 atom. Thus, a new stereogenic center forms at the original carbonyl carbon atom, and two configurations are possible. If the hydroxyl group of the hemiacetal projects below the plane, the compound is α-D-glucopyranose. If it projects above the plane, the compound is β-D-glucopyranose.

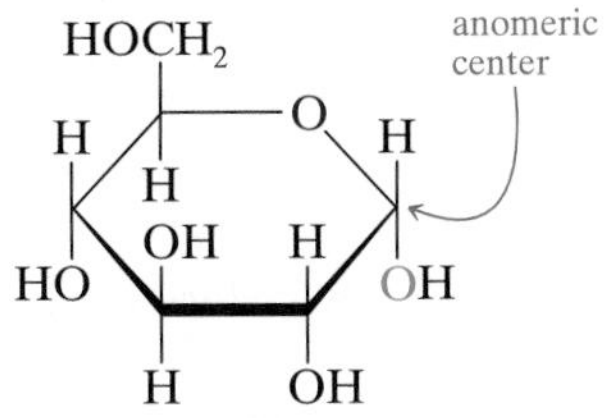

α-D-glucopyranose (a hemiacetal)

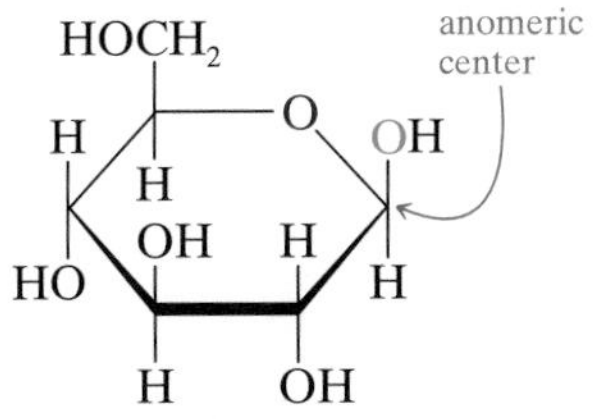

β-D-glucopyranose (a hemiacetal)

The α and β forms of D-glucose are diastereomers that differ in configuration at one stereogenic center. Diastereomers whose configurations differ only at the hemiacetal center are called **anomers.** The stereogenic carbon atom at the hemiacetal center is called the **anomeric carbon atom.**

Aldohexoses are most stable as pyranoses, but substantial quantities of furanoses exist in equilibrium with pyranoses. Now let's consider the cyclic hemiketals of the ketohexose D-fructose. D-Fructose cyclizes in aqueous solution to give a mixture containing 40% of a mixture of α- and β-D-fructofuranose and 60% of a mixture of α- and β-D-fructopyranose. The furanose isomers form when the C-5 OH group adds to the carbonyl carbon atom of the C-2 keto group. A ring of four carbon atoms and one oxygen atom results.

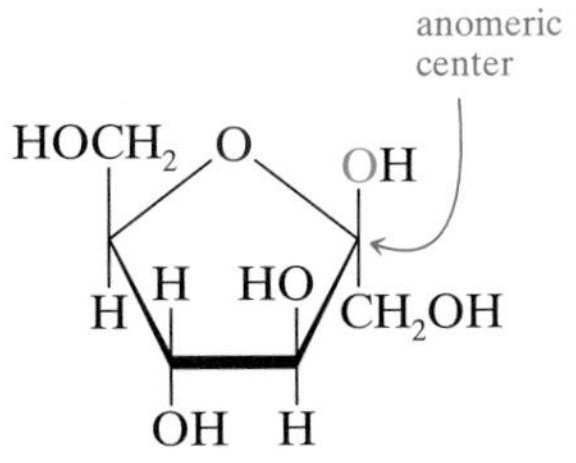

β-D-fructofuranose (a hemiketal)

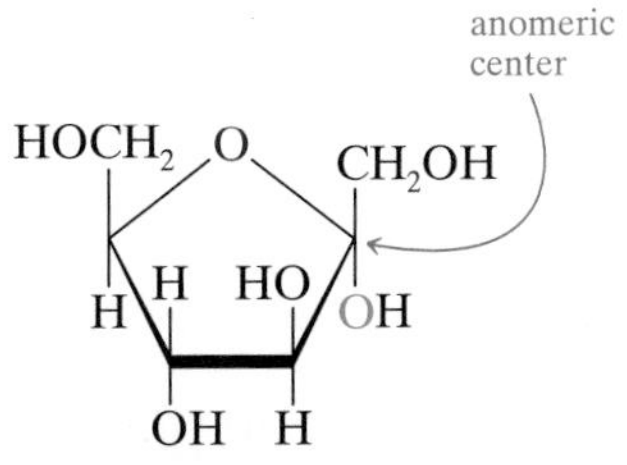

α-D-fructofuranose (a hemiketal)

The pyranose isomers form when the C-6 OH group adds to the C-2 carbonyl carbon atom. Again, α and β designate the configuration of the hydroxyl group at the anomeric carbon atom.

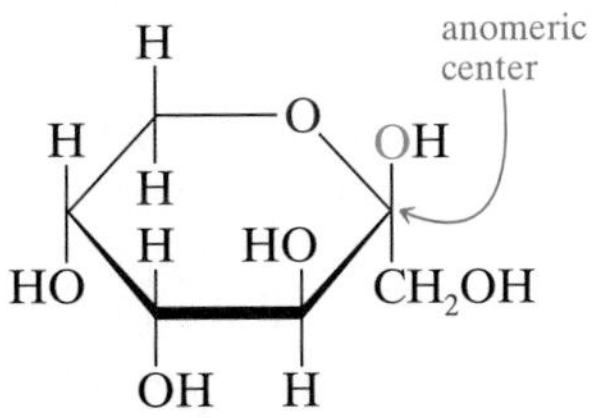

β-D-fructopyranose (a hemiketal)

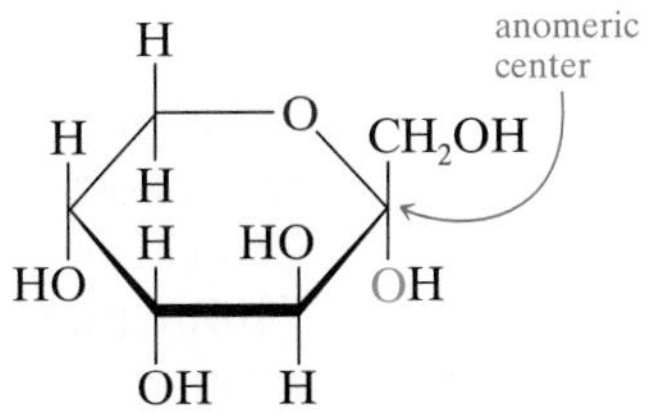

α-D-fructopyranose (a hemiaketal)

EXAMPLE 21.2

Draw the Haworth projection of the α anomer of the pyranose form of D-galactose.

Solution

Since the pyranose form is a six-membered ring, draw a ring with five carbon atoms and one oxygen atom.

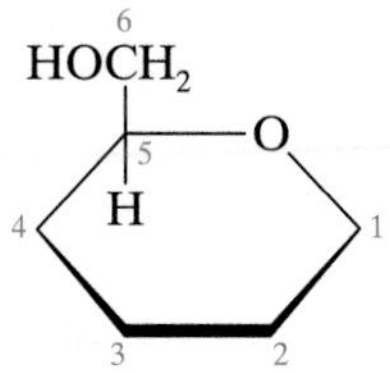

For the D configuration the CH_2OH group is above the plane of the ring. Now place the hydroxyl groups and hydrogen atoms at carbon atoms 2, 3, and 4. An atom or group on the right in the Fischer projection is below the ring of the Haworth projection, and an atom or group on the left is above the ring.

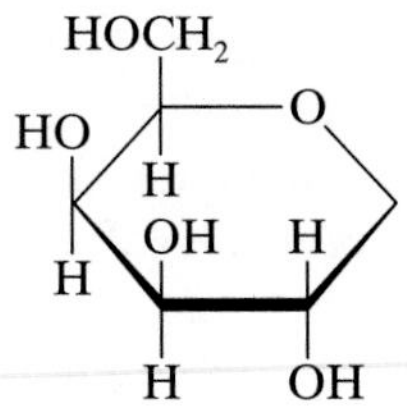

Finally, the α anomer must have a hydroxyl group below the plane of the ring at the anomeric carbon atom, which is the number 1 carbon atom in this case.

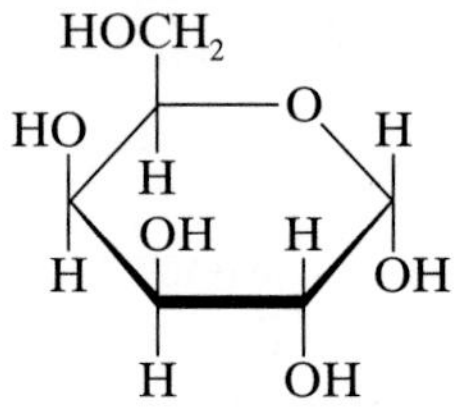

PROBLEM 21.2

Draw the Haworth projection formula of the α anomer of the furanose of D-arabinose.

Mutarotation

Now let's consider the experimental consequences of the formation of anomers of monosaccharides using glucose as an example. Crystalline α-D-glucose melts at 146°C and has $[\alpha]_D = +112.2$. The β anomer melts at 150°C and has $[\alpha]_D = +18.7$. If either anomer is dissolved in water, the specific rotation of the solution slowly changes to an equilibrium value of +54. A gradual change in rotation to an equilibrium point is known as **mutarotation** (from the Latin *mutare*, "to change"). Mu-

FIGURE 21.5 Equilibrium Between Anomeric Forms of Glucose

α-D-glucose (36%) $[\alpha]_D = +112.2$ ⇌ open-chain ⇌ β-D-glucose (64%) $[\alpha]_D = +18.7$

tarotation results from the equilibration of the cyclic hemiacetals with the open-chain form in solution. Ring opening followed by ring closure can form either the α or β anomer (Figure 21.5). Both anomers then exist in equilibrium. A solution of glucose at equilibrium contains 36% of the α anomer and 64% of the β anomer; less than 0.01% of the open-chain form is present.

$$\underset{(36\%)}{\alpha\text{-D-glucose}} \rightleftharpoons \underset{(0.01\%)}{\text{open-chain glucose}} \rightleftharpoons \underset{(64\%)}{\beta\text{-D-glucose}}$$

21.6 Glycosides

Foxglove contains digitalis, which is a glycoside containing a steroidal alcohol as the aglycone. It is used as a medicine in the treatment of heart disease. (Photo Researchers, Inc.)

In Section 17.7 you learned that acid catalyzes the reaction of hemiacetals and hemiketals with alcohols to yield acetals and ketals, respectively. In this reaction an —OR′ group replaces the —OH group. The hemiacetals and hemiketals of monosaccharides also react with alcohols to form acetals and ketals called **glycosides.** The carbon–oxygen bond formed is called a **glycosidic bond.** The non-carbohydrate group bonded to the anomeric carbon atom of a glycoside is an **aglycone.** Since the α- and β-forms of hemiacetals or hemiketals exist in equilibrium, two possible glycosides may form. Glycosides are named by citing the alkyl group first and then replacing the *-ose* ending of the carbohydrate with *-oside.* The configuration at the glycosidic carbon atom must be indicated.

methyl-α-D-glucoside   methyl-β-D-glucoside

Anomeric glycosides are diastereomers and have different physical properties. Glycosides are common in nature and are stable in neutral or basic solution. In acidic solution, glycosides hydrolyze to form a carbohydrate and an aglycone.

EXAMPLE 21.3

Determine the component functional groups of the following compound. From what compounds can the substance be formed?

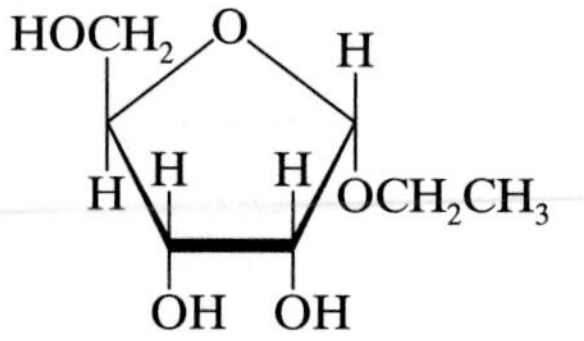

Solution

The compound is a furanose form of a carbohydrate, since there are four carbon atoms and one oxygen atom in a five-membered ring. The ring carbon atom on the right is an acetal because one hydrogen atom and two —OR groups are bonded to it. The acetal has the α configuration, and the alcohol used to form the acetal is ethanol.

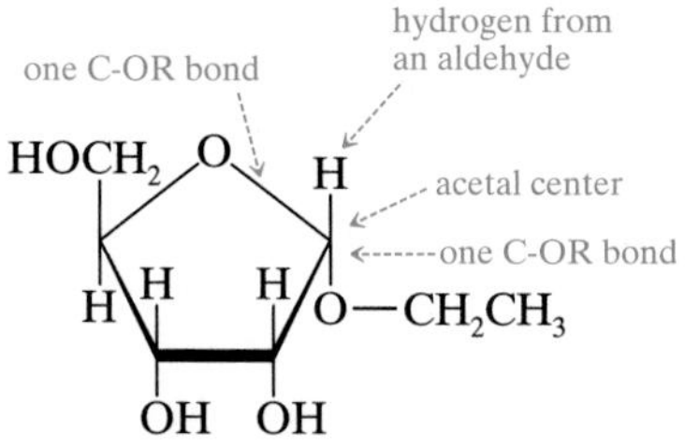

The carbohydrate has the D configuration because the CH_2OH group is "up" in the Haworth projection. The other two chiral carbon atoms of the pentose have hydroxyl groups "down," which corresponds to the right in the Fischer projection. The compound is ribose.

PROBLEM 21.3

Determine the component functional groups of the following compound. From what compounds can the substance be formed?

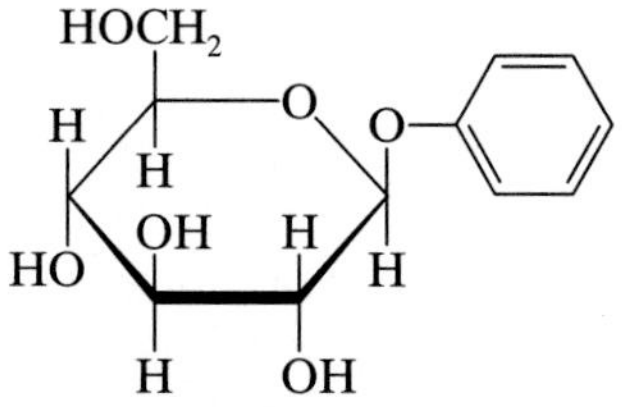

21.7 Oxidation of Monosaccharides

In Section 17.4 you learned that aldehydes are oxidized by Tollens' reagent, Benedict's solution, and Fehling's solution. These reagents also react with aldehyde groups in carbohydrates. For example, Benedict's solution can detect glucose in urine. If no glucose is present in the urine, the Benedict's solution added to it remains blue, but the presence of glucose results in a red precipitate. Depending on the amount of glucose present, the mixture of the precipitate and solution may vary from green to yellow to orange to red with increasing glucose concentration.

Oxidation of an aldose by Benedict's solution forms a glyconic acid. In the specific case of D-glucose, the glyconic acid is called D-gluconic acid. In a basic solution, the conjugate base D-gluconate is formed.

D-glucose (pyranose form) ⇄ D-glucose (open form) $\xrightarrow{Cu^{2+}}$ D-gluconate (anion)

Although aldoses exist as hemiacetals, there is some of the open-chain form with an aldehyde group. As the aldehyde reacts, more of the open-chain form is produced, as expected on the basis of Le Châtelier's principle.

Ketoses also react with Benedict's solution. You would not expect this, because ketones are not oxidized by Benedict's solution. However, an α-hydroxy ketone can isomerize to an α-hydroxy aldehyde (Section 17.10). A ketose forms an enediol that may not only revert to the α-hydroxy ketone but may also form an isomeric α-hydroxy aldehyde.

α-hydroxy ketone ⇌ enediol ⇌ α-hydroxy aldehyde

Movement of hydrogen from the C-2 hydroxyl group to the C-1 atom in the enediol regenerates the original ketose. However, movement of hydrogen from the C-1 hydroxyl group to the C-2 atom forms an aldose. In basic solution, then, the aldose formed reacts with Benedict's solution, and more ketose is converted into the aldose; the equilibrium shifts as predicted by Le Châtelier's principle.

Carbohydrates that react with Benedict's solution are called **reducing sugars.** The term *reducing* refers to the effect of the carbohydrate on Benedict's solution. The carbohydrate is oxidized, but the Benedict's solution is reduced. Both aldoses and ketoses are reducing sugars. Glycosides are not reducing sugars, because they are not hydrolyzed to form an aldose in the basic Benedict's solution.

Other hydroxyl groups of aldoses may be oxidized. For example, the terminal $-CH_2OH$ group can be enzymatically oxidized in cells without oxidation of the aldehyde group. The product is a uronic acid. Oxidation of D-glucose gives D-glucuronic acid, a component in the heteropolysaccharide hyaluronic acid, which is found in the vitreous humor of the eye.

carboxyl group

D-glucuronic acid

EXAMPLE 21.4

Is ribulose (Figure 21.2) a reducing sugar?

Solution

This ketose exists in equilibrium with an enediol intermediate, which is also in equilibrium with ribose.

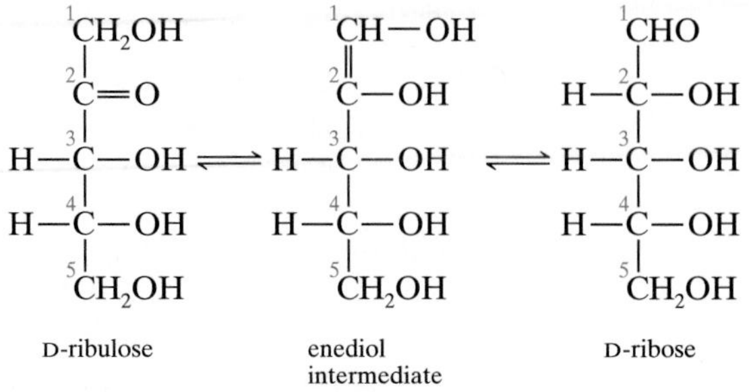

D-ribulose ⇌ enediol intermediate ⇌ D-ribose

Since ribose contains an aldehyde group, ribulose will give a reaction with Benedict's solution and is classified as a reducing sugar.

PROBLEM 21.4

Is the following substance a reducing sugar?

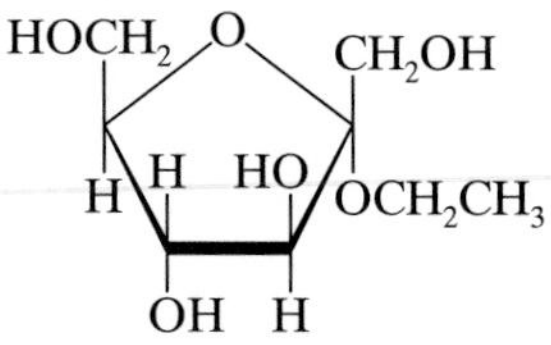

21.8 Reduction of Monosaccharides

Although pentoses and hexoses exist predominately as hemiacetals or hemiketals, they react with reducing agents. Sodium borohydride reduces aldoses and ketoses to polyalcohols called **alditols.** The reduction occurs via the aldehyde group in the small amount of the open-chain form of the aldose in equilibrium with its cyclic hemiacetal. As the aldehyde is reduced, the equilibrium shifts to produce more aldehyde until eventually all of the aldose is reduced.

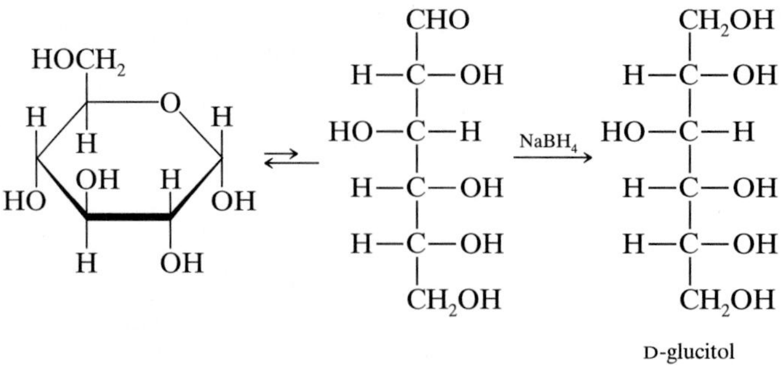

$NaBH_4$

D-glucitol

D-Glucitol occurs in some fruits and berries. Produced and sold commercially as a sugar substitute, it is also called sorbitol.

21.9 Disaccharides

Disaccharides are glycosides of two monosaccharides; one serving as the hemiacetal or hemiketal and the other as the alcohol. One common bonding links the C-1 atom of the hemiacetal to the C-4 atom of the second monosaccharide, which serves as the alcohol. Such bonds are designated ($1 \rightarrow 4$), indicating the location and configuration of oxygen at the glycosidic bond. Maltose, cellobiose, and lactose have $1 \rightarrow 4$ disaccharide bonds. Sucrose has a $1 \rightarrow 2$ disaccharide bond.

Maltose

Maltose is produced by hydrolysis of starch by the enzyme amylase. Further hydrolysis, catalyzed by the enzyme maltase, gives two molecules of D-glucose. The glycosidic oxygen atom of one glucose is α, and it is bonded to the C-4 atom of another glucose. Maltose then is $\alpha(1 \rightarrow 4)$ linked. The right-hand ring of maltose is a hemiacetal and can be either an α- or β-hemiacetal.

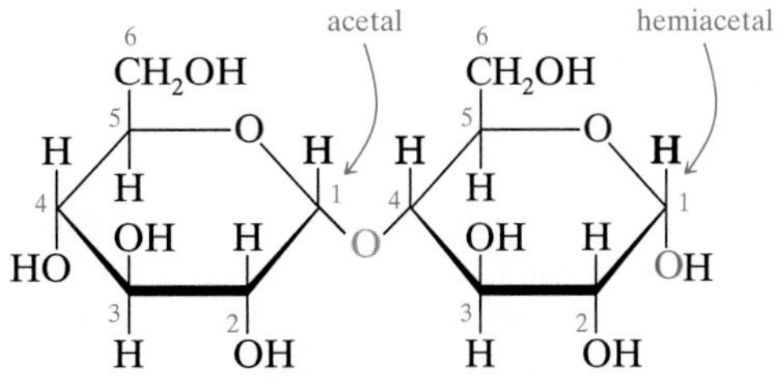

acetal hemiacetal
6 CH_2OH 5 O H 4 H OH H HO 3 H 2 OH 1 O 6 CH_2OH 5 O H 4 H OH H 3 H 2 OH 1 H OH

α-maltose

The prefixes α and β refer to the configuration at the hemiacetal center. This designation should not be confused with the glycosidic bond at the acetal center, which is always α in maltose.

Maltose mutarotates because two isomeric hemiacetals are in equilibrium with an open-chain compound. For the same reason, maltose is a reducing sugar. The aldehyde group formed by ring opening of the hemiacetal can react with Benedict's solution.

Cellobiose

Cellobiose is also composed of two molecules of D-glucose. However, the glycosidic linkage is $\beta(1 \rightarrow 4)$. The hemiacetal can be either α or β. Like maltose, cellobiose will mutarotate and is a reducing sugar. In solution, the two forms of cellobiose exist in equilibrium. Again, do not confuse the hemiacetal center with the acetal center or glycosidic bond, which is always β in cellobiose.

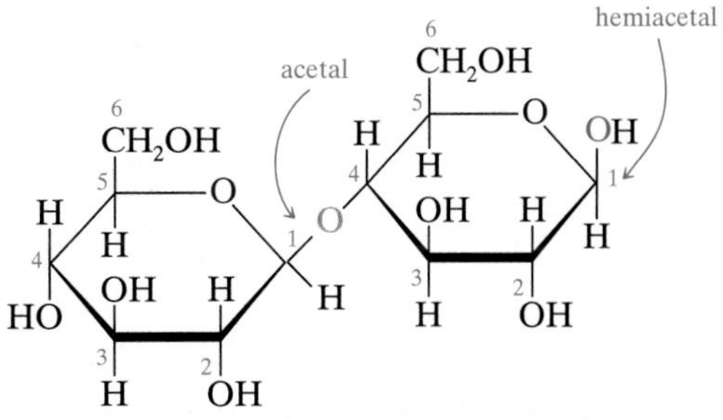

hemiacetal acetal
6 CH_2OH 5 O H 4 H OH H HO 3 H 2 OH 1 H O 6 CH_2OH 5 O H 4 H OH H 3 H 2 OH 1 OH H

β-cellobiose

Cellobiose is produced by hydrolysis of cellulose, a homopolysaccharide. Humans do not have an enzyme to hydrolyze cellobiose. Simple differences in configuration at the $(1 \rightarrow 4)$ linkage result in remarkable changes in the chemical reactivity of biomolecules and the enzymes required to catalyze their reactions.

Lactose

Lactose, often called milk sugar, is a disaccharide found in the milk of mammals, including both humans and cows. In lactose the C-1 atom of galactose is joined by a β-glycosidic linkage to the C-4 atom of glucose.

Lactose Intolerance

People with lactose intolerance lack the enzyme lactase needed to hydrolyze lactose and should not eat food containing lactose. If they ingest food that contains lactose, a high level of unhydrolyzed lactose in their intestinal fluids draws water from tissues by osmosis. The result is abdominal distention, cramping, and diarrhea. Although lactose intolerance is not life-threatening, it is unpleasant.

The level of the enzyme lactase in humans varies with both age and race. Most humans have sufficient lactase for the early years of life, when milk is a major part of their diet. However, in adulthood the enzyme level decreases, and lactose intolerance results. This trait shows remarkable genetic variations. For example, the majority of northern Europeans have high lactase levels, as do several nomadic pastoral tribes in Africa (see the figure). The ability to digest milk as adults may be the result of an evolutionary process in societies that consumed large amounts of milk and milk products such as cheese. Those individuals with the enzyme necessary to digest milk may have had an adaptational advantage.

Some peoples, such as the Thai and the Chinese, have a high lactose intolerance. Similarly, the Ibo and Yoruba of Nigeria cannot tolerate lactose as adults. The Fula and Hausa of the Sudan have quite different lactose intolerances. The Fula raise and milk a breed of cattle called fulani, whereas the Hausa, who show lactose intolerance, do not raise cattle. The Tussi, a cattle-owning class of the Rundi of east Africa, also can digest lactose.

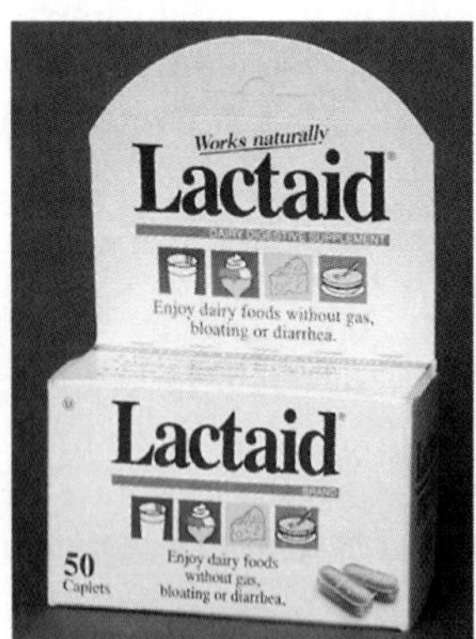

Lactaid, which is available in grocery stores in the dairy section, does not require a prescription. (Peter Arnold, Inc.)

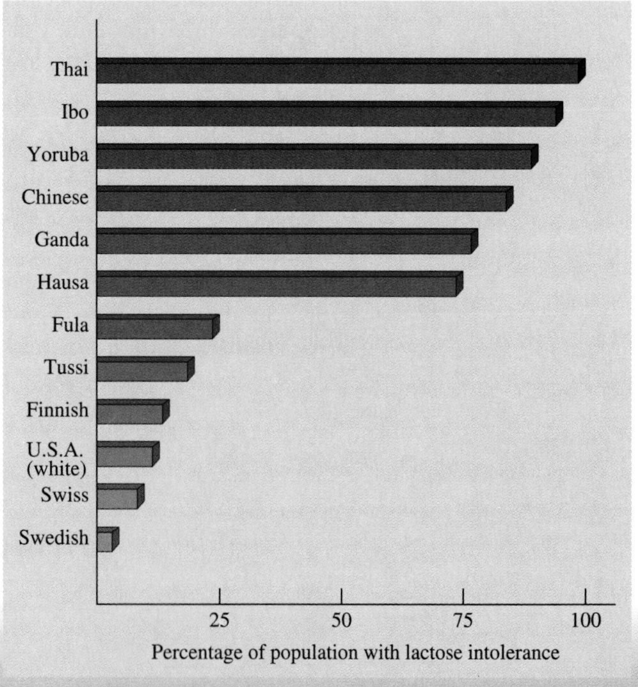

Lactose Intolerance

β-glycosidic bond
CH_2OH H O OH H 4 H OH H H HO CH_2OH O H 1 O OH H H H OH H H H OH

β-lactose

The lactose content of milk varies with species; cow's milk contains about 5% lactose, whereas human milk contains about 7%. The enzyme lactase, present in the small intestine, catalyzes hydrolysis of lactose to form glucose and galactose. Galactose is then isomerized into glucose in a reaction catalyzed by the enzyme UDP-galactose-4-epimerase.

Some infants suffer from a genetic disease called *galactosemia.* These children lack the enzyme necessary to convert galactose into glucose. The children require a diet that excludes galactose. Otherwise the ingestion of milk or other foods containing galactose causes vomiting, diarrhea, cataracts, and even mental retardation.

Sucrose

Sucrose, common table sugar, is composed of glucose and fructose. It is a quite different disaccharide from those discussed to this point. The glycosidic linkage is α at the C-1 atom of glucose and β at the C-2 atom of fructose. Sucrose has both an acetal and a ketal functional group. As a result sucrose cannot mutarotate, nor is it a reducing sugar.

glucose: $HOCH_2$ O H H H OH H HO H OH O
α linkage
fructose: $HOCH_2$ O H H HO CH_2OH OH H
β linkage

sucrose

EXAMPLE 21.5

Describe the structure of the following disaccharide.

CH_2OH O H H OH H HO H H OH O CH_2 O H H H OH H HO OH H OH

Solution

The hemiacetal center located on the ring at the right has a hydroxyl group below the plane of the ring and has the α configuration. The glycosidic bond is from the C-1 atom of the ring on the left to the C-6 atom of the ring on the right. Furthermore, the oxygen bridge is β to the ring on the left. The bridge is $\beta(1 \rightarrow 6)$. Both monosaccharide components are glucose.

PROBLEM 21.5

Describe the structure of the following disaccharide.

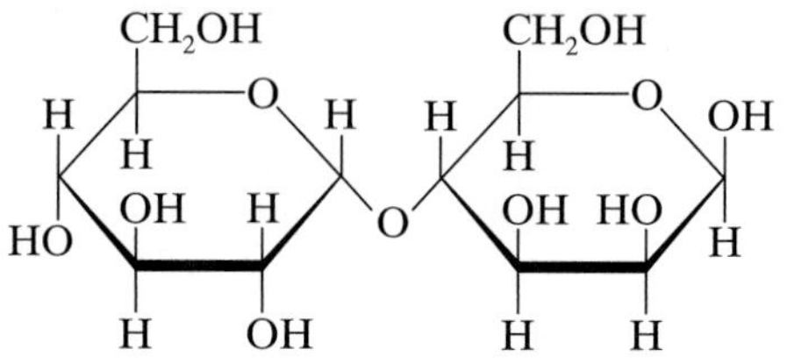

21.10 Polysaccharides

Polysaccharides are polymers of monosaccharides linked to each other by glycosidic bonds similar to those described for disaccharides. Starch and cellulose are homopolysaccharides that contain only glucose. Among the heteropolysaccharides are hyaluronic acid, found in the vitreous humor of the eye; heparin, an anticoagulant in blood; and chondroitin, a component of cartilage and tendons. Since the

FIGURE 21.6 Structures of Amylose and Cellulose

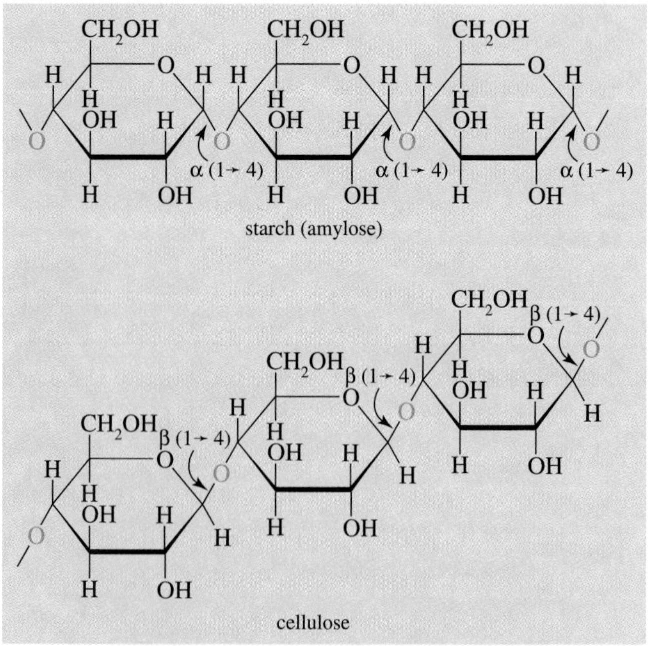

structures of heteropolysaccharides are more complex than those of homopolysaccharides, only homopolysaccharides will be considered in this section.

The glycosidic linkage of polysaccharides (Figure 21.6) is of great biological importance. Only α-linked polysaccharides can be digested by the majority of animals, because they lack the enzyme required to hydrolyze β-linked polysaccharides. However, some bacteria have the enzyme to hydrolyze β-linked polysaccharides. Cattle and other herbivores have these microorganisms in their digestive tracts. Thus these bacteria perform the hydrolysis so that cattle can use the resulting monosaccharides.

Starch is available in potatoes, rice, wheat, and other cereal grains. The molecular weight and structure of starch depend on its source. Amylose and amylopectin are two types of starch. Amylose is a linear polymer (Figure 21.6). The molecular weight of amylose may be between 40,000 and 400,000 amu, corresponding to between 200 and 2000 glucose units. Amylopectin contains chains similar to those in amylose, but only about 25 glucose units occur per chain. Branches of glucose-containing chains are interconnected by a glycosidic linkage between the hydroxyl group of a C-6 of one chain and a C-1 of another glucose chain (Figure 21.7). The molecular weight of amylopectin may be as high as 1 million amu. Since each chain has an average molecular weight of 3000 amu, there may be as many as 300 interconnected chains.

Glycogen is synthesized by animals as a storage form of glucose and is similar in structure to amylopectin. It has a molecular weight greater than 3 million amu. This molecular weight is due not to longer chains but rather to more branching of chains. The average chain length in glycogen is 12 glucose units.

During periods of diminished food intake, animals draw on glycogen stores for energy. Although glycogen is found throughout the body, the largest amounts are in the liver. An adult has enough glycogen for about 15 hours of normal activity.

FIGURE 21.7 Branched Polysaccharide Structure of Amylopectin or Glycogen

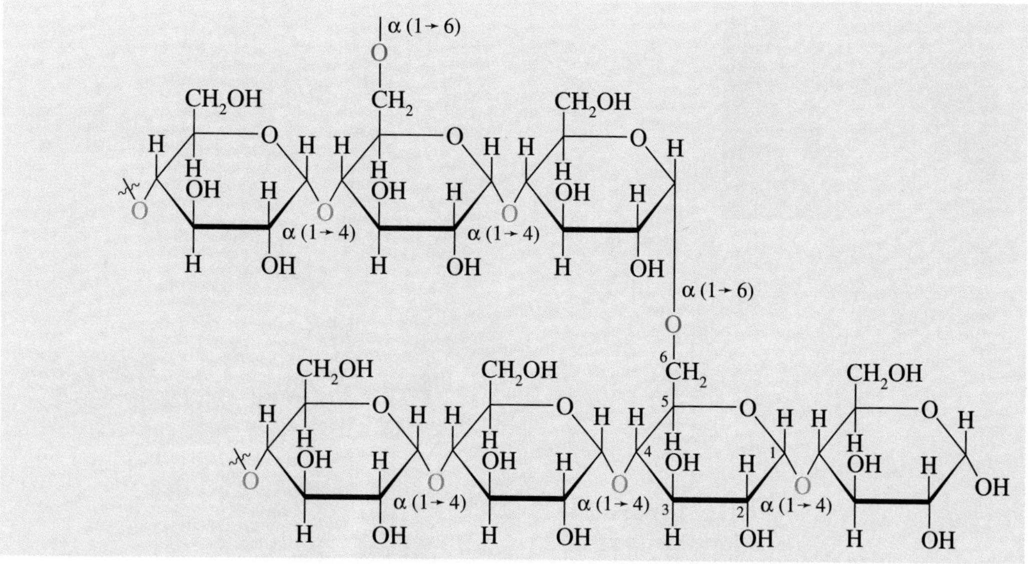

Carbohydrates and Human Blood Groups

Complex carbohydrates coat the surfaces of nearly all human cells, acting as markers that identify the cell. Some of the markers flag the cell as "self." These molecules enable the immune system to avoid attacking the body's own cells, instead recognizing and destroying foreign cells, whether from a transplant or a parasite.

Human blood cells contain surface markers that divide blood into three major classes, designated A, B, and O. Many minor blood groups are also known, but we will not consider them. The classification of blood groups relies on differences in the structures of oligosaccharides bonded to a protein called glycophorin that is embedded in the membrane of red blood cells.

The blood-group oligosaccharides contain several different monomers: galactose (Gal), *N*-acetylgalactosamine (GalNAc), and *N*-acetylglucosamine (GlcNAc). They also contain 6-deoxy-α-L-galactose. This sugar has the common name α-L-fucose.

The α-L-fucose moiety of each blood group is attached to a trisaccharide in blood groups A and B and to a disaccharide in blood group O (see the figure). In each oligosaccharide the β-galactose residue is attached to α-L-fucose by an α-(1 → 2′)-glycosidic bond. The sugar at the reducing end of these oligosaccharides is

N-acetylgalactosamine (GalNAc)

N-acetylglucosamine (GlcNAc)

6-deoxy-α-L-galactose (α-L-fucose)

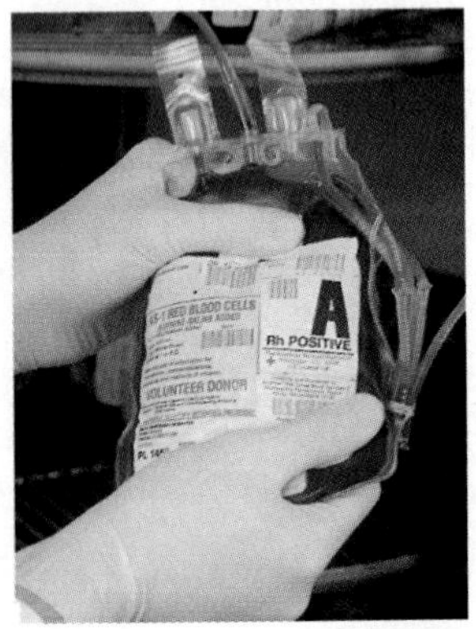

Blood in blood banks is typed and stored for use in transfusions. (Rainbow)

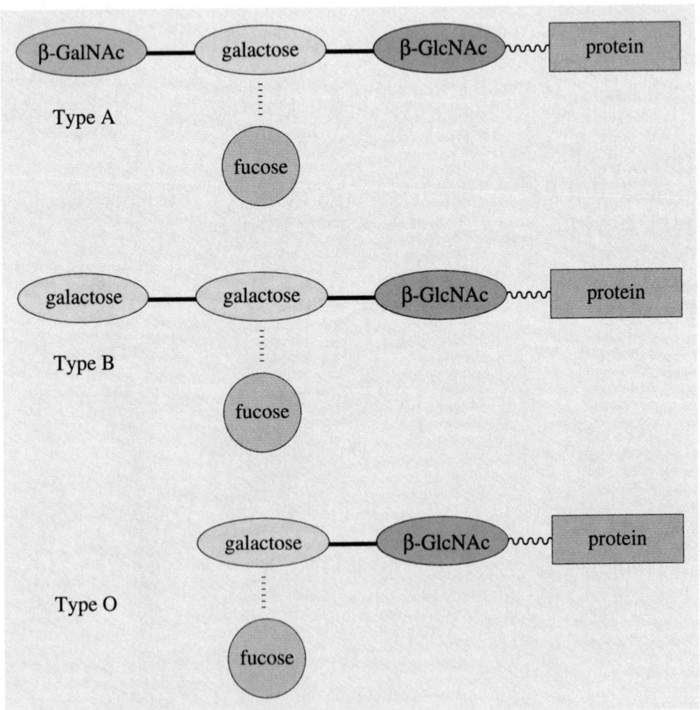

Carbohydrates in Blood Groups

linked to glycophorin by an α-glycosidic bond to the hydroxyl group of a serine residue in the protein.

Each blood group is further subdivided into two types of chains that differ in their glycosidic linkages. In a type 1 chain, the β-Gal moiety is linked to β-GlcNAc by a $(1 \rightarrow 4')$-glycosidic bond. In a type 2 chain, the β-Gal moiety is linked to β-GlcNAc by a $(1 \rightarrow 3')$-glycosidic bond.

The carbohydrates are the *antigenic determinants* of their groups. A person with type A blood makes antibodies that "attack" type B blood, forming clumps of type B cells. Similarly, a person with type B blood makes antibodies that "attack" type A blood. However, persons who are type A or type B do not make antibodies against type O blood, so type O persons are called "universal donors." They are not, however, universal acceptors because they produce antibodies against both type A and type B blood.

Summary

The simplest carbohydrates, known as **monosaccharides,** are polyhydroxy aldehydes and polyhydroxy ketones. The aldehydes are called **aldoses,** whereas the ketones are called **ketoses.** The configuration at the chiral carbon atom farthest from the aldehyde or ketone group determines the D or L configuration of the compound.

Aldoses and ketoses form cyclic hemiacetals and hemiketals. The five-membered rings are **furanoses,** whereas the six-membered rings are **pyranoses.** A **Haworth projection** is used to represent the hemiacetals and hemiketals. The additional chiral center formed in the hemiacetal or hemiketal is called the **anomeric** center. This center is responsible for the phenomenon of **mutarotation,** in which the optical rotation of a mixture of **anomers** attains an equilibrium value.

Monosaccharides can react with alcohols to form acetals and ketals, which are called **glycosides.** If the alcohol is another monosaccharide, a **disaccharide** results.

Oligosaccharides are small polymers containing from 2 to 20 linked monosaccharides. **Polysaccharides** contain many monosaccharide units joined by glycosidic bonds. The most common polysaccharides in plants are **starch** and **cellulose.** In starch there are $\alpha(1 \rightarrow 4)$ glycosidic bonds between glucose units, whereas in cellulose there are $\beta(1 \rightarrow 4)$ glycosidic bonds. In animals, **glycogen** is a branched polymer of glucose containing $\alpha(1 \rightarrow 4)$ and $\alpha(1 \rightarrow 6)$ glycosidic bonds.

Reducing sugars react with Benedict's solution to give a red precipitate. Monosaccharides are reducing sugars. Disaccharides with a hemiacetal or hemiketal center are also reducing sugars.

When glucose 6-phosphate is abundant in a cell, or if the ATP level is high, **glycogenesis** (glycogen creation) occurs. The reverse of this process, conversion of glycogen into glucose 6-phosphate, is called **glycogenolysis** (glycogen hydrolysis). **Glycolysis** is a sequence of 10 reactions that convert glucose into pyruvic acid and simultaneously form ATP. Conversion of pyruvic acid into glucose is called **gluconeogenesis.** The conversion of glucose 6-phosphate by the cell into ribulose 5-phosphate occurs in the **pentose phosphate pathway.** In this pathway ribulose 5-phosphate is converted into ribose 5-phosphate, which is required for the synthesis of ribonucleic acids.

Infants who lack the enzyme necessary to convert galactose into glucose suffer from a genetic disease called **galactosemia. Lactose intolerance** is a condition in which people lack the enzyme lactase (also known as β-galactosidase) needed to hydrolyze lactose.

Exercises

Classification of Monosaccharides

21.1 What is an aldose? How does it differ from a ketose?

21.2 To what carbon atom do the letters D and L refer in monosaccharides?

21.3 To which category does each of the following monosaccharides belong?

(a)
CHO
H—|—OH
H—|—OH
H—|—OH
CH_2OH

(b)
CHO
HO—|—H
HO—|—H
H—|—OH
H—|—OH
CH_2OH

(c)
CH_2OH
=O
H—OH
H—OH
H—OH
CH_2OH

21.4 To which category does each of the following monosaccharides belong?

(a)
CHO
H—OH
HO—H
H—OH
CH_2OH

(b)
CHO
HO—H
HO—H
H—OH
HO—H
CH_2OH

(c)
CH_2OH
=O
H—OH
HO—H
HO—H
CH_2OH

21.5 Draw the Fischer projection formula of D-mannose.

21.6 Draw the Fischer projection formula of D-ribose.

21.7 Classify each of the following monosaccharides as D or L.

(a)
CH_2OH
=O
H—OH
H—OH
CH_2OH

(b)
CHO
HO—H
H—OH
H—OH
H—OH
CH_2OH

(c)
CH_2OH
=O
HO—H
CH_2OH

21.8 Classify each of the following monosaccharides as D or L.

(a)
CH_2OH
=O
HO—H
HO—H
CH_2OH

(b)
CH_2OH
HO—H
=O
HO—H
H—OH
CH_2OH

(c)
CH_2OH
=O
HO—H
CH_2OH

Glucose and Metabolism

21.9 What is the role of glucose 6-phosphatase in the cell? Which organs contain this enzyme?

21.10 Distinguish between glycogenolysis and gluconeogenesis.

21.11 Which organs synthesize and store glucose? Which organs can export glucose?

21.12 In what way can the liver form glucose for release to other organs?

Haworth Projection Formulas

21.13 Draw the Haworth projection of the pyranose form of each of the following compounds.

(a) α-D-mannose (b) β-D-galactose
(c) α-D-glucose (d) α-D-galactose

21.14 Draw the Haworth projection of the furanose form of each of the following compounds.

(a) α-D-fructose (b) β-D-fructose
(c) α-D-ribulose (d) β-D-xylulose

21.15 Identify each of the following structures by the name of the monosaccharide.

(a)
CH_2OH
O
HO H
H
OH H
H OH
H OH

(b)
CH_2OH
O
H OH
H
OH HO
HO H
H H

(c)
H
O
H H
H
H H
HO OH
OH OH

21.16 Identify each of the following structures by the name of the monosaccharide.

(a)

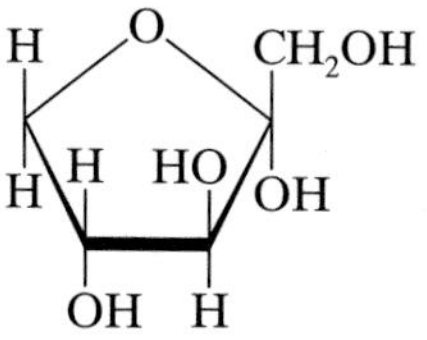

(b)

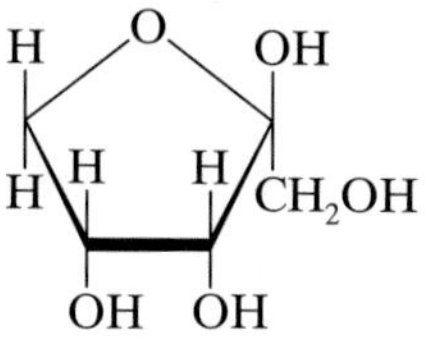

(c)

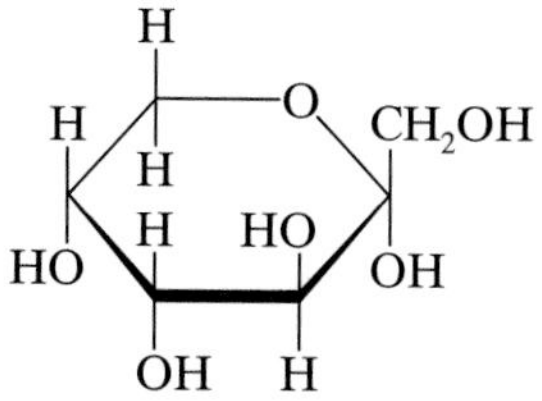

(II)

21.17 Which of the compounds in Exercises 21.15 and 21.16 are pyranoses, and which are furanoses?

21.18 What is the configuration at the anomeric carbon atom of each compound in Exercises 21.15 and 21.16?

Mutarotation

21.19 Can all aldopentoses and aldohexoses mutarotate?

21.20 Will L-glucose mutarotate?

21.21 Name two disaccharides that will mutarotate.

21.22 What disaccharide will not mutarotate?

21.23 Which of the following compounds can mutarotate?

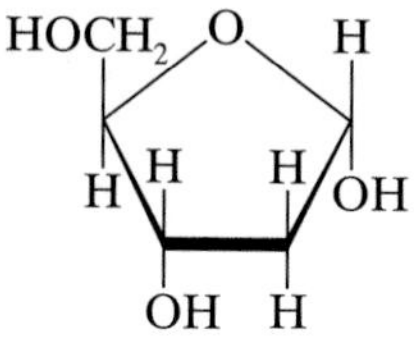

I

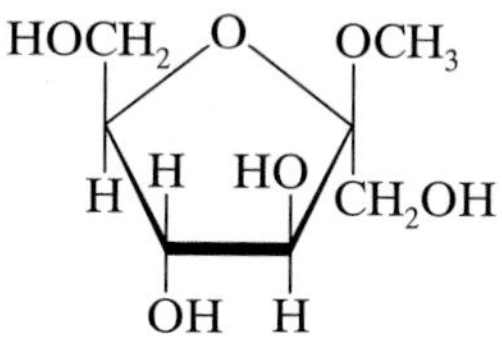

II

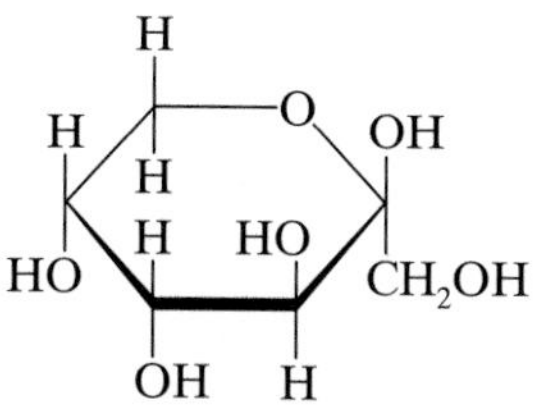

III

21.24 Which of the following compounds can mutarotate?

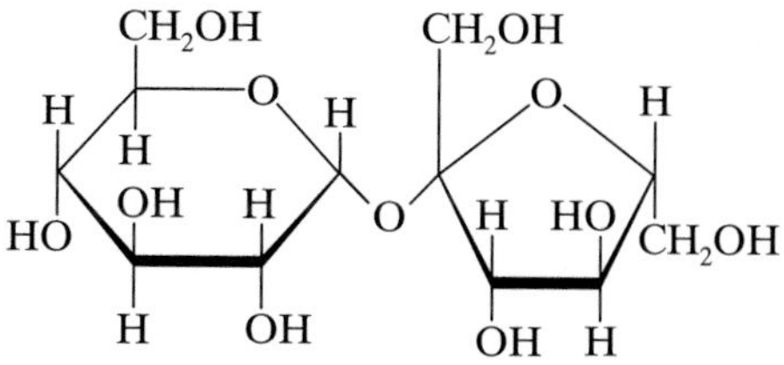

(I)

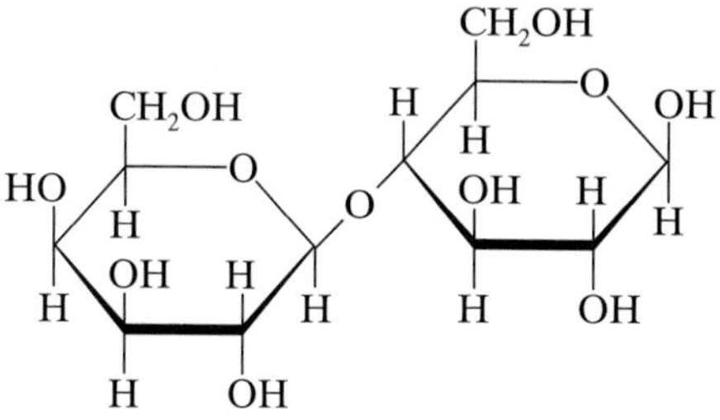

Glycosides

21.25 Draw the Haworth projection formulas of the two glycosides formed from the pyranose form of glucose and methyl alcohol.

21.26 Draw the Haworth projection formulas of the two glycosides formed from the furanose form of fructose and ethyl alcohol.

21.27 Identify the following compound by its components, the configuration at the acetal or ketal center, and the ring form.

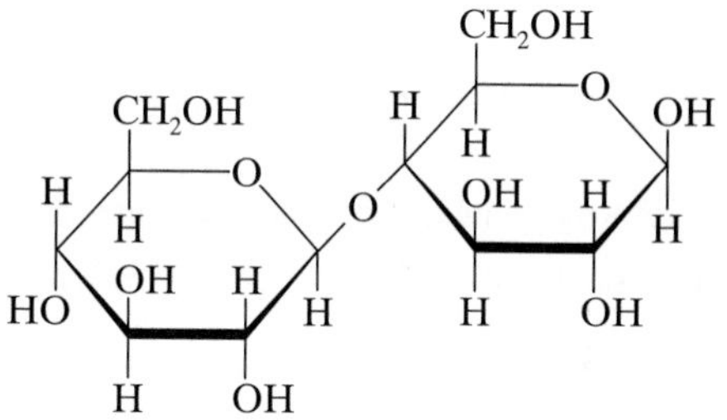

21.28 Identify the following compound by its components, the configuration at the acetal or ketal center, and the ring form.

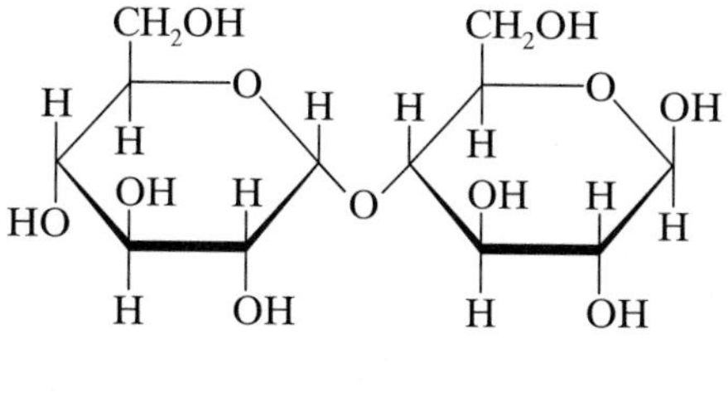

Disaccharides

21.29 What monosaccharides are produced from the hydrolysis of lactose?

21.30 What monosaccharides are produced from the hydrolysis of sucrose?

21.31 Describe the type of glycosidic linkage in the following compound.

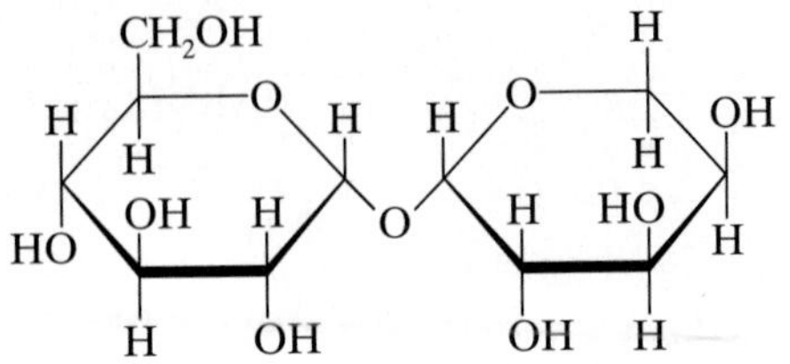

21.32 Describe the type of glycosidic linkage in the following compound.

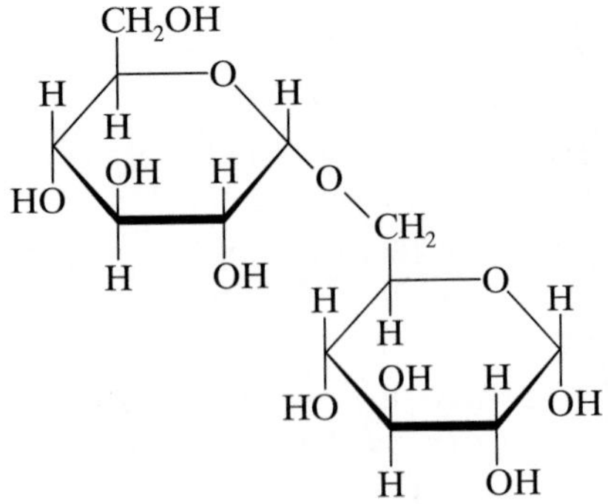

Polysaccharides

21.33 What are the structural similarities between amylose and amylopectin?

21.34 What are the structural differences between amylose and cellulose?

21.35 Explain why humans can't digest cellulose.

21.36 What is the biological importance of glycogen?

Isomerization of Monosaccharides

21.37 Write the structure of the aldose and ketose that can exist in equilibrium with D-glucose in basic solution.

21.38 Write the structure of the aldose and ketose that can exist in equilibrium with D-galactose in basic solution.

21.39 Write the structure of one aldose and one ketose that can exist in equilibrium with D-ribose in basic solution.

21.40 Write the structure of two aldoses that can exist in equilibrium with D-xylulose in basic solution.

Oxidation of Carbohydrates

21.41 Ketones do not react with Benedict's solution, but fructose does. Explain this difference.

21.42 Are all aldoses reducing sugars?

21.43 Why is the ketose D-ribulose a reducing sugar?

21.44 Which common disaccharides are reducing sugars?

21.45 Is the disaccharide in Exercise 21.31 a reducing sugar?

21.46 Is the disaccharide in Exercise 21.32 a reducing sugar?

Reduction of Monosaccharides

21.47 D-Xylitol is used as a sweetener in some chewing gums. Deduce the structure of D-xylitol from its name.

21.48 Identify the two compounds that result from the reduction of D-fructose by $NaBH_4$. Why are two compounds formed?

22

Lipids

These King penguins have layers of fat that allow them to live in Antarctica. (Photo Researchers, Inc.)

22.1 Classification of Lipids
22.2 Terpenes
22.3 Steroids
22.4 Waxes
22.5 Fatty Acids
22.6 Triacylglycerols
22.7 Glycerophospholipids
22.8 Sphingophospholipids
22.9 Glycosphingolipids
22.10 Biological Membranes
22.11 Transport across Membranes

Essays
- *Whale Oil*
- *Eicosanoids*

22.1 Classification of Lipids

Unlike carbohydrates and proteins, lipids are not as easily characterized by their structural features. Lipids encompass a range of molecular structures and, as a result, an extraordinary range of biochemical functions. As the molecules making up dietary fat, certain lipids provide a major source of metabolic energy. As components of biological membranes, lipids provide an insoluble partition between a cell and its watery environment. Lastly, as hormones, lipids regulate a wide spectrum of cellular activities.

Some of the functions of lipids are performed by individual molecules, such as steroidal hormones. Other functions are the result of aggregates of lipids that function as units, like the glycerophospholipids and sphingolipids in biological membranes. The composition of biological membranes and the highly specialized functions they serve are a major focus of this chapter.

Lipids are relatively nonpolar compounds, and they can be separated from more polar cellular substances by their solubility in nonpolar organic solvents. In fact, lipids were historically classified as compounds of biological origin that are soluble in organic solvents. The term lipid is sometimes used as a synonym for fat (from the Greek *lipos,* "fat"). However, fat is only one of the various types of lipids.

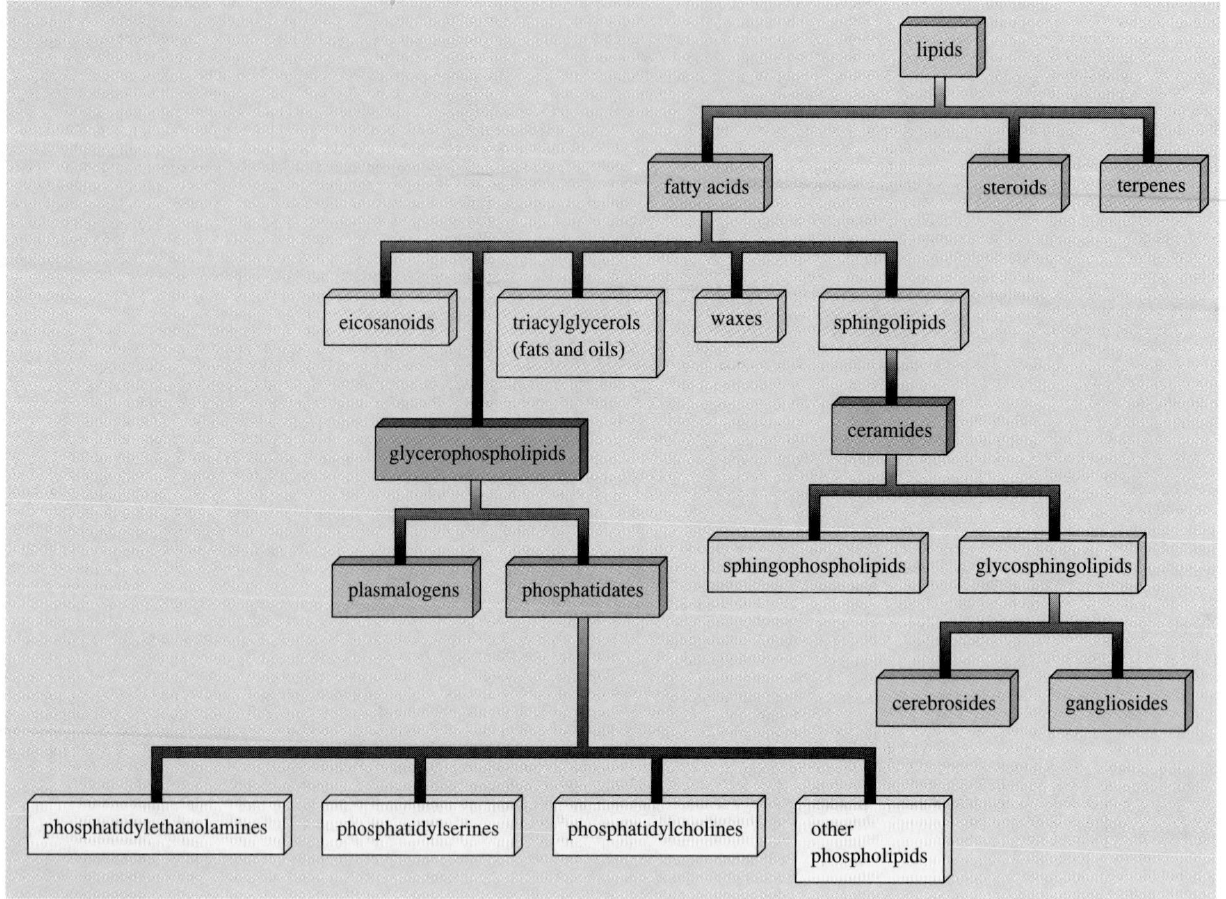

FIGURE 22.1
Classification of Lipids

The major classes of lipids and their relationship to one another are listed in Figure 22.1. Terpenes, produced in plants, and steroids, which are important hormones in animals, are not discussed in much detail in this chapter. The emphasis of this chapter is on the remaining lipids, all of which can be hydrolyzed by aqueous basic solution. The major hydrolysis products of these lipids are the long-chain carboxylic acids (with an even number of carbon atoms) called *fatty acids.* The other components resulting from the hydrolysis of a lipid determine its subclass.

1. **Waxes** are esters of long-chain alcohols and fatty acids.
2. **Triacylglycerols,** also known as *triglycerides,* are esters of glycerol and long-chain fatty acids.
3. **Glycerophospholipids** are composed of glycerophosphate (an ester of glycerol and phosphoric acid), long-chain fatty acids, and certain low molecular weight alcohols.
4. **Sphingolipids** are composed of sphingosine, a long-chain fatty acid present as an amide, and at least one other component. There are two major subclasses of sphingolipids: *Sphingophospholipids* contain a phosphate unit and choline and thus structurally resemble glycerophospholipids. *Glycosphingolipids* contain carbohydrate units but not a phosphate unit.

Waxes, which are located on the surfaces of plants and some animals, provide a protective barrier against water. Triacylglycerols function as a long-term reserve source of energy for animals as well as providing thermal insulation. Glycerophospholipids and sphingolipids are components of cell membranes.

22.2 Terpenes

Terpenes are abundant in the oils of plants and flowers. They are responsible for the odors of pine trees and for the colors of carrots and tomatoes. Vitamins A, E, and K are terpenes.

Terpenes consist of two or more *isoprene* units joined together, usually head to tail. These compounds may have different degrees of unsaturation and a variety of functional groups. Nevertheless, the isoprene units are usually easy to identify.

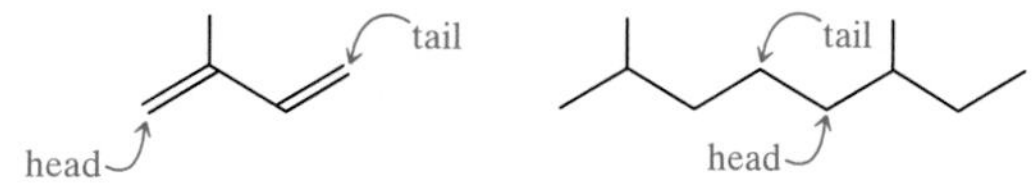

tail
head
tail
head

isoprene
2-methyl-1,3-butadiene

two isoprene units
linked head to tail

Farnesol, which is part of the coating of apples, contains three isoprene units joined head to tail. Dashed lines indicate where the three units are joined. Carvone, found in oil of caraway, contains a ring that consists of two isoprene units.

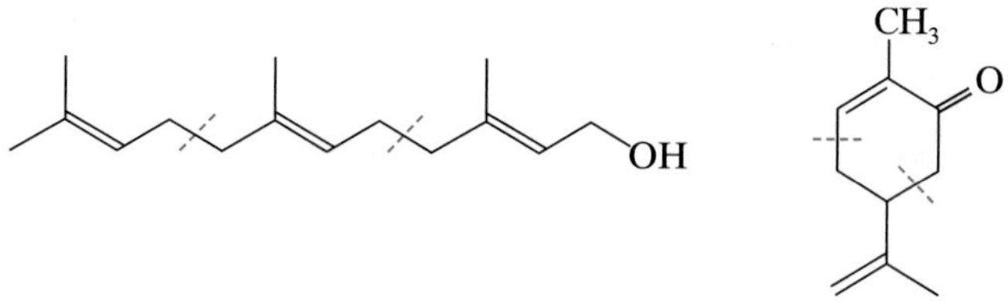

OH
CH_3
O

isoprene units in farnesol

isoprene units in carvone

Terpenes are classified by the number of isoprene units they contain. The **monoterpenes,** the simplest terpene class, contain two isoprene units, and **sesquiterpenes** have three isoprene units. Examples of these structures are shown in Figure 22.2 using bond-line structures. **Diterpenes, triterpenes,** and **tetraterpenes** contain 4, 6, and 8 isoprene units, respectively.

22.3 Steroids

Steroids are tetracyclic compounds containing three six-membered rings and a five-membered ring. These compounds contain a variety of functional groups such as hydroxyl, carbonyl, and carbon–carbon double bonds. Each ring is assigned a letter, and the carbon atoms are numbered by a standard system.

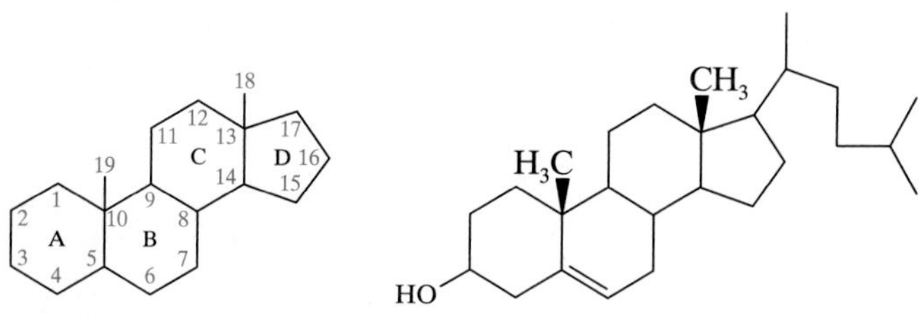

18 12 11 13 17 C D 16 19 14 15 1 9 2 10 8 A B 3 5 7 4 6

steroid ring system

CH_3
H_3C
HO

cholesterol

FIGURE 22.2 Classification of Terpenes

monoterpene

α-phellandrene (eucalyptus)

sesquiterpene

α-selinene (celery)

diterpene

OH

vitamin A (present in mammalian tissue and fish oil)

triterpene

squalene (shark liver oil)

tetraterpene

β-carotene (present in carrots and other vegetables)

The Role of Cholesterol

This weight lifter from Russia depends on androgens for his muscle development. (Photo Researchers, Inc.)

Cholesterol is a steroid that is synthesized in most animals. It occurs in amounts up to 25% in cell membranes; the higher the percentage of cholesterol, the more rigid is the cell membrane (Section 22.10). In humans the central nervous system has a high cholesterol content, and the brain is about 10% cholesterol. Human plasma contains about 50 mg of free cholesterol per 100 mL and about 170 mg of cholesterol esterified with various fatty acids.

There is debate about the role of dietary sources of cholesterol in heart disease. However, about 800 mg of cholesterol per day is synthesized by the liver and to some extent by the intestine of an adult on a low-cholesterol diet. When cholesterol is absorbed from dietary sources, the synthesis process is repressed.

Arteriosclerosis or "hardening of the arteries" is one of the problems of aging. When accompanied by a buildup of cholesterol deposits on the inner surfaces of arteries, the condition is called atherosclerosis. Although not all the reasons for the deposition of cholesterol are known, one is the body's decreased ability to metabolize fat as one grows older. Therefore, the cholesterol content of the tissues and blood increases.

Cholesterol is a precursor for the formation of several steroid hormones. **Hormones** provide a communication pathway between tissues in multicellular organisms. Cholesterol is converted to progesterone by shortening the chain attached at the C-17 position (ring D). Progesterone is converted to corticosteroids and sex hormones (Figure 22.3).

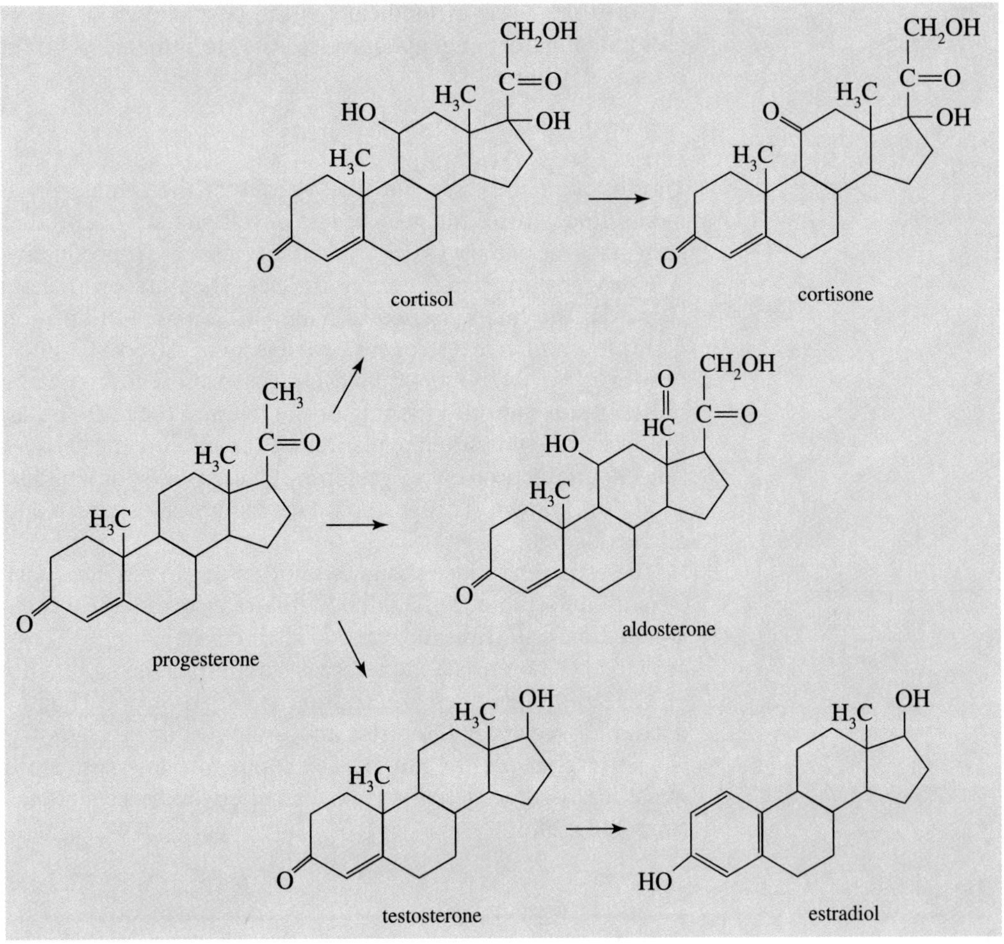

FIGURE 22.3 Progesterone and Derived Steroid Hormones

Corticosteroids

Steroidal hormones produced in the adrenal cortex are called **corticosteroids.** They are of two types: *glucocorticoids* and *mineralocorticoids.*

Glucocorticoids, along with insulin, control the glucose balance in the body. Glucocorticoids such as cortisol promote gluconeogenesis and the formation of glycogen, whereas insulin facilitates the use of glucose by cells. These two substances must be in balance to properly metabolize the body's sugar.

The adrenal cortex secretes 25 mg of cortisol per day. This steroid depresses the synthesis of protein in muscle tissue. As a consequence, amino acids are available for conversion to pyruvic acid and then to glucose by gluconeogenesis and to glycogen by glycogenesis. Thus, cortisol increases the supply of carbohydrates at the expense of proteins.

Mineralocorticoids affect the electrolyte balance of body fluids and hence the water balance. Aldosterone, secreted by the adrenal cortex, is the most active mineralocorticoid. This hormone causes the kidney tubules to reabsorb Na^+, Cl^-, and HCO_3^-. About 0.2 mg of aldosterone is secreted each day by the average adult.

Cortisone, used in medicine, is the best known of the corticosteroids. It is an anti-inflammatory agent used in the treatment of rheumatoid arthritis and bronchial asthma.

Sex Hormones

The testes of the male and the ovaries of the female produce steroidal sex hormones that control the growth and development of reproductive organs, the development of secondary sex characteristics, and the reproductive cycle.

Estrogens are female sex hormones. They are produced in the adrenal cortex. However, the major source of male and female sex hormones is the gonads. Progesterone and two estrogens, estrone and estradiol, which are produced in the ovaries, control the menstrual cycle. Estrogen secretion causes growth of the lining of the uterus and the ripening of the ovum. Progesterone secretion prevents other ova from ripening after ovulation and maintains the fertilized egg after implantation. If production of progesterone decreases, spontaneous abortion occurs. This condition, known as progesterone deficiency, is responsible for patterns of repeated miscarriages.

The estrogens cause the growth of tissues in female sexual organs. Although estrogens are secreted in childhood, the rate increases 20-fold after puberty. The fallopian tubes, uterus, and vagina all increase in size. The estrogens also initiate growth of the breasts and the breasts' milk-producing ductile system.

Male sex hormones are **androgens.** Testosterone (Figure 22.3) stimulates production of sperm by the testes and promotes the growth of the male sex organs. It is also responsible for muscle development. Thus, testosterone has two biological roles: androgenic activity (sex characteristic-determining) and anabolic activity (muscle-building).

22.4 Waxes

A wax coats this large leaf in Loma Loma, Fiji. (Robert J. Ouellette)

Waxes are esters of fatty acids and long-chain alcohols containing an even number of carbon atoms. They can be represented by a block diagram—a method that we shall use for other classes of lipids as well.

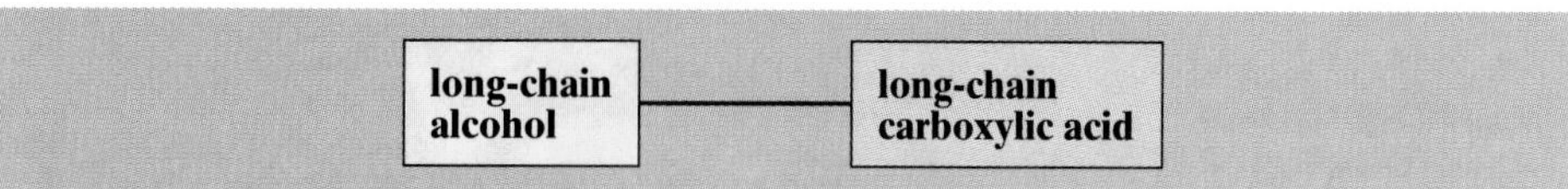

Waxes are low-melting solids that coat the surfaces of plant leaves and fruits and also coat the hair and feathers of some animals. The coatings on plants and fruits prevent loss of water and provide some protection against microorganisms. On the surfaces of feathers, hair, or skin of animals, waxes provide a water-repellent surface. If the wax is removed from the feathers of an aquatic bird as a result of an oil spill, the feathers become wet, and the bird loses its buoyancy.

Waxes are usually mixtures of esters. Lanolin, the wax from sheep's wool, is a mixture of esters derived from 32 different alcohols and 36 different carboxylic acids. Waxes also contain small amounts of other compounds such as alcohols and hydrocarbons. Beeswax, secreted by bees and used as a structural material for the beehive, contains about 20% hydrocarbon. On hydrolysis, beeswax yields several fatty acids and a mixture of alcohols having 24 to 36 carbon atoms. One example of an ester found in beeswax contains a carboxylic acid with 14 carbon atoms and an alcohol with 26 carbon atoms.

Whale Oil

Whale oil isn't an oil—that is, it isn't a triglyceride. It is actually a mixture of waxes! As much as 4 tons of whale "oil" is contained in the head of a sperm whale. The whale uses this mixture to control its buoyancy. One of the compounds is an ester of a 16-carbon acid and a 16-carbon alcohol. Note that this compound has fewer carbon atoms than the esters in beeswax. Hence, the compound has a lower melting point.

$$CH_3(CH_2)_{14}-C(=O)-O-CH_2(CH_2)_{14}CH_3$$

whale oil

Most solids have a greater density than liquids. Therefore, the volume of a substance generally decreases when it is converted from a liquid to a solid. Whales take advantage of this feature of liquids and solids. The oil tends to freeze at the depths of the ocean where the whales feed. As a consequence the density of the whale increases, and it is able to stay submerged without expending energy. The whale controls the amount of liquid and solid oil, and therefore its average density, by passing cold sea water through chambers in its head or by increasing the circulation of warm blood in the same area.

$$CH_3(CH_2)_{12}-C(=O)-O-CH_2(CH_2)_{24}CH_3$$

an ester found in beeswax

EXAMPLE 22.1

Half of the dry weight of a copepod that lives in the waters of British Columbia is the compound $C_{36}H_{62}O_2$. Hydrolysis of this compound yields an unbranched acid, $C_{20}H_{30}O_2$, and a straight-chain alcohol, $C_{16}H_{34}O$. Hydrogenation of the acid yields $C_{20}H_{40}O_2$. Describe the structure of the $C_{36}H_{62}O_2$ compound.

Solution

Since the compound yields an acid and an alcohol when hydrolyzed, it must be an ester. The formula $C_{16}H_{34}O$ corresponds to that of a saturated alcohol ($C_nH_{2n+2}O$). The acid, $C_{20}H_{30}O_2$, must contain five double bonds, since hydrogenation results in the addition of 10 hydrogen atoms. The location of the double bonds cannot be determined from the information given. A general representation of the wax is

$$C_{19}H_{29}C(=O)-O-(CH_2)_{15}CH_3$$

PROBLEM 22.1

One of the components of carnauba wax, found on the leaves of palm trees, is an ester of a saturated unbranched carboxylic acid containing 26 carbon atoms and a saturated unbranched alcohol containing 30 carbon atoms. Write its structure.

22.5 Fatty Acids

Triglycerides, glycerophospholipids, sphingophospholipids, and glycosphingolipids all yield fatty acids on hydrolysis. **Fatty acids** are long-chain carboxylic acids containing an even number of carbon atoms. Over 100 different fatty acids are found in lipids in amounts that vary with the organism. However, those with 14 to 22 car-

TABLE 22.1 Formulas and Melting Points of Fatty Acids

Name	Formula	Melting Point (°C)
Saturated		
lauric	$CH_3(CH_2)_{10}CO_2H$	44
myristic	$CH_3(CH_2)_{12}CO_2H$	58
palmitic	$CH_3(CH_2)_{14}CO_2H$	63
stearic	$CH_3(CH_2)_{16}CO_2H$	71
arachidic	$CH_3(CH_2)_{18}CO_2H$	77
Unsaturated		
oleic	$CH_3(CH_2)_7CH{=}CH(CH_2)_7CO_2H$	13
linoleic	$CH_3(CH_2)_4CH{=}CHCH_2CH{=}CH(CH_2)_7CO_2H$	−5
linolenic	$CH_3CH_2CH{=}CHCH_2CH{=}CHCH_2CH{=}CH(CH_2)_7CO_2H$	−11
arachidonic	$CH_3(CH_2)_4CH{=}CHCH_2CH{=}CHCH_2CH{=}CHCH_2CH{=}CH(CH_2)_3CO_2H$	−50

bon atoms are prevalent in mammals, and the 16- and 18-carbon acids are the most abundant.

The degree of unsaturation and the location of the double bonds is another structural variable. The 18-carbon oleic acid, which is the most abundant unsaturated fatty acid, has the cis configuration at C-9. Mammals cannot synthesize the polyunsaturated fatty acids linoleic and linolenic acids, which contain two and three carbon–carbon double bonds, respectively. For this reason polyunsaturated fatty acids are required in the diet. However, mammals can synthesize other polyunsaturated fatty acids from these two fatty acids.

The formulas and melting points of some fatty acids are given in Table 22.1. The melting points of the saturated fatty acids increase with increasing chain length because the London forces increase with increasing chain length.

The cis unsaturated acids are "bent" molecules because of the geometry about the double bonds. These "bends" hinder efficient packing, so the London forces between molecules are weaker. As a result, cis-unsaturated acids have lower melting points than saturated acids. Increasing the number of cis double bonds lowers the melting point further. Space-filling models of oleic, linoleic, and stearic acids are shown in Figure 22.4.

FIGURE 22.4 Space-Filling Models of Fatty Acids

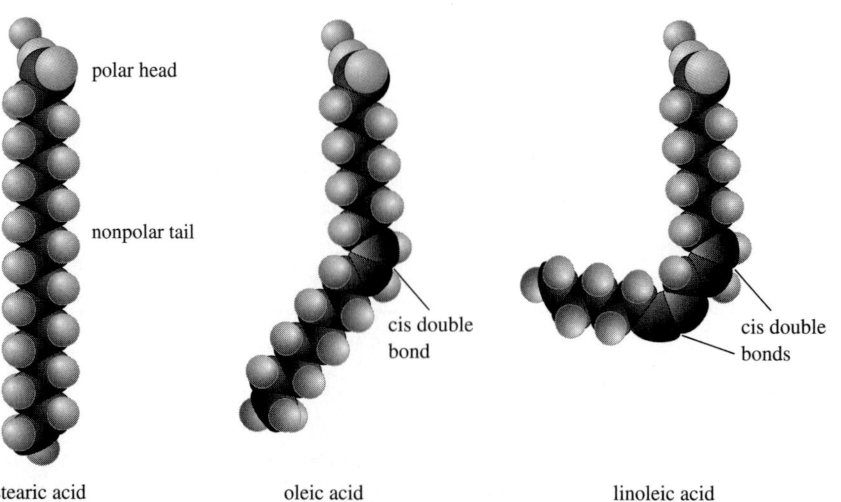

Eicosanoids

Long-chain fatty acids are widely distributed in natural products. Some have profound effects on biological processes even at extremely low concentrations. Prostaglandins, leukotrienes, and thromboxanes make up a class of hormones called *eicosanoids,* all of which derive from arachidonic acid, a 20-carbon unsaturated acid (see the figure). However, unlike many hormones, the eicosanoids are not produced by glands and secreted into the blood to affect cells throughout the body. Instead eicosanoids act locally, in the cells that produce them or in neighboring cells. They regulate normal processes, such as smooth muscle contraction, and affect cellular processes that result from diseases. For example, prostaglandins contribute to the symptoms of an illness such as swelling, nausea, vomiting, and pain.

Prostaglandins are 20-carbon fatty acids that contain a trans-substituted, five-membered ring. They are classified as PGA through PGI based on the number and type of their functional groups, such as hydroxyl groups and ketone groups. A subscript on the letter indicates the number of double bonds.

The prostaglandins were originally isolated from the prostate gland. But all nucleated cells produce them, and they can affect any cell type, even at very low concentrations. They can also alter the effects of many hormones. For example, hormone-sensitive enzymes that hydrolyze lipids respond to insulin and to other hormones that regulate the concentration of blood glucose. PGE_1 inhibits these enzymes at a concentration of 10 nM (10^{-8} M).

Prostaglandins affect virtually every aspect of reproduction. They regulate menstruation and control fertility and contraception. PGE_2 stimulates smooth muscle contraction in the uterus. It has been used clinically to induce labor and to abort pregnancies prematurely.

A class of steroid hormones called corticosteroids also inhibits prostaglandin biosynthesis. These steroids act in the following way. Arachidonic acid does not occur free in cells, but is esterified to a phospholipid and is incorporated in this form into cell membranes. Arachidonic acid is released by enzymes called phospholipases, and corticosteroids inhibit these enzymes. Corticosteroids such as cortisone are used to reduce inflammation.

Much research aims to develop synthetic prostaglandins for use as therapeutic drugs. Natural prostaglandins cannot be taken orally because they are rapidly degraded and do not survive long enough for effective action. Thus, one research goal is to develop modified prostaglandins that can be administered orally.

Eicosanoids

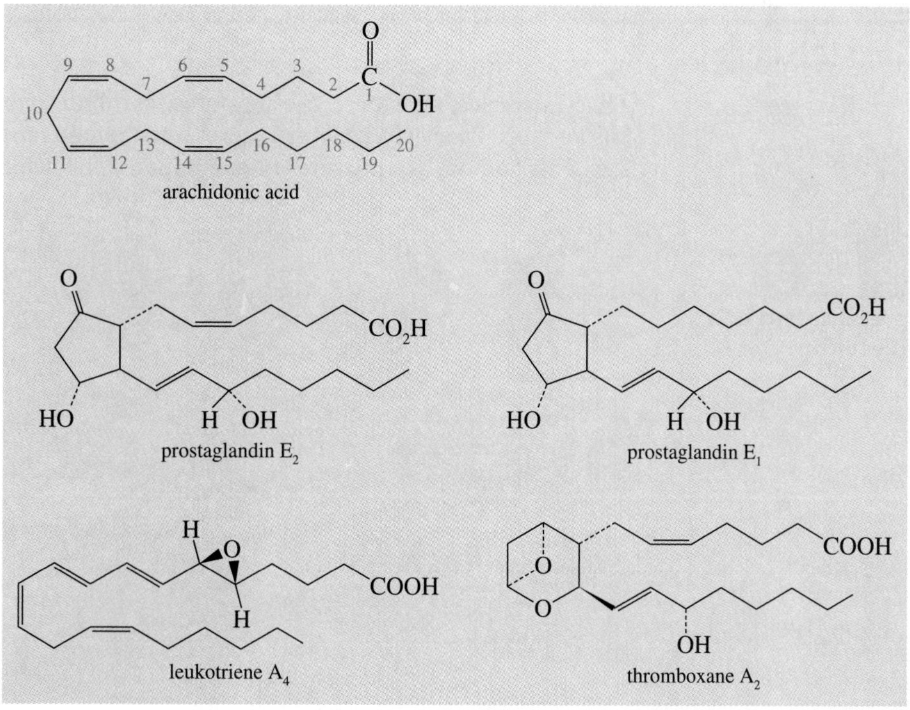

arachidonic acid

prostaglandin E_2

prostaglandin E_1

leukotriene A_4

thromboxane A_2

Leukotrienes act as mediators in inflammatory disorders such as asthma, ulcerative colitis, rheumatoid arthritis, and psoriasis. Leukotriene A_4 is a long-acting bronchoconstrictor released in the lungs during severe asthmatic reactions. Drugs that can modulate the effects of leukotrienes could prevent or moderate such hypersensitivity. Researchers are currently studying ways to intervene at the site of the biosynthesis of the leukotrienes or at the receptor sites for them.

Thromboxane A_2 is produced in blood platelets and causes constriction of smooth muscle of arterial walls, reducing blood flow.

EXAMPLE 22.2

The melting point of palmitoleic acid (*cis*-9-hexadecenoic acid) is -1 °C. Compare this melting point with that of palmitic acid and explain the difference.

$$\underset{\text{H}}{\overset{CH_3(CH_2)_5}{}}\!\!\!\!\diagdown\!\!\!\! C{=}C \!\!\!\!\diagup\!\!\!\! \overset{(CH_2)_7CO_2H}{\underset{\text{H}}{}}$$

palmitoleic acid

Solution

The melting point of palmitic acid is 63 °C, which is 64° higher than the melting point of palmitoleic acid. This difference is due to the presence of a cis double bond in palmitoleic acid, which prevents close approach of the hydrocarbon chains and results in weaker London forces.

PROBLEM 22.2

Using data in Table 22.1, estimate the melting point of $CH_3(CH_2)_{22}CO_2H$.

22.6 Triacylglycerols

Triacylglycerols, which are triesters of glycerol and fatty acids, are also known as fats and oils. Triacylglycerols are represented by the block diagram shown in Figure 22.5. Fats and oils are mixtures of compounds. The acid fraction of these molecules

FIGURE 22.5 Representation of a Triglyceride

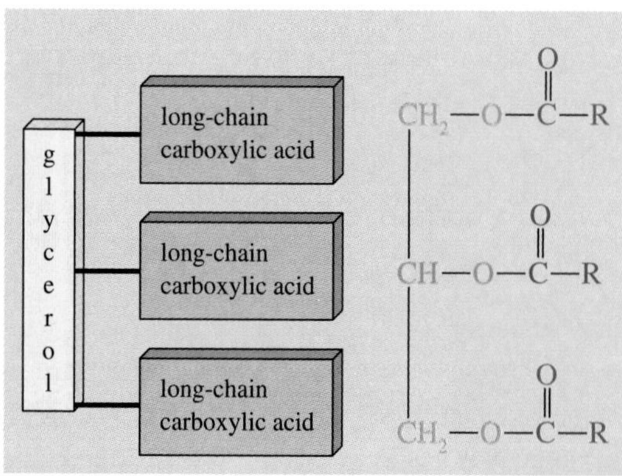

TABLE 22.2 Composition of Fats and Oils

	Melting Point (°C)	Saturated Fatty Acids (%)				Unsaturated Fatty Acids(%)		
		Myristic	Palmitic	Stearic	Arachidic	Oleic	Linoleic	Linolenic
Animal fats								
butter	32	11	29	9	2	27	4	—
lard	30	1	28	12	—	48	6	—
human fat	15	3	24	8	—	47	10	—
Plant oils								
corn	−20	1	10	3	—	50	34	—
cottonseed	−1	1	23	1	1	23	48	—
linseed	−24	—	6	2	1	19	24	47
olive	−6	—	7	2	—	84	5	—
peanut	3	—	8	3	2	56	26	—
soybean	−16	—	10	2	—	29	51	6

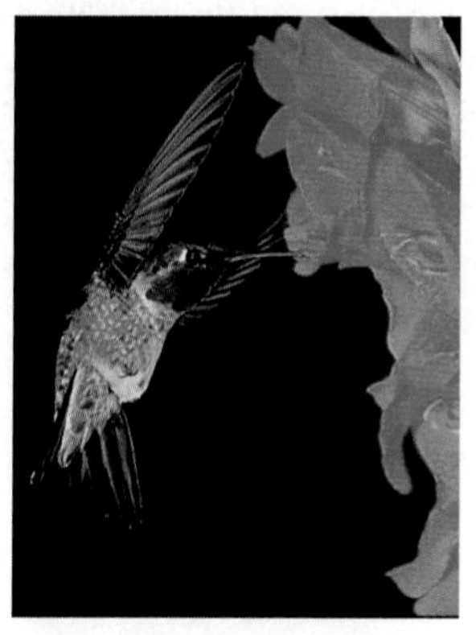

Hummingbirds store large quantities of triglycerides to provide energy for their metabolic processes. (Photo Researchers, Inc.)

varies in chain length and degree of unsaturation. A single molecule of a fat or oil may contain up to three different acid residues. **Fats** have a high percentage of saturated acids, whereas **oils** have a high percentage of unsaturated acids. Fats are solids or semisolids and are usually obtained from animals. The important acids found in these sources are myristic, palmitic, stearic acids and arachidic (Table 22.1). The unsaturated acids found in oils are oleic, linoleic, and linolenic acid; all contain 18 carbon atoms, but they differ in their degree of unsaturation (Table 22.1).

Oils are obtained from vegetable sources such as olives, peanuts, corn, and soybeans. The unsaturated acid residues in the molecules of oils lower their melting points, and they are usually liquids.

Animals accumulate fat (adipose tissue) when their intake of food exceeds their energy output requirements. (The metabolism of fats is discussed in Chapter 27.) Vital organs such as kidneys are enclosed in adipose tissue, which provides a protective cushion against a blow. A subcutaneous layer of fat helps insulate the animal against heat loss. Although plants do not generally store fats and oils for energy requirements, some (such as peanuts and olives) produce triglycerides in abundance.

The composition of triglycerides varies, as indicated in Table 22.2. The relationship between consumption of saturated fats and arterial disease has been the object of extensive medical research. Unsaturated fats are beneficial in preventing arterial deposits.

EXAMPLE 22.3

Soybean oil is 51% linoleic acid. Draw a structure for one of the possible components of soybean oil.

Solution

Although all oils are mixtures of triglycerides, this percentage of linoleic acid means that there must be large amounts of triglycerides containing two or three units of linoleic acid.

linoleic acid

$$\begin{array}{l} CH_2-O-\overset{O}{\overset{\|}{C}}-(CH_2)_7-CH=CHCH_2CH=CH-(CH_2)_4-CH_3 \\ | \\ CH-O-\overset{O}{\overset{\|}{C}}-(CH_2)_7-CH=CHCH_2CH=CH-(CH_2)_4-CH_3 \\ | \\ CH_2-O-\overset{O}{\overset{\|}{C}}-(CH_2)_7-CH=CHCH_2CH=CH-(CH_2)_4-CH_3 \end{array}$$

PROBLEM 22.3

How many moles of hydrogen gas would react with an oil containing one unit each of oleic, linoleic, and linolenic acid? What would be the product?

22.7 Glycerophospholipids

This swiss chocolate contains lecithins as emulsifying agents to keep the components from separating. (The Image Bank)

Glycerophospholipids are also known as *phosphoglycerides*. They consist of a phosphatidic acid that is esterified with certain small alcohols. Phosphatidic acids are formed by esterification of two hydroxyl groups of glycerol with carboxylic acids, and one hydroxyl group with phosphoric acid. (At physiological pH, phosphatidic acids exist in an ionized phosphatidate form.) The C-1 atom of glycerol in a phosphatidyl ester is usually esterified with a saturated acid containing 12 to 20 carbon atoms. The acid at C-2 is usually unsaturated. The phosphate group is always bonded to the C-3 atom.

$$\begin{array}{l} \qquad\qquad\quad CH_2-O-\overset{O}{\overset{\|}{C}}-R \\ \qquad\qquad\quad | \\ R-\overset{O}{\overset{\|}{C}}-O-CH \\ \qquad\qquad\quad | \\ \qquad\qquad\quad CH_2-O-\overset{O}{\overset{\|}{P}}-O-H \\ \qquad\qquad\qquad\qquad\quad | \\ \qquad\qquad\qquad\qquad\quad OH \end{array}$$

phosphatidic acid

Phosphatidic acids combined with alcohols such as ethanolamine, choline, serine, or inositol are called phosphatidyl esters or phosphatides. Examples of the various types of glycerophospholipids are listed in Table 22.3. Block diagrams of a glycerophospholipid and two other lipids in membranes are given in Figure 22.6.

$$\begin{array}{l} \qquad\qquad\qquad\qquad\qquad\qquad\qquad\quad CH_2-O-\overset{O}{\overset{\|}{C}}-(CH_2)_{14}CH_3 \\ \qquad\qquad\qquad\qquad\qquad\qquad\qquad\quad | \\ CH_3(CH_2)_7CH=CH(CH_2)_7-\overset{O}{\overset{\|}{C}}-O-CH \\ \qquad\qquad\qquad\qquad\qquad\qquad\qquad\quad | \\ \qquad\qquad\qquad\qquad\qquad\qquad\qquad\quad CH_2-O-\overset{O}{\overset{\|}{P}}-O-CH_2CH_2NH_2 \\ \qquad\qquad\qquad\qquad\qquad\qquad\qquad\qquad\qquad\quad | \\ \qquad\qquad\qquad\qquad\qquad\qquad\qquad\qquad\qquad\quad OH \end{array}$$

unsaturated fatty acid; saturated fatty acid; phosphate group; ethanolamine

phosphatidylethanolamine

TABLE 22.3. Composition of Glycerophospholipids

General Structure	X =	X—OH Name	Type of Glycerophospholipid
$CH_2-O-C(=O)-R$ $CH-O-C(=O)-R'$ $X-O-P(=O)(O^-)-O-CH_2$	$H-$	Water	Phosphatidic acid
	$H_3\overset{+}{N}CH_2CH_2-$	Ethanolamine	Phosphatidylethanolamine
	$(CH_3)_3\overset{+}{N}CH_2CH_2-$	Choline	Phosphatidylcholine
	$^-O_2CCHCH_2-$ (with NH_3^+ on CH)	Serine	Phosphatidylserine
	OH OH O OH HO OH	Inositol	Phosphatidylinositol

The phosphatides exist as anions at physiological pH (approximately 7). The dissociation of the proton on the phosphate oxygen atom is essentially complete at pH 7.0. Phosphatidylethanolamine is protonated at the nitrogen atom at pH 7.0. Phosphatidylcholine has a positive charge at the nitrogen atom because choline contains a quaternary ammonium ion. As a result, both phosphatidylcholine and phosphatidylethanolamine are dipolar but have no net charge. Phosphatidylserine has three charged sites at pH 7.0. The phosphate oxygen atom bears a negative

FIGURE 22.6 Representation of Components of Membrane Lipids

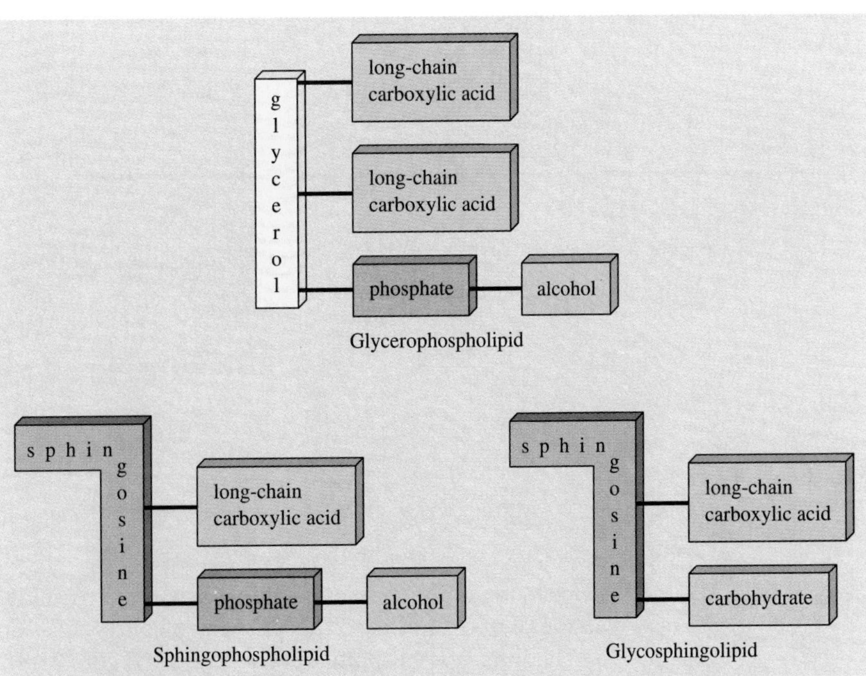

FIGURE 22.7
Space-Filling Model of a Glycerophospholipid

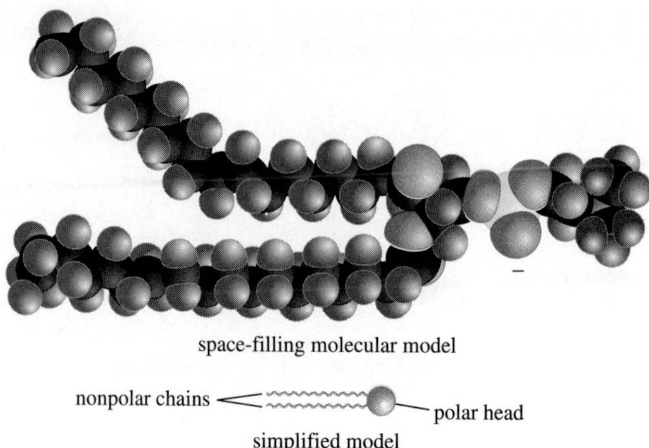

charge because the proton is dissociated at pH 7.0. In addition, the carboxyl group of serine exists as the carboxylate ion, and the amine group of serine is protonated. Thus, phosphatidylserine bears a net negative charge.

The older name of phosphatidylcholines is **lecithins.** The phosphatidylcholines are found in micelles of the protoplasm of body cells. These micelles are emulsifying agents and transport fat molecules from one tissue to another. The nonpolar fat molecules dissolve within the micelle much as grease does in a soap micelle.

Phosphatidylethanolamines used to be called **cephalins.** These lipids are found in the heart and liver and in high concentration in brain tissue. Phosphatidylethanolamines are also essential to the blood-clotting mechanism.

A molecular model of a phosphatidylcholine containing one unsaturated and one saturated acid is shown in Figure 22.7. All glycerophospholipids have polar sites, which are represented by a circle in the simplified symbol. The two nonpolar hydrocarbon chains of the fatty acids are represented with wavy lines or "tails" attached to the polar "head." These structural features constitute a "head-and-tail" model that is used in representing the structure of cell membranes (Section 22.10).

22.8 Sphingophospholipids

Sphingophospholipids contain sphingosine, a long-chain unsaturated amino alcohol, rather than glycerol.

$$\begin{array}{l} CH_3(CH_2)_{12}-CH{=}\overset{4}{C}H-\overset{3}{C}H-OH \\ \qquad\qquad\qquad\quad H-\underset{\displaystyle H}{\underset{|}{N}}-\overset{2}{C}H \\ \qquad\qquad\qquad\qquad\qquad\ {}_1CH_2-O-H \end{array}$$

sphingosine

Conversion of the amine of sphingosine into an amide with a fatty acid yields a ceramide. Esterification of the primary alcohol of a ceramide with phosphoric acid and choline gives a sphingophospholipid (Figure 22.8).

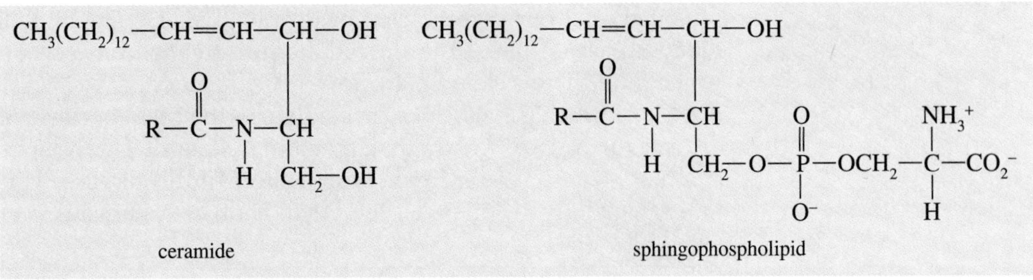

ceramide

sphingophospholipid

FIGURE 22.8 Structure of a Ceramide and a Sphingophospholipid

Compare the block diagrams of a sphingophospholipid and a glycerophospholipid (Figure 22.6). Although the components are different, the overall structures are similar. Both compounds have a polar head and two nonpolar tails. However, there are significant chemical differences. Sphingophospholipids have a single amide group and are stable to hydrolysis; glycerophospholipids have two carboxylic esters that are easily hydrolyzed.

Sphingomyelins, found in the myelin sheath surrounding nerve fibers, are sphingophospholipids. Sphingomyelins have acid residues that are 20–26 carbon atoms long. These long chains form a very stable coating for nerve fibers. In individuals with some genetic diseases, the carbon chains are shorter, which results in defects in the myelin sheath. Gaucher's disease, Niemann–Pick disease, multiple sclerosis, and leukodystrophy are all the result of unstable myelin membranes.

22.9 Glycosphingolipids

Glycosphingolipids are similar to sphingophospholipids; they contain sphingosine and a fatty acid residue bonded as an amide as in a ceramide. However, glycosphingolipids contain no phosphate. A carbohydrate is bonded via a glycosidic linkage to the primary alcohol oxygen atom of the ceramide. A block-diagram representation of a glycosphingolipid is shown in Figure 22.6.

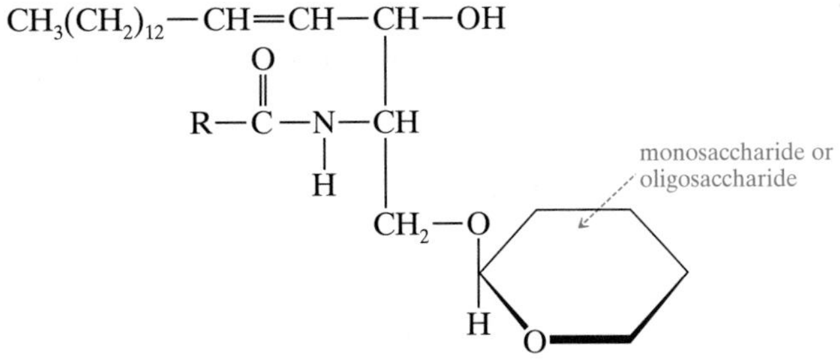

monosaccharide or oligosaccharide

glycosphingolipid

Cerebrosides and gangliosides are glycosphingolipids. Cerebrosides contain only glucose or galactose, whereas gangliosides contain an oligosaccharide. Both are found in the myelin sheath. Cerebrosides are in the white matter of the central nervous system; gangliosides occur in the gray matter of the brain.

Gangliosides are synthesized or degraded by sequential addition or removal of monosaccharide units. The degradation of a ganglioside occurs inside lysosomes, which have the necessary enzymes for the reaction. Tay–Sachs disease is a genetic disease in which gangliosides accumulate in the brain because of a deficiency of the enzymes necessary for their degradation. Retarded development, blindness, and death at an early age are the consequences of this disease, which occurs 100 times more frequently in American Jews than in other Americans.

EXAMPLE 22.4

Plasmalogens are a class of lipids having the following general structure. Identify the components of the lipid and determine which class of lipids it most closely resembles.

```
      O       CH2—O—CH=CH—R
      ‖       |
R—C—O—CH           O
              |          ‖
              CH2—O—P—O—CH2CH2NH2
                         |
                         OH
```

Solution

The center portion of the structure is derived from glycerol. A phosphate ester of ethanolamine appears at the bottom right. An ester of a carboxylic acid occurs at C-2 of glycerol. Only the residue at the top right is different from the other classes of lipids. Instead of an ester, the residue is an ether. This structure most closely resembles a phosphatidylethanolamine.

PROBLEM 22.4

Identify the components of the following compound and classify it.

```
                              O
                              ‖
             CH3(CH2)14—C—O—CH2
                              O      |
                              ‖      |
CH3(CH2)7C=C(CH2)7—C—O—CH          O              NH3+
           |   |                     |          ‖              |
           H   H                     CH2—O—P—OCH2—C—CO2−
                                            |              |
                                            O−             H
```

22.10
Biological Membranes

We recall that long-chain carboxylate ions associate in micelles that have the nonpolar hydrocarbon chain in the interior and the polar head on the surface. Phospholipids have two nonpolar hydrocarbon chains and do not fit well in micelles. However, they do readily form lipid bilayers as depicted in Figure 22.9 that are about 5 nm thick. The hydrophobic hydrocarbon chains point toward the interior of the bilayer. The hydrophilic heads are in contact with aqueous solutions on each side of the two layers.

Lipid bilayers are one part of biological membranes that separate cells from their environment as well as organelles from one another within the cytoplasm of the cell. The properties of the membrane are determined by the type of fatty acid

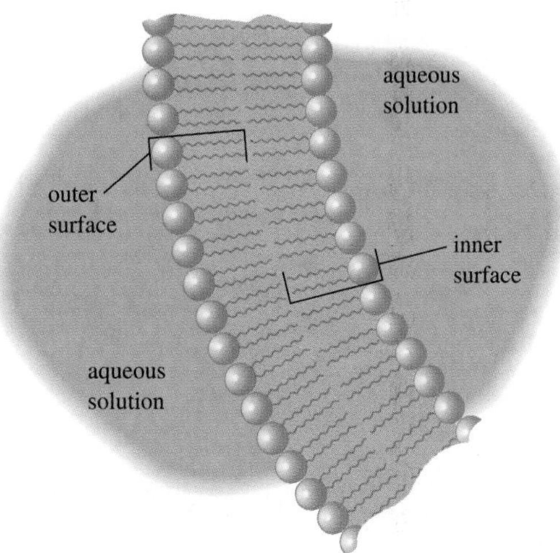

FIGURE 22.9 Cross-Section of a Lipid Bilayer
The circles represent the polar head of the lipid components of the bilayer that extend toward the aqueous media. The two hydrophobic tails of each lipid point to the interior.

and polar group in the phospholipids. Cell membranes allow nutrients to enter the cell and substances produced within the cell to get out.

Composition of Membranes

Cell membranes consist largely of lipids and proteins with smaller amounts of carbohydrates. These additional components can increase the thickness of the membrane to as much as 10 nm. The fraction of each component is related to the function of the membrane, but the average of 40% lipid and 60% protein by mass is representative. However, the myelin covering of nerve fibers is about 80% lipid. This covering is very nonpolar and serves a protective function.

The polar proteins of cell membranes participate in the transport process that occurs across the cell membrane. Mitochondrial membranes are only about 20% lipid, and proteins are the major part of the cell membrane. The organelle plays an important role in energy conversions within cells, and many molecules must cross the mitochondrial membrane.

Each membrane has characteristic amounts of the various types of lipids. Brain tissue has large amounts of phosphatidylserines. Phosphotidylethanolamines make up 50% of the membrane of red blood cells.

The distributions of phospholipids between the inner and outer surfaces of the lipid bilayer are not the same. Sphingomyelins and phosphatidylcholines make up half of the outer layer of the plasma membrane of human erythrocytes but only 20% of the inner layer. Phosphatidylethanolamines and phosphatidylserines are the major components of the inner layer.

Proteins in Membranes

Protein molecules are present in the bilayer in three different ways (Figure 22.10). Proteins extending from the edge of the bilayer into the interior of the membrane are **integral proteins.** Some of these proteins are embedded in only one side of the membrane, but most extend through to the other side. Proteins that extend across the membrane are called **transmembrane proteins.** Integral proteins interact with the interior of the membrane by hydrophobic forces between certain nonpolar portions of a protein and the tails of the lipid.

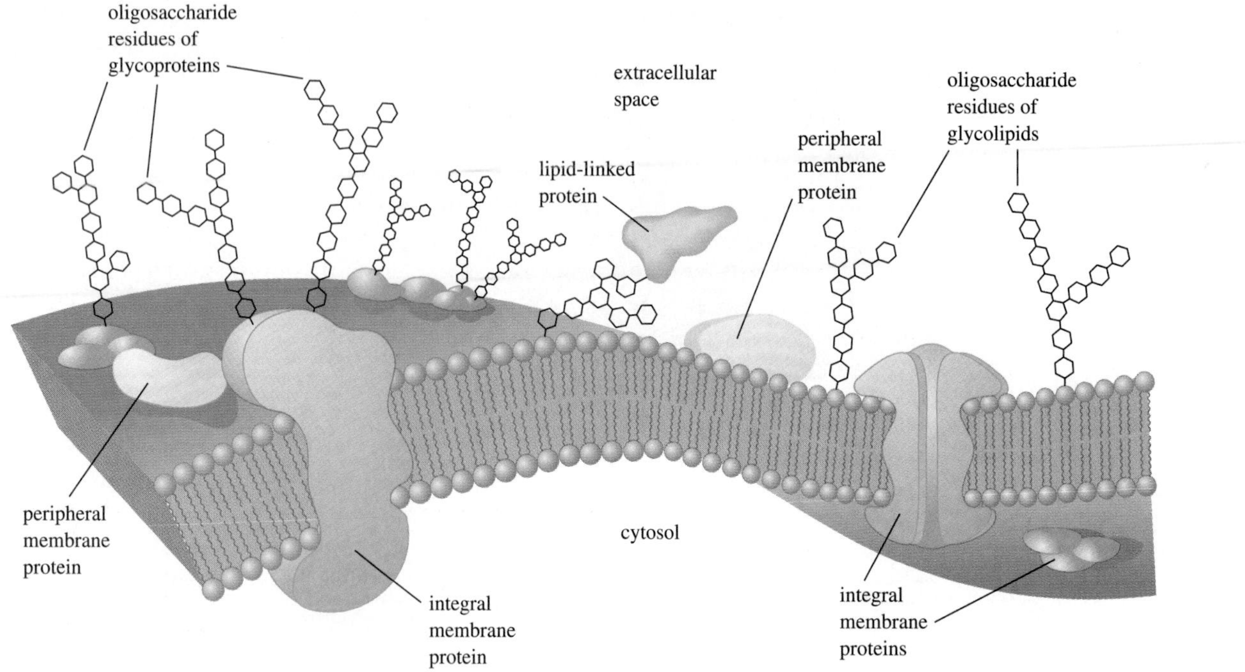

FIGURE 22.10 Proteins and the Structure of a Eukaryotic Membrane

Proteins that are associated with only the surface of the bilayer are called **peripheral proteins.** Peripheral proteins may be bound to the membrane by electrostatic forces or hydrogen bonds. As a result, these proteins may easily be removed from the membrane.

Lipid-linked proteins are covalently bonded to phosphatidylinositol incorporated in the cell membrane. Thus, these proteins are less easily removed than peripheral proteins.

Carbohydrates in Membranes

Membranes contain carbohydrates combined with lipids in glycolipids and combined with proteins in glycoproteins. In mammals these glycolipids and glycoproteins are located on the outer surface of the cell membrane. The carbohydrate portion is hydrophilic and remains directed toward water in the external environment. The protein is anchored in the membrane. The arrangement provides some intercellular recognition. It not only allows the grouping of cells to form tissue but also aids the immune system in recognizing foreign cells.

Membrane Fluidity

The London forces that hold membranes together depend on both the length and geometry of the hydrocarbon chain. The longer the chain, the stronger the London forces and the more rigid the membrane. Unsaturation affects the flexibility of the membrane because unsaturated fatty acids have bends in the chain, and they do not pack together efficiently in the bilayer. As a result membranes with a high degree of unsaturation are more flexible.

The lipids and proteins in membranes can move laterally (on one side) in the membrane. A phospholipid can move about 10^{-4} cm/s, which means that it can move from one end of a bacterium to the other in about a second. In contrast lipid movement from one side of a membrane to the other is very slow. Such transverse diffusion or "flip-flop" of the polar head from one side to the other occurs at a rate only 10^{-9} times as fast as lateral movement. Thus, any difference in the lipid composition on each side of the membrane is maintained.

The lateral mobility of proteins varies. Some proteins are as mobile as lipids, whereas others are essentially immobile. The protein rhodopsin, which is responsible for sight, is very mobile. Other proteins that are anchored to the cytoskeleton remain fixed in place. Proteins, which are much more polar than lipids, do not undergo transverse diffusion.

22.11 Transport Across Membranes

Molecules and ions pass through a biological membrane to provide food for the cell and to release waste products from the cell. Smaller hydrophobic molecules such as O_2 as well as polar but uncharged molecules such as H_2O, urea, and ethanol diffuse across the lipid part of the membrane relatively rapidly. However, large polar molecules and most ions such as Na^+ and K^+ diffuse very slowly. In order to maintain cellular processes, molecules move across membranes by specialized transport mechanisms called facilitated diffusion and active transport.

Facilitated Diffusion

Facilitated diffusion occurs without the expenditure of cellular energy and in a direction from high concentration to low concentration. Each substance moves across the membrane faster than in simple diffusion because "carriers" known as channel proteins facilitate the process. These carriers are transmembrane proteins with molecular weights in the range of 9000 to 40,000 amu. There are a variety of carriers, each specific for certain molecules or ions.

A carrier protein meets a specific molecule or ion at one surface of the membrane and forms a complex. Formation of the complex causes a conformational change in the protein that allows the molecule to slip through a "channel" to the other side of the membrane. Once the molecule is released, the protein returns to its original conformation (Figure 22.11). Glucose enters cells in this manner.

The transport of anions in human erythrocytes occurs by facilitated diffusion. A protein exchanges the bicarbonate ion inside the cell for the chloride ion outside the cell. Carbon dioxide produced by cellular metabolism dissolves in aqueous media as the bicarbonate ion. Its concentration within the cell is higher than outside the cell, and it is transported spontaneously to the region of lower concentration. To maintain charge balance, the chloride ion flows into the cell.

Active Transport

Active transport also involves a specific interaction between a component of the cell and the molecule to be transported. However, active transport occurs against the "natural" flow expected from concentration differences. Thus, material moves from a region of low concentration to one of high concentration. Cellular ATP provides the energy needed for active transport.

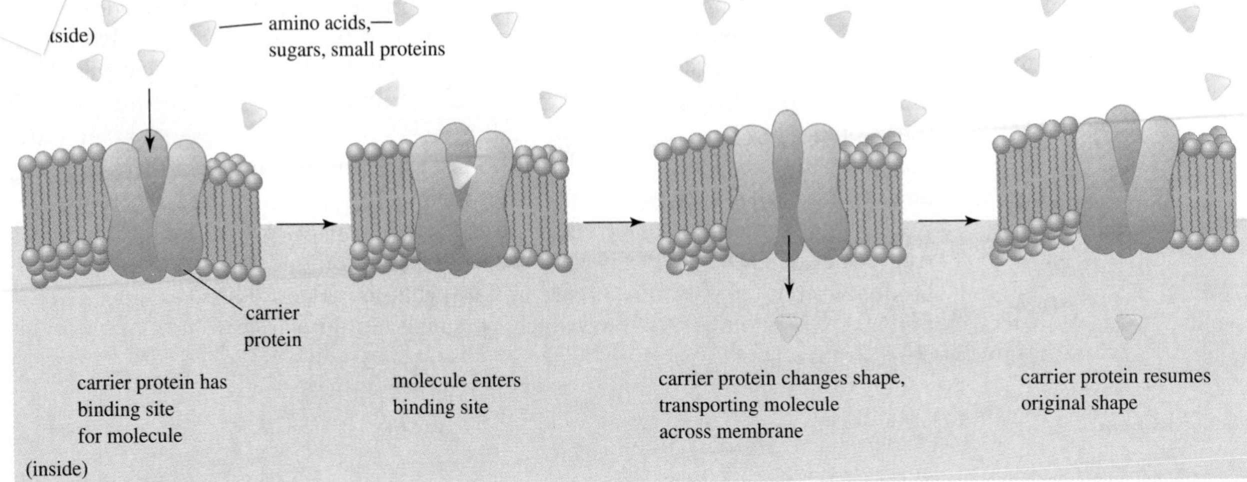

FIGURE 22.11 Faciliated Diffusion by A Carrier Protein

Carrier proteins change their conformation and allow specific molecules to pass to the other side of the membrane.

Active transport proteins have two sites responsible for the transport. One is the site on one face of the membrane, either the inner or outer face depending on the direction of transport required. It recognizes the particular ion or molecule to be transported. The second site is within the membrane and binds ATP to provide energy to the protein, which must change its conformation to move the molecule or ion across the membrane. Because energy is required, such active transport proteins are called *pumps.* They require energy much like a water pump uses energy to pump water uphill.

One of the most important active transport processes in animal cells is the sodium ion–potassium ion transport system known as the *sodium pump.* Sodium ions are pumped out of the cell to maintain a concentration of 0.1 M within the cell, while the extracellular sodium ion concentration is 0.14 M. At the same time, potassium ions are pumped into the cell from an extracellular concentration of 0.05 M to provide an intracellular concentration of about 0.15 M. The entire process is controlled because any imbalance in the total concentration of ions would cause a change in the osmotic pressure. The cell would swell if the osmotic pressure increased or shrink if the osmotic pressure decreased.

Summary

Lipids are a broad class of nonpolar materials that are grouped because of their solubility in nonpolar organic solvents. Terpenes and steroids are not hydrolyzed by base. Waxes, triglycerides (triacylglycerols), phosphoglycerides, sphingophospholipids, and glycosphingolipids are hydrolyzed to yield fatty acids and other components characteristic of the class of lipids.

Terpenes contain multiple isoprene units, whereas **steroids** contain a fused set of four rings. Two important classes of steroids are the **corticosteroids** (**glucocorticoids** and **mineralocorticoids**), produced by the adrenal cortex, and the **sex hormones** (**androgens** and **estrogens**), produced by the gonads. Estrogens and androgens are female and male sex hormones, respectively .

Waxes are esters of fatty acids and long-chain alcohols, both of which contain an even number of carbon atoms. **Tri-**

acylglycerols, or **triglycerides,** are esters of glycerol and **fatty acids.** The degree of unsaturation determines the classification of a triacylglycerol as a **fat** or an **oil.**

Phosphate groups occur in both glycerophospholipids and sphingophospholipids. The **glycerophospholipids** consist of one unit of glycerol, phosphate, an alcohol, and two units of a fatty acid. **Sphingophospholipids** have one unit of sphingosine, phosphate, choline, and a fatty acid in an amide linkage.

Glycosphingolipids contain one unit of sphingosine, a fatty acid bonded as an amide, and a sugar. Glycosphingolipids contain no phosphate. **Cerebrosides** contain glucose or galactose, whereas **gangliosides** contain an oligosaccharide.

In water, the phosphate-containing lipids form a **lipid bilayer.** The assembly is maintained by **hydrophobic** forces between nonpolar hydrocarbon chains in the middle of the bilayer. The **hydrophilic** heads of the phospholipids interact with water on both sides of the bilayer. Other components such as proteins, glycolipids, and glycoproteins are also part of the bilayer. These components control the passage of nutrients into the cell. **Integral, peripheral,** and **lipid-linked** proteins are associated with the bilayer.

Molecules and ions may pass through a biological membrane by **facilitated diffusion** and **active transport.** Active transport requires energy from the hydrolysis of ATP. The sodium pump is an important active transport system that maintains the proper concentration of ions in a cell.

Exercises

Classification of Lipids

22.1 Which of the following lipids undergo hydrolysis reactions?

(a) waxes (b) steroids (c) terpenes (d) cephalins

22.2 Which of the following contain a fatty acid as part of their structure?

(a) terpene (b) waxes (c) steroids (d) oils

22.3 What are the differences and similarities in the structures of glycerophospholipids and sphingophospholipids?

22.4 What are the differences and similarities in the structures of glycerophospholipids and glycosphingolipids?

22.5 Classify the following terpene, which is the sex attractant secreted by the dorsal gland of the springbok, a South African gazelle. Indicate the separation into isoprene units.

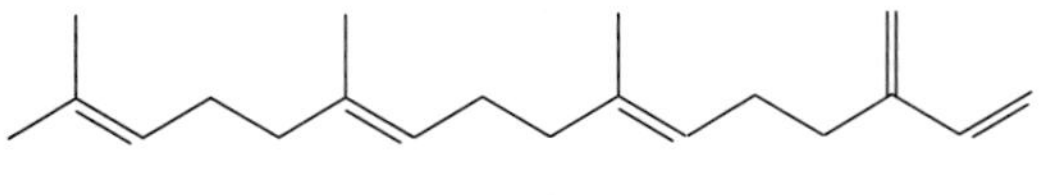

22.6 Classify the following terpene and divide it into isoprene units.

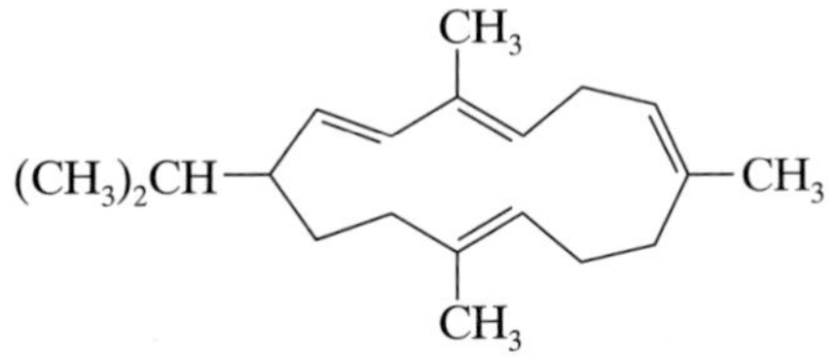

CH_3

$(CH_3)_2CH$— —CH_3

CH_3

Steroids

22.7 Identify the functional groups in cortisol.

22.8 Identify the functional groups in aldosterone.

Waxes

22.9 One of the components of carnauba wax used in floor polish is an ester containing a saturated unbranched carboxylic acid with 20 carbon atoms and a saturated unbranched alcohol with 32 carbon atoms. Write a condensed structural formula for this wax.

22.10 The wax of a particular copepod is unsaturated. This species lives in cold water and uses the wax as a source of metabolic energy. Explain the benefit of the unsaturation in the acid portion of this ester.

Fatty Acids

22.11 Cod liver oil is a triglyceride containing palmitoleic acid. Suggest a structure for the acid.

22.12 Estimate the melting point of $CH_3(CH_2)_{20}CO_2H$.

22.13 Steareolic acid is named 9-octadecynoic acid by the IUPAC method. The molecular formula is $C_{18}H_{32}O_2$. Write its structure.

22.14 A compound called hypogeic acid is prepared in the laboratory and is now named 7-hexadecenoic acid. Its melting point is 33 °C. What is the likely geometry at the double bond?

22.15 Why does linoleic acid have a lower melting point than oleic acid?

22.16 The melting point of elaidic acid (*trans*-9-octadecenoic acid) is 45 °C. Compare this value with the melting points of stearic acid and oleic acid and explain the differences.

Triacylglycerols

22.17 Write a balanced equation for the hydrolysis of a fat molecule using a base.

22.18 A sample of one oil is hydrolyzed to produce 50% oleic acid, 35% linoleic acid, and 15% saturated acids. A second oil produces 25% oleic acid and 50% linoleic acid and 25% saturated acids. Which oil is more unsaturated?

22.19 Is the following compound more likely classified as a fat or an oil?

$$\begin{array}{l} CH_2-O-\overset{O}{\overset{\|}{C}}-C_{17}H_{31} \\ | \\ CH-O-\overset{O}{\overset{\|}{C}}-C_{17}H_{33} \\ | \\ CH_2-O-\overset{O}{\overset{\|}{C}}-C_{17}H_{29} \end{array}$$

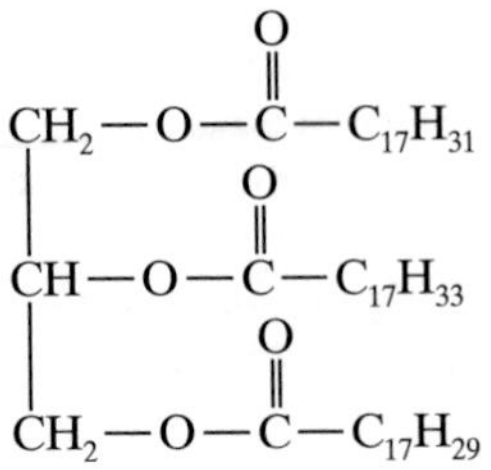

22.20 Is the following compound more likely classified as a fat or an oil?

$$\begin{array}{l} CH_2-O-\overset{O}{\overset{\|}{C}}-(CH_2)_{14}CH_3 \\ | \\ CH-O-\overset{O}{\overset{\|}{C}}-(CH_2)_{10}CH_3 \\ | \\ CH_2-O-\overset{O}{\overset{\|}{C}}-(CH_2)_7CH=CH(CH_2)_7CH_3 \end{array}$$

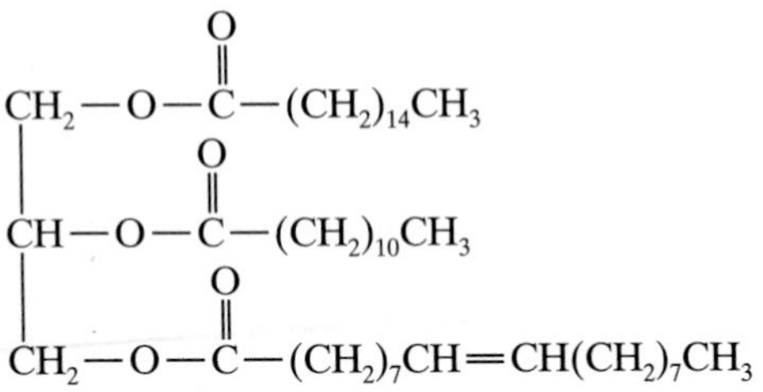

22.21 Identify the fatty acids in the compound of Exercise 22.19.

22.22 Identify the fatty acids in the compound of Exercise 22.20.

22.23 Draw the structure of a triacylglycerol containing palmitic acid as an ester at the secondary carbon atom and stearic acid as esters at the two primary carbon atoms of glycerol. Can this compound exist in an optically active form?

22.24 Should the following compound be optically active?

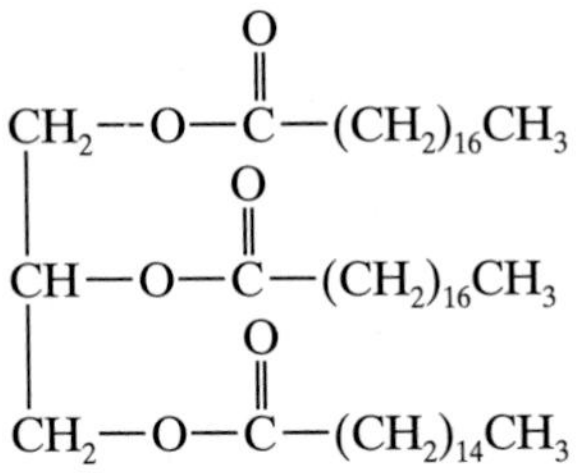

$$\begin{array}{l} CH_2-O-\overset{O}{\overset{\|}{C}}-(CH_2)_{16}CH_3 \\ | \\ CH-O-\overset{O}{\overset{\|}{C}}-(CH_2)_{16}CH_3 \\ | \\ CH_2-O-\overset{O}{\overset{\|}{C}}-(CH_2)_{14}CH_3 \end{array}$$

22.25 State whether each of the following is likely to be found as an ester in a naturally occurring compound.

(a) $CH_3(CH_2)_{15}CO_2H$
(b) $CH_3(CH_2)_{20}CO_2H$
(c) $(CH_3)_2CH(CH_2)_{14}CO_2H$

22.26 Hydrolysis of an optically active triacylglycerol gives 1 mole each of glycerol and oleic acid and 2 moles of stearic acid. Write a structure for the triacylglycerol.

Phosphoglycerides

22.27 What products result from hydrolysis of a glycerophospholipid?

22.28 Draw the structures of the alcohols found in glycerophospholipids.

22.29 Identify the components of the following glycerophospholipid.

$$\begin{array}{l} CH_2-O-\overset{O}{\overset{\|}{C}}-(CH_2)_{16}CH_3 \\ | \\ CH-O-\overset{O}{\overset{\|}{C}}-(CH_2)_7CH=CH(CH_2)_7CH_3 \\ | \\ CH_2-O-\underset{O^-}{\underset{|}{\overset{O}{\overset{\|}{P}}}}-OCH_2CH_2\overset{+}{N}(CH_3)_3 \end{array}$$

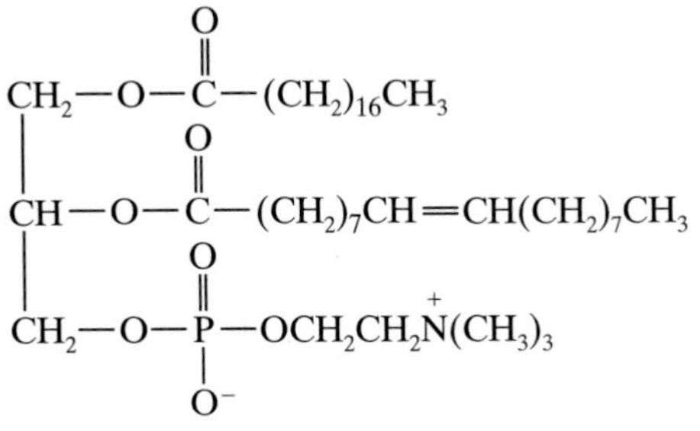

22.30 What are the hydrolysis products of the following glycerophospholipid?

$$\begin{array}{l} CH_2-O-\overset{O}{\overset{\|}{C}}-(CH_2)_{14}CH_3 \\ | \\ CH-O-\overset{O}{\overset{\|}{C}}-(CH_2)_7CH=CH(CH_2)_7CH_3 \\ | \\ CH_2-O-\underset{O^-}{\underset{|}{\overset{O}{\overset{\|}{P}}}}-OCH_2-\underset{H}{\underset{|}{\overset{NH_3^+}{\overset{|}{C}}}}-CO_2^- \end{array}$$

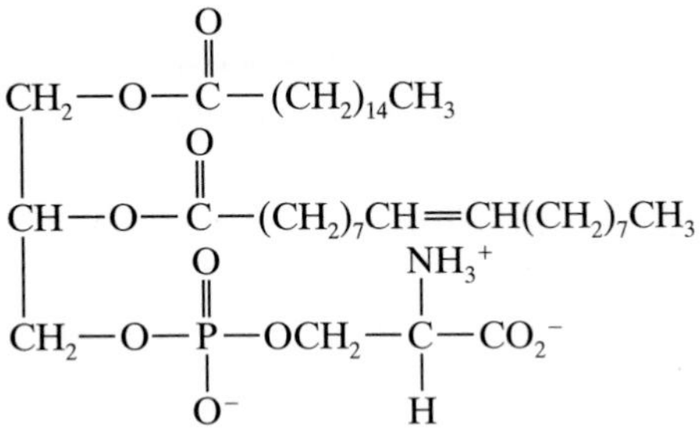

22.31 What charges exist on the polar head of each of the types of glycerophospholipid at physiological pH?

22.32 How many ionizable hydrogen atoms are there in a phosphatidic acid?

Sphingophospholipids

22.33 How do sphingophospholipids differ from glycerophospholipids?

22.34 Why are sphingophospholipids not hydrolyzed as readily as glycerophospholipids?

22.35 Sphingophospholipids are said to have two nonpolar tails. One is a fatty acid residue. What is the structure of the second chain?

22.36 What are sphingomyelins? What structural feature allows them to serve an important biological function?

Glycosphingolipids

22.37 How are glycosphingolipids similar to sphingophospholipids? In what ways do the two types of compounds differ?

22.38 What is the difference between a cerebroside and a ganglioside? Where are these compounds found?

22.39 What type of bond joins the sugar unit to the sphingosine part of a glycosphingolipid?

22.40 From the structure of glycosphingolipids, predict whether these molecules are more stable in acidic or basic solution.

Biological Membranes

22.41 How does the structure of the fatty acid affect the rigidity of a cell membrane?

22.42 How are proteins incorporated in a cell membrane?

22.43 What kind of forces hold a cell membrane together?

22.44 What relationship exists between the protein content and the permeability of a cell membrane?

22.45 Peripheral proteins can be removed from a membrane by washing it with a detergent solution. Explain why.

22.46 Where is the sugar portion of a glycolipid located in a cell membrane?

Diffusion and Transport

22.47 Describe two ways in which materials cross cell membranes.

22.48 Why is an active transport system necessary to maintain the potassium ion content of a cell?

This mountain sheep in Banff National Park (Alberta, Canada) has horns and a coat of wool that both consist of protein.
(Robert J. Ouellette)

23 Amino Acids and Proteins

23.1 The Biological Role of Proteins
23.2 Amino Acids
23.3 Acid–Base Properties of Amino Acids
23.4 Isoionic Point
23.5 Peptides
23.6 Biological Activity of Peptides
23.7 Bonding in Proteins
23.8 Structure of Proteins
23.9 Denaturation of Proteins
23.10 Hydrolysis of Proteins

Essays
- *Lipoproteins and Blood Cholesterol*
- *Straight and Curly Hair*
- *Complete Proteins in the Diet*

23.1 The Biological Role of Proteins

From amoebas to zebras, the proteins of all organisms contain amino acids bonded together by amide bonds between amino and carboxylic acid functional groups. In proteins, amide bonds are called **peptide bonds.**

peptide bonds

$$\sim\!\text{NHCH}(\text{R})\overset{\text{O}}{\overset{\|}{\text{C}}}\text{—NHCH}(\text{R}')\overset{\text{O}}{\overset{\|}{\text{C}}}\text{—NHCH}(\text{R}'')\overset{\text{O}}{\overset{\|}{\text{C}}}\!\sim$$

Each protein has a unique amino acid composition, and all the amino acids are strung together in a specific order that is required for its specialized function.

After water, proteins are the most abundant component of the organelles. The name *protein* is derived from the Greek *proteios,* meaning "preeminence" or "holding first place," which reflects the role that they play in virtually all cellular processes. The name was suggested in 1839 by the Dutch chemist Gerardus Johannes Mulder, who could not have known how prophetic his suggested name would be. The scope of the biological functions of proteins includes the following:

1. ***Enzymatic catalysis.*** The enzymes that catalyze virtually all reactions in cells are proteins. From a simple reaction such as the hydration of carbon dioxide to give carbonic acid to very complex reactions such as the replication of a DNA molecule, enzymes are indispensable. They increase reaction rates by at least a millionfold. Without this remarkable rate enhancement, life would not be possible.
2. ***Transport of material.*** Without the specific carrier proteins involved in both facilitated diffusion and active transport of molecules across cell membranes, the cell would die. Other proteins carry substances throughout the body, as in the case of hemoglobin, which transports oxygen in erythrocytes. Serum albumin carries fatty acids from adipose tissues to other tissues for catabolic reactions that release the stored energy in these reduced molecules. The protein ferrin transports iron from the liver to the bone marrow, where heme is synthesized for hemoglobin.
3. ***Immune protection.*** Antibodies are proteins that recognize and combine with foreign substances, such as viruses, that have entered the blood or tissue. They play their role by distinguishing between self—those things that belong—and nonself—those things that do not belong. The immune system fashions antibodies that attack and destroy specific foreign invaders.
4. ***Communication and regulation.*** Not all hormones are proteins, but those that are function by allowing communication between cells via the endocrine system. Insulin, a protein which regulates blood glucose levels, is a hormone. Human growth hormone is also a protein.
5. ***Coordination of motion.*** Contractile proteins, as found in muscles, coordinate motion by stretching and contracting. The heart, our most important muscle, contracts and expands because of the action of the proteins actin and myosin. The propulsion of sperm by the motion of their flagella is another example of the role of proteins in motion.
6. ***Mechanical support.*** Collagen is a protein that provides mechanical support of animals in the form of bones, tendons, and skin. Our hair and fingernails contain the protein keratin. The multicellular structure of higher animal forms would collapse without the structure provided by proteins.

Classification by Solubility

One of the earliest classifications of proteins was based on solubility. Two major classes are **globular proteins,** which are soluble in water or salt solutions, and **fibrous proteins,** which are insoluble in water.

Globular proteins have tightly folded peptide chains that are roughly spherical. These proteins serve mobile or dynamic functions in a living organism. Albumins and globulins are two types of globular proteins. They include ovalbumin in egg white, lactalbumin in milk, and serum albumin in the blood. Albumins in blood are carriers of some molecules that otherwise would not be soluble in blood. Globulins also occur in blood serum and are part of the body's defense against disease.

Fibrous proteins contain polypeptide chains arranged in extended parallel layers or sheetlike structures. They have connective, structural, or protective functions. Examples of classes of fibrous proteins are collagens, elastins, myosins, and keratins. Skin and tendons, which are connective tissues, contain collagen. The elastic tissues in ligaments and arteries contain elastins. Myosins occur in muscles; keratins occur in hair, wool, nails, and hooves.

Lipoproteins and Blood Cholesterol

Lipoproteins are complex compounds composed of proteins and lipids that regulate the cholesterol level in the blood. Blood serum is 0.5–1.0% lipoprotein. The lipoproteins are classified according to their density, which is related to the fractions of lipid and protein in them (see chart).

Very low density lipoproteins (VLDLs) have densities in the 1.006–1.018 g/mL range and consist of about 5% protein. Low-density lipoproteins (LDLs) have densities of 1.019–1.063 g/mL and consist of about 25% protein. High-density lipoproteins (HDLs) have densities of 1.063–1.21 g/mL and consist of about 50% protein.

The VLDLs are the principal carriers of triacylglycerols, whereas LDLs are the carriers of 80% of the cholesterol. High-density lipoproteins carry the remaining cholesterol. The functions of LDLs and HDLs in carrying cholesterol are quite different. Low-density lipoproteins carry cholesterol to cells for their use, whereas HDLs carry excess cholesterol away from cells to the liver for processing and excretion from the body. Individuals with high HDL levels have an efficient means of removing unneeded cholesterol and avoiding heart disease. If the cholesterol carried to the cells exceeds the cell's requirements and the HDL level is too low, the excess cholesterol will be left as a deposit on the walls of the arteries.

The average concentration of HDL is 45 mg/100 mL for men and 55 mg/100 mL for women. Some scientists feel that the HDL level explains why proportionately fewer women than men have heart attacks. Exercising also is thought to increase the HDL concentration. Male long-distance runners may have as much as 75 mg HDL/100 mL.

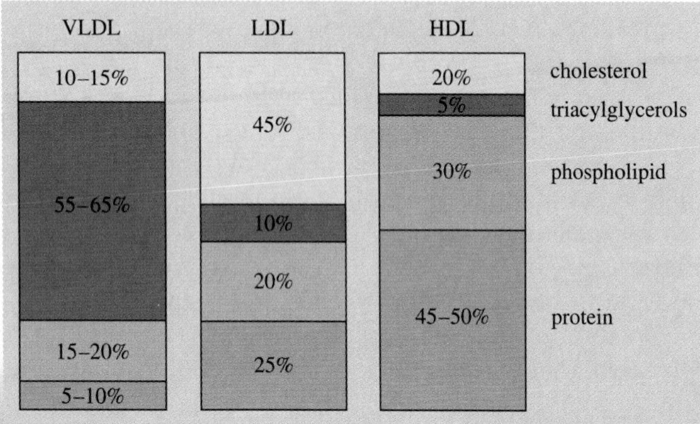

The density of liproproteins increases as the percent protein increases: VLDL < LDL < HDL.

Classification by Composition

Proteins that contain other compounds in addition to amino acids are **conjugated proteins.** The non-amino acid unit is called a **prosthetic group.** There are five types of conjugated proteins.

1. When the prosthetic group is a carbohydrate, the protein is a **glycoprotein.** Heparin, which inhibits the clotting of blood, and mucin, a component of saliva, are both glycoproteins.
2. Proteins combined with nucleic acids in ribosomes and some viruses are **nucleoproteins** (Chapter 29).
3. **Phosphoproteins** contain phosphate esters. Vitellin of egg yolk and casein of milk are both phosphoproteins.
4. When the prosthetic group is colored, the complex protein is called a **chromoprotein.** The red pigment heme occurs in the hemoglobin of the blood.
5. **Lipoproteins** are proteins that contain lipid molecules.

23.2

Amino Acids

About 250 amino acids are found in natural sources; however, only about 20 of them occur in large amounts in proteins. The amino acids of proteins in plants and animals are **α-amino acids;** they have an amino group attached to the carbon atom adjacent to the carboxylic acid group. The Fischer projection formula for an α-amino acid is

$$\begin{array}{c} CO_2H \\ | \\ H_2N-C-H \\ | \\ R \end{array}$$

an α-amino acid

The R of this general formula may be hydrogen, an alkyl group, an aromatic ring, or a heterocyclic ring. Except for glycine, in which R is hydrogen, naturally occurring amino acids contain at least one chiral center and are of the L configuration.

Classification of Amino Acids

The amino acids in proteins are listed in Table 23.1, where they are divided into neutral, acidic, and basic amino acids. The neutral amino acids are further divided by the polarity of the R group. The three-letter abbreviations are used to describe the sequence of amino acids in proteins.

Neutral amino acids have one amino group and one carboxyl group. The simplest neutral amino acids have an alkyl group at the α-carbon atom. Proline has a secondary rather than a primary amino group. Two of the neutral amino acids, serine and threonine, are also alcohols. Phenylalanine, tyrosine, and tryptophan are neutral amino acids with aromatic rings. Two of the neutral amino acids, cysteine and methionine, have a sulfur atom.

Amino acids that have an additional amino or carboxyl group are classified as **basic** or **acidic amino acids,** respectively. The three basic amino acids are lysine, arginine, and histidine. The acidic amino acids are aspartic acid and glutamic acid. Conversion of the second carboxylic acid group of the acidic amino acids into an amide yields asparagine and glutamine, which are neutral amino acids.

The properties of amino acids depend on the nature of the R group attached to the α-carbon atom. These R groups are classed as hydrophilic (water loving) or hydrophobic (water hating). The amino and carboxylic acid groups in basic and acidic amino acids are hydrophilic, as are the hydroxyl groups of serine, threonine, and tyrosine. Hydrophobic amino acids have alkyl or aromatic R groups, which do not hydrogen bond to water.

Essential Amino Acids

Adequate amounts of about 10 of the 20 amino acids are synthesized in the body. The remaining amino acids must be obtained from food because either there is no biochemical pathway to produce them or the pathway produces inadequate amounts. Amino acids that must be obtained from food are called **essential amino acids.** Of course, all amino acids are necessary, but the term *essential* is reserved for those amino acids that must be obtained in the diet. The essential amino acids and their estimated minimum daily requirements are listed in Table 23.2.

TABLE 23.1 Common α-Amino Acids in Proteins

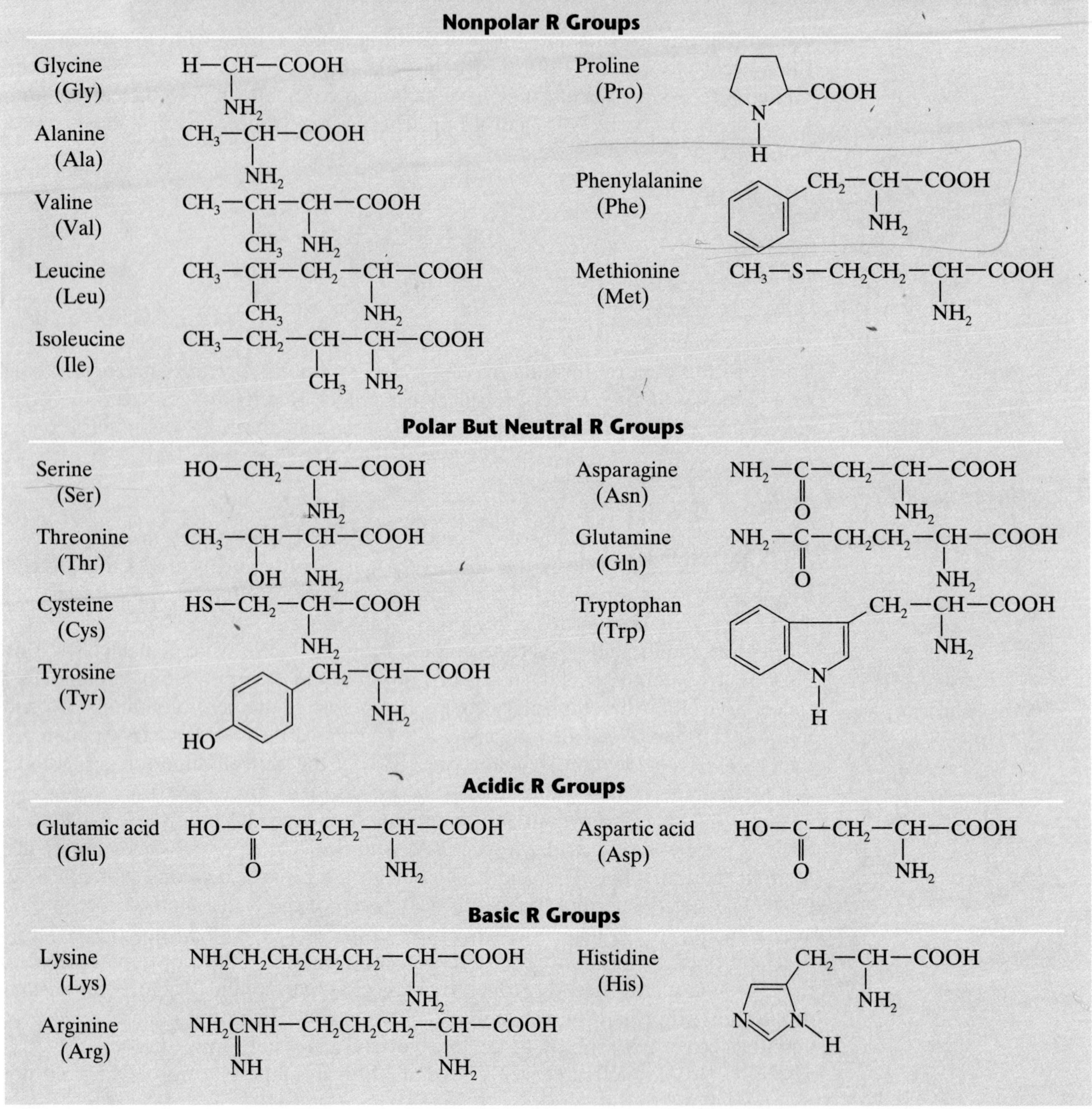

Nonpolar R Groups			
Glycine (Gly)	$H-\underset{NH_2}{CH}-COOH$	Proline (Pro)	$\underset{H}{N}$ (ring) $-COOH$
Alanine (Ala)	$CH_3-\underset{NH_2}{CH}-COOH$	Phenylalanine (Phe)	$CH_2-\underset{NH_2}{CH}-COOH$
Valine (Val)	$CH_3-\underset{CH_3}{CH}-\underset{NH_2}{CH}-COOH$	Methionine (Met)	$CH_3-S-CH_2CH_2-\underset{NH_2}{CH}-COOH$
Leucine (Leu)	$CH_3-\underset{CH_3}{CH}-CH_2-\underset{NH_2}{CH}-COOH$		
Isoleucine (Ile)	$CH_3-CH_2-\underset{CH_3}{CH}-\underset{NH_2}{CH}-COOH$		
Polar But Neutral R Groups			
Serine (Ser)	$HO-CH_2-\underset{NH_2}{CH}-COOH$	Asparagine (Asn)	$NH_2-\underset{O}{\overset{}{C}}-CH_2-\underset{NH_2}{CH}-COOH$
Threonine (Thr)	$CH_3-\underset{OH}{CH}-\underset{NH_2}{CH}-COOH$	Glutamine (Gln)	$NH_2-\underset{O}{C}-CH_2CH_2-\underset{NH_2}{CH}-COOH$
Cysteine (Cys)	$HS-CH_2-\underset{NH_2}{CH}-COOH$	Tryptophan (Trp)	$CH_2-\underset{NH_2}{CH}-COOH$; ring $\underset{H}{N}$
Tyrosine (Tyr)	HO (ring) $CH_2-\underset{NH_2}{CH}-COOH$		
Acidic R Groups			
Glutamic acid (Glu)	$HO-\underset{O}{C}-CH_2CH_2-\underset{NH_2}{CH}-COOH$	Aspartic acid (Asp)	$HO-\underset{O}{C}-CH_2-\underset{NH_2}{CH}-COOH$
Basic R Groups			
Lysine (Lys)	$NH_2CH_2CH_2CH_2CH_2-\underset{NH_2}{CH}-COOH$	Histidine (His)	$CH_2-\underset{NH_2}{CH}-COOH$; ring N, $N-H$
Arginine (Arg)	$NH_2\underset{NH}{C}NH-CH_2CH_2CH_2-\underset{NH_2}{CH}-COOH$		

Tyrosine is not listed as an essential amino acid because phenylalanine is converted into tyrosine by the body. Histidine is essential for growth in infants and may be essential for adults as well. The rate of synthesis of histidine cannot meet the needs of a growing body, so histidine may or may not be considered an essential amino acid depending on the age and state of health of the individual.

TABLE 23.2 Essential Amino Acids and Daily Requirements

Amino Acid	Minimum Daily Requirement (g) Women	Men
isoleucine	0.45	0.7
leucine	0.6	1.1
lysine	0.5	0.8
methionine	0.55	1.0
phenylalanine	1.1	1.4
threonine	0.3	0.5
tryptophan	0.15	0.25
valine	0.65	0.8

23.3 Acid–Base Properties of Amino Acids

The properties of amino acids resemble those of salts rather than those of the uncharged molecules shown in Table 23.1. Consider ethylamine, acetic acid, and glycine. Ethylamine is a gas and acetic acid is a liquid at room temperature, whereas glycine is a solid.

$$\underset{\text{mp} = -84\ °\text{C}}{CH_3CH_2NH_2} \qquad \underset{\text{mp} = 16\ °\text{C}}{CH_3CO_2H} \qquad \underset{\text{mp} = 232\ °\text{C}}{NH_2CH_2CO_2H}$$

The amino acids have low solubility in organic solvents, but they are moderately soluble in water, unlike the majority of organic compounds. The physical properties of amino acids resemble those of ionic compounds.

Amino Acids Exist as Zwitterions

The acidic carboxylic acid group and the basic amino group of amino acids transfer protons in acid–base reactions. The amino group accepts a proton, forming an ammonium group. A proton is lost from the carboxylic acid to form a carboxylate group. A double ion, called a **zwitterion,** results.

$$\overset{+}{N}H_3\text{—}\underset{\underset{R}{|}}{\overset{\overset{CO_2^-}{|}}{C}}\text{—}H$$

a zwitterion

The ammonium and carboxylate sites of the zwitterion act as an acid and a base, respectively. As a result, amino acids can act as buffers in the cell. A conjugate base is formed in basic solution; the proton is lost from the ammonium group giving an amine group. The conjugate base has a negative charge due to the charge of the carboxylate ion.

$$^{+}NH_3-\underset{H}{\overset{H}{|}}C-CO_2^- + OH^- \longrightarrow NH_2-\underset{H}{\overset{H}{|}}C-CO_2^- + H_2O$$

zwitterion conjugate base

The conjugate acid is formed in acidic solution when the carboxylate group is protonated. The conjugate acid is positive because of the charge on the ammonium ion.

$$^{+}NH_3-\underset{H}{\overset{H}{|}}C-CO_2^- + H_3O^+ \longrightarrow {}^{+}NH_3-\underset{H}{\overset{H}{|}}C-CO_2H + H_2O$$

zwitterion conjugate acid

Amino Acids and pH

The form in which the amino acids exist depends on the pH of the solution. To understand this phenomenon we have to examine the pK_a values of the functional groups. For glycine the pK_a values of the $-CO_2H$ and $-NH_3^+$ groups are 2.35 and 9.78, respectively (Table 23.3). The pK_a($-CO_2H$) indicates the acidity of the carboxylic acid which is the conjugate acid of the zwitterion. The pK_a ($-NH_3^+$) indicates the acidity of the ammonium ion of the zwitterion.

TABLE 23.3 pK_a Values of Amino Acids at 25 °C

Amino Acid	pK_a ($-CO_2H$)	pK_a ($-NH_3^+$)	pK_a (side chain)
glycine	2.35	9.78	
alanine	2.35	9.87	
valine	2.29	9.72	
leucine	2.33	9.74	
isoleucine	2.32	9.76	
methionine	2.17	9.27	
proline	1.95	10.64	
phenylalanine	2.58	9.24	
tryptophan	2.43	9.44	
serine	2.19	9.44	
threonine	2.09	9.10	
cysteine	1.89	10.78	8.38
tyrosine	2.20	9.11	10.07
asparagine	2.02	8.80	
glutamine	2.17	9.13	
aspartic acid	1.99	10.00	3.96
glutamic acid	2.13	9.95	4.32
lysine	2.16	9.20	10.80
arginine	1.82	8.99	12.48
histidine	1.81	9.15	6.00

$$^{+}NH_3-\underset{H}{\overset{H}{\underset{|}{\overset{|}{C}}}}-CO_2H + H_2O \rightleftharpoons {}^{+}NH_3-\underset{H}{\overset{H}{\underset{|}{\overset{|}{C}}}}-CO_2^- + H_3O^+ \qquad K_a = 5.0 \times 10^{-3} \quad pK_a = 2.35$$

$$^{+}NH_3-\underset{H}{\overset{H}{\underset{|}{\overset{|}{C}}}}-CO_2^- + H_2O \rightleftharpoons NH_2-\underset{H}{\overset{H}{\underset{|}{\overset{|}{C}}}}-CO_2^- + H_3O^+ \qquad K_a = 1.6 \times 10^{-10} \quad pK_a = 9.78$$

When an amino acid dissolves in an aqueous solution, several species exist. When the pH of a solution is equal to the pK_a of an acid, the concentrations of an acid and its conjugate base are equal. At pH = 2.35, the concentrations of the conjugate acid of glycine and the zwitterion are equal. At pH = 9.78, the concentrations of the conjugate base of glycine and the zwitterion are equal. At pH values between 2.35 and 9.78, the zwitterion is the major form of the amino acid. The zwitterion is the major form in biological fluids because their pH values are close to 7.

23.4 Isoionic Point

The **isoionic point,** abbreviated pH_i is the pH at which the concentration of the zwitterion is a maximum. At a pH above the isoionic point, the conjugate base predominates. At a pH below the isoionic point, the conjugate acid exists. The isoionic points of some amino acids are given in Table 23.4. For neutral amino acids, the pH_i is the average of the two pK_a values.

The calculation of the isoionic points of acidic and basic amino acids is somewhat complicated because there are several acid–base equilibria to consider. However, for an acidic amino acid the isoionic point is less than 7, whereas for a basic amino acid, with the exception of histidine, the isoionic point is significantly greater than 7.

The isoionic point of an amino acid is measured by observing its behavior in an electric field. A charged molecule migrates to the electrode of opposite charge. At low pH an amino acid has a net positive charge and migrates toward the negative electrode. At high pH an amino acid has a net negative charge and migrates toward the positive electrode. At the isoionic point the zwitterion predominates, and it does not migrate toward either electrode.

Proteins Have Isoionic Points

Proteins have isoionic points characteristic of their component amino acids (Table 23.5). At their isoionic point, proteins have no net charge and are the least soluble in aqueous solution. As a consequence, a protein tends to precipitate from solution at the isoionic point. For example, casein is least soluble at pH 4.7. The pH of cow's milk is 6.3, and casein has a net negative charge and is soluble at this pH. If the milk is acidified, the casein is protonated and then precipitates. Casein, which is used in making cheese, is obtained by adding an acid to milk or by bacterial action that produces lactic acid and causes a decrease in the pH of milk.

TABLE 23.4 Isoionic Points of Amino Acids

Amino Acid	pK_i
glycine	6.06
alanine	6.10
valine	5.96
leucine	5.98
isoleucine	6.02
methionine	5.74
proline	6.30
phenylalanine	5.48
tryptophan	5.89
serine	5.68
threonine	5.60
cysteine	5.07
tyrosine	5.66
asparagine	5.41
glutamine	5,65
aspartic acid	2.77
glutamic acid	3.22
lysine	9.74
arginine	10.76
histidine	7.59

Separation of Amino Acids and Proteins

Mixtures of amino acids can be separated and identified by a process called **electrophoresis,** which is based on the charge of an amino acid at a given pH (Figure 23.1). A paper strip saturated with a buffer solution at a selected pH bridges two vessels containing the buffer. A sample of the amino acid mixture is placed at the center of the paper, and an electric potential is applied between the two vessels. If the buffer pH equals the isoionic point of an amino acid, the zwitterion predominates, and does not migrate. An amino acid having a negative charge at that pH migrates toward the positive electrode, whereas an amino acid having a positive

TABLE 23.5 Isoionic Points of Proteins

Protein	Isoionic Point
pepsin (enzyme)	1.1
casein (milk protein)	4.6
egg albumin	4.7
urease (enzyme)	5.0
insulin (hormone)	5.3
hemoglobin	6.8
ribonuclease (enzyme)	9.5
chymotrypsin (enzyme)	9.5

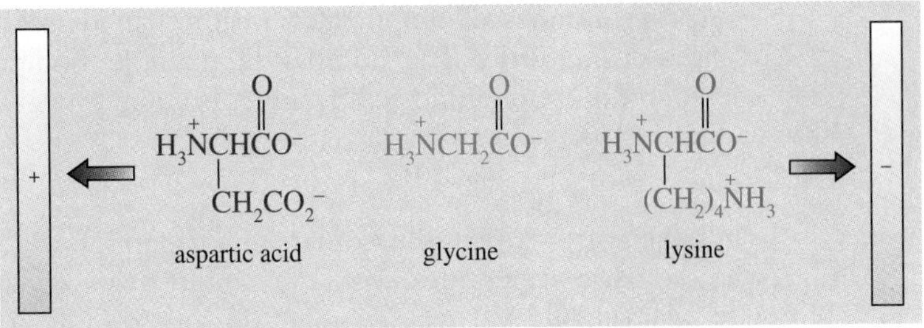

FIGURE 23.1 Separation of Amino Acids by Electrophoresis
At pH 6.0 glycine exists largely as the zwitterion and does not migrate toward either electrode. Aspartic acid is largely deprotonated and has a net negative charge at this pH; it migrates toward the positive electrode. Lysine is largely protonated and has a net positive charge at this pH; it migrates toward the negative electrode.

charge at that pH migrates toward the negative electrode. After a period of time the original "spot" of the amino acid sample is separated into spots corresponding to each of the amino acids present.

Electrophoretic separation of proteins is used in clinical laboratories. Because proteins have different charges and molecular weights, they move at different rates in the electrophoresis apparatus. Electrophoresis is used to analyze blood serum. For example, the identification of certain enzymes in the blood is diagnostic for myocardial infarction.

23.5 Peptides

When the α-amino group of one amino acid is linked to the carboxyl group of a second amino acid by an amide bond, the product is called a **peptide.** Each amino acid in a peptide is called an amino acid residue. A prefix indicates the number of amino acids in a peptide. If the peptide contains two amino acid units, it is a **dipeptide;** if it contains three amino acids, it is a **tripeptide;** and so forth. But a peptide that contains, say, 14 amino acids is more likely to be called a 14-peptide than a tetradecapeptide. Peptides that contain only a "few" amino acids are called **oligopeptides.**

A peptide has two ends: the end with a free α-amino group is called the **N-terminal amino acid residue.** The end with the free carboxyl group is called the **C-terminal amino acid residue.** Peptides are named from the N-terminal amino acid to the C-terminal amino acid. Two examples of this nomenclature for isomeric dipeptides containing glycine and alanine are shown.

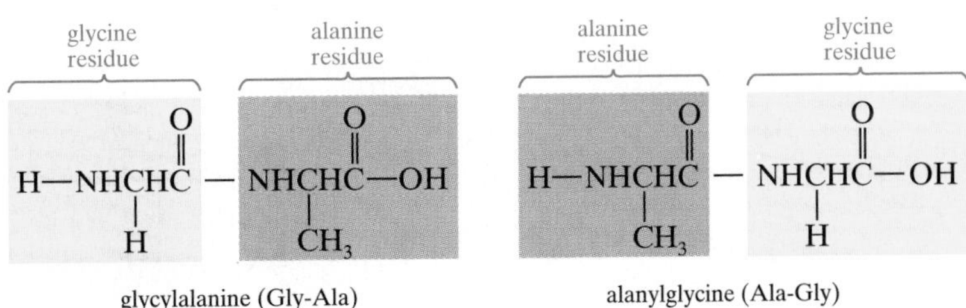

As the number of amino acid residues increases, the number of isomers increases dramatically. The number of possible isomeric peptides containing one each of n different amino acid residues is equal to $n!$, where

$$n! = 1 \times 2 \times 3 \times \ldots \times (n - 1) \times n$$

Thus, there are six possible isomers of a tripeptide with three different amino acids. The isomeric tripeptides with the amino acids glycine, alanine, and valine are Gly-Ala-Val, Gly-Val-Ala, Val-Gly-Ala, Val-Ala-Gly, Ala-Gly-Val, and Ala-Val-Gly. For a peptide with one each of 20 different amino acid residues there are 2,432,902,008,176,640,000 isomers!

Proteins with 100 to 300 amino acid residues are common in the thousands of proteins found in cells. These proteins contain two or more residues of the same amino acid, in which case the given formula to calculate the number of isomers does not apply. However, the number of isomers is still astronomically large. It is this structural diversity that makes possible the many functions of proteins presented in Section 23.1.

The assembly of amino acids into peptides and proteins is directed by ribonucleic acids (Chapter 29). How a single required structure of an enzyme can be assembled to the exclusion of the billions of billions of isomers is one of the marvels of life. Yet, each species assembles not just one such protein but thousands of different proteins required for their myriad functions. A eukaryotic cell may contain several thousand different proteins, and the human body is thought to contain at least 10^5 different proteins.

EXAMPLE 23.1

Identify the terminal amino acids of tuftsin, a tetrapeptide that stimulates and promotes the destruction of tumor cells. Write the amino acid sequence using three-letter abbreviations for the amino acids. Also write the complete name without abbreviations.

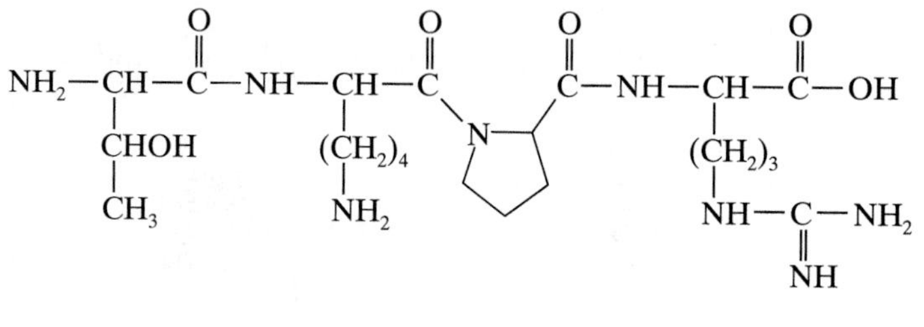

Solution

The N-terminal and C-terminal amino acids are threonine and arginine, respectively. The amino acid sequence is Thr-Lys-Pro-Arg. The name is threonyllysylprolylarginine.

PROBLEM 23.1

Identify the components of aspartame, an artificial sweetener.

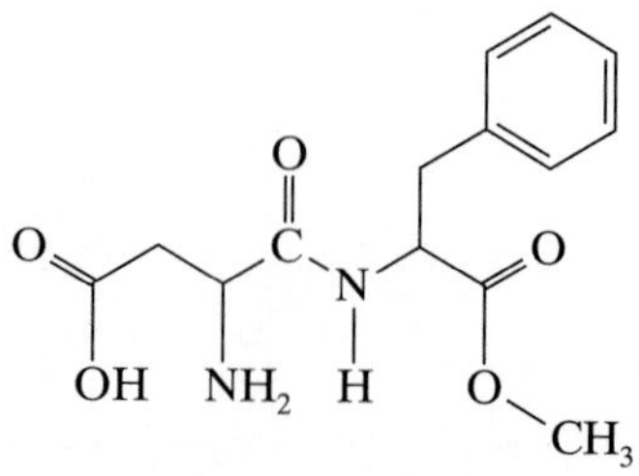

23.6 Biological Activity of Peptides

Cells contain many small peptides that have diverse functions (Table 23.6). Some act as hormones with physiological functions, such as pain relief and control of blood pressure. These oligopeptides are produced and released in small amounts. They are rapidly metabolized, but their physiological action is necessary for only a short time. For example, the 14-peptide somatostatin, which inhibits the release of other hormones such as insulin, glucagon, and secretin, has a biological half-life (the time in which one-half of a compound is metabolized) of less than 4 minutes.

Enkephalins are peptides that bind specific receptor sites in brain cells to reduce pain. The enkephalin receptor sites have a high affinity for opiates, including morphine, heroin, and other structurally similar substances. Hence, enkephalin receptors are commonly called *opiate receptors.* Opiates mimic the enkephalins normally present in the body to mitigate pain.

Tyr-Gly-Gly-Phe-Leu

an enkephalin

Peptides are produced in many tissues. For example, kidney cells secrete angiotensin II, which increases blood pressure by constricting blood vessels. The production of excess angiotensin II is responsible for some forms of hypertension.

Asp-Arg-Val-Tyr-Ile-His-Pro-Phe

angiotensin II

Oxytocin, a peptide hormone produced in the pituitary gland, causes the contraction of smooth muscle, such as that of the uterus. It is used clinically to induce

TABLE 23.6 Peptide Hormones

Hormone	Amino Acid Residues	Molecular Weight	Function
tuftsin	4	501	stimulates phagocytosis
met-enkephalin	5	645	analgesic activity
angiotensin II	8	1031	affects blood pressure
oxytocin	8	986	affects uterine contractions
vasopressin	8	1029	an antidiuretic
bradykinin	9	1069	produces pain
somatostatin	14	1876	inhibits release of other hormones
gastrin	17	2110	promotes pepsin secretion
secretin	27	2876	stimulates pancreatic secretions
glucagon	29	3374	stimulates glucose production from glycogen
calcitonin	32	3415	decreases calcium level in blood
relaxin	48	5500	relaxation of uterine muscles
insulin	51	5700	affects blood sugar level

labor or to increase the effectiveness of uterine contractions. Vasopressin, another pituitary peptide hormone, regulates the excretion of water by the kidneys and affects blood pressure. The structures of oxytocin and vasopressin differ by only two amino acids. They are both cyclic peptides that result from a disulfide bond (—S—S—) between what would otherwise be the N-terminal amino acid cysteine and another cysteine five amino acid residues away. The C-terminal amino acid exists as an amide in both compounds.

S———S (between Cys-1 and Cys-6)

Cys-Tyr-**Ile**-Gln-Asn-Cys-Pro-**Leu**-Gly-NH_2

oxytocin

S———S (between Cys-1 and Cys-6)

Cys-Tyr-**Phe**-Gln-Asn-Cys-Pro-**Arg**-Gly-NH_2

vasopressin

The structural difference between oxytocin and vasopressin may seem small at first glance. But when we compare oxytocin and vasopressin, we see that residue 3 in oxytocin is isoleucine and that residue 3 in vasopressin is phenylalanine. This change has relatively little effect: Both residues are nonpolar and about the same size. However, residue 8 in oxytocin is leucine, a nonpolar amino acid with a *sec*-butyl side chain, whereas residue 8 in vasopressin is arginine, an amino acid with a strongly basic side chain and a positive charge at pH 7. Because of this difference in charge, the receptor for oxytocin has a weak affinity for vasopressin, and the receptor for vasopressin has a very low affinity for oxytocin. These peptides therefore bind different receptors and have different functions.

23.7 Bonding in Proteins

There are four types of bonds in proteins—the peptide bond, the disulfide bond, the hydrogen bond, and the ionic bond. In addition, there are hydrophobic interactions. The characteristics of these bonds will be used to describe the structure and conformation of proteins in Section 23.8.

The Peptide Bond

The peptide bond is the strongest bond in a protein. There is restricted rotation about the carbon–nitrogen bond of amides (Section 19.8). This restricted rotation is the result of some double-bond character shown in the dipolar resonance form. Peptides tend to exist in trans conformations about the carbon–nitrogen bond.

Although rotation is somewhat limited about the peptide bond, the bond between the α-carbon atom and the carbonyl carbon atom is a rotationally free single bond. Similarly, the single bond between the nitrogen atom and the α-carbon atom of the next amino acid is also rotationally free. Free rotation also occurs around the bonds between the α-carbon atoms and the R groups. Thus, a protein chain consists of rigid peptide units connected to one another by freely rotating single bonds.

The Disulfide Bond

The **disulfide bond** is a covalent bond between two sulfur atoms. A disulfide bond results from the oxidation of the —SH (sulfhydryl) groups of two cysteine molecules to form cystine.

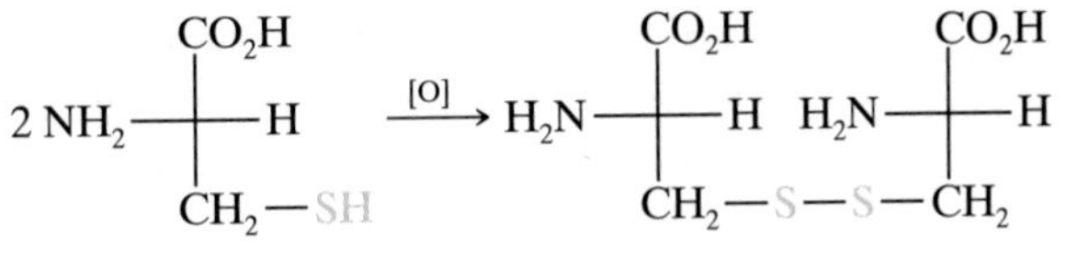

$$2\,NH_2\text{—}\underset{CH_2\text{—}SH}{\overset{CO_2H}{C}}\text{—}H \xrightarrow{[O]} H_2N\text{—}\underset{CH_2\text{—}S\text{—}S\text{—}CH_2}{\overset{CO_2H}{C}}\text{—}H \quad H_2N\text{—}\overset{CO_2H}{C}\text{—}H$$

cystine

Many proteins—especially relatively small ones containing fewer than 100 amino acid residues—have a high cysteine content. Each of these cysteine residues has a sulfhydryl group that can be oxidized to form a disulfide bond, as in cystine.

Disulfide bonds form after a protein has folded into its biologically active conformation. Once formed, the protein conformation is much less flexible. Intrachain disulfide bonds occur in small peptides such as oxytocin and vasopressin. Disulfide bonds can also link a cysteine residue in one polypeptide chain with a cysteine residue in another polypeptide chain, as in insulin (Section 23.8).

Hydrogen Bonds

Proteins contain many functional groups that can form hydrogen bonds. Although individual hydrogen bonds are weaker than peptide and disulfide bonds, there are many hydrogen bonds and they help stabilize the folded conformation of proteins. Intramolecular hydrogen bonding between the amide hydrogen atom of one peptide unit and the carbonyl oxygen atom of another peptide unit within a molecule is very common.

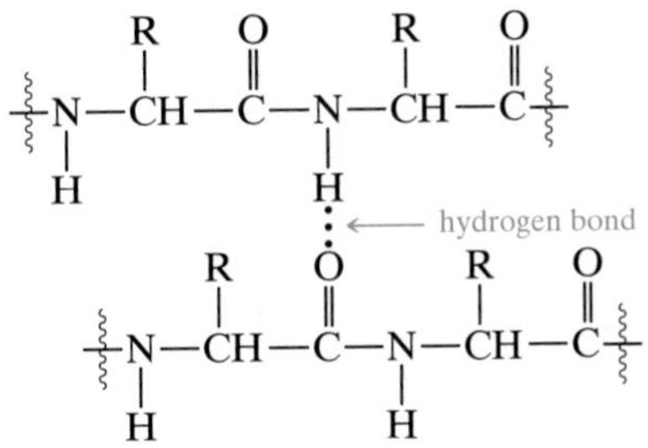

$$\text{—}\underset{H}{N}\text{—}\overset{R}{CH}\text{—}\overset{O}{\overset{\|}{C}}\text{—}\underset{H}{N}\text{—}\overset{R}{CH}\text{—}\overset{O}{\overset{\|}{C}}\text{—}$$

hydrogen bond

$$\text{—}\underset{H}{N}\text{—}\overset{R}{CH}\text{—}\overset{O}{\overset{\|}{C}}\text{—}\underset{H}{N}\text{—}\overset{R}{CH}\text{—}\overset{O}{\overset{\|}{C}}\text{—}$$

Hydrogen bonds also form with amino acid side chains such as the —OH groups of serine, threonine, and tyrosine, as well as the —NH group in the ring of tryptophan.

Ionic Bonds

At physiological pH, some of the R groups attached to the polypeptide chain are charged. Ionic attractive forces between the carboxylate groups and the ammonium groups pull portions of chains together. An intrachain ionic bond between acidic and basic amino acids is called a **salt bridge.**

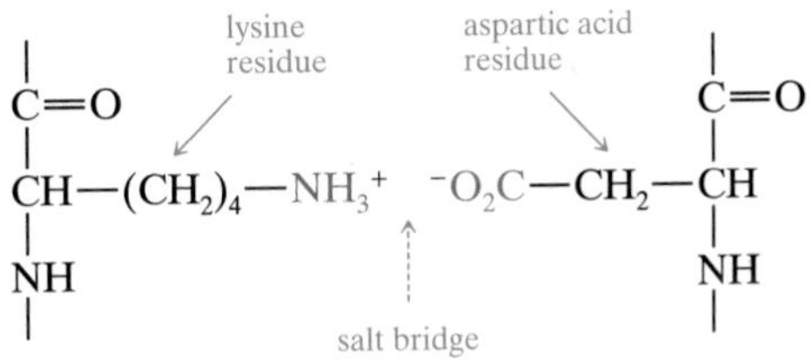

lysine residue — aspartic acid residue

$$\underset{NH}{\overset{C=O}{CH}}\text{—}(CH_2)_4\text{—}NH_3^+ \quad {}^-O_2C\text{—}CH_2\text{—}\underset{NH}{\overset{C=O}{CH}}$$

salt bridge

Hydrophobic Interactions

Proteins contain many nonpolar side chains. These side chains are repelled by water and tend to associate with one another on the "inside" of a folded protein molecule, out of contact with water. The tendency of nonpolar side chains to avoid the solvent is called the **hydrophobic effect.** The hydrophobic interactions in proteins are similar to those in the micelle of a soap (Section 18.8) or the bilayer of lipids in membranes (Section 22.10). Hydrophobic interactions among nonpolar side chains in proteins are weak, but abundant, and are primarily responsible for maintaining the folded conformation of a protein. The aromatic rings of amino acids are very polarizable and form hydrophobic interactions.

23.8 Structure of Proteins

The biological activity of a protein depends on the three-dimensional shape of the molecule called its **native state** or **native conformation.** Any alteration of structure caused by breaking any bond destroys the function of the protein. Protein structure is described at four levels: primary, secondary, tertiary, and quaternary. Each of these divisions is arbitrary because it is the total structure of the protein that controls function. Nevertheless, it is useful to consider the levels of structure individually. An overview of the four levels of protein structure is shown in Figure 23.2.

Primary Structure

The substantial strength of spider webs is due to proteins. (Peter Arnold, Inc.)

The linear sequence of amino acids in a protein and the location of disulfide bonds is called its **primary structure.** For example, insulin consists of two peptide chains, called the A chain and the B chain, linked by two disulfide bonds. The A chain has 21 amino acids, and the B chain has 30 amino acids (Figure 23.3). There is also an intrachain disulfide bond within the A chain. Insulins from different animals have slightly different amino acid sequences, as noted in Figure 23.3. Because the sequence within the cyclic portion of the shorter chain does not affect the physiological function of the insulin, diabetic individuals who became allergic to one type of insulin were often given insulin from another animal source. However, this problem has been eliminated by the synthesis of insulin using recombinant DNA technology in which bacteria produce human insulin with virtually no allergens.

Human hemoglobin consists of heme, which is a complex heterocyclic compound, and globin. Globin is composed of four protein chains: two α chains and two β chains. There are 141 and 146 amino acids in the α and β chains, respectively. In some people the sixth amino acid from the N-terminal end of the β chain is valine rather than glutamic acid. This difference of a single amino acid out of 146 causes the cells to sickle, and their passage through blood vessels is restricted. The associated circulatory problems are known as *sickle cell anemia.*

Differences in hemoglobin in various animals do not affect its oxygen-carrying capacity. The β chains of gorilla and human hemoglobins differ only at position 104. In gorillas the amino acid is lysine, whereas in humans it is arginine, another basic amino acid. The pig β chain differs from human hemoglobin at 17 sites, and that of the horse at 26 sites. In spite of the different composition of hemoglobin in animals, there are nine positions that contain the same amino acids in all hemoglobin molecules. These amino acid residues are important to the oxygen-binding function of hemoglobin.

Secondary Structure

The specific spatial arrangement of the amino acid residues close to one another in the polypeptide chain is called its **secondary structure** (Figure 23.2). Recall that the bonds separating the planar and rigid peptide units can rotate freely. These peptide units are oriented at angles to each other and can exist in a variety of conforma-

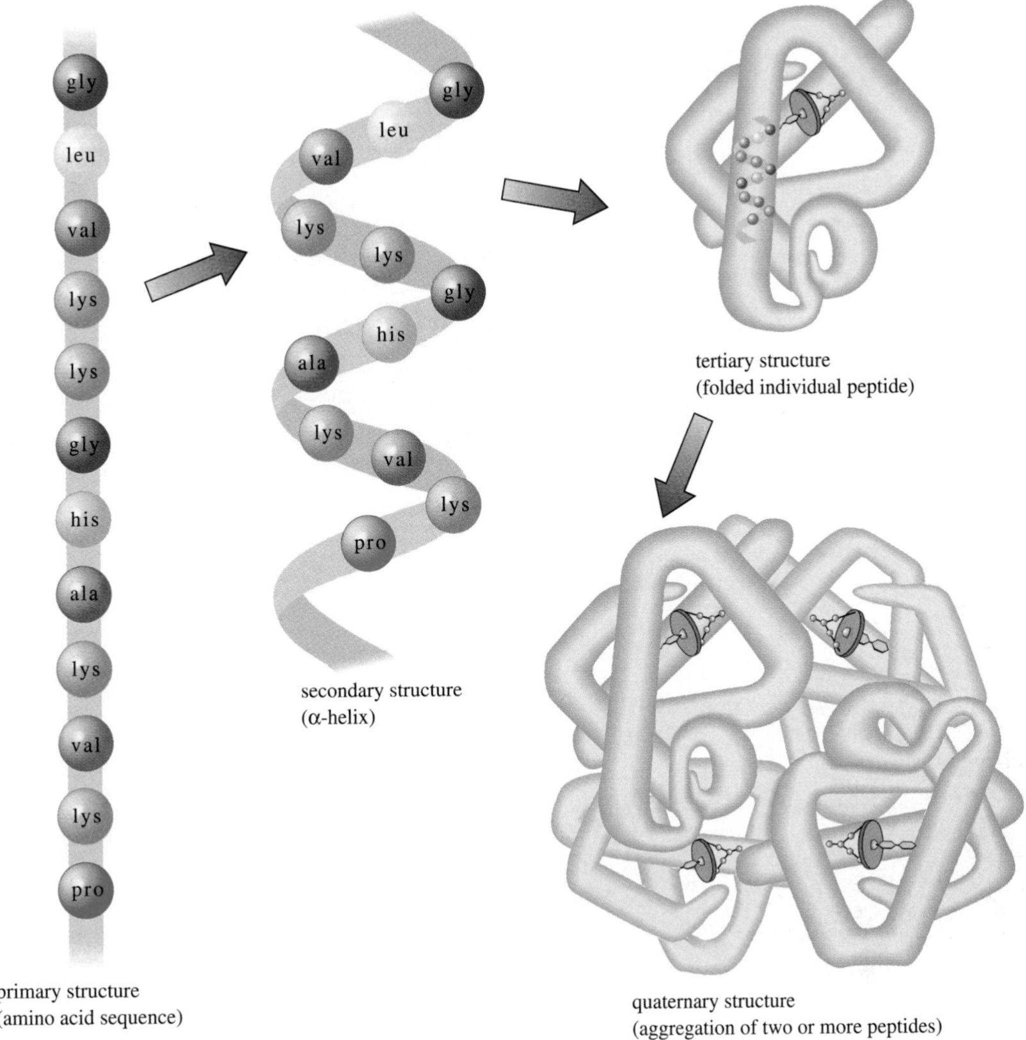

FIGURE 23.2
The Four Levels of Protein Structure
The levels of protein structure are determined by the primary structure. The secondary structure is the result of hydrogen bonding. The folded protein chain shown in the tertiary structure is the result of numerous interactions but primarily hydrophobic bonding. Quaternary structure is not found in all proteins. In hemoglobin each of the four chains enfold a heme containing iron that transports oxygen.

tions. Many proteins exist as chains coiled into a spiral known as a helix. The helix could be either right- or left-handed, but for proteins consisting of L-amino acids the right-handed (or α) helix is more stable than the left-handed helix. The spiral is held together by hydrogen bonds between the proton of the N—H group of one amino acid and the oxygen atom of the C═O group of another amino acid in the next turn of the helix (Figure 23.4). You are familiar with at least one object that is helix-shaped, the coiled telephone cord.

In addition to the intrachain hydrogen bonding that occurs in the α helix, there are proteins with interchain hydrogen bonding. The β-pleated sheet consists of extended protein chains arranged in an alternating sequence with chains running in opposite directions (Figure 23.4.) Interchain hydrogen bonds occur in fibrin (the

Animal	Positions 8	9	10
sheep	Ala	Gly	Val
cow	Ala	Ser	Val
pig	Thr	Ser	Ile
horse	Thr	Gly	Ile

Ser–Leu–Tyr–Gln–Leu–Glu–Asn–Tyr–Cys–Asn

Gly–Ile–Val–Glu–Gln–Cys–Cys

Phe–Val–Asn–Gln–His–Leu–Cys–Gly–Ser–His–Leu–Val–Glu

Val Leu Tyr Leu Ala

Cys Gly Glu Arg Gly Phe Phe Tyr Thr Pro Lys Ala

FIGURE 23.3 Primary Structure of Insulin

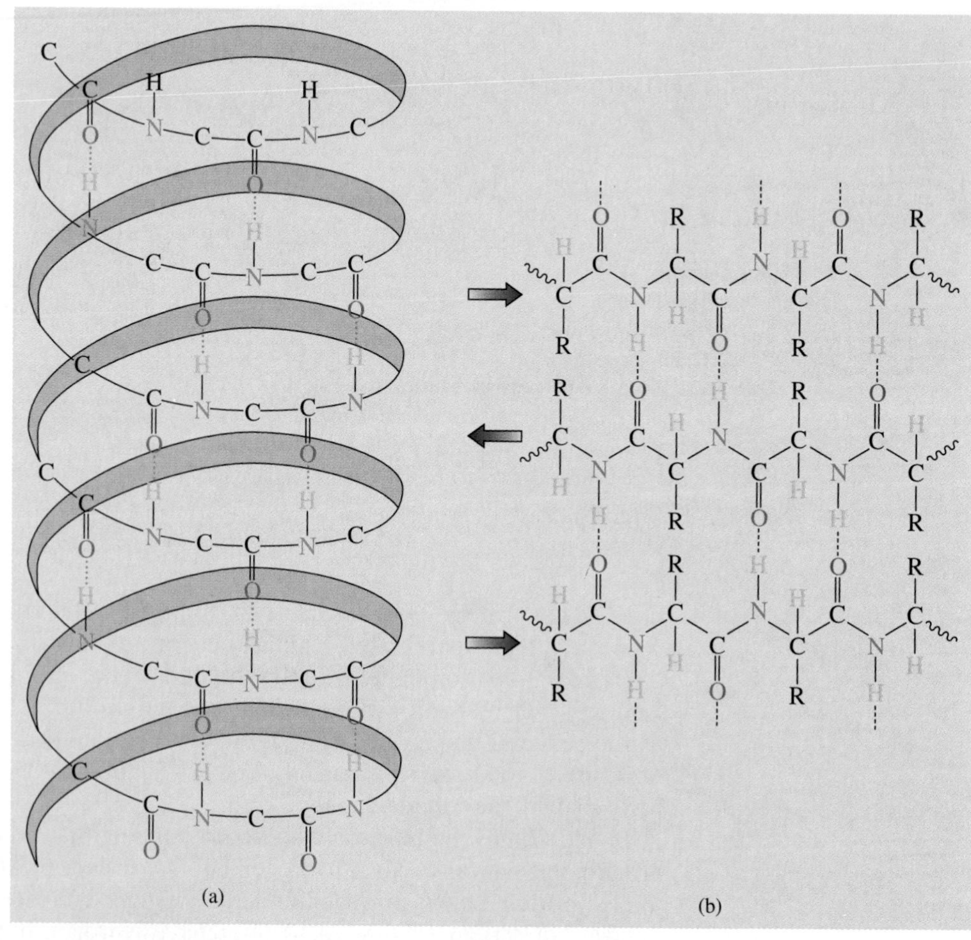

FIGURE 23.4 Hydrogen Bonding in Proteins
(a) The intramolecular hydrogen bonds between coils of an α-helix are shown only on the "front." (b) The intermolecular hydrogen bonds between the chains of proteins in a β-pleated sheel occur between chains oriented in opposing directions as shown by the arrows.

blood-clotting protein), myosin (a protein of muscle), keratin (the protein of hair), and fibroin (the protein of silk).

Only amino acids with small side chains can form a β-pleated sheet. For example, silk contains approximately 45% glycine, 30% alanine, and 12% serine. The primary structure consists of glycine separated by either alanine or serine units.

-Gly-Ala-Gly-Ser-Gly-Ala-Gly-

The hydrogen atoms and side chains of the amino acids extend above and below the plane of the sheet. The chains can closely associate without the side chains bumping into each other. Furthermore, the sheets can stack closely together when the groups are small.

Tertiary Structure

The three-dimensional shape of the protein is its **tertiary structure.** This spatial arrangement brings together a number of amino acid residues that are far apart in the polypeptide chain. The proximity of amino acids in the tertiary structure is responsible for the activity of enzymes.

The three-dimensional folded shape of proteins (Figure 23.2) depends on their primary and secondary structures, which together make possible long-range interaction between amino acids. The forces of attraction between amino acids include ionic bonds, hydrogen bonds, and hydrophobic interactions.

The hydrophobic portions of the protein associate within the interior of the folded structure. Polar or charged (hydrophilic) groups are located at the surface near water molecules. Thus, there is no single reason for a specific shape of a globular protein. Each shape is unique and is the result of both composition and the sequence of the polypeptide chain.

Quaternary Structure

The **quaternary structure** of a protein is the association of several protein chains or subunits into a closely packed arrangement (Figure 23.2). Each of the subunits has its own primary, secondary, and tertiary structure. The subunits fit together because of their shape and are held together by forces other than covalent bonds.

The subunits in a quaternary structure must be specifically arranged for the entire protein to function properly. Any alteration in the structure of the subunits or how they associate causes marked changes in biological activity. A list of some proteins that have quaternary structure is given in Table 23.7.

TABLE 23.7 Proteins with Quaternary Structure

Protein	Molecular Weight	Number of Subunits	Biological Function
alcohol dehydrogenase	80,000	4	enzyme for alcohol fermentation
aldolase	150,000	4	enzyme for glycolysis
fumarase	194,000	4	enzyme in the citric acid cycle
hemoglobin	65,000	4	oxygen transport in blood
insulin	11,500	2	hormone regulating metabolism of glucose

Hemoglobin consists of two pairs of different proteins, each protein enfolding a molecule of heme. The two identical α-chains and two identical β-chains are arranged tetrahedrally in a three-dimensional structure (Figure 23.2). These units are held together by hydrophobic interactions, hydrogen bonding, and salt bridges. The four protein subunits of hemoglobin do not behave independently. When one heme molecule binds O_2, the conformation of the surrounding protein chain is slightly altered. Changes in conformation at one site caused by a change at a spatially separated site of a protein molecule are called **allosteric effects.** As a result of allosteric effects, each heme in the other subunits can then bind more easily to additional oxygen molecules. As each oxygen binds, there are further conformational changes in the other protein chains that enhance their binding capability. As a consequence, once oxygenation occurs at one heme, there is cooperation at all other sites in hemoglobin, which can then carry four oxygen molecules.

23.9 Denaturation of Proteins

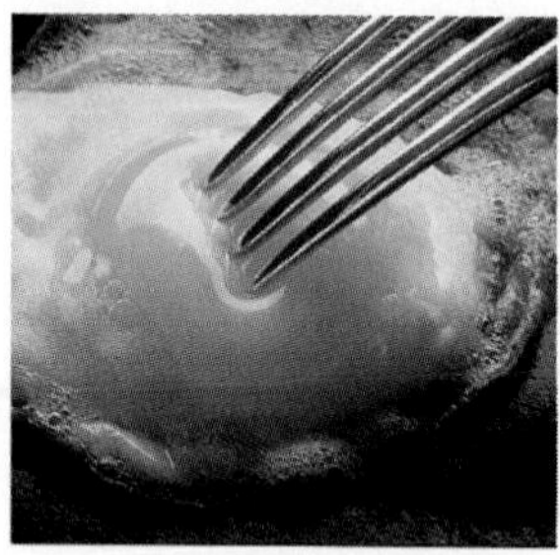

Fried eggs are different than raw eggs because the protein is denatured by heat. (The Image Bank)

Denaturation is the destruction of the native conformation of a protein, which destroys the biochemical function of the molecule. In denaturation the hydrogen bonds, disulfide bonds, ionic bonds, and hydrophobic interactions responsible for secondary, tertiary, and quaternary structures are disrupted. When there is quaternary structure, denaturation first affects the forces holding the subunits together. Next, the tertiary structure is disrupted, and the protein unfolds. Finally, the hydrogen bonds are broken, and a random-coil protein results.

Denaturation may be reversible or irreversible. The formation of a native protein from a denatured protein is called **renaturation** (Figure 23.5). When the enzyme aldolase is treated with 4 M urea, the tetramer separates into its four subunits, and these protein chains unfold to some degree. However, when the protein is separated from the urea, the chains refold and the subunits rejoin. This denaturation process is reversible. Conversely, if the disulfide bonds in insulin are cleaved by reduction and then reoxidized, denaturation is irreversible.

Denaturation may be caused by heat or radiation, acids or bases, oxidizing or reducing agents, hydrogen-bonding solvents, and heavy-metal salts. We now consider the action of each of these denaturation agents.

Heat and Radiation

If a protein is heated or irradiated, the added energy causes increased vibration and motion within the molecule. Hydrophobic interactions and hydrogen bonds are disrupted, and the protein is denatured. The most common example of denaturation by heat is the frying of an egg. Most proteins are denatured when heated above 50°C, a temperature above the normal body temperature of 37°C. Burns, as well as sunburn, cause denaturation of protein in the skin. Surgical cauterization or heating of body protein is used to seal small blood vessels.

FIGURE 23.5 Denaturation and Renaturation of Proteins

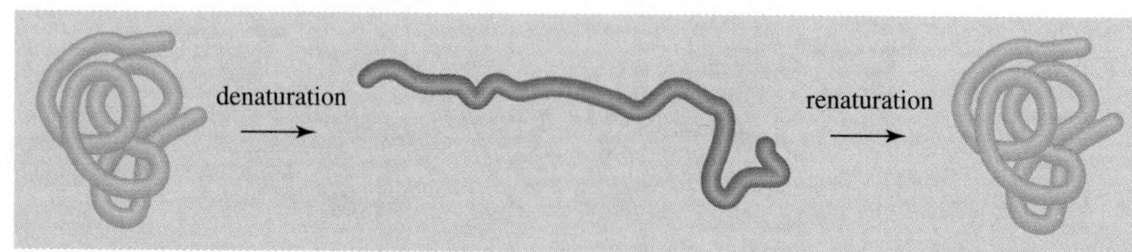

Acids and Bases

When the pH of a solution of a protein is changed, the carboxyl group in an acidic amino acid may exist as the acid or as the carboxylate ion. Similarly, the amino group of basic amino acids may exist as the amine or as an ammonium ion. As a consequence, the ionic bonds in the protein are disrupted. The degree of destruction of the protein structure depends on how much the pH is changed. If the change is slight, then restoring the pH can result in complete regeneration of structure (Figure 23.6).

Oxidizing and Reducing Agents

Reducing agents convert disulfide bonds to sulfhydryl groups and the structure of proteins is disrupted (Figure 23.6). Oxidizing agents can convert sulfhydryl groups back into disulfide bonds, but the structure of the protein may be different from that of the native protein.

Hydrogen-Bonding Solvents

Hydrogen-bonding solvents denature proteins by disrupting the hydrogen bonds within the protein molecule. The —OH group of ethanol effectively competes with hydrogen bonds within the protein. A 70% solution of ethanol is able to pass across the cell wall of bacteria. Once inside the cell, it denatures the bacterial proteins. As a result, ethanol is a good antiseptic.

FIGURE 23.6
Types of Denaturation Processes
(a) The denaturation of a protein by acid or base may be reversible if the pH changes are small and the degree of unfolding of the chain is small. (b) Reduction of disulfide bonds followed by oxidation may not regenerate the native protein.

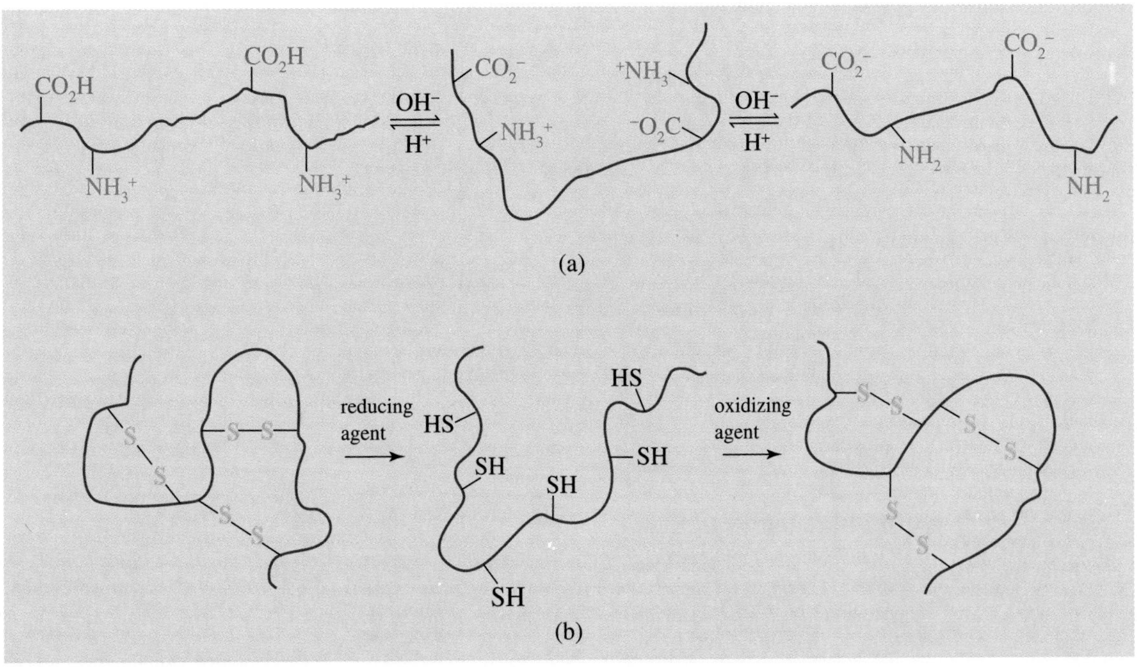

Straight and Curly Hair

All human hair contains the protein keratin, but whether you have straight or curly hair is genetically specified. Keratin has a high percentage of cysteine, and it is the disulfide bridges that are primarily responsible for the shape of hair. The locations of the disulfide bridges in straight hair are at the same level and maintain the chains in a "straight" arrangement (see Figure). In curly hair there are disulfide bonds between cysteines of the same chain, which causes some kinks. As a result, the disulfide bonds between chains are not at the same level.

You can do something about your hair but it takes a little chemistry. For example, straight hair can be changed by a "permanent wave"—a misnomer because it isn't permanent. First, a lotion that is a reducing agent is applied to the hair. The disulfide bonds are cleaved and converted to sulfhydryl groups. Then the hair is shaped on curlers, and a second solution is applied. This solution contains an oxidizing agent that oxidizes the sulfhydryl groups back to disulfide bonds. The positions of the bridges that form between the protein strands are different from those in the original strands because the relative positions of the sulfhydryl groups have been moved. In addition, some disulfide bonds within a single strand may result. Alas, while the hair is now curly, the situation is not permanent. As new hair eventually grows in, the strands are the same genetically determined shape that you wanted to change.

Hair permanents depend on doing redox reactions that can be carried out in the home. (The Stock Market)

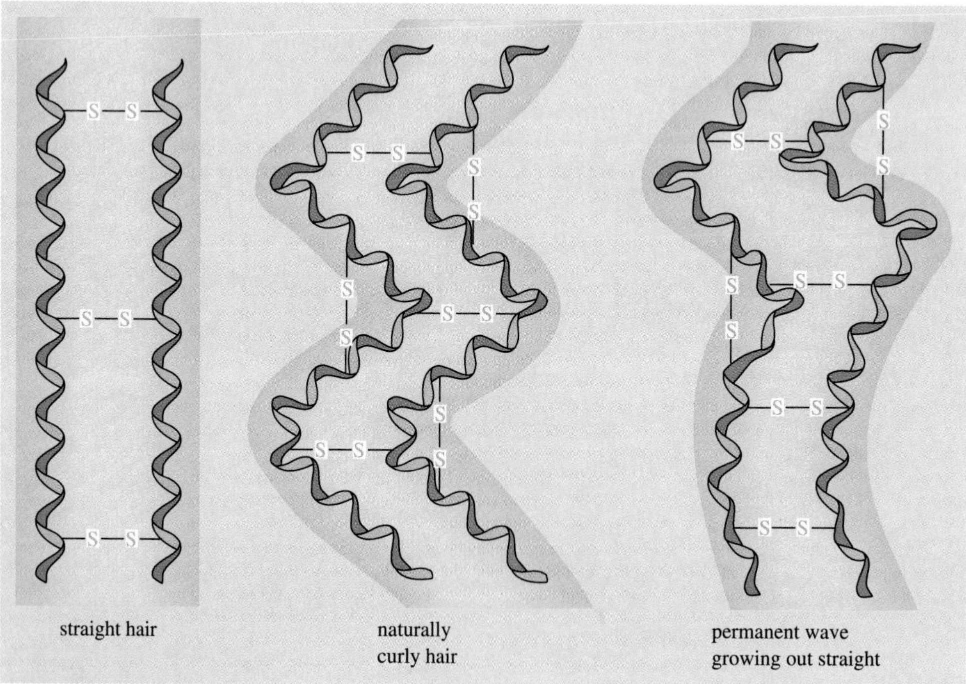

Disulfide Bonds in Hair

If disulfide bonds join protein chains at the same level, the hair strands are straight. Disulfide bonds connecting different levels and intramolecular disulfide bonds occur in curly hair. A permanent wave breaks and forms new disulfide bonds, but the new hair that grows will still be straight.

Heavy-Metal Salts

Heavy-metal ions such as Ag^+, Hg^{2+}, and Pb^{2+} denature proteins by irreversibly reacting with sulfhydryl bonds to form metal–sulfur bonds. The metal ions also combine with the carboxylate ions in the side chains of acidic amino acids.

23.10 Hydrolysis of Proteins

Peptide bonds are quite unreactive toward hydrolysis. They can be hydrolyzed by heating for hours in 6 M HCl at 100 °C. Complete hydrolysis produces the constituent amino acids of the peptide or protein. However, the digestion of proteins in the body must occur at a faster rate and at much lower acid concentration. The stomach contains only 0.1 M HCl.

Enzymes that catalyze the hydrolysis of proteins are called **proteases.** They are used in the laboratory to determine the structure of peptides and proteins. The proteases pepsin, chymotrypsin, and trypsin preferentially attack certain types of amide groups in a protein. Pepsin hydrolyzes the peptide bond at the nitrogen end of the aromatic amino acids phenylalanine, tyrosine, or tryptophan. Chymotrypsin hydrolyzes the peptide bond of the same three amino acids but at the carboxyl end of the amino acid. Trypsin hydrolyzes peptide bonds of the basic amino acids lysine and arginine at the carboxyl end of the molecule.

$$\text{Lys-Glu-Tyr-Leu} \xrightarrow{\text{pepsin}} \text{Lys-Glu} + \text{Tyr-Leu}$$

$$\text{Lys-Glu-Tyr-Leu} \xrightarrow{\text{chymotrypsin}} \text{Lys-Glu-Tyr} + \text{Leu}$$

$$\text{Lys-Glu-Tyr-Leu} \xrightarrow{\text{trypsin}} \text{Lys} + \text{Glu-Tyr-Leu}$$

Some enzymes hydrolyze peptide bonds from one end of the molecule. **Carboxypeptidases** sequentially hydrolyze peptides only from the end that has the free carboxyl group. **Aminopeptidases** sequentially hydrolyze peptides only from the end containing the free amino group. By determining the amino acids produced by a carboxypeptidase or an aminopeptidase at timed intervals, the sequence of amino acids can be determined. When the tetrapeptide Arg-Glu-Lys-Trp reacts with a carboxypeptidase, it first liberates tryptophan. The remaining tripeptide, Arg-Glu-Lys, then yields lysine. When an aminopeptidase is used, the sequential formation of amino acids is arginine followed by glutamic acid.

EXAMPLE 23.2

Predict the products of the chymotrypsin-catalyzed hydrolysis of the following enkephalin.

Tyr-Gly-Gly-Phe-Leu

Solution

Chymotrypsin catalyzes hydrolysis at the carboxyl end of aromatic amino acids. The enkephalin contains both phenylalanine and tyrosine. Tyrosine is the N-terminal amino acid, and hydrolysis at its carboxyl end results in free tyrosine.

$$\text{Tyr-Gly-Gly-Phe-Leu} \longrightarrow \text{Tyr} + \text{Gly-Gly-Phe-Leu}$$

Phenylalanine is bonded to the C-terminal amino acid leucine. Hydrolysis at the carboxyl group of phenylalanine frees leucine. A tripeptide results.

$$\text{Gly-Gly-Phe-Leu} \longrightarrow \text{Gly-Gly-Phe} + \text{Leu}$$

PROBLEM 23.2

Predict the products of the pepsin-catalyzed hydrolysis of the enkephalin of Example 23.2.

Complete Proteins in the Diet

The composition of dietary protein must provide the proper mix of essential amino acids and other amino acids to supply those processes that form body protein. If one or more amino acids are not available at the time of synthesis of a protein, then the protein is not made. Protein synthesis is an "all or nothing" event.

Dietary proteins are rated in terms of biological value on a percentage scale. Complete proteins have a high biological value—they supply all of the amino acids in the amounts required for normal growth. A list of protein sources and their biological value is given in the table. Note that hen's eggs, cow's milk, and fish provide proteins of high biological value. Plant proteins vary more in biological value than animal proteins. However, not all plant proteins are deficient in the same amino acids. Gliadin, a wheat protein, is low in lysine; zein, a corn protein, is low in both lysine and tryptophan. Societies that eat large amounts of corn or wheat products must have other sources of lysine.

Biological Value of Dietary Proteins

Food	Biological Value (%)
whole hen's egg	94
whole cow's milk	84
fish	83
beef	73
soybeans	73
white potato	67
whole grain wheat	65
whole grain corn	59
dry beans	58

Vegetarians must carefully choose their food so that all the essential amino acids are available on a daily basis. For example, wheat is low in lysine, but beans are high in lysine as well as tryptophan. On the other hand, wheat is high in cysteine and methionine, whereas beans are low in these two amino acids. By eating both beans and wheat, the vegetarian increases the percentage of usable proteins. Some societies and ethnic groups have developed diets that provide a good nutritional supply of proteins even without the benefit of nutritionists. The American Indian thrived on a diet that included both corn and beans—a mixture that we call succotash. Rice and black-eyed peas of the South as well as corn tortillas and beans in Mexico provide a reasonable balance of amino acids.

Mixtures of various beans are an important dietary source of proteins for vegetarians. (Tony Stone Images)

The diets in some areas of the world today fall below the minimum daily requirement of protein owing to economic conditions and, in some cases, social and religious customs. As income decreases, the more costly animal protein is replaced by cereal grains and other incomplete protein sources. If a variety of plant proteins are not available, a number of diseases in young children result. Kwashiorkor is a protein deficiency disease that develops in young children after weaning, when their diet is changed to starches. The disease is characterized by bloated bellies and patchy skin. After a certain point, death is inevitable. Some forms of mental retardation also result from incomplete nutrition.

Summary

Proteins play crucial roles in many biological processes, and the functions of proteins are used to classify them. Proteins provide structural support, transport material, act as chemical messengers, cause motion, serve as catalysts, and provide protection.

Proteins are classified into two broad categories, **globular** and **fibrous,** based on their solubility. Proteins that contain other compounds in addition to amino acids are **conjugated proteins.** Hydrolysis of proteins yields amino acids, but other materials are also formed from conjugated proteins. The non-amino acid unit is called a **prosthetic group.**

The 20 **amino acids** in proteins all have an α-amino group and, with the exception of the optically inactive glycine, are of the L configuration. The amino acids are classified as **neutral, acidic,** or **basic.** Amino acids that must be obtained from food are called **essential amino acids.**

Amino acids act as both acids and bases. The form of the amino acid in aqueous solution depends on the pH. The pH at which the amino acid exists as a **zwitterion** is the **isoionic point.** The isoionic points of neutral amino acids are in the 5.5–6.5 range. Basic amino acids have higher isoionic points, and acidic amino acids have lower isoionic points. **Electrophoresis** is used to separate amino acids and proteins based on their electric charge in a chosen buffer solution.

The structure of proteins is described in four parts. The **primary structure** is the sequence of the amino acids in the protein formed by **peptide bonds. Secondary structure** is the arrangement or conformation about the peptide backbone. Hydrogen bonding is the principal contributor to secondary structure. **Tertiary structure** refers to the overall folding and bending of the peptide backbone. **Quaternary structure** is the arrangement of subunits in those proteins that have several protein chains.

Denaturation is the loss of biological activity of a protein caused by changes in the secondary, tertiary, or quaternary structure. Heat, changes in pH, oxidizing or reducing agents, certain metal ions, and some hydrogen-bonding solvents denature proteins. The re-formation of a protein in its native state from a denatured protein is called **renaturation.**

Enzymes that catalyze the hydrolysis of proteins are called **proteases. Carboxypeptidases** sequentially hydrolyze peptides only from the end that has the free carboxyl group. **Aminopeptidases** sequentially hydrolyze peptides only from the end containing the amino group.

Exercises

Amino Acids

23.1 D-Glutamic acid is found in bacterial cell walls. Draw a projection formula of the amino acid.

23.2 Earthworms have some D-serine. Draw the projection formula of this amino acid.

23.3 The following compound is an unusual amino acid found in collagen. From what amino acid could this compound be derived by metabolic reactions?

$$\underset{}{NH_2CH_2}\underset{OH}{\underset{|}{C}}HCH_2CH_2\underset{NH_2}{\underset{|}{C}}HCO_2H$$

23.4 The following antibacterial agent is contained in garlic. From what amino acid might it be derived by metabolic reactions?

$$CH_2{=}CHCH_2\overset{\overset{O}{\|}}{S}CH_2\underset{NH_2}{\underset{|}{C}}HCO_2H$$

23.5 The following compound is an unusual amino acid that functions as a neurotransmitter. Classify this amino acid, and determine its IUPAC name.

$$NH_2CH_2CH_2CH_2CO_2H$$

23.6 The following compound is one of the amino acids formed in the biosynthesis of penicillin. Classify this amino acid, and determine its common name.

$$HO_2C\underset{NH_2}{\underset{|}{C}}HCH_2CH_2CH_2CO_2H$$

23.7 Give the name and write the structure of each of the following amino acids.

(a) Gly (b) Ala (c) Phe (d) Pro

23.8 Give the name and write the structure of each of the following amino acids.

(a) Tyr (b) Lys (c) Ser (d) Cys

Essential Amino Acids

23.9 Tyrosine can be synthesized by the body. Why, then, is tyrosine considered an essential amino acid?

23.10 Why is histidine not listed as an essential amino acid in Table 23.2?

23.11 How do the amino acids in fish, meat, and cereal grains compare in terms of biological value?

23.12 Why is the soybean useful as a substitute for beef or as an extender of beef products?

23.13 What precautions must vegetarians take to provide a balanced diet?

23.14 Ovo-lacto vegetarians eat eggs and drink milk. Do these individuals have as many dietary concerns as total vegetarians?

Acid–Base Properties

23.15 Draw the structures of alanine at pH = 1 and pH = 12.

23.16 Draw the structures of glutamic acid at pH = 1 and pH = 10.

23.17 Write the structure of the zwitterion of serine.

23.18 Write the structure of the zwitterion of valine.

Isoionic Points

23.19 What structure of alanine predominates at the isoionic point? What structure becomes more important at pH values higher than the isoionic point?

23.20 Describe the values for the isoionic points of neutral, acidic, and basic amino acids.

23.21 Estimate the isoionic points of the following tripeptides.

(a) Ala-Val-Gly (b) Ser-Val-Asp (c) Lys-Ala-Val

23.22 Estimate the isoionic points of the following tripeptides.

(a) Glu-Val-Ala (b) Arg-Val-Gly (c) His-Ala-Val

23.23 Examine the structures of oxytocin and vasopressin (Section 23.6). Which should have the higher isoionic point?

23.24 Examine the structure of the enkephalin in Section 23.6 and estimate its isoionic point.

23.25 The isoionic point of chymotrypsin is 9.6. What does this value indicate about the composition of chymotrypsin?

23.26 The isoionic point of pepsin is 1.1. What does this value indicate about the composition of pepsin?

Peptides

23.27 Write the complete formula and the condensed formula of alanylserine.

23.28 How does glycylserine differ from serylglycine?

23.29 Which amino acids can form peptides with carboxyl groups or carboxylate groups at internal positions in the peptide chain?

23.30 Which amino acids can form peptides with amino groups or ammonium groups at internal positions in the peptide chain?

23.31 Identify the amino acids contained in the following tripeptide. Name the compound.

H O $CH(CH_3)_2$
NH_2 N C N O
O $HSCH_2$ H OH

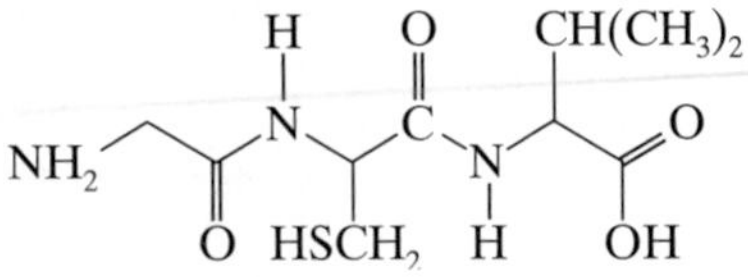

23.32 Identify the amino acids contained in the following tripeptide. Name the compound.

$HOCH_2$ H O CH_3
NH_2 N C N O
O H OH

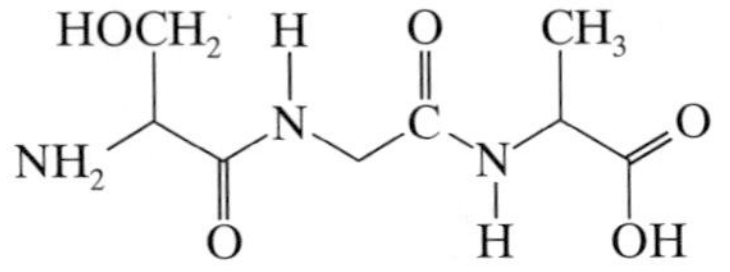

Proteins

23.33 Which of the following amino acids tend to exist in the interior of a globular protein in an aqueous solution?

(a) glycine (b) phenylalanine
(c) glutamic acid (d) arginine

23.34 Which of the following amino acids tend to exist in the interior of a globular protein in an aqueous solution?

(a) proline (b) cysteine
(c) glutamine (d) aspartic acid

23.35 If a globular protein is embedded in a lipid bilayer, which of the amino acids listed in Exercise 23.33 will be in contact with the interior of the bilayer?

23.36 If a globular protein is embedded in a lipid bilayer, which of the amino acids listed in Exercise 23.34 will be in contact with the interior of the bilayer?

23.37 Noting that proline is a secondary amine, explain how proline can disrupt the α-helix of a protein.

23.38 Examine the structures of valine and glutamic acid and suggest a reason why human hemoglobin is affected by the substitution of valine for glutamic acid at position 6 in the β-chain.

Denaturation of Proteins

23.39 What is meant by the term denaturation?

23.40 What would happen to the structure of insulin if it was treated with a reducing agent?

23.41 Why does cooking an egg denature the egg proteins?

23.42 Explain how egg whites can be used to treat lead poisoning.

23.43 Explain how pH changes could change the tertiary structure of a protein that has a large number of salt bridges as part of its tertiary structure.

23.44 What would be the effect of increasing the pH in a solution of a protein containing a large amount of lysine?

Enzymatic Hydrolysis

23.45 What is the result of treating Lys-Gly-Tyr-Leu with pepsin?

23.46 What is the result of treating Lys-Gly-Tyr-Leu with chymotrypsin?

23.47 Will each of the following tripeptides be cleaved by trypsin? If so, name the products.

(a) Arg-Gly-Tyr (b) Glu-Asp-Gly
(c) Phe-Trp-Ser (d) Ser-Phe-Asp

23.48 Will each of the following tripeptides be cleaved by trypsin? If so, name the products.

(a) Asp-Lys-Ser (b) Lys-Tyr-Cys
(c) Asp-Gly-Lys (d) Arg-Glu-Ser

23.49 Indicate which of the tripeptides in Exercise 23.47 will be cleaved by pepsin and name the products.

23.50 Indicate which of the tripeptides in Exercise 23.47 will be cleaved by chymotrypsin and name the products.

23.51 List the products formed from the following peptide when it is treated with pepsin.

Trp-Cys-Ala-Lys-Leu-Gly-Phe-Pro-Cys

23.52 List the products formed from the following peptide when it is treated with trypsin.

Trp-Cys-Ala-Lys-Leu-Gly-Phe-Pro-Cys

23.53 What amino acid is formed first when Gly-Ala-Phe is treated with an aminopeptidase?

23.54 What amino acid is formed first when Gly-Ala-Phe is treated with a carboxypeptidase?

The enzymes of yeast are responsible for the chemical action that makes bread rise. (The Stock Market)

24 Enzymes

24.1 Biological Catalysts
24.2 Classification of Enzymes
24.3 Models of Enzyme Action
24.4 Active Sites
24.5 Cofactors
24.6 Kinetics of Enzyme Reactions
24.7 Enzyme Inhibition
24.8 Regulatory Enzymes: Feedback Inhibition
24.9 Zymogens

Essays
- *Hot Spas for Bacteria*
- *Mercury and Lead Poisoning*

24.1 Biological Catalysts

The chemical reactions in biological systems are similar to those discussed in the organic chemistry chapters. Dehydration of alcohols and the related hydration of double bonds, oxidation of alcohols and reduction of carbonyl compounds, and nucleophilic acyl substitution reactions of acid derivatives such as esters and amides are all examples of reactions that also occur in biological systems. The differences lie in the complexity of the structures of biomolecules and the extraordinary speed of these reactions in a cell.

Unlike most chemistry in the laboratory, life processes occur only in aqueous solutions, usually at a pH near 7, and at a relatively low temperature of 37 °C. Moreover, the reactions occur at extraordinary rates and with incredible efficiency, producing desired products in high yield, meaning without byproducts. Yet the "reagents" required for biological reactions are not the strong acids and bases or strong oxidizing agents usually associated with reactions in a laboratory.

Another notable feature of biological reactions is the control exercised over which reactions occur, where they occur, and in what sequence. Almost all biological reactions are controlled by biological catalysts known as enzymes.

The ΔG° for an enzyme-catalyzed reaction is the same as for an uncatalyzed reaction carried out in a laboratory. In other words, enzymes can't change an energetically unfavorable reaction into a favorable one or change the equilibrium constant for a reaction. They do, however, decrease the time required for the reaction to occur. An enzyme provides a pathway with a lower activation energy than that of the uncatalyzed reaction, which allows the substrate molecules to react at a faster rate.

Enzymes are extraordinary catalysts because they increase the rate of a reaction by a factor of 10^6 or more. Reactions that otherwise would require years at neutral pH and at 37 °C occur in seconds in the presence of an enzyme. For example, carbonic anhydrase catalyzes the hydration of carbon dioxide.

$$H_2O(l) + CO_2(aq) \xrightarrow{\text{carbonic anhydrase}} H_2CO_3(aq)$$

The catalyzed reaction occurs 10^7 times as fast as the uncatalyzed reaction. Without the catalyst, the transfer of CO_2 from the tissues into the blood would occur too slowly to maintain life processes.

Enzymes are Specific

The reactant in a biological reaction catalyzed by an enzyme is called a **substrate.** Each enzyme has a characteristic **specificity,** meaning that it recognizes and reacts with certain substrates or even just one substrate. Because enzymes are specific, every organism has thousands of enzymes in a single cell to catalyze all the reactions required to maintain life.

Some enzymes are **linkage specific,** which means they react with a group of substrates with certain functional groups. For example, a *lipase* hydrolyzes the ester linkage of triacylglycerols regardless of which carboxylic acid residue is present. The hydrolysis of the amide linkage in proteins requires different enzymes called *proteases.*

Some enzymes are more specific in their reactivity. Enzymes with **group specificity** act on the functional group of structurally similar molecules. For example, each of the proteases pepsin, trypsin, and chymotrypsin is limited in the type of peptide linkages it hydrolyzes.

Enzymes that act on one substrate and no others have **absolute specificity.** Succinate dehydrogenase is an enzyme of absolute specificity in the citric acid cycle. It catalyzes the oxidation of succinic acid to fumaric acid by the oxidizing agent flavine adenine dinucleotide (FAD).

$$\text{FAD} + HO_2CCH_2CH_2CO_2H \xrightarrow{\text{succinate dehydrogenase}} (H)(HO_2C)C{=}C(CO_2H)(H) + \text{FADH}_2$$

succinic acid — fumaric acid

Enzymes usually are **stereospecific,** meaning they act on only one enantiomer. Amylase catalyzes the hydrolysis of the α-glycosidic linkage of starch but does not affect the β-glycosidic linkage of cellulose. Aspartase catalyzes the conversion of fumaric acid into L-aspartic acid.

$$(H)(HO_2C)C{=}C(CO_2H)(H) + NH_4^+ \xrightarrow{\text{aspartase}} NH_2{-}CH(CO_2H)(CH_2CO_2H) + H^+$$

fumaric acid — L-aspartic acid

24.2 Classification of Enzymes

The first discovered enzymes were given common names such as pepsin, trypsin, and chymotrypsin, which reflected their source or some specific feature of the enzyme. The name *pepsin* for the enzyme that catalyzes the hydrolysis of proteins in the digestive juices of the stomach is derived from the Greek *pepsis,* meaning "digestion." A system for naming enzymes was achieved by adding the suffix *-ase* to the name of the substrate on which the enzyme acts. Thus, a *lipase* is an enzyme that catalyzes the hydrolysis of the ester bonds of lipids.

In 1961 the International Union of Biochemistry decided to name enzymes systematically by indicating their function with a stem to which the suffix *-ase* is attached. Enzymes are classified into the six groups listed in Table 24.1 and are further categorized into subclasses. For example, the hydrolases that catalyze the hydrolysis of peptides are known as *peptidases,* whereas those that catalyze the hydrolysis of esters are *esterases.*

Oxidoreductases

Oxidoreductases are enzymes that catalyze oxidation–reduction reactions. They are often designated by names such as *oxidase, peroxidase, oxygenase, dehydrogenase,* or *reductase.* The oxidation of lactic acid to pyruvic acid catalyzed by lactate dehydrogenase occurs by the removal of two hydrogen atoms.

$$CH_3-\underset{H}{\overset{OH}{C}}-CO_2H \xrightarrow{\text{lactate dehydrogenase}} CH_3-\overset{O}{\overset{\|}{C}}-CO_2H + 2\,H^+ + 2\,e^-$$

lactic acid — pyruvic acid

An oxidation reaction must be accompanied by a simultaneous reduction reaction. Thus, an oxidoreductase that catalyzes an oxidation is combined with an oxidizing agent such as NAD^+ or FAD that becomes reduced. These processes are discussed in Chapter 25.

Transferases

Transferases are enzymes that catalyze the transfer of groups of atoms from one compound to another. The enzymes that catalyze the transfer of amino groups from amino acids to keto acids are called *aminotransferases* or *transaminases.*

TABLE 24.1 Classification of Enzymes

Enzyme Class	Type of Reaction Catalyzed
oxidoreductase	oxidation–reduction
transferase	transfer of a group from one compound to another
hydrolase	hydrolysis
lyase	nonhydrolytic addition or removal of groups
isomerase	conversion of a substance into one of its isomers
ligase	synthesis of one large molecule from two smaller molecules

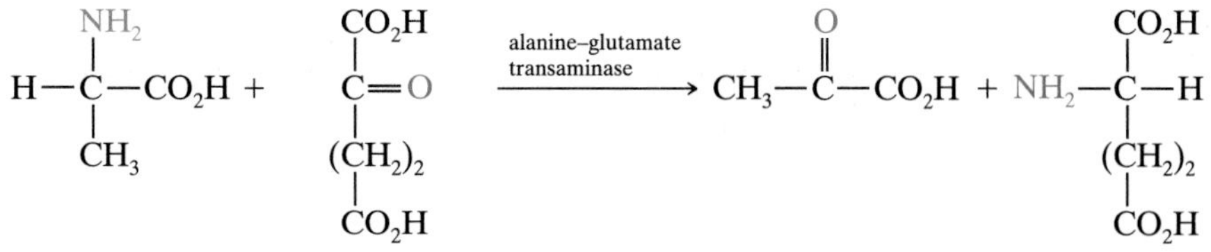

$$\underset{\text{alanine}}{H-\underset{CH_3}{\overset{NH_2}{C}}-CO_2H} + \underset{\alpha\text{-ketoglutaric acid}}{HO_2C-(CH_2)_2-C(=O)-CO_2H} \xrightarrow{\text{alanine–glutamate transaminase}} \underset{\text{pyruvic acid}}{CH_3-\overset{O}{\overset{\|}{C}}-CO_2H} + \underset{\text{glutamic acid}}{NH_2-\underset{(CH_2)_2CO_2H}{\overset{CO_2H}{C}}-H}$$

Hydrolases

Hydrolases catalyze the hydrolysis of esters, glycosidic bonds in carbohydrates, and peptide bonds in proteins. They are important in digestive processes. The digestive enzymes pepsin, trypsin, and chymotrypsin (Chapter 23) are all hydrolases.

$$\text{-Lys-Glu-Tyr-Leu-} \xrightarrow[H_2O]{\text{pepsin}} \text{-Lys-Glu} + \text{Tyr-Leu-}$$

Isomerases

Isomerases catalyze isomerization reactions, which rearrange the functional groups within a molecule. One example is the conversion of glucose 6-phosphate to fructose 6-phosphate in glycolysis reactions (Chapter 26).

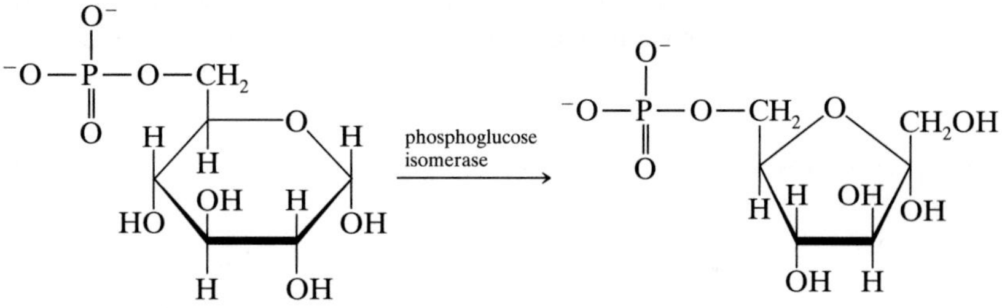

$$\text{glucose 6-phosphate} \xrightarrow{\text{phosphoglucose isomerase}} \text{fructose 6-phosphate}$$

glucose 6-phosphate   fructose 6-phosphate

Lyases

Lyases (from the Greek *lain,* "to break") catalyze the addition of small molecules such as water, ammonia, or carbon dioxide to double bonds such as C=C and C=N, or the reverse elimination reactions to form double bonds. The addition of water to fumaric acid, one of the steps in the citric acid cycle, is catalyzed by fumarate hydratase (fumarase).

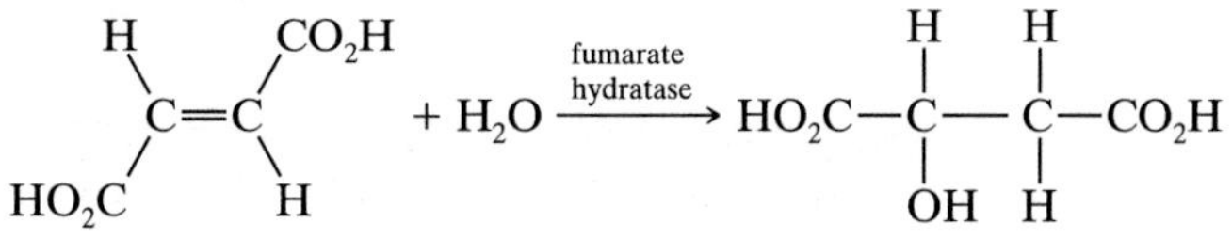

$$\underset{\text{fumaric acid}}{HO_2C-CH=CH-CO_2H} + H_2O \xrightarrow{\text{fumarate hydratase}} \underset{\text{malic acid}}{HO_2C-\underset{OH}{\overset{H}{C}}-\underset{H}{\overset{H}{C}}-CO_2H}$$

Ligases

Ligases (from the Latin *ligare,* "to tie together"), also called **synthetases,** catalyze the synthesis of large molecules from two smaller substrate molecules. These nonspontaneous reactions require ATP as a source of energy. The conversion of glutamic acid into glutamine catalyzed by glutamate synthetase is one example.

$$\text{ATP} + NH_3 + HO{-}\overset{\overset{\displaystyle O}{\|}}{C}{-}(CH_2)_2{-}\overset{\overset{\displaystyle NH_2}{|}}{CH}{-}\overset{\overset{\displaystyle O}{\|}}{C}{-}OH \xrightarrow{\text{glutamate synthetase}}$$

glutamic acid

$$H_2N{-}\overset{\overset{\displaystyle O}{\|}}{C}{-}(CH_2)_2{-}\overset{\overset{\displaystyle NH_2}{|}}{CH}{-}\overset{\overset{\displaystyle O}{\|}}{C}{-}OH + H_2O + \text{ADP} + P_i$$

glutamine

EXAMPLE 24.1

What type of enzyme is required for the following reaction? Suggest a common name for the enzyme that catalyzes the reaction.

$$\text{sucrose} + H_2O \longrightarrow \text{glucose} + \text{fructose}$$

Solution

The reaction, which involves hydrolysis of an acetal of glucose and a ketal of fructose (Section 21.9), requires a hydrolase. The common name sucrase indicates that sucrose is the substrate, but it does not indicate which reaction of sucrose is catalyzed.

PROBLEM 24.1

Phosphate esters play important biological roles. These esters are formed by passing a phosphate group from a higher energy phosphate ester to another alcohol to yield a lower energy phosphate ester. What type of enzyme catalyzes this reaction?

24.3 Models of Enzyme Action

A model to explain how enzymes function must account for their specificity and provide a mechanism for how a reaction is accelerated. Both features of enzyme-catalyzed reactions result from the formation of an enzyme–substrate complex that subsequently reacts. The steps in the mechanism are depicted as follows:

$$\underset{\text{enzyme}}{E} + \underset{\text{substrate}}{S} \rightleftharpoons \underset{\text{enzyme–substrate complex}}{ES}$$

$$ES \longrightarrow \underset{\text{enzyme}}{E} + \underset{\text{product}}{P}$$

Thus, it is not the substrate that reacts but the enzyme–substrate complex. Although not shown in the simplified mechanism, the complex reacts with additional molecules. If the enzyme is a hydrolase, water is required. In the case of a dehydrogenase, an oxidizing agent is required.

The binding in the enzyme–substrate complex may be the result of an ionic bond, a hydrogen bond, hydrophobic interactions, or a combination of these bonds. For example, an ionic bond may form between an ammonium ion of the lysine residue in the enzyme and a negative site in the substrate. The binding energy for enzyme–substrate interactions is usually in the range of 3 to 12 kcal/mole.

Lock-and-Key Theory

The German chemist Emil Fischer suggested in 1890 that enzymes combine with substrates that have complementary molecular geometry. The **lock-and-key theory** pictures an enzyme as a large, irregularly shaped molecule with a crevice in the

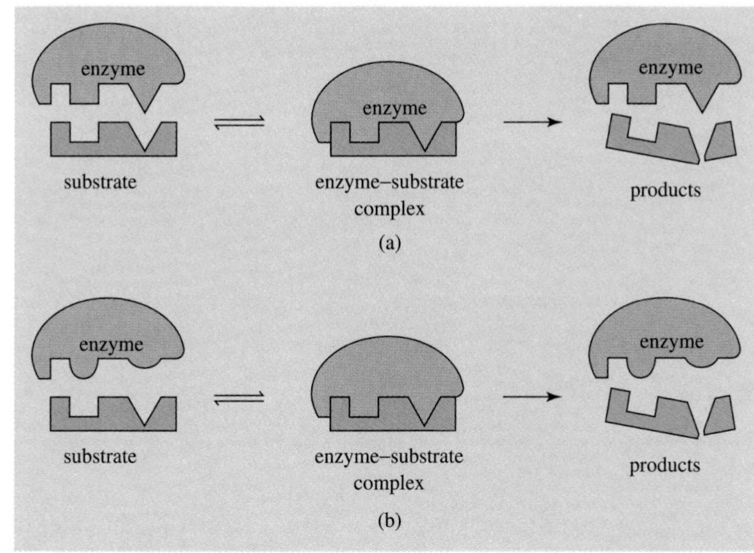

FIGURE 24.1 Models for Enzyme Action
In the lock-and-key model (a) the enzyme and the substrate have fixed complementary shapes. In the induced-fit model (b) the enzyme changes conformation to fit the substrate.

folded protein chain suitable for binding a substrate. The binding occurs only if the enzyme and substrate fit like a lock and key (Figure 24.1). Small changes in the substrate or in the enzyme prevent chemical reaction because the two molecular surfaces no longer have the complementary shapes required to fit together.

Induced-Fit Hypothesis

As the structures of enzymes were established, it became clear that not all enzymes have the specific shapes required by the lock-and-key theory. To account for the action of some enzymes, Daniel F. Koshland Jr., an American biochemist, suggested an alternative model, the induced-fit hypothesis.

The **induced-fit hypothesis** pictures an enzyme as a molecule that can change its conformation as the substrate approaches and starts to bind. The substrate induces a change in the shape of the enzyme, so that the enzyme–substrate complex achieves the proper complementary fit (Figure 24.1). Both carboxypeptidase A, a protein-digesting enzyme, and hexokinase, an enzyme that catalyzes the conversion of glucose into glucose 6-phosphate, operate by induced-fit pathways.

Chirality of Substrates

Enzymes have many chiral centers and have a "handedness" that allows them to fit and bind with a particular molecule but not with its enantiomer (Figure 24.2). For example, the formation of L-tryptophan from L-serine and indole is catalyzed by tryptophan synthetase. The enantiomer, D-serine, does not bond to the enzyme and is not converted into tryptophan.

$$\text{indole (N–H)} + HOCH_2{-}\underset{\substack{|\\H}}{\overset{\substack{NH_2\\|}}{C}}{-}CO_2H \xrightarrow{\text{tryptophan synthetase}} \text{indol-3-yl (N–H)}{-}CH_2{-}\underset{\substack{|\\H}}{\overset{\substack{NH_2\\|}}{C}}{-}CO_2H$$

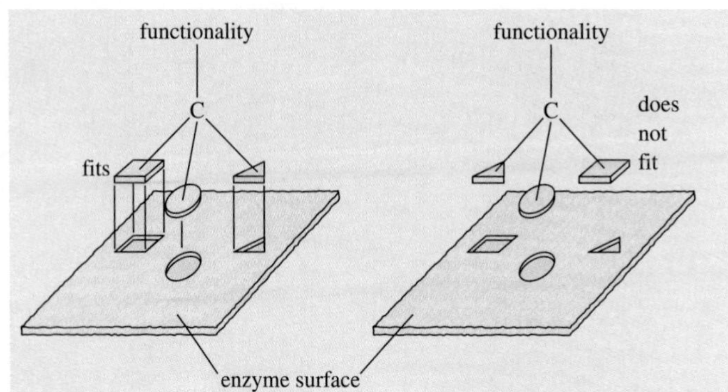

FIGURE 24.2 Specificity of an Enzyme for an Enantiomer
The enantiomer on the left fits into the template created by the enzyme, and the functionality is available for reaction. The mirror-image enantiomer on the right does not fit the enzyme, and its reactions are not catalyzed.

24.4 Active Sites

Two of the reasons why enzymes accelerate the rates of reactions are the high local concentration of reactants and their specific alignment within the enzyme structure. When the reactants are brought together at the enzyme, the concentration is much higher than in a dilute solution. The reactants are held in a precisely defined alignment required for reaction. In solution, reactions occur only if the reactant molecules are properly aligned when they collide.

The third reason for the high rates of enzyme-catalyzed reactions is a structural feature called the **active site.** Functional groups within the enzyme form the active site that is responsible for the catalysis. An active site is only a small portion of the enzyme molecule. It consists of a few amino acids out of a chain of hundreds of amino acids that are in proximity as a result of the tertiary structure of the enzyme. For example, the active site in lysozyme, an enzyme that cleaves polysaccharides in the cell walls of bacteria, consists of amino acid residues at positions 35, 52, 62, 63, and 101 in the sequence of 129 amino acids. The tertiary structure of the enzyme places these amino acids close in space to create the active site. These amino acids participate in the reaction and provide a catalyzed reaction pathway with a low activation energy. The remainder of the protein shapes the site required to select and bind the substrate.

24.5 Cofactors

Some enzymes are composed only of protein. Other enzymes require one or more nonprotein components, called **cofactors,** to be biologically active. A cofactor may be a metal ion, or it may be an organic molecule called a **coenzyme.** The enzyme–cofactor complex is called a **holoenzyme;** the inactive enzyme alone is called an **apoenzyme.**

apoenzyme + metal ion ⟶ holoenzyme
apoenzyme + coenzyme ⟶ holoenzyme

Metal Ion Cofactors

Usually only one metal ion per molecule of enzyme is required to form the holoenzyme, which is sometimes called a **metalloenzyme.** Aminopeptidases contain manganese or magnesium, whereas carboxypeptidases contain zinc. The weight percent

FIGURE 24.3 Models of Combination of Apoenzyme and Coenzyme
(a) A cosubstrate forms part of the active site. (b) A prosthetic group may be near the active site and may provide sites for binding and reactions such as proton transfer.

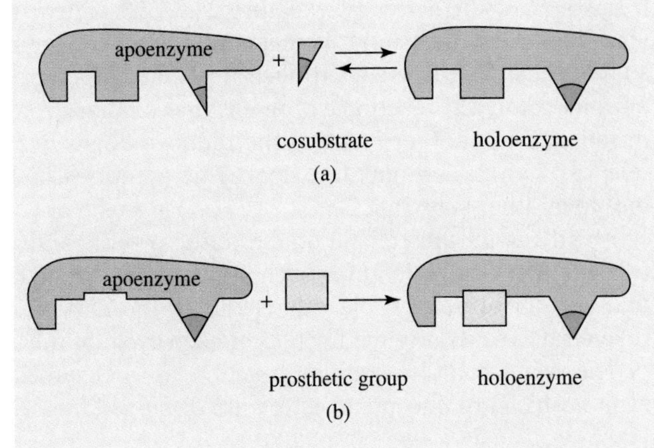

of the metal in the holoenzyme is in the range of tenths of a percent because the molecular weights of apoenzymes are very large. For this reason only small amounts of certain metal ions are required in our diet.

Coenzymes

Coenzymes have lower molecular weights than apoenzymes. The coenzyme joins with the apoenzyme to form the features required for enzymatic activity. The two types of coenzymes are called *cosubstrates* and *prosthetic groups.* A cosubstrate

FIGURE 24.4 Structures of Some Coenzymes

Coenzyme	Related vitamin
thiamine pyrophosphate	B_1
pyridoxal phosphate	B_6
riboflavin phosphate	B_2

forms part of the active site (Figure 24.3). It is altered in the reaction and dissociates from the active site. It is regenerated in a different enzyme-catalyzed reaction so that it can be recycled. Cosubstrates shuttle metabolic groups between different enzyme-catalyzed reactions. If an enzyme oxidizes a substrate, the coenzyme is reduced, and in another reaction the reduced coenzyme is oxidized. If the substrate releases a phosphate unit, the coenzyme accepts it and later transfers a phosphate unit to another substrate.

Prosthetic groups remain bound to the protein and are located at or near the active site. They have features such as nucleophilic sites or sites for proton transfer that are not present in the side chains of the enzyme. The prosthetic site must be regenerated to its original form after each cycle of the catalyzed reaction.

The water-soluble vitamins function as coenzymes (Figure 24.4). These coenzymes, which are phosphate esters, are discussed further in Chapter 25.

24.6 Kinetics of Enzyme Reactions

In most chemical reactions the rate of formation of product increases with increasing concentration of the reactant. In enzyme-catalyzed reactions, the rate increases with increasing substrate concentration, but only until a limiting rate is reached (Figure 24.5). This maximum rate or velocity (V_{max}) is reached because essentially all the enzyme has been converted to the reactive enzyme–substrate complex. The enzyme is then "saturated," and the reaction occurs only as fast as the enzyme–substrate complex reacts.

$$E + S \rightleftharpoons E - S$$

Enzymes are present in low concentrations, and as expected from Le Châtelier's principle, the large substrate concentration shifts the equilibrium reaction toward the enzyme–substrate complex.

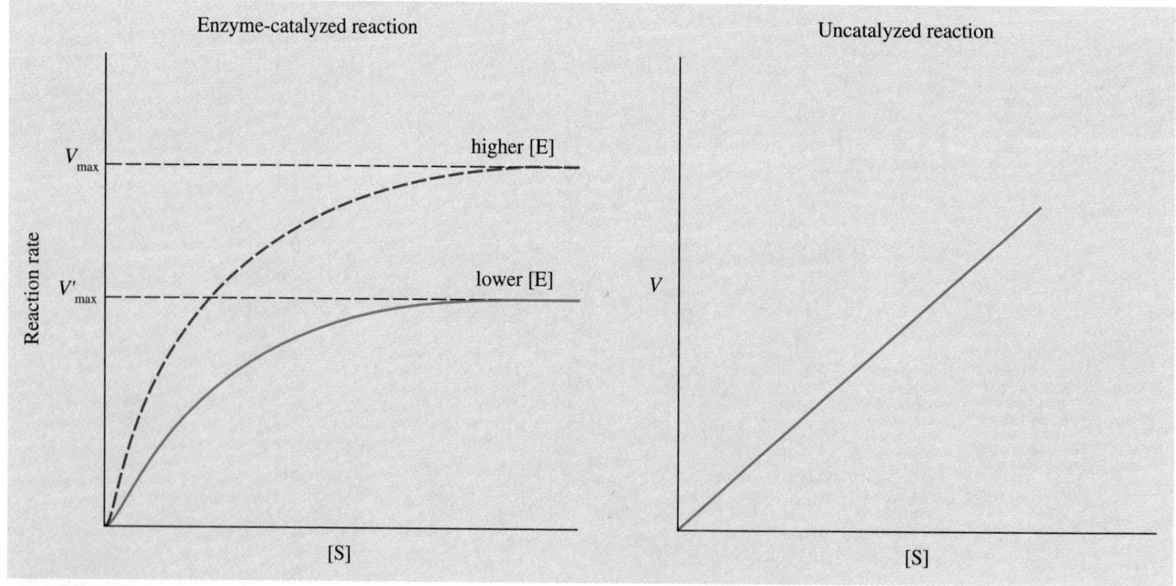

FIGURE 24.5
Comparison of Rates of Enzyme-Catalyzed and Uncatalyzed Reactions

The role of an enzyme and the limit placed on the rate of reaction by saturation may be better understood using an analogy in human terms. Assume that you can get to a rock concert from a parking lot only by taking a shuttle bus. The number of people arriving at the concert is not affected by how many people are waiting at the parking lot. The number depends on the frequency of the bus trips and on the number of buses available. The buses can hold only so many people. Once "saturated" they deliver their passengers. The concert goers who can't get on the bus have to wait until the next trip.

The V_{max} for an enzyme-catalyzed reaction is directly proportional to the concentration of the enzyme. Thus, increasing the amount of enzyme increases the concentration of the enzyme–substrate complex. (In our bus analogy this result corresponds to putting more buses in service).

Each enzyme has an intrinsic activity, which is the ratio of V_{max} to [E] and is known as the turnover number.

$$\textbf{turnover number} = \frac{V_{max}}{[E]}$$

The **turnover number** is the number of substrate molecules converted into product per minute for each molecule of enzyme under saturation conditions. The turnover numbers of enzymes are very high (Table 24.2).

Stability of Enzyme–Substrate Complexes

The equilibrium constant for formation of the enzyme–substrate complex can be experimentally determined. However, the quantity K_m is usually tabulated for the reverse reaction, that is, the dissociation of the complex.

$$E-S \rightleftharpoons E + S$$

$$K_m = \frac{[E][S]}{[ES]}$$

A large K_m, or large dissociation constant for the enzyme–substrate complex, indicates weak binding. A small K_m means that the enzyme and substrate are strongly bound. A list of K_m values is given in Table 24.3.

The smaller the K_m, the higher is the fraction of the enzyme–substrate complex. This means that the conversion of the enzyme–substrate complex into product can

TABLE 24.2 Turnover Numbers of Some Enzymes

Enzyme	Turnover Number (min^{-1})
carbonic anhydrase	36,000,000
ketosteroid isomerase	17,100,000
fumarase	1,200,000
β-amylase	1,100,000
β-galactosidase	12,500
phosphoglucomutase	1,240
succinate dehydrogenase	1,150

TABLE 24.3 Stability of Enzyme–Substrate Complexes

Enzyme	Substrate	K_m (M)
chymotrypsin	acetyl-L-tryptophanamide	5×10^{-3}
lysozyme	hexa-*N*-acetylglucosamine	6×10^{-6}
β-galactosidase	lactose	4×10^{-3}
threonine deaminase	threonine	5×10^{-3}
carbonic anhydrase	CO_2	8×10^{-3}
pyruvate carboxylase	pyruvate	4×10^{-4}

occur at low substrate concentrations. A large K_m means that the enzyme–substrate complex tends to dissociate into enzyme and substrate. Thus product can form only when the substrate concentration is large and drives the equilibrium toward the enzyme–substrate complex.

Effect of Temperature on Reaction Rate

The rates of ordinary chemical reactions increase as the temperature increases. The rates of enzyme-catalyzed reactions behave similarly, but a maximum rate, V_{max}, occurs at an optimum temperature (Figure 24.6). For humans this optimum temperature is near 37 °C, the normal body temperature. Above the optimum temperature, the rate of reaction decreases sharply because enzymes are denatured by heat and cannot form the enzyme–substrate complex.

The effect of temperature on the reaction rate of enzymes has several practical consequences. Surgical instruments are sterilized by heating them with steam in an autoclave at 110 °C to denature the protein of bacteria. Although low temperatures don't kill bacteria, the rate of their activity is slowed by decreasing the temperature. Thus, we are advised to store our cooked turkey in the refrigerator rather than to keep it on the kitchen counter on Thanksgiving Day.

FIGURE 24.6 Effect of Temperature on an Enzyme-Catalyzed Reaction

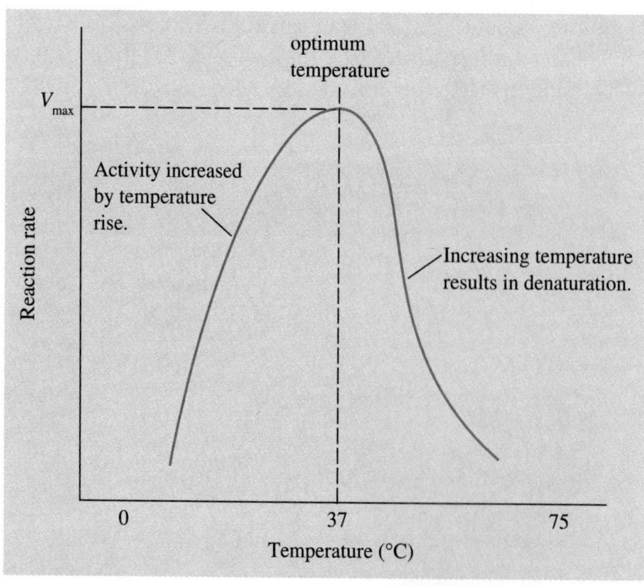

Hot Spas for Bacteria

Not all bacteria are deactivated by heating. In fact, some bacteria have evolved in very hot places. In the hot springs at Yellowstone National Park there are bacteria that function normally at temperatures near that of boiling water. There are also animals such as tube worms that thrive in deep, hot vents on the ocean floor where cracks in the Earth's crust generate superheated water. Under the high pressure of the ocean, the boiling point of water is about 110 °C. The animal species found near the deep, hot vents also have bacteria associated with them. Apparently all these life forms have proteins that are not easily denatured and whose activity depends on temperatures that would be a death sentence for most bacteria. Yet, these bacteria live in a hot spa that is to their liking. However, if they were to be transported to cooler climes, such as in your body, they would not fare well. Their metabolic rates are dependent on high temperature, and we might expect them to be as lethargic as their land-based bacterial cousins whose activity decreases when cooled in a refrigerator.

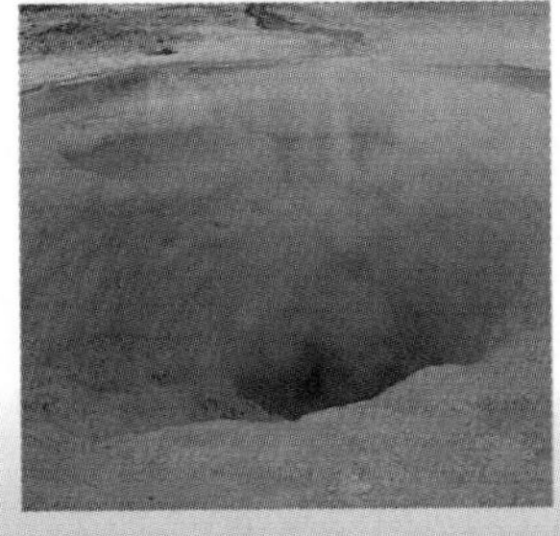

Temperature tolerant bacteria exist in the hot springs in Yellowstone National Park. (Robert J. Ouellette)

Effect of pH on Reaction Rate

Most enzymes have a maximum activity at a specific or optimum pH. Many operate at maximum efficiency near pH 7, the pH of the cellular cytosol. However, there are environments in the body where normal chemical activity requires pH values far from 7. In the stomach the pH is approximately 2, and enzymes responsible for hydrolysis of proteins must survive to be efficient catalysts at this pH. Pepsin is an example; it is most active near pH 2 (Figure 24.7). Trypsin, another protease, is most active near pH 8, the pH of the intestine.

There are two reasons why pH affects rate. First, the tertiary structure depends in part on ionic bonds, which are affected by pH. Second, the acidic and basic amino acid side chains that may be part of the active site can lose or gain protons depending on pH.

FIGURE 24.7 Effect of pH on Activity of Enzymes

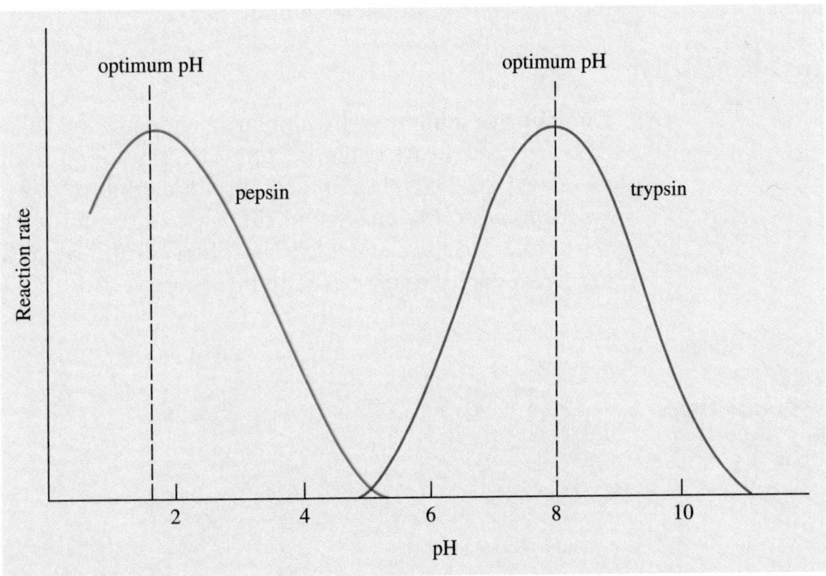

EXAMPLE 24.2

The catalytic activity of lysozyme is maximum at pH = 5.0. The active site involves the side chains of aspartic acid and glutamic acid, whose pK_a values are 3.96 and 4.32, respectively. In what form does each of these side chains exist at maximum catalytic activity?

Solution

At the pH equal to the pK_a of the side chain, there is an equal concentration of the carboxylic acid and the carboxylate ion. Since the pH of maximum activity is higher than 3.96 and 4.32, the solution is less acidic (more basic). As a consequence, a higher fraction of each side chain will exist as the carboxylate ion. Since the side chain of aspartic acid is more acidic than that of glutamic acid, there will be a larger amount of the conjugate base of the aspartic acid side chain than of the glutamic acid side chain.

PROBLEM 24.2

The optimum pH for chymotrypsin is 7.8. The active site contains serine and histidine. In what form is histidine at the optimum pH?

24.7 Enzyme Inhibition

The acceleration of reaction rates is only part of the chemistry of cells. Cells must also slow the rate of reactions and sometimes even shut off some reactions. Otherwise, the cell would use up all available substrate and would form larger amounts of product than could be conveniently used at a specific time.

Enzyme inhibitors deactivate and regulate the reactivity of enzymes. There are two type of inhibitors, competitive and noncompetitive. Taking a cue from these natural processes, chemists have developed drugs that inhibit bacterial enzymatic activity but that don't affect human cells.

In subsequent sections we will examine two other methods that control biological reactions. Some products of biological reactions inhibit the enzymes that are required for their formation in a process called **feedback inhibition** (Section 24.8). Another method of reaction control is based on the formation of inactive forms of enzymes called **zymogens** that are activated as needed (Section 24.9).

Competitive Inhibition

Competitive inhibitors are similar in shape to the substrate, so they can occupy the enzyme's active site (Figure 24.8). The binding of the inhibitor may be irreversible or reversible. In *irreversible* competitive inhibition the inhibitor forms a strong covalent bond to the enzyme at the active site, and further enzymatic activity is prevented because the active site is "plugged up." Many toxins and nerve gas poisons are irreversible competitive inhibitors.

FIGURE 24.8 Model for Competitive Inhibition

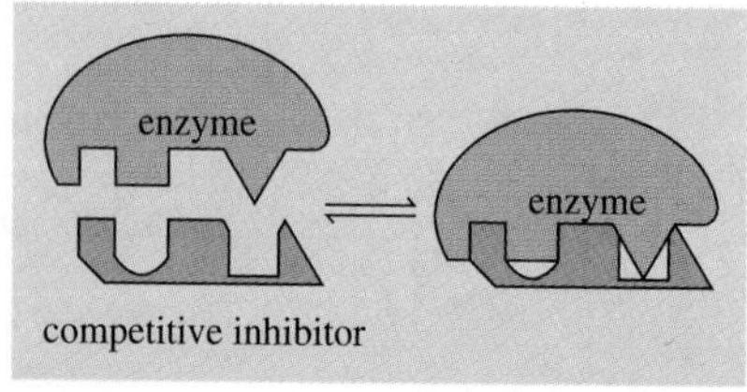

Reversible competitive inhibitors form a complex in equilibrium with the enzyme. Malonic acid is a reversible competitive inhibitor. This compound structurally resembles succinic acid, the substrate for succinate dehydrogenase. It binds with the enzyme to form an enzyme–substrate complex that does not lead to product. The succinic acid, which must compete with malonic acid for the active site, cannot be as rapidly converted to product in the presence of malonic acid.

$$FAD + \underset{\text{succinic acid}}{HO_2C-CH_2-CH_2-CO_2H} \xrightarrow{\text{succinate dehydrogenase}} (HO_2C)(H)C{=}C(H)(CO_2H) + FADH_2$$

$$FAD + \underset{\text{malonic acid}}{HO_2C-CH_2-CO_2H} \xrightarrow{\text{succinate dehydrogenase}} \text{no reaction}$$

The products of enzyme–catalyzed reactions also may be reversible competitive inhibitors. For example, 2,3-diphosphoglycerate, formed from 1,3-diphosphoglycerate, is itself a reversible competitive inhibitor for diphosphoglycerate mutase.

$$CH_2(-O-PO_3^{2-})-CH(OH)-CH_2-O-PO_3^{2-} \xrightarrow{\text{diphosphoglycerate mutase}} CH_2(OH)-CH(-O-PO_3^{2-})-CH_2-O-PO_3^{2-}$$

Thus, as the product forms, enzyme activity decreases. Additional details of such enzyme regulation will be given in the next section.

Noncompetitive Inhibition

Noncompetitive inhibition occurs when an inhibitor binds with some site other than the active site. The conformation of the enzyme is changed so that the catalytic properties of the active site are altered and the turnover number is decreased (Figure 24.9). Noncompetitive inhibitors may bind at some point far removed from the active site, but the resulting conformational change is drastic enough to alter the catalytic activity at the binding site. These changes are called *allosteric effects.* In many cases noncompetitive inhibition is reversible.

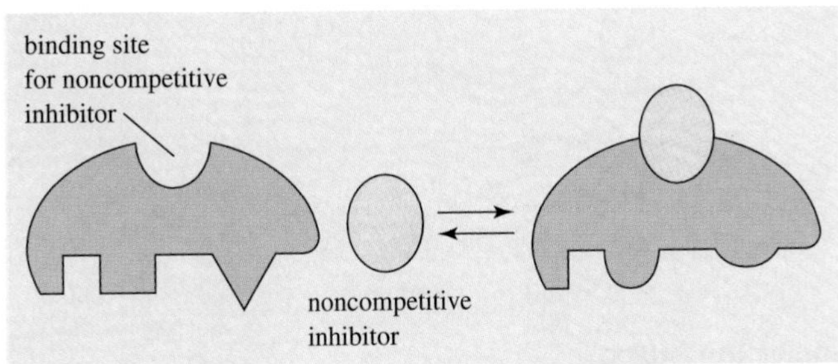

FIGURE 24.9 Model for Noncompetitive Inhibition

Mercury and Lead Poisoning

Salts of mercury and lead are irreversible noncompetitive inhibitors of many enzymes. Metal ions such as mercury(II) combine with sulfhydryl groups. The mercury–sulfur bonds can form between cysteine units within the same protein or between cysteines of two proteins. Recall that the primary structure of the protein, and as a result the tertiary structure, depends on the number and location of free sulfhydryl groups and disulfide linkages. When a reaction occurs with free sulfhydryl groups, the conformation of the protein changes. The mercury–sulfur bonds are very strong, and as a result the enzyme activity is irreversibly inhibited.

Mercury poisoning can occur from mercury that was dumped in streams or lakes decades ago. Although elemental mercury is unreactive, some microorganisms can convert the metal into organic mercury compounds such as the $CH_3—Hg^+$ ion. These compounds enter the food chain as they are ingested by fish. The dose of the mercury compound increases when we eat the fish.

Paint chips containing lead on old wood may poison a child. (Photo Edit)

Detection of Inhibition

Competitive and noncompetitive inhibitors affect V_{max} differently. In competitive inhibition the inhibitor and the substrate compete for the binding site on the enzyme. When the inhibitor occupies binding sites, some of the substrate molecules cannot react. However, for a given enzyme concentration, the substrate may effectively compete for binding sites by increasing its concentration. As the substrate concentration is increased, the enzyme sites are increasingly occupied by substrate. The V_{max} value eventually reaches the same value as in the absence of the competitive inhibitor (Figure 24.10).

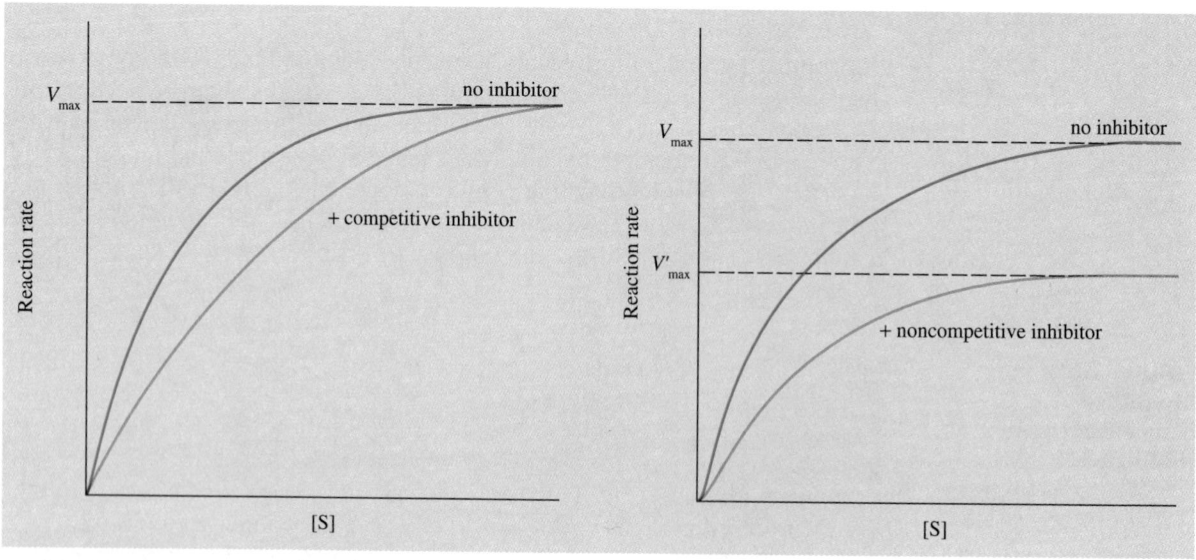

FIGURE 24.10
Determination of Type of Inhibition

The decrease in reaction rate caused by a noncompetitive inhibitor cannot be overcome by increasing the concentration of substrate. The enzyme molecules combined with the noncompetitive inhibitor are no longer active. As a result, fewer enzyme molecules are available to catalyze the reaction with the substrate and V_{max} decreases to V'_{max} (Figure 24.10).

EXAMPLE 24.3

Elastase, a protease, has valine and serine at its binding site. The following compound is an inhibitor of elastase. Which amino acid of the binding site is likely to be involved in forming a covalent bond with the aldehydic carbon atom?

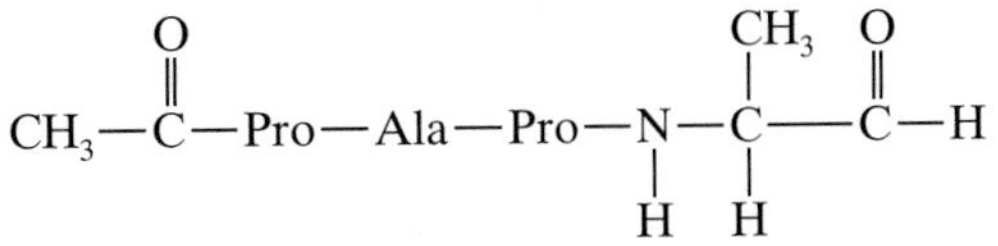

$$CH_3-\overset{\overset{O}{\|}}{C}-Pro-Ala-Pro-\underset{\underset{H}{|}}{N}-\underset{\underset{H}{|}}{\overset{\overset{CH_3}{|}}{C}}-\overset{\overset{O}{\|}}{C}-H$$

Solution

Valine has no functional groups in the side chain, whereas serine has a hydroxyl group. An alcohol and an aldehyde react to form a hemiacetal. The covalent bond is depicted as forming between the carbonyl carbon atom of the elastase inhibitor and the oxygen atom of the hydroxyl group of serine.

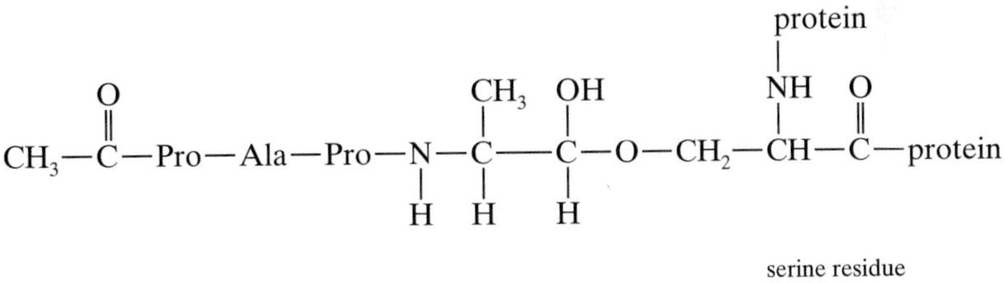

$$CH_3-\overset{\overset{O}{\|}}{C}-Pro-Ala-Pro-\underset{\underset{H}{|}}{N}-\underset{\underset{H}{|}}{\overset{\overset{CH_3}{|}}{C}}-\underset{\underset{H}{|}}{\overset{\overset{OH}{|}}{C}}-O-CH_2-\overset{\overset{\overset{protein}{|}}{NH}}{\overset{|}{CH}}-\overset{\overset{O}{\|}}{C}-protein$$

serine residue

PROBLEM 24.3

Oxalate ion ($^-O_2C{-}CO_2^-$) is a competitive inhibitor of pyruvate kinase. How might the oxalate ion bind to the active site of the enzyme?

24.8 Regulatory Enzymes: Feedback Inhibition

Cells can't waste vital resources. Thus the activity of enzymes within cells is initiated or shut off according to the demands of life processes. When a vital compound is absent or in short supply, an enzyme catalyzes its formation. Once the concentration of the necessary compound is sufficient, its production is terminated and the enzyme activity is switched off.

Complex control mechanisms are required for an organism to effectively use its enzymes. Two of the most important control mechanisms are the operation of regulatory enzymes in feedback inhibition and the formation of enzymes from zymogens. The chemistry of zymogens is the subject of the next section.

Feedback inhibition involves the control of a reaction sequence by a negative modulator and a regulatory enzyme. A **regulatory enzyme** catalyzes the first reaction of a sequence of reactions and is inhibited by the end product of the sequence. A **negative modulator** is a product at the end of a series of reactions that inhibits an earlier step in the same reaction series. The negative modulator "feeds back" to inhibit some preceding step. Consider the sequence of reactions of threonine via intermediates labeled A, B, and C and ultimately forming isoleucine (Figure 24.11). The steps are catalyzed by enzymes labeled E_1 through E_4.

FIGURE 24.11
Feedback Inhibition

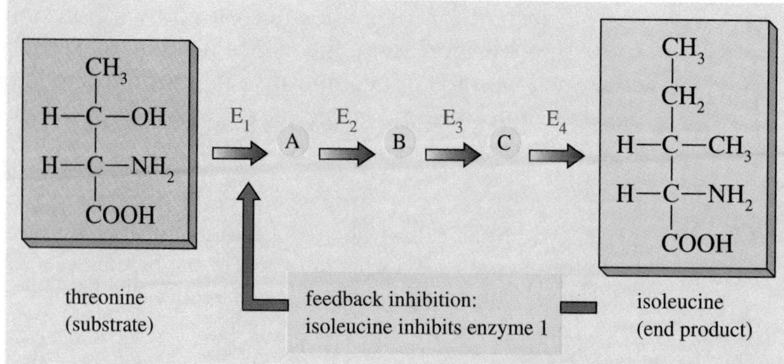

In this sequence of steps the final product cannot be formed unless the product C and, in turn, the product B are first formed. Formation of these requires the appropriate enzymes. If the final product is a competitive inhibitor of E_1, the rate of formation of A is decreased, and as a consequence the rates of formation of B and C are decreased. The reaction rate is controlled by the equilibrium constant for the enzyme–product complex.

$$E_1 + \text{product} \rightleftharpoons E_1\text{–product}$$

As long as the product is present in sufficient quantity, the concentration of E_1 is diminished by formation of the complex. As the product is used up by the organism, the equilibrium shifts, E_1 is released, and the entire sequence of reactions commences again. In this example the product is the negative modulator, and E_1 is the regulatory enzyme.

The first step in the conversion of threonine into isoleucine is catalyzed by threonine deaminase. Isoleucine occupies a site on threonine deaminase different from the one that binds to threonine. This second site is called the *allosteric site* because when isoleucine binds, the conformation of the enzyme changes. The binding characteristics are altered so that threonine now binds poorly. Regulation of enzyme activity by noncompetitive inhibition involving allosteric changes is called **allosteric regulation.**

24.9 Zymogens

Some enzymes are synthesized and stored in an inactive form called a **zymogen** or proenzyme. They are transformed into active enzymes by hydrolysis when the organism needs them as catalysts. The peptidases trypsin and chymotrypsin are both produced and stored in the pancreas as the zymogens trypsinogen and chymotrypsinogen, respectively. This is a necessary protection for the pancreas, which makes them. If the peptidases were produced in active form, the pancreas would be destroyed by its own protein-digesting enzymes. Premature release of protein-digesting enzymes is called *pancreatitis,* a condition that can result from injury to the pancreas.

There are two additional safety features for the pancreas. The zymogens produced in the pancreas travel from the endoplasmic reticulum to the Golgi apparatus, where they are surrounded by membranes made of protein and lipid. These zymogen granules are then secreted into a duct to the duodenum when stimulated by a hormone. The pancreas also contains an inhibitor that prevents conversion of the zymogen into an active enzyme.

Summary

Reactant molecules in enzyme-catalyzed reactions are **substrates. Enzymes** are proteins that are biological catalysts. Many have **absolute specificity.** Names of enzymes are based on the name of the substrate and the type of reaction catalyzed. Enzymes are classified as **oxidoreductases, transferases, hydrolases, lyases, isomerases,** and **ligases.**

The concentrations of products at equilibrium are not affected by the enzyme. Enzymes change reaction rates by providing a reaction pathway with a lower activation energy. The reaction pathway with an enzyme involves formation of an **enzyme–substrate complex** in which the substrate and enzyme bind because of **complementary shapes.** Two models account for the formation of the complex: the **lock-and-key theory** and the **induced-fit hypothesis.** The lock-and-key theory pictures enzymes and substrates as rigid molecules, whereas the induced-fit hypothesis depicts a flexible enzyme that can assume a shape induced by the substrate. Enzymatic activity occurs at a small portion of the enzyme called the **active site,** created by the tertiary structure, which places several amino acid side chains near one another.

Many enzymes require a cofactor for catalytic activity. The complete catalyst, or **holoenzyme,** consists of the protein part, or **apoenzyme,** and the **cofactor.** Cofactors in **metalloenzymes** are metal ions. Organic cofactors are **coenzymes.** Most water-soluble vitamins or their derivatives are coenzymes.

The kinetics of enzyme reactions are described by a turnover number and an equilibrium constant. The **turnover number** is the number of substrate molecules converted into product per minute for each molecule of enzyme under **saturation** conditions. A K_m **value** gives the stability of the enzyme–substrate complex.

Enzymatic reactions have a maximum velocity at a characteristic **optimum temperature** and **optimum pH.** The optimum temperature for most enzymes in the human body is 37 °C. Higher temperatures decrease the rate of reaction as a result of alteration of the enzyme structure and eventual denaturation. Changes in pH alter the tertiary structure of the enzyme and affect its binding efficiency.

The activity of enzymes is hindered by molecules called **inhibitors. Competitive inhibitors** compete with the substrate for binding at the active site and are usually structurally similar to the natural substrate. **Noncompetitive inhibitors** bind with the enzyme at some site other than the active site. Generally, noncompetitive inhibitors have no structural relationship to the substrate.

Regulatory enzymes are required for economical use of the energy and materials of the cell. **Allosteric enzymes** provide **feedback inhibition** by means of a negative modulator. The **negative modulator** produced at the end of a series of steps is a reversible inhibitor for an earlier reaction. A **zymogen** is an inactive form of an enzyme that is activated when required by a cell.

Exercises

Enzyme Specifity

24.1 How do enzymes differ from ordinary catalysts? In what way are they similar?

24.2 What is meant by the term specificity? What is absolute specificity?

Classification of Enzymes

24.3 What type of enzyme would catalyze the conversion of glucose into fructose?

24.4 What type of enzyme would catalyze the conversion of alanylserine into alanine and serine?

24.5 What substrates should react when catalyzed by a phosphatase?

24.6 What substrates should react when catalyzed by a peptidase?

24.7 What type of enzyme is required for the following reaction?

$$\begin{array}{c} CO_2^- \\ | \\ H{-}C{-}OH \\ | \\ CH_2OPO_3^{2-} \end{array} \longrightarrow \begin{array}{c} CO_2^- \\ | \\ H{-}C{-}OPO_3^{2-} \\ | \\ CH_2OH \end{array}$$

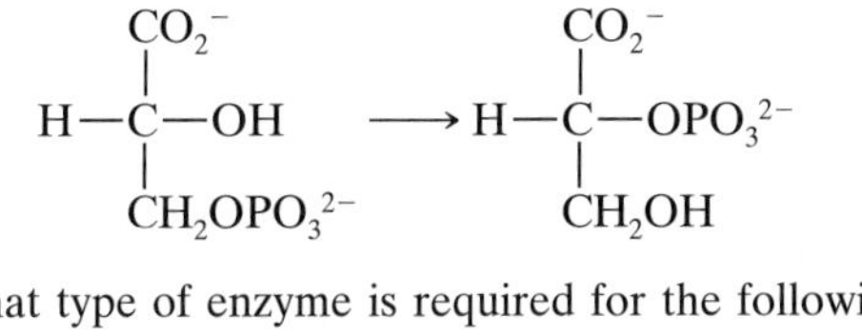

24.8 What type of enzyme is required for the following reaction?

$$\begin{array}{c} CO_2H \\ | \\ C{=}O \\ | \\ CH_3 \end{array} \longrightarrow CH_3CHO + CO_2$$

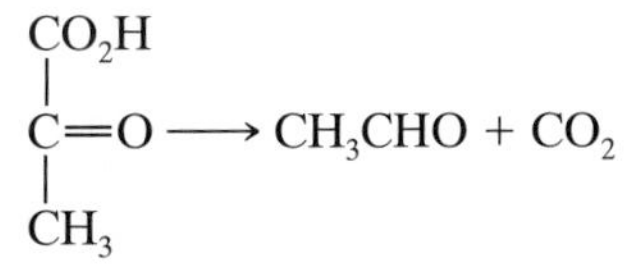

24.9 What type of enzyme is required for the following reaction, which is part of the pentose phosphate pathway?

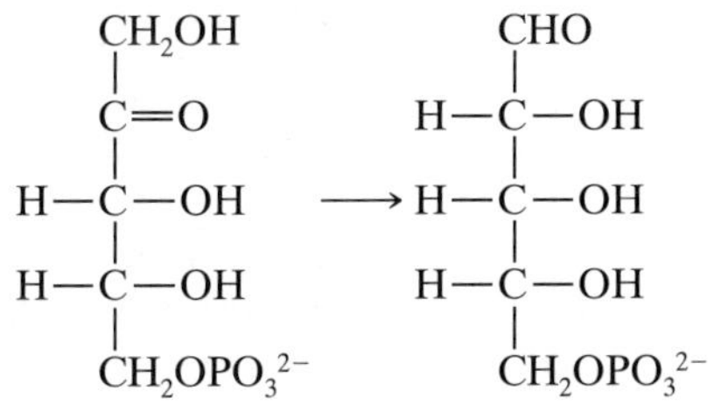

$$\begin{array}{c} CH_2OH \\ | \\ C{=}O \\ | \\ H{-}C{-}OH \\ | \\ H{-}C{-}OH \\ | \\ CH_2OPO_3^{2-} \end{array} \longrightarrow \begin{array}{c} CHO \\ | \\ H{-}C{-}OH \\ | \\ H{-}C{-}OH \\ | \\ H{-}C{-}OH \\ | \\ CH_2OPO_3^{2-} \end{array}$$

24.10 What type of enzyme is required for the following reaction, which occurs in the citric acid cycle?

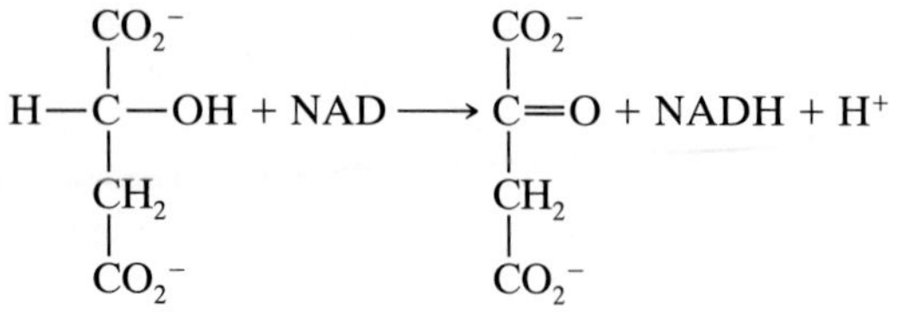

$$\mathrm{H{-}\overset{\displaystyle CO_2^-}{\underset{\displaystyle \underset{\displaystyle CO_2^-}{|}\;\; CH_2}{C}}{-}OH} + \mathrm{NAD} \longrightarrow \mathrm{\overset{\displaystyle CO_2^-}{\underset{\displaystyle \underset{\displaystyle CO_2^-}{|}\;\; CH_2}{C}}{=}O} + \mathrm{NADH} + \mathrm{H^+}$$

Models of Enzyme Action

24.11 What two models are used to describe the specificity of enzymes?

24.12 Define each of the following terms.

(a) complementary shape
(b) enzyme–substrate complex
(c) lock-and-key theory

24.13 Why are enzymes specific for only one enantiomer?

24.14 How can two amino acids separated by many other amino acids still combine in an active site?

24.15 Name two oxidizing agents that act as coenzymes with oxidoreductases.

24.16 Distinguish among the terms apoenzyme, holoenzyme, and metalloenzyme.

24.17 Why are higher concentrations of coenzymes required as compared with enzyme concentrations?

24.18 Why are only low concentrations of metals required nutritionally even though the metals ions are cofactors?

Enzyme Activity

24.19 The enzyme acetylcholinesterase contains serine, histidine, and aspartic acid at the active site. The enzyme is inactive in acidic solution but becomes active as the pH is increased. Explain this result.

24.20 The enzyme pepsin catalyzes the hydrolysis of dietary protein in the stomach but not in the intestine, where the pH is greater than 7. Explain this difference in the activity of pepsin.

24.21 What is meant by saturation in enzyme reactions?

24.22 What is the significance of the turnover number of enzymes?

24.23 What do K_m values of enzymes indicate about the rate of enzyme-catalyzed reactions?

24.24 Why do most enzymes lose their catalytic activity when heated to 50 °C?

24.25 Histidine is part of the active site in many enzymes that catalyze reactions involving protonation and deprotonation. Explain the suitability of histidine on the basis of one of its pK_a values.

24.26 Hydrogen bonding occurs between the hydroxyl group of tyrosine in an enzyme and the carbonyl oxygen atom of a substrate. What would happen to the rate of the reaction above pH 8?

24.27 The cancer cells in some forms of leukemia require asparagine, and the cancer may be suppressed by intravenous administration of asparaginase, which catalyzes the hydrolysis of asparagine to aspartic acid. The K_m value of asparaginases from various sources differ. Would an asparaginase with a low or high K_m be most effective?

24.28 The receptor protein for progesterone is active at pH 7.0. The side chains of the protein form hydrogen bonds with the two carbonyl groups of progesterone. At this pH what amino acids might be involved?

Enzyme Inhibition

24.29 Malonic acid ($HO_2CCH_2CO_2H$) is a competitive inhibitor for the enzyme succinate dehydrogenase. Explain how malonic acid acts as an inhibitor.

24.30 Explain how a competitive inhibitor can be experimentally distinguished from a noncompetitive inhibitor.

24.31 Why is Pb^{2+} a poison even at low concentrations?

24.32 The arsenate ion (AsO_4^{3-}) is toxic. Explain how arsenate might act as a competitive inhibitor in enzyme-catalyzed reactions requiring transfer of phosphate groups.

24.33 Iodoacetamide reacts with an enzyme's sulfhydryl group by the following reaction. Is this an example of reversible or irreversible inhibition?

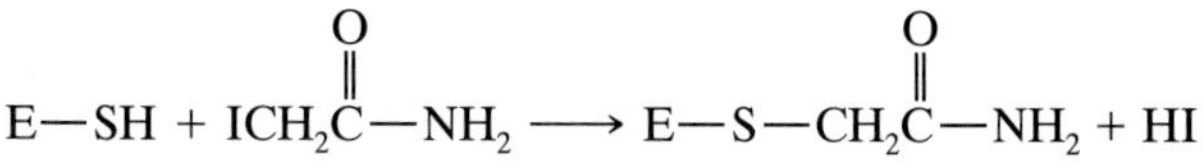

$$\mathrm{E{-}SH + ICH_2\overset{\displaystyle O}{\overset{\|}{C}}{-}NH_2 \longrightarrow E{-}S{-}CH_2\overset{\displaystyle O}{\overset{\|}{C}}{-}NH_2 + HI}$$

24.34 Diisopropylphosphofluoridate is a nerve gas poison that reacts with the hydroxyl group of serine at the active site of acetylcholinesterase. A mole of HF is produced in the reaction. Draw the structure of the product.

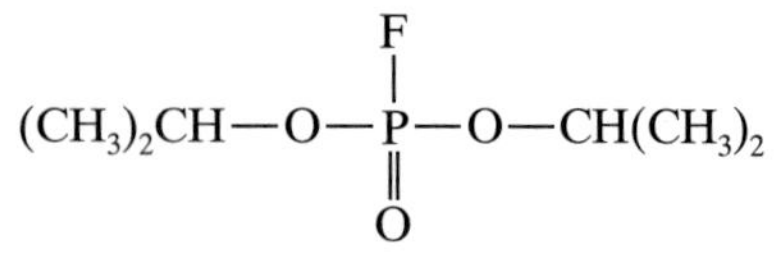

$$\mathrm{(CH_3)_2CH{-}O{-}\overset{\displaystyle F}{\overset{|}{\underset{\displaystyle \underset{\displaystyle O}{\|}}{P}}}{-}O{-}CH(CH_3)_2}$$

diisopropylphosphofluoridate

24.35 Consider the following rates of an enzyme-catalyzed reaction in the presence and absence of an inhibitor. Is the inhibitor competitive or noncompetitive? ([S] = substrate concentration)

[S] (μ mole/L)	Rate (μ mole/min) No Inhibitor	Rate (μ mole/min) Inhibitor (1×10^{-4} M)
3	10.5	2.0
5	14.5	3.0
10	22.5	4.5
30	34.0	7.0
90	40.5	8.0

24.36 Consider the following rates of an enzyme-catalyzed reaction in the presence and absence of an inhibitor. Is the inhibitor competitive or noncompetitive? ([S] = substrate concentration)

[S] (μ mole/L)	Rate (μ mole/min) No Inhibitor	Rate (μ mole/min) Inhibitor (1×10^{-4} M)
3	10.0	4.0
5	14.5	6.5
10	22.5	11.5
30	34.0	22.5
90	40.5	34.5

Regulatory Enzymes and Zymogens

24.37 Explain the action of a modulator.

24.38 Explain how allosteric enzymes function.

24.39 Trypsinogen is formed in the pancreas. Explain why the pancreas contains a trypsin inhibitor.

24.40 How does a peptidase convert a zymogen into an active enzyme?

24.41 Acute pancreatitis is a severe and usually fatal disease in which the zymogens produced by the pancreas are activated while still inside the pancreas. Explain why this condition may be fatal.

24.42 Clotting of blood occurs by a series of zymogen activations. Explain why the enzymes for blood clotting are produced as zymogens.

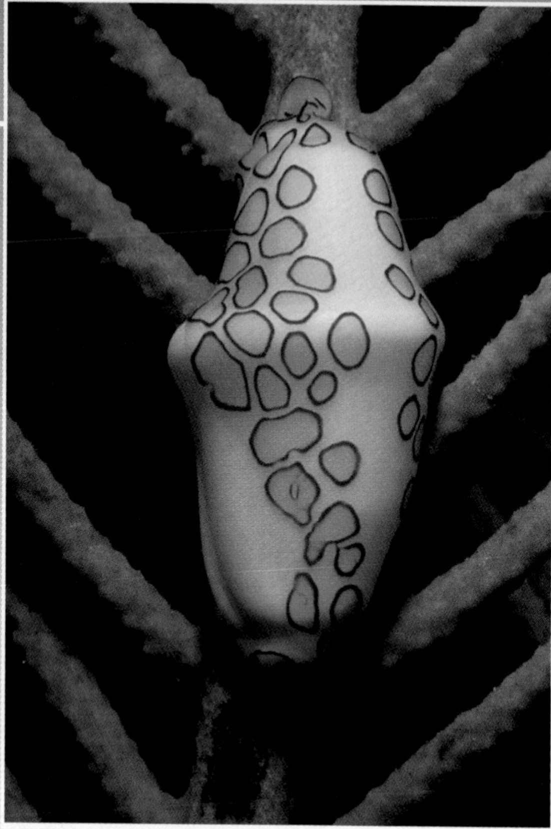

All life forms, including this flamingo tongue feeding on a gorgonian, depend on food to maintain themselves. (Robert Ouellette)

25

Biochemical Energy

25.1 Food and Free Energy
25.2 Free Energy and Oxidation State
25.3 Metabolic Reactions and ATP
25.4 Coupling of Reactions
25.5 Substrate-Level Phosphorylation
25.6 Energy and Redox Reactions
25.7 Cellular Respiration
25.8 Citric Acid Cycle
25.9 The Mitochondria

25.1 Food and Free Energy

The law of conservation of energy applies to all reactions, including biological reactions that occur in the cell. The energy of all living things is ultimately derived from the radiant energy of the sun. Plants use light energy in photosynthesis to produce compounds in a reduced form. In this process, energy is stored in the compounds formed. Animals that eat those plants convert the compounds to oxidized forms and release energy that is used for mechanical work, growth and repair of cells, and for storage of energy in compounds that can provide that energy at a later date. This later release of energy occurs when the cells oxidize energy-storage compounds such as glycogen in catabolic reactions. In this and the next three chapters we will learn how energy is stored and how it is used.

Stages of Catabolic Reactions

The release of energy via catabolic reactions of foodstuffs occurs in three stages (Figure 25.1). The first stage, digestion, is a hydrolytic process that converts large molecules into smaller molecules to be absorbed by the body. Glycosidic bonds in

FIGURE 25.1 Stages of Conversion of Food into Energy

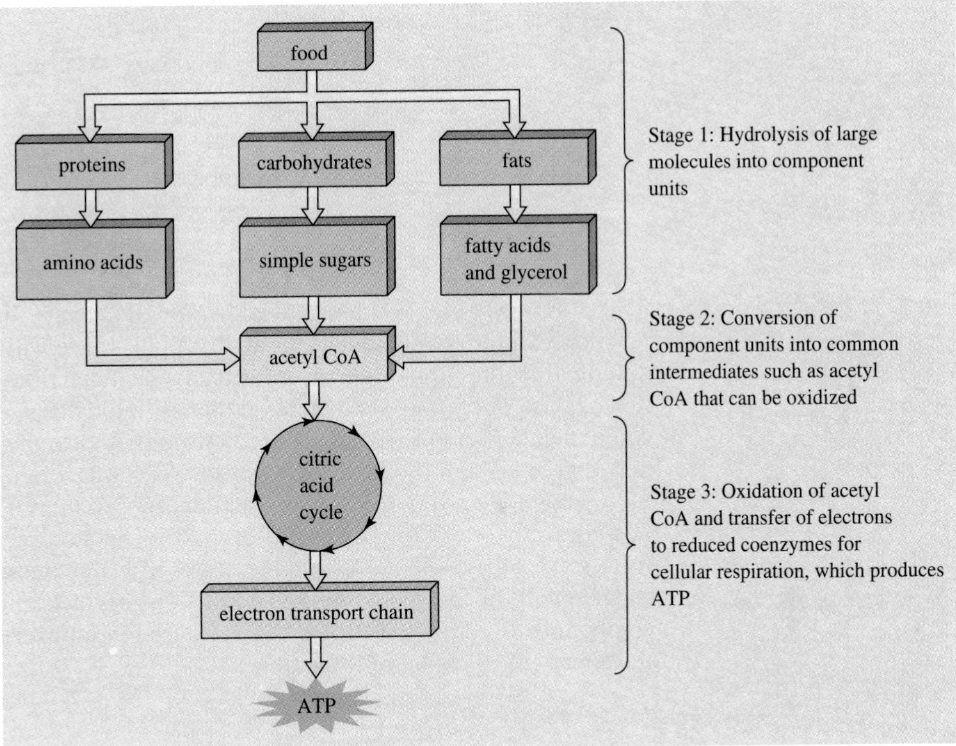

disaccharides and polysaccharides are hydrolyzed to form monosaccharides. Ester bonds of fats and oils are hydrolyzed to form glycerol and fatty acids. Amide bonds of proteins are hydrolyzed to form amino acids. Each hydrolysis step is catalyzed by specific enzymes. However, the small amount of energy released in these hydrolysis reactions is not saved in storage compounds.

In the second stage, hydrolysis products of digestion are degraded to the few simple compounds that are central to metabolism. Each of these compounds in turn is converted into the acetyl group of acetyl CoA. In this stage some ATP is generated, but the amount of energy stored in ATP is only a small fraction of the amount available from further oxidation in stage 3.

In the third stage, cells conserve and store most of the energy contained in foodstuffs. Acetyl CoA is oxidized in the citric acid cycle to yield carbon dioxide (Section 25.8). Electrons are transferred to the coenzymes NAD^+ and FAD to form NADH and $FADH_2$ (Section 25.6). The electrons are then transferred to oxygen via the electron transport chain, which generates the largest quantity of ATP in a process called *cellular respiration* (Section 25.7).

Free Energy and Biochemical Reactions

You learned in Section 10.3 that spontaneous chemical reactions occur in the direction of a negative free energy change ($\Delta G^\circ < 0$). These principles apply to biochemical reactions as well. Cells extract energy from food by breaking down molecules in **catabolic reactions.** These reactions are spontaneous; the free energy of the products is less than the free energy of the reactants. Some energy is released, and the reaction is **exergonic.** The more energy released, the more complete is the reaction in that direction and the larger is the equilibrium constant.

exergonic reaction:

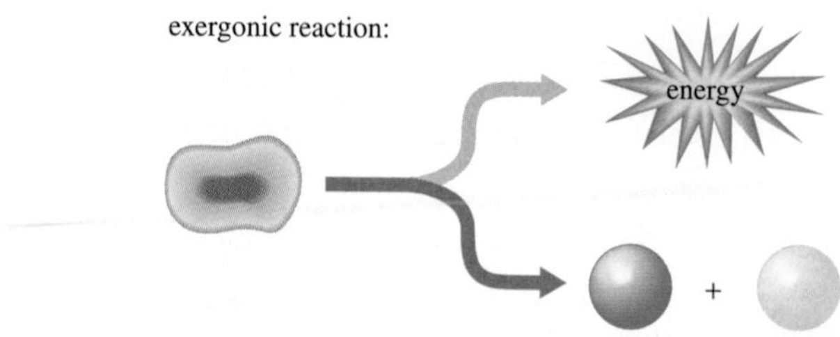

The flow of water is a useful analogy for the change in free energy in chemical reactions. Water spontaneously flows downhill, so we will use the term *downhill* to describe the direction of energy flow in chemical reactions that are exergonic.

Cells use energy in **anabolic** reactions to build the molecules required for cell components. Anabolic reactions are nonspontaneous; the free energy of the products is greater than the free energy of the reactants. Energy must be absorbed, and the reaction is **endergonic.** Such reactions run uphill. Of course, as in the case of flowing water, we know that such a change can occur only if energy is supplied from some other source. A gasoline-powered water pump could be used to move water uphill. In the cell, endergonic reactions occur because stored free energy in compounds is converted to ATP and released in an exergonic reaction to provide the energy for the anabolic reaction.

endergonic reaction:

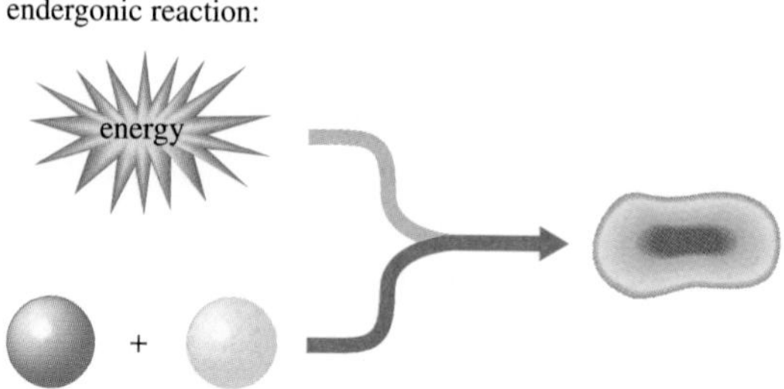

$\Delta G°$ Can Be Calculated

The $\Delta G°$ for a reaction is the same whether the end product is obtained in one step or in a series of steps. Consider the analogy of water flowing downhill. We know that the water tends to go downhill spontaneously to a position of lower potential energy. The energy released does not depend on the actual path that the water takes but only on the vertical distance between the top and bottom of the hill. In other words it doesn't matter whether the water flows down a smooth incline, passes over several cascades in rapids, or drops over a single waterfall. The same principle applies to free energy changes. The $\Delta G°$ for the conversion of a specific reactant to a specific product is the same regardless of whether the reaction occurs in a single step or in multiple steps. Thus, the $\Delta G°$ for a chemical reaction can be calculated from the $\Delta G°$ of several other reactions. Consider the combustion (oxidation) of methane.

$$CH_4 + 2\,O_2 \longrightarrow CO_2 + 2\,H_2O \qquad \Delta G° = -191.4 \text{ kcal/mole}$$

The same overall reaction is obtained by adding the following three reactions.

$$\begin{array}{rll} CH_4 \longrightarrow C + 2\,H_2 & \Delta G^\circ_1 = & +12.2 \text{ kcal/mole} \\ C + O_2 \longrightarrow CO_2 & \Delta G^\circ_2 = & -94.3 \text{ kcal/mole} \\ 2\,H_2 + O_2 \longrightarrow 2\,H_2O & \Delta G^\circ_3 = & -109.3 \text{ kcal/mole} \\ \hline CH_4 + 2\,O_2 \longrightarrow CO_2 + 2\,H_2O & \Delta G^\circ = & -191.4 \text{ kcal/mole} \end{array}$$

The sum of the individual ΔG° values for each reaction equals ΔG° for the combustion of methane because the net reaction is the same. The three individual reactions shown do not actually occur when methane is burned. They were chosen only to show how to calculate the desired ΔG° for the oxidation of methane.

The additivity of ΔG° values for reactions is useful in discussing series of biochemical reactions. Some individual reactions may be endergonic, whereas others are exergonic. The net ΔG° for the series of reactions is the sum of the individual ΔG° values.

Metabolic Pathways

All of the chemical reactions in metabolism occur in sequences of steps. In most metabolic pathways the product of one reaction is the reactant of the next. Such pathways are shown as linear arrays of intermediates.

$$\text{initial reactant} \longrightarrow A \longrightarrow B \longrightarrow C \longrightarrow \text{final product}$$

Several central metabolic processes, like the citric acid cycle, are cyclic. These cycles ultimately regenerate the first member of the series of reactions. The net change occurs in the reactant that combines with the first member of the series and is brought into the cycle. At one or several points along the way the various components of the reactant are released in bits and pieces as products.

Keeping Score

In the next three chapters we will keep an energy "scoreboard" to track how much energy from food is converted into ATP. The flow of energy in metabolic reactions is not always in one direction. Usually there is an "investment" phase in which some of the energy "currency" of ATP is used (Figure 25.2). For example, ATP may be consumed in a reaction required to activate a substrate for subsequent reactions. In subsequent reactions some energy "recovery" occurs, and ATP is formed. Eventually, the metabolic reactions enter a "profit" phase to produce additional

FIGURE 25.2 Energy Flow in Metabolic Reactions

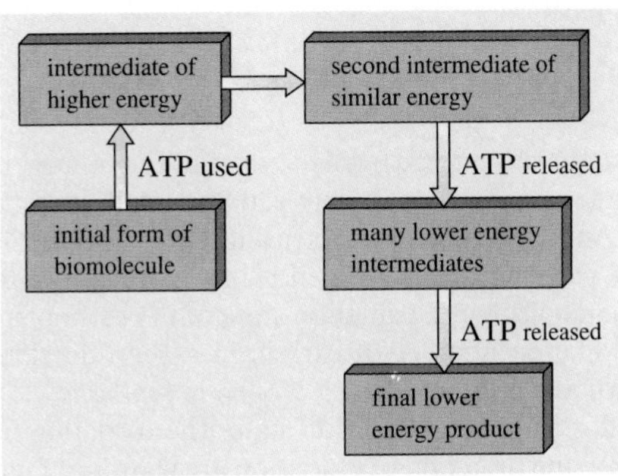

ATP. The oxidation of an acetyl group in the citric acid cycle followed by cellular respiration is an example of a profitable sequence of reactions.

EXAMPLE 25.1

The $\Delta G°$ for the oxidation of ethanol to acetic acid is -106.7 kcal/mole. The $\Delta G°$ for the oxidation of acetic acid to carbon dioxide is -204.0 kcal/mole. What is the $\Delta G°$ for the oxidation of ethanol to carbon dioxide?

Solution

First, write the balanced equation for each reaction.

$$\begin{array}{lll} C_2H_5OH + O_2 \longrightarrow CH_3CO_2H + H_2O & & \Delta G° = -106.7 \text{ kcal/mole} \\ \underline{CH_3CO_2H + 2\,O_2 \longrightarrow 2\,CO_2 + 2\,H_2O} & & \underline{\Delta G° = -204.0 \text{ kcal/mole}} \\ C_2H_5OH + 3\,O_2 \longrightarrow 2\,CO_2 + 3\,H_2O & & \Delta G° = ? \end{array}$$

The equation for the oxidation of ethanol to carbon dioxide and water is the sum of the other two equations. Thus, the $\Delta G°$ for the final equation is calculated to be $-106.7 + (-204.0) = -310.7$ kcal/mole.

PROBLEM 25.1

The $\Delta G°$ for the hydrogenation of acetylene to form ethane is -57.9 kcal/mole; the $\Delta G°$ for the hydrogenation of ethylene to form ethane is -24.2 kcal/mole. Calculate the $\Delta G°$ for the reaction of acetylene with hydrogen to produce ethylene.

25.2 Free Energy and Oxidation State

This isolated birch tree in Vermont is storing the energy of sunlight by forming molecules containing more energy than carbon dioxide and water. (Robert Ouellette)

Photosynthesis is an endergonic reaction because plant molecules have a higher free energy than the compounds from which they were formed.

$$6\,CO_2 + 6\,H_2O \longrightarrow C_6H_{12}O_6 + 6\,O_2 \qquad \Delta G° = +686 \text{ kcal}$$

The reaction is a multistep process. However, the free energy change of all of the steps equals the difference in the free energies of the reactants and the products. When glucose is oxidized, either by burning in air or in metabolic reactions, 686 kcal/mole of energy is released. Note that this reaction is the reverse of the photosynthesis reaction.

$$C_6H_{12}O_6 + 6\,O_2 \longrightarrow 6\,CO_2 + 6\,H_2O \qquad \Delta G° = -686 \text{ kcal}$$

The energy stored in carbon compounds is related to the oxidation state of carbon. Plants synthesize relatively reduced molecules that have a high free energy content. Animals eat plants, releasing energy stored in the bonds of plant molecules. The carbon dioxide formed is in a more oxidized state. In both photosynthesis and metabolism, the oxidation state of the carbon atoms changes.

The oxidation number of carbon in carbon dioxide is +4, its maximum value. When carbon combines with hydrogen in methane, its oxidation state decreases to −4, which is its lowest oxidation state. The oxidation numbers of one-carbon compounds and the direction of oxidation are shown in Figure 25.3.

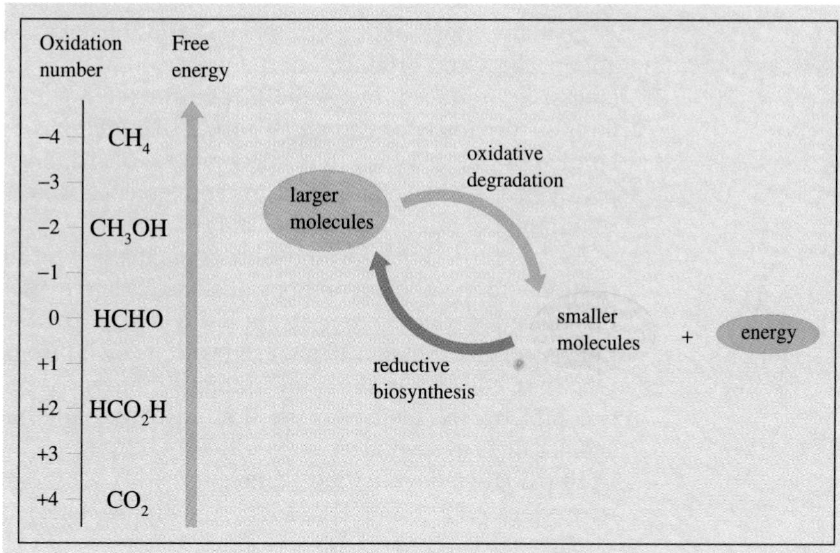

FIGURE 25.3 Oxidation Number and Free Energy of Carbon Compounds

Photosynthesis yields glucose, a "reduced molecule" compared with carbon dioxide. Reduced molecules are "energy rich," and they release energy to produce oxidized molecules that are "energy poor." Catabolic reactions in animals break down reactants in food into energy-poor products in higher oxidation states by **oxidative degradation reactions.** Anabolic reactions, which synthesize energy-rich products in a lower oxidation state—that is, a more reduced state—are **reductive biosynthetic reactions.**

EXAMPLE 25.2

Fats have more stored energy than carbohydrates on a per-gram basis. Why?

Solution

Fats are esters of glycerol and long-chain fatty acids. The majority of the carbon atoms in a fat are bonded to two hydrogen atoms. Carbohydrates, such as glucose, have oxygen atoms bonded to most of the carbon atoms and are more oxidized. Fats are more reduced than carbohydrates and therefore have more stored energy.

PROBLEM 25.2

Examine the structure of proteins and determine if their stored energy is similar to that of carbohydrates or fats.

25.3 Metabolic Reactions and ATP

Although the same equation represents both the combustion and oxidative degradation of glucose, the mechanisms differ. In addition, the manner in which energy is released differs. The energy liberated in the combustion reaction is more than the organism needs at one time. The human body needs steadily produced energy for muscle movement, active transport, and transmission of nerve impulses, as well as for anabolic reactions.

Oxidative degradation occurs in a series of steps that release energy in small usable packets and produce intermediate compounds called **metabolites.** Some heat energy is produced, but some steps produce adenosine triphosphate (ATP), the body's storehouse of energy (Figure 25.4). ATP has the heterocyclic base adenine bonded to ribose, which in turn is bonded to a triphosphate group. At physiological pH, ATP has a −4 charge, but for convenience we often use the shorthand representation that doesn't show the charge.

Let's return to the flow of water as an analogy for the comparison of direct combustion and metabolic reactions. If a single dam is used to generate electric power, the energy produced comes from a single step. This is analogous to direct combustion. A series of smaller dams, each with its own turbines, is analogous to metabolic reactions. Ultimately, the same amount of energy is released per gallon of water as it completes the course to the bottom of the hill, but the energy is produced in smaller units at several sites.

In oxidative degradation, some energy is released as heat, but some is also used to produce ATP from ADP (adenosine diphosphate).

FIGURE 25.4 Structure and Hydrolysis of ATP

Either of the two anhydride bonds linking phosphate units can be hydrolyzed and release energy.

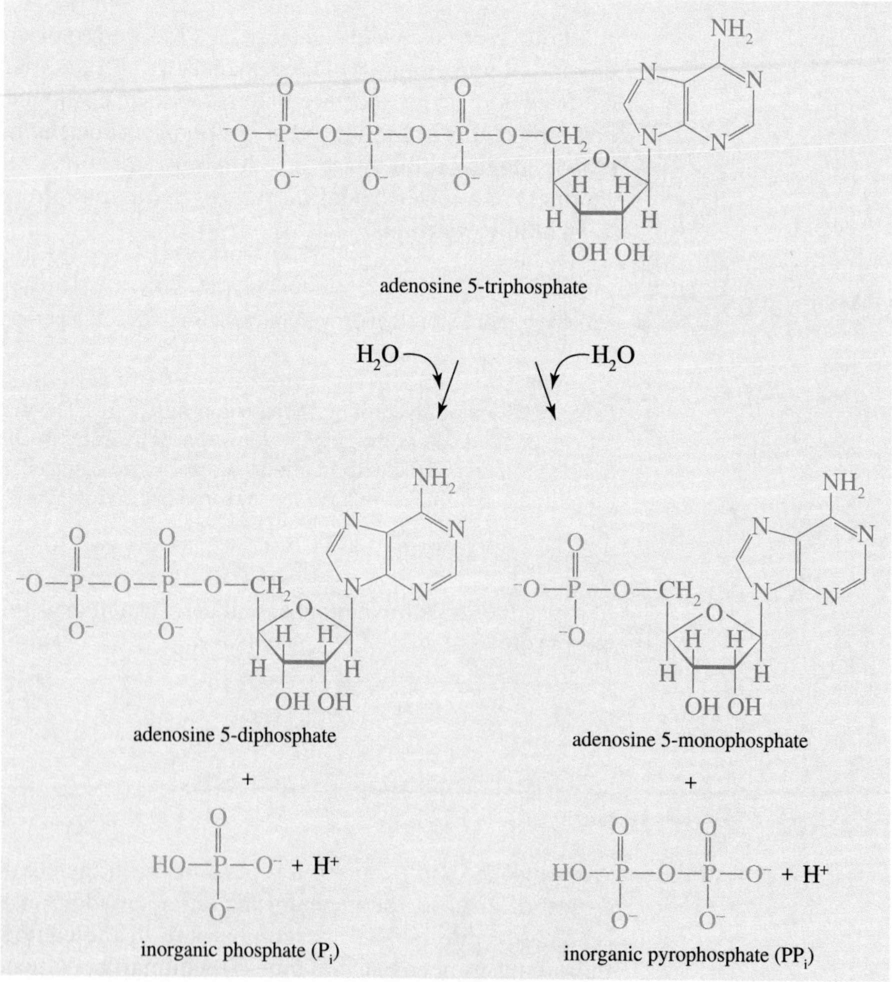

adenosine 5-triphosphate

H_2O H_2O

adenosine 5-diphosphate + inorganic phosphate (P_i) $HO-P(O)(O^-)-O^- + H^+$

adenosine 5-monophosphate + inorganic pyrophosphate (PP_i) $HO-P(O)(O^-)-O-P(O)(O^-)-O^- + H^+$

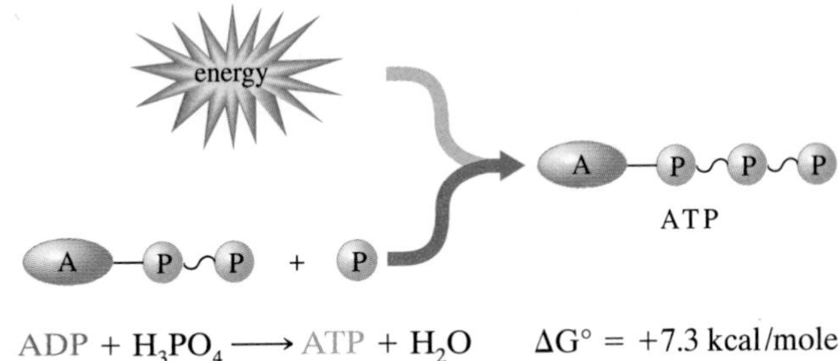

$$ADP + H_3PO_4 \longrightarrow ATP + H_2O \qquad \Delta G° = +7.3 \text{ kcal/mole}$$

Thus, ATP has free energy of 7.3 kcal/mole stored in one of its phosphoric acid anhydride bonds. This energy is released to the organism by hydrolysis when needed. In the shorthand equation P_i represents inorganic ions of phosphoric acid that exist at physiological pH.

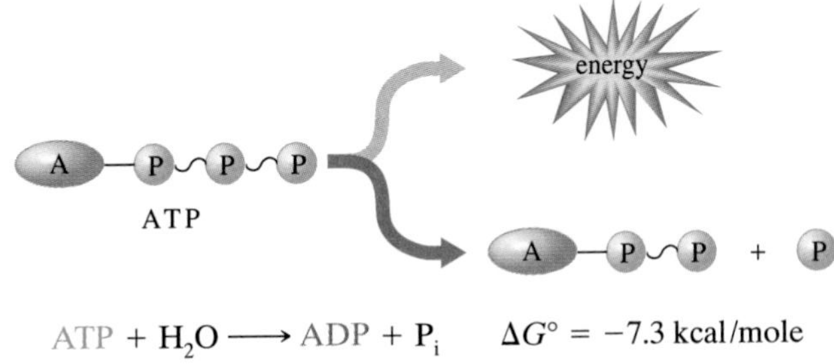

$$ATP + H_2O \longrightarrow ADP + P_i \qquad \Delta G° = -7.3 \text{ kcal/mole}$$

The anhydride bond in ADP can also be hydrolyzed to produce adenosine monophosphate (AMP).

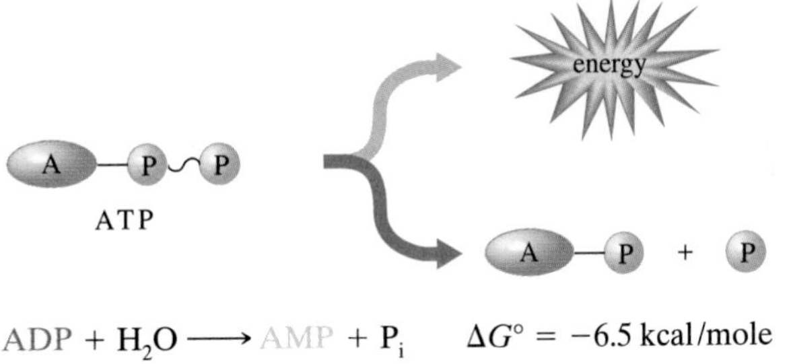

$$ADP + H_2O \longrightarrow AMP + P_i \qquad \Delta G° = -6.5 \text{ kcal/mole}$$

In the metabolism of fatty acids (Chapter 27), ATP is hydrolyzed to AMP and pyrophosphoric acid. In biological systems, pyrophosphoric acid exists as the pyrophosphate ion, PP_i.

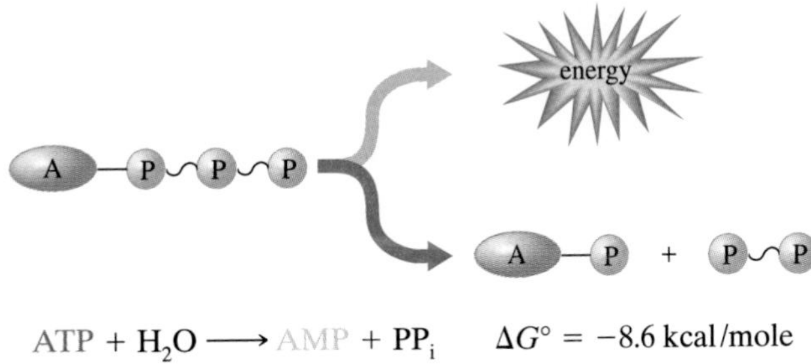

$$ATP + H_2O \longrightarrow AMP + PP_i \qquad \Delta G° = -8.6 \text{ kcal/mole}$$

Phosphorylation, the reverse of the hydrolysis reactions of ATP, can occur only if energy is available. Without phosphorylation, all ATP would eventually be converted to ADP or AMP. However, free energy stored in food molecules is used to convert ADP or AMP back to ATP, and cells maintain a reservoir of ATP. Excess ATP is used in anabolic reactions to synthesize compounds such as glycogen and fats, which also store free energy. When food is not available, the ATP requirements for mechanical work and transport are obtained from the stored compounds, which undergo catabolism and produce ATP.

EXAMPLE 25.3

The ΔG° for the formation of glucose 6-phosphate from glucose and a source of phosphate is +3.3 kcal/mole. What is the ΔG° for the hydrolysis of glucose 6-phosphate?

$$\text{glucose} + H_3PO_4 \longrightarrow \text{glucose 6-phosphate} + H_2O \qquad \Delta G^\circ = +3.3\text{ kcal/mole}$$

Solution

The hydrolysis of glucose 6-phosphate is the reverse of the reaction that forms it. Thus, for this hydrolysis, the $\Delta G^\circ = -3.3$ kcal/mole.

$$\text{glucose 6-phosphate} + H_2O \longrightarrow \text{glucose} + H_3PO_4 \qquad \Delta G^\circ = -3.3\text{ kcal/mole}$$

PROBLEM 25.3

The ΔG° for the hydrolysis of L-glycerol 3-phosphate to give glycerol and phosphoric acid is −2.2 kcal/mole. What is the ΔG° for the esterification of glycerol with phosphoric acid?

25.4 Coupling of Reactions

Virtually all metabolic processes involve a sequence of steps. In some cases an endergonic reaction occurs because a subsequent exergonic reaction uses the product of the first reaction as a reactant. The oxidation of malic acid by the coenzyme nicotinamide adenine dinucleotide (NAD^+) is one step in the citric acid cycle. The ΔG° is positive, and this single reaction is not spontaneous.

$$\text{malic acid} + NAD^+ \longrightarrow \text{oxaloacetic acid} + NADH + H^+ \qquad \Delta G^\circ = +6.1\text{ kcal/mole}$$

At equilibrium there would be little oxaloacetic acid. The conversion of oxaloacetic acid into citric acid is a subsequent step in the citric acid cycle. This reaction is favorable and uses the oxaloacetic acid formed in the preceding step.

$$\text{oxaloacetic acid} + \text{acetyl CoA} \longrightarrow \text{citric acid} + \text{CoA} \qquad \Delta G^\circ = -7.7\text{ kcal/mole}$$

As the oxaloacetic acid is used in the second reaction, more malic acid is converted into oxaloacetic acid by the first reaction. This is another example of Le Châtelier's principle. The second reaction "pulls" the first reaction by removing one of its products. The net result of the two reactions is the release of 1.6 kcal/mole.

Unlike consecutive reactions, a **coupled reaction** is a single reaction that can be viewed as two hypothetical separate reactions. One hypothetical step may be an energy-releasing reaction and the other an energy-consuming reaction. This analysis allows us to keep track of which reactants have the stored energy and where it goes.

As part of the metabolism of glucose (Chapter 26), the monosaccharide is converted to a phosphate ester in a coupled reaction. Consider as a hypothetical step the phosphorylation reaction of glucose with inorganic phosphate as a reagent. This reaction is not spontaneous. Consider also the hydrolysis of ATP, which is spontaneous. The sum of the two processes corresponds to an exergonic reaction.

$$\begin{array}{ll} \text{glucose} + P_i \longrightarrow \text{glucose 6-phosphate} + H_2O & \Delta G^\circ = +3.3 \text{ kcal/mole} \\ \text{ATP} + H_2O \longrightarrow \text{ADP} + P_i & \Delta G^\circ = -7.3 \text{ kcal/mole} \\ \hline \text{ATP} + \text{glucose} \longrightarrow \text{ADP} + \text{glucose 6-phosphate} & \Delta G^\circ = -4.0 \text{ kcal/mole} \end{array}$$

The two separate reactions do not occur as written. In other words, ATP does not have to hydrolyze to form P_i, which then phosphorylates glucose. Instead, a coupled reaction occurs in which the phosphate is transferred from ATP to glucose at the active site of the enzyme hexokinase (Figure 25.5). The reaction occurs spontaneously, with $\Delta G^\circ = -4.0$ kcal/mole, because the sum of the free energies of glucose and ATP is larger than the sum of the free energies of glucose 6-phosphate and ADP.

In biochemical equations it is convenient to use the following representation for coupled reactions to save space:

$$\text{glucose} \xrightarrow[\text{ATP} \quad \text{ADP}]{} \text{glucose 6-phosphate}$$

The curved arrow represents a simultaneous reaction that occurs in a number of reactions in biochemistry. Thus, ATP and ADP are usually not written on the same line as the other reactants and products.

25.5 Substrate-Level Phosphorylation

Substrate-level phosphorylation is a transfer of a phosphate group of ATP or some other phosphorylating agent to a substrate. The direction of the transfer is governed by ΔG°. A compound with high free energy transfers its phosphate to a compound of lower free energy. Several phosphorylated compounds found in cells are listed in Table 25.1 with the free energy change for their hydrolysis. Compounds to-

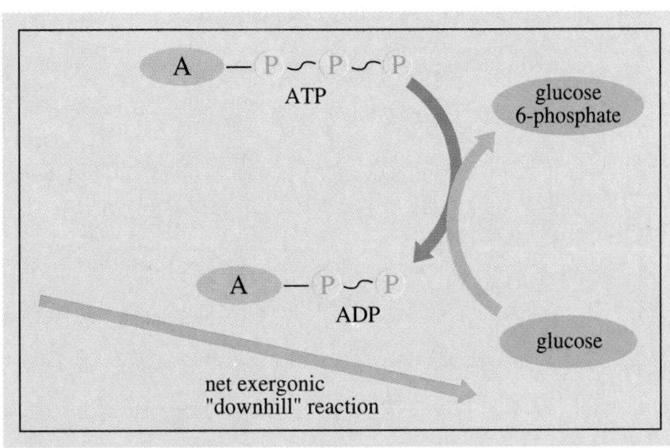

FIGURE 25.5 Coupled Phosphate Transfer Reaction

TABLE 25.1 Free Energy of Hydrolysis of Phosphorylating Agents

Compound	$\Delta G°$ at pH 7.0 (kcal)
phosphoenolpyruvate (PEP)	−14.8
1,3-diphosphoglycerate	−11.8
phosphocreatine	−10.3
acetyl phosphate	−10.1
ATP to AMP and PP_i	−8.6
pyrophosphate (PP_i)	−8.0
acetyl CoA	−7.5
ATP to ADP and P_i	−7.3
ADP	−6.5
glucose 1-phosphate	−5.0
fructose 6-phosphate	−3.8
glucose 6-phosphate	−3.3
L-glycerol 3-phosphate	−2.2

ward the top of the table release larger amounts of energy in hydrolysis. These compounds can transfer phosphate to ADP and convert it to ATP.

$$\begin{array}{ll} \text{PEP} + H_2O \longrightarrow \text{pyruvic acid} + P_i & \Delta G° = -14.8 \text{ kcal/mole} \\ \text{ADP} + P_i \longrightarrow \text{ATP} + H_2O & \Delta G° = +7.3 \text{ kcal/mole} \\ \hline \text{PEP} + \text{ADP} \longrightarrow \text{pyruvic acid} + \text{ATP} & \Delta G° = -7.5 \text{ kcal/mole} \end{array}$$

The phosphates below ATP in the table have a smaller tendency to transfer phosphate. ATP is a better phosphorylating agent than those substances below it.

$$\begin{array}{ll} \text{ATP} + H_2O \longrightarrow \text{ADP} + P_i & \Delta G° = -7.3 \text{ kcal/mole} \\ \text{glycerol} + P_i \longrightarrow \text{L-glycerol 3-phosphate} + H_2O & \Delta G° = +2.2 \text{ kcal/mole} \\ \hline \text{ATP} + \text{glycerol} \longrightarrow \text{ADP} + \text{L-glycerol 3-phosphate} & \Delta G° = -5.1 \text{ kcal/mole} \end{array}$$

ATP is an effective carrier of phosphate groups. It is formed when ADP accepts phosphate from substances of high free energy and it can transfer phosphate to substances of low free energy in coupled reactions (Figure 25.6).

FIGURE 25.6 ATP—The Central Phosphate Transfer Compound

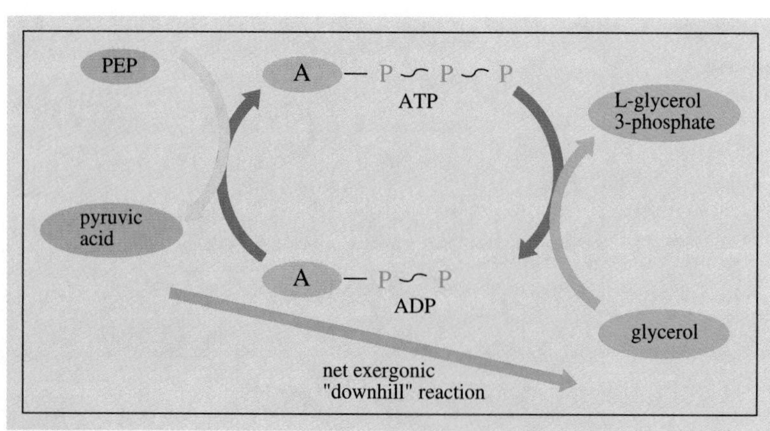

EXAMPLE 25.4

The $\Delta G°$ for the phosphorylation of fructose to give fructose 6-phosphate is +3.8 kcal/mole. Is a coupled reaction of fructose with ATP spontaneous?

Solution

The phosphorylation of fructose can occur with the transfer of a phosphate group from ATP. The hydrolysis of ATP to ADP releases more free energy than that required to phosphorylate fructose. The coupled reaction is spontaneous.

$$\begin{array}{lr} \text{fructose} + P_i \longrightarrow \text{fructose 6-phosphate} + H_2O & \Delta G° = +3.8 \text{ kcal/mole} \\ \text{ATP} + H_2O \longrightarrow \text{ADP} + P_i & \Delta G° = -7.3 \text{ kcal/mole} \\ \hline \text{fructose} + \text{ATP} \longrightarrow \text{fructose 6-phosphate} + \text{ADP} & \Delta G° = -3.5 \text{ kcal/mole} \end{array}$$

PROBLEM 25.4

The $\Delta G°$ for the hydrolysis of phosphocreatine is −10.3 kcal/mole. Can this reaction be coupled to convert ADP into ATP?

25.6 Energy and Redox Reactions

As in the case of organic compounds, the oxidation of biochemical compounds is most readily seen as a gain of oxygen or a loss of hydrogen. Oxidation in cells occurs by the removal of hydrogen atoms as protons as well as by the removal of electrons present in carbon–hydrogen bonds.

Oxygen is an oxidizing agent brought into tissues by respiration, but it does not directly oxidize metabolites to produce CO_2 and H_2O. The oxidation process occurs in many electron-transfer steps. A small group of oxidizing agents is used in biochemical reactions to remove hydrogen. The structures of the coenzymes NAD^+ and flavin adenine dinucleotide (FAD), which are oxidizing agents, are shown in Figure 25.7.

NAD⁺ (Nicotinamide Adenine Dinucleotide)

The portion of NAD^+ that is the oxidizing agent is nicotinamide (niacin or vitamin B_3). Only this portion of the molecule is shown when it is represented as an oxidizing agent.

$$\text{(pyridinium ring with H, } -\overset{O}{\overset{\|}{C}}-NH_2\text{, } N^+-R) + H^+ + 2e^- \rightleftharpoons \text{(dihydropyridine ring with H, H, } -\overset{O}{\overset{\|}{C}}-NH_2\text{, } \ddot{N}-R)$$

NAD^+ gains two electrons and is reduced. A substrate that is oxidized by NAD^+ loses two electrons. The substrate loses two protons, but only one proton combines with NAD^+. The reaction for the oxidation of ethanol to acetaldehyde is

$$CH_3-\underset{H}{\overset{H}{\underset{|}{\overset{|}{C}}}}-OH + NAD^+ \longrightarrow CH_3-\overset{O}{\overset{\|}{C}}-H + NADH + H^+$$

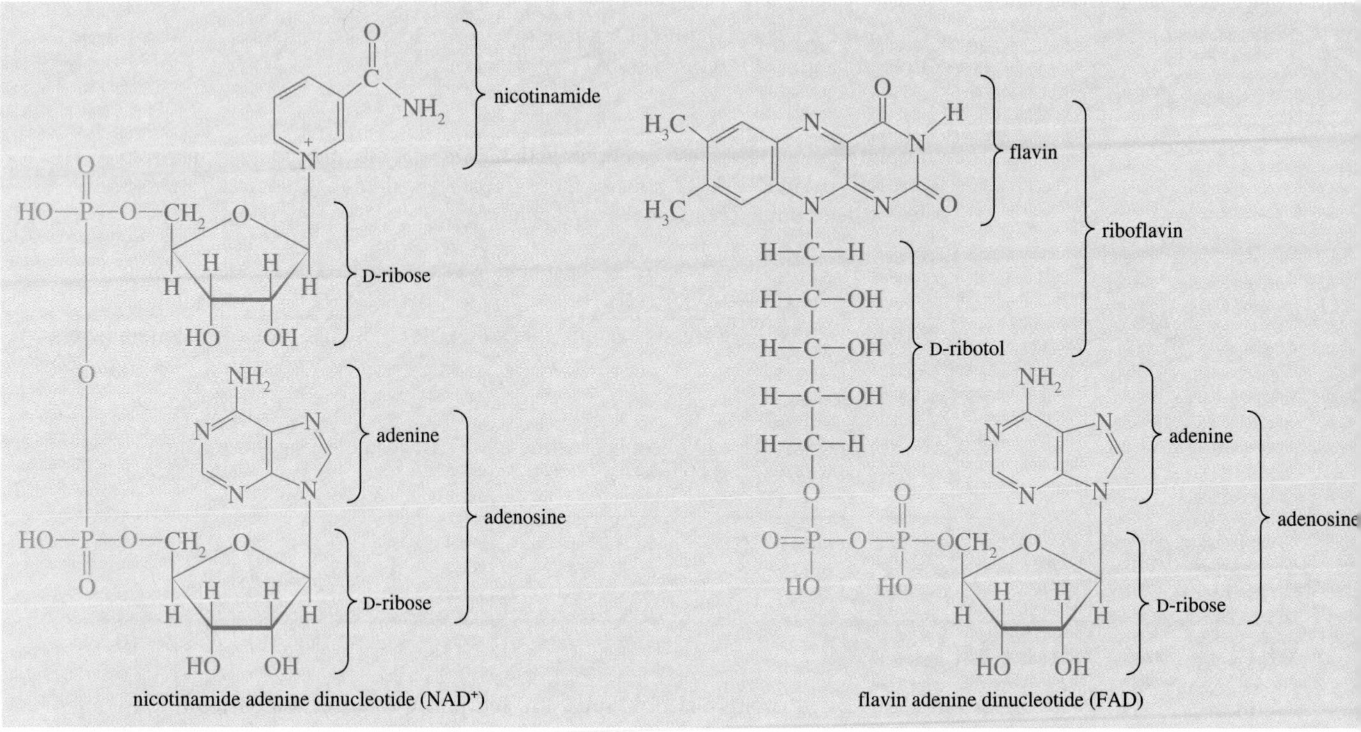

FIGURE 25.7 Structure of NAD$^+$ and FAD

nicotinamide adenine dinucleotide (NAD$^+$)

flavin adenine dinucleotide (FAD)

NAD$^+$ oxidizes aldehydes to carboxylic acids by the addition of an oxygen atom from water.

$$CH_3-\overset{\overset{\displaystyle O}{\|}}{C}-H + NAD^+ + H_2O \longrightarrow CH_3-\overset{\overset{\displaystyle O}{\|}}{C}-OH + NADH + H^+$$

As substrates are oxidized, their free energy decreases and some energy is saved by the reduction of NAD$^+$. The reduced form, NADH, can then either reduce other substrates or reduce oxygen to water in cellular respiration via the electron transport chain (Section 25.7).

FAD (Flavin Adenine Dinucleotide)

FAD contains riboflavin, vitamin B_2. The flavin portion of the molecule is reduced by accepting two electrons and two protons.

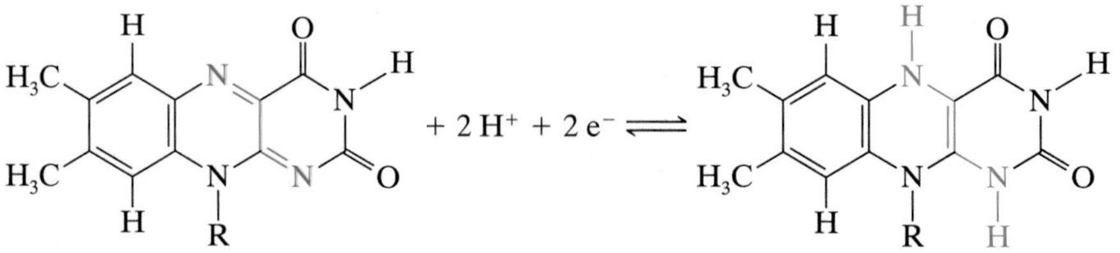

$$+ 2\,H^+ + 2\,e^- \rightleftharpoons$$

FAD dehydrogenates $-CH_2CH_2-$ units to form a double bond. For example, in the metabolism of fatty acids, one step is the removal of hydrogen atoms from the

α and β positions of a coenzyme A–activated compound. The trans isomer is formed exclusively.

$$RCH_2CH_2-\overset{\overset{\displaystyle O}{\|}}{C}-SCoA + FAD \longrightarrow \underset{H}{\overset{R}{>}}C=C\underset{\underset{\underset{\displaystyle O}{\|}}{C}-SCoA}{\overset{H}{<}} + FADH_2$$

Coenzymes Must Be Reoxidized

Reduction of the coenzymes NAD^+ and FAD cannot continue indefinitely, because they are available in limited amounts. For coenzymes to continue to act as oxidizing agents, their reduced forms must be reoxidized. Oxidation in biological systems indirectly involves oxygen. Consider the oxidation of NADH by O_2.

$$NADH + H^+ + \tfrac{1}{2}O_2 \longrightarrow NAD^+ + H_2O \qquad \Delta G° = -53 \text{ kcal/mole}$$

If this reaction occurred in a single step, the energy released would be in a larger "packet" than the cell could use. Oxidation of NADH or $FADH_2$ occurs in several steps in the electron transport chain so that the energy is stored in ATP.

25.7 Cellular Respiration

In cellular respiration, NADH and $FADH_2$ are oxidized to NAD^+ and FAD, respectively, in a series of steps called the **electron transport chain.** The oxidation of NADH occurs in six steps in which the electrons removed are eventually passed along to oxygen via electron carriers.

The electron transport process is like an old-time bucket brigade. Water obtained from the water source is handed in a bucket from one person to another until at the end of the line the water is thrown on the fire. In the electron transport chain, each substance receives electrons from the preceding substance. Then the reduced material is reoxidized by passing electrons along to the next substance in the chain.

The first step in the electron transport chain is the oxidation of NADH by a flavin mononucleotide, FMN.

$$NADH + H^+ + FMN \xrightarrow[\text{ADP} \quad \text{ATP}]{} NAD^+ + FMNH_2$$

One mole of ATP is produced in a coupled reaction, so some of the energy stored in NADH is saved. In the second step, $FMNH_2$ is oxidized by coenzyme Q (CoQ).

$$FMNH_2 + CoQ \longrightarrow FMN + CoQH_2$$

Coenzyme Q is a lipid-soluble molecule that shuttles between FMN and the cytochromes, which are the next carriers in the chain.

Cytochromes are proteins that contain heme as a prosthetic group. Individual cytochromes differ in the proteins surrounding the heme. Since heme contains iron, the net oxidation–reduction reaction is

$$Fe^{2+} \underset{\text{reduction}}{\overset{\text{oxidation}}{\rightleftharpoons}} Fe^{3+} + e^-$$

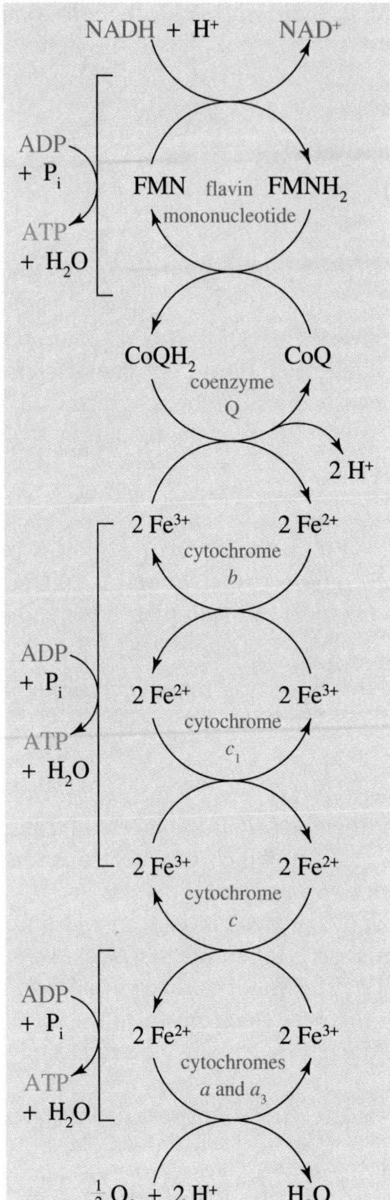

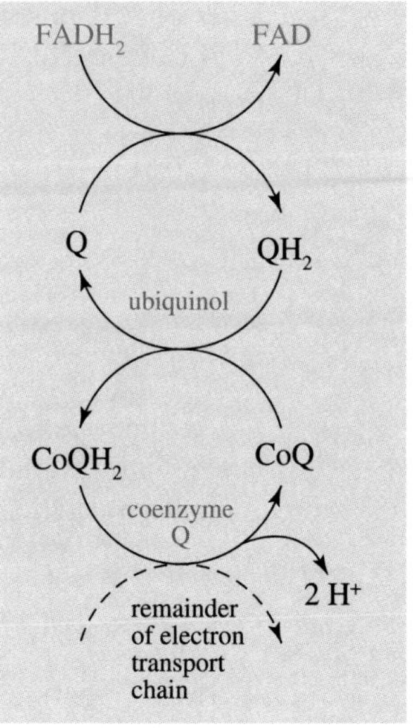

FIGURE 25.8 Coupled Reaction of NADH and $FADH_2$ in the Electron Transport Chain

Each step of the cytochrome reactions passes along the electrons, which eventually reduce oxygen to form water (Figure 25.8). ATP is formed in two of the steps. Since 1 mole of ATP is formed in the reduction of FMN, the oxidation of 1 mole of NADH produces 3 moles of ATP. Thus, some of the energy stored in NADH from the oxidation of substrate is saved in coupled reactions in which ATP is formed.

Electrons are passed between the members of the chain, but the protons are not passed to the cytochromes. According to the stoichiometry of the reaction, protons are required to combine with oxygen to form water. The method by which protons are released from the substrate will be discussed in Section 25.9.

$FADH_2$ also is oxidized and transfers two electrons in the electron transport chain, but $FADH_2$ enters the chain differently than does NADH. $FADH_2$ cannot reduce FMN—it reduces ubiquinol (Q) to QH_2. There is another difference in this step. ATP is not produced in a coupled reaction because the reaction is not sufficiently exergonic. QH_2 then reduces CoQ to $CoQH_2$, and the remaining steps are identical to those in the series described for NADH (Figure 25.8). In summary, $FADH_2$ passes two electrons along a chain to oxygen and releases two protons, but only 2 moles of ATP are formed per mole of $FADH_2$.

Oxidative Phosphorylation

Energy is saved in cellular respiration by the phosphorylation of ADP. Three of the steps in the oxidation of NADH transfer energy to ADP and form ATP. In the oxidation of $FADH_2$ two steps produce ATP. The equations representing these two processes follow.

$$\text{NADH} + 3\,\text{ADP} + \tfrac{1}{2}\,\text{O}_2 + 3\,\text{P}_i + \text{H}^+ \longrightarrow \text{NAD}^+ + 3\,\text{ATP} + \text{H}_2\text{O}$$
$$\text{FADH}_2 + 2\,\text{ADP} + \tfrac{1}{2}\,\text{O}_2 + 2\,\text{P}_i \longrightarrow \text{FAD} + 2\,\text{ATP} + \text{H}_2\text{O}$$

The formation of ATP in these coupled reactions of the electron transport chain is called **oxidative phosphorylation.** The ATP is produced under aerobic conditions. *Aerobic* means with or requiring oxygen; in the case of the electron transport chain it applies to the last step of the chain.

The mechanism by which free energy is transferred to ATP in the electron transport chain is still a subject of research. However, each component of the electron transport chain is anchored in the inner membrane of a mitochondrion (Section 25.9), and the individual components are arranged for efficient transfer of electrons.

The free energy change for the direct oxidation of NADH by oxygen is -53 kcal/mole. The energy stored in ATP by the electron transport chain is $+21.9$ kcal/mole, since 3 moles of ATP are produced. For each mole of NADH oxidized the fraction of the energy saved by ATP that can be used as an energy source in other reactions is about 42%.

$$\frac{3 \times 7.3\ \text{kcal/mole ATP}}{53.0\ \text{kcal/mole NADH}} \times 100 = 42\%\ \text{stored}$$

25.8 Citric Acid Cycle

In the **citric acid cycle** a two-carbon-atom group called the acetyl group is oxidized to CO_2 and H_2O. The citric acid cycle is also called the *Krebs cycle* or *tricarboxylic acid cycle* (TCA). The details of the sources of the acetyl group will be described in Chapters 26 and 27. Entry into the cycle occurs via acetyl CoA.

$$\underset{\text{acetyl group}}{\text{CH}_3{-}\overset{\overset{\displaystyle\text{O}}{\|}}{\text{C}}{-}} \qquad \underset{\text{acetyl coenzyme A}}{\text{CH}_3{-}\overset{\overset{\displaystyle\text{O}}{\|}}{\text{C}}{-}\text{S}{-}\text{CoA}}$$

Some of the energy released in the oxidation of the acetyl group is saved by forming NADH and $FADH_2$, both of which yield ATP in the electron transport chain when they are oxidized.

The reactions of the citric acid cycle are somewhat like a chemical merry-go-round. The product of the last of a series of reactions is a reactant for the first reaction in another trip around the cycle (Figure 25.9). For convenient reference the steps of the cycle are numbered. The acids exist as anions at physiological pH.

FIGURE 25.9 Reactions of the Citric Acid Cycle

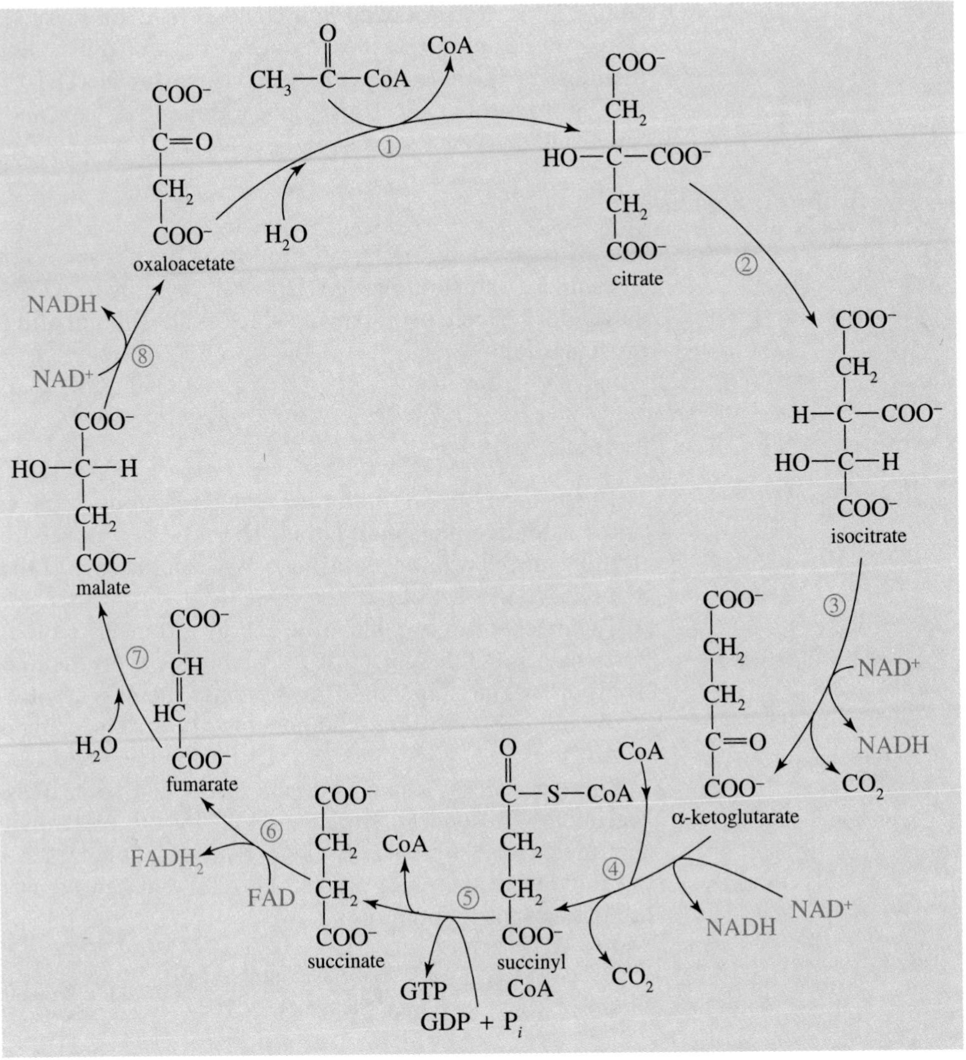

In step 1 a six-carbon atom molecule of citric acid forms from the four-carbon atom molecule of oxaloacetate and acetyl CoA. The reaction is an aldol type of condensation (Section 17.11) in which the α-carbon atom of the acetyl CoA adds to the carbonyl carbon atom of oxaloacetate. Each subsequent step can be explained based on the chemistry of functional groups. However, the objective of this section is only to summarize what occurs in the citric acid cycle. During the cycle the citric acid is converted to a five-carbon compound in step 3 and to a four-carbon compound in step 4. The carbon atoms are released as carbon dioxide.

Oxidations that occur in the cycle are dehydrogenation reactions, but no oxygen is required. The hydrogen atoms and electrons are donated to NAD^+ and FAD. The reduced coenzymes are then reconverted to oxidized forms in the electron transport chain.

In the citric acid cycle, NADH is produced in coupled reactions 3, 4, and 8, and $FADH_2$ is produced in reaction 6. In step 5, a mole of guanosine triphosphate (GTP) is formed from guanosine diphosphate (GDP). The GTP has a somewhat higher free energy than ATP. As a consequence, GTP transfers a phosphate group to ADP by substrate-level phosphorylation. The net reaction of the citric acid cycle is

$$\text{acetyl CoA} + 3\,NAD^+ + FAD + GDP + P_i + 2\,H_2O \longrightarrow 2\,CO_2 + 3\,NADH + FADH_2 + GTP + 2\,H^+ + CoA$$

The 3 moles of NADH produce 9 moles of ATP in the electron transport chain, and the 1 mole of $FADH_2$ produces 2 moles of ATP. Since 1 mole of ATP is formed from GTP in the cycle, a total of 12 moles of ATP is produced by the oxidation of 1 mole of an acetyl group.

The citric acid cycle does not directly involve molecular oxygen. However, the cycle can occur only under aerobic conditions because NAD^+ and FAD are required. These two coenzymes are available only if they are formed in the electron transport chain where electrons are ultimately transferred to molecular oxygen as the oxidizing agent.

Control of the Citric Acid Cycle

The rate of oxidation of acetyl groups in the citric acid cycle is governed by the ATP needs of the cell. The citric acid cycle requires a supply of NAD^+ and FAD, which are regenerated from NADH and $FADH_2$ by oxidation in the electron transport chain. Along with the oxidation of NADH and $FADH_2$ there is an accompanying formation of ATP. If ATP is needed and is formed, then NADH and $FADH_2$ are reconverted into NAD^+ and FAD, which in turn are used in the citric acid cycle. Thus, ATP is responsible for the control of the citric acid cycle. This control is exerted at three points in the cycle.

The synthesis of citrate from oxaloacetate and acetyl CoA in step 1 is an important control point. Unless the acetyl unit enters the cycle, there are no subsequent reactions. ATP is an allosteric inhibitor of the enzyme citrate synthetase required in this step. As the concentration of ATP increases, the enzyme is less effective in binding acetyl CoA, and so the rate of the citric acid cycle is decreased.

A second control point of the citric acid cycle occurs at step 3. Both ADP and NAD^+ bind at the enzyme isocitrate dehydrogenase and allosterically enhance the activity of the enzyme for isocitrate. Thus, as the ATP concentration decreases and the ADP concentration increases, the citric acid cycle is stimulated. In addition, NADH acts as a competitive inhibitor of the enzyme by displacing the required NAD^+. Thus, NADH generated in the citric acid cycle itself inhibits the cycle. If ATP is required by the cell, oxidation of NADH occurs in the electron transport chain. The citric acid cycle then occurs because the concentration of the competitive inhibitor is reduced.

Control of the citric acid cycle also occurs at step 4, which is inhibited by NADH. The product of step 4, succinyl CoA, is a competitive inhibitor of the reaction. However, its concentration is decreased by the subsequent reactions.

In summary, the citric acid cycle is controlled through a number of complementary chemical reactions. The introduction of the acetyl unit into the cycle, as well as its rate of conversion, are decreased when the ATP level of the cell is high.

25.9 The Mitochondria

Now let's consider how the metabolic reactions of stages 2 and 3 (Section 25.1) occur, because this affects their interrelationships. Glycolysis and the pentose phosphate pathway occur in the cytoplasm. The citric acid cycle, oxidative phosphoryla-

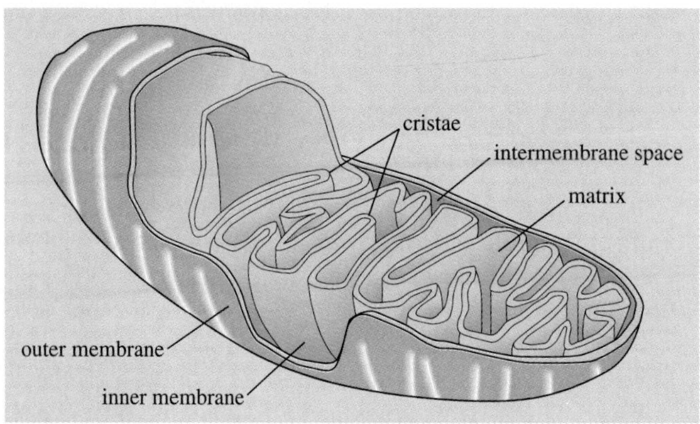

FIGURE 25.10 Structure of the Mitochondrion

tion, and oxidation of fatty acids occur in the mitochondria. Finally, gluconeogenesis and the urea cycle (Chapter 28) depend on both the cytoplasm and the mitochondria. Thus, the interrelationships among the various metabolic reactions depend on the flow of compounds across the mitochondrial membranes.

The mitochondria in the cell are called the power plants because they obtain energy from oxidation and produce ATP. A mitochondrion consists of an outer membrane with a convoluted inner membrane (Figure 25.10). The inner membrane creates a structure called a **crista.** The space between the outer membrane and the inner membrane is the **intermembrane space.** The space contained within the crista is the **matrix.**

The reactions of the electron transport chain occur on the inner membrane. The citric acid cycle occurs in the matrix, where the necessary enzymes are located. The reduced coenzymes NADH and $FADH_2$ from the citric acid cycle in the matrix are oxidized in the electron transport chain on the inner membrane. Thus, mitochondria are factories for the conversion of substrates into stored energy.

Each of the membranes of a mitochondrion contains lipid and protein molecules. However, the membranes have different permeabilities. The outer membrane is permeable to most small molecules. The inner membrane is impermeable to nearly all ions and most molecules, but oxygen, carbon dioxide, and water can cross. Large molecules such as fatty acids and charged materials such as ADP and ATP must be actively transported across the inner membrane, and energy is required.

The flow of ADP and ATP across the inner mitochondrial membrane is controlled. As ADP enters the matrix via a carrier, ATP exits via a carrier. The controlled flow in both directions across a membrane is **facilitated exchange diffusion.** The carrier is ATP–ADP translocase, a protein that constitutes about 6% of the membrane.

How is the oxidation of NADH coupled to the phosphorylation of ADP? These two processes are coupled as a result of a difference in proton (or hydronium ion) concentration across the inner mitochondrial membrane. The protons generated in the electron transport chain are pumped from the matrix to the intermembrane space by carriers of the inner mitochondrial membrane. The proton concentration of the intermembrane space is higher; the pH is 1.4 units lower than that of the matrix. Therefore, not only are there "pumps" in the inner membrane, but also the inner membrane must be impermeable to proton flow in the opposite direction. The difference in proton concentration causes a voltage difference of 0.22 V between the matrix and the intermembrane space. It is the energy difference corresponding

FIGURE 25.11
Structure of NADPH

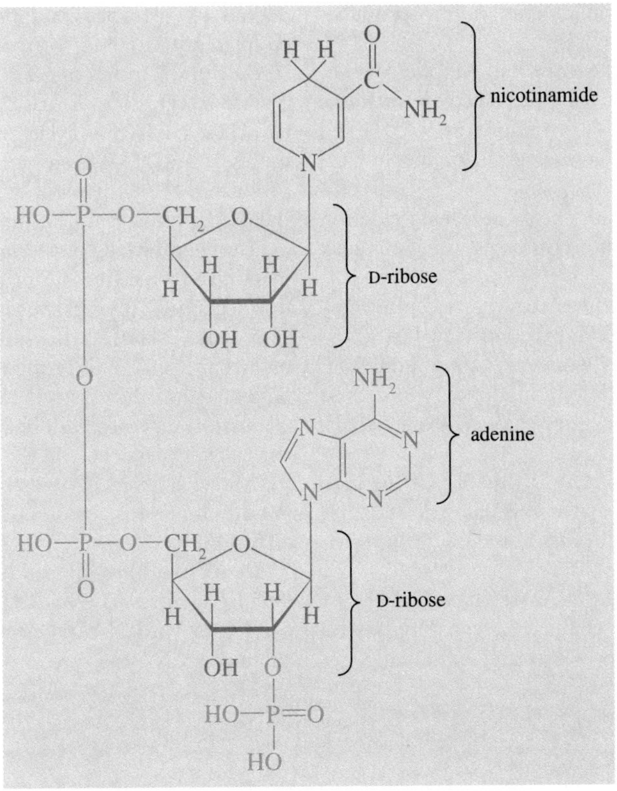

to this voltage difference that drives the synthesis of ATP in an enzyme-catalyzed reaction.

Biosynthesis and NADPH

In biosynthesis, energy provided by ATP converts low-energy reactants into high-energy products. Since the products are more reduced than the reactants, a reducing agent is required. Although NADH and $FADH_2$ are available in cells from oxidative degradation, these compounds are used primarily for the generation of ATP in the electron transport chain.

Nicotinamide adenine dinucleotide phosphate (NADPH) is the reducing agent of reductive biosynthesis. It is similar to NADH but one of the hydroxyl groups of ribose is esterified with phosphate (Figure 25.11). The extra phosphate group is discerned by enzymes, and NADPH is the source of reducing power for reductive biosynthetic reactions. NADPH is generated as glucose 6-phosphate is oxidized to ribose 5-phosphate in the pentose phosphate pathway (Section 26.8). The activity of this process is low in muscle tissue but is very high in adipose (fat storage) tissue, where NADPH is used to synthesize fatty acids.

Summary

Free energy change, $\Delta G°$, represents the energy from a chemical reaction that can be used to do work. An **endergonic** reaction occurs with an increase in free energy, whereas an **exergonic** reaction occurs with a decrease in free energy. Free energy changes are additive and may be used to calculate the $\Delta G°$ of a chemical reaction.

The oxidation state of carbon determines the energy content of biological molecules. Photosynthesis produces reduced, or energy-rich, compounds. Metabolism produces oxidized, or energy-poor, compounds.

ATP is a widely distributed carrier and supplier of biochemical energy that is produced by the **electron transport chain** in reactions coupled with the oxidation of NADH and $FADH_2$. ATP is the source of free energy for mechanical work, active transport, and biosynthesis.

Coupled reactions combine an energy-releasing reaction and an energy-consuming reaction. Both reactions join as a single reaction at an active site.

Substrate-level phosphorylation occurs between a phosphate donor and an acceptor. The reaction transfers phosphate from a high-energy to a low-energy compound. ATP is an intermediate energy carrier. Compounds whose free energy of hydrolysis is more negative than ATP's are high-energy compounds.

Oxidation of biochemicals occurs with NAD^+ to give NADH and H^+ or with FAD to give $FADH_2$. The reoxidation of the reduced coenzymes occurs in the electron transport chain.

Cellular respiration oxidizes NADH and $FADH_2$ in steps of the **electron transport chain.** Electrons are passed among the components of the chain and eventually accepted by oxygen. **Oxidative phosphorylation** occurs via coupled reactions to form 3 moles of ATP from 1 mole of NADH and 2 moles of ATP from 1 mole of $FADH_2$.

An acetyl group is oxidized to carbon dioxide in the **citric acid cycle (Krebs cycle** or **tricarboxylic acid cycle**). Three moles of NADH, 1 mole of $FADH_2$, and 1 mole of GTP are produced. The equivalent of 12 moles of ATP is produced in the oxidation of 1 mole of an acetyl group.

The membranes of the **mitochondria** have selective permeability to control the production of energy from oxidation. The inner membrane creates a structure called a **crista.** The space between the outer membrane and the inner membrane is the **intermembrane space.** The space contained within the crista is the **matrix.** The reactions of the electron transport chain occur on the inner membrane. The citric acid cycle occurs in the matrix. As ADP enters the matrix via a carrier, ATP exits via a carrier. The controlled flow in both directions across a membrane is **facilitated exchange diffusion.**

Reductive biosynthesis requires ATP as well as the reducing agent NADPH. The extra phosphate of NADPH compared with NADH distinguishes the two as reducing agents.

Exercises

Free Energy and Oxidation

25.1 Why is carbon dioxide an energy-poor molecule?

25.2 Does methylamine or methanamide have carbon in the higher oxidation state?

25.3 What changes in the oxidation state of carbon occur in photosynthesis?

25.4 What changes in the oxidation state of carbon occur in the metabolism of glucose?

25.5 Is the conversion of ethanol into acetaldehyde an oxidation or a reduction reaction?

25.6 Is the conversion of acetic acid into acetaldehyde an oxidation or a reduction reaction?

ATP

25.7 What structural units make up ATP?

25.8 Why is ATP called the energy currency in biological systems?

25.9 What is the charge on ADP at physiological pH?

25.10 What is the charge on AMP at physiological pH?

25.11 Using symbolic formulas, write an equation for the hydrolysis of ATP to ADP.

25.12 Using symbolic formulas, write an equation for the hydrolysis of ATP to AMP.

Coupled Reactions

25.13 Can two endergonic reactions be coupled?

25.14 Explain how the hydrolysis of ATP can be part of a coupled reaction.

25.15 What convention is used in equations to represent common biochemical conversions such as ATP $\rightarrow$ ADP in coupled reactions?

25.16 Explain why a series of endergonic reactions can occur when a strongly exergonic reaction occurs at the end of a series.

Substrate Level Phosphorylation

25.17 Does PEP have a higher or lower free energy of hydrolysis than ATP?

25.18 Does glucose 6-phosphate have a higher or lower free energy of hydrolysis than ATP?

25.19 Using the data in Table 25.1, determine if phosphoenol pyruvate should phosphorylate ADP.

25.20 Using the data in Table 25.1, determine if L-glycerol 3-phosphate should phosphorylate ADP.

25.21 Determine the ΔG° for the isomerization of glucose 6-phosphate into glucose 1-phosphate.

25.22 Determine the ΔG° for the isomerization of glucose 6-phosphate into fructose 6-phosphate.

Cellular Respiration

25.23 What is the first step in the electron transport chain starting with NADH?

25.24 What is the first step in the electron transport chain starting with $FADH_2$?

25.25 How many moles of ATP are produced in coupled reactions of NADH in the electron transport chain?

25.26 How many moles of ATP are produced in coupled reactions of $FADH_2$ in the electron transport chain?

25.27 What substances cause the electron transport chain to produce ATP?

25.28 Why does $FADH_2$ yield less ATP than NADH does in the electron transport chain?

25.29 Are the protons from NADH passed along the electron transport chain?

25.30 Are the protons from $FADH_2$ passed along the electron transport chain?

25.31 How does $FADH_2$ enter the electron transport chain?

25.32 How many steps in the electron transport chain produce ATP?

Citric Acid Cycle

25.33 What group is oxidized in the citric acid cycle?

25.34 What are the carbon products of the citric acid cycle?

25.35 How many steps in the citric acid cycle involve oxidation?

25.36 What reduced forms of coenzymes are formed in the citric acid cycle?

25.37 How many moles of ATP are produced from 1 mole of acetyl group in the citric acid cycle?

25.38 Why are the citric acid cycle and the electron transport chain interrelated processes?

25.39 Explain how ATP controls the rate of conversion of acetyl groups into CO_2 in the citric acid cycle.

25.40 Explain how NAD^+ controls the rate of conversion of acetyl groups into CO_2 in the citric acid cycle.

25.41 Malonic acid is a competitive inhibitor in the citric acid cycle. What enzyme does it inhibit?

25.42 Why is succinyl coenzyme A a competitive inhibitor of the citric acid cycle?

Mitochondria

25.43 The number of mitochondria in a cell varies according to the function of the cell. Why do liver cells have a large number of mitochondria?

25.44 The mitochondria have transport carriers for substances such as citric acid and α-ketoglutaric acid. Explain why.

25.45 How is the flow of ADP and ATP across the mitochondrial membrane controlled?

25.46 Across what cellular membrane are the protons of the respiration chain pumped?

Cows can convert the carbohydrate cellulose which we cannot digest into milk containing lactose that we can digest. (The Stock Market)

26

Metabolism of Carbohydrates

26.1 Digestion of Carbohydrates
26.2 Metabolic Pathways of Glucose
26.3 Glycolysis
26.4 Metabolism of Pyruvic Acid
26.5 The Yield of ATP from Glucose
26.6 Gluconeogenesis
26.7 The Pentose Phosphate Pathway
26.8 Glycogenolysis and Glycogenesis
26.9 Regulation of Carbohydrate Metabolism

Essay
- *Fermentation—An Anaerobic Process*

26.1 Digestion of Carbohydrates

Carbohydrates provide 50% of the daily energy needs of most Americans. The digestion of starch, the principal dietary carbohydrate, starts in the mouth, where the enzyme amylase contained in saliva catalyzes hydrolysis of some of the glycosidic bonds of starch (Figure 26.1). However, the extent of hydrolysis is limited because contact between the saliva and the food is inefficient due to the large size of the chewed food particles. A mixture of lower molecular weight carbohydrates called dextrins and some maltose is produced.

Hydrolysis of starch and dextrins continues in the esophagus but is dramatically slowed in the acid fluid of the stomach, where enzymatic activity of amylase is low. The enzyme is a protein and it is denatured by the HCl in the stomach. There are no other enzymes in the stomach that can catalyze the hydrolysis of glycosidic linkages, but the acid of the stomach does to a limited extent catalyze some hydrolysis of carbohydrates.

The carbohydrates in the **chyme,** the partially digested contents of the stomach, then enter the small intestine. The chyme is acidic but its pH rises to the 7–8 range as the result of reaction with bicarbonate ion as the chyme mixes with the secretions of the pancreas. The hydrolysis of glycosidic bonds starts again because the

FIGURE 26.1 Digestion of Carbohydrates

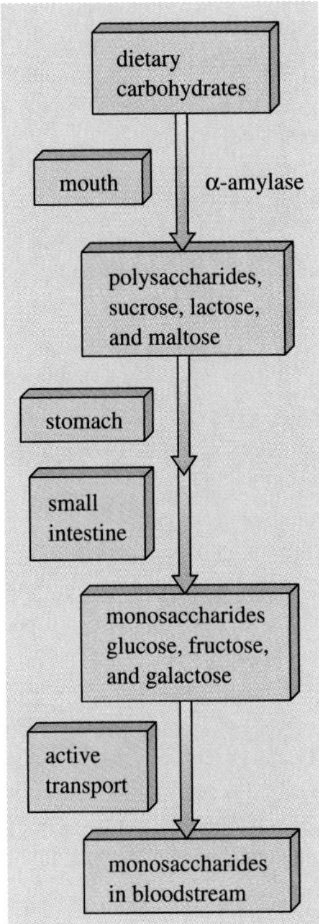

pancreas also secretes amylase, which is an effective catalyst in near-neutral solutions. Maltose and some glucose are produced.

Further hydrolysis of maltose is catalyzed by the enzyme maltase contained in the epithelial cells of the small intestine. Hydrolysis of the dietary disaccharides lactose and sucrose also occurs. These reactions are catalyzed by the enzymes lactase and sucrase, which also are located in the epithelial cells of the small intestine.

Glucose is the major monosaccharide produced in digestion, but fructose and galactose are also formed. All these monosaccharides are absorbed by the epithelial cells of the small intestine by an active transport process and are then secreted into the capillaries of the small intestine. The bloodstream then distributes the monosaccharides to all cells of the body.

Because glucose is the major monosaccharide produced by the hydrolysis of carbohydrates, it is appropriate to consider what happens to glucose in the body. Fructose and galactose are converted into glucose or intermediates that are derived from glucose.

26.2 Metabolic Pathways of Glucose

The glucose level of the blood and the amount of glycogen in the liver are regulated by interrelated metabolic pathways. These carefully balanced sequences of reactions produce energy, store chemical energy, or produce intermediates required for biosynthesis (Figure 26.2).

FIGURE 26.2 Metabolic Pathways of Glucose

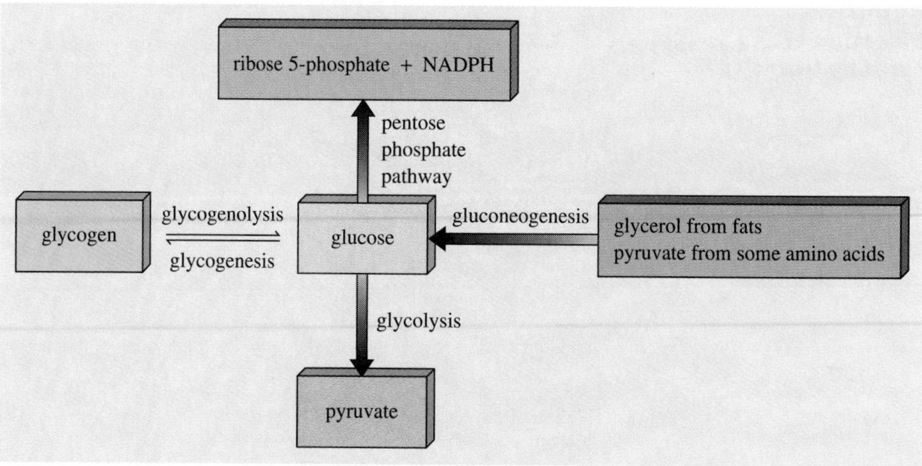

Gluconeogenesis and Glycolysis

Gluconeogenesis is the synthesis of glucose from pyruvate provided by noncarbohydrate compounds such as glycerol, lactate, and some amino acids. This process occurs when glucose is not available from other sources.

In glycolysis, also known as the **Embden–Meyerhof pathway,** 1 mole of glucose is converted into 2 moles of pyruvic acid, 2 moles of NADH, and 2 moles of ATP. The fates of pyruvic acid and NADH depend on the availability of oxygen in the cell. With sufficient oxygen, NADH is oxidized in the electron transport chain in the mitochondria, and pyruvic acid is converted into acetyl CoA, which is then oxidized in the citric acid cycle. If the supply of oxygen is insufficient, pyruvic acid is reduced to lactic acid by NADH. The lactic acid is converted in the liver back into pyruvic acid and then into glucose via gluconeogenesis. The majority of this chapter is devoted to an analysis of glycolysis and gluconeogenesis.

Pentose Phosphate Pathway

The pentose phosphate pathway converts glucose as its 6-phosphate into NADPH, used in reductive biosynthesis, and ribose 5-phosphate, used in the synthesis of RNA and DNA. Since many cells require much more NADPH than ribose 5-phosphate, the carbohydrate is converted into intermediates of the glycolysis pathway. The fates of NADPH and ribose 5-phosphate are discussed in Section 26.7.

Glycogenesis and Glycogenolysis

Glycogen, a branched polymer of glucose, is an easily mobilized store of chemical energy. When the glucose level in the blood drops, glycogen in the liver is hydrolyzed in a series of steps called *glycogenolysis.* If there is too much glucose in the blood, *glycogenesis* occurs to store the glucose as glycogen. These two processes regulate the blood glucose level. The degradation reactions and synthesis of glycogen, which occur by different pathways, are considered in Section 26.8.

26.3 Glycolysis

Glycolysis is a series of reactions that convert glucose to pyruvate with the simultaneous formation of ATP. In aerobic organisms, glycolysis and the subsequent reactions of the citric acid cycle and the electron transport chain store some of the energy derived from glucose. Pyruvate is oxidized to form carbon dioxide and water.

$$\underset{\text{glucose}}{C_6H_{12}O_6} \xrightarrow{\text{glycolysis}} \underset{\text{pyruvate}}{2CH_3-\overset{\overset{\displaystyle O}{\|}}{C}-CO_2^-} \xrightarrow{O_2} CO_2 + H_2O$$

The Organization of Glycolysis

To understand what is going on in glycolysis, it will be necessary to focus on changes in molecules that have many functional groups. You can understand the reactions if you relate them to the reactions of simpler molecules discussed in the organic chemistry chapters of this text.

All intermediates in glycolysis have either six or three carbon atoms, and each is phosphorylated. The six-carbon molecules are either glucose or fructose derivatives; the three-carbon molecules are derivatives of dihydroxyacetone, glyceraldehyde, glycerate, or pyruvate. It is important to recognize that two three-carbon units are derived from one six-carbon unit. Thus, any ATP or reduced enzymes that are formed with a three-carbon unit must be doubled to account for the total products derived from the original six-carbon unit.

The 10 reactions of glycolysis are summarized in Figure 26.3. However, it is useful to group several reactions together to focus on the major changes that occur. Four groups of reactions are considered.

Formation of Fructose 1,6-Diphosphate

The first three reactions convert glucose into fructose 1,6-diphosphate. Energy is required for the sum of the three reactions.

Step 1

HOCH$_2$ / H, H, OH, H, HO, OH, H, OH

glucose

+ ATP $\xrightarrow{\text{hexokinase}}$

$^{2-}O_3POCH_2$ / H, H, OH, H, HO, OH, H, OH

glucose 6-phosphate

+ ADP

Step 2

$^{2-}O_3POCH_2$ / H, H, OH, H, HO, OH, H, OH

glucose 6-phosphate

$\underset{\text{isomerase}}{\overset{\text{phosphoglucose}}{\rightleftharpoons}}$

$^{2-}O_3POCH_2$ O CH$_2$OH / H, H, HO, OH, OH, H

fructose 6-phosphate

FIGURE 26.3
Glycolysis Reactions

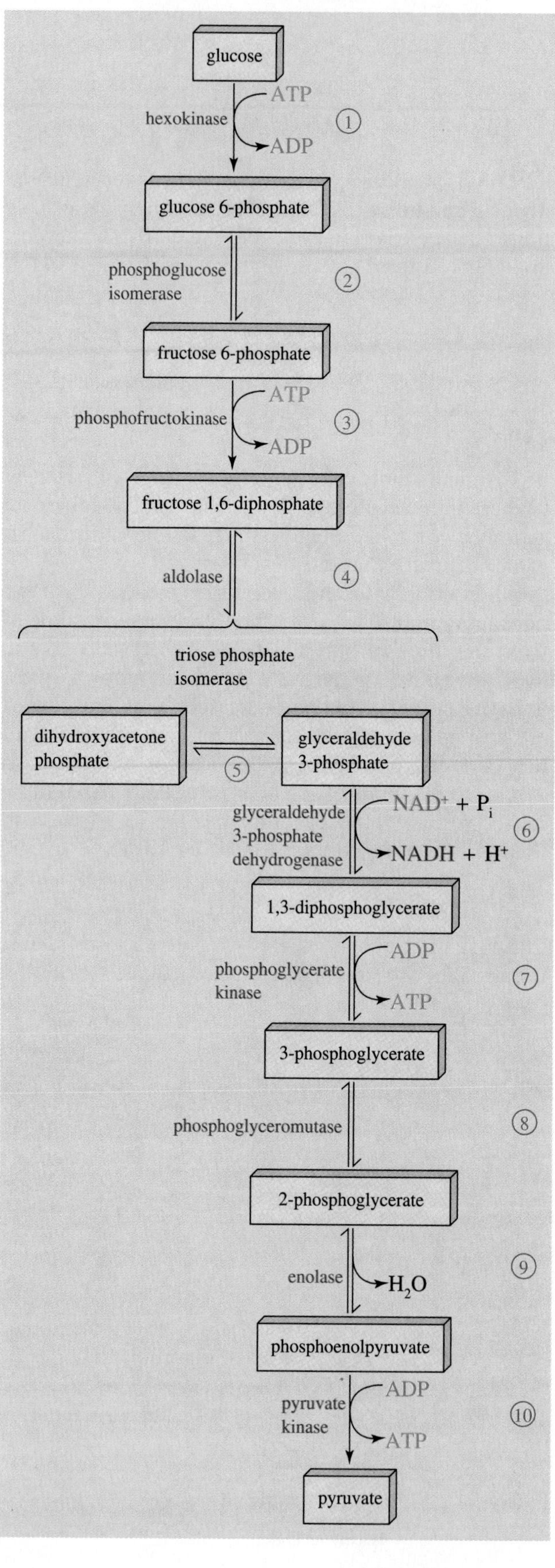

Step 3

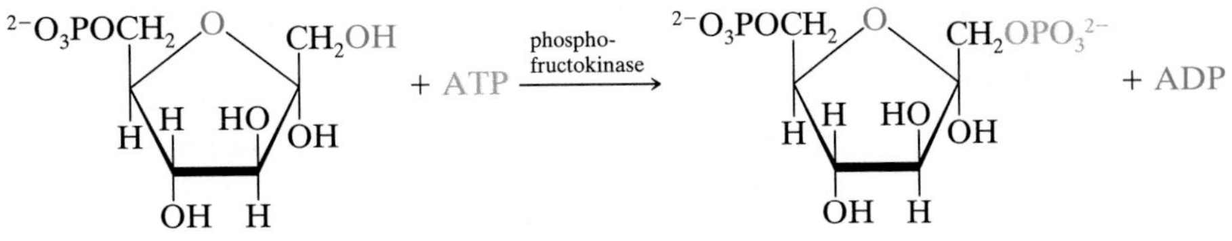

fructose 6-phosphate + ATP —(phosphofructokinase)→ fructose 1,6-diphosphate + ADP

In step 1 the glucose in a cell is phosphorylated by ATP to produce glucose 6-phosphate. The isomerization of glucose 6-phosphate to fructose 6-phosphate in step 2 occurs via the enediol intermediate discussed in Section 21.7. The pyranose ring of glucose 6-phosphate exists in equilibrium with an open-chain form, which isomerizes to the open-chain form of fructose 6-phosphate followed by ring closure to give a furanose. A second phosphorylation by ATP in step 3 converts fructose 6-phosphate into fructose 1,6-diphosphate.

Formation of Glyceraldehyde 3-Phosphate

The next sequence in the glycolysis reactions consists of two reactions that convert the six-carbon glucose into two three-carbon glyceraldehyde 3-phosphate units.

Step 4

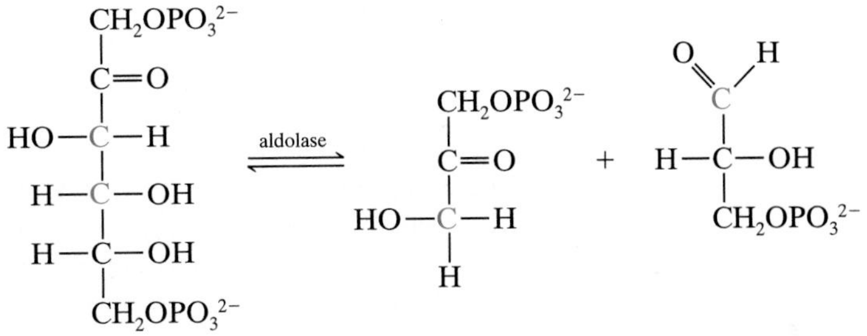

fructose 1,6-diphosphate ⇌(aldolase) dihydroxyacetone phosphate + glyceraldehyde 3-phosphate

Step 5

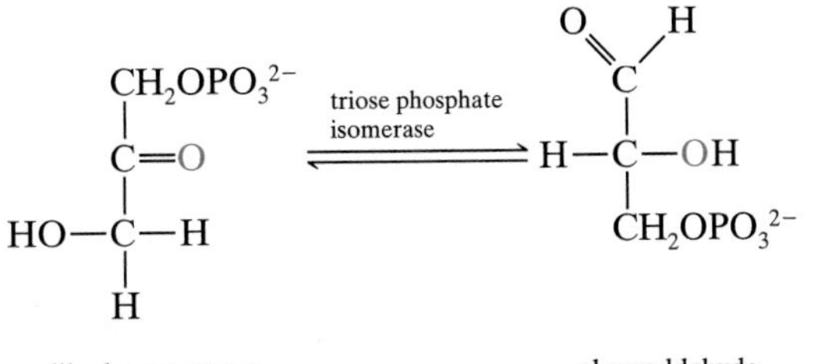

dihydroxyacetone phosphate ⇌(triose phosphate isomerase) glyceraldehyde 3-phosphate

Step 4 is written using the open form of fructose 1,6-diphosphate to show the reverse aldol reaction. The two products are tautomers and are interconverted via an enediol intermediate in step 5. However, only glyceraldehyde 3-phosphate reacts in the subsequent steps of glycolysis. Although at equilibrium 96% of the mixture of two isomers exists as dihydroxyacetone phosphate, the isomerization reactions occur because the equilibrium position is shifted by the removal of glyceraldehyde phosphate. Thus, we can consider the net result of steps 4 and 5 to be the formation of two units of glyceraldehyde 3-phosphate. After this point the products of all reactions of three-carbon units must be doubled to account for all the carbon atoms of the original glucose.

Energy Conservation Steps

Up to step 5, 2 moles of ATP has been invested per mole of glucose, and none of the free energy of glucose has been stored. The next two steps extract some of the energy of glyceraldehyde 3-phosphate.

Step 6

$$\underset{\text{glyceraldehyde 3-phosphate}}{\mathrm{HC(=O)\!-\!CH(OH)\!-\!CH_2OPO_3^{2-}}} + NAD^+ + P_i \xrightleftharpoons{\text{glyceraldehyde 3-phosphate dehydrogenase}} \underset{\text{1,3-diphosphoglycerate (1,3-DPG)}}{\mathrm{{}^{2-}O_3PO\!-\!C(=O)\!-\!CH(OH)\!-\!CH_2OPO_3^{2-}}} + NADH + H^+$$

Step 7

$$\underset{\text{1,3-diphosphoglycerate}}{\mathrm{{}^{2-}O_3PO\!-\!C(=O)\!-\!CH(OH)\!-\!CH_2OPO_3^{2-}}} + ADP \xrightleftharpoons{\text{phosphoglycerate kinase}} \underset{\text{3-phosphoglycerate}}{\mathrm{{}^{-}O\!-\!C(=O)\!-\!CH(OH)\!-\!CH_2OPO_3^{2-}}} + ATP$$

Step 6 is the oxidation of the aldehyde group of glyceraldehyde 3-phosphate into a phosphorylated acid. The product, 1,3-diphosphoglycerate, has a high phosphate transfer potential (Section 25.5). In step 7 a phosphate is transferred to give ATP and 3-phosphoglycerate. As a consequence of these two reactions, one NADH and one ATP are formed, both of which contain stored energy. Energy in excess of the original invested in steps 1 and 3 has now been obtained, because NADH generates ATP in the electron transport chain.

Formation of Pyruvate and ATP

The final three steps of glycolysis convert 3-phosphoglycerate into pyruvate and ATP.

Step 8

$$\underset{\text{3-phosphoglycerate}}{\mathrm{{}^{-}O\!-\!C(=O)\!-\!CH(OH)\!-\!CH_2OPO_3^{2-}}} \xrightleftharpoons{\text{phosphoglyceromutase}} \underset{\text{2-phosphoglycerate}}{\mathrm{{}^{-}O\!-\!C(=O)\!-\!CH(OPO_3^{2-})\!-\!CH_2OH}}$$

Step 9

$$\underset{\text{2-phosphoglycerate}}{\mathrm{{}^{-}O\!-\!C(=O)\!-\!CH(OPO_3^{2-})\!-\!CH_2OH}} \xrightleftharpoons{\text{enolase}} \underset{\text{phosphoenolpyruvate}}{\mathrm{{}^{-}O\!-\!C(=O)\!-\!C(OPO_3^{2-})\!=\!CH_2}} + H_2O$$

TABLE 26.1 Reactions That Use and Produce ATP in Formation of Pyruvate

Reaction	ATP Produced or Consumed per Glucose
glucose $\longrightarrow$ glucose 6-phosphate	−1
fructose 6-phosphate $\longrightarrow$ fructose 1,6-diphosphate	−1
2 1,3-diphosphoglycerate $\longrightarrow$ 2 3-phosphoglycerate	+2
2 phosphoenolpyruvate $\longrightarrow$ 2 pyruvate	+2
	Net +2

Step 10

$$\underset{\text{phosphoenolpyruvate}}{CH_2{=}C(OPO_3^{2-})COO^-} + ADP \xrightarrow{\text{pyruvate kinase}} \underset{\text{pyruvate}}{CH_3C(=O)COO^-} + ATP$$

Step 8 is an isomerization reaction that transfers a phosphate group from one oxygen atom to another in the molecule. Dehydration of an alcohol occurs in step 9. The unsaturated compound produced is a phosphate of an enol and has a high phosphate transfer potential. Transfer of a phosphate to ADP in step 10 does not yield an enol but rather the more stable keto compound, pyruvate.

Energy Yield of Glycolysis

The net reaction for the conversion of glucose into pyruvate is

$$\text{glucose} + 2\,P_i + 2\,ADP + 2\,NAD^+ \longrightarrow 2\ \text{pyruvate} + 2\,ATP + 2\,NADH + 2\,H^+ + 2\,H_2O$$

The four steps of glycolysis in which ATP is generated or consumed are summarized in Table 26.1. Recall that 2 moles of three-carbon units are involved in each step after the formation of 1 mole of fructose 1,6-diphosphate.

26.4 Metabolism of Pyruvic Acid

In Chapter 20 it was stated that pyruvic acid, glucose, and acetyl CoA are three key compounds in metabolic reactions. The centrality of pyruvic acid is shown in Figure 26.4. The interconversion of pyruvic acid and alanine by transamination links the metabolism of carbohydrates and amino acids. Pyruvic acid is also linked to the citric acid cycle. It is converted into oxaloacetic acid inside the mitochondria and by this means replenishes the citric acid cycle intermediates. In addition, oxaloacetic acid is converted into glucose 6-phosphate in gluconeogenesis. Pyruvic acid can also be converted into acetyl CoA if additional ATP is needed by the cell.

In this section we consider the fate of pyruvic acid under aerobic and anaerobic conditions. Lactic acid and pyruvic acid are interrelated by an oxidation–reduction

FIGURE 26.4 Metabolic Pathways of Pyruvic Acid

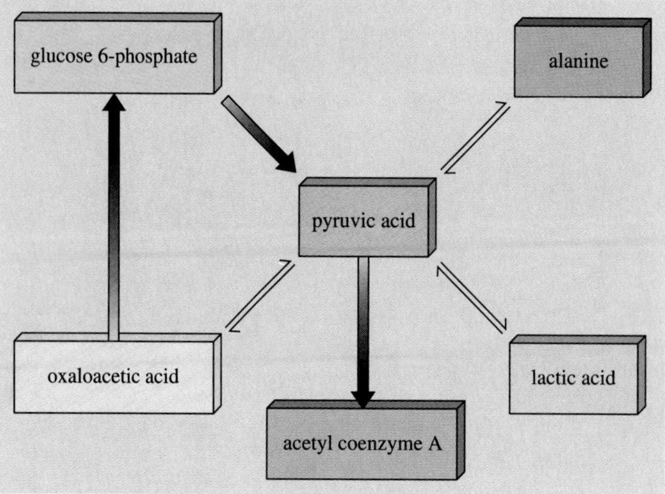

reaction. Under aerobic conditions, pyruvic acid is converted to acetyl CoA, which is further oxidized in the citric acid cycle. Under anaerobic conditions, such as in strenuous exercise, pyruvic acid is reduced to lactic acid, which then is reoxidized to pyruvic acid in the liver.

Fate of Pyruvic Acid Under Aerobic Conditions

Glycolysis converts glucose into pyruvate and stores some of the released energy in ATP. To further release the energy originally stored in glucose, pyruvic acid must be further oxidized. In addition, it is necessary that the NADH produced in glycolysis be processed to release its stored energy. In fact, if NADH were not processed to form NAD^+, the glycolysis process itself would slow down and eventually stop.

The oxidation of both pyruvic acid and NADH can occur under aerobic conditions. Pyruvate reacts with CoA to produce acetyl CoA. This step, which is the link between glycolysis and the citric acid cycle, occurs in the matrix of the mitochondria. Two moles of NADH are produced for each original mole of glucose.

$$CH_3-\overset{\overset{O}{\|}}{C}-CO_2H + CoA \xrightarrow[NAD^+ \;\to\; NADH + H^+]{} CH_3-\overset{\overset{O}{\|}}{C}-CoA + CO_2$$

The NADH formed in this reaction as well as that from the glycolysis reaction is oxidized in the electron transport chain when there is sufficient oxygen present—that is, under aerobic conditions. As previously stated, the NAD^+ formed is required for glycolysis, but as we now see, it is also needed for the formation of acetyl CoA for the citric acid cycle.

Fate of Pyruvic Acid Under Anaerobic Conditions

If you exercise too strenuously, your lungs and circulatory system cannot deliver enough oxygen to operate the electron transport chain, and an anaerobic condition is created. Because NADH cannot be efficiently oxidized, the supply of NAD^+

Fermentation—An Anaerobic Process

We function best under aerobic conditions, but we also can switch to anaerobic pathways for a period of time. Simpler organisms such as yeasts, called **faculative anaerobes,** are more adaptive and can function efficiently either aerobically or anaerobically. Some bacteria, such as those responsible for botulism or gangrene, function only anaerobically and in fact are poisoned by oxygen. Such organisms are **strict anaerobes.**

To regenerate NAD^+ under anaerobic conditions, yeast has a two-step metabolic pathway available that is substantially different from the Cori cycle in humans. The process that forms ethyl alcohol is called *fermentation.* In the first step pyruvic acid is decarboxylated in an irreversible process to give acetaldehyde. Both Mg^{2+} as a cofactor and thiamine pyrophosphate as a coenzyme are required. The CO_2 formed is the reason why yeast makes baked products such as bread rise.

$$\underset{\text{pyruvic acid}}{CH_3-\overset{\overset{\displaystyle O}{\|}}{C}-CO_2H} \xrightarrow{\text{decarboxylase}} \underset{\text{acetaldehyde}}{CH_3-\overset{O}{\overset{\|}{C}}-H} + CO_2$$

The acetaldehyde is reduced by NADH to give ethanol. This process regenerates the NAD^+ required for glycolysis. Zinc ion is a required cofactor for the reaction.

$$CH_3-\overset{\overset{\displaystyle O}{\|}}{C}-H \xrightarrow[NADH + H^+ \quad NAD^+]{\text{dehydrogenase}} CH_3-\underset{\underset{\displaystyle H}{|}}{\overset{\overset{\displaystyle OH}{|}}{C}}-H$$

When bread bakes, the ethanol formed evaporates. When fermentation of yeast occurs in an aqueous solution of grape juice, the ethanol formed is retained. Usually the CO_2 formed in the first step escapes and a "still" wine results. If the final stages of fermentation are carried out in a corked bottle, the CO_2 builds up, and a sparkling wine such as champagne results. The gas formed during fermentation is released when the cork is "popped." The net reaction for fermentation is

$$C_6H_{12}O_6 + 2\,ADP + 2\,P_i \longrightarrow 2\,CH_3CH_2OH + 2\,CO_2 + 2\,ATP$$

Carbon dioxide plays an important role during fermentation. It provides an atmosphere in the fermentation vat that excludes oxygen. Without this protective cover, aerobic processes would occur. One of these converts ethanol to acetic acid, and the product is vinegar, not ethyl alcohol.

When fermentation by yeast cells occurs initially the alcohol formed is dilute, and the cells can continue their anaerobic processes. However, in a limited environment such as a wine vat, the alcohol concentration continues to increase. In the 12–13% range the alcohol kills the yeast cells. This is a case in which a metabolic product kills the organism. In more complex life forms such as humans, the metabolic products usually do not accumulate to the extent that death occurs. For example, under conditions of extreme exercise, the resultant muscle fatigue makes us stop the exercise.

Certain bacteria that are strict anaerobes cause gangrene when they are located in deep wounds away from oxygen. They anaerobically produce products such as butyric acid, acetic acid, butanol, ethyl alcohol, and acetone. The carbon dioxide formed in the fermentation process is retained in the wound and prevents oxygen from entering the area. Eventually the bacteria produce substances that when retained locally cause tissue death.

Treatment of gangrene is difficult and usually requires removal of the dying tissue. In some cases, placing the patient in a hyperbaric oxygen chamber helps. Under higher pressure, oxygen is forced into the tissues and kills the bacteria, which cannot function aerobically.

Fermentation processes are central to the manufacture of wine. (The Image Bank)

needed as an oxidizing agent in glycolysis, in the citric acid cycle, and for the conversion of pyruvic acid into acetyl CoA cannot be replenished. In order for cellular activities to continue, the muscle cells must derive energy by an anaerobic pathway.

Under anaerobic conditions, pyruvic acid is reduced to lactic acid by NADH. This process generates NAD^+, which allows glycolysis to continue producing two ATP molecules per molecule of glucose.

$$\underset{\text{pyruvate}}{CH_3-\overset{\overset{O}{\|}}{C}-CO_2H} \xrightarrow[NADH\ +\ H^+\ \ NAD^+]{\text{lactate dehydrogenase}} \underset{\text{L-lactate}}{CH_3-\overset{OH}{\underset{H}{C}}-CO_2H}$$

Multiplying this reaction by 2 and adding it to the net reaction of glycolysis gives the net reaction for anaerobic glycolysis:

$$\text{glucose} + 2\ ADP + 2\ P_i \longrightarrow 2\ \text{lactate} + 2\ ATP + 2\ H_2O$$

Note that this reaction is not an oxidation or a reduction. The oxidation in glycolysis is balanced by the reduction of pyruvic acid.

The anaerobic process generates some ATP, but only a small fraction of the chemical energy of glucose has been released. There is also a penalty to the body under anaerobic conditions. If the muscle tissues cannot obtain enough oxygen, the lactic acid concentration increases in the muscles and blood, which causes fatigue. The accumulated lactic acid constitutes an **oxygen debt.** After stopping the exercise that caused the oxygen debt, we continue to breathe hard. The liver takes up the lactic acid from the blood and oxidizes it back to pyruvic acid, and the aerobic oxidation of pyruvic acid starts again.

Another metabolic pathway occurs under anaerobic conditions to oxidize the lactic acid to pyruvic acid so that it can be metabolized. Lactic acid diffuses out of the muscle into the blood and is carried to the liver, where the Cori cycle occurs. In the **Cori cycle,** lactic acid is oxidized to pyruvic acid, which is then converted into glucose via gluconeogenic processes. The net result of the Cori cycle is a shift of the metabolic burden from the muscles to the liver (Figure 26.5).

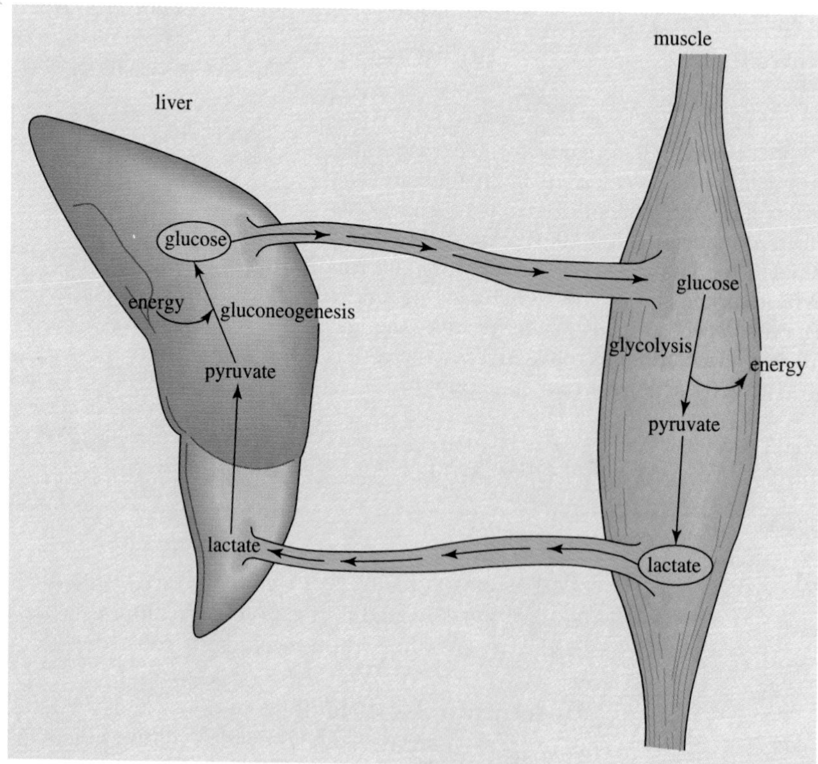

FIGURE 26.5
The Cori Cycle
Lactic acid formed in the muscle under anaerobic conditions is carried to the liver to re-form glucose. Since gluconeogenesis requires energy, the metabolic burden is shifted to the liver.

26.5 The Yield of ATP from Glucose

At this point we can see how the biological system can really "profit" from the oxidation of glucose under aerobic conditions. In total, 1 mole of glucose produces 2 moles of ATP, 4 moles of NADH, and 2 moles of acetyl CoA. Acetyl CoA then produces NADH, $FADH_2$, and ATP in the citric acid cycle. Oxidation of the NADH and $FADH_2$ forms ATP in the electron transport chain.

There is one point to be considered before calculating the energy derived from the metabolism of glucose. Glycolysis enzymes are in the cytoplasm of the cell, and 2 moles of NADH per mole of glucose (step 6) are released in the cytoplasm. However, the 2 moles of NADH per mole of glucose produced in converting pyruvic acid into acetyl CoA are released in the mitochondria.

Since the reactions of the citric acid cycle and those of the electron transport chain occur in the mitochondria, it is necessary to transport cytoplasmic NADH into them. However, mitochondria are impermeable to NADH and NAD^+. Thus, the electrons from NADH are shuttled across the mitochondrial membrane, but not by NADH itself. The two shuttles for electrons are called the *glycerol 3-phosphate shuttle* and the *malate–aspartate shuttle.*

In the glycerol 3-phosphate shuttle, the reduced carrier diffuses across the mitochondrial membrane and is oxidized by FAD (Figure 26.6). The dihydroxyacetone phosphate then diffuses out of the mitochondria into the cytoplasm. The net reaction is

$$\underset{\text{(cytoplasmic)}}{\text{NADH}} + H^+ + \underset{\text{(mitochondrial)}}{\text{FAD}} \longrightarrow \underset{\text{(cytoplasmic)}}{NAD^+} + \underset{\text{(mitochondrial)}}{FADH_2}$$

As a consequence, only 2 moles of ATP are available from the oxidation of $FADH_2$ rather than 3 moles of ATP that could be formed for each NADH if it could have entered the mitochondria. Thus, a price is paid for this transport, but without such a shuttle none of the energy stored in cytoplasmic NADH would be available.

In the malate–aspartate shuttle, electrons from NADH in the cytoplasm are transferred to malate, which crosses the inner mitochondrial membrane. This elec-

FIGURE 26.6 The Glycerol 3-Phosphate Shuttle

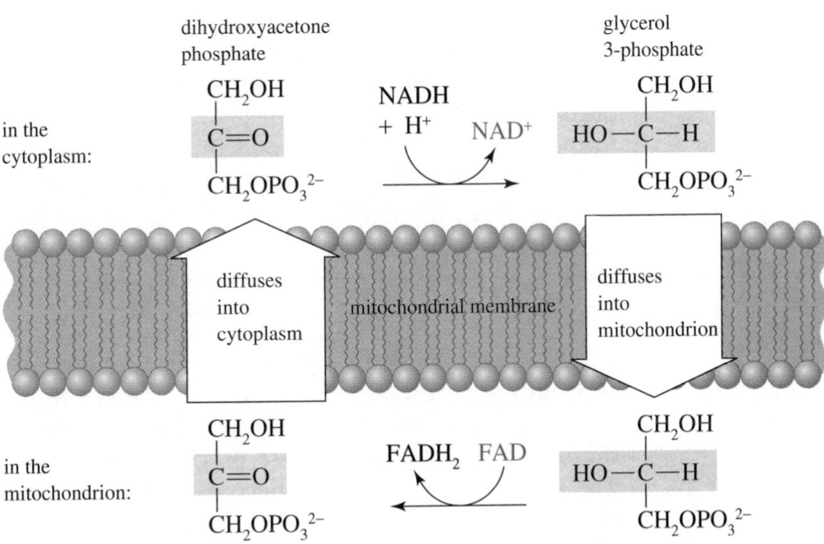

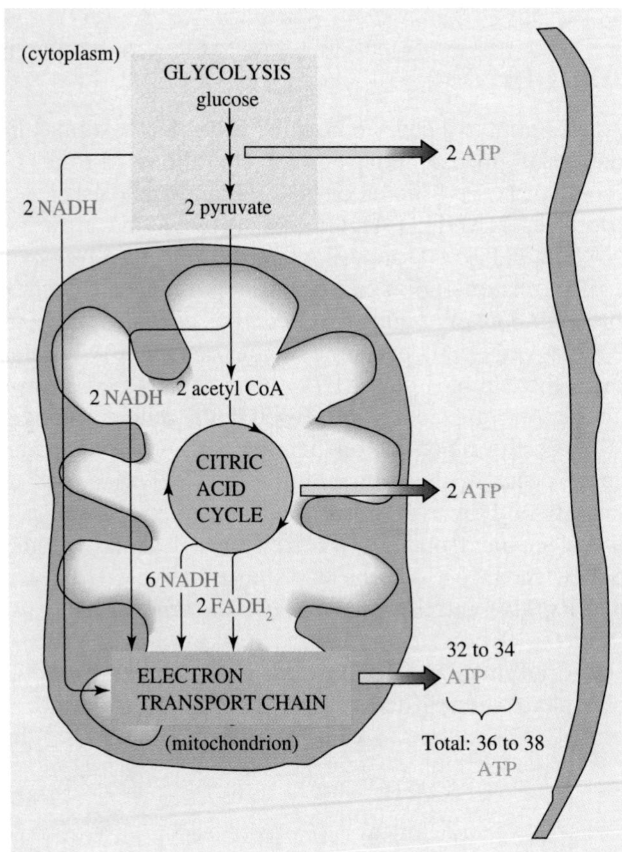

FIGURE 26.7 Formation of ATP from Metabolism of Glucose

tron carrier is then oxidized by NAD^+ within the mitochondrial matrix to form NADH. The net reaction of the shuttle is

$$\underset{\text{(cytoplasmic)}}{NADH} + \underset{\text{(mitochondrial)}}{NAD^+} \longrightarrow \underset{\text{(cytoplasmic)}}{NAD^+} + \underset{\text{(mitochondrial)}}{NADH}$$

Thus, the NADH produced in the cytoplasm yields 3 moles of ATP in the electron transport chain in the mitochondria.

Now, let us summarize the steps in the metabolism of 1 mole of glucose in terms of the ATP produced. Using the glycerol 3-phosphate shuttle, 36 moles of ATP are formed. (If the malate–aspartate shuttle were used, a total of 38 moles of ATP rather than 36 would be formed.) A summary is illustrated in Figure 26.7.

26.6 Gluconeogenesis

The liver has a limited supply of glycogen for export. Thus, the brain, which is dependent on glucose as a fuel, would be starved if food intake were restricted for even one day. In the absence of carbohydrates, the necessary glucose can be synthesized from noncarbohydrate precursors by a process called **gluconeogenesis.** The sources of material for glucose production are lactic acid, amino acids, and glycerol.

The pain evidenced by this runner in a Texas state competition is the result of depleted glycogen stores. (The Image Works)

The steps in gluconeogenesis are listed in Figure 26.8. At first glance, these reactions look like the reverse of the glycolysis reactions (Figure 26.3), but some are not. Steps 1, 3, and 10 of glycolysis are too exergonic to run in reverse. Alternative steps are required to get around this energy bottleneck. The other seven steps of glycolysis can run in reverse and are catalyzed by the same enzymes.

The reactions of glycolysis that cannot be reversed are

$$\text{glucose} + \text{ATP} \xrightarrow{\text{hexokinase}} \text{glucose 6-phosphate} + \text{ADP}$$

$$\text{fructose 6-phosphate} + \text{ATP} \xrightarrow{\text{phosphofructokinase}} \text{fructose 1,6-diphosphate} + \text{ADP}$$

$$\text{phosphoenolpyruvate} + \text{ADP} \xrightarrow{\text{pyruvate kinase}} \text{pyruvate} + \text{ATP}$$

These reactions—steps 1, 3, and 10 in glycolysis—must be bypassed by different reactions in gluconeogenesis. Steps 1, 8, and 10 in gluconeogenesis correspond to the "reverse" of steps 10, 3, and 1 of glycolysis, respectively.

Pyruvate is converted to phosphoenolpyruvate (steps 1 and 2 in Figure 26.8) by way of oxaloacetate, which is not part of the glycolysis pathway. Two reactions fueled by ATP and GTP overcome the unfavorable free energy difference between pyruvic acid and phosphoenolpyruvate.

$$\underset{\text{pyruvate}}{CO_2^- \text{—} C(=O) \text{—} CH_3} + CO_2 + \text{ATP} + H_2O \xrightleftharpoons{\text{pyruvate carboxylase}} \underset{\text{oxaloacetate}}{CO_2^- \text{—} C(=O) \text{—} CH_2 \text{—} CO_2^-} + \text{ADP} + P_i$$

$$\underset{\text{oxaloacetate}}{CO_2^- \text{—} C(=O) \text{—} CH_2 \text{—} CO_2^-} + \text{GTP} \xrightleftharpoons{\text{phosphoenolpyruvate carboxykinase}} \underset{\text{phosphoenolpyruvate}}{CO_2^- \text{—} C(\text{—}OPO_3^{2-}) \text{=} CH_2} + CO_2 + \text{GDP}$$

Conversion of fructose 1,6-diphosphate into fructose 6-phosphate (step 8 in Figure 26.8) is catalyzed by fructose 1,6-diphosphatase, a different enzyme from the one required for the reverse glycolysis reaction. Hydrolysis of glucose 6-phosphate into glucose (step 10 in Figure 26.8) is catalyzed by glucose 6-phosphatase, also a different enzyme from the one required for the reverse glycolysis reaction. The stoichiometry of the reaction for gluconeogenesis is

$$2\ \text{pyruvate} + 4\ \text{ATP} + 2\ \text{GTP} + 2\ \text{NADH} + 2\ H^+ + 2\ H_2O \longrightarrow \text{glucose} + 4\ \text{ADP} + 2\ \text{GDP} + 6\ P_i + 2\ \text{NAD}^+$$

The free energy change for gluconeogenesis is −9 kcal/mole. Since $\Delta G°$ for glycolysis is −20 kcal/mole, a direct reverse of the glycolysis pathway would require $\Delta G°$ of +20 kcal/mole. Thus, gluconeogenesis cannot proceed by the reverse of the steps of glycolysis reactions. To reconvert pyruvic acid to glucose in gluconeogenesis, six molecules of high-energy phosphates (4 ATP + 2 GTP) are required. The gluconeogenesis pathway then can occur only at the expense of other ATP-producing metabolic processes.

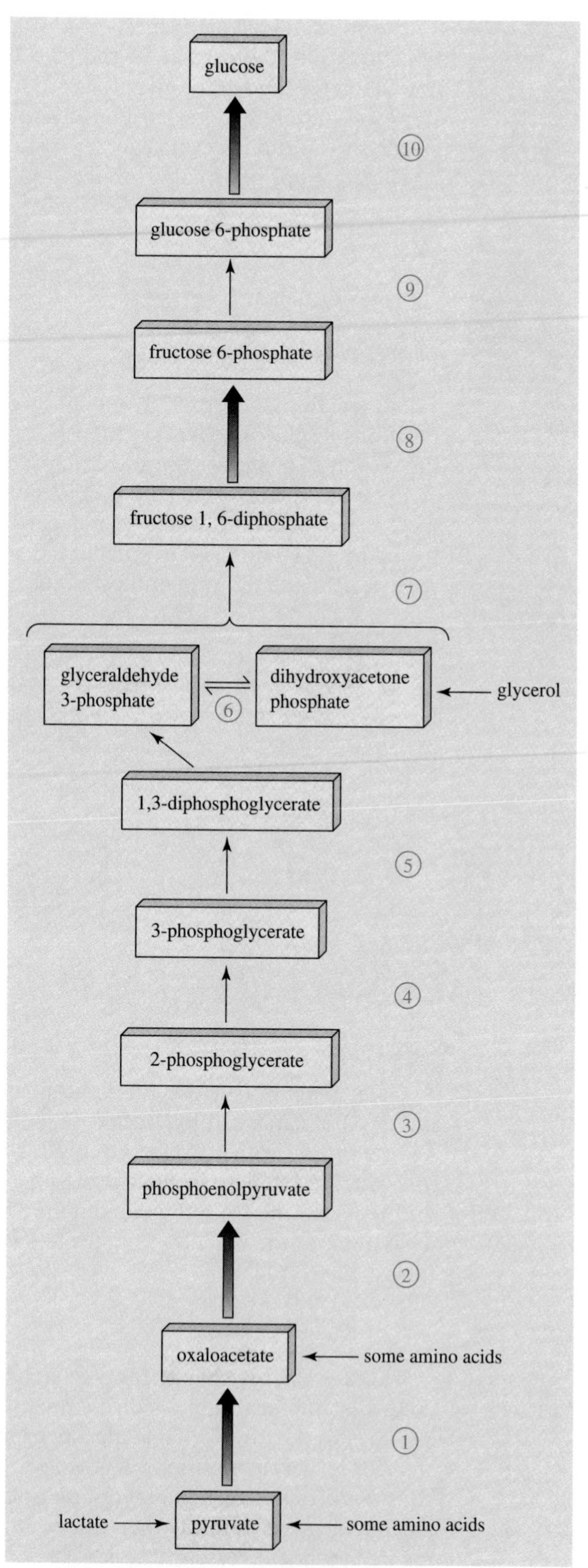

FIGURE 26.8 Gluconeogenesis Pathway

The distinctive reactions of gluconeogenesis are indicated by large arrows. The other reactions are the reverse of glycolysis. Entry points of substrates are indicated.

26.7 The Pentose Phosphate Pathway

In the pentose phosphate pathway, glucose 6-phosphate is oxidized to ribose 5-phosphate and NADPH, thereby producing two important intermediates for other metabolic processes.

$$\text{glucose 6-phosphate} + 2\,\text{NADP}^+ + \text{H}_2\text{O} \longrightarrow \text{ribose 5-phosphate} + 2\,\text{NADPH} + 2\,\text{H}^+ + \text{CO}_2$$

The ribose 5-phosphate is used to form nucleotides. The NADPH, a coenzyme, is required for reductive biosynthetic reactions such as the synthesis of lipids.

The pentose phosphate pathway is usually discussed as a series of two separate sequences of reactions known as the *oxidative* and *nonoxidative* stages. In the oxidative stage, NADPH is formed as glucose 6-phosphate is converted to ribulose 5-phosphate.

$$\text{glucose 6-phosphate} + 2\,\text{NADP}^+ + \text{H}_2\text{O} \longrightarrow \text{ribulose 5-phosphate} + 2\,\text{NADPH} + 2\,\text{H}^+ + \text{CO}_2$$

Ribulose 5-phosphate reacts to form two isomeric compounds. One reaction catalyzed by phosphopentose isomerase yields ribose 5-phosphate via an enediol intermediate. The second reaction forms the epimer xylulose 5-phosphate. Both isomers are intermediates in the nonoxidative stage of the pentose phosphate pathway. In addition, a small amount of ribose 5-phosphate is used to synthesize ribonucleotides.

The Pentose Phosphate Pathway and Glycolysis Are Linked

A nonoxidative section of the pentose phosphate pathway catalyzes the interconversion of three-, four-, five-, six-, and seven-carbon carbohydrate derivatives. The reactions are complex and involve details whose discussion is not appropriate in this text. Enzymes link three reactions that are depicted using only the number of carbon atoms in the carbohydrate derivatives.

$$\text{C}_5 + \text{C}_5 \xrightleftharpoons{\text{transketolase}} \text{C}_3 + \text{C}_7$$

$$\text{C}_3 + \text{C}_7 \xrightleftharpoons{\text{transaldolase}} \text{C}_4 + \text{C}_6$$

$$\text{C}_5 + \text{C}_4 \xrightleftharpoons{\text{transketolase}} \text{C}_3 + \text{C}_6$$

The sum of these three reactions is the formation of two hexoses and a triose from three pentoses. Both products are intermediates in the glycolysis pathway.

$$3\ \text{ribose 5-phosphate} \longrightarrow 2\ \text{fructose 6-phosphate} + \text{glyceraldehyde 3-phosphate}$$

In the balanced equation given for the oxidative section of the pentose phosphate pathway, 2 moles of NADPH are produced per mole of ribose 5-phosphate. However, there are times when the biological needs for these two substances are not in a 2:1 ratio.

If more ribose 5-phosphate than NADPH is required, it is formed from C_3 and C_6 glycolysis intermediates by reversal of the reactions of the nonoxidative branch of the pentose phosphate pathway. If more NADPH is required than ribose

5-phosphate, the C_5 carbohydrate is converted into glycolysis intermediates, and they are oxidized.

26.8 Glycogenolysis and Glycogenesis

Glycogenolysis, the release of glucose from glycogen, occurs by a sequence of three reactions. The first is the removal of a glucose unit from the nonreducing end of glycogen catalyzed by glycogen phosphorylase to produce glucose 1-phosphate and a glycogen molecule with one less glucose unit.

$$\text{glycogen } (n \text{ residues}) + P_i \xrightarrow{\text{glycogen phosphorylase}} \text{glucose 1-phosphate} + \text{glycogen } (n-1 \text{ residues})$$

glucose 1-phosphate

These marathon runners in the New York marathon depend on large stores of muscle glycogen at the start of the race. (The Image Bank)

The ΔG° for the reaction is small because the energy required to hydrolyze the glycosidic bond is nearly equal to that provided by the phosphorylation of glucose. The glucose 1-phosphate is isomerized to glucose 6-phosphate, catalyzed by phosphoglucomutase. The ΔG° of -1.6 kcal/mole for this reaction is also very small.

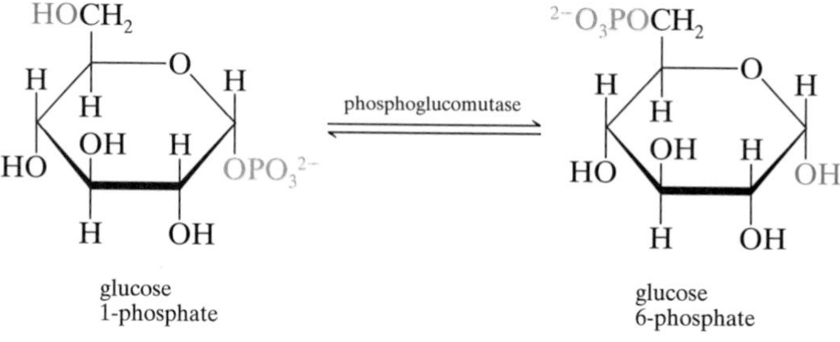

Glucose 6-phosphate, in contrast with glucose, cannot leave a cell. Thus, the third step in glycogenolysis is the hydrolysis of glucose 6-phosphate. The enzyme glucose phosphatase, which is present only in the kidneys, intestine, and liver, catalyzes the hydrolysis of this phosphate.

$$\text{glucose 6-phosphate} + H_2O \xrightarrow{\text{glucose phosphatase}} \text{glucose} + P_i$$

The liver not only stores large amounts of glycogen but can also release glucose for use by other organs. Organs such as muscles and the brain lack this enzyme. Thus, the glucose 6-phosphate is retained in muscles and the brain, which need it as fuel for the formation of ATP.

Glycogenesis, the synthesis of glycogen, requires a separate pathway from that for glycogenolysis. In glycogenesis, glycogen is elongated by formation of additional glycosidic bonds with glucose in a three-step reaction sequence. The ΔG° of -0.5 kcal/mole for the reaction is very small.

In step 1, glucose 6-phosphate is isomerized to glucose 1-phosphate by phosphoglucomutase. This reaction is the reverse of step 2 of the glycogenolysis pathway.

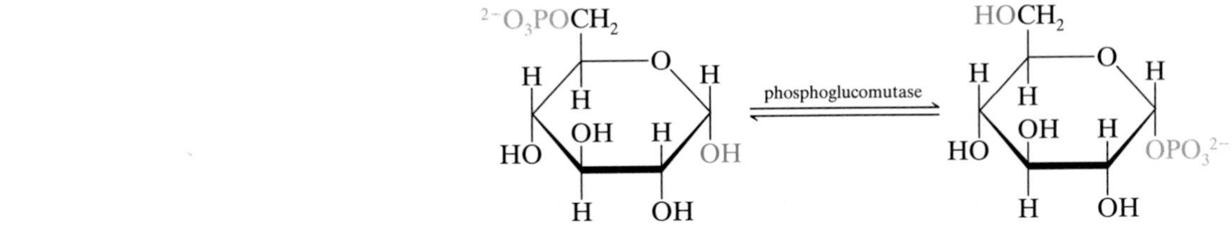

glucose 6-phosphate

glucose 1-phosphate

Glucose 1-phosphate is then activated by forming a uridine diphosphate derivative. The UDP-glucose is formed from the reaction of glucose 1-phosphate and uridine triphosphate (UTP) in a reaction catalyzed by UDP-glucose pyrophosphorylase (Figure 26.9). Note that one phosphate bond is formed and one is cleaved. Thus, the $\Delta G°$ for the reaction is essentially zero—that is, the reaction is readily reversible. However, the pyrophosphate formed is rapidly hydrolyzed in an essentially irreversible reaction. The summation of these two steps provides UDP-glucose.

$$\begin{array}{r}
\text{glucose 1-phosphate} + \text{UTP} \longrightarrow \text{UDP-glucose} + \text{PP}_i \\
\text{PP}_i + H_2O \longrightarrow 2\ \text{P}_i \\
\hline
\text{glucose 1-phosphate} + \text{UTP} + H_2O \longrightarrow \text{UDP-glucose} + 2\ \text{P}_i
\end{array}$$

Uridine diphosphate glucose (UDP-glucose), an activated form of glucose, is a donor of glucose. It transfers a glucose molecule to the growing glycogen chain at the nonreducing end (Figure 26.9). The reaction is catalyzed by glycogen synthetase.

26.9 Regulation of Carbohydrate Metabolism

The various metabolic pathways described in this chapter, which consist of hundreds of reactions, must be coordinated for biosynthetic or degradative sequences to occur. The amounts of enzymes are controlled by the rate of protein synthesis—a feature that will not be further discussed. However, the activities of those enzymes will be discussed. Allosteric interactions provide important controls over the activity of certain enzymes in many of the metabolic pathways of carbohydrates.

Glycolysis and Gluconeogenesis

Glycolysis and gluconeogenesis pathways are reciprocally regulated, so that while one is active the other is relatively inactive. The major control point in glycolysis is the conversion of fructose 6-phosphate into fructose 1,6-diphosphate. This phosphofructokinase-catalyzed step is inhibited by ATP but is allosterically stimulated by AMP. The major control point of gluconeogenesis is the conversion of fructose 1,6-diphosphate into fructose 6-phosphate, which is catalyzed by fructose 1,6-diphosphatase. This enzyme is inhibited by AMP, the reverse of the effect on phosphofructokinase in the glycolysis pathway. Thus, the two pathways are controlled so that they are not active at the same time.

Citrate also exerts some control over glycolysis and gluconeogenesis pathways. It inhibits phosphofructokinase but stimulates fructose 1,6-diphosphatase. When the citrate level is high, gluconeogenesis occurs rather than glycolysis.

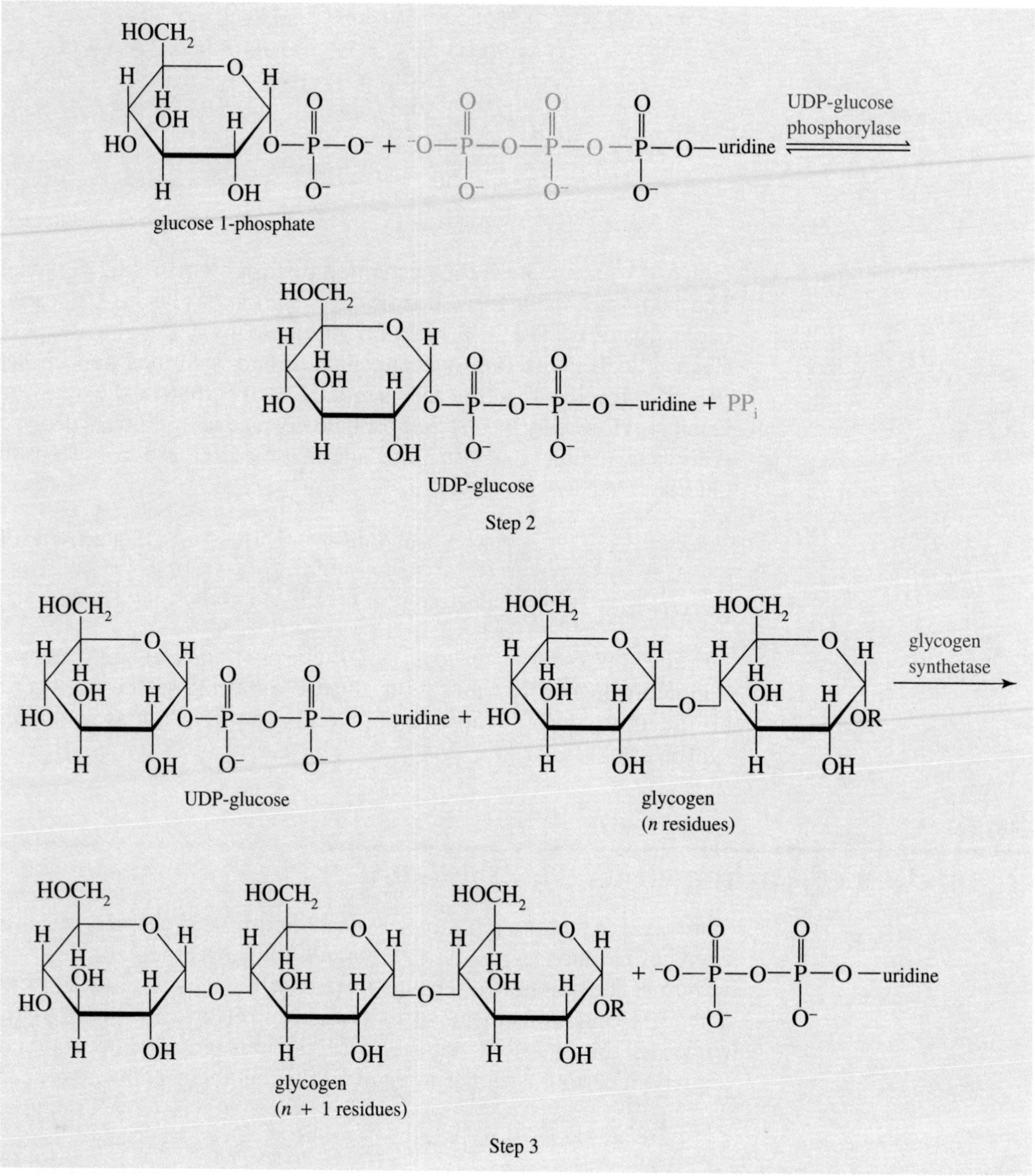

glucose 1-phosphate

UDP-glucose phosphorylase

UDP-glucose

Step 2

glycogen synthetase

UDP-glucose

glycogen
(n residues)

glycogen
(n + 1 residues)

Step 3

FIGURE 26.9
Glycogenesis

Glycogenesis and Glycogenolysis

The synthesis and degradation reactions of glycogen are also coordinated. While one process is active the other is inactive. Both processes are strongly affected by hormones. Insulin, a pancreatic hormone, binds to protein receptors in cell membranes and facilitates entry of glucose into the cell. The release of insulin is triggered by high glucose levels. Insulin also increases the rate of synthesis of glycogen.

The pancreatic hormones glucagon and epinephrine have an effect opposite that of insulin. They stimulate the breakdown of glycogen. Glucagon is active in the liver, and epinephrine in the muscles. Although the series of reactions affected by the hormones is complex, the net result is activation of glycogen phosphorylase required for glycogenolysis and deactivation of glycogen synthetase required for glycogenesis.

Summary

Dietary carbohydrates are digested to the monosaccharides glucose, fructose, and galactose. The latter two compounds are converted into intermediates that are metabolized by the glucose pathways.

Because the cells need glucose, the body has a series of interrelated metabolic pathways to maintain the requisite glucose level. Formation of glycogen and its subsequent hydrolysis provide glucose during periods of low carbohydrate intake or when glucose is being rapidly used.

Glycolysis (Embden–Meyerhof pathway) is a series of enzyme-catalyzed reactions that produce 2 moles each of pyruvic acid, ATP, and NADH from 1 mole of glucose. Under **anaerobic** conditions, NADH reduces pyruvic acid to lactic acid. Lactic acid is oxidized back to pyruvic acid in the liver. The accumulated lactic acid constitutes an **oxygen debt.** Under **aerobic** conditions, pyruvic acid is converted to acetyl coenzyme A, which is oxidized in the citric acid cycle.

Gluconeogenesis, which occurs in the liver, converts pyruvic acid into glucose, which then is released into the blood to be transported to other organs. Transport of lactic acid to the liver and its conversion into pyruvic acid is called the **Cori cycle.**

The **pentose phosphate pathway** forms ribose 5-phosphate and NADPH. The pentose can be interconverted with glycolysis intermediates. NADPH is used in reductive biosynthesis.

If there is too much glucose in the blood, **glycogenesis** occurs to store the glucose as glycogen. When the glucose level in the blood drops, glycogen in the liver is hydrolyzed in a series of steps called **glycogenolysis.**

Carbohydrate metabolism is regulated by a combination of enzymes and hormones. The principal enzyme controls are phosphofructokinase in glycolysis and fructose 1,6-diphosphatase in gluconeogenesis. The hormones epinephrine and glucagon inactivate glycogen synthetase and activate glycogen phosphorylase. Insulin facilitates the entry of glucose into cells.

Exercises

Digestion of Carbohydrates

26.1 Where does the enzymatic hydrolysis of starch first occur?

26.2 Why is the enzymatic hydrolysis of starch slow in the stomach?

26.3 How are the metabolisms of fructose and of galactose related to the metabolism of glucose?

26.4 Suggest a way in which mannose may be converted to glucose in a metabolic reaction.

Glucose Metabolism

26.5 Distinguish between glycogenolysis and glycogenesis.

26.6 Distinguish between glycolysis and gluconeogenesis.

26.7 What organ uses glucose as its sole fuel?

26.8 What is the role of glycogen in controlling the blood glucose level? What two processes involve glycogen formation and hydrolysis?

26.9 What is the fate of pyruvic acid under anaerobic and under aerobic conditions?

26.10 What substances are produced in the pentose phosphate pathway?

Glycolysis

26.11 How many moles of pyruvic acid are produced from glucose by glycolysis?

26.12 What is the first step in glycolysis, and what enzyme catalyzes the reaction?

26.13 What is the end product of glycolysis?

26.14 Which coenzyme functions as the oxidizing agent in glycolysis?

26.15 How is NADH oxidized under aerobic conditions?

26.16 How is NADH used in glycolysis under anaerobic conditions?

26.17 How many moles of ATP and NADH are produced from 1 mole of glucose in glycolysis?

26.18 How many steps in glycolysis require ATP? How many steps in glycolysis produce NAD^+?

26.19 What type of enzyme catalyzes the following reaction?

glucose 6-phosphate ⟶ fructose 6-phosphate

26.20 What type of enzyme catalyzes the following reaction?

fructose 6-phosphate ⟶ fructose 1,6-diphosphate

Pyruvic Acid

26.21 What reactions occur with pyruvic acid under anaerobic conditions?

26.22 Where is lactic acid oxidized to pyruvic acid?

26.23 Outline the Cori cycle.

26.24 What reaction occurs with pyruvic acid under aerobic conditions?

Energy from Metabolism of Glucose

26.25 How many molecules of acetyl coenzyme A are formed from one molecule of glucose? How many molecules of CO_2 are formed in this reaction?

26.26 How does the energy derived from the oxidation of the NADH formed in glycolysis differ from that of the NADH formed in the citric acid cycle?

26.27 Why are shuttles required to make use of the electrons of NADH produced in the cytoplasm?

26.28 What two shuttles operate in the mitochondria to transport reducing power across the membrane?

Gluconeogenesis

26.29 What are the sources of compounds for gluconeogenesis?

26.30 What portion of a fat can be used for gluconeogenesis? What portion cannot be used?

26.31 How many steps of glycolysis must be bypassed in gluconeogenesis?

26.32 What intermediate do gluconeogenesis and the citric acid cycle have in common?

Pentose Phosphate Pathway

26.33 How does NADPH differ from NADH in biological function?

26.34 What compound contains the carbon atom lost from glucose in its conversion to ribose?

26.35 How many moles of NADPH are produced per mole of ribose?

26.36 Why does the first enzyme in the pentose phosphate pathway use $NADP^+$ rather than NAD^+?

26.37 What happens to the ribose that is produced because of a high NADPH requirement but is not itself required by the organism?

26.38 How is ribose produced when NADPH is not required?

Glycogenolysis and Glycogenesis

26.39 What reaction determines whether glucose formed by glycogenolysis can leave a cell?

26.40 Which organs contain the enzyme glucose 6-phosphatase?

26.41 What reaction makes glycogenesis essentially irreversible?

26.42 What triphosphate is involved in the first step of glycogenesis?

Control of Carbohydrate Metabolism

26.43 Which reaction in glycolysis is a control point?

26.44 Which reaction in gluconeogenesis is a control point?

27

Metabolism of Triglycerides

The humps of this camel on the steppe grasslands in Mongolia are a source of stored chemical energy in the form of triglycerides. (Photo Researchers, Inc.)

27.1 Digestion of Triglycerides
27.2 Catabolic Reactions of Triglycerides
27.3 Energy Production from Fatty Acids
27.4 Ketosis
27.5 Biosynthesis of Fatty Acids

Essays

- *Fats, Exercise, and Diet*

27.1 Digestion of Triglycerides

Although lipids are relatively insoluble in water, our body have several compounds that complex with lipids and form micelles that are efficiently transported in body fluids. Emulsifying agents in the digestion process act like soap toward the globules of fat and break the fat into smaller globules which make them more accessible to hydrolysis and hence to their metabolism. A class of compounds called *lipoproteins* transports lipids in the blood.

Triacylglycerols pass unchanged from the mouth to the stomach, where the churning action of the stomach breaks the material into smaller droplets of fat (Figure 27.1). These droplets mix with partially digested foods giving chyme which then leaves the stomach. Pancreatic lipases, which catalyze the hydrolysis of lipids, are released as the chyme mixture enters the small intestine. Bile salts, secreted from the gallbladder, increase the solubility of the lipids by forming micelles. Only then can digestion of triglycerides begin. However, the hydrolysis is not complete, and a mixture of monoglycerides and diglycerides forms along with fatty acids.

The mixture of hydrolyzed products is taken up by the villi that project into the lumen of the small intestine (Figure 27.2). The surface of the villi consists of epithelial cells and microvilli. This structure provides a large surface area that allows the

FIGURE 27.1 Digestion of Triacylglycerols

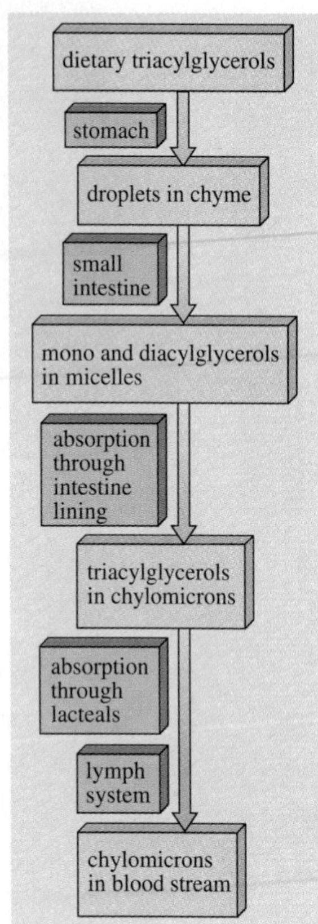

hydrolysis products to diffuse into the villi. Surprisingly, after this passage, these fragments are reassembled into triglycerides. Then the triglycerides join with an aggregate of cholesterol, cholesterol esters, phospholipids, and proteins to give **chylomicrons,** one of the several classes of lipoproteins. The chylomicrons pass through the lymphatic system and into the bloodstream. They are delivered via the vascular system to the liver and adipose tissue as well as to other organs such as the heart and lungs. The triglycerides ultimately are distributed as tissue lipids or as storage fat called *depot fat.* This rather circuitous route leaves the triglycerides in essentially the same form as when ingested.

Lipid Distribution in the Body

The **tissue lipids** contained in cell membranes are glycerophospholipids, phosphosphingolipids, and glycosphingolipids. These lipids are used to maintain cell membranes and the membranes of cell organelles such as the mitochondria. Except in extreme cases of starvation, the tissue lipids are not used to provide energy.

Depot fat, also called **adipose tissue,** is where most of the body's triglyceride supply is stored. This tissue contains fat cells called *adipocytes.* They have large fat droplets that constitute nearly the entire volume of the cell. Adipose tissue occurs as a subcutaneous (under the skin) layer, where it insulates against temperature changes. Substantial quantities of adipose tissue surround vital organs such as the kidneys and the spleen and provide protection against physical shock as well as insulation. However the primary function of adipose tissue is as a reserve of stored

FIGURE 27.2 Villi of the Small Intestine

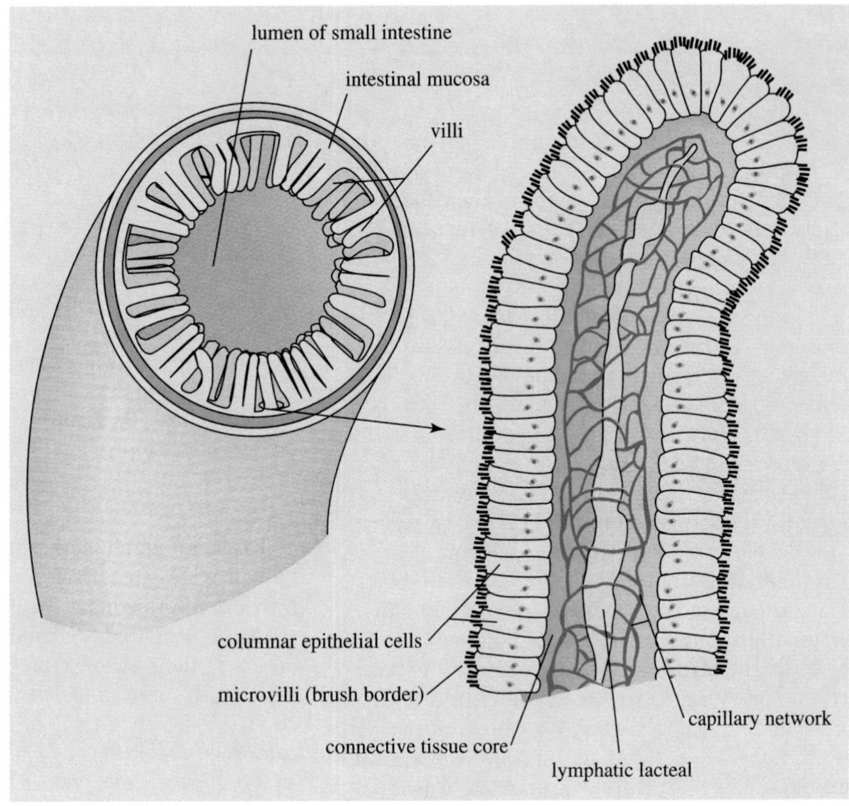

chemical energy. Given sufficient water, the average adult can sustain life for approximately 40 days from adipose tissue.

Lipids are Mobile

There is no way to excrete excess lipid material from the body; lipids are removed only by oxidation. As a consequence, with no exercise, overeating will cause obesity. In most cases excessive fat arises not from ingested fat but from carbohydrates that are converted to fat. When food is continually ingested and exercise is frequent, the adipose tissue is constantly changing. Newly absorbed fats form new adipose tissue, and old adipose tissue is metabolized.

The release of fatty acids from adipose tissue is called **fatty acid mobilization.** In this process ester bonds are hydrolyzed by enzymes called **lipases.** The lipases differ depending on their location in the body and how they are regulated.

$$\begin{array}{l} R-CO_2-CH_2 \\ \quad\quad\quad\quad | \\ R-CO_2-CH \\ \quad\quad\quad\quad | \\ R-CO_2-CH_2 \end{array} + 3\,H_2O \xrightarrow{\text{lipase}} 3\,R-CO_2H + \begin{array}{l} CH_2-OH \\ | \\ CH\ -OH \\ | \\ CH_2-OH \end{array}$$

Free fatty acids are transported by serum albumin to various parts of the body where they enter cells. Unlike the polar glucose molecule, which requires insulin and a membrane transport system to enter the cells, the nonpolar fatty acid molecules easily cross the lipid bilayer of cell membrane. Thus, most cells will oxidize fatty acids in preference to glucose. As a consequence, the body's glycogen supply is conserved so that circulating glucose remains available for the brain.

Fats, Exercise, and Diet

Fats are digested more slowly than carbohydrates and stay in the digestive system longer. They are also easily stored in the body, much to the chagrin of many individuals. However, these fat deposits keep the body running after the glycogen stores are depleted. Fats provide a long-term energy storage, whereas carbohydrates provide a short-term energy storage.

Health authorities recommend that less than 30% of total consumer calories be derived from fat, but about 35–40% of the American diet consists of fats. Fats in the diet cannot be converted into glucose; they are converted into other fats in the body. In addition, after the glycogen stores are full, any excess carbohydrates ingested are also converted into fat.

Judging by articles listed on the covers of many magazines displayed at the supermarket checkout counter, weight is the most important issue in American society. These magazines list ways to lose a stated amount of weight (usually too much) in a stated amount of time (usually too short). However, we can't escape the law of conservation of energy. If we take in more calories than we use up in everyday activities, we will store that energy as fat. Conversely, if we limit our caloric intake and increase our activity, we will use stored fat to make up the energy deficit. These facts are fundamental and can't be changed by some fad "diet." Only by regular exercise and by changing our eating habits can we maintain our weight at a desirable level. The caloric expenditure for various forms of exercise, based on a 150-lb person, is given in the table. Note that a hamburger (without french fries or a shake) at most fast-food chains contains 500–600 kcal, and a large fraction of that is fat. It would take about an hour of strenuous exercise to prevent the potential weight gain from that hamburger.

One way or the other, there is no substitute for exercise to maintain your weight. (Photo Researchers, Inc.)

Exercise and Caloric Expenditure

Activity	kcal/hour
bicycling (10 mph)	420
calisthenics (heavy)	600
cross-country skiing (5 mph)	700
ice skating (rapid)	700
rope skipping (vigorous)	800
running (6 mph)	750
swimming (50 yd/min)	750
walking (4 mph)	420

Fat is not an altogether undesirable substance. It not only provides a compact form of stored energy but also provides intermediates for many important biosynthetic processes required to maintain life. Men and women differ in their storage of fats. The average male has about 12% body fat, whereas the average female has about 20% body fat. Of course, percent body fat values can range either lower or higher. Amounts significantly above these average values are used to classify levels of obesity. A conditioned male marathon runner may have as little as 6% body fat. There is a minimum required amount of fat for continued good health. For males it is about 3% and for females about 10%.

One pound of adipose tissue is equivalent to 3500 kcal. Thus, if we reduce our caloric intake by 500 kcal a day, we can lose a pound by the end of the week. For most people, this requires a significant change in eating habits. Thus, it is understandable that doctors warn against trying to lose more than about 2 lb a week.

Many quick-weight-loss diets or diet pills do little more than change the water balance in the body by dehydration. In a short time the weight is regained as the body rehydrates. Low-carbohydrate regimens cause the body to draw down its glycogen reserves. Because 1 g of glycogen is associated with about 2 g of hydrogen-bonded water molecules, the loss of glycogen does result in an enhanced weight loss. Loss of a pound of glycogen decreases body weight by 3 lb. However, the loss is not where we want it. The fat remains, and our body has efficient mechanisms to regain the lost glycogen and of course the water as well. Ingested fat goes exactly where we probably don't want it, and it is not processed until large amounts of glycogen are used. Therefore, the best way to prevent fat buildup is to reduce fat in the diet. Our body will produce the fat we need from other dietary sources. Only a decrease in caloric intake and an increase in exercise will decrease the fat stores.

27.2 Catabolic Reactions of Triglycerides

Lipases catalyze the hydrolysis of triglycerides to produce glycerol and fatty acids. Glycerol contains only a small fraction of the carbon atoms of the triglyceride. Thus, the major source of energy from triglycerides is from the fatty acids. Glycerol is phosphorylated in the liver to form L-glycerol 3-phosphate, which is subsequently oxidized to dihydroxyacetone phosphate. The dihydroxyacetone phosphate is isomerized to glyceraldehyde 3-phosphate and enters the glycolysis pathway for glucose.

$$\underset{\text{glycerol}}{HO{-}CH(CH_2OH)_2} \xrightarrow[\text{ATP} \to \text{ADP}]{\text{glycerol kinase}} \underset{\text{L-glycerol 3-phosphate}}{HOCH_2{-}CH(OH){-}CH_2OPO_3^{2-}} \xrightarrow[\text{NAD}^+ \to \text{NADH} + \text{H}^+]{\text{glycerol phosphate dehydrogenase}} \underset{\text{dihydroxyacetone phosphate}}{HOCH_2{-}C({=}O){-}CH_2OPO_3^{2-}}$$

One mole of ATP is consumed, but one mole of NADH is produced. Recall that NADH yields 3 moles of ATP in the electron transport chain. Thus a net of 2 moles of ATP are formed.

The fatty acids formed in hydrolysis must cross the cell membrane of cells requiring energy. Three general processes must occur. These are:

1. Activation of the fatty acid by formation of an acyl CoA.
2. Transport of the acyl CoA into the mitochondrial matrix.
3. Oxidation of the acyl CoA to give acetyl CoA and the reduced coenzymes NADH and $FADH_2$.

Activation of Fatty Acids

Fatty acids are activated by conversion to a thioester (Chapter 18) formed with coenzyme A (CoA). Energy is required as an "investment" so that the resulting fatty acid derivatives can release energy in subsequent steps. The activation occurs in two steps. First, a mixed anhydride is formed as the carboxyl group displaces pyrophosphate (PP_i) from ATP in a nucleophilic substitution reaction. The sulfhydryl group of coenzyme A then attacks the acyl adenylate in a second nucleophilic substitution reaction to form acyl SCoA and AMP.

$$R{-}\overset{O}{\overset{\|}{C}}{-}O^- + ATP \rightleftharpoons R{-}\overset{O}{\overset{\|}{C}}{-}AMP + PP_i$$

$$R{-}\overset{O}{\overset{\|}{C}}{-}AMP + HS{-}CoA \rightleftharpoons R{-}\overset{O}{\overset{\|}{C}}{-}S{-}CoA + AMP$$

The equilibrium constant for the sum of the two reactions is close to 1, so the reaction is readily reversible.

$$RCO_2^- + CoA + ATP \rightleftharpoons \text{acyl CoA} + AMP + PP_i$$

The energy released in the conversion of ATP into AMP and PP_i is approximately equal to the energy required to form the thioester linkage in the acyl CoA. However, the reaction is driven in the forward direction because pyrophosphate is rapidly hydrolyzed. Thus the sum of two reactions makes the overall reaction irre-

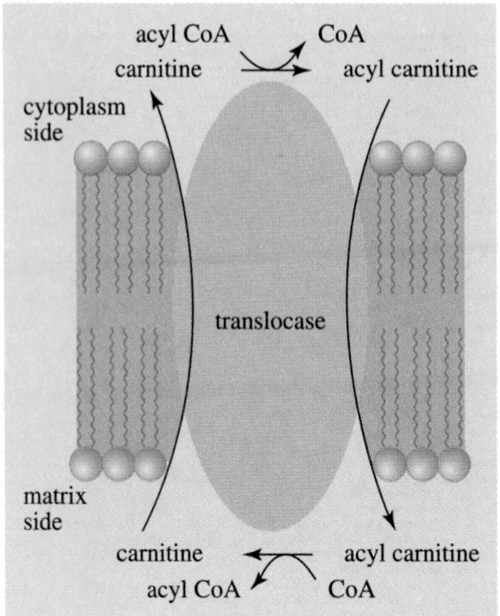

FIGURE 27.3
Transport of Acyl CoA
Translocase allows transport of acyl carnitine into the mitochondrial matrix. Carnitine returns in exchange for acyl carnitine.

versible. In total the energy equivalent of two phosphoric acid anhydride bonds is released.

$$\begin{aligned} \mathrm{RCO_2^- + CoA + ATP} &\rightleftharpoons \mathrm{acyl\ CoA + AMP + PP_i} \\ \mathrm{PP_i + H_2O} &\longrightarrow \mathrm{2\,P_i + 2\,H^+} \\ \hline \mathrm{RCO_2^- + CoA + ATP + H_2O} &\longrightarrow \mathrm{acyl\ CoA + AMP + 2\,P_i + 2\,H^+} \end{aligned}$$

Transport of Activated Fatty Acids into the Mitochondrial Matrix

The activation of fatty acids occurs on the outer mitochondrial membrane, but oxidation occurs in the mitochondrial matrix. However, the acyl CoA cannot directly traverse the inner mitochondrial membrane. Therefore, the acyl group is transferred from the sulfur atom of CoA to the hydroxyl group of carnitine, a zwitterionic compound derived from lysine. This more polar acyl carnitine compound is shuttled across the membrane (translocated) by a transmembrane protein called **translocase.** In the matrix, the acyl group is transferred to CoA, and the carnitine then returns to the outer membrane in exchange for an entering acyl carnitine (Figure 27.3).

$$\underset{\text{acyl CoA}}{\mathrm{R{-}\overset{\overset{\displaystyle O}{\|}}{C}{-}SCoA}} + \underset{\text{carnitine}}{\mathrm{HO{-}\underset{\displaystyle CH_2{-}CO_2^-}{\underset{|}{\overset{\overset{\displaystyle N(CH_3)_2}{|}\\ \overset{\displaystyle CH_2}{|}}{CH}}}}} \rightleftharpoons \underset{\text{acyl carnitine}}{\mathrm{R{-}\overset{\overset{\displaystyle O}{\|}}{C}{-}O{-}\underset{\displaystyle CH_2{-}CO_2^-}{\underset{|}{\overset{\overset{\displaystyle \overset{+}{N}(CH_3)_2}{|}\\ \overset{\displaystyle CH_2}{|}}{CH}}}}} + \mathrm{HSCoA}$$

Degradation of Fatty Acids into Two-Carbon Units

The acyl CoA is degraded in the mitochondrial matrix by a sequence of four reactions. In each sequence the chain length of the fatty acid is shortened, and two-carbon-atom units of acetyl CoA are formed. The acetyl CoA is subsequently oxidized in the citric acid cycle. The degradation sequence involves four steps: dehydrogenation, hydration, oxidation, and thiolytic cleavage (Figure 27.4). We now look at the details of each of these four steps.

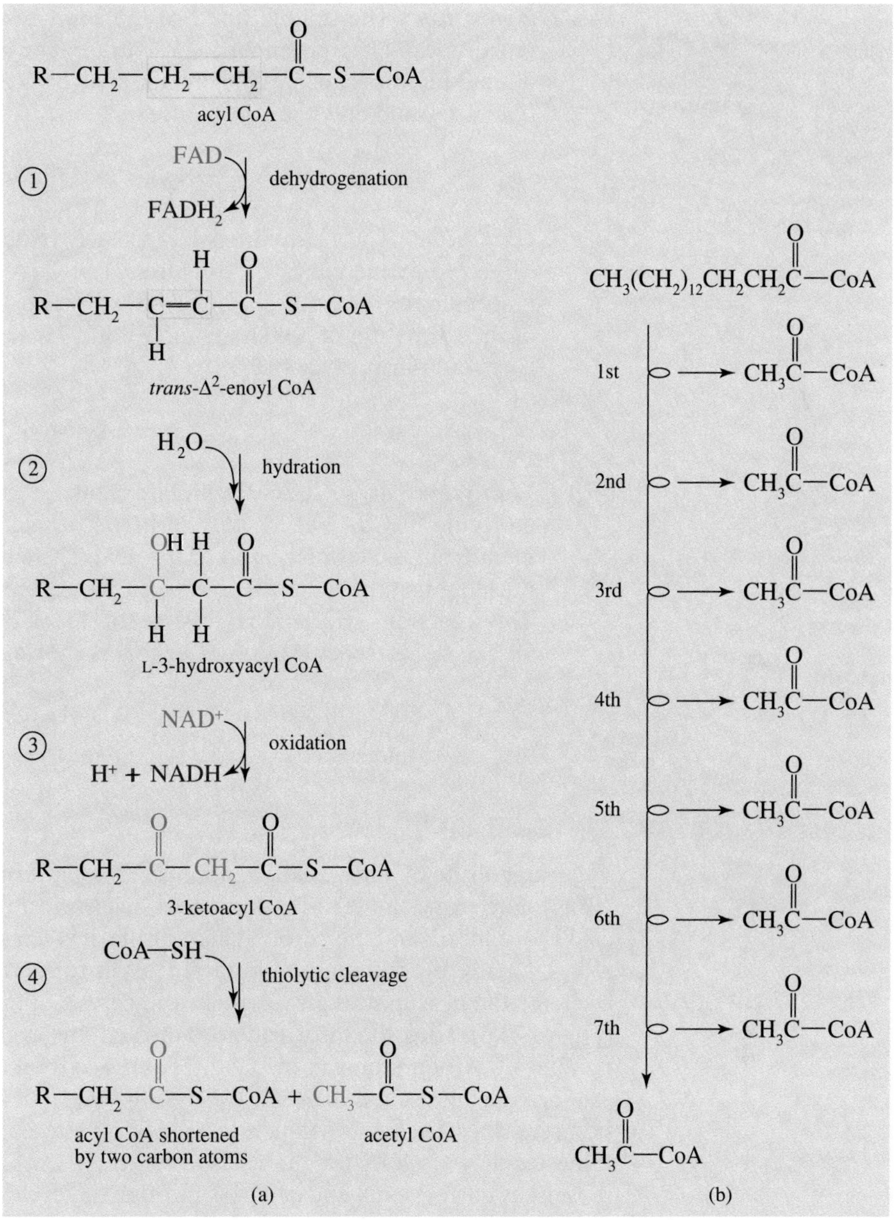

FIGURE 27.4
The Fatty Acid Cycle
(a) The four steps in one cycle. (b) The sequence of cycles for the oxidation of palmitic acid. Each loop represents one cycle.

$R{-}CH_2{-}CH_2{-}CH_2{-}C({=}O){-}S{-}CoA$

acyl CoA

① FAD → $FADH_2$ dehydrogenation

$R{-}CH_2{-}CH{=}CH{-}C({=}O){-}S{-}CoA$

trans-Δ^2-enoyl CoA

② H_2O hydration

$R{-}CH_2{-}CH(OH){-}CH_2{-}C({=}O){-}S{-}CoA$

L-3-hydroxyacyl CoA

③ NAD^+ → $H^+ + NADH$ oxidation

$R{-}CH_2{-}C({=}O){-}CH_2{-}C({=}O){-}S{-}CoA$

3-ketoacyl CoA

④ CoA—SH thiolytic cleavage

$R{-}CH_2{-}C({=}O){-}S{-}CoA + CH_3{-}C({=}O){-}S{-}CoA$

acyl CoA shortened by two carbon atoms; acetyl CoA

(a)

$CH_3(CH_2)_{12}CH_2CH_2C({=}O){-}CoA$

1st → $CH_3C({=}O){-}CoA$

2nd → $CH_3C({=}O){-}CoA$

3rd → $CH_3C({=}O){-}CoA$

4th → $CH_3C({=}O){-}CoA$

5th → $CH_3C({=}O){-}CoA$

6th → $CH_3C({=}O){-}CoA$

7th → $CH_3C({=}O){-}CoA$

$CH_3C({=}O){-}CoA$

(b)

Step 1: ***Dehydrogenation*** The enzyme acyl CoA dehydrogenase catalyzes the dehydrogenation of the activated acid to form a trans double bond between the α- and β-carbon atoms of the acid.

$$\text{acyl CoA} + \text{FAD} \longrightarrow trans\text{-}\Delta^2\text{-enoyl CoA} + FADH_2$$

The Δ^2 refers to the position of the double bond. There are four different enzymes that catalyze this reaction, and each requires FAD as an oxidizing agent. Each specific enzyme catalyzes the dehydrogenation of acids most efficiently for a certain range of chain lengths. Since 1 mole of $FADH_2$ yields 2 moles of ATP in the electron transport chain, this step provides energy from the metabolism of fatty acids.

Step 2: ***Hydration*** The double bond of the *trans*-Δ^2-enoyl CoA is hydrated in a reaction catalyzed by enoyl CoA hydrase. The addition of water to the trans double bond occurs specifically; the hydroxyl group is placed on the β-carbon atom, and only the L isomer results.

$$trans\text{-}\Delta^2\text{-enoyl CoA} + H_2O \longrightarrow \text{L-3-hydroxyacyl CoA}$$

Step 3: ***Oxidation*** The L-3-hydroxyacyl CoA is oxidized by NAD^+ in a reaction catalyzed by the enzyme L-3-hydroxyacyl CoA dehydrogenase. The enzyme is specific for the L stereoisomer but works for any chain length. As a result of the formation of NADH, 3 moles of ATP are eventually produced in the electron transport chain.

$$\text{L-3-hydroxyacyl CoA} + NAD^+ \longrightarrow \text{3-ketoacyl CoA} + NADH + H^+$$

Step 4: ***Thiolytic cleavage*** β-Ketothiolase catalyzes a thiolytic cleavage in which a molecule of CoA reacts with acyl CoA to cleave a two-carbon-atom fragment from the carboxyl end of the acid. The sulfur atom of CoA serves as a nucleophile, attacking the 3-keto carbon atom and displacing acetyl CoA. This reaction, which is a reverse of the Claisen condensation (Section 18.11), yields a coenzyme A ester that is shorter by two carbon atoms.

$$\underset{(n \text{ carbon atoms})}{\text{3-ketoacyl CoA}} + \text{CoA} \longrightarrow \underset{(n\text{-}2 \text{ carbon atoms})}{\text{acyl CoA}} + \text{acetyl CoA}$$

The Process Is Repeated

The catabolism of fatty acids continues in a pattern called the **fatty acid cycle** (FAC). Each time the series of four steps occurs, 1 mole of acetyl CoA per mole of acid is produced, and the carbon chain of the acyl CoA is shortened by two carbon atoms. Finally, the four-carbon butyryl CoA is formed, and it proceeds around the cycle for the last time to give 2 moles of acetyl CoA per mole of the original acid (Figure 27.6). Thus, the fatty acid goes through the cycle one less time than half the number of carbon atoms in the acid. For an n-carbon acid, $n/2$ units of acetyl CoA are formed, but the number of cycles is only $n/2 - 1$. Note that the original activation step does not occur again; it is needed only to convert the original acid to the thioester acyl CoA. After the release of acetyl CoA, the shortened acid chain is present as an acyl CoA and can pass through the cycle again.

EXAMPLE 27.1

How many moles of acetyl CoA are produced from the metabolism of 1 mole of $CH_3(CH_2)_{12}CO_2H$? How many times must the fatty acid cycle occur?

Solution

The compound contains 14 carbon atoms, so seven acetyl CoA units will be produced. Since the last cycle of metabolism produces two acetyl groups from a four-carbon acid, the cycle occurs only six times.

PROBLEM 27.1

How many moles of acetyl CoA are produced from the metabolism of 1 mole of stearic acid? How many times must the fatty acid cycle occur?

27.3
Energy Production from Fatty Acids

Based on the reactions in the previous section, let's calculate the total energy released by the metabolism of palmitic acid. An overall accounting is given in Figure 27.5.

Each cycle of the sequence of reactions yields 1 mole each of NADH and $FADH_2$. Each reduced coenzyme is oxidized in the electron transport chain, and ATP is produced. Recall that 3 moles of ATP are produced from NADH and 2 moles from $FADH_2$. Thus, the product of one turn through the fatty acid cycle is 5 moles of ATP.

$$\frac{1\text{ NADH}}{\text{turn}} \times \frac{3\text{ ATP}}{1\text{ NADH}} + \frac{1\text{ FADH}_2}{\text{turn}} \times \frac{2\text{ ATP}}{1\text{ FADH}_2} = \frac{5\text{ ATP}}{\text{turn}}$$

Since the number of turns of the cycle is $n/2 - 1$, the number of moles of ATP generated from oxidation of reduced coenzymes in the electron transport chain is $5(n/2 - 1)$.

The acetyl CoA produced is oxidized in the citric acid cycle. Recall that this cycle produces 3 moles of NADH, 1 mole of $FADH_2$, and 1 mole of ATP. Subsequent oxidation of the reduced coenzymes in the electron transport chain yields 11 moles of ATP.

$$\frac{3\text{ NADH}}{\text{cycle}} \times \frac{3\text{ ATP}}{1\text{ NADH}} + \frac{1\text{ FADH}_2}{\text{cycle}} \times \frac{2\text{ ATP}}{1\text{ FADH}_2} = \frac{11\text{ ATP}}{\text{cycle}}$$

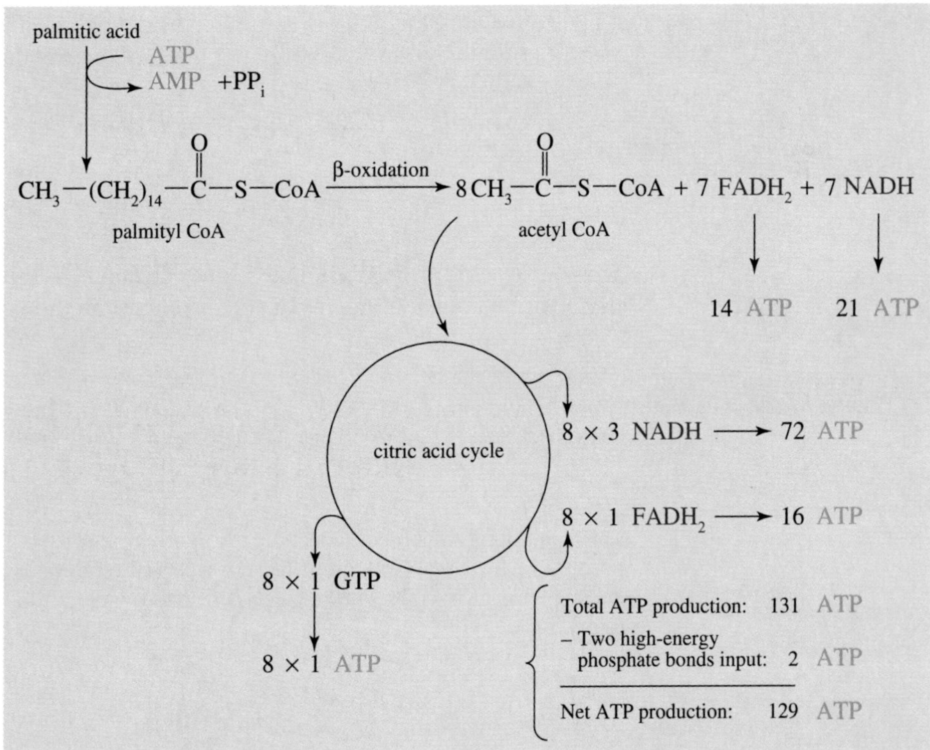

FIGURE 27.5
ATP Production from Oxidation of Palmitic Acid

Thus, a total of 12 moles of ATP is produced in the citric acid cycle per mole of acetyl CoA.

In the reaction of palmitic acid the equivalent of 2 moles of ATP is required to activate the acid because ATP is converted to AMP. The acid then goes through the fatty acid cycle seven times.

activation step		−2 ATP
seven turns in fatty acid cycle	7 × 5 ATP =	35 ATP
		33 ATP

Each mole of acetyl CoA produced in the seven turns of the fatty acid cycle then produces 12 moles of ATP by oxidation in the citric acid cycle, followed by oxidation of the reduced coenzymes in the electron transport chain.

ATP from the fatty acid cycle		33 ATP
ATP from the citric acid cycle and electron transport	8 × 12 ATP =	96 ATP
		129 ATP

The oxidation of palmitic acid liberates 2340 kcal/mole. The amount of energy stored in the formation of ATP is obtained by multiplying the number of moles of ATP by 7.3 kcal/mole. Thus, 942 kcal is stored per mole of palmitic acid. The biological storage process is 40% efficient.

$$\frac{942 \text{ kcal/mole saved in ATP}}{2340 \text{ kcal/mole released}} \times 100 = 40\%$$

It is instructive to compare the amount of energy obtained from the same mass of glucose and a fatty acid. A total of 36 moles of ATP is produced per mole of glucose (molar mass = 180 g). For palmitic acid (molar mass = 256 g), 129 moles of ATP are formed. Dividing the number of moles of ATP by the molar mass gives a measure of the amount of energy per gram of each compound.

$$\frac{36 \text{ moles ATP}}{180 \text{ g glucose}} = 0.20 \text{ mole ATP/g glucose}$$

$$\frac{129 \text{ moles ATP}}{256 \text{ g palmitic acid}} = 0.50 \text{ mole ATP/g palmitic acid}$$

Thus, fatty acids provide about 2.5 times as much ATP as glucose on a per gram basis and are an efficient means of storing energy in the body.

EXAMPLE 27.2

How many moles of ATP are produced from the metabolism of 1 mole of $CH_3(CH_2)_{12}CO_2H$?

Solution

This compound, which contains 14 carbon atoms, must pass through the cycle six times to produce seven acetyl CoA units. The six cycles result in the formation of six units of $FADH_2$ and six units of NADH. The reduced coenzymes yield a total of 30 ATP by oxidative phosphorylation.

$$6 \text{ NADH} \times \frac{3 \text{ ATP}}{1 \text{ NADH}} + 6 \text{ FADH}_2 \times \frac{2 \text{ ATP}}{1 \text{ FADH}_2} = 30 \text{ ATP}$$

The seven acetyl CoA units yield 84 ATP in the citric acid cycle.

$$7 \text{ acetyl CoA} \times \frac{12 \text{ ATP}}{1 \text{ acetyl CoA}} = 84 \text{ ATP}$$

Two ATP must be deducted for the activation of the acid. The total number of moles of ATP produced per mole of the acid is 30 + 84 − 2 = 112.

PROBLEM 27.2

How many moles of ATP are produced from the metabolism of 1 mole of lauric acid?

27.4 Ketosis

There are abnormal physiological conditions in which fat metabolism is the principal source of energy. In starvation, diabetes, and low-carbohydrate diets stored adipose tissue becomes the sole energy source. In each case the cells don't receive sufficient glucose, and the rate of oxidation of fatty acids increases. In diabetes, a condition in which carbohydrate metabolism is impaired, fat becomes the alternative energy source. Any condition that decreases the availability of glucose causes a decrease in the level of oxaloacetic acid required for the citric acid cycle. Oxaloacetic acid is normally produced by the carboxylation of pyruvic acid, which in turn is produced in the glycolysis of carbohydrates. Thus, if fat and carbohydrate oxidation are not in balance, there is a shortage of oxaloacetic acid. As a result, acetyl CoA cannot be processed by the citric acid cycle.

When acetyl CoA cannot enter the citric acid cycle, it combines with itself in the liver mitochondria, a condition called **ketosis** (Figure 27.6). The first reaction is a

FIGURE 27.6 Formation of Ketone Bodies

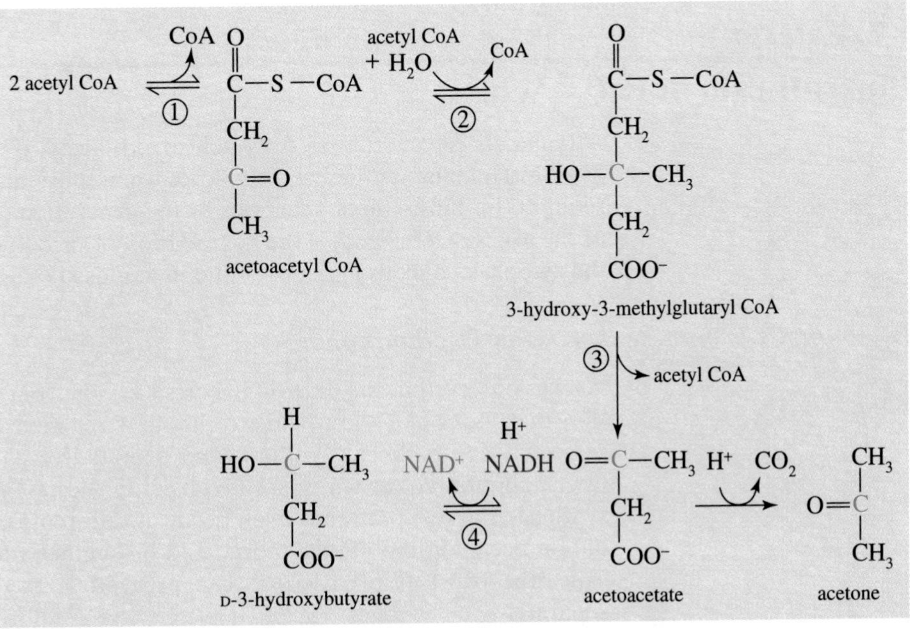

2 acetyl CoA ⇌(① −CoA) $CH_3COCH_2C(=O){-}S{-}CoA$ (acetoacetyl CoA) ⇌(② + acetyl CoA + H_2O, −CoA) $^-OOCCH_2C(OH)(CH_3)CH_2C(=O){-}S{-}CoA$ (3-hydroxy-3-methylglutaryl CoA)

③ → − acetyl CoA: $O{=}C(CH_3)CH_2COO^-$ (acetoacetate)

acetoacetate → (H⁺, −CO_2) $CH_3C(=O)CH_3$ (acetone)

acetoacetate + NADH + H⁺ ⇌(④) NAD⁺ + $HO{-}CH(CH_3)CH_2COO^-$ (D-3-hydroxybutyrate)

Claisen condensation of 2 moles of acetyl CoA. The acetoacetyl CoA formed then reacts with acetyl CoA and water to give 3-hydroxy-3-methylglutaryl CoA. In the third step, 3-hydroxy-3-methylglutaryl CoA gives acetoacetate in a reverse Claisen reaction. The sum of the three reactions is

$$2 \text{ acetyl CoA} + H_2O \longrightarrow \text{acetoacetate} + 2 \text{ CoA} + H^+$$

Reduction of acetoacetate in the mitochondrial matrix forms D-3-hydroxybutyrate. Acetoacetate also can decarboxylate (lose CO_2) to give acetone. Acetoacetate, D-3-hydroxybutyrate, and acetone are collectively known as **ketone bodies.**

In normal metabolism the concentration of ketone bodies in the blood is usually less than 1 mg/100 mL. They can be used as metabolic fuel and to generate acetyl CoA. **Ketosis** occurs when the rate of formation of ketone bodies exceeds the rate of their oxidation. This condition occurs when the blood level exceeds 3 mg/100 mL. A level of ketone bodies in the blood greater than 20 mg/100 mL is called **ketonemia.** With continued production, the ketone bodies may exceed the amount that tissue cells can use. When the blood level reaches 70 mg/100 mL, the renal threshold for ketone bodies, the excess ketone bodies are excreted in the urine. The condition of "ketones in the urine" is called **ketonuria.** At still higher ketone body concentrations, **acetone breath** results from acetone excretion via the lungs.

The conditions of ketonemia, ketonuria, and acetone breath are collectively called *ketoacidosis.* The excess acids (acetoacetic acid and D-3-hydroxybutyric acid) in the blood lower the blood pH. The pH of blood is normally maintained by bicarbonate ions controlled by the kidneys.

$$H_3O^+ + HCO_3^- \rightleftharpoons H_2CO_3 + H_2O$$

However, when ketosis occurs, there is insufficient bicarbonate ion to neutralize the acid. A low blood pH makes hemoglobin less efficient in transporting oxygen. Brain cells become oxygen starved; coma and death may result from severe ketosis.

27.5 Biosynthesis of Fatty Acids

All animals can synthesize fatty acids from acetyl CoA, a process known as **lipogenesis.** Humans synthesize fatty acids when more nutrients are digested than are required for biosynthetic reactions or the generation of energy. The primary source of the excess acetyl CoA is the degradation of carbohydrates. Lipogenesis occurs in the cytoplasm of cells in the liver and in adipose tissue.

Biosynthesis Is Not the Reverse of Degradation

As was observed in the case of glycolysis and gluconeogenesis, a biochemical reaction that looks like the reverse of another reaction often does not occur by the same reaction pathway. Synthetic and degradative pathways are always quite distinct, and different enzymes are involved. In the case of fatty acid degradation and synthesis, the two pathways even occur in different cellular compartments. Degradation occurs in the mitochondria, and biosynthesis takes place in the cytoplasm. Thus, the two sets of reactions can proceed at the same time, each controlled separately.

The following statements outline the differences between oxidative degradation and biosynthesis of fatty acids.

1. The intermediates of fatty acid synthesis are bonded to the sulfhydryl group of a protein called **acyl carrier protein (ACP).** In oxidative degradation, the intermediates are bonded to the sulfhydryl group of coenzyme A.
2. All the enzymes required in reductive biosynthesis are joined in a single polypeptide chain called a **fatty acid synthetase** system. The intermediates remain with the system until all reactions are completed. Degradative enzymes in the mitochondria are not associated, and the steps occur independently.
3. The reduction reactions in biosynthesis occur with the coenzyme NADPH. Oxidative degradation uses NAD^+. These two enzymes differ by the extra phosphate group in NADPH (Section 25.9).
4. The growing fatty acid chain in biosynthesis receives its two-carbon units from malonyl ACP, which is formed from acetyl ACP. The two-carbon units generated in oxidative degradation are CoA derivatives.

Formation of Malonyl CoA Starts the Process

An irreversible preliminary step is required to start the series of fatty acid synthesis reactions. This step converts acetyl CoA into malonyl CoA by reaction with bicarbonate and is catalyzed by acetyl CoA carboxylase.

$$\underset{\text{acetyl CoA}}{CH_3-\overset{O}{\overset{\|}{C}}-S-CoA} + ATP + HCO_3^- \xrightarrow{\text{acetyl CoA carboxylase}}$$

$$\underset{\text{malonyl CoA}}{{}^-O-\overset{O}{\overset{\|}{C}}-CH_2-\overset{O}{\overset{\|}{C}}-S-CoA} + ADP + P_i + H^+$$

The reaction proceeds only if the ATP level is high. This situation occurs when the ATP/ADP ratio of the cell is high, and the level of acetyl CoA produced from nutrients is high. Acetyl CoA carboxylase is an allosteric enzyme that requires the vitamin biotin as a cofactor. The key allosteric activator is citrate. Palmitoyl CoA, the end product of fatty acid synthesis, inhibits this first step.

The Chain-Elongation Steps

Fatty acid synthesis, which occurs by a repeating series of steps, is preceded by conversion of CoA derivatives into ACP derivatives. The malonyl ACP required for each turn of the biosynthetic cycle is formed by an acyl transfer reaction catalyzed by malonyl transacylase. A similar conversion of acetyl CoA into acetyl ACP is catalyzed by acetyl transacylase.

$$\text{acetyl CoA} + \text{ACP} \longrightarrow \text{acetyl ACP} + \text{CoA}$$
$$\text{malonyl CoA} + \text{ACP} \longrightarrow \text{malonyl ACP} + \text{CoA}$$

Once malonyl ACP and acetyl ACP are formed, a series of four steps occurs in the multienzyme fatty acid synthetase system to increase the length of the growing fatty acid chain by two carbon atoms (Figure 27.7). Let's now examine each step separately.

FIGURE 27.7 Biosynthesis of Fatty Acids

(a) The four steps in one cycle. (b) The sequence of cycles for synthesis of palmitic acid. Each loop represents one cycle.

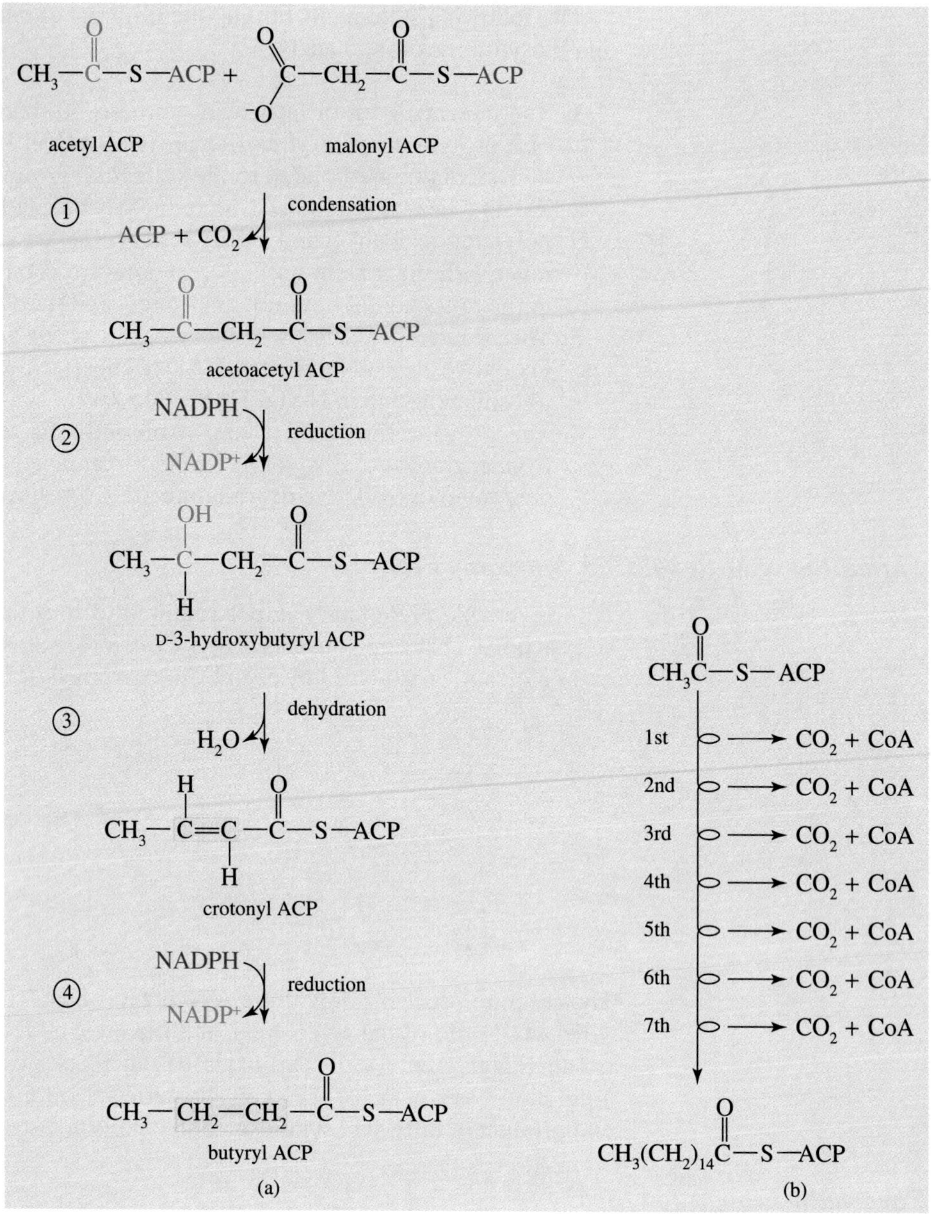

$$CH_3-\overset{O}{\overset{\|}{C}}-S-ACP + {}^{-}O-\overset{O}{\overset{\|}{C}}-CH_2-\overset{O}{\overset{\|}{C}}-S-ACP$$

acetyl ACP, malonyl ACP

① condensation ($\rightarrow ACP + CO_2$)

$$CH_3-\overset{O}{\overset{\|}{C}}-CH_2-\overset{O}{\overset{\|}{C}}-S-ACP$$

acetoacetyl ACP

② reduction (NADPH $\rightarrow$ $NADP^+$)

$$CH_3-\underset{H}{\overset{OH}{C}}-CH_2-\overset{O}{\overset{\|}{C}}-S-ACP$$

D-3-hydroxybutyryl ACP

③ dehydration ($\rightarrow H_2O$)

$$CH_3-\overset{H}{C}=\underset{H}{C}-\overset{O}{\overset{\|}{C}}-S-ACP$$

crotonyl ACP

④ reduction (NADPH $\rightarrow$ $NADP^+$)

$$CH_3-CH_2-CH_2-\overset{O}{\overset{\|}{C}}-S-ACP$$

butyryl ACP

(a)

$$CH_3\overset{O}{\overset{\|}{C}}-S-ACP$$

1st $\rightarrow CO_2 + CoA$
2nd $\rightarrow CO_2 + CoA$
3rd $\rightarrow CO_2 + CoA$
4th $\rightarrow CO_2 + CoA$
5th $\rightarrow CO_2 + CoA$
6th $\rightarrow CO_2 + CoA$
7th $\rightarrow CO_2 + CoA$

$$CH_3(CH_2)_{14}\overset{O}{\overset{\|}{C}}-S-ACP$$

(b)

Step 1: **Condensation** A carbon–carbon bond forms via the biological equivalent of a Claisen condensation. A two-carbon unit and a three-carbon unit join to produce a four-carbon unit and CO_2. Condensation of 2 moles of acetyl ACP to produce acetoacetyl ACP would be unfavorable. However, with malonyl ACP as a reactant, the release of CO_2 provides an important decrease in free energy that drives the reaction. Note that malonyl ACP provides the two additional carbon atoms to the growing fatty acid chain. In subsequent turns of the cycle, malonyl ACP adds two carbon atoms to the acyl ACP derivative of an elongated chain.

Step 2: **Reduction** The ketone is reduced to an alcohol by NADPH. This reaction may look like the reverse of the oxidation of an alcohol to a ketone in the degradation of fatty acids. However, there are two differences. In biosynthesis the D isomer is formed, whereas in oxidative degradation the L isomer is oxidized. NADPH is the reducing agent, whereas NAD^+ is the oxidizing agent in the related fatty acid degradation step.

Step 3: **Dehydration** The alcohol is dehydrated forming the trans isomer. In the first series of reactions, the product is crotonyl ACP. In repeating sequences the general product is a *trans*-Δ^2-enoyl ACP of an elongated unsaturated fatty acid.

Step 4: **Reduction** The alkene is reduced, forming butyryl ACP in the first cycle. NADPH is again the reducing agent. In subsequent turns of the cycle, the product is an acyl ACP containing two additional carbon atoms for each turn.

The Process Is Repeated

The product of step 4 in the series of four steps is the reactant in the next turn of this cycle. In the first cycle, acetyl ACP is converted into butyryl ACP. Butyryl ACP then reacts with malonyl ACP to eventually give a six-carbon acid. Continued cycles produce compounds having 8, 10, . . . , carbon atoms, but elongation ceases at the 16-carbon carboxylic acid, palmitic acid. Further elongation of the chain occurs by an alternative enzyme system that will not be discussed.

Biosynthesis of fatty acids occurs in the fatty acid synthetase system with all intermediates present as ACP derivatives. However, the necessary intermediates are supplied to the system as CoA derivatives. Thus, we can determine the overall stoichiometry of the reaction sequence using CoA derivatives. To produce palmitic acid, the equivalent of eight acetyl CoA precursors are required, and the series of steps must occur seven times. Two of the carbon atoms are provided by acetyl CoA. The remaining 14 carbon atoms are supplied by 7 moles of malonyl CoA. The stoichiometry is

$$\text{acetyl CoA} + 7\text{ malonyl CoA} + 14\text{ NADPH} + 7\text{ H}^+ \longrightarrow \text{palmitate} + 7\text{ CO}_2 + 14\text{ NADP}^+ + 8\text{ CoA} + 6\text{ H}_2\text{O}$$

Malonyl CoA is supplied by adding one carbon atom to acetyl CoA. The equation for the synthesis of the 7 moles of malonyl CoA required for the preceding reaction is

$$7\text{ acetyl CoA} + 7\text{ CO}_2 + 7\text{ ATP} \longrightarrow 7\text{ malonyl CoA} + 7\text{ ADP} + 7\text{ P}_i + 7\text{ H}^+$$

We obtain the overall stoichiometry by adding the two equations.

$$8\text{ acetyl CoA} + 7\text{ ATP} + 14\text{ NADPH} \longrightarrow \text{palmitate} + 14\text{ NADP}^+ + 8\text{ CoA} + 6\text{ H}_2\text{O} + 7\text{ ADP} + 7\text{ P}_i$$

Note that the synthesis of palmitate requires the input of 8 acetyl CoA, 7 ATP, and 14 NADPH. Because each NADPH is equivalent to 3 ATP, the number of ATP required for the biosynthesis is $7 + (3 \times 14) = 49$ ATP.

How Are the Necessary Acetyl Groups Provided?

Two questions can be raised about the biosynthesis of palmitic acid which requires eight molecules of acetyl CoA. First, where do the acetyl CoA molecules come from? Second, how do they get to the site of biosynthesis?

Acetyl CoA is formed in the mitochondria from pyruvate but must be transported to the cytoplasm where it is needed for biosynthesis. However, the mitochondrial membrane is not readily permeable to acetyl CoA. The transport problem is solved because citrate, which is easily transported to the cytoplasm, is formed by condensation of acetyl CoA with oxaloacetate (Figure 27.8). The citrate is then cleaved in the cytoplasm to form the required acetyl CoA. The oxaloacetate formed in this step cannot cross directly back to the mitochondria. Instead, it is reduced to malate and then decarboxylated to give pyruvate.

$$\text{oxaloacetate} + \text{NADH} + \text{H}^+ \rightleftharpoons \text{malate} + \text{NAD}^+$$
$$\text{malate} + \text{NADP}^+ \longrightarrow \text{pyruvate} + \text{CO}_2 + \text{NADPH}$$

The pyruvate then diffuses across the mitochondrial membrane and is carboxylated to form oxaloacetate, which can react with acetyl CoA to give citrate. The net result is a transport of acetyl CoA from the mitochondria to the cytoplasm.

$$\text{pyruvate} + \text{CO}_2 + \text{ATP} + \text{H}_2\text{O} \longrightarrow \text{oxaloacetate} + \text{ADP} + \text{P}_i + 2\,\text{H}^+$$

Sources of NADPH

Where do the 14 moles of NADPH required for the biosynthesis of palmitic acid come from? Part of the answer is provided by summing the preceding three equations representing the reaction of and formation of oxaloacetate. The net equation is

$$\text{NADP}^+ + \text{NADH} + \text{ATP} + \text{H}_2\text{O} \longrightarrow \text{NADPH} + \text{NAD}^+ + \text{ADP} + \text{P}_i + \text{H}^+$$

Thus, one NADPH is formed for every acetyl CoA that is transferred from the mitochondria to the cytoplasm. In the formation of palmitic acid, the required eight molecules of acetyl CoA are formed along with eight molecules of NADPH. The remaining six NADPH molecules required in the balanced equation for the formation of palmitic acid are supplied from the pentose phosphate pathway (Section 26.8).

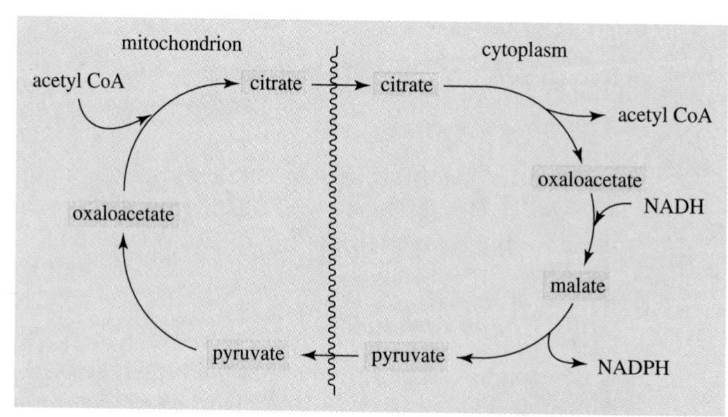

FIGURE 27.8 Transport of Acetyl CoA

Acetyl CoA is transferred as citrate from the mitochondria to the cytoplasm. NADPH is formed in the cytoplasm where it is required for fatty acid biosynthesis.

Summary

About 35% of the American diet consists of fats. Fatty acids in fats cannot be converted into glucose. Fats, which are more reduced than carbohydrates, have a higher caloric content than carbohydrates, which are hydrated.

Fats are digested in the intestine and are then reassembled after passage through the intestinal wall. The fatty acids are used to form tissue lipids and storage fats. **Depot fat,** also called **adipose tissue,** is connective tissue in the body that contains cellular fat in cells called adipocytes. The depot fat serves as a long-term source of chemical energy. The release of fatty acids from adipose tissue is called **fatty acid mobilization.** The reaction involves hydrolysis of the ester bonds by enzymes called **lipases.**

Tissue lipids are contained in cell membranes; they consist of glycerophospholipids, phosphosphingolipids, and glycosphingolipids. These lipids are used to maintain cell membranes and the membranes of cell organelles like the mitochondria. Except in extreme cases of starvation, the tissue lipids are not used to provide energy.

Fatty acids are degraded two carbon atoms at a time and form acetyl CoA. There are four repeating reactions in the **fatty acid cycle.** In each turn of the cycle one molecule of acetyl CoA and a fatty acid with two fewer carbon atoms are produced. A turn of the cycle also produces one molecule each of NADH and $FADH_2$. The reduced coenzymes are oxidized in the electron transport chain, and energy is stored as ATP. The acetyl CoA is oxidized in the citric acid cycle, which, when followed by the oxidation of reduced coenzymes, produces more ATP.

Acetoacetate, D-3-hydroxybutyrate, and acetone are **ketone bodies** that are produced as a result of the excessive oxidation of fatty acids. Accumulation of ketone bodies results in **ketosis,** which affects the pH of blood and therefore the oxygen-carrying efficiency of hemoglobin. A level of ketone bodies in the blood greater than 20 mg/100 mL is called **ketonemia.** The condition of "ketones in the urine" is called **ketonuria.** At still higher ketone body concentrations, **acetone breath** results from excretion of acetone via the lungs.

Fatty acid biosynthesis occurs in the cytoplasm of the cell and is catalyzed by a special enzyme complex called **fatty acid synthetase.** The intermediates in the reaction are bonded to an **acyl carrier protein.** In the first step, a bond is formed between acetyl ACP and malonyl ACP. Subsequent steps involve reduction of a carbonyl group, dehydration of an alcohol, and reduction of an alkene. The reducing agent in fatty acid biosynthesis is NADPH.

Exercises

Digestion of Triglycerides

27.1 What is depot fat?

27.2 What is fatty acid mobilization?

27.3 Where does the digestion of triglycerides begin?

27.4 What are the products of the digestion of triglycerides?

27.5 What happens to the products of lipid digestion after they pass through the intestinal wall?

27.6 How are lipids carried in the blood?

27.7 In what forms are lipids distributed in the body?

27.8 How do tissue lipids differ from adipose tissue?

27.9 How can carbohydrates be converted into fat?

27.10 What process releases fatty acids from adipose tissue?

Catabolic Reactions of Triglycerides

27.11 Why is the metabolic pathway of fatty acids called a cycle?

27.12 Why is ATP hydrolyzed to AMP rather than ADP in the activation of a fatty acid?

27.13 What is the configuration of the unsaturated fatty acid formed by dehydrogenation?

27.14 What is the configuration of the 3-hydroxy fatty acid?

27.15 How many moles of acetyl CoA are produced in the metabolism of stearic acid?

27.16 How many moles of acetyl CoA are produced in the metabolism of palmitic acid?

27.17 How many times does the fatty acid cycle occur for stearic acid?

27.18 How many times does the fatty acid cycle occur for palmitic acid?

27.19 How many moles of acetyl CoA would be produced in the oxidation of decanoic acid?

27.20 How many moles of acetyl CoA would be produced in the oxidation of myristic acid?

27.21 The fatty acid cycle is also called β-oxidation. Explain why.

27.22 Why are fatty acids not convertible into glycogen?

27.23 How is FAD recovered from the $FADH_2$ produced in the fatty acid cycle?

27.24 How is NAD^+ re-formed from the NADH produced in the fatty acid cycle?

27.25 Fats can provide a small quantity of glucose. What part of the fats can undergo gluconeogenic reactions?

27.26 What fraction of the carbon atoms in glyceryl tristearate can undergo gluconeogenic reactions?

27.27 What carboxylic acid is produced in the last turn of the fatty acid cycle if 10-phenyldecanoic acid is introduced into a cell?

27.28 What carboxylic acid is produced in the last turn of the fatty acid cycle if 9-phenylnonanoic acid is introduced into a cell?

27.29 What are ketone bodies?

27.30 What reaction produces ketone bodies?

27.31 Define the term ketonemia.

27.32 Define the term ketonuria.

27.33 What is the condition of acetone breath?

27.34 What is ketosis?

27.35 Why can ketosis result in death?

27.36 What conditions cause the formation of ketone bodies?

Biosynthesis of Fatty Acids

27.37 What binds the intermediates in fatty acid biosynthesis?

27.38 What special feature serves to catalyze the reactions of fatty acid biosynthesis?

27.39 Where does fatty acid biosynthesis occur?

27.40 How is acetyl CoA converted into malonyl CoA?

27.41 How many kcal/mole are required to place each two-carbon unit in a fatty acid chain?

27.42 What types of reactions occur in each of the four steps of fatty acid biosynthesis?

27.43 How many moles of acetyl CoA are required for the biosynthesis of myristic acid?

27.44 How many moles of NADPH and ATP are required to synthesize myristic acid?

27.45 How many moles of ATP are required for the biosynthesis of palmitic acid?

27.46 How many moles of NADPH are required for the biosynthesis of myristic acid?

27.47 How does the NAD^+/NADH ratio affect the oxidation of fatty acids?

27.48 How does the $NADP^+$/NADPH ratio affect the biosynthesis of fatty acids?

28 Metabolism of Proteins

This brown bear in Alaska obtains some protein from salmon caught from the stream. (Peter Arnold, Inc.)

28.1 Digestion of Proteins
28.2 Overview of Amino Acid Metabolism
28.3 Deamination of Amino Acids in Catabolism
28.4 The Urea Cycle
28.5 The Fate of the Carbon Skeleton
28.6 Decarboxylation of Amino Acids
28.7 Biosynthesis of Amino Acids

Essay
- *Diet, Amino Acid Supplements, and Muscles*
- *Nitrogen Balance, Starvation, and Kwashiorkor*
- *Phenylketonuria*

28.1 Digestion of Proteins

Digestion of proteins begins in the stomach and continues in the small intestine. Proteins cannot be absorbed across intestinal membranes, but hydrolysis of their peptide bonds produces the constituent amino acids, which can be transported.

$$\text{—NH}\underset{\underset{R_1}{|}}{\text{CH}}\overset{\overset{O}{\|}}{\text{C}}\text{—NH}\underset{\underset{R_2}{|}}{\text{CH}}\overset{\overset{O}{\|}}{\text{C}}\text{—NH}\underset{\underset{R_3}{|}}{\text{CH}}\overset{\overset{O}{\|}}{\text{C}}\text{—} \xrightarrow{\text{hydrolysis}}$$

(peptide bonds: the two C—N bonds between residues)

$$\text{—NH}\underset{\underset{R_1}{|}}{\text{CH}}\overset{\overset{O}{\|}}{\text{C}}\text{—OH} + \text{NH}_2\underset{\underset{R_2}{|}}{\text{CH}}\overset{\overset{O}{\|}}{\text{C}}\text{—OH} + \text{NH}_2\underset{\underset{R_3}{|}}{\text{CH}}\overset{\overset{O}{\|}}{\text{C}}\text{—}$$

The digestion process is summarized in Figure 28.1. About 10% of the peptide bonds in proteins are hydrolyzed in the stomach. Freshly secreted gastric juice, which is about pH 1, contains HCl. Mixing of the stomach contents with the gastric juice increases the pH of the mixture to about 2. The acid of the stomach denatures

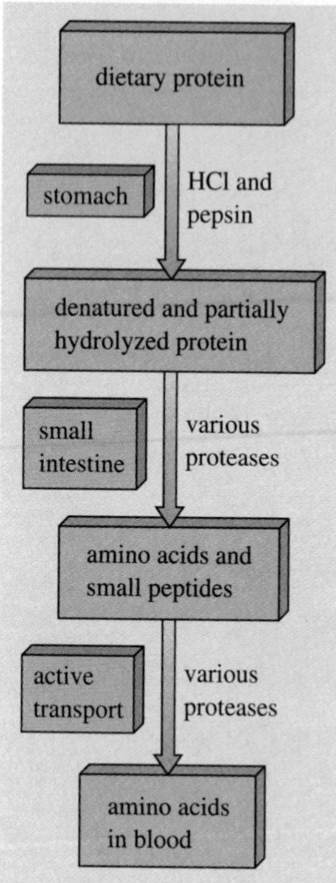

FIGURE 28.1 Digestion of Dietary Protein

the proteins, unfolding the molecules and making the peptide bonds more susceptible to enzymatic hydrolysis. The major enzyme in the stomach is *pepsin,* which catalyzes the hydrolysis of peptide bonds at the nitrogen end of aromatic amino acids (Section 23.10).

Protein digestion is largely completed in the small intestine, where the hydrolysis is catalyzed by the digestive enzymes of the pancreas. These pancreatic secretions contain the zymogens trypsinogen, chymotrypsinogen, and procarboxypeptidase, which form their corresponding enzymes in a series of cascading reactions. The intestinal cell wall releases an enzyme that hydrolyzes trypsinogen to give trypsin. Trypsin then activates chymotrypsinogen to give chymotrypsin and procarboxypeptidase to give carboxypeptidase. Once released in active form these three enzymes and aminopeptidase, which is also secreted in intestinal juice, have a broad range of hydrolytic capabilities. Chymotrypsin attacks the carboxyl end of aromatic acids; trypsin attacks the carboxyl end of the basic amino acids lysine and arginine. Carboxypeptidase is an exopeptidase—that is, it hydrolyzes peptides sequentially from the carboxyl end of the peptide chain. An aminopeptidase is an exopeptidase that hydrolyzes peptides sequentially from the amino end of the peptide chain.

About 60 g of proteins per day are hydrolyzed to amino acids. The free amino acids formed are actively transported across the intestinal wall into the portal vein, where they are absorbed into the blood and are carried to the liver for metabolism. The liver is the site for both the degradation and synthesis of amino acids. The amino acids obtained directly from the diet as well as those synthesized by the liver are transported to the cells of the body to form the various necessary proteins. Amino acids in excess of normal requirements either are converted to other nonprotein nitrogen-containing molecules or are degraded.

28.2 Overview of Amino Acid Metabolism

The role of amino acids in metabolic reactions is distinctly different from those of the components of fats and carbohydrates. The primary role of amino acids is to build proteins, not to supply energy to cells. Amino acids constantly undergo turnover in cells; they are in dynamic equilibrium. Proteins are continually being degraded, rebuilt, and replaced. Proteins in a specific location appear to stay the same, but "old" proteins are hydrolyzed and "new" proteins are synthesized. About 300–400 g of protein in an adult human body are replaced every day. This process requires a collection of free amino acids throughout the body called the **amino acid pool** (Figure 28.2). The amino acids circulate in blood and other fluids throughout the body, and the composition of the pool changes as some amino acids leave while others enter. About 100 g of free amino acids are in circulation. Glutamic acid and the related glutamine account for about 50 g; about 10 g is a mixture of the **essential amino acids,** which can be obtained only from dietary sources. The amino acids are used in four major ways:

1. protein synthesis
2. synthesis of nonprotein nitrogen compounds
3. synthesis of other nonessential amino acids
4. production of ATP, glycogen, or fatty acids

Diet, Amino Acid Supplements, and Muscles

The myth that you must eat a lot of meat to form the protein required to build muscles is just that—a myth! Proteins are not transported whole and placed in your muscles. The body degrades unneeded amino acids and converts some of them to carbohydrates but most of them to fats. Thus, overeating meat can actually make you fat unless you exercise enough to burn the larger number of calories that result from this diet. Muscles are built slowly and only through exercise. The required amino acids are obtained in a balanced diet that includes some protein. However, the amino acids can also be synthesized by the body from other food sources.

A lot of money is wasted by some of the public who purchase amino acid supplements to help them bulk up by taking large doses of these compounds. Others buy them because there are reputed medical benefits to ingesting large amounts of certain amino acids. Leucine and phenylalanine are touted as pain relievers. Tryptophan and tyrosine are said to cure insomnia and depression. Even if these benefits are realized, the ingestion of large doses of amino acid supplements is dangerous. Studies in animals show that the body cannot handle abnormal amounts of amino acids, and imbalances can cause a variety of other ailments. This result is not unexpected. Our bodies have been genetically conditioned to work with reasonable amounts of biomolecules as produced by normal metabolic processes. In other words, if it isn't broken, don't try to fix it.

The excessive use of L-tryptophan caused several hundred cases of a blood disorder (eosinophilia–myalgia syndrome) in 1989. Hospitalization was required in about half the cases, some of which were very serious. However it turned out that the cause of the disorder was not L-tryptophan but an impurity in the commercial source. Nevertheless, if large doses had not been taken, the disorder would not have occurred.

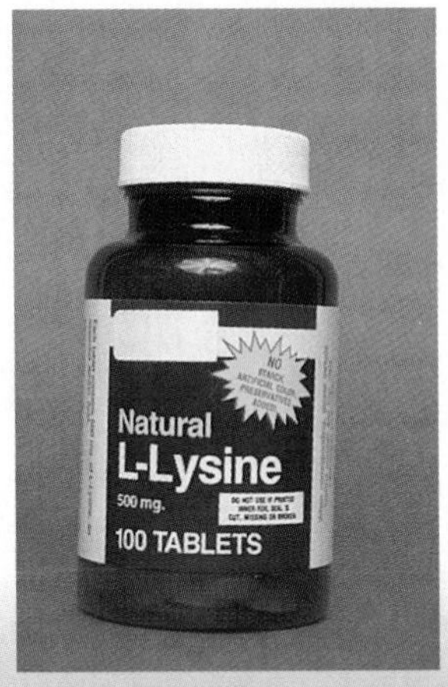

L-Lysine, one of the essential amino acids can be purchased. (Peter Arnold, Inc.)

Protein Synthesis

About 75% of the amino acids in the amino acid pool are used to build proteins. The time required to degrade a protein and replace it is the **protein turnover rate.** This value is a half-life, the time required to replace half of a specific protein. The time in residence as a protein varies considerably based on the function of the protein. Peptide hormones and enzymes have turnover rates of a few minutes. For liver and blood proteins the turnover rate is about 6 days. Muscle proteins have a turnover rate of about 6 months. Collagens, the structural proteins, are replaced in about 3 years. The synthesis of proteins is under the control of a group of ribonucleic acids called mRNA (Chapter 29).

Nitrogen Balance

Amino acids are the principal source of the nitrogen atoms needed to synthesize a variety of nitrogen-containing biomolecules. Examples include neurotransmitters, some hormones, the heme of hemoglobin, the choline and ethanolamine of glycerophospholipids, and the purines and pyrimidines of nucleic acids. These varied processes are beyond the scope of this text.

There is no "permanent" storage form for amino acids as there is for carbohydrates and lipids. Thus, the nitrogen atoms derived from the metabolism of proteins must be either used or excreted. Only a small fraction of the nitrogen atoms of the

FIGURE 28.2
Amino Acid Pool

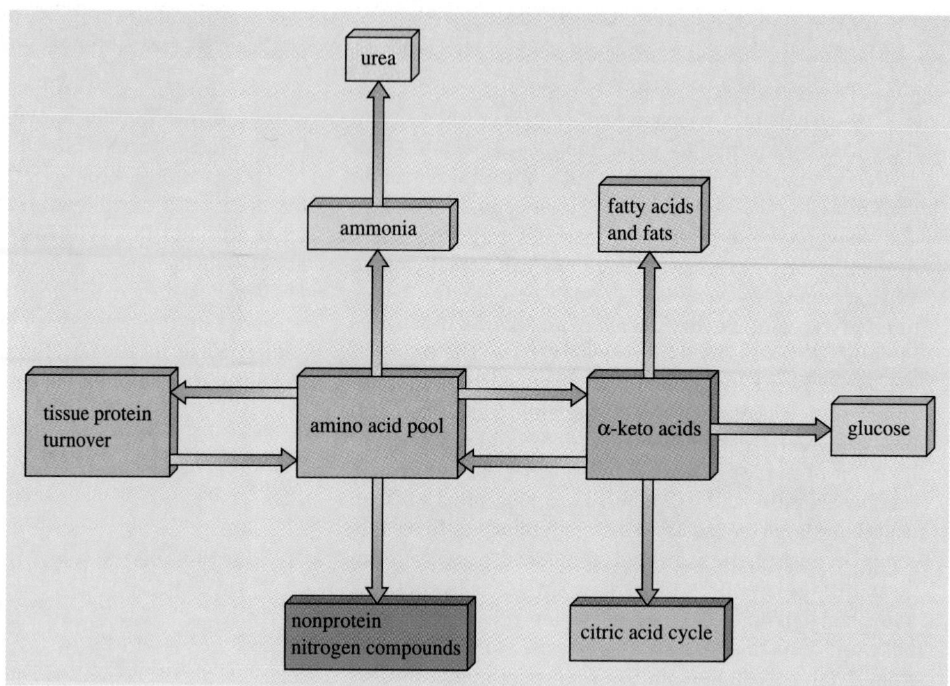

body are temporarily stored in the amino acid pool, largely as glutamic acid and the related glutamine. The **nitrogen balance** is controlled by the daily intake of nitrogen compounds and the loss of nitrogen compounds by excretion. During periods of growth, the nitrogen balance must be positive (intake must exceed excretion) so that proteins and other nitrogen-containing compounds can be synthesized. During fasting (excretion of nitrogen exceeds intake), a negative balance results.

Synthesis of Nonessential Amino Acids

Of the set of 20 amino acids, humans must obtain a group of 9 from their diets. These **essential amino acids** are so named because they either cannot be synthesized from other amino acids or other nitrogen compounds or cannot be synthesized at a sufficiently rapid rate for biosynthesis of proteins. The "nonessential" amino acids are equally needed by the body but can be produced by biosynthetic reactions to be discussed in this chapter. Most of the nonessential amino acids can be made in 1–3 steps. The essential amino acids require 7–10 steps. It is interesting that plants and many microorganisms can make all the amino acids, but humans cannot. Thus, we have to rely on dietary proteins to maintain ourselves. One important process that we and other animals use to synthesize amino acids is called **transamination.** This reaction, which transfers an α-amino group from an α-amino acid to an α-keto acid, is discussed in Section 28.3. The reactions are catalyzed by enzymes called transaminases or aminotransferases.

$$\mathrm{R{-}\overset{\displaystyle H}{\overset{|}{\underset{\underset{\displaystyle NH_2}{|}}{C}}}{-}CO_2H} + \mathrm{R'{-}\overset{\displaystyle O}{\overset{\|}{C}}{-}CO_2H} \longrightarrow \mathrm{R{-}\overset{\displaystyle O}{\overset{\|}{C}}{-}CO_2H} + \mathrm{R'{-}\overset{\displaystyle H}{\overset{|}{\underset{\underset{\displaystyle NH_2}{|}}{C}}}{-}CO_2H}$$

Nitrogen Balance, Starvation, and Kwashiorkor

The fate of amino acids obtained from dietary sources depends on the requirements of the organism at that time. A normal diet provides sufficient quantities of amino acids for biosynthesis, and any excess is either used for fuel or stored as glycogen or fat. An adult requires only about 0.8 g of dietary protein per kilogram of body weight. Thus, a 130-lb (60-kg) individual requires only 48 g of dietary protein. Although the required amount of protein is small, there are areas of the world where even this amount is not available or where the dietary sources have a poor balance of the essential amino acids. Children and pregnant women suffer the most because they require a positive nitrogen balance to form protein.

You undoubtedly have seen pictures of malnourished children in the tropics of Africa who have distended stomachs. This condition results from a disease called kwashiorkor. After children are weaned and no longer have a source of milk, they may have available only high-carbohydrate, low-protein foods. With insufficient protein, they not only are underdeveloped but also develop edema.

Wasted individuals, who are so emaciated that their bones show through the skin, are shown in news releases from areas of war or famine. These starving people have reached a stage at which they can no longer call on stored glycogen or fats. The glycogen stores are rapidly depleted, and then the fat reserves are used. After fat is depleted from vital organs such as the kidneys, heart, and even bone marrow, the body then metabolizes body protein. Thus, the body literally eats itself. A condition in which the body must rely on structural protein to provide intermediates for energy production is not one that can be sustained for long. Not only does the individual become more susceptible to disease but the organs start to fail. Blindness can result, and eventually circulatory failure occurs because the heart is so weakened.

Temporary starvation conditions are more tolerable for adults than for children. The developing organs and especially the brain require balanced diets. Thus, children who experience starvation not only may show stunted growth but may be retarded as well.

Oxidation of Amino Acids

Each of the 20 amino acids found in proteins has a different oxidative degradation pathway. Although these pathways are diverse in their complexity, there are some common features that can be described. It is convenient to divide these related degradation processes into two parts, the reactions of the carbon skeleton and the reactions involving the α-amino groups (Figure 28.3). The metabolic pathways involving the carbon atoms provide intermediates for the citric acid cycle. You already know that the citric acid cycle is a source of considerable biochemical energy. In addition to providing intermediates for the citric acid cycle, some of the carbon atoms of amino acids are stored as glycogen or fat; however, amino acids provide only about 15% of the total energy needs of an adult.

The metabolic reactions of amino acids that use nitrogen atoms to form nonprotein nitrogen-containing compounds are very complex, and we will not consider them here. We will however consider the ultimate fate of nitrogen atoms as they are converted into urea in the urea cycle (Section 28.4).

28.3 Deamination of Amino Acids in Catabolism

At some point, all amino acids that are not incorporated in proteins are deaminated—the α-amino groups are removed from the original amino acid. The majority of the amino acids are deaminated by **transamination,** which converts them into α-keto acids. In transamination the nitrogen atom of one amino acid is retained in a second amino acid. The second common mechanism of nitrogen removal is **oxidative deamination.** In this process the nitrogen atoms are lost as materials that are excreted by the organism. Two amino acids, serine and threonine, undergo **direct deamination** by a dehydration–hydration process.

Transamination—The Role of Glutamate

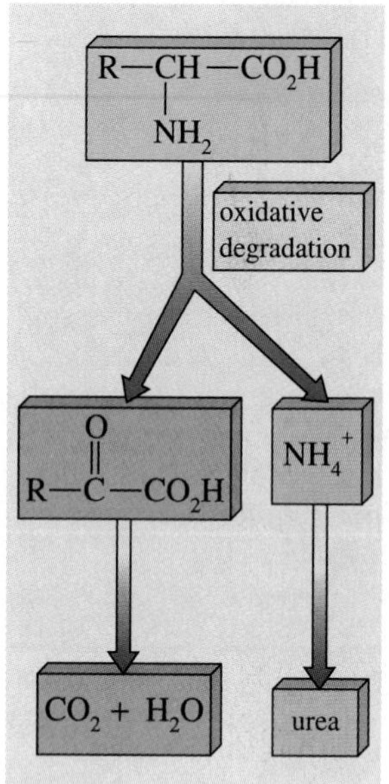

FIGURE 28.3 Oxidation of Amino Acids

Transamination exchanges nitrogen atoms between amino acids and one of three α-keto acids—pyruvate, α-ketoglutarate, or oxaloacetate. The general equation is

$$R-\underset{\displaystyle NH_3^+}{\underset{|}{CH}}-CO_2^- + R'-\overset{\displaystyle O}{\overset{\|}{C}}-CO_2^- \xrightleftharpoons{\text{transaminase}} R-\overset{\displaystyle O}{\overset{\|}{C}}-CO_2^- + R'-\underset{\displaystyle NH_3^+}{\underset{|}{CH}}-CO_2^-$$

Although the transamination reaction appears to be a simple transfer of nitrogen atoms between two molecules, the reaction actually involves several steps and coenzyme intermediates that carry the nitrogen atom. Pyridoxal phosphate and pyridoxamine phosphate, two forms of vitamin B_6, react as outlined in Figure 28.4. Pyridoxal phosphate reacts with amino acids to give α-keto acids and pyridoxamine phosphate. In the "reverse" reaction α-keto acids react with pyridoxamine phosphate to give amino acids and pyridoxal phosphate.

CHO, HO, $CH_2OPO_3^{2-}$, CH_3, N — pyridoxal phosphate

CH_2NH_2, HO, $CH_2OPO_3^{2-}$, CH_3, N — pyridoxamine phosphate

There are at least 50 transaminases, but each requires pyridoxal phosphate. Each transaminase has a twofold specificity and catalyzes a specific reaction. The primary specificity is for the α-keto acid, which is the amino group acceptor. To a lesser extent transaminases have some specificity for the α-amino acid that donates the amino group. The majority of transaminases are specific for α-ketoglutarate as the acceptor α-keto acid, which is converted into glutamate. Transaminases that use pyruvate and oxaloacetate yield alanine and aspartate, respectively. Ultimately even the alanine and aspartic acid react via a transamination reaction with α-ketoglutarate to give glutamate. For example, aspartate aminotransferase catalyzes the transfer of the amino group from aspartate to α-ketoglutarate.

$$\underset{\text{aspartate}}{HO_2CCH_2\underset{\displaystyle NH_3^+}{\underset{|}{CH}}CO_2^-} + \underset{\alpha\text{-ketoglutarate}}{HO_2CCH_2CH_2\overset{\displaystyle O}{\overset{\|}{C}}CO_2^-} \rightleftharpoons \underset{\text{oxaloacetate}}{HO_2CCH_2\overset{\displaystyle O}{\overset{\|}{C}}CO_2^-} + \underset{\text{glutamate}}{HO_2CCH_2CH_2\underset{\displaystyle NH_3^+}{\underset{|}{CH}}CO_2^-}$$

Another transaminase, alanine aminotransferase, catalyzes the transfer of the amino group from alanine to α-ketoglutarate.

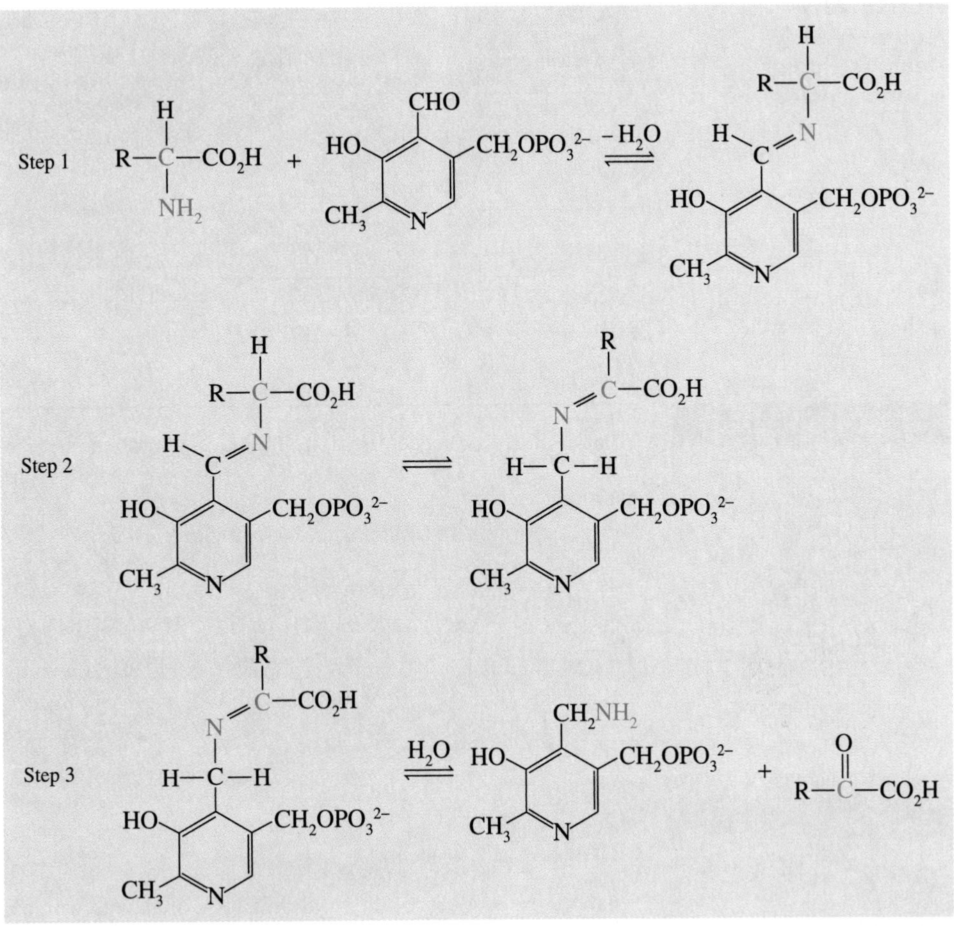

FIGURE 28.4 Mechanism of a Transamination Reaction

Transamination requires two forms of vitamin B_6 and occurs in three steps. In step 1, an imine of the amino acid forms. In step 2, an isomerization occurs. In step 3, hydrolysis of an imine gives the α-keto acid and the second form of vitamin B_6. By the reverse of these steps an α-keto acid is converted into an amino acid.

$$\underset{\text{alanine}}{CH_3CH(NH_3^+)CO_2^-} + \underset{\alpha\text{-ketoglutarate}}{HO_2CCH_2CH_2C(=O)CO_2^-} \rightleftharpoons \underset{\text{pyruvate}}{CH_3C(=O)CO_2^-} + \underset{\text{glutamate}}{HO_2CCH_2CH_2CH(NH_3^+)CO_2^-}$$

Whatever the pathway of transamination, the final acceptor of an amino group from many amino acids is α-ketoglutarate, which produces glutamate.

$$\underset{\alpha\text{-amino acid}}{RCH(NH_3^+)CO_2^-} + \underset{\alpha\text{-ketoglutarate}}{HO_2CCH_2CH_2C(=O)CO_2^-} \rightleftharpoons \underset{\alpha\text{-keto acid}}{RC(=O)CO_2^-} + \underset{\text{glutamate}}{HO_2CCH_2CH_2CH(NH_3^+)CO_2^-}$$

The glutamate produced in this reaction and as a result of a series of other transaminations can be oxidatively deaminated to give α-ketoglutarate and NH_4^+ (Figure 28.5). The α-ketoglutarate is used in the citric acid cycle. The ammonium ion is eliminated in the urea cycle. Glutamate is also the major donor of amino groups in the biosynthesis of amino acids from α-keto acids in reactions catalyzed by glutamate transaminase.

FIGURE 28.5 Summary of Deamination of α-Amino Acids

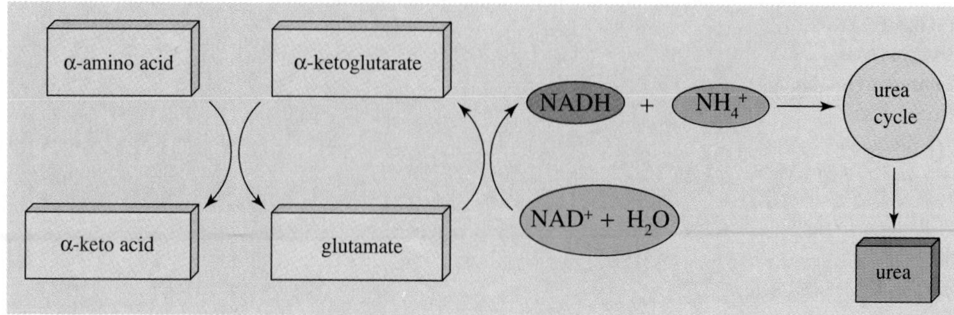

EXAMPLE 28.1

What are the products of the transamination of cysteine with α-ketoglutaric acid?

Solution

Draw the formulas of cysteine and α-ketoglutaric acid.

$$HSCH_2\overset{\overset{\displaystyle NH_2}{|}}{C}HCO_2H \qquad HO_2CCH_2CH_2\overset{\overset{\displaystyle O}{\|}}{C}CO_2H$$

Now, interchange the amino group of cysteine with the keto group of α-ketoglutaric acid, making sure that the proper number of bonds is present in each product.

$$HSCH_2\overset{\overset{\displaystyle O}{\|}}{C}CO_2H \qquad HO_2CCH_2CH_2\overset{\overset{\displaystyle NH_2}{|}}{C}HCO_2H$$

PROBLEM 28.1

What are the products of the transamination of alanine with oxaloacetic acid?

Oxidative Deamination

Oxidative deamination is the conversion of an amino acid into a keto acid with the release of an ammonium ion. We saw that in transamination the α-amino group is transferred to α-ketoglutaric acid to form glutamic acid. Oxidative deamination of the glutamate formed in transaminations unloads the nitrogen of amino acids as ammonium ions, which are then converted to urea in the urea cycle (Section 28.4). The oxidative deamination is catalyzed by glutamate dehydrogenase; either NADP$^+$ or NAD$^+$ can be the coenzyme for the reaction. In the first step the amine is dehydrogenated to form an imine, which is then hydrolyzed to the keto acid.

$$\begin{array}{c} CO_2^- \\ | \\ {}^+H_3N{-}CH \\ | \\ CH_2 \\ | \\ CH_2 \\ | \\ CO_2^- \end{array} \underset{NAD^+ \quad NADH + H^+}{\rightleftharpoons} \begin{array}{c} CO_2^- \\ | \\ {}^+H_2N{=}C \\ | \\ CH_2 \\ | \\ CH_2 \\ | \\ CO_2^- \end{array} \underset{H_2O \quad NH_4^+}{\rightleftharpoons} \begin{array}{c} CO_2^- \\ | \\ O{=}C \\ | \\ CH_2 \\ | \\ CH_2 \\ | \\ CO_2^- \end{array}$$

glutamate (an iminium ion) α-ketoglutarate

The activity of glutamate dehydrogenase is allosterically regulated. Both ATP and GTP (guanosine triphosphate) inhibit the enzyme, whereas ADP and GDP (guanosine diphosphate) activate it. Thus, a decrease in the amount of ATP in the cell accelerates the oxidative deamination of amino acids. This means that amino acids are used as an energy source when fatty acids and carbohydrates are not available.

The net reaction of transamination of an α-amino acid followed by oxidative deamination is

$$\alpha\text{-amino acid} + NAD^+ + H_2O \longrightarrow \alpha\text{-keto acid} + NH_4^+ + NADH + H^+$$

When the α-amino acid is glutamic acid, the product, which is α-ketoglutarate, enters the citric acid cycle and is oxidized.

The conversion of glutamic acid into α-ketoglutarate by deamination is reversible, and it is a major link in the metabolism of proteins, fats, and carbohydrates. The reverse reaction, which converts α-ketoglutarate to glutamate, is one of the few ways to convert nitrogen in the form of ammonium ions into amino acids. Once glutamate forms, it can transfer the nitrogen atom to keto acids, giving other amino acids. It is necessary only that the organism have the appropriate α-keto acids. This availability determines the difference between essential and nonessential amino acids.

Direct Deamination of Serine and Threonine

Both serine and threonine are unique amino acids that have a β-hydroxyl group. They are dehydrated by serine dehydrase and threonine dehydrase, respectively. Loss of water from serine gives aminoacrylate. This compound, which is the nitrogen analog of an enol (Section 17.9), can tautomerize to an imine, which is the nitrogen analog of the keto form. Hydrolysis of this reactive compound gives a ketone. Threonine goes through a similar series of steps.

$$^+H_3N{-}CH(CO_2^-){-}CH_2OH \xrightarrow[-H_2O]{\text{serine dehydrase}} {}^+H_3N{-}C(CO_2^-){=}CH_2 \rightleftharpoons {}^+H_2N{=}C(CO_2^-){-}CH_3 \xrightarrow[H_2O \;\; NH_4^+]{} O{=}C(CO_2^-){-}CH_3$$

serine — aminoacrylate — pyruvate

EXAMPLE 28.2

What is the product of the oxidative deamination of aspartic acid?

Solution

First, write the formula for aspartic acid. Next, remove the amino group and a hydrogen atom from the atom with the amino group. Then add a carbonyl oxygen atom.

$$HO_2CCH_2CH(NH_2)CO_2H$$

$$HO_2CCH_2C(=O)CO_2H$$

PROBLEM 28.2

What is the product of the oxidative deamination of phenylalanine?

28.4
The Urea Cycle

This anemone fish in Fiji releases excess nitrogen directly into its environment as ammonium ions. (Robert J. Ouellette)

Nitrogen atoms not used for biosynthesis are converted into ammonia. Animals differ in the way that they handle excess ammonia. All marine animals excrete the NH_4^+ directly into their aqueous environment. Birds, terrestrial reptiles, and insects convert NH_4^+ into uric acid, which is excreted. Terrestrial vertebrates convert NH_4^+ into urea in the urea cycle, and the urea is then excreted. The average adult excretes about 30 g of urea daily, but this amount can be much higher for a high-protein diet.

urea   uric acid

The synthesis of urea, which occurs in the liver, involves some of the intermediates in the citric acid cycle and two amino acids that do not occur in proteins—ornithine and citrulline. The cycle is summarized in Figure 28.6. The reactions will be discussed in the sequence in which they occur. One of the nitrogen atoms of urea is transferred from aspartate; the other is derived from NH_4^+. In order to keep track of the nitrogen atoms involved in the net reaction to form urea, these atoms are colored in each structure.

Carbamoyl phosphate is synthesized from NH_4^+, CO_2, ATP, and water in step 1. The reaction, which is essentially irreversible, consumes 2 moles of ATP. Carbon dioxide and one of the nitrogen atoms are brought into the cycle in this step.

FIGURE 28.6
The Urea Cycle

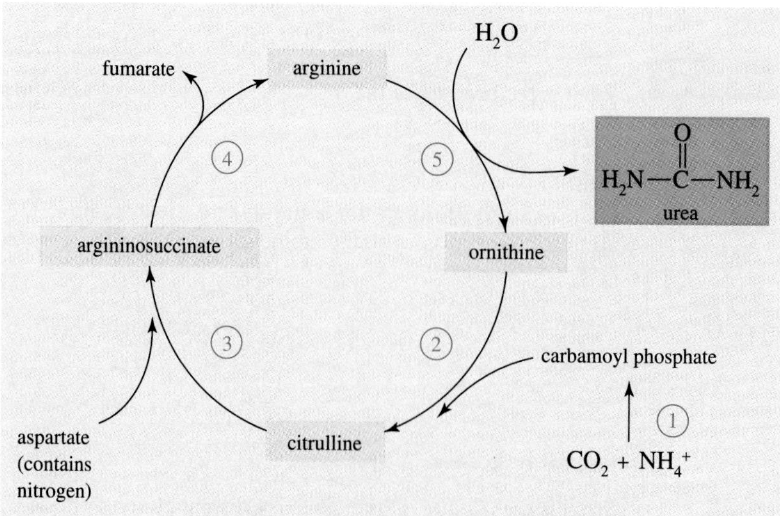

Step 1

$$CO_2 + NH_4^+ + 2\,ATP + 2\,H_2O \longrightarrow {}^+NH_2{-}\overset{\displaystyle O}{\overset{\|}{C}}{-}O{-}\overset{\displaystyle O}{\overset{\|}{P}}(O^-){-}O^- + 2\,ADP + P_i + 3\,H^+$$

carbamoyl phosphate

Ornithine is the carrier of the carbon and nitrogen atoms needed to form urea. In step 2 ornithine combines with the carbamoyl phosphate to form citrulline.

Step 2

$$\underset{\text{ornithine}}{NH_3^+{-}(CH_2)_3{-}CH(NH_3^+){-}CO_2^-} + \underset{\text{carbamoyl phosphate}}{NH_2{-}\overset{\displaystyle O}{\overset{\|}{C}}{-}OPO_3^{2-}} \longrightarrow \underset{\text{citrulline}}{NH_2{-}\overset{\displaystyle O}{\overset{\|}{C}}{-}NH{-}(CH_2)_3{-}CH(NH_3^+){-}CO_2^-} + P_i$$

Citrulline condenses with aspartate in step 3. Note that the amino group of aspartate bonds to the carbon atom brought to citrulline in step 2. Two nitrogen atoms are bonded to a common carbon atom. This unit will eventually be released as urea. This reaction is driven by cleavage of ATP to AMP and the subsequent rapid hydrolysis of pyrophosphoric acid (PP_i).

Step 3

$$\underset{\text{citrulline}}{NH_2{-}\overset{\displaystyle O}{\overset{\|}{C}}{-}NH{-}(CH_2)_3{-}CH(NH_3^+){-}CO_2^-} + \underset{\text{aspartate}}{{}^+H_3N{-}CH(CO_2^-){-}CH_2{-}CO_2^-} \xrightarrow[\text{ATP}\ \ \text{AMP} + PP_i]{} \underset{\text{argininosuccinate}}{{-}O_2C{-}CH_2{-}CH(CO_2^-){-}NH{-}\overset{\displaystyle NH_2^+}{\overset{\|}{C}}{-}NH{-}(CH_2)_3{-}CH(NH_3^+){-}CO_2^-}$$

In step 4, argininosuccinate is cleaved into arginine and fumarate. The carbon atoms of aspartate are now in the fumarate. The fate of fumarate and the relationship between the citric acid cycle and the urea cycle will be discussed shortly.

Step 4

$$\underset{\text{argininosuccinate}}{{}^-O_2C{-}CH_2{-}CH(CO_2^-){-}NH{-}\overset{\displaystyle NH_2^+}{\overset{\|}{C}}{-}NH{-}(CH_2)_3{-}CH(NH_3^+){-}CO_2^-} \longrightarrow \underset{\text{arginine}}{NH_2{-}\overset{\displaystyle NH_2^+}{\overset{\|}{C}}{-}NH{-}(CH_2)_3{-}CH(NH_3^+){-}CO_2^-} + \underset{\text{fumarate}}{(H)(CO_2^-)C{=}C(H)(CO_2^-)}$$

Finally, in step 5, arginine is hydrolyzed to release urea and ornithine. The cycle is now complete, and the ornithine is available for another turn around the cycle.

Step 5

$$\underset{\text{arginine}}{\text{H}-\underset{\underset{CO_2^-}{|}}{\overset{\overset{\text{NH}-\overset{\overset{NH_2^+}{\|}}{C}-NH_2}{|}}{\overset{(CH_2)_3}{\overset{|}{C}}}}-NH_3^+} \xrightarrow{H_2O} \underset{\text{ornithine}}{\text{H}-\underset{\underset{CO_2^-}{|}}{\overset{\overset{NH_3^+}{|}}{\overset{(CH_2)_3}{\overset{|}{C}}}}-NH_3^+} + \underset{\text{urea}}{O{=}\overset{\overset{NH_2}{|}}{C}-NH_2}$$

The reactions of the urea cycle occur at two different sites. The formation of NH_4^+ from amino acids, conversion into carbamoyl phosphate (step 1), and synthesis of citrulline (step 2) occur in the mitochondrial matrix; steps 3, 4, and 5 occur in the cytoplasm. The urea cycle requires three molecules of ATP per molecule of urea formed. Two ATP are used in step 1, and one ATP is used in step 3. However, since ATP is converted to AMP instead of ADP in step 3, and the PP_i is rapidly hydrolyzed, the energy equivalent of four ATP molecules is expended. The stoichiometry of the urea cycle is

$$HCO_3^- + NH_4^+ + 2\,H_2O + \text{aspartate} + 3\,\text{ATP} \longrightarrow \text{urea} + \text{fumarate} + 2\,\text{ADP} + 2\,P_i + \text{AMP} + PP_i$$

The Urea Cycle Is Linked to the Citric Acid Cycle

The fumarate released in step 4 of the urea cycle is also a component of the citric acid cycle, where it is hydrated to malate, which in turn is oxidized to oxaloacetate. The oxaloacetate can be transaminated to form the aspartate necessary for continued operation of the urea cycle. The linkage of the urea cycle, the three components of the citric acid cycle, and the transamination step are shown in Figure 28.7. Alternatively, the oxaloacetate can be converted into glucose by gluconeogenesis or into citrate by reaction with acetyl CoA as part of the citric acid cycle.

Hyperammonemia

The ammonia formed in oxidative deamination is toxic even at low concentrations and must be excreted. Levels of 5 mg per 100 mL of blood in humans are toxic. Un-

FIGURE 28.7 Linkage of the Urea and Citric Acid Cycles
Fumarate from argininosuccinate in the urea cycle is incorporated into the citric acid cycle. Oxaloacetate from the citric acid cycle is transaminated to aspartate, which is incorporated into the urea cycle.

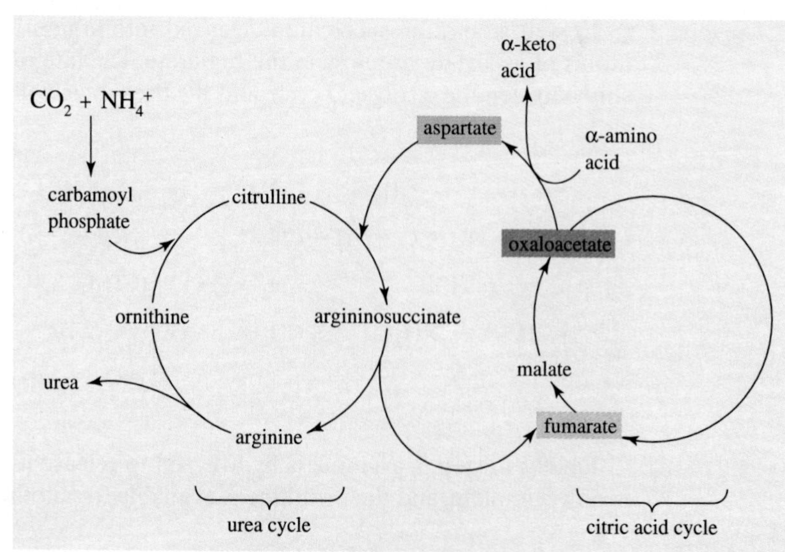

der normal circumstances the concentration of ammonia is in the 1–3 μg/100 ml range. An increase in ammonia concentrations causes tremors, blurred vision, and slurred speech. Still higher concentrations, as occur in a condition called **hyperammonemia,** can cause mental retardation or coma and death.

Hyperammonemia results from a deficiency of any of the enzymes of the urea cycle. High levels of NH_4^+ are toxic because a shift occurs in the equilibrium catalyzed by glutamate dehydrogenase, which converts α-ketoglutarate to glutamate. This depletion of α-ketoglutarate, which is a vital intermediate of the citric acid cycle, reduces the rate of formation of ATP. Hyperammonemia is diagnosed by the elevated level of glutamine in the cerebrospinal fluid.

28.5 The Fate of the Carbon Skeleton

The removal of α-amino groups from any of the 20 amino acids gives α-keto acids. In this section we consider the fate of the carbon skeleton of these α-keto acids. Each amino acid goes through multistep reactions requiring numerous specialized enzymes. Although the details of these steps are beyond the scope of this book, they all eventually converge in steps that lead to the citric acid cycle (Figure 28.8). The 20 amino acids are converted into seven compounds: pyruvate, acetyl CoA, acetoacetyl CoA, α-ketoglutarate, succinyl CoA, fumarate, and oxaloacetate.

Amino acids that are degraded to pyruvate, α-ketoglutarate, succinyl CoA, fumarate, or oxaloacetate are classified as **glucogenic.** The compounds formed from these amino acids can be converted into phosphoenolpyruvate and into glucose via gluconeogenesis.

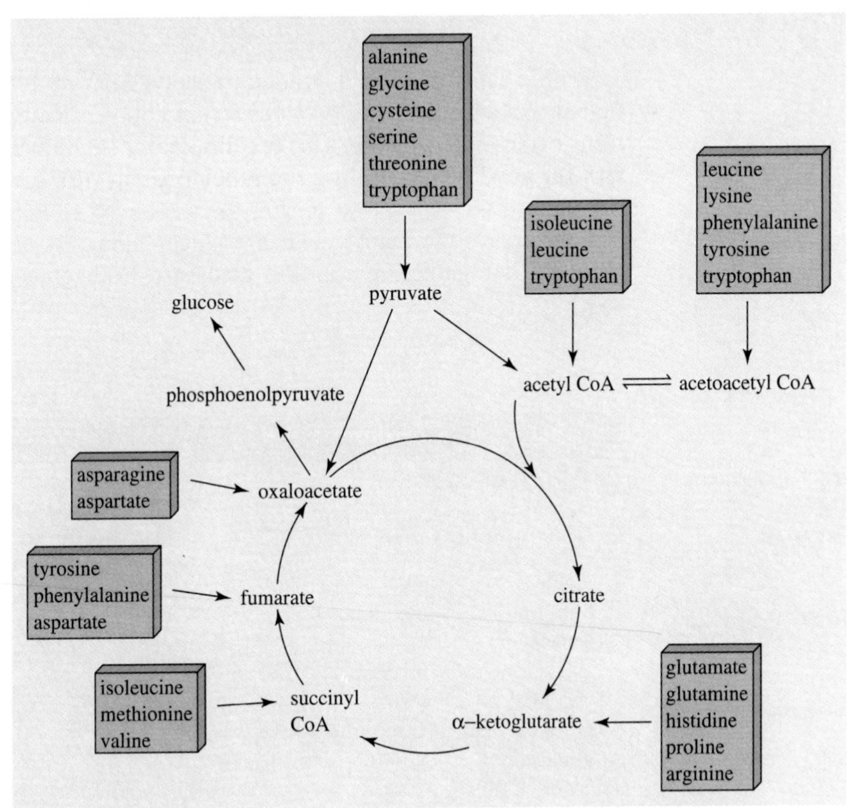

FIGURE 28.8 Fate of Carbon Skeletons of Amino Acids
The α-keto acids produced from amino acids are major metabolic intermediates that can be converted into glucose or oxidized in the citric acid cycle. Entry points of intermediates into the citric acid cycle are indicated for each amino acid.

Phenylketonuria

The degradation of phenylalanine involves conversion into tyrosine catalyzed by the enzyme phenylalanine-4-monoxygenase. In individuals lacking this enzyme, phenylalanine undergoes a transamination reaction to yield phenylpyruvic acid.

$$C_6H_5-CH_2-\underset{\displaystyle NH_2}{\underset{|}{CH}}-CO_2H$$

phenylalanine

$$C_6H_5-CH_2-\overset{\displaystyle O}{\overset{\|}{C}}-CO_2H$$

phenylpyruvic acid

The phenylpyruvic acid accumulates in the body and inhibits the conversion of pyruvic acid to acetyl CoA, which deprives the cells of the energy normally derived via glycolysis of glucose. Because the glycolysis reaction is the most important one for the brain, which depends on glucose as its primary source of energy, mental retardation will develop in newborn infants who lack the enzyme. This genetic disease, which can be detected early by the amount of phenylpyruvic acid in the urine, is called **phenylketonuria** or **PKU.**

Blood tests of newborn children, which are required in the United States, can detect elevated levels of phenylpyruvic acid. Restriction of phenylalanine from the diet for the first 6 years of life allows the brain to develop normally. A normal diet is usually resumed when the brain is fully developed.

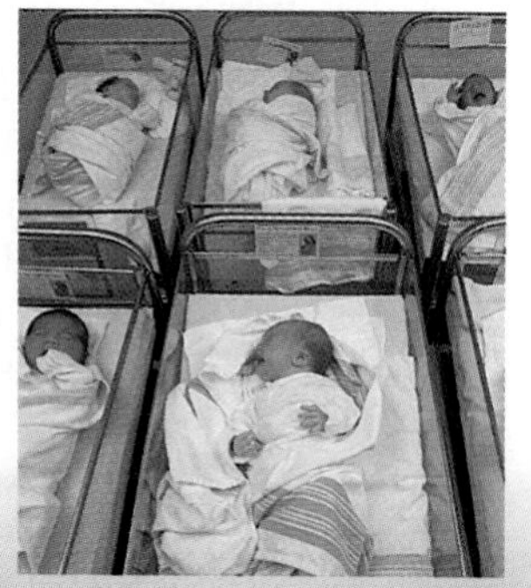

These newborn babies in Danbury, Connecticut have been tested to determine whether they have the genetic disease phenylketonuria. (Stock Market)

Amino acids that are degraded to acetyl CoA or acetoacetyl CoA are called **ketogenic** because they can produce ketone bodies. Ketogenic amino acids cannot be transformed into glucose. This fact illustrates that a balanced protein diet is important for good health. Eating too much protein with too little carbohydrate may result in ketosis.

Fourteen of the amino acids are glucogenic; only one amino acid, leucine, is exclusively ketogenic. Five amino acids are both glucogenic and ketogenic (Table 28.1).

TABLE 28.1 Glucogenic and Ketogenic Amino Acids

Glucogenic		Glucogenic and Ketogenic	Ketogenic
alanine	glycine	isoleucine	leucine
arginine	histidine	lysine	
asparagine	methionine	phenylalanine	
aspartic acid	proline	tryptophan	
cysteine	serine	tyrosine	
glutamic acid	threonine		
glutamine	valine		

28.6 Decarboxylation of Amino Acids

Although most amino acids transfer their amino groups by transamination or are oxidatively deaminated to form α-keto acids, some also are decarboxylated. Decarboxylation is the loss of CO_2 (Section 19.6). The reaction is represented by the general equation

$$R{-}\underset{\displaystyle |}{\overset{\displaystyle NH_2}{CH}}{-}CO_2H \longrightarrow R{-}\overset{\displaystyle NH_2}{\underset{}{CH}}{-}H + CO_2$$

The amino acids arginine, lysine, histidine, and tyrosine are decarboxylated by bacterial decarboxylases that occur in intestinal bacteria. The product amines, called *ptomaines,* are dangerous if they accumulate.

EXAMPLE 28.3

What is the structure of the amine produced by the decarboxylation of lysine?

Solution

First, draw the structure of lysine. Then remove the elements of CO_2 and add a hydrogen atom to the carbon atom that lost the CO_2. This process maintains the stoichiometry and the correct number of bonds to carbon.

$$NH_2CH_2CH_2CH_2CH_2\overset{\displaystyle NH_2}{CH}CO_2H$$

$$NH_2CH_2CH_2CH_2CH_2\overset{\displaystyle NH_2}{CH_2}$$

This compound, known as cadaverine (Chapter 19), is a product of animal decay.

PROBLEM 28.3

What amino acid, when decarboxylated, would yield the following amine?

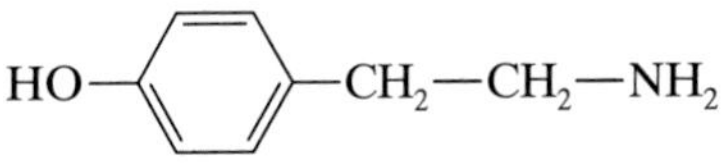

$$HO{-}C_6H_4{-}CH_2{-}CH_2{-}NH_2$$

Histamine and Antihistamines

The decarboxylation of the amino acid histidine produces histamine. Histamine that is released in tissues stimulates nerve endings and produces allergic reactions that cause pain and itching.

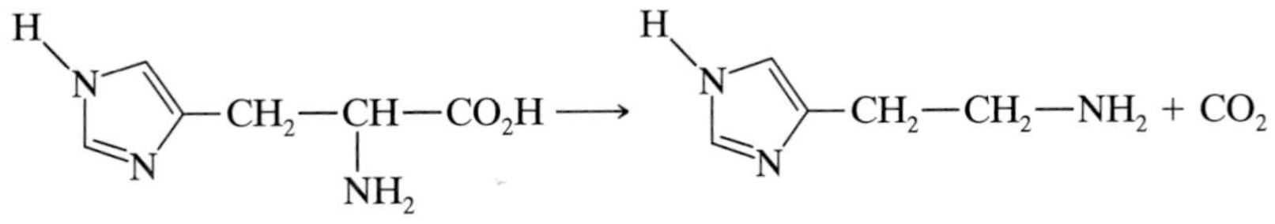

$$\text{(imidazolyl)}{-}CH_2{-}\underset{NH_2}{CH}{-}CO_2H \longrightarrow \text{(imidazolyl)}{-}CH_2{-}CH_2{-}NH_2 + CO_2$$

histidine histamine

Antihistamines are nonprescription drugs that relieve the allergy symptoms and prevent excess histamine production. Like histamine, each of the three structures shown contains a $-CH_2CH_2N-$ unit. The antihistamines are competitive inhibitors that occupy receptor sites in nerves. They produce no direct action, but by preventing histamine from occupying sites, they act as antagonists. *Antagonists* are drugs that prevent the physiological effect of another substance.

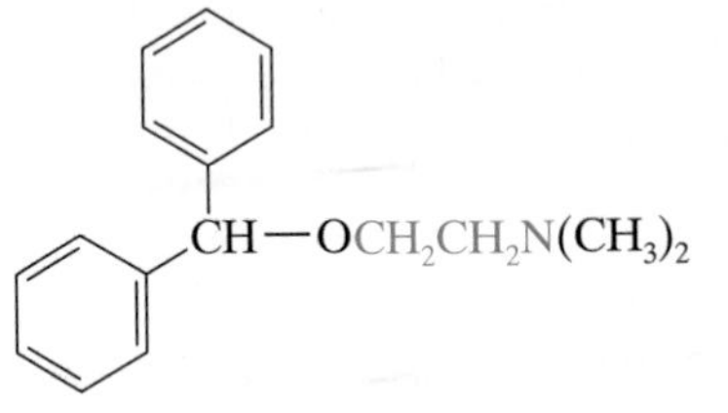

$(C_6H_5)_2CH-OCH_2CH_2N(CH_3)_2$

diphenhydramine

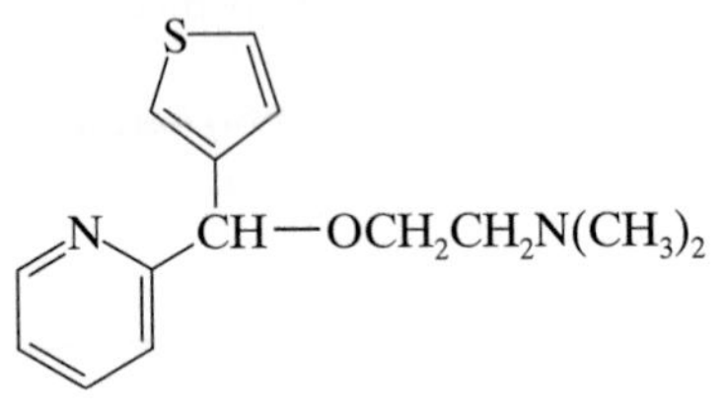

methapyrilene

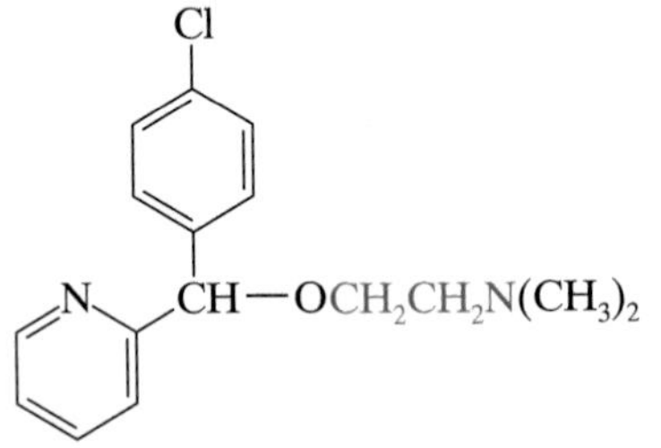

chlorpheniramine

28.7 Biosynthesis of Amino Acids

The ability of animals to synthesize amino acids varies with the species. Recall that humans cannot synthesize nine of the amino acids required for protein formation. The essential amino acids, which must be obtained from the diet, cannot be synthesized because our bodies have neither the correct carbon skeletons available nor the enzymes necessary to produce these skeletons. Most of the biosynthetic pathways that other animal species or plants use to produce these amino acids are exceedingly complex, and discussion of them is beyond the scope of this text.

The biosynthesis of the essential amino acids usually involves many more steps than are required for the formation of nonessential amino acids. For example, the aromatic amino acids phenylalanine and tryptophan are derived from chorismate in three- and five-step processes, respectively. However the chorismate must be made in a seven-step process from phosphoenolpyruvate and erythrose 4-phosphate. The energy requirements for forming the essential amino acids as a class are higher than those for forming the nonessential amino acids.

The synthetic pathways for some of the nonessential amino acids are the reverse of transamination and oxidative deamination. Thus, glutamic acid, alanine, and aspartic acid are all formed by transamination reactions from corresponding α-keto acids that are all intermediates in the citric acid cycle. For example, aspartate is formed from oxaloacetate by transfer of an amino group from glutamate, the universal donor of amino groups.

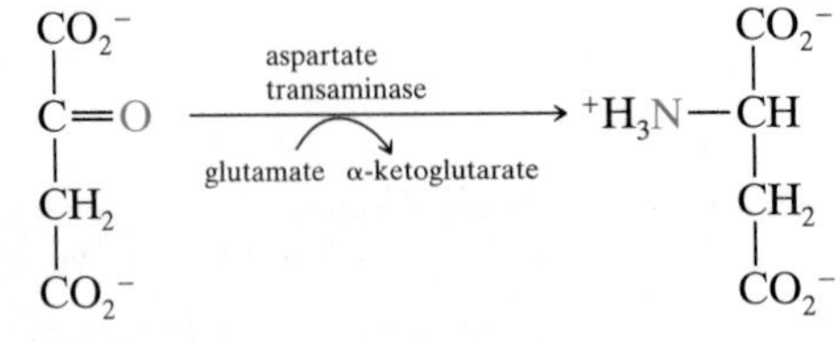

$$^{-}O_2C-C(=O)-CH_2-CO_2^{-} \xrightarrow[\text{glutamate} \rightarrow \alpha\text{-ketoglutarate}]{\text{aspartate transaminase}} {}^{-}O_2C-CH(NH_3^{+})-CH_2-CO_2^{-}$$

oxaloacetate → aspartate

A few amino acids, such as serine and proline, are prepared by specialized pathways from common metabolic intermediates. Serine is formed in the following three steps from 3-phosphoglycerate:

1. oxidation of a secondary alcohol by NAD^+
2. transamination of the ketone using glutamate
3. hydrolysis of a phosphate ester

$$\mathrm{H{-}C(CO_2^-)(OH){-}CH_2OPO_3^{2-}} \xrightarrow[\mathrm{NAD^+ \;\to\; NADH + H^+}]{\text{3-phosphoglycerate dehydrogenase}} \mathrm{O{=}C(CO_2^-){-}CH_2OPO_3^{2-}}$$

3-phosphoglycerate → 3-phosphohydroxypyruvate

$$\xrightarrow[\text{glutamate} \;\to\; \alpha\text{-ketoglutarate}]{\text{phosphoserine transaminase}} \mathrm{{}^+H_3N{-}CH(CO_2^-){-}CH_2OPO_3^{2-}} \xrightarrow[\mathrm{H_2O \;\to\; P_i}]{\text{3-phosphoserine phosphatase}} \mathrm{{}^+H_3N{-}CH(CO_2^-){-}CH_2OH}$$

3-phosphoserine → serine

Proline is formed from glutamate in the following steps:

1. reduction of a carboxylic acid to an aldehyde
2. cyclic addition to produce an imine
3. reduction of the imine to give proline

$$\mathrm{{}^+H_3N{-}C(CO_2^-)(H){-}(CH_2)_2{-}CO_2^-} \xrightarrow[\mathrm{ATP \;\to\; ADP}]{\gamma\text{-glutamate kinase}} \mathrm{{}^+H_3N{-}C(CO_2^-)(H){-}(CH_2)_2{-}C({=}O)OPO_3^{2-}} \xrightarrow[\mathrm{NADPH + H^+ \;\to\; NADP^+ + P_i}]{\text{glutamate }\gamma\text{-semialdehyde dehydrogenase}} \mathrm{{}^+H_3N{-}CH(CO_2^-){-}CH_2{-}CH_2{-}C(H){=}O}$$

glutamate → γ-glutamate phosphate → glutamate γ-semialdehyde

glutamate γ-semialdehyde ⇌ ($-H_2O$) Δ^1-pyrroline 5-carboxylate

$$\Delta^1\text{-pyrroline 5-carboxylate} \xrightarrow[\mathrm{NADPH + H^+ \;\to\; NADP^+}]{\Delta\text{-pyrroline 5-carboxylate reductase}} \text{proline}$$

proline ← Δ^1-pyrroline 5-carboxylate

Summary

The enzymes pepsin, chymotrypsin, trypsin, carboxypeptidase, and aminopeptidase catalyze the hydrolysis of proteins into amino acids. The amino acids enter the **amino acid pool,** which is a reserve of free amino acids in a state of equilibrium. Each protein in the body is constantly being degraded and rebuilt. The time required to degrade a protein and replace it is the **protein turnover rate.** Unneeded amino acids are degraded in the liver to produce either carbohydrates or fats or are oxidized directly to carbon dioxide. Nitrogen atoms in excess of immediate biosynthetic needs are converted to ammonium ion and urea and are excreted in the urine. The **nitrogen balance** relates the daily intake of nitrogen compounds and the loss of nitrogen compounds by excretion.

The α-amino groups of many amino acids are transferred to α-ketoglutarate in **transamination reactions. Oxidative deamination** of glutamate yields ammonium ion, which then is converted to urea in the **urea cycle.**

The carbon skeletons of all 20 amino acids are eventually converted into seven compounds: pyruvate, α-ketoglutarate, succinyl CoA, fumarate, oxaloacetate, acetyl CoA, and acetoacetyl CoA. Several amino acids produce either acetyl CoA or acetoacetyl CoA, which can be transformed into acetyl CoA.

The amino acids that can be converted into pyruvate, α-ketoglutarate, succinyl CoA, fumarate, and oxaloacetate can eventually produce glucose. These amino acids are **glucogenic.** Amino acids that are degraded to acetyl CoA or acetoacetyl CoA are **ketogenic.** Fourteen amino acids are glucogenic, and only leucine is exclusively ketogenic. Five amino acids are both glucogenic and ketogenic. Some amino acids undergo **decarboxylation** to produce amines.

Glutamate provides the α-amino group required for the synthesis of some nonessential amino acids by transamination of α-keto acids. Independent pathways exist for the syntheses of serine and proline.

Exercises

Digestion of Proteins

28.1 What is the amino acid pool?

28.2 What is meant by protein turnover?

28.3 What percentage of digested amino acids is used to build proteins?

28.4 What percentage of human energy needs is provided by amino acids?

28.5 How does the turnover rate for enzymes differ from that of muscle protein?

28.6 Why must there be a positive nitrogen balance in a period of active growth?

28.7 What happens to the nitrogen balance during a fast?

28.8 What happens to the carbon atoms of unneeded amino acids?

28.9 What is the major enzyme for protein digestion in the stomach?

28.10 What digestive enzymes are provided by the pancreas?

Transamination

28.11 Explain the function of the transamination reaction in the metabolism of amino acids.

28.12 Write an equation for the transamination of alanine and α-ketoglutaric acid.

28.13 The following α-keto acid can substitute for an essential amino acid in the diet. What is the amino acid? How can the keto acid substitute for an essential amino acid?

$$CH_3-\underset{\displaystyle CH_3}{\underset{|}{CH}}-CH_2-\overset{\displaystyle O}{\overset{\|}{C}}-CO_2H$$

28.14 The following α-keto acid can substitute for an essential amino acid in the diet. What is the amino acid? How can the α-keto acid substitute for an essential amino acid?

$$CH_3-CH_2-\underset{\displaystyle CH_3}{\underset{|}{CH}}-\overset{\displaystyle O}{\overset{\|}{C}}-CO_2H$$

Oxidative Deamination

28.15 What is oxidative deamination?

28.16 Write the structure of the product of oxidative deamination of glutamic acid.

28.17 What coenzyme is required for oxidative deamination?

28.18 What is the fate of the α-ketoglutaric acid formed in oxidative deamination?

Urea Cycle

28.19 What two amino acids enter the urea cycle?

28.20 Where does the synthesis of urea occur?

28.21 What is hyperammonemia?

28.22 Why are high levels of ammonium ion toxic?

28.23 What substance reacts and is re-formed in the urea cycle?

28.24 What is the fate of the fumaric acid formed in the urea cycle?

28.25 What steps in the urea cycle occur in the mitochondrial matrix?

28.26 What steps in the urea cycle occur in the cytoplasm?

Degradation of Amino Acids

28.27 What compounds are produced from glucogenic amino acids?

28.28 What compounds are produced from ketogenic amino acids?

28.29 What amino acid is exclusively ketogenic?

28.30 Name the five amino acids that are both glucogenic and ketogenic.

28.31 Threonine undergoes a dehydration reaction followed by hydration to form a keto acid. Write the structures of the intermediates and the final product.

28.32 Cysteine is degraded via the following series of intermediates. What types of transformations occur in each step? The last intermediate loses SO_2 to give a keto acid. What is that keto acid?

$$\begin{array}{c} CO_2^- \\ | \\ {}^+H_3N{-}C{-}H \\ | \\ CH_2SH \end{array} \longrightarrow \begin{array}{c} CO_2^- \\ | \\ {}^+H_3N{-}C{-}H \\ | \\ CH_2SO_2^- \end{array} \longrightarrow \begin{array}{c} CO_2^- \\ | \\ C{=}O \\ | \\ CH_2SO_2^- \end{array}$$

Decarboxylation of Amino Acids

28.33 Bacterial enzymes catalyze the decarboxylation of ornithine, an amino acid in the urea cycle. What is the structure of the product?

28.34 What amino acid produces the neurotransmitter 4-aminobutanoic acid via a decarboxylation reaction?

Biosynthesis of Amino Acids

28.35 What is the major pathway for the synthesis of nonessential amino acids?

28.36 Outline how proline is formed from glutamic acid.

28.37 What substance might be used to synthesize tyrosine?

28.38 What substance is used to synthesize serine?

29 Nucleic Acids

This young kangaroo will eventually leave its mother's pouch but will maintain its genetic heritage for life. (Photo Researchers, Inc.)

29.1 The Control of Life Processes
29.2 Nucleic Acids Are Polymers
29.3 Polynucleotides
29.4 The Double Helix of DNA
29.5 DNA Replicates Itself
29.6 The Flow of Genetic Information
29.7 DNA Produces Several Types of RNA
29.8 The Genetic Code
29.9 Protein Synthesis and Translation
29.10 Gene Mutations and Genetic Disease
29.11 Recombinant DNA and Genetic Engineering
29.12 Viruses

Essays
- Nucleosides in Medicine
- A Gene Defect May Cause Osteoarthritis
- Recombinant DNA and Agriculture
- Gene Therapy in Cystic Fibrosis

29.1 The Control of Life Processes

In the preceding five chapters we have touched on but a few of the thousands of biochemical reactions essential for life. Virtually all of them are catalyzed by enzymes. Each enzyme has a unique primary, secondary, tertiary, and sometimes a quaternary structure whose precise shape is ideally suited for a specific task. How does an organism make each of the thousands of enzymes that control life processes? How does an organism pass on information between generations on how to reproduce these enzymes and other body proteins? In this chapter we will learn that deoxyribonucleic acid (DNA) has an informational code that is responsible for the synthesis of all proteins. In addition, DNA is the molecule that passes on that information from one generation to another.

The Molecular Basis of Heredity

Even in ancient times, the Greeks recognized that organisms pass on traits to their offspring. However it wasn't until the 1860s that the experiments of Gregor Mendel firmly established the fundamental principles of heredity. It took almost a hundred years for the molecular basis of heredity to be established.

There are probably few in a modern society who haven't heard of DNA and how it identifies each of us even more than do our fingerprints. DNA matching is becoming a common method for proving that an individual was at a crime scene. The identity of an individual can be established using microscopic quantities of cellular evidence left behind, such as blood or semen.

The experimental verification that DNA is the molecule that contains the genetic information that is passed between generations and also determines the uniqueness of each individual was not easily established. DNA was discovered in 1869 by Friedrich Meischer shortly after the work of Mendel, but its role in life processes was not recognized until the middle of the next century.

Biologists learned that the cell nucleus contains structures called *chromosomes* that are responsible for heredity traits and that these chromosomes are passed along in cellular division because the chromosomes are "doubled" in the division process. Chromosomes consist of protein and DNA, but it was thought more likely that the genetic information was encoded in the proteins. DNA is a high molecular weight polymer of nucleotides that consist of deoxyribose, phosphate, and four possible nitrogen-containing bases. This number of simple units was regarded as insufficient to contain the molecular information for life processes. Proteins, on the other hand, contain 20 amino acids that could be combined to give a variety of structures that could be the carriers of genetic information. However, as we will see, the four bases of DNA are arranged in a sequence to form three-letter "words" or triplets placed along a backbone of deoxyribose and phosphate. The triplets provide the specific "language" used to pass on hereditary information.

In 1928, based on work with several species of bacteria and how they infected mice, Frederick Griffith, a British medical officer, showed that it was not the protein of the bacteria that was responsible for assimilation of genetic information. It took until 1944 before the American bacteriologist Oswald Avery showed that it was DNA that was responsible for Griffith's observations. Although Avery's work was treated with skepticism, over the next decade a number of scientists obtained additional evidence implicating DNA as the genetic material of the cell. Studies of viruses that infect bacteria closed the case. Viruses are essentially only DNA with a protective coat of protein. Studies using radioisotopes to alternately label the protein and the DNA showed that it is the DNA that enters the bacterium and takes over its metabolic machinery. Thus, it is the DNA that contains the genetic information for its replication. Eventually it became clear that not just viruses but all life forms contain DNA that is species specific and that the molecular diversity of life forms can be explained even though DNA contains only four types of nitrogen-containing bases in the form of nucleotides.

DNA and Replication

Self-replication is the central process for the continuance of life. Each organism transmits hereditary information using **chromosomes** located in the nucleus of the cell. Chromosomes contain all the primary genetic information that determines the characteristics of the living organism. Thus, carrot seeds produce carrots and not cabbages. Mice give birth to mice and not elephants. The number of chromosomes and their molecular structure is unique for each species. The number of chromosomes is always an even number; they occur as pairs within the nucleus.

Each chromosome contains a series of hundreds to thousands of genes, which in turn contain information for a limited function, such as for the synthesis of a particular enzyme. The location of genes on the various chromosomes of a species can be "mapped." The mapping process is not a simple task but one with considerable possible benefits. Genes are responsible for some 5000 known genetic diseases, and the location of only about 250, such as the gene for cystic fibrosis, have been identified. Technology is being developed to artificially alter genes to change genetic directions. It remains to be seen how successful medical intervention at the level of the gene will be.

DNA can make exact copies of itself in a process called *replication.* DNA also directs the synthesis of three types of ribonucleic acids (RNA) in a process called *transcription.* The three types of RNA molecules, ribosomal RNA, messenger RNA, and transfer RNA, control the synthesis of all proteins in a process called *translation.*

We begin this chapter by examining the structures of nucleotides, which are components of nucleic acids. Then we will discuss the structures of the nucleic acids and their role in the control of all life processes.

29.2 Nucleic Acids Are Polymers

The nucleic acids DNA and RNA are polymers of nucleotides. Nucleotides consist of three units: a sugar, a nitrogen base, and a phosphate group. The sugar in RNA is the β-anomer of D-ribose, which accounts for the name ribonucleic acid. The sugar in DNA is the β-anomer of D-2-deoxyribose, hence the name deoxyribonucleic acid. D-2-Deoxyribose lacks a hydroxyl group at the C-2 atom.

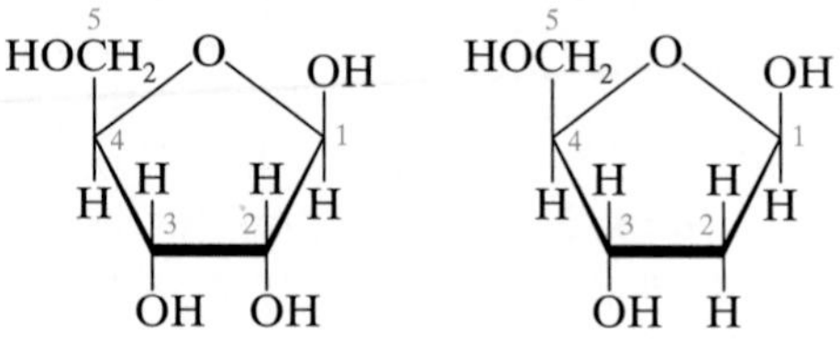

β-D-ribose

β-D-2-deoxyribose

Two purine bases—adenine (A) and guanine (G)—are in both DNA and RNA. The capital letters are shorthand notations used to represent the bases.

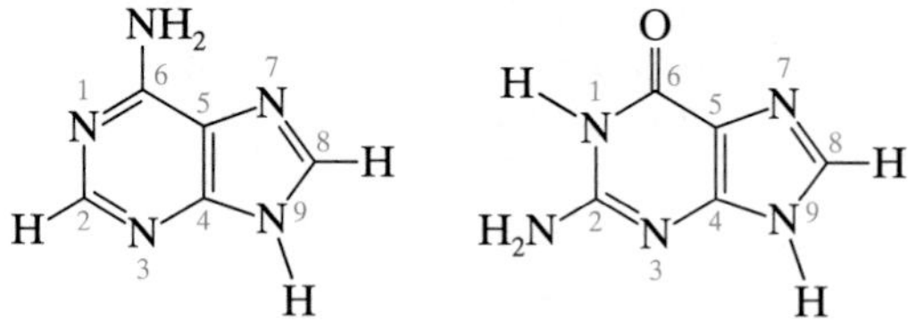

adenine (A)

guanine (G)

Three pyrimidine bases—cytosine (C), thymine (T), and uracil (U)—are found in nucleic acids. Cytosine appears in both RNA and DNA. Thymine is found exclusively in DNA; uracil is found only in RNA. Note that thymine and uracil differ only by a methyl group at the 5-position.

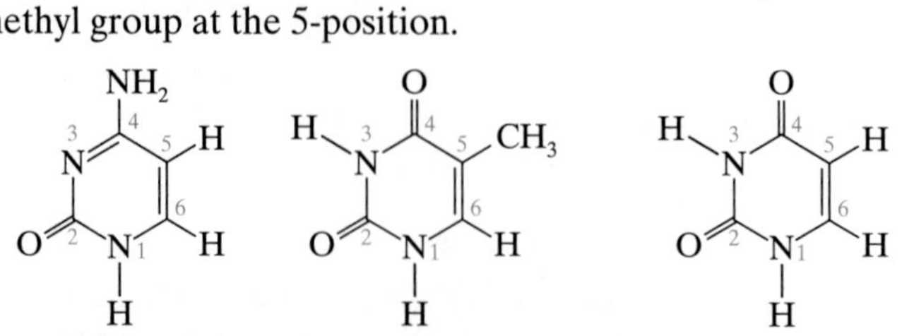

cytosine (C)

thymine (T)
(found in DNA)

uracil (U)
(replaces thymine in RNA)

In summary, DNA and RNA contain four bases each: two purines and two pyrimidines. In DNA the bases are A, G, C, and T; in RNA the bases are A, G, C, and U.

Nucleosides Contain a Sugar and a Base

A **nucleoside** is a nitrogen glycoside formed when the hemiacetal center of the sugar and an —NH of the base join with the elimination of a molecule of water.

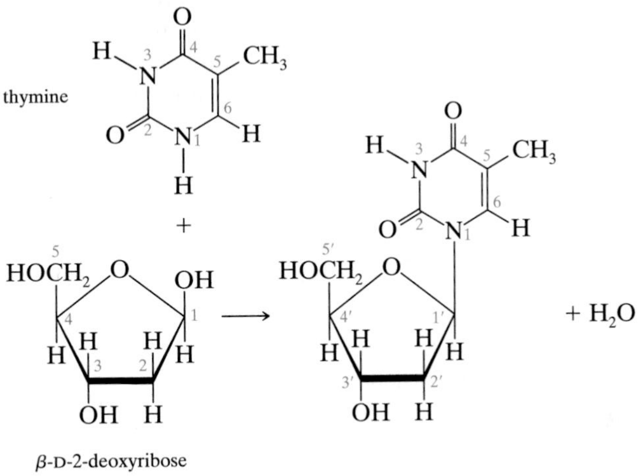

thymine

+

β-D-2-deoxyribose

$\longrightarrow$ $+ H_2O$

Nucleosides in Medicine

Medical intervention at the nucleoside level can result in profound physiological changes because the treatment goes to the very heart of what controls all life processes. Some treatments for both cancer and AIDS (acquired immune deficiency syndrome) use chemically modified nucleosides.

Rapidly dividing cancer cells require a large supply of DNA. Thymidine, one of the components of DNA, is synthesized in the cell by methylation of uridine. Administration of fluorouracil as part of chemotherapy produces fluorouridine, which is an inhibitor of the enzyme that catalyzes the conversion of uridine into thymidine. Since cancer cells grow at a more rapid rate than normal cells, the growth of the tumor is more adversely affected. However, fluorouracil also interferes with DNA synthesis in normal cells and causes side effects that must be monitored. During chemotherapy the weakened body is more susceptible to bacterial infections.

Azidothymidine (AZT) is a nucleoside used in the treatment of persons who are HIV-positive and those who have AIDS. It is not a cure for this disease, but there is some retardation in its debilitative effects. AZT has been approved for limited use in the United States, but it causes severe anemia. Blood transfusions may be needed and close medical supervision is required.

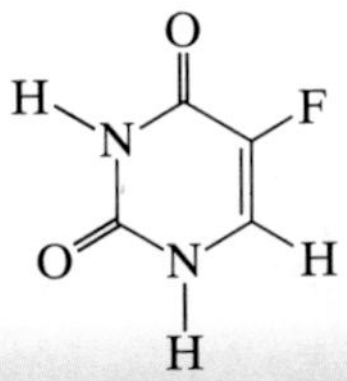

fluorouracil

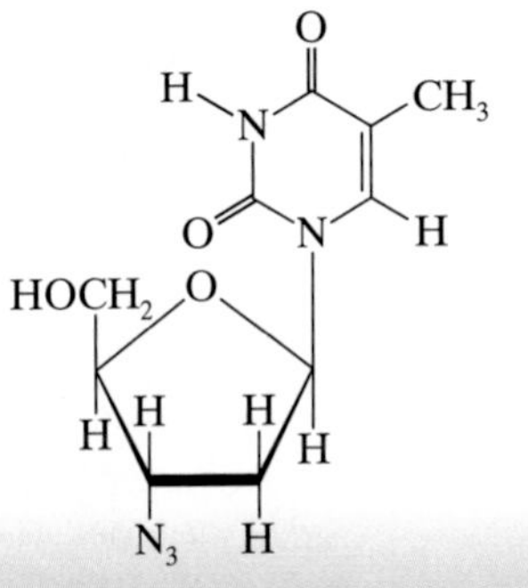

azidothymidine

The configuration is β at the anomeric carbon atom of the sugar in both DNA and RNA. The ring atoms in the base are numbered; primed numbers are used for the atoms of the carbohydrate. The nitrogen base is always attached to the 1′ carbon atom of the carbohydrate, and there is a primary hydroxyl group located at the 5′ carbon atom. Ribonucleosides have secondary hydroxyl groups at the 2′ and 3′ carbon atoms, whereas deoxyribonucleosides have a secondary hydroxyl group only at the 3′ carbon atom.

The names of the nucleosides are listed in Table 29.1. Ribonucleosides are named by modifying the name of the nitrogen base. Purine derivatives have the ending *-osine;* pyrimidine derivatives end in *-idine.* The same convention is used for deoxyribonucleosides, along with the prefix *deoxy-*.

Nucleotides Are the Monomers of Nucleic Acids

A **nucleotide** is an ester of a nucleoside and phosphoric acid. The hydroxyl group at the 5′ carbon atom acts as the alcohol to form the ester.

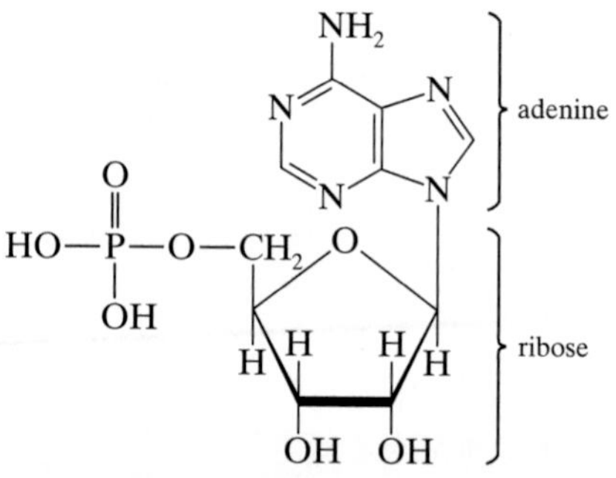

NH_2
N N
adenine
O
N N
HO—P—O—CH_2 O
OH
H H H H
ribose
OH OH

adenylic acid

The two protons of the monophosphate ester are ionized at physiological pH, and the ester exists as an ion with a −2 charge in solution. The names of the nucleotides are listed in Table 29.1.

TABLE 29.1 Names of Nucleosides and Nucleotides in DNA and RNA

Base	Nucleoside	Nucleotide
	DNA	
adenine (A)	deoxyadenosine	deoxyadenylic acid
guanine (G)	deoxyguanosine	deoxyguanylic acid
thymine (T)	deoxythymidine	deoxythymidylic acid
cytosine (C)	deoxycytidine	deoxycytidylic acid
	RNA	
adenine (A)	adenosine	adenylic acid
guanine (G)	guanosine	guanylic acid
uracil (U)	uridine	uridylic acid
cytosine (C)	cytidine	cytidylic acid

EXAMPLE 29.1

Classify the following structure, identify its components, and name it.

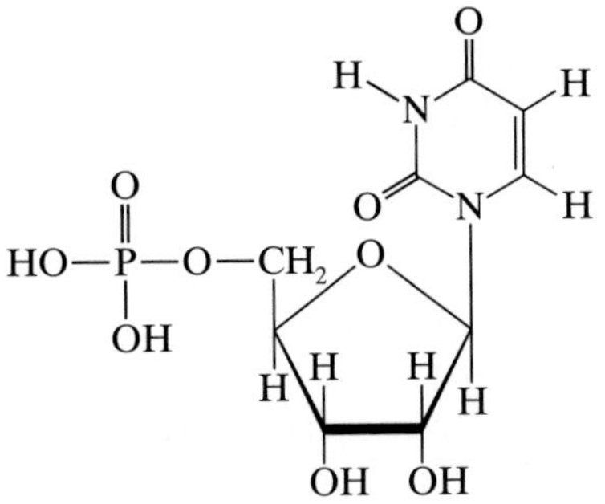

Solution

The compound contains a phosphate group and is a nucleotide. The sugar is ribose, which means that the compound is a ribonucleotide. The base is uracil, which can be found only in ribonucleotides. From Table 29.1 we find that the name of the compound is uridylic acid.

PROBLEM 29.1

Classify the following structure, identify its components, and name it.

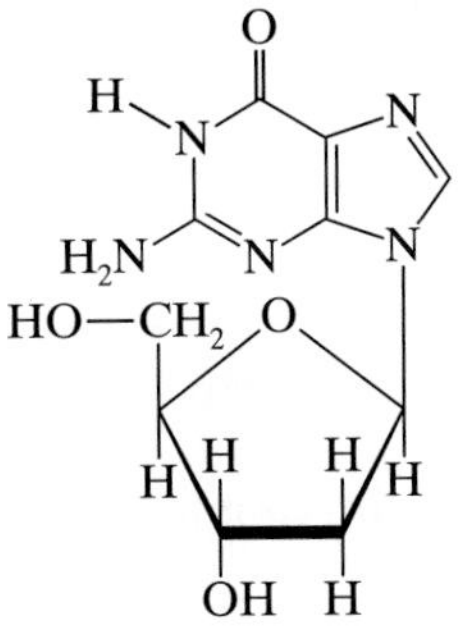

29.3 Polynucleotides

The composition and structure of molecules have been a focus throughout this text. Although nucleic acids are complex structures, some of the same considerations presented earlier are important to understanding their chemical reactions. As in the case of proteins, it is convenient to discuss the structure of polynucleotides in terms of primary and secondary features.

Composition of Polynucleotides

The amount of A, T, G, and C in the DNA of several animals, plants, and microorganisms was first reported by the biochemist Erwin Chargaff in 1947. DNA composition is species specific, and thus the idea that diversity of life forms results from differences in DNA became better established. But a more important result of Chargaff's analysis is the interesting pattern that exists between the quantities of the bases (Table 29.2). Regardless of the species, the amount of C is equal to the

TABLE 29.2 Composition of DNA of Various Species

	Mole %					
Species	**A**	**T**	**G**	**C**	**A/T**	**G/C**
Escherichia coli	26.0	23.9	24.9	25.2	1.09	0.99
yeast	31.3	32.9	18.7	17.1	0.97	1.05
sea urchin	32.8	32.1	17.7	18.4	1.02	0.96
herring	27.8	27.5	22.2	22.6	1.01	0.98
salmon	29.7	29.1	20.8	20.4	1.02	1.02
ox	29.0	29.1	20.7	20.7	1.01	1.00
pig	29.8	29.1	20.7	20.7	1.02	1.00
rat	28.6	28.4	21.4	21.5	1.01	1.00
human	30.9	29.4	19.9	19.8	1.01	1.00

amount of G within experimental error; the amount of A is equal to the amount of T. These equalities are known as Chargaff's rules.

$$\% \text{C} = \% \text{G} \qquad \% \text{A} = \% \text{T}$$

Although the amount of each of the pairs is different from one life form to another, the sum of the percentages of A and G, the two purine bases, equals the sum of the percentages of C and T, the two pyrimidine bases. The relationship is

$$\% \text{A} + \% \text{G} = \% \text{C} + \% \text{T}$$

We will return to these relationships in DNA later in the next section. There are no simple relationships in the content of the four bases of the three types of RNA.

Primary Structure of Polynucleotides

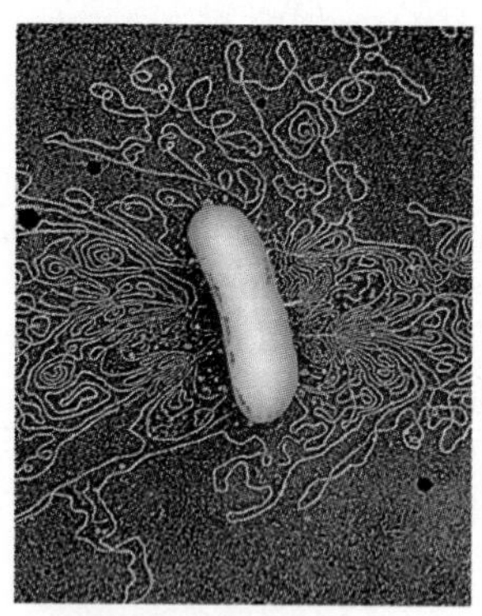

This damaged E. coli is spewing forth its DNA. (Photo Researchers, Inc.)

Nucleotides, bonded through phosphodiester linkages, make up the primary structure of both RNA and DNA molecules (Figure 29.1). Phosphodiesters are formed between the 3′ hydroxyl group of one nucleotide and the 5′ phosphate ester of another nucleotide. Thus, DNA and RNA both have a backbone of alternating sugar and phosphate units. Note that each phosphodiester has one acidic hydrogen atom, hence the name *nucleic acids.* However, at physiological pH the nucleic acids are ionized. The differences in structure are in the sugar (deoxyribose in DNA and ribose in RNA) and in the order of the bases. Recall that thymine exists only in DNA, and uracil exists only in RNA.

As in the case of protein structures, a convention is used to list the sequence of nucleotides. The direction is from the 5′ end, which exists as a phosphate ester, to the 3′ end, which has the free hydroxyl group. For RNA the convention is

$$5' \xrightarrow[\text{RNA}]{\text{A-C-G-...-U}} 3'$$

The same process is used for DNA, but a lowercase d is placed at the left of the first base to indicate the deoxyribose in the backbone. The presence of thymine rather than uracil also indicates that the nucleic acid is deoxyribonucleic acid.

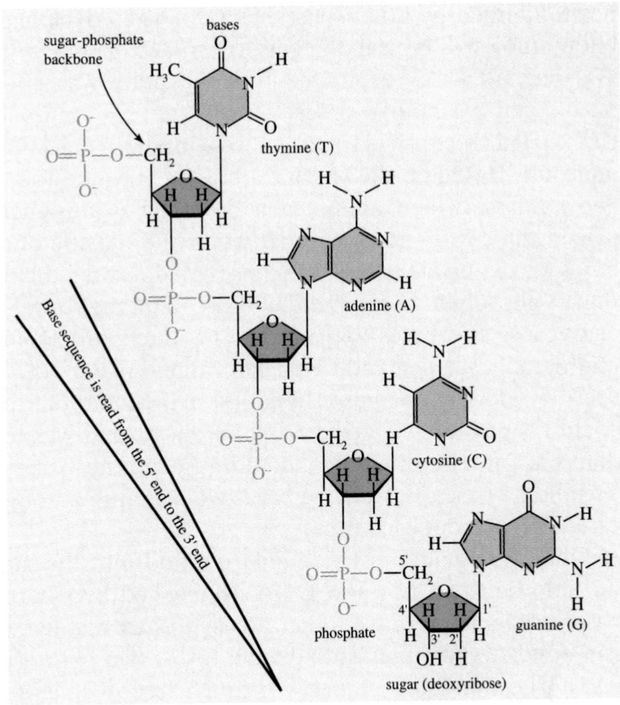

FIGURE 29.1 Structure of a Polynucleotide

$$5' \xrightarrow[\text{DNA}]{\text{dT-C-G-...-A}} 3'$$

When discussing the structure of DNA, its replication, transcription into RNA, and translation into protein we will use some structural symbols for the bases. The significance of their shapes will become apparent shortly. A sequence of nucleotides is represented by a linear series of these symbols.

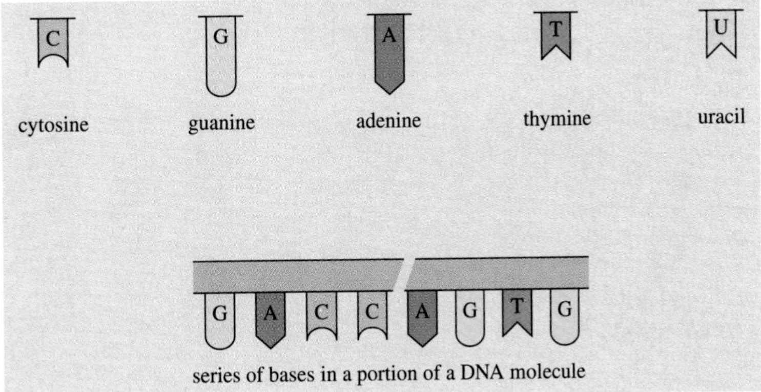

29.4 The Double Helix of DNA

The % C = %G and % A = % T relationship reported by Chargaff was used by J. D. Watson and F. H. C. Crick in 1952 to postulate the secondary structure of DNA. Their model was based on a crude X-ray "photograph" of DNA that had

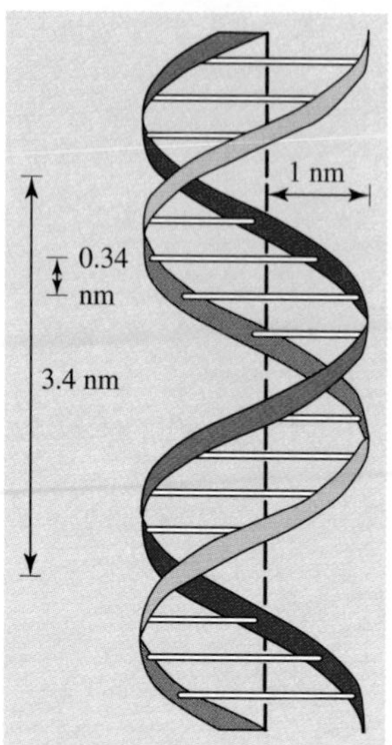

FIGURE 29.2
The Double Helix
The 2-deoxyribose-phosphate backbones represent the two strands of DNA that run in opposite directions. The bridges between the two strands are the paired nitrogen-containing bases. These bridges are 0.34 nm apart. One complete turn of the helix occurs every 3.4 nm.

been obtained by other scientists working to decipher the structure of DNA. This photograph was typical of a pattern expected for a uniform helix-shaped substance. Watson and Crick estimated that the helix was about 2 nm wide and that the purine and pyrimidine bases were about 0.34 nm apart. At this point Watson and Crick tried to construct various wire models to account for the dimensions of the molecule. Based on the width of the structure they suggested that DNA consists of two strands twined about each other. They designed a satisfactory model that placed the sugar–phosphate chain on the outside and the purine and pyrimidine bases on the inside. The model resembled a twisted ladder, with the sides being the sugar–phosphate chain and the bases being the rungs (Figure 29.2).

Because the bases appeared to be stacked 0.34 nm apart based on the X-ray photograph, Watson and Crick determined that 10 layers of bases (rungs) occur for each 360° turn of the helix. In addition, they felt that the uniform width of the helix could result only if a purine base in one strand was located across from a pyrimidine base in the other strand. Because purine bases are about twice as wide as pyrimidine bases, two purine bases would cause a bulge in the helix if they were located across from each other.

Finally, Watson and Crick incorporated into the model the stoichiometric relationship between bases in DNA established by Chargaff. They proposed that the bases across from one another are hydrogen bonded. Cytosine forms three hydrogen bonds to guanine, and adenine forms two hydrogen bonds to thymine (Figure 29.3). The matching of bases through hydrogen bonds is called **complementary base pairing.** For each C molecule in DNA there is a molecule of the complementary base G; for each A molecule there is a complementary base T. Note that this requirement also means that for each purine base there is a pyrimidine base. Thus, the Watson–Crick model is consistent with the experimentally observed relationship in the bases of DNA.

FIGURE 29.3
Hydrogen Bonding in DNA
The pairing of a purine with a pyrimidine base by the proper number of hydrogen bonds not only stabilizes the structure but also results in a helix with a constant diameter. Because the strands run in opposite directions, the deoxyriboses and their phosphodiesters in the two strands are oriented differently.

The individual strands of polydeoxyribonucleotides run in opposite directions in the double helix and are said to be antiparallel. As a result, the ribose units face in opposite directions, and the phosphodiester linkages are "above" the ribose unit in one strand but "below" in the other strand (Figure 29.3). The following diagram uses shorthand representations that summarize the stoichiometric relationships between bases.

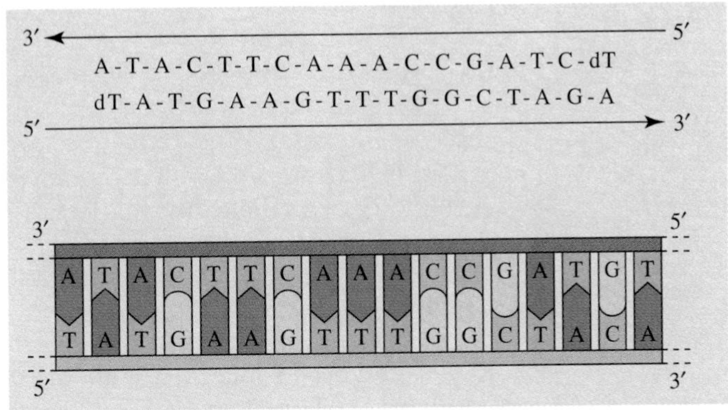

The strength of a hydrogen bond is approximately 5 kcal/mole. Although this is less than the energy of a covalent bond, the very large number of hydrogen bonds between the two DNA strands gives a very stable structure.

29.5 DNA Replicates Itself

Replication is a copying process by which DNA is supplied to the new cells formed by cell division. DNA is replicated in the cell nucleus. As new cells are formed, replication of DNA is controlled by the hydrogen-bonding properties of DNA. Each strand of DNA serves as a pattern or template for producing a new complementary strand. This new complementary strand is identical to the original complementary strand. Since DNA is a double-stranded molecule, each strand replicates its own complementary strand.

Studies using radioisotopes have established that replication of DNA occurs by a **semiconservative process;** that is, each new DNA molecule has one strand of the parent DNA molecule, half the DNA of the previous generation. The second strand is formed in the replication process (Figure 29.4).

Although the replication process is complex in detail, some general features can be described. In replication, DNA must unwind so that the new complementary strand can form. The enzyme DNA helicase works its way in between the two strands and then moves along the helix, temporarily separating the strands as it goes. This process exposes the bases at a point in the double helix called the **replication fork** (Figure 29.5). The replication fork is not a single site but moves as DNA unravels and creates a place for synthesis of the new strand of DNA.

The enzyme DNA polymerase, which catalyzes the replication, binds to each unwound strand. This enzyme recognizes each base in the DNA chain and matches it with a free nucleotide according to the base pairing rules. Then it binds the nucleotides into a growing strand by bonding the 5′ position of one unit to the 3′ position of another unit to form the backbone. This process is known as 5′ → 3′ replication. Note that two DNA polymerases must enter the fork and move in opposite directions on each strand in order to replicate DNA.

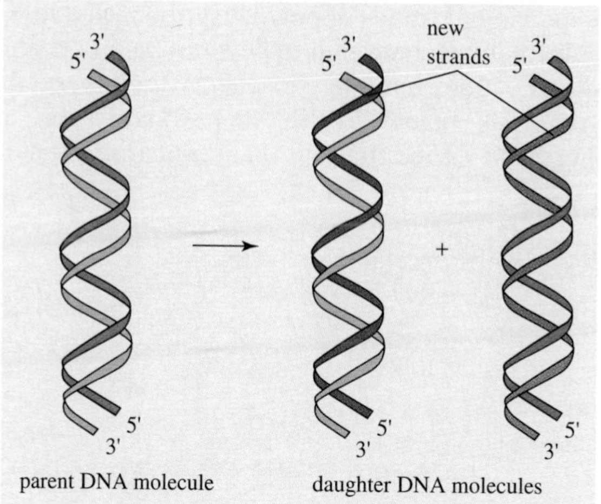

FIGURE 29.4 Semiconservative Replication of DNA

Replication occurs at a rate of about 100 nucleotides per second. The process is not error free. About 1 in 10,000 is incorrectly paired, but the DNA polymerase checks each nucleotide as it is added to the strand. If an incorrect nucleotide is discovered, the polymerase "backs up," removes it, and places the correct nucleotide before continuing. As a result, errors in the completed DNA molecule are fewer

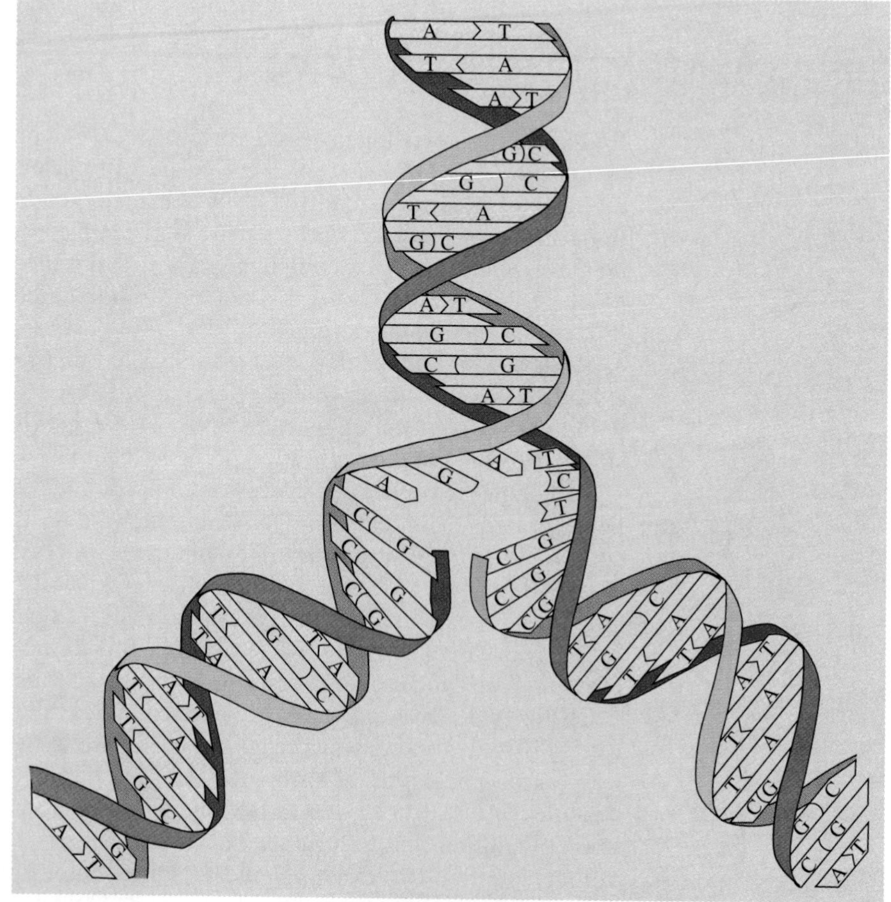

FIGURE 29.5 Model for Replication of DNA

The two strands have been separated at the replication fork by DNA helicase (not shown). The base sequences in the two double-stranded DNA molecules are identical to those in the original DNA molecule. The sequence of base pairs is shown using geometric representations of the bases.

than 1 in 1 billion. Even this small incidence of error can be corrected by DNA repair enzymes. These enzymes also repair and replace sections of DNA that have been damaged by chemicals or radiation. The repair enzymes cut out sections of a strand and fill in the gap with the proper nucleotides.

EXAMPLE 29.2

Part of the nucleotide sequence in one chain of DNA follows. Write a representation of the complementary DNA chain.

$$5' \xrightarrow{\text{dA-T-C-G}} 3'$$

Solution

The complementary base pairs in DNA are A with T and G with C. For each A in one chain there is a T in the other. For each C in one chain there is a G in the other. The complementary chain is T-A-G-dC.

PROBLEM 29.2

Part of the nucleotide sequence in one chain of DNA follows. Write a representation of the complementary DNA chain.

dC-T-A-G-C

29.6 The Flow of Genetic Information

Experiments have established that genes, which are segments of DNA, are responsible for control of the synthesis of proteins. In the following sections we will learn how a nucleotide sequence in DNA codes for formation of an amino acid sequence in a protein. In addition we will see how DNA in the nucleus controls protein synthesis that occurs in the cytoplasm. The "message" is translated from one molecular language into another by an intermediate molecule, mRNA, that is formed in the nucleus and moves to the cytoplasm.

The flow of genetic information is commonly called the **central dogma.** We summarize the tenets of this dogma before considering details of protein synthesis.

1. DNA stores and transmits all hereditary information, including the instructions for the synthesis of all proteins.
2. DNA is replicated in the cell nucleus when cells are ready to divide. **Replication** is a copying process by which DNA is supplied to new cells formed by cell division.
3. Information for protein synthesis is passed to messenger RNA in a process called **transcription.**
4. mRNA directs the synthesis of protein by a process called **translation** using a triplet sequence of nucleotides.

The series of steps in protein synthesis as outlined by the central dogma is shown in Figure 29.6. In subsequent sections we will examine the transcription process and the code used to translate nucleotide information into a series of amino acids in a protein chain.

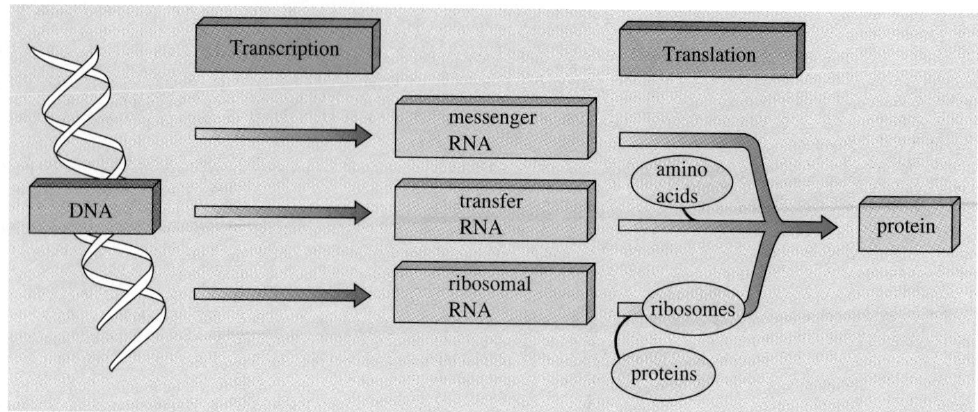

FIGURE 29.6 Information Flow in a Cell *Information flows from DNA to mRNA in transcription and then to proteins in translation.*

29.7 DNA Produces Several Types of RNA

Protein synthesis requires three types of RNA, all synthesized by DNA. These are messenger RNA (mRNA), transfer RNA (tRNA), and ribosomal RNA (rRNA). All three types of RNA are formed by a transcription process in which DNA is a template for base pairing. Transcription copies only certain selected genes into RNA. Moreover, the information contained in those genes resides on only one strand of the double helix. The strand that contains the code that is transcribed into mRNA is the **sense strand.**

Transcription is distinguished from replication in several ways. First, RNA polymerase rather than DNA polymerase is the catalyst. Second, the polymerase incorporates uracil rather than thymine in RNA. Finally, because only one strand of DNA is transcribed, the RNA molecules formed are single strands rather than double strands. This process is represented in Figure 29.7. A shorthand representation of transcription follows.

$$5' \xrightarrow[\text{dA-T-C-G}]{\text{DNA strand}} 3'$$

$$3' \xleftarrow[\text{RNA strand}]{\text{U-A-G-C}} 5'$$

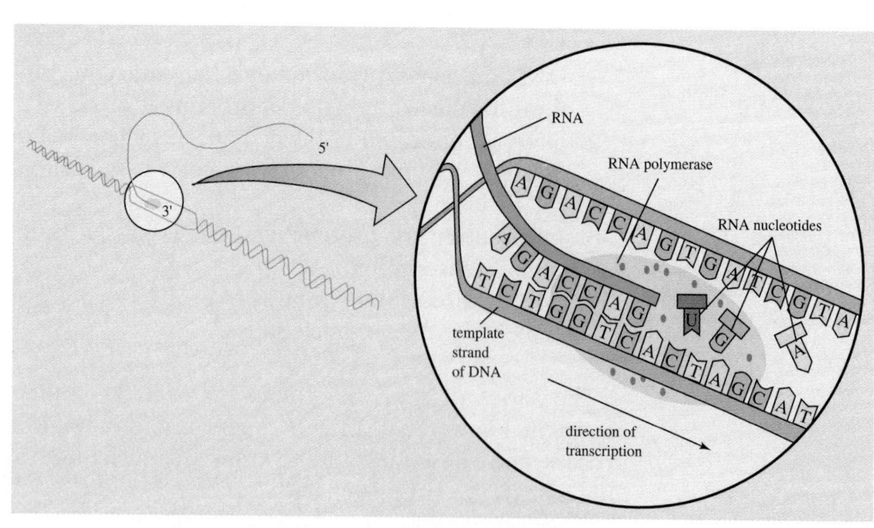

FIGURE 29.7 Transcription and Formation of mRNA *One strand of DNA serves as a template to form mRNA. Note that uracil is placed in mRNA by adenine, whereas the base pair of adenine in DNA is thymine. RNA polymerase as shown is much smaller than the actual molecule, which encompasses about 50 nucleotides at a time.*

Steps in Transcription

RNA polymerase locates the proper position in DNA by recognizing a sequence of nucleotides called the **promoter region.** This region precedes that of the nucleotide sequence to be transcribed. RNA polymerase occupies a region of approximately 50 nucleotides, so that it is associated with the promoter region and about a dozen base triplets.

RNA polymerase forces open the DNA double helix and moves along the sense strand in the 3′ to 5′ direction. It then pairs the appropriate nucleotide from the nucleotide pool in the nucleus to its complement in the DNA sense strand. The nucleotides are joined in the growing RNA chain. After about 10 nucleotides are added, that portion is separated and "dangles" while the remaining RNA continues to be synthesized.

Eventually, transcription is completed, and the RNA chain is separated from DNA. This event as well as the release of the RNA polymerase occurs because the enzyme reaches a sequence of nucleotides beyond those required for transcription that triggers the chemical reactions required for release.

Ribosomal RNA

Ribosomal RNA combines with about 50 proteins to form complex structures called **ribosomes,** whose molecular weight is about 3 million. Ribosomes are the sites of protein synthesis. The ribosomes are like protein factories, but they cannot manufacture products without directions and workers. Each ribosome consists of two subunits. In a eukaryotic cell the smaller unit consists of one molecule of rRNA and about 30 proteins. The larger unit consists of three rRNA molecules and about 50 proteins. It is this larger unit that contains the enzymes for the synthesis of peptide chains from amino acids as well as the sites where synthesis occurs.

Messenger RNA

Messenger RNA carries instructions or directions for protein synthesis in a code or **codon** composed of a series of nitrogen bases. It is a long, single-stranded molecule, and its molecular weight depends on the length of the protein whose synthesis it directs. There are three nucleotides in the code for every amino acid in the protein to be formed. However, the identities of the RNA initially formed by transcription from DNA and that which eventually translates genetic information in protein synthesis are not the same. The pre–messenger RNA is modified in eukaryotic cells. The mRNA then enters the cytoplasm through the pores of the nuclear envelope prepared to serve as a translator of the nucleotide language into protein sequences.

Transfer RNA

Most tRNA molecules have fewer than 100 nucleotides. **Transfer RNA** binds and delivers individual amino acids from the amino acid pool to the site of protein synthesis. A specific tRNA carries one type of amino acid. However, several different tRNA molecules can transport the same amino acid (Section 29.6).

Each tRNA has an amino acid attachment site and a template recognition site. Each 3′ end has a -C-C-A sequence that is the amino acid attachment site. The amino acid bonds to adenylic acid. The base sequence of the template recognition site is discussed in Section 29.9.

In 1965 R. Holley of Cornell University determined the nucleotide sequence of a tRNA for alanine. He suggested that the single-stranded molecule forms intrachain hydrogen bonds with some complementary bases. He postulated a "clover-

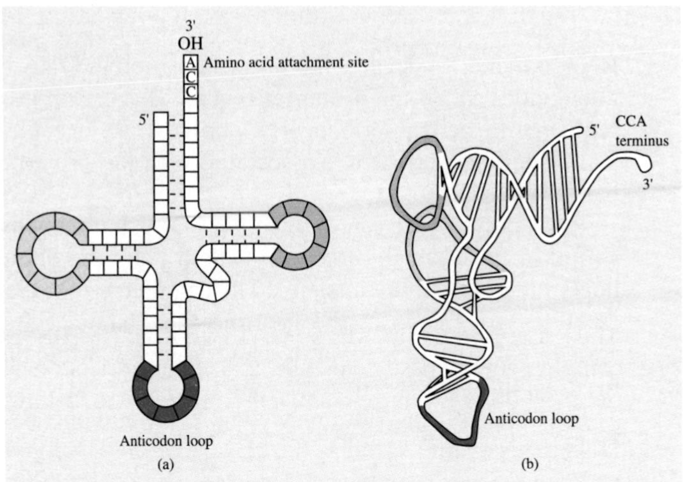

FIGURE 29.8
Structure of a tRNA
(a) A two-dimensional simplified "cloverleaf" structure in which the squares represent nucleotides. Base pairing occurs between a number of bases to give the characteristic loops. (b) The actual three-dimensional structure. The three loops are indicated in color; the anticodon loop is at the bottom of the structure. The 3′ terminus of every tRNA contains the sequence CCA.

leaf" model for the tRNA of alanine (Figure 29.8) having the maximum number of hydrogen bonds. The base pairings have been confirmed experimentally, and the shape of the three-dimensional molecule has been established.

One outside bend of a loop of transfer RNA contains a three-nucleotide sequence called an **anticodon,** which is the complement of a codon located in mRNA. The interaction of the codon of mRNA and the anticodon of tRNA is described in Section 29.8.

EXAMPLE 29.3

What portion of an mRNA chain will be produced from the following portion of a DNA strand?

$$5' \xrightarrow{\text{dG-C-A-T}} 3'$$

Solution

Base pairing in RNA will result in C in RNA opposite G in DNA, and G in RNA opposite C in DNA. The base T in DNA will pair with A in RNA, but A in DNA will pair with U in RNA.

$$3' \xleftarrow[\text{C-G-U-A}]{} 5'$$

PROBLEM 29.3

What sequence of bases in DNA is required to produce the following section of a messenger RNA chain?

$$5' \xrightarrow{\text{G-C-A-U}} 3'$$

29.8 The Genetic Code

We have been referring to DNA as a storehouse of information that contains a code that can be translated into information for protein synthesis. Both replication and transcription have been described only in terms of a series of nucleotides with-

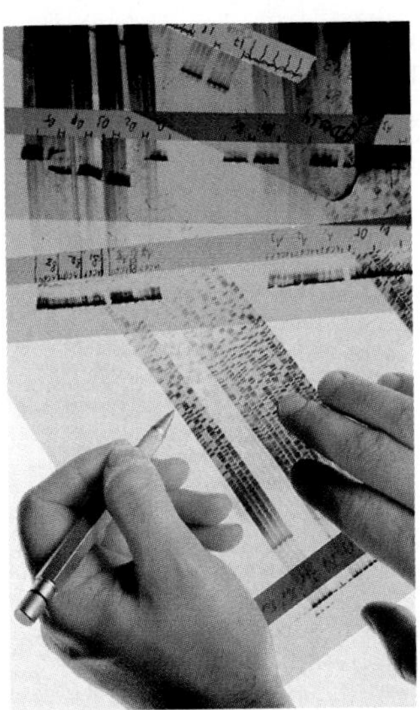

The sequence of DNA provides a molecular "fingerprint" that is unique for each of us. The determination of the base sequence is now a standard procedure used in the prosecution of criminals. (Tony Stone Images)

out regard for any relationship between neighboring nucleotides. However, the nucleotide sequence itself is a code that contains information much like the simple series of dots and dashes of the Morse code represent words. All genetic information is coded with only four bases.

If a series of two letters were the basis for forming "words," then $4^2 = 16$ words are possible, and there would be too few code words to specify the 20 different amino acids. George Gamow, of the University of Colorado, suggested that the four bases in mRNA are read as three-letter code words or codons. With three-letter words there are $4 \times 4 \times 4 = 64$ possible codons (Table 29.3).

Codons are the three-base sequences present only in mRNA. Transfer RNA, which associates with mRNA in the ribosomes in the translation process, contains a complementary base sequence that is an anticodon. In 1962 the relationship between codons in mRNA and amino acids was determined. *Escherichia coli* ribosomes were bound to synthetic mRNA molecules containing only uridylic acid. The ribosomes associated only with the tRNA of phenylalanine and formed polyphenylalanine. Therefore, the base sequence U-U-U specifies phenylalanine.

The relationship between each codon and the amino acid incorporated into a protein has been established (Table 29.3). There is more than one codon for most amino acids, so the genetic code is redundant or degenerate. A group of codons referred to as **synonyms** can specify the same amino acid. The reason for the existence of synonymous codons is not clear, but there must be advantages to the cell in having multiple ways of specifying its need for a particular amino acid.

There are some interesting relationships between codons that specify a common amino acid. Most amino acids are specified by two or four, and a few by six, synonyms. The sets of two or four synonyms differ only in the third letter. For example, cysteine is specified by UGC and UGU; alanine is specified by GCA, GCC,

TABLE 29.3 Codons and Specified Amino Acids

First Base	Second Base: U		C		A		G		Third Base
U	UUU	Phe	UCU	Ser	UAU	Tyr	UGU	Cys	U
	UUC	Phe	UCC	Ser	UAC	Tyr	UGC	Cys	C
	UUA	Leu	UCA	Ser	UAA		UGA		A
	UUG	Leu	UCG	Ser	UAG		UGG	Trp	G
C	CUU	Leu	CCU	Pro	CAU	His	CGU	Arg	U
	CUC	Leu	CCC	Pro	CAC	His	CGC	Arg	C
	CUA	Leu	CCA	Pro	CAA	Gln	CGA	Arg	A
	CUG	Leu	CCG	Pro	CAG	Gln	CGG	Arg	G
A	AUU	Ile	ACU	Thr	AAU	Asn	AGU	Ser	U
	AUC	Ile	ACC	Thr	AAC	Asn	AGC	Ser	C
	AUA	Ile	ACA	Thr	AAA	Lys	AGA	Arg	A
	AUG	Met	ACG	Thr	AAG	Lys	AGG	Arg	G
G	GUU	Val	GCU	Ala	GAU	Asp	GGU	Gly	U
	GUC	Val	GCC	Ala	GAC	Asp	GGC	Gly	C
	GUA	Val	GCA	Ala	GAA	Glu	GGA	Gly	A
	GUG	Val	GCG	Ala	GAG	Glu	GGG	Gly	G

GCG, and GCU. Arginine, leucine, and serine have sets of six synonyms: four related codons plus a set of two codons not related to the other four.

Because the codons are automatically read three nucleotides at a time there is no need for the equivalent of spacing similar to words in a sentence or for punctuation marks such as commas. However, it is still necessary to know where the code starts and stops in the instructions to make a particular protein.

The codons UAA, UAG, and UGA are involved in termination of the protein chain and are like the period at the end of a sentence. This feature of the genetic code is discussed in the next section, as is the initiation process.

EXAMPLE 29.4

What are the codons for alanine? How are they related?

Solution

Table 29.3 shows the codons GCU, GCC, GCA, and GCG for alanine. These four codons are related by identical letters in the first two positions. Only the third letter differs.

PROBLEM 29.4

What are the codons for cysteine? How are they related?

29.9 Protein Synthesis and Translation

Amino acids attached to transfer RNA (tRNA) are brought to ribosomal RNA (rRNA), the site of protein synthesis. The translation of the codons of mRNA into a sequence of amino acids requires the following substances.

1. amino acids
2. ATP to activate the amino acids
3. transfer RNA
4. messenger RNA
5. ribosomal RNA
6. guanosine triphosphate (GTP)

Amino Acids Are Activated for Reaction

To prepare an amino acid for protein synthesis, ATP reacts with the carbonyl carbon atom to yield an activated acyl derivative in which adenosine monophosphate (AMP) is bonded to the amino acid. The product is called an aminoacyl adenosine monophosphate or an aminoacyl adenylate (Figure 29.9). The reaction is catalyzed by enzymes known as aminoacyl synthetases, and a specific enzyme is required to activate each amino acid.

Esterification of tRNA

Aminoacyl adenosine monophosphates of amino acids react with a molecule of tRNA. Although each tRNA has a unique structure, they all terminate in the sequence CCA. The carbonyl carbon atom of the amino acid forms an ester with a hydroxyl group of the terminal adenylic acid.

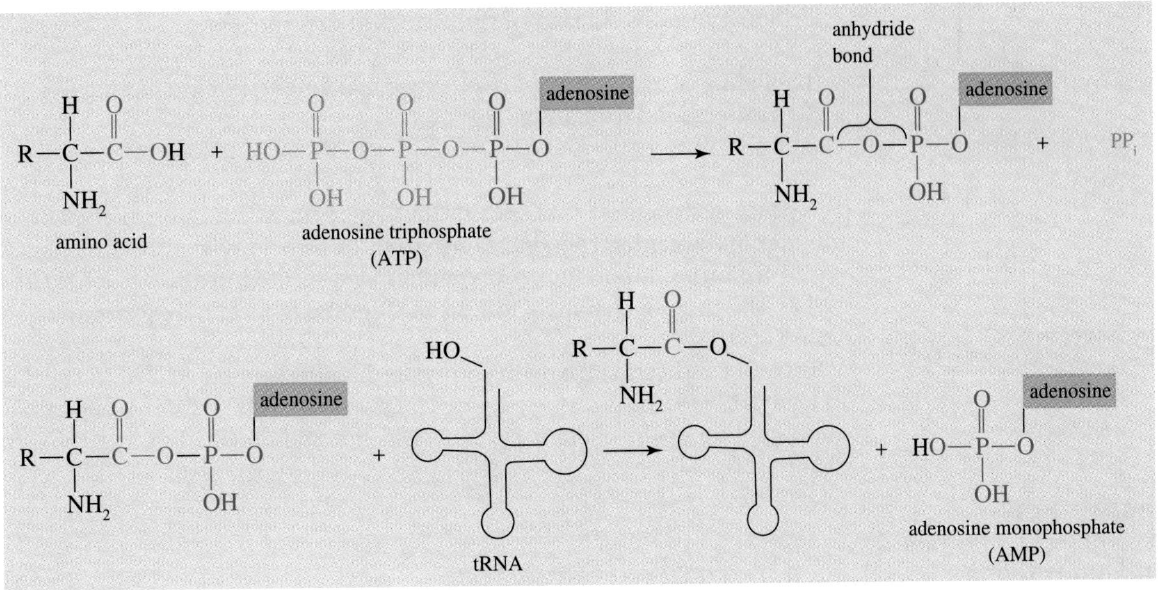

FIGURE 29.9
Formation of Activated Amino Acid and Activated tRNA

After formation of the proper aminoacyl RNA molecule, the anticodon in tRNA associates with the codon of mRNA. Thus the correct amino acid is brought to the codon site in mRNA, where base pairing occurs.

Site of Protein Synthesis

The two ribosome units that provide the site for protein synthesis are not bound together until they are assembled at the mRNA. Messenger RNA and the two ribosome units associate starting at the 5′ end, and the assembled ribosome then traverses the mRNA chain to the 3′ end. The larger ribosome unit contains the site of attachment of tRNA with its associated amino acid. However, the complete ribosome is required for the specific coupling of tRNA anticodons with the mRNA codons. After one ribosome starts its journey, other ribosomes start theirs. Thus, several ribosomes simultaneously translate the same mRNA molecule. Each ribosome is at a different position along the mRNA chain, and several proteins are in various stages of growth (Figure 29.10).

FIGURE 29.10
Ribosomes and mRNA in Protein Formation
The assembled ribosomes move from the 5′ to the 3′ end of mRNA and function independently as they synthesize a series of protein chains.

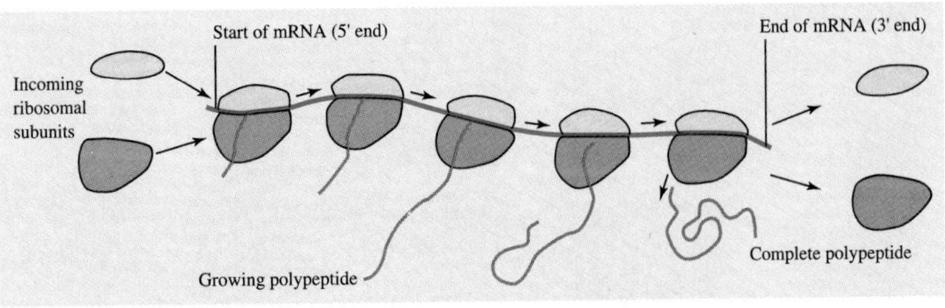

Protein synthesis consists of three steps at the ribosome:

1. binding of an aminoacyl tRNA by complementary base pairing
2. peptide bond formation
3. translocation

The processes occur at two sites in the larger ribosome unit called the **donor (P) site** and the **acceptor (A) site.** During synthesis, a growing protein chain is transferred from the donor site to the amino acid located at the acceptor site (Figure 29.11). The protein chain is synthesized from the N-terminal toward the C-terminal amino acid.

Protein synthesis starts with formylmethionine combined with a special tRNA. In Figure 29.11a, the growing peptide chain with a terminal fMet occupies a P site. At the other end of the chain is cysteine, which is still attached to its tRNA. In Figure

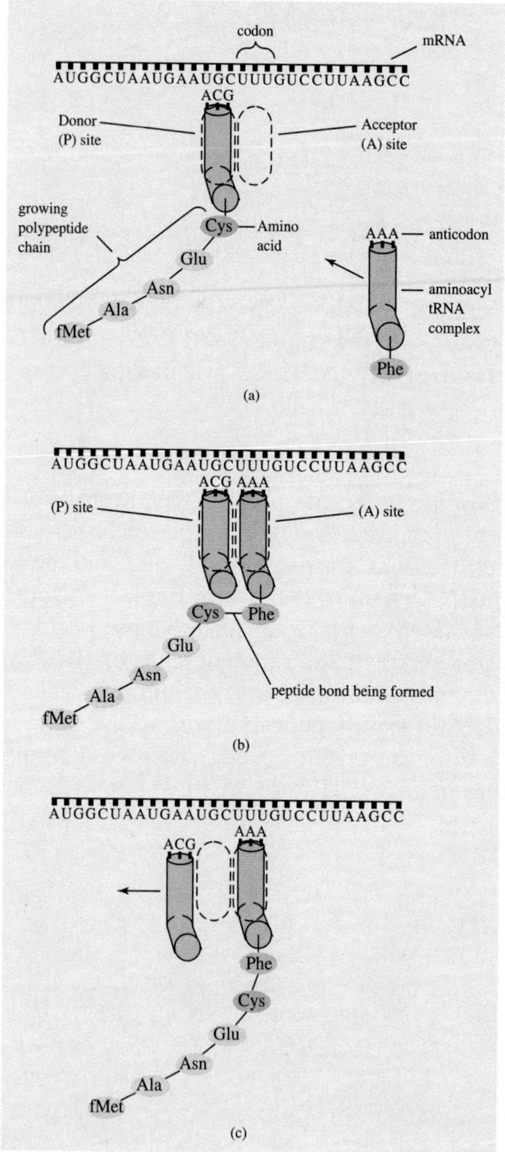

FIGURE 29.11 Synthesis of a Protein at Sites within a Ribosome

(From Hill, Fiegl, Baum–page 624.)

29.11b, an aminoacyl tRNA for phenylalanine has entered the A site, and a peptide bond is being formed between cysteine and phenylalanine. Because the peptide chain is now attached to the tRNA in the A site, the free tRNA for cysteine leaves the P site (Figure 29.11c). The ribosome then moves to the next trinucleotide sequence to the right. This process is called **translocation** and is catalyzed by translocase, with the expenditure of energy provided by GTP. As a result, the tRNA with its attached protein chain is shifted from the A site to the P site. Another amino acyl tRNA then enters the A site, and the entire sequence is repeated until the protein is complete. Note that bonding is always from the carboxyl group of the last peptide in the growing peptide chain to the amino group of the next amino acid.

EXAMPLE 29.5

What anticodon exists in the template recognition site of an aminoacyl tRNA to bind at the site having the codon sequence AGU in mRNA?

Solution

The anticodon must consist of complementary bases ((U For A, C for G, A for U) in order to bind aminoacyl tRNA to the codon sequence by base pairing. The base sequence must be UCA.

PROBLEM 29.5

What portion of an mRNA will associate with an aminoacyl tRNA having the sequence CAU in the template recognition site?

EXAMPLE 29.6

What peptide is synthesized from the following base sequence in mRNA?

-G-C-U-G-A-A-U-G-G

Solution

Reading the letters as sequences of three-letter words, we have the codons GCU, GAA, and UGG. These codons specify alanine, glutamic acid, and tryptophan, respectively. The peptide will be Ala-Glu-Trp.

PROBLEM 29.6

What codon sequence is required to synthesize the amino acid sequence Phe-Ile-Gly in a protein?

Initiation and Termination

Protein synthesis requires both initiation and termination steps. **Initiation** is the start of the growth of the protein chain. **Termination** is the conclusion of the synthesis of the protein chain and its release from the ribosome.

The process of initiation is complex. However, for a number of mRNA molecules, the first amino acid in the chain is a modified methionine called formylmethionine, and it is specified by the codon AUG. Termination of the protein synthesis is indicated by the termination codons UAA, UGA, or UAG. There are no tRNA molecules with the anticodon to pair with these trinucleotide sequences. The codons instead bind some proteins called *release factors.* Binding of these factors activates the enzyme needed to hydrolyze the peptide from the tRNA and the P site. The peptide chain is then released from the ribosome.

Protein Modification

Most proteins formed by translation of mRNA are not the actual proteins required by the cell and are modified after release from the ribosome. One such modification is the hydrolysis of the formyl group of formylmethionine or of methionine itself. In some cases the protein chain is cleaved, as in the conversion of proinsulin into insulin. Other modifications include joining the protein to another group such as a lipid or a carbohydrate.

29.10 Gene Mutations and Genetic Disease

A **mutation** of a gene is a change of one or more of the nucleotides in DNA. Nucleotides may be substituted, added, or deleted. Before considering the consequences of mutation for protein synthesis, let's see what happens to a sentence when letters are substituted, added, or deleted. Since codons are three-letter words, our sentence is

THE BIG DOG SAW THE CAT

Substitution of one letter by another in the message might or might not make sense depending on the letter substituted and the location in the word. For example, consider

The first two words and the last three make sense, but DOH does not make sense. Some of the message is understandable. Replacement of a single nucleotide in DNA is called a **point mutation.**

Now consider a "point mutation" of our sentence when H is substituted for the first letter of a word.

THE BIG HOG SAW THE CAT (sense)

The meaning of the message is changed, but it does make sense. However, if you want to send information about a dog, the sentence with the word hog is flawed.

Consider adding a letter to the sentence. If the words are still read three letters at a time, a major part of the sentence becomes nonsense if we add E after the G of DOG.

THE BIG DOG ESA THE ECA T (ESA THE ECA T: nonsense)

Deletion of a letter such as the G of DOG also results in nonsense.

THE BIG DOS AWT HEC AT (DOS AWT HEC AT: nonsense)

Both additions and deletions in DNA are examples of **frame shift errors,** which cause a different series of words to be read.

Now let us consider the effect of substitution, addition, and deletion on codons. If DNA is mutated, then mRNA is mutated as well. As a consequence, protein synthesis may be flawed. The effect of defective proteins on biological function depends on which base is changed.

The consequences of changing the base corresponding to the third letter in a codon may not be serious, because many amino acids have several codons that differ only in the third letter. For example, glycine is coded by GGA, GGG, GGU, and GGC. Replacement of C by U in the third letter will still place the correct amino acid in the protein (Figure 29.12).

A point mutation that results in replacement of the same type of amino acid such as one neutral amino acid by another neutral amino acid at some point away from the active site often is of no consequence. If the replacement amino acid has a similar shape, it may not significantly change the conformation of the polypeptide nor the resulting activity of an enzyme. Thus, although changing the first base of the codon GGC for glycine from G to A gives the codon for serine, the amino acids are both "small" neutral amino acids (Figure 29.12).

A single amino acid may significantly change the conformation of a protein to alter biological function. In normal adult hemoglobin, glutamic acid is the sixth amino acid in the chain. The codons for glutamic acid are GAA and GAG. In sickle cell hemoglobin, valine is in the sixth position. The codons for valine are GUA, GUG, GUU, and GUC. Substitution of U for A, the second letter in the glutamic acid codon, gives a codon for valine. The replacement of glutamic acid by valine changes the tertiary structure of hemoglobin because glutamic acid is an acidic amino acid, whereas valine is a neutral amino acid.

Some point mutations can cause serious biological damage, as when an improper amino acid is located at the active site of an enzyme. Consider the codon UCG for serine. Mutation to give UUG substitutes leucine for serine in the protein. Since serine is required at active sites in trypsin and chymotrypsin, the resulting enzyme cannot hydrolyze proteins.

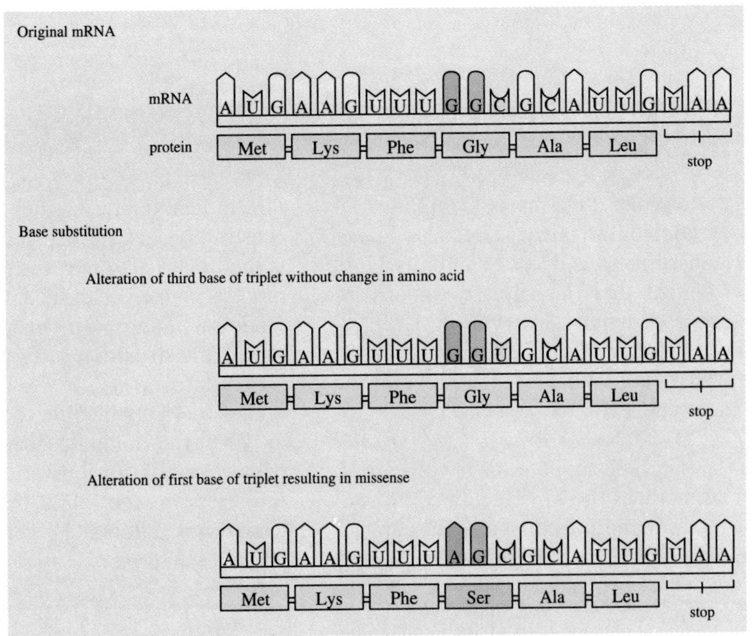

FIGURE 29.12 Result of Point Mutations in mRNA

FIGURE 29.13 Result of a Frame Shift Mutation

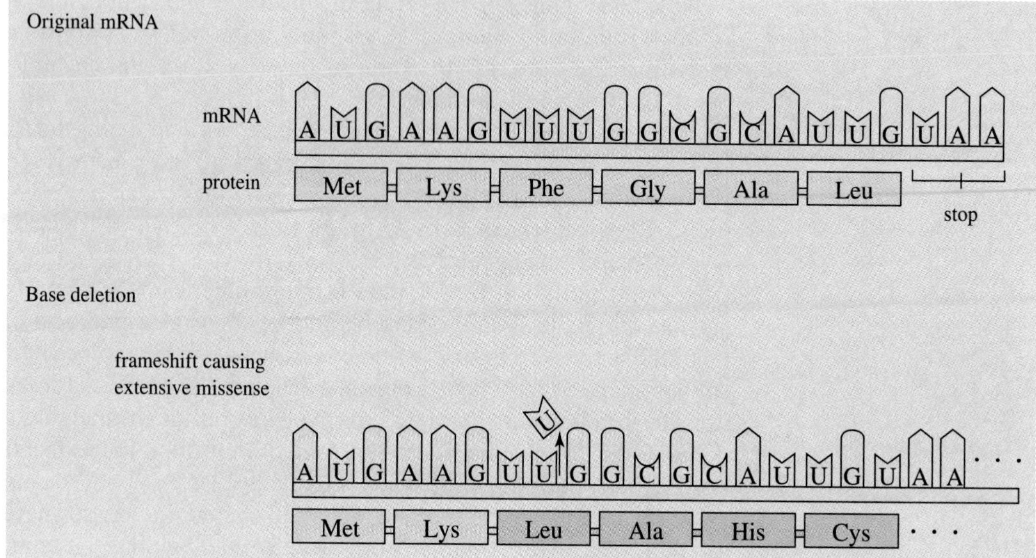

There are also serious consequences if a mutation changes a codon for an amino acid into a termination codon. One of the codons for lysine is AAG. If U is substituted for the first A, the termination codon UAG results. As a consequence, when this codon is reached, the synthesis of the protein stops and a useless protein results.

A deletion or addition mutation also has serious consequences. Beyond the point of change, the instructions lead to the formation of an unsuitable sequence of amino acids. The effect is shown for deletion of the third U required to code for phenylalanine (Figure 29.13). By reading G, the former first letter of the codon for glycine, the frame shifted codon now specifies leucine. Not only is leucine's shape quite different from phenylalanine's, but all subsequent amino acids are "wrong."

A Gene Defect May Cause Osteoarthritis

Osteoarthritis disables millions of people. It involves the disintegration of the cartilage between bones, which results in pain, stiffness, and loss of mobility. Until recently it was thought that this disease was an inevitable consequence of aging. However, in 1990, researchers at the Jefferson Medical College in Philadelphia found a genetic mutation that is responsible for the production of the weak cartilage associated with osteoarthritis.

There are more than 1000 amino acids in collagen II, the protein that strengthens the cartilage. In some people the gene for the formation of collagen II has an "incorrect" base sequence. It directs the placement of arginine rather than cysteine, the proper amino acid. The substitution has serious consequences because cysteine is the amino acid that forms disulfide bridges. Furthermore, arginine is a basic amino acid, whereas cysteine is a neutral amino acid. This one substitution significantly affects the structural integrity of collagen II and causes it to disintegrate.

Doctors may be able to screen people from families in which the mutation occurs. They could advise these people to look for jobs or choose lifestyles to minimize the stress on the joints that can eventually cause osteoarthritis. Ultimately, gene therapy may be able to prevent the disease.

EXAMPLE 29.7

What are the consequences of a genetic mutation that results in replacement of CGC in mRNA by CCC?

Solution

The change is in the second letter and is expected to result in the specification of a different amino acid in protein synthesis. The codon CGC specifies the basic amino acid arginine, whereas the codon CCC specifies the secondary amino acid proline. Since these amino acids differ substantially in structure and acid–base properties, the tertiary structure and function of the protein may be seriously altered.

PROBLEM 29.7

What are the consequences of a genetic mutation that results in replacement of AGC by AGG in mRNA?

29.11 Recombinant DNA and Genetic Engineering

Now that we understand how nature puts together nucleic acids, we might ask if chemists could do the same thing in the laboratory. Of course, they would need a reason to do so because nature does such a good job—or does it? A number of genetic diseases result from faulty genes that prevent the normal production of the enzymes required for "normal" life. Intervention in the fundamental process of human life does pose serious moral questions. However, there are many who feel that improvement of the quality of life and prevention of death by the alteration of genes are valid considerations that are on a higher moral ground. The construction and alteration of DNA to make genes is called **genetic engineering.**

The DNA of Bacteria Is Easily Changed

In addition to the DNA in its nuclear region, the *E. coli* bacterium has some DNA in circular, coiled molecules called **plasmids.** These plasmids have only a few genes and they replicate independently of the chromosomes in the nuclear region.

Plasmids can be removed from *E. coli* and "snipped" open by special enzymes called **restriction enzymes.** An opened plasmid has two "sticky ends" with nucleotides that can base pair with other polynucleotides if they can find them (Figure 29.14). Chemists have also learned how to remove portions of human genes that are responsible for desirable biological functions, by cutting them at two places using restriction enzymes. This section of human DNA also has two "sticky ends." The human DNA then inserts into the open bacterial plasmid using an enzyme called DNA ligase. The modified plasmid is called **recombinant DNA.** After the modified plasmid is reinserted into the bacterium, the bacterium multiplies and reproduces the new plasmid. The recombinant DNA has properties and functions characteristic of both the original DNA and the inserted DNA.

Using recombinant DNA techniques, *E. coli* produces proteins that it normally does not make. For example, it produces human insulin when spliced with a DNA fragment that transcribes the messenger RNA for human insulin. The insulin can be "harvested" from the *E. coli* colony, which is a miniature chemical factory. Dia-

FIGURE 29.14
Recombinant DNA
In the first step a plasmid from a prokaryotic cell is cleaved by a restriction enzyme. The plasmid vector is then joined to a section of "foreign" DNA obtained by use of a restriction enzyme on the DNA of a eukaryotic cell.

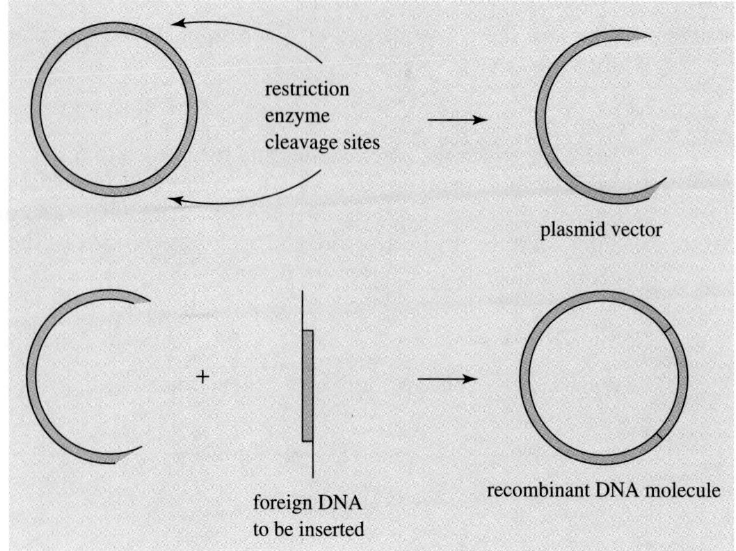

betic persons formerly used insulins obtained from animal sources that differ only slightly from human insulin. However, allergic reactions sometimes occurred, and another insulin had to be used. Since *E. coli* makes human insulin, there is no allergic reaction to this insulin.

The production of human growth hormone is another example of the benefits of recombinant DNA techniques. Children who suffer from pituitary dwarfism may grow no taller than four feet. Growth hormones are species specific, which means that the growth hormone for humans cannot be obtained from any other animal. The only source of human growth hormone used to be cadavers. This scarce supply allowed some children to be treated and to grow to a more "normal" height. However, the cost of obtaining the hormone was high. In addition, there was increased concern that viruses lurking in the cadavers could be transmitted to the children via the hormone sample. A plentiful safe supply became available from recombinant DNA technology in 1985.

In 1989 recombinant DNA technology produced erythropoietin (EPO)—a human protein hormone produced by the kidneys. It stimulates the production of red blood cells in the bone marrow. EPO is used to treat anemia, especially as a result of kidney damage.

Can Recombinant DNA Be Used in Human Genes?

Some abnormal genes can be detected in chromosomes. Genetic counseling allows prospective parents to prevent perpetuation of inherited diseases. The procedure is voluntary. It is unlikely that most societies would tolerate government control over such individual decisions as having children.

Intervention at the level of the DNA in the nucleus of the cell could correct a genetic abnormality. The defective gene could be removed and replaced by a "normal" gene. For example, if the gene responsible for the genetic disease known as hemophilia could be changed, the risk of bleeding to death would be eliminated for the 1 in 10,000 males that are born with this condition. However, the ethical implications and potentials for abuse of gene manipulation are serious concerns. It is

Recombinant DNA and Agriculture

Intervention using recombinant DNA to change the characteristics of plants is far easier than in higher animals because many plants can be grown from a single cell. Moreover, the ethical questions are less onerous because geneticists have been developing new plants for a long time by selective breeding. However, the task of improving the characteristics of a plant by recombinant DNA techniques is a continuing challenge. First, the gene of interest must be located that may improve the plant, increase its yield, decrease its susceptibility to pathogens, or improve its shelf life in the market.

Some crop plants have been developed that have increased resistance to herbicides such as Roundup. This herbicide is nontoxic to animals and is short-lived in the environment. As a result of increased resistance of the crop plant, the herbicide can more selectively destroy weeds and not affect the crop yield. Advances have also been made in increasing the resistance of plant roots to pests such as corn rootworm.

Other projects have met with mixed success because of public concern about genetically altered products. One example is the development of tomatoes that are less resistant to rot and can be picked when fully developed rather than while green. Until this concern abates, there may be fewer attempts made to improve the quality of our food supply. The possibilities are almost unlimited. One potential benefit would be the development of plants that provide a more complete protein source. The majority of societies in the world have a predominately vegetarian diet. The most common form of malnutrition is protein deficiency. Improving the quality of the protein in crops is important in developing or underdeveloped countries that have minimum sources of the more complete animal proteins.

Perhaps the most exciting prospect for recombinant DNA technology is the development of plants that could fix nitrogen. A limited number of plants convert atmospheric nitrogen into nitrogen-containing compounds that are useful to the plant. Most plants must be fertilized to increase the crop yield. Fertilizers are expensive because they are produced in chemical plants by methods that consume large quantities of fossil fuel. As a consequence, fertilizers are not available to poorer nations. In addition, if plants such as corn could be genetically engineered to fix nitrogen with the efficiency of legumes, the world would benefit immensely. Not only would crop yields increase but the soil would be rejuvenated simultaneously.

The genetically altered tomato on the left is larger than an ordinary tomato on the right. (Peter Arnold, Inc.)

Gene Therapy in Cystic Fibrosis

Cystic fibrosis is caused by genetic dysfunction of a chloride channel in the membrane of airway and other epithelial cells. Loss of function of this channel causes abnormal salt transport, which results in an excessive accumulation of mucus in the lungs and gastrointestinal tract. The lungs are gradually destroyed by progressive damage of the respiratory epithelium as a result of chronic infections. Correction of the condition would require introduction of a gene directing formation of functional chloride channels.

Active research in animals as well as several human studies in clinical trials have been carried out. The most advanced studies have used adenoviruses as vectors for gene delivery. They can be efficiently introduced into epithelial cells. In 1993, three persons with cystic fibrosis were treated, and the protein that forms the chloride ion channel as well as the mRNA for the protein were detected in one of the patients, and the protein was detected in the bronchial epithelia of another patient. However, it is known that adenoviruses cause dose-dependent inflammation in the airways of primates. Continued research in the use of adenoviruses may improve the delivery of the gene at high doses without inflammatory side effects.

In clinical trials in 1994 the possibility of using a liposome-mediated delivery to the nasal epithelium was examined as a way to avoid the side effects associated with adenoviruses. Some restoration of salt transport was observed. However, the efficiency of the gene transfer is low. Continued research is directed toward greater efficiency of cell targeting and gene transfer.

one thing to cure a disease, but it is quite another to allow individuals to start custom-designing their children to have "desirable" physical characteristics.

29.12 Viruses

Viruses, which cause a variety of diseases in plants and animals including smallpox, measles, mumps, and influenza in humans, are intermediate between living and nonliving material. They are composed of either DNA or RNA, but not both, surrounded by a protein coat. Most viruses are RNA, not DNA, based. Viruses do not have the necessary nucleic acids to replicate themselves or the amino acids and enzymes necessary to support life. Because they must have a host cell to reproduce, viruses are not considered "living."

Viruses are very small (0.05 to 0.2 μm) and come in a variety of shapes. Both the shape of the virus and its protein coat affect how it functions in infecting cells. Viruses attack cells and take over the chemical machinery. The protein coat contains an enzyme that breaks down the cell membrane of the host cell. The viral DNA or RNA then "squirts" into the cell. Each virus has a unique membrane-dissolving enzyme, so that it attacks only selected types of cells. Thus, there are viruses that will attack only blood cells, only heart muscle, or only the spinal cord.

Once inside the cell, the viral nucleic acids can become a "silent gene" and remain for long periods of time before becoming active. Alternatively, they may immediately take over the operation of the cell, which then stops making its own DNA, RNA, and protein. The cell starts to replicate the nucleic acids of the virus as well as the viral protein. Many copies of the virus are made by a single cell, which is eventually destroyed because it no longer produces the materials necessary for its own survival. The new viruses are released and attack other cells. If the process is not stopped, the organism gets sick and may die.

A virus can infect only a limited number of types of cells. Apparently it recognizes certain receptor sites on the surface of the cell. Unfortunately for that cell, the lock-and-key arrangement between it and the virus leads to the cell's demise. Some viruses are less specific and can infect a range of species. For example, swine flu virus can infect humans as well as hogs. The rabies virus infects a range of mammals, including squirrels, raccoons, dogs, and humans.

How Does RNA Take Over a Cell?

Those few viruses that are DNA based take over cells by simply using the nucleotides and enzymes. However, if the virus is RNA based, the process is a bit more complicated. Remember that RNA is produced in cells under the direction of DNA. A host cell has no directions or enzymes for making the RNA of the virus. Thus, the virus also has to have some conspirator molecules to use the cell. There are two ways in which an RNA virus can replicate itself. One process involves a protein enzyme called *RNA replicase* that catalyzes the manufacture of RNA from directions encoded on the viral RNA. Thus, some RNA molecules can reproduce themselves within a cell. Some viruses carry the necessary RNA transcriptase, whereas others use the enzyme in the host cell. Both the polio virus and the rabies virus operate by using RNA transcriptase.

The second mechanism occurs in viruses whose RNA directs the synthesis of a protein enzyme called *DNA polymerase* or *reverse transcriptase.* As a consequence,

RNA is responsible for the synthesis of DNA, which then forms more RNA. Viruses that operate by this mechanism are called **retroviruses.**

Are There Medicines Against Viruses?

Antibiotics are ineffective against diseases induced by viruses. There are some virucidal agents, but their effectiveness is limited. They must stop the reproduction of viral nucleic acids inside host cells but not prevent the normal replication of the host's DNA. Vidarabine, sold as Vira-A, is similar to adenosine but contains arabinose rather than ribose. It is a viricidal agent used only to fight life-threatening diseases, because it is toxic if dosages are not carefully controlled. It may also cause chromosomal damage.

Many viral infections can be prevented by immunization with a vaccine. The vaccine contains an active or inactive viral form that stimulates the body's natural immune system to produce antibodies against the virus. These antibodies deactivate invading active viruses before they can enter host cells.

Smallpox has been eradicated by a worldwide program of vaccination. Similarly, the incidence of measles and polio has been greatly reduced in the United States and many other countries by use of vaccines.

Summary

Chromosomes located in the nucleus of the cell contain the genetic information—the **genetic code—**that determines the characteristics of the living organism. The **nucleic acids** contain three types of components: a sugar, a heterocyclic nitrogen base, and a phosphate group. **Ribonucleic acids** contain β-D-ribose, whereas **deoxyribonucleic acids** contain β-D-2-deoxyribose. The **purine** and **pyrimidine** bases in ribonucleic acids are **adenine (A), uracil (U), cytosine (C)** and **guanine (G),** whereas the bases in deoxyribonucleic acids are **adenine, thymine (T), cytosine,** and **guanine.**

Nucleosides are N-glycosides of a sugar and one of the heterocyclic nitrogen bases. **Nucleotides** are phosphate esters of nucleosides. Nucleic acids are polymers of nucleotides. **Polynucleotides** contain **phosphodiester** linkages between the 5′-hydroxyl group of one sugar and the 3′-hydroxyl group of another sugar unit.

DNA consists of two **antiparallel strands** of polynucleotide chains coiled about an axis to form a **double helix.** The purine and pyrimidine bases, which are pointed toward the interior of the helix, form **complementary** hydrogen bonds to each other. Cytosine forms three hydrogen bonds to guanine, and adenine forms two hydrogen bonds to thymine.

The **replication** of the double helix of DNA occurs by a **semiconservative** process at a **replication fork,** catalyzed by **DNA polymerase.** The helix partially uncoils, and each chain serves as a **template** for the synthesis of its **complementary chain.** Each daughter DNA has one strand of the original DNA and one new strand.

RNA is a single-stranded molecule that contains phosphodiester linkages between the 5′-hydroxyl group of ribose of one nucleotide unit and the 3′-hydroxyl group of another nucleotide. There are three types of RNA: **ribosomal RNA (rRNA), transfer RNA (tRNA),** and **messenger RNA (mRNA).**

DNA produces RNA by synthesis of a complementary strand in a process called **transcription** catalyzed by **RNA polymerase.** Information contained in **codons** in mRNA is used in a **translation** process to form proteins at a site in the ribosome. tRNA contains an **anticodon** that associates with the codon of mRNA in the **ribosomes.**

Protein synthesis requires both **initiation** and **termination** steps. The chain-building processes occur at two sites in the ribosomes called the **donor (P) site** and the **acceptor (A) site.** During synthesis a growing protein chain is transferred by **translocation** from the donor site to the amino acid located at the acceptor site.

A **mutation** of a gene is a change of one or more of the nucleotides in DNA. In **point mutation** one base is replaced by another at a single site. Both additions and deletions are examples of **frame shift errors.**

The construction of DNA to make genes is called **genetic engineering.** The process uses DNA in circular, coiled molecules from *E. coli* called **plasmids.** They are "snipped" open by **restriction enzymes** and modified. The modified plasmid is called **recombinant DNA.**

Exercises

DNA and RNA

29.1 How are all DNA molecules structurally alike?

29.2 How do DNA molecules differ?

29.3 What group bridges the sugars in DNA and RNA?

29.4 DNA and RNA are acidic materials. Explain why.

29.5 Indicate whether each of the following is a purine or pyrimidine base.

(a) uracil (b) cytosine
(c) adenine (d) thymine

29.6 Which bases are found in DNA? Which bases are found in RNA?

Nucleosides and Nucleotides

29.7 Write the structural formula for adenosine monophosphate.

29.8 Write a structural formula for a nucleoside containing thymine and deoxyribose.

29.9 What is the configuration of the N-glycoside bond of nucleotides?

29.10 What hydroxyl group of ribonucleosides is phosphorylated to give ribonucleotides?

29.11 How many free hydroxyl groups exist in a polynucleotide of deoxyribose?

29.12 How many free hydroxyl groups exist in a polynucleotide of ribose?

29.13 How many chiral centers are there in a ribonucleoside?

29.14 How many chiral centers are there in a deoxyribonucleoside?

29.15 Identify the components of the following substance.

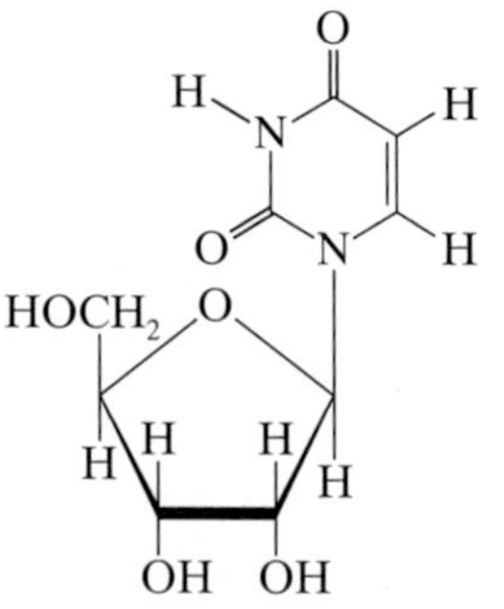

29.16 Identify the components of the following substance.

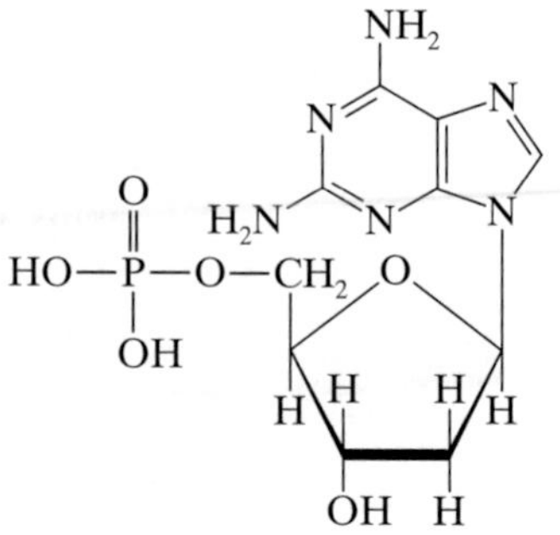

The Double Helix

29.17 Does each of the following pairs occur in the DNA molecule? If not, explain why.

(a) G-A (b) G-T (c) C-U

29.18 Does each of the following pairs occur in the DNA molecule? If not, explain why.

(a) A-A (b) C-T (c) A-C

29.19 What forms the backbone structure of each strand of the DNA double helix?

29.20 Where are the base pairs located in the double helix?

Replication

29.21 Describe the replication process.

29.22 What is the replication fork?

29.23 What is the complementary strand for a DNA strand with the sequence dG-C-A-T-C-A-G?

29.24 What is the complementary strand for a DNA strand with the sequence dA-T-C-A-G-T-A?

Transcription

29.25 What mRNA would be formed from the base sequence in Exercise 29.23?

29.26 What mRNA would be formed from the base sequence in Exercise 29.24?

29.27 What is the complementary base in mRNA for A in DNA?

29.28 What base in mRNA is the complement of T in DNA?

29.29 A portion of DNA has the sequence C-C-C-T-G-T-A-C-A-C-C-T. What base sequence will form in mRNA? What peptide will be formed?

29.30 A portion of DNA has the sequence G-G-G-T-G-C-A-G-A-C-C-A. What base sequence will form in mRNA? What peptide will be formed?

Translation

29.31 How is a codon involved in the translation process?

29.32 How is an anticodon involved in the translation process?

29.33 What amino acid does each of the following codons in mRNA place in a protein chain?

(a) GUU (b) CCC (c) UUU (d) ACG

29.34 What amino acid does each of the following codons in mRNA place in a protein chain?

(a) UGU (b) GCA (c) AGG (d) CUA

29.35 How are acceptor and donor sites in the ribosome used in protein synthesis?

29.36 What amino acid initiates protein synthesis?

29.37 What anticodons in tRNA will base-pair with the codons of Exercise 29.33?

29.38 What anticodons in tRNA will base-pair with the codons of Exercise 29.34?

29.39 What aromatic amino acids have similar codons?

29.40 Do basic amino acids have similar codons?

29.41 What base sequence in mRNA is necessary to form Val-Asp-Ala-Gly?

29.42 What base sequence in mRNA is necessary to form Ser-Glu-Pro-Phe?

Codons and Mutations

29.43 What are the genetic consequences of a frame shift?

29.44 Why are substitution mutations generally less serious than deletion or addition mutations?

29.45 Describe two types of substitution mutations that will cause seriously defective proteins.

29.46 What differences are there between the codons for acidic amino acids and those for basic amino acids?

Recombinant DNA

29.47 What are restriction enzymes?

29.48 Where is a plasmid found? What is it made of?

29.49 What are the sources of the DNA used in producing recombinant DNA?

29.50 Why might it be more difficult to use recombinant DNA techniques in human cells?

Viruses and Bacteria

29.51 How can we chemically fight bacterial infections?

29.52 How are viral infections different from bacterial infections?

29.53 Plants are susceptible to some viral infections. Why don't these viruses affect humans?

29.54 Viruses could be called parasites. Why?

29.55 What enzyme must a virus have in order to produce RNA from viral RNA?

29.56 What enzyme must a virus have in order to produce DNA from viral RNA?

A phlebotomist drawing a blood sample from a patient. (The Stock Market)

30 Body Fluids

30.1 Solutions in the Body
30.2 pH of Body Fluids
30.3 Solutes in Body Fluids
30.4 Digestive Juices
30.5 Blood
30.6 Blood Cells
30.7 Blood Gases
30.8 The Immune System
30.9 Blood Clotting
30.10 Urine
30.11 The Kidneys

Essays
- *Bacteria and Ulcers*
- *Bone Marrow Transplants*

30.1 Solutions in the Body

Virtually all the life processes discussed in this book occur in aqueous solutions collectively known as **body fluids.** The solutes include simple ions, small neutral covalent molecules, and some charged covalent molecules as well as some aggregates of large molecules that are similar to micelles. The total volume of water in a 70-kg person is approximately 40 L, or about 60% of total body weight. A newborn infant may contain as much as 75% water. The average decreases as we age. In obese persons the average may be as low as 45%.

Approximately 25 L, or 62%, of body fluids is present as **intracellular fluid,** the fluid within the 100 trillion cells of the body. About 2 L of intracellular fluid is found in red blood cells. There are 15 L, or 38%, of water in the body fluids called **extracellular fluid,** the fluid in the spaces outside of cells. About 3 L of extracellular fluids is in **blood plasma;** the remaining 12 L are largely **interstitial fluids,** which lie in the spaces between cells. There are also small quantities of cerebrospinal fluid, intraocular fluid, and fluids of the gastrointestinal tract.

All the organs of the body and their associated chemical reactions depend on the aqueous environment of the extracellular fluids and the exchange of those liquids with the external environment. In addition, the exchange of gases with the external environment depends on the internal aqueous environment. The major organs that control this equilibrium are the lungs and the kidneys, both subjects of this chapter. In addition, the skin and the alimentary tract are important parts of the body that are also considered.

In this chapter we will consider body fluids according to their major physiological function. Digestive juices, blood, and urine will be discussed, but the properties of blood will be considered in the most detail. Digestive juices prepare and convert food into compounds that the body can metabolize for growth and energy. Blood provides a transportation network for nutrients from the digestive tract, oxygen, carbon dioxide, waste products, enzymes, antibodies, and hormones. Urine eliminates some waste products from the body and is important in maintaining the body's water and electrolyte balance.

30.2 pH of Body Fluids

Regulation of pH is one of the most important functions of the body, because most of the rates of reaction in body fluids are pH dependent. The average hydronium ion concentration of all extracellular fluid is 4×10^{-8} M, which is pH 7.4. This is also the pH of the blood. The pH of intracellular fluids varies between 7.0 and 7.2.

There are three major buffer systems that control the pH of body fluids, namely, the bicarbonate buffer, the phosphate buffer, and the protein buffer. The protein buffer in blood may be considered in two parts: plasma protein and hemoglobin.

The Bicarbonate Buffer

The bicarbonate buffer, first discussed in Section 11.10, consists of the bicarbonate ion and carbonic acid. Because the pK_a of carbonic acid is 6.1, the ratio of bicarbonate to carbonic acid is about 20 to 1 at pH 7.4—the pH of body fluids. As a result, this buffer is effective only in countering excess acidity. The bicarbonate ion reacts with the hydronium ions to form carbonic acid.

$$HCO_3^- + H_3O^+ \rightleftharpoons H_2CO_3 + H_2O$$
$$H_2CO_3 \rightleftharpoons H_2O + CO_2$$

The bicarbonate buffer of blood is regulated by two of the major organs. The lungs expire carbon dioxide formed by the decomposition of carbonic acid. The kidneys eliminate excess bicarbonate ion in urine.

The Phosphate Buffer

The phosphate buffer consists of HPO_4^{2-} and $H_2PO_4^-$. The pK_a of $H_2PO_4^-$ is 6.8, a value closer to the pH of body fluids than is the 6.1 of the bicarbonate buffer. Because the ratio $H_2PO_4^-/HPO_4^{2-}$ is about 4 to 1, the phosphate buffer is effective against both acid and base. However, its concentration in extracellular fluids is much smaller than that of the bicarbonate buffer. Therefore, the phosphate buffer has less capacity.

$$HPO_4^{2-} + H_3O^+ \rightleftharpoons H_2PO_4^- + H_2O$$

The phosphate buffer is important in intracellular fluids, where the concentration of phosphate species is higher than in extracellular fluid. In addition, the pH of intracellular fluid is closer to the pK_a of the phosphate buffer.

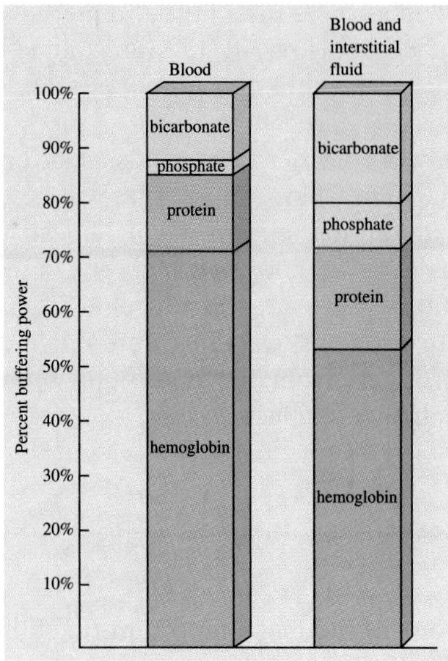

FIGURE 30.1 Relative Buffering Power of Buffers
The relative buffering power is shown as a percentage of the total for the body, both for the blood and for all interstitial fluid.

The Protein Buffer

Recall that amino acids are amphoteric and exist as zwitterions (Section 23.2). Therefore, amino acids are effective buffers themselves. However, the majority of amino acids in the body are present in the form of protein. Nevertheless, the large number of $—CO_2^-$ and $—NH_3^+$ groups in side chains of the protein can also exist as $—CO_2H$ and $—NH_2$ groups. Thus, proteins can also act as buffers. The isoionic point of hemoglobin is 6.8, so at the pH of blood the major form of hemoglobin is its anion. The mechanism of the hemoglobin buffer will be examined in Section 30.7, where we consider the transport of blood gases.

Relative Buffering Power

The relative buffering power of the buffer systems of the body depends not only on their pK_a values but also on their concentrations. For the blood, the percentage of the buffering action is greatest for hemoglobin (Figure 30.1). Considering the blood and all interstitial fluids, the buffering action of hemoglobin is a smaller fraction of the total because of the larger quantity of interstitial fluids.

If all tissue cellular fluids are considered, then 75% of the buffering power of body fluids is within the cells. The phosphate buffer and the buffer of intracellular protein together dominate the total buffer capacity.

30.3 Solutes in Body Fluids

The solutes in body fluids include gases, small ions, small polar organic molecules, and large organic molecules with polar groups on their surface. Our emphasis will be on small ions and ionized biomolecules, mainly proteins.

Electrolytes is a term used to describe the ions in body fluids. Recall that the concentrations of these ions are given in milliequivalents per liter (Section 9.8). The concentrations of the cations and anions of the body are shown in Figure 30.2 for interstitial fluid, plasma, and intracellular fluid. Note that the proteins are listed

FIGURE 30.2 Electrolyte Composition of Body Fluids

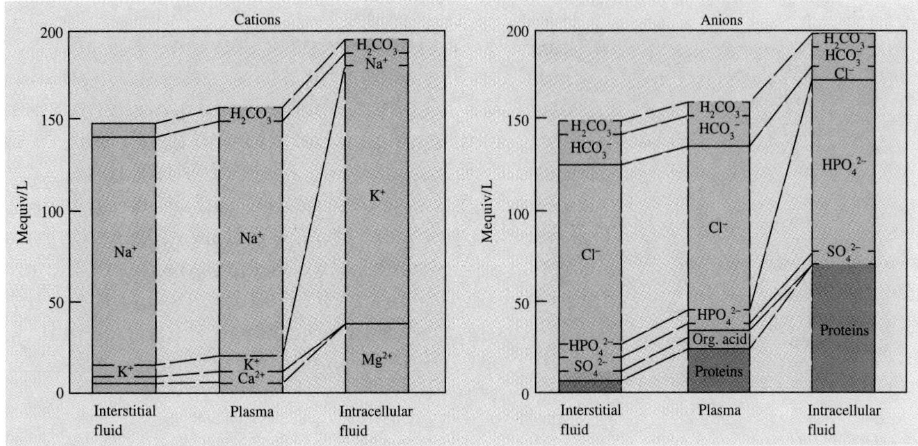

as anions because their isoionic points average 6.8, and the pH of most fluids is higher (more basic).

The representation of the composition of body fluids using milliequivalents per liter provides a different picture from one based on percent by mass. According to concentration, the major components appear to be ions. However, the molecular weight of proteins is much higher than the formula weight of any of the ions. Taking into account this fact and determining the composition of body fluids based on mass gives a very different picture. By mass, the proteins and the nonelectrolytes make up 90% of the solutes in plasma, about 60% in interstitial fluid, and 97% in intracellular fluid.

Extracellular fluids contain high concentrations of sodium and chloride ions and a reasonably large concentration of bicarbonate ion. There are only small concentrations of potassium, calcium, magnesium, phosphate, sulfate, and organic anions. The concentration of this fluid, which constitutes the environment in which the cells are located, must be carefully regulated. Much as we depend on the atmosphere about us for our survival, the cells can function only if the proper aqueous environment within our bodies is maintained.

Compared with extracellular fluids, intracellular fluids are substantially different in composition. There are only small concentrations of sodium and chloride ions and virtually no calcium ions in intracellular fluids. However, there are large concentrations of potassium and phosphate ions as well as moderate concentrations of magnesium and sulfate ions. The cell membrane maintains and controls this difference in concentrations by facilitated diffusion and active transport mechanisms (Section 22.11).

Osmotic Pressure

The diffusion of water but not dissolved solutes across a selectively permeable membrane is a phenomenon called *osmosis* (Section 9.13). Recall that as a result of differences in osmotic pressure, water moves from the compartment with the lower concentration of solute toward the compartment with the higher concentration of solute. Recall also that the osmotic pressure depends only on the molar concentration of the solute, not on the identity of the solute or its molecular (formula) weight. Thus, the chloride ion, the glucose molecule, and albumin—whose molecular weight is 70,000—have the same effect on osmotic pressure at the same concentration.

The total osmotic pressure of a solution is the sum of the contributions of the osmolarity of all solutes. Approximately 80% of the osmotic pressure of interstitial fluid and blood plasma is due to sodium and chloride ions. In intracellular fluid approximately 50% of the osmotic pressure is due to potassium ions.

Small changes in concentration on either side of the membrane can cause considerable diffusion of water (Section 9.13). In the case of red blood cells, diffusion out of the cell causes crenation, and diffusion into the cell may cause hemolysis. The transfer of water through cell membranes is so rapid that there is essentially always an equilibrium between the two sides of the membrane. Thus, the amount of fluid within cells can be changed by altering the composition of extracellular fluid. If a hypertonic solution is added to the extracellular fluid, water will move out of the cell. Conversely, if a hypotonic solution is added, water moves into the cells.

Water Balance

The kidneys (Section 30.11) are the major organ responsible for maintaining the balance of not only water in the body but also the concentration of the dissolved electrolytes. The hypothalamus, which monitors the concentration of solutes in blood plasma, adjusts hormone secretions. For example, vasopressin (Section 23.6) is an antidiuretic hormone. If the osmotic pressure increases, a decrease in the vasopressin level results, and less water is excreted. As a consequence, the concentration of solutes in the urine increases.

Some abnormal conditions cause excess secretion of vasopressin. The water content of the body then increases, and the concentrations of electrolytes may decrease to dangerous levels. The reverse problem occurs in diabetes insipidus, in which there is insufficient secretion of vasopressin. Excessive amounts of dilute urine are eliminated, but the problem can be controlled by administration of synthetic hormones.

30.4 Digestive Juices

In digestion, large and complex molecules of food are broken down into smaller and simpler molecules that are then absorbed through the intestinal walls and into the blood. The digestive process occurs at many points throughout the digestive tract of the human body. Each organ provides chemicals called digestive juices. These juices include saliva, gastric juice, pancreatic juice, intestinal juice, and bile. All the digestive juices with the exception of bile contain enzymes.

Saliva

Chewed food in the mouth is moistened and physically ground into smaller pieces. This important process exposes more surfaces of the food for more rapid attack by the digestive juices. Saliva, which is 99% water, is secreted into the mouth and moistens the food, making swallowing easier. The remaining 1 percent consists of mucin, a lubricating glycoprotein, and α-amylase (ptyalin), an enzyme that hydrolyzes starch. The saliva is buffered by inorganic salts to maintain the pH in the 5.8–7.0 range.

An average person secretes about 1 L of saliva per day from the parotid, sublingual, and submaxillary glands. Chewing activates the flow of saliva, as does the thought, sight, or smell of food.

The partial hydrolysis of starch into lower molecular weight polysaccharides and maltose is the principal chemical reaction that occurs in saliva. Further hydrolysis of starch occurs farther along the digestive tract but not in the stomach, where acid inactivates the α-amylase. Proteins and lipids pass unchanged through the mouth.

Gastric Juice

Swallowed food is passed through the esophagus into the stomach, where it is mixed by peristaltic action (churning) with gastric juice. The gastric juice, secreted by glands in the walls of the stomach, is regulated by the hormone *gastrin.*

About 2 to 3 L of gastric juice is secreted daily. This clear, pale yellow fluid is approximately 98% water and 0.5% hydrochloric acid (approximately 0.1 M), which participates in hydrolysis. The enzymes pepsin and gastric lipase in gastric juice hydrolyze proteins and fats, respectively.

Prior to secretion, the enzyme pepsin is present as the inactive zymogen *pepsinogen,* which is converted into the active enzyme by stomach acid. The molecular weight of pepsinogen is 40,400 and that of pepsin is 32,700 after 42 amino acids are removed from the N-terminal end of the zymogen. Most of the 16 basic amino acids in pepsinogen are removed in hydrolysis. The active sites of pepsin, which hydrolyzes peptides containing phenylalanine, tyrosine, and tryptophan, are exposed only after hydrolysis of the zymogen. Otherwise, this protein-digesting enzyme would digest the very glands that produce it.

About 15% of the peptide bonds in proteins are hydrolyzed by pepsin in the stomach. Although the duodenum and small intestine do not contain pepsin, they have other very efficient catalysts for hydrolysis of peptides.

Gastric lipase catalyzes the hydrolysis of the ester linkages of fats and oils at an optimum rate near pH 7. Because the enzymatic activity is substantially less effective at the high acidity of gastric juice, very little hydrolysis of fats and oils occurs in the stomach.

Chyme, the liquid mixture of food in the stomach, is released into the duodenum, the first part of the lower digestive tract (Figure 30.3) through the pyloric valve.

Intestinal Juice

The glands in the mucosa of the duodenum secrete an enzyme-rich intestinal juice when chyme enters from the stomach. The enzymes in intestinal juice provided by the pancreas are listed in Table 30.1.

The major digestion of carbohydrates occurs in the duodenum, where pancreatic α-amylase, which has a higher activity than salivary α-amylase, is secreted. Within minutes, carbohydrates are converted into oligosaccharides that are then further hydrolyzed by enzymes in the brush border of the duodenal epithelium.

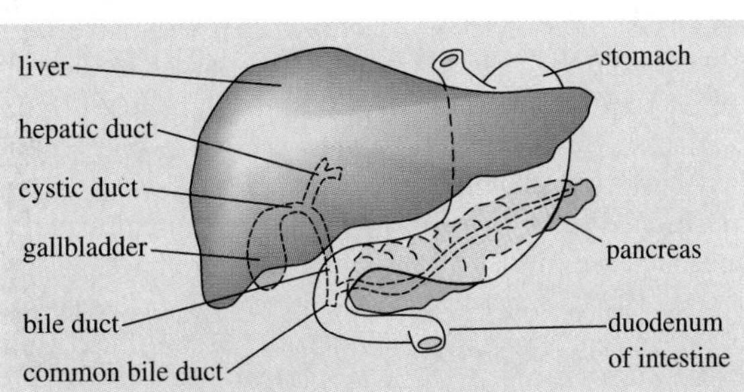

FIGURE 30.3 Relationship of the Liver, Gallbladder, and Pancreas

TABLE 30.1 Enzymes of the Pancreas

Enzyme	Function
Endopeptidases	
trypsin	cleaves peptides at C-terminal position of arginine or lysine
chymotrypsin	cleaves peptides at C-terminal position of aromatic amino acids
elastase	cleaves peptide bonds at C-terminal position of glycine or alanine
Exopeptidases	
carboxypeptidase	cleaves amino acids one at a time from C-terminal position
aminopeptidase	cleaves amino acids one at a time from N-terminal position
Amylolytic	
α-amylase	cleaves α-1,4-glycosidic linkages of amylose
Lipases	
lipase	cleaves ester bonds at 1- and 3-positions of triacylglycerols
phospholipase	cleaves ester bond at 2-position of glycerophospholipids
Nucleolytic	
ribonuclease	cleaves ribonucleic acids into nucleotides
deoxyribonuclease	cleaves deoxyribonucleic acids into nucleotides

Intestinal lipase catalyzes the partial hydrolysis of simple lipids (Section 27.1). The more complex phospholipids are hydrolyzed by lecithinases and phosphatases. The fatty acids and 2-monoglycerides readily diffuse across the brush border membrane, after which the components are reassembled.

Enteropeptidase, the most important enzyme in intestinal juice, is secreted by the mucosa of the duodenum. It catalyzes the hydrolysis of one bond of trypsinogen, the inactive pancreatic zymogen. A hexapeptide is removed, and the resulting trypsin is activated as a result of a small change in its tertiary structure.

$$\text{trypsinogen} \xrightarrow{\text{enteropeptidase}} \text{trypsin} + \text{val-asp-asp-asp-asp-lys}$$

Bacteria and Ulcers

For a very long time the peptic ulcers characterized by lesions in the gastroduodenal area were said to be the result of excessive amounts of stomach acid. It has been established that ulcers are not the result of deficiencies in the mucosal defense mechanisms but rather are caused by the larger-than-normal secretion of HCl. Thus, cures have been directed at decreasing the release of that acid. Medical treatment often required continued therapy using drugs.

It has recently been established, however, that the majority of ulcers are the result of a *Helicobacter pylori* infection. The question of how the bacteria survive in the acidic environment is one of interest. It is known that the bacteria produce urease, the enzyme that hydrolyzes urea to produce ammonia. The ammonia may neutralize acid in the gastric lumen and protect the bacteria.

How this infection by *Helicobacter pylori* affects the acidity of the stomach is not well established. It is known that the bacteria increase the secretion of gastrin. Whether this release is responsible for a chain of reactions that causes the ulcer is not known. However, what is clear is that eliminating the bacteria also stops ulcers. The treatment is rather straightforward: A combination of ameprazole and amoxicillin taken orally eradicates the bacteria. The benefit of this treatment is that once the bacterial infection is eliminated there is no need for continued medication.

In addition to hydrolyzing peptides, trypsin also converts other zymogens secreted in pancreatic juice into their active enzymes. About 50% of the ingested protein is digested and absorbed in the duodenum.

Aminopeptidase is one of the peptidases secreted by the brush border of the small intestine. It hydrolyzes proteins and peptides from the N-terminal end. In addition, both dipeptidases (which cleave dipeptides to amino acids) and dipeptidyl aminopeptidases (which cleave a dipeptide from the N-terminal end of a peptide) are released. Several carboxypeptidases are also released in pancreatic juice as the zymogens.

Pancreatic Juice

The liberation of the hormone secretin by the duodenal mucosa is stimulated by the entry of acidic chyme into the duodenum. This hormone stimulates the discharge of digestive juices by the pancreas into the duodenum. When the flow of chyme ceases, the production of secretin stops, and in turn, the pancreas no longer secretes its juice. The pancreas also produces a secretion containing the hormone insulin, which goes directly into the bloodstream.

Bile

Bile is an alkaline fluid that does not contain any digestive enzymes. It is produced in the liver and stored in the gallbladder (Figure 30.3). Bile salts, bile pigments, and cholesterol are the important constituents of bile. Bile is forced out of the gallbladder through the bile duct into the duodenum by a contraction of that organ that is stimulated by the hormone cholecystokinin. The hormone secretin, involved in the flow of pancreatic juice, also stimulates the liver to produce more bile. Bile salts emulsify lipids, allowing more effective hydrolysis of fats and oils by the enzyme steapsin. In addition, bile salts emulsify the fatty acid products for better absorption out of the digestive tract.

30.5 Blood

Within the circulatory system, which consists of 10^5 km of blood vessels, blood transports and interchanges substances between the body and its environment. Oxygen, minerals, and organic compounds are transported to the cells of the body by the blood, and carbon dioxide and waste products are carried away. In addition, blood carries antibodies, enzymes, and hormones to sites within the body.

Blood regulates the environment of the body. For example, blood regulates the temperature of the body in part by transport of heat energy from the interior to surface capillaries as the blood circulates. Because water has a high heat capacity, regulation of the water balance also helps regulate the temperature. In addition, control of the water balance affects the concentration of electrolytes. The blood system in conjunction with the kidneys maintains the fluid and electrolyte balance. Finally, the buffers in blood also regulate pH (Section 30.2), thus allowing chemical reactions to occur without substantial changes in the acid–base balance.

Blood is the medium responsible for much of the defense of the body against infection. It carries blood cells and plasma proteins that together constitute the immune system (Section 30.8). One of the simplest defense systems is the clotting of blood described in Section 30.9.

Properties of Blood

There are 5 to 6 L of blood in an average adult; the actual amount depends on body size. In addition, the volume is altered by conditions such as diarrhea and hemorrhaging, which decrease blood volume, and pregnancy and fever, which increase blood volume. Although the specific gravity of blood ranges from 1.055 to 1.065, it is about 3.5 to 5.5 times as viscous as water. Hence, the saying "blood is thicker than water" refers to its viscosity. The viscosity increases as the total number of cells increases.

The color of blood depends on its oxygen content. Highly oxygenated blood is bright red, but as oxygen is removed, it turns dark purple. This color change is reversible as the oxygen content changes.

Components of Blood

Some of the components of blood are listed in Table 30.2. However, the composition of blood is usually considered in terms of the liquid and the suspended cells, which together are **whole blood.** The liquid, called **plasma,** constitutes about 55% of blood and is about 92% water. Electrolytes and proteins are the principal substances dissolved or colloidally suspended in plasma. The concentrations of the electrolytes in interstitial fluid, plasma, and intracellular fluid have already been discussed (Figure 30.2).

TABLE 30.2 Composition of Blood Plasma

Class	Substance	Normal Concentration Range
cations	sodium (Na^+)	136–145 mequiv/L
	potassium (K^+)	3.5–5.0 mequiv/L
	calcium (Ca^{2+})	4.3–5.3 mequiv/L
	magnesium (Mg^{2+})	1.2–1.8 mequiv/L
	iron (Fe^{3+})	60–160 μg/L
	copper (Cu^{2+})	70–155 μg/L
anions	chloride (Cl^-)	98–106 mequiv/L
	bicarbonate (HCO_3^-)	23–28 mequiv/L
	lactate	0.7–1.8 mequiv/L
	sulfate (SO_4^{2-})	0.9–1.1 mequiv/L
	phosphate ($HPO_4^{2-}/H_2PO_4^-$)	3.0–4.5 mg/dL
protein	total	6–8 g/dL
	albumin	3.4–4.0 g/dL
	globulin	2.2–4.0 g/dL
fats	cholesterol	150–200 mg/L
	phospholipids	150–220 mg/L
	triglycerides	145–250 mg/L
carbohydrates	glucose	80–120 mg/dL
vitamins	vitamin B_{12}	200–800 pg/mL
	vitamin A	0.15–0.6 μg/mL
	vitamin C	0.4–1.5 mg/dL
others	creatinine	62–133 μmole/L
	uric acid	0.15–0.48 mmole/L
	bilirubin	0.1–1.2 mg/dL
	aldosterone	3–10 ng/dL
	cortisol	5–18 μg/dL
	ketone bodies	0.2–2.0 mg/dL

Plasma proteins consist of 55% albumins, 38% globulins, and 7% fibrinogen. The albumins transport substances such as fatty acids that are insoluble in water. There are three types of globulins. The α-and β-globulins form complexes with ions such as Cu^{2+} and Fe^{2+} that would otherwise be insoluble in the slightly basic blood plasma. The γ-globulins contain antibodies that fight against infections (Section 30.8).

Separated from cellular material by centrifugation, plasma can be maintained for medical use for a longer time than can whole blood. Blood plasma used for laboratory analysis is separated from whole blood by adding a clotting agent and centrifuging the collected sample to move the clot to the bottom of the sample tube.

The cellular material of blood, which constitutes 45% of whole blood and consists of red blood cells, white blood cells, and platelets, will be discussed in the next section. However, much of our focus on the chemistry of blood centers on the role of red blood cells (erythrocytes) in the transport of blood gases. By cell count, erythrocytes constitute approximately 95% of all of the cells in the blood under normal conditions.

30.6 Blood Cells

There are three types of blood cells, each with specific roles. They are *erythrocytes* (red blood cells), *leukocytes* (white blood cells), and *thrombocytes* (platelets). These three types of cells are collectively called the **formed elements** (Table 30.3). A microliter (10^{-6} L) contains 4–6 million erythrocytes, 4000–10,000 leukocytes, and a few hundred thousand platelets. (A cubic millimeter (mm^3) is an older concentration unit that is equal to a microliter.)

Erythrocytes

The 5–6 L of blood in an average person contains 3×10^{13} red blood cells. A severe shortage of these cells is called **anemia;** a large excess is called **polycythemia.** Anemia may result from an inadequate rate of formation of erythrocytes **(aplastic anemia),** excessive rate of destruction of erythrocytes **(hemolytic anemia),** or loss of erythrocytes due to hemorrhaging.

TABLE 30.3 Blood Cells

Blood Cell Type	Normal Range (cells/μL)
Erythrocytes	
men	$4.5–6.5 \times 10^6$
women	$3.8–5.8 \times 10^6$
Leukocytes	4000–10,000
neutrophils	4000–7000
lymphocytes	2500–5000
monocytes	100–1000
eosinophils	0–500
basophils	0–100
Platelets	150,000–350,000

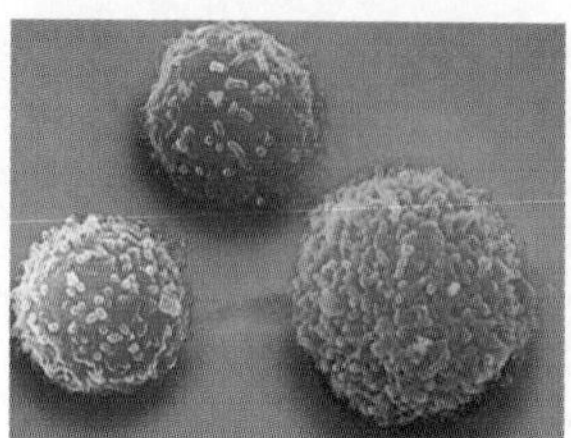

Color enhanced scanning electron microphotograph of two lymphocytes and a macrophage. (Photo Researchers, Inc.)

The amount of erythrocytes in blood is reported as a **hematocrit value,** which is the volume percent of the cells in a blood sample. It is determined by centrifugation, which packs the cells at the bottom of a centrifuge tube. A typical normal value is 47 $\pm$ 5% for men and 42 $\pm$ 5% for women. The hematocrit value is simply the percentage of the tube occupied with cells.

Red blood cells are disc shaped with a depression in the center—much like a doughnut without the hole. This shape has sufficient surface area to efficiently exchange blood gases. Red blood cells are about 7 μm in diameter and 2.5 μm thick. The shape of erythrocytes is maintained by a "skeleton" that consists of a mesh of fibrous proteins attached inside the plasma membrane. Yet the cell has sufficient flexibility to pass through small curved blood vessels.

Erythrocytes do not contain a nucleus or mitochondria and cannot reproduce. The number of these cells in the bloodstream is maintained by new cells formed in the bone marrow. Under normal circumstances, the bone marrow forms enough new red blood cells to replace the old ones that wear out. Human red blood cells have an average lifetime of 4 months. Some hemolyze, and others are cleaned out of the system by macrophages. About 3 million of the approximately 30 trillion "age" and are degraded each second. The iron released is recycled in new red blood cells.

Leukocytes

Leukocytes differ from erythrocytes in that they have a nucleus. There are several varieties of leukocytes, known as *basophils, eosinophils, lymphocytes, monocytes,* and *neutrophils.* Their primary roles are as defenders against foreign organisms.

Neutrophils are the most prevalent of the various leukocytes. They are the rapid deployment force, responding instantly to microbial infection by detecting foreign protein. Any defect or dysfunction of neutrophils results in massive infections.

Eosinophils are present only in small amounts, but their concentration can increase as they respond to infection. They are effective against parasites and are involved in some allergic reactions.

Basophils are granules that contain heparin, an anticoagulant, and histamine, a vasodilator. These two substances apparently increase blood flow and facilitate the transport of other leukocytes to combat infection.

Monocytes are larger than the other leukocytes. They are transformed into macrophages, which are structures that ingest matter such as microorganisms, dead cells, and denatured protein. In this process they produce material that is recognized by lymphocytes. In addition they secrete the agent interleukin-1 that promotes the maturation of T lymphocytes, one of the several subclasses of lymphocytes.

Lymphocytes are a group of plasma cells that include T cells and B cells. The T cells, which constitute 60% of the lymphocytes, are vital to the immune defense mechanism. These cells do not depend on antibodies for their function. There are several subtypes of T cells. One known as the helper T cell is the one attacked by the AIDS virus. B cells, which constitute 30% of lymphocytes, have immunoglobulin and antibodies on their surface. In the immune system, T cells are associated with cellular response, whereas B cells are associated with a humoral response.

The concentration of white blood cells as a group can be as high as 20,000/μL in appendicitis. However, the value can be below 5000/μL when they are killed fighting viral infections such as chicken pox, measles, or mumps.

Individuals with leukemia have high concentrations of white blood cells because their production is out of control. However, these cells are of little help in destroy-

Bone Marrow Transplants

Patients with a bone marrow disease such as leukemia or aplastic anemia may be cured with a bone marrow transplant. The cells may be obtained from a donor who is a close relative. The donor's and the relative's blood leukocytes are screened to determine if the antigenic patterns match. A match is required so that the transplanted cells engraft successfully. Otherwise the patient's adaptive immune system will reject the tissue. In addition, the functional lymphocytes of the graft may attack the host tissue of the patient and could destroy every organ of the patient.

Advances have been made in decreasing the possibility of rejection by using immunosuppressive agents such as steroids, cyclosporine, and anti–T cell antiserum. Other techniques include the complete removal of the patient's bone marrow, often accompanied by radiation to completely eliminate any traces of the original bone marrow. Using such techniques, bone marrow transplants have been done using the bone marrow of unrelated individuals. Nevertheless, the likelihood of a successful transplant increases the more closely the donor's leukocyte antigens match those of the patient.

Because there are thousands of combinations of leukocyte antigens, the chance of finding a perfect match is about 1 in 1 million. However, even with such small odds there are enough potential donors in the United States to guarantee that all patients could be treated. The problem is finding a willing donor. The National Marrow Donor Program has started a registry to list leukocyte antigen types of potential donors. As of May 1995 the database contained 1.6 million registrants. The goal is to make it possible to identify living donors for every patient requiring a bone marrow transplant.

ing invaders because they are not fully mature. As a result, individuals with leukemia have a low resistance to infection.

Thrombocytes

Thrombocytes, or platelets, are small cells without a nucleus that contain large amounts of phospholipids. The platelets, which number 150,000 to 350,000/μL, are involved in the mechanism of blood clotting. If they are not produced in sufficient number or are destroyed internally, such as by a viral infection, severe bleeding occurs when a blood vessel is damaged.

30.7 Blood Gases

To generate the energy necessary to maintain an organism, oxygen is required to oxidize food. The reactions that liberate this energy in cellular respiration were discussed in Section 25.7. Approximately 600 L of oxygen are required each day. Because air is 20% oxygen, the daily air intake of the lungs must be about 3000 L. Oxidation of organic materials generates 480 L of carbon dioxide per day.

The Structure of the Lungs

A large contact area is needed for air and blood to exchange oxygen and carbon dioxide. The windpipe, or trachea, branches into two bronchi that successively branch into an increasing number of smaller tubules called *bronchioles.* Ultimately these tubules terminate in about 300 million alveoli in the lungs. Each alveolus consists of a tiny blind sac surrounded by a bundle of fine capillaries of the circulatory system (Figure 30.4). These capillaries are nearly 2400 km (1500 miles) in length and have an exposed surface area of approximately 74 m^2 (800 ft^2). This large surface area allows passage of oxygen from the air in the alveoli into the blood, which flows through these capillaries at a rate of about 5 L/min. Thus, virtu-

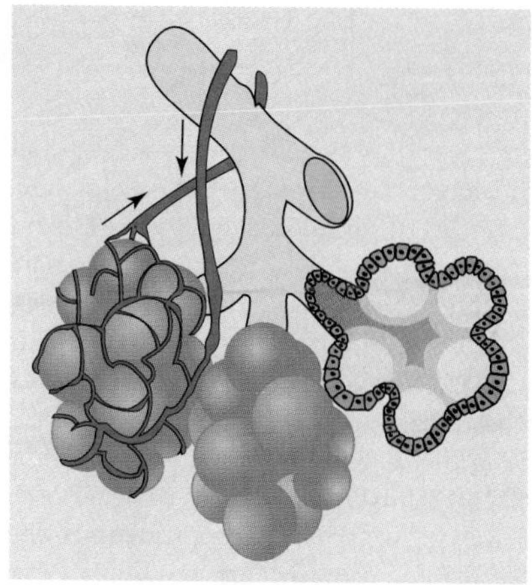

FIGURE 30.4 Alveoli of the Lungs and Associated Capillaries

ally the entire blood supply passes through the lungs once each minute. In periods of extreme exercise the blood flow rate may rise to 20 L/min.

Oxygen Transport

A hemoglobin molecule contains four heme units, each surrounded by a polypeptide chain, as previously shown in Figure 23.2. Each hemoglobin molecule can carry four oxygen molecules. As a result of conformational changes induced by the successive binding of oxygen molecules (an allosteric effect) the heme-protein subunits have a greater affinity for oxygen. Thus, if one site accepts oxygen, the remaining sites are rapidly filled.

There is 15 g of hemoglobin per 100 mL of blood, which combine with 20 mL of gaseous oxygen to give oxyhemoglobin. Without hemoglobin the solubility of oxygen is only 0.3 mL/100 mL—a quantity insufficient to deliver oxygen to cells.

The percentage of heme molecules that carry oxygen is the **percent saturation,** a quantity that depends on the partial pressure of oxygen. Because the equilibrium relationship between oxygen and the hemoglobin molecule depends on complex allosteric effects, a graph of percent oxygen saturation (oxyhemoglobin) versus the partial pressure of oxygen is an S-shaped curve (Figure 30.5).

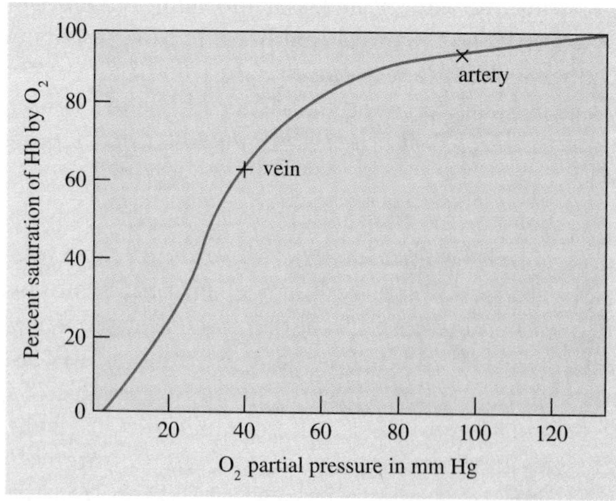

FIGURE 30.5 Oxygen Saturation Curve of Hemoglobin

The partial pressure of oxygen in the lungs is 100 mm Hg, which almost completely saturates arterial blood as the oxygen migrates from the lungs into the bloodstream. The oxyhemoglobin is carried to cells, where the partial pressure of oxygen is in the 25–40 mm Hg range. Oxygen is transferred from the blood to the tissues. In venous blood, where the partial pressure of oxygen is about 40 mm Hg, the percent saturation is still about 65%. When the blood returns to the lungs, the oxygen that was transferred to the cells is replaced. Note that for situations where the partial pressure of oxygen is lower than 40 mm Hg, the percent saturation falls off rapidly. The percent saturation drops by a factor of about 3 for a change from 40 mm Hg to 20 mm Hg.

The effect of percent oxygen saturation is evident at high altitudes where, as the result of decreased air pressure, the partial pressure of oxygen is insufficient to saturate the blood. The consequence of this oxygen insufficiency is a decreased ability to perform normal functions. Anyone who has climbed to high altitudes knows how easy it is to tire and that a high breathing rate is necessary. Above 18,000 ft, oxygen deficiency is so serious that unconsciousness would result if airplanes were not pressurized.

In Denver—the Mile High City—the effect of altitude is hardly noticeable because the atmospheric pressure is 640 mm Hg, and the partial pressure of oxygen is 134 mm Hg. However, in the Andes the atmospheric pressure and the partial pressure of oxygen are 380 and 80 mm Hg, respectively. People who live in the Andes adapt to the decreased oxygen level by producing a higher concentration of red blood cells. On Mount Everest, the atmospheric pressure and the partial pressure of oxygen are 253 and 53 mm Hg, respectively. High-altitude mountain climbers in the Himalayas "adapt" by using portable oxygen tanks.

Blood volume varies for different species, and even the concentration of hemoglobin may be different. Seals can dive to great depths in the ocean, where they can remain for considerable periods of time, because they have about twice the average blood volume for their weight and a high hemoglobin concentration. These factors are the reason why they can store sufficient oxygen for their dives. They also have a diving reflex that permits them to cut off blood flow to most of the body except the brain, heart and those muscles required to dive.

Effect of Acid on Oxygen Transport

Although four oxygen molecules can be carried per molecule of hemoglobin, we will show only one molecule in equations to simplify the representation. In addition, we will use HHb to represent hemoglobin and to emphasize the fact that there are acidic hydrogen atoms in the molecule. For the oxygenation of hemoglobin we write the following equation.

$$\mathrm{HHb} + \mathrm{O_2} \rightleftharpoons \mathrm{HbO_2^-} + \mathrm{H^+}$$

Because hydrogen ions are formed, this equilibrium can be shifted by changes in pH. At higher acidity (lower pH) the oxygenation of hemoglobin would be repressed. However, the bicarbonate ion of the bicarbonate buffer can neutralize the acid formed in the oxygenation of hemoglobin. The resulting carbonic acid decomposes to give carbon dioxide.

$$\mathrm{HCO_3^-} + \mathrm{H^+} \longrightarrow \mathrm{H_2CO_3}$$

$$\mathrm{H_2CO_3} \xrightarrow{\text{carbonic anhydrase}} \mathrm{H_2O} + \mathrm{CO_2}$$

Not only is the acid formed in the oxygenation of blood neutralized, but carbon dioxide, the product of metabolic oxidation, is removed by expiration from the lungs. The net equation for oxygenation of hemoglobin is obtained by summing the three equations.

$$HHb + O_2 + HCO_3^- \rightleftharpoons HbO_2^- + CO_2 + H_2O \quad \text{(net oxygenation reaction)}$$

Carbon Dioxide Transport

Carbon dioxide can be transported more easily than oxygen because it can be converted into other compounds. Carbon dioxide combines with water to form carbonic acid, which in turn can form bicarbonate ion. Carbon dioxide can also react with hemoglobin, but not at the same site where oxygen binds. It reacts with an amino group of the protein to form a carbamate called *carbaminohemoglobin.* This reaction is illustrated as follows, where an $—NH_2$ group contained within the molecule is appended to the HHb symbol for emphasis.

$$HHb—NH_2 + CO_2 \longrightarrow HHb—NH—CO_2^- + H^+$$

Most of the carbon dioxide is converted to carbonic acid, which is rapidly converted into bicarbonate ion.

$$H_2O + CO_2 \rightleftharpoons H_2CO_3$$
$$H_2CO_3 \longrightarrow HCO_3^- + H^+$$

Note that when carbon dioxide forms, the hydrogen ion concentration increases as a result of the formation of both carbaminohemoglobin and carbonic acid. The pH is controlled by the reaction of oxyhemoglobin, a component of the hemoglobin buffer.

$$HbO_2^- + H^+ \rightleftharpoons HHb + O_2$$

The net equation for removal of carbon dioxide is obtained by summing the three equations. Not only is the carbon dioxide removed, but the acid generated is neutralized as well.

$$HbO_2^- + CO_2 + H_2O \rightleftharpoons HHb + O_2 + HCO_3^- \quad \text{(net deoxygenation reaction)}$$

This reaction is the reverse of the one for the oxygenation of hemoglobin. This relationship illustrates the effect of the hemoglobin buffer mentioned in Section 30.2. The symbols HHb and HbO_2^- represent the acid and base forms of the buffer. In addition, we see how the two interrelated reactions of oxygen transport and carbon dioxide transport are mutually controlled. Under a high partial pressure of oxygen, the oxygenation reaction occurs. Deoxygenation occurs when oxygen is required by a cell and carbon dioxide from metabolic reactions must be removed.

30.8 The Immune System

The immune system is exceedingly complex and can be discussed in only a cursory fashion in this text. Although the processes are interrelated, the immune system can be considered in two parts—the *innate* immune system and the *adaptive* immune system. We will consider the role of each of these systems in combating **pathogens,** which is a general term for any invader that the body does not recognize as "self." The **innate immune system** comprises those responses that are neither specific nor inducible. That is, the response is not directed against any specific pathogen, nor is it based on any previous encounter with that pathogen. The **adaptive immune system** has a "memory" of previous invasions, and it has stored long-lived cells that can be reactivated to specifically fight that pathogen when it enters at a future date. This system provides antibodies that respond to specific foreign molecules called **antigens.** Antigens may be glycoproteins on the membrane surface of specific bacteria or the protein coat of a virus. The specific antigen is recognized by noncovalent interactions between proteins of the antigen and those of the antibody.

Inflammation

Invaders that penetrate body tissue trigger a response that you see as an inflammation. This response is initiated by plasma protein and blood cells as they meet the invader. A relatively rapid increase in blood flow to the infected area occurs and results in swelling and redness. In addition, some components of the blood cells release mediators that increase capillary permeability, allowing diffusion of molecules into the infected site.

Although the inflammatory response is vital to defending the system, it may get out of hand if sustained and may cause tissue damage. Some anti-inflammatory agents control the inflammation without blocking the efficiency of action against the microbial invaders. Histamine is one of the chemical messengers that directs the inflammatory response. It is formed by decarboxylation of the amino acid histidine (Section 28.6). Antihistamines are drugs that relieve inflammation and prevent excess histamine production.

The Innate Immune System

Our skin is the first line of defense against infection. If it is breached as a result of an abrasion or puncture wound, the invaders that enter can find their way into the circulatory system. Natural openings to the body are protected by secretions such as mucus that contains immunoglobulins and antibacterial enzymes such as lysozyme. Any pathogen that gets past these defenses usually does not enter the bloodstream directly. It first must lodge in an organ and then enter the circulatory system. However, the vasculature of organs and spaces called sinusoids is lined with macrophages. In simplest terms, macrophages eat invaders. Inhaled microbes are destroyed by mobile alveolar macrophages that circulate in but are confined to the pulmonary capillaries. Other major organs such as the liver, spleen, and bone marrow have fixed macrophages.

The Adaptive Immune System

The specificity of the adaptive immune system depends on an immunological memory that responds to an antigen of the pathogen. This specificity is associated with the structure of antibody molecules as well as with the receptors on T and B lymphocytes. There are hundreds of thousands of antigens that are recognized by the immune system. The body does not store very many of any one kind of antibody nor does it have many molecules of any one type of the lymphocyte required to fight a specific pathogen. However, once these few circulating lymphocytes recognize the pathogen, a process called **clonal selection** generates a large number of the specific required T or B cells. The importance of this process is the reason why the AIDS virus, which attacks the T cells, is so devastating to the immune system.

TABLE 30.4 Properties of Antibodies

Antibody	Molecular Weight	Units/Molecule	Concentration (mg/dL)
IgG	150,000	1	600–1500
IgA	400,000	2	85–300
IgM	900,000	5	50–400
IgD	180,000	1	15
IgE	190,000	1	0.02

The memory of this portion of the immune system is the basis for vaccination against infectious agents. Immunity results when killed or mutant forms of the virus are injected. Although not infectious, the sample contains an antigen such as the protein coat of a virus. The immune system learns to cope with and devises a battle plan to rid the body of that foreign substance. Once the plan is devised, the immune system is ready to take on a real invader in the future because it remembers its past experience with the antigen.

Antibodies

Antibodies are glycoproteins which are present in both plasma and other body fluids that can specifically bind to an antigen. There are five major classes of antibodies, called *immunoglobulin G* (IgG), *IgA, IgD, IgE,* and *IgM.* IgG is the most prevalent in blood serum and is responsible for induced immunity to many microorganisms. The molecular weights of the various immunoglobulins are in the 150,000–900,000 range (Table 30.4).

Within each class there are thousands of different antibodies, but each one consists of two "heavy" protein chains and two "light" protein chains held together in a Y-shaped molecule. Each polypeptide chain within a class of antibodies has an extensive region with an identical amino acid sequence. The diversity of antibodies is accounted for by the regions of variable amino acid sequences.

There can be too much of a good thing when it comes to immunoglobulins. Asthma and some allergies are caused by an excess of IgE. There are also a number of **autoimmune diseases** that result from attack on cells that should be recognized as "self." The most common is rheumatoid arthritis, in which immunoglobulins attack connective tissue at joints. Lupus erythematosus is the result of attack on nucleic acids and blood components.

30.9 Blood Clotting

The formation of blood clots is important for preventing the loss of blood when a blood vessel is injured. The blood clot is also the first line of defense against bacterial attack through the skin. Blood clots contain blood cells enmeshed in an insoluble fibrous protein called *fibrin.* Fibrin is a monomer-like substance that polymerizes to form threads that entangle platelets and seal off the damaged area to prevent further loss of blood.

As vital as the clotting process is to the maintenance of the organism, the formation of blood clots (thrombi) within a blood vessel, known as **thrombosis,** is life-threatening. The conditions that cause such events are beyond the scope of this

TABLE 30.5 Factors of Blood Clotting

Name	Common Name
Factor I	fibrinogen
Factor II	prothrombin
Factor III	thromboplastin
Factor IV	Ca^{2+}
Factor V	proaccelerin
Factor VI	
Factor VII	proconvertin
Factor VIII	antihemophilic factor
Factor IX	Christmas factor
Factor X	Stuart factor
Factor XI	plasma thromboplastin antecedent
Factor XII	Hageman factor

text. However, the consequences are almost self-evident. Blood clots within a large blood vessel that break off from vascular attachment then move in the blood until they lodge in blood vessels too small to allow their passage. The result is an embolism, the most common being a pulmonary embolism. Venous thrombi in the leg are a common source of embolisms.

The coagulation process for blood clot formation involves numerous cascading reactions, and factors (Table 30.5). The product of one reaction catalyzes a subsequent reaction. Calcium ion, factor IV, is the simplest of these substances. Several of the processes involve the conversion of a zymogen into an active enzyme. The zymogens fibrinogen and prothrombin are factors I and II, respectively. Tissue thromboplastin is factor III. Hemophilia A, the most common form of the disease, is the result of a defect in the gene responsible for formation of factor VIII. Hemophilia B is the result of a deficiency of factor IX.

Damaged tissue cells release thromboplastin, which is an enzyme for the conversion of the zymogen prothrombin into thrombin. Several cofactors are also required, the most important being Ca^{2+}. Thrombin in turn is a catalyst that converts fibrinogen into fibrin. The reactions are summarized as follows:

$$\text{prothrombin} \xrightarrow[\text{Ca}^{2+}]{\text{thromboplastin}} \text{thrombin}$$

$$\text{fibrinogen} \xrightarrow{\text{thrombin}} \text{fibrin}$$

Anticoagulants

The body has anticoagulants that inhibit one or more of the series of reactions that cause clotting of blood. Heparin, a heteropolysaccharide, is the principal agent secreted into the circulatory system to prevent blood clots within this vital transport system. It acts by blocking the catalytic activity of both thromboplastin and thrombin. This compound is also administered by injection to prevent thrombosis.

Sodium fluoride, as well as salts of the organic acids citric acid or oxalic acid, are used as anticoagulants for blood samples collected for analysis. The fluoride, citrate, and oxalate ions all form strong complexes with calcium ion, which then cannot serve as a factor in the activation of prothrombin. The blood remains whole rather than clotting. Citrate salts can be used to prevent clotting in blood that is to be returned to the patient, because citric acid is a normal part of the metabolism of the body. Neither fluoride nor oxalate can be so used because both are poisonous to the body in the quantities used to preserve the blood.

Vitamin K has a strong effect on the clotting process. It is a coenzyme required to convert a glutamate side chain of prothrombin into a form suitable for binding calcium ion during its conversion of prothrombin to thrombin. Thus, any chemical agent that affects this process or any of the other steps in the formation of clots can serve as an anticoagulant. Potential anticoagulants are evaluated based on potency, the time required for anticoagulation to occur, and the time required to recover the coagulation function.

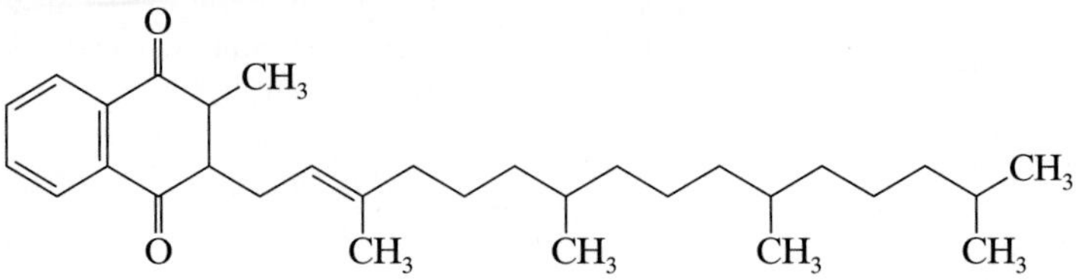

O CH$_3$ CH$_3$ O CH$_3$ CH$_3$ CH$_3$ CH$_3$

vitamin K$_1$

Dicumarol, which has been used for a long time as an anticoagulant, acts by repressing prothrombin formation and may also inhibit the enzyme for which vitamin K is the coenzyme. Dicumarol is used medically as an anticoagulant to prevent further blood clots in patients who have had heart attacks. The coagulant activity of blood is reduced to 50% of normal by the end of 12 hours and to 20% by 24 hours. In other words, the blocking process does not occur immediately. Normal coagulation times return in 1 to 3 days after termination of therapy.

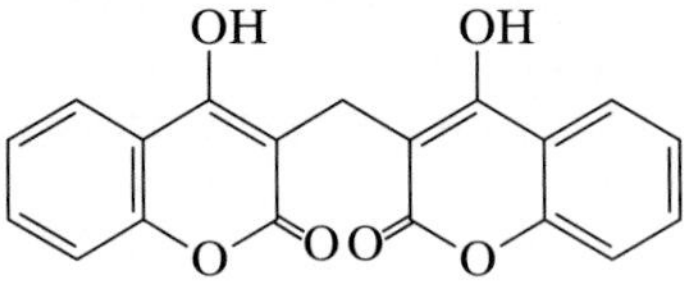

OH OH O O O O

Dicumarol

A number of other synthetic anticoagulants, with varying degrees of potency, are available for use in treating patients. Warfarin is a very powerful drug whose dosage must be carefully monitored. It can be administered orally, intravenously, or intramuscularly.

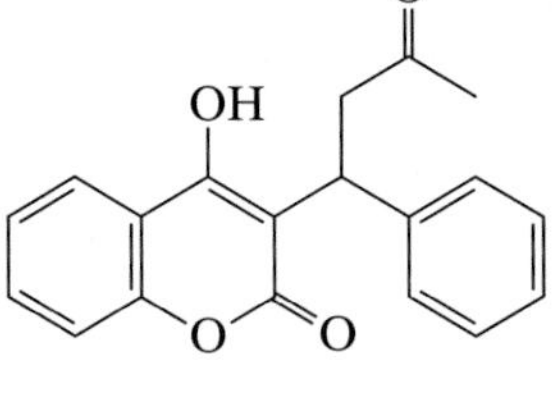

O OH O O

Warfarin

30.10 Urine

Waste products are excreted by the human body through the lungs, the skin, the intestines, and the kidneys. The lungs eliminate both water and carbon dioxide. The skin eliminates water in perspiration, which contains small amounts of salts. The intestines excrete feces, which contain undigested and undigestible material along with some water and salts. The principal excretory organs of the body are the two kidneys, which control water balance and the concentrations of most of the constituents of body fluids.

Properties of Urine

Beavers take in large amounts of water that they excrete as very dilue urine. (Photo Researchers, Inc.)

The specific gravity of urine is normally in the range of 1.003 to 1.030. Diabetes mellitus (insulin deficiency) causes an increase in the specific gravity due to the high concentration of sugar in the urine. If large amounts of water are excreted, as in the case of diabetes insipidus (vasopressin deficiency), the specific gravity may be close to 1.000.

Urine is normally acidic, with an average pH value of 6.0. However, the pH may range from 4.7 to 8.0, depending on diet. High-protein diets increase the acidity because of formation of sulfates and phosphates. Fever causes similar changes in acidity, as does acidosis. Diets high in vegetables and fruits result in more basic urine.

The color of urine is normally pale yellow or amber. The color intensity depends on the volume of urine produced and the amount of waste products. As the volume of water excreted increases, the color of urine becomes lighter. The color of urine is due to urobilinogen and urobilin, which are metabolic products derived from hemoglobin. Colors other than pale yellow usually indicate malfunctions in the body. A red color may indicate the presence of blood in the urine. A greenish-brown color may indicate the presence of bile in the urine. Bile normally is excreted in the feces. However, an intestinal obstruction may cause bile in the urine.

Constituents of Urine

Approximately 50 to 60 g of dissolved solids are excreted in urine daily. About 55% of those solids are organic compounds; inorganic salts constitute the remaining 45%. The organic compounds are end products of metabolic reactions.

About 25 g of urea is excreted daily. Urea is the end product of the metabolism of proteins. Uric acid is the product of metabolism of purines from nucleoproteins. Uric acid output is increased in liver disease, leukemia, and gout. Deposits of uric acid or its salts in joints and tissue are characteristics of gout. When uric acid crystallizes in kidneys, kidney stones result.

uric acid

Creatine is produced from three amino acids—arginine, glycine, and methionine. It is found in all tissues, especially in muscles, blood, and the brain. *Creatinuria* is an abnormal condition in which there is excessive creatine in the urine. It occurs in diabetes mellitus, muscular dystrophy, prolonged fevers, and during starvation. Creatinuria may also occur in pregnancy and during menstruation.

$$\mathrm{NH{=}C(NH_2){-}N(CH_3){-}CH_2{-}CO_2H}$$

creatine

Creatinine is an amide metabolically produced by cyclization of creatine to prepare for its excretion. The number of milligrams of creatinine excreted in the urine in a 24-hour period per kilogram of body weight is called the **creatinine coefficient.** The normal creatinine coefficient is 20 to 26 for males and 14 to 22 for females.

H, O, N, HN, N, CH_3

creatinine

The amount of inorganic salts excreted daily varies with food intake and the body's requirements. Large amounts of phosphates result from diets that are high in phosphates such as nucleoproteins and phospholipids. Excessive excretion of phosphates occurs in **hyperparathyroidism** and in some bone diseases. Sulfates in the urine are derived from the metabolism of sulfur-containing proteins, which is influenced by diet.

30.11 The Kidneys

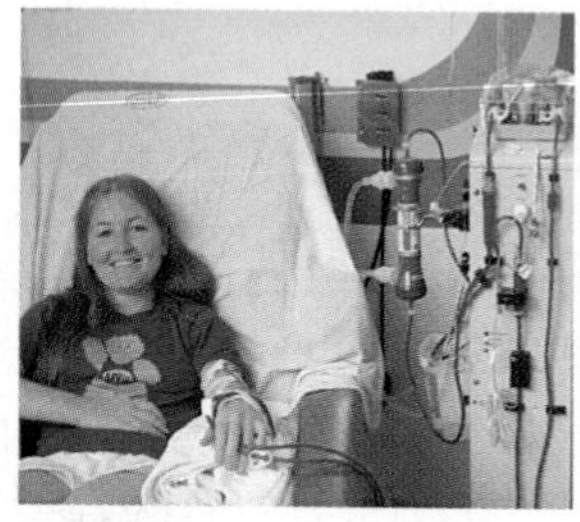

When the kidneys fail, dialysis is required to remove waste products in order to maintain life. (Rainbow)

About 20% of the blood pumped by the heart goes directly to the kidneys. The rate of flow is about 1.2 L/min, which means that on average the entire blood supply passes through the kidneys in about 4 min. This blood supply is not required for the metabolic reactions of the kidney itself but rather passes through the kidney to permit removal of undesirable substances from body fluids.

Each of our two kidneys consists of about 1 million nephrons, each of which independently forms urine. Thus, we need consider only one nephron rather than the entire kidney. Blood reaches the kidneys via the renal arteries, which branch into the arterioles and then into capillaries. The capillaries of an arteriole form a rounded ball called a *glomerulus,* which is contained in a structure called *Bowman's capsule* (Figure 30.6). The Bowman's capsule is connected by a tube to the bladder, so that as urine forms it is transmitted and stored until it is excreted. The blood continues its flow through the arteriole and then through a capillary network surrounding the tubules of the nephron. It finally passes back into the veins.

Water and water-soluble compounds can diffuse across membranes in the millions of glomeruli contained in each kidney. The cells, platelets, and protein material in the blood remain in the capillaries because they cannot pass through the membrane. Thus, the glomerular filtrate has essentially the same composition as the blood with the exception of proteins. The filtrate enters the Bowman's capsule and passes through the proximal tubule and then through Henle's loop. Then the liquid goes through the distal tubule and finally into the collecting tubule to the bladder for excretion. As the fluid flows through the tubules most of the water and some of the solutes are reabsorbed by the blood in the peritubular capillaries that surround the tubules. Thus, the function of the kidneys is to retain unwanted

FIGURE 30.6 A Kidney Nephron and Its Circulatory System

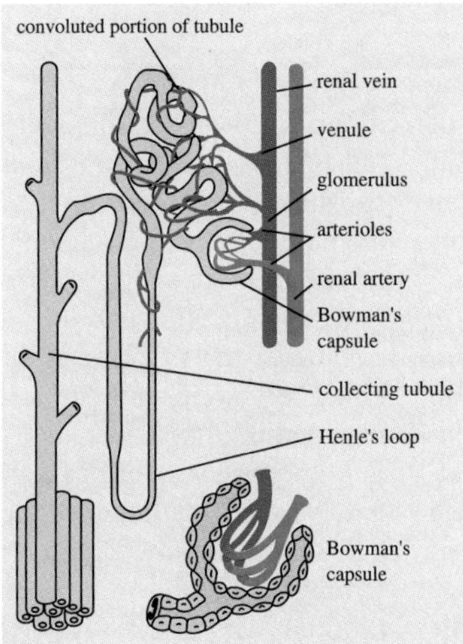

solutes of the blood and to return the important electrolytes to the blood. Under normal circumstances, only unwanted portions of the blood are excreted as urine.

Approximately 125 mL/min of glomerular filtrate passes through the kidneys from the 1.2 L of blood processed by the kidneys every minute. On a daily basis about 180 L of filtrate is processed, and about 1.4 L of urine is excreted. Thus, more than 99% of the filtrate is reabsorbed into the blood. This process is critically important in maintaining fluid balance. Vasopressin is the hormone that allows us to intake varying quantities of water and yet maintain our fluid balance.

The return of solutes to the bloodstream requires more than just diffusion processes. As in transport across any other membrane, both facilitated diffusion and active transport are required to maintain the electrolyte balance of the blood. Some of these processes are controlled by hormones. For example, aldosterone controls the level of sodium ion in blood.

Acid–Base Balance

As discussed in Section 30.2, the phosphate buffer is one of several buffers that control the pH of body fluids. It is the principal buffer of the kidneys and controls the amount of hydronium ion that is excreted. About 50 to 100 mequiv of acid must be excreted each day. The urine carries away some of that acid as the dihydrogen phosphate ion. It also eliminates some acid as the ammonium ion by reaction with ammonia produced by oxidative deamination of glutamate.

$$HPO_4^{2-} + H_3O^+ \longrightarrow H_2PO_4^- + H_2O$$
$$NH_3 + H_3O^+ \longrightarrow NH_4^+ + H_2O$$

In the process of removing acid, the kidneys also assist in removal of CO_2 and restoring HCO_3^- in the body fluids. The hydrogen ion that is eliminated via the phosphate buffer is produced from carbon dioxide by reactions that we have seen several times in this chapter.

$$H_2O + CO_2 \xrightarrow{\text{carbonic anhydrase}} H_2CO_3$$
$$H_2CO_3 \longrightarrow HCO_3^- + H^+$$

As a direct consequence, the bicarbonate that reacted in the bicarbonate buffer of the blood is regenerated and returned to the blood by the kidneys.

Summary

About 25 L, or 62%, of body fluids is **intracellular fluid,** of which about 2 L is found in red blood cells. **Extracellular fluid** includes **interstitial fluid** (12 L) and **blood plasma** (3 L). Digestive juices, blood, and urine are three examples of body fluids.

The pH of body fluids is controlled by the **bicarbonate buffer,** the **phosphate buffer,** and the **protein buffer,** with the protein buffer having the largest buffering power. The osmotic pressure of body fluids is controlled by the concentration of solutes, not the identity of the solutes. Water balance of the body is controlled by daily intake and excretion in urine. **Vasopressin** is the antidiuretic hormone responsible for control of water balance.

Digestive juices include **saliva, gastric juice, pancreatic juice, intestinal juice,** and **bile.** With the exception of bile, all other digestive juices contain enzymes, many of which are formed from zymogens.

The three major roles of blood are transport, regulation, and defense. Hemoglobin transports oxygen and carbon dioxide in erythrocytes. The amount of oxygen carried depends on the partial pressure of the gas. Hemoglobin is allosterically regulated and can transport four oxygen molecules.

The **formed elements** are broadly classified as **erythrocytes** (red blood cells), **leukocytes** (white blood cells), and **thrombocytes** (platelets). The subclasses of leukocytes are **basophils, eosinophils, lymphocytes, monocytes,** and **neutrophils.**

The combination of oxygen with hemoglobin in respiration produces a proton that is neutralized by bicarbonate ion to form carbon dioxide, which is exhaled. When the partial pressure of carbon dioxide increases in a cell, some bicarbonate ion and protons are formed. The proton shifts the oxyhemoglobin–hemoglobin equilibrium and causes deoxygenation to supply oxygen to the cell.

Plasma protein and the various types of white blood cells constitute the immune system, which comprises the innate immune system and the adaptive immune system. The **innate immune system** is not specific or inducible. The **adaptive immune system** has a memory that can be activated to fight specific pathogens. Noncovalent bonds between **antigen** and **antibody** initiate the adaptive immune response. Lymphocytes recognize specific pathogens and generate additional lymphocytes by **clonal selection.** There are five classes of **immunoglobulins;** IgG is the most prevalent in serum. **Autoimmune diseases** are the result of attack of the body's cells by the immunoglobulins.

Blood clotting occurs by a series of steps requiring activators called **factors.** Two of the steps are the conversion of prothrombin to thrombin and fibrinogen to fibrin. Fibrin forms threads that entangle blood cells to form the clot. **Anticoagulants,** both natural and synthetic, prevent clot formation.

The kidneys control the water balance, pH, and electrolyte concentration by a filtration process of blood followed by return of the majority of the filtrate to the blood. Undesired constituents are eliminated as urine.

Exercises

Body Fluids

30.1 What are the two major divisions of body fluids? Which constitutes the major amount?

30.2 Which specific body fluid is present in the largest amount?

30.3 Which of the body fluids functions to cool the body?

30.4 What body fluid rids the body of wastes?

30.5 What are the three buffers that control the pH of body fluids?

30.6 Which buffers are important in intracellular fluid?

30.7 Which buffer has the largest buffering power in blood?

30.8 What organs control the pH of body fluids?

30.9 What cation is found in the highest concentration in interstitial fluid and in intracellular fluid, respectively?

30.10 Which fluid, interstitial or intracellular, contains the higher concentration of Mg^{2+}?

30.11 What anion is found in the highest concentration in interstitial fluid?

30.12 Which fluid, interstitial or intracellular, contains the higher concentration of SO_4^{2-}?

Digestive Juices

30.13 Which body fluids have digestive enzymes or related zymogens?

30.14 Name the enzymes that catalyze the digestion of each of the following.

(a) amylose (b) triacylglycerols (c) polypeptides

30.15 Suggest a reason why cancer of the pancreas is the most serious form of cancer and why patients with this cancer have a drastically shortened life.

30.16 Pancreatitis is a disease that occurs when ducts of the pancreas are blocked and trypsinogen accumulates. At high levels, the trypsinogen overcomes the trypsin inhibitor.

What is the expected result and why can acute cases of pancreatitis destroy the pancreas?

30.17 Why might enteropeptidase be considered the on-off switch for the digestive enzymes of intestinal juice?

30.18 Name three enzymes that operate in intestinal juice.

Blood

30.19 What is whole blood and what are its two components?

30.20 What are the three classes of globulins and what are their functions?

30.21 Crushing injuries that release the contents of cells into plasma cause a condition called hyperkalemia. Recalling the origin of the symbols of the elements from other languages (Table 1.2), deduce what this name indicates.

30.22 A condition called hyponatremia can occur as a result of vomiting, diarrhea, and burns. Recalling the origin of the symbols of the elements from other languages (Table 1.2), deduce what this name indicates.

Blood Cells

30.23 What is the hematocrit value?

30.24 What term describes a shortage of erythrocytes? What term describes an excess of erythrocytes?

30.25 Name two subclasses of lymphocytes and describe the differences in their structures.

30.26 Which blood cell is attacked by the AIDS virus?

Blood Gases

30.27 How many oxygen molecules bond to a hemoglobin molecule?

30.28 How do hemoglobin and oxyhemoglobin differ in color?

30.29 Suggest a reason why fetuses have a modified form of hemoglobin called fetal hemoglobin, which has a larger affinity for oxygen.

30.30 How is the percent saturation affected if the pH of blood decreases?

30.31 How is the affinity of hemoglobin for oxygen affected by a high partial pressure of carbon dioxide?

30.32 What mechanisms remove carbon dioxide from tissues?

Immune System

30.33 How are immune responses similar to the formation of an enzyme–substrate complex?

30.34 What types of cells in association with antibodies are part of the adaptive immune response?

30.35 What is the physical cause of inflammation?

30.36 How might an antihistamine stop the formation of histamine from histidine?

30.37 Which part of the immune system has a memory function?

30.38 What is clonal selection and why is it important in the immune system?

Blood Clotting

30.39 What are two reasons for the formation of blood clots?

30.40 What is thrombosis and what are its dangers?

30.41 What factors are required for the conversion of prothrombin to thrombin?

30.42 What factor is required for the conversion of fibrinogen to fibrin?

30.43 Why are the enzymes required for blood clotting secreted as zymogens?

30.44 How does sodium citrate function in preventing blood clots?

30.45 Based on their medical use, are synthetic anticoagulants reversible or irreversible enzyme inhibitors?

30.46 What three features are considered in the choice of an anticoagulant?

30.47 Consider the structures of Dicumarol and Warfarin and suggest why they can function as competive inhibitors of enzymes using vitamin K as a coenzyme.

30.48 What structural feature of phenprocouman makes this substance an anticoagulant?

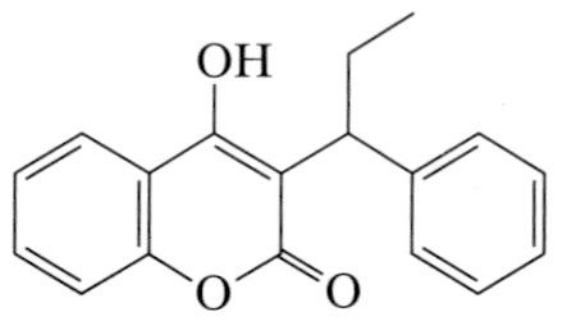

OH
O O

phenprocouman

Urine and the Kidneys

30.49 What percentage of blood passes through the glomeruli of the kidneys?

30.50 What percentage of the glomerular filtrate is reabsorbed by the blood?

30.51 Which major component of blood is not transferred into the glomerular filtrate?

30.52 What mechanisms are required to return desired components from the glomerular filtrate to blood?

30.53 What is the average pH of urine?

30.54 What buffer controls the pH of urine? Which excreted component removes the excess acid?

APPENDIX A

Scientific Notation and Significant Figures

A.1

Calculations with Significant Figures

In this text you may have to convert a measured quantity in one unit into its equivalent in another unit, or you may have to calculate one physical quantity from one or more other measured physical quantities. In either case, the accuracy implied in the "final" answer cannot be any more certain than the least certain quantity used in the calculation. Your calculator may display eight numbers, but you have to decide which numbers to keep in the answer. The accuracy implied by the calculated quantity is expressed by the number of significant figures. That number of significant figures depends on the number of significant figures in each measured quantity and on the type of mathematical operation. Two different procedures are used, one for addition and subtraction and another for multiplication and division.

Rounding Off Nonsignificant Figures

The process of rounding off numbers discards the parts of a number that are not significant. The rules for rounding off nonsignificant figures are

1. If the first nonsignificant figure is less than 5, drop it and all other nonsignificant figures.
2. If the first nonsignificant figure is more than 5 or is 5 followed by digits other than all zeros, drop all nonsignificant figures and increase the last significant figure by 1.
3. If the first nonsignificant figure is 5 alone or is 5 followed by only zeros, drop all nonsignificant figures and increase the last significant figure by 1 if it is odd but leave it alone if it is even.

If 34.51 is expressed to three significant figures, the answer is 34.5 (rule 1), as the 1 in the hundredth place is less than 5. However, if the number 34.51 is expressed to two significant figures, the answer is rounded off to 35 (rule 2), as the nonsignificant figures are 5 followed by digits other than zero.

Addition and Subtraction

When numbers are added or subtracted, the answer should contain no more digits to the right of the decimal point than are in the quantity that has the smallest number of digits to the right of the decimal point. After the mathematical operation is completed, express the answer to that position after properly rounding off the number.

Consider the measured lengths 2.2 m and 13.31 m, which contain two and four significant figures, respectively. However, if the two quantities are added to calculate the sum of the lengths, we have to consider the number of digits to the right of the decimal point, not the number of significant figures. The quantity 2.2 m has one digit to the right of the decimal point—it is expressed to the tenths place. The quantity 13.31 m has two numbers to the right of the decimal point—it is expressed to the hundredths place. The sum of the numbers 2.2 m and 13.31 m displayed on a calculator is 15.51. However, the proper answer is 15.5 m. Only the tenths place is common to both numbers. The hundredths place is not given in 2.2 because the measurement was made only to the nearest tenth of a meter. The fact that the other quantity was determined to the nearest hundredth of a meter is unimportant in calculating the sum of the two lengths.

The same rule based on the place value to the right of the decimal point applies also to subtraction. The difference between the two measured lengths 12.2 m and 3.31 m displayed on a calculator 8.89. However, the proper answer is 8.9, expressed using only one place to the right of the decimal point and after rounding the 8 in the tenths place to 9 because the value of the hundredths place is greater than 5.

Example A.1

What is the volume of a sample obtained by adding 25.1 mL, 15 mL, and 14.15 mL? Express the answer to the proper number of significant figures.

SOLUTION

A calculator displays the result 54.25. The quantity 15 ml has no numbers to the right of the decimal point, and the place common to all numbers is the units place. Thus, the answer must be expressed as 54 after the numbers to the right of the decimal point are discarded. This procedure can be demonstrated by placing a line just to the left of the decimal point in the following column of numbers. None of the numbers to the right can be retained according to the rule expressed for addition or subtraction.

25 | .1
15 |
14 | .15

54 | .25 ml = 54 ml

Example A.2

An 8.14-g sample is removed from a larger sample whose mass is 51.1256 g. What is the mass of the remaining sample? Express the answer to the proper number of significant figures.

SOLUTION

Subtracting using a calculator gives the display 42.9856, but the answer can be expressed only to the nearest hundredth of a gram because the mass of the 8.14-g sample is known only to the hundredth place. The answer is 42.99 after rounding up based on the numbers 5 and 6 found in the next two places to the right.

Multiplication and Division

In multiplication or division of measured numbers, the answer must not contain more significant figures than are found in the quantity with the least number of significant figures in the measurements. Thus, an area calculated by multiplying 0.201 m $\times$ 3.2 m on a calculator is 0.6432. The answer must be expressed to only two significant figures because 3.2 m has only two significant figures. The proper answer is 0.64 m^2.

The quotient of two numbers may be required, as in the calculation of the density of a sample whose mass is 60.31 g and volume is 3.2 mL. The quotient 60.31/3.2 obtained from a calculator is 18.846875. However, only two significant figures may be retained because there are only two significant figures in 3.2 mL. The correct answer is 19 g/mL after rounding up.

Example A.3

You fill the gas tank of your car and travel 276.1 miles and then refill the gas tank with 9.8 gallons. What is the calculated gas mileage of your car?

SOLUTION

Division using your calculator gives 28.173469.

$$\frac{276.1}{9.8} = 28.173469$$

However, since the divisor contains only two significant figures, the quotient may have only two significant figures. The answer is 28 mi/gal.

Example A.4

A cylinder with a cross section of 6.65 cm^2 is filled with a liquid to a height of 5.5 cm. What is the volume of the liquid?

SOLUTION

The volume calculated by multiplying 6.65 $cm^2 \times$ 5.5 cm on a calculator is 36.57. The answer is not 36.57 cm^3 because the height is known to only two significant figures. The correct answer is 37 cm^3.

$$(6.65\ \text{cm}^2)\ (5.5\ \text{cm}) = 36.57 = 37\ \text{cm}^3$$

Exercises

1 Perform each of the following calculations for common sets of units, and give the answer to the proper number of significant figures.

(a) 340 + 3.4 + 0.4589 (b) 124.2 + 4.2 + 32.22
(c) 3.99 + 121.455 + 0.05 (d) 0.123 + 3.43 + 101.2

2 Perform each of the following calculations for common sets of units, and give the answer to the proper number of significant figures.

(a) 5.0035 − 4.49 (b) 445.56 − 0.195
(c) 0.1246 − 0.0003 (d) 0.3544 − 0.04

3 Perform each of the following calculations, and express the answer to the proper number of significant figures.

(a) 6.0035 $\times$ 2.49 (b) 145.22 $\times$ 0.195
(c) 0.1246 $\times$ 0.02 (d) 9.3522 $\times$ 0.050

4 Perform each of the following calculations, and express the answer to the proper number of significant figures.

(a) 7.0035/1.49 (b) 145.22/0.292
(c) 0.1122/0.02 (d) 3.3522/0.20

A.2 Exact Values in Calculations

Conversion factors are used to convert a measured quantity expressed in one unit into another unit. Within the metric system, these conversions are all exact (as are those within the English system). A liter is exactly 1000 mL; a yard is exactly 3 ft.

In any calculation using exact quantities such as conversions factors, the final answer is based only on the accuracy of the measured quantity. The conversions factors are exact, and their values correspond to an infinite number of significant figures. The conversion factor often looks as if it has fewer significant figures than the measured quantity. However, the number of significant figures in the measured quantity is used to determine the number of significant figures in your answer. The equivalent distance of 23 yards is 69 feet even though the conversion factor 3 ft/yd appears to have only one significant figure. Similarly, in converting 0.3468 m into centimeters we use a conversion factor based on the fact that there are 100 centimeters in a meter. Because the conversion factor is an exact quantity, the answer has the same number of significant figures as the starting quantity.

$$0.3468\ \text{m} \times \frac{100\ \text{cm}}{1\ \text{m}} = 34.68\ \text{cm}$$

A.3

Multiple Mathematical Operations

When calculations involve two or more steps, rounding should be done only on the final answer. Rounding at each step in a calculation can cause an error in the final result. With a calculator, you can perform multiple arithmetic operations in succession and round the final answer to the correct number of significant figures based on the quantities used in the calculation. For example, for the product of 1.23 m × 4.003 m × 3.22 m, the answer must have three significant figures. When you multiply 1.23 m by 4.003 m, the calculator displays 4.92369. Then multiplying by 3.22 produces 15.8542818. At this point, round off the answer to 15.9 m^3, which contains three significant figures.

When intermediate answers are given for each step in the solution to the examples in this text, one digit beyond the required number of significant figures is retained. The extra digit is underlined. Correct rounding of the final answer is based on the proper number of significant figures. In the multiplication of 1.23 by 4.003, the calculator display is recorded as 4.923 even though only three significant figures are permitted.

Multiplication of 4.923 by 3.22 gives 15.85206. Note that this quantity is different from the 15.8542818 obtained by successive multiplications on the calculator, but the difference is beyond the significant figures. With either number, rounding to three significant figures gives the same result, 15.9.

A.4

Scientific Notation

Physical quantities such as the mass of an electron as well as calculated quantities are often expressed as a number called the *coefficient* multiplied by 10 raised to some power known as the *exponent.*

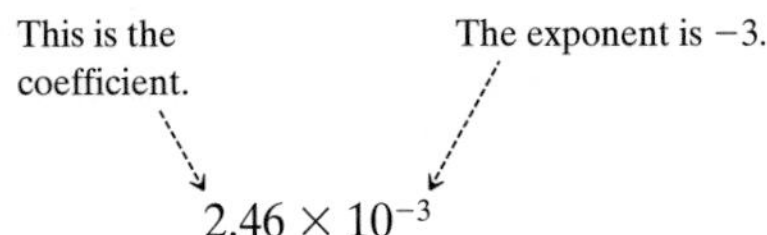

Numbers written in this way are called *exponential numbers.* If the number preceding the exponential portion of the quantity is restricted to quantities between 1 and 10, the expression is called *scientific notation.*

Converting Exponential Numbers into Decimal Numbers

A positive exponent of 10 indicates the number of times the preceding number must be multiplied by 10 to obtain the quantity as a nonexponential number. For example, 1.326×10^3 means

$$1.326 \times 10 \times 10 \times 10 = 1326$$

Each time you multiply by 10, the decimal point is located one more place to the right. Thus, you can convert a quantity in scientific notation with a positive exponent to its numerically equivalent nonexponential number by "moving" the decimal point to the right the number of places corresponding to the exponent.

A negative exponent of 10 indicates the number of times the preceding number must be divided by 10 to obtain the quantity as a nonexponential number. For example, 1.3×10^{-4} means

$$\frac{1.3}{10 \times 10 \times 10 \times 10} = 0.00013$$

Each time you divide by 10 the decimal point is located one more place to the left. Thus, you can convert a quantity in scientific notation with a negative exponent to its numerically equivalent nonexponential number by "moving" the decimal point to the left by a number of places corresponding to the exponent.

Converting Decimal Numbers into Scientific Notation

In order to express a calculated quantity with the proper number of significant figures it is usually best to use scientific notation. The number preceding the 10 raised to a power is selected to retain the correct number of significant figures.

To change a number greater than 1 into its equivalent in scientific notation, move the decimal point to the left n places until the number is greater than 1 but less than 10. This number is the coefficient. The number of places that the decimal is moved becomes the exponent of 10, as in 10^n.

The exponent 2 corresponds to the number of places that the decimal point was shifted.

$$357.8 = 3.578 \times 10^2$$

The decimal point is moved to the left by two places to obtain a number between 1 and 10.

You can always check to see that the exponential number is properly expressed by remembering that moving the decimal point in the indicated direction makes the number smaller. This movement of the decimal point corresponds to dividing the number by 10 each time the decimal moves one place. If you divide the number by 10^3, then you must multiply the resulting quantity by 10^3 to make the value of the exponential number equal to the decimal number.

To change a number less than 1 into its equivalent in scientific notation, move the decimal point n places to the right until the number is greater than 1 but less than 10. The number of places is the negative coefficient in 10^{-n}.

The exponent −4 corresponds to the number of places that the decimal point was shifted.

$0.00023 = 2.3 \times 10^{-4}$

The decimal point is moved to the right by four places to obtain a number between 1 and 10.

This movement of the decimal point corresponds to multiplying the number by 10 each time the decimal moves one place. If you multiply the number by 10^4, then you must divide the resulting quantity by 10^4—which is equivalent to multiplying by 10^{-4}—to make the value of the exponential number equal to the decimal number.

EXAMPLE A.5

Do the numbers 2462.89 and 2.46289×10^4 represent the same value?

SOLUTION

Either express the first number in scientific notation or convert the second number into a decimal number. To express 2462.89 in scientific notation, move the decimal point 3 places to the left to obtain a number that is greater than 1 but less than 10.

$$2462.89 = 2.46289 \times 10^3$$

This number does not have the same exponent and therefore is not the same value as 2.46289×10^4.

To convert 2.46289×10^4 into a decimal number, move the decimal point four places to the right.

$$2.46289 \times 10^4 = 24{,}628.9$$

Thus, by either method you can show that 2462.89 and 2.46289×10^4 do not represent the same value.

Exercises

5 Express the following numbers in scientific notation.
(a) 244.89 (b) 0.0476 (c) 4143.8
(d) 0.000455 (e) 41.6469 (f) 0.000056

6 The following numbers are expressed in scientific notation. Write the numerical equivalent of each as a decimal number.
(a) 5.02×10^4 (b) 1.256×10^3 (c) 3.15×10^{-4}
(d) 1.2×10^{-6} (e) 2.100×10^{-5} (f) 2.01×10^{-3}

Adding and Subtracting Exponential Numbers

All exponential numbers must be expressed to the same power to add or subtract them. If the exponential numbers have the same exponent, add or subtract the coefficients and express the answer in scientific notation, changing the exponent if necessary. The rules for retaining significant figures in the addition of exponential numbers are the same as for decimal numbers.

$$\begin{array}{rl} 7.5\,|\times 10^{-12}\text{ mole} & \\ +4.2\,|\,6\times 10^{-12}\text{ mole} & \\ \hline 11.76\ \ \times 10^{-12}\text{ mole} & = 11.8\times 10^{-12}\text{ mole} \\ & = 1.18\times 10^{-11}\text{ mole} \end{array}$$

If the exponential numbers have different exponents, rewrite the quantities so that they all have the same exponent. As a result, one or more of the numbers may not be in scientific notation.

Consider the subtraction of the quantity 5.5×10^{22} atoms from the quantity 1.3×10^{23} atoms. We change the coefficient and the exponent of 5.5×10^{22} to obtain the equivalent quantity 0.55×10^{23}. The subtraction can now be done.

$$\begin{array}{rl} 1.3\,|\times 10^{23}\text{ atoms} & \\ -0.5\,|\,5\times 10^{23}\text{ atoms} & \\ \hline 0.7\,|\,5\times 10^{23}\text{ atoms} & = 0.8\times 10^{23}\text{ atoms} \\ & = 8\times 10^{22}\text{ atoms} \end{array}$$

Multiplying and Dividing Exponential Numbers

Two steps are required to multiply numbers in exponential form. First, multiply the coefficients and then multiply the exponential portion of the numbers. The multiplication of the exponential portion is equivalent to the algebraic addition of the exponents.

$$\begin{aligned} (2.0 \times 10^{-2}\text{ mole})\,(6.0 \times 10^{23}\text{ atoms/mole}) &= (2.0 \times 6.0)(10^{-2} \times 10^{23}) \\ &= (12) \times (10^{(-2+23)}) \\ &= 12 \times 10^{21} = 1.2 \times 10^{22}\text{ atoms} \end{aligned}$$

To divide numbers in exponential form, divide the coefficients and then divide the exponential portion of the number. The division of the exponential portion is equivalent to subtracting the exponent of the quantity in the denominator from the exponent of the quantity in the numerator.

$$\frac{3.6 \times 10^{-2}\text{ mole}}{1.5 \times 10^{2}\text{ L}} = \frac{3.6}{1.5} \times \frac{10^{-2}}{10^{2}} = 2.4 \times (10^{-2-(2)}) = 2.4 \times 10^{-4}\text{ mole/L}$$

Exercises

7 Do the following calculations.
(a) $(2.0 \times 10^{-3}) + (3.1 \times 10^{-2})$
(b) $(2.0 \times 10^{2}) + (3.1 \times 10^{3})$
(c) $(2.0 \times 10^{-2}) - (3.1 \times 10^{-3})$
(d) $(2.0 \times 10^{3}) - (3.1 \times 10^{2})$

8 Do the following calculations.
(a) $(2.0 \times 10^{-3}) \times (3.0 \times 10^{-2})$
(b) $(4.0 \times 10^{2}) \times (3.1 \times 10^{-3})$
(c) $(2.4 \times 10^{-2}) \div (4.8 \times 10^{-3})$
(d) $(3.6 \times 10^{3}) \div (1.2 \times 10^{-2})$

Answers to Problems

1 (a) 344 (b) 160.6 (c) 125.50 (d) 104.8

2 (a) 0.51 (b) 445.36 (c) 0.1243 (d) 0.31

3 (a) 14.9 (b) 28.3 (c) 0.002 (d) 0.47

4 (a) 4.70 (b) 497 (c) 6 (d) 17

5 (a) 2.4489×10^2 (b) 4.76×10^{-2} (c) 4.1438×10^3 (d) 4.55×10^{-4} (e) 4.16469×10^1 (f) 5.6×10^{-5}

6 (a) 50,200 (b) 1256 (c) 0.000315 (d) 0.0000012 (e) 0.00002100 (f) 0.00201

7 (a) 3.3×10^{-2} (b) 3.3×10^3 (c) 1.7×10^{-2} (d) 1.7×10^3

8 (a) 6.0×10^{-5} (b) 1.2 (c) 5.0 (d) 3.0×10^5

APPENDIX B

GLOSSARY

absolute specificity The ability of an enzyme to catalyze the reaction of only one substrate.
absolute zero The lowest temperature attainable; −273.15 °C.
accuracy The degree to which a measurement represents the true value of what is measured.
acetal A compound formed from the reaction of 1 mole of an aldehyde and 2 moles of an alcohol, $RCH(OR)_2$.
acetone breath Condition that results from release of acetone by the lungs during ketosis.
acetyl coenzyme A Acetyl derivative of coenzyme A; a central intermediate for all metabolic processes that provides acetyl groups for oxidation.
acetyl group The CH_3CO group.
acetylcholine An ester of choline that serves as a chemical messenger in transmitting messages from nerve cells to muscles.
achiral compound A compound that can be superimposed on its mirror image.
acid A proton donor.
acid anhydride A compound formed by loss of water in the reaction of two molecules of an acid.
acid ionization constant The equilibrium constant (K_a) for the ionization of a weak acid.
acidic amino acid An amino acid that has more than one carboxyl group.
acidic oxide An oxide of a nonmetal that combines with water to form an acid.
acidic solution A solution with a higher concentration of hydronium ions than exists in pure water.
acidosis Condition in which the blood pH is below 7.36.
actinides Elements 90–103, located in a row outside the main periodic table.
activation energy The minimum energy required for a reaction to occur.
active site The region in an enzyme that has a unique arrangement of amino acid side chains required for catalytic activity.
active transport The movement of material across membranes from low concentration to high concentration.
acyclic Compound that contains no rings.
acyl carrier protein The carrier of acyl groups in fatty acid biosynthesis.
acyl chloride Carboxylic acid derivative with the RCOCl functional group.
acyl group The carbonyl-containing portion of a compound such as an ester or amide.
addition polymer Polymer that forms from the addition of monomers to each other, usually involving double bonds.
addition reaction Addition of general reagent XY to a multiple bond.
adipose tissue Another name for depot fat found in connective tissue.
adenine A purine base found in both RNA and DNA.
adenosine diphosphate (ADP) Hydrolysis product of adenosine triphosphate consisting of ribose, adenine, and two phosphate groups.
adenosine monophosphate (AMP) Hydrolysis product of adenosine triphosphate consisting of ribose, adenine, and one phosphate group.
adenosine triphosphate (ATP) High-energy phosphate compound consisting of ribose, adenine, and three phosphate groups.
adipocyte A fat cell.
aerobic In the presence of oxygen (air).
AIDS Acquired immune deficiency syndrome; a condition with a variety of symptoms resulting from infection of T cells by a retrovirus.
alcohol A compound containing a hydroxyl group bonded to a saturated carbon atom.
alcohol dehydrogenase An enzyme required to oxidize alcohol to an aldehyde.
aldehyde A carbonyl compound whose carbonyl carbon atom is bonded to one hydrogen atom and either an alkyl or aryl group.
aldol reaction Reaction between two molecules of an aldehyde to give a β-hydroxy aldehyde.
aldolase An enzyme required for the reverse aldol reaction, which cleaves fructose 1,6-diphosphate into two three-carbon intermediates.
aldose A carbohydrate containing an aldehyde group.
alkali metals The elements of Group IA of the periodic table.
alkaline earth metals The elements of Group IIA.
alkaline solution An aqueous solution with pH > 7.
alkalosis Condition in which the blood pH rises above 7.4.
alkane Hydrocarbon having only carbon–carbon single bonds and the molecular formula C_nH_{2n+2}
alkene Hydrocarbon having a carbon–carbon double bond.
alkoxide ion Anion (RO^-) produced by loss of a proton from an alcohol.
alkoxy group The RO— group present in ethers.
alkyl ammonium ion Derivative of ammonium ion in which one or more alkyl groups replace hydrogen, as in RNH_3^+.
alkyl group A group of carbon and hydrogen atoms that resembles an alkane but has one less hydrogen atom.
alkyl halide A compound with an alkyl group bonded to a halogen atom.
alkyne Hydrocarbon having a carbon–carbon triple bond.
allosteric effect A change in conformation at one site caused by a change in conformation at a second, spatially separated, site.

allosteric regulation Noncompetitive inhibition that causes conformational changes.
alloy A solid solution of metals.
α-helix Secondary structure of a protein in which the coil is stabilized by hydrogen bonds.
alpha particle The nucleus of the helium atom produced in radioactive decay.
alpha rays Radiation consisting of alpha particles.
amide A compound whose functional group is a nitrogen atom bonded to a carbonyl carbon atom.
amine A derivative of ammonia in which one or more hydrogen atoms are replaced by an alkyl or aryl group.
amino acid An amino-substituted carboxylic acid. In proteins the amino group is at the α-position.
amino acid pool A reserve of free amino acids in the body.
amino acid residue The amino acid unit as it exists in a peptide chain.
amino group The functional group $-NH_2$.
aminoacyl synthetase An enzyme that covalently links an amino acid to the correct tRNA molecule for use in protein synthesis.
aminopeptidase Enzyme that sequentially hydrolyzes a peptide from the end that has the free amino group.
aminotransferase Also known as transaminase; an enzyme that catalyzes the transfer of an amino group from one molecule to another.
amphetamines A family of arylethyl amines that are physiologically active.
amphoteric Capable of acting as either an acid or a base.
amylopectin A component of starch that has branched chains of glucose.
amylose A component of starch that has a linear arrangement of glucose.
anabolic steroids Synthetic substances related to testosterone that are used to promote muscle development.
anabolism Metabolic reactions that require energy to build complex molecules from smaller molecules.
anaerobic processes Processes that do not require oxygen.
androgens Male sex hormones; testosterone is one example.
anesthetic A drug that decreases or eliminates sensation in any part of the body or in the whole body.
angular (bent) molecule A planar molecule with three atoms arranged at an angle other than 180°.
anion Negatively charged ion.
anomers Stereoisomers that differ in configuration at the anomeric center, as in a monosaccharide.
antagonist Drug that opposes the effect of another compound.
antibody Protein that recognizes and combines with a foreign substance.
anticodon A series of three bases of tRNA that base-pair with the codon of mRNA.
antihistamine Drug used as an inhibitor of histamine.
antiparallel strands Series of polynucleotides in DNA that run in opposite directions when read from the 5′ to the 3′ end.
antiseptic An agent used to destroy or slow growth of microorganisms on the body.
apoenzyme An inactive enzyme without the cofactor required to form the holoenzyme.
aqueous solution A homogeneous mixture of a solute in water.
arachidonic acid An unsaturated fatty acid that is the precursor of prostaglandins.
arene Hydrocarbon with an aromatic ring.
aromatic compound An unusually stable ring compound represented by a Lewis structure containing alternating single and double bonds.
artificial transmutation Formation of radioisotopes by bombarding an element with atomic particles.
aryl group A substituent derived from an aromatic compound.
-ate The ending used to designate that ion in a pair of polyatomic ions that contains the larger number of oxygen atoms. In organic chemistry, the ending used to designate an ester.
atherosclerosis Deposition of excess cholesterol on the walls of the arteries.
atmosphere A unit of pressure equal to the pressure exerted by the atmosphere under standard conditions, which supports a column of mercury 760 mm high.
atom Smallest particle of which elements and compounds are made.
atomic mass unit A quantity equal to 1.6605×10^{-24} g used as the unit in describing atomic mass.
atomic number A number equal to the number of protons in the nucleus of an atom of the element.
atomic orbital A region in space about a nucleus in which one or two electrons may be located.
atomic radius The radius of an atom, given in nanometers.
atomic spectrum Spectrum of light emitted by an excited atom, characteristic of each element.
atomic symbol A one- or two-letter symbol for an element or an atom of an element.
atomic weight The weighted average of the masses of the isotopes of an element present in natural abundance.
Avogadro's hypothesis Equal volumes of gases under the same conditions contain the same number of atoms or molecules.
Avogadro's number The number of structural units of a substance in a mole; 6.02×10^{23}.
AZT A nucleoside used to inhibit the replication of the human immunodeficiency virus.
background radiation Radiation coming from naturally occurring isotopes or cosmic rays.
balancing Adjustment of coefficients in a chemical equation.
barometer A device to measure atmospheric pressure.
basal metabolic rate Energy required to maintain the body.
base A proton acceptor.
base pair Hydrogen-bonded pair of nitrogen bases; G to C and A to T, or A to U.
basic amino acid Amino acid that has an extra amino group.
basic oxide A metal oxide that combines with water to form a base.
basic solution A solution with a lower concentration of hydronium ions than exists in pure water.
Benedict's solution An alkaline solution of complexed cupric ion used as a test reagent for aldehydes.
benzene Aromatic compound with the molecular formula C_6H_6.
beta particle High energy particle identical with an electron.
beta rays Rays consisting of beta particles.
bilayer *See* lipid bilayer.
bile Fluid that emulsifies fats to aid in digestion.
binary compound A compound containing only two elements.
biochemistry The study of the composition, structure, and reactions of substances in living systems.
biological value Rating of proteins based on essential amino acid content.

biomolecule A naturally occurring molecule that plays a role in life processes.
biosynthesis The synthesis of one biomolecule from another by one or more reactions in a cell.
blood–brain barrier Cells in the brain that limit the types of molecules that can enter.
blood sugar The amount of glucose present in blood.
boiling A process that occurs when the vapor pressure of a liquid is equal to atmospheric pressure.
boiling point The temperature at which the vapor pressure of a liquid equals atmospheric pressure.
bond angle The angle between two covalent bonds at a common atom in a molecule.
bond energy Energy required to break a chemical bond in a molecule.
bond length Distance between nuclei of two covalently bonded atoms.
bond-line structure A shorthand representation of a molecule using lines and omitting most carbon and hydrogen atoms.
bonding A description of how atoms are held or fastened together in a molecule.
bonding electrons The electrons shared between atoms.
branched alkane An alkane with an alkyl group bonded to a parent alkane.
Boyle's law The volume of a fixed amount of a gas at constant temperature varies inversely with pressure.
branched chain A sequence of bonded atoms that have additional atoms attached to points within the chain.
Bronsted–Lowry theory Acid–base theory describing acids as proton donors and bases as proton acceptors.
buffer A solution that resists a change in pH and consists of a weak acid and a salt of its conjugate base.
C-terminal amino acid Amino acid with a free carboxyl group at the end of a peptide.
calorie A unit of energy used in chemistry to describe the change in energy in a chemical reaction or a physical process.
carbanion A negative carbon ion with three bonds and an electron pair on a carbon atom.
carbocation A carbon ion with three bonds and a positive charge on a carbon atom.
carbohydrate A polyhydroxy aldehyde or ketone or a compound that can be hydrolyzed to produce a polyhydroxy aldehyde or ketone.
carbonyl A group consisting of a carbon atom and an oxygen atom bonded by a double bond.
carboxyl group The $—CO_2H$ group; the functional group of carboxylic acids.
carboxylate group The anion formed by loss of a proton from a carboxylic acid, represented by $—CO_2^-$.
carboxylic acid Organic compound with the general molecular formula RCO_2H.
carboxypeptidases Enzymes that sequentially hydrolyze peptides only from the end that has the free carboxyl group.
carcinogen Substance that produces cancer.
catabolism Metabolic reactions that release energy and degrade larger molecules into smaller molecules.
catalysis The increase in the speed of a reaction in the presence of a substance called a catalyst.
catalyst A substance that increases the speed of a chemical reaction.
catalytic site Position within an enzyme that provides the catalytic function.
cathode The negative electrode, to which positive ions (cations) are attracted.
cation Positively charged atomic particle that results from the loss of one or more electrons.
cell membrane The lipid bilayer surrounding a cell.
cellulose A polysaccharide of glucose with β-1,4 linkages.
Celsius scale Temperature scale on which the melting point and the boiling point of water are 0 °C and 100 °C, respectively.
centimeter A unit of length equal to $\frac{1}{100}$ meter.
central dogma A series of theories about the transmission of hereditary information and protein synthesis.
cephalin A phosphatidylethanolamine.
ceramide An amide of a fatty acid and sphingosine.
cerebroside A glycosphingolipid containing only glucose or galactose, found in the brain.
chain reaction Reaction in which one product initiates a new reaction, as in nuclear fission or the chlorination of alkanes.
Charles's law The volume of a fixed amount of a gas at constant pressure is directly proportional to the absolute temperature.
chemical bond The attractive force that holds atoms together in compounds.
chemical change A process in which the composition or structure of one or more substances is altered.
chemical energy The energy stored in substances that can be released during a chemical reaction.
chemical equation Representation of reactants and products of a chemical reaction.
chemical equilibrium Condition in which the rate of the forward reaction equals the rate of the reverse reaction.
chemical formula Representation of a chemical compound using atomic symbols and subscripts.
chemical properties Characteristics of matter that govern changes in chemical composition during chemical reactions.
chemical reaction The conversion of reactants into products.
chemical symbol *See* atomic symbol.
chiral The property of an object of not being superimposable on its mirror image.
chiral compound A compound that cannot be superimposed on its mirror image.
cholesterol A steroid that is the precursor of steroid hormones.
chromosomes Units contained in somatic cells that possess genetic information.
chylomicron Tiny fat droplet in the lymph fluid and blood.
chymotrypsin A protease produced in the pancreas and secreted into the small intestine.
cis isomer An isomer that has two groups of atoms oriented on the same side of a structural feature such as a cycloalkane ring.
citrate synthase The enzyme that condenses an acetyl group with oxaloacetate in the first step of the citric acid cycle.
citric acid cycle A series of reactions that oxidize an acetyl group to carbon dioxide and water and save stored energy in NADH, $FADH_2$, and ATP.
Claisen condensation Reaction joining two ester molecules to form a β-ketoester.
codon Three bases on mRNA that specify an amino acid in, or termination of, a protein chain.
coefficient The number of units of a substance relative to other reactants and products in a chemical equation.
coenzyme A cofactor that is an organic molecule.

coenzyme A A thiol ester that transfers acyl groups in acetyl CoA.
cofactor A nonprotein material that is an essential part of some enzymes.
collagen Protein component of connective tissue.
colligative property A property of a solution that depends only on the number of dissolved particles.
combination reaction The direct union of two or more substances to produce one new substance.
combined gas law Law relating the pressure, volume, and temperature of a gas sample.
combustion A rapid chemical reaction of a substance with oxygen.
competitive inhibitor A compound similar to a substrate that binds to the active site of an enzyme.
common names Arbitrary names, that are unrelated to composition, but have been used for a long time.
complementary base pairing The matching of bases in DNA and RNA through hydrogen bonding.
complementary shapes Two structures that fit together to form a unit.
complete protein A protein source that contains all the essential and nonessential amino acids.
composition The identity and amount of the components of a sample of matter.
compounds Pure substances composed of elements joined together by forces called bonds.
concentrated solution A solution that has a high concentration of solute.
concentration A measure of the amount of solute in a given amount of solvent or solution.
condensation Conversion of a gas into a liquid.
condensation polymer A polymer made by reacting monomers to give a polymer and some small molecule such as water.
condensation reaction Reaction combining two molecules and forming water as a second product.
condensed formula A simplified structural formula in which some of the bonds are not shown but implied.
configuration The spatial arrangement of atoms.
conformers Different spatial arrangements of atoms in space as a result of rotation about single bonds.
conjugate acid The acid formed when a base gains a proton.
conjugate base The base formed when an acid loses a proton.
conjugated protein Substance consisting of a protein combined with a nonprotein material.
conversion factor A multiplier having two or more units associated with it that is used to convert a quantity in one unit into its equivalent in another unit.
coordinate covalent bond The bond between two atoms formed by the contribution of a pair of electrons from just one of the atoms.
core An atomic nucleus and all electrons other than those in the valence shell.
Cori cycle Process in which lactic acid is oxidized to pyruvic acid in the liver and then converted into glucose.
corticosteroids Steroids produced by the adrenal cortex; include glucocorticoids and mineralocorticoids.
coupled reactions Reactions that occur together. One is energy releasing, and the other is energy consuming.
covalent bond A bond formed by the sharing of a pair of electrons between two atoms.
covalent compound Compound of discrete molecules joined by covalent bonds.
crenation The shrinkage of red blood cells in contact with a hypertonic solution as the result of differences in osmotic pressure.
crista A structure created by the inner membrane of the mitochondrion.
curie Radioactivity unit equal to 3.7×10^{10} disintegrations per second.
cycloalkane A hydrocarbon that contains a ring of carbon atoms bonded by single covalent bonds.
cytochrome Iron-containing compound that functions as an electron carrier in the electron transport chain.
cytosine Pyrimidine base found in both RNA and DNA.
cytosol Liquid medium of the cytoplasm minus the organelles.
D isomer Compound with a configuration related to that of D-glyceraldehyde.
Dalton's atomic theory Elements are composed of indestructible atoms that combine to form compounds.
Dalton's law of partial pressures The sum of the partial pressures of the gases in a mixture equals the total pressure of the mixture.
data Facts obtained from an experiment.
daughter nucleus Isotope formed in a nuclear decay reaction.
deamination The loss of an amino group from a molecule such as an amino acid.
decarboxylation The loss of carbon dioxide from a carboxyl group, as in a carboxylic acid.
decomposition reaction A reaction that converts a single substance into two or more simpler substances.
degenerate code Related triplet codes that specify the same amino acid.
dehydration The removal of water from a molecule.
dehydrogenation reaction Reaction in which the reactant loses hydrogen.
denaturation The loss or destruction of the native conformation of a protein.
density The ratio of mass per unit volume.
deoxyadenosine A nucleoside containing the purine adenine and 2-deoxyribose.
deoxycytidine A nucleoside containing the pyrimidine cytosine and 2-deoxyribose.
deoxyguanosine A nucleoside containing the purine guanine and 2-deoxyribose.
deoxyribonucleic acid (DNA) A polynucleotide containing deoxyribose, phosphate, and a mixture of adenine, thymine, guanine, and cytosine.
deoxyribose Aldopentose in which the —OH group attached to C-2 of ribose is replaced by hydrogen.
deoxythymidine A nucleoside containing the pyrimidine thymine and 2-deoxyribose.
depot fat The cellular fat found in connective tissues.
detergent Compound with polar and nonpolar groups that can form micelles.
dextrorotatory Capable of rotating the plane of polarized light in a clockwise direction.
diabetes Disease caused by a deficiency of insulin.
diastereomers Stereoisomers that are not mirror images of each other. *See also* enantiomers.
diatomic molecule A molecule composed of two atoms, either alike or different.
dicarboxylic acid Compound with two carboxylic acid groups, as in succinic acid.
digestion Hydrolysis of large food molecules into smaller molecules.

dilute solution A solution that has a low concentration of solute.
dilution A decrease in concentration as a result of addition of solvent.
dipeptide Two amino acids connected by a peptide (amide) linkage.
dipole A pair of opposite charges of equal magnitude at a distance from each other in a molecule.
dipole–dipole attraction Attraction between polar molecules.
diprotic acid An acid that can transfer two protons.
disaccharide A sugar (carbohydrate) formed from two monosaccharides joined by an acetal or ketal bond.
disinfectant Agent used to destroy or slow growth of microorganisms on inanimate objects.
displacement reaction Reaction in which an atom or group of atoms replaces an atom or group of atoms in a reactant.
dissociation The separation of ions in water.
distillation A physical method used to separate liquids based on their boiling points.
disulfide A group represented by R—S—S—R.
disulfide bridge The S—S bond between cysteine residues in peptide chains.
double bond The bond formed by the sharing of two pairs of electrons between two atoms.
DNA *See* deoxyribonucleic acid.
DNA polymerase The enzyme that catalyzes the replication of DNA.
double helix An antiparallel association of two polydeoxyribonucleotides; the structure of DNA.
double replacement reaction A reaction in which two compounds exchange atoms or groups of atoms to produce two different compounds.
driving force The tendency of a chemical system to proceed toward a minimum free energy.
dynamic equilibrium A condition in which two or more opposing processes occur at the same time and are in balance.
eicosanoid Derivative of a 20-carbon fatty acid, found in compounds such as prostaglandins.
elastin A fibrous protein found in tendons and arteries.
electrode A material suspended in a medium through which electricity passes.
electrolytes Ionic compounds or compounds that form ions when dissolved in water.
electron A subatomic particle with a mass of 9.109×10^{-28} g and a charge of -1.60×10^{-19} C (represented by a relative charge of -1).
electron carrier Oxidizing agent involved in the transport of electrons in an electron transport chain.
electron configuration A description of the arrangement of the electrons in the atom by shells, subshells, and orbitals.
electron-dot structure A structure that uses dashes for covalent bonds and pairs of dots for lone-pair electrons.
electron-dot symbol A symbol giving the number of valence electrons as dots located around the elemental symbol.
electron shell A name for principal energy levels designated by integers 1 to *n*.
electron spin A property of the electron; may be either clockwise or counterclockwise.
electron transport chain A series of reations in which electron carriers remove electrons from coenzymes and transfer them to oxygen.
electronegativity Number that indicates the electron-attracting tendency of an atom.
electrophile An electron-deficient species that tends to attack electron-rich centers.
electrophoresis A method of separating amino acids or proteins according to their net charge at a selected pH.
element Pure substance that cannot be decomposed into any simpler substance(s) by ordinary chemical reactions.
Embden–Meyerhof pathway Anaerobic metabolic pathway for the conversion of glucose into pyruvic acid.
empirical formula The simplest whole-number ratio of the atoms present in a compound.
enantiomers Stereoisomers that are mirror images of each other. *See also* diastereomers.
endergonic Energy-absorbing reaction with products of higher free energy than that of reactants; $\Delta G° > 0$.
endoplasmic reticulum A network of membrane-bound channels in a cell.
endothermic reaction A process that absorbs heat energy from the surroundings.
endpoint The point of neutralization in an acid–base titration.
enediol An intermediate in the conversion of isomeric α-hydroxy carbonyl compounds.
energy Ability to do work; expressed in calories or joules.
energy of activation The minimum energy required in a molecular collision to initiate a reaction between reactants.
enkephalins Peptides that bind at receptor sites in the brain to reduce pain.
enthalpy A quantity symbolized by $H°$. $\Delta H°$ is the energy difference between two states or substances.
entropy A quantity symbolized by $S°$. $\Delta S°$ is a measure of the change in the degree of disorder in a system.
enzyme A biochemical catalyst that is predominantly protein.
enzyme–substrate complex Combination of an enzyme and a substrate in an enzyme-catalyzed reaction.
equilibrium A state in which opposing processes are in balance.
equilibrium constant A numerical quantity reflecting the relationship between the concentrations of reactants and products at equilibrium.
equivalent Quantity of an acid containing 1 mole of protons. Quantity of a base that reacts with 1 mole of protons. Abbreviated equiv.
essential amino acids Amino acids that cannot be synthesized by the body and must be obtained from food.
essential fatty acids Unsaturated fatty acids that cannot be synthesized by the body and must be obtained from food.
ester A compound formed by the reaction of an alcohol and an acid, represented by RCO_2R.
esterase A hydrolase that catalyzes the hydrolysis of esters.
esterification Ester formation, as in the reaction of an alcohol and a carboxylic acid.
estrogens Female sex hormones; examples are estrone and estradiol.
ether Compound with the R—O—R structure.
eukaryote Cell that contains a true nucleus.
evaporation Conversion of a liquid into a vapor by escape of surface molecules.
exergonic reaction Energy-releasing reaction with products of lower free energy than that of reactants.
exothermic reaction A process that releases heat energy to the surroundings.
extensive property A property that depends on the amount of a sample.

extracellular fluid Body fluid located outside the cells.
facilitated diffusion A transport process that occurs without cellular energy requirements.
fact Something observed or experienced that can be confirmed by another observer.
factor unit method The use of conversion factors and cancellation of units in the solution of problems.
fat-soluble vitamins Nonpolar vitamins such as vitamins A, D, and E.
family A set of chemically related elements located in a group of the periodic table.
fats Esters of glycerol and long-chain saturated carboxylic acids.
fatty acid A long-chain carboxylic acid containing an even number of carbon atoms.
fatty acid biosynthesis A series of reactions that occur in the cytoplasm of the cell and convert two-carbon-atom acetyl units into fatty acids.
fatty acid cycle A sequence of reactions that successively degrade a fatty acid chain by two-carbon-atom acetyl units.
fatty acid mobilization The release of fatty acids by hydrolysis from adipose tissue.
fatty acid synthetase A complex of enzymes used in fatty acid biosynthesis.
feedback inhibition A method of controlling an enzyme level in which a product of a series of reactions is an inhibitor of an enzyme required for an earlier step.
Fehling's solution An alkaline solution of complexed cupric ion used as a test reagent for aldehydes.
fermentation The anaerobic metabolism of glucose.
fibrin An insoluble blood protein of a blood clot.
fibrous protein A protein arranged in a sheetlike structure.
Fischer projection formula Planar representation of compound with stereogenic centers.
fission The splitting of a nucleus into two lighter nuclei.
flavin adenine dinucleotide (FAD) Reducing agent that accepts two hydrogen atoms in biochemical reactions.
formula Representation of the combination of elements present in a compound.
formula weight The sum of the atomic weights of the atoms indicated by the formula unit of an ionic compound.
fossil fuels Coal, natural gas, and petroleum.
free energy change A quantity symbolized by ΔG° that measures the spontaneity of a chemical reaction.
free radical Reactive species containing one unpaired electron.
freezing A process in which a liquid is converted to a solid.
freezing point depression Difference between freezing point of a solvent and that of a solution.
fructose Ketohexose found in fruits.
functional group An atom or group of atoms that confers specific properties to an organic molecule.
fused-ring compound A compound in which two or more aromatic rings are joined through two or more carbon atoms.
fusion The combination of two nuclei into a larger nucleus.
galactose An aldohexose that is a component of lactose.
gamma rays High-energy radiation from a radioactive atom.
ganglioside A glycosphingolipid containing a higher saccharide as the sugar unit.
gas A state of matter that has no characteristic shape or volume.
Gay–Lussac's law At constant volume, the pressure of a fixed amount of a gas is directly proportional to the absolute temperature.
Geiger counter A device used to detect and measure radiation.
gene Segment of DNA containing information for synthesis of a protein.
genetic code Information in DNA needed to maintain the identity of an organism.
genetic engineering The use of recombinant DNA to synthesize proteins.
geometric isomers Isomers that have the same sequence of atoms but whose orientation in space differs.
globular proteins Tightly folded proteins that are roughly spherical.
glucocorticoid Steroids produced by the adrenal cortex and involved in the control of glucose levels in the body.
glucogenic amino acids Amino acids that can be converted to pyruvate, oxaloacetate, succinyl CoA, α-ketoglutarate acid, or fumarate.
gluconeogenesis The conversion of pyruvic acid into glucose by the liver.
glucose An aldohexose that is a component of starch, cellulose, lactose, and sucrose.
glucose 6-phosphatase The enzyme that dephosphorylates glucose 6-phosphate in the last step of gluconeogenesis.
glucose tolerance A measure of the ability of the body to handle dietary glucose.
glucoside A glycoside of glucose.
glutamate dehydrogenase An enzyme that catalyzes the oxidative deamination of glutamate.
glyceride A lipid that has glycerol as the backbone of the structure.
glycerophospholipid A lipid with glycerol linked by ester bonds to two fatty acids and a phosphate group that is also bonded to an amino alcohol.
glycogen A storage form of glucose in animals.
glycogen synthetase The enzyme that forms glycosidic bonds between glucose and the growing glycogen molecule.
glycogenesis The conversion of glucose into glycogen.
glycogenolysis The conversion of glycogen into glucose.
glycolipid A covalent molecule containing both a sugar and a lipid unit.
glycolysis The conversion of glucose into pyruvic acid.
glycoprotein A protein containing carbohydrates on the surface, as in cell surfaces.
glycoside An acetal or ketal of a carbohydrate.
glycosidic linkage Acetal or ketal bond between the anomeric carbon atom and a hydroxyl group of a second monosaccharide.
glycosphingolipid A substance consisting of sphingosine, a fatty acid, and a carbohydrate.
Golgi complex A stack of flattened membrane sacs in a cell.
gram A mass equal to $\frac{1}{1000}$ of a kilogram.
group Vertical column in the periodic table; also known as a family.
guanine Purine base found in both DNA and RNA.
guanosine A nucleoside composed of the purine guanine and ribose.
half-life Time required for one-half of a radioisotope to decay.
halogens The elements of Group VIIA.
Haworth projection A planar representation of a cyclic carbohydrate.
HDL (high-density lipoprotein) Substance that transports cholesterol via the blood away from the body tissues to the liver.
heat A form of energy that can be transferred between two objects of different temperature.

heat of fusion The amount of energy released as 1 g of a solid is converted into a liquid at its melting point.
heat of reaction The energy difference between the products and the reactants.
heat of vaporization The amount of energy required to convert 1 g of liquid to a gas at its boiling point.
α-helix Secondary structure of polypeptide chains arranged in a spiral.
heme An iron-containing group present in hemoglobin.
hemiacetal A compound formed by the reaction of 1 mole each of an aldehyde and an alcohol.
hemiketal A compound formed by the reaction of 1 mole each of a ketone and an alcohol.
hemoglobin Protein in the blood that transports oxygen.
hemolysis The bursting of a red blood cell in contact with a hypotonic solution as the result of differences in osmotic pressure.
Henry's law The solubility of a gas in a liquid is proportional to its partial pressure.
heteroatom An atom other than a carbon atom that is part of the backbone of an organic molecule.
heterocyclic compound A compound having one or more atoms other than carbon in the ring.
heterogeneous mixture A mixture that is not uniform throughout.
high-density lipoprotein See (HDL).
histamine Compound formed by the decarboxylation of histidine.
holoenzyme An active enzyme; consists of an apoenzyme and a cofactor.
homogeneous mixture A mixture that is uniform throughout.
homologous series A series of compounds in which adjacent members differ by a repeating unit.
hormone Chemical messengers that are produced by endocrine glands. Some hormones are proteins.
Hund's rule Electrons of equal energy locate singly in different orbitals of a subshell before pairing occurs.
hybrid orbitals The result of mixing of two or more orbitals, such as *s* and *p*, to form orbitals suitable for bonding.
hybridization Combination of two or more atomic orbitals to form orbitals for bonding.
hydration Addition of water to a molecule.
hydrocarbon A compound containing only carbon and hydrogen.
hydrogen bond An intermolecular attraction between an electropositive hydrogen atom and a nonbonded electron pair of an electronegative atom of a neighboring molecule.
hydrogenation Reaction with hydrogen that converts a multiple bond into a single bond.
hydrolysis Reaction of a salt with water that sometimes results in a solution with a pH different from 7.
hydronium ion The ion H_3O^+; the principal form in which protons are found in aqueous solution.
hydrophilic Water-attracting; a term used to describe the surface of lipid bilayers and micelles.
hydrophobic Water-repelling; a term used to describe the interactions in the interior of lipid bilayers and micelles.
hydrophobic bonding A term describing the London forces between nonpolar hydrocarbon chains.
hydroxyl group The —OH group.
hyperammonemia Condition of elevated level of NH_4^+ in the blood.
hyperglycemia Condition of high blood sugar level.
hyperthermia A substantial increase in the body temperature for a period of time.
hypertonic solution A solution with a higher osmotic pressure than that of the solution within a cell.
hyperventilation Condition of breathing faster than normal.
hypervitaminosis A condition resulting from ingestion of excessive amounts of vitamins.
hypo- The prefix used for the polyatomic ion containing a smaller number of oxygen atoms than a similar polyatomic ion named with an -ite ending.
hypoglycemia Condition of low blood sugar.
hypothermia A substantial decrease in the body temperature for a period of time.
hypothesis A tentative model.
hypotonic solution Solution with a lower osmotic pressure than that of the solution within a cell.
hypoventilation Condition of breathing more slowly than normal.
-ic The ending formerly used to designate the higher oxidation state of metal ions in compounds.
ideal gas A gas whose behavior is predicted by the ideal gas law and that conforms to the kinetic molecular theory.
ideal gas law equation An equation relating the pressure, volume, temperature, and number of moles of a gas.
ideal gas constant $R = 0.0821$ L atm mole^{-1} K^{-1}.
immune system Groups of cells that defend the body against infection.
incomplete protein Protein that does not contain sufficient amounts of all the essential and nonessential amino acids.
indicator An acidic or basic substance that changes color when pH changes.
induced dipole A separation of charge within a molecule caused by a temporary dipole in the vicinity.
induced-fit hypothesis A model picturing an enzyme as conformationally flexible so it can change to accept a substrate.
inert gas Nonreactive elements located in the rightmost column of the periodic table.
inhibitors Compounds that destroy or deactivate enzymes.
initiation The start of growth of the protein chain.
initiation step First step in a free radical mechanism.
inner transition elements The elements in two rows outside the main periodic table.
inorganic compounds Compounds that do not contain carbon, with the exception of carbonates and cyanides.
insensible perspiration Evaporation of water from the skin and from air in the lungs.
insulin Protein hormone that controls blood sugar level.
integral proteins Proteins that extend from the surface into the interior of a membrane.
intensive property A property that is not dependent on the amount of sample.
intermembrane space The space between the outer and inner membranes in the mitochondrion.
intermolecular forces Forces of attraction between separate molecules.
International System of Units (SI) Fundamental units used in the sciences.
interstitial fluid Fluid in tissues but not inside cells.
inverse square law Intensity of radiation decreases as the square of the distance increases.
ion An electrically charged atom or molecular particle in which the number of electrons is not equal to the number of protons.
ion product constant The product of the molar concentration of hydronium ions and hydroxide ions in water.

ionic bond The bond resulting from the transfer of electrons from a metal to a nonmetal.
ionic compound A neutral collection of oppositely charged ions.
ionic equation *See* net ion equation.
ionic radius The radius of a spherical ion present in an ionic compound.
ionization The removal of an electron from an atom. The formation of ions on dissolution in water. The transfer of a proton to water.
ionization energy The energy required to remove the electron with the highest energy from an atom and form an ion.
ionizing radiation High-energy radiation that removes electrons from molecules.
irreversible inhibitor A substance that deactivates an enzyme by formation of a covalent bond.
irreversible reaction A reaction that does not proceed appreciably in the reverse direction.
isoionic point The pH at which there is no net charge on an amino acid or protein.
isomerism The existence of two different compounds with the same molecular formula.
isomers Substances having the same molecular formula but different structures.
isopropyl group The alkyl group $(CH_3)_2CH—$.
isotonic solution Solution with the same osmotic pressure as normal body fluid.
isotope Any of two or more forms of an element that have the same number of protons but differ in the number of neutrons.
-ite The ending used to designate that ion in a pair of polyatomic ions that contains the smaller number of oxygen atoms.
IUPAC Acronym for the International Union of Pure and Applied Chemistry.
IUPAC rules Set of rules for naming compounds.
Joule The SI unit of energy.
Kelvin scale Temperature scale also called the absolute temperature scale.
keratin The fibrous protein of hair, fur, and fingernails.
ketal A compound formed from the reaction of 1 mole of a ketone and 2 moles of an alcohol, $R_2C(OR)_2$.
ketoacidosis The decrease in pH resulting from formation of ketone bodies.
ketogenic amino acids Amino acids that can be converted into acetoacetyl CoA or acetyl CoA.
ketohexose A ketose with six carbon atoms.
ketone A compound whose carbonyl carbon atom is bonded to two alkyl or aryl groups or an alkyl group and an aryl group.
ketone bodies Acetoacetic acid, β-hydroxybutyric acid, and acetone.
ketonemia Condition of high concentration of ketone bodies in the blood.
ketonuria Condition of high concentration of ketone bodies in the urine.
ketose A carbohydrate with a ketone functional group.
ketosis Condition of high concentrations of ketone bodies in the body.
kilocalorie A quantity of heat equal to 1000 calories.
kilogram Base unit of mass in the SI system.
kilometer A length equal to 1000 meters.
kinetic energy The energy associated with a moving object.
kinetic molecular theory A model that describes gas behavior based on the concept of an average kinetic energy that is directly proportional to the absolute temperature.
kinetics The study of the rates of chemical reactions.
Krebs cycle The citric acid cycle.
L-sugar Monosaccharide with an OH group on the left at the stereogenic center farthest from the carbonyl group in the Fischer projection formula.
lactase A hydrolase that catalyzes the hydrolysis of lactose.
lactate dehydrogenase Enzyme that catalyzes conversion of lactate into pyruvate.
lactose Disaccharide containing galactose and glucose.
lactose intolerance A condition in which lactose cannot be hydrolyzed for further metabolic reactions.
lanthanides Elements 58–71, located in a row outside the main periodic table.
law An explicit statement of fact obtained by observation or experimentation.
law of conservation of energy Energy can be converted from one form to another but cannot be either created or destroyed in a chemical reaction.
law of conservation of matter Matter is neither created nor destroyed in a chemical reaction but may change in form.
law of partial pressures The sum of the partial pressures of all gases in a mixture is equal to the total pressure.
LDL (low-density lipoproteins) Substances that transport cholesterol to tissues via the blood.
Le Châtelier's principle If a stress is applied to a system at equilibrium, the system will adjust to reduce the stress.
lecithin A phosphatidylcholine.
levorotatory Capable of rotating the plane of polarized light in a counterclockwise direction.
Lewis base An electron-pair donor.
Lewis octet rule A rule referring to the eight electrons in the valence shell of an atom or ion in a compound.
Lewis structure A structural formula showing all valence electrons.
ligase An enzyme that catalyzes the joining of two molecules.
limiting reactant Reactant present in an amount that determines the amount of product that can be obtained.
linear molecule A molecule in which all the atoms are arranged along a common axis.
linkage specific The property of an enzyme that allows reaction of only one type of bond in a substrate.
lipase Enzyme that catalyzes the hydrolysis of lipid ester bonds.
lipid bilayer Two layers of lipid molecules arranged to form a membrane.
lipids A class of biomolecules that includes fats.
lipogenesis Synthesis of fatty acids from acetyl CoA.
lipoproteins Lipid-containing proteins.
liquid A state of matter that has a definite volume but no characteristic shape.
liter Volume unit equal to 1000 cm^3.
lock-and-key theory A model that pictures an enzyme as conformationally rigid.
London forces Intermolecular forces involving temporary and induced dipoles.
lone-pair electrons Valence-shell electrons associated with an atom but not involved in bonding.
low-density lipoprotein See (LDL).
lyase An enzyme that catalyzes the removal of groups and forms products containing double bonds.
lysosomes Organelles that contain granules of protein enzymes.
macroscopic Large enough to be observed directly.
maltase A hydrolase that catalyzes the hydrolysis of maltose.
maltose A disaccharide containing two units of glucose.

Markovnikov's rule Rule predicting the product of addition to a double bond.
mass Quantity of matter.
mass number A number equal to the sum of the numbers of protons and neutrons in the nucleus of the atom.
matter Anything that occupies space and has mass.
measurement A comparison with a standard measuring device.
melting The conversion of a solid into a liquid.
melting point The temperature at which a solid is converted into a liquid.
membrane Material that separates cells from their environment as well as organelles from each other.
mercaptan A compound containing the sulfhydryl group, —SH.
meso compound A compound with chiral centers that is symmetrical and is not optically active.
messenger RNA (mRNA) A polynucleotide with a series of codons required to synthesize proteins.
meta Prefix specifying the 1,3 relation of substituents on a benzene ring.
metabolism Chemical reactions in living systems.
metabolites Intermediate and final products in metabolism.
metal An element that is a good conductor of electricity.
metalloid Element with properties between those of a metal and a nonmetal.
meter Base unit of length in the metric and SI systems.
micelle An aggregate of molecules or ions assembled so that hydrophobic portions are in the interior and hydrophilic portions are on the surface.
microgram (μg) A mass equal to 10^{-6} gram or 10^{-3} milligram.
microliter (μl) A volume equal to 10^{-6} liter or 10^{-3} milliliter.
milliequivalent One thousandth of an equivalent. Abbreviated mequiv.
milligram A mass equal to 10^{-3} gram.
milliliter Volume equal to 10^{-3} liter.
millimeter A length equal to 10^{-3} meter.
millimeter of mercury (mm Hg) A unit of pressure. 760 mm Hg is equal to 1 atmosphere.
mineralocorticosteroid A steroid produced by the adrenal cortex that is involved in the control of ions in the body, which in turn affects water balance.
mirror image Reversed image found by reflection in a mirror.
miscible Term describing liquids that can dissolve in each other in all proportions.
mitochondria The oval organelles that store chemical energy by producing ATP.
mixture A sample of matter that can be separated into two or more components by physical methods.
mobilization Hydrolysis of triacylglycerols in adipose tissue to release fatty acids into bloodstream.
model An idea that may correspond to what is responsible for a natural phenomenon.
molar mass The mass in grams of 1 mole of that substance.
molar volume 22.4 L/mole for gases at STP.
molarity The ratio of moles of solute per liter of solution.
mole Avogadro's number of structural units of matter.
mole ratio A ratio of the coefficients of two substances given by a balanced equation.
molecular compound A compound whose smallest representative unit is a molecule.
molecular formula A representation of a molecule indicating the number and type of each atom present in the molecule.
molecular orbital A region of space about two or more atoms where pairs of electrons are shared.
molecular weight The sum of the atomic weights of the component atoms of the molecule.
molecule A combination of atoms held together by covalent bonds in discrete units.
monatomic ion Ion resulting from gain or loss of one or more electrons from an atom.
monomer Small molecule that combines with similar small molecules to give a polymer.
monoprotic acid An acid that can transfer only one proton.
monosaccharide A simple carbohydrate that cannot be further hydrolyzed.
mRNA *See* messenger RNA.
multiple bond A bond with more than one shared pair of electrons.
mutarotation The change in optical rotation due to equilibrium between anomeric forms.
mutation A change in one or more nucleotides in DNA.
NAD^+ (nicotinamide adenine dinucleotide) An oxidized coenzyme used in catabolic reactions.
NADPH (nicotinamide adenine dinucleotide phosphate) A reduced coenzyme used in biosynthesis.
native state The natural conformation of a protein.
natural gas Mixture of alkanes that is primarily methane.
negative modulator A product at the end of a series of reactions that is an inhibitor for an earlier step.
net ionic equation An equation giving only the species actually participating in a chemical reaction.
neutral amino acid An amino acid containing only one amino group and one carboxyl group.
neutral solution A solution in which the pH is 7.
neutralization reaction Reaction of an acid or acidic oxide with a base or a basic oxide.
neutron A neutral subatomic particle that is found in the nucleus of an atom. It has a mass of 1 amu.
nicotinamide adenine dinucleotide See NAD^+.
nicotinamide adenine dinucleotide phosphate See NADP.
nitrogen balance The relationship between the daily intake of nitrogen compounds and the quantity of nitrogen compounds excreted.
nitrogen bases Heterocyclic compounds of purine and pyrimidine found in DNA and RNA.
nitrogen fixation Conversion of gaseous nitrogen into nitrogen compounds.
noble gases The elements of Group VIIIA of the periodic table; also known as rare gases or inert gases.
nomenclature A system of names or rules for naming compounds.
nonbonding electrons Valence-shell electrons associated with an atom but not involved in bonding.
noncompetitive inhibition A combination of an inhibitor with an enzyme at a point other than the active site.
nonelectrolyte A neutral covalent compound that does not form ions in solution and does not conduct electricity.
nonessential amino acid An amino acid that can be synthesized by the body.
nonmetal An element that is not a metal; usually a poor conductor of electricity.
nonpolar bond A covalent bond in which electrons are shared equally.
nonpolar molecule A molecule in which any polarity of bonds cancels.
nonreducing sugar A sugar that is not oxidized by Benedict's or Fehling's solution.

nonspontaneous reaction A reaction requiring continuous addition of energy to occur.
normal boiling point The temperature at which the vapor pressure of a liquid equals 1 atm.
normal hydrocarbon A hydrocarbon molecule without branches.
normality Concentration unit of a solution indicating the number of equivalents of acid or base in a liter of solution.
nuclear equation An equation giving both atomic numbers and mass numbers of reactants and products for nuclear reactions.
nuclear fission The splitting of an atom into two smaller atoms.
nuclear fusion The combination of two atoms to form a larger atom.
nucleic acids Compounds composed of pyrimidine and purine bases, ribose or deoxyribose, and phosphoric acid.
nucleophile An electron-rich species that reacts with electron-deficient centers.
nucleoside A pentose and a heterocyclic nitrogen base combined as an *N*-glycoside.
nucleotide The product of phosphorylation of a nucleoside.
nucleus The region in the center of an atom that contains the protons and neutrons. Also the control center of a cell, directing chemical reactions of the cell as well as its reproduction.
nylon A polyamide made from a diamine and a dicarboxylic acid.
obligate anaerobe An organism that cannot live in an environment with molecular oxygen.
octane number A rating scale of the burning efficiency of hydrocarbons.
octet rule Rule stating that atoms tend to have eight outer shell electrons about them in compounds.
oil A triglycerol containing a high percentage of unsaturated fatty acids.
oligosaccharide An intermediate-sized saccharide.
open-chain sugar Noncyclic form of a monosaccharide.
optical activity The ability of a substance to rotate plane-polarized light.
optical isomers Compounds that are nonsuperimposable mirror images of each other.
optical rotation The amount and direction of rotation of plane-polarized light caused by a chiral compound.
optimum pH The pH at which an enzyme-catalyzed reaction occurs at the maximum rate.
optimum temperature The temperature at which the reaction rate of an enzyme-catalyzed reaction is maximum.
orbital A region in space about the nucleus of an atom where no more than two electrons may be found.
orbital overlap The interpenetration of one atomic orbital by another to form a molecular orbital.
organelles Structural components of a cell such as the nucleus, endoplasmic reticulum, and mitochondria.
organic chemistry The chemistry of the compounds of carbon.
organic compounds Compounds that contain carbon atoms.
ortho Prefix specifying the 1,2 relationships of substituents on a benzene ring.
osmosis The net flow of solvent molecules through a semipermeable membrane.
osmotic pressure The pressure required to stop the net transfer of a solvent across a semipermeable membrane.
osteoporosis A disease of calcium loss from bones.
-ous The ending formerly used to designate the lower oxidation state of metal ions in compounds.
oxidase An enzyme that catalyzes the oxidation of a compound.
oxidation The loss of electrons by a substance.
oxidation number A positive or negative integer assigned to describe an element as a free atom, an ion, or as part of a polyatomic ion or molecule.
oxidation–reduction reaction A reaction in which the oxidation numbers of two or more atoms are changed. Also referred to as a redox reaction.
oxidative deamination The conversion of an amino acid into a keto acid with the release of an ammonium ion.
oxidative phosphorylation The coupled reactions of formation of ATP in the electron transport chain.
oxidizing agent The substance that gains electrons and is reduced in a redox reaction.
oxygen debt A condition of high lactic acid concentration in the blood.
oxytocin Peptide hormone that regulates smooth muscle contractions.
pH A logarithmic expression of the hydronium ion concentration, equal to $-\log [H_3O^+]$.
P_i A representation of inorganic phosphate ions in biochemical reactions.
pK_a A measure of the acidity of an acid, equal to $-\log K_a$.
pK_b A measure of the basicity of a base, equal to $-\log K_b$.
paired electrons Two electrons of opposite spin in the same orbital.
para Prefix specifying the 1,4 relationship of substituents on a benzene ring.
parent chain The longest continuous chain identified for the purposes of deriving the IUPAC name of an organic compound.
partial pressure The pressure exerted by a single gas in a gaseous mixture.
parts per million (ppm) A unit of concentration; the number of parts of a component per million parts of solution.
pentose phosphate pathway A series of reactions that convert glucose 6-phosphate into ribulose 5-phosphate and NADPH in the cell.
pepsin An enzyme that cleaves peptides at the nitrogen end of the aromatic amino acid groups.
pepsinogen A zymogen that is converted to pepsin.
peptidase A hydrolase that catalyzes the hydrolysis of peptide bonds.
peptide bond The amide bond in a polypeptide or protein.
per- The prefix used for the polyatomic ion containing a larger number of oxygen atoms than a similar polyatomic ion named with an -ate ending.
percent composition The mass of each element present in the compound divided by the total mass of all elements in the compound and then multiplied by 100.
percent yield Percentage of theoretical yield that is produced in a reaction.
period A horizontal row in the periodic table.
periodic law The periodic recurrence of properties of elements considered in order of increasing atomic number.
periodic table An arrangement of elements, with elements of similar properties grouped together.
peripheral proteins Proteins attached by ionic forces to the surface of a bilayer.
perspective formulas Structural formulas written in two dimensions that impart some three-dimensional aspects to the representation.
pH scale A numerical scale used to describe the acidity or basicity of a solution.

phenol A compound in which a hydroxyl group is bonded to a carbon atom of an aromatic ring.
phenyl group Substituent derived by removal of a hydrogen atom from benzene.
phenylketonuria Genetic disease caused by the absence of an enzyme required to convert phenylalanine into tyrosine.
pheromone A compound released by a species to communicate with other members of the species.
phosphate ester Compound formed by reaction of an alcohol with phosphoric acid.
phosphatidate A glycerol derivative esterified at C-1 and C-2 with fatty acids and at C-3 with a phosphoryl group.
phosphatidylcholine A glycerophospholipid in which the phosphoryl group is esterified with the hydroxyl group of the amino alcohol choline. Also known as lecithin.
phosphatidylethanolamine A phosphoglyceride in which the phosphoryl group is esterified with the hydroxyl group of the amino alcohol ethanolamine. Also known as cephalin.
phosphodiester The phosphate ester unit between ribose units of nucleotides in polynucleotides.
phosphoglycerides Molecules consisting of one unit each of glycerol, phosphate, and alcohol and two units of fatty acids.
phosphorylation Reaction combining a phosphate unit with a substrate.
photosynthesis Conversion of CO_2 and H_2O into glucose by plants.
physical change A process that occurs without a change in the composition of matter.
physical properties Characteristics of matter that do not involve changes in the chemical composition of substances.
pi (π) bond Side-by-side overlap of two *p* orbitals.
plane-polarized light Light consisting of waves vibrating in a single plane.
plasmid Circular molecule of DNA contained in prokaryotes.
β-pleated sheet The structure resulting from intermolecular hydrogen bonds between alternating series of polypeptide chains with small side chains.
point mutation The substitution of a single base in a codon.
polar covalent bond A bond formed by sharing electrons between two atoms of unequal electronegativity.
polar molecule Molecule in which the bond polarities do not cancel.
polarimeter An instrument used to measure the optical activity of molecules.
polarizability The ease with which an electron cloud can be distorted by nearby charges.
polyamide A polymer resulting from formation of amide bonds between monomer units, as in nylon.
polyatomic ion A group of bonded atoms that bears a positive or negative charge.
polyester A polymer resulting from formation of ester bonds between monomer units, as in Dacron.
polymer Large molecule made up of repeating units called monomers.
polypeptides Molecules consisting of α-amino acids linked by peptide (amide) bonds.
polysaccharides Polymers consisting of many monosaccharides linked by glycosidic bonds.
positron Positively charged particle with the mass of an electron.
ppm *See* parts per million.
precipitate An insoluble solid deposited from a solution.
precipitation reaction A process in which a precipitate forms by the combination of ions in solution.
precision The degree of reproducibility of a measurement.
precursor The molecule from which a product is immediately derived in a synthetic or biological reaction.
pressure Force per unit area. In gases the pressure is measured in terms of the height of a column of mercury that can be supported by the gas.
primary alcohol A carbon compound with a hydroxyl group bonded to a primary carbon atom.
primary amine An amine with a single hydrocarbon group in place of one hydrogen atom of ammonia.
primary carbon atom A carbon atom that is directly bonded to only one other carbon atom.
primary protein structure The sequence of amino acids in a polypeptide or protein.
products The substances produced in a chemical reaction.
progesterone A female sex hormone responsible for maintaining pregnancy.
prokaryote A cell that does not contain a nucleus.
proof A concentration unit used for alcohol solutions; proof is twice the concentration in volume percent.
propagation steps Repeating consecutive steps in a free radical reaction.
property A characteristic of a material that can be used to identify it.
prostaglandins Biological derivatives of arachidonic acid that occur in low concentrations in body tissue and have a wide range of physiological activities.
prosthetic group A nonprotein material combined in a conjugated protein.
protease An enzyme that catalyzes the hydrolysis of a protein.
protein A polymer of amino acids joined by peptide bonds.
protein turnover rate The time required for the degradation and replacement of proteins in the cells of the body.
proton A subatomic particle with a mass of 1 amu and a charge of 1.60×10^{-19} C (represented by a relative charge of +1).
pure substance Matter that has uniform properties, has a definite composition, and cannot be separated into components by physical means.
purine bases Adenine and guanine; components of nucleic acids.
pyramidal molecule A nonplanar molecule with three groups bonded to a common central atom.
pyranose Cyclic form of a sugar containing six atoms in the ring.
pyridoxal phosphate A form of a coenzyme derived from vitamin B_6 having an aldehyde group.
pyridoxine phosphate A form of a coenzyme derived from vitamin B_6 having an amino group.
pyrimidine bases Cytosine, thymine, and uracil; components of nucleic acids.
qualitative Not associated with a number.
quantitative Associated with a number.
quaternary ammonium ion An ammonium ion having four hydrocarbon groups bonded to nitrogen and bearing a positive charge.
quaternary protein structure The manner in which protein subunits (chains) are assembled to give the whole protein.
racemic mixture An equimolar mixture of enantiomers.
rad Radiation absorbed dose; a measure of radiation exposure.
radiation The emission of alpha, beta, or gamma rays.
radiation sickness Symptoms that result from exposure to varying amounts of radiation.
radical A chemical species with an unpaired electron.
radioactive decay The change of radioactive isotopes into other isotopes by emission of radiation.

radioactive disintegration series A series of nuclear reactions in which the product in each reaction is the reactant in the next.
radioisotopes Atoms that undergo nuclear decay.
rate of reaction The speed at which a reactant is converted into a product.
ratio Two quantities expressed as a quotient.
RBE Relative biological effectiveness; a measure of the effect of radiation on an organism.
reactants The substances that enter into a chemical reaction.
reaction mechanism Description of the sequence of steps of bond formation and cleavage in a chemical reaction.
reactions Changes that occur in the composition and structure of matter.
reagent A compound or mixture of compounds used to carry out a chemical test.
real gas Any gaseous substance that actually exists.
receptor A site on a cell surface that binds a molecule and affects the operation of the cell.
recombinant DNA A chemical combination of DNA of a plasmid with DNA from an animal source.
redox reaction *See* oxidation–reduction reaction.
reducing agent A substance that loses electrons and is oxidized in a redox reaction.
reducing sugar A carbohydrate that causes the reduction of Benedict's solution (or Tollens' reagent).
reductase An enzyme that catalyzes a reduction reaction.
reduction The gain of electrons by a substance.
reductive amination The conversion of a carbonyl group to an amine.
regulatory enzyme An enzyme that catalyzes the first reaction of a sequence of reactions and is inhibited by the end product of the sequence.
rem Roentgen equivalent in man; a unit of radiation exposure.
renaturation The re-formation of a native protein from a denatured protein.
replication A duplication process to generate DNA for cell division.
replication fork The site of replication of DNA.
representative elements The elements in which the *s* and *p* subshells of the highest energy level are being filled.
resonance structure Two or more plausible Lewis structures used when no single structure can accurately represent the molecule.
respiration The intake and use of oxygen in oxidation reactions in the body.
respiration chain The transfer of electrons from oxidation reactions to biochemical intermediates and eventually to oxygen.
retinal The aldehyde form of vitamin A that binds to the protein opsin in the eye.
retrovirus A virus that can convert genetic information in RNA into DNA.
reverse transcriptase An enzyme that can synthesize DNA from RNA.
reversible reaction Reaction that proceeds in both forward and reverse directions.
riboflavin A vitamin that is a component of the coenzyme flavine adenine dinucleotide.
ribonucleic acid (RNA) Substance containing ribose, phosphate, and one of the bases adenine, uracil, guanine, or cytosine.
ribose A pentose present in ribonucleic acids.
ribosomal RNA (rRNA) RNA used to produce ribosomes.
ribosomes Material consisting of ribosomal RNA and protein; the sites of protein synthesis.
RNA *See* ribonucleic acid.
RNA polymerase The enzyme required to produce RNA by a transcription process.
roentgen A measure of radiation based on ions produced in air by the radiation.
salt Another term for an ionic compound.
salt bridge An ionic bond between parts of a protein chain.
saponification The hydrolysis of an ester bond by a strong base.
saturated fat Fat consisting mainly of saturated fatty acids.
saturated hydrocarbon A hydrocarbon that has only carbon–carbon single bonds.
saturated solution A solution in which additional solute will not dissolve.
science The observation, description, experimental investigation, and explanation of natural phenomena.
scientific method A sequence of steps consisting of observation, experimentation, and formulation of laws and theories that leads to scientific knowledge.
scientific notation A number expressed as a product of a coefficient between 1 and 10 multiplied by 10 raised to a power.
secondary alcohol A hydrocarbon compound with a hydroxyl group bonded to a secondary carbon atom.
secondary amine An amine with two hydrocarbon groups in place of two hydrogen atoms of ammonia.
secondary carbon atom A carbon atom bonded to two other carbon atoms.
secondary protein structure The spatial arrangement of amino acid residues that are close to each other in the polypeptide chain.
semipermeable membrane A material that allows water to pass through it.
sex hormones Steroids produced predominantly by the gonads (ovaries in females and testes in males).
shell A description of the electrons in a space about the nucleus. The shells are designated by integers.
side chain A group of atoms appended to a main chain, as in hydrocarbons or peptides.
significant figures The figures in a measured number that are known with certainty plus the first estimated number.
single bond A shared pair of electrons between atoms.
single-replacement reaction Reaction in which one element substitutes for or replaces another element in a molecule.
soap A salt of long-chain carboxylic acids.
solid A state of matter that has a definite volume and shape.
solubility A measure of the amount of solute that can dissolve in a given amount of solvent.
solute The minor component of a solution.
solution A homogeneous mixture of two or more substances.
solvent The major component of a solution.
***sp* hybrid orbital** One of two orbitals produced by mixing one *s* and one *p* orbital.
sp^2 hybrid orbital One of three orbitals produced by mixing one *s* and two *p* orbitals.
sp^3 hybrid orbital One of four orbitals produced by mixing one *s* and three *p* orbitals.
specific gravity The ratio of the density of a substance to the density of water at the same temperature.
specific heat The heat energy required to raise the temperature of 1 gram of a substance by 1 °C.

specificity The selectivity of enzymes for the individual substrates and reactions that they catalyze.
spectator ions Ions present in solution that are not involved in a reaction.
sphingolipid A lipid derived from the amino alcohol sphingosine rather than glycerol.
sphingomyelin A sphingolipid found in the sheath surrounding nerves.
sphingophospholipids Lipids that consist of one unit each of sphingosine, an amide of a fatty acid, phosphate, and choline.
spontaneous reaction Reaction that occurs without an outside source of energy.
starch A polymer of glucose containing α linkages between glucose units.
stereochemical specificity A property of an enzyme that allows it to catalyze the reaction of a single enantiomer.
stereochemistry Description of the processes occurring as a result of the arrangement of groups about a stereogenic center.
stereogenic center Carbon atom containing four different groups.
stereoisomers Isomers with the same structure but different configurations.
steroids Lipids containing a characteristic system of four fused rings of carbon atoms.
stoichiometry The numerical relationship among the quantities of reactants and products in a chemical reaction.
STP (standard temperature and pressure) 0 °C (273 K) and 1 atm.
straight chain A sequence of carbon atoms bonded to each other without intervening atoms or side chains.
strong acid A substance that essentially completely transfers its proton to water.
strong base A substance that essentially completely removes the proton of an acid.
strong electrolyte Substance that dissociates or ionizes completely in water.
structural formula Formula representing the spatial arrangement of atoms and bonds using lines to represent bonds.
structure The arrangement of the components of a substance.
subatomic particle An electron, proton, or neutron.
sublimation The change of a substance from a solid into a gas without passing through the liquid state.
subscript Number placed to the lower right of an atomic symbol in a chemical formula.
subshell A part of a shell characterized by a shape according to type. The subshells are labeled *s*, *p*, *d*, and *f*.
substances The materials in a sample of matter.
substituent An atom or group of atoms attached to a skeleton of carbon atoms.
substitution reaction Reaction in which one atom or group of atoms replaces another atom or group of atoms in a molecule.
substrate A reactant in an enzyme-catalyzed reaction.
substrate-level phosphorylation The phosphorylation of a low-energy compound by a high-energy phosphate.
sucrose A disaccharide containing glucose and fructose.
sulfa drug An antibacterial drug derived from sulfanilamide.
sulfhydryl group The —SH group.
superimposition The simultaneous blending of all atoms in a model with atoms in another model to show identity.
surface tension A property of liquids in which a membrane-like surface results from intermolecular forces.
synthesis The chemical conversion of one molecule into another.
systematic name IUPAC name containing information about the composition of a substance.
tautomer One of a pair of isomers that differ because of a shift of a hydrogen atom.
temperature A measure of the degree of "hotness" of a material.
template A pattern used to guide the synthesis of compounds with complementary bases.
temporary dipole A separation of charge produced momentarily in an otherwise nonpolar substance.
termination The conclusion of synthesis of the protein chain.
termination codons Codons that signal that the construction of a protein is complete.
tertiary alcohol A hydrocarbon compound with a hydroxyl group bonded to a tertiary carbon atom.
tertiary amine An amine with hydrocarbon groups in place of all three hydrogen atoms of ammonia.
tertiary carbon atom A carbon atom bonded to three other carbon atoms.
tertiary protein structure The spatial arrangement of amino acid residues that are far apart in the polypeptide chain.
testosterone A male sex hormone.
tetrahedral molecule A molecule that has an atom located in the center of a tetrahedron and four atoms bonded to the central atom located at the corners of the tetrahedron.
theoretical yield Calculated amount of product formed from a given amount of reactant.
theory A concept or model used to explain a law.
thioesters Esters of thiols.
thiol A compound containing a sulfhydryl group bonded to a carbon atom.
thymine Pyrimidine base contained in DNA.
tissue lipids Lipid materials in cell membranes.
titration A procedure to determine the amount of an acid or base by neutralization.
Tollens' reagent An alkaline solution of $[Ag(NH_3)_2]^+$ that is used as a test for aldehydes.
torr Pressure unit equal to 1 mm Hg.
trace element Any element required by the body in only small amounts.
trans isomer An isomer that has two groups of atoms oriented on opposite sides of a structural feature such as a cycloalkane ring.
transamination An interconversion process between keto compounds and amino compounds.
transcription The means by which information for protein synthesis is passed from DNA to messenger RNA.
transfer RNA (tRNA) RNA molecules having an amino acid attachment site and a template recognition site needed to bring the proper amino acid to messenger RNA.
transferase An enzyme that catalyzes the transfer of a group such as a phosphate from one molecule to another.
transition elements The elements in which the *d* or *f* subshells are in the process of being filled.
translation The transfer of information from a sequence of nucleotides to form a sequence of amino acids in a peptide chain.
transmembrane proteins Proteins that extend across the membrane.
transmutation The change of an isotope of one element into an isotope of another element in a nuclear reaction.
transport protein A protein that transports material across a cell membrane or between tissues in the body.

triacylglycerols A newer name for triglycerides.
tricarboxylic acid (TCA) cycle The citric acid cycle.
triglycerides Esters of glycerol and fatty acids.
trigonal planar molecule A molecule with three atoms arranged around a central atom, with all atoms in a common plane and all bond angles at 120°.
trigonal pyramidal molecule A molecule with the central atom bonded to three other atoms so that a three-sided pyramid is formed. The three atoms bonded to the central atom are in a common plane.
triple bond A bond formed by the sharing of three pairs of electrons between two atoms.
triprotic acid An acid that can transfer three protons.
tRNA *See* transfer RNA.
turnover number The number of substrate molecules converted into product per minute per enzyme molecule under saturated conditions.
unit A quantity of measure adopted relative to a standard.
universal gas constant The proportionality constant *(R)* of the ideal gas law equation.
unsaturated fatty acid A long-chain carboxylic acid having one or more double bonds.
unsaturated hydrocarbon A hydrocarbon that has double and/or triple bonds.
unsaturated solution A solution in which more solute could be dissolved.
unshared electron pair A pair of valence-shell electrons associated with an atom but not involved in bonding.
uracil Pyrimidine base found in RNA.
urea cycle A metabolic pathway in which ammonia is converted into urea.
valence-shell electron-pair repulsion (VSEPR) theory A theory relating the shape of molecules to the distribution of electron pairs about a central atom.
valence shell electrons Electrons of the *s* and *p* subshell in the highest occupied energy level.
vapor Gaseous phase produced from a liquid or solid.
vapor pressure The pressure of a vapor in equilibrium with its liquid form.
vaporization The process of converting molecules from the liquid to the gaseous state.
vasopressin A peptide hormone that regulates water balance.
very low density lipoproteins (VLDL) A plasma protein that carries lipids synthesized by the liver to adipose tissue.
virus A species containing a polynucleotide, usually RNA, and a surrounding protein coat that can enter host cells.
viscosity A measure of resistance to flow.
vitamin An organic substance that is required in trace amounts in the diet.
VLDL (very low density lipoproteins) The principal carriers of triglycerides.
water-soluble vitamins Vitamin C and the B group of vitamins.
wax An ester of a fatty acid and a long-chain alcohol.
weak acid An acid that only partially ionizes in water.
weak electrolyte Substance that does not dissociate or ionize completely in water.
weight The result of the gravitational attraction between an object and the Earth.
zwitterion An electrically neutral ion resulting from transfer of a proton from an acidic to a basic site in a molecule.
zymogen An inactive storage form of an enzyme.

APPENDIX C

Answers to Problems

CHAPTER 1

1.1 three hydrogen atoms and one nitrogen atom combined in a molecule

1.2 An oxygen atom is bonded to two carbon atoms. Each of these carbon atoms is bonded to two hydrogen atoms and another carbon atom, which is bonded to three hydrogen atoms.

1.3 Nitrogen and hydrogen are reactants; ammonia is the product.
nitrogen + hydrogen $\longrightarrow$ ammonia

1.4 gas

1.5 Chemical change

CHAPTER 2

2.1 4, 3, 3

2.2 2×10^2 nm

2.3 2.5×10^3 mL

2.4 3.9×10^{-6} g

2.5 6602 °F

2.6 248 °F

2.7 22.5 g/cm^3

2.8 2.3×10^2 g

2.9 6.1 cm^3

2.10 5.0×10^2 cal

CHAPTER 3

3.1 39.9 amu

3.2 146

3.3 53 electrons, 53 protons, 78 neutrons

3.4 35.46 amu

3.5 5th period, Group IB; Cu and Au

3.6 −30 °C

3.7 2

3.8 5

3.9 $1s^2 2s^2 2p^6 3s^2 3p^6 4s^2 3d^{10} 4p^6 5s^2$

3.10 $1s^2 2s^2 2p^6 3s^2 3p^6 4s^2 3d^{10} 4p^4$; 2 unpaired electrons

3.11 $1s^2 2s^2 2p^6 3s^2 3p^6 4s^2 3d^{10} 4p^6 5s^2 4d^{10} 5p^2$

3.12 Bi > Br

3.13 Cl has higher ionization energy; As has greater tendency to lose electrons

3.14 Ge is less electronegative than S

CHAPTER 4

4.1 30 protons, 28 electrons

4.2 $Ca^{2+} < 0.197$ nm; $S^{2-} > 0.104$ nm

4.3 Mg^{2+} and O^{2-} in 1 : 1 ratio

4.4 +3

4.5 $(+1) + (+3) + 2(-1) + (-2) = 0$

4.6 ammonium carbonate

CHAPTER 5

5.1
```
:F̤—N̈—F̤:
     |
    :F̤:
```

5.2
```
:S̤=C=S̤:
```

5.3
```
     H   H
     |   |
  H—C—C—C̈l:
     |   |
     H   H
```
nonpolar C—C and C—H bonds but polar C—Cl bond

5.4
```
        :Ö:
         |
  H—Ö—P—Ö—H
         |
        :Ö—H
```

5.5 The N—O single and double bonds can be interchanged in resonance forms.

5.6 Both are tetrahedral about the carbon atom. Methanol is angular about oxygen; methylamine is pyramidal about nitrogen.

5.7
```
     H   H   H
     |   |   |
  H—C—C—C—OH
     |   |   |
     H   H   H
```

CHAPTER 6

6.1 $C_2H_5OH + 3\ O_2 \longrightarrow 2\ CO_2 + 3\ H_2O$

6.2 $Ca_3(PO_4)_2 + 3\ H_2SO_4 \longrightarrow 3\ CaSO_4 + 2\ H_3PO_4$

6.3 $Pb^{2+} + CrO_4^{2-} \longrightarrow PbCrO_4$

6.4 −2

6.5 +3

6.6 O_2 is the oxidizing agent; NH_3 is the reducing agent

6.7 reduction; reducing agent

CHAPTER 7

7.1 183.1 amu

7.2 144.0 amu

7.3 29.2% N, 8.3% H, 12.5% C, 50.0% O

7.4 56.8 moles

7.5 4.89×10^{21} ions

7.6 $C_5H_{14}N_2$

7.7 $C_{10}H_8$

7.8 3 moles H_2/1 mole WO_3; 1 mole WO_3/3 moles H_2

7.9 0.10 mole

7.10 92 g

7.11 3.2 kcal

CHAPTER 8

8.1 7.6×10^3 cm Hg

8.2 6.0 atm

8.3 93.8 mL
8.4 156 °C
8.5 146 mL
8.6 120 amu
8.7 111 torr
8.8 The forces are greater in $GeBr_4$.
8.9 They are essentially equal.
8.10 CO is polar
8.11 No hydrogen bonding occurs in the —S—H group.

CHAPTER 9

9.1 22 g
9.2 12 mL
9.3 1.5×10^{-5} g
9.4 8.9×10^{-8} M
9.5 350 mL
9.6 10 mg
9.7 positive
9.8 5.7×10^3 g/mole

CHAPTER 10

10.1 Combustion would occur faster in the oxygen atmosphere.
10.2 Ripening is a reaction that occurs faster at room temperature.
10.3 $K = \dfrac{[SO_3]^2}{[SO_2]^2\,[O_2]}$
10.4 1.9 M
10.5 The reaction will shift to the left.
10.6 The equilibrium would shift to the left.
10.7 The equilibrium constant increases.

CHAPTER 11

11.1 It is an acid.
11.2 $H_2N{-}CH_2{-}C(=O){-}O^-$ $H_3N^+{-}CH_2{-}C(=O){-}OH$
11.3 1.1×10^{-10} M; basic
11.4 2×10^{-4}
11.5 9.94
11.6 0.1m CN^-
11.7 2.12
11.8 0.98 g
11.9 0.0250 N

CHAPTER 12

12.1 $^{14}_{7}N$
12.2 2 mg

CHAPTER 13

13.1 ether to the left of the ring; amide to the right of the ring
13.2 Yes, they are isomers.
13.3 Br^- is a base; carbocation is an acid.
13.4 reduction; reducing agent
13.5 elimination followed by reduction

CHAPTER 14

14.1 $C_{100}H_{202}$
14.2 four primary carbon atoms bonded to a central quaternary carbon atom
14.3 3,3,4-trimethylhexane
14.4 two *tert*-butyl groups near the —OH group; methyl group
14.5 $C_{10}H_{20}O$
14.6 [structure: bicyclic ring with O, CH_3, H, and CH_3CH_2 substituents]
14.7 (a) isobutylcyclopentane
(b) propylcyclobutane
(c) *cis*-1,5-dibromocyclodecane
14.8 one; five

CHAPTER 15

15.1 $C_{15}H_{26}$
15.2 cis, cis, trans going from right to left
15.3 (a) [cyclohexadiene with CH_3] (b) [cyclohexadiene with CH_3]
15.4 $CH_3{-}CH(CH_3){-}CH_2{-}CH(OH)CH_2CH(CH_3)CH_2{-}CH_3$
15.5 1-bromo-1-ethylcyclobutane
15.6 $CH_2{=}CHCH{=}CHC{\equiv}CC{\equiv}CC{\equiv}CCH{=}CH_2$
15.7 reduction of 11-tetradecyn-1-ol using sodium and ammonia
15.8 *trans*-5-phenyl-3-heptene

CHAPTER 16

16.1 *trans*-6-nonen-1-ol
16.2 tertiary
16.3 The second compound will react faster with the Lucas reagent.
16.4 [structure: $H_2C{=}C(OPO_3^{2-}){-}C(=O)OH$]
16.5 Jones reagent gives

cyclohexyl–CH_2CO_2H cyclohexyl–$C(=O)CH_3$

PCC gives

cyclohexyl–CH_2CHO cyclohexyl–$C(=O)CH_3$

CHAPTER 17

17.1 $C_6H_5{-}C(=O){-}C{\equiv}C{-}C{\equiv}C{-}CH_3$
17.2 $CH_3CH_2CH_2CH_2CHO$, $(CH_3)_2CHCH_2CHO$, $(CH_3)_3CCHO$, $CH_3CH_2CH(CH_3)CHO$

17.3 OH
$CH-CH_3$ (on cyclohexane ring)

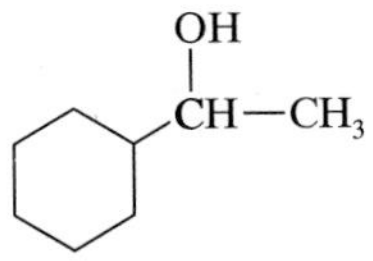

17.4 ketal

17.5 2-methylpropanal

CHAPTER 18

18.1 ether on right benzene ring; carboxylic acid on heterocylic ring; amide between heterocyclic ring and left benzene ring; chloro group

18.2 3,5-dihydroxy-3-methylpentanoic acid

18.3 oxaloacetic acid; right carboxyl group

18.4 $NH_2CH_2CH_2CH_2CH_2CH_2NH_2$

18.5 $H-\overset{O}{\overset{\|}{C}}-O-CH_2-\overset{CH_3}{\overset{|}{C}H}-CH_3$

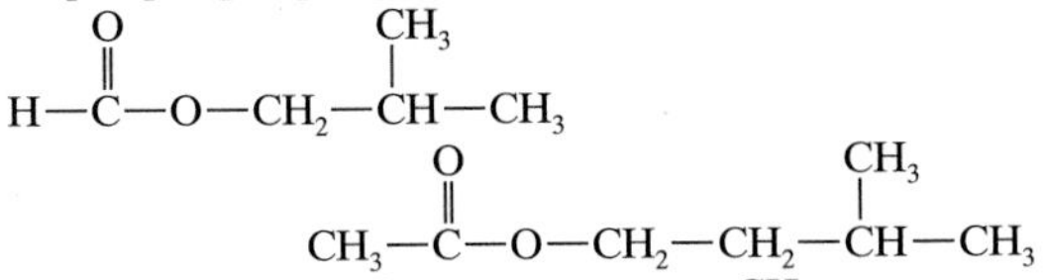

$CH_3-\overset{O}{\overset{\|}{C}}-O-CH_2-CH_2-\overset{CH_3}{\overset{|}{C}H}-CH_3$

18.6 $CH_3-\overset{CH_3}{\overset{|}{C}H}-CH_2-\overset{O}{\overset{\|}{C}}-O-CH_2CH_2-\overset{CH_3}{\overset{|}{C}H}-CH_3$

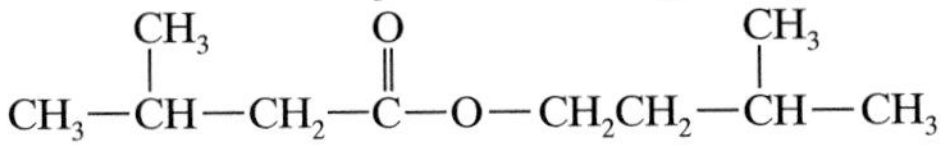

3-methylbutyl 3-methylbutanoate

CHAPTER 19

19.1 amide; amine

19.2 secondary

19.3 CH_3O, CH_3O-, CH_3O (on benzene ring) $-CH_2CH_2NH_2$

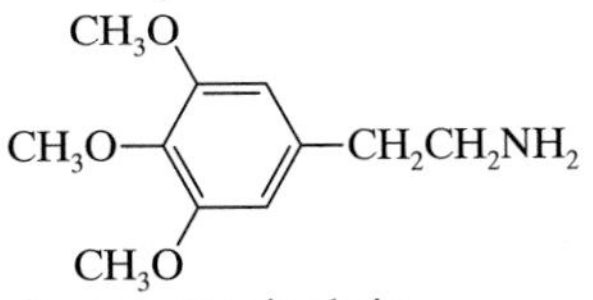

19.4 nitrogen atom in chain

19.5 only trimethylamine does not react; only ethylmethylamine gives a nitrosoamine

19.6 3-methylbenzoic acid and diethylamine

19.7 CH_3, CH_3 (on benzene ring) $-\overset{H}{\overset{|}{N^+}}-H$ with H below; $HO-\overset{O}{\overset{\|}{C}}-$ (piperidine ring with N–CH_3)

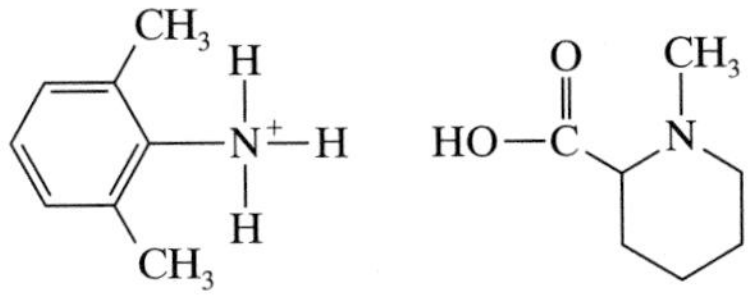

CHAPTER 20

20.1 -10.3 kcal/mole

20.2 Yes, the carbon atom of the five-membered ring bonded to the aromatic ring is chiral.

20.3

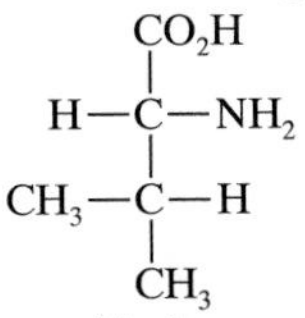

$$\begin{array}{c} CO_2H \\ | \\ H-C-NH_2 \\ | \\ CH_3-C-H \\ | \\ CH_3 \end{array}$$

20.4 two chiral centers gives 4 stereo isomers

20.5

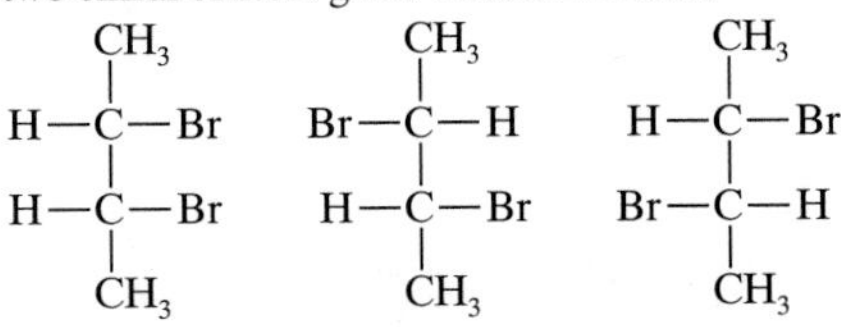

$$\begin{array}{c} CH_3 \\ | \\ H-C-Br \\ | \\ H-C-Br \\ | \\ CH_3 \end{array} \quad \begin{array}{c} CH_3 \\ | \\ Br-C-H \\ | \\ H-C-Br \\ | \\ CH_3 \end{array} \quad \begin{array}{c} CH_3 \\ | \\ H-C-Br \\ | \\ Br-C-H \\ | \\ CH_3 \end{array}$$

CHAPTER 21

21.1 I and II are diastereomers; I and III are enantiomers.

21.2 $HOCH_2$, O, H, H, H, HO, OH, OH, H (furanose ring)

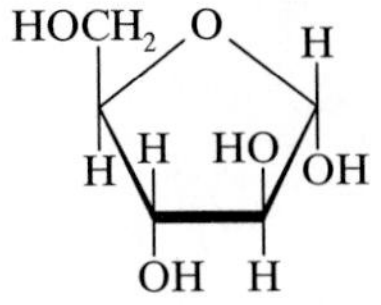

21.3 It is an acetal of glucose and phenol.

21.4 No, it is a ketal of fructose.

21.5 α-glucose bonded $1 \rightarrow 4$ to β-mannose.

CHAPTER 22

22.1 $CH_3(CH_2)_{24}CO_2CH_2(CH_2)_{28}CH_3$

22.2 85 °C

22.3 six; glyceryl tristearate

22.4 palmitic acid, oleic acid, glycerol, phosphoric acid, serine; it is a phosphatidylserine

CHAPTER 23

23.1 aspartic acid, phenylalanine

23.2 Try-Gly-Gly and Phe-Leu

CHAPTER 24

24.1 transferase

24.2 conjugate base

24.3 The carboxylate ion will probably form an ionic bond.

CHAPTER 25

25.1 -33.7 kcal/mole

25.2 They are more oxidized and are more similar to carbohydrates.

25.3 $+2.2$ kcal/mole

25.4 Yes, the $\Delta G° = -3.0$ kcal/mole.

CHAPTER 27

27.1 9 moles, 8 times

27.2 95 moles/mole acid

CHAPTER 28

28.1 pyruvic acid and aspartic acid

28.2 phenylpyruvic acid

28.3 tyrosine

CHAPTER 29

29.1 nucleoside; deoxyribose and guanine; deoxyguanosine

29.2 G-A-T-C-dG

29.3 C-G-T-A

29.4 UGU and UGC; they differ in the third base.

29.5 GUA

29.6 UUUAUUGGU is one of many possibilities.

29.7 Arginine will replace serine, which will be a serious mutation.

APPENDIX D

Answers to Odd-Numbered Exercises

CHAPTER 1

1.1 Fish also lay eggs.
1.3 Push "on" button again, check power cord.
1.5 The facts must be verified.
1.7 To allow other people to check the facts.
1.9 C_3H_8
1.11 $C_2H_6O_2$
1.13 one atom of sulfur and two atoms of oxygen per molecule
1.15 19
1.17 $C_{16}H_{18}N_2O_5S$
1.19

```
H—S
  |
  H
```

1.21

```
  H   H
  |   |
H—N—N—H
```

1.23 same composition; structures differ in order of connection of atoms.
1.25 (a) CH_4O (b) CH_5N
1.27 sulfur + oxygen ⟶ sulfur dioxide
1.29 hydrogen peroxide is reactant; water and oxygen are products
1.31 liquid
1.33 gas
1.35 (a) physical (b) physical (c) chemical (d) physical (e) chemical (f) physical
1.37 All foods are mixtures of chemicals.
1.39 The symbols are Cu, Ag, and Au; all are located in a single column labeled IB.

CHAPTER 2

2.1 student 1, accurate and precise; student 2, not accurate but precise; student 3, accurate but not precise
2.3 (a) 5 (b) 3 (c) 5 (d) 3 (e) 6 (f) 2 (g) 4 (h) 6
2.5 (a) 3 (b) 4 (c) 3 (d) 2 (e) 4 (f) 3
2.7 (a) mL (b) pg (c) nm (d) dm (e) cL (f) kg
2.9 (a) milligram (b) picometer (c) centiliter (d) kilogram (e) milliliter (f) kilometer
2.11 (a) 5.9 cm (b) 1.53×10^{-3} km (c) 0.348 L (d) 5238 mL (e) 56.6 dm (f) 3.53 m
2.13 2.5 mm
2.15 500
2.17 1.5×10^6 nm
2.19 2×10^{-3} g/L
2.21 0.65 g
2.23 3.25 g/L
2.25 −35 °C
2.27 −129.3 °F
2.29 106 °F
2.31 4.4 K
2.33 7.9 g/mL
2.35 1.3×10^2 g
2.37 1.28×10^3 g
2.39 5.0×10^3 mL
2.41 1.10
2.43 11 g
2.45 11
2.47 40 °C
2.49 11 kg
2.51 3.8 Cal
2.53 72 g

CHAPTER 3

3.1 6.02×10^{23} atoms O
3.3 1.21×10^{-1} nm
3.5 number of electrons and protons is the same.
3.7 atomic number is the number of protons; as a subscript to the left of the symbol
3.9 (a) 8 protons, 8 electrons, 8 neutrons
(b) 11 protons, 11 electrons, 12 neutrons
(c) 13 protons, 13 electrons, 14 neutrons
(d) 16 protons, 16 electrons, 16 neutrons
(e) 18 protons, 18 electrons, 22 neutrons
(f) 20 protons, 20 electrons, 20 neutrons
3.11 (a) ${}^{19}_{9}F$ (b) ${}^{30}_{14}Si$ (c) ${}^{28}_{14}Si$ (d) ${}^{31}_{15}P$
3.13 The isotopes have 0, 1, and 2 neutrons, respectively.
3.15 ${}^{131}_{53}I$ isotope has 78 neutrons; ${}^{127}_{53}I$ isotope has 74 neutrons
3.17 (a) 4, VIIIB (b) 5, VIIIB (c) 6, IIB (d) 4, IB (e) 5, IIB (f) 6, VIIIB
3.19 C, S, Br
3.21 (a) 2 in $n = 1$, 5 in $n = 2$
(b) 2 in $n = 1$, 8 in $n = 2$, 4 in $n = 3$
(c) 2 in $n = 1$, 3 in $n = 2$
(d) 2 in $n = 1$, 7 in $n = 2$
(e) 2 in $n = 1$, 8 in $n = 2$, 6 in $n = 3$
(f) 2 in $n = 1$
3.23 (a) 5 (b) 4 (c) 3 (d) 7 (e) 6 (f) 2
3.25 (a) 2 (b) 4 (c) 3
3.27 (a) $2s,2p$ (b) $3s,3p$ (c) $2s$ (d) $2s,2p$ (e) $3s,3p$ (f) $2s$
3.29 (a) 1 (b) 3 (c) 5 (d) 1 (e) 3 (f) 5
3.31 (a) 2 (b) 2 (c) 2 (d) 2
3.33 (a) $1s^22s^22p^4$ (b) $1s^22s^22p^63s^23p^6$ (c) $1s^22s^2$ (d) $1s^22s^22p^2$ (e) $1s^22s^22p^63s^23p^3$ (f) $1s^22s^1$

3.35 (a) $1s^22s^22p^3$ (b) $1s^22s^22p^63s^23p^2$ (c) $1s^22s^22p^1$
(d) $1s^22s^22p^5$ (e) $1s^22s^22p^63s^23p^4$ (f) $1s^2$

3.37 (a) 4 (b) 1 (c) 1
(d) 3 (e) 3 (f) 4

3.39 (a) 3 (b) 2 (c) 1
(d) 1 (e) 2 (f) 0

3.41 (a) :Ö· (b) :Är: (c) :Be
(d) ·Ċ: (e) ·P: (f) Li·

3.43 (a) K· (b) :B̈r· (c) Mg:
(d) Ca: (e) ·G̈e· (f) :K̈r:

3.45 (a) 2, VA (b) 3, IA
(c) 3, VIIA (d) 4, VIIB

3.47 (a) 5, IVB (b) 4, IVA
(c) 6, VIB (d) 3, VA

3.49 (a) VA (b) IIIB
(c) VIIA (d) IVA

3.51 (a) 2 (b) 3 (c) 1 (d) 2

3.53 Ca, Rb, Ag, Ni

3.55 (a) Mg > Si (b) Ge > Br
(c) Se > S (d) Sn > Si

3.57 (a) 0.089 nm (b) 3.66 g/cm^3

3.59 (a) Cl (b) Si (c) K
(d) C (e) Sb (f) Pb

3.61 (a) F (b) Li (c) Cl
(d) S (e) Ge (f) Li

3.63 (a) F (b) S (c) Li
(d) N (e) Cl (f) P

CHAPTER 4

4.1 Se^{2-} has 36 electrons; P^{3-} has 18 electrons

4.3 Ce^{2+} has 55 electrons; Ce^{4+} has 54 electrons

4.5 (a) O^{2-} (b) S^{2-} (c) I^-
(d) Br^- (e) N^{3-} (f) F^-

4.7 (a) sulfide ion (b) iodide ion (c) oxide ion
(d) fluoride ion (e) nitride ion (f) bromide ion

4.9 (a) Mg (b) K (c) Al
(d) Br^- (e) S^{2-} (f) N^{3-}

4.11 (a) SO_4^{2-} (b) PO_4^{3-} (c) OH^-
(d) NH_4^+ (e) CN^- (f) CO_3^{2-}
(g) SO_3^{2-} (h) HCO_3^-

4.13 (a) 1 tungsten atom, 4 oxygen atoms, and an excess of 2 electrons
(b) 1 hydrogen atom, 1 phosphorus atom, 4 oxygen atoms, and an excess of 2 electrons
(c) 2 phosphorus atom, 7 oxygen atoms, and an excess of 4 electrons
(d) 1 vanadium atom, 2 oxygen atoms, and a deficiency of 1 electron

4.15 (a) $FeCl_3$ (b) NaOH (c) $Mg(OH)_2$
(d) CdS (e) MnF_2 (f) Cs_3N

4.17 (a) LiF (b) $MgBr_2$ (c) Li_2O
(d) MgSe (e) AlF_3 (f) Na_2S

4.19 PO_4^{3-}; $Pb_3(AsO_4)_2$

4.21 $Na_4P_2O_7$; $Ca_2P_2O_7$

4.23 (a) calcium hydroxide (b) lithium perchlorate
(c) sodium phosphate (d) potassium sulfate
(e) sodium nitrate (f) ammonium nitrite

4.25 (a) barium hydroxide (b) potassium chlorite
(c) cesium phosphate (d) potassium sulfite
(e) sodium nitrite (f) ammonium nitrate

4.27 (a) LiF (b) ZnO (c) $MgCO_3$
(d) KNO_3 (e) NaCN (f) Al_2S_3

4.29 (a) FeF_3 (b) Cu_2O (c) $FeSO_4$
(d) $Cu(CN)_2$ (e) Fe_2O_3 (f) $Cu(NO_3)_2$

4.31 (a) $Ca(ClO)_2$ (b) $Al(OH)_3$ (c) $CuSO_4$

CHAPTER 5

5.1 (a) H—H (b) :Ï—Ï: (c) :F̈—F̈:
(d) :B̈r—B̈r: (e) :C̈l—C̈l: (f) :N≡N:

5.3 (a) H (b) Br (c) H
(d) O (e) I (f) Si

5.5 (a) H:Ö:S:Ö:H with :Ö: above and :Ö: below S (b) H—P̈—H with H below P
(c) :F̈—C—F̈: with :F̈: above and :F̈: below C (d) H—S̈e—H
(e) :B̈r—C—B̈r: with :B̈r: above and :B̈r: below C (f) :C̈l—Si—C̈l: with :C̈l: above and :C̈l: below Si

5.7 (a) :S̈—H$^-$ (b) H—P$^+$—H with H above and H below P (c) H—Ö$^+$—H with H below O
(d) :C≡N:$^-$ (e) :Ö—S—Ö: with :Ö: above and :Ö: below S, $2-$ (f) H—N$^+$—H with H above and H below N

5.9 (a) :F̈—C—F̈: with :F̈: above and :F̈: below C (b) H—C—Ö: with H above C, H below C and H below O
(c) H—C—N̈—H with H above, H above and H below (d) (:C̈l)(:C̈l:)C=Ö:
(e) H—C≡N: (f) (H)(H)C=N̈—Ö—H

5.11 Ö=N—Ö:$^-$ (with :Ö: below N) ⟷ $^-$:Ö—N=Ö: (with :Ö: below N) ⟷ :Ö—N—Ö:$^-$ (with :O double-bonded below N)

5.13 (a) tetrahedral (b) trigonal pyramidal
(c) angular (d) angular

5.15 (a) tetrahedral (b) tetrahedral
(c) trigonal pyramidal (d) angular

5.17 (a) 180° (b) 109°
(c) 109° (d) 180°

5.19 (a) polar (b) polar (c) nonpolar
(d) polar (e) nonpolar (f) nonpolar

5.21 4 single bonds; 1 double bond and 2 single bonds; 1 triple bond and 1 single bond; 2 double bonds

5.23 (a) Br—CH_2—CH_2—Br
(b) CH_3—CH_2—CH_2—CH_2—CH_3

(c) $CH_3—CH_2—CH_2—SH$
5.25 (a) CH_2BrCH_2Br
(b) $CH_3CH_2CH_2CH_2CH_3$
(c) $CH_3CH_2CH_2SH$
5.27 (a) $C_6H_{12}S$ (b) $C_{12}H_{20}O$
5.29 (a) CH_3Cl (b) CH_2Br_2 (c) CH_3F

CHAPTER 6

6.1 Coefficients are given in order of appearance:
(a) 2,1,2 (b) 2,3,2
(c) 1,1,2 (d) 2,2,1
(e) 1,1,1,2 (f) 4,1,4,3
6.3 Coefficients are given in order of appearance:
(a) 1,6,1,2 (b) 1,3,3,1 (c) 1,12,4,3
(d) 2,1,1,4 (e) 1,3,1,3 (f) 1,2,1,1
6.5 Coefficients are given in order of appearance:
(a) 2,7,4,6 (b) 1,8,5,6 (c) 2,15,10,10
(d) 1,3,2,3 (e) 2,15,10,12 (f) 1,7,5,5
6.7 (a) $2\ Ba + O_2 \longrightarrow 2\ BaO$
(b) $H_2 + I_2 \longrightarrow 2\ HI$
(c) $2\ Li + S \longrightarrow Li_2S$
(d) $2\ Al + 3\ Cl_2 \longrightarrow 2\ AlCl_3$
(e) $Ca + Cl_2 \longrightarrow CaCl_2$
(f) $3\ Mg + N_2 \longrightarrow Mg_3N_2$
6.9 (a) $2\ HgO \longrightarrow 2\ Hg + O_2$
(b) $SrCO_3 \longrightarrow SrO + CO_2$
(c) $2\ Au_2O_3 \longrightarrow 4\ Au + 3\ O_2$
(d) $2\ Ag_2O \longrightarrow 4\ Ag + O_2$
6.11 (a) $2\ KBr + Cl_2 \longrightarrow 2\ KCl + Br_2$
(b) $SnO_2 + 2\ H_2 \longrightarrow Sn + 2\ H_2O$
(c) $4\ Mg + Fe_3O_4 \longrightarrow 3\ Fe + 4\ MgO$
(d) $2\ Al_2O_3 + 3\ C \longrightarrow 4\ Al + 3\ CO_2$
6.13 (a) $Zn + 2\ HCl \longrightarrow ZnCl_2 + H_2$
(b) $SiCl_4 + 2\ Mg \longrightarrow 2\ MgCl_2 + Si$
(c) $Br_2 + 2\ NaI \longrightarrow I_2 + 2\ NaBr$
(d) $2\ ZnO + C \longrightarrow CO_2 + 2\ Zn$
6.15 Only AgBr is insoluble.
6.17 Only AgI is insoluble.
6.19 (a) $Pb(NO_3)_2 + 2\ HCl \longrightarrow PbCl_2 + 2\ HNO_3$
(b) $FeCl_2 + 2\ NaOH \longrightarrow Fe(OH)_2 + 2\ NaCl$
(c) $Cd(NO_3)_2 + H_2S \longrightarrow CdS + 2\ HNO_3$
(d) $AgNO_3 + CsCl \longrightarrow AgCl + CsNO_3$
6.21 (a) $MnSO_4 + (NH_4)_2S \longrightarrow MnS + (NH_4)_2SO_4$
(b) $CaCO_3 + H_2SO_4 \longrightarrow CaSO_4 + H_2O + CO_2$
(c) $CdSO_4 + 2\ NaOH \longrightarrow Cd(OH)_2 + Na_2SO_4$
(d) $3\ ZnCO_3 + 2\ H_3PO_4 \longrightarrow Zn_3(PO_4)_2 + 3\ H_2O + 3\ CO_2$
6.23 (a) $Pb^{2+} + 2\ I^- \longrightarrow PbI_2$
(b) $Cd^{2+} + S^{2-} \longrightarrow CdS$
(c) $Ag^+ + Br^- \longrightarrow AgBr$
(d) $Ba^{2+} + SO_4^{2-} \longrightarrow BaSO_4$
6.25 all are $Ag^+ + Br^- \longrightarrow AgBr$
6.27 (a) $H_3PO_4 + 3\ KOH \longrightarrow 3\ H_2O + K_3PO_4$
(b) $2\ Fe(OH)_3 + 3\ H_2SO_4 \longrightarrow Fe_2(SO_4)_3 + 6\ H_2O$
(c) $2\ H_3PO_4 + 3\ Ca(OH)_2 \longrightarrow 6\ H_2O + Ca_3(PO_4)_2$
(d) $Zn(OH)_2 + 2\ HNO_3 \longrightarrow Zn(NO_3)_2 + 2\ H_2O$
6.29 (a) Al oxidized, Co reduced
(b) Cu oxidized, Br reduced
(c) K oxidized, Br reduced
6.31 (a) Sb_2O_3 is the oxidizing agent; Fe is the reducing agent
(b) Sb_2O_5 is the oxidizing agent; C is the reducing agent
(c) H_2SO_4 is the oxidizing agent; NaBr is the reducing agent
6.33 (a), (b), and (c) will react
6.35 Oxidation because the molecule has gained an oxygen atom

CHAPTER 7

7.1 (a) 28.0 (b) 44.0 (c) 64.1 (d) 80.1
7.3 394.0
7.5 (a) 62.0 (b) 110.3 (c) 92.8 (d) 291.6
7.7 (a) 142.1 (b) 138.2 (c) 93.9 (d) 164.1
7.9 (a) %C = 7.8%; %Cl = 92.2%
(b) %Si = 27.0%; %F = 73.0%
(c) %Ge = 18.5%; %Br = 81.5%
(d) %P = 11.5%; %Br = 88.5%
7.11 (a) 43.4% Na; 11.3% C; 45.3% O
(b) 12.6% Li; 29.2% S; 58.2% O
(c) 55.3% K 14.6% P; 30.1% O
(d) 36.7% Fe; 21.1% S; 42.1% O
7.13 (a) 40.9% C; 4.5% H; 54.5% O
(b) 83.9% C; 10.5% H; 5.6% O
(c) 55.8% C; 6.5% H; 14.5% N; 2.3% P; 4.4% Co; 16.5% O
7.15 (a) 2.00 (b) 0.498 (c) 0.010 (d) 1.00
7.17 (a) 10.0 (b) 0.010 (c) 2.0 (d) 0.100
7.19 (a) 0.0100 (b) 0.0500 (c) 1.00 (d) 0.500
7.21 2.56×10^{-4} mole
7.23 1.3×10^{-4} mole
7.25 (a) 6.00×10^{22} (b) 6.02×10^{24}
(c) 3.01×10^{23} (d) 6.02×10^{21}
7.27 (a) 6.0×10^{22} (b) 3.0×10^{24}
(c) 6.0×10^{23} (d) 1.5×10^{22}
7.29 (a) 1.20×10^{22} (b) 1.51×10^{23}
(c) 6.02×10^{22} (d) 3.01×10^{23}
7.31 (a) 2×10^{23} (b) 4×10^{22} (c) 6×10^{23}
7.33 9×10^{15}
7.35 CH_3O
7.37 GaP
7.39 (a) CH_2 (b) CH_3 (c) CH
7.41 $C_6H_{10}O_4$
7.43 $C_8H_{10}N_4O_2$
7.45 (4 moles Al)/(3 moles MnO_2)
7.47 (2 moles AgCl)/(1 mole $BaCl_2$)
7.49 (6 moles SO_2)/(2 moles Bi_2S_3)
7.51 2.4
7.53 0.04
7.55 2.5
7.57 53.2
7.59 7.19
7.61 5.1×10^2
7.63 -1.3×10^2 kcal
7.65 87.1%
7.67 27 g
7.69 7.16

CHAPTER 8

8.1 There is less air pushing down on a surface at a high altitude.

8.3 (a) 0.500 atm (b) 38.0 cm Hg
(c) 76.0 torr (d) 19.0 cm Hg
(e) 0.918 atm (f) 51 cm Hg
8.5 0.292 atm
8.7 8.75×10^3 L
8.9 10 L
8.11 501 mL
8.13 112 mL
8.15 −9 °C
8.17 4.3 atm
8.19 2.3×10^8 L
8.21 18 mL
8.23 (a) 2.24 L; 0.100 mole (b) 2.24 L; 0.100 mole
(c) 1.68 L; 0.075 mole
8.25 6.02×10^{23}; 2.7×10^{22}
8.27 42 g/mole
8.29 1.7 moles
8.31 4.3×10^{15}
8.33 77 torr
8.35 0.6 atm
8.37 5 atm
8.39 $SbCl_3$
8.41 SnI_4
8.43 Stronger intermolecular forces exist between $SiCl_4$ molecules.
8.45 CCl_4, 8.26×10^3; CBr_4, 1.09×10^4 cal/mole; Br_4
8.47 −3 °C
8.49 mercury freezes at −39 °C
8.51 It will melt at a higher temperature.
8.53 The larger number of sites for hydrogen bonding increase the intermolecular forces and hence the surface tension.
8.55 They both have similar London forces.
8.57 CCl_4 has higher London forces.
8.59 ethylene glycol can form more hydrogen bonds
8.61 Trimethylamine does not have any N—H bonds with which to form hydrogen bonds.

CHAPTER 9

9.1 unsaturated
9.3 (a) 1.00% (b) 4.00% (c) 3.0%
9.5 (a) 20.0% (b) 5% (c) 0.3%
9.7 1 mL SO_2
9.9 0.002 mL
9.11 0.2 ppm
9.13 (a) 0.20 M (b) 0.50 M (c) 4.18 M
9.15 (a) 45.0 g (b) 85.5 g (c) 15.0 g
9.17 (a) 0.40 L (b) 10.0 L (c) 0.10 L
9.19 6.11×10^{-3} M
9.21 10 M
9.23 0.19 moles
9.25 4.2×10^{-3} M
9.27 (a) 0.57 M (b) 1.5 M (c) 3.6 M
9.29 (a) 250 mL (b) 2500 mL (c) 500 mL
9.31 $HBr + H_2O \longrightarrow H_3O^+ + Br^-$
9.33 0.158
9.35 reduction of the CO_2 pressure lowers the solubility of CO_2
9.37 decrease
9.39 strong attractive forces between NH_3 and H_2O molecules due to hydrogen bonding
9.41 soluble in water but not CCl_4
9.43 1-pentanol
9.45 water soluble
9.47 Water will pass into the cells by osmosis.
9.49 1.2 mm Hg
9.51 6.49×10^4

CHAPTER 10

10.1 negative $\Delta H°$ for exothermic reaction
10.3 one that occurs without an outside source of energy
10.5 (a) not spontaneous at any T
(b) may be spontaneous at low T
10.7 One reaction is the reverse of the other in terms of the products and reactants.
10.9 The reaction will go faster at a higher temperature.
10.11 The chemical reactions associated with metabolism go faster at higher temperatures.
10.13 The chemical reaction associated with ripening goes faster at higher temperatures, slower at lower temperatures.
10.15 B with the lower E_a
10.17 not a correct statement; the activation energy is constant
10.19 $CO + 2\,H_2 \rightleftharpoons CH_3OH$
$K = \dfrac{[CO]\,[H_2]^2}{[CH_3OH]}$
10.21 (a) $K = \dfrac{[O_3]^2}{[O_2]^3}$ (b) $K = \dfrac{[NH_3]^2}{[N_2]\,[H_2]^3}$
(c) $K = \dfrac{[CH_3Cl]\,[HCl]}{[CH_4]\,[Cl_2]}$
10.23 (a) $K = \dfrac{[CH_4]\,[H_2S]^2}{[CS_2]\,[H_2]^4}$ (b) $K = \dfrac{[HCl]^4\,[O_2]}{[Cl_2]^2\,[H_2O]^2}$
(c) $K = \dfrac{[NO_2]^2}{[N_2O_4]}$
10.25 61.0
10.27 3.11×10^{-2} M
10.29 (a) shift right (b) shift left
(c) shift left (d) shift right
10.31 (a) shift right (b) shift left (c) shift left
10.33 (a) shift left (b) shift right
(c) shift left (d) shift right
10.35 shift to right
10.37 only in (a)
10.39 none will
10.41 only (b)
10.43 low pressure and high temperature
10.45 only (b)

CHAPTER 11

11.1 yes, if it can lose or gain a proton such as in H_2O, HCO_3^-, and HSO_4^-
11.3 $H_2SO_3 + H_2O \rightleftharpoons H_3O^+ + HSO_3^-$
$HSO_3^- + H_2O \rightleftharpoons H_3O^+ + SO_3^{2-}$
11.5 The acid loses H^+ to produce the conjugate base.
11.7 (a) $HClO_4$ (b) HCl (c) HNO_3
(d) HCO_3^- (e) CH_3CO_2H (f) H_2SO_4
11.9 $C_8H_{11}NO_2O_3^-$
11.11 H—C(H)(H)—C(=O)—C(=O)—O⁻ or $C_3H_3O_3^-$
11.13 $C_{18}H_{21}NO_3H^+$
11.15 The predominant species in solution is NH_3 (aq).
11.17 (a) weak (b) strong (c) strong
(d) weak (e) strong (f) weak

11.19 $K_{a1} = \dfrac{[H_3O^+][H_2PO_4^-]}{[H_3PO_4]}$

$K_{a2} = \dfrac{[H_3O^+][HPO_4^{2-}]}{[H_2PO_4^-]}$

$K_{a3} = \dfrac{[H_3O^+][PO_4^{3-}]}{[HPO_4^{2-}]}$

11.21 5.87×10^{-6}
11.23 (a) $[H_3O^+] = 0.1$ M; $[OH^-] = 1 \times 10^{-13}$ M
(b) $[H_3O^+] = 0.01$ M; $[OH^-] = 1 \times 10^{-12}$ M
(c) $[OH^-] = 1 \times 10^{-3}$ M; $[H_3O^+] = 1 \times 10^{-11}$ M
(d) $[H_3O^+] = 0.1$ M; $[OH^-] = 1 \times 10^{-13}$ M
(e) $[OH^-] = 1 \times 10^{-4}$ M; $[H_3O^+] = 1 \times 10^{-10}$ M
(f) $[OH^-] = 0.01$ M; $[H_3O^+] = 1 \times 10^{-12}$ M
11.25 (a) 3.0 (b) 2.7 (c) 12.0 (d) 11.5
11.27 (a) 3.7 (b) 6.7 (c) 11.7 (d) 1.0
11.29 CNO^-
11.31 5.0×10^{-4}
11.33 2.0×10^{-3} M
11.35 They contain a natural indicator that is one color in acid and another color in base.
11.37 pink at pH 3; yellow at pH 7
11.39 high ratio decreases pH; low ratio increases pH
11.41 HCl is a strong acid
11.43 8.37×10^{-7} M
11.45 9.01
11.47 increase pH; decrease pH
11.49 (a) 0.50 meq HCl (b) 0.50 meq HNO_3
(c) 1.0×10^3 meq HCl
11.51 (a) 0.20 N NaOH (b) 0.1 ON HCl
(c) 1.0 N H_2SO_4
11.53 0.22 g $C_3H_6O_3$
11.55 15 mL NaOH
11.57 0.0417 N $HClO_4$
11.59 3.75 g

CHAPTER 12

12.1 ^1_1H
12.3 (a) $^{21}_{9}F \longrightarrow {}^{0}_{-1}e + {}^{21}_{10}Ne$
(b) $^{31}_{14}Si \longrightarrow {}^{0}_{-1}e + {}^{31}_{15}P$
(c) $^{28}_{12}Mg \longrightarrow {}^{0}_{-1}e + {}^{28}_{13}Al$
12.5 (a) $^{212}_{84}Po \longrightarrow {}^{4}_{2}He + {}^{208}_{82}Pb$
(b) $^{240}_{96}Cm \longrightarrow {}^{4}_{2}He + {}^{236}_{94}Pu$
(c) $^{252}_{99}Es \longrightarrow {}^{4}_{2}He + {}^{248}_{97}Bk$
12.7 (a) $^{239}_{94}Pu$ (b) $^{218}_{84}Po$
(c) $^{230}_{92}U$ (d) $^{241}_{95}Am$
12.9 (a) $^{20}_{9}F$ (b) $^{230}_{90}Th$
(c) $^{226}_{88}Ra$ (d) $^{27}_{13}Al$
12.11 (a) $^{18}_{8}O$ (b) $^{11}_{5}B$ (c) $^{15}_{7}N$
12.13 (a) $^{26}_{12}Mg$ (b) $^{41}_{19}K$ (c) $^{7}_{3}Li$
12.15 17190 years old
12.17 1.0 mg
12.19 14.8 days
12.21 $^{55}_{27}Co$
12.23 (a) $^{23}_{11}Na + {}^{2}_{1}H \longrightarrow {}^{24}_{11}Na + {}^{1}_{1}H$
(b) $^{10}_{5}B + {}^{4}_{2}He \longrightarrow {}^{13}_{7}N + {}^{1}_{0}n$
(c) $^{59}_{27}Co + {}^{1}_{0}n \longrightarrow {}^{56}_{25}Mn + {}^{4}_{2}He$
12.25 (a) $^{237}_{94}Pu$ (b) $^{30}_{15}P$ (c) $^{131}_{53}I$
12.27 (a) $^{1}_{0}n$ (b) $^{4}_{2}He$ (c) $^{1}_{0}n$
12.29 (a) $^{256}_{101}Md$ (b) $^{263}_{106}Sg$ (c) $^{247}_{99}Es$ (d) $^{249}_{100}Fm$
12.31 strips electrons from molecules, producing reactive ions
12.33 Radiation produces ions and electrons in the low-pressure gas chamber.
12.35 3.7×10^{10} Bq
12.37 16 m
12.39 short because higher dosage results
12.41 emit gamma radiation so the radiation can be detected outside the body
12.43 isotopes for therapy are alpha or beta emitters

CHAPTER 13

13.1 Only compounds in (c) are isomers.
13.3 (a) CH_3CHBr_2 and CH_2BrCH_2Br
(b) CH_3CH_2OH and CH_3OCH_3
(c) $CH_3CHBrCl$ and CH_2BrCH_2Cl
13.5 isomers; differ in location of both chlorine atom and fluorine atoms
13.7 Pairs in (a) and (b) represent the same compound; the pair in (c) are isomers.
13.9 (a) ester (b) ether (c) alkene
13.11 (a) alkene and ketone
(b) benzene ring, alkene, and ester
(c) alkene, aldehyde
13.13 (a) homolytic C—H cleavage; homogenic H—Cl formation
(b) homolytic Cl—Cl cleavage; homogenic C—Cl formation
13.15 (a) H—C(H)=Ö$^+$—H (b) CH_3—N$^+$(H)(H)—CH_3
(c) CH_3—C≡N$^+$—H
13.17 ether is the Lewis base; BF_3 is the Lewis acid
13.19 (a) reduction (b) oxidation (c) oxidation
13.21 reduction
13.23 elimination; $ZnCl_2$
13.25 first step is substitution and byproduct is HCl; second step is addition and byproduct is HCl; third step is elimination and byproduct is HCl

CHAPTER 14

14.1 $C_{31}H_{64}$; $CH_3(CH_2)_{29}CH_3$
14.3 (a) $CH_3CH_2CH_2CH_3$ (b) $CH_3CH_2CH_2CH(CH_2CH_3)CH_2CH_3$
(c) $CH_3CH_2CH(CH_3)CH_2CH_3$
14.5 I and II are the same; III and IV are the same
14.7 (a) propyl (b) *sec*-butyl (c) isobutyl
14.9 (a) 2 primary and 3 secondary
(b) 3 primary, 2 secondary, 1 tertiary
(c) 4 primary, 1 secondary, 1 quaternary
14.11 (a) 4 primary, 1 quaternary
(b) 3 primary, 2 secondary, 1 tertiary
(c) 3 primary, 2 secondary, 1 tertiary
14.13 (a) 2-methylbutane (b) 3-methylhexane
14.15 (c) 2-methylpentane (d) 3-methylheptane
(a) CH_3—CH_2—CH(CH_3)—CH_2—CH_3
(b) CH_3—CH_2—CH(CH_3)—CH(CH_3)—CH_2—CH_3

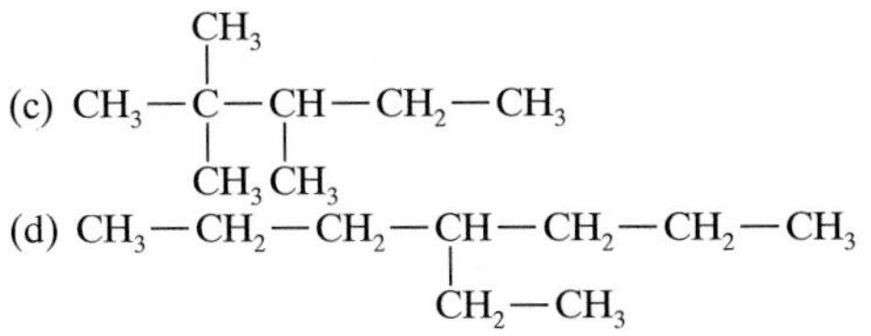

(c) $CH_3-\underset{CH_3}{\overset{CH_3}{\underset{|}{\overset{|}{C}}}}-\underset{CH_3}{\underset{|}{CH}}-CH_2-CH_3$

(d) $CH_3-CH_2-CH_2-\underset{CH_2-CH_3}{\underset{|}{CH}}-CH_2-CH_2-CH_3$

14.17 2-methylhexane and 3-methylhexane

14.19 (a) cyclopropyl–Cl (b) 1,1-dichlorocyclobutane (Cl, Cl on one ring carbon) (c) cyclooctane ring

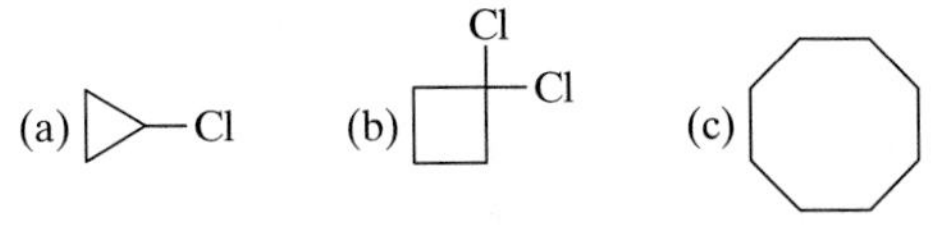

14.21 (a) 1,1-dimethylcycloheptane
(b) cyclodecane
(c) *trans*-1,2-dichlorocyclohexane
(d) 1,1-dichlorocyclohexane

14.23 (a) $C_{10}H_{18}$ (b) $C_8H_{12}O$
(c) $C_7H_{13}N$ (d) C_6H_6

14.25 It is combustible.

14.27 octane; 2,2,3,3-tetramethylbutane

14.29 (a) 2 (b) 2 (c) 2 (d) 3

14.31

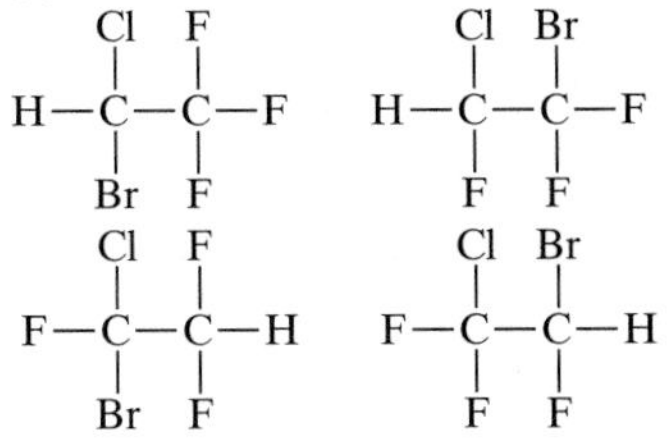

$H-\underset{Br}{\overset{Cl}{C}}-\underset{F}{\overset{F}{C}}-F \qquad H-\underset{F}{\overset{Cl}{C}}-\underset{F}{\overset{Br}{C}}-F$

$F-\underset{Br}{\overset{Cl}{C}}-\underset{F}{\overset{F}{C}}-H \qquad F-\underset{F}{\overset{Cl}{C}}-\underset{F}{\overset{Br}{C}}-H$

CHAPTER 15

15.1 (a) $C_{10}H_{16}$ (b) C_6H_{12}
(c) $C_{10}H_{16}$ (d) C_8H_{12}

15.3 $CH_3CH_2CH{=}CH_2$ 1-butene
$(CH_3)_2C{=}CH_2$ 2-methyl-1-propene

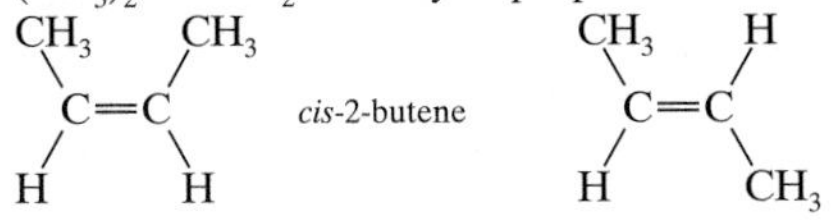

$\begin{array}{ccc} CH_3 & & CH_3 \\ & C{=}C & \\ H & & H \end{array}$ *cis*-2-butene $\qquad \begin{array}{ccc} CH_3 & & H \\ & C{=}C & \\ H & & CH_3 \end{array}$ *trans*-2-butene

15.5

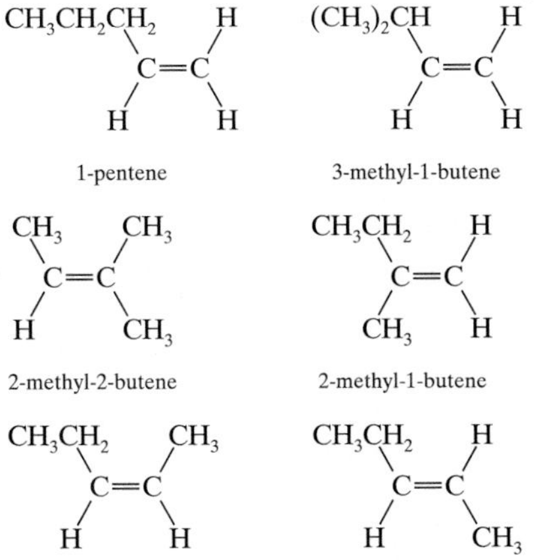

$\begin{array}{ccc} CH_3CH_2CH_2 & & H \\ & C{=}C & \\ H & & H \end{array}$ 1-pentene $\qquad \begin{array}{ccc} (CH_3)_2CH & & H \\ & C{=}C & \\ H & & H \end{array}$ 3-methyl-1-butene

$\begin{array}{ccc} CH_3 & & CH_3 \\ & C{=}C & \\ H & & CH_3 \end{array}$ 2-methyl-2-butene $\qquad \begin{array}{ccc} CH_3CH_2 & & H \\ & C{=}C & \\ CH_3 & & H \end{array}$ 2-methyl-1-butene

$\begin{array}{ccc} CH_3CH_2 & & CH_3 \\ & C{=}C & \\ H & & H \end{array}$ *cis*-2-pentene $\qquad \begin{array}{ccc} CH_3CH_2 & & H \\ & C{=}C & \\ H & & CH_3 \end{array}$ *trans*-2-pentene

15.7

$\begin{array}{ccc} CH_3 & & Cl \\ & C{=}C & \\ H & & H \end{array}$ *cis*-1-chloropropene $\qquad \begin{array}{ccc} CH_3 & & H \\ & C{=}C & \\ H & & Cl \end{array}$ *trans*-1-chloropropene

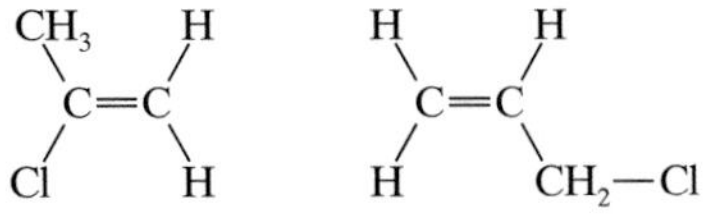

$\begin{array}{ccc} CH_3 & & H \\ & C{=}C & \\ Cl & & H \end{array}$ 2-chloropropene $\qquad \begin{array}{ccc} H & & H \\ & C{=}C & \\ H & & CH_2-Cl \end{array}$ 3-chloropropene

15.9 (b) and (c) exist as cis–trans isomers

15.11 cis, trans

15.13 (a) 2-methyl-1-propene
(b) 2,3-dimethyl-2-butene
(c) 2-methyl-2-butene
(d) *trans*-2,3-dichloro-2-pentene

15.15 (a) 1,2-dimethylcyclohexene
(b) 1,3-dimethylcyclohexene
(c) cyclodecene
(d) 6-ethyl-1-methylcyclohexene

15.17 (a) $\begin{array}{ccc} CH_3CH_2 & & CH_3 \\ & C{=}C & \\ H & & CH_3 \end{array}$ (b) $\begin{array}{ccc} CH_3CH_2CH_2CH_2 & & H \\ & C{=}C & \\ H & & H \end{array}$

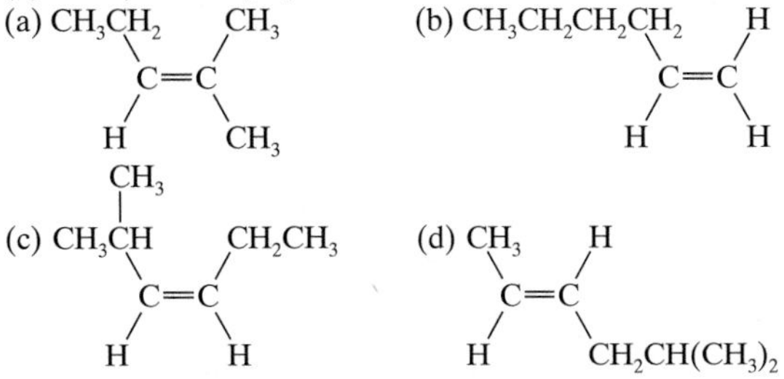

(c) $\begin{array}{ccc} CH_3CH(CH_3) & & CH_2CH_3 \\ & C{=}C & \\ H & & H \end{array}$ (d) $\begin{array}{ccc} CH_3 & & H \\ & C{=}C & \\ H & & CH_2CH(CH_3)_2 \end{array}$

15.19 (a) cyclohexene ring (b) cyclopentene ring with $-CH_3$ on the double-bond carbon (c) cyclohexene ring with Br on each double-bond carbon (d) cyclohexene ring with CH_3 and CH_3 on one carbon

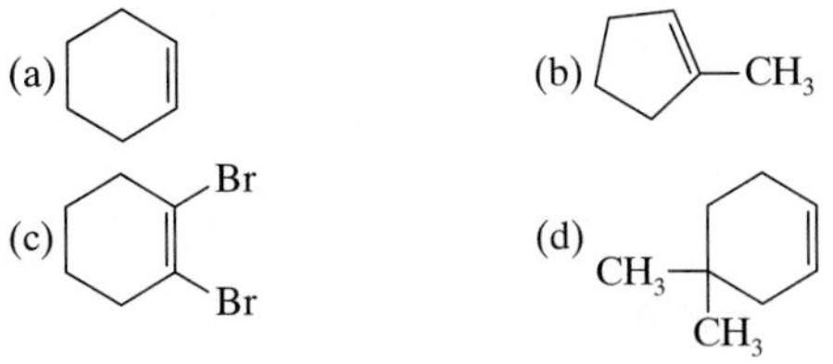

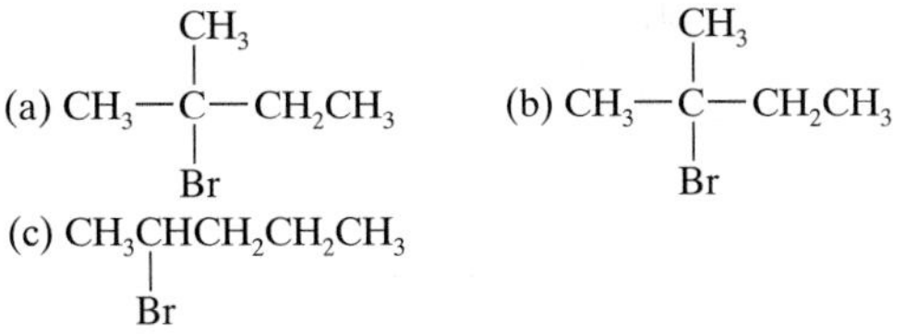

15.21 The color of the bromine is lost as it reacts with 1-hexene.

15.23 1835 $\quad BrCH_2-\underset{Br}{\overset{CH_3}{C}}-CH_2CH_3$

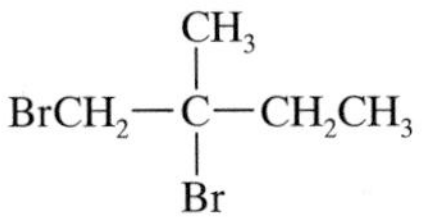

15.25 (a) $CH_3-\underset{Br}{\overset{CH_3}{C}}-CH_2CH_3$ (b) $CH_3-\underset{Br}{\overset{CH_3}{C}}-CH_2CH_3$
(c) $CH_3\underset{Br}{CH}CH_2CH_2CH_3$

15.27 $(-CH_2-\overset{CN}{CH}-)_n$

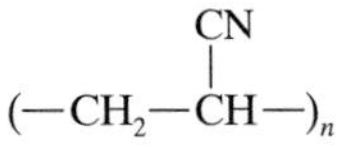

15.29 (a) 1-pentyne
(b) 2,2-dimethyl-3-hexyne
(c) 4-methyl-2-hexyne

15.31 (a) $CH_3C{\equiv}CCH_2CH_2CH_3$
(b) $HC{\equiv}CCH(CH_3)CH_2CH_3$
(c) $CH_3CH_2C{\equiv}C-\underset{CH_2-CH_3}{CH}CH_2CH_2CH_3$

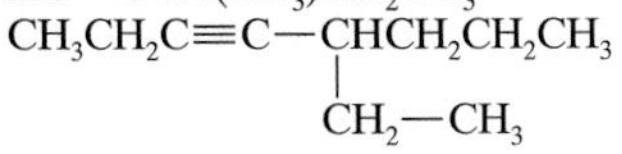

15.33 (a) $CH_3CH_2CH_3$ (b) $CH_3CBr_2CHBr_2$
(c) $CH_3CH_2CBr{=}CH_2$ (d) $CH_3CBr_2CH_2CH_3$

15.35 (a) 3 (b) 4 (c) 4

15.37 (a) para (b) ortho (c) meta

15.39 (a) ethylbenzene (b) isopropylbenzene
(c) *p*-diethylbenzene (d) 1,3,5-trimethylbenzene

15.41 4-chloro-3,5-dimethylphenol

15.43 (a) $CH_3CH_2CHCH_2CH_2CH_2CH_3$ with C_6H_5 on C-3 (b) $CH_3CH{-}C_6H_5$ with C_6H_5
(c) CH_3CHCH_2Cl with C_6H_5

15.45 para isomer

15.47 (a) $C_{18}H_{12}$ (b) $C_{18}H_{12}$

15.49 Br Br

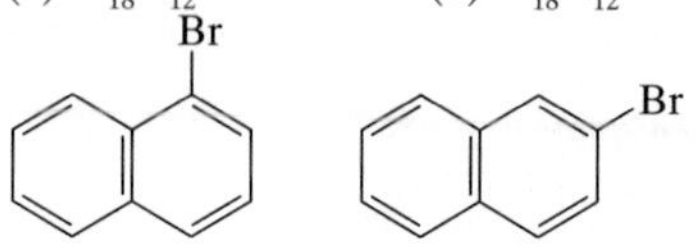

15.51 N N N N N N

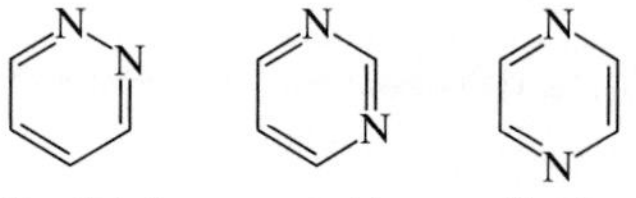

The third compound has no dipole moment.

15.53 thiophene

CHAPTER 16

16.1 (a) $(CH_3)_2CCH_2CH_2CH_3$ with OH (b) $CH_3CH_2CHCH_2OH$ with CH_3
(c) $(CH_3)_2CHCHCH_2OH$ with CH_3

16.3 (a) 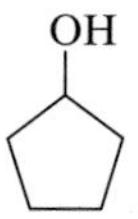 OH (b) 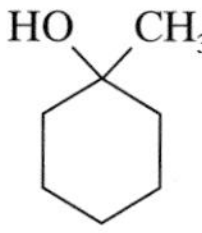 HO CH_3
(c) 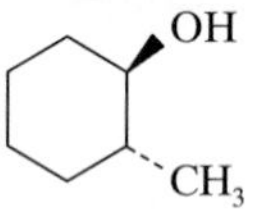 OH CH_3 (d) 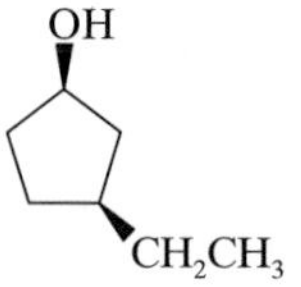 OH CH_2CH_3

16.5 (a) 3-ethyl-2-hexanol
(b) 4,7-dimethyl-5-decanol
(c) 4-methyl-3-hexanol

16.7 (a) primary (b) primary

16.9 Two hydroxyl groups per molecule that form hydrogen bonds increase the solubility.

16.11 more carbon atoms in 1-butanol; it has more hydrocarbon character

16.13 rate of reaction is I > III > II.

16.15 (a) 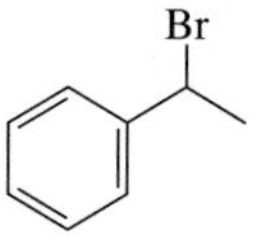 Br (b) 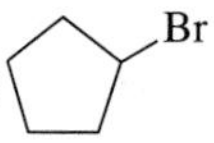 Br
(c) 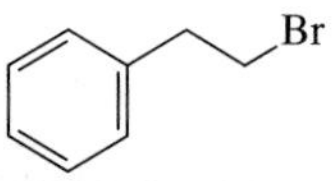 Br

16.17 (a) $CH_3CH{=}CH_2$
(b) $(CH_3)_2C{=}CH_2$
(c) $CH_3CH_2CH_2CH{=}CH_2$
(d) $CH_3CH_2CH{=}CH_2$ and
cis and trans $CH_3CH{=}CHCH_3$

16.19 $CH_3(CH_2)_6CO_2H$ from 1-octanol is acidic and soluble in base; $CH_3(CH_2)_5COCH_3$ from 2-octanol is a ketone.

16.21 (a) $CH_3(CH_2)_2C{\equiv}CCH_2\overset{O}{\overset{\|}{C}}CH_3$
(b) $CH_2{=}CH{-}CH{=}CH{-}CH_2CHO$
(c) 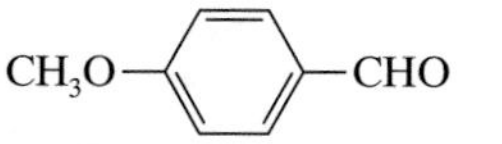 CH_3O–(benzene ring)–CHO
(d) 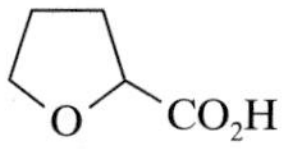 (ring) O CO_2H

16.23 CH_3CH_2, H, C=C, H, $CH_2CH_2CH_2CH_2CHO$

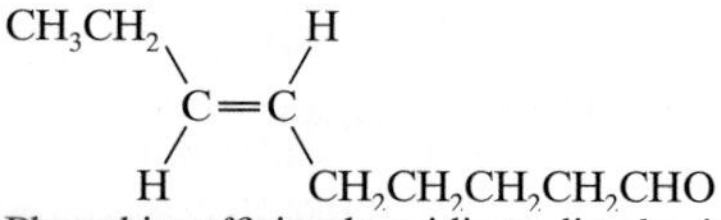

16.25 Phenol is sufficiently acidic to dissolve in basic solution.

16.27 4-chloro-2-phenylphenol

16.29 $CH_3OCH_2CH_2CH_3$; $CH_3OCH(CH_3)_2$; $CH_3CH_2OCH_2CH_3$

16.31 (a) dicyclopentyl ether
(b) phenyl propyl ether
(c) cyclopentyl propyl ether

16.33 (a) 2-methoxypentane (b) 2-methoxy-4-methylpentane
(c) 3-ethoxyhexane

16.35 $CH_3CH_2CH_2SH$ and $CH_3CH(SH)CH_3$

16.37 $CH_3CH_2CH_2S^-Na^+$ forms and is not volatile

16.39 (a) $CH_3CH_2CH_2SH$ (b) $(CH_3)_2CHCH(SH)CH_2CH_3$
(c) 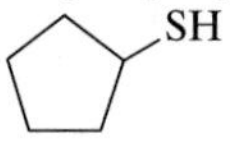 SH

CHAPTER 17

17.1 (a) $CH_3CH_2CH(CH_3)CHO$
(b) $(CH_3CH_2)_2CHCH_2CHO$
(c) $CH_3CH_2CH_2CHBrCHO$
(d) $CH_3CH_2CH_2CH_2CH(CH_3)CH(CH_3)CH_2CHO$

17.3 (a) butanal (b) 3,3-dimethylbutanal
(c) 2-methylpropanal (d) 2-ethyl-3-methylpentanal

17.5 (a) 4-chloro-2,3-dimethylheptanal
(b) 6-ethyl-3-methyl-2-nonanone
(c) 4-ethyl-2,5-dimethylheptanal
(d) 8-methyl-4-nonanone

17.7 (a) $CH_3{-}C(CH_3)_2{-}CHO$
(b) 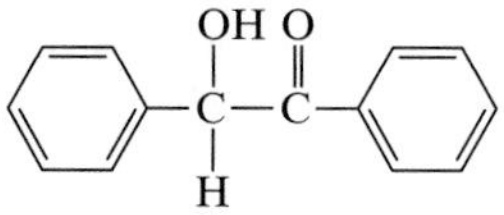 OH O, C–C, H
(c) $CH_2{=}CH{-}CHO$
(d) $CH_3{-}\overset{CH_3}{C}{=}CH{-}\overset{O}{\overset{\|}{C}}{-}CH_3$

17.9 There is no hydrogen atom bonded to an electronegative atom.

17.11 They do not have a hydroxyl group to hydrogen bond to water.

17.13 a red precipitate; a silver mirror

17.15 (a) 2-propanol (b) 3-pentanol (c) 1-hexanol

17.17 (a) 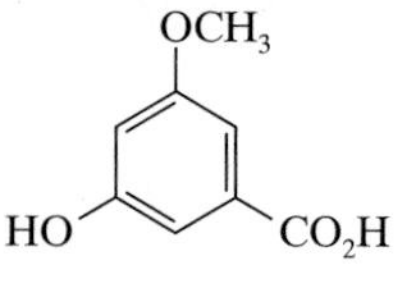 OCH_3, HO, CO_2H (b) 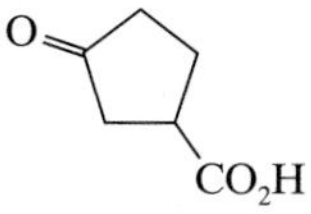 O, CO_2H
(c) 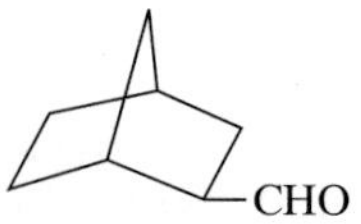 CHO

17.19 The carbonyl carbon atom has no groups attached that would block attack of the nucleophile.
17.21 (a) acetal (b) ketal (c) acetal (d) hemiacetal
17.23 (a) acetal (b) hemiacetal
(c) hemiketal (d) ketal
17.25 primary alcohol, secondary alcohol and a ketal
17.27 (a) $CH_3CH{=}NCH_3$ $(CH_3)_2C{=}NCH_2CH_3$
$(CH_3CH_2)_2C{=}N{-}OH$
17.29 There is only one geometric isomer possible for cyclohexanone but two for the cyclopentanecarbaldehyde derivative.
17.31 It has no α-carbon atom.
17.33 (a) CH_3, H / C=C / HO, H (b) H, OH / C=C / CH_3CH_2, H
(c) OH (d) H, OH / C=C / H, H
17.35 (a) CH_3, O (b) O, CH_3
17.37 $CH_3CH(CH_3)CH(OH)C(CH_3)(CH_3)CHO$

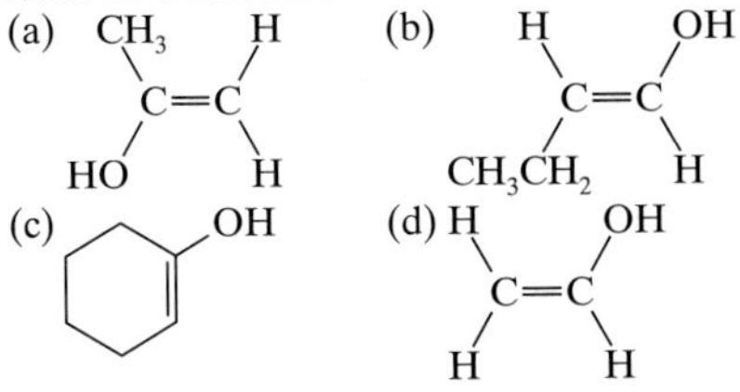

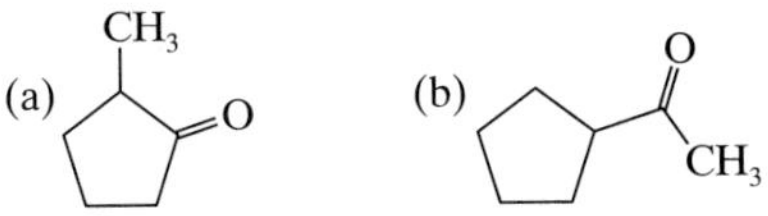

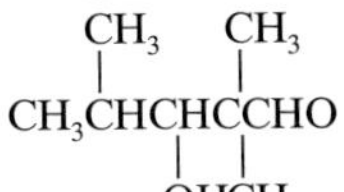

CHAPTER 18

18.1 (a) 3-methylpentanoic acid
(b) 2,4-dimethylpentanoic acid
(c) 3-methylhexanoic acid
(d) 3-ethyl-5,5-dimethylhexanoic acid
18.3 $CH_3{-}CH(CH_3){-}CH_2{-}C_6H_4{-}CH(CH_3){-}CO_2H$
18.5 fewer sites to form hydrogen bonds in the alcohol
18.7 longer chain has higher London forces
18.9 anion is more stable because of resonance
18.11 formic acid
18.13 The oxygen atom of the methoxy group is electronegative and inductively withdraws electron density from the carboxyl group.
18.15 The nitrogen atom of the heterocyclic ring is electronegative and inductively withdraws electron density from the carboxyl group.
18.17 $CH_3CH(OH)CO_2H + NaOH \longrightarrow CH_3CH(OH)CO_2Na + H_2O$
18.19 (a) sodium butyrate; sodium butanoate
(b) potassium stearate; potassium octadecanoate
(c) calcium propionate; calcium propanoate
18.21 (a) $CH_3CH_2CH_2CO_2Na + HCl \longrightarrow CH_3CH_2CH_2CO_2H + NaCl$
(b) $CH_3(CH_2)_{16}CO_2K + HCl \longrightarrow CH_3(CH_2)_{16}CO_2H + KCl$
(c) $(CH_3CH_2CO_2)_2Ca + 2\,HCl \longrightarrow 2\,CH_3CH_2CO_2H + CaCl_2$
18.23 $C_nH_{2n}O_2$
18.25 two; $CH_3CH_2CH_2CO_2H$ and $(CH_3)_2CHCO_2H$
18.27 (a) $CH_3CO_2CH_2(CH_2)_6CH_3$ (b) $HCO_2C(CH_3)_3$
(c) $CH_3CH_2CH_2CO_2CH_2CH_3$
(d) $CH_3(CH_2)_3CO_2CH_2CH_2CH_3$
18.29 (a) ethyl methanoate (b) methyl butanoate
(c) octyl ethanoate
18.31 (a) butyl butyrate (b) pentyl acetate
(c) octyl acetate
18.33 The carboxylate anion hydrogen bonds to water.
18.35 same molecular weight and same functional groups
18.37 (a) acetic acid and phenol
(b) ethanol and benzoic acid
(c) ethanol and formic acid
18.39 (a) CH_3CO_2Na and C_6H_5ONa
(b) CH_3CH_2OH and $C_6H_5CO_2Na$
(c) $CH_3CH_2OH + HCO_2Na$
18.41 (a) $CH_3CH_2CH_2CO_2CH_2CH_2CH_2CH_3$
(b) $CH_3CO_2CH_2CH_2CH_2CH_2CH_3$
(c) $CH_3CO_2CH_2(CH_2)_6CH_3$
18.43 (a) ethanol and formic acid
(b) 1-pentanol and acetic acid
(c) ethanol and acetic acid
18.45 mandelic acid $C_6H_5{-}CH(OH){-}CO_2H$
18.47 adipic acid, glutaric acid, succinic acid
18.49 glutaric acid and 1,4-butanediol
18.51 (a) and (c)
18.53 $CH_3CH_2CO_2CH_3 + CH_3CH_2SH$
18.55 $CH_3{-}O{-}PO_3H_2$; two
18.57 one, two; two
18.59 (a) methyl ethanoate (b) ethyl propanoate
18.61 $CH_3{-}C(=O){-}CH_2{-}C(=O){-}O{-}CH_3$

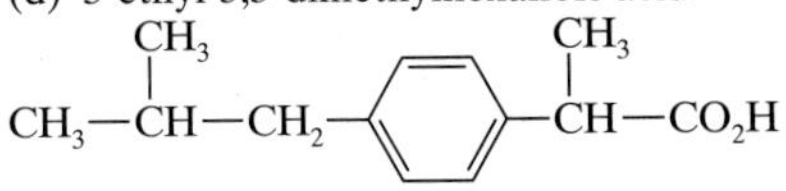

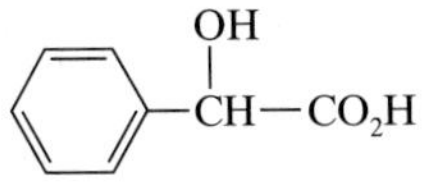

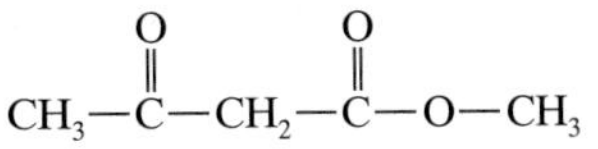

CHAPTER 19

19.1 from left to right in the structure; amide, ketone, ether, alcohol, amine
19.3 from left to right; amine, amide
19.5 tertiary
19.7 (a) and (c) are secondary; (b) is tertiary
19.9 (a) tertiary (b) secondary
(c) tertiary (d) secondary
19.11 (a) 3-hexanamine
(b) *N,N*-dimethyl-1-butanamine
(c) 4-methyl-2-pentanamine
19.13 (a) *N*-ethylcyclohexanamine
(b) cyclopentanamine
(c) *N,N*-dimethylaniline
(d) *N,N*-dipropyl-1-pentanamine
19.15 H, NH_2, H
19.17 (a) CH_2CH_3, N, H (b) Br, N

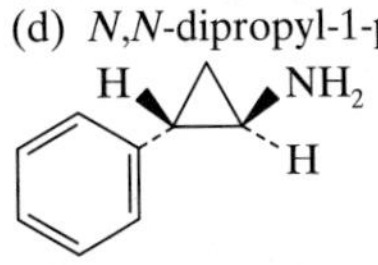

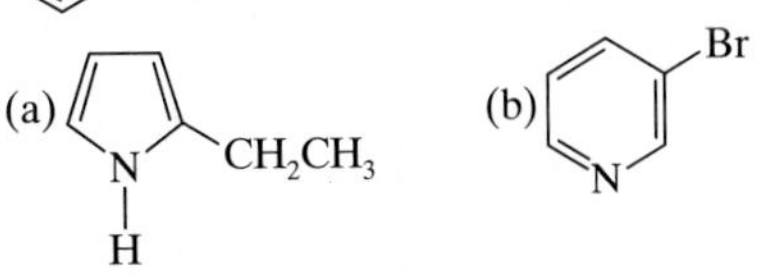

(c) 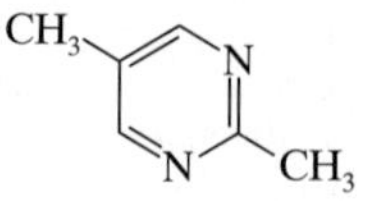 [pyrimidine ring with CH_3 at C-5 and CH_3 at C-2]

(d) 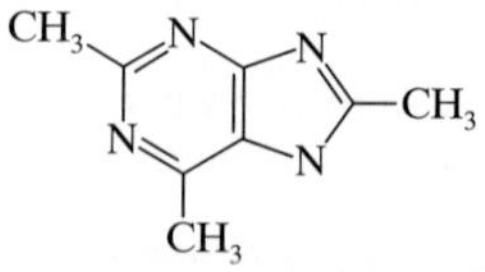 [purine ring with CH_3 groups at C-2, C-6 and C-8]

19.19 two; dimethylamine and ethylamine
19.21 four isomers; butylamine, *sec*-butylamine, isobutylamine, *tert*-butylamine
19.23 Trimethylamine cannot form hydrogen bonds.
19.25 The amines in fish are protonated to produce nonvolatile salts.
19.27 (a) and (d) should be in the 10^{-3} to 10^{-4} range;
(b) and (c) should be about 10^{-10}.
19.29 triethylamine
19.31 methylpropylamine, methylisopropylamine, or diethylamine
19.33 (a) and (c) give nitrogen gas;
(b) does not give a visible reaction.
19.35 (a) *N,N*-dimethylpropanamide
(b) *N*-ethyl-*N*-methylmethanamide
(c) *N*-methyl-3-methylbutanamide
(d) *N*-methylbenzamide
19.37 two electron pairs of the oxygen atom; hydrogen atom bonded to the nitrogen atom; electron pair on nitrogen atom
19.39 The ester cannot form hydrogen bonds, but the amide can.
19.41 (a) $C_6H_5CO_2^- + CH_3CH_2NH_2$
(b) $CH_3CH_2CO_2H + (CH_3)_2NH_2^+$
(c) $HO_2C—C_6H_4—CO_2H + 2\ NH_4^+$
19.43
(a) 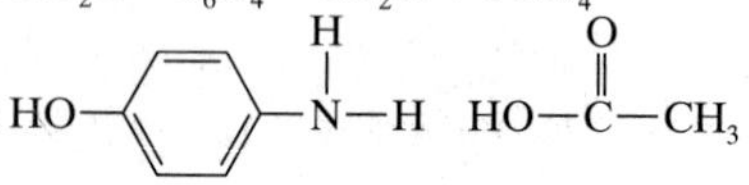 HO—C_6H_4—NH—H HO—C(=O)—CH_3
(b) 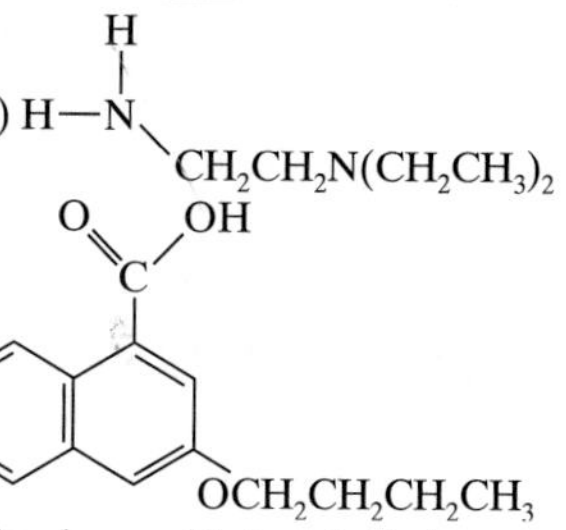 H—N(H)—$CH_2CH_2N(CH_2CH_3)_2$ [naphthalene]—C(=O)OH with $OCH_2CH_2CH_2CH_3$
19.45 Nylon has amide bonds that are more resistant to hydrolysis than ester bonds.

CHAPTER 20

20.1 carbon, hydrogen, and oxygen
20.3 sodium, potassium, calcium, and chloride ions.
20.5 positive
20.7 −10.3 kcal/mole
20.9 site of ATP formation
20.11 lipids and proteins
20.13 (a) and (c) have one chiral center each and are optically active
20.15 (a) has a plane of symmetry perpendicular to the plane of the ring and passing through the C-1 to C-2 bond.
20.17 (a) 1 (b) 2
20.19 6
20.21 They are enantiomers.
20.23 configuration is still D; sign of rotation cannot be predicted
20.25 left to right −19.6 and +14.8
20.27 [Fischer projections] CO_2H / H—C—Br / CH_3 and CO_2H / NH_2—C—H / CH_3
20.29 [Fischer projection] CO_2H / NH_2—C—H / CH_2OH
20.31 8
20.33 Ribitol has a plane of symmetry passing through the C-3 atom.
20.35 One of the compounds is meso because it has a plane of symmetry passing through the C-3 atom.
20.37 4
20.39 Biochemicals are chiral and provide the environment to favor formation of one enantiomer.
20.41 The receptors are chiral and interact differently with each enantiomer.

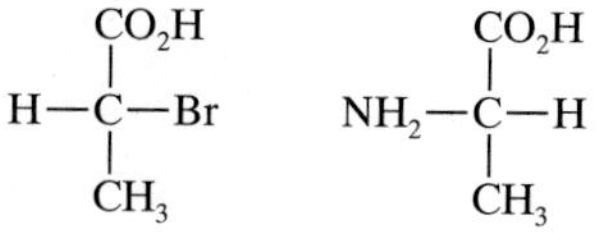

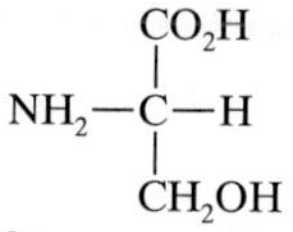

CHAPTER 21

21.1 aldoses contain aldehydes; ketoses contain ketones
21.3 (a) aldopentose (b) aldohexose (c) ketohexose
21.5 [Fischer projection] CHO / HO—C—H / HO—C—H / H—C—OH / H—C—OH / CH_2OH
21.7 (a) D (b) D (c) L
21.9 It hydrolyzes glucose 6-phosphate to glucose.
21.11 muscles and liver; liver
21.13 [Haworth projections (a)–(d) of pyranoses with $HOCH_2$, H, OH and HO substituents]
21.15 (a) α-D-galactose (b) β-D-mannose (c) α-D-ribose
21.17 (a), (b), and (c) of 21.15 and (c) of 21.16 are pyranoses.
21.19 yes
21.21 lactose and maltose
21.23 I and III; II is a ketal

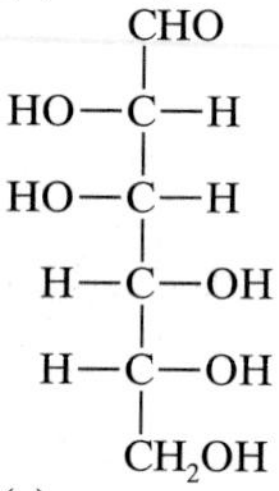

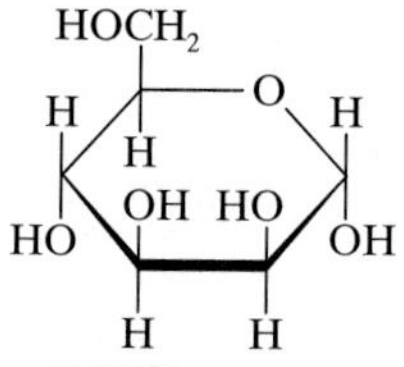

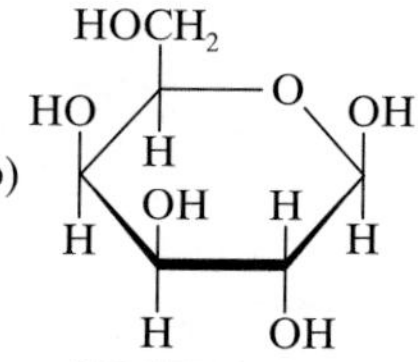

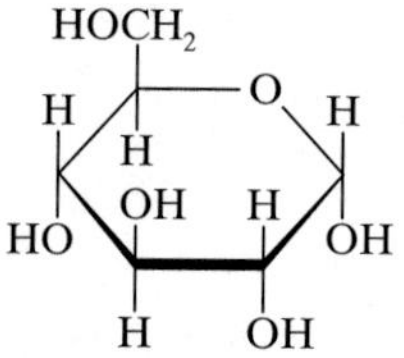

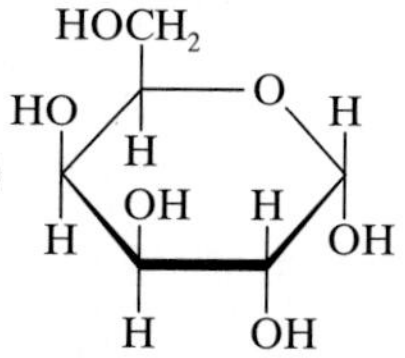

21.25

```
    HOCH2                              HOCH2
 H /|----O\  H                      H /|----O\  O—CH3
  / H      \/                        / H      \/
 \ OH   H  /                         \ OH   H  /
HO \______/ O—CH3                  HO \______/  H
    |     |                            |     |
    H     OH                           H     OH
```

21.27 glucose with β-(1 —> 4) link to glucose
21.29 glucose and galactose
21.31 α; 1,1
21.33 Both are α-linked polymers of glucose.
21.35 They do not have the enzyme to hydrolyze β-linked glucose.
21.37 D-mannose, D-fructose
21.39 arabinose and ribulose
21.41 It can form glucose via an enediol intermediate.
21.43 It can form ribose via an enediol intermediate.
21.45 No, it has only acetal centers.
21.47 It is an alditol related to xylose

CHAPTER 22

22.1 (a) and (d)
22.3 glycerol in phosphoglycerides; sphingosine in sphingophospholipids; phosphate and fatty acid in both
22.5 diterpene
22.7 alkene, two ketones; primary, secondary and tertiary alcohols
22.9 $CH_3(CH_2)_{18}CO_2CH_2(CH_2)_{30}CH_3$
22.11 It is a 16-carbon unsaturated acid with one double bond.
22.13 $CH_3(CH_2)_7C{\equiv}C(CH_2)_7CO_2H$
22.15 Linoleic has more double bonds; it has lower London forces.
22.17

```
RCO2—CH2                                 CH2—OH
     |                                   |
RCO2—CH  + 3 OH⁻ ⟶ 3 RCO2⁻ +  CH—OH
     |                                   |
RCO2—CH2                                 CH2—OH
```

22.19 an oil
22.21 linoleic, oleic, and linolenic acids
22.23 The center carbon atom is not chiral.

```
CH3(CH2)16CO2—CH2
              |
CH3(CH2)14CO2—CH
              |
CH3(CH2)16CO2—CH2
```

22.25 Only (b) can occur in an ester.
22.27 glycerol, fatty acids, and phosphoric acid
22.29 stearic acid, oleic acid, glycerol, phosphoric acid, choline
22.31 Phosphatidyl choline and phosphatidyl ethanolamine both have a positive and a negative charge; phosphatidyl serine has two negative charges and a positive charge phosphatidyl inositol has a negative charge.
22.33 The former contains sphingosine; the latter contains glycerol.
22.35 the unsaturated carbon chain present in sphingosine
22.37 Both contain sphingosine. The former contains a carbohydrate, whereas the latter contains phosphate bonded to various alcohols.
22.39 a glycosidic bond
22.41 Saturated acids make the membrane more rigid.
22.43 London forces
22.45 It is on the membrane surface.
22.47 Active transport requires energy and occurs against concentration differences. Facilitated diffusion requires no energy and is concentration dependent.

CHAPTER 23

23.1 the mirror image of L-glutamic acid places the amino group on the right.
23.3 lysine
23.5 a neutral amino acid; 4-aminobutanoic acid
23.7 (a) glycine (b) alanine (c) phenylalanine (d) proline
23.9 It can be synthesized only if phenylalanine is available.
23.11 decrease in the order fish> meat> cereal grain
23.13 balanced diet of vegetables for complete protein
23.15

```
CH3—CH—CO2H     CH3—CH—CO2⁻
    |               |
    NH3⁺            NH2
```

23.17

```
HOCH2—CH—CO2⁻
      |
      NH3⁺
```

23.19 The zwitterion exists at the isoionic point. At higher pH the carboxylate salt predominates.
23.21 (a) 7 (b) less than 7 (c) greater than 7
23.23 vasopressin, because of arginine
23.25 It has a larger number of basic amino acids than acidic amino acids.
23.27

```
            O
            ‖
NH2—CH—C—NH—CH—CO2H     Ala-Ser
    |           |
    CH3         CH2—OH
```

23.29 acidic amino acids
23.31 glycine, cysteine, valine; glycylcysteylvaline
23.33 (a) and (b)
23.35 (a) and (b)
23.37 It does not have an N—H bond for hydrogen bonding.
23.39 loss of the native conformation of a protein
23.41 The secondary structure is disrupted by heat.
23.43 protonation of carboxylate group; loss of proton from ammonium group
23.45 Lys-Gly and Tyr-Leu are formed.
23.47 Only (a) will react, giving Arg and Gly-Tyr.
23.49 (a) Arg-Gly and Tyr (b) no reaction (c) Phe, and Trp-Ser (d) Ser and Phe-Asp
23.51 Trp-Cys-Ala-Lys-Leu-Gly + Phe-Pro-Cys
23.53 Gly

CHAPTER 24

24.1 Both accelerate chemical reactions, but enzymes are specific.
24.3 isomerase
24.5 phosphate esters
24.7 isomerase
24.9 isomerase
24.11 lock-and-key model and induced-fit hypothesis
24.13 Enzymes are chiral.
24.15 NAD^+ and FAD
24.17 Coenzymes react with the substrate and are used up.
24.19 At high pH histidine is not protonated; the aspartic acid side chain is in the form of the carboxylate ion.
24.21 The enzyme is completely converted to an enzyme–substrate complex.

24.23 strong binding of substrate
24.25 isoionic point of histidine is near 7; protons transferred easily at neutral pH
24.27 low K_m
24.29 It resembles succinic acid and can occupy the active site.
24.31 The ion reacts with sulfhydryl groups.
24.33 irreversible because of covalent bond formation
24.35 noncompetitive
24.37 It is an end product in a reaction series and inhibits an earlier step
24.39 The inhibitor prevents the conversion of the zymogen into an active enzyme, which would destroy the pancreas.
24.41 The active enzyme catalyzes the hydrolysis of the proteins of the pancreas.

CHAPTER 25

25.1 It is completely oxidized.
25.3 Its oxidation state is decreased.
25.5 oxidation
25.7 adenine, ribose, and three phosphate units
25.9 −3
25.11 $ATP + H_2O \longrightarrow P_i + ADP$
25.13 No, one reaction must be exergonic in order to provide the energy for endergonic reactions.
25.15 curved arrow above or below the reaction arrow
25.17 Higher—its free energy of hydrolysis (−14.8 kcal/mole) is larger than that of ATP (−7.3 kcal/mole).
25.19 Yes, the ΔG° would be $(-14.8) - (-7.3) = -7.5$ kcal/mole.
25.21 1.7 kcal/mole
25.23 Oxidation of NADH and reduction of FMN occurs in the first step.
25.25 3 moles of ATP
25.27 FMN and two cytochromes
25.29 No, the protons are released.
25.31 $FADH_2$ reduces ubiquinol
25.33 acetyl group
25.35 steps 3, 4, 6, and 8
25.37 12 moles
25.39 ATP inhibits two steps in the citric acid cycle; ADP stimulates one step.
25.41 succinate dehydrogenase
25.43 The liver must do a lot of biochemical work and requires energy.
25.45 A carrier, ATP–ADP translocase, is located in the membrane.

CHAPTER 26

26.1 in the mouth
26.3 Fructose and galactose must be converted into glucose.
26.5 Glycogenolysis is the hydrolysis of glycogen; glycogenesis is the formation of glycogen from glucose.
26.7 the brain
26.9 Under anaerobic conditions pyruvic acid is reduced to lactic acid; under aerobic conditions it is converted to acetyl coenzyme A.
26.11 2
26.13 pyruvic acid
26.15 in the electron transport chain
26.17 2 moles of each
26.19 phosphoglucose isomerase
26.21 reduction by NADH to lactate
26.23 Lactic acid formed in muscles is oxidized in the liver to pyruvic acid and then to glucose.
26.25 Two molecules of acetyl coenzyme A; 1 molecule of acetyl coenzyme A is converted into 2 molecules of carbon dioxide.
26.27 NADH reducing power must pass across the mitochondrial membrane for oxidation in the electron transport chain
26.29 lactic acid, some amino acids, and glycerol
26.31 three steps
26.33 NADPH has an additional phosphate group.
26.35 2 moles of NADPH
26.37 It is converted into glycolysis intermediates.
26.39 hydrolysis of the glucose 6-phosphate
26.41 hydrolysis of pyrophosphate
26.43 the phosphofructokinase-catalyzed step

CHAPTER 27

27.1 connective tissue that contains cellular fat
27.3 in the intestines
27.5 They are reassembled into triglycerides.
27.7 tissue lipid and depot fat
27.9 Acetyl coenzyme A formed in the aerobic reaction of pyruvic acid is converted into fatty acids.
27.11 Each turn of the cycle results in formation of a compound of two fewer carbon atoms.
27.13 trans isomer
27.15 9 moles
27.17 8 times
27.19 5 moles
27.21 Oxidation occurs at the beta position of the acid.
27.23 oxidation in the electron transport chain
27.25 glycerol of fat
27.27 phenylacetic acid
27.29 acetone, 3-hydroxybutyrate, acetoacetate
27.31 a condition of ketone bodies in the blood
27.33 presence of acetone in the breath at higher concentrations
27.35 Blood pH is affected, and oxygen transport is hindered.
27.37 acyl carrier proteins
27.39 in the cytoplasm of liver cells
27.41 51 kcal/mole
27.43 7 moles
27.45 7 moles
27.47 The NAD^+ level must be high.

CHAPTER 28

28.1 a reserve of amino acids in a state of equilibrium
28.3 75%
28.5 very rapid for enzymes; very slow for muscle protein
28.7 The nitrogen balance is negative.
28.9 pepsin
28.11 convert amino acids to α-ketoacids
28.13 leucine
28.15 conversion of an amino group in an amino acid into a carbonyl group
28.17 NAD^+
28.19 ornithine and citrulline
28.21 a condition of elevated NH_4^+ in the blood.
28.23 ornithine
28.25 steps 1 and 2
28.27 pyruvate, α-ketoglutarate, succinyl CoA, and oxaloacetate
28.29 leucine

28.31 $$^{+}H_3N-\underset{\underset{\displaystyle CH_3}{|}}{\overset{\overset{\displaystyle CO_2^-}{|}}{C}}{=}CH \longrightarrow {}^{+}H_2N{=}\overset{CO_2^-}{C}-CH_2-CH_3 \longrightarrow O{=}\overset{CO_2^-}{C}-CH_2-CH_3$$

28.33 putrescine; $NH_2(CH_2)_4NH_2$
28.35 transamination reaction produces the nonessential amino acids
28.37 phenylalanine

CHAPTER 29

29.1 All contain heterocyclic bases, deoxyribose, and phosphate esters.
29.3 phosphodiesters
29.5 Uracil, cytosine, and thymine are pyrimidines.
29.7 NH_2; N N N N (adenine); $HO-\overset{O}{\overset{\|}{P}}(OH)-O-CH_2$; O; H H H H; OH OH

29.9 beta
29.11 none
29.13 four
29.15 ribose, uracil
29.17 None of the pairs exist in DNA. Too much space is required for G-A; G-T doesn't have proper number of hydrogen bonds; uracil is not found in DNA
29.19 deoxyribose and phosphodiester linkages
29.21 DNA unravels to form a replication fork, and new DNA strands are formed.
29.23 -dC-G-T-A-G-T-dC
29.25 C-G-U-A-G-U-C
29.27 U is the complement in mRNA for A in mRNA.
29.29 GGGACAUGUGGA; Gly-Thr-Cys-Gly
29.31 Codons base-pair with anticodons of tRNA.
29.33 (a) Val (b) Pro (c) Phe (d) Thr
29.35 protein is transferred from donor site to amino acid at acceptor site
29.37 (a) CAA (b) GGG (c) AAA (d) UGC
29.39 tyrosine and phenylalalanine
29.41 GUU GAU GCU GGU is one possible sequence.
29.43 Very defective proteins are formed.
29.45 substitution to give termination codons, and substitution for amino acids at the active site
29.47 snip open plasmids and DNA molecules
29.49 DNA is removed from human genes.
29.51 Membrane-dissolving enzymes are specific
29.53 RNA replicase

CHAPTER 30

30.1 intracellular and extracellular
30.3 blood
30.5 bicarbonate, phosphate, protein
30.7 hemoglobin
30.9 Na^+; K^+
30.11 Cl^-
30.13 saliva, gastric juice, intestinal juice, pancreatic juice
30.15 Many of the necessary enzymes are produced by the pancreas.
30.17 It hydrolyzes trypsinogen to trypsin, which in turn activates other zymogens.
30.19 Whole blood is plasma and blood cells.
30.21 K^+ is released from the damaged cell.
30.23 the percent by volume of red blood cells in whole blood
30.25 T cells and B cells; B cells are associated with antibodies
30.27 four
30.29 to allow a larger amount of oxygen to be transported to tissue that is rapidly growing
30.31 decreased
30.33 A noncovalent interaction between both components occurs.
30.35 increased fluid in the area of infection
30.37 adaptive immune system
30.39 stop blood flow and decrease possibility of infection
30.41 thromboplastin and Ca^{2+}
30.43 to avoid inadvertent clotting
30.45 reversible
30.47 the two rings in both compounds structurally resemble those in vitamin K
30.49 about 20%
30.51 protein and cells
30.53 6.0

Index

Absolute configuration, 501
Absolute specificity, 585
Absolute zero, 35
Acceptor site, 702
Accuracy, 24
Acetal , 418
    in carbohydrates, 521
    cyclic, 419
Acetaldehyde, 408
Acetaminophen, 369
Acetate ion, 266
    hydrolysis of, 274
Acetic acid, 258, 433
    acid ionization constant, 268
    ionization, 267
    reaction with hydroxide, 267
    reaction with water, 266
    structure, 258
Acetone, 408
Acetone breath, 658
Acetophenone, 371, 408
Acetyl coenzyme A
    and citric acid cycle, 619
    energy of hydrolysis, 614
    in fatty acid cycle, 653
    in fatty acid biosynthesis, 659
    formation of ketone bodies, 658
Acetyl group, 430
    in citric acid cycle, 619
    in thioesters, 449
Acetylcholine, 449
    and muscle relaxants, 470
Acetylcholinesterase, 470
Acetylene, 313, 355
    bonding in, 110
    molecular model, 354
    reaction with hydrogen, 367
Acetylsalicylic acid, 443
Achiral, 493
Acid, 127, 255
    Brønsted-Lowry concept, 257
    carboxylic, 429
    conjugate, 262
    diprotic, 259
    equivalents, 280
    equivalent weight of, 280
    fatty, 541
    ionization constant, 268
    Lewis, 325
    monoprotic, 258
    percent ionization, 268
    properties of, 255
    strength, 266
    strong, 257
    triprotic, 259
    weak, 258
Acid anhydride, 432
    reactivity, 450
Acid-base reaction, 260
    of alcohols, 389
    of amino acids, 563
    in organic chemistry, 325
Acid chloride, 432
    reactivity, 448
Acid ionization constant, 268
    inorganic acids, 268
    carboxylic acids, 438
Acid rain, 256
Acidic amino acid, 561
Acidic solution, 264
    pH, 270
Acidic oxides, 127
Acidic salt, 274
Acidity
    of alcohols, 389
    of amino acids, 564
    of ammonium salts, 466
    of carboxylic acids, 437
    of α-hydrogen atoms, 422
    of phenols, 390
    of thiols, 400
Acidosis, 278
Aconitase, 393
ACP (see acyl carrier protein)
Acrilan, 366
Actinide, 54
Activated complex, 241
Activation energy, 240
Active site, 590
Active transport, 553
Acyclic, 333
Acyl carrier protein, 659
Acyl carnitine, 652
Acyl group, 430
Acyl coenzyme A, 651
Adaptive immune system, 729
Addition polymerization, 365
Addition reaction, 328
    alkenes, 362
    alkynes, 368
    carbonyl compounds, 416
    mechanism, alkenes, 364
    mechanism, carbonyl compounds, 416
    of nitrogen compounds, 420
    in vision, 421
Adenine, 686
    in DNA, 690
Adenosine diphosphate, 610
Adenosine monophosphate, 610
Adenosine triphosphate
    activation of amino acids, 701
    activation of fatty acids, 651
    active transport, 553
    biochemical reactions, 487
    energy of hydrolysis, 611
    from fatty acids, 655
    in gluconeogenesis, 639
    from glucose metabolism, 638
    hydrolysis, 610
    phosphorylation of glycerol, 651
    protein synthesis, 700
    structure, 451
    in urea cycle, 676
    yield from glucose, 638
Adipic acid, 431
    in nylon, 475
Adipose tissue, 648
ADP (see adenosine diphosphate)
Adrenaline, 465
Aerobic, 618
Aglycone, 521
Air, composition of, 182
Alanine, 562
    isoionic point, 566
    pK_a, 564
Alanine aminotransaminase, 587
Alcohol, 317, 381
    acid-base reactions, 389
    addition to carbonyl compounds, 417
    boiling points, 388
    classification, 385
    dehydration, 392
    density, 41
    formation of esters, 441
    hydrogen bonding in, 388
    nomenclature, 385
    oxidation of, 394
    proof, 206
    reaction with sodium, 382
    solubility, 389
    structure, 382
    substitution reaction, 390
Alcohol dehydrogenase, 396
    quaternary structure, 575
Aldehyde, 317, 407
    addition of alcohols, 417
    addition of nitrogen compounds, 420
    aldol condensation, 423
    boiling points, 411
    formation of acetals, 418
    nomenclature, 408
    oxidation, 412
    physical properties, 410
    reduction, 414
    solubility, 411
Alditol, 524
Aldohexose 513
Aldol, 423
Aldol condensation, 423
Aldopentose, 513
Aldose, 512
Aldosterone, 539
    in blood, 722
Aldotetrose, 513
Alkali metal, 54
Alkalosis, 278
Alkaline earth metal, 54
Alkane, 333
    boiling point, 345
    branched, 335
    chlororination, 347
    classification of, 336
    effect on skin, 344
    mechanism of chlorination, 347
    nomenclature of, 337
    normal, 334
    octane number, 346
    oxidation, 345
    physical properties, 344
    solubility, 344
Alka-Seltzer, 262
Alkene, 317, 353
    addition reactions, 362
    boiling points, 356
    geometric isomers, 356
    hydrogenation, 361
    nomenclature, 359
    oxidation, 360
    polymerization, 365
    properties, 355
    reduction, 360
Alkoxide, 389
Alkoxy group, 398
Alkyl group, 335, 337
Alkyne, 317, 353
    boiling points, 356
    nomenclature, 366
    properties, 355
    reactions, 367
Allose, 514
Alloy, 203
Allosteric effect, 576
Allosteric regulation, 600
Altrose, 514
Aluminum
    chemical symbol, 16
    electron configuration, 63
    electronegativity, 71
    ionization energy, 70
    radius, 69
    in solid propellant, 124
    specific heat, 41
Alpha particle, 289
Amalgam, 204
Amide, 318, 431
    bonding in, 472
    classification, 472
    formation, 474
    hydrogen bonding, 473
    hydrolysis, 476
    nomenclature, 473
    properties, 473
    in proteins, 458
    resonance, 473
Amine, 318
    basicity, 466
    boiling points, 463
    classification, 459
    heterocyclic, 467
    hydrogen bonding, 464
    nomenclature, 461
    physical properties, 463
    reactions with nitrous acid, 470
    solubility, 464
    solubility in acid, 468
    structure, 459
Amino group, 460
Amino acid, 561
    acid-base properties, 563
    acidic, 561
    activation of, 701
    basic, 561
    biosynthesis, 680
    classification, 561
    codons for, 699
    C-terminal, 567
    decarboxylation, 679
    electrophoresis, 566
    essential, 561, 666
    glucogenic, 677

isoionic point, 566
ketogenic, 678
neutral, 561
N-terminal, 567
nitrogen balance, 668
oxidative deamination, 672
pK_a values, 564
structures of, 562
transamination, 670
zwitterion form, 563
Amino acid pool, 666
Aminopeptidase, 579
Ammonia
in aqueous solution, 260
basicity, 270
boiling point, 12, 188, 463
formation of, 251
heat of vaporization, 189
hydrogen bonding in, 194
K_b, 466
melting point, 12
polarity, 107
production in U.S.A., 153
pK_b, 466
reaction with acid, 260
reaction with water, 260
shape, 104
structure, 8
specific heat, 41
VSEPR theory, 103
Ammonium ion, 86
hydrolysis of, 274
in urea cycle, 674
Ammonium nitrate
in cold packs, 214
decomposition, 128
Amphetamine, 465
Amphoteric, 263
amu (see atomic mass unit)
β-Amylase, 593
Amylopectin, 528
Amylose, 528
Anabolic reaction, 487, 606
Anaerobic exercise, 634
Androgen, 540
Anemia, 723
Angular molecule, 103
Anhydride
of carboxylic acid, 432
of phosphoric acids, 451
Aniline, 371
boiling point, 463
K_b, 466
Anion, 78
list of, 86
Anisole, 371
Anode, 209
Anomer, 519
Anomeric carbon atom, 519
Antacid, 262
Anthracene, 375
Antibodies, 559, 730
Anticodon, 698
Antigens, 729
Antihistamine, 679, 729
Antimony
electronegativity, 71
elemental symbol, 16
ionization energy, 70
radius, 69
Aplastic anemia, 723
Apoenzyme, 590
Aqueous solution, 118, 203
Arabinose, 514
Arachidic acid, 542
Arachidonic acid, 354, 542
Arene, 369
Arginine, 562
isoionic point, 566
pK_a, 564
in urea cycle, 675
Arginosuccinate, 675
Argon
boiling point, 189
electron configuration, 63
heat of vaporization, 189
ionization energy, 70
radius of, 69
specific heat, 41
Aromatic hydrocarbon, 353, 369
carcinogenic, 374
polycyclic, 374
nomenclature, 371
Aromaticity, 369
Arrhenius, Svante, 209
Arsenic
electronegativity, 71
elemental symbol, 16
ionization energy, 70
radius, 69
Arteriosclerosis, 538
Aryl group, 373
Asparagine, 562
Aspartase, 585
Aspartate aminotransferase, 670
Aspartic acid, 562
from fumaric acid, 585
isoionic point, 566
pK_a, 564
in urea cycle, 675
Aspirin, 369
-ate, 87
Atmosphere, 173
partial pressures, 182
Atmospheric pressure, 173
Atom, 6, 46
Bohr model of, 56
Dalton's theory of, 46
sizes of, 47
subatomic particles of, 48
Atomic bomb, 297
Atomic dating, 294
Atomic mass unit, 47
Atomic number, 49
and periodicity, 53
Atomic orbital, 57
Atomic radii, 48
periodic trends, 69
Atomic theory
Bohr's, 56
Dalton's, 46
Atomic weight, 51
and periodic table, 53
ATP (see adenosine triphosphate)
Atto-, 30
Avogadro, Amadeo, 149
Avogadro's hypothesis, 180
Avogadro's number, 149
AZT (azidothymidine), 687
Balancing
equations, 119
nuclear equations, 291
Barium
electronegativity, 71
ionization energy, 70
radius, 69
Barium ion, 85
Barometer, 172
Basal metabolic rate, 41
Base, 255
Brønsted-Lowry, 257
conjugate, 262
Lewis, 325
properties of, 255
strong, 260
Base ionization constant, 269
Base pairing, 692
Basic amino acid, 561
Basic oxide, 127
reaction with acid, 132
reaction with water, 127
Basic salt, 274
Basic solution, 259, 265
and pH, 270
Basophils, 724
Beeswax, 541
Bends, 217
Benedict's solution
effect on aldehydes, 412
effect on carbohydrates, 522
test for diabetes, 413
Bent molecule, 103
Benzaldehyde, 371, 408
Benzalkonium chloride, 469
Benzene, 316
delocalization, 370
Benzethonium chloride, 470
Benzoic acid, 371
Benzophenone, 408
Benzpyrene, 374
Benzyl, 373
Beryllium
electron configuration, 61
electron dot symbol, 65
electronegativity, 71
Hund's rule, 64
ionization energy, 70
radius, 69
Beta particle, 289
Bicarbonate buffer, 278, 715
Bicarbonate ion, 86
in antacid, 262
in blood, 212, 278, 722
in body fluids, 715
in fatty acid synthesis, 659
K_a, 270
in seawater, 203
Bile, 721
Biochemical reactions, 487
free energy, 605
Biochemistry, 482
Biological value, 580
Biosynthesis,
of amino acids, 680
of fatty acids, 658
Bisulfate ion, 86
ionization, 259
Bisulfite ion, 86
ionization, 259
K_b, 270
Blood, 721
alcohol, 383
buffer, 278
gases, 725
glucose in, 413
ketone bodies, 658
lipoproteins, 560
pH, 278
plasma, 714
properties, 722
solubility of gases, 217
Blood clotting, 730
BMR (see basic metabolic rate)
Body fluids, 714
electrolytes, 717
pH, 715
solutes, 716
Bohr, Niels, 56
Bohr model of the atom, 56
Boiling point, 11, 187
alcohols, 388
aldehydes, 410
alkanes, 345
alkenes, 356
alkynes, 356
amines, 463
carboxylic acids, 436
effect of altitude, 188
normal, 188
solutions, 220
Bombykol, 492
Bond, 7
coordinate covalent, 99
covalent, 93
double, 95
hydrogen, 191
ionic, 81
length, 94
pi, 109
polar covalent, 98
in proteins, 570
sigma, 108
triple, 95
types of, 78
Bond energy, 95
Bond length, 94
Bond line structures, 112
Bonding electrons, 94
Boron
electron configuration, 62
electron dot symbol, 65
electronegativity, 71
Hund's rule, 64
ionization energy, 70
radius, 69
Boyle's law, 174
Boyle, Robert, 174
Branched alkane, 335
Breathing, 176
Bridged ring compounds, 340
Bromide ion, 86
radius, 80
in seawater, 203
Bromine
addition to alkene, 363
addition to alkyne, 369
density, 38
electronegativity, 71
chemical symbol, 16
in human body, 483
ionization energy, 70
radius, 69
specific heat, 41
Brønsted-Lowry concept, 257
Buffer, 276
in blood, 278
in body fluids, 715
phosphate, 278
Buret, 33, 280
Butane
boiling point, 345
octane number, 346
structural formula, 314
Butyl group, 339
sec-Butyl group, 339
t-Butyl group, 339
Butyric acid, 433
Cadaverine, 464
Cadmium ion, 85
in phosphate rock, 88
radius, 81
Cal (kcal), 41
cal (calorie), 40
Calcium
and bones, 84
in diet, 40
electron configuration, 63
electronegativity, 71
chemical symbol, 16
ionization energy, 70
radius, 69
specific heat, 41
Calcium carbonate
in antacids, 262

Calcium carbonate *(cont.)*
decomposition, 127
reaction with acid, 261
use on soil, 276
Calcium chloride,
in hot packs, 214
Calcium ion, 85
in blood, 722
in blood clotting, 731
in body fluids, 717
in hard water, 447
in plasma, 212
radius, 80
in seawater, 203
Calcium oxide, 153
Calorie, 40
Capric acid, 433
Caproic acid, 433
Caprylic acid, 433
Captive zero, 26
Carbamoyl phosphate, 675
Carbanion, 324
Carbocation, 324
in addition to alkenes, 364
stability, 365
Carbocyclic, 333
Carbohydrate, 484, 510
blood groups, 530
classification, 511
digestion, 626
in E. coli, 484
glycosides, 521
in glycosphingolipids, 549
in membrane, 552
mutarotation, 520
oxidation, 522
Carbon
bonding, 95, 312
electron configuration, 62
electron dot symbol, 65
electronegativity, 70
chemical symbol, 16
in human body, 483
Hund's rule, 64
ionization energy, 70
isotopes, 51
and organic compounds, 95, 311
radius, 69
reaction with oxygen, 126
Carbon-14 dating, 294
Carbon dioxide
and carbonic acid, 129
dipole moment, 107
shape, 104
solubility in water, 217
Carbon monoxide
boiling point, 12
density, 38
and hemoglobin, 100
melting point, 12
reaction with hydrogen, 240
solubility in water, 217
structure, 237
Carbon tetrachloride, 347
Lewis structure, 97
from methane, 347
polarity, 107
Carbonate buffer, 278
Carbonate ion, 86
K_b, 270
reaction with acid, 261
Carbonic acid, 129
ionization, 259
Carbonic anhydrase, 585
turnover number, 593
K_m, 594
Carbonyl carbon atom, 317, 405
Carbonyl oxygen atom, 317, 405
Carbonyl group, 317, 405
addition reactions, 416
hydrogen bonding in water, 411
keto-enol forms, 422
oxidation reactions, 412
reduction reactions, 414
Carbonyl oxygen atom, 317, 405
Carboxylate, 437
Carboxyl group, 429
Carboxylic acid, 318
acidity, 437
boiling points, 436
Claisen condensation, 452
decarboxylation, 440
formation of esters, 442
hydrogen bonding, 436
nomenclature, 433
physical properties, 435
reaction with base, 439
salts, 439
solubility, 437
Carboxypeptidase, 579
Carnitine,
β-Carotene, 355, 538
Carvone, 504
Casein,
isoionic point, 566
Cathode, 209
Catabolic reaction, 487, 605
of amino acids, 669
and free energy, 605
of trigylcerides, 651
Catabolism, 487
Cation, 70, 78
names of, 85
Catalyst, 119
biological, 584
in living systems, 239
and reaction path, 243
specific, 240, 585
Cell
membrane, 489
organization, 490
and osmosis, 224
Cellobiose, 525
Cellular respiration, 617
Cellulose, 528
Celsius scale, 34
Centi-, 30
Central dogma, 695
Cephalin, 548
Ceramide, 549
Cerebrosides, 549
Cetyltrimethylammonium chloride, 469
Chain reaction
in chlorination, 347
nuclear, 297
Chargaff, Erwin, 689
Charge
of electron, 48
of ions, 70, 79
of proton, 49
Charles's, J.A.C., 176
Charles' law, 177
Chemical bond
coordinate covalent, 99
covalent, 93
ionic, 81
polar covalent, 98
Chemical change, 13
Chemical energy, 18
and biochemistry, 487
Chemical equation, 9, 117
balancing, 119
net ionic, 131
nuclear, 288
symbols used in, 118
Chemical equilibrium, 244
effect of concentration, 248
effect of pressure, 248
effect of temperature, 250
Chemical family, 54
Chemical formula, 7
ionic compounds, 87
molecules, 95
Chemical properties, 11
and periodic table, 67
Chemical reactions, 13
types of, 125
Chemical symbol, 16
Chemistry, 5
nuclear, 287
Chiral, 493
Chiral center 494
Chlorate ion, 86
Chloride ion,
in blood, 212, 722
in body fluids, 717
formation of, 79
oxidation number, 134
radius of, 80
in seawater, 203
Chlorination, 347
Chlorine
boiling point, 12
electron configuration, 63
electronegativity, 71
chemical symbol, 16
in human body, 483
ionization energy, 70
isotopes, 50
melting point, 12
oxidation number, 134
production in U.S.A., 153
radius of, 69
reaction with alkanes, 347
specific heat, 41
Chlorite ion, 86
Chloroform, 347
boiling point, 12
from methane, 347
melting point, 12
specific heat, 41
Chloromethane, 347
Lewis structure, 96
Cholesterol, 537
and heart disease, 538
and lipoproteins, 560
structure, 343
Choline, 449, 470
in glycerophospholipids, 547
Chromatin, 491
Chromium
in alloy, 204
chemical symbol, 16
Chromoprotein, 560
Chromosome, 685
Chylomicrons, 648
Chyme, 626, 719
Chymotrypsin
from chymotrypsinogen, 600
isoionic point, 566
hydrolysis of proteins, 579
K_m, 594
Chymotrypsinogen, 600
Cicutoxin, 356
Cimetidine, 375
Cis-isomer
in alkenes, 357
in cycloalkanes, 341
Citric acid, 431
Citric acid cycle, 619
control of, 621
and urea cycle, 676
Citrulline, 675
Claisen condensation, 452
fatty acid biosynthesis, 660
Classification, 3
alcohols, 385
amides, 472
amines, 459
amino acids, 561
carbohydrates, 511
carbon atom, 335
elements, 17
enzymes, 586
hydrocarbons, 333, 353
lipids, 535
mixtures, 14
monosaccharides, 511
of organic reactions, 327
proteins, 559
Clonal selection, 729
Codon, 697, 699
table of, 699
termination, 703
Coefficients
in scientific notation, 27
in equations, 118
Coenzyme, 590
structures, 591
Coenzyme A, 449
in biochemical Claisen condensation, 452
in citric acid cycle, 620
in fatty acid cycle, 651
Coenzyme Q, 618
Cofactor, 590
Cold packs, 214
Collagen, 559
and osteoarthritis, 706
turnover rate, 667
Colligative property, 219
boiling point elevation, 220
freezing point depression, 220
osmotic pressure, 222
Combination reactions, 126
Combined gas law, 179
Competitive inhibition, 596
Complementary base pairing, 692
Complementary shape, 588
Complete protein, 580
Composition, 6
of biochemicals, 483
of compounds, 148
of E. coli, 484
of fats and oils, 545
of human body, 483
of lipids, 536
of membranes, 551
percent, 148
polynucleotides, 689
proteins, 558
Compound, 15
comparison with mixtures, 15
empirical formula of, 154
molecular, 93
ionic, 81
percent composition, 148
Concentrated solution, 204
Concentration of solution, 204
effect of, on equilibrium, 248
molarity, 207
normality, 280
parts per million, 207
percent concentration, 205
volume/volume %, 206
weight/volume %, 205
weight/weight %, 205
Concerted reaction, 323
in substitution reaction, 392
Condensation, 11, 171
of aldehydes, 423
Condensation polymerization, 444
Condensation reaction, 328
Condensed structural formula, 110
Conductivity, of solutions, 209

Configuration, 492
  of amino acids, 561
  of monosaccharides, 512
  of nucleosides, 688
Conformations, 316
  of ethane, 315
  of peptides, 576
Conjugate acid, 262
Conjugated protein, 560
Conjugate base, 262
  table of, 267
Conservation of energy, 18
Conservation of mass, 13, 46
Contractile protein, 559
Control rods, 298
Conversion factor, 28
Coordinate covalent bond, 99
  in hemoglobin, 100
Copper
  boiling point, 12
  chemical symbol, 16
  in human body, 483
  melting point, 12
  specific heat, 41
Core, 65
Cori cycle, 636
Corn oil, 545
Corticosteroids, 539
Cortisol, 539
Cortisone, 540
Cottonseed oil, 545
Coupled reaction, 612
  in electron transport chain, 617
  in substrate level phosphorylation, 613
Covalent bond, 93
Creatine, 733
Creatinine coefficient, 734
Crenation, 224
Cresol, 385
Crick, F. H.C., 691
Crista, 622
Critical mass, 297
Cryogens, 153
Cupric ion, 85
  in Benedict's solution, 412
Cuprous ion, 85
Curie, 301
Cyanide ion, 86
  K_b, 270
Cycloalkane, 333, 340
  geometric isomers, 341
  in nature, 343
  nomenclature, 342
  oxidation of, 345
  physical properties, 345
Cyclobutane, 340
Cyclohexane, 340
Cyclopentane, 340
Cyclopropane, 340
Cysteine, 562
Cystine, 401
Cytochromes, 617
Cytoskeleton, 489
Cytoplasm, 489
Cytosine, 686
  in DNA, 690
Cytosol, 490
Dacron, 444
Dalton, John
  atomic theory, 46
  law of partial pressures, 181
Data, 4
Dating, carbon-14, 294
Daughter nucleus, 288
D configuration, 500
Dead Sea scrolls, 295
Deamination
  direct, 673
  oxidative, 672
Decarboxylation, 440
  of amino acids, 679
  of oxalosuccinate, 441
  of pyruvic acid, 635
Decay, nuclear, 288
Deci-, 30
Decomposition reaction, 127
Definite proportions, law of, 46
Dehydration
  of alcohols, 392
  of citric acid, 393
  mechanism of, 394
Dehydrogenase, 586
Dehydrogenation
  in citric acid cycle, 620
  by NAD^+, 615
Deka-, 30
Delocalization, 102
  in benzene, 370
Denaturation, 576
Denatured alcohol, 383
Density, 37
  of gases, 38
  of liquids and solids, 38
  of lipoprotein, 560
  urine, 733
  water, 196
Deoxyribonucleic acid (see DNA)
Deoxyribose, 686
Depot fat, 648
Detergent, 446
Dextrorotatory, 498
Diastereomers, 503
Diatomic molecules, 46
1,1-Dichloroethane, 322
1,2-Dichloroethane, 322
Dichloromethane, 347
Dicumarol, 732
Diethyl ether, 397
  anesthetic, 400
  structure, 8
  vapor pressure, 187
Diffusion
  in membranes, 553
Digestion
  carbohydrates, 626
  proteins, 665
  triglycerides, 647
Digestion juices, 718
Dihydroxyacetone, 515
Dihydroxyacetone phosphate, 424
  in glycolysis, 631
  glycerol 3-phosphate shuttle, 637
Dilute solution, 204
Dilution, 209
Dimethyl ether, 382
Dipeptide, 567
1,3-Diphosphoglycerate, 632
Diphosphoglylcerate mutase, 597
Dipole, 98
  induced, 191
  moment, 106
  temporary, 191
Dipole-dipole forces, 192
Diprotic acid, 259
Disaccharide, 525
Dissociation, 210
Distillation, 14
  of alcohol, 383
Disulfide bond, 401
  in proteins, 570
Diterpene, 537
DNA (deoxyribonucleic acid), 684
  composition, 689
  hydrogen bonding, 692
  polymerase, 710
  primary structure, 690
  recombinant, 707
  replication, 685, 693
  secondary structure, 691
DNA polymerase, 710
Donor site, 702
Dopamine, 458
Double bond, 95
  in carbonyl group, 405
  in ethylene, 109, 354
Double helix, 691
Double replacement reaction, 129
Dynamic equilibrium, 171
E. coli
  composition, 484
  and recombinant DNA, 707
Eicosanoids. 543
Electrode, 209
Electrolytes, 209
  in body fluids, 210, 716
  ionization of, 210
  strong, 210
  weak, 210
Electron, 48
  charge, 48
  energy level diagram, 62
  in covalent bonds, 94
  in ionic bonds, 81
  lone pair, 94
  mass, 48
  nonbonding, 94
  orbitals, 59
  paired, 60
  spin, 59
  unpaired, 50
  unshared pair, 94
  valence, 65
Electron configuration, 61
  and pairing, 60
  periodic table and, 65
Electron density, 59
Electron dot symbol, 65
Electronegativity
  and covalent bonds, 98
  effect on acids, 438
  table of values 71
Electron shells, 57
Electron transfer
  in ionic compounds, 81
  in redox reactions, 134
Electron transport chain, 617
Electrophile, 324
Electrophoresis, 566
Element, 15
  atomic weights, 51
  classification, 17
  diatomic molecules, 46
  electron dot symbols, 65
  metals, 17
  nonmetals, 17
  periodic table, 17, 54
  periodic properties, 67
  representative, 54
  semimetals, 18
  symbols of, 16
  transition, 54
  transmutation of, 295
  transuranium, 296
Elimination reaction, 327
Empirical formula, 154
  calculation of, 155
  and molecular formula, 156
Enantiomer, 494
  properties of, 496
  representation of, 500
Endergonic, 487
Endoplasmic reticulum, 491
Endothermic reaction, 18, 161
Endpoint, 280
Enediol, 423
  in glycolysis, 631
  in isomerization of ketose, 523
  in pentose phosphate pathway, 641
Energy, 18
  activation, 240
  biochemical reaction, 605
  chemical, 18
  in chemical reactions, 229
  conservation of, 18
  from fatty acids, 655
  food and, 604
  ionization, 70
  and oxidation state, 608
  units of, 40
Energy level, 56
Enflurane, 400
Enkephalin, 569
Enol, 422
Enolase, 632
Enthalpy, 19, 230
  and free energy, 233
Entropy, 19, 232
Entropy change, 230
  changes in state, 172
  chemical reactions, 232
  in solutions, 214
Enzyme, 9
  absolute specificity, 585
  active site, 590
  allosteric regulation, 600
  chirality, 589
  classification, 586
  coenzymes, 590
  cofactors, 590
  competitive inhibition, 596
  effect of pH, 595
  effect of temperature, 594
  feedback inhibition, 599
  induced fit hypothesis, 589
  inhibition, 596
  kinetics, 592
  lock and key theory, 588
  models, 588
  noncompetitive inhibition, 597
  regulatory, 599
  specificity, 585
  substrate complex, 593
  turnover number, 593
  and zymogens, 600
Enzyme-substrate complex, 592
  stability, 593
Eosinophils, 724
Epinephrine, 458, 465
Equation
  balancing, 119
  combination, 126
  decomposition, 127
  double replacement, 129
  ideal gas, 181
  net ionic, 131
  nuclear, 291
  oxidation-reduction, 134
  precipitation, 125
  symbols in, 118
Equilibrium, 229
  dynamic, 171
  between liquid and vapor, 186
  between liquid and solid, 189
  chemical, 244
  effect of concentration changes, 248
  effect of pressure changes, 248
  effect of temperature changes, 250
  and Le Chatelier's principle, 247
  and reaction completeness, 246
  self ionization of water, 264
  in saturated solution, 213
  static, 171

Equilibrium constant, 244
acid ionization, K_a, 268
effect of temperature, 247
of water, K_w, 264
Equivalent, 211, 280
Equivalent weight, 211, 280
Erythrocytes, 723
Erythrose, 514
Erythrulose, 515
Essential amino acid, 561
Essential fatty acid, 561, 666
Esters, 318, 430, 441
Claisen condensation, 452
formation, 442
hydrolysis of, 445
of inorganic acids, 441
nomenclature, 442
odors, 443
of phosphoric acids, 391, 450
saponification, 445
Estradiol, 539
Estrogen, 540
Ethane, 95
boiling point, 345
conformation, 315
Ethanol, 383
antidote for methanol poisoning, 396
boiling point, 389
Ethanolamine
in glycerophospholipids, 546
Ether
anesthetic, 400
density, 38
vapor pressure, 187
Ethers, 317, 382, 397
nomenclature, 398
Ethoxide, 389
Ethyl acetoacetate, 452
Ethyl alcohol, 383
solubility in water, 389
structure, 8
vapor pressure, 187
Ethyl group, 337
Ethylene, 95
addition reactions, 362
bonding in, 109
molecular model, 354
production in U.S.A., 153
Ethylene glycol, 384
in polyesters, 444
Eukaryote, 488
Eukaryotic cell, 489
Evaporation, 185
Exa-, 30
Exact number, 25
Exergonic, 487
Exothermic reaction, 18, 161
Exponents, 27
Extensive property, 37
Extracellular fluids, 714
Facilitated exchange diffusion, 553
Factor unit method, 28
Faculative anaerobe, 635
FAD/$FADH_2$ (flavin adenine dinucleotide)
in citric acid cycle, 619
in electron transport chain, 617
in mitochondria, 637
structure, 616
yield from fatty acids, 655
Fahrenheit scale, 34
Family, 54
Fat, 545
and exercise, 650
Fatty acid, 541
ATP from, 655
biosynthesis, 658
dehydrogenation, 653
melting points, 542
metabolism, 651
transport, 652
Fatty acid cycle, 654
Fatty acid mobilization, 649
Fatty acid synthetase, 659
Feedback inhibition, 596, 599
Fehling's solution, 412
Femto-, 30
Fermentation, 635
Ferric ion, 85
Ferrous ion, 85
Fetal alcohol syndrome, 384
Fibrinogen, 731
Fibrous protein, 559
Film badge, 301
Fischer projection formula, 499
in monosaccharides, 512
Fission, nuclear, 297
Fleming, Alexander, 10
Fluoride ion,
in seawater, 203
radius, 80
Fluorine
chemical symbol, 16
electron configuration, 62
electron dot symbol, 65
electronegativity, 71
Hund's rule, 64
ionization energy, 70
radius, 71
Fluorouracil, 687
Formaldehyde, 408
bonding, 406
effect on vision, 421
shape, 104
VSEPR theory, 104
Formic acid, 433
bonding, 430
physical properties, 436
Formula, 7
condensed structural, 110
empirical, 154
of ionic compounds, 87
molecular, 7
structural, 109, 314
Formula unit, 83
Formula weight, 147
Frame shift error, 704
Free energy, 233
and biological processes, 235
and oxidation state, 608
Free energy change, 605
Freezing, 11, 171
Freezing point, 11, 171
Freezing point depression, 220
Freon, 348
Fructose, 515
Haworth projection formula, 519
in sucrose, 527
Fructose 1,6-diphosphate, 424
in gluconeogenesis, 639
in glycolysis, 629
Fructose 6-phosphate
in glycolysis, 629
in gluconeogenesis, 639
Fumarase,
molecular weight, 575
turnover number, 593
Fumarate hydratase, 587
Fumaric acid, 431
and aspartic acid, 585
in citric acid cycle, 620
from succinic acid, 585
in urea cycle, 675
Functional groups, 316
Furan, 375
Furanose, 517
Fused ring compound, 340
Fusion
of carbon rings, 374
heat of, 189
nuclear, 300
Galactose, 514
in lactose, 527
Galactosemia, 527
β-Galactosidase, 593
Galvanize, 139
Gamma ray, 290
Ganglioside, 549
Gas, 11
Boyle's law, 174
Charles's law, 177
combined gas law, 179
Dalton's law of partial pressure, 181
Gay-Lussac's law, 178
ideal, 170
ideal gas equation, 181
kinetic theory of, 170
molar volume of, 181
noble, 17
pressure, 172
real, 170
Gas constant, universal, 181
Gastric acid, 719
denaturation of protein, 666
pH, 665
Gay-Lussac's law, 178
Geiger counter, 301
Gene, 685
genetic engineering, 707
mutation, 704
therapy, 709
Genetic code, 698
Genetic disease, 704
Genetic effects, 303
Geometric isomerism
in alkenes, 356
in cycloalkane, 341
and pheromones, 358
and vision, 421
Germanium
electronegativity, 71
ionization energy, 70
radius, 69
semimetal, 68
Giga-, 30
Globular protein, 559
Globulins, 559
Glucagon, 569
Glucocorticoid, 539
Glucogenic amino acid, 677
Gluconeogenesis, 517, 628,
comparison with glycolysis, 628
control of, 643
net reaction, 638
Glucose
and ATP, 638
in blood, 413, 722
in cellobiose, 525
in cerebrosides, 549
from gluconeogenesis, 517, 628, 638
from glycogenolysis, 516, 628
glycolysis, 516, 628
Haworth projection formula, 518
hemiacetal, 518
in lactose, 527
in maltose, 525
metabolic pathways, 516, 627
mutarotation, 520
from starch, 529
structure, 514
in sucrose, 527
Glucose 1-phosphate, 516, 642
Glucose 6-phosphate, 516
coupled reaction, 613
in glycogenolysis, 642
in glycolysis, 629
pentose phosphate pathway, 641
Glutamate transaminase, 671
Glutamic acid, 652
conversion to glutamine, 235
isoionic point, 566
from α-ketoglutaric acid, 671
oxidation deamination, 672
pK_a, 564
Glutamine, 562
Glutaric acid, 431
Glyceraldehyde
in carbohydrates, 514
configuration, 500
Glyceraldehyde 3-phosphate, 424
in glycerol 3-phosphate shuttle, 637
in glycolysis, 631
Glycerol, 384
phosphorylation, 614
reaction with nitric acid, 441
Glycerol kinase, 651
Glycerol 3-phosphate, 614, 651
Glycerol phosphate dehydrogenase, 651
Glycerol 3-phosphate shuttle, 637
Glycerophospholipid, 546
composition, 547
Glycine, 562
isoionic point, 566
melting point, 563
pK_a, 564
Glycogen, 529
formation from glucose, 642
structure, 529
Glycogen phosphorylase, 642
Glycogenesis, 516, 628
control of, 644
Glycogenolysis, 516, 628
control of, 644
Glycolysis, 516, 628
and ATP, 638
comparison with gluconeogenesis, 628
control of, 643
net reaction, 633
steps in, 630
Glyconic acid, 522
Glycoprotein, 560
Glycoside, 521
in disaccharides, 525
Glycosidic linkage, 511, 521
Glycosphingolipid, 549
Gold
density, 38
chemical symbol, 16
specific heat, 41
Golgi complex, 491
Graduated cylinder, 33
Gram, 34
Group, 17, 54
periodic table, 54
GTP (guanosine triphosphate)
from citric acid cycle, 620
in protein synthesis, 700
Guanine, 686
in DNA, 690
Gulose, 514
Half-life, 292
table of, 293
use in dating, 294
Halogen, 54
Halogenation, 346
Haworth projection formula, 518
HDL (high-density lipoprotein), 560

Heat
  of fusion, 189
  of reaction, 161, 230
  of solution, 213
  specific, 41
  of vaporization, 188
Hecto-, 3
Heimlich maneuver, 176
Helium
  chemical symbol, 16
  density, 38
  and diving, 217
  electron configuration, 61
  Hund's rule, 64
  ionization energy, 70
  solubility in water, 217
α-Helix, 573
Hematocrit value, 724
Hemiacetal, 417
  in carbohydrates, 518
  cyclic, 419
Hemiketal, 417
  in carbohydrates, 518
Hemoglobin
  association with oxygen, 249
  carbon dioxide transport, 728
  effect of carbon monoxide, 100
  isoionic point, 566
  molecular weight, 575
  percent saturation, 726
  primary structure, 572
  quaternary structure, 575
  and sickle cell anemia, 572
Hemolysis, 224
Henry's law, 216
Heparin, 560, 731
Heredity, 684
Heteroatom, 334
Heterocyclic aromatic amine, 461
Heterocyclic compound, 334
Heterogeneous mixture, 14
Heterogenic process, 324
Heterolytic cleavage, 323
Heteropolysaccharide, 511
Hexane, 335
  boiling point, 345
  octane number, 346
Hexachlorophene, 385
Hexamethylenediamine, 475
Hexokinase, 629
Hexylresorcinol, 385
Histamine, 679, 729
Histidine, 562
  and histamine, 679
  isoionic point, 566
  pK_a, 564
Holoenzyme, 590
Homeostasis, 36
Homogeneous mixture, 14
Homogenic process, 324
Homologous series, 335
Homopolysaccharide, 511
Homolytic cleavage, 323
Hormone, 538
Hot packs, 214
Human growth factor, 708
Hund's rule, 60
Hyaluronic acid, 511
Hybrid orbital
  sp, 110, 314
  sp_2, 109, 314
  sp_3, 108, 313
Hybridization, 108, 313
Hydrocarbon, 333
  physical properties, 344
  saturated, 333
  unsaturated, 333, 353
Hydrochloric acid, 258
  equivalent weight, 281
  in gastric juice, 665, 719
Hydrogen
  chemical symbol, 16
  covalent bond in, 94
  density, 38
  diatomic molecule, 46
  electron configuration, 61
  electronegativity, 71
  chemical symbol, 16
  in human body, 483
  Hund's rule, 64
  ionization energy, 70
  isotopes of, 51
  molecule, 46
  mass of, 47
  oxidation number, 134
  radius, 69
  in rockets, 124
Hydrogenation, 361
α-Hydrogen atom, 422
Hydrogen bomb, 300
Hydrogen bond, 191
  in alcohols, 388
  in amides, 473
  in amines, 464
  in ammonia, 194
  in carboxylic acids, 436
  of carbonyl compounds in water, 411
  in DNA, 692
  in proteins, 571
  in water, 194
Hydrogen chloride
  polarity of, 98
  reaction with water, 258
Hydrocyanic acid, 258
Hydrogen fluoride
  acid ionization constant, 268
  and hydrogen bonding, 194
Hydrogen ion, 257
  and pH, 270
  and hydronium ion, 257
Hydrolase, 587
Hydrolysis
  of amides, 476
  of ATP, 610
  of cellulose, 526
  of esters, 445
  of lipids, 536
  of metal ions, 274
  of phosphorylating agents, 614
  of proteins, 579
  of salts, 273
Hydrolysis reaction, 328
Hydronium ion, 257
  in acidic solution, 264
  in basic solution, 265
  reaction with hydroxide, 260
  and pH, 270
Hydrophilic, 446
Hydrophobic, 446
  effect, 572
  interactions in proteins, 571
Hydroxide ion, 86
  in acidic solution, 264
  in basic solution, 259
  reaction with hydronium ion, 260
Hydroxyl group, 317, 381
Hyperammonemia, 677
Hyperparathyroidism, 734
Hyperthermia, 36
Hypertonic, 224
Hypervitaminosis, 216
Hypo-, 87
Hypochlorite ion, 86
Hypothermia, 36
Hypothesis, 4
Hypotonic, 224
Ibuprofen, 369
Ice
  density of, 38, 196
  equilibrium with water, 190
Ichthrothereol, 356
Ideal gas, 170
Ideal gas law, 181
Idose, 514
Imidazole, 467
Imine, 318, 420
  in biosynthesis of proline, 681
  in vision, 421
Immune system, 728
Immunoglobulin, 730
Indicator, 256, 272
Indole, 461
Induced dipole, 191
Induced fit hypothesis, 589
Inductive effect, 437
Inhibitor, 596
Initiation
  of protein synthesis, 703
  step in halogenation, 347
Inner transition element, 54
Inorganic compounds, 310
Insulin
  from E. coli, 708
  control of metabolism, 644
  isoionic point, 566
  molecular weight, 569
  primary structure, 574
  quaternary structure, 575
Integral protein, 551
Intensive property, 37
Intermembrane space, 622
Intermolecular forces, 169
International Bureau of Weights and Measures, 31
International System of Units, 31
International Union of Pure and Applied Chemistry, 337
Interstitial fluid, 714
Intestinal juice, 719
Intracellular fluid, 714
Invert soap, 469
Iodine
  electronegativity, 71
  in human body, 483
  ionization energy, 70
  radioisotope, 306
  radius of , 69
Ion,
  acid-base properties, 274
  anion, 78
  cation, 70, 78
  in E. coli, 484
  hydronium, 257
  hydroxide, 86
  names, 84
  polyatomic, 86
  radii, 80
  spectator, 131
Ionic bond, 81
  in protein, 571
Ionic compound, 81, 87
  medical applications, 90
  names, 89
  solubilities, 130
  structure, 83
Ionic radii,
  and enzyme reactions, 81
  table of, 80
Ionization, 210
  of acids, 257
  of atoms, 70
  of electrolytes, 210
  of water, 263
Ionization constant
  acid, 268
  base, 269
Ionization energy, 70
Ionization reaction 257
Ionizing radiation, 301
Ion product constant, 264
Ionophore, 399
Iron
  in blood, 722
  bond with carbon monoxide, 100
  chemical symbol, 16
  in cytochrome, 617
  in human body, 483
  oxidation of, 139
  specific heat, 41
Isobutane, 321
Isobutyl group, 339
Isocitric acid, 620
Isoflurane, 400
Isoionic point, 565
Isoleucine, 562
  biosynthesis, 600
  isoionic point, 566
  pK_a, 564
Isomerase, 587
Isomers, 311, 320
  aromatic compounds, 372
  butane, 321
  in cycloalkanes, 341
  dichloroethanes, 322
  functional groups, 321
  in peptides, 568
  positional, 322
  skeletal, 321
Isoprene, 537
Isopropyl group, 339
Isotonic, 224
Isotope, 50
  in biochemistry, 52
-ite, 87
IUPAC rules, 337
Jones reagent, 395
Joule, 40
K electron capture, 291
K, equilibrium, 244
K_a, acid ionization constant, 268
  of acetone, 422
  of alcohols, 390
  of ammonium salts, 466
  of carboxylic acids, 438
  of phenols, 390
  of thiols, 400
K_b, base ionization constant, 269
  of amines, 466
  table of, 270
K_w, ion product constant, 264
Kekule, August, 370
Kelvin scale, 35
Keratin, 559
Ketal, 418
Keto form, 422
Ketogenic amino acid, 678
α-Ketoglutaric acid, 431
  in citric acid cycle, 620
  in transamination, 670
Ketone, 317, 407
  addition of alcohols, 417
  addition of nitrogen compounds, 420
  boiling points, 411
  formation of ketals, 418
  nomenclature, 409
  physical properties, 40
  reduction, 414
  solubility, 411
Ketone bodies, 658
Ketonemia, 658
Ketonuria, 658
Ketose, 513
  and Benedict's solution, 523

Ketosis, 657
Kevlar, 475
Kidneys, 734
Kilo-, 30
Kilocalorie, 41
Kilogram, 34
Kinetic molecular theory, 170
Kinetics, 228, 236
 effect of catalysis, 239, 243
 effect of concentration, 237
 effect of reactants, 236
 effect of temperature, 238
 of enzyme reactions, 592
Krypton
 boiling point, 189
 electron configuration, 63
 heat of vaporization, 189
 ionization energy, 70
 radius, 69
Krebs cycle, 61
Lactalbumin, 559
Lactam, 432
Lactase, 527
Lactate dehydrogenase, 586, 636
Lactic acid, 431
 and anaerobic exercise, 635
 configuration, 501
 from pyruvic acid, 636
Lactone, 432
Lactose, 527
Lactose intolerance, 526
Lanthanide, 54
Lauric acid, 433, 542
Law, 5
 Boyle's, 174
 Charles's, 177
 combined gas, 179
 conservation of energy, 18
 conservation of mass, 13, 46
 Gay-Lussac's, 178
 gravity, 5
 ideal gas, 181
 partial pressures, 181
 periodic, 53
L-configuration, 500
LDH (see lactate dehydrogenase)
LDL (low density lipoprotein), 560
Lead
 chemical symbol, 16
 denaturation of protein, 579
 density, 38
 electronegativity, 71
 ionization energy, 70
 poisoning, 598
 radius, 69
Leading zero, 26
Leaving group, 391
 in nucleophilic acyl substitution, 448
Le Chatelier's principle, 189
 in acetal formation, 418
 boiling point and, 190
 chemical equilibrium, 247
 concentration and equilibrium, 248
 in ester formation, 442
 in ester hydrolysis, 445
 freezing point and, 190
 pressure and equilibrium, 248
 temperature and equilibrium, 250
Lecithin, 548
Leucine, 562
 isoionic point, 566
 K_a, 564
Leukocytes, 723
 types of, 724
Levorotatory, 498
Lewis, G. N., 78
Lewis acid, 325
Lewis octet rule, 78
Lewis structure, 82, 95
 rules, 96
Ligase, 587
Line spectra, 56
Linear molecule, 103
Linoleic acid, 542
Linolenic acid, 354, 542
Lipase, 649, 719
Lipid, 486
 classification, 535
 in E. coli, 484
 in membrane, 550
 tissue, 648
Lipogenesis, 658
Lipoprotein, 560
Liquid, 11, 183
 boiling point of, 187
 characteristics, 11
 compressibility, 183
 effect of pressure on, 183
 equilibria, 186, 189
 evaporation, 185
 heat of vaporization, 188
 surface tension, 184
 and vapor pressure, 186
 viscosity, 184
Liter, 33
Lithium
 chemical symbol, 16
 electron configuration, 61
 electron dot symbol, 65
 electronegativity, 71
 Hund's rule, 64
 ionization energy, 70
 radius of, 69
Lithium hydroxide, 132
Lithium ion,
 radius of, 80
Lock and key theory, 588
London forces, 191
 in alkanes, 344
 in alkenes and alkynes, 356
 in membranes, 552
Lone pair electrons, 94
Lucas test, 390
Lucite, 366
Lymphocytes, 724
Lyase, 587
Lysine, 562
 isoionic point, 566
 pK_a, 564
Lysol, 384
Lysosome, 492
Lysozyme, 594
Lyxose, 514
Macroscopic, 6
Magnesium
 chemical symbol, 16
 electron configuration, 63
 electronegativity, 71
 in enzymes, 590
 in human body, 483
 ionization energy, 70
 isotopes, 51
 radius, 69
Magnesium carbonate, 129
Magnesium ion
 in blood, 722
 in body fluids, 717
 in hard water, 447
 in plasma, 212
 in sea water, 203
Magnesium hydroxide, 262
Malate-aspartate shuttle, 637
Maleic acid, 431
Malic acid, 431
 in citric acid cycle, 612, 620
 from fumaric acid, 587
 oxidation of, 397
Malonic acid, 431
Malonyl CoA, 659
Malleable, 18
Maltose, 525
Mannose, 514
Manganese
 chemical symbol, 16
 in human body, 483
Markovnikov's rule, 363
Mass, 6
 atomic, 47
 conservation of, 13, 46
 of electron, 48
 molar, 181
 of neutron, 48
 of proton, 48
Mass number, 50
Matrix, 622
Matter, 5
 chemical properties, 13
 compounds, 15
 density, 37
 elements, 15
 heterogeneous, 14
 homogeneous, 14
 intensive properties, 37
 kinetic molecular theory of, 170
 physical properties, 11
 mixtures of, 14
 states of, 11
Measurement, 23
 length, 32
 mass, 34
 metric prefixes, 30
 metric system, 29
 pressure, 172
 significant figures in, 25
 temperature, 34
 uncertainty in, 24
 volume, 33
Mechanisms, 323
 acetal formation, 420
 addition to alkenes, 364
 addition to carbonyl compounds, 46
 chlorination of alkanes, 347
 dehydration of alcohols, 394
 nucleophilic acyl substitution, 448
 nucleophilic substitution, 391
Mega-, 30
Melting, 11, 171
Melting point, 11
 of carboxylic acids, 436
 of fatty acids, 542
Membrane, 550
 carbohydrate in, 552
 composition of, 551
 diffusion in, 553
 fluidity of, 552
 in mitochondria, 551
 protein in, 551
 structure of, 551
 transport across, 553
Mendel, Gregor, 684
Mendeleev, Dmitri, 53
Mercaptan, 400
Mercuric ion, 85
 denaturation of protein, 579
Mercurous ion, 85
Mercury
 barometer, 172
 chemical symbol, 16
 density, 38
 ions, 85
 poisoning, 598
 specific heat, 41
Meso compound, 505
Messenger RNA, 697
 in protein synthesis, 701
Meta-, 372
Metabolic acidosis, 278
Metabolism, 487
 amino acids, 666
 and ATP, 609
 carbohydrates, 626
 fatty acid, 651
 glucose, 516, 627
 proteins, 666
 stages, 605
 triglycerides, 651
Metabolite, 487
Metal, 17
 alkali, 54
 alkaline earth, 54
 and periodic table, 68
 transition, 54
Metalloid, 18
 and periodic table, 68
Metalloenzymne, 590
Meter, 32
Methamphetamine, 465
Methane
 boiling point, 12, 188
 density of, 38
 halogenation, 347
 heat of combustion, 346
 hybridization in, 108, 313
 melting point, 12
 shape, 104
 specific heat, 41
 structure, 104
 VSEPR theory, 105
Methanol
 boiling point, 389
 Lewis structure, 96
 poisoning, 383
 solubility, 389
 structure, 382
 uses, 383
Methionine, 562
 isoionic point, 566
 pK_a, 564
Methoxyamphetamine, 465
Methotrexate, 375
Methyl alcohol (see methanol)
Methyl group, 337
Methyl radical, 347
Methyl salicylate, 369, 443
Methylamine, 459
 boiling point, 463
 K_b, 466
 Lewis structure, 96
 structure, 459
Methylene chloride, 347
Metric prefixes, 30
Metric system, 31
Micelle, 446
Micro-, 30
Microcurie, 301
Microgram, 34
Milk of Magnesia, 262
Milli-, 30
Milliequivalent, 211
Milliliter, 33
Millimeter, 32
Mineralocorticoid, 539
Mirror image, 492
Miscible, 215
Mitochondria, 491, 621
 and ATP, 638
 and acyl transfer, 652
 and glycerol 3-phosphate shuttle, 637

and malate-aspartate shuttle, 637
membrane, 622
Mixture, 14
mm Hg, 173
Model, 4
of atom, 46
of competitive inhibition, 596
of enzyme action, 588
of membrane, 555
of noncompetitive inhibition, 597
Moderator, 297
Molar heat of vaporization, 188
Molar mass, 149, 151
Molar volume, 181
Molarity, 207
Mole, 149
Avogadro's number, 150
calculations using, 152
and molar mass, 150
and stoichiometry, 158
Mole method, 158
Mole ratio, 158
Molecular formula, 7
from empirical formula, 177
Molecular geometry, 103
Molecular model, 8, 104, 113
Molecular structure, 8
Molecular weight, 147
Molecule, 7
diatomic, 93
of elements, 46
molecular weights of, 147
polarity of, 106
Monocytes, 724
Monomer, 365
Monoprotic acid, 258
Monosaccharide, 511
classification, 511
configuration, 512
cyclic form, 517
mutarotation, 520
oxidation, 522
reduction, 524
Monoterpene, 537
Muriatic acid, 257
Muscle relaxant, 470
Mutarotation, 520
Multiple bonds, 95
Mutation
and genetic disease, 704
from radiation, 303
Myelin sheath, 549
Myosins, 559
Myristic acid, 433, 542
NAD^+/NADH (nicotinamide adenine dinucleotide), 396
and electron transport chain, 617
and citric acid cycle, 619
and glycerol 3-phosphate shuttle, 637
and malate-aspartate shuttle, 638
structure, 616
yield from fatty acid, 655
yield from glycolysis, 638
NADPH (nicotinamide adenine dinucleotide phosphate)
in biosynthesis, 623
fatty acid biosynthesis, 659
pentose phosphate pathway, 641
structure, 616
Nano-, 30
Nanogram, 34
Naphthalene, 375
Native state, 572
Negative modulator, 599
Neon
electron configuration, 62
Hund's rule, 64
ionization energy, 70
radius, 69
Net ionic equation, 131
Neutral amino acid, 561
Neutralization point, 280
Neutralization reaction, 132, 260
Neutral solution, 264
Neutron, 49
Newton, Sir Isaac, 5
Niacin, 375
Nickel
chemical symbol, 16
catalyst in hydrogenation, 361
Nickel ion, 185
Nicotinamide adenine dinucleotide, (see NAD^+)
Nicotinamide adenine dinucleotide phosphate (see NADPH)
Nitrate ion, 86
Nitric acid, 127, 258
Nitric oxide, 126
Nitride ion, 86
Nitrile, 318
Nitrite ion, 86
and cancer, 471
K_b, 270
Nitrogen
abundance in air, 182, 203
and bends, 217
bonding in, 95
chemical symbol, 16
density, 38
diatomic, 46
electron configuration, 62
electron dot symbol, 65
electronegativity, 70
in human body, 483
ionization energy, 70
Hund's rule, 64
in organic compounds, 457
partial pressure in air, 182
radius, 69
reaction with hydrogen, 127
reaction with oxygen, 126
solubility in water, 217
specific heat, 41
Nitrogen balance, 667
Nitrogen dioxide, 126
Nitrogen fixation, 239
Nitroglycerin, 441
N-Nitrosoamines, 470
Nitrous acid
acid ionization constant, 268
reaction with amines, 470
Noble gases, 17
and periodic table, 54
Nomenclature
alcohols, 385
aldehydes, 408
alkanes, 337
alkenes, 359
alkynes, 366
amides, 473
amines, 461
aromatic hydrocarbons, 371
carboxylic acids, 433
cycloalkanes, 342
esters, 442
ethers, 398
ketones, 409
thiols, 400
Nomex, 475
Nonactin, 399
Nonbonding electrons, 94
Noncompetitive inhibition, 596
Nonelectrolyte, 210
Nonmetal, 17
and periodic table, 68
Nonspontaneous reaction, 231
Norepinephrine, 465
Normal alkane, 334
Normal alkyl group, 339
Normal boiling point, 188
Normality, 280
Nuclear chemistry, 287
Nuclear envelope, 491
Nuclear fission, 297
Nuclear fusion, 300
Nuclear power plant, 297
Nuclear region, 489
Nuclear transmutation, 295
Nucleic acid, 486
in E. coli, 484
Nucleolus, 491
Nucleophile, 324
Nucleophilic acyl substitution, 448
Nucleoprotein, 560
Nucleoside, 687
in medicine, 687
names, 688
Nucleotide, 688
Nucleus
of atom, 49
in cell, 491
daughter, 288
Nylon, 475
Observations, 2
Octane
boiling point, 12
melting point, 12
octane number, 346
specific heat, 41
structure, 8
Octane number, 346
Octet rule, 78
Oils, 545
Oil of wintergreen, 369
Oleic acid, 542
Oligopeptide, 567
Oligosaccharide, 511
Olive oil, 545
Optical activity, 497
Optical isomer, 498
Optimum pH, 595
Optimum temperature, 594
Orbital, 57, 59
formation of sigma bond, 108
formation of pi bond, 109
hybrid, 108
shape of p, 60
shape of s, 59
Organelles, 482, 489
Organic compound, 95, 310
in daily life, 312
Orlon, 366
Ornithine, 675
Ortho-, 372
Osmolarity, 223
Osmosis, 221
and cells, 224, 717
Osmotic pressure, 222
Osteoarthritis, 706
-ous, 85
Oxalic acid, 431
Oxaloacetic acid, 431
in citric acid cycle, 620
in gluconeogenesis, 639
in transamination, 670
in urea cycle, 676
Oxidation, 136
of alcohols, 394
of aldehydes, 412
of alkanes, 345
of alkenes, 360
in biochemistry, 608
of monosaccharides, 522
by NAD^+, 141, 615
of thiols, 401
Oxidation number, 134
Oxidation-reduction reactions, 134
in organic chemistry, 326
Oxidation state, carbon, 609
Oxidative deamination, 670, 672
Oxidative degradation reaction, 609
Oxidative phosphorylation, 618
Oxide ion, 86
radius, 80
Oxidizing agent, 136
strong, 138
weak, 138
Oxidoreductase, 586
Oxygen
abundance in air, 203
chemical symbol, 16
density, 38
diatomic, 46
electron configuration, 62
electron dot symbol, 65
electronegativity, 71
in human body, 483
Hund's rule, 64
ionization energy, 70
isotopes, 51
partial pressure in air, 182
radius, 69
in rockets, 124
solubility in water, 217
specific heat, 41
transport in blood, 727
Oxygen debt, 636
Oxytocin, 570
Ozone, 101
Ozone layer, 348
Paired electrons, 60
Palladium
catalyst in hydrogenation, 361
chemical symbol, 16
Palmitic acid, 433, 542
Pancreatic juice, 721
Para-, 372
Paraffin, 345
Parkinson's disease, 458
Partial pressure, 181
gas solubility, 216
in medicine, 182
Parts per million, 207
Pathogens, 728
Penicillin, 10
Pentane
boiling point, 345
octane number, 346
Pentose phosphate pathway, 517, 628
and glycolysis, 641
net reaction, 641
Pepsin
digestion of proteins, 666
hydrolysis of proteins, 579
isoionic point, 566
Pepsinogen, 719
Peptide, 567
hormone, 569
isomerism, 568
Peptide bond, 558
PeptoBismol, 262
per, 87
Percent composition, 148
Percent concentration, 205
Percent ionization, 268
Percent saturation, 726
Percent yield, 162
Perchlorate ion, 86
Perchloric acid, 258

Period, 17
Periodic law, 53
Periodic table, 17, 53
  and chemical properties, 67
  and electron configuration, 65
  family, 54
  groups in, 17, 54
  periods in, 17
  and properties, 67
Peripheral protein, 552
Permanganate ion, 86
Perspective structural formula, 113, 313
Peta-, 30
pH, 269
  acid solutions, 270
  in acidosis, 278
  basic solutions, 270
  body fluids, 715
  enzyme reactions, 595
  in ketosis, 658
  measurement, 272
  meter, 272
  in respiratory acidosis, 279
  in saliva, 718
  urine, 733
Phenanthrene, 375
Phenol, 381
  acidity, 390
  germicide, 384
Phenol coefficient, 385
Phenolphthalein, 280
Phenylalanine, 562
  isoionic point, 566
  pK_a, 564
Phenylketonuria, 678
Phenyl group, 373
Phenylpropanolamine, 465
Phosphate buffer, 278, 715
Phosphate ion, 86
  in fertilizer, 88
  K_b, 270
Phosphatide, 546
Phosphatidic acid, 546
Phosphatidylcholine, 547
  in membrane, 551
Phosphatidylethanolamine, 546
  in membrane, 551
Phosphatidylserine, 547
  in membrane, 551
Phosphide ion, 86
Phosphodiester, 690
Phosphoenolpyruvate, 632
Phosphoenolpyruvate carboxykinase, 639
Phosphofructokinase, 631
2-Phosphoglycerate, 632
3-Phosphoglycerate, 632
Phosphoglycerate kinase, 632
Phosphoglucomutase
  in glycolysis, 642
  in glycogenesis, 643
Phosphoglucose isomerase, 587
  in glycolysis, 629
  turnover number, 593
Phosphoglyceride, 546
Phosphoglyceromutase, 632
3-Phosphohydroxypyruvate, 681
Phosphoprotein, 560
Phosphoric acid
  acid ionization constant, 259
  equivalent weight of, 281
  esters, 450
  ionization of, 259
  production in U.S.A., 153
Phosphorus
  chemical symbol, 16
  electron configuration, 63
  electronegativity, 71
  elemental form, 46
  in human body, 483
  ionization energy, 70
  isotopes in biochemistry, 52
  radius, 69
3-Phosphoserine, 681
Phosphorylating agent, 614
Phosphorylation, 613
Photomultiplier, 301
Photosynthesis, 510
Phthalic acid, 434
Physical properties, 11
  of alcohols, 388
  of aldehydes, 410
  of alkanes, 344
  of alkenes, 355
  of amines, 463
  of carboxylic acids, 435
  of esters, 443
  of ketones, 410
  and periodic table, 67
Phytoplankton, 510
Pi bond, 109, 353
  in acetylene, 110
  in benzene, 370
  in carbonyl group, 406
  in carboxyl group, 430
  in ethylene, 109
Pico-, 30
Piperidine, 460
Pipet, 33
pK_a
  amino acids, 564
  ammonium salts, 466
  carboxylic acids, 438
pK_b
  of amines, 466
PKU (see phenylketonuria)
Plane-polarized light, 497
Plane of symmetry, 493
Plasma, 722
Platinum
  chemical symbol, 16
  reduction of carbonyl compounds, 414
β-Pleated sheet, 573
Plexiglass, 366
Point mutation, 704
Polar covalent bond, 98
Polarimeter, 497
Polarity
  of carbonyl group, 406
  and electronegativity, 99
  and intermolecular attraction, 192
  of molecules, 106
Polarizability, 191
Polarizing filter, 497
Polyamide, 475
Polyatomic ion, 86
Polycyclic aromatic hydrocarbons, 374
Polycythemia, 723
Polyene, 354
Polyesters, 444
Polyethylene, 365
Polymer, 365
  table of, 366
Polymerization
  addition, 365
  condensation, 444
Polynucleotides, 689
Polypropylene, 366
Polyprotic, 259
Polysaccharide, 528
Polyunsaturated, 354, 542
Polyvinyl chloride (PVC), 366
p orbital, 59
Positional isomer, 322
Positron, 290
Positron emission tomography, 305
Potassium
  chemical symbol, 16
  electron configuration, 63
  electronegativity, 71
  in human body, 483
  ionization energy, 70
  radius of, 69
Potassium dichromate
  oxidation of alcohols, 394
Potassium ion, 85
  in blood, 212, 722
  in body fluids, 717
  radius, 80
  in seawater, 203
Potassium permanganate,
  oxidation of alkenes, 360
ppm, 207
Precipitate, 125
Precipitation reaction, 125, 129
Precision, 24
Pressure, 172
  and altitude, 173
  atmospheric, 172
  effect on equilibrium, 248
  effect on solubility, 216
  measurement of, 172
  in ocean, 172, 217
  osmotic, 222
  partial, 181
  vapor, 186
  and volume, 175
  units of, 173
Primary
  alcohol, 387
  amine, 459
  carbon atom, 336
  structure of protein, 572
  structure of polynucleotides, 690
Product, 9
  chemical equation, 118
Progesterone, 539
Prokaryote, 489
Proline, 562
  biosynthesis of, 681
  isoionic point, 566
  pK_a, 564
Promoter region, 697
Proof, 206
Propagation step, 347
Propane, 334
  boiling point, 345
Properties, 3,11
  of acids and bases, 255
  chemical, 13
  colligative, 219
  extensive, 37
  intensive, 37
  of metals and nonmetals, 17
  and periodic table, 67
  physical, 11
  of subatomic particles, 48
Propionic acid, 433
Propyl group, 339
Prostaglandin, 543
Prosthetic group, 560
Protease, 579, 585
Protein, 485
  biological function, 558
  biological value, 580
  bonding in, 558
  classification, 559
  complete, 580
  conjugated, 560
  denaturation, 576
  digestion of, 665
  disulfide bond in, 570
  in E. coli, 484
  electrophoresis, 566
  essential amino acids, 561
  fibrous, 559
  globular, 559
  hydrogen bonding in, 571
  hydrolysis of, 579
  hydrophobic interactions, 571
  integral, 551
  ionic bond, 571
  isoionic point, 566
  in membranes, 551
  peripheral, 552
  primary structure, 572
  quaternary structure, 575
  renaturation, 576
  secondary structure, 572
  structure, 572
  tertiary structure, 575
  transcription, 695
  transmembrane, 551
Protein buffer, 716
Protein hormones, 559
Protein synthesis
  initiation, 703
  mutation and, 705
  termination, 703
  translation, 703
Protein turnover rate, 667
Prothrombin, 731
Proton, 49
  acceptor, 257
  from acid, 257
  charge, 49
  donor, 257
  mass of, 49
psi, 172
psicose, 515
Pure substance, 15
Purine, 458, 461
Purine bases, 686
Putrescine, 464
Pyranose, 517
Pyridine, 375, 460
  basicity, 467
Pyridoxal phosphate, 591, 670
Pyridoxamine phosphate, 670
Pyridoxine, 375
Pyrimidine, 458, 461
Pyrimidine bases, 686
Pyrophosphate, 391
  in fatty acid cycle, 651
Pyrophosphoric acid, 391, 450
Pyrrole, 375
  basicity, 467
Pyrrolidine, 460
Pyruvate carboxylase, 594
Pyruvic acid, 431
  from glycolysis, 633
  in gluconeogenesis, 639
  metabolism, 633
  in transamination, 671
Qualitative, 4
Quantitative, 4
Quantized, 56
Quaternary
  ammonium salt, 469
  carbon atom, 336
  structure of protein, 575
R (universal gas constant), 181
Racemic mixture, 506
Rad, 302
Radical
  in chlorination reaction, 347
  from heterolytic cleavage, 323
  in ozone layer, 348
  from radiation, 302

Radii
of atoms, 69
of ions, 80
of proton, 48
Radioactivity, 289
detection of, 301
genetic effects of, 303
half life, 292
inverse square law, 304
units of, 301
Radiation safety, 304
Radiation sickness, 303
Radioisotopes, 288
use in diagnosis, 304
use in therapy, 306
Radon
chemical symbol, 16
electron configuration, 63
Raney nickel, 361
Rare gas, 54
Rate of reaction, 228
effect of catalyst, 239
effect of concentration, 237
effect of pressure, 237
effect of reactant, 236
effect of temperature, 238
Rate determining step, 323
RBE (relative biological effectiveness), 302
Reactant, 9
in chemical equation, 118
Reaction, 9
acid-base, 260
activation energy, 240
addition, 327
anabolic, 487, 606
catabolic, 487, 605
chain, 297, 347
combination, 126
complete, 246
condensation, 328
decomposition, 127
double replacement, 129
effect of catalyst on, 239
effect of reactants on, 236
effect of temperature on, 238, 250
elimination, 327
endothermic, 18, 161
enthalpy of, 161, 230
entropy of, 232
exothermic, 18, 161
net ionic, 131
neutralization, 132
nonspontaneous, 231
oxidation-reduction, 134
precipitation, 129
rates, 228, 236
rearrangement, 328
redox, 134
single replacement, 128
spontaneous, 19, 231
substitution, 327
types of, 125
Reaction coordinate diagram, 241
Real gas, 170
Recombinant DNA, 707
and agriculture, 709
Redox reaction, 134
in biochemistry, 141
in organic chemistry, 140
Reducing agent, 136
Reducing sugar, 523
Reductase, 586
Reduction, 136
of alkenes, 360
of carbonyl compounds, 414
of monosaccharides, 524
Reductive biosynthetic reaction, 609
Regulatory enzyme, 599
Rem, 302
Renaturation, 576
Replication fork, 693
Replication, 685, 693
Representative element, 54
Resonance structure, 101
in amides, 472
in anilines, 467
benzene, 370
of carboxylates, 437
in ozone, 101
in phenoxide ion, 390
Respiratory acidosis, 279
Respiratory alkalosis, 279
Restriction enzyme, 707
Retinal, 408, 421
Retrovirus, 711
Reverse osmosis, 223
Reverse transcriptase, 710
Rhodopsin, 421
Riboflavin phosphate, 591
Ribonuclease
isoionic point, 566
Ribonucleic acid (see RNA)
Ribose, 514
in nucleic acids, 686
Ribose 5-phosphate, 628, 641
Ribosomal RNA, 697
Ribosome, 489
in endoplasmic reticulum, 491
in protein synthesis, 701
Ribulose, 515
Ribulose 5-phosphate, 641
Ritalin, 465
RNA (ribonucleic acid), 686
messenger, 697
primary structure, 698
ribosomal, 697
transfer, 697
RNA replicase, 710
Rockets, 124
Roentgen, 302
Rolaids, 262
Rust, 139
Rutherford, Ernest, 49
Salicylic acid, 434
Salt
boiling point, 12
of carboxylic acids, 439
as electrolyte, 210
from acid and base, 132
hydrolysis, 273
melting point, 12
solubility, 218
Salt bridge, 571
Saponification, 445
Saturated
fatty acids, 542
hydrocarbon, 333
solution, 204, 213
Science, 2
Scientific law, 5
Scientific method, 3
Scientific notation, 27
Scintillation counter, 301
Seawater
ions, 203
osmotic pressure, 224
Secondary
alcohol, 387
amine, 459
carbon atom, 336
structure of DNA, 691
structure of proteins, 572
Selenide ion, 86
Selenium
electronegativity, 71
ionization energy, 70
radius, 69
Self-ionization of water, 263
Semimetal, 18, 68
Semiconservative replication, 693
Semipermeable membrane, 221
Sense strand, 696
Serine, 562
biosynthesis, 681
deamination, 673
isoionic point, 566
in glycerophospholipids, 547
pK_a, 564
Serotonin, 458
Sesquiterpene, 537
Sex hormones, 540
Shell, 57
SI system, 31
Sickle cell anemia, 572
Sigma bond, 108
Significant figures, 25
Silicon
chemical symbol, 16
electronegativity, 71
ionization energy, 60
radius, 69
Silver
chemical symbol, 16
specific heat, 41
Silver ion, 85
in Tollens' reagent, 412
Single replacement reactions, 128
Skeletal isomer, 321
Soap, 446
Sodium
chemical symbol, 16
electron configuration, 63
electronegativity, 71
in human body, 483
ionization energy, 70
radius, 69
reaction with alcohols, 382
reaction with chlorine, 68
reaction with water, 68
Sodium acetate, 274
Sodium bicarbonate, 127
Sodium chloride
bonding in, 82
crystal structure, 83
dissolution of, 213
melting point, 78
osmotic pressure, 222
solubility, 218
Sodium hydroxide, 260
equivalent weight, 281
production in U.S.A., 153
Sodium ion, 85
in blood, 212, 722
in body fluids, 717
in seawater, 203
radius, 80
Sodium oxide, 82
Sodium pump, 554
Sodium stearate, 446
Solid, 11, 189
density of, 38
equilibrium with liquid, 189
heat of fusion, 189
Solubility, 204
of alcohols, 389
of amines, 464
of carbon dioxide, 216
of carboxylate salts, 439
of carboxylic acids, 437
effect of pressure on, 216
effect of temperature on, 217
of esters, 443
of gases, 216
of ionic compounds, 130
of oxygen, 217
of sodium chloride, 218
Solute, 202
Solution, 14
aqueous, 203
boiling points of, 220
colligative properties of, 219
concentrated, 204
concentration of, 204
dilute, 204
dilution of, 209
formation of, 212
freezing point, 220
hypertonic, 224
hypotonic, 224
isotonic, 224
miscible, 215
molarity, 207
neutral, 264
normality, 280
osmosis, 221
percent concentration, 205
saturated, 204, 213
unsaturated, 204
Solvent, 202
s orbital, 59
Somatic effects, 303
Sorbose, 515
Soybean oil, 545
Specificity, 585
Specific gravity, 40
Specific heat, 41
Specific rotation, 498
Spectator ion, 131
Spin, electron, 59
Sphingomyelin, 549
Sphingophospholipid, 548
Sphingosine, 548
Spirocyclic, 340
sp hybrid orbital, 110, 314
sp^2 hybrid orbital, 109, 314
in amides, 472
in carbonyl group, 406
in carboxylic acid, 430
in ethylene, 109, 313
sp^3 hybrid orbital, 108, 313
in alcohols, 381
in amines, 459
in methane, 108, 313
Spontaneous reaction, 19, 231
and free energy, 233
Squalene, 538
Standard temperature and pressure, 180
Starch, 528
States of matter, 11
symbols in equations, 118
Stearic acid, 433, 542
Stereogenic center, 494
Stereoisomer, 492
Stereospecificity, 497
in enzymes, 585
Steroid, 537
Stoichiometry, 158
STP, (see standard temperature and pressure)
Strict anaerobes, 635
Strong acid, 257
Strong base, 260
Strong electrolyte, 210
Strontium
in bones, 55
electronegativity, 71
ionization energy, 70
radioactive decay, 293
radius, 69
in seawater, 203

Structural formula, 109, 314
  condensed, 110
  perspective, 113, 313
Structural models, 113
Structure, 7, 8
  of membranes, 551
  of peptides, 567
  of polynucleotides, 691
  of proteins, 572
  of tRNA, 698
Styrene, 371
Styrofoam, 366
Subatomic particles, 48
Sublimation, 12
Submicroscopic, 6
Subscript, 86, 88, 116
Subshell, 58
  order of energy, 62
  and periodic table, 66
Substance, 3
Substitution reaction, 327
  alcohols, 390
  alkanes, 347
  nucleophilic acyl, 448
Substrate, 585
  and chirality, 589
Substrate level phosphorylation, 613
Succinate dehydrogenase, 585
  inhibition, 597
  turnover number, 593
Succinic acid, 431
  in citric acid cycle, 620
  dehydrogenation, 585
Sucrose, 527
Sulfate ester, 441
Sulfate ion, 86
  in blood, 722
  in body fluids, 717
  in plasma, 212
  in seawater, 203
Sulfhydryl group, 400
Sulfide ion, 86
Sulfite ion, 86
Sulfur
  in carbon compounds, 400
  chemical symbol, 16
  electron configuration, 63
  electronegativity, 71
  elemental form, 46
  in human body, 483
  ionization energy, 70
  radius, 69
  reaction with oxygen, 126
Sulfur dioxide, 126
Sulfur trioxide
  reaction with water, 127
Sulfuric acid, 259
  in acid rain, 256
  dehydration of alcohols, 392
  equivalent weight, 281
  ionization, 259
  production in U.S.A., 153
Superimposable, 493
Surface tension, 184
Symbol
  chemical, 16
  in chemical equations, 118
  electron dot, 65
  of ions, 79
Symmetrical reagent, 362
Synthetases, 587
Tagatose, 515
Talose, 514
Tartaric acid, 504
Tautomer, 423
Teflon, 366
Tellurium
  electronegativity, 71
  ionization energy, 70
  radius, 69
Temperature, 34
  Celsius, 34
  effect on enzyme reactions, 595
  effect on equilibrium, 247
  effect on free energy, 234
  effect on organisms, 238
  effect on reaction rates, 238
  effect on solubility, 217
  effect on vapor pressure, 186
  Fahrenheit, 34
  Kelvin, 35
Temporary dipole, 191
Tera-, 30
Terephthalic acid, 434
Termination codon
Terpene, 537
Tertiary
  alcohol, 387
  amine, 459
  carbon atom, 336
  structure of protein, 575
Testosterone, 539
Tetrachloromethane, 347
Tetrahedral molecule, 103
Tetraterpene, 537
Theoretical yield, 162
Theory, 4
  atomic, 46
  kinetic molecular, 170
  lock and key, 588
  VSEPR, 104
Thermal pollution, 219
Thiamine, 375
Thiamine pyrophosphate, 591
Thioether, 319
Thioester, 432
  in nature, 449
Thiol, 319, 400
  acidity, 400
  esters, 449
  oxidation, 401
Thomson, J. J., 48
Threonine, 562
  deamination, 673
  isoionic point, 566
  pK_a, 564
Threose, 514
Thrombin, 731
Thrombocytes, 723, 725
Thrombosis, 730
Thymine, 686
  in DNA, 690
Tissue lipid, 648
Tin
  chemical symbol, 16
  electronegativity, 71
  ionization energy, 70
  radius, 69
Titration, 279
Tollens' reagent, 412
Toluene, 371
Torr, 173
Torricelli, Evangelista, 172
Trailing zeros, 26
Trans isomer
  in alkenes, 357
  in cycloalkanes, 341
Transaldolase, 641
Transaminase, 586
Transamination, 668
  mechanism, 671
Transcription, 686, 695
Transferase, 586
Transfer RNA, 697
  structure, 698
Transition metals, 54
Transketolase, 641
Translation, 695
  and protein synthesis, 700
Translocase, 703
Translocation, 703
Transmembrane protein, 551
  in mitochondrial matrix, 622
Transmutation, 295
Transport protein, 559
Transuranium element, 296
Triacylglycerol, 544
Tricarboxylic acid cycle, 619
Trichloromethane, 347
Triglyceride, 544
  catabolic reaction, 651
  composition, 536
  digestion, 647
Trigonal planar molecule, 103
Trigonal pyramidal molecule, 103
Triose phosphate isomerase, 631
Tripeptide, 567
Triple bond, 95
Triphosphate, 391
Triphosphoric acid, 450
Triprotic acid, 259
Tristearin, 446
Triterpene, 537
Trypsin
  hydrolysis of protein, 579
  optimum pH, 595
  from trypsinogen, 600
Trypsinogen, 600
Tryptophan, 562
  isoionic point, 566
  pK_a, 564
  from serine, 589
Tryptophan synthetase, 589
Tums, 262
Turnover number, 593
Tyrosine, 562
  isoionic point, 566
  pK_a, 564
UDP-glucose, 643
Ulcers, 720
Unit, 24
  of energy, 40
  of length, 32
  of mass, 34
  metric, 29
  of pressure, 173
  of temperature, 35
  of volume, 33
Universal gas constant, 181
Unpaired electron, 60
Unsaturated
  fatty acids, 542
  hydrocarbon, 353
  solution, 204
Unshared electron pairs, 94
Unsymmetrical reagent, 362
Uracil, 686
Uranium
  decay, 293
  density, 38
Urea, 674, 733
Urea cycle, 674
  and citric acid cycle, 676
Urease
Uric acid, 205, 674, 733
Uridine diphosphate glucose, 643
Urine, 732
  density, 733
  pH, 733
Valence, 96
Valence electrons, 65
  in electron dot symbols, 65
  in Lewis structures, 94
Valence shell, 65
Valence-shell electron-pair repulsion, 103
Valeric acid, 433
Valine, 562
  isoionic point, 566
  pK_a, 564
van der Waals forces, 190
Vaporization, 171
Vaporization, heat of, 188
Vapor pressure, 186
Vasopressin, 570
  and water balance, 718, 735
Vidarabine, 711
Vinyl chloride, 366
Virus, 710
Viscosity, 184
Vision, 421
Vitamin A, 355
  solubility, 216
Vitamin C
  solubility, 216
Vitamin K_1, 732
VLDL (very low-density lipoprotein), 560
Volume
  molar, 181
  units of, 33
Volumetric analysis, 282
Volumetric flask, 33
VSEPR theory, 104
Warfarin, 732
Water
  amphoteric, 263
  boiling point, 188
  density, 38, 196
  in E. coli, 484
  equilibrium constant, K_w, 264
  equilibrium with ice, 196
  freezing point, 11
  heat of vaporization, 189
  hydrogen bonding, 194
  normal boiling point, 188
  melting point, 11
  in organisms, 484
  pH of, 270
  polarity, 107
  reaction with acidic oxides, 127
  reaction with basic oxides, 127
  self-ionization, 263
  shape, 104
  specific heat, 41
  surface tension, 184
  vapor pressure, 187
Watson, J.D., 691
Waxes, 540
Weak acid, 258
Weak base, 267
Weak electrolyte, 210
Weight, 6
Whale oil, 541
Xenon
  boiling point, 189
  electron configuration, 63
  heat of vaporization, 189
  ionization energy, 70
  radius, 69
Xylose, 514
Xylulose, 515
Zinc
  activity, 128
  chemical symbol, 16
  in enzymes, 81, 590
  galvanized steel, 139
  in human body, 483
  reaction with HCl, 118
  specific heat, 41
Zinc ion, 85
Zwitterion, 563
Zymogen, 596
  in blood clotting, 731

Atomic Numbers and Atomic Masses of the Elements

Atomic masses are based on ${}^{12}_{6}C$. Numbers in parentheses are the mass numbers of the most stable isotopes of radioactive elements.

Element	Symbol	Atomic Number	Atomic Mass	Element	Symbol	Atomic Number	Atomic Mass
Actinium	**Ac**	89	227.0278	Erbium	**Er**	68	167.26
Aluminum	**Al**	13	26.98154	Europium	**Eu**	63	151.96
Americium	**Am**	95	(243)	Fermium	**Fm**	100	(257)
Antimony	**Sb**	51	121.760	Fluorine	**F**	9	18.998403
Argon	**Ar**	18	39.948	Francium	**Fr**	87	(223)
Arsenic	**As**	33	74.9216	Gadolinium	**Gd**	64	157.25
Astatine	**At**	85	(210)	Gallium	**Ga**	31	69.723
Barium	**Ba**	56	137.33	Germanium	**Ge**	32	72.59
Berkelium	**Bk**	97	(247)	Gold	**Au**	79	196.9665
Beryllium	**Be**	4	9.01218	Hafnium	**Hf**	72	178.49
Bismuth	**Bi**	83	208.9804	Hahnium	**Ha**	105	(262)
Boron	**B**	5	10.811	Hassium	**Hs**	108	(265)
Bromine	**Br**	35	79.904	Helium	**He**	2	4.002602
Cadmium	**Cd**	48	112.41	Holmium	**Ho**	67	164.9304
Calcium	**Ca**	20	40.078	Hydrogen	**H**	1	1.00794
Californium	**Cf**	98	(251)	Indium	**In**	49	114.818
Carbon	**C**	6	12.011	Iodine	**I**	53	126.9045
Cerium	**Ce**	58	140.12	Iridium	**Ir**	77	192.217
Cesium	**Cs**	55	132.9054	Iron	**Fe**	26	55.845
Chlorine	**Cl**	17	35.453	Krypton	**Kr**	36	83.80
Chromium	**Cr**	24	51.996	Lanthanum	**La**	57	138.9055
Cobalt	**Co**	27	58.9332	Lawrencium	**Lr**	103	(262)
Copper	**Cu**	29	63.546	Lead	**Pb**	82	207.2
Curium	**Cm**	96	(247)	Lithium	**Li**	3	6.941
Dysprosium	**Dy**	66	162.50	Lutetium	**Lu**	71	174.967
Einsteinium	**Es**	99	(252)	Magnesium	**Mg**	12	24.305
Element 110	—	110	(271)	Manganese	**Mn**	25	54.9380
Element 111	—	111	(272)	Meitnerium	**Mt**	109	(266)